Fuel Science and Technology Handbook

CHEMICAL INDUSTRIES

A Series of Reference Books and Textbooks

Consulting Editor
HEINZ HEINEMANN
Heinz Heinemann, Inc.,
Berkeley, California

Volume 1: Fluid Catalytic Cracking with Zeolite Catalysts, *Paul B. Venuto and E. Thomas Habib, Jr.*

Volume 2: Ethylene: Keystone to the Petrochemical Industry, *Ludwig Kniel, Olaf Winter, and Karl Stork*

Volume 3: The Chemistry and Technology of Petroleum, *James G. Speight*

Volume 4: The Desulfurization of Heavy Oils and Residua, *James G. Speight*

Volume 5: Catalysis of Organic Reactions, *edited by William R. Moser*

Volume 6: Acetylene-Based Chemicals from Coal and Other Natural Resources, *Robert J. Tedeschi*

Volume 7: Chemically Resistant Masonry, *Walter Lee Sheppard, Jr.*

Volume 8: Compressors and Expanders: Selection and Application for the Process Industry, *Heinz P. Bloch, Joseph A. Cameron, Frank M. Danowski, Jr., Ralph James, Jr., Judson S. Swearingen, and Marilyn E. Weightman*

Volume 9: Metering Pumps: Selection and Application, *James P. Poynton*

Volume 10: Hydrocarbons from Methanol, *Clarence D. Chang*

Volume 11: Foam Flotation: Theory and Applications, *Ann N. Clarke and David J. Wilson*

Volume 12: The Chemistry and Technology of Coal, *James G. Speight*

Volume 13: Pneumatic and Hydraulic Conveying of Solids, *O. A. Williams*

Volume 14: Catalyst Manufacture: Laboratory and Commercial Preparations, *Alvin B. Stiles*

Volume 15: Characterization of Heterogeneous Catalysts, *edited by Francis Delannay*

Volume 16: BASIC Programs for Chemical Engineering Design, *James H. Weber*

Volume 17: Catalyst Poisoning, *L. Louis Hegedus and Robert W. McCabe*

Volume 18: Catalysis of Organic Reactions, *edited by John R. Kosak*

Volume 19: Adsorption Technology: A Step-by-Step Approach to Process Evaluation and Application, *edited by Frank L. Slejko*

Volume 20: Deactivation and Poisoning of Catalysts, *edited by Jacques Oudar and Henry Wise*

Volume 21: Catalysis and Surface Science: Developments in Chemicals from Methanol, Hydrotreating of Hydrocarbons, Catalyst Preparation, Monomers and Polymers, Photocatalysts and Photovoltaics, *edited by Heinz Heinemann and Gabor A. Somorjai*

Volume 22: Catalysis of Organic Reactions, *edited by Robert L. Augustine*

Volume 23: Modern Control Techniques for the Processing Industries, *T. H. Tsai, J. W. Lane, and C. S. Lin*

Volume 24: Temperature-Programmed Reduction for Solid Materials Characterization, *Alan Jones and Brian McNicol*

Volume 25: Catalytic Cracking: Catalysts, Chemistry, and Kinetics, *Bohdan W. Wojciechowski and Avelino Corma*

Volume 26: Chemical Reaction and Reactor Engineering, *edited by J. J. Carberry and A. Varma*

Volume 27: Filtration: Principles and Practices, second edition, *edited by Michael J. Matteson and Clyde Orr*

Volume 28: Corrosion Mechanisms, *edited by Florian Mansfeld*

Volume 29: Catalysis and Surface Properties of Liquid Metals and Alloys, *Yoshisada Ogino*

Volume 30: Catalyst Deactivation, *edited by Eugene E. Petersen and Alexis T. Bell*

Volume 31: Hydrogen Effects in Catalysis: Fundamentals and Practical Applications, *edited by Zoltán Paál and P. G. Menon*

Volume 32: Flow Management for Engineers and Scientists, *Nicholas P. Cheremisinoff and Paul N. Cheremisinoff*

Volume 33: Catalysis of Organic Reactions, *edited by Paul N. Rylander, Harold Greenfield, and Robert L. Augustine*

Volume 34: Powder and Bulk Solids Handling Processes: Instrumentation and Control, *Koichi Iinoya, Hiroaki Masuda, and Kinnosuke Watanabe*

Volume 35: Reverse Osmosis Technology: Applications for High-Purity-Water Production, *edited by Bipin S. Parekh*

Volume 36: Shape Selective Catalysis in Industrial Applications, *N. Y. Chen William E. Garwood, and Frank G. Dwyer*

Volume 37: Alpha Olefins Applications Handbook, *edited by George R. Lappin and Joseph L. Sauer*

Volume 38: Process Modeling and Control in Chemical Industries, *edited by Kaddour Najim*

Volume 39: Clathrate Hydrates of Natural Gases, *E. Dendy Sloan, Jr.*

Volume 40: Catalysis of Organic Reactions, *edited by Dale W. Blackburn*

Volume 41: Fuel Science and Technology Handbook, *edited by James G. Speight*

Additional Volumes in Preparation

Oxygen in Catalysis, *A. Bielanski and Jerzy Haber*

Industrial Drying Equipment: Selection and Application *Kees van't Land*

Fuel Science and Technology Handbook

edited by

James G. Speight
Western Research Institute
Laramie, Wyoming

Marcel Dekker, Inc. New York and Basel

ISBN 0-8247-8171-6

This book is printed on acid-free paper.

MARCEL DEKKER, INC.
270 Madison Avenue, New York, New York 10016

Current printing (last digit):
10 9 8 7 6 5 4 3 2 1

PRINTED IN THE UNITED STATES OF AMERICA

Preface

The last two decades have seen perturbations of energy supply systems that are unlikely to be settled within the two following decades. Disruptions of oil supply and quantum leaps (as well as quantum decreases!) in oil prices have, to say the least, sent shudders through the industry and to the consumers that are still being felt and discussed.

The optimists live for the moment that oil prices are relatively low whereas the realists look to the time that stability of liquid fuel supplies in the United States will be a reality with assured supplies from domestic sources.

At the same time, it is recognized that supplies of crude petroleum in North America are dwindling, and, unless new fields are found, current estimates put a lifetime of North American petroleum reserves at less than fifty years. It is therefore quite evident that if the United States and Canada are to have any degree of self-sufficiency in the supplies of fuels other sources must be found.

Such sources other than "conventional" petroleum do indeed exist and are spread throughout North America. These sources are heavy oils, tar sand bitumen, coal, oil shale, and natural gas. These resources represent tremendous potential since they also encourage fulfillment of the goal of moving toward a degree of energy self-sufficiency. However, the mere existence of these resources is not the complete answer to recognizing the national supplies of fuels.

The major emphasis in most energy scenarios is on the production of liquid fuels from the various energy resources. Although much work has been carried out in the last two decades, there still remain many unanswered questions relative to the conversion of tar sand bitumen, coal, and oil shale to liquid fuels. The operation of two commercial tar sand plants in the Athabasca region of Canada has been a success. The commercialization of plants for the conversion of coal and oil shale to liquid fuels has had only minimal success. On the other hand, Fischer-Tropsch conversion of natural gas to liquid fuels is seeing some success.

The use of the energy resources as combustible fuels is another issue that must be addressed. The use of coal as a combustible fuel has been known for centuries and the use of oil shale as a supplementary fuel at plants such as the Rohrbach Zement plant near Rottweil in West Germany is only a beginning in the realization of the full potential of these fossil fuel resources.

On the issue of the combustion of fossil fuels, it is necessary to consider the gradual rise in temperature of the earth's atmosphere that has occurred over the last several decades. This so-called greenhouse effect arises because of the increased concentration of carbon dioxide in the earth's atmosphere. The effects of global warming are not yet fully understood, but the general prognosis is not good (Moran, J. M., Morgan, M. D., and Wiersma, J. H. 1986. *Introduction to Environmental Science.* W. H. Freeman and Company, New York; Keepin, W. 1986. *Annual Review of Energy 11*:357).

The emissions of carbon dioxide into the atmosphere are known to come from the combustion of fossil fuels—particularly the heavier fuels such as coal and heavy oils. It is believed that there is a strong need to move to the higher hydrogen/carbon fossil fuels to combat carbon dioxide production. Thus, there is the need not only to promote the use of natural gas as a fuel for combustion but also to understand the chemistry and engineering of the combustion of the heavier fossil fuels.

It is the purpose of this book to outline the current methods and known technologies that will aid in the development of the unconventional sources of liquid fuels. Petroleum is used as the benchmark since it is the fossil fuel resource to which all others are compared not only from the point of view of technology but also price. The various aspects of petroleum technology are presented as a lead-in to the potential (and limitations) of heavy oil tar sand bitumen, coal, oil shale, and natural gas.

Other sections deal in some detail with newer, or emerging, technical aspects of these fossil fuel resources. Examples are heavy oil and bitumen processing, influence of porphyrins on processing, catalytic gasification of coal, mild gasification of coal as well as updates on the various technological aspects of oil shale processing, and the use of natural gas as a clean fuel and chemical feedstock. In addition, material relating to the environmental aspects of heavy oil and coal processing is also included.

In summary, this book provides a ready, easy-at-hand reference source to compare the scientific and technological aspects of the major fossil fuel resources of the United States, Canada, and many other parts of the world.

James G. Speight

Editor's Note

For the sake of simplicity, illustrations contained in the text are usually line drawings, and no attempt has been made to illustrate the myriad of valves, heat exchangers, and the like that may occur within a processing sequence.

In addition, training in the engineering disciplines is more likely to employ the Fahrenheit temperature scale whereas many of the other scientific disciplines employ the Celsius (Centigrade) temperature scale. For the sake of clarity, the text contains temperatures in both scales but it should be noted that exact conversion is not always possible. Accordingly, the interconversion of these scales in this text is often to the nearest 5° where such license (especially in describing processing temperatures) would not cause serious errors or misconceptions. In all other cases, conversion is as close as possible.

The sections relating to the testing of fossil fuels and fossil fuel products contain references to the relevant standard test method. In some cases, reference is given to older methods as well as to the current methods. Even though some of the older test methods are no longer in use insofar as they have been replaced by newer methods, it is considered useful to refer to these methods as they have played an important role in the evolution of the newer methods. Indeed, some of the older methods are still preferred by many laboratories, hence their inclusion in this text.

Contents

Preface *iii*
Editor's Note *vii*
Contributors *xi*

Part I PETROLEUM

1 Origin, Occurrence, and Recovery 1
2 Terminology and Classification 51
3 Composition and Properties 71
4 Fractionation 121
5 Refining: Distillation 149
6 Refining: Thermal Methods 165
7 Refining: Treatment Methods 203
8 Refining Heavy Feedstocks 229
9 Catalysts 251
10 Petroleum Products 261
11 Chemicals from Petroleum 277
References to Part I 307

Part II TAR SANDS

12 Origin, Occurrence, and Recovery 317
13 Tar Sand Composition and Properties 375
14 Bitumen Composition and Properties 387

15	Bitumen Upgrading	411
16	Current and Future Development	419
	References to Part II	431

Part III COAL

17	Origin, Occurrence, and Recovery	437
18	Terminology and Classification	543
19	Composition	573
20	Coking and Carbonization	645
21	Combustion	655
22	Gasification	661
23	Liquefaction	735
24	Chemicals from Coal	763
	References to Part III	775

Part IV OIL SHALE

25	Origin, Occurrence, and Recovery	795
26	Characterization, Testing, and Classification of Oil Shales	839
27	Physical/Chemical Properties	855
28	Chemical Structure of Kerogen	877
29	Retorting	911
30	Shale Oil Refining	1005
31	Chemicals from Shale Oil	1031
	References to Part IV	1039

Part V NATURAL GAS

32	Origin, Occurrence, and Recovery	1055
33	Definitions and Terminology	1087
34	Composition and Properties	1095
35	Processing: General Concepts	1099
36	Processing: Specific Processes	1113
37	Chemicals from Natural Gas	1153
	References to Part V	1165

Index *1169*

Contributors

R. Terry K. Baker, A.R.I.C., Ph.D. D.Sc. Professor, Chemical Engineering Department, Auburn University, Auburn, Alabama (*Chapter 22, Section VI*)

Keith D. Bartle, Ph.D. Senior Lecturer, School of Physical Chemistry, University of Leeds, Leeds, England (*Chapter 19*)

Jan F. Branthaver, B.A., Ph.D. Senior Research Chemist, Exploratory and New Products Division, Western Research Institute, Laramie, Wyoming (*Chapter 3, Section II.C and II.D, Chapter 8, Section III*)

Rita K. Hessley, Ph.D Professor, Department of Chemistry, Western Kentucky University, Bowling Green, Kentucky (*Part III, Chapters 17-18, 20-24*)

Nelly M. Rodriguez, B.S., Ph.D. Research Associate, Chemical Engineering Department, Auburn University, Auburn, Alabama (*Chapter 22, Section VI*)

Richard H. Schlosberg, Ph.D. Research Associate, Exxon Chemical Company, Annandale, New Jersey (*Chapter 11, 24*)

Charles S. Scouten, Ph.D. Senior Scientist, Research and Development Department, Amoco Oil Company, Naperville, Illinois (*Part IV, Chapter 25-30*)

Colin E. Snape, Ph.D. Professor, Department of Pure and Applied Chemistry, University of Strathclyde, Glasgow, Scotland (*Chapter 19*)

James G. Speight, B.Sc., Ph.D. Chief Scientific Officer/Executive Vice President, Western Research Institute, Laramie, Wyoming (*Parts I, II, V, Chapter 22, Section VII, Chapter 31*)

Fuel Science and Technology Handbook

Part I

Petroleum

1

Origin, Occurrence, and Recovery

I	Introduction	2
II	Origin	4
	A. Establishment of Source Beds	5
	B. Nature of the Source Material	5
	C. Transformation of Source Material into Petroleum	12
	D. Accumulation in Sediments	12
	E. In Situ Transformation	15
III	Occurrence	16
IV	Primary Recovery	21
	A. Exploration	21
	B. Drilling	26
	C. Well Completion	26
	D. Production	30
	E. Natural Gas	30
V	Secondary Recovery	32
VI	Enhanced Oil Recovery	34
	A. Chemical Methods	36
	B. Miscible Methods	39
	C. Thermal Methods	42
	D. Product Quality	46
VII	Transportation	46

I. INTRODUCTION

Petroleum is a mixture of gaseous, liquid, and solid hydrocarbon-type chemical compounds that occur in sedimentary rock deposits throughout the world (Speight, 1980). In the crude state petroleum has minimal value, but when refined it provides high-value liquid fuels, solvents, lubricants, and many other products (Purdy, 1957). The fuels derived from petroleum contribute approximately one-third to one-half of the world's total supply of energy and are used not only for transportation fuels (i.e., gasoline, diesel fuel, aviation fuel, etc.) but also to heat buildings. Petroleum products are used to lubricate machines and a once-maligned byproduct—asphalt—is used to provide highway surfaces and roofing materials and is now of premium value.

Crude petroleum is a mixture of compounds boiling at different temperatures that can be separated into a variety of different generic fractions (Table 1). Since there is wide variation in the properties of crude petroleum (Table 2), the proportions in which the different fractions occur vary with origin (Gruse and Stevens, 1960). Thus,

Table 1 Crude Petroleum is a Mixture of Compounds That Can Be Separated into Different Generic Boiling Fractions

Fraction	Boiling (°C)	Range[a] (°F)
Light naphtha	-1–150	30–300
Gasoline	-1–180	30–355
Heavy naphtha	150–205	300–400
Kerosene	205–260	400–500
Stove oil	205–290	400–550
Light gas oil	260–315	400–600
Heavy gas oil	315–425	600–800
Lubricating oil	>400	>750
Vacuum gas oil	425–600	800–1100
Residuum	>600	>1100

[a]For convenience, boiling ranges are interconverted to the nearest 5°.

Table 2 There is a Wide Variation in the Properties of Different Crude Petroleums

Origin	Specific gravity (water = 1.000)	Approximate physical composition (%)			
		Gasoline and gas	Kerosene	Gas oil	Residuum (1000°F+)
California	0.858	36.6	4.4	36.0	23.0
Pennsylvania	0.800	47.4	17.0	14.3	21.3
Oklahoma	0.816	47.6	10.8	21.6	20.0
Texas	0.864	44.9	4.2	23.2	27.1
Iraq	0.844	45.3	15.7	15.2	23.8
Iran	0.836	45.1	11.5	22.6	20.8
Kuwait	0.860	39.2	8.3	20.6	31.9
Bahrain	0.861	26.1	13.4	34.1	26.4
Saudi Arabia	0.840	34.5	8.7	29.3	27.5

some crude oils have higher proportions of the lower boiling components whereas others have higher proportions of residuum (asphaltic components).

Petroleum is by far the most commonly used liquid fuel source. Nevertheless, dwindling supplies of this resource make it essential that other sources of liquid fuels be found. Such sources (as described elsewhere in this text: see Parts II, III, and IV) represent enormous potential that must yet be realized. For example, liquid fuels from petroleum are only a fraction of those that could ultimately be produced from heavy oils, tar sand bitumens, coal, and oil shale (Figure 1). In fact, the occurrence of petroleum and its uses as sources of much-needed liquid fuels may be likened to the "tip of the iceberg" when compared to the aforementioned resources of fossil fuels.

Continuing development of the various resources is obviously needed to expand the availability of techniques by which liquid fuels can be produced from the fossil fuel resources which will ultimately lead to their most beneficial use.

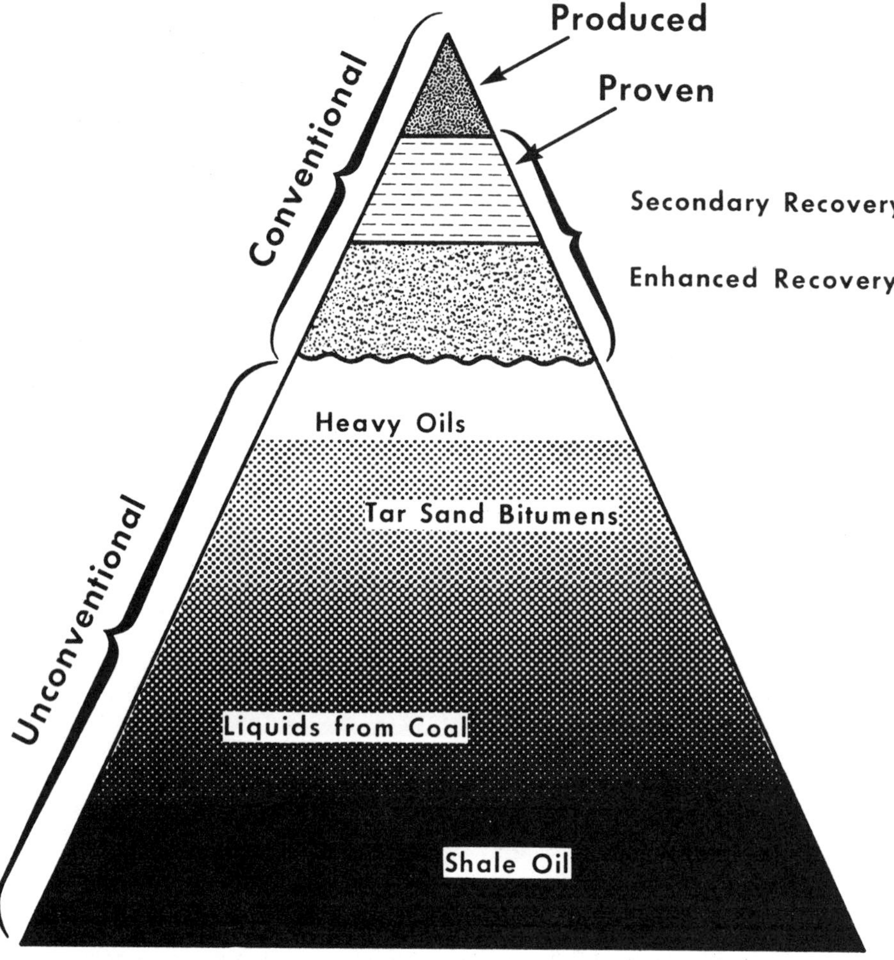

Figure 1 Simplified representation of the use of fossil fuel resources as sources of liquid fuels.

II. ORIGIN

The issue of petroleum genesis has long been a topic of research interest and it is now generally accepted that petroleum formation is associated with the development of fine-grained sedimentary rocks, deposited in a marine or near-marine environment (Brooks and Welte, 1984). In addition, petroleum is the product of plant and animal debris incorporated in those sediments at the time of deposition. However, the details of this transformation and the mechanism by which petroleum is expelled from the source sediment and accumulates in the reservoir rock are still unclear (Hobson and Tiratsoo, 1975; Nagy and Colombo, 1967).

A. Establishment of Source Beds

Flowing water can erode particles from rocks and transport them to sites where the current is less strong. The water also contains suspended organic matter which will settle with the minerals and, hence serve as the oil-generating substance. There are arguments in favor of marine plankton as the prominent source material but local conditions may favor the accumulation of other organisms such as marine algae, the remains of larger marine animals, or even material from terrestrial sources. Both the nature and the quantity of the organic and inorganic settling matter may vary and, as a consequence, it is often assumed that the characteristics of the petroleum found in different deposits are related not only to the nature of the source material but also to the relative amounts of the different constituents in the source material as well as to the local conditions in the area where the petroleum was formed.

The debris that settles at the bottom of the sea is attacked by bacteria and any material converted into water-soluble material or into gases dissipates in the sea. The remaining material is partly transformed by bacteria and buried under the steadily increasing cover of sediments. As the pressure in the sediment increases, the water content diminishes from 70–80% to 10% or less (depending on the depth and type of sediment), and anaerobic bacterial decomposition probably continues for a considerable length of time during which biochemical transformations occur (Eglington and Murphy, 1969).

B. Nature of the Source Material

Relatively few clues to the character of the source materials can be derived from the chemical composition of crude oils. However, porphyrins, which are complex derivatives of the basic material porphine (Chap. 8, Sec. III), are among the nitrogen-containing constituents of the crude oil and these compounds are recognized as the degradation products of the chlorophylls (photosynthetic pigments of plants

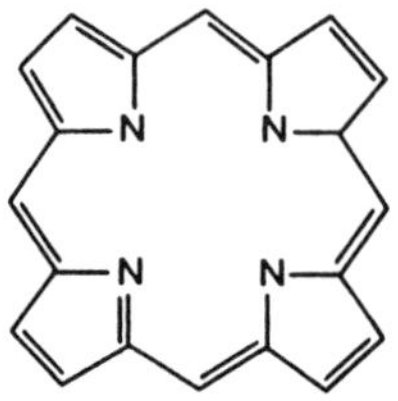

Porphine

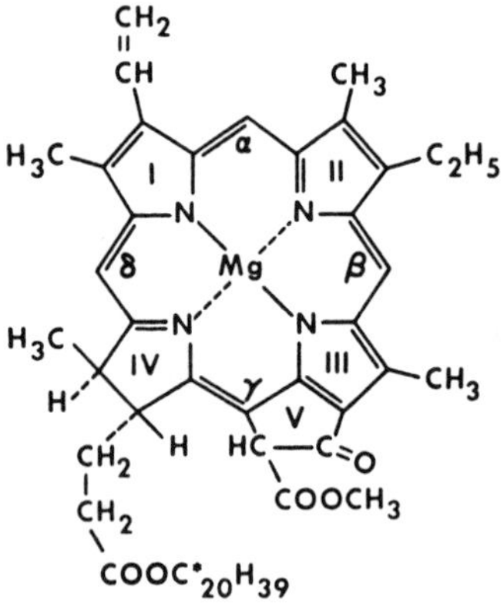

Chlorophyll a

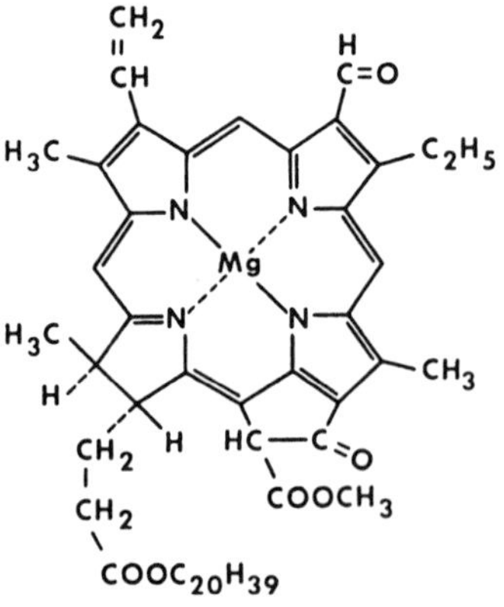

Chlorophyll b

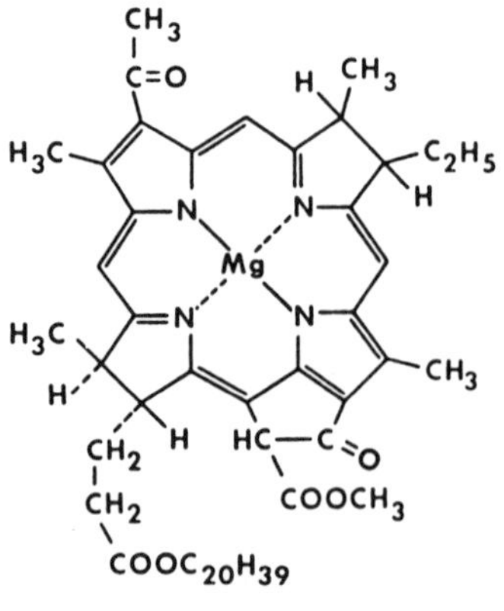

Bacteriochlorophyll

and some bacteria) and of the hemes and hematins (respiratory pigments of both plants and animals). Most, if not all, of the porphyrin material in crude oils is complexed with a metal, of which vanadium is the most important, followed by nickel; iron and copper may also be present. Vanadium contents of crude oils range from a few parts per million for the paraffinic petroleums to a thousand or so parts per million for certain heavy, asphaltic crudes (Gruse and Stevens, 1960; Tissot and Welte, 1978).

Hydrocarbons have also contributed to the formation of crude oil as they are widely distributed in living plants in concentrations up to several tenths of a percent on a dry weight basis (Atkinson and Zuckerman, 1981). Paraffinic and naphthenic hydrocarbon components undoubtedly survive after burial to become ultimately incorporated in any crude oil formed. The polyenes, which comprise the terpene hydrocarbons and the carotenoid pigments,

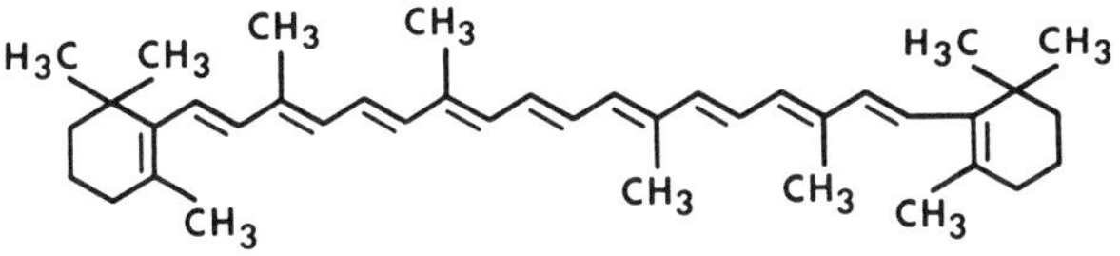

β-carotene

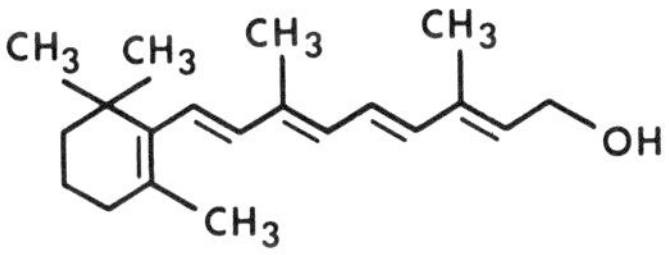

Vitamin A

also survive for a time.

Fatty acids have long been regarded as source materials for petroleum because they might be converted to hydrocarbons by the simple elimination of carbon dioxide (Speight, 1980);

$$\underset{\text{Fatty Acid}}{R.CO_2H} \longrightarrow \underset{\text{Hydrocarbon}}{R.H} + \underset{\text{Carbon Dioxide}}{CO_2}$$

The natural fatty acids (Table 3) are largely of the straight-chain type; many are unsaturated and exhibit wide variations in the arrangement of the double bonds. However, the fats of the simplest and most primitive organisms are usually composed of very complex mixtures of fatty acids, in contrast to the higher plants and animals for which the component acids are few in number.

Table 3 Fatty Acids and Their Natural Source (Speight, 1980).

Acid	Structure	Natural source
Butyric	$CH_3CH_2CH_2CO_2H$	Cow butterfat
Hexanoic (caproic)	$CH_3(CH_2)_4CO_2H$	Goat butterfat
Lauric	$CH_3(CH_2)_{10}CO_2H$	Laurel oil
Myristic	$CH_3(CH_2)_{12}CO_2H$	Nutmeg
Palmitic	$CH_3(CH_2)_{14}CO_2H$	Palm oil
trans-Crotonic	$CH_3CH{=}CHCO_2H$	Croton oil
Oleic	$CH_3(CH_2)_7CH{=}CH(CH_2)_7CO_2H$	Olive oil
Tariric	$CH_3(CH_2)_{10}C{\equiv}C(CH_2)_4CO_2H$	Lichens
Sorbic	$CH_3CH{=}CHCH{=}CHCO_2H$	Mountain ash berries
Linoleic	$CH_3(CH_2)_4CH{=}CHCH_2CH{=}CH(CH_2)_7CO_2H$	Cottonseed oil
Linolenic	$CH_3CH_2(CH{=}CHCH_2)_3(CH_2)_6CO_2H$	Linseed oil
Ricinoleic	$CH_3(CH_2)_5CH(OH)CH_2CH{=}CH(CH_2)_7CO_2H$	Castor oil
Licanic	$CH_3(CH_2)_{13}CO(CH_2)_2CO_2H$	Oiticica oil
Eleostearic	$CH_3(CH_2)_3(CH{=}CH)_3(CH_2)_7CO_2H$	Tung oil
Sterculic	$CH_3(CH_2)_7C{=}C(CH_2)_7CO_2H$ \ / CH_2	Lichens
Mycomycin	$HC{\equiv}C{-}C{\equiv}C{-}CH{=}C{=}CHCH{=}CHCH{=}CHCH_2CO_2H$	Lichens
Tuberculostearic	$CH_3(CH_2)_7CH(CH_3)(CH_2)_8CO_2H$	Tubercle bacillus

Fats are mixtures of various glycerides and are made up not only of the symmetrical compounds, i.e., glycerides of which the three fatty acid radicals are the same:

$$\begin{array}{ll} CH_2OCOC_{15}H_{31} & CH_2OCOC_{17}H_{35} \\ | & | \\ CHOCOC_{15}H_{31} & CHOCOC_{17}H_{35} \\ | & | \\ CH_2OCOC_{15}H_{31} & CH_2OCO\ C_{17}H_{35} \end{array}$$

tripalmitin **tristearin**

but are also of mixed compounds which contain two or three different acyl radicals in the molecule:

$$\begin{array}{ll} CH_2OCOC_{17}H_{35} & CH_2OCOC_{17}H_{35} \\ | & | \\ CHOCOC_{17}H_{35} & CHOCOC_{17}H_{35} \\ | & | \\ CH_2OCO\ C_{15}H_{31} & CH_2OCOC_{17}H_{33} \end{array}$$

palmitodistearin **oleodistearin**

Fats from aquatic organisms are largely of the unsaturated type and, on a weight basis, constitute an important part of both freshwater and marine organisms; it is presumed that they represent a significant contribution to the total organic fraction of sediments.

Carbohydrates form an important part of all plants and also will contribute to the total organic content of the sediments, particularly in the near-shore area. Carbohydrates are quite varied in character, ranging from the simple sugars:

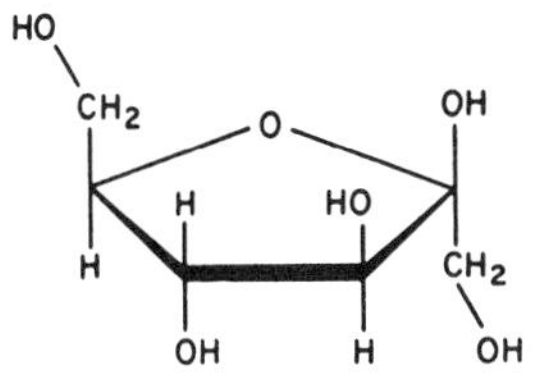

Fructose 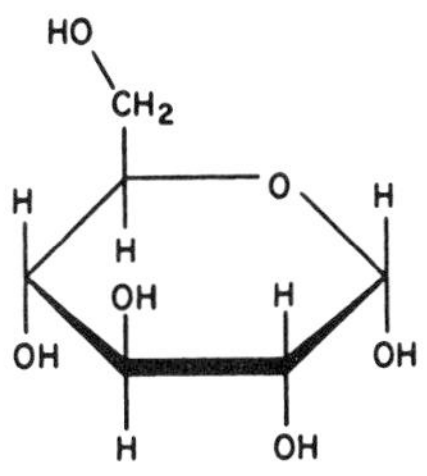 **α-Glucose**

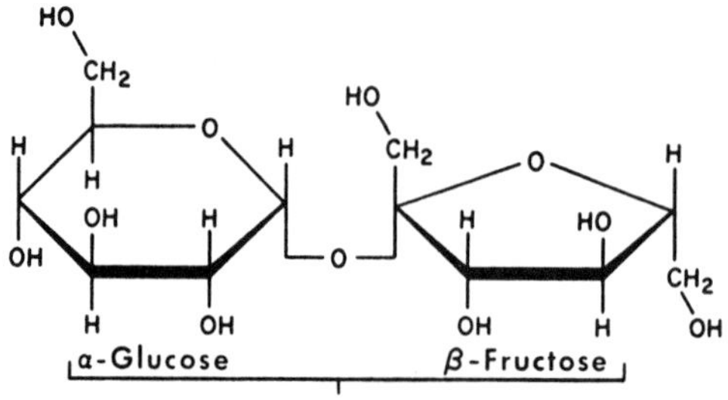

α-Glucose β-Fructose

Sucrose

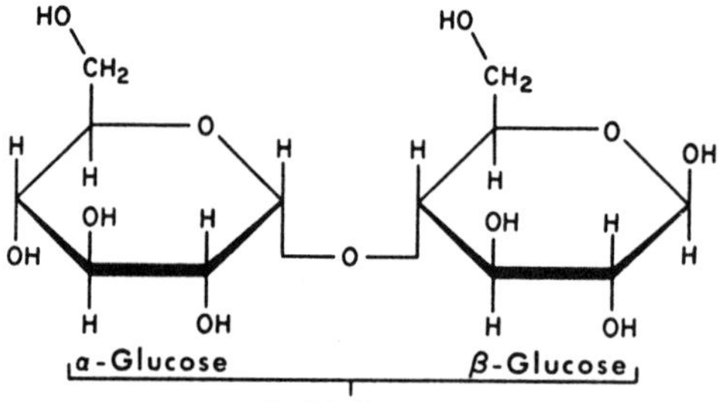

α-Glucose β-Glucose

β-Maltose

to polysaccharides such as glycogen, the starches, and the celluloses:

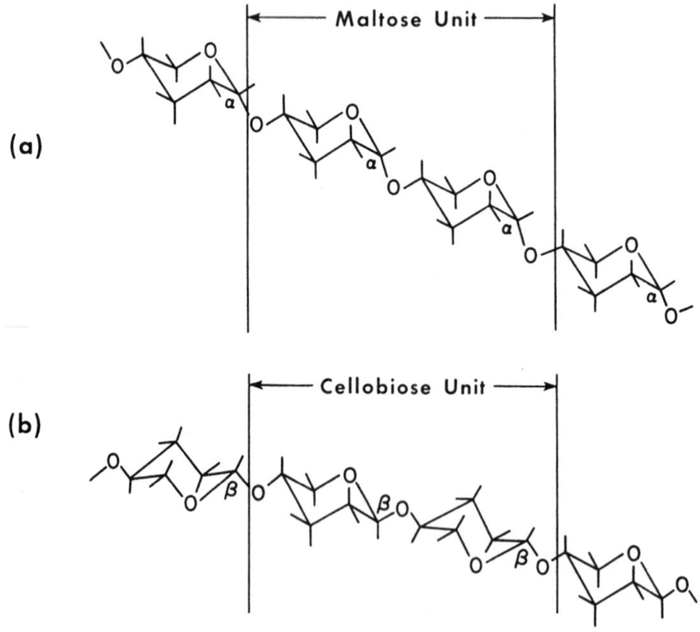

(a) Maltose Unit

(b) Cellobiose Unit

The simpler members of the carbohydrates are water-soluble and hence form a highly acceptable substrate for bacteria. Thus, part of the total carbohydrate is destroyed by oxidative organisms; another part is consumed by animal forms and converted into tissue; and finally, some of it is entrapped in the sediments. The fraction which escapes immediate destruction is probably largely composed of the water-soluble and biochemically more resistant compound types.

Proteins:

$$H_2N-\underset{\underset{\textstyle O}{\|}}{\overset{\overset{\textstyle R'}{|}}{CH}-C}\left[-NH-\overset{\overset{\textstyle R''}{|}}{CH}-\underset{\underset{\textstyle O}{\|}}{C}-\right]_n NH-\overset{\overset{\textstyle R'''}{|}}{CH}-C\begin{matrix}\nearrow O \\ \searrow OH\end{matrix}$$

where $R' \neq R'' \neq R'''$

are another important and highly complex component of all living matter. They are polymeric substances composed of one or more of some 25 amino acids linked to one another through the carboxyl carbon and the nitrogen, and chemically the proteins are very reactive.

In the native states, proteins are hydrolyzed by either alkaline or acidic media to yield water-soluble products, and as for the carbohydrates, it is likely that they are consumed in the aquatic environment either by oxidative destruction or by consumption as food by animal life. Native protein is rare outside the living cell owing to the ease with which it loses structural organization (denaturation). In many instances, solubility is greatly diminished after denaturation and incorporation in the sediment is distinctly favored.

Lignin is a mixture of complex, high molecular weight, amorphous substances and forms the cell wall structure of plants, particularly those of woody type. Despite the extensive work on the lignin of the higher land plants, its structure has not been defined except that it is probably built up of phenylpropane units containing methoxyl and hydroxyl groups. In sediments it is frequently found as a complex with protein or carbohydrate, but so little is known about the lignin of marine plants that the ligneous substances in sediments cannot definitely be typed as to source.

Sterols, which are based on a saturated, condensed ring structure, are widely distributed in animals and plants and are also believed to be source material for petroleum, although very little is known of their natural degradation reactions.

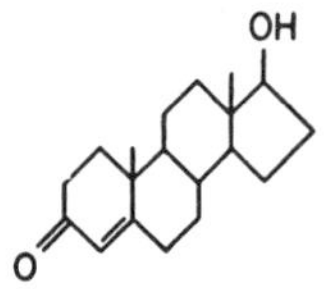

Testosterone 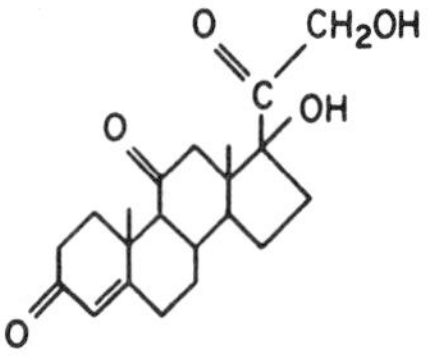 Cortisone

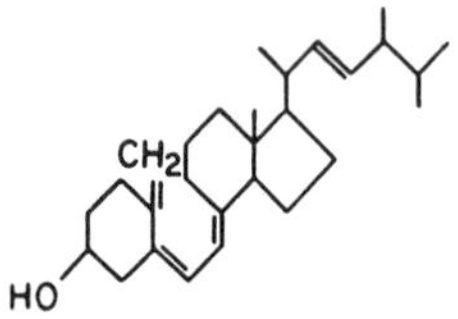

Vitamin D 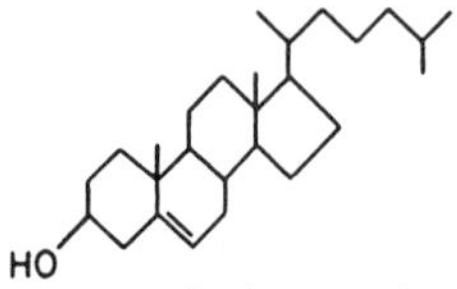 Cholesterol 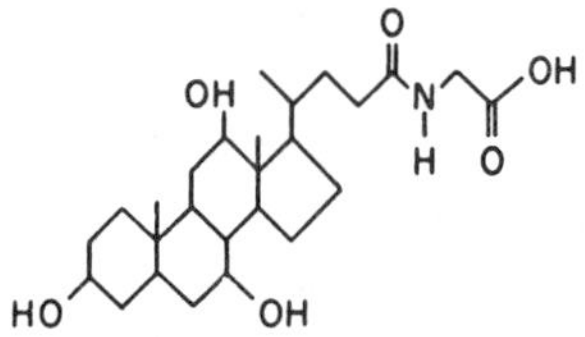 Glycocholic Acid

In general, the organic detritus of a sediment will vary with the environment just as the aquatic organisms vary in their chemical make-up. This is very important in understanding the differences in petroleum types; gross differences and structural variations of petroleum constituents are now believed to arise because of required variations in the source materials and also to variations in the conditions (geophysical) under which the petroleum was formed.

C. Transformation of Source Material into Petroleum

The least well understood phase of petroleum genesis is the chemistry of the transformation of the organic matter into petroleum. The discovery of prolific bacterial growths in recent sediments suggested that biological activity might possibly be a means of generating crude oil. Bacteria are reputed to attack carbohydrates and proteins readily but their effect on certain lipids, proteins, and lignins is relatively unknown (Brooks and Welte, 1984).

D. Accumulation in Sediments

It is generally recognized that petroleum has usually not been formed in the reservoir rock from which it is obtained but rather migrated into the reservoir rock (Speight, 1980). The predominant theory assumes that as the sedimentary layers superimposed in the source bed became thicker, the pressure increased and the compression of the source bed caused liquid organic matter to migrate to sediments with a higher permeability, which as a rule are sands or porous limestones. It is believed that during migration, petroleum does not move through the bulk of the nonsource rock and shale bodies but through faults and fractures which may be in the form of a channel network which permits leakage of oil and gas from one zone to another. During this migration, the composition of the oil may be changed due to physical causes such as filtration, adsorption, and interaction with minerals such as elemental sulfur or even with sulfur-containing minerals (e.g., sulfates) and the like.

Once the oil has accumulated in the reservoir rock, gravitational forces are presumed to be dominant thereby causing the oil, gas, and water to segregate according to their relative densities in the upper parts of the reservoir (Landes, 1959).

Any porous and permeable stratum will suffice as the reservoir and a very common reservoir rock is a porous or fractured limestone, especially of the reef (bioherm) type. Most reservoir rocks are sedimentary rocks and almost always the coarser grained of the sedimentary rocks: sand, sandstones, limestones, and dolomites.

A less common reservoir is a fractured shale or even igneous or metamorphic rock. Many geologists are of the opinion that oil found in at least some reef structures is indigenous because of the large concentration of organisms in reefs; moreover, there is often no other obvious source of the oil.

The cap rock and basement rock, which have a far lower permeability than the reservoir rock, act as a seal to prevent the escape of oil and gas from the reservoir rock. Typical cap and basement rocks are clays and shales, i.e., strata in which the pores are much finer than those of reservoir rocks. Rocks such as marls and dense limestones can also serve as cap and basement rocks provided that any pores are very small, and there are cases where evaporites (salt, anhydrite, gypsum) act as effective sealants.

The distribution of the fluids in a reservoir rock is dependent on the densities of the fluids as well as on the properties of the rock. If the pores are of uniform size and evenly distributed, there will be an upper zone where the pores are filled mainly by gas (the gas cap), and a middle zone in which the pores are occupied principally by oil with gas in solution, and a lower zone with its pores filled by water. Such accumulations usually occur at suitable locations in the earth's strata (Figure 2) and may be several miles in length. A certain amount of water (approx. 10–30%) occurs along with the oil in the middle zone.

There is a transition zone from the pores occupied entirely by water to pores occupied mainly by oil in the reservoir rock, and the thickness of this zone depends on the densities and interfacial tension of the oil and water as well as on the sizes of the pores. Similarly, there is some water in the pores in the upper gas zone, which

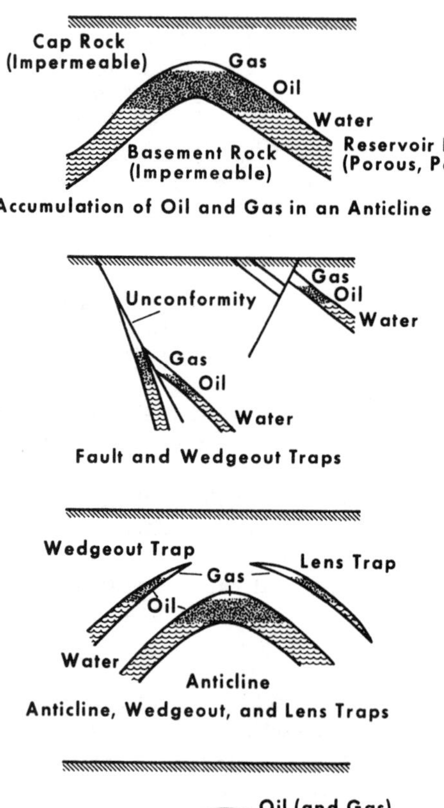

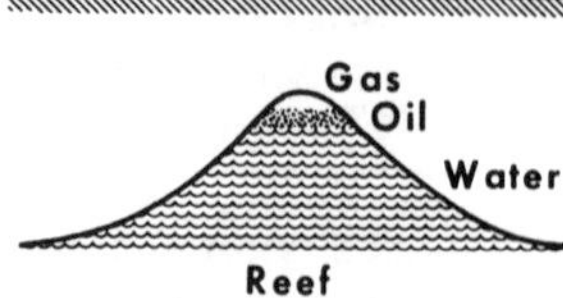

Figure 2 Geologic features with the capability of holding oil can vary widely in type.

has at its base a transition zone from pores occupied largely by gas to pores filled mainly by oil.

The water found in the oil and gas zones is known generally as interstitial water and usually occurs as collars around grain contacts, as a filling of pores with unusually small throats connecting with adjacent pores, or, to a much smaller extent, as wetting films on the surface of the mineral grains when the rock is preferentially wet by water (Figure 3).

E. In Situ Transformation

There are many ways in which a petroleum can be gradually altered in composition (Figure 4). The most common is probably exposure of the oil to the weathering action of the atmosphere, as might occur at an outcrop of a petroleum-bearing stratum, and a heavy asphaltic oil at the exposure will give way to lighter and more volatile crude within the formation. A less evident case of oxidation, which is caused by movement of oxygen-bearing waters through the permeable stratum, might occur within the reservoir at the oil-water contact. It has been suggested that the alteration of petroleum by oxygen-bearing groundwaters can be used to explain the origin of certain asphalt deposits and that the light components were removed by the percolating waters. Bacteria carried into the formation might also cause considerable alteration of the crude under the prevailing oxidizing conditions.

Gas deasphalting may also occur; this process is analogous to propane deasphalting in the refinery (Chap. 9, Sec. II.E) and to the routine laboratory procedure for the separation of asphaltenes from

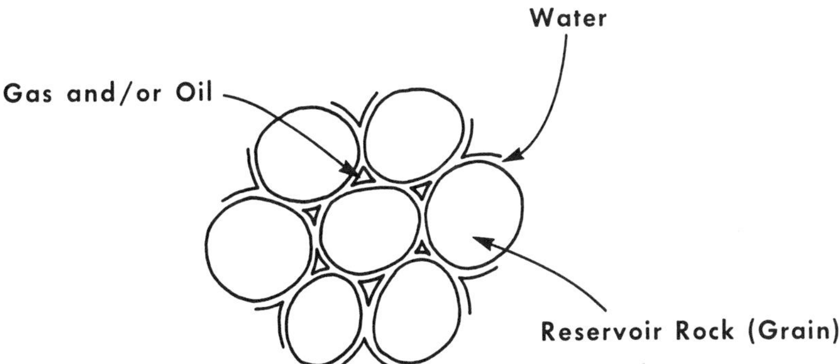

Figure 3 Representation of oil and gas accumulations within a reservoir.

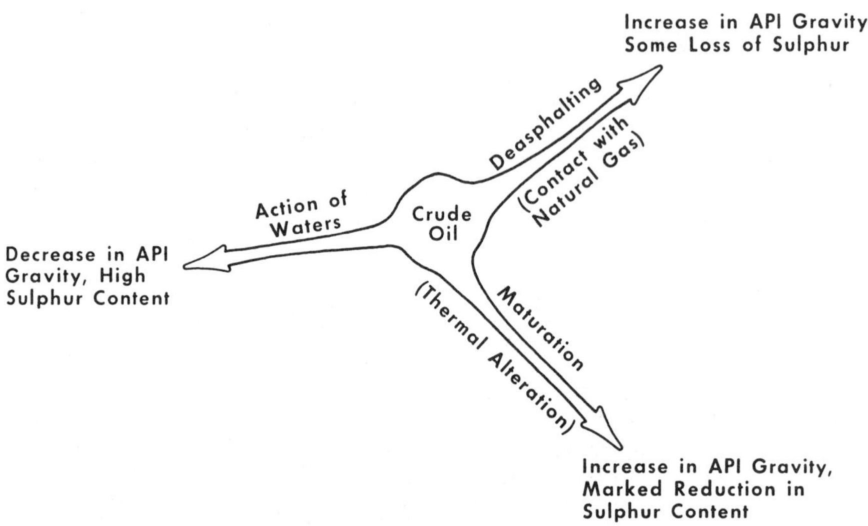

Figure 4 Schematic representation of the in situ transformation of crude oil.

petroleum by the addition of low-boiling liquid hydrocarbons (Chap. 4, Sec. II). If there is an increase in pressure as well as in temperature due to increased depth of burial, the result could be an increased solubility of the gases generated during maturation in the crude oil. The result is a change in the phase equilibria (Figure 5) and, consequently, the separation of asphaltic material.

On this basis it is conceivable that young, heavy oils might be converted into lighter crudes over the course of geologic time by mild thermal cracking, but if such changes are to occur, catalysis may be necessary.

III. OCCURRENCE

The use of petroleum or derived materials such as bitumen is not new; rather, it is a very old art (Abraham, 1945; Forbes, 1958). However, the petroleum industry is essentially a twentieth century industry, although in order to understand the evolution of the industry it is essential to have an understanding of the first uses of petroleum (Table 4).

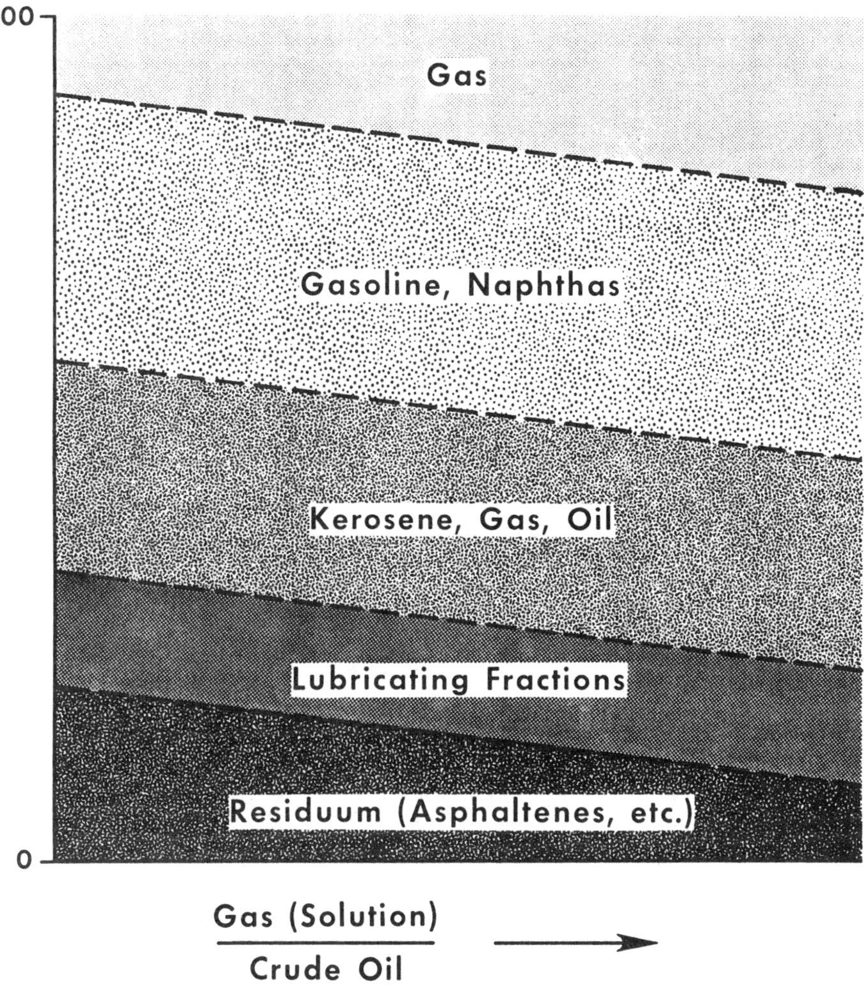

Figure 5 Effect of dissolved gas on the composition of the reservoir fluid.

The first recorded use of petroleum dates back over 5000 years ago when it was recognized that the heavier derivatives of petroleum (asphalts) could be used for caulking and as an adhesive for jewelry or for construction purposes. There is also documented use of asphalt for medicinal purposes. Approximately 2000 years ago, Arabian scientists developed methods for the distillation of petroleum which were introduced into Europe by way of Spain.

Table 4 Petroleum and Derivatives Such as Asphalt Have Been Known and Used for Almost 6000 Years

3800 B.C.	First documented use of asphalt for caulking reed boats.
3500 B.C.	Asphalt used as cement for jewelry and for ornamental applications.
3000 B.C.	Documented use of asphalt as a construction cement by Sumerians; also believed to be used as a road material; asphalt used to seal bathing pool or water tank at Mohenjo Daro.
2500 B.C.	Documented use of asphalt and other petroleum liquids (oils) in the embalming process; asphalt believed to be widely used for caulking boats.
1500 B.C.	Documented use of asphalt for medicinal purposes and (when mixed with beer) as a sedative for the stomach; continued reference to use of asphalt liquids (oil) as illuminant in lamps.
1000 B.C.	Documented use of asphalt as a waterproofing agent by lake dwellers in Switzerland.
500 B.C.	Documented use of asphalt mixed with sulfur as an incendiary device in Greek wars; also use of asphalt liquid (oil) in warfare.
350 B.C.	Documented occurrence of flammable oils in Persia.
300 B.C.	Documented use of asphalt and liquid asphalt as incendiary device (Greek fire) in warfare.
250 B.C.–250 A.D.	Documented occurrences of asphalt and oil seepages in several areas of the "fertile crescent" (Mesopotamia); repeated documentation of the use of liquid asphalt (oil) as an illuminant in lamps.
750 A.D.	First documented use in Italy of asphalt as a color in paintings.

(continued)

Table 4 (Cont.)

950–1000 A.D.	Report of destructive distillation of asphalt to produce an oil; reference to oil as nafta (naphtha).
1100 A.D.	Documented use of asphalt for covering (lacquering) metalwork.
1200 A.D.	Continued use of asphalt and naphthas as an incendiary device in warfare; use of naphtha as an illuminant/incendiary material.
1500–1600 A.D.	Documentation of asphalt deposits in the Americas; first attempted documentation of the relationship of asphalts and naphtha (petroleum).
1600–1800 A.D.	Asphalt used for a variety of tasks; relationship of asphalt to coal and wood tar studied; asphalt studied; used for paving; continued documentation of the use of naphtha as an illuminant and the production of naphtha from asphalt; importance of naphtha as fuel realized.
1859	Discovery of petroleum in North America; birth of modern day petroleum science and refining.

Interest in petroleum was also documented in China since petroleum was encountered when drilling for salt and appears on documents of the third century A.D. The Baku region of northern Persia was also reported (by Marco Polo in 1271–1273) to have a commercial petroleum industry. The interest in petroleum continued up to the modern times with an increasing interest in nafta (naphtha) when it was discovered that this material could be used as an illuminant and as a supplement to asphaltic incendiaries for use during warfare. The nafta of that time was obtained from shallow wells or by the destructive distillation of asphalt. This can perhaps be equated to the modern day coking operations (Chap. 6, Sec. I) where the overall objective of the process is to convert residua into liquid fuels.

The modern oil industry began in 1859 with the discovery and subsequent commercialization of petroleum in Pennsylvania (Bell, 1945). After completion of the first well (by Edwin Drake), the surrounding areas were immediately leased and extensive drilling took place. Crude oil output in the United States increased from approximately 2000 barrels (bbl) (1 bbl = 42 U.S. gal = 33.6 Imperial gal = 5.61 ft^3 = 158.8 liters) in 1859 to nearly 3 million bbl in 1863 and approximately 10 million bbl in 1874.

In 1861 the first cargo of oil, contained in wooden barrels, was sent across the Atlantic to London, and by the 1870s refineries, tank cars, and pipelines had become characteristic features of the industry. Throughout the remainder of the nineteenth century the United States and Russia were the two areas in which the most striking developments took place. The end of this period was characterized by the opening up of the oil fields in the Far East.

At the outbreak of World War I in 1914, the two major producers were the United States and Russia but supplies of oil were also being obtained from Indonesia, Rumania, and Mexico. During the 1920s and 1930s, attention was also focused on other areas for oil production such as the United States, the Middle East, and Indonesia. At this time European and African countries were not considered to be major oil-producing areas.

In the post-1945 era, Middle Eastern countries continued to rise in importance because of new discoveries of vast reserves. The United States, though continuing to be the biggest producer, was also the major consumer and thus was not a major exporter of oil. At this time, oil companies began to roam much farther in the search for oil, and resulted in significant discoveries in Europe, Africa, and Canada.

At the present time, many countries are recognized as having reserves of crude oil (Table 5; *BP Stat. Rev. World Energy*, June 1987). At current rates of production, proven oil reserves are sufficient for only 30 years and more reserves need to be discovered to replace those being consumed. For example, oil production from the Organization of Petroleum-Exporting Countries (OPEC) grew by more than 13% during 1986 with Saudi Arabia's production increasing by some 45% during the year. In the non-OPEC countries, the decline in production was realized predominantly by the United States, where output decreased by 3%.

In fact, according to these data, the United States has less than 10 years of proven reserves at the current rates of output compared with more than 90 years of output in Saudi Arabia or an average of some 85 years of output at current production rates for the whole of the Middle Eastern countries. In addition, the United States is one of the largest importers of petroleum (Table 6; *BP Stat. Rev. World Energy*, June 1987).

In summary, world oil consumption increased by 2.5% in 1986 and reached 59.9 million bbl/day. This is the highest rate of growth since 1978.

IV. PRIMARY RECOVERY

As already noted, crude oil accumulates over geologic time in porous underground rock formations called reservoirs, where it has been trapped by overlying and adjacent impermeable rock. Oil reservoirs sometimes exist with an overlying gas "cap," or in communication with aquifers, or both. The oil resides together with water, and sometimes free gas, in very small holes (pore spaces) and fractures. The size, shape, and degree of interconnection of the pores vary considerably from place to place in an individual reservoir. Thus, the anatomy of a reservoir is complex, both microscopically and macroscopically.

Properties of crude oil and formation water in different parts of an individual reservoir generally vary only slightly, although there are notable exceptions. For different reservoirs, crude oils display a wide spectrum of properties. Some crude oils are thinner than water, while others are thicker than cold molasses. Crude oils all contain dissolved gas in varying amounts. Most crude oils are less dense than water. The formation waters in different reservoirs vary widely in salinity and hardness.

Because of the various types of accumulations and the existence of wide ranges of both rock and fluid properties, reservoirs respond differently and must be treated individually.

The exploration, production, and transportation of petroleum are very much recent technologies; the majority of the innovations involved in these technologies have occurred since the 1920s (Speight, 1980). Advances in engineering and mechanical disciplines have been mainly responsible for realizing the vast potential of petroleum.

A. Exploration

The exploration for petroleum originated in the latter part of the nineteenth century when geologists began to map land features to search out favorable places to drill for oil (Landes, 1959; Hobson and Tiratsoo, 1975). Of particular interest to the geologists were the outcrops, which provided evidence of alternating layers of porous and impermeable rock. The porous rock (typically a sandstone, limestone, or dolomite) provides the reservoir for the petroleum while the impermeable rock (typically clay or shale) acts as a trap and prevents the migration of the petroleum from the reservoir.

By the early 1900s, most of the areas where surface structural characteristics offered the promise of oil had been investigated. In

Table 5 Estimated Reserves of Crude Oil in Various Countries

	Thousand million tonnes	Thousand million barrels	Share of total (%)	R/P ratio
North America				
USA	4.1	32.5	4.6	8.5
Canada	1.0	7.9	1.1	12.3
Total North America	5.1	40.4	5.7	9.0
Latin America				
Argentina	0.3	2.3	0.3	14.4
Brazil	0.3	2.3	0.3	10.1
Ecuador	0.2	1.7	0.2	16.7
Mexico	7.6	54.7	7.8	56.3
Venezuela	3.6	25.0	3.6	38.7
Others	0.4	2.9	0.4	10.9
Total Latin America	12.4	88.9	12.6	37.7
Western Europe				
Norway	1.4	10.5	1.5	31.2
United Kingdom	0.7	5.3	0.8	5.5
Others	0.3	2.4	0.3	13.3
Total Western Europe	2.4	18.2	2.6	12.2
Middle East				
Abu Dhabi	4.1	31.0	4.4	80.6
Dubai	0.2	1.4	0.2	9.7
Iran	6.7	48.8	6.9	71.1
Iraq	6.3	47.1	6.7	74.7
Kuwait	12.7	91.9	13.1	*
Neutral zone	0.7	5.2	0.7	43.3
Oman	0.6	4.0	0.6	19.8
Qatar	0.4	3.2	0.5	25.0
Saudi Arabia	22.7	166.6	23.7	90.3
Syria	0.2	1.4	0.2	19.2
Others	0.2	1.4	0.2	31.3
Total Middle East	54.8	402.0	57.2	85.5
Africa				
Algeria	1.1	8.8	1.3	26.9
Angola	0.2	1.2	0.2	11.7
Egypt	0.5	3.6	0.5	12.1

(continued)

Table 5 (Cont.)

	Thousand million tonnes	Thousand million barrels	Share of total (%)	R/P ratio
[Africa]				
Libya	2.8	21.3	3.0	55.1
Nigeria	2.2	16.0	2.3	30.2
Tunisia	0.2	1.8	0.3	46.1
Others	0.3	2.5	0.4	13.6
Total Africa	7.3	55.2	8.0	29.3
Asia and Australasia				
Japan		0.1		14.0
Brunei	0.2	1.4	0.2	25.6
Indonesia	1.1	8.3	1.2	16.7
Malaysia	0.4	2.8	0.4	15.0
Other South-East Asia		0.2		2.8
India	0.6	4.2	0.6	18.0
Other South Asia		0.2		6.0
Australia	0.2	1.7	0.2	6.8
New Zealand		0.2		21.2
Total Asia and Australasia	2.5	19.1	2.6	14.3
Centrally-Planned Economies (CPEs)				
China	2.4	18.4	2.6	18.5
USSR	8.0	59.0	8.4	13.1
Others	0.3	1.9	0.3	11.0
Total CPEs	10.7	79.3	11.3	14.0
Total world	95.2	703.1	100.0	32.5

Proved reserves of oil are generally taken to be those quantities which geological and engineering information indicate with reasonable certainty can be recovered in the future from known reservoirs under existing economic and operating conditions.
Reserves/Production (R/P) ratio: If the reserves remaining at the end of any year are divided by the production in that year, the result is the length of time that those remaining reserves would last if production were to continue at the then-current level.
Source: The estimates contained in this table are those published by the *Oil and Gas Journal* in its "Worldwide Oil" issue of 29th December 1986, plus an estimate of natural gas liquids for North America. Reserves of shale oil and tar sands are not included. UK reserves data are taken from the UK Department of Energy's Brown Book, published in May 1987.

Table 6 Imports and Exports of Crude Oil for Various Countries

	Million tonnes				Thousand barrels daily			
	Crude imports	Produce imports	Crude exports	Product exports	Crude imports	Product imports	Crude exports	Product exports
USA	208.8	88.3	7.5	29.8	4205	1840	145	620
Canada	15.0	6.1	25.7	7.3	305	125	525	150
Latin America	61.7	18.5	114.6	61.1	1235	385	2325	1275
Western Europe	346.9	88.5	29.5	24.1	7015	1845	590	500
Middle East	0.9	8.7	470.2	69.1	15	180	9440	1440
North Africa	1.0	4.1	98.1	23.9	20	85	2020	495
West Africa	0.2	2.2	96.8	1.6	5	45	1945	35
East and Southern Africa	20.9	1.5	—	0.2	420	30	—	5
South Asia	15.8	5.4	—	2.8	315	110	—	60
South-East Asia	81.0	16.5	52.8	23.4	1630	345	1065	485
Japan	165.8	38.6	—	1.0	3335	805	—	20
Australasia	4.5	4.8	1.3	3.9	90	100	25	80
USSR, E. Europe & China[a]	50.0	5.8	84.6	63.6	1015	120	1695	1325
Destination not known[b]	8.6	22.8	—	—	170	475	—	—
Total world	981.1	311.8	981.1	311.8	19,775	6490	19,775	6490

[a]Excludes intraregional trade within the Communist bloc.
[b]Includes changes in the quantity of oil in transit, transit losses, minor movements not otherwise shown, unidentified military use, etc.

the early 1920s, the era of subsurface exploration for oil began (Forbes, 1958). New geologic and geophysical techniques were developed for areas where the strata were not sufficiently exposed to permit surface mapping of the subsurface characteristics.

In the 1960s, the development of geophysics provided methods for exploring below the surface of the earth. The principles used are basically magnetism (magnetometer), gravity (gravimeter), and sound waves (seismograph).

The magnetometer is a specially designed magnetic compass which detects minute differences in the magnetic properties of rock formations, thus helping to find structures that might contain oil, such as the layers of sedimentary rock that may lie on top of the much denser igneous, or basement, rock.

The data give clues to places that might conceal anticlines or other oil-favorable structures (Figure 2). Of even more value is the determination of the approximate total thickness of the sedimentary rock, which can save unwarranted expenditures later or more costly geophysics or even the drilling of a well in which the sediment might not contain sufficient oil to warrant further investigation. Most magnetometer surveys used now are performed by air, which permits very large-scale surveys to be made rapidly and permits surveys over regions which might otherwise be inaccessible.

The gravimeter detects differences in gravity and gives an indication of the location and density of underground rock formations. Differences from the normal can be caused by geologic and other influences, and such differences provide an indication of subsurface structural formations.

The seismograph measures the shock waves from explosions initiated by triggering small, controlled charges of explosives in the bottom of shallow holes in the ground. The formation depth is determined by measuring the time elapsed between the explosion and detection of the reflected wave at the surface. Seismic geophysical work is also carried out on the water, greatly aiding the search for oil on the continental shelves and other areas covered by water. A marine seismic project moves continually, with detectors being towed behind the boat at a constant speed and a fairly constant depth. Explosive charges are detonated at a position and time determined by the speed of the boat, so that a continuous survey of the reflecting horizons can be obtained.

Another valuable exploration method is geophysical borehole logging, which involves drilling a well and then employing instruments to log or make measurements at various levels in the hole by such means as electrical resistivity, radioactivity, acoustics, or density. In addition, formation samples (cores) are taken for physical and chemical tests.

B. Drilling

Drilling for oil is a complex operation that has evolved considerably over the past 100 years. The older cable-tool method (Figure 6) used almost extensively until 1900, involves raising and dropping a heavy bit and drill stem attached by cable to a cantilever arm at the surface. It pulverizes the rock and earth, gradually forming the hole. The cable-tool system is generally preferred only for penetrating hard rock at shallow depths and when oil reservoirs are expected at the shallow depths. The weight of the column is usually enough to attain penetration, but it can be augmented by a hydraulic pressure cylinder at the surface.

Although there are many variations in design, all modern rigs have essentially the same components (Figure 7). The hoisting, or draw works, raises and lowers the drill pipe and casing, which can weigh as much as 200 tons (200,000 kg). The height of the derrick depends on the numbers of joints of drill pipe necessary for the drilling operation. The drill column is rotated by the rotary table, located in the middle of the rig floor above the borehole. The table imparts rotary motion to the drill stem through the kelly attached to the upper end of the column. The kelly fits into a shaped hole in the center of the rotary table.

The drilling bit is connected to drill collars at the bottom of the stem. These are thick steel cylinders, 6–9 m long; as many as 10 may be screwed together. They concentrate weight at the bottom of the column and exert tension on the more flexible pipe above, reducing the tendency of the hole to go off-line and the drill pipe to fracture. Drill bits have many designs (Figure 8); variations include number of blades, type of metal, and shape of cutting or abrading components.

A mud-circulating system is one of the most important parts of a rig. This system maintains the mud in proper condition, free of rock cuttings or other abrasive materials it might bring up from the hole, as well as retaining proper physical and chemical characteristics.

A recent innovation is the concept of directional drilling, which is used to reach formations not located directly below the penetration point. Wells are also drilled in water using marine platforms but because these platforms are capital-intensive, directional drilling is used on a widely spaced pattern, from a single-platform anchorage. Directional drilling has also been used, when the occasion permits, to drill from land under the surface of a body of water.

C. Well Completion

Completing a well and preparing for the production of oil involves insertion of a casing, which comprises one or more strings of tubing and which is carried out in part during drilling. The casing provides (1)

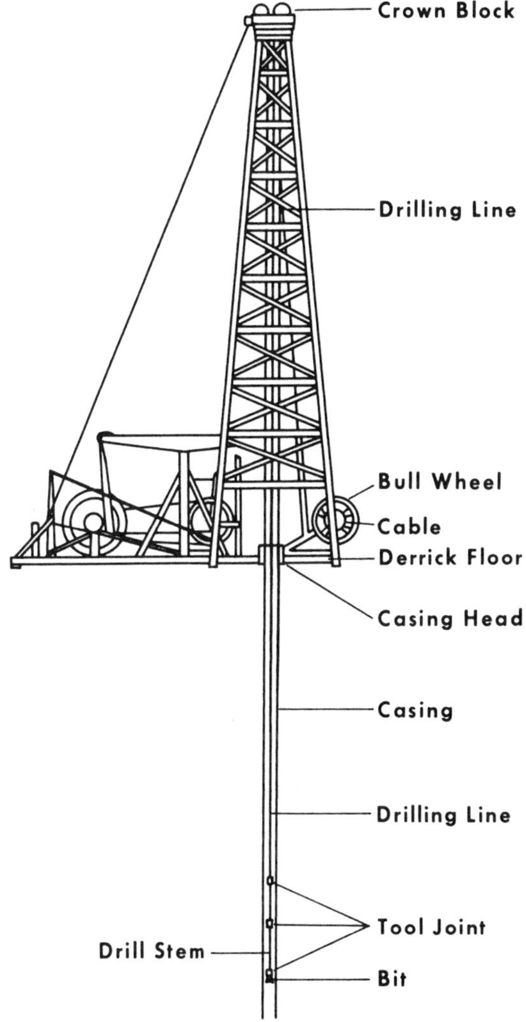

Figure 6 The older cable-tool method of drilling for oil involved raising and dropping a heavy drill bit and drill stem on a cantilever arm to pulverzie the rock and earth.

a permanent wall to the borehole to prevent cave-ins and inflow of unwanted water; (2) a return passage for the mud stream; and (3) control of the well during production.

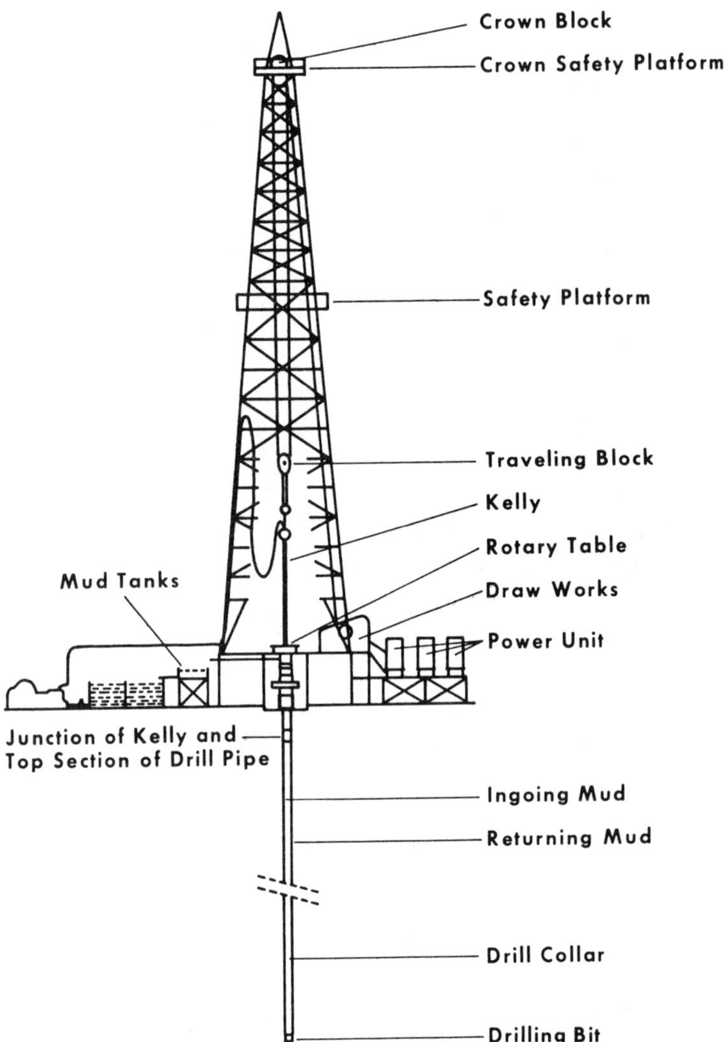

Figure 7 Modern methods of drilling use the rotary drill system.

Access to producing strata is achieved through holes in the casing wall and a cement sheath is then injected between the casing and the borehole wall to add strength. An assembly of valves, known as the "Christmas tree" (Figure 9), is installed above a master valve at the casing head if oil is expected to flow naturally or by gas or airlift; if mechanical lift is anticipated, the Christmas tree assembly is not used.

Figure 8 The design of drill bits can vary widely depending on the nature of the earth formations.

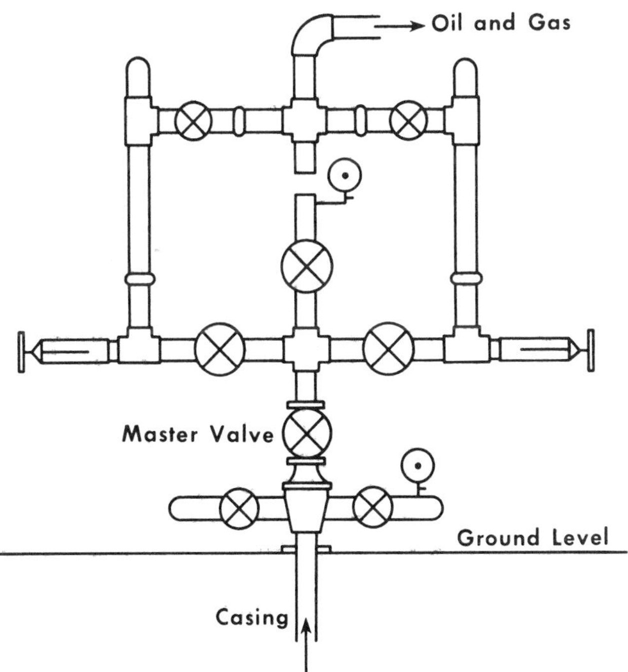

Figure 9 Wells must be capped to control the flow of oil; such caps are referred to as "Christmas trees."

D. Production

Primary oil recovery depends on natural reservoir energy to drive the oil through the complex pore network to producing wells. The driving energy may be derived from liquid expansion and evolution of dissolved gases from the oil as reservoir pressure is lowered during production, expansion of free gas or a gas cap, influx of natural water, gravity, or combinations of these effects. The recovery efficiency for primary production is generally low when liquid expansion and solution gas evolution are the driving mechanisms. Much higher recoveries are associated with reservoirs having water and gas cap drives, and with reservoirs where gravity effectively promotes drainage of the oil from the rock pores. Eventually, the natural drive energy is dissipated. When this occurs, energy must be added to the reservoir to produce significant amounts of additional oil.

In dissolved gas drive, the propulsive force is the gas in solution in the oil which tends to come out of solution because of the pressure release at the point of penetration of a well. Dissolved gas drive is the least efficient type of natural drive because of rapid drops in the bottom hole pressure; the total recovery may be less than 20%.

If gas overlies the oil in the reservoir, it can be utilized to drive the oil into wells situated at the bottom of the oil-bearing zone. By producing oil only from below the gas cap, it is possible to maintain a high gas cap pressure until the end of the life of the pool. However, if the oil deposit is not systematically developed and bypassing of the gas occurs, an undue proportion of oil will be left behind. The usual recovery of oil in a gas cap field is 40--50% of the oil in place.

The most efficient propulsive force for recovering oil is natural water drive, in which the pressure of the water forces the lighter recoverable oil out of the reservoir into the producing wells. In a water drive field it is essential that the removal rate be so adjusted that the water moves up evenly as space is made available for it by the removal of the hydrocarbons. The recovery in water drive pools may run as high as 80%. The force behind the water drive may be either hydrostatic pressure, expansion of the reservoir water, or a combination of both.

Gravity drive is an important factor when oil columns of several thousands of feet exist, as they do in some North American fields. Furthermore, the last bit of recoverable oil is produced in many pools by gravity drainage of the reservoir.

E. Natural Gas

Natural gas (see also Part V) is a combustible gas that occurs in porous rock in the earth's crust and is found with, or near to, crude oil reservoirs. However, it may occur alone in separate reservoirs. More commonly, it forms a gas cap entrapped between liquid petroleum and an impervious rock layer (cap rock) in a petroleum reservoir. Under conditions of greater pressure the gas will be intimately mixed with, or dissolved in, crude oil.

Typical natural gas consists of hydrocarbons of very low boiling point. Methane (CH_4); the first member of the paraffin series, with a boiling point of -159°C (-254°F), makes up approximately 85% of the gas. Ethane (C_2H_6), with a boiling point of -89°C (-128°F), may be present in amounts up to 10%. Propane (C_3H_8) has a boiling point of -42°C (-44°F) (Table 7).

Table 7 Approximate Composition Data for "Wet" and "Dry" Natural Gas (Calorific Value: 900–1100 Btu/ft^3)

Constituents	Composition (vol %)		
Hydrocarbons	"wet"	←(range)→	"dry"
Methane	84.6		96.0
Ethane	6.4		2.0
Propane	5.3		0.6
Isobutane	1.2		0.18
n-Butane	1.4		0.12
Isopentane	0.4		0.14
n-Pentane	0.2		0.06
Hexanes	0.4		0.10
Heptanes	0.1		0.08
Nonhydrocarbons			
Carbon dioxide		0--5	
Helium		0–0.5	
Hydrogen sulfide		0–5	
Nitrogen		0–10	
Argon		0–0.05	
Radon, krypton, xenon		Traces	

Types of natural gas vary according to composition and can be dry or lean gas (mostly methane), wet gas (considerable amounts of so-called higher hydrocarbons), sour gas (much hydrogen sulfide), sweet gas (little hydrogen sulfide), residue gas (higher paraffins having been extracted), and casinghead gas (derived from an oil well by extraction at the surface). Natural gas has no distinct odor; its main use is for fuel but it can also be used to make chemicals, and liquefied petroleum gas (see Part V).

There are also large quantities of natural gas trapped in reservoirs "too tight" to permit recovery by conventional technology. Twenty basins in the United States contain significant amounts of gas in such low-permeability formations. The leading recovery technology for tight gas reservoirs involves massive hydraulic fracturing. The purpose of this technique is to create artificial fractures in the reservoir to provide a conduit for gas to flow to the well. Fractures are created by pumping fluid into the formation until the pressure breaks the rock. A propping agent, usually sand, is mixed with the fluid and is thus carried into the fracture.

Some natural gas wells also produce helium, which can occur in commercial quantities; nitrogen and carbon dioxide are also found in some natural gases. Gas is usually separated at as high a pressure as possible, reducing compression costs where the gas is to be used for gaslift or delivered to a pipeline. After gas removal, lighter hydrocarbons and hydrogen sulfide are removed as necessary to obtain a crude oil of suitable vapor pressure for transport, yet retaining most of the natural gasoline constituents.

V. SECONDARY RECOVERY

Secondary oil recovery involves the introduction of energy into a reservoir by injecting gas or water under pressure. Separate wells are usually used for injection and production. The injected fluids maintain reservoir pressure, or repressure the reservoir after primary depletion, and displace a portion of the remaining crude oil to production wells (Figure 10).

The most common method of producing oil from nonflowing wells is by means of a pump that provides a mechanical lift to the fluids in the reservoir. A pump barrel is lowered into the well on a string of solid steel rods known as sucker rods. Up-and-down movement of the sucker rods forces the oil up the tubing to the surface. This vertical movement may be supplied by a walking beam powered by a nearby engine,

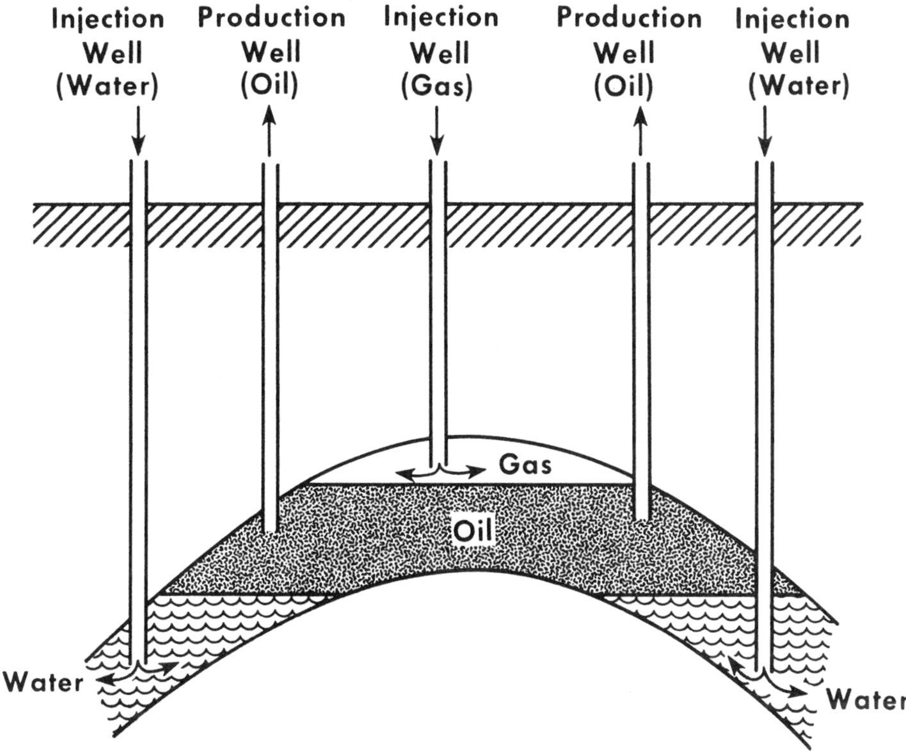

Figure 10 Petroleum recovery is a complex operation that relies on natural phenomena (gas-cap drive) or applied methods such as water flood.

or it may be brought about through the use of a pump jack, which is connected with a central power source by means of pull rods.

During withdrawal of fluids from a well, it is usual practice to maintain pressures in the reservoir near the original levels by pumping gas or water into the reservoir as the hydrocarbons are withdrawn. This practice has the advantage not only of maintaining the production of individual wells and increasing considerably the ultimate yield, but also it may bring about the conservation of gas that otherwise would be wasted.

In the older fields, it was not the usual practice to maintain the reservoir pressure, and it is now necessary to obtain petroleum from these fields by means of secondary recovery projects. Several methods have been developed to obtain oil from reservoirs where previous economic policies dictated that ordinary production systems were no longer viable.

Water injection (Figure 10) is predominantly a secondary recovery process (water flood). The principal reason is: reservoir formation water is ordinarily unavailable in volume during the early years of an oil field, and pressure maintenance water from outside the field may be injected back into the reservoir for reservoir pressure maintenance.

The viscosity of petroleum (Chap. 3, Sec. III) is an important factor that must be taken into account when the heavier oils are recovered from a reservoir or, perhaps what is more pertinent, when these materials have to be transported from one point in a refinery to another. The high viscosity of certain residua may dictate that the materials only be transported short distances in heated pipes or that the residuum be diluted with an aromatic naphtha to be recovered after the mix reaches its destination.

Certain reservoir types, such as those with very viscous crude oils and some low-permeability carbonate (limestone, dolomite, chert) reservoirs, respond poorly to conventional secondary recovery techniques. In these reservoirs it is desirable to initiate enhanced oil recovery (EOR) operations as early as possible. This may mean abbreviating considerably the conventional secondary recovery operations or bypassing them altogether.

VI. ENHANCED OIL RECOVERY

Conventional primary and secondary recovery processes are ultimately expected to produce about one-third of the original oil discovered. Recoveries from individual reservoirs can range from less than 5% to as high as 80% of the original oil in place. This broad range of recovery efficiency is a result of variations in the properties of the specific rock and fluids involved from reservoir to reservoir, as well as the kind and level of energy that drives the oil to producing wells where it is captured.

The oil remaining after conventional recovery operations is retained in the pore space of reservoir rock at a lower concentration than originally existed. The produced oil is replaced by gas and/or water in the pores. In portions of the reservoir that have been contacted or swept by the injection fluid, the residual oil remains as droplets (or ganglia) trapped in either individual pores or clusters of pores. It may also remain as films partly coating the pore walls. Entrapment of this residual oil is predominantly due to capillary and surface forces and to pore geometry.

In the pores of those volumes of reservoir rock that were not well swept by displacing fluids, the oil continues to exist at higher concentrations and may exist as a continuous phase. This macroscopic bypassing of the oil occurs because of reservoir heterogeneity, the placement of wells, and the effects of viscous, gravity, and capillary forces, which act simultaneously in the reservoir. The resultant effect depends on conditions at individual locations. The higher the mobility of the displacing fluid relative to that of the oil (the higher the mobility ratio), the greater the propensity for the displacing fluid to bypass oil. Due to fluid density differences, gravity forces cause vertical segregation of the fluids in the reservoir so that water tends to underrun, and gas to override, the oil-containing rock. These mechanisms can be controlled or utilized to only a limited extent in primary and secondary recovery operations.

The intent of enhanced oil recovery is to increase the effectiveness of oil removal from pores of the rock (displacement efficiency) and to increase the volume of rock contacted by injected fluids (sweep efficiency). EOR processes use thermal, chemical, or fluid phase behavior effects to reduce or eliminate capillary forces that trap oil within pores, to thin the oil or otherwise improve its mobility, or to alter the mobility of the displacing fluids. In some cases, the effects of gravity forces, which ordinarily cause vertical segregation of fluids of different densities, can be minimized or even used to advantage.

The degree to which EOR methods are applicable in the future will depend on the development of improved process technology; on improved understanding of fluid chemistry, phase behavior, and physical properties; and on the accuracy of geology and reservoir engineering in characterizing the physical nature of individual reservoirs.

Enhanced oil recovery is defined as the incremental ultimate oil that can be economically recovered from a petroleum reservoir over oil that can be economically recovered by conventional primary and secondary methods. Since the early 1950s, a significant amount of laboratory research and field testing has been devoted to developing EOR methods as well as defining the requirements for a successful recovery and the limitations of the various methods (Table 8; Meyer, 1977).

Chemical, miscible, and thermal methods are the three categories of EOR processes generally recognized as most promising. The various EOR processes differ considerably in complexity, the physical mechanisms responsible for oil recovery, and the amount of experience that has been derived from field application.

Table 8 Generalized Screening Criteria for Oil Recovery Methods

Recovery method	Requirements	Limitations
Water flood	Reservoir uniformity	Fissuring, dismembering
Cyclic stimulation	Hydrophylity, nonuniformity	Dismembering, hydrophobity
Surfactant flood	Low watercut, sand content	Clay content, fissuring
Polymer flood	High permeability	Fissuring
Gas-water injection	Low permeability, anisotropic	Isotropic, hydrophobity
CO_2 flood	Steep dip, anisotropic	Salinity, fissuring
Micellar flood	Low salinity	Fissuring
Combustion	Thin reservoir	Fissuring
Steam injection	Thick reservoir	Depth more than 1000 m
Caustic flood	Naphthenic acids, asphaltenes	Saline
Steam cyclic simulation	Asphaltenes, paraffin	Hydrophobity, water saturation

A. Chemical Methods

Chemical methods include polymer flooding, surfactant (micellar/polymer, microemulsion) flooding, and alkaline flood processes. Polymer flooding is conceptually simple and inexpensive, and its commercial use is increasing despite relatively small potential incremental oil production. Surfactant flooding is complex and requires detailed laboratory testing to support field project design. As demonstrated by recent field tests, it has excellent potential for improving the recovery of low- to moderate-viscosity oils. Surfactant flooding is expensive and has been used in few large-scale projects. Alkaline flooding has been used only in those reservoirs containing specific types of high-acid-number crude oils.

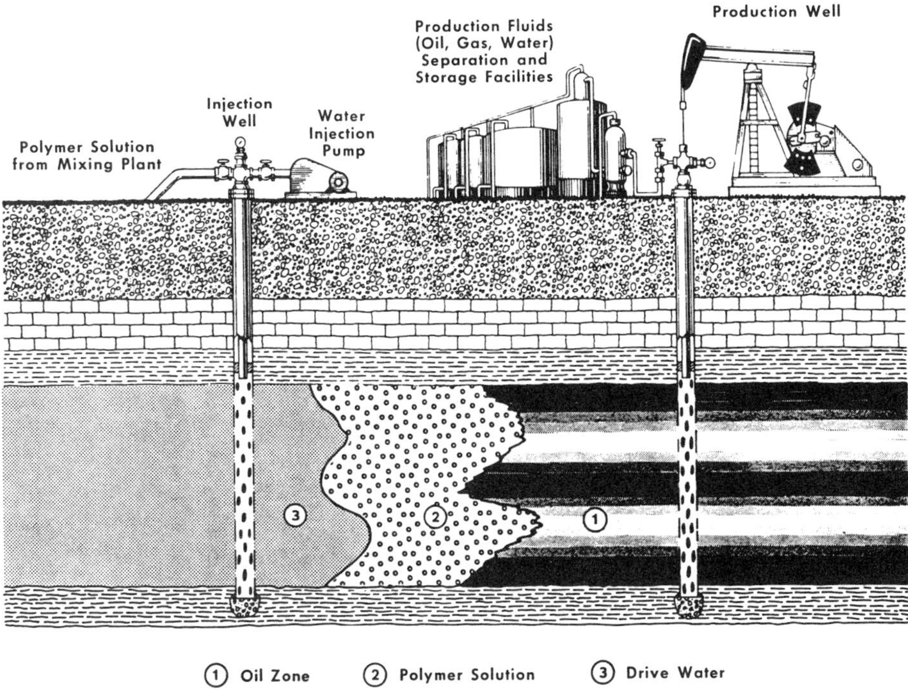

Figure 11 Schematic representation of a polymer flooding operation.

1. Polymer Flooding

Conventional water flooding can often be improved by the addition of polymers to injection water (Figure 11) to improve the mobility ratio between the injected and in-place fluids. The polymer solution affects the relative flow rates of oil and water, and sweeps a larger fraction of the reservoir than water alone, thus contacting more of the oil and moving it to production wells. Polymers currently in use are produced both synthetically (polyacrylamides) and biologically (polysaccharides). The polymers may also be crosslinked in situ to form highly viscous fluids that will divert the subsequently injected water into different reservoir strata.

Polymer flooding has its greatest utility in heterogeneous reservoirs and those that contain moderately viscous oils. Oil reservoirs with adverse water flood mobility ratios have potential for increased oil recovery through better areal sweep efficiency. Heterogeneous reservoirs may respond favorably as a result of improved vertical sweep efficiency. Because the microscopic displacement efficiency is not affected, the increase in recovery over water flood will likely be modest and limited to the extent that sweep efficiency is improved but the incremental cost is also moderate. Currently, polymer flooding is being used in a significant number of commercial field projects. The process may be used to recover oils of higher viscosity than those for which a surfactant flood might be considered.

2. Surfactant Flooding

Surfactant flooding (Figure 12) is a multiple-slug process involving the addition of surface-active chemicals to water. These chemicals reduce the capillary forces that trap the oil in the pores of the rock.

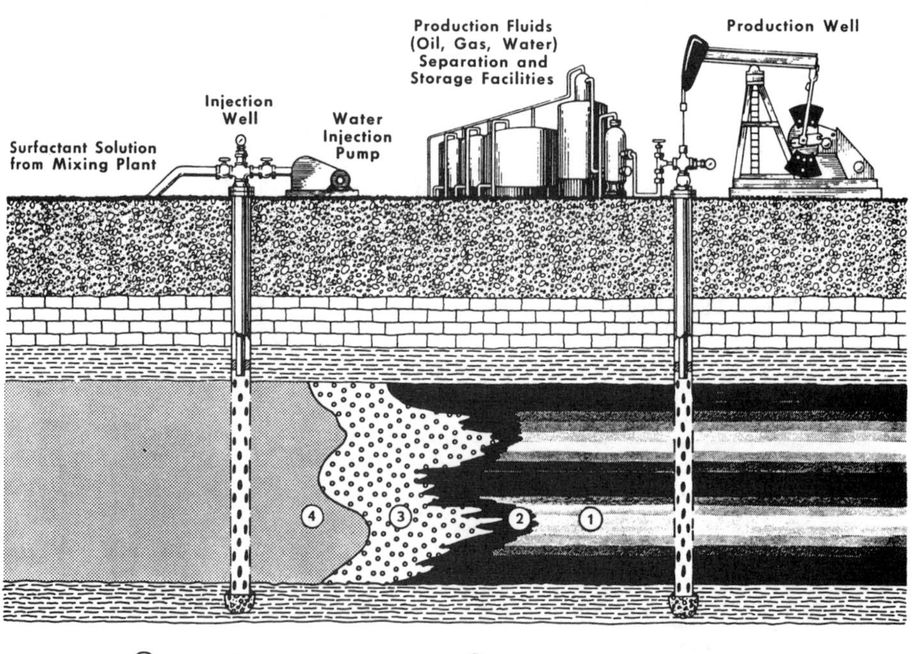

Figure 12 Schematic representation of a surfactant flooding operation.

The surfactant slug displaces the majority of the oil from the reservoir volume contacted, forming a flowing oil/water bank that is propagated ahead of the surfactant slug. The principal factors that influence the surfactant slug design are interfacial properties, slug mobility in relation to the mobility of the oil/water bank, the persistence of acceptable slug properties and slug integrity in the reservoir, and cost.

The surfactant slug is followed by a slug of water containing polymer in solution. The polymer solution is injected to preserve the integrity of the more costly surfactant slug and to improve sweep efficiency. Both of these goals are achieved by adjusting the polymer solution viscosity, in relation to the viscosity of the surfactant slug, in order to obtain a favorable mobility ratio. The polymer solution is then followed by injection of drive water, which continues until the project is completed.

Each reservoir has unique fluid and rock properties and specific chemical systems must be designed for each individual application. The chemicals used, their concentrations in the slugs, and the slug sizes will depend on the specific properties of the fluids and the rocks involved and on economic considerations.

3. Alkaline Flooding

Alkaline flooding (Figure 13) adds inorganic alkaline chemicals such as sodium hydroxide, sodium carbonate, or sodium orthosilicates to floodwater to enhance oil recovery by one or more of the following mechanisms: interfacial tension reduction, spontaneous emulsification, or wettability alteration. These mechanisms rely on the in situ formation of surfactants during the neutralization of petroleum acids in the crude oil by the alkaline chemicals in the displacing fluids.

Although emulsification in alkaline flooding processes decreases injection fluid mobility to a certain degree, emulsification alone may not provide adequate sweet efficiency. Sometimes polymer is included as an ancillary mobility control chemical in an alkaline water flood to augment any mobility ratio improvements due to alkaline-generated emulsions.

B. Miscible Methods

Miscible floods using carbon dioxide, nitrogen, or hydrocarbons as miscible solvents have their greatest potential for enhanced recovery of low-viscosity oils. Commercial hydrocarbon miscible floods have been operated since the 1950s. CO_2-miscible flooding on a large scale is relatively recent and is expected to make the most significant contribution to miscible enhanced recovery in the future.

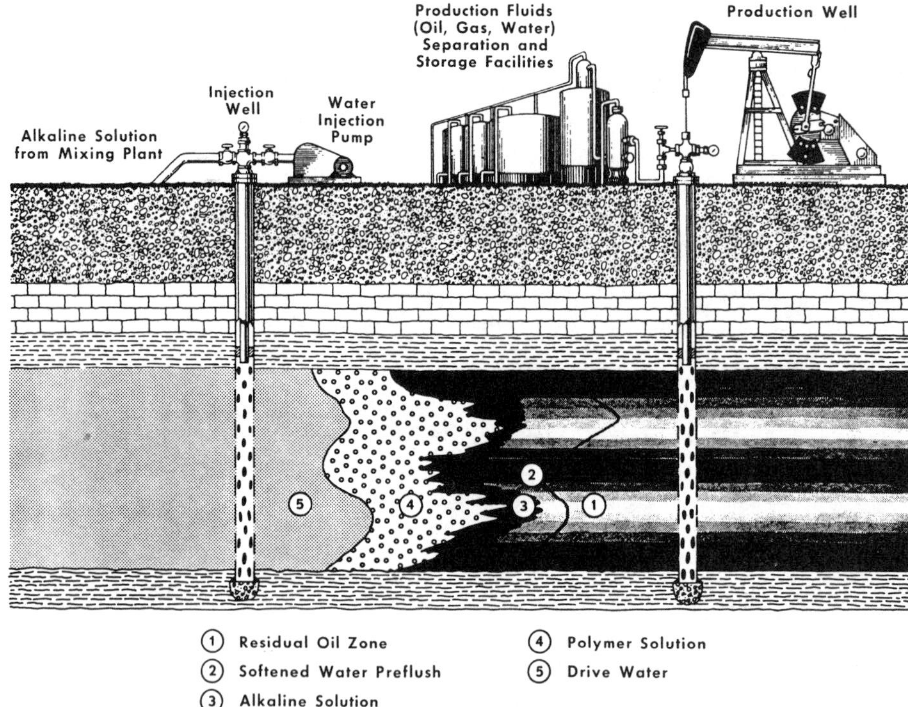

Figure 13 Schematic representation of an alkaline flooding operation.

1. CO_2 Miscible Flooding

CO_2 is capable of miscibly displacing many crude oils, thus permitting recovery of most of the oil from the reservoir rock that is contacted (Figure 14). The CO_2 is not miscible with the oil initially. However, as CO_2 contacts the in situ crude oil, it extracts some of the hydrocarbon constituents of the crude oil into the CO_2, and CO_2 is dissolved into the oil. Miscibility is achieved at the displacement front when no interfaces exist between the hydrocarbon-enriched CO_2 mixture and the CO_2-enriched in situ oil. Thus, by a dynamic (multiple-contact) process involving interphase mass transfer, miscible displacement overcomes the capillary forces that otherwise trap oil in pores of the rock.

The reservoir operating pressure must be kept at a high enough level to develop and maintain a mixture of CO_2 and extracted hydrocarbons that at reservoir temperature will be miscible with the crude oil. Impurities in the CO_2 stream, such as nitrogen or methane, increase the pressure required for miscibility. Dispersive mixing due to reservoir heterogeneity and diffusion tends to locally alter and destroy the miscible composition, which must then be regenerated by additional extraction of hydrocarbons. In field applications, there may actually be both miscible and near-miscible displacements proceeding simultaneously in different parts of the reservoir.

The volume of CO_2 injected is specifically chosen for each application, and usually ranges from 20 to 40% of the reservoir pore volume. In the later stages of the injection program, CO_2 may be driven through the reservoir by water or a lower cost inert gas. To achieve higher sweep efficiency, water and CO_2 are often injected in alternate cycles.

In some applications, particularly in carbonate (limestone, dolomite, chert) reservoirs where it is likely to be used most frequently, CO_2 may prematurely break through to producing wells. When this occurs, remedial action using mechanical controls in injection and production wells may be taken to reduce CO_2 production. However, substantial CO_2 production is considered normal. Generally, this produced CO_2 is reinjected, often after processing, to recover valuable light hydrocarbons.

For some reservoirs, miscibility between CO_2 and oil cannot be achieved but CO_2 can still be used to recover additional oil. CO_2 swells crude oils, thus increasing the volume of pore space occupied by the oil and reducing the quantity of oil trapped in the pores. It also reduces oil viscosity. Both effects improve the mobility of the oil. CO_2-immiscible flooding has been demonstrated in both pilot and commercial projects, but overall it is expected to make a relatively small contribution to enhanced oil recovery.

2. Other Processes

Hydrocarbon gases and condensates have been used for over 100 commercial and pilot miscible floods. Depending on the composition of the injected stream and the reservoir crude oil, the mechanism for achieving miscibility with reservoir oil can be similar to that obtained with CO_2 (dynamic or multiple-contact miscibility), or the miscible solvent and in situ oil may be miscible initially (first contact miscibility). Except in special circumstances, these light hydrocarbons are generally too valuable to be used commercially.

Nitrogen and flue gases have also been used for commercial miscible floods. Minimum miscibility pressures for these gases are usually higher than for CO_2, but in high-pressure, high-temperatures reservoirs where miscibility can be achieved, these gases may be a cost-effective alternative to CO_2.

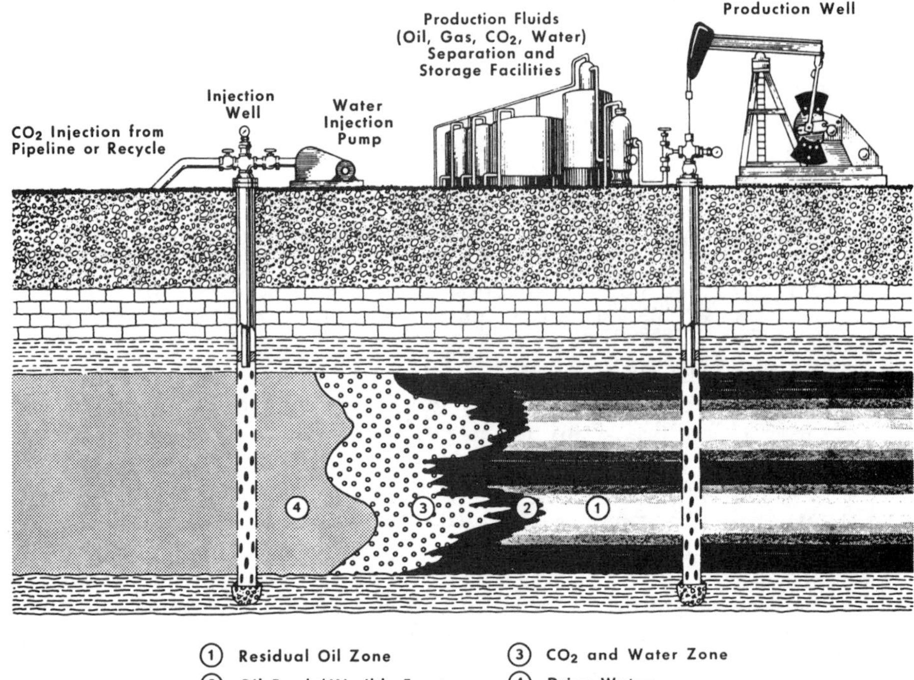

Figure 14 Schematic representation of a CO_2-miscible flooding operation.

C. Thermal Methods

Thermal EOR processes add heat to the reservoir to reduce oil viscosity and/or to vaporize the oil. In both instances, the oil is made more mobile so that it can be more effectively driven to producing wells. In addition to adding heat, these processes provide a driving force (pressure) to move oil to producing wells.

Thermal recovery methods include cyclic steam injection, steam flooding, and in situ combustion. The steam processes are the most advanced of all EOR methods in terms of field experience, and thus have the least uncertainty in estimating performance, provided that a good reservoir description is available. Steam processes are most often applied in reservoirs containing viscous oils and tars, usually in place of, rather than following, secondary or primary methods. Commercial application of steam processes has been underway since the early 1960s. In situ combustion has been field-tested under a wide variety of reservoir conditions, but few projects have proved economic and advanced to commercial scale.

1. Steam Injection

Steam injection has been commercially applied since the early 1960s and generally occurs in two steps: (1) steam stimulation of production wells (i.e., direct steam stimulation; Figure 15) and (2) steam drive by steam injection to increase production from other wells (i.e., indirect steam stimulation; Figure 16).

In cases where there is some natural reservoir energy, steam stimulation normally precedes steam drive. In steam stimulation, heat is

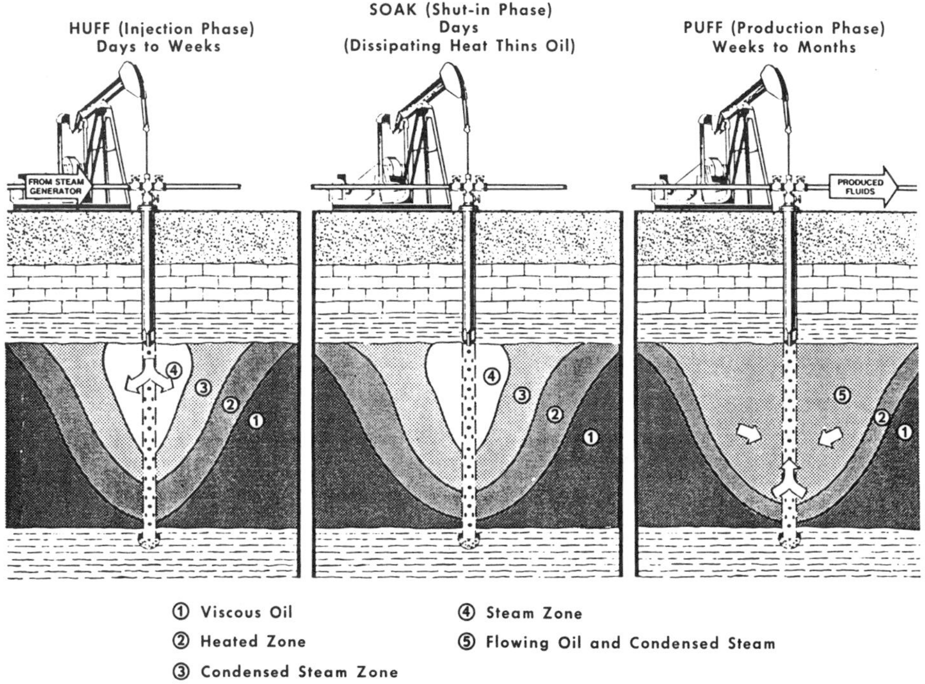

Figure 15 Schematic representation of a cyclic steam stimulation operation.

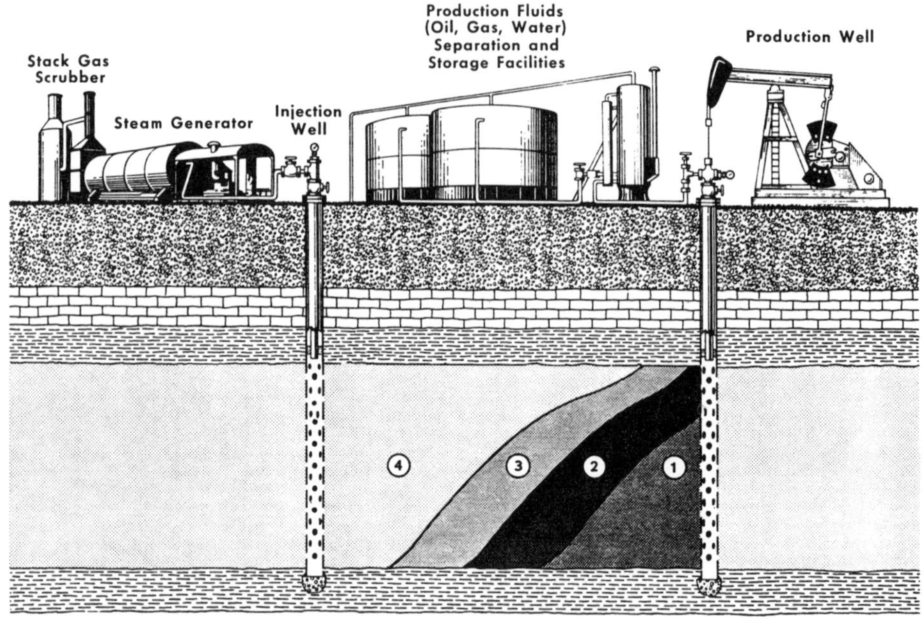

Figure 16 Schematic representation of a steam flooding operation.

applied to the reservoir by the injection of high-quality steam into the production well. This cyclic process, also called "huff and puff" or "steam soak," uses the same well for both injection and production. The period of steam injection is followed by production of reduced viscosity oil and condensed steam (water). One mechanism aiding production is the flashing of hot water (originally condensed from steam injected under high pressure) back to steam as pressure is lowered when a well is put back on production.

When natural reservoir drive energy is depleted and productivity declines, most cyclic steam injection projects are converted to steam drives. In some projects, producing wells are periodically steam-stimulated to maintain high production rates. Normally, steam drive projects are developed on relatively close well spacing to achieve thermal communication between adjacent injection and production wells. To date, steam methods have been applied almost exclusively in relatively thick reservoirs containing viscous crude oils.

2. In Situ Combustion

In situ combustion is normally applied to reservoirs containing low-gravity oil but has been tested over perhaps the widest spectrum of conditions of any EOR process. Heat is generated within the reservoir by injecting air and burning part of the crude oil. This reduces the oil viscosity and partially vaporizes the oil in place. The oil is driven forward by a combination of steam, hot water, and gas drive. The relatively small portion of the oil that remains after these displacement mechanisms have acted becomes the fuel for the in situ combustion process. Production is obtained from wells offsetting the injection locations. In some applications, the efficiency of the total in situ combustion operation can be improved by alternating water and air injection (see Figure 17). The injected water tends to improve

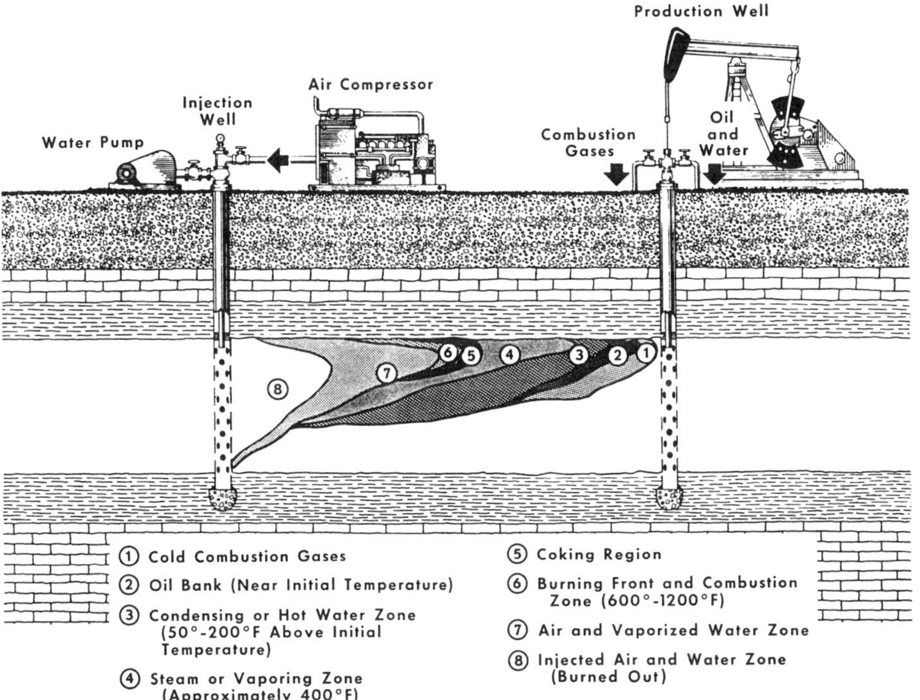

Figure 17 Schematic representation of an in situ combustion operation.

the utilization of heat by transferring heat from the rock behind the combustion zone to the rock immediately ahead of the combustion zone.

The performance of in situ combustion is predominantly determined by four factors: the quantity of oil that initially resides in the rock to be burned; the quantity of air required to burn the portion of the oil that fuels the process; the distance to which vigorous combustion can be sustained against heat losses; and the mobility of the air or combustion product gases. In many otherwise viable field projects, the high gas mobility has limited recovery through its adverse effect on the areal or vertical sweep efficiency of the burning front. Due to the density contrast between air and reservoir liquids, the burning front tends to override the reservoir liquids. To date, combustion has been most effective for the recovery of viscous oils in moderately thick reservoirs where reservoir dip and continuity promote effective gravity drainage, or where operational factors permit close well spacing.

D. Product Quality

It must also be remembered that in any field where primary production is followed by a secondary or enhanced production method, there will be noticeable differences in properties between the produced fluids (Table 9). While the differences in elemental composition may not reflect these differences to any great extent, more significant differences will be evident from an inspection of the physical properties. One issue that arises from the physical property data is that such oils may be outside the range of acceptability for refining techniques other than thermal options. In addition, overloading of thermal process units will increase as the proportion of the heavy oil in the refinery feedstock increases. Obviously, there is the need for more and more refineries to accept larger proportions of heavy crude oils as the refinery feedstock *and* have the capabilities of processing such materials.

Several such processing options are under investigation/development (Chap. 8, Sec. IV) that may well give the refineries this capability. It is to be hoped that their inclusion on-stream will be timely.

VII. TRANSPORTATION

Crude oil and its products are usually transported into and out of a refinery by pipeline, although tankers also play an important role in the movement of oil from the well to the refinery; railroad cars and motor vehicles are also used to a large extent for the transportation of petroleum products (Speight, 1980). In most instances, serious attempts are made to remove extraneous material from the crude oil prior to transportation.

Fluids produced from a well are seldom pure crude oil; in fact, a variety of materials may be produced by oil wells in addition to liquid and gaseous hydrocarbons. The natural gas itself may contain as impurities one or more nonhydrocarbon substances. The most abundant of these impurities is hydrogen sulfide, which imparts a noticeable odor to the gas.

By far the most abundant extraneous material is water. Many wells, especially during their declining years, produce vast quantities of saltwater, and disposing of it is both a serious and expensive problem. Furthermore, the brine may be corrosive, which necessitates frequent replacement of casing, pipe, and valves, or it may be saturated so that the salts tend to precipitate upon reaching the surface. If the reservoir rock is poorly cemented sandstone, large quantities of sand are produced along with the oil and gas and can cause considerable damage to pipes and fittings.

The pipeline used for petroleum transportation may be anywhere from 2 to 36 (or even larger) inches in diameter and may cover many thousands of miles. The tanker fleet that is owned, or used, by the world's oil companies is also responsible for the movement of a considerable portion of the world's crude oil. Indeed, we have recently entered the age of the "supertanker" where weights in excess of 250,000 tons are not uncommon. It is, however, well known that these ships received the majority of their attention because of the accidents, which appear to plague vessels of these sizes, and the resultant spillage of many thousands or even millions of gallons of crude oil to the detriment of the surrounding environment.

One of the first steps to be taken in the preparation of crude oil for transportation is the removal of excessive quantities of water. Crude oil at the wellhead usually contains emulsified water in proportions that may reach amounts approaching 80–90%. It is generally required that crude oil to be transported by pipeline contain substantially less water than may appear in the crude at the wellhead. In fact, water contents of 0.5–2.0% have been specified as the maximum tolerable amount in a crude oil that is to be moved by pipeline. It is therefore necessary to remove the excess water from the crude oil prior to transportation.

A crude oil-water emulsion may be either a dispersion of small globules of water in the oil or a dispersion of small globules of oil in the water. Whatever the form of the emulsion, it must be broken to cause an effective separation of the two components.

Table 9 Differences Between Primary and Secondary/Enhanced Recovered Oils (Thomas et al., 1987)

Property	Mobil steam flood		Santa Fe steam flood		Santa Fe fire flood	
	Primary	Thermal	Primary	Thermal	Primary	Thermal
Carbon, wt%	85.4	85.4	86.9	86.9	85.7	85.8
Hydrogen	11.2	11.2	11.7	11.4	11.1	11.4
Nitrogen	1.2	0.9	0.8	0.9	1.0	0.9
Sulfur	2.5	2.1	1.2	1.3	2.0	2.2
Oxygen (by diff.)	—	0.4	—	—	0.2	—
H/C ratio	1.56	1.56	1.60	1.57	1.54	1.58
Nickel, ppm	51	58	49	66	69	30
Vanadium, ppm	127	104	22	37	119	125

Iron, ppm	8	12	30	29	17	80
Gravity, °API	10.4	24.5	15.0	12.9	10.4	11.2
Specific gravity, 16°C/16°C	0.997	0.907	0.966	0.980	0.997	0.992
Viscosity, cp						
16°C	777,000	210,000	7,250	69,600	536,000	258,000
38°C	30,200	2,210	786	8,650	20,500	12,300
60°C	2,830	1,210	174	541	1,950	1,430
Pour point, °C	18	13	-4	10	13	13
Acid number, mg KOH/g	3.79	3.39	3.20	2.48	3.41	3.30
Conradson carbon residue, wt%	10.08	17.08	9.39	16.00	15.17	17.50
Ash, wt%	<0.01	<0.01	<0.01	<0.01	<0.01	<0.01
Water, wt%	<0.1	<0.1	<0.2	<0.1	<0.1	<0.2

In an emulsion, the globules of one phase are usually surrounded by a thin film of an emulsifying agent which prevents them from congregating into large droplets and, thereafter, settling. In the case of an oil-water emulsion, the emulsifying agent may be part of the heavier (asphaltic) more polar constituents. The film may be broken either mechanically, electrically, or by the use of demulsifying agents (Chap. 5, Sec. I). Thus, the proportion of water in the oil can be reduced to the specified amounts, thereby rendering the crude oil suitable for transportation.

The transportation of crude oils may be further simplified by blending crude oils from several wells and thereby homogenizing the feedstock to the refinery. It is, however, usual practice to blend crude oils of similar characteristics although fluctuations in the properties of the individual crude oils may cause significant variations in the properties of the blend over a period of time. However, the technique of blending several crudes prior to transportation, or even after transportation but prior to refining, may eliminate the frequent need to change the processing conditions that would perhaps be required to process each of the crudes individually.

2

Terminology and Classification

I	Nomenclature and Terminology	51
	A. Native Materials	53
	B. Manufactured Materials	58
	C. Derived Materials	62
II	Classification	63
	A. Compound Type	64
	B. Correlation Index	66
	C. Density	67
	D. Viscosity-Gravity Constant	68
	E. Characterization Factor	69

I. NOMENCLATURE AND TERMINOLOGY

Petroleum is an extremely complex mixture of hydrocarbon compounds and other compounds containing nitrogen, oxygen, and sulfur as well as metallic (porphyrinic) constituents. The actual boundaries of petroleum can only be arbitrarily defined in terms of boiling point and carbon number (Figure 1). The material is so diverse that petroleum from different sources will exhibit different boundary limits. Thus, it is not surprising that petroleum has been difficult to define precisely.

The nomenclature of petroleum and related materials is of such an extensive and diverse character that many schemes have been

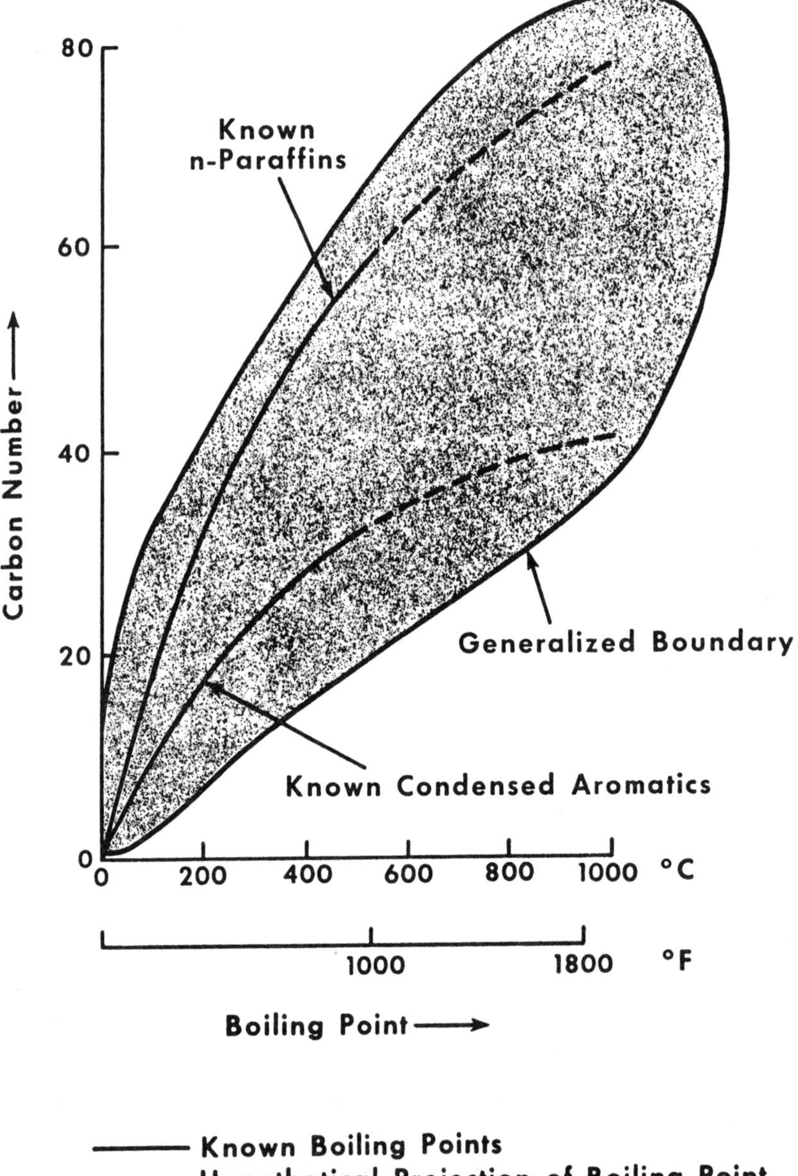

Figure 1 The physical boundaries for petroleum can be arbitrarily defined using carbon number and boiling point data.

suggested but, fortunately, only a few individual terms have survived. For the most part, it is preferable to place petroleum and related materials into two major classes, i.e., those that are of natural origin and those that are manufactured, with a third class of materials being those that are derived from the natural or manufactured products (Table 1; Speight, 1980).

A. Native Materials

1. Petroleum

Petroleum, and the equivalent term crude oil, covers a wide assortment of materials consisting of mixtures of hydrocarbons and other

Table 1 Petroleum and Related Materials Can be Divided into Various Class Subgroups

Natural materials	Manufactured materials	Derived materials
Petroleum	Wax	Oils
Heavy oil	Residuum	Resins
Mineral wax	Asphalt	Asphaltenes
Bitumen (native asphalt)	Tar	Carbenes
Asphaltite	Pitch	Carboids
Asphaltoid		
Bituminous rock		
Bituminous sand		

Notes:
1. Tar sands—a misnomer; tar is a product of coal processing.
2. Oil sands—also a misnomer but equivalent to usage of "oil shale."
3. Bituminous sands—more correct; bitumen is a naturally occurring asphalt.
4. Asphalt—a product of a refinery operation; usually made from a residuum.
5. Residuum—the nonvolatile portion of petroleum and often further defined as "atmospheric" (bp >350°C; >660°F) or vacuum (bp > 565°C; >1050°F).

compounds containing variable amounts of sulfur, nitrogen, and oxygen which may vary widely in volatility, specific gravity, and viscosity (Tables 2 and 3). The combined metals, notably those of vanadium and nickel, usually occur in the more viscous crude oils in amounts up to several thousand parts per million (Gruse and Stevens, 1960).

Table 2 Characteristics of Various Residua Produced from Tia Juana (Venezuela) Light Crude Oil

Characteristic			Boiling range		
		°F	>650	>950	>1050
		°C	>345	>510	>656
Yield on crude, vol %	100.0		48.9	23.8	17.9
Gravity, °API	31.6		17.3	9.9	7.1
Specific gravity	0.8676		0.9509	1.007	1.0209
Sulfur, wt %	1.08		1.78	2.35	2.59
Carbon residue (Conradson), wt %	—		9.3	17.2	21.6
Nitrogen, wt %	—		0.33	0.52	0.60
Pour point, °F	-5		45	95	120
Viscosity:					
Kinematic, cST	10.2		890	—	—
@100°F	10.2		890	—	—
@210°F	—		35.0	1010	7959
Furol (SFS) sec					
@122°F	—		172	—	—
@210°F	—		—	484	3760
Universal (SUS) sec @210°F	—		165	—	—
Metals:					
Vanadium, ppm	—		185	—	450
Nickel, ppm	—		25	—	64
Iron, ppm	—		28	—	48

Table 3 Generalized Ranges for the Bulk Fractions in Crude Petroleum, Heavy Oil, and Residua

	Range of composition (w/w %)			Carbon residue (w/w %)
	Asphaltenes	Resins	Oils	
Petroleum	<0.1–12.0	3–22	67–97	0.2–10.0
Heavy oil	11–45	14–39	24–64	10.0–22.0
Residua	11–29	29–39	?--49	18.0--32.0

Note: See also Chap. 4, Secs. II and III.

2. Heavy Oil

The definition of a heavy oil is quite arbitrary but the term is generally applied to a petroleum that has an API gravity of less than 20° and usually, but not always, has a sulfur content higher than 2% by weight. Furthermore, in contrast to conventional crude oils, heavy oils are darker in color and may even be black (Speight, 1981).

The term "heavy oil" is often arbitrarily used to describe both the heavy oils, which require thermal stimulation of recovery from the reservoir, and bituminous sands (tar sands), wherein the heavy bituminous product is recovered by a mining operation (Figure 2).

3. Mineral Wax

Mineral waxes occur naturally as yellow-to-dark brown, solid substances and are composed largely of paraffins; fusion points vary from 60°C (140°F) to as high as 95°C (205°F). They are usually found in association with considerable mineral matter as a filling in veins and fissures or as an interstitial material in porous rocks. The similarity in character of these native products is substantiated by the fact that, with minor exceptions where local names have prevailed, the original term ozokerite (ozocerite) has served without notable ambiguity for mineral wax deposits (Gruse and Stevens, 1960).

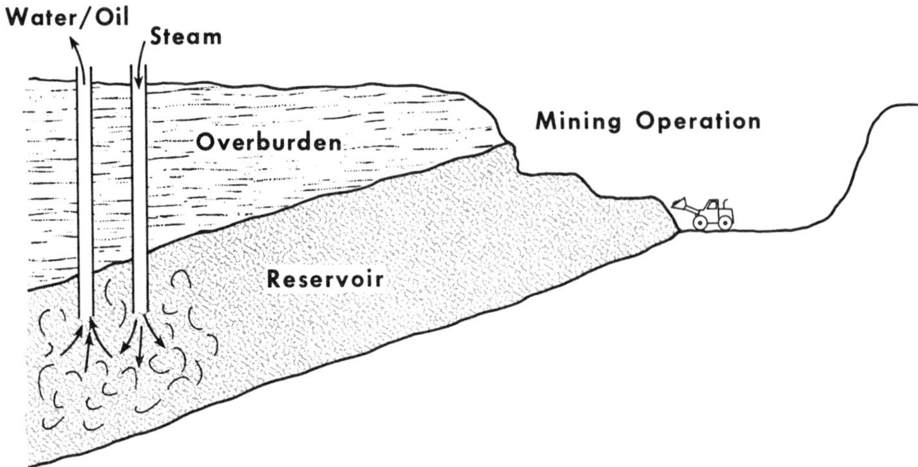

Figure 2 Heavy oils may be recovered by thermal methods such as steam stimulation whereas tar sand bitumens are currently recovered commercially by mining operations.

4. Native Asphalt (Bitumen)

Native asphalts (bitumens) include a wide variety of reddish brown-to-black materials of semisolid, viscous-to-brittle character which can exist in nature with no mineral impurity or with 50% or more mineral matter. Frequently, bitumen is found filling pores and crevices of sandstones, limestones, or argillaceous sediments, in which case the organic and associated mineral matrix is known as rock asphalt (Abraham, 1945; Hoiberg, 1960).

It is incorrect to refer to native bituminous materials as tar and/or pitch. Although the word "tar" is descriptive of the black, heavy, bituminous material involved, it is best to avoid its use with respect to natural materials and to restrict the meaning to the volatile, or near-volatile, products produced in the destructive distillation of bituminous or other organic substances. Similarly, pitch is the distillation residue of the various types of tars.

5. Asphaltite

Asphaltites are a variety of naturally occurring, dark brown-to-black, solid, nonvolatile bituminous substances that are differentiated from native asphalts primarily by their high content of material insoluble in *n*-pentane (asphaltene) or other liquid hydrocarbons (Chap. 4, Sec. II). The resultant high-temperature fusion (approximate range: 115–330°C; 240–625°F) is characteristic. The names applied to the two rather distinct types included in this group are now accepted and used for the most part without ambiguity. Gilsonite (originally known as uintaite from the locations of the deposits in the Uinta Basin of western Colorado and eastern Utah) is characterized by a bright luster, conchoidal fracture, and with a fixed carbon in the range 10–20%. The second type in this category is grahamite, which is very much like gilsonite in external characteristics but is distinguishable by its black streak, relatively high fixed carbon value (35–55%), and high temperature of fusion which is accompanied by a characteristic intumescence (Speight, 1980).

A third, but rather broad, category of asphaltites includes a group of materials known as glance pitch which physically resemble gilsonite but have certain of the properties of grahamite. They have been referred to as intermediates between the two but it is considered more likely that they represent an intermediate stage (or stages) between native asphalt and grahamite.

6. Asphaltoid

Asphaltoides are a group of brown-to-black, solid materials that are differentiated from the asphaltites by their infusibility and low solubility in carbon disulfide. These substances have also been variously referred to as asphaltic pyrobitumens since under isothermal conditions they decompose to bitumen-like materials. The term pyrobitumen is less correct than the term asphaltoids, since these materials closely resemble the asphaltites (Speight, 1980).

There is still some confusion regarding the classification of asphaltoids although the existence of four types is recognized: elaterite, wurtzilite, alberltite, and impsonite (in order of increasing density and fixed carbon content).

7. Bituminous Rock

Bituminous rocks are those rocks in which the bituminous material is found impregnating relatively shallow sand, sandstone, and limestone strata, or as a filling in veins and fissues in fractured rocks. The deposits contain as much as 20% bituminous material and, if the organic material in the rock matrix is a natural asphalt, it is usual to refer to the deposit as a rock asphalt, to distinguish it from those native asphalts that are relatively mineral-free. If the material is of the asphaltite or asphaltoid type, the corresponding terms should be used, i.e., rock asphaltite or rock asphaltoid (Speight, 1980).

B. Manufactured Materials

1. Wax

The term paraffin wax is restricted to the colorless, translucent, highly crystalline material obtained from the light-lubricating fractions of paraffinic crude oils (wax distillates). The commercial products melt in the approximate range 50–65°C (120–150°F). Dewaxing of heavier fractions leads to semisolid materials known as petrolatums and solvent deoiling of the petroleum or of heavy, waxy residua results in dark-colored waxes of sticky, plastic-to-hard nature. The waxes are composed of fine crystals and contain, in addition to *n*-paraffins, appreciable amounts of isoparaffins and long-chain cyclics. Melting points of the commercial grades are in the 70–90°C (160–195°F) range (Brooks et al., 1954).

Highly paraffinic waxes are also produced from peat, lignite, or shale oil tar and paraffin waxes, known as ceresins, also may be prepared from ozocerite, which is quite similar to the waxes from petroleum.

2. Residuum (Residua)

A residuum is the residue obtained from petroleum after nondestructive distillation has removed all of the volatile materials. The temperature of the distillation is usually maintained below 350°C (660°F) since the rate of thermal decomposition of petroleum constituents is minimal below this temperature but the rate of thermal decomposition of petroleum constituents is substantial above 350°C (Speight, 1981).

Residua are black, viscous materials and are thus obtained by atmospheric or vacuum distillation of a crude oil (Figure 3). They may be liquid at room temperature (generally atmospheric residua) or almost solid (generally vacuum residua) depending upon the nature of the crude oil (Figure 4). When a residuum is obtained from a crude oil but thermal decomposition has commenced, it may be more correct to refer to this product as pitch.

The differences between a parent petroleum and the residua are due to the relative amounts of various constituents present (Figure 5). These constituents (Chap. 5) are removed or remain by virtue

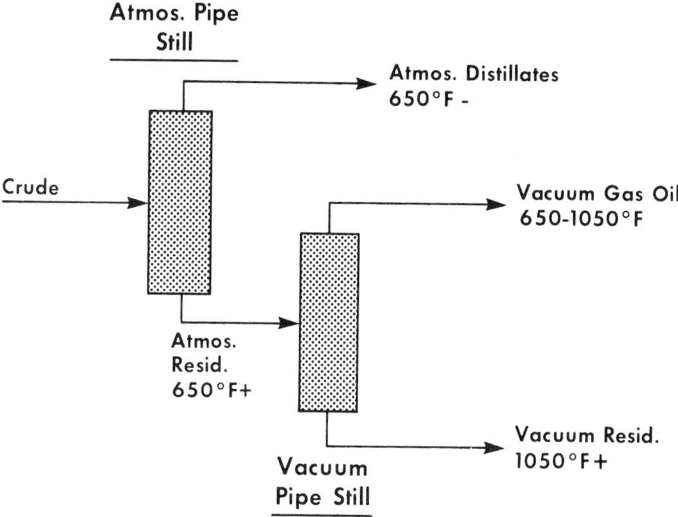

Figure 3 Residua are obtained by removal of the volatile constituents of the feedstock at atmospheric pressure or at reduced pressure, but the properties of the residua differ considerably with the "end point" (see Table 2 in this chapter).

of their relative volatility. In the simplest sense, an asphalt (propane) consists of the majority of the nonvolatile materials originally in the petroleum.

3. Asphalt

Asphalts are prepared from petroleum (Figure 6) and these products resemble the native asphalts closely. It is recommended to distinguish between the two types of asphalts by use of the qualifying terms native and petroleum—except, of course, in those occasional instances in which the source is other than petroleum, e.g., wurtzilite asphalt (Bland and Davidson, 1967).

When the asphalt has been produced simply by distillation of an asphaltic crude, the product can be referred to as residual, or straight run, petroleum asphalt. If the asphalt is prepared by solvent extraction of residua or by light hydrocarbon (propane) precipitation, or if blown or otherwise treated, the term should be modified accordingly to the quality of the product.

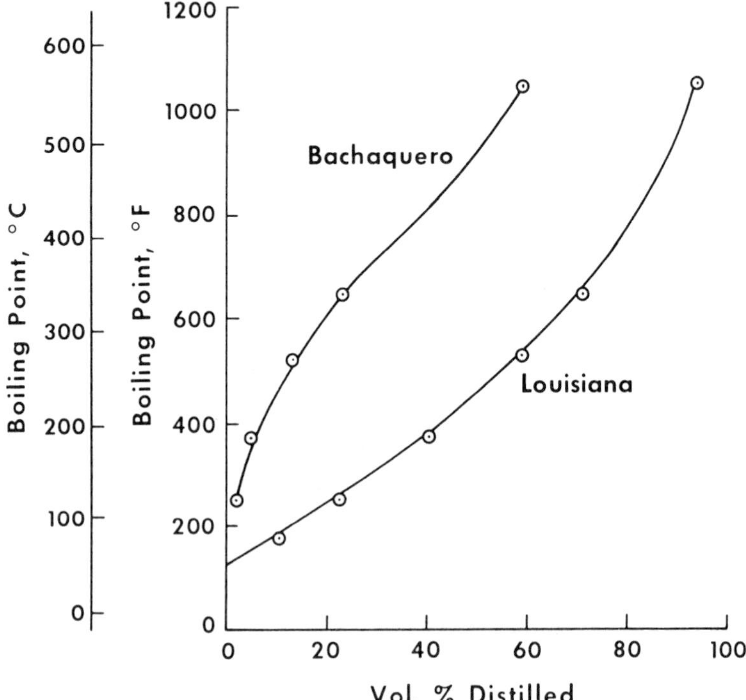

Properties of the 1050°F + Residua:

	Louisiana	Bachaquero
Gravity, API	12.1	2.8
Sulfur, wt. %	0.50	3.71
Nitrogen, wt. %	0.42	0.60
MNI, wt. %	10.5	20.0
Con. Carbon, wt. %	15.8	27.5
Nickel, ppm	26	100
Vanadium, ppm	19	900
Pour Point, °F	-	130

Figure 4 Different crude oils contain different amounts of residua and the properties of specific "cut point" residua may be different.

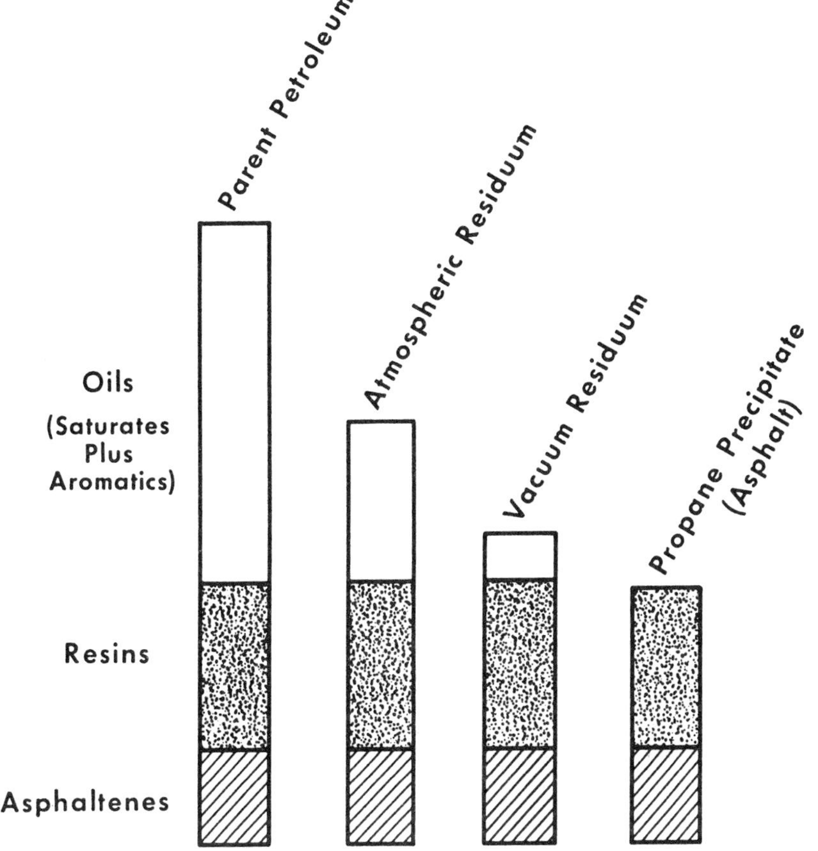

Figure 5 Schematic representation of the relationship between a crude oil, two residua from the crude oil, and the propane asphalt (see also Table 3 in this chapter).

4. Tar and Pitch

Tars are the result of the destructive distillation of many organic materials and are brown-to-black, oily, viscous liquids. Tar is most commonly produced from bituminous coal and is generally understood to refer to the coal product, but it is advisable to specify coal tar if there is the possibility of ambiguity. The most important factor in determining the yield and character of the coal tar is the carbonizing temperature (Speight, 1983; Chap. 20, Sec. III).

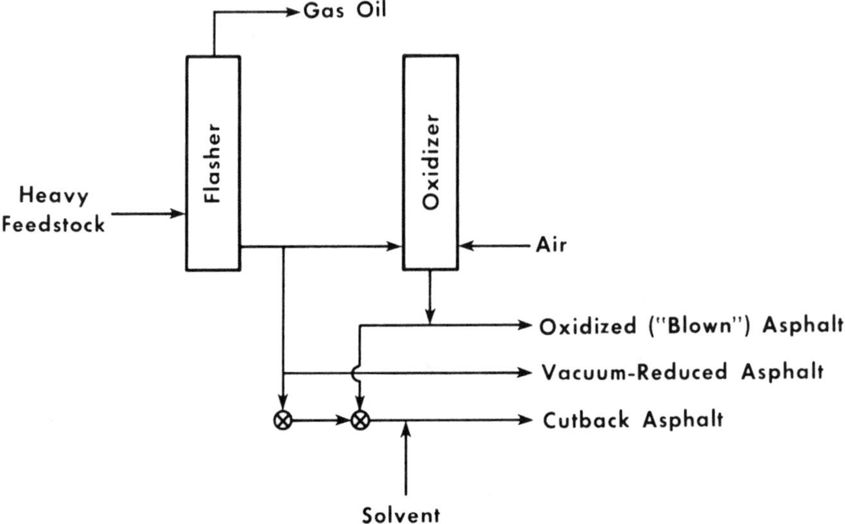

Figure 6 Simplified representation of the preparation of asphalts from petroleum residua.

C. Derived Materials

1. Asphaltenes, Carbenes, Carboids

All of the petroleum-related materials mentioned in the preceding sections are capable of being separated by solvents into several fractions which are sufficiently distinct in character to warrant the application of individual names (Pfeiffer, 1950).

Treatment of petroleum, petroleum residua, or bituminous materials with a low-boiling liquid hydrocarbon will result in the separation of brown-to-black "powdery" materials known as asphaltenes. The reagents for effecting this separation are *n*-pentane and *n*-heptane, although other light paraffins have been used (vanNes and Westen, 1951; Speight, 1981).

Asphaltenes separated from crude oils and their residua and most bitumens (native asphalts) dissolve readily in benzene, carbon disulfide, chloroform, or other chlorinated hydrocarbon solvents. However, in the case of the more complex native materials, or petroleum residua which have been heated intensively or for prolonged periods, the *n*-pentane-insoluble (or *n*-heptane-insoluble) fraction may not dissolve completely. The asphaltene fraction has therefore been restricted to that of the insoluble portion which will dissolve in these solvents. The benzene insoluble fraction is collectively referred to as carbene and carboid and the fraction soluble in carbon disulfide (or pyridine) but insoluble in benzene is defined as carbene.

2. Resins and Oils

The portion of a petroleum soluble in, for example, pentane or haptane is often referred to as maltene, and can be further subdivided by percolation through any surface-active material, such as fuller's earth or alumina, to yield an oil fraction and a more strongly adsorbed, deep red-to-brown, semisolid material known as resin. Several other ways have been proposed for separating the resin fraction, e.g., a common procedure involves precipitation by liquid propane (Speight, 1980).

The fraction precipitated by propane may also contain acidic material, often referred to as asphaltic or asphaltogenic acids. These acids can probably be regarded simply as cyclic and noncyclic organic acids of high molecular weight. These acids usually appear in the resin fraction, but if they have been removed or are absent, the resins are said to be neutral. Removal of the resins from maltenes leaves the oils fraction.

The resins and oils (maltenes) have also been referred to as petrolenes thereby adding further confusion to this "system" of nomenclature. However, it has been accepted by many workers in petroleum chemistry that the term *petrolenes* be applied to that part of the pentane-soluble material which is low boiling (<300°C/<570°F/760 mm) and can be distilled without thermal decomposition. Consequently, the term *maltenes* is now arbitrarily assigned to the pentane-soluble portion of petroleum which is relatively high boiling (>300°C/760 mm; Speight, 1980).

II. CLASSIFICATION

On a variety of terms, petroleum can be classified as a naturally occurring hydrocarbon mixture (Figure 7) but for the present purpose it is necessary to attempt to classify the various petroleums precisely.

Many attempts have been made to classify petroleums but it is unfortunate that no successful universal method of classification has yet evolved. The original methods of classification arose because of commercial interests and were a means of providing refinery operators with a rough guide to processing conditions. Therefore, systems based on a superficial inspection of a physical property such as specific gravity or API (Baumé) gravity (Chap 3, Sec. III) are easily

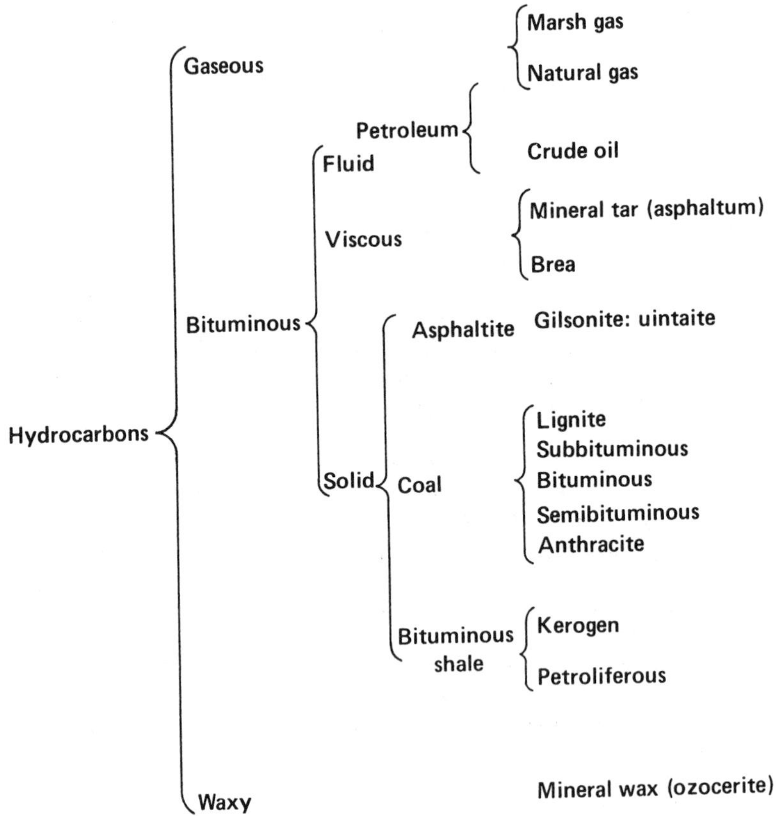

Figure 7 Generalized classification scheme for and interrelationship of hydrocarbon resources.

applied and are actually used to a large extent in expressing the quality of crude oils.

A. Compound Type

The oldest and most widely used classification of petroleum distinguishes between oils either on a paraffin or on an asphalt base (Figure 8) and arose because some oils separate paraffin wax on cooling but other oils show no separation of paraffin wax on cooling. The

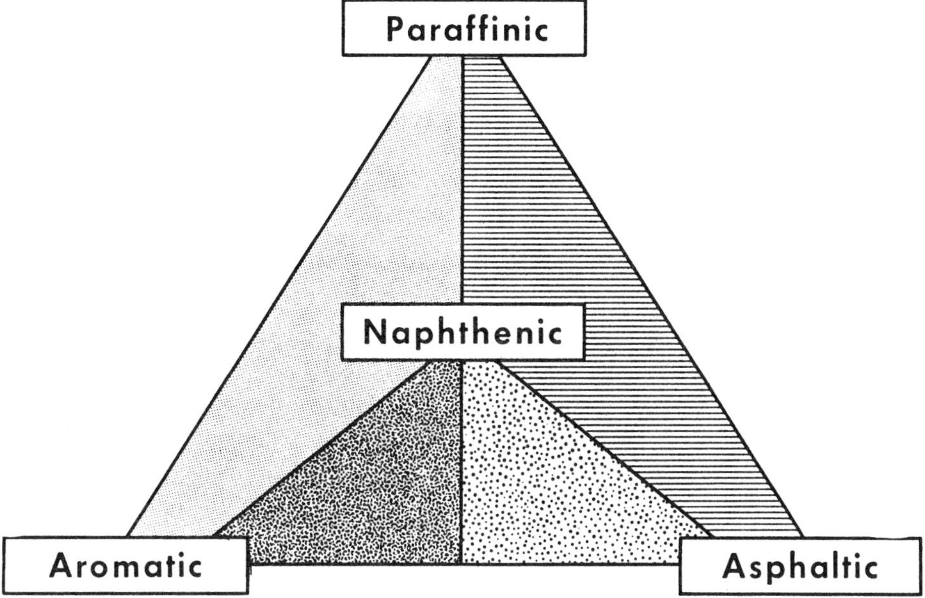

Figure 8 Simplified composition diagram for crude oils showing the different types of crude oils and a means of classification.

terms paraffin base and asphalt base petroleum were introduced and have remained in common use (vanNes and Westen, 1951).

The presence of paraffin wax is usually reflected in the paraffinic nature of the constituent fractions and a high asphaltic content corresponds with the so-called naphthenic properties of the fractions. As a result, the misconception that paraffin base crudes consist mainly of paraffins and asphalt base crudes mainly of cyclic (or naphthenic) hydrocarbons has arisen. In addition to paraffin and asphalt base oils, a mixed base had to be introduced for those oils which leave a mixture of asphaltic bitumen and paraffin wax as a residue by nondestructive distillation.

In practice, a distinction is often made between light and heavy crude oils (indicating the proportion of low-boiling material present) and which, in combination with the above distinction (paraffinic, asphaltic, etc.), doubles the number of possible classes.

An early attempt to give the classification system a quantitative basis suggested that a crude should be called asphaltic if the distillation residue contained less than 2% of wax and paraffinic if it contained more than 5%. A division according to the chemical composition of the 250–300°C (480–570°F) fraction was also suggested (Table 4) but the difficulty in using such a classification is that in the fractions boiling above 200°C (390°F), the molecular constituents can no longer be placed in one group because most of them are of a typically mixed nature.

B. Correlation Index

The correlation index (CI) developed by the U.S. Bureau of Mines is based on the plot of specific gravity versus the reciprocal of the boiling point in degrees Kelvin (°K = °C + 273) for pure hydrocarbons on which the line described by the constants of the individual members of the normal paraffin series is given a value of CI = 0 and a parallel line passing through the point for the values of benzene is given as CI = 100 (Figure 9), thus:

$$CI = 473.7d - 456.8 + \frac{48640}{K}$$

in which K, in the case of a petroleum fraction, is the average boiling point determined by the standard Bureau of Mines distillation method and d is the specific gravity (Gruse and Stevens, 1960).

Values for the index between 0 and 15 indicate a predominance of paraffinic hydrocarbons in the fraction; from 15 to 20 a predominance

Table 4 Petroleum Classification According to Chemical Composition

	Comp. of 250–300°C fraction				
Class of crude	% par.	% naphth.	% arom.	% wax	% asphalt
Paraffinic	46–61	22–32	12–25	1.5–10	0–6
Paraffinic-naphthenic	42–45	38–39	16–20	1–6	0–6
Naphthenic	15–26	61–76	8–13	Trace	0–6
Paraffinic-naphthenic-aromatic	27–35	36–47	26–33	0.5–1	0–10
Aromatic	0–8	57–78	20–25	0–0.5	0–20

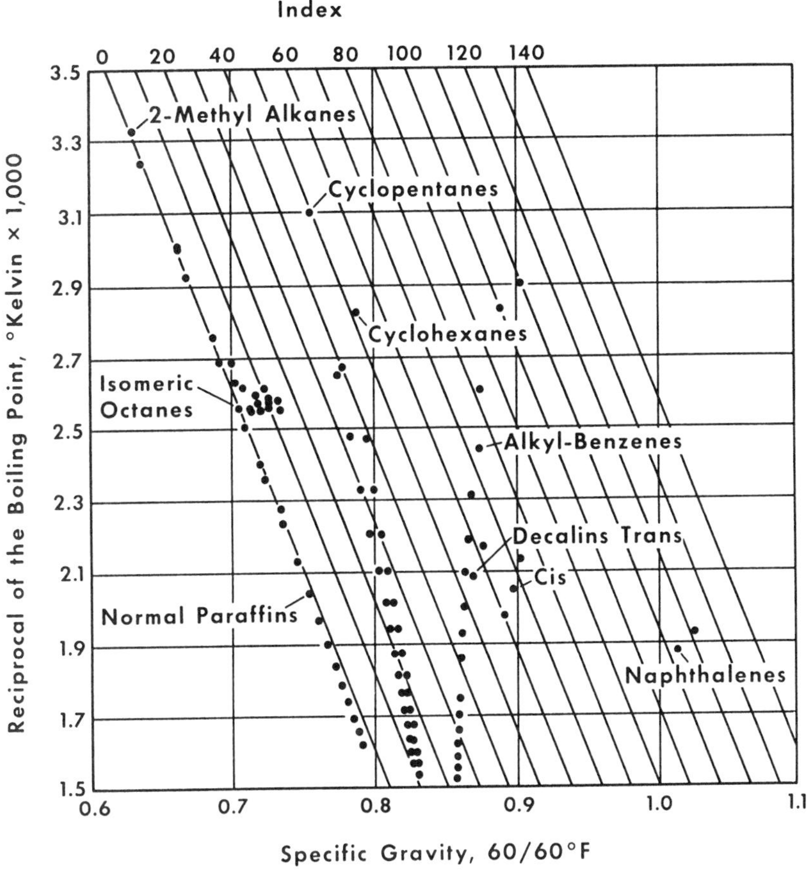

Figure 9 Reference data for the correlation index.

either of naphthenes or of mixtures of paraffins, naphthenes, and aromatics; an index value above 50 indicates predominantly aromatic character.

C. Density

Density (specific gravity) has been, since the early years of the industry, the principal and often the only specification of crude oil products and was taken as an index of the proportion of gasoline and, particularly, kerosene present (Gruse and Stevens, 1960). As long as only one kind of petroleum was in use the relations were approximately true, but as crude oils having other properties were discovered and came into use, the significance of density measurements disappeared. Nevertheless, crude oils of particular types are still rated by gravity. The use of density values has been advocated for application based on the American Petroleum Institute (API) gravity.

For many years, petroleum and heavy oils were very generally defined in terms of physical properties. For example, heavy oils were considered to be those petroleum-type materials that had gravity somewhat less than 20° API with the "heavier" oils falling into the API gravity range 10–15° (e.g., Cold Lake crude oil = 12° API) and tar sand bitumens falling into the 5–10° API range (e.g., Athabasca bitumen = 8° API). Residua would vary depending on the temperature at which distillation was terminated but usually vacuum residua were in the range 2–8° API (Speight, 1981).

A more formal classification system has been proposed through the United Nations Institute for Training and Research (UNITAR) and which depends on gravity and viscosity (Figure 10). This system affords a better classification of petroleum, heavy oils, and bitumen; the scale can also be used for residua or other heavy feedstocks.

D. Viscosity-Gravity Constant

The viscosity-gravity constant (vgc) parameter has been used to some extent as a means of classifying crude oils (Gruse and Stevens, 1960). This constant was one of the early indexes proposed to characterize (or classify) oil types:

$$\mathrm{vgc} = \frac{10d - 1.0752 \log (v - 38)}{10 - \log (v - 38)}$$

where d is the specific gravity and v is the Saybolt viscosity at 38°C (100°F). For oils so viscous that the low-temperature viscosity is difficult to measure, an alternative formula:

$$\mathrm{vgc} = \frac{d - 0.24 - 0.022 \log (v - 35.5)}{0.755}$$

was proposed in which the 99°C (210°F) Saybolt viscosity is used. The viscosity-gravity constant is of particular value in indicating a predominantly paraffinic or cyclic (naphthenic) composition. The lower the index number, the more paraffinic the stock, e.g., naphthenic lubricating oil distillates have vgc 0.876 while the raffinate obtained by solvent extraction of lubricating oil distillate has vgc $\sim$0.840.

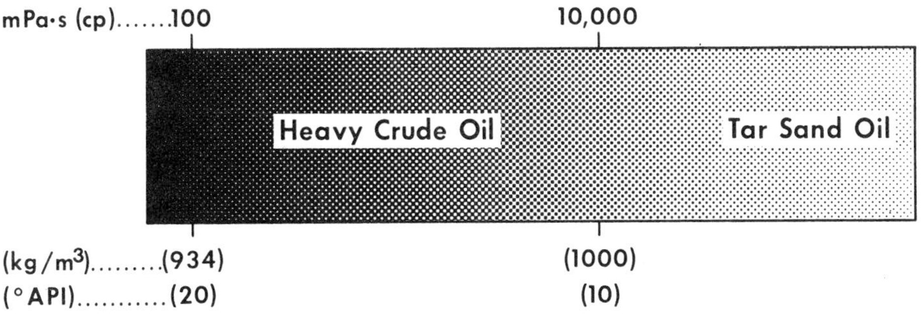

Type of crude	Characteristics
1. Conventional or "light" crude oil	Density-gravity range less than 934 kg/gm^3 (>20° API)
2. "Heavy" crude oil	Density-gravity range from 1000 kg/m^3 to more than 934 kg/m^3 (10° API to <20° API) Maximum viscosity of 10,000 mPa.s (cp)
3. "Extra-heavy" crude oil; may also include atmospheric residua (bp >340°C; >650°F)	Density-gravity greater than 1000 kg/m^3 (<10° API) Maximum viscosity of 10,000 mPa.s (cp)
4. Tar sand bitumen or natural asphalt; may also include vacuum residua (bp >510°C; >950°F)	Viscosity greater than 10,000 mPa.s (cp) Density-gravity greater than 1000 kg/m^3 (<10° API)

Figure 10 Classification of crude oils by density-gravity.

E. Characterization Factor

This factor evolved through the work of the Universal Oil Products organization and is defined by the formula:

$$K = \frac{\sqrt[3]{T_B}}{d}$$

where T_B is the average boiling point in degrees Rankine (degrees Fahrenheit + 460) and d is the specific gravity. This factor has been shown to be additive on a weight basis. It was originally devised to show the thermal cracking characteristics of viscous oils; thus, highly paraffinic oils have $K \sim 12.5-13.0$ while cyclic (naphthenic) oils have $K \sim 10.5-12.5$ (Gruse and Stevens, 1960).

3

Composition and Properties

I	Ultimate Composition	72
II	Chemical Composition	72
	A. Hydrocarbon Components	72
	B. Nonhydrocarbon Components	75
	C. Porphyrins	83
	D. Oil Shales and Coal	96
III	Physical Properties	97
	A. Density and Specific Gravity	99
	B. Sulfur Content	100
	C. Viscosity	100
	D. Volatility	109
	E. Carbon Residue	111
	F. Metals Content	111
	G. Aniline Point	114
	H. Liquefaction and Solidification	115
	I. Surface and Interfacial Tension	116
	J. Specific Heat	117
	K. Conductivity	117
	L. Refractive Index	117
IV	Use of the Data	118

Petroleum is not a uniform material. In fact, its composition can vary not only with the location and age of the oil field but also with the depth of the individual well; two adjacent wells may produce petroleum with markedly different characteristics. On a molecular basis, petroleum is a complex mixture of hydrocarbons plus organic compounds of sulfur, oxygen, and nitrogen as well as compounds containing metallic constituents, particularly vanadium, nickel, iron, and copper.

I. ULTIMATE COMPOSITION

The ultimate analysis (elemental composition) of petroleum is not reported to the same extent as for coal. Nevertheless, from the data available (Speight, 1980) is appears that the proportions of the elements in petroleum vary only slightly over narrow limits:

Carbon	83.0–87.0%
Hydrogen	10.0–14.0%
Nitrogen	0.1–2.0%
Oxygen	0.05–1.5%
Sulfur	0.05–6.0%

in spite of the wide variation in physical properties from the lighter, more mobile crude oils at one extreme to the heavier asphaltic crude oils at the other extreme (Figure 1).

II. CHEMICAL COMPOSITION

The high proportion of carbon and hydrogen indicate that hydrocarbons are the major constituents of petroleum. Elements other than carbon and hydrogen, i.e., nitrogen, oxygen, and sulfur, appear as organic compounds (Table 1) and, while they are usually a relatively minor part of petroleum, they can, in the heavier asphaltic crude oils, constitute a considerable percentage of the whole.

A. Hydrocarbon Components

The hydrocarbon components of petroleum are composed of paraffinic, naphthenic, and aromatic groups. Olefinic groups are not usually found in crude oils and acetylenic hydrocarbons are very rare indeed. It is

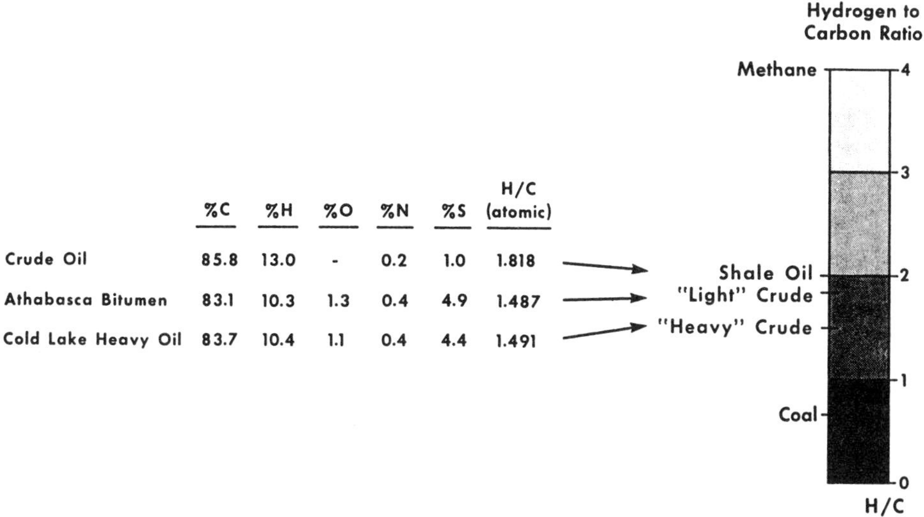

	%C	%H	%O	%N	%S	H/C (atomic)
Crude Oil	85.8	13.0	-	0.2	1.0	1.818
Athabasca Bitumen	83.1	10.3	1.3	0.4	4.9	1.487
Cold Lake Heavy Oil	83.7	10.4	1.1	0.4	4.4	1.491

Figure 1 Variation in the ultimate (elemental) composition and H/C (atomic) ratios of fossil fuels.

convenient therefore to divide the hydrocarbon components of petroleum into the following three categories (Rossi et al., 1953):

1. Paraffins—saturated hydrocarbons with straight or branched chains, but without any ring structure:

$CH_3(CH_2)_nCH_3$ Straight-Chain Paraffin

$CH_3CH_2CH_2(CH_2)_mCH_2CH(CH_3)CH_3$ Branched-Chain Paraffin

2. Naphthenes—saturated hydrocarbons containing one or more rings, each of which may have one or more paraffinic sidechains (more correctly known as "alicyclic" hydrocarbons):

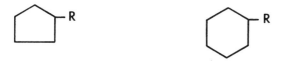

Alkylcyclopentane Alkylcyclohexane

Table 1 Petroleum is a Complex Mixture of Various Chemical Types (See Chap. 3, Sec. II)

1. Hydrocarbons	2. Sulfur Compounds	3. Nitrogen Compounds
$CH_3(CH_2)_nCH_3$	R-S-R'	
$CH_3(CH_2)_nCH(CH_3)CH_3$		

4. Oxygen Compounds	5. Metallic Constituents
$R\text{-}CO_2H$	Porphyrinic
OH	Nonporphyrinic
$\overset{O}{\overset{\|}{-C}}\text{-O-R}$	

3. Aromatics—hydrocarbons containing one or more aromatic nuclei, such as benzene, naphthalene, phenanthrene ring systems which may be linked up with (substituted) naphthene rings and/or paraffinic side chains:

Benzene

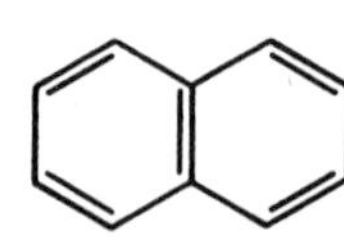

Naphthalene

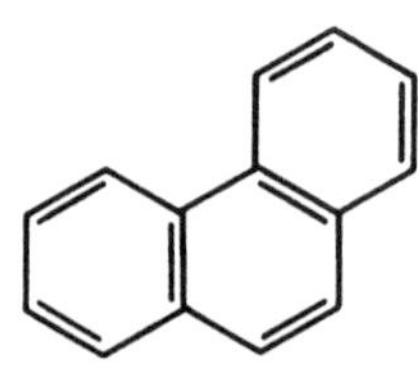

Phenanthrene

The proportion of paraffins in crude oils varies with the type of crude, but within any one crude oil the proportion of paraffinic hydrocarbons usually decreases with increasing molecular weight (Figure 2).

The relationship between the various hydrocarbon constituents of crude oils is one of hydrogen addition or hydrogen loss:

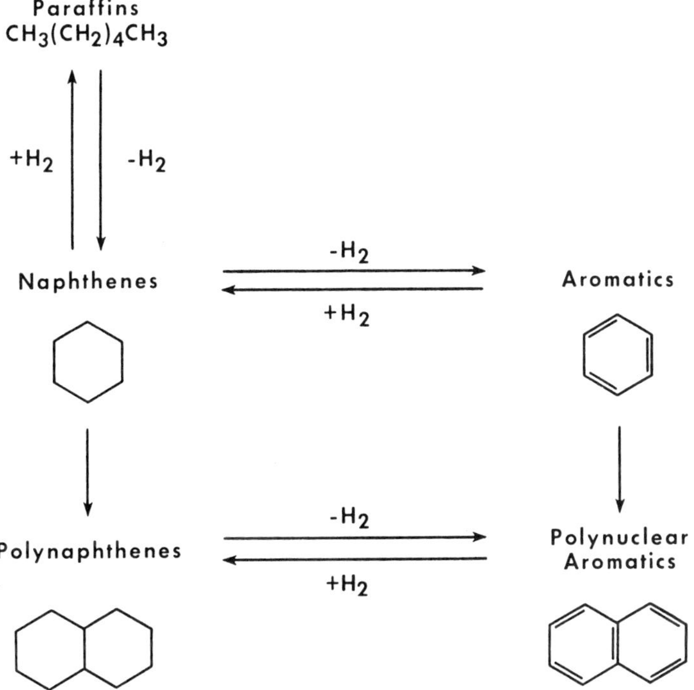

and there is no reason to deny the occurrence of these interconversion schemes during the formation, maturation, and in situ alteration of petroleum.

B. Nonhydrocarbon Components

Inclusion of organic compounds of nitrogen, oxygen, and sulfur only serves to present crude oil as an even more complex mixture than was originally conceived. As the boiling point of the petroleum fraction increases, not only the number of the constituents but also the molecular complexity of the constituents increases (Figure 2; Speight, 1981).

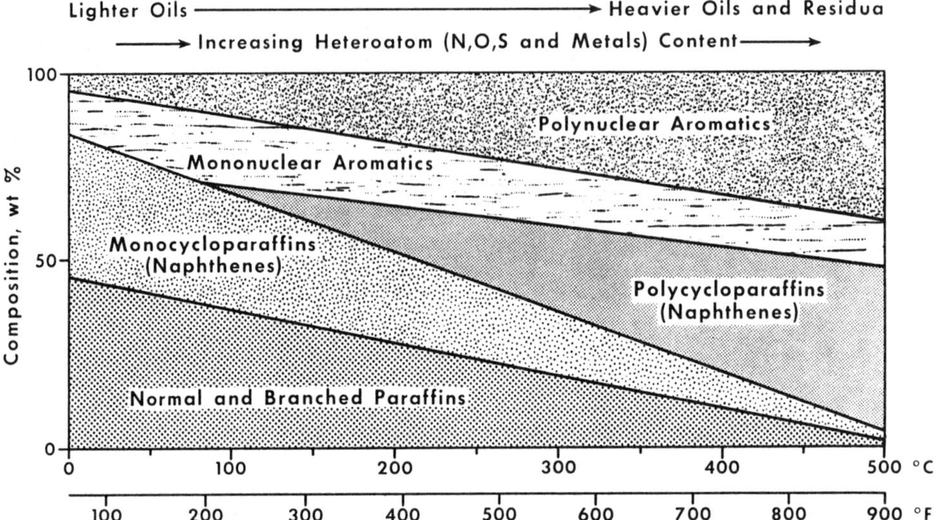

Figure 2 Distribution of the various compound types throughout the boiling range of petroleum.

Crude oils contain appreciable amounts of organic nonhydrocarbon constituents, mainly sulfur-, nitrogen-, and oxygen-containing compounds and, in smaller amounts, organometallic compounds in solution and inorganic salts in colloidal suspension. These tend to concentrate mainly in the nonvolatile residues.

Although their concentration in certain fractions may be quite small, their influence is important. For example, the deposition of inorganic salts suspended in the crude can cause serious breakdowns in refinery operations; the thermal decomposition of deposited inorganic chlorides with evolution of free hydrochloric acid can give rise to serious corrosion problems in the distillation equipment. The presence of organic acidic components, i.e., mercaptans and acids, can also promote metallic corrosion. In catalytic operations, passivation and/or poisoning of the catalyst can be caused by deposition of traces of metals (vanadium, nickel) or by chemisorption of nitrogen-containing compounds on the catalyst, thus necessitating the frequent regeneration or replacement of the catalyst (Bland and Davidson, 1967).

1. Sulfur Compounds

Sulfur compounds of one type or another are present in all crude oils; in general, the higher the density of the crude oil (or the lower the

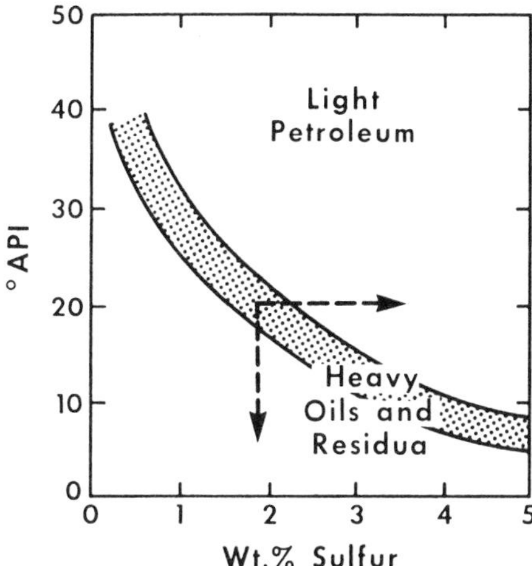

Figure 3 Generalized relationship of API gravity to sulfur content.

API gravity of the crude oil, the higher the sulfur content (Figure 3). The total sulfur in the crude can vary from perhaps 0.04% for a light paraffin oil to about 5.0% for a heavy crude oil but the sulfur content of crude oils produced from broad geographic regions will vary with time, depending on the composition of newly discovered fields, particularly those in different geologic environments (Gruse and Stevens, 1960).

Sulfur compounds in petroleum vary from the thiols (R–SH) to the polycyclic sulfides, such as compounds with five- and six-membered rings (Table 2):

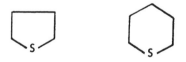

Various types of thiophenes, e.g.:

2,3,4-trimethyl-thiophene	2,3,-dimthyl-4-ethyl-thiophene	2,4,5-tetramethyl-thiophene

Table 2 Representative Sulfur-containing Compounds in Crude Oils

RSH	Thiols (Mercaptans)
RSR'	Sulfides
 [ring structure containing S]	Cyclic Sulfides
RSSR'	Disulfides
 [ring structure containing S]	Thiophene
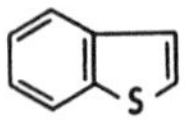 [ring structure containing S]	Benzothiophene
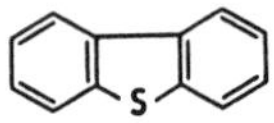 [ring structure containing S]	Dibenzothiophene
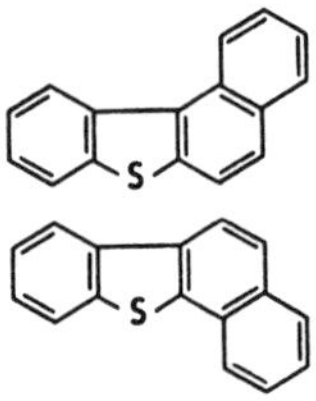 [ring structures containing S]	Naphthobenzothiophene

have also been isolated from crude oils while benzothiophene derivatives are usually present in the higher boiling petroleum fractions. Disulfides are not regarded as being generally present in petroleum but may arise by oxidation of thiols during processing:

$$\text{R-SH} \xrightarrow{[O]} \text{R-S-S-R}$$

2. Oxygen Compounds

The total oxygen content of petroleums is usually less than 2% although larger amounts have been reported. However, in cases where the oxygen content is phenomenally high it may be that the oil has suffered prolonged exposure to the atmosphere either during or after production. The oxygen content of petroleum does increase with the boiling point of the fractions examined (Gruse and Stevens, 1960).

Carboxylic acids in petroleum with less than eight carbon atoms per molecule are almost entirely aliphatic in nature; monocyclic acids begin at C_6 and predominate above C_{14}. This indicates that the structures of the carboxylic acids correspond with those of the hydrocarbons with which they are associated in the crude oil, i.e., in the range where paraffins are the prevailing type of hydrocarbon, the aliphatic acids may be expected to predominate; similarly, in the ranges where the monocycloparaffins and dicycloparaffins prevail, one may expect to find principally monocyclic and dicyclic acids, respectively.

In addition to the carboxylic acids and phenolic compounds, the presence of ketones, esters, ethers, and anhydrides has been claimed for a variety of petroleums. However, the precise identification of these compounds is difficult as most of them occur in the higher molecular weight nonvolatile residua.

3. Nitrogen Compounds

The nitrogen content of petroleum is low and generally falls within the range 0.1–0.9%, although crude oils with no detectable nitrogen or even trace amounts are not uncommon. However, in general, the "heavier" the oil the higher the nitrogen content (Figure 4). The presence of nitrogen in petroleum is of great significance in refinery operations since these compounds are responsible for the poisoning of cracking catalysts and they also contribute to gum formation in such petroleum products. The use of heavy oils as catalytic cracking feedstocks has accentuated the harmful effects of the nitrogen compounds, which are more prevalent in these higher boiling crude oils (Speight, 1981).

Basic nitrogen compounds of relatively low molecular weight can be extracted with dilute mineral acids, but strong bases of higher molecular weight may remain unextracted because of unfavorable partition between the oil and aqueous phases. Nitrogen compounds extractable with dilute mineral acids from petroleum distillates were found to consist of pyridines, quinolines, and isoquinolines carrying alkyl substituents, as well as for a few pyridines in which the substituent was a cyclopentyl or cyclohexyl group. The compounds which cannot be

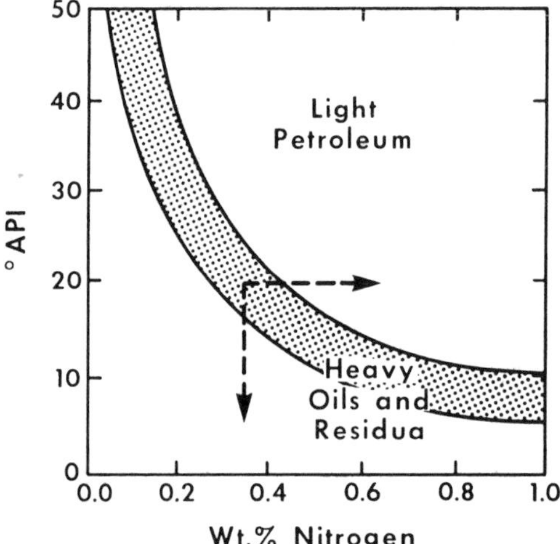

Figure 4 Generalized relationship of API gravity to nitrogen content.

extracted with dilute mineral acids contain the greater part of the nitrogen in petroleum and are generally of the carbazole, indole, and pyrrole types (Table 3).

Porphyrins (Chap. 3, Sec. II.C) are also constituents of petroleum and usually occur in the nonbasic portion of the nitrogen-containing concentrate. The simplest porphyrin is porphine and consists of four pyrrole molecules joined by methine (−CH=) bridges. The methine bridges establish conjugated linkages between the component pyrrole nuclei forming a more extended resonance system. Although the resulting structure retains much of the inherent character of the pyrrole components, the larger conjugated system gives inreased aromatic character to the porphine molecule (Falk, 1964).

4. Metallic Compounds

The occurrence of metallic constituents in crude oil is of considerably greater interest to the petroleum industry than might be expected from the very small amounts present. Even minute amounts of iron, copper, and particularly nickel and vanadium in the feedstocks for catalytic cracking affect the activity of the catalyst and result in increased gas and coke formation and reduced yields of gasoline.

The ash residue left after burning of a crude oil is due to the presence of these metallic constituents, part of which occur as inorganic

Table 3 Representative Nitrogen-Containing Compounds in Crude Oils

Nonbasic		
Pyrrole	C_4H_5N	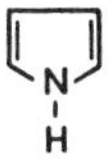
Indole	C_8H_7N	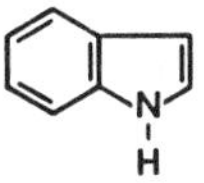
Carbazole	$C_{12}H_9N$	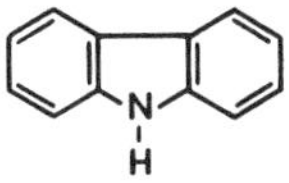
Benzo(a)carbazole	$C_{16}H_{11}N$	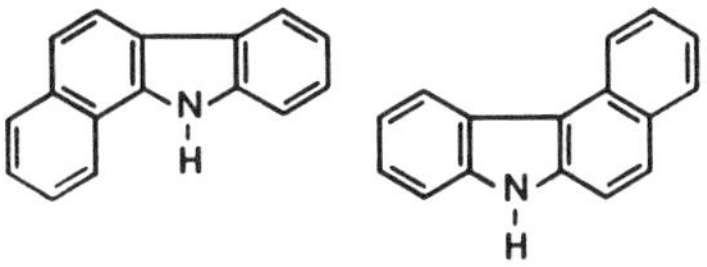
Basic		
Pyridine	C_5H_5N	
Quinoline	C_9H_7N	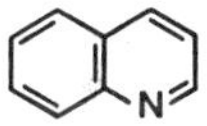
Indoline	C_8H_9N	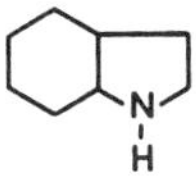
Benzo(f)quinoline	$C_{13}H_9N$	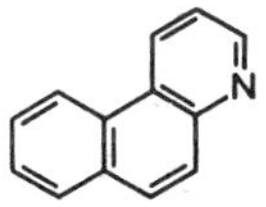

water-soluble salts (mainly chlorides and sulfates of sodium, potassium, magnesium, and calcium) and which occur in the water phase of crude oil emulsions. These are removed in the desalting operations, either by evaporation of the water and subsequent water washing, or by breaking the emulsion, thereby causing the original mineral content of the crude to be substantially reduced. Other metals are present in the form of oil-soluble organometallic compounds either as complexes, metallic soaps, or in the form of colloidal suspensions, and the total ash from desalted crudes is of the order 0.1–100 mg/liter (Speight, 1980).

Evidence for the presence of several other metals in oil-soluble form has been produced and thus zinc, titanium, calcium, and magnesium compounds have been identified in addition to vanadium, nickel, iron, and copper (Table 4).

Distillation concentrates the metallic constituents in the residues (Speight, 1981), although some can appear in the higher boiling distillates. However, the latter may be due in part to entrainment.

Table 4 Ranges of Principle Trace Elements Found in Petroleum

Element	Range in petroleum (ppm)
Cu	0.2--12.0
Ca	1.0–2.5
Mg	1.0–2.5
Ba	0.001–0.1
Sr	0.001--0.1
Zn	0.5–1.0
Hg	0.03--0.1
Ce	0.001--0.6
B	0.001--0.1
Al	0.5–1.0
Ga	0.001–0.1
Ti	0.001–0.4
Zr	0.001--0.4
Si	0.1–5.0
Sn	0.1–0.3
Pb	0.001–0.2
V	5.0–1500
Fe	0.04–120
Co	0.001–12
Ni	3.0–120

Nevertheless, there is evidence that a portion of the metallic constituents may occur in the distillates by volatilization or entrainment of the organometallic compounds present in the petroleum. As the percentage overhead product obtained by vacuum distillation of a reduced crude is increased, the amount of metallic constituents in the overhead oil is also increased. The majority of the vanadium, nickel, iron, and copper in residual stocks may be precipitated along with the asphaltenes by hydrocarbon solvents.

C. Porphyrins

Porphyrins are naturally occurring chemical species that have been known to exist in petroleum for more than 50 years. They are given separate consideration in this section because of their uniqueness. Furthermore, they are not always discussed along with other nitrogen-containing constituents of petroleum but they are considered to be a metallo-containing organic material that will also occur in some crude oils. Thus, they are given consideration here as separate and distinct chemical entities.

1. Properties

Porphyrins are cyclic, conjugated compounds with a basic structural unit consisting of four pyrrole groups connected by four methine groups. The dimensions of porphine (Figure 5) and other porphyrin molecules are such that bonding of the four nitrogen atoms with a large number of metal ions can occur to form highly stable metal chelates. The two hydrogen atoms bonded to the pyrrolic nitrogen atoms, called imino hydrogens, are replaced in this process (Figure 5).

Metal chelates are molecules in which a metal ion is linked by one or more coordinate bonds to nonmetal atoms. The nonmetal portion of a chelate molecule is referred to as a ligand.

Porphyrins chelated with metal ions are classified as metalloporphyrins, a term which also includes some nonchelates. For example, the metals sodium and potassium can replace the slightly acidic imino hydrogens of some porphyrins. However, the products of this reaction are not chelates because there is no coordinate bond formed between sodium or potassium ions and the porphyrin nitrogen atoms.

Porphyrins not bonded to metals are often referred to as free base porphyrins. The term metal complex is often used in place of the term metal chelate.

A large number of different porphyrin compounds exist in nature or have been synthesized. Most of these compounds have substituents

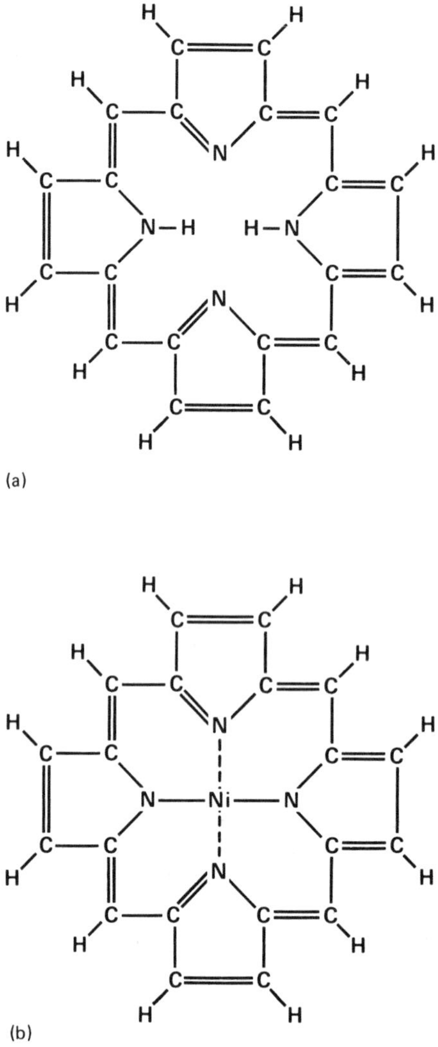

(a)

(b)

Figure 5 Structure of (a) porphine and (b) the nickel chelate of porphine.

other than hydrogen on many of the ring carbons. The nature of the substituents on porphyrin rings determines the classification of a specific porphyrin compound into one of various types according to one common system of nomenclature. These porphyrin types have well-known trivial names or acronyms, which will be used in this work. The formal nomenclature of porphyrins has been reviewed by Bonnett (1978; see Figure 6).

When one or two double bonds of a porphyrin macrocycle are hydrogenated, a chlorin or a phlorin is formed (Figure 7). Chlorins are components of chlorophylls and are of great biochemical importance. Chlorins in chlorophylls have the unique feature of possessing an isocyclic ring, formed by two methylene groups bridging a pyrrolic carbon to a methine carbon. Geologic porphyrins which contain this structural feature are assumed to be derived from chlorophylls. VO-DPEP has an isocyclic ring, designated as ring E (Figure 6). Geoporphyrins having this five-membered isocyclic ring are known as DPEP types. Recently other geoporphyrins have been discovered that have isocyclic rings with more than five carbons. This has led to use of the term cycloalkanoporphyrin to designate any porphyrin having an isocyclic ring structural element. A DPEP-type porphyrin therefore is a cycloalkanyl porphyrin having a five-membered isocyclic ring.

Etioporphyrins (Figure 8) also are commonly found in geologic materials. Porphyrins of this type are distinguished by having no substituents other than hydrogen on methine carbons.

Benzoporphyrins and tetrahydrobenzoporphyrins (Figure 9) also have been identified in geologic materials. These compounds have

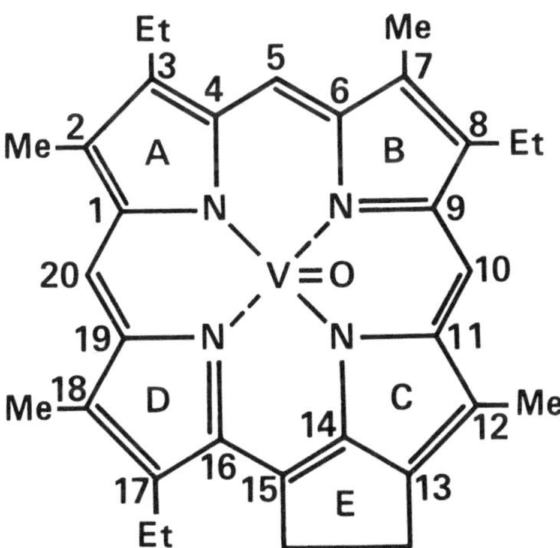

Figure 6 Structure of vanadyl desoxyphylloerythroetioporphyrin (VO-DPEP) with formalized numbering of the porphyrin system.

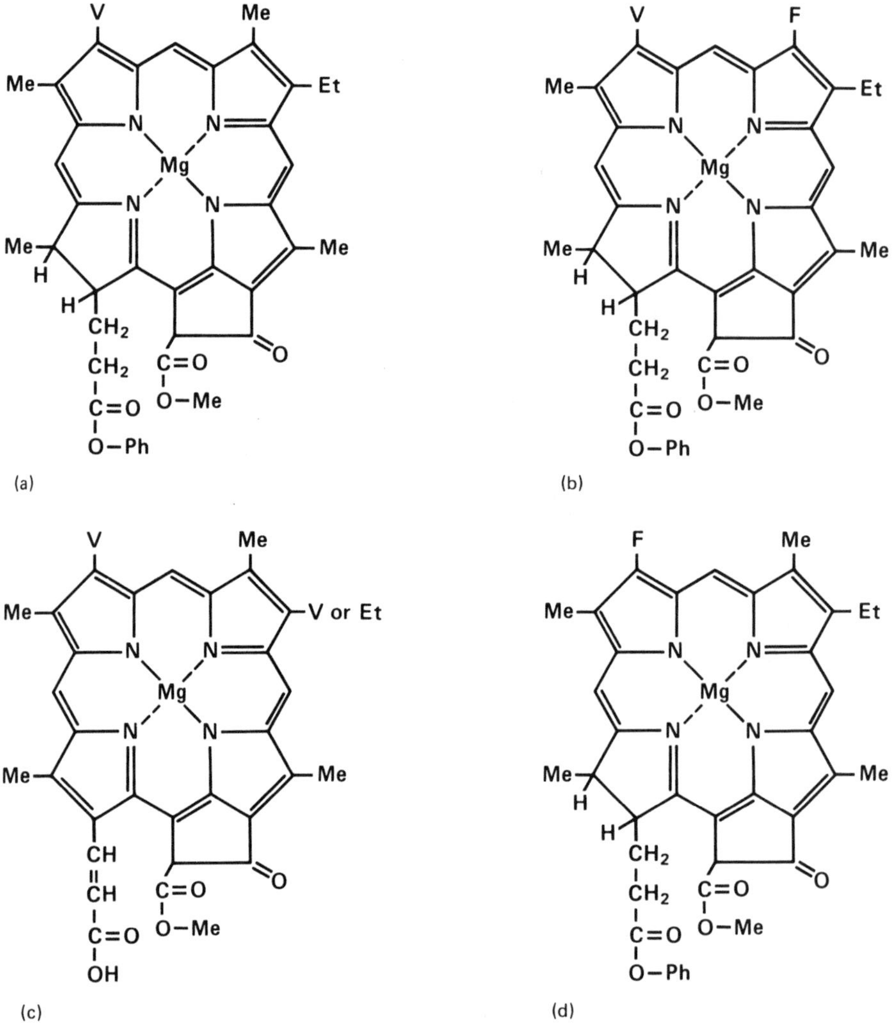

(a) (b)

(c) (d)

Figure 7 Structures of chlorins, as represented by chlorophylls a, b, the two chlorophyll c's, and chlorophyll d. In these structures, Me = methyl; Et = ethyl; V = vinyl; F = formyl; pH = phytyl; Fr = farnesyl.

either a benzene ring or a hydrogenated benzene ring fused onto a pyrrole unit. Porphyrins having combined features characteristic of DPEP, benzo, and tetrahydrobenzo types are also known. The

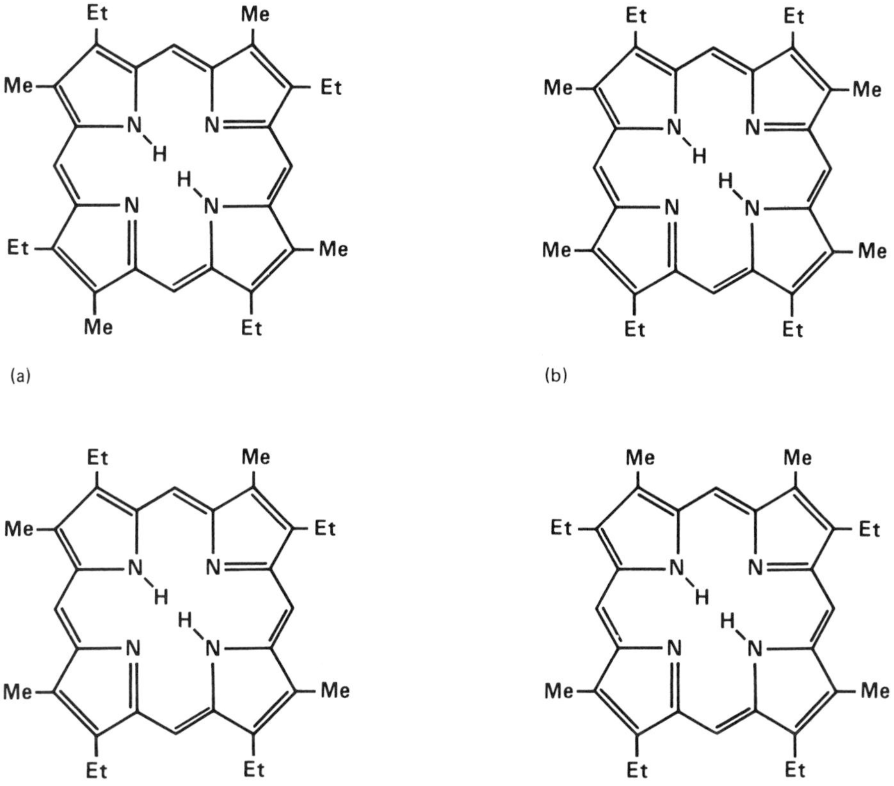

(a)

(b)

(c)

(d)

Figure 8 Structures of etioporphyrin. I-(a); etioporphyrin II-(b); etioporphyrin III-(c); etioporphyrin IV-(d).

structures of the vanadyl chelates of two synthetic porphyrins, octaethylporphine (OEP) and tetraphenylporphine (TPP), are illustrated (Figure 9) for reference.

The ultraviolet (UV)-visible spectra of porphyrins are characterized by several peaks, some with very large extinction coefficients (Smith, 1975). Solutions of free base porphyrins usually exhibit UV-visible spectra consisting of five broad peaks in the 350- to 700-nm range, of which the one at 390–420 nm, called the Soret peak, is always the largest. The other peaks are in the 490- to 700-nm range (Figure 10). The relative heights of the four non-Soret peaks can provide information about the nature of substituents on a porphyrin macrocyclic ring. Metalloporphyrin UV-visible spectra normally

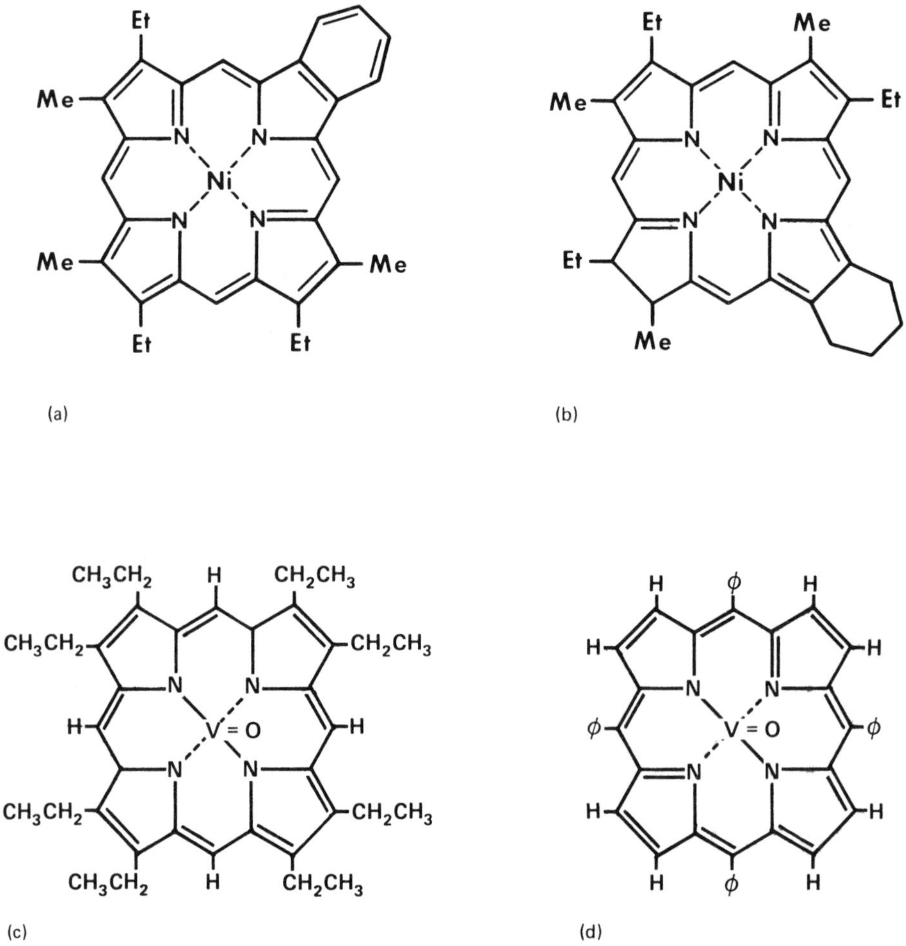

(a) (b)

(c) (d)

Figure 9 Structures for (a) a benzoporphyrin; (b) a tetrahydrobenzoporphyrin; (c) vanadyl octaethylporphine; and (d) vanadyl tetraphenylporphine.

exhibit two peaks in addition to the Soret peak. The positions of the maxima of these peaks often serve to identify which metal is chelated with a given porphyrin (Figure 11). The distinctive nature of the UV-visible spectra of porphyrin solutions and the large extinction coefficients of some of the peaks enable porphyrins to be detected in small concentrations, sometimes even when they are part of complex mixtures (Sugihara and Bean, 1962).

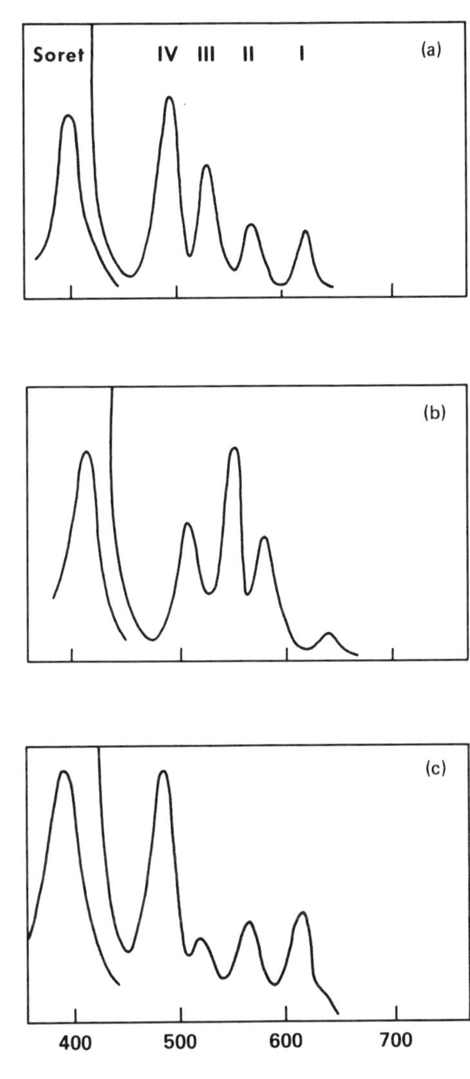

Figure 10 UV-visible spectra of various porphyrins; an etioporphyrin (a); a benzoporphyrin (b); a DEPE porphyrin (c).

Most porphyrins are sufficiently volatile that they can be studied by mass spectrometry. Low-voltage mass spectra exhibit prominent parent ion peaks and doubly charged ion peaks. Because of the large size of porphyrin molecules, isotope peaks are prominent in their mass

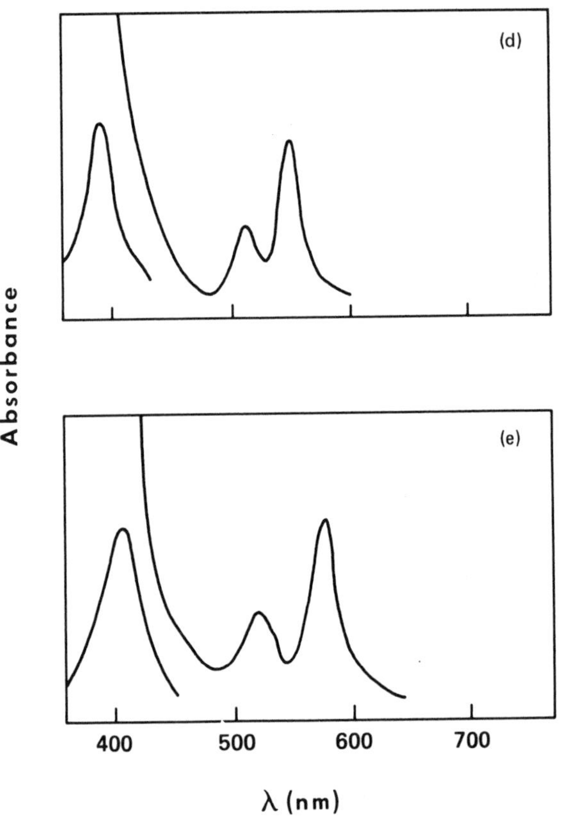

Figure 11 UV-visible spectra of metal porphyrins; a nickel etioporphyrin (d); a vanadyl etioporphyrin (e).

spectra. Cracking patterns are similar to those of other substituted aromatic molecules. If ethyl or higher alkyl substituents are present, P-15 (parent ion minus methyl substituent), P-29 (parent ion minus ethyl substituent), or other peaks are observed in the mass spectra that correspond to cleavage β to the aromatic ring. The presence of four nitrogen atoms results in a mass defect, as does the presence of a metal. Therefore high-resolution mass spectrometry is a useful method for porphyrin analysis. Mass spectrometric studies have been important in elucidating the structures of geoporphyrins, which usually occur as extended series of methylene homologs (Baker, 1969; Figures 12 and 13). Each porphyrin type consists of a series of compounds ranging over several carbon numbers. Each mass number in the mass spectra may correspond to several compounds with the same carbon

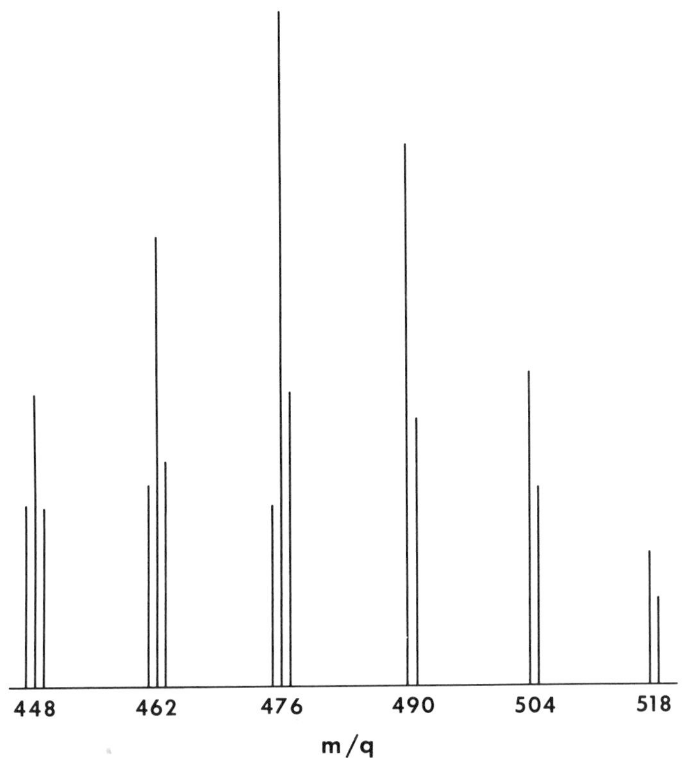

Figure 12 Mass spectrum of DPEP-type porphyrin concentrate from Boscan crude oil.

number. Porphyrins of carbon number higher than 35 (which corresponds to the peqks at m/q = 518 and 519 in the mass spectrum of the DPEP concentrate) are found in Boscan crude (not shown in Figures 12 and 13). The peak intensities form a roughly Gaussian distribution, which is typical of the mass spectra of porphyrins from many geologic materials. In the mass spectrum of the Boscan DPEP concentrate, the peak at m/q = 476 is highest in intensity. This peak corresponds to a porphyrin or porphyrins of carbon number 32. In most crude oil porphyrin mass spectra, such maxima are usually observed at peaks corresponding to porphyrins having carbon numbers of 30, 31, or 32.

The Boscan porphyrins originally existed largely as vanadyl $(V = 0)^{2+}$ chelates. They were demetallized and then the porphyrin types were separated from each other. It is not necessary to demetallize

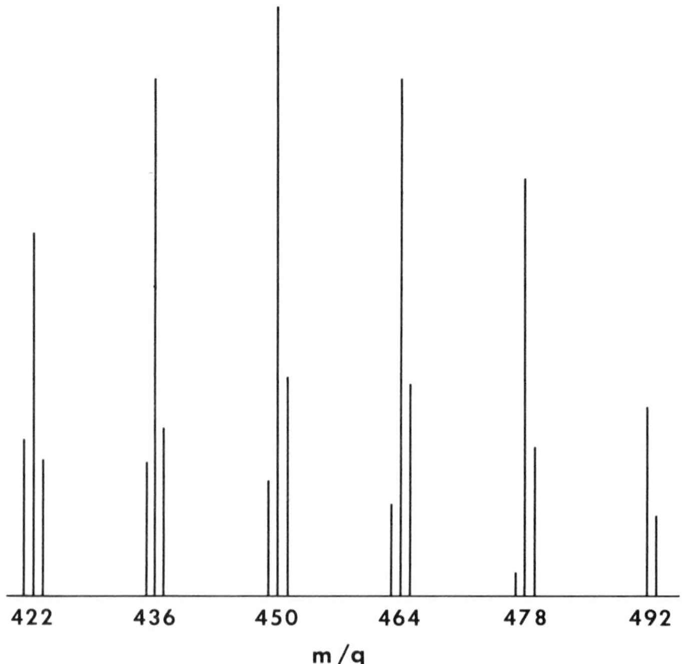

Figure 13 Mass spectrum of etioporphyrin concentrate from Boscan crude oil.

porphyrins to obtain mass spectra, as metalloporphyrins also give good mass spectra.

Most porphyrin suites in geologic materials are complex and are usually present in small concentrations. Isolating porphyrins from these materials and separating the individual compounds is almost impossibly tedious. Unless this is done, the usefulness of some analytical techniques is limited in porphyrin investigations. This is not true for mass spectrometry, which can be used to study samples that have not undergone extensive workup.

Nuclear magnetic resonance (NMR) methods have been used to determine exact structures of individual geoporphyrins. Nuclear Overhauser effect (nOe) studies have been of particular value. This technique can be used to determine the exact arrangement of substituents around a porphyrin ring (Sanders et al., 1978). In order to use this technique, porphyrins must be isolated from the matrices in which they occur and must be substantially purified. This may be accomplished by high-performance liquid chromatography (HPLC), size exclusion chromatography (SEC), and other separation techniques. The isolation of porphyrins from geologic materials has been reviewed by Quirke (1987a).

Chemical properties of porphyrins are determined by various structural features. Free base porphyrins react with metal salts and other metal chelates under certain conditions to incorporate metal ions into the middle of the porphyrin ring. The reverse reaction can be made to occur, particularly in strongly acidic media such as hydrofluoric, sulfuric, or sulfonic acids. Free base porphyrins protonate in acidic media and these protonated porphyrins may be water-soluble. This property aids in the isolation of porphyrins from bitumens.

Both free base porphyrins and metalloporphyrins can be hydrogenated. One or more molecules of hydrogen may be added until all double bonds are saturated. The reduced products are not as strong chelating agents as the parent porphyrins.

Similar to other aromatic molecules, porphyrins readily undergo electrophilic substitution reactions.

2. Precursors

Petroleum, tar sand bitumen, oil shale kerogen, and coal are known to be largely derived from biological sources, a fact which studies of geoporphyrins helped establish (Treibs, 1934). Algae and bacteria are believed to be the most important precursors of petroleum, tar sand bitumen, and some oil shale kerogens, while coal and other oil shale kerogens are derived largely from woody plants.

The organic matter of plants growing in and around most bodies of water is consumed by various grazing animals, fungi, and bacteria. These organisms recycle plant organic carbon to carbon dioxide. In bodies of water depleted in oxygen, this recycling of carbon is not efficient. Much dead plant matter persists in sediments on the bottom of these waters. This plant matter would be consumed by bottom-feeding organisms in more oxygenated waters. Unoxidized plant material in anoxic sediments becomes more deeply buried as sediments accumulate. Eventually the sediments either lithify or, if they are composed of woody plant remains, coalify. If they contain sufficiently large amounts of organic matter, the lithified sediments are good source rocks for petroleum generation. Upon deeper burial, the source rocks will yield petroleum, which migrates into reservoirs. During these processes, chlorophylls present in the original plant material become converted into porphyrins.

3. Occurrence

The study of porphyrins in fossil fuels began with the discovery of vanadyl porphyrins in an Austrian oil shale by Alfred Treibs (1934). Subsequently, Treibs found vanadyl porphyrins and what he believed to be iron porphyrins in crude oils. Treibs's iron porphyrins later were shown to be nickel porphyrins (Glebovskaya and Volkenshtein, 1948). The work of Treibs is believed by some authorities to have founded the science of organic geochemistry.

As a result of Treibs's investigation of geoporphyrins there arose the concept of the biomarker, a key concept of organic geochemistry. Biomarkers are of interest because they establish a linkage between compounds found in the geosphere and their presumed biological precursors. Information can be obtained from the study of biomarkers about the chemical transformations that occur during the conversion of biological materials to the organic matter of fossil fuels. Thus a great deal of information about porphyrins has been accumulated by organic geochemists. Processing scientists should be familiar with some of this information because of the troublesome nature of porphyrins in the refining of fossil fuels.

Almost all crude oils and tar sand bitumens contain detectable amounts of vanadyl (2+) and nickel (2+) porphyrins. More mature, lighter crudes usually contain only small amounts of these compounds. Heavy oils may contain large amounts of vanadyl and nickel porphyrins. Vanadium concentrations of over 1000 ppm are known for some crudes, and a substantial amount of the vanadium in these crudes is chelated with porphyrins. In high-sulfur crudes of marine origin, vanadyl porphyrins are more abundant than nickel porphyrins. Low-sulfur crudes of lacustrine origin usually contain more nickel porphyrins than vanadyl porphyrins. Of all the metals in the periodic table, only vanadium and nickel have been proved definitely to exist as chelates in significant amounts in a large number of crude oils and tar sand bitumens. The existence of iron porphyrins in some crudes has been claimed (Franceskin et al., 1986). Geochemical reasons for the absence of substantial quantities of porphyrins chelated with metals other than nickel and vanadium in most crude oils and tar sand bitumens have been advanced by Quirke (1987b). Neither are free base porphyrins found in petroleum or tar sand bitumens.

The metalloporphyrin suites in crude oils and tar sand bitumens are composed of several compound types. Cycloalkanyl porphyrins with five-membered isocyclic rings, the DPEP types, usually predominate. Cycloalkanyl porphyrins with isocyclic rings having more than five carbons may be present in some samples in small amounts. Certain crude oils are characterized by porphyrin suites in which etioporphyrins predominate. Benzoporphyrins and tetrahydrobenzoporphyrins have been found in many heavy crudes and tar sand bitumens, but not as the dominant compound type.

In petroleum and tar sand bitumen, the several porphyrin types exist as extended series of methylene homologs. These homologous series may be "fingerprinted" by mass spectrometry, as was discussed previously. The series usually appear as Gaussian distributions of peaks in mass spectra. Porphyrins of carbon numbers 28–33 are almost always the most prominent members of the series. The homologous series extend to much higher carbon numbers in some materials, but these higher homologs are relatively involatile and are apparently present at relatively low concentrations. Some of the high-carbon-number porphyrins have extended alkyl side chains (Quirke et al., 1980). Each of the mass numbers in the mass spectra of the homologous series may correspond to several isomers.

It is evident that the total number of individual porphyrin compounds present in a crude oil can be quite large. Both vanadyl and nickel porphyrin suites may consist of varying amounts of several different porphyrin types. Each of these porphyrin types may itself consist of a large number of methylene homologs covering a wide range of carbon numbers. Many isomers of the same carbon number may exist. If all these geoporphyrins arose from a limited number of biological chlorophylls, then it is of great interest to determine what geochemical conditions were responsible for the transformation of the chlorophylls to geoporphyrins. This topic has been reviewed by several authors (Hodgson et al., 1967; Baker, 1969; Baker and Palmer, 1978; Baker and Louda, 1986; Filby and Van Berkel, 1987).

If vanadium and nickel contents of crudes are measured and compared with porphyrin concentrations (usually estimated by measuring areas under peaks in UV-visible spectra), it is usually found that not all the metal content can be accounted for as porphyrins. In some crudes, as little as 10% of total metals appear to be chelated with porphyrins. Only rarely can all measured nickel and vanadium in a crude oil be accounted for as porphyrinic (Erdman and Harju, 1963). Currently, most investigators believe that much of the vanadium and nickel in crude oils is chelated with ligands that are not classical porphyrins. These metal chelates are referred to as nonporphyrin metal chelates or complexes. A substantial amount of evidence for the existence of such compounds has been reported, but as yet no isolations of specific compounds or unequivocal structure proofs have been performed (Crouch et al., 1983; Fish et al., 1984; Reynolds et al., 1987). It has been suggested that sulfur is involved in the coordination spheres of these compounds (Sugihara et al., 1965; Dickson et al., 1972). Naphthenic acid salts of nickel have been suggested to account for some of the nonporphyrin metal complexes (Fish et al., 1987a). A model in which metal ions occupy holes in aromatic sheets has been proposed by Yen (1975).

Other investigators question the existence of nonporphyrin metal chelates in petroleum (Goulon et al., 1984). It is claimed that in systems such as heavy crudes, in which intermolecular associations are important, UV-visible spectra do not provide reliable measurements of porphyrin concentrations, invariably greatly undercounting them. The physicochemical methods employed by Goulon et al. should not be affected by such matrix effects and these methods could not detect coordination sites involving metals in petroleum that were not porphyrinic. In the remainder of this chapter it will be assumed that nonporphyrin chelates exist in fossil fuels, but it should be borne in mind that minority views sometimes prevail in science. By nonporphyrin metals will be meant that amount of metal content of a bituminous material not accounted for as porphyrinic when the bituminous material is analyzed for porphyrin content by methods such as UV-visible spectrometry.

Metalloporphyrins and nonporphyrin metal chelates become concentrated in asphaltenes when tar sand bitumen or petroleum is treated with *n*-pentane, *n*-hexane, *n*-heptane, or other alkanes. The deasphaltened oils (DAO; maltenes) contain smaller concentrations of porphyrins than the parent materials and usually very small concentrations of nonporphyrin metals. Size exclusion chromatography has been used to separate crude oils and their asphaltenes into fractions in which metals are chelated almost entirely with porphyrins and fractions in which the metals appear to be largely nonporphyrinic (Kowanko et al., 1978; Fish et al., 1987a; Reynolds et al., 1987).

Some other metals appear to be organically associated in petroleum and tar sand bitumens but are usually present in small amounts compared with nickel and vanadium. Iron is present in substantial amounts in some bituminous materials, but it has been suggested that some of it is anthropogenic. It is not known whether or not the other metals are chelated with porphyrins.

D. Oil Shales and Coal

Most of the organic matter in coals and oil shales is insoluble in organic solvents under normal conditions of temperature and pressure. Thus it is not so easily demonstrated that metals bonded to organic ligands exist in these substances. Coals and oil shales contain finely divided mineral particles that are disseminated throughout insoluble, macromolecular organic matrices. What might appear to be a metal with consistent organic affinities among a number of coals and oil shales may be an inorganically bonded metal occurring as a minor component in finely divided clays or other minerals. Porphyrins are found in the fractions of coals and oil shales that are soluble in organic solvents.

Curiously, porphyrins in a large variety of coals of both recent and ancient origin are chelated with gallium (Bonnet et al., 1987). Iron porphyrins also have been identified in coals. Porphyrin contents of coals are much lower than those of heavy crude oils, based on the concentrations of porphyrins found in the soluble fractions of coals. The homologous series of coal porphyrins seem to consist of members having even numbers of carbons and are predominantly of the etio type. Other metals in coals probably exist chelated with organic ligands or as salts of phenols or organic acids. Titanium may be one of the metals in coal that is organically bonded in one of these ways (Narain et al., 1986).

Some oil shales are very rich in metalloporphyrins (Van Berkel and Filby, 1987). Vanadium and nickel porphyrins and nonporphyrins are the predominant metal chelates in oil shale organic matter. The large Green River oil shale deposits contain nickel porphyrins of both the DPEP and etio types (Morandi and Jensen, 1966). Organoarsenic compounds also have been identified as components of these shales (Fish et al., 1987b). Iron and magnesium salts of organic acids are present in Green River shales (Vandergrift et al., 1980). In the organic matter of the large Devonian oil shale deposits of the eastern United States, vanadium porphyrins predominate over nickel porphyrins.

III. PHYSICAL PROPERTIES

In general terms, refinery processes can be conveniently divided into three different types (Speight, 1980):

1. Separation, i.e., division of the feedstock into various streams (or fractions) depending on the nature of the crude material
2. Conversion, i.e., the production of salable materials from the feedstock by skeletal alteration, or even by alteration of the chemical type, of the feedstock constituents, and
3. Finishing, i.e., purification of the various product streams by a variety of processes which essentially remove impurities from the product

The separation and finishing processes may involve distillation or treatment with a "wash" solution while the conversion processes are usually regarded as those processes which change the number of carbon atoms per molecule, alter the molecular hydrogen/carbon ratio, or even change the molecular structure of the material (isomerization) without affecting the number of carbon atoms per molecule (Figure 14).

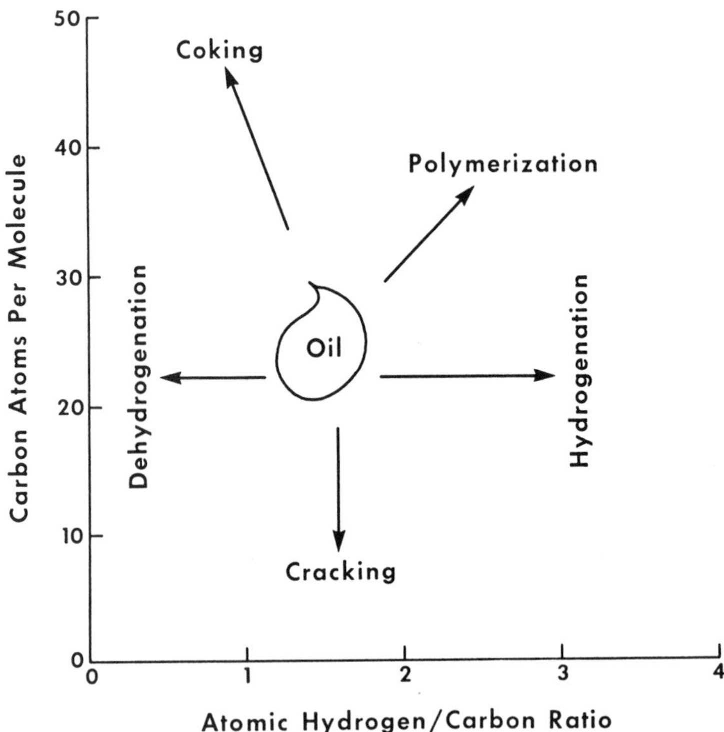

Figure 14 Simplified representation of the chemistry of refinery processes.

Thus, although it is possible to classify refinery operations in the general terms outlined briefly above, the behavior of various feedstocks in these refinery operations is not so simple. For example, the physical and chemical composition of a feedstock play a large part not only in determining the nature of the products that arise from the refining operations but also in determining the means by which a particular feedstock should be processed (Nelson, 1958; Gary and Handwerk, 1975; Speight, 1981).

It is apparent, therefore, that the judicious choice of a crude oil to produce any given product is just as important as the selection of the product for any given purpose. Thus, initial inspection of the crude oil—whether or not this be conventional examination of the physical properties from which may come deductions about the most logical means of refining—is of the utmost importance. In fact, evaluation of crude oils from physical property data as to which refining sequences should be employed for any particular crude oil is a predominant part of the initial examination of any material that is destined for use as a refinery feedstock.

The evaluation of crude oil and any heavy oil or residuum usually involves an examination of one or more of the physical properties of the material. As a result, a set of basic characteristics can be obtained which can be correlated with processability. This also requires comparison of the observed data with those data obtained for other materials of known processability as data obtained without any reference point are of limited value.

Feedstocks can be assessed in terms of sulfur content, carbon residue, nitrogen content, metal content, and such properties as the API (American Petroleum Institute) gravity and viscosity, which help the refinery operator to gain an understanding of the nature of the material to be processed. Thus, the products from high-sulfur feedstocks often require extensive treatment to remove (or alter) the corrosive sulfur compounds while nitrogen compounds and the various metals that occur in crude oils will cause serious loss of catalyst life; the carbon residue presents an indication of the amount of coke that may be formed in the process to the detriment of the liquid products.

To satisfy requirements with regard to determining the processability of a crude oil or a heavier feedstock, most refiners have developed their own methods of analysis and evaluation. Although many of these methods are proprietary and, as a result, not normally available, there are various standards organizations, such as the American Society for Testing and Materials (ASTM) in North America and the Institute of Petroleum (IP) in Britain, which have devoted considerable time and effort to the correlation and standardization of methods for the inspection and evaluation of refinery feedstocks and products. Accordingly, brief mention is made here of the various tests that might be applied to the evaluation of petroleum, heavy oils, and residua, and for convenience reference is made to the corresponding ASTM test.

A. Density and Specific Gravity (ASTM D287, D1298, D941, D1217, and D1555)

In the early years of the petroleum industry, density was the principal specification for feedstocks and refinery products, and was particularly used to give an estimate of the most desirable product, i.e., kerosene, in crude oil. Specific gravity, which approximates closely (but not exactly) density, is the most common term at present, but both factors suffer from the disadvantage that only a very narrow range of values must apply to a wide range of crude oils, heavy oils, bitumens, and asphalts (Table 5).

In order to combat this extremely narrow range of values and in an attempt to inject a more meaningful relationship between the physical properties and processability of the various crude oils, the American Petroleum Institute devised a measurement of gravity based on the Baumé scale for industrial liquids, which is an inverse of the specific gravity scale:

$$°\text{API} = \frac{141.5}{\text{sp. gr. } 60°/60°\text{F}} - 131.5$$

Thus, the API gravity of a lighter crude oil may be of the order of 45° API while a much heavier asphaltic crude oil could have a gravity of the order of 10–20° API (Table 5) with some residua and bitumens having gravities of order of 5–10° API. This is in keeping with the general trend of increased aromaticity leading to decreased API gravity (or, more correctly, increased specific gravity).

It is also possible to recognize certain preferred trends between the API gravity of petroleums, bitumens, and residua and one or more of the other physical parameters. For example, correlations exist between the API gravity and sulfur content (Figure 3), nitrogen content (Figure 4), viscosity (Figure 15), carbon residue (Figure 15), and asphaltene plus resin content (Figure 16).

B. Sulfur Content (ASTM D129, D1552, D4294)

Sulfur content and API gravity are the two properties which have had the greatest influence on the value of petroleum as a feedstock. The sulfur content varies from about 0.1% to about 3% by weight for the more conventional crude oils to as much as 5–6% for some of the heavier oils and bitumens (Table 5, Figure 3). Depending on the sulfur content of the crude oil feedstock residua may be of the same order or even have a substantially higher sulfur content. Indeed, the very nature of the distillation process by which residua are produced, i.e., removal of distillate without thermal decomposition, dictates that the majority of the sulfur, which is predominantly in the higher molecular weight fractions, be concentrated in the residuum (Figure 17).

C. Viscosity (ASTM D445, D88, D2161, D341, and D2270)

Viscosity is an important fluid characteristic and indicates the relative mobility of various crude oils. There are several methods for the

Table 5 Some Properties of Crude Oils, Heavy Oils (Bitumens), and Residua[a]

Era	Specific gravity	API gravity	Carbon residue, w/w %	Sulfur, w/w T	Physical composition (w/w %)		
					Asphal-tenes	Resins	Oils
Conventional (Alberta) crude oils[a]							
Upper Cretaceous	0.851	34.8	2.92	0.21	0.9	12.2	86.9
	0.842	36.5	2.98	0.31	0.3	13.0	86.7
	0.850	34.9	2.56	0.25	<0.1	11.6	88.4
	0.830	39.0	1.29	0.16	0.2	9.4	90.4
Lower Cretaceous	0.856	33.8	2.91	0.82	6.9	16.0	77.1
	0.917	22.8	9.32	0.25	11.7	16.5	71.8
	0.908	24.3	8.56	1.71	10.4	16.1	73.5
	0.880	29.3	5.72	1.56	7.2	13.0	79.8
	0.922	22.0	4.45	2.63	15.6	16.7	67.7
	0.810	43.2	0.28	0.08	<0.1	9.4	90.6
	0.837	37.6	1.36	0.15	<0.1	13.7	86.3
	0.851	34.1	2.86	0.19	2.5	6.3	91.2
	0.804	41.4	0.34	0.03	<0.1	6.9	93.1
	0.837	37.5	1.53	0.13	<0.1	9.3	90.7
Mississippian Jurassic	0.894	26.8	4.61	1.05	3.1	12.6	84.3
	0.924	21.6	8.78	2.03	10.9	21.4	67.7
	0.863	32.4	6.25	1.44	3.8	8.4	87.8
	0.849	35.2	0.73	0.46	∿0.2	12.5	87.3
	0.822	40.5	1.56	0.22	<0.1	9.1	90.9

(continued)

Table 5 (cont.)

Era	Specific gravity	API gravity	Carbon residue, w/w %	Sulfur, w/w %	Physical composition (w/w %) Asphaltenes	Resins	Oils
[Mississippian Jurassic]	0.905	24.8	6.46	1.80	4.9	10.0	85.1
	0.846	35.7	2.79	0.45	4.9	9.9	85.2
Upper Devonian	0.854	34.2	4.21	1.17	3.4	16.9	79.7
	0.828	39.3	2.59	0.26	0.8	13.7	85.5
	0.829	39.1	1.55	0.35	12.5	12.4	75.1
	0.827	39.8	0.29	0.21	<0.1	3.4	96.6
	0.820	41.1	1.10	0.15	<0.1	5.7	94.3
	0.809	43.4	0.44	0.14	~0.4	10.8	88.8
	0.840	36.9	1.40	0.38	<0.1	7.9	92.1
	0.827	39.7	1.23	0.19	<0.1	6.3	93.7
Lower Devonian	0.890	27.4	6.28	0.88	6.4	14.3	79.3
	0.899	25.8	0.91	1.54	8.1	14.4	77.5
	0.835	37.9	2.77	0.15	2.8	11.1	86.1
	0.825	40.1	2.74	0.24	2.0	9.9	88.1
	0.851	34.7	1.87	0.90	~0.2	12.6	87.2
Heavy oils and natural asphalts (bitumens)							
Cold Lake (Alberta)	0.999	10.1	13.6	4.4	15.7	28.7	55.6
Lloydminister (Alberta)	0.966	15.0	11.8	4.3	12.9	38.4	48.7
Qayarah (Iraq)	0.964	15.3	15.6	8.4	20.4	36.1	43.5
Boscan (Venezuela)	0.998	10.3	10.4[b]	5.6	11.9	24.1	64.0

Athabasca (Canada)	1.030	5.9	18.5	4.9	16.9	34.1	49.0
Trinidad Lake	1.070	0.7	10.8[c]	6.2	33.3	29.4	31.9[d]
Bermudez Lake	1.058	2.2	13.4[c]	4.0	35.3	14.4	39.6[d]
Neuchatel (Switzerland)	1.016	7.8	—	—	19.9	32.2	42.2[d]
Selenitza (Albania)	1.080	—	—	6.9	45.2	18.8	24.5
P.R. Springs (Utah)	0.998	10.3	12.5[b]	0.8	16.0	84.0	
Asphalt Ridge (Utah)	0.986	12.0	9.1[b]	0.4	3.4	96.6	
Tar Sand Triangle	0.992	11.1	21.6[b]	4.4	26.0	74.0	
Residua							
Wilmington (>485°C; >905°F)	1.010	8.6	18.0[b]	2.1	12.8	87.2	
Mexico[e]	1.015	7.9	28.9[c]	6.4	23.2	76.8	
California[e]	1.065	1.4	24.8[c]	1.2	28.8	71.2	
Gulf Coast[e]	1.022	6.9	25.7[c]	0.8	27.8	72.2	
Venezuela	1.012	8.3	32.0	3.1	11.5	29.3	46.1[d]
Kuwait (>565°C; >1050°F)	1.033	5.5	23.1	5.5	11.1	39.4	49.5

[a] *Source*: Reproduced from Koots, J. A., and J. G. Speight, *Fuel*, 1975, *54*, 180, by permission of the publisher, IPC Business Press Ltd. ©.
[b] Ramsbottom carbon residue.
[c] Residue on ignition—reported as fixed carbon (mmf).
[d] The remainder appears as asphaltic acids and asphaltic anhydrides.
[e] Cut point not given.

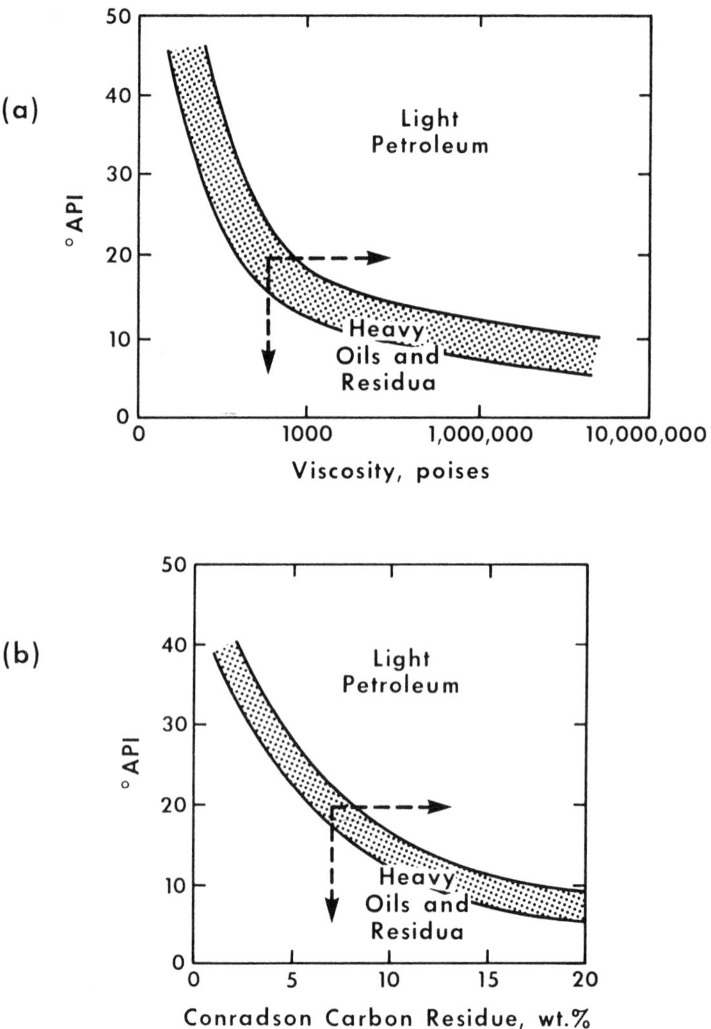

Figure 15 Generalized relationship between API gravity and (a) viscosity; (b) carbon residue.

determination of viscosity, with the units of viscosity varying depending on the method. However, viscosity is often expressed as "stokes" or "poises" (poises = stokes × specific gravity) and the viscosity of crude oils may vary from several stokes (or poises) for the lighter crude oils to several thousand stokes (or poises) for the

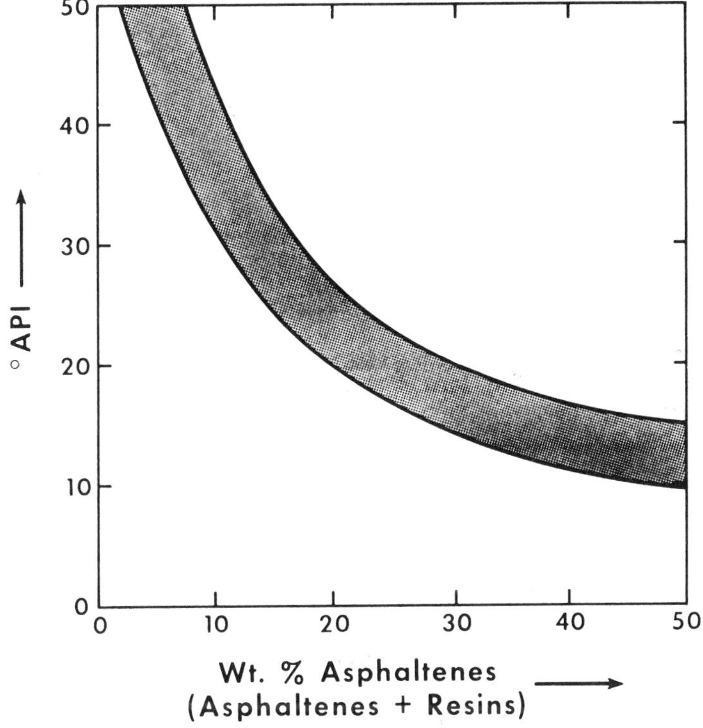

Figure 16 Generalized relationship between API gravity and asphaltics (asphaltenes plus resins) content of crude oils.

heavier crude oils even to several million stokes (or poises) for the bitumens and residua (Figure 15). Viscosity data may also be reported as Saybolt universal seconds (SUS) or as Saybolt fural seconds (SFS) which, by means of conversion factors, can be reported as stokes.

The Saybolt universal viscosity (ASTM D88) is the time in seconds required for the flow of 60 ml of petroleum from a container, at constant temperature, through a calibrated orifice. The Saybolt fural viscosity (ASTM D88) is determined in a similar manner except that a larger orifice is employed.

It is possible to interconvert the several viscosity scales (Figure 18), especially Saybolt to kinematic viscosity (ASTM D2161):

$$\text{Kinematic viscosity} = a \times \text{Saybolt sec} + \frac{b}{\text{Saybolt sec}}$$

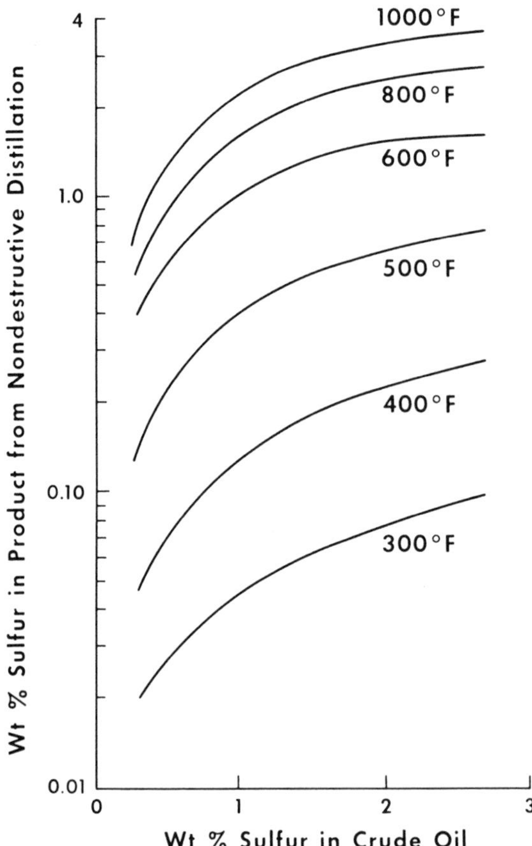

Figure 17 Generalized relationship of sulfur content of distillates to sulfur content of the original crude oil.

where a and b are constants. The Saybolt universal viscosity equivalent to a given kinematic viscosity varies slightly with the temperature at which the determination is made because the temperature of the calibrated receiving flask used in the Saybolt method is not the same as that of the oil. Conversion of kinematic viscosities from 2 to 70 centistokes at 100 and 210°F to equivalent Saybolt universal viscosities in seconds is achieved using tabular data (Table 6).

Appropriate multipliers are listed to convert kinematic viscosities over 70 centistokes:

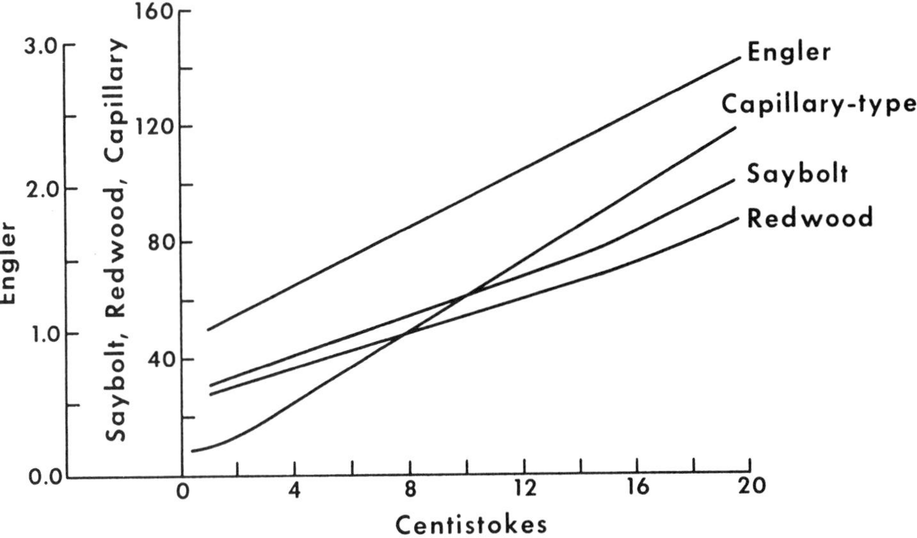

Figure 18 Interrelationships of various viscosimeter scales.

Table 6 Conversion of Kinematic Viscosities (2–70 Centistokes) to Saybolt Viscosities at 100 and 210°F (ASTM D2161)

Kinematic viscosity (centistokes)	Saybolt sec	
	100°F (38°C)	210°F (99°C)
2	32.6	32.9
10	58.9	59.3
20	97.8	98.5
30	140.3	142.3
40	186.3	187.6
50	232.1	233.8
60	278.3	280.2
70	324.4	326.7

Saybolt sec at 100°F (38°C) = cs × 4.635

Saybolt sec at 210°F (99°C) = cs × 4.667

The viscosity-temperature coefficient of a lubricating oil is an important expression of its suitability. A convenient number to express this property is very useful and, hence, a viscosity index (ASTM D2270) was derived:

$$\text{Viscosity index} = \frac{L - U}{L - H} \times 100$$

where L and H are the viscosities of the zero and 100 index reference oils [both having the same viscosity at 210°F (99°C)] and U is that of the unknown, all at 100°F (38°). Originally the viscosity index was calculated from Saybolt viscosity data, but subsequently figures were provided for kinematic viscosities.

The classification of lubricating oils by viscosity is a matter of some importance. A useful system is that of the Society of Automotive Engineers (SAE). Each oil class carries an index designation (Table 7). For those classes designated by letter and number, maximum and minimum viscosities are specified at 0°F (-18°C); those designated by number only are specified in viscosities at 210°F (99°C). Viscosity is also used in specifying several grades of fuel oils and in setting the requirements for kerosenes and insulating oils.

Table 7 Viscosity Values (Saybolt Universal Sec) for Lubricating Oils

SAE designation	Viscosity (SUS)			
	At 0°F		At 210°F	
	Min.	Max.	Min.	Max.
5W		4,000		
10W	6,000	Less than 12,000		
20W	12,000	48,000		
20			45	Less than 58
30			58	Less than 70
40			70	Less than 85
50			85	110

D. Volatility

The volatility of a liquid or liquefied gas may be defined as its tendency to vaporize, i.e., to change from the liquid to the vapor or gaseous state. Because one of the three essentials for combustion in a flame is that the fuel must be in the gaseous state, volatility is a primary characteristic of liquid fuels. Thus, their vaporizing tendencies are the basis for the general characterization of liquid petroleum fuels such as liquefied petroleum gas, natural gasoline, motor and aviation gasolines, naphthas, kerosene, gas oils, diesel fuels, and fuel oils (ASTM D2715).

The flash point of petroleum or a petroleum product is the temperature to which the product must be heated under the specified conditions of the method to give off sufficient vapor to form a mixture with air that can be ignited momentarily by a specified flame (ASTM D56, D92, D93), whereas the fire point is the temperature to which the product must be heated under the prescribed conditions of the method to burn continuously when the mixture of vapor and air is ignited by a specified flame (ASTM D92).

From the viewpoint of safety flash points are of most significance at or slightly above the maximum temperatures (30–60°C) that may be encountered in storage, transportation, and use of liquid petroleum products, either in closed or in open containers.

One other aspect of volatility that receives considerable attention is the vapor pressure of petroleum and its constituent fractions. The vapor pressure is the force exerted on the walls of a closed container by the vaporized portion of a liquid. Conversely it is the force which must be exerted on the liquid to prevent it from vaporizing further (ASTM D323). The vapor pressure increases with temperature for any given gasoline, liquefied petroleum gas, or other product. The temperature at which the vapor pressure of a liquid, either a pure compound or a mixture of many compounds, equals 1 atmosphere (14.7 lb/in.^{2}, absolute) is designated as the boiling point of the liquid.

Distillation involves the general procedure of vaporizing the petroleum liquid in a suitable flask either at atmospheric pressure (ASTM D86, D216, D285, D447, and D2892) or at reduced pressures (ASTM D1160), and the data are reported in terms of one or more of the following items (see Figure 4 in Chap. 2 for a simple distillation curve).

(1) *Initial boiling point* is the thermometer reading in the neck of the distillation flask when the first drop of distillate leaves the tip of the condenser tube. This reading is materially affected by a number of test conditions, namely, room temperature, rate of heating, condenser temperature, and several others.

(2) *Distillation temperature* is usually observed when the level of the distillate reaches each 10% mark on the graduated receiver, with

the temperatures for the 5 and 95% marks often included. Conversely, the volume of the distillate in the receiver, i.e., the percent recovered, is often observed at specified thermometer readings.

(3) *End point or maximum temperature* is the highest thermometer reading observed during distillation. In most cases it is reached when all of the sample has been vaporized. If a liquid residue remains in the flask after the maximum permissible adjustments in heating rate are made, the fact is recorded as indicative of the presence of very high-boiling compounds.

(4) *Dry point*, for special purposes such as solvents and relatively pure hydrocarbons, is the thermometer reading at the instant the flask becomes dry. For these purposes dry point is considered to be more indicative of the final boiling point then end-point or maximum temperature.

(5) *Recovery* is the total volume of distillate recovered in the graduated receiver.

(6) *Residue* is the liquid material, mostly recondensed vapors, left in the flask after it has been allowed to cool at the end of distillation. The residue is measured by transferring it to an appropriate small graduated cylinder. Low or abnormally high residues indicate the absence of presence, respectively, of high-boiling components.

(7) *Total recovery* is the sum of the liquid recovery and residue.

(8) *Distillation loss* is determined by subtracting the total recovery from 100%. It is, of course, the measure of the portion of the vaporized sample which does not condense under the conditions of the test. Like the initial boiling point, distillation loss is affected materially by a number of test conditions, namely, condenser temperature, sampling and receiving temperatures, barometric pressure, heating rate in the early part of the distillation, etc. Provisions are made for correcting high distillation losses for the effect of low barometric pressure because of the practice of including distillation loss as one of the items in some specifications for motor gasolines.

(9) *Percent evaporated* at a specific thermometer reading or other distillation temperature, or the converse, is often reported instead of, or in addition to, percent recovered. The amounts that have been evaporated are usually obtained by plotting observed thermometer readings against the corresponding observed recoveries plus, in each case, the distillation loss. The initial boiling point is plotted with the distillation loss as the percent evaporated. Distillation data, in terms of percent evaporated, are considerably more reproducible, particularly for the more volatile products. As examples, distillation specifications for a conventional petroleum and a heavy crude oil are reproduced in Table 8.

E. Carbon Residue

The carbon residue of a crude oil is a measure of the thermal coke-forming propensity and bears a general relationship to the sulfur content, nitrogen content, asphaltene (or asphaltene plus resin–asphaltics) content, and viscosity of the oil (Figure 19).

There are two methods for determining the carbon residue—the Conradson method (ASTM D189) and the Ramsbottom method (ASTM) D524)—and both are equally applicable to the high-boiling fractions of crude oils which decompose to volatile material and coke when distilled at a pressure of 1 atmosphere. Heavy oils and residua which contain metallic constituents (and distillation of crude oils concentrates these constituents in the residua) will have erroneously high carbon residues. Thus, the metallic constituents must first be removed from the oil or they can be estimated as ash by complete burning of the coke after carbon residue determination.

Although both methods of carbon residue determination involve burning of the sample in a limited amount of oxygen, the techniques employed are quite different, but nevertheless it is possible to interconnect the data (Figure 20). Indeed, it may be possible, for any one particular heavy feedstock and one particular refining operation, to make a rough correlation between carbon residue and product yield.

To overcome the extremely small values of carbon residue obtained by the Conradson and Ramsbottom methods when applied to the lighter distillate fuel oils, it is customary to distill such products to 10% residual oil and determine the carbon residue thereof.

F. Metals Content (ASTM D4075)

The heavier oils and residua contain relatively high proportions of metals either in the form of salts or as organometallic constituents (such as the metalloporphyrins), which are extremely difficult to remove from the feedstock. Indeed, the nature of the process by which residua are produced virtually dictates that all of the metals in the original crude oil will be concentrated in the residuum (Speight, 1981). The exception here is those metallic constituents which may actually volatilize under the distillation conditions and appear in the higher boiling distillates. The deleterious effect of metallic constituents on the catalyst are known and serious attempts have been made to develop catalysts that can tolerate a high concentration of metals without serious loss in catalyst activity or catalyst life.

Table 8 Distillation Specifications for (a) A Light (Leduc) Crude Oil and (b) A Heavy (Athabasca) Crude Oil (Bitumen)

Distillation (U.S. Bureau of Mines Routine Method)

(a) Fraction number	Cut at °C	Cut at °F	%	Sum (%)	Sp. gr 60°F	°API 60°F	CI	SUV 100°F	Cloud test (°F)
Stage 1: Distillation at atmospheric pressure, 758 mm Hg									
1	50	122	2.8	2.8	0.656	82.4	-		
2	75	167	3.3	6.1	0.677	77.5	11		
3	100	212	5.1	11.2	0.713	67.0	18		
4	125	257	6.8	18.0	0.741	59.5	22		
5	150	302	6.2	24.2	0.761	54.4	24		
6	175	347	5.9	30.1	0.779	50.1	26		
7	200	392	5.0	35.1	0.794	46.7	27		
8	225	437	5.0	40.1	0.809	43.4	28		
9	250	482	5.1	45.2	0.821	40.9	29		
10	275	527	6.1	51.3	0.833	38.4	30		
Stage 2: Distillation continued at 40 mm Hg pressure									
11	200	392	5.2	56.5	0.849	35.2	33	40	10
12	225	437	4.8	61.3	0.858	33.4	34	46	30
13	250	482	5.0	66.3	0.870	31.1	36	57	50
14	275	527	4.7	71.0	0.882	28.9	39	85	70
15	300	572	4.8	75.8	0.891	27.3	40	155	90
Residuum			20.2	96.0	0.946	18.1			

Carbon residue of residuum: 6.0%

Distillation (U.S. Bureau of Mines Routine Method)

(b) Fraction number	Cut at °C	Cut at °F	%	Sum (%)	Sp. gr. 60°F	°API 60°F	CI	SUV 100°F	Cloud test (°F)
Stage 1: Distillation at atmospheric pressure, 762 mm Hg									
1	50	122							
2	75	167							
3	100	212							
4	125	257							
5	150	302	0.9	0.9					
6	175	347	0.8	1.7	0.809	43.4			
7	200	392	1.1	2.8	0.823	40.4	41		
8	225	437	1.1	3.9	0.848	35.4	47		
9	250	482	4.1	8.0	0.866	31.9	50		
10	275	527	11.9	19.9	0.867	31.7	46		
Stage 2: Distillation continued at 40 mm Hg pressure									
11	200	392	1.6	21.5	0.878	20.7	47	36	Below 0
12	225	437	3.2	24.7	0.929	20.8	67	66	Below 0
13	250	482	6.1	30.8	0.947	17.9	73	118	Below 0
14	275	527	6.4	37.2	0.958	16.2	75	178	Below 0
15	300	572	10.6	47.8	0.972	14.1	78	508	Below 0
Residuum			49.5	97.3					

Carbon residue of residuum: 39.6%
Carbon residue of crude: 19.6%

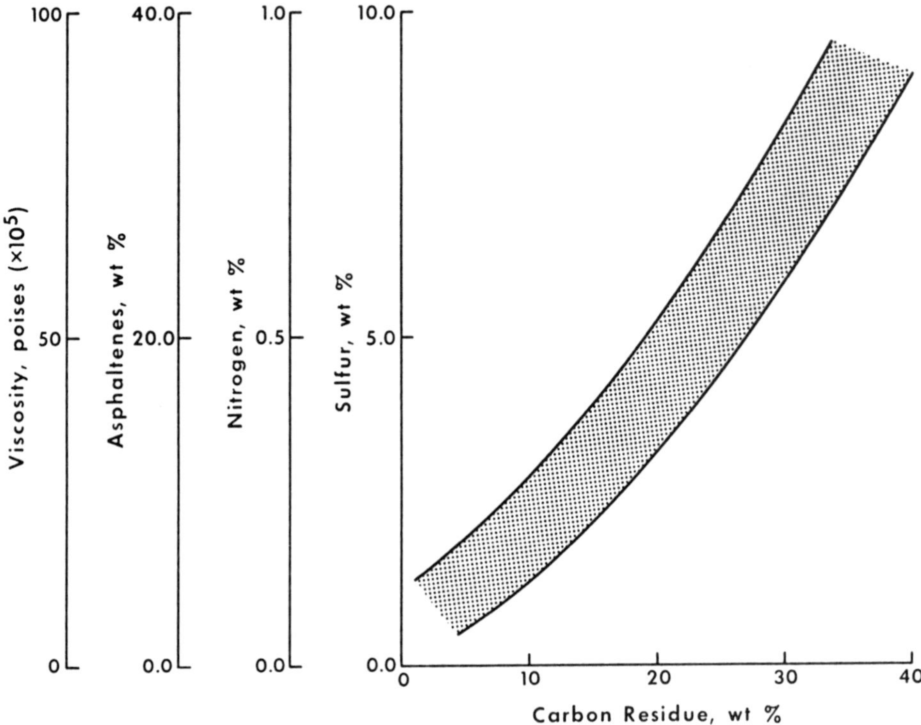

Figure 19 Relationship of crude oil carbon residue to other crude oil parameters.

G. Aniline Point

The aniline point of a liquid was originally defined as the consolute or critical solution temperature of the two liquids, i.e., the minimum temperature at which they are miscible in all proportions. The term is now most generally applied to the temperature at which exactly equal parts of the two are miscible.

Although an arbitrary index (ASTM D611, D801), the aniline point is of considerable value in the characterization of petroleum products. For oils of a given type, it increases slightly with molecular weight, while for those of given molecular weight, it increases rapidly with paraffinicity. As a consequence, it was one of the first properties proposed for the group analysis of petroleum products with respect to aromatic and naphthene content.

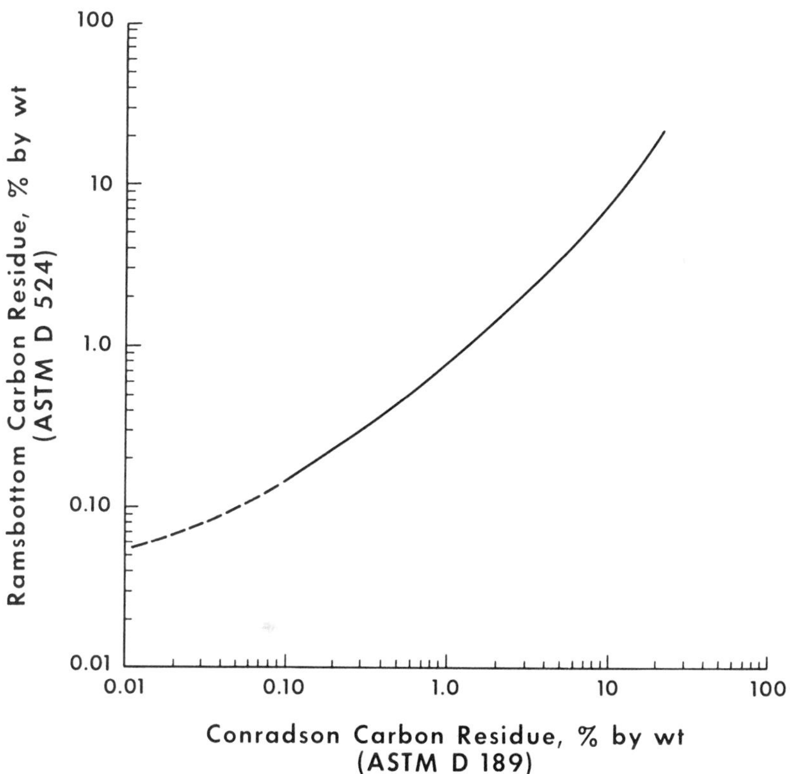

Figure 20 Interconversion of Ramsbottom and Conradson carbon residue data.

H. Liquefaction and Solidification

Petroleum and the majority of its products are liquids at ambient temperature and problems which may arise from solidification during normal use are not common. Nevertheless, the melting point is a test (ASTM D87, D127) that is widely used by suppliers of wax and by the wax consumers; it is particularly applied to the highly paraffinic or crystalline waxes.

Petroleum oils become more or less plastic solids when cooled to sufficiently low temperatures. This is due to the congealing of the various hydrocarbons which constitute the oil. The cloudpoint of a petroleum oil is the temperature at which paraffin wax or other solidifiable compound present in the oil is chilled under definitely prescribed conditions (ASTM D2500, D3117). As cooling is continued, all petroleum oils become more and more viscous and flow becomes slower and slower. The pour point of a petroleum oil is the lowest temperature at which the oil will pour or flow under definitely prescribed conditions when it is chilled without disturbance at a standard rate (ASTM D97).

The solidification characteristics of a petroleum product depend on its grade or kind. For plastic solids such as greases, the temperature of interest is that at which fluidity occurs, commonly known as the dropping point; the dropping point of a grease is the temperature at which the grease passes from a plastic solid to a liquid state and begins to flow under the conditions of the test (ASTM D566, D2265). For another type of plastic solid, including petrolatum and microcrystalline wax, both melting point and congealing point are of interest. The melting point of a wax is the temperature at which the wax becomes sufficiently fluid to drop from the thermometer; the congealing point is the temperature at which melted petrolatum ceases to flow when allowed to cool under definitely prescribed conditions (ASTM D938).

For another type of solid, paraffin wax, the solidification temperature is of interest. For such purposes, melting point is defined as the temperature at which the melted paraffin wax begins to solidify, as shown by the minimum rate of temperature change, when cooled under prescribed conditions.

I. Surface and Interfacial Tension

The surface tension of petroleum and its products has been studied for many years, but the narrow range of values (around 24–38 dynes/cm) for such widely diverse materials as gasoline (26 dynes/cm), kerosene (30 dynes/cm), and the lubricating fractions (34 dynes/cm) has rendered the surface tension of little value for any attempted characterization.

On the other hand, the interfacial tension of petroleum, and especially of petroleum products, against aqueous solutions provides valuable information (ASTM D971). The interfacial tension of petroleum is subject to the same constraints, i.e., differences in composition, molecular weight, etc., as is the surface tension. Where oil-water systems are involved, the pH of the aqueous phase influences the tension at the interface; the change is small for highly refined oils, but increased pH will cause a rapid decrease for poorly refined, contaminated, or slightly oxidized oils.

The interfacial tension between oil and distilled water provides an indication of compounds in the oil that have an affinity for water. The measurement of interfacial tension has received special attention because of its use in assessing when an oil in constant use will reach the limit of its serviceability. For example, the interfacial tension of turbine oil against water is lowered by the presence of oxidation products, impurities from the air or rust particles, and certain antirust compounds intentionally blended in the oil. A depletion of the antirust additive may cause an increase of interfacial tension, whereas the formation of oxidation products or contamination with dust and rust lowers the interfacial tension.

J. Specific Heat

Specific heat is defined as the quantity of heat required to raise a unit mass of material through 1° of temperature (ASTM D2766). Specific heats are extremely important engineering quantities in refinery practice because they are used in all calculations on heating and cooling petroleum products.

K. Conductivity

The electrical conductivity of hydrocarbon oils is also exceedingly small (ASTM D3114), being of the order of 10^{-19}–10^{-12} ohm/cm^{-1}. Available data indicate that the observed conductivity is frequently more dependent on the method of measurement and the presence of trace impurities than on the chemical type of the oil. Conduction through oils is not ohmic, i.e., the current is not proportional to field strength; in some regions it is observed to increase exponentially with the latter. Time effects are also observed, the current being at first relatively large and decreasing to a smaller steady value. This is due partly to electrode polarization and partly to ions removed from the solution. Most oils increase in conductivity with rising temperatures.

L. Refractive Index

The refractive index is the ratio of the velocity of light in a vacuum to the velocity of light in the substance. The measurement of the refractive index is very simple (ASTM D1218), requires small quantities of material, and, consequently, has found wide use in the characterization of hydrocarbons and petroleum samples. For closely separated fractions of similar molecular weight, the values increase in the order paraffin, naphthene, and aromatic; those for polycyclic naphthenes and polycyclic aromatics are usually higher than for the corresponding monocyclics. For a series of hydrocarbons of essentially the same type, the refractive index increases with molecular weight, especially in the paraffin series. Thus, the refractive index can be used to provide valuable information about the composition of hydrocarbon (petroleum) mixtures; as with density, low values indicate paraffinic materials and higher values indicate the presence of aromatic compounds. However, the combination of refractive index and density may be used to provide even more definite information about the nature of a hydrocarbon mixture and, hence, the use of the refractivity intercept (n - d/2) (ASTM D2159).

IV. USE OF THE DATA

The data derived from any one or more of the evaluation techniques described above (see, for example, Table 9) can be employed to give the refiner an indication of the means by which the crude feedstock should be processed. However, it must be emphasized that to proceed from the raw evaluation data to full-scale production is not the preferred step; further evaluation of the processability of the feedstock is usually necessary through the use of a pilot scale operation. To take the evaluation of a feedstock one step further, it may then be possible to develop correlations between the data obtained from the actual plant operations (as well as the pilot plant data) with one or more of the physical properties determined as part of the initial feedstock evaluation. For example, it may be possible to calculate yields for conventional delayed coking operations by using the carbon residue and the API gravity (Table 10) of the feedstock.

However, it is essential that when such data are derived the parameters employed be carefully specified. For example, the data presented in the tables were derived on the basis of straight-run residua having API gravities less than 18°, the gas oil end point was of the order of 470–495°C (875–925°F), the gasoline end point was 205°C (400°F), and the pressure in the coke drum was standardized at 35–45 psi. Obviously, there are benefits to the derivation of such specific data but the numerical values, while only representing an approximation, may vary substantially when applied to different feedstocks (Table 10).

Table 9 Properties of Different Cut Point Residua Produced From Tia Juana Crude Oil

Property	Unit									
Boiling range	°F	Whole crude	>430	>565	>650	>700	>750	>850	>950	>1050
	°C		>220	>295	>345	>370	>400	>455	>510	>565
Yield on crude	Vol. %	100.0	70.2	57.4	48.9	44.4	39.7	31.2	23.8	17.9
Gravity	°API	31. 6	22.5	19.4	17.3	16.3	15.1	12.6	9.9	7.1
Specific gravity		0.8676	0.9188	0.9377	0.9509	0.9574	0.9652	0.9820	1.007	1.0209
Sulfur	Wt.%	1.08	1.42	1.64	1.78	1.84	1.93	2.12	2.35	2.59
Carbon residue (Conradson)	Wt.%		6.8	8.1	9.3	10.2	11.2	13.8	17.2	21.6
Nitrogen	Wt.%				0.33	0.36	0.39	0.45	0.52	0.60
Pour point	°F	-5	15	30	45	50	60	75	95	120
Viscosity										
Kinematic @ 100°F	cs	10.2	83.0	315	890	1590	3100			
@ 210°F			9.6	19.6	35.0	50.0	77.0	220	1010	7959
Furol @ 122°F	(SFS)sec			70.6	172	292	528			
@ 210°F						25.2	37.6	106	484	3760
Universal @ 210°F	(SUS)sec		57.8	96.8	165	234	359	1025		
Metals:										
Vanadium	ppm				185					450
Nickel	ppm				25					64
Iron	ppm				28					48

Table 10 Examples of Evaluation Data for the Estimation of Product Yields from the Delayed Coking of (a) Wilmington and (b) East Texas Crude Oil Residua

(a)

Coke, wt.% $= 39.68 - 1.60 \times {}^{\circ}\mathrm{API}$

Gas ($\leq C_4$), wt.% $= 11.27 - 0.14 \times {}^{\circ}\mathrm{API}$

Gasoline, wt.% $= 20.5 - 0.36 \times {}^{\circ}\mathrm{API}$

Gas oil, wt.% $= 28.55 + 2.10 \times {}^{\circ}\mathrm{API}$

Gasoline, vol.% $= \left(\dfrac{186.5}{131.5 + {}^{\circ}\mathrm{API}}\right)$ (gasoline, wt.%)

Gas oil vol.% $= \left(\dfrac{155.5}{131.5 + {}^{\circ}\mathrm{API}}\right)$ (gas oil, wt.%)

(b)

Coke, wt.% $= 45.76 - 1.78 \times {}^{\circ}\mathrm{API}$

Gas ($<C_4$), wt.% $= 11.92 - 0.16 \times {}^{\circ}\mathrm{API}$

Gasoline, wt.% $= 20.5 - 0.36 \times {}^{\circ}\mathrm{API}$

Gasoline, vol.% $= \left(\dfrac{186.5}{131.5 + {}^{\circ}\mathrm{API}}\right)$ (gasoline, wt.%)

Gas oil, vol.% $= \left(\dfrac{155.5}{131.5 + {}^{\circ}\mathrm{API}}\right)$ (gas oil, wt.%)

4

Fractionation

I	Distillation	122
	A. Atmospheric Pressure	122
	B. Reduced Pressure	122
II	Solvent Treatment	124
	A. Influence of Solvent Type	124
	B. Influence of Temperature	129
	C. Influence of Degree of Dilution	129
III	Adsorption	130
	A. USBM-API Method	131
	B. SARA Method	132
	C. Other Methods	134
IV	Use of Composition Data	140

The fractionation methods available to the petroleum industry are varied but do allow a reasonably effective degree of separation. However, the problem is to separate the petroleum constituents without any alteration to their molecular structure and the procedures employed are those which segregate the constituents according to molecular size and type (Pfeiffer, 1950; van Nes and Westen, 1951).

It is necessary to understand that the names of the various fractions arise from the operational means by which the fractions were derived. Thus, when referring to a particular fraction, the name must be qualified for clarity, e.g., pentane-asphaltenes, heptane-asphaltenes, clay-resins, etc. (Speight, 1981).

The order in which the several fractionation methods are used is determined not only by the nature and/or composition of the crude oil but also by the effectiveness of a particular process and its compatability with the other separation procedures that will be employed. Even though there are wide variations in the nature of petroleum (Figure 1), there have been many attempts to devise standard methods of petroleum fractionation (Hoiberg, 1960).

I. DISTILLATION

A. Atmospheric Pressure

Distillation has found wide applicability in petroleum chemistry but it is generally recognized that the fractions separated by distillation are only rarely, if at all, suitable for designation as a petroleum product. Each usually requires some degree of refining which, of course, varies with the impurities in the fraction and the desired properties of the finished product (Chaps. 6 and 7). Nevertheless, distillation is the most important fractionating process for the separation of petroleum hydrocarbons; it is an essential part of any refinery operation (Nelson, 1958; Bland and Davidson, 1967).

Insofar as petroleum is a complex mixture, there is little emphasis commercially on the isolation of the individual compounds. The aim of the distillation is predominantly the separation into several fractions of substantially broad boiling ranges.

The initial fractionation of crude oil essentially involves distillation of the material into various fractions (Table 1 in Chap. 5) which vary depending on the nature and composition of the crude oil.

The kerosene (stove oil) and light gas oil fractions are often referred to as "middle distillates" and usually represent the last fractions to be separated by distillation at atmospheric pressure. This leaves the fractions from the heavy gas oil and higher boiling material, which are collectively called atmospheric residuum or "reduced crude."

B. Reduced Pressure

Separation of the reduced crude into the constituent fractions requires that the distillation be carried out under reduced pressure because of the tendency of these higher boiling materials to participate

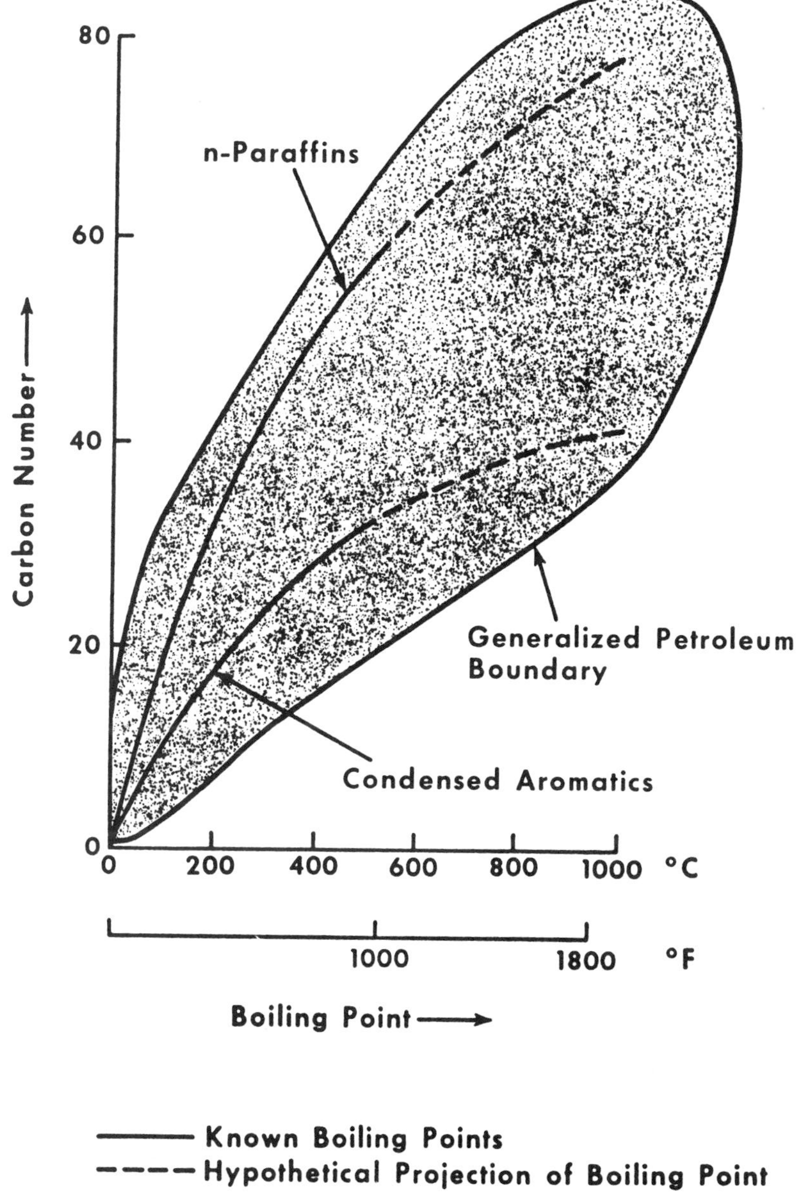

Figure 1 The constituents of petroleum fall into a molecular weight and boiling point range where very little is known about model compounds.

in rearrangement, condensation, or decomposition reactions at temperatures above 350°C (660°F).* The lowering of the pressure is accompanied by a corresponding decrease in the boiling points of the individual constituents. For example, a material boiling near 350°C at 1 atmosphere (760 mm Hg) may boil over 100° lower (around 230°C; 445°F) at 25–30 mm Hg and the potential for thermal decomposition is markedly reduced.

II. SOLVENT TREATMENT

Fractionation of petroleum by distillation is an excellent means by which the more volatile constituents can be isolated and studied. However, the nonvolatile (vacuum) residuum, which may actually constitute from 1 to 50% of the petroleum, cannot be fractionated by distillation without the possibility of thermal decomposition and, as a result, alternate methods of fractionation have been developed.

A. Influence of Solvent Type

If chosen carefully, solvents effect a separation between the constituents of residua, bituminous materials, and virgin petroleum according to differences in molecular weight and aromatic character.

On the basis of the solubility in a variety of solvents, it has become possible to distinguish between the various operational fractions of petroleum and other feedstocks (Figure 2; Pfeiffer, 1950; vanNes and Westen, 1951; Traxler, 1961; Speight, 1980, 1981):

1. Carboids–insoluble in carbon disulfide (or pyridine); insoluble in benzene; insoluble in low molecular weight paraffins
2. Carbenes–soluble in carbon disulfide (or pyridine); insoluble in benzene; insoluble in low molecular weight paraffins
3. Asphaltenes–soluble in carbon disulfide (or pyridine); soluble in benzene; insoluble in low molecular weight paraffins
4. Maltenes–soluble in carbon disulfide (or pyridine); soluble in benzene; soluble in low molecular weight paraffins

---

*It is generally recognized that reactions initiated by thermal means become significant above 350°C but are usually slow or inconsequential at temperatures below 350°C. For this reason, 350°C is often referred to as the "cracking" (thermal decomposition) temperature.

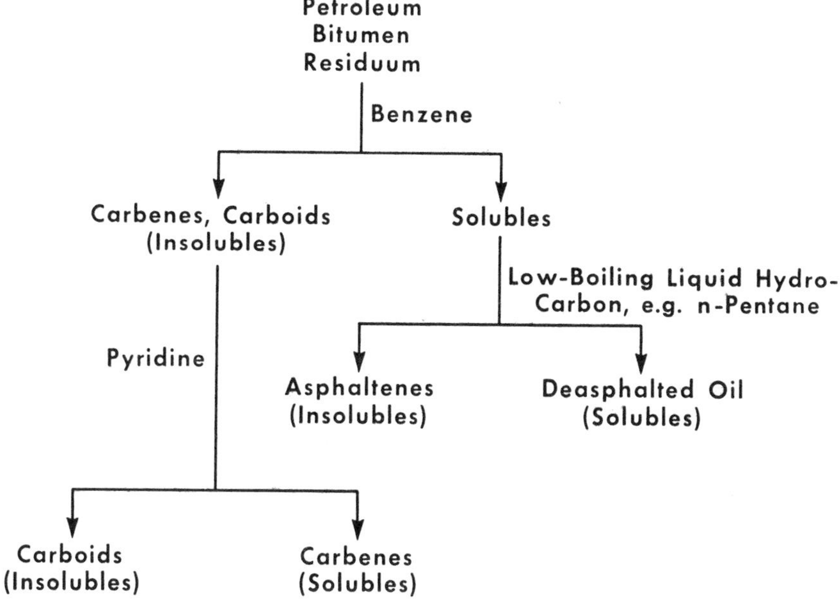

Figure 2 Schematic representation of the fractionation of crude oil into higher molecular weight fractions.

Most petroleums contain practically no carboids and carbenes but residua from the cracking processes may contain 2% by weight or more. Some of the highly paraffinic crude oils may contain only small portions of asphaltenes.

The separation of crude oils into two fractions (asphaltenes and maltenes) is qualitatively and quantitatively reproducible and is conveniently achieved using low molecular weight paraffinic hydrocarbons. These liquids have selective solvency for hydrocarbons and simple, relatively low molecular weight hydrocarbon derivatives. The more complex higher molecular weight compounds are precipitated by addition of $\geq$40 volumes of n-pentane or n-heptane in the methods generally preferred at present.

Variation in solvent type also causes significant changes in asphaltene yield. For example, branched chain paraffins or terminal olefins do not precipitate the same amount of asphaltenes as do the corresponding normal paraffins (Figure 3). The solvent power of the solvents (i.e., the ability of the solvents to dissolve asphaltenes) increases in the order 2-methylparaffin < n-paraffin < terminal olefin (Mitchell and Speight, 1973).

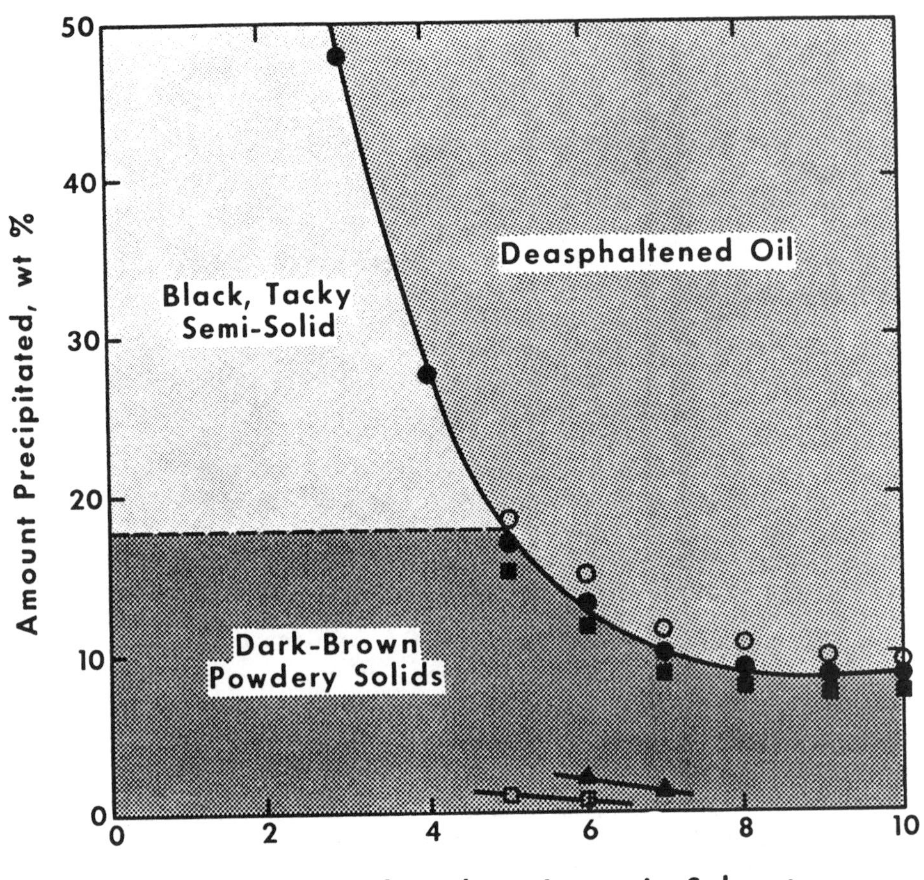

Figure 3 Precipitate yields from a natural bitumen using various solvent types.

Cycloparaffins (naphthenes) have a remarkable effect on asphaltene yield and give results totally unrelated to those from any other nonaromatic solvent (Figure 3). For example, when cyclopentane, cyclohexane, or their methyl derivatives are employed as precipitating media, only about 1% of the material remains insoluble.

Further fractionation of petroleum is also possible by variation of the solvent. For example, there are procedures for solvent treatment at low temperatures (-4°C to -20°C; 4–25°F) which will bring about fractionation of the maltenes. The hydrocarbon solvents pentane and hexane have been claimed adequate for this purpose but may not be successful for feedstocks.

Other fractionation procedures (Hoiberg, 1960) involving the use of solvents include a procedure using polar solvents for separation into five fractions by the following stepwise extractions: (1) asphaltenes, insoluble in hexane; (2) "hard" resins, insoluble in 80:20 isobutyl alcohol-cyclohexane mixture; (3) waxes, insoluble in 1:2 mixture of acetone and methylene chloride at 0°C (-18°F); (4) "soft" resins, insoluble in isobutyl alcohol; and (5) oils, the balance of the sample which remains soluble in insobutyl alcohol (Figure 4).

A fractionation procedure was also devised using *n*-butanol and acetone as the solvents (Figure 5) but the preliminary separation of the asphaltenes was incomplete. The method consisted of (1) separation of the "asphaltics" by *n*-butanol and (2) separation of the butanol-soluble portion into "paraffinics" and "cyclics" by chilling an acetone solution of the two, but has the unfortunate result that all three fractions obtained may contain asphaltenes.

A further method of fractionation (Figure 6) consisted of stepwise separation into the following fractions: (1) asphaltenes, precipitated by *n*-pentane; (2) resins, precipitated with propane and subdivided by fractionation with aniline into (a) soft resins and (b) hard resins; (3) wax, precipitated with methylisobutyl ketone; (4) oils, remaining fraction separated with acetone into (a) paraffinic oils and (b) naphthenic oils. A particularly interesting feature of this method is the subdivision of the resin fraction by solubility in aniline and the subdivision of the paraffinic fraction into three components: wax, paraffinic oils, and naphthenic oils.

Another all-solvent procedure (Figure 7) involves the use of acetone, which discharges a resin fraction, and dimethyl formamide, which then discharges a saturates fraction.

Fractionation by solvent treatment is often preferred because the components of the feedstock are not contacted by any external chemical other than the solvent and, hence, do not have the opportunity to undergo any chemical changes.

Such methods also allow the fractionation of feedstocks to be achieved without loss of material (on an absorbent) and produce fractions of varying polarity. The obvious benefit of such a technique

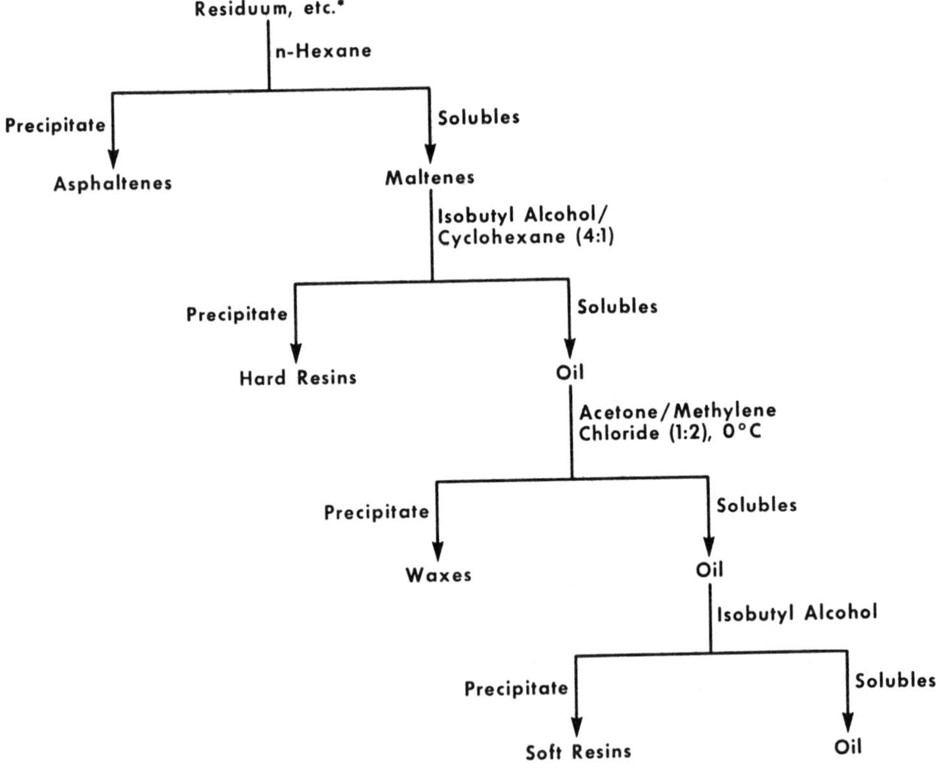

Figure 4 Fractionation using various solvents.

is the complete recovery of material thereby allowing a more quantitative and qualitative examination of the feedstocks.

The disadvantages of an all-solvent separation technique are that, first, in some instances, low temperatures (e.g., 0°C to -10°C and the like) are advocated as a means of effecting oil fractionation with solvents (Rostler, 1965; Speight, 1979). Such requirements may cause inconveniences in a typical laboratory operation by requiring a permanently cool temperature during the separation. Second, it must be recognized that large volumes of solvent may be required to effect a reproducible separation in the same manner as the amounts required for consistent asphaltene separation (American Society for Testing and Materials, 1984, ASTM D2006-70, D2007-80, D4124-84,

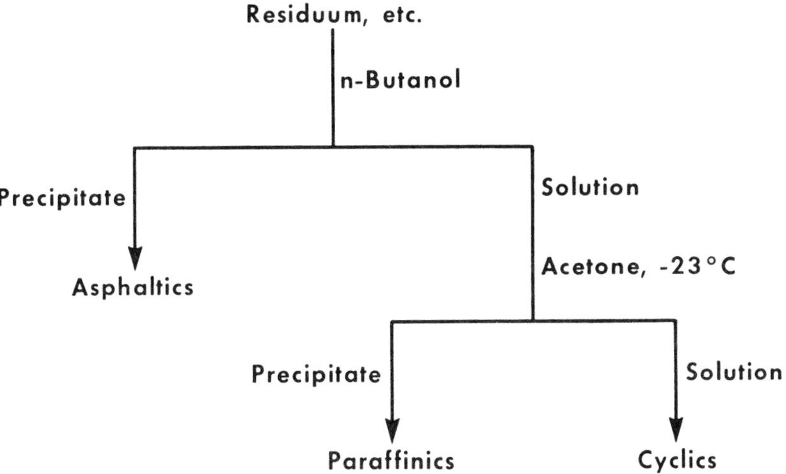

Figure 5 Fractionation using *n*-butanol and acetone.

D893-80; Institute of Petroleum, 1982, IP 143/82). Finally, it is also essential that the solvent be of sufficiently low boiling point that complete removal of the solvent from the product fraction can be effected.

Although not specifically included in the three main disadvantages of the all-solvent approach, it should also be recognized that the solvent must not react with the feedstock constituents.

B. Influence of Temperature

When the precipitating solvent is used in a large excess, the quantity of precipitate increases with increasing temperature (Pfeiffer, 1950). For example, at ambient temperatures (approx. 21°C; 70°F) Athabasca bitumen affords 17% by weight asphaltenes with pentane but at 35°C (95°F) 22.5% by weight "asphaltenes" are produced using the same bitumen/pentane ratio. Similar effects have been noted with other hydrocarbon solvents at temperatures up to 70°C (160°F).

C. Influence of Degree of Dilution

At constant temperature, the quantity of precipitate first increases with increasing ratio of solvent to feedstock and then reaches a maximum (Figure 8; Mitchell and Speight, 1973).

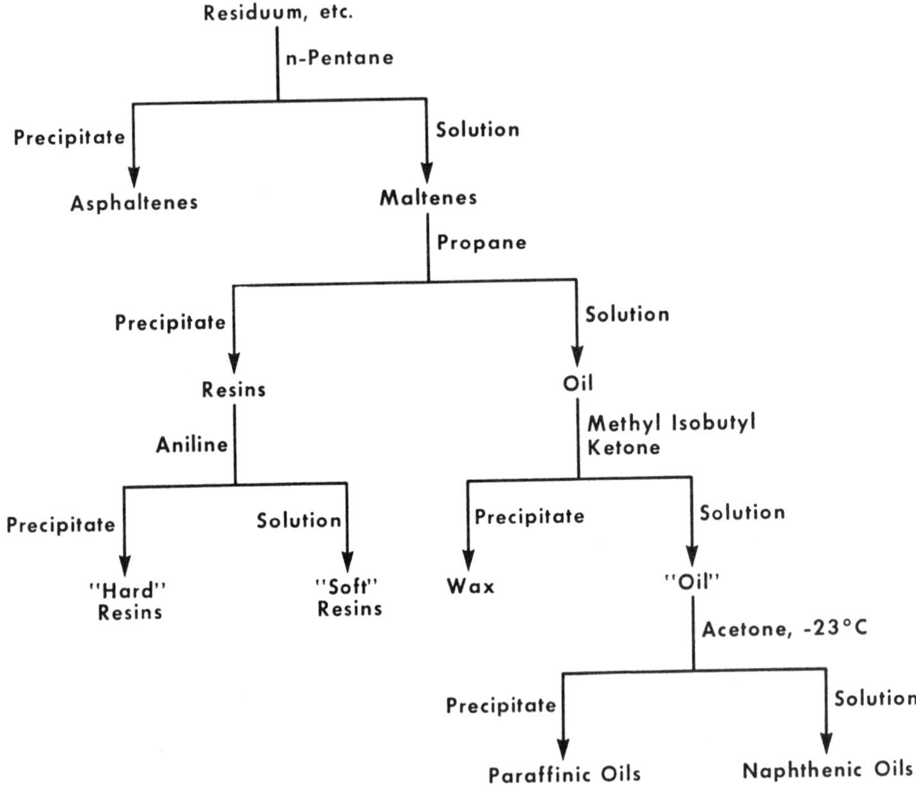

Figure 6 Fractionation using various solvents.

III. ADSORPTION

Two procedures have received considerable attention over the years: (1) the United States Bureau of Mines—American Petroleum Institute (USBM-API) method and (2) the saturates-aromatics-resins-asphaltenes (SARA) method. This latter method is often also called the saturates-aromatics-polars-asphaltenes (SAPA) method. It is these two methods that will represent the standard methods of petroleum fractionation. Other methods will also be noted, especially where the method has added further meaningful knowledge to compositional studies.

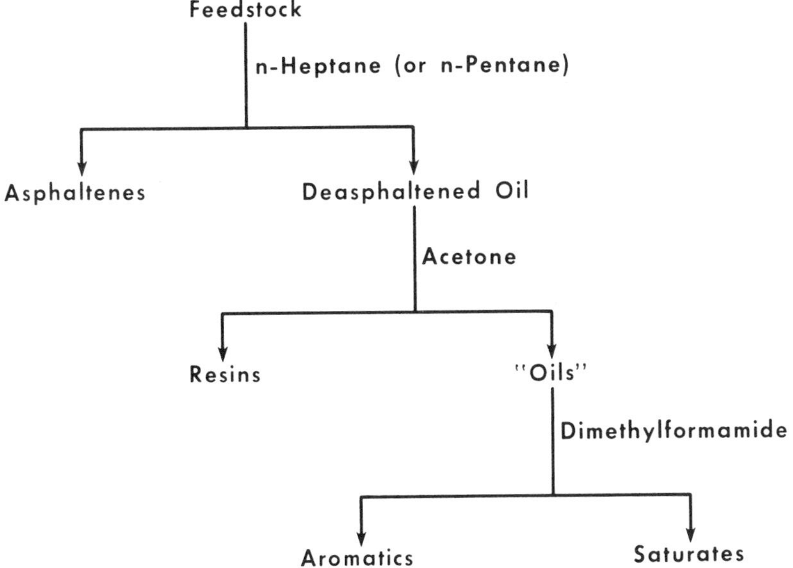

Figure 7 Schematic representation of an all-solvent fractionation procedure.

A. USBM-API Method

The USBM-API method (Figure 9) employs ion exchange chromatography and coordination chromatography with adsorption chromatography to separate heavy oils and residua into seven broad fractions: acids, bases, neutral nitrogen compounds, saturates, and mono-, di-, and polyaromatic compounds. The acid and base fractions are isolated by ion exchange chromatography, the neutral nitrogen compounds by complexation chromatography using ferric chloride, the saturates and aromatics by adsorption chromatography on activated alumina (Jewell et al., 1972a) or on a combined silica-alumina column (Hirsch et al., 1972; Jewell et al. 1972b).

The feedstock sample can be separated into chemically significant fractions that are suitable for analysis according to compound type (Jewell et al., 1972b; McKay et al., 1975). Although originally conceived for the separation of distillates, this method has been successfully applied to determining the composition of heavy oils, tar sand bitumens, and shale oils (Cummins et al., 1975; McKay et al.,

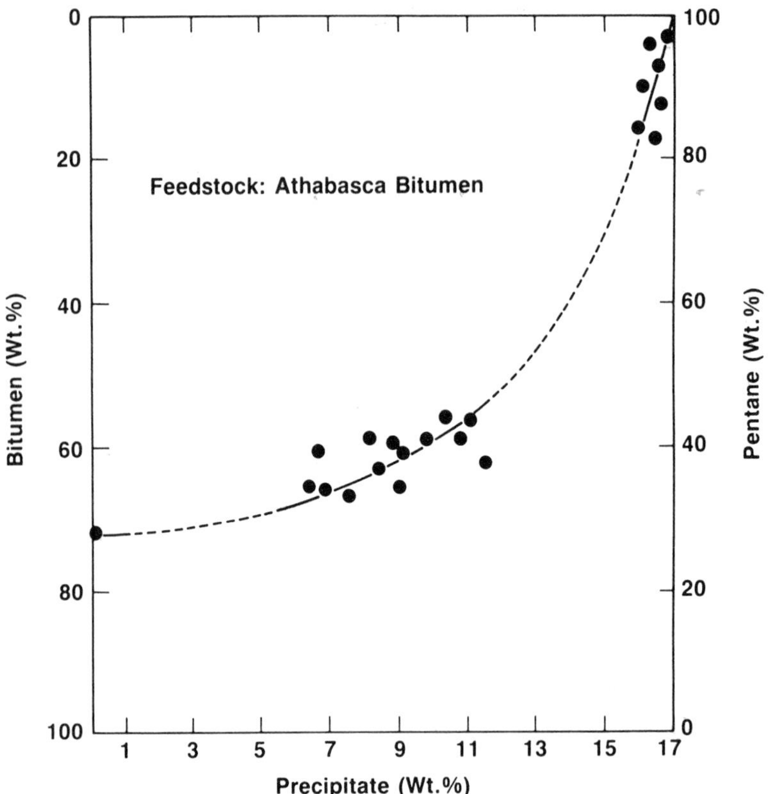

Figure 8 Variation of precipitate yield with the amount of solvent added.

1976; Bunger, 1977). Originally, the method required distillation of a feedstock into narrow boiling point fractions but whole feedstocks such as the bitumens from Utah or Athabasca tar sands (Selucky et al., 1977) have been separated into classes of compounds without prior distillation or without prior removal of asphaltenes. The latter finding is supported by the separation of asphaltenes using ion exchange materials (McKay et al., 1977; Francisco and Speight, 1984).

B. SARA Method

The SARA method (Jewell et al., 1974) is essentially an extension of the API method which allows more rapid separations by placing the

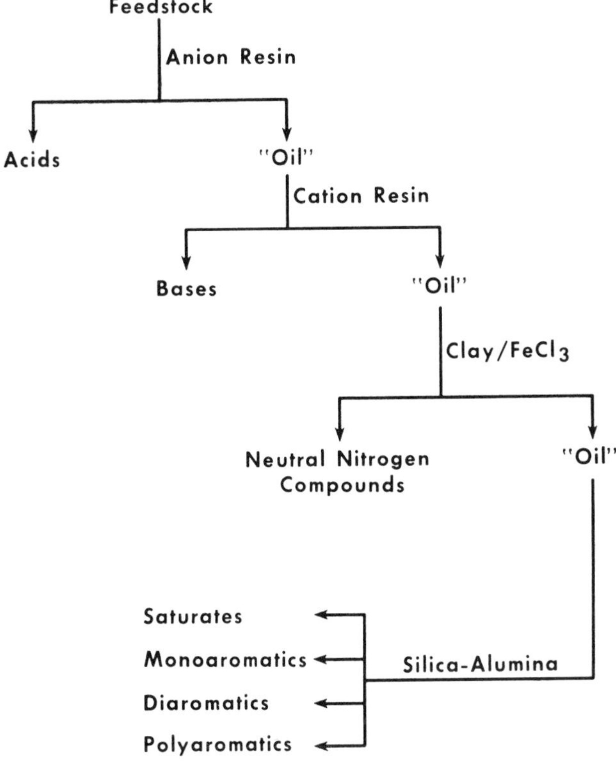

Figure 9 Schematic representation of the USBM-API fractionation procedure.

two ion exchange resins and the $FeCl_3$–clay–anion exchange resin packing into a single column. The adsorption chromatography of the nonpolar part of the same is still performed in a separation operation. Since the asphaltene content of petroleum (and synthetic fuel) feedstocks is often an important aspect of processability, an important feature of the SARA method is that the asphaltenes are separated as a group. Perhaps more important is that the method is reproducible and applicable to a large variety of the most difficult feedstocks, such as residua, tar sand bitumen, shale oil, and coal liquids.

Both the USBM-API and SARA methods require some caution if the asphaltenes are first isolated as a separate fraction. For example, asphaltene yield varies with the hydrocarbon used for the separation and with other factors (Girdler, 1965; Mitchell and Speight, 1973; Speight et al., 1984). An inconsistent separation technique can give rise to problems resulting from residual asphaltenes in the deasphaltened oil which can undergo irreversible adsorption on the solid adsorbent.

C. Other Methods

While the USBM-API and SARA methods are widely used separation schemes for studying the composition of heavy petroleum fractions and other fossil fuels, there are several other schemes which have also been used successfully and which have found common usage in investigations of feedstock composition. For example, a simple alternative (Figure 10) to the SARA sequence is the chromatographic preseparation of a deasphaltened sample on deactivated silica or alumina with pentane (or hexane) into saturated materials followed by elution with benzene for aromatic materials and with benzene/methanol for polar materials (resins). This allows the sample to be chromatographed further into narrower (more similar) fractions without mutual interference on the adsorbent.

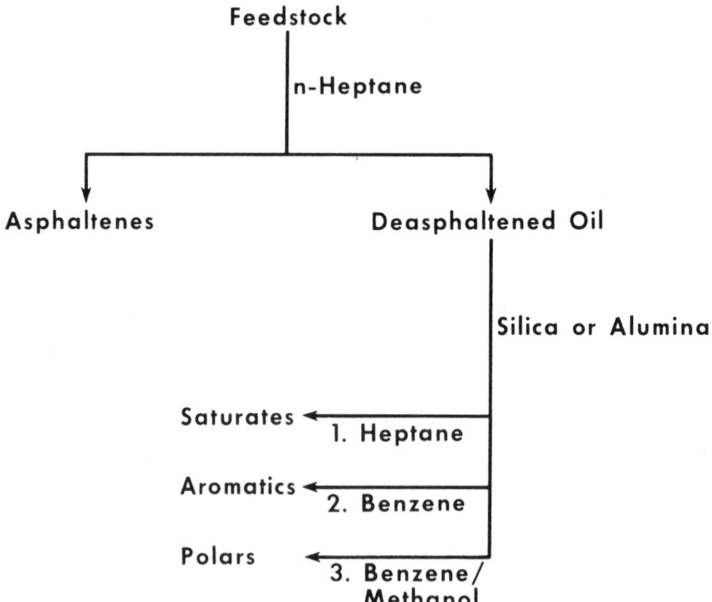

Figure 10 Schematic representation of a simplified alternate procedure to the USBM-API and SARA procedures.

The selection of any separations procedure depends primarily on what information is desired regarding the feedstock. For example, separation into multiple fractions to examine the minute details of feedstock composition will require a complex sequence of steps. One example of such a separation scheme (Figure 11) involves the fractionation of Athabasca tar sand bitumen into four gross fractions and subfractionation of these four fractions (Boyd and Montgomery, 1963).

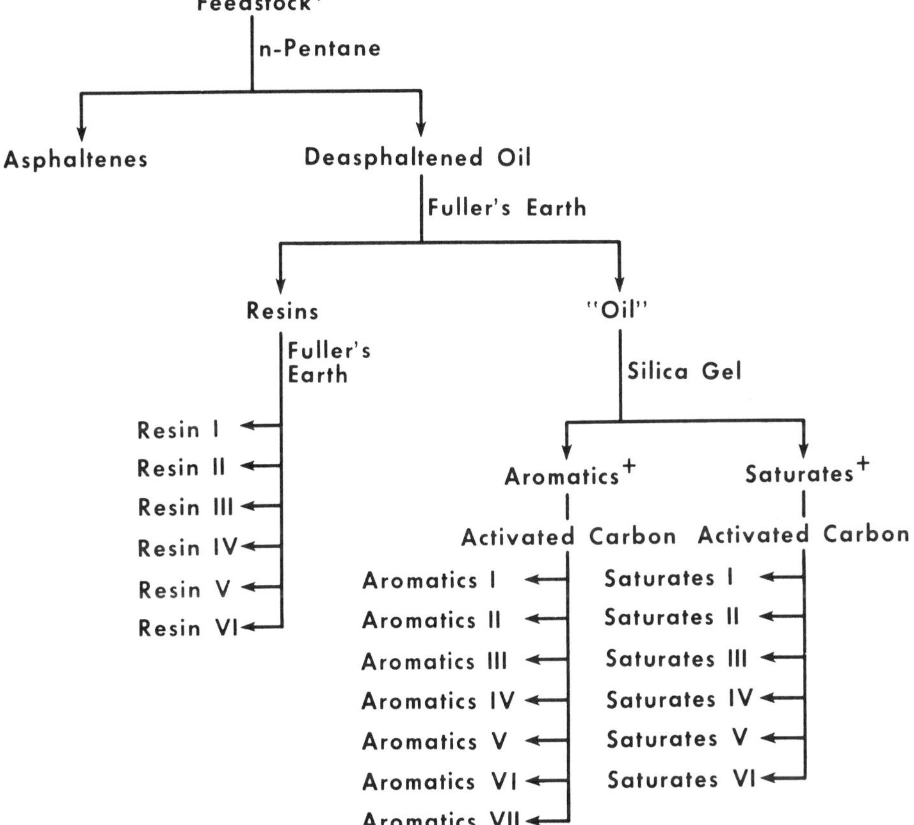

* In this case the feedstock was whole tar sand; direct treatment of the bitumen is recommended by the current author because of solvent/sand interference.

+Other low yield fractions were also produced, but not shown here.

Figure 11 Multiple subfractionation of major fractions can provide more information about feedstock character.

This allowed the investigators to study the distribution of the functional types within the bitumen.

Other investigators (Oudin, 1970) reported that a combination of chromatography using alumina and silica gel is suitable for deasphaltened oils (Figure 12). A more complex scheme, also involving the use of silica, resulted in the successful separation of hexane-deasphalted crude oils (Seifert, 1975). There is also a report (Al-Kashab and Neumann, 1976) of the direct fractionation of hydrocarbon and heteroatom compounds from deasphaltened residua on a dual alumina-silica column with subsequent treatment of the polar fraction with cation and anion exchange resins into basic, acidic, and neutral materials. The method (Figure 13) also includes chromatography of the

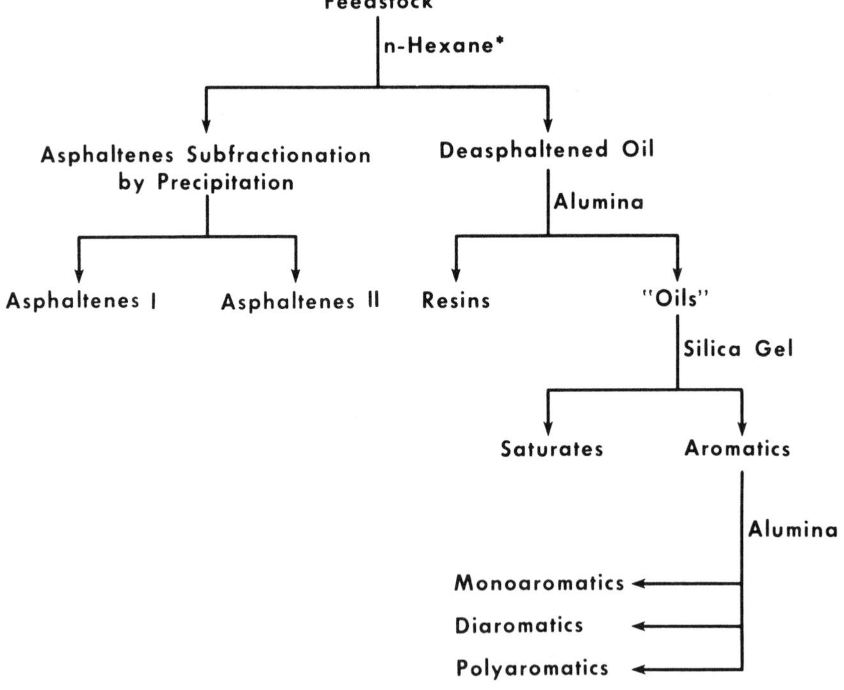

Figure 12 Schematic representation of a general procedure for fractionation/subfractionation of feedstocks.

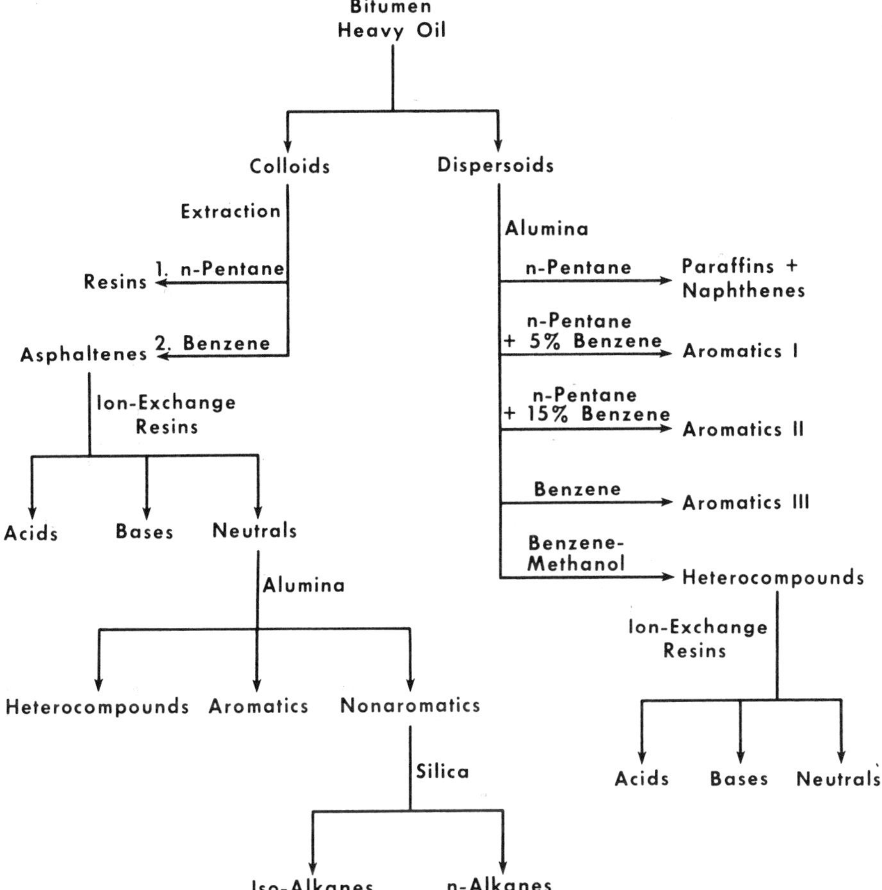

Figure 13 Schematic representation of a more complex scheme for the fractionation of heavy feedstocks.

asphaltenes but only with highly polar asphaltene samples are the basic and acid compounds first removed with ion exchange resins. The remainder of the feedstock is separated into saturates, aromatics, and hetero compounds using alumina-silica adsorption. Separation of the saturates into *n*-alkanes and isoalkanes plus cycloalkanes is achieved by use of urea and thiourea.

One of the problems of such a fractionation scheme is the initial separation of the feedstocks into two ill-defined fractions (colloids and dispersant) without first removing the asphaltenes. As already noted, asphaltenes are specifically defined by the method of separation. They are less well-defined using liquids such as ethyl acetate in place of the more often used hydrocarbons such as pentane and heptane (Speight, 1979). The use of ethyl acetate undoubtedly leads to asphaltene material in the "dispersant" and nonasphaltene material in the "colloids." Application of a more standard deasphalting technique would undoubtedly improve this method and provide an excellent insight into feedstock composition.

One of the common problems of any of the fractionation schemes described above is the nature of the adsorbent. In the early reports of petroleum fractionation (Pfeiffer, 1950), "clays" often appeared as an adsorbent to effect separation of the feedstock into various constituent fractions. However, "clay" (Fuller's earth, attapulgus clay, and the like) is often difficult to define with any degree of precision from one batch to another. Variations in the nature and properties of the clay can and will cause differences not only in the yields of composite fractions but also in the distribution of the compound types in those fractions. In addition, irreversible adsorption of the more polar constituents onto the clay can be a serious problem when further investigations of the constituent fractions are planned.

One option for resolving this problem has been the use of more "standard" adsorbents such as alumina and silica. These materials are more easy to define and are often accompanied by guarantees of composition and type by various manufacturers. They also tend to irreversibly adsorb less of the feedstock than a clay. Once the nature of the adsorbent is guaranteed, reproducibility becomes a reality. Without reproducibility the analytical method does not have credibility.

It is, of course, worth noting at this point that there are indeed ASTM standard methods for the separation of petroleum into constituent fractions. In these methods, not only are the procedures specified in great detail but so are the materials and equipment employed. In keeping with its goals, the ASTM requires reproducibility and precision (ASTM D2006-70, D2007-80, D4124-84).

These three ASTM methods essentially provide for the separation of a feedstock into four or five constituent fractions (Figures 14–16). It is interesting to note that as the methods have evolved there has been a change from the use of pentane (ASTM D2006-70 and ASTM D2007-80) to heptane (ASTM D4124-84) to separate asphaltenes. This is, in fact, in keeping with the production of a more consistent fraction that represents these higher molecular weight; more complex constituents of petroleum (Girdler, 1965; Speight et al., 1984).

Two of the methods (ASTM D2007-80 and ASTM D4124-84) use adsorbents to fractionate the deasphaltened oil but the third method (ASTM D2006-70) advocates the use of various grades of sulfuric

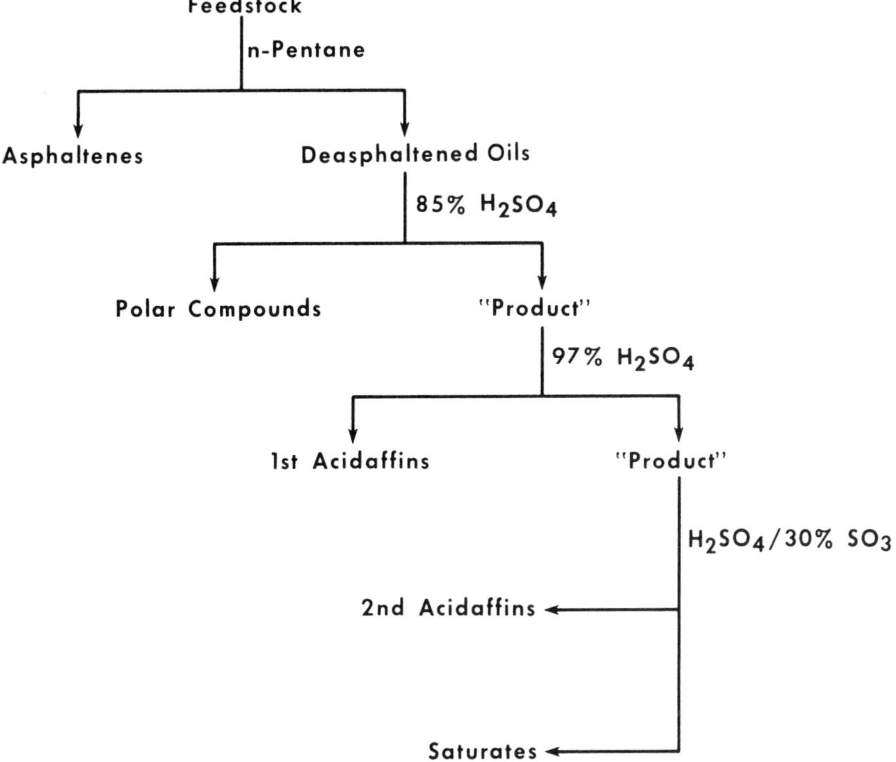

Figure 14 Schematic representation of the ASTM D2006-70 procedure.

acid to separate the material into compound types. Caution is advised in the application of this method since the method does not work well with all feedstocks. For example, when the sulfuric acid method (ASTM D2006-70) is applied to the separation of Athabasca bitumen and similar feedstocks, complex emulsions can be produced.

Obviously, there are precautions that must be taken when attempting to separate heavy feedstocks or polar feedstocks into their constituent fractions. The disadvantages of using ill-defined adsorbents are that adsorbent performance will differ with the same feed and in certain instances may even cause chemical and physical modification of the feed constituents. The use of a chemical reactant such as sulfuric acid should only be advocated with caution since feeds will react differently and may even cause irreversible chemical changes and/or emulsion formation. These advantages may be of little consequence

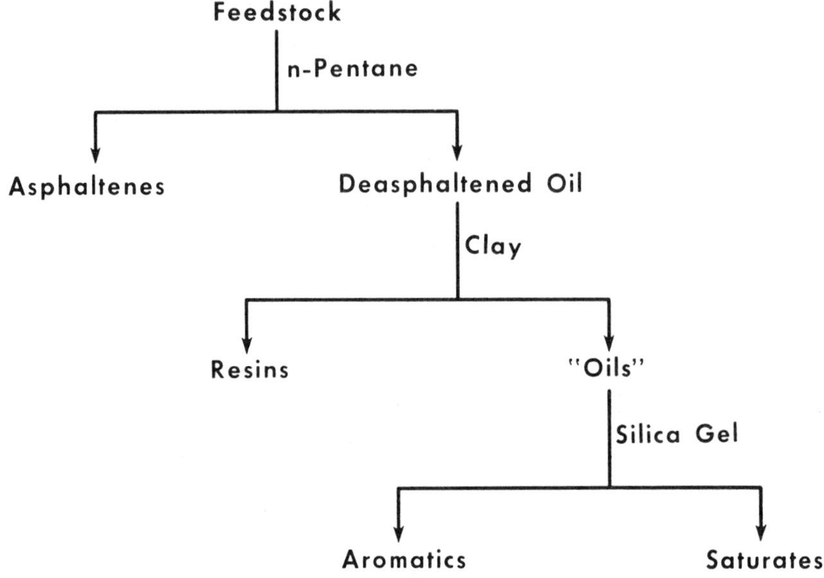

Figure 15 Schematic representation of the ASTM D2007-80 procedure.

where it is not, for various reasons, the intention to recover the various product fractions *in toto* or in the original state but in terms of the compositional evaluation of different feedstocks the disadvantages are very real.

IV. USE OF COMPOSITION DATA

In the simplest sense, petroleum can be considered to be a composite of four major operational fractions (Figure 17). However, it must never be forgotten that the nomenclature of these fractions lies within the historical development of petroleum science, and that the fraction names are operational and are released more to the general characteristics than to the identification of specific compound types. Nevertheless, once a convenient fractionation technique has been established, it is possible to compare a variety of different feedstocks varying from a conventional petroleum to a propane asphalt (Corbett and Petrossi, 1978).

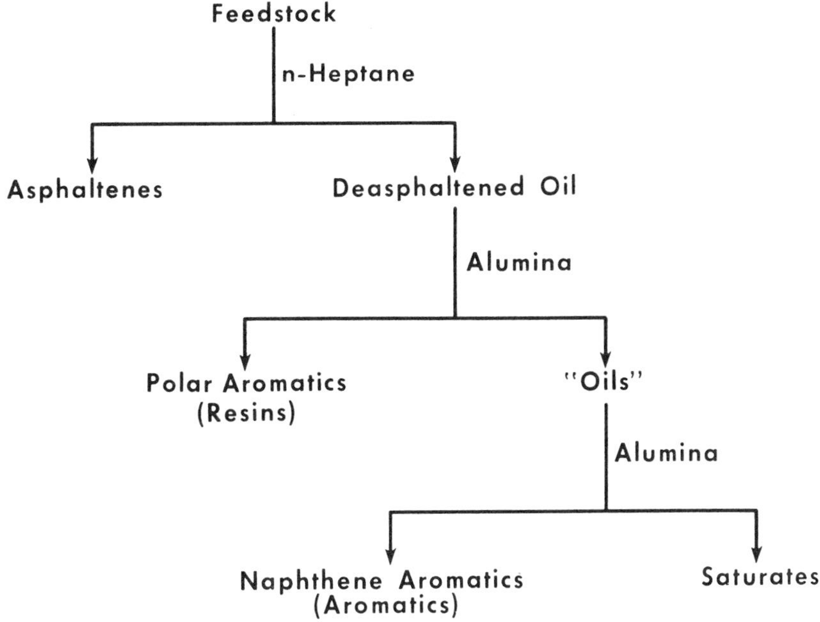

Figure 16 Schematic representation of the ASTM D4124-84 procedure.

It is noteworthy here that throughout the history of studies related to petroleum composition, there has been considerable attention paid to the asphaltic constituents (i.e., the asphaltenes and resins). This is due in no small part to the tendency of the asphaltenes to be responsible for high yields of thermal coke and also be responsible for shortened catalyst lifetimes in refinery operations (Figure 18). In fact, it is the unknown character of the asphaltenes that has also been responsible for drawing the attention of investigators for the last five decades (Speight, 1984). In addition, just as residua contain the majority of all of the potential coke-forming constituents and catalyst poisons that were originally in the crude oil because the distillation process is essentially a concentration process and most of the coke formers and catalyst poisons are nonvolatile, the asphaltene fraction contains most of the coke-forming constituents and catalyst poisons that are originally present in a heavy oil or residuum.

One of the early findings of composition studies was that the behavior and properties of any material are dictated by composition (Speight, 1981). Although the early studies are primarily focused

Heavy Oil
Bitumen, Etc.

n-Pentane (40 Volumes)

Asphaltenes
(Precipitate)

Deasphalted Oil
(Solubles)

Adsorbent, Elute
With (A) n-Pentane
and (B) Pyridine

Oils
(Pentane Eluate)

Resins
(Pyridine Eluate)

Adsorb on Silica Gel,
Elute With (A) n-Pentane
(B) Benzene

Saturates
(Pentane Eluate)

Aromatics
(Benzene Eluate)

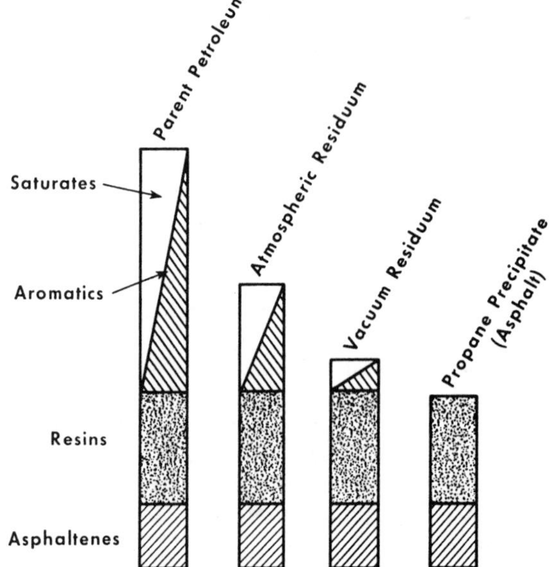

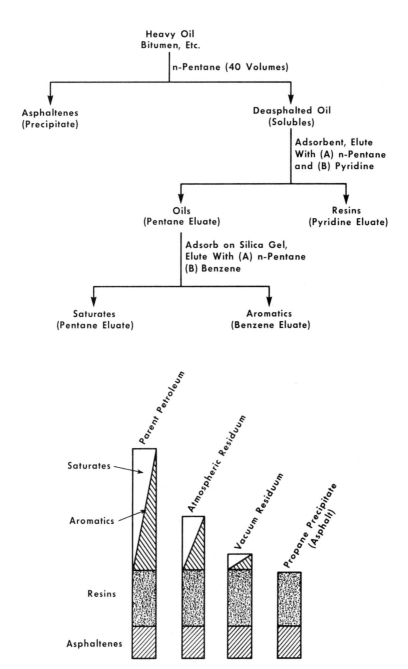

Figure 17 Schematic representation of the separation of petroleum into four major fractions. It should be noted that the names are operational definitions and may not be truly representative of the chemical nature of the fractions.

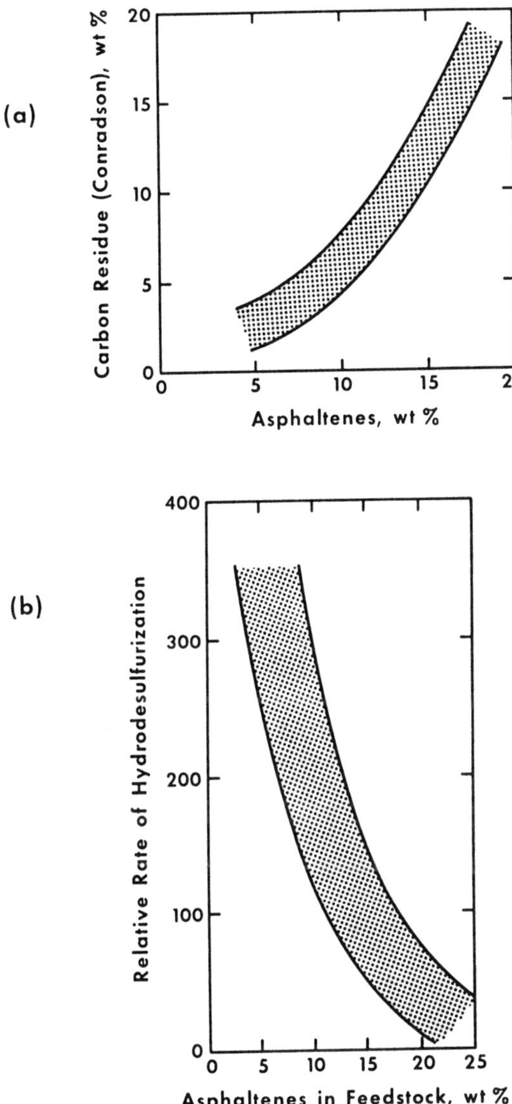

Figure 18 Asphaltenes cause high yields of thermal coke (carbon residue) and large decreases in rates of catalytic hydrodesulfurization.

on the composition and behavior of asphalt, the techniques developed for those investigations have provided an excellent means of studying heavy feedstocks (Tissot, 1984). Later studies have focused not only on the composition of petroleum as major operational fractions but on further fractionation which allows different feedstocks to be compared on a relative basis to provide a very simple but convenient feedstock "map."

However, such a map does not give any indication of the complex interrelationships of the various fractions (Koots and Speight, 1975); although predictions of feedstock behavior is possible using such data. It is necessary to take the composition studies one step further using subfractionation of the major fractions to obtain a more representative indication of petroleum composition.

Thus, by careful selection of an appropriate technique, it is possible to obtain an overview of petroleum composition which can be used for behavioral predictions. By taking the approach one step further and by assiduous collection of various subfractions, it becomes possible to develop the petroleum map and add an extra dimension to compositional studies (Figures 19-21). Petroleum and heavy feedstocks then appear more as a continuum rather than as four specific fractions.

It must be understood that such representations will vary for different feedstocks. For example, conventional petroleum will contain a higher proportion of volatile saturates and aromatics while coal liquids and shale oil will contain a higher proportion of polar aromatics but not necessarily a higher proportion of asphaltenes.

Such a concept has also been applied to the asphaltene fraction of petroleum (Long, 1981) in which asphaltenes are considered to be a complex state of matter based on molecular weight and polarity (Figure 22). The advantage of such a concept is that it can be used to explain differences in asphaltene yield with different hydrocarbons (pentane and heptane) as well as differences in the character of asphaltenes from petroleum and from coal liquids.

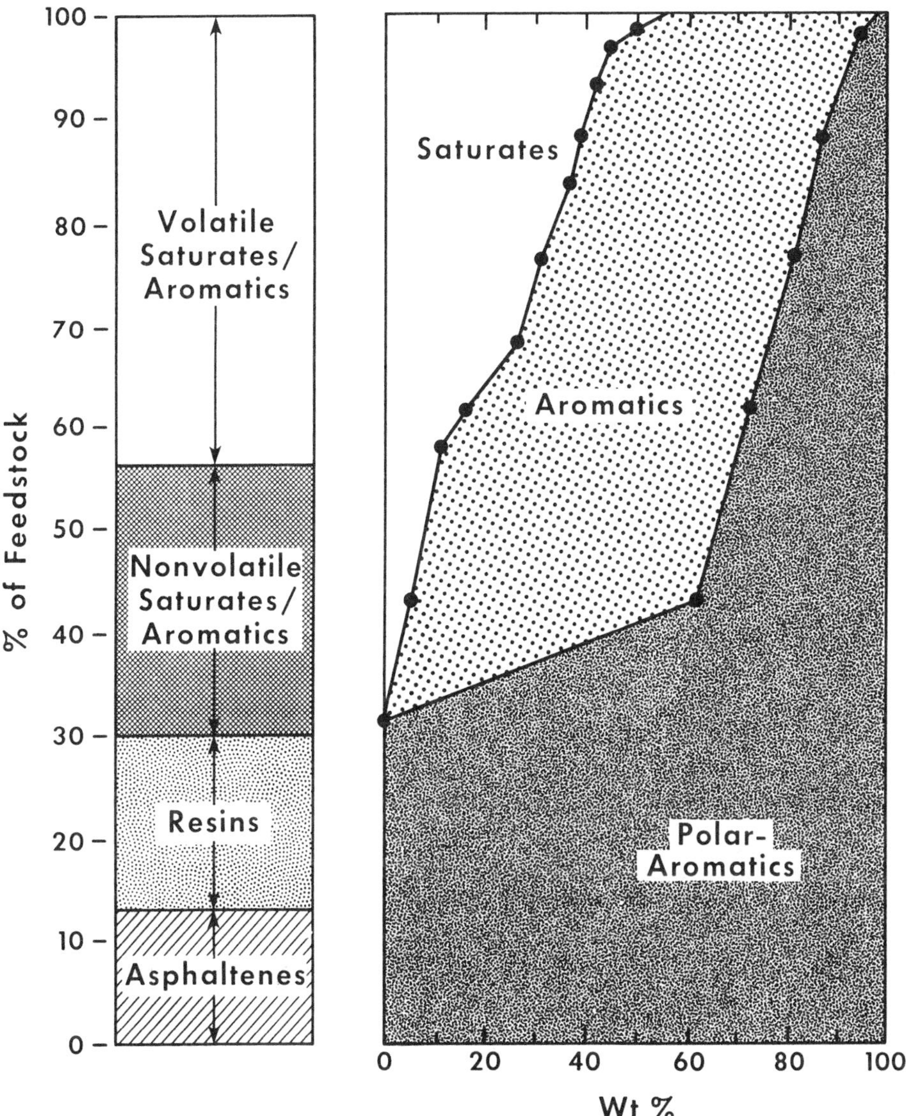

Figure 19 A carefully constructed "map" provides an understanding of feedstock composition that is not evident from the more general separation procedures.

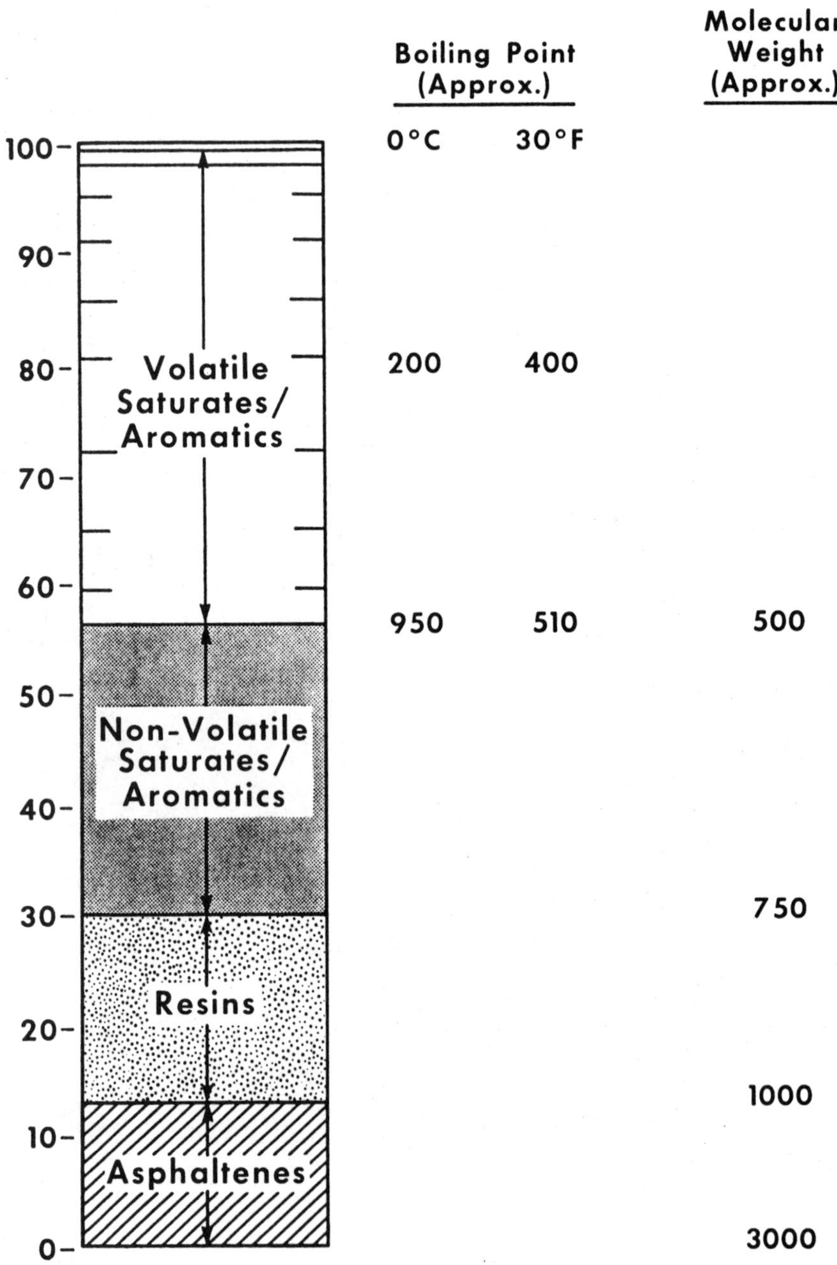

Figure 20 Illustration of the variation of composition and molecular weight with boiling range.

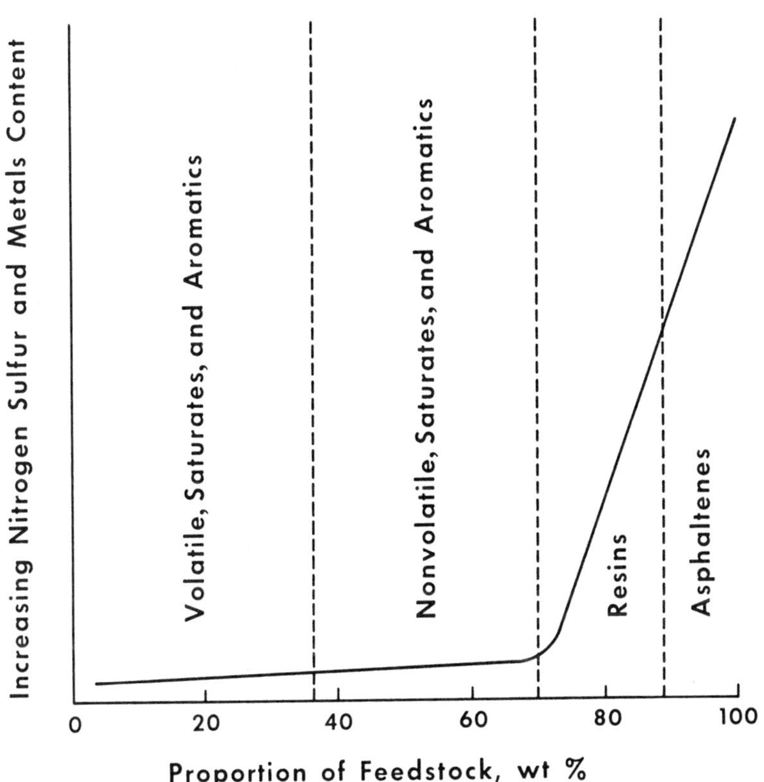

Figure 21 Simplified representation of the variation of the nitrogen, sulfur, and metals content with crude oil composition.

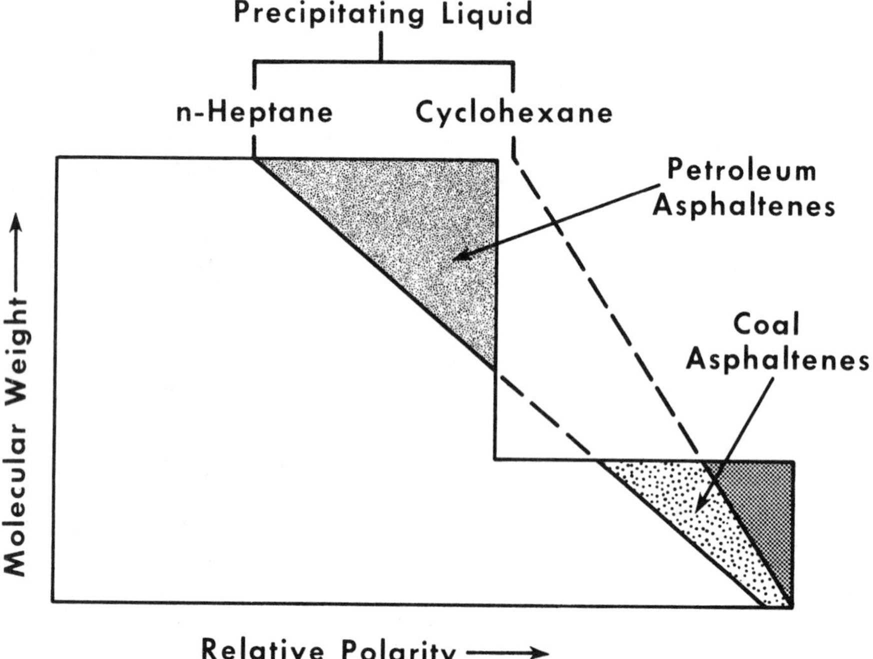

Figure 22 Schematic representation of asphaltene composition using general "molecular weight" and "polarity" parameters.

5

Refining: Distillation

I	Pretreatment	152
II	Early Processes	152
III	Modern Processes	156
	A. Atmospheric Distillation	156
	B. Vacuum Distillation	158
IV	Other Processes	159
	A. Stripping	159
	B. Rerunning	159
	C. Stabilization and Light End Removal	159
	D. Superfractionation	162
	E. Azeotropic and Extractive Distillation	163

Refining consists of a series of steps by which the original crude oil is eventually converted into salable products in the amounts dictated by the market (Kobe and McKetta, 1958).

A refinery is a variety of processes (Figure 1) which vary in number depending on the crude oil and the products required. The processes must be selected and products manufactured to give a balanced operation. A refinery must be flexible and have the ability to change operations as needed (Speight, 1981).

A refining installation must also include all necessary nonprocessing facilities; adequate tankage for storing crude oil, intermediate, and finished products; a dependable source of electric power; materials-

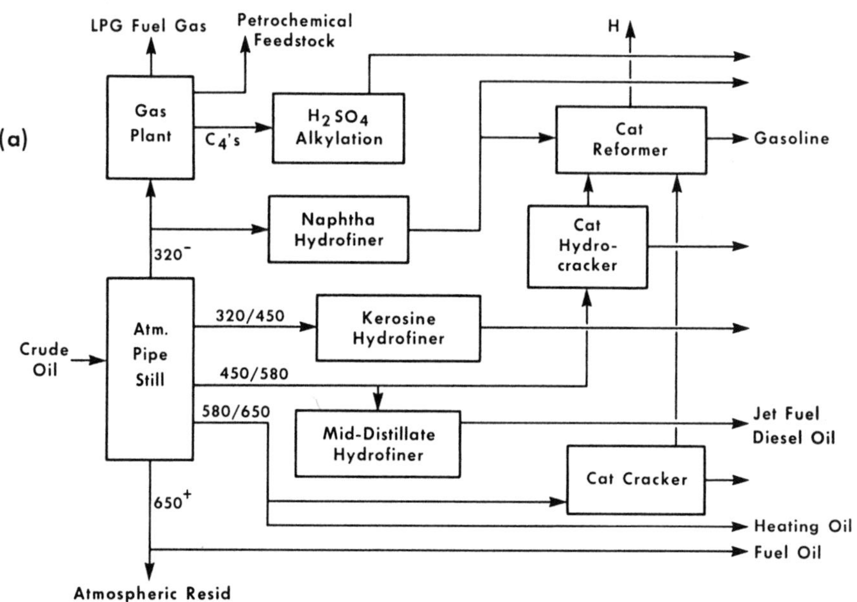

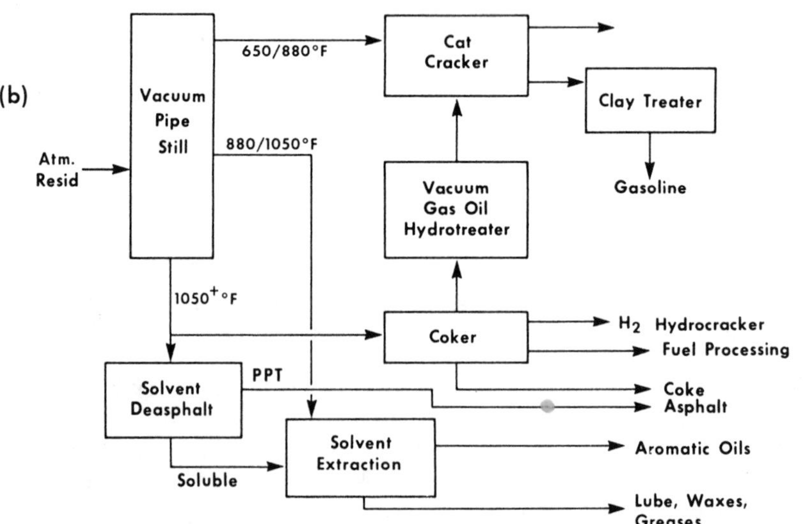

Figure 1 Schematic representation of the interrelationship of the various refinery processes.

handling equipment, workshops and supplies for maintaining continuous 24-hr-a-day, 7-days-a-week operation; waste disposal and water treatment equipment; and product-blending facilities (Nelson, 1958).

In the early stages of refinery development, when illuminating and lubricating oils were the main products, distillation was the major, and often only, refinery process. As the demand for gasoline increased, conversion processes were developed because distillation could no longer supply the necessary quantities (Bell, 1945).

Distillation has remained a major refinery process and it is a process to which just about every crude which enters the refinery is subjected. The most important and primary function of distillation in the refinery is its use for the separation of crude oil into component fractions (Table 1).

It is possible to obtain products which range from low-boiling materials to a heavy "nonvolatile" residue or "bottoms" with lighter materials taken off at intermediate points. The bottoms (or reduced crude) may then be processed by vacuum distillation in order to separate the high-boiling lubricating oil fractions without the danger of

Table 1 Nomenclature and Boiling Ranges of the Various Distillation Fractions of Petroleum

	Boiling range[a]	
Fraction	(°C)	(°F)
Light naphtha	-1–150	30–300
Heavy naphtha	150–205	300–400
Gasoline	-1–180	30–355
Kerosene	205–260	400–500
Stove oil	205–290	400–550
Light gas oil	260–315	400–600
Heavy gas oil	315–425	600–800
Lubricating oil	>400	>750
Vacuum gas oil	425–600	800–1100
Residuum	>600	>1100

[a]For convenience, boiling ranges are interconverted to the nearest 5°.

decomposition which occurs at high temperatures (>350°C; >660°F). Atmospheric distillation can be terminated with a lower boiling fraction ("cut") if it is felt that vacuum or steam distillation will yield a better quality product or if the process appears to be economically more favorable (Guthrie, 1960).

I. PRETREATMENT

Even though distillation is the first step in crude oil refining, crude oil which is contaminated by saltwater must first be treated to remove the emulsion. If saltwater is not removed, the materials of construction of the heater tubes and column intervals will be exposed to chloride ion attack and to the corrosive action of hydrogen chloride which will be formed at the temperature of the column feed.

Three general approaches have been taken to the desalting of crude petroleum (Figure 2) but the selection of a particular process depends on the type of salt dispersion and the properties of the crude oil. For example, simple brine suspensions may be removed from crude oil by heating under pressure sufficient to prevent vapor loss [90–150°C (200–300°F)/50–250 psi], then allowing the material to settle in a large vessel. Alternatively, coalescence is aided by passage through a tower packed with sand, gravel, or similar material.

Emulsions may also be broken by addition of treating agents such as soaps, fatty acids, sulfonates, and long-chain alcohols. When a chemical is used for emulsion breaking during desalting, it may be added at one or more of three points in the system: (1) to the oil before it is mixed with freshwater; (2) to the freshwater before mixture with the oil; and (3) to the mixture of oil and water. A high potential field across the settling vessel will also aid coalescence and break emulsions, in which case dissolved salts and impurities are removed with the water.

Corrosion in the distillation unit can be reduced by "flashing" the crude oil feed. The temperature of the feed is raised (by heat exchange with the products from the distillation stages) and fed to a flash unit at a pressure of approximately 2–3 atmospheres. Dissolved hydrogen sulfide will be removed prior to introducing the feedstock to the atmospheric column to diminish the corrosive attack of hydrogen sulfide which occurs at elevated temperatures in the presence of steam.

II. EARLY PROCESSES

Distillation has been used as a means of refining petroleum for many years. The original petroleum still was a cast iron vessel mounted on

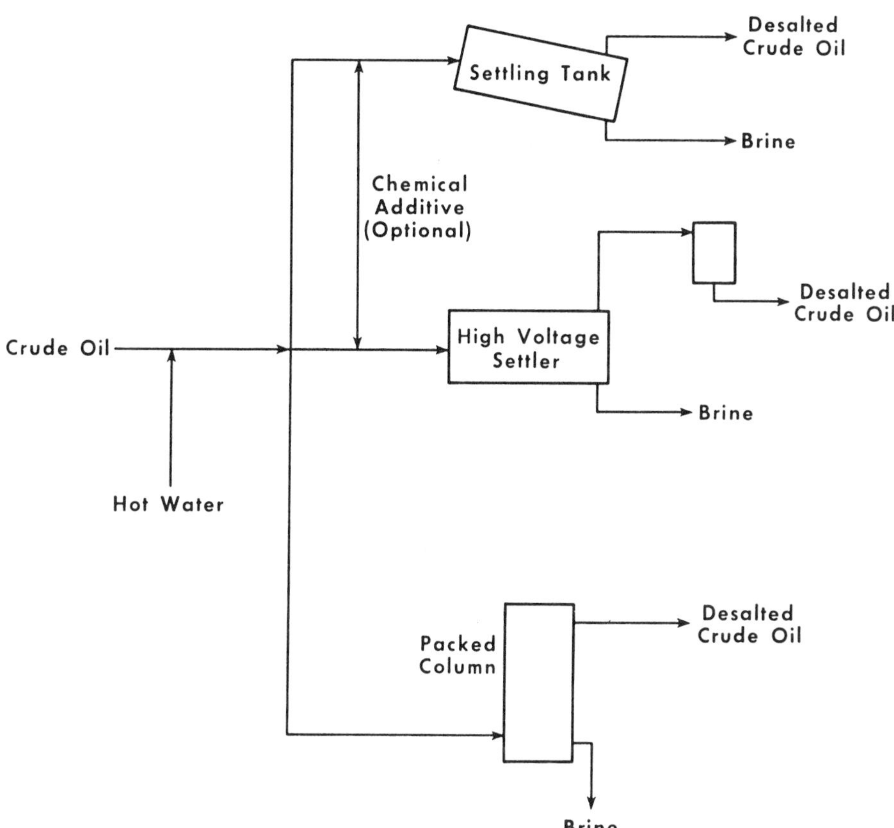

Figure 2 General methods for desalting crude oil.

brickwork over a fire. The heat caused the petroleum in the still to boil and the vapors passed through a pipe or gooseneck which leads from the top of the still to a condenser (Figure 3). The latter was a coil or worm of pipe immersed in a tank of running water. Heating a batch of crude petroleum caused the more volatile, lower boiling components to vaporize and then be condensed in the worm to form naphtha. As the distillation progressed, the higher boiling components became vaporized and were condensed to produce kerosene, which at the time was regarded as the major petroleum product. At the conclusion of the distillation, i.e., when all the possible kerosene had been obtained, the residue was removed from the still and discarded. The still was then refilled with petroleum and the operation repeated.

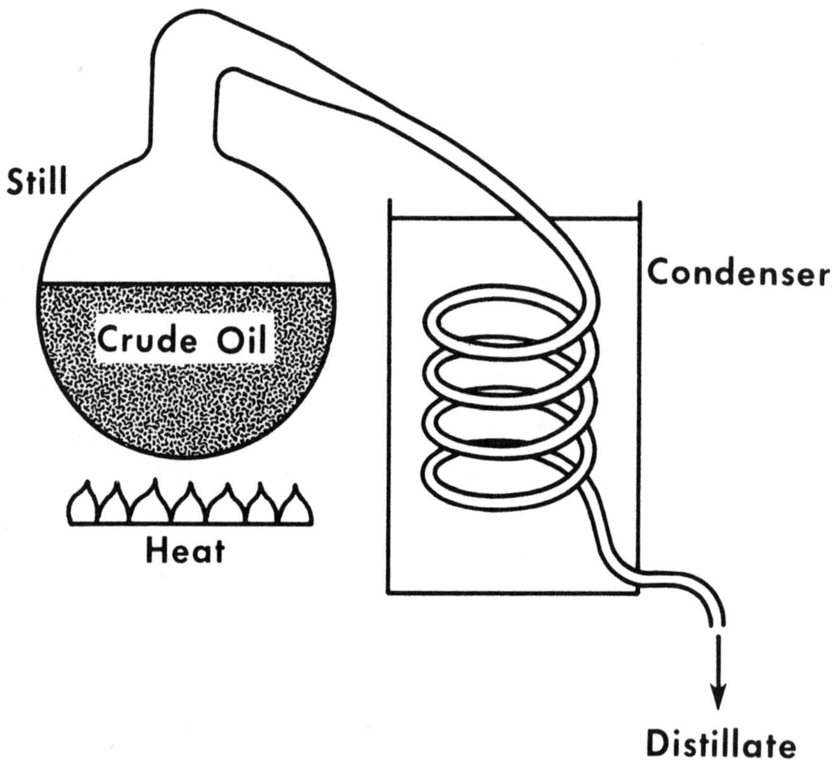

Figure 3 Simplified representation of an early petroleum still.

The capacity of the stills at that time was usually several barrels* of petroleum and it often required three or more days to run (distill) a batch of crude oil (Speight, 1980).

The 1880s saw the introduction of the continuous distillation of petroleum, and the method was used to separate naphtha and kerosene from the remainder of the crude petroleum which was used as a residual fuel oil. The method employed a number of stills coupled together

*The unit of measure in the petroleum industry is the barrel (bbl). One barrel = 42 U.S. gallons = 33.6 Imperial gallons = 5.61 ft^3 = 158.8 liters.

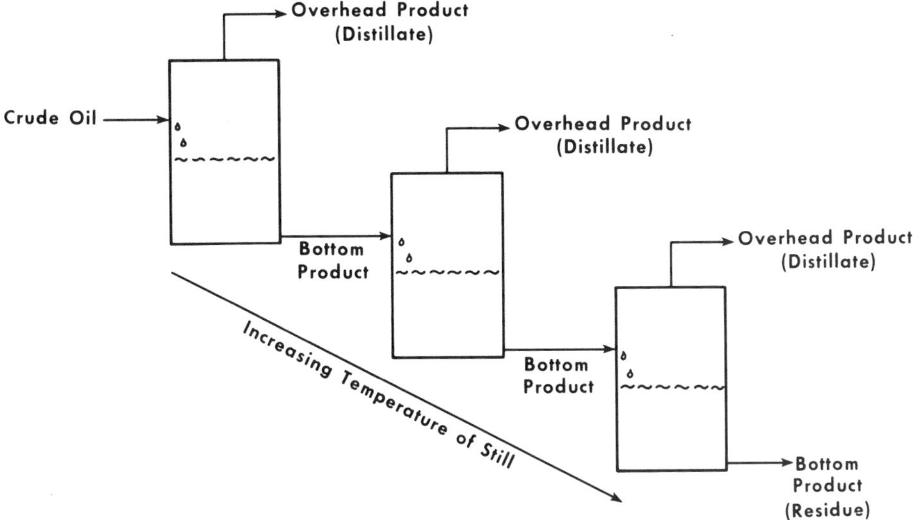

Figure 4 Schematic representation of a battery still as used in the early days of the petroleum industry.

in a row (battery). Each still was heated separately and was hotter than the preceding one, and the stills were arranged so that oil flowed by gravity from the first to the last (Figure 4).

In the early 1900s, a method of partial (or selective) condensation was developed that allowed a more exact separation of petroleum fractions. A partial condenser (van Dyke tower; Figure 5) was inserted between the still and the conventional water-cooled condenser. The lower section of the tower was packed with stones and insulated with brick. The heavier vapors entering the tower condensed on the stones and drained back into the still. Noncondensed material then passed into a section containing air-cooled tubes on which heavier vapors from the stone-packed section condensed; the condensate was withdrawn as a petroleum fraction. The overhead (noncondensable) material from the air-cooled section then entered a second tower that also contained air-cooled tubes and often produced a second fraction. Finally, residual vapors were condensed in the conventional water-cooled condenser to yield a third fraction.

The van Dyke tower was one of the first stages in a series of improvements which ultimately lead to the distillation units found in modern refineries and which separate petroleum fractions by fractional distillation.

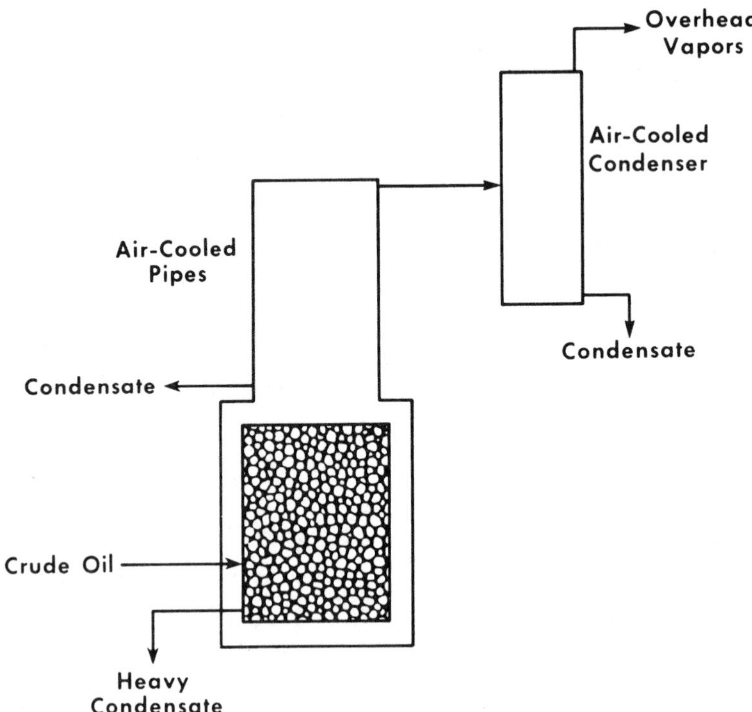

Figure 5 Schematic representation of a van Dyke tower.

III. MODERN PROCESSES

A. Atmospheric Distillation

The present day petroleum distillation unit is, in fact, a collection of distillation units which enable a fairly efficient degree of fractionation to be achieved. In contrast to the early units, which consisted of separate stills, a tower is used in the modern day refinery.

The feed to a fractional distillation tower is heated by flow-through pipe arranged within a large furnace (pipe still heater or pipe still furnace); hence the frequent use of the term "pipe still" for the distillation unit (Figure 6). The furnace heats the feed to a predetermined temperature—usually a temperature at which a calculated portion of the feed will change into vapor. The vapor is held under pressure in the pipe in the furnace until it discharges as a foaming stream into the fractional distillation tower. Here the unvaporized or liquid portion of the feed descends to the bottom of the tower to be pumped

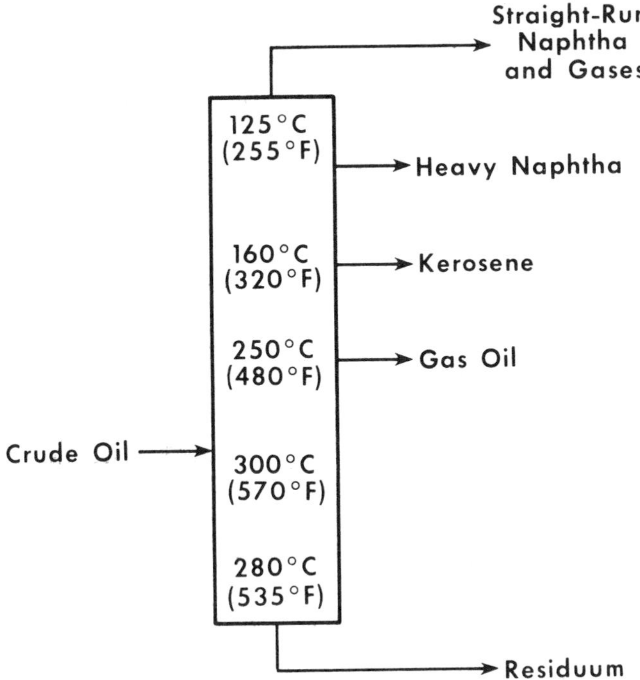

Figure 6 A pipe still furnace (heater) and distillation (fractionating or bubble) tower showing the variation of temperature within the tower.

away as a bottom product while the vapors pass up the tower to be fractionated into gas oils, kerosene, and naphthas.

Heat exchangers are also used to preheat the feed to the furnace. These exchangers are bundles of tubes arranged within a shell so that a stream passes through the tubes in the opposite direction to a stream passing through the shell. Thus, cold crude oil is heated by passage through a series of heat exchangers where, at the same time, hot products from the distillation tower are cooled. This results in an overall saving of energy and is a contributing factor in the economical operation of a refinery.

The tower is divided into a number of horizontal sections by metal trays or plates and each one is the equivalent of a still. The more trays, the more redistillations and hence the better the fractionation or separation of the mixture fed into the tower. A tower for fractionating crude petroleum may be 13 ft in diameter and 85 ft high, while a tower stripping unwanted volatile material from a gas oil may be only 3 or 4 ft in diameter and 10 ft high. Towers concerned with the distillation of liquefied gases are only a few feet in diameter but may be up to 200 ft in height. A tower used in the fractionation of crude petroleum may have from 16 to 28 trays while one used in the fractionation of liquefied gases may have 30 to 100 trays. The feed to a typical tower enters the vaporizing or flash zone—an area without trays. The majority of the trays are usually located above this area.

Liquid collects on each tray to a depth of, say, several inches and the depth is controlled by a dam or weir. As the liquid level rises, excess liquid spills over the weir into a channel (downspout) which carries the liquid to the tray below.

The temperature of the trays is progressively cooled from bottom to top (Figure 6). The bottom tray is heated by the incoming heated feedstock although, in some instances, a steam coil (reboiler) is used to supply additional heat. As the hot vapors pass upward in the tower, condensation occurs on the trays until a refluxing (simultaneous boiling of a liquid and condensing of the vapor) occurs on the trays. Vapors continue to pass upward through the tower whereas the liquid on any particular tray will spill onto the tray below and so on until the heat at one particular point is too intense for the material to remain liquid. It then becomes vapor and joins the other vapors passing upward through the tower. The whole tower thus simulates a collection of several (or many) stills with the composition of the liquid at any one point or on any one tray remaining fairly consistent. This allows part of the refluxing liquid to be tapped off at various points as "sidestream" products.

Of all the units in a refinery, the distillation unit is required to have greatest flexibility in terms of variable quality of feedstock and range of product yields. The maximum permissible temperature of the feedstock in the vaporizing furnace is the factor limiting the range of products in a single-stage (atmospheric) column. Thermal decomposition or cracking of the hydrocarbons begins as the temperature of the oil approaches 360°C (680°F) and is undesirable because the carbon is deposited on the tubes with consequent formation of hot spots and eventual failure of the affected tubes.

B. Vacuum Distillation

The boiling point of the heaviest cut obtainable at atmospheric pressure is limited by the temperature at which the residue starts to decompose or crack. If the stock is required for the manufacture of lubricating oils, further fractionation without cracking may be desirable, and this may be achieved by distillation under vacuum conditions (<100 mm Hg). Volumes of vapor at these pressures are large and pressure drops must be small so that vacuum units will be of larger diameter than the accompanying atmospheric unit.

Then trays similar to those used in the atmospheric column are used in vacuum distillation, the column diameter may be extremely high, i.e., up to 45 ft. In order to maintain low-pressure drops across the trays the liquid seal must be minimal. The low holdup and relatively high viscosity of the liquid limits the tray efficiency, which tends to be much lower than in the atmospheric column.

IV. OTHER PROCESSES

A. Stripping

Stripping is a fractional distillation operation carried out on each sidestream product immediately after it leaves the main distillation tower. The purpose of stripping is to remove the more volatile components and thus reduce the flash point of the sidestream product.

B. Rerunning

Rerunning is a general term covering the redistillation of any material and indicating, usually, that a large part of the material is distilled overhead. For example, by separating the wide-cut fraction into a light and heavy naphtha, the rerun tower acts, in effect, as an extension of the crude distillation tower.

C. Stabilization and Light End Removal

The gaseous and more volatile liquid hydrocarbons produced in a refinery are collectively known as the light hydrocarbons or light ends (Table 2). Removal of these volatile components is known as stabilization. The simplest stabilization process is a stripping process.

An example of more precise stabilization is the means by which the mixture of hydrocarbons produced by cracking is treated. The overhead products from the cracked mixture consists of light ends and cracked gasoline with light ends dissolved in it. To prevent excessive pressure buildup during storage, the cracked gasoline and the dissolved gases are pumped to a stabilizer maintained under a pressure

Table 2 Hydrocarbon Constituents of "Light Ends"

Hydrocarbon	Carbon atoms	Molecular weight	Boiling point °C	Boiling point °F	Uses
Methane	3	16	-182	-296	Fuel gas
Ethane	2	30	-89	-128	Fuel gas
Ethylene	2	28	-104	-155	Fuel gas, petrochemicals
Propane	3	44	-42	-44	Fuel gas, LPG
Propylene	3	42	-48	-54	Fuel gas, petrochemicals polymer gasoline
Isobutane	4	58	-12	11	Alkylate, motor gasoline
n-Butane	4	58	-1	31	Motor gasoline
Isobutylene	4	56	-7	20	Synthetic rubber and chemicals, polymer gasoline, alkylate, motor gasoline
Butylene-1[a]	4	56	-6	21	Synthetic rubber and chemicals, alkylate, polymer gasoline, motor gasoline
Butylene-2[a]	4	56	1	34	
Isopentane	5	72	28	82	Motor and aviation gasolines
n-Pentane	5	72	36	97	Motor and aviation gasolines
Pentylenes	5	70	30	86	Motor gasolines
Isohexane	6	86	61	141	Motor and aviation gasolines
n-Hexane	6	86	69	156	Motor and aviation gasolines

[a]Numbers refer to the positions of the double bond. For example, butylene-1 (or butene-1 or but-1-ene) is $CH_3CH_2CH{=}CH_2$ while butylene-2 (or butene-2 or but-2-ene) is $CH_3CH{=}CHCH_3$.

of approximately 100 psi and operated with reflux. This fractionating tower makes a cut between the lightest liquid component (pentane) and the heaviest gas (butane). The bottom product is thus a liquid free of all gaseous components boiling lower than pentane; hence the fractionating tower is known as a debutanizer (Figure 7).

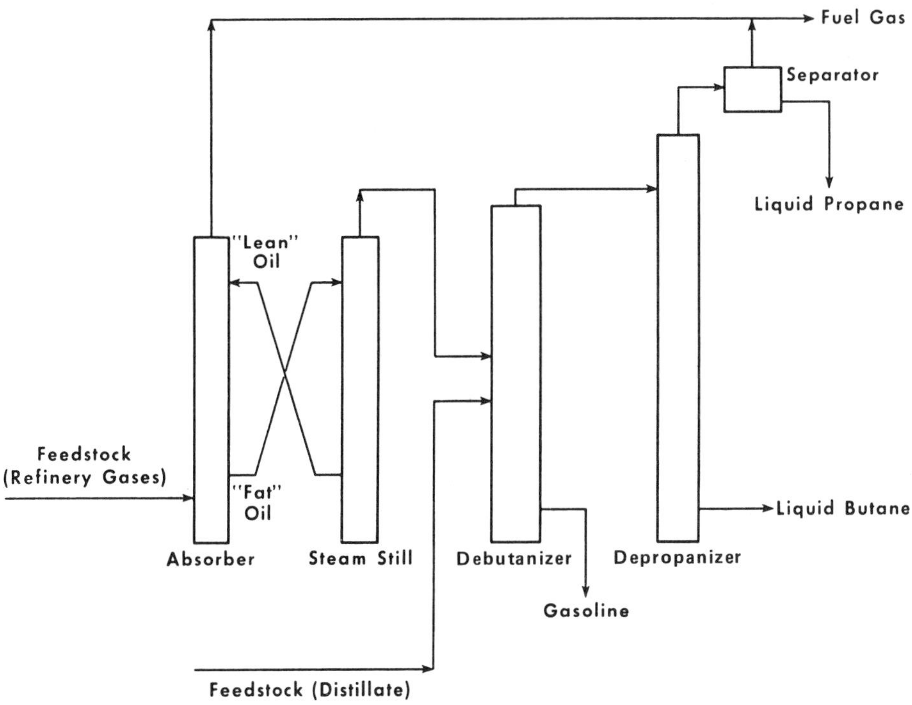

Figure 7 A "light ends" plant.

A depropanizer is very similar to the debutanizer except that it is smaller in diameter due to the smaller volume being distilled and it is taller because of the larger number of trays required to make a sharp distinction between the butane and propane fractions. Since the normally gaseous propane must exist as a liquid in the tower, a pressure of 200 psi is maintained. The bottom product, known as the butane fraction, stabilizer bottoms, or refinery casinghead, is a high-vapor-pressure material that must be stored in refrigerated tanks or pressure tanks. The depropanizer overhead, consisting of propane and lighter gases, is normally used as refinery fuel gas.

A depentanizer is a fractional distillation tower that removes the pentane fraction from a debutanized (butane-free) fraction. Depentanizers are similar to debutanizers and were recently introduced to segregate the pentane fractions from selected cracked gasolines and reformates. The pentane fraction when added to a premium gasoline makes this gasoline extraordinarily responsive to the demands of an engine accelerator.

The gases produced as overhead products from crude distillation, stabilization, and depropanization units may be delivered to a gas absorption plane for the recovery of small amounts of butane and heavier hydrocarbons. The gas absorption plant consists essentially of two towers. One tower is the absorber where the butane and heavier hydrocarbons are separated from the lighter gases. This is done by spilling a light oil, called lean oil, down the absorber over trays similar to those in a fractional distillation tower. The gas mixture enters at the bottom of the tower and rises to the top. As it does this, it contacts the lean oil which absorbs the butane and heavier hydrocarbons but not the lighter hydrocarbons. The latter leave the top of the absorber as dry gas. The lean oil, enriched with butane and heavier hydrocarbons, becomes fat oil. This is pumped from the bottom of the absorber into the second tower where fractional distillation separates the butane and heavier hydrocarbons as an overhead fraction and the oil, once again lean oil, as the bottom product. The condensed butane and heavier hydrocarbons are included with the refinery casinghead or stabilizer bottoms. The dry gas is frequently used as fuel gas for refinery furnaces. It contains, however, propane and propylene, which may be required for liquefied petroleum gas or for the manufacture of polymer gasoline or petrochemicals. Separation of the propane fraction (propane and propylene) from the lighter gases is accomplished by further distillation.

D. Superfractionation

Fractional distillation does not completely separate one petroleum fraction from another. One product overlaps another depending on the efficiency of the distillation unit, which depends on the number of trays in the tower, the amount of reflux used, and the rate of distillation.

Complete separation is not required for the ordinary uses of these materials but certain materials such as solvents (e.g., hexane, heptane, benzene, toluene, xylene) are required at a high degree of purity. This requires highly efficient fractional distillation towers specially designed for the purpose and referred to as superfractionators. Several towers with 50–100 trays and operated with high reflux ratios may be required to separate a single compound at the necessary purity.

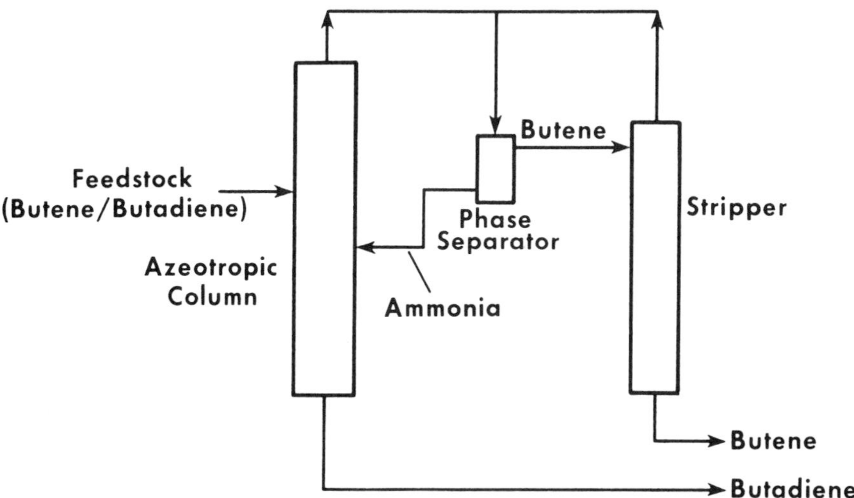

Figure 8 Schematic representation of the separation of butene and butadiene by azeotropic distillation.

E. Azeotropic and Extractive Distillation

The similarities and differences between azeotropic and extractive distillation may be illustrated by two processes that have been proposed for the separation of C_4 hydrocarbons (Figures 8 and 9). Thus, butadiene and butene may be separated by the use of liquid ammonia, which forms an azeotrope with butene. The more volatile ammonia-butene azeotrope is condensed, cooled, and allowed to separate into a butene layer and a heavier ammonia layer. The butene layer is fed to a second column where any remaining ammonia is removed as a butene-ammonia azeotrope and the butene is recovered as bottom product. The ammonia layer is returned to the lower section of the first azeotropic distillation column. Butadiene is recovered as bottom product from this column.

Using acetone-water as an extractive solvent for butanes and butenes, butene is retained as a bottom product from the extractive distillation column with the acetone/water. The butene and the extractive solvent is fed to a second column where the butene is removed as overhead. The acetone/water solvent from the base of this column is recycled to the first column.

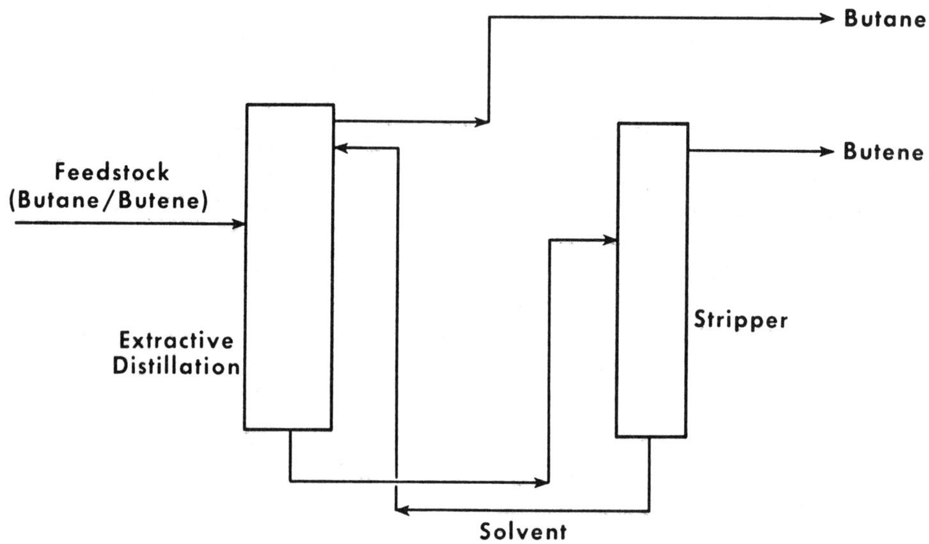

Figure 9 Extractive distillation as used to separate butane from butene.

6

Refining: Thermal Methods

I	Thermal Cracking	166
	A. Early Processes	167
	B. Commercial Processes	169
II	Catalytic Cracking	174
	A. Commercial Processes	174
III	Hydroprocessing	178
	A. Commercial Processes	180
IV	Thermal Reforming	182
	A. Commercial Processes	183
V	Catalytic Reforming	184
	A. Commercial Processes	184
VI	Isomerization	187
	A. Commercial Processes	188
VII	Alkylation	189
	A. Commercial Processes	190
VIII	Polymerization	193
	A. Commercial Processes	193
IX	Hydrogen Production	195
	A. Commercial Processes	196

Thermal conversion processes are designed to increase the yield of liquid products (e.g., gasoline) obtainable from petroleum either directly (by means of the production of suitable components from the heavier feedstock) or indirectly (by production of olefins and the like, which are precursors to the liquid fuel components). These processes may also be characterized by the physical state, i.e., liquid and/or vapor phase, in which the decomposition occurs. The state depends on the nature of the feedstock as well as conditions of pressure and temperature (Bland and Davidson, 1967).

Heavier feedstocks contain considerable amounts of asphaltic materials and it may be necessary to maintain the proportion of these coke-forming constituents at a minimum both for catalytic and noncatalytic processes to avoid catalyst poisoning or reduction in catalyst activity where it is essential that as much of the nitrogen and metals (such as vanadium and nickel) as possible be removed from the feedstock (Speight, 1981).

There are a number of thermal processes such as tar separation (flash distillation), vacuum flashing, visbreaking, and coking which are directed at upgrading feedstocks by removal of the asphaltic fraction. However, there is also the method of deasphalting with liquid hydrocarbon gases such as propane, butane, or isobutane, which is very effective in the preparation of residua for cracking feedstocks. The process is often referred to as propane decarbonizing or solvent decarbonizing (Chap. 7, Sec. II.E).

In practice, liquid propane is contacted countercurrently with descending heavy oil in the deasphalting tower. The propane is separated from the deasphalted oil by evaporation and steam stripping. The asphalt-propane mixture is heated, flash-distilled, and stripped.

The thermal cracking of petroleum fractions is a very complex process but is often visualized as a series of simple thermal conversion (Figure 1). The reactions involve the formation of transient free radical species that may react further in several ways to produce the observed product slate. It is for this reason that the slate of products from thermal cracking is considered to be difficult to predict (Germain, 1969).

I. THERMAL CRACKING

One of the earliest processes used after distillation is the noncatalytic conversion of higher boiling petroleum stocks into lower boiling products and is known as thermal cracking. Petroleum hydrocarbons undergo cracking when subjected to temperatures over 350°C (660°F). Sufficiently high temperatures will convert oils entirely to gases and

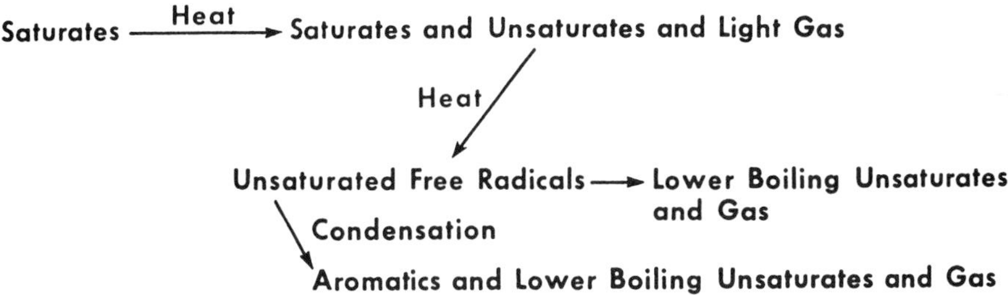

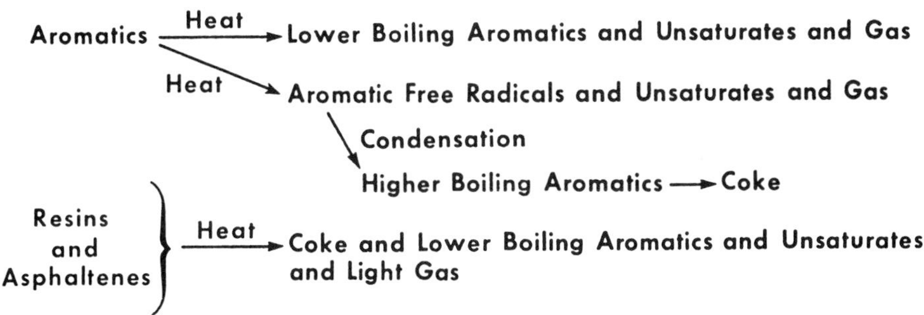

Figure 1 Simplified representation of the thermal decomposition of the various constituents of petroleum.

coke but cracking conditions are controlled to produce as much as possible of the desired product, which is usually gasoline but which may be gases (for petroleum use) or a lower viscosity oil for use as a fuel oil. The feedstock (cracking stock) may be almost any fraction obtained from crude petroleum but the greatest amount of cracking is carried out on gas oils—a term which refers to the portion of crude petroleum that boils between the fuel oils (kerosene and/or stove oil) and the residuum. Reduced crudes are also cracked by the processes are somewhat different from those used for gas oils.

A. Early Processes

Cracking was used commercially in the production of oils from coal and shales before the petroleum industry began, and before the discovery that the heavier products could be decomposed to lighter oils was used to increase the production of kerosene and called cracking distillation (Kobe and McKetta, 1958).

As the need for gasoline arose, the necessity of prolonging the cracking process became apparent and a process known as pressure cracking evolved. Pressure cracking was a batch operation in which some 200 bbl of gas oil was heated to about 425°C (800°F) in stills (shell stills) especially reinforced to operate at pressure as high as 95 psi. The gas oil was held under maximum pressure for 24 hr while the temperature was maintained. Distillation was then started and during the next 48 hr 70–100 bbl of a lighter distillation was obtained which contained the gasoline components (Stephens and Spencer, 1956).

The large-scale production of cracked gasoline was first developed by Burton in 1912. The process employed batch distillation in horizontal shell stills and operated at about 400°C (about 750°F) and 75–95 psi (Figure 2). Heating a batch volume of oil was soon considered cumbersome and during the years 1914–1922 a number of successful continuous cracking processes were developed. The tube-and-tank cracking process (Figure 3) is typical of the early continuous cracking processes. The cracking reactions formed coke which in the

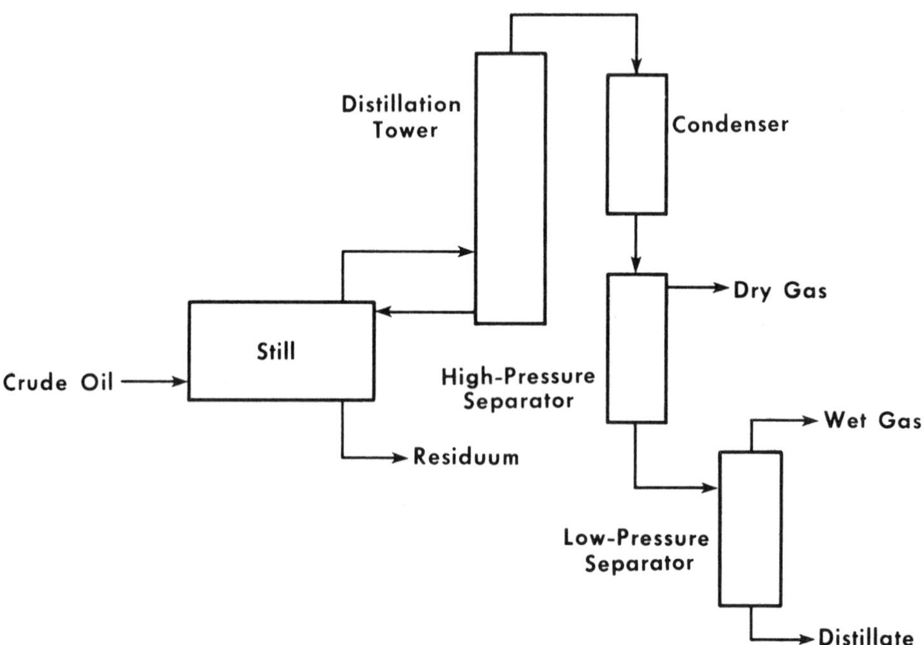

Figure 2 Schematic representation of the Burton cracking process.

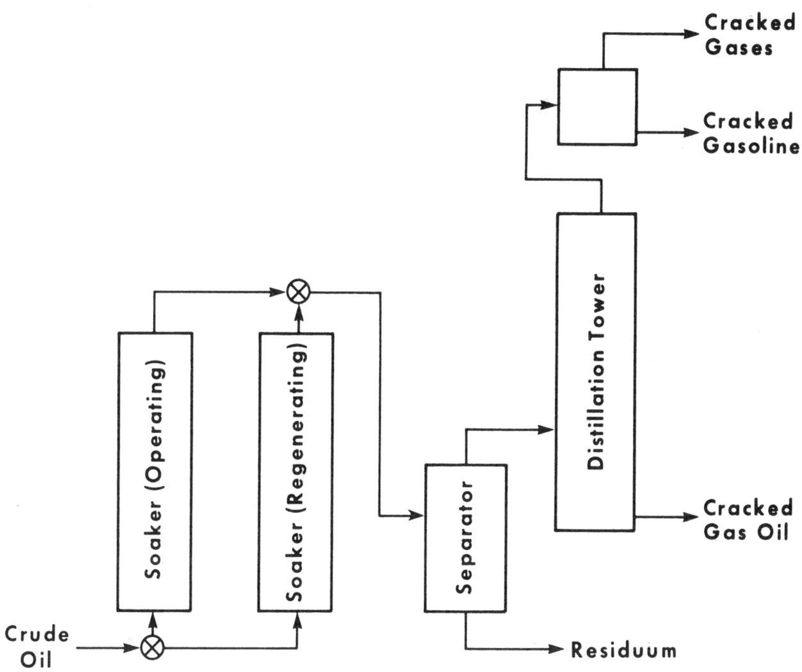

Figure 3 Schematic representation of the tube-and-tank cracking process.

course of several days filled the soaker. The gas oil stream was then switched to a second soaker and the first soaker was cleaned out by drilling operations similar to those used to drill an oil well.

The cracked material (other than coke) left the on-stream soaker to enter an evaporator (tar separator) maintained under a much lower pressure than the soaker where, because of the lower pressure, all of the cracked material except the tar became vaporized. The vapor left the top of the separator where it was distilled into separate fractions: gases, gasoline, and gas oil. The tar which deposited in the separator was pumped out for use as an asphalt or as a heavy fuel oil.

B. Commercial Processes

Thermal cracking of higher boiling materials to produce motor liquid fuels is now becoming an obsolete process. New units are rarely installed, but a few refineries still operate thermal cracking units built in previous years. There are, however, processes that can be regarded as having evolved from the original concept of thermal cracking (Table 1). These are visbreaking and the various coking processes.

1. Visbreaking

Viscosity breaking, or visbreaking, is a mild thermal cracking operation used to reduce the viscosity of residua.

Conditions range from 455 to 510°C (850 to 950°F) and from 50 to 300 psi at the heating coil outlet. Liquid phase cracking takes place at these low-severity conditions. In addition to the major product, fuel oil, material in the gas oil and gasoline boiling range is produced.

The residuum is passed through a furnace where it is heated to a temperature of 480°C (895°F) under an outlet pressure of about 100 psi (Figure 4). The cracked products are then passed into a flash-distillation chamber. The overhead material from this chamber is then

Table 1 Summary of the Various Cracking Processes Employed in Modern Refineries

Visbreaking
Mild (880–920°F) heating at 50–200 psig
Reduce viscosity of fuel oil
Low conversion (10%) to 430°F$^-$
Heated coil or drum
Delayed Coking
Moderate (900–960°F) heating at 90 psig
Soak drums (845–900°F) coke walls
Coked until drum solid
Coke (removed hydraulically) 20–40% on feed
Yield 430°F$^-$, 30%
Fluid coking
Severe (900–1050°F) heating at 10 psig
Oil contacts refractory coke
Bed fluidized with steam-even heating
Higher yields of light ends (C_5^-)
Less coke make

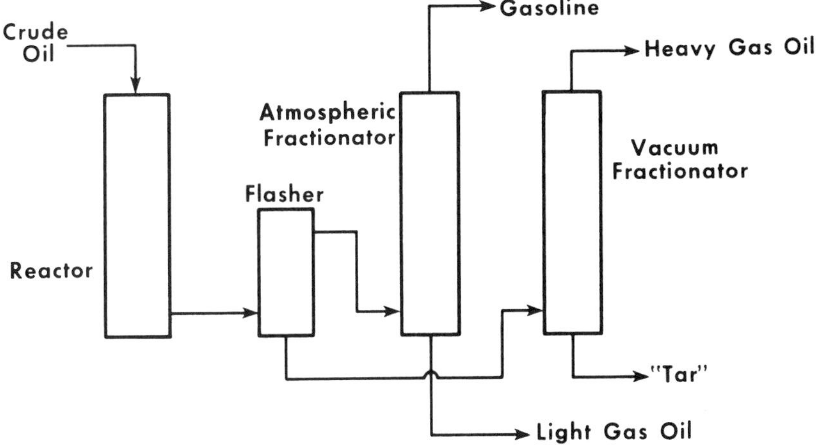

Figure 4 Schematic representation of the visbreaking process.

fractionated to produce a low-quality gasoline as an overhead product and light gas oil as bottoms. The liquid products from the flash chamber are cooled with a gas oil flux and then sent to a vacuum fractionator. This yields a heavy gas oil distillate and a residual tar of reduced viscosity.

2. Coking

Coking is a thermal process for the continuous conversion of heavy, low-grade oils into lighter products. The increased use of heavy oils in refineries has sparked renewed interest in coking operations.

Coking processes generally utilize longer reaction times than thermal cracking processes. To accomplish this, drums or chambers (reaction vessels) are employed, but it is necessary to use two or more such vessels in order that decoking can be accomplished in those vessels not on-stream without interrupting the semicontinuous nature of the process.

Delayed coking: Delayed coking is a semicontinuous process in which the heated feedstock enters one of a pair of coking drums where the cracking reactions continue (Figure 5). The cracked products leave as overheads and coke deposits form on the inner surface of the drum. To give continuous operation, two drums are used; while one is on-stream, the other is being cleaned. The temperature in the coke

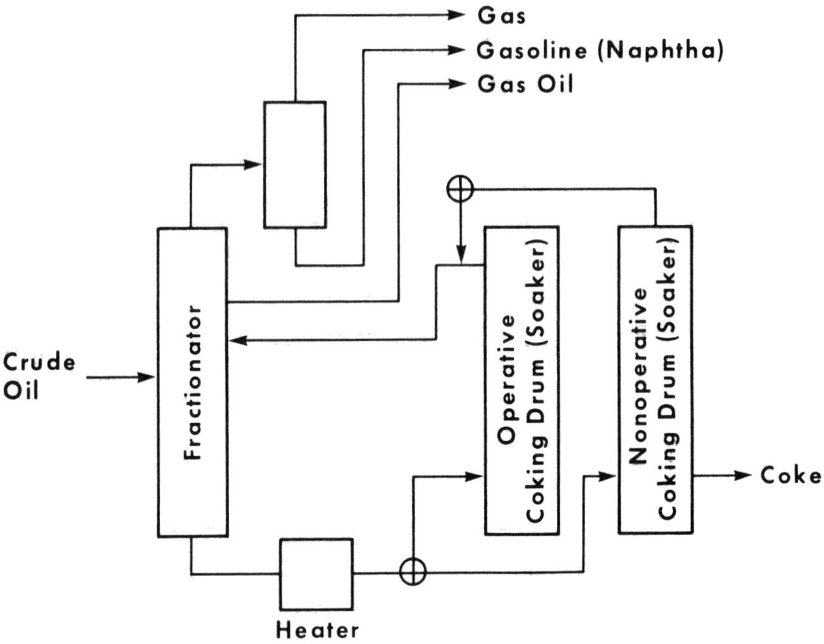

Figure 5 Schematic representation of the delayed coking process.

drum ranges from 415 to 450°C (780 to 840°F) at pressures from 15 to 90 psi.

The coke drum is usually on-stream for about 24 hr before becoming filled with porous coke and the following procedure is used to remove the coke: (1) the coke deposit is cooled with water; (2) one of the heads of the coking drum is removed to permit the drilling of a hole through the center of the deposit; and (3) a hydraulic cutting device, which uses multiple high-pressure water jets, is inserted into the hole and the wet coke removed from the drum.

Fluid coking: Fluid coking is a continuous process which uses the fluidized solids technique to convert residua and other heavy feedstocks under conditions which result in decreased yields of coke and increased yields of liquid products.

Fluid coking uses two vessels—a reactor and a burner. Coke particles are circulated between these to transfer heat (generated by burning a portion of the coke) to the reactor (Figure 6). The reactor holds a bed of fluidized coke particles and steam is introduced at the bottom of the reactor to fluidize the bed. The pitch feed coming

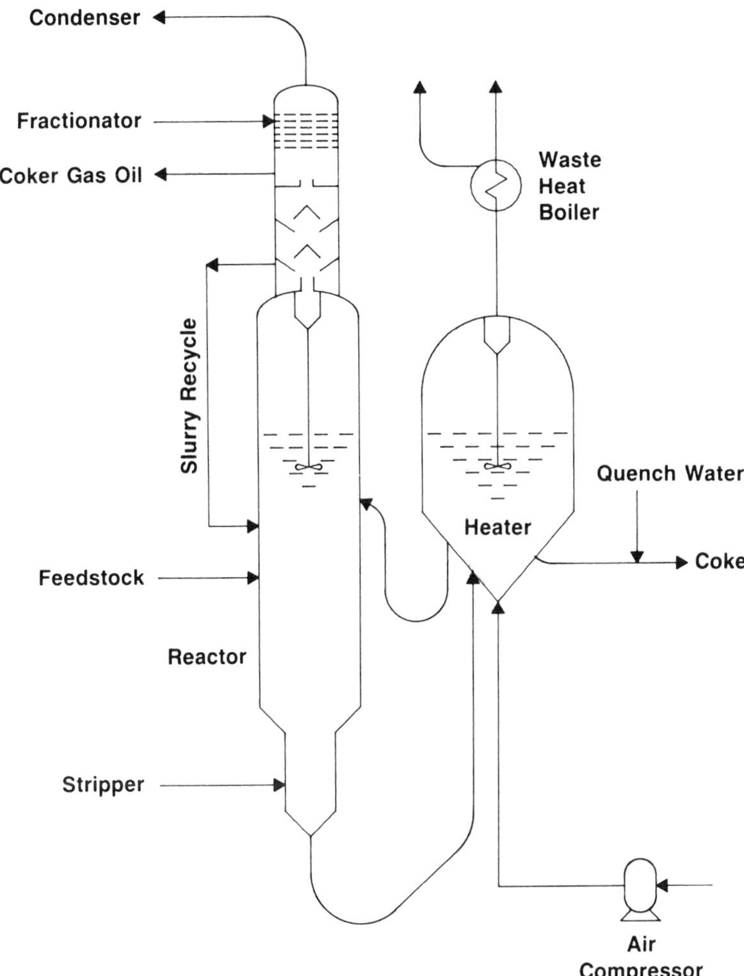

Figure 6 Schematic representation of the fluid coking process.

from the bottom of a vacuum tower at, for example, 260–370°C (500–700°F) is injected directly into the reactor. The temperature in the coking vessel ranges from 480 to 565°C (900 to 1050°F) and the pressure is substantially atmospheric, so the incoming feed is partly vaporized and partly deposited on the fluidized coke particles.

In the reactor, the coke particles flow down through the vessel into a stripping zone at the bottom. Steam displaces the product vapors between the particles and the coke then flows into a riser which leads to the burner. Steam is added to the riser to reduce the solids loading and to induce upward flow. The average bed temperature in the burner is 590–650°C (1095–1200°F) and air is added as needed to maintain the temperature by burning part of the product coke. The pressure in the burner may range from 5 to 25 psi.

Coke is one of the products of the process and it must be withdrawn from the system in order to keep the solids inventory from increasing. The net coke produced is removed from the burner bed through a quench elutriator drum where water is added for cooling and cooled coke is withdrawn and sent to storage. During the course of the coking reaction, the particles tend to grow in size. The size of the coke particles remaining in the system is controlled by a grinding system within the reactor.

II. CATALYTIC CRACKING

The original incentive to develop cracking processes arose from the need to increase gasoline supplies. Since cracking could virtually double the volume of gasoline from a barrel of crude oil, this purpose of cracking was wholly justified (Germain, 1969).

Catalytic cracking (Table 2) is basically the same as thermal cracking but it differs in the use of a catalyst, which is not (in theory) consumed in the process, to direct the course of the cracking reactions to produce more of the desired higher octane hydrocarbon products.

In general, catalytic cracking may be regarded as the modern method for converting high-boiling petroleum fractions, such as gas oil, into gasoline and other low-boiling fractions. Thus, catalytic cracking in the usual commercial process involves contacting a gas oil fraction with an active catalyst under suitable conditions of temperature, pressure, and residence time so that a substantial part (>50%) of the gas oil is converted into gasoline and lower boiling products, usually in a single-pass operation (Bland and Davidson, 1967).

However, during the cracking reaction, carbonaceous material is deposited on the catalyst, which markedly reduces its activity; therefore, removal of the deposit is very necessary. This is usually accomplished by burning in the presence of air until catalyst activity is reestablished.

A. Commercial Processes

The several processes presently employed in catalytic cracking differ mainly in the method of catalyst handling although there is overlap with regard to catalyst type and the nature of the products.

Table 2 Generalized Summary of Catalytic Cracking Operations

Conditions
- Solid acidic catalyst (silica-alumina, zeolite, etc.)
- 900–1000°F (solid/vapor contact)
- 10–20 psig

Feeds
- Virgin naphthas to atmospheric residua
- Pretreated to remove salts (metals)
- Pretreated to remove asphalts

Products
- Lower MW components
- C_3–C_4 gases > C_2^- gases
- Isoparaffins
- Coke → heat

Variations
- Fixed bed
- Moving bed
- Fluidized bed

The catalyst, which may be an activated natural or synthetic material, is employed in bead, pellet, or microspherical form and can be used as a fixed bed, moving bed, or fluid bed. The fixed-bed process was the first process to be used commercially; it uses a static bed of catalyst in several reactors which allows a continuous flow of feedstock to be maintained (Bland and Davidson, 1967; Speight, 1980).

The major process variables are temperature, pressure, catalyst/oil ratio (ratio of the weight of catalyst entering the reactor per hour to the weight of oil charged per hour), and space velocity (weight or volume of the oil charged per hour per weight or volume of catalyst in the reaction zone). Wide flexibility in product distribution and quality is possible through control of these variables along with the extent of internal cycle. Increased conversion can be obtained by (1) higher temperature, (2) higher pressure, (3) lower space velocity, and (4) a higher catalyst/oil ratio.

1. Fixed-Bed Processes

Fixed-bed processes were the first of the modern catalytic cracking processes. In the Houdry fixed bed, the catalyst in the form of

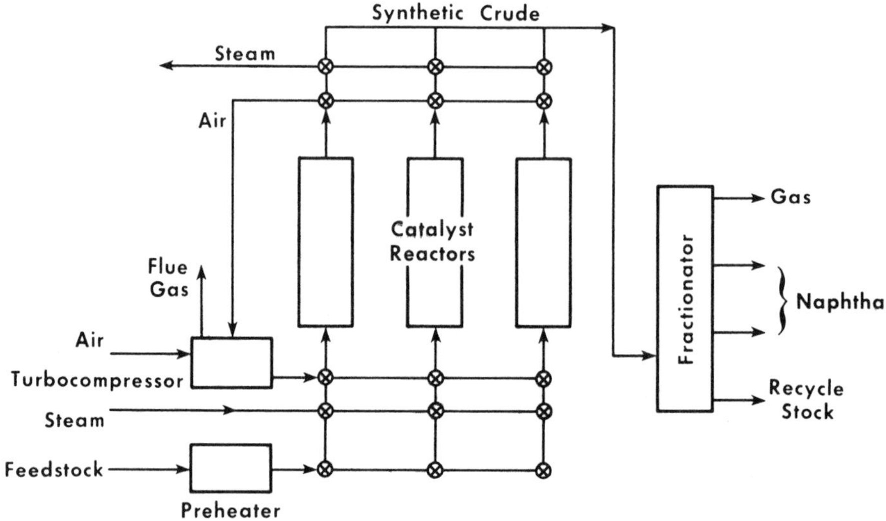

Figure 7 Schematic representation of the Houdry fixed-bed catalytic cracking process.

small lumps or pellets is made up of layers or beds in several (four or more) catalyst-containing drums called converters (Figure 7). Feedstock vaporized at about 450°C (840°F) and under 7–15 psi passed through one of the converters where the cracking reactions took place. After a short time, deposition of coke renders the catalyst ineffective and, using a synchronized valve system, the feedstream is turned into a neighboring converter while the catalyst in the first converter was regenerated by carefully burning the coke deposits with air.

2. Fluid-Bed Processes

The fluid catalytic cracking process is the most widely used process of all and is characterized by the use of a finely powdered catalyst which is moved through the processing unit (Figure 8). The catalyst particles are of such a size that when "aerated" with air, or hydrocarbon vapor, the catalyst behaves like a liquid and can be moved through pipes. Vaporized feedstock and fluidized catalyst flow together into a reaction chamber where the catalyst, still dispersed in the hydrocarbon vapors, forms beds in the reaction chamber and the cracking reactions take place. The cracked vapors pass through

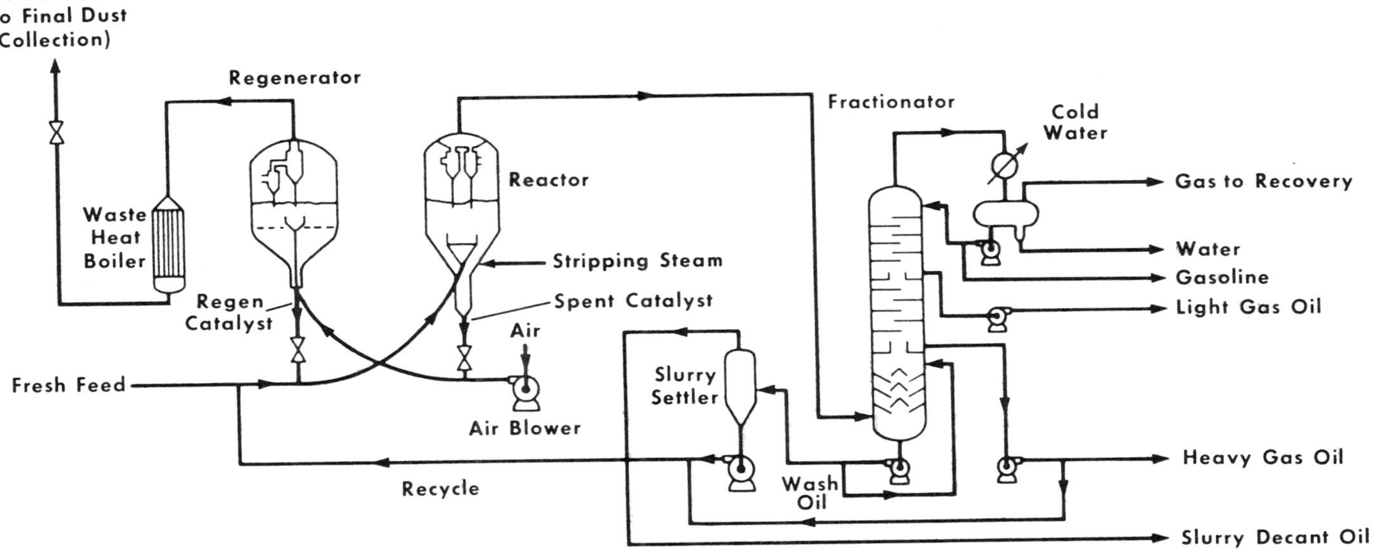

Figure 8 Schematic representation of a fluid bed catalytic cracking process.

cyclones located in the top of the reaction chamber and the catalyst powder is thrown out of the vapors by centrifugal force. The cracked vapors then enter the bubble towers where fractionation into light and heavy cracked gas oils, cracked gasoline, and cracked gases takes place.

Since the catalyst in the reactor becomes contaminated with coke, the catalyst is continuously withdrawn from the bottom of the reactor and lifted by means of a stream of air into a regenerator where the coke is removed by controlled burning. The regenerated catalyst then flows to the fresh feed line where the heat in the catalyst is sufficient to vaporize the fresh feed before it reaches the reactor where the temperature is about 510°C (950°F).

3. Moving-Bed Processes

The Houdriflow process employs a continuous moving-bed process in an integrated single vessel for the reactor and regenerator kiln. The charge stock, sweet or sour, can be any fraction of the crude boiling between naphtha and soft asphalt. The catalyst is transported from the bottom of the unit to the top in a gas lift employing compressed flue gas and steam (Figure 9).

The reactor feed and catalyst pass concurrently through the reactor zone to a disengager section, in which vapors are separated and directed to a conventional fractionation system. The spent catalyst, which has been steam-purged of residual oil, flows to the kiln for regeneration after which steam and flue gas are used to transport the catalyst to the reactor.

III. HYDROPROCESSING

Hydrogenation processes for the conversion of petroleum fractions and products may be classified as destructive and nondestructive. Destructive hydrogenation (hydrogenolysis or hydrocracking; Table 3) is characterized by the cleavage of carbon-to-carbon linkages accompanied by hydrogen saturation of the fragments to produce lower boiling products. Such treatment requires severe processing conditions and the use of high hydrogen pressures to minimize polymerizations and condensations leading to coke formation.

Nondestructive or simple hydrogenation is generally used for the purpose of improving product quality without appreciable alteration of the boiling range. Mild processing conditions are employed so that only the more unstable materials are attacked. Thus, nitrogen, sulfur,

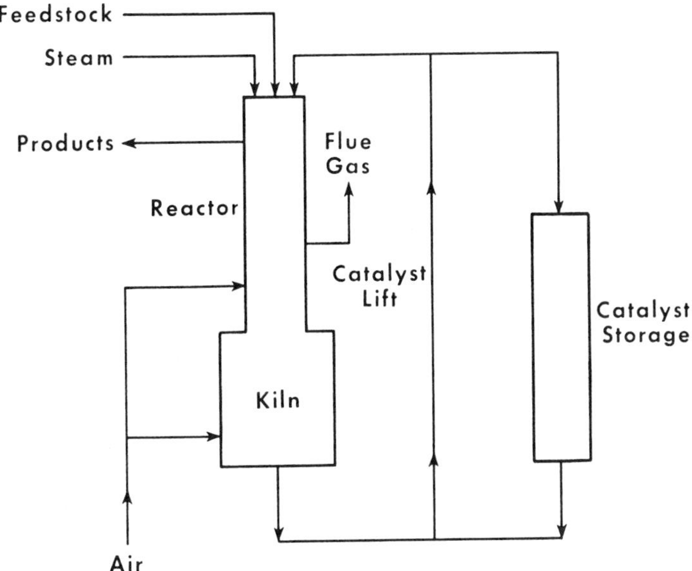

Figure 9 Schematic representation of the Houdriflow moving-bed catalytic cracking process.

Table 3 Generalized Summary of Hydrocracking Process Operations

Conditions
Solid acid catalyst (silica-alumina with rare earth, etc.)
500–800°F (solid/liquid contact)
1000–2000 psig H_2
Feeds
Refractory (aromatic) stream
Most S, N, metals, and H_2O removed
Coker oils, cycle oils
Products
Lower MW isoparaffins
Some C_4^- gases
Residual tar (recycle)
Variations
Fixed bed
Ebullating bed

and oxygen compounds undergo hydrogenolysis to split out ammonia, hydrogen sulfide, and water, respectively. Olefins are saturated and unstable compounds, such as diolefins, which might lead to the formation of gums or insoluble materials are converted to more stable compounds (Scott and Bridge, 1971; Aalund, 1975).

A. Commercial Processes

Hydrocracking is similar to catalytic cracking with hydrogenation superimposed and with the reactions taking place either simultaneously or sequentially. The purpose of hydrocracking is to convert high-boiling feedstocks to lower boiling products by cracking the hydrocarbons in the feed and hydrogenating the unsaturated materials in the product streams. The sulfur and nitrogen atoms are converted to hydrogen sulfide and ammonia, but a probably more important role of the hydrogenation is to hydrogenate the coke precursors rapidly and prevent their conversion to coke.

A comparison of hydrocracking with another important refinery process, hydrotreating, to which it is closely related, is useful in assessing the parts played by these two processes in refinery operations. Hydrotreating of distillates may be defined simply as the removal of sulfur and nitrogen compounds by selective hydrogenation. The hydrotreating catalysts are usually cobalt plus molybdenum or nickel plus molybdenum (in the sulfide) forms and impregnated on an alumina base. The hydrotreated operating conditions are such that appreciable hydrogenation of aromatics will not occur: 1000–2000 psi hydrogen and about 370°C (700°F). The desulfurization reactions are invariably accompanied by small amounts of hydrogenation and hydrocracking, the extent of which depends on the nature of the feedstock and the severity of desulfurization.

Hydrocracking is an extremely versatile process which can be utilized in many different ways. One of its advantages is its ability to break down high-boiling aromatic stocks which are produced by catalytic cracking or coking. To take full advantage of hydrocracking the process must be integrated in the refinery with other process units.

The commercial processes for treating, or finishing, petroleum fractions with hydrogen all operate in essentially the same manner. The feedstock is heated and passed with hydrogen gas through a tower or reactor filled with catalyst pellets. The reactor (Figure 10) is maintained at a temperature of 260–425°C (500–800°F) at pressures of 100–1000 psi depending on the particular process, the nature of the feedstock, and the degree of hydrogenation required. After leaving the reactor, excess hydrogen is separated from the

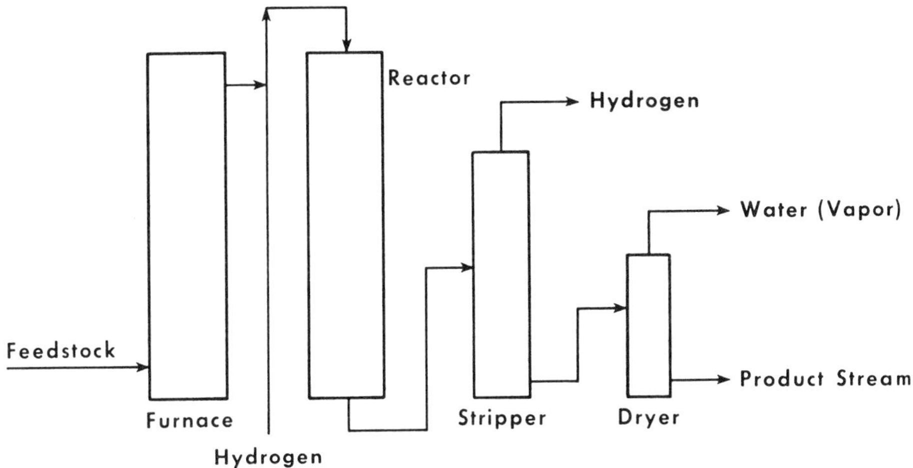

Figure 10 A hydrofiner unit.

treated product and recycled through the reactor after removal of hydrogen sulfide. The liquid product is passed into a stripping tower where steam removes dissolved hydrogen and hydrogen sulfide and, after cooling, the product is run to finished product storage or, in the case of feedstock preparation, pumped to the next processing unit.

1. H-Oil

The H-oil process is basically a catalytic hydrogenation technique whereby during the reaction considerable hydrocracking takes place (Figure 11). The process is used to upgrade heavy sulfur-containing crudes and residual stocks to high-quality sweet distillates, thereby reducing fuel oil yield. A modification of H-oil called Hy-C cracking will convert heavy distillates to middle distillates and kerosene.

Oil and hydrogen are fed upward through the reactors as a liquid-gas mixture at a velocity such that catalyst is in continuous motion. Catalyst of small particle size can be used, giving efficient contact among gas, liquid, and solid with good mass and heat transfer. Part of the reactor effluent is recycled back through the reactors for temperature control and to maintain the requisite liquid velocity. The entire bed is held within a narrow temperature range, which provides

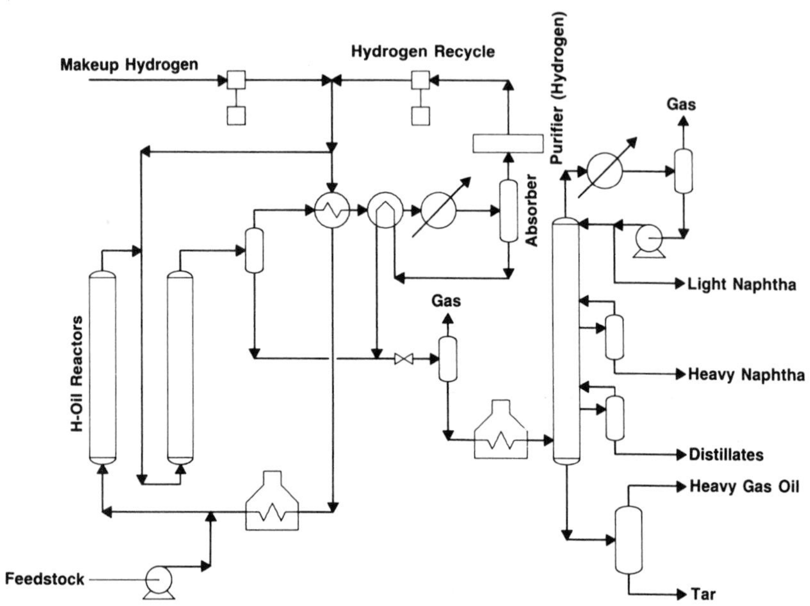

Figure 11 The H-oil process.

essentially an isothermal operation with an exothermic process. Because of the movement of catalyst particles in the liquid-gas medium, deposition of tar and coke is minimized and fine solids entrained in the feed will not lead to reactor plugging. The catalyst can also be added and withdrawn from the reactor without destroying the continuity of the process.

The reactor effluent is cooled by exchange and separates into vapor and liquid. After scrubbing in lean oil absorber, hydrogen is recycled and the liquid product is either stored directly or fractionated prior to storage and blending.

IV. THERMAL REFORMING

When the demand for higher octane gasolines developed during the early 1930s, attention was directed to ways and means of improving the octane number of fractions within the boiling range of gasoline. Straight-run gasolines, for example, frequently had very low octane numbers and any process that would improve them would aid in meeting the demand for more octane numbers. Such a process—called thermal reforming—was developed and used widely but to a much less extent than thermal cracking. Thermal reforming was a natural development from thermal cracking since reforming is also a decomposition reaction due to heat. Cracking converts heavier oils into gasoline; reforming converts or reforms gasolines into higher octane gasolines. The equipment for thermal reforming is essentially the same as for thermal cracking but higher temperatures are used (Nelson, 1958).

A. Commercial Processes

Thermal reforming (Figure 12) is in general less effective and less economical than catalytic processes and has been largely supplanted. As practiced, a single-pass operation was employed at temperatures in the range 540–760°C (1000–1140°F) and pressures of about 500–1000 psi. The degree of octane number improvement depended on the extent of conversion but was not directly proportional to the extent of crack per pass. Not only is the octane level changed by the depth of the cracking but the gasolines produced increase in volatility; the distillation curves approach limiting values for volume over at given temperatures. At very deep conversions the production of coke and gas became prohibitively high. The gases produced, though high in methane, were quite olefinic and the process was generally accompanied by either a separate gas polymerization operation or one in which C_3–C_4

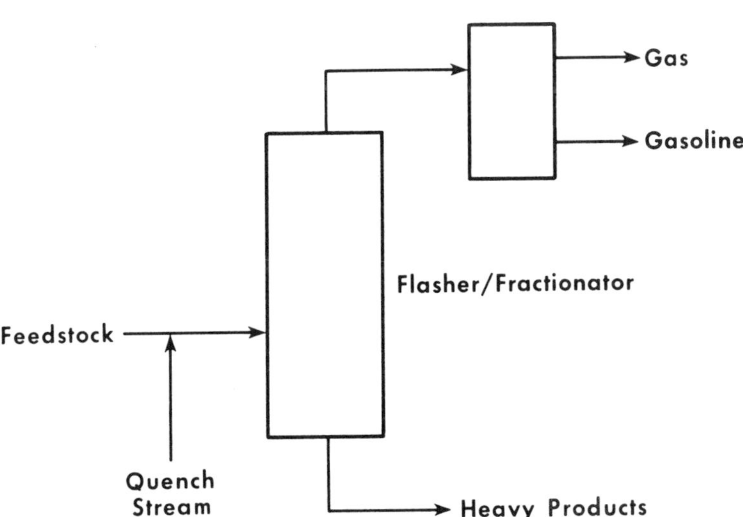

Figure 12 Schematic representation of the thermal reforming process.

gases, autogenous as well as extraneous, were added back to the reforming system.

V. CATALYTIC REFORMING

Like thermal reforming, catalytic reforming converts low-octane gasolines into high-octane gasolines (reformates). Where thermal reforming could produce reformates with research octane numbers of 65–80 depending on the yield, catalytic reforming produces reformates with octane numbers of the order of 90–95. Catalytic reforming is conducted in the presence of hydrogen over hydrogenation-dehydrogenation catalysts which may be supported on alumina or silica-alumina.

A. Commercial Processes

The commercial processes available for use can be broadly classified as of the fixed-bed, moving-bed, and fluid bed types. The fluid and moving-bed processes used mixed nonprecious metal oxide catalysts in units equipped with separate regeneration facilities. Fixed-bed processes use predominantly platinum-containing catalysts in units equipped for cyclic, occasional, or no regeneration.

Dehydrogenation is a main chemical reaction in catalytic reforming and, consequently, hydrogen gas is produced in large quantities. The hydrogen is recycled through the reactors where the reforming takes place to provide the atmosphere necessary for the chemical reactions and also prevents carbon from being deposited on the catalyst, thus extending its operating life. An excess of hydrogen above whatever is consumed in the process is produced and, as a result, catalytic reforming processes are unique in that they are the only petroleum refinery processes to produce hydrogen as a byproduct.

Catalytic reforming usually is carried out by feeding a naphtha (after pretreating with hydrogen if necessary) and hydrogen mixture to a furnace where the mixture is heated to the desired temperature (450–520°C; 840–965°F) and then passed through fixed-bed catalytic reactors at hydrogen pressures of 100–1000 psi. Normally two or more reactors are used in series, and reheaters are located between adjoining reactors in order to compensate for the endothermic reactions taking place. Sometimes as many as four or five are kept on-stream in series while one or more is being regenerated. The on-stream cycle of any one reactor may vary from several hours to many days, depending on the feedstock and reaction conditions.

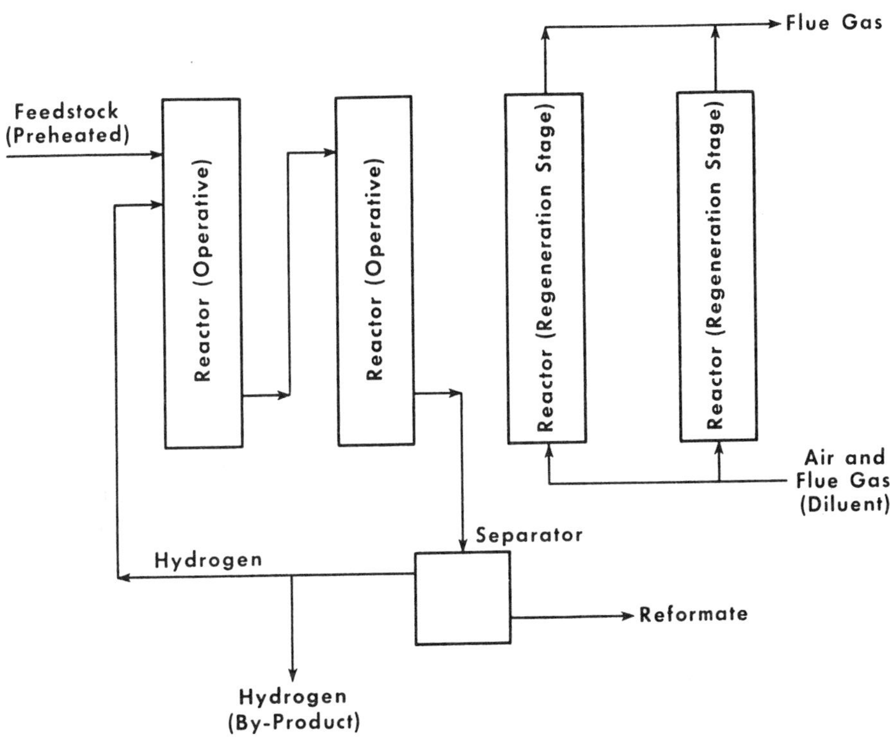

Figure 13 A fixed-bed hydroforming unit.

1. Fixed-Bed Processes

An example of a fixed-bed process is the hydroforming process (Figure 13) which makes use of molybdena-alumina catalyst pellets arranged in fixed beds. The process utilizes four reaction vessels or catalyst cases, two of which are regenerated while the other two are on the process cycle. Naphtha feed is preheated to 480–540°C (900–1000°F) and passed in series through the two catalyst cases under a pressure of 150–300 psi. Gas containing 70% hydrogen produced by the process was passed through the catalyst cases with the naphtha. The material leaving the final catalyst case entered a four-tower system where fractionation distillation separated hydrogen-rich gas, a reformate suitable for motor gasoline, and an aromatic polymer boiling above 205°C (400°F).

After 4–16 hr on process cycle, the catalyst is regenerated by burning carbon deposits from the catalyst at a temperature of 565°C

(1050°F) by blowing air diluted with flue gas through the catalyst. The air also reoxidized the reduced catalyst (9% molybdenum oxide on activated alumina pellets) and removes sulfur from the catalyst.

2. Moving-Bed Processes

Hyperforming is a moving-bed reforming process (Figure 14) which uses catalyst pellets of cobalt molybdate with a silica-stabilized alumina base. In operation, the catalyst moves downward through the reactor by gravity flow and is returned to the top by means of a solids-conveying technique (Hyperflow) which moves the catalyst at low velocities and with minimum attrition loss. Feedstock (naphtha vapor) and recycle gas flow upward, countercurrent to catalyst and regeneration of catalyst is accomplished in either an external vertical lift line or a separate vessel. Operating conditions in the reactor are 400 psi and 425–480°C (800–900°F) the higher temperature being employed for a straight-run naphtha feedstock while catalyst regeneration takes place at 510°C (950°F) and 415 psi.

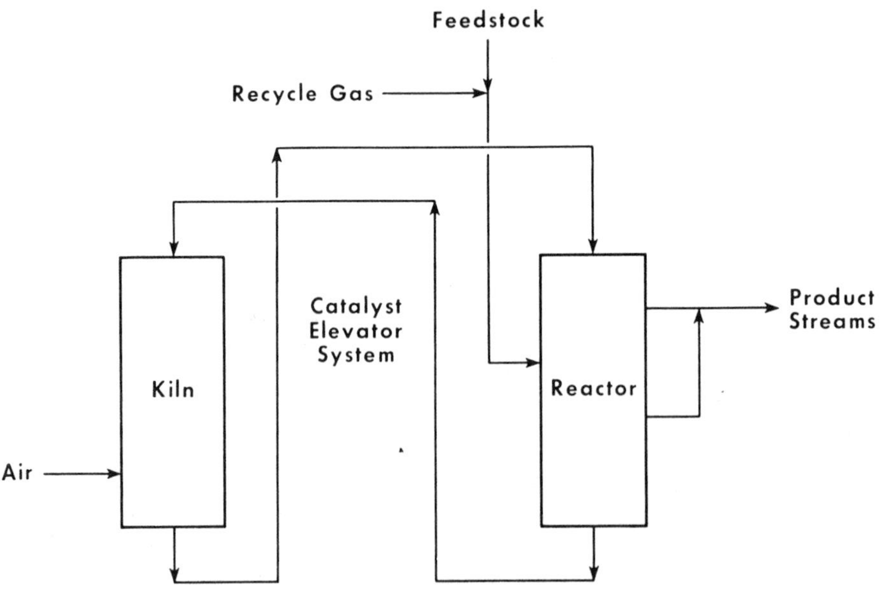

Figure 14 Moving-bed catalytic reforming.

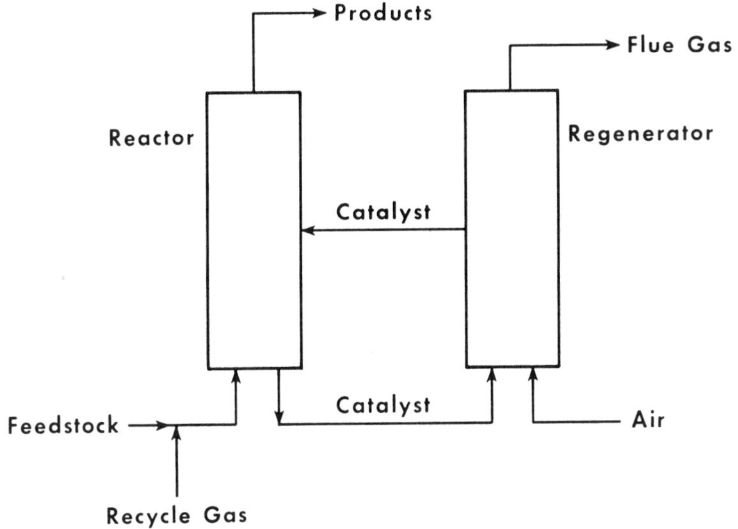

Figure 15 Fluid bed reforming.

3. Fluid-Bed Processes

In catalytic reforming processes using a fluidized-solids-catalyst-bed (Figure 15) continuous regeneration with a separation or integrated reactor is practiced to maintain catalyst activity by coke and sulfur removal. Cracked or virgin naphthas are charged with hydrogen-rich recycle gas to the reactor. A molybdena (10.0%) on alumina catalyst—not materially affected by normal amounts of arsenic, iron, nitrogen, or sulfur—is used. Operating conditions in the reactor are about 200–300 psi and 480–510°C (900–950°F).

Fluid bed operation with its attendant excellent temperature control prevents over- and underreforming operations, resulting in more selectivity in the conditions needed for optimum yield of the desired product.

VI. ISOMERIZATION

Catalytic reforming processes provide high-octane constituents in the heavier gasoline fraction, but the normal paraffin components of the lighter gasoline fraction, especially butane (C_4) to hexane (C_6), have poor octane ratings. The conversion of these normal paraffins to their isomers (isomerization), e.g.,

$$\underset{n\text{-pentane}}{CH_3 \cdot CH_2 \cdot CH_2 \cdot CH_2 \cdot CH_3} \rightarrow \underset{\text{isopentane}}{CH_3 \cdot CH_2 \cdot \underset{\displaystyle CH_3}{\underset{|}{CH}} \cdot CH_3}$$

yields gasoline components of high octane rating in this lower boiling range. Conversion is obtained in the presence of a catalyst (aluminum chloride activated with hydrochloric acid) and it is essential to inhibit side reactions such as cracking and olefin formation.

Various companies have developed and operated isomerization processes which increased the octane numbers of light naphthas from, way, 70 or less to more than 80. In a typical process, a naphtha is passed over an aluminum chloride catalyst at 120°C (250°F) and under a pressure of about 800 psi to produce the isomerate.

A. Commercial Processes

The earliest important process (Figure 16) was the formation of isobutane, which is required as an alkylation feed, and the isomerization may take place in the vapor phase, with the activated catalyst supported on a solid phase, or in the liquid phase with a dissolved catalyst. Thus a pure butane feed is mixed with hydrogen (to inhibit olefin formation) and passed to the reactor at 110–170°C (230–340°F) and 200–300 psi. The product is cooled and the hydrogen separated; the cracked gases are then removed in a stabilizer column. The stabilizer bottom product is passed to a superfractionator and the normal and isobutanes are separated. With pentanes, the equilibrium is favorable at higher temperatures, and operating conditions of 300–1000 psi and 240–500°C (465–930°F) may be used.

Present isomerization applications in petroleum refining are to provide additional feedstock for alkylation units or high-octane fractions for gasoline blending. Straight-chain paraffins (*n*-butane, *n*-pentane, *n*-hexane) are converted to respective iso compounds by continuous catalytic (aluminum chloride, noble metals) processes. Natural gasoline or light straight-run gasoline can provide feed by first fractionating as a preparatory step. High volumetric yields (>95%) and 40–60% conversion per pass are characteristic of the isomerization reaction.

Nonregenerable aluminum chloride catalyst is employed with various carriers in a fixed-bed or liquid contactor. Platinum or other metal

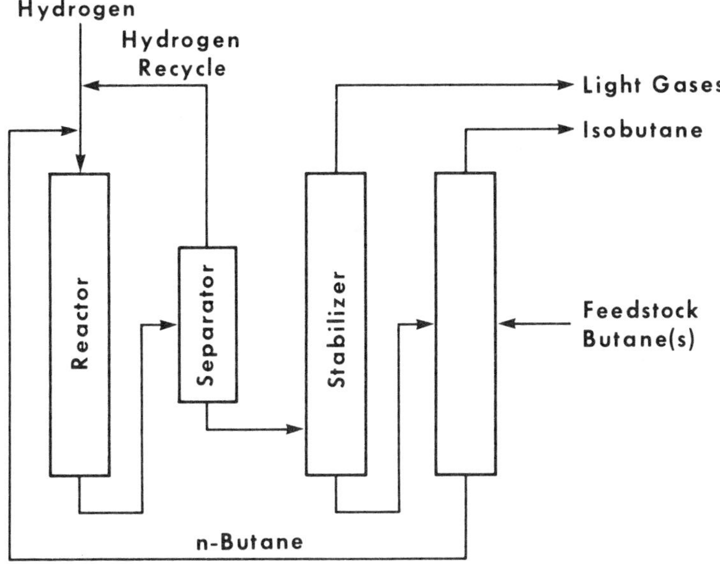

Figure 16 Schematic representation of the fixed-bed isomerization of butane.

catalyst processes utilize fixed-bed operation and can be regenerable or nonregenerable (Figures 17 and 18). The reaction conditions vary widely depending on the particular process and feedstock, 40–480°C (100–900°F) and 150–1000 psi; residence time in the reactor is 10–40 min.

VII. ALKYLATION

The combination of olefins with paraffins to form higher isoparaffins is termed alkylation:

$$\underset{\text{Paraffin}}{R{-}H} + \underset{\text{olefin}}{CH_2{=}CH_2} \rightarrow \underset{\text{"alkylate" (paraffin)}}{R{\cdot}CH_2{\cdot}CH_3}$$

Since olefins are reactive (hence, unstable) and are responsible for exhaust pollutants, their conversion to high-octane isoparaffins is desirable where possible. In refinery practice, only isobutane is

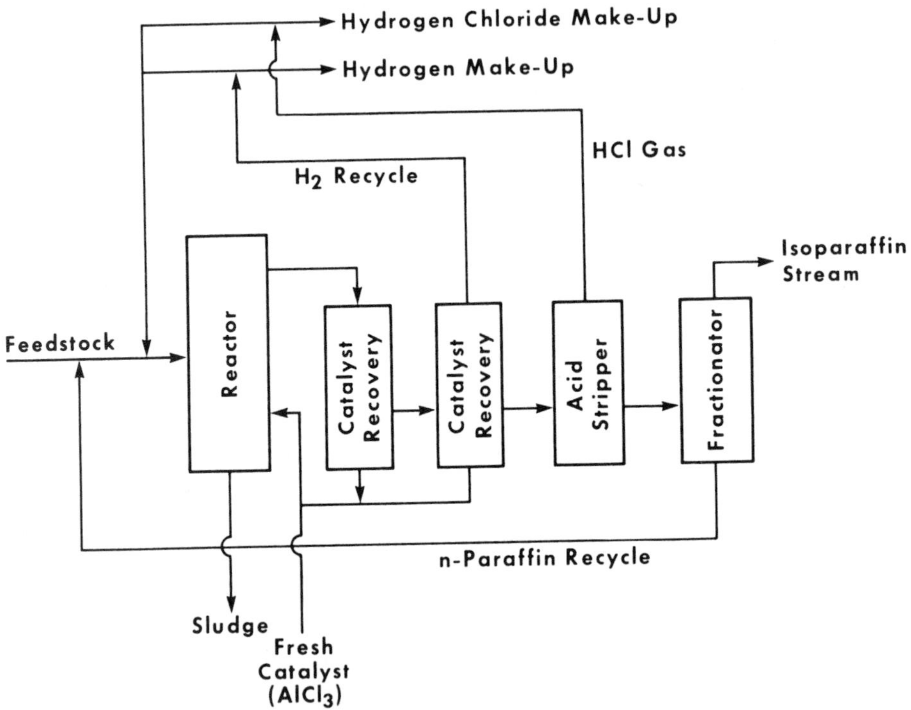

Figure 17 Schematic representation of the isomerization process using aluminum chloride.

alkylated, by reaction with iso- or normal butene, and isooctane is the product. Although alkylation is possible without catalysts, commercial processes use aluminum chloride, sulfuric acid, or hydrogen fluoride as catalyst, when the reactions can take place at low temperatures, minimizing undesirable side reactions such as polymerization of olefins.

Alkylate is composed of a mixture of isoparaffins which have octane numbers that vary with the olefins from which they were made. Butylenes produce the highest octane numbers, propylene the lowest, and pentylenes the intermediate. However, all alkylates have high octane numbers (>87) and are particularly valuable because of this.

A. Commercial Processes

The alkylation reaction as practiced in petroleum refining is the union, through the agency of a catalyst, of an olefin (ethylene, propylene,

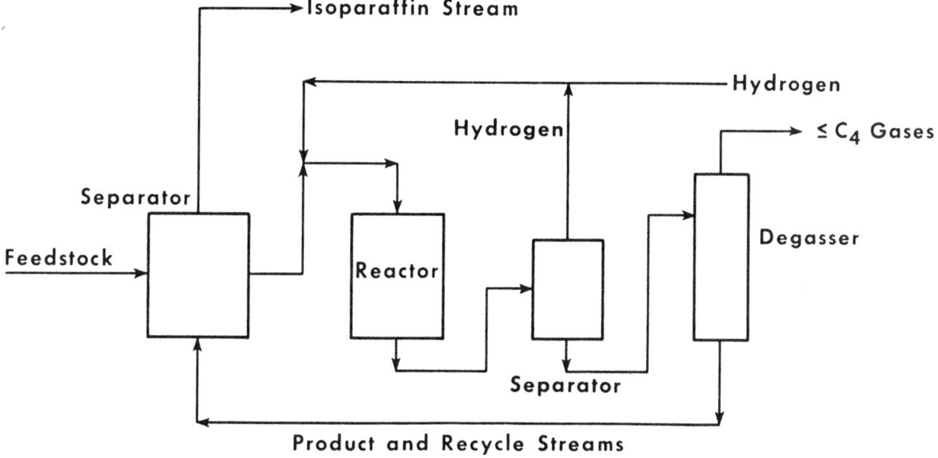

Figure 18 Isomerization with noble metal catalysts.

butylene, and amylene) with isobutane to yield high-octane, branched chain hydrocarbons in the gasoline boiling range. Olefin feedstock is derived from the gas make of a catalytic cracker while isobutane is recovered from refinery gases or produced by catalytic butane isomerization.

In thermal catalytic alkylation, ethylene or propylene is combined with isobutane at 50–280°C (125–450°F) and 300–1000 psi in the presence of metal halide catalysts such as aluminum chloride. Conditions are less stringent in catalytic alkylation; olefins (C_3, C_4, and C_5) are combined with isobutane in the presence of an acid catalyst (sulfuric or hydrofluoric) at low temperatures and pressures [-1 to 40°C (30–105°F)] and atmosphere to 150 psi.

1. Cascade Sulfuric Acid Process

This is a low-temperature process (Figure 19) employing concentrated sulfuric acid catalyst to react olefins with isobutane to produce high-octane aviation or motor fuel blend stock. The olefin feed is split into equal streams and charged to the individual reaction zones of the cascade reactor. Isobutane-rich recycle and refrigerant streams are introduced in the front of the reactor and pass through the reaction zones. The olefin is contacted with the isobutane and acid in the reaction zones, which operate at 2–7°C (35–45°F) and 5–15 psi after which vapors are withdrawn from the top of the reactor, compressed,

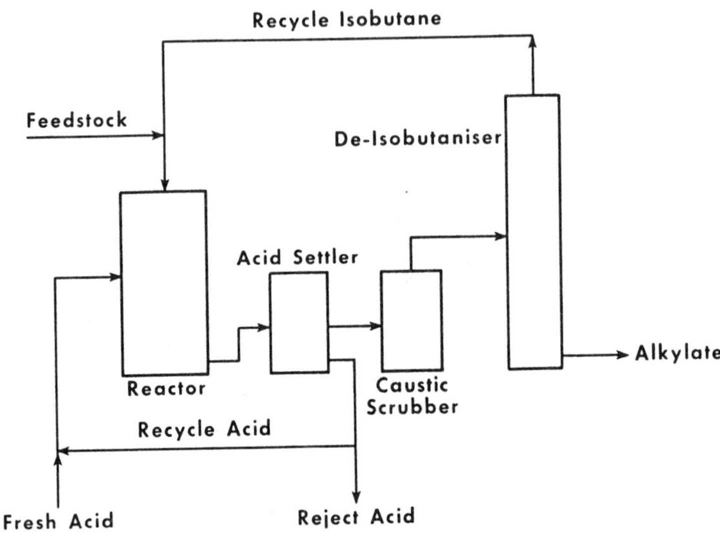

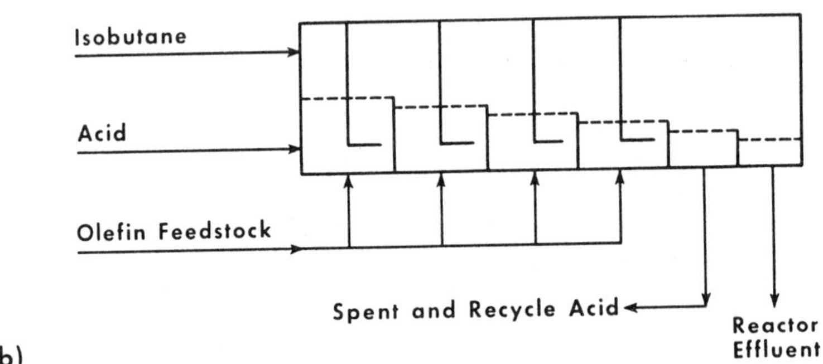

Figure 19 Schematic representation of the different process configurations for alkylation using sulfuric acid: (a) Conventional reactor. (b) Cascade reactor.

and condensed. Part of this stream is sent to a depropanizer to control propane concentration in the unit. Depropanizer bottoms and the remainder of the stream are combined and returned to the reactor. Spent acid is withdrawn from the bottom of the settling zone, while hydrocarbons spill over a baffle into a special withdrawal section and are hot-water-washed with caustic addition for pH control before being successively depropanized, deisobutanized, and debutanized. Alkylate can then be taken directly to motor fuel blending or be rerun to produce aviation grade blend stock.

VIII. POLYMERIZATION

In the petroleum industry, polymerization is the process by which olefin gases are converted to liquid condensation products which may be suitable for gasoline (hence polymer gasoline) or other liquid fuels. The feedstock usually consists of propylenes and butylenes from cracking processes or may even be selective olefins for dimer, trimer, or tetramer production:

$CH_2{=}CH_2$	$\text{-}(CH_2.CH_2)_2\text{-}$	$\text{-}(CH_2.CH_2)_3\text{-}$	$\text{-}(CH_2.CH_2)_4\text{-}$
Olefin	Dimer	Trimer	Tetramer

A. Commercial Processes

Polymerization may be accomplished thermally or in the presence of a catalyst at lower temperatures. Thermal polymerization is regarded as not being as effective as catalytic polymerization but has the advantage that it can be used to "polymerize" saturated materials that cannot be induced to react by catalysts. The process consists essentially of vapor phase cracking of, say, propane and butane followed by prolonged periods at the high temperature (510–595°C; 950–1100°F) for the reactions to proceed to near-completion.

Olefins can also be conveniently polymerized by means of an acid catalyst. Thus, the treated, olefin-rich feedstream is contacted with a catalyst (sulfuric acid, copper pyrophosphate, phosphoric acid) at 150–220°C (300–425°F) and 150–1200 psi, depending on feedstock and product requirement. The reaction is exothermic and temperature is usually controlled by heat exchange. Stabilization and/or fractionation systems separate saturated and unreacted gases from the product. In both thermal and catalytic polymerization processes, the feedstock is usually pretreated to remove sulfur and nitrogen compounds.

1. Thermal Polymerization

Thermal polymerization converts butanes and lighter gases into liquid condensation products. Olefins are produced by thermal decomposition

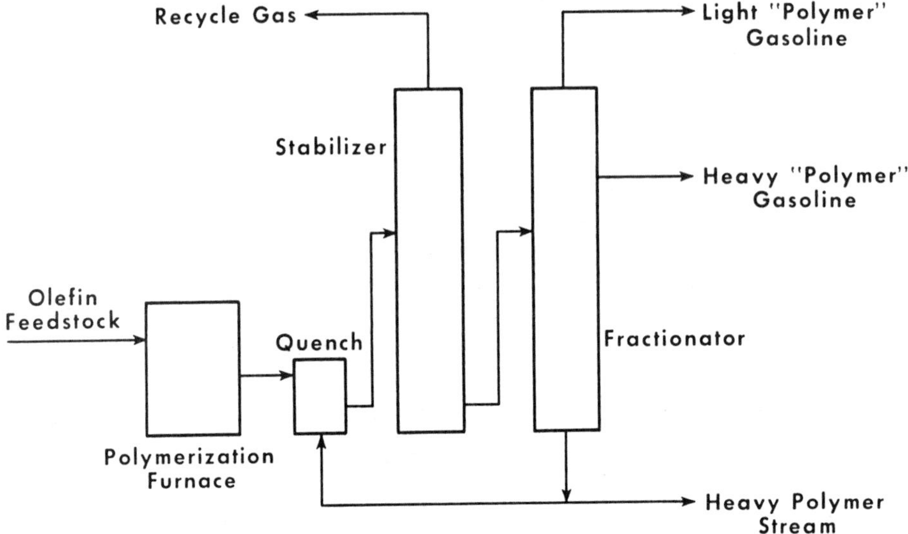

Figure 20 Thermal polymerization.

and polymerized by heat and pressure (Figure 20). Thus liquid feed under a pressure of 1200–2000 psi is pumped to a furnace heated to 510–595°C (950–1100°F) from which the various streams are separated by fractionation.

2. Phosphoric Acid Polymerization

This process (Figure 21) converts propylene and/or butylene to high-octane gasoline or petrochemical polymers. The catalyst, pelleted kieselguhr impregnated with phosphoric acid, is used in a chamber or tubular reactor. The exothermic reaction temperature is controlled by using saturates (separated from the effluent as recycle to the feed) as a quench liquid between the catalyst chamber beds. Tubular reactors are temperature-controlled by water or oil circulation around the catalyst tubes.

Reaction temperatures and pressures are 175–225°C (350–435°F) and 400–1200 psi. Olefins and aromatics may be united by alkylation for special applications at 205–315°C (400–600°F) and 400–900 psi and a rerun column is required in addition to the usual fractionating.

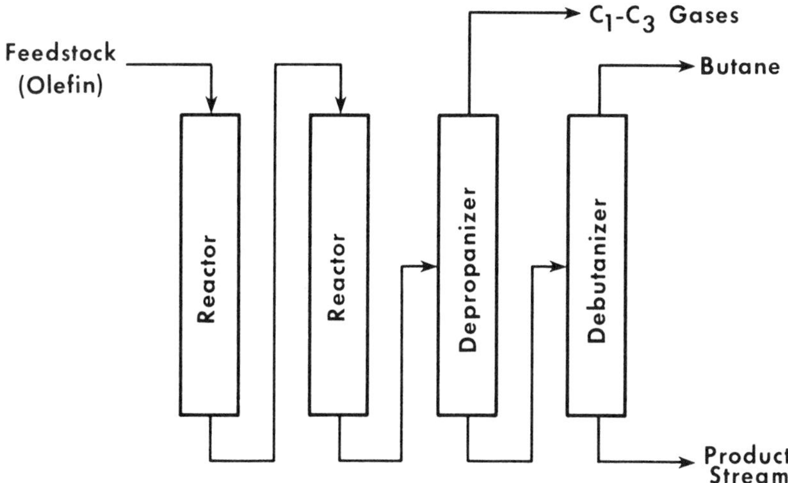

Figure 21 Schematic representation of the catalytic polymerization process.

3. Bulk Acid Polymerization

This is a process to produce high-octane polymer gasoline from all types of light olefin feed and the olefin concentration can be as high as 95%; liquid phosphoric acid is used as the catalyst.

The olefin feed is washed (caustic and water) and then contacted thoroughly by liquid phosphoric acid in a small reactor. The effluent stream and the acid are separated in a settler, and acid is returned to the reactor through a cooler. Gasoline is first stabilized and washed with caustic prior to storage. The heat of reaction is removed by circulation through an exchanger prior to contact with the olefin feed and catalyst activity is maintained by continuous addition of fresh acid and withdrawal of spent acid.

IX. HYDROGEN PRODUCTION

The trend to increased hydrogenation (hydrocracking and/or hydrotreating) processes in refineries coupled with the need to process the heavier oils, which require substantial quantities of hydrogen for upgrading, has resulted in vastly increased demands for this gas. Part of the hydrogen needs can be satisfied by hydrogen recovery from catalytic reformed product gases but other external sources are required.

A. Commercial Processes

Most of the external hydrogen is manufactured either by steam-methane reforming or by oxidation processes. However, other processes such as steam-methanol interaction or ammonia dissociation may also be used as sources of hydrogen. Electrolysis of water will produce high-purity hydrogen but the power costs may be prohibitive.

1. Steam Methane Reforming

Steam-methane reforming is a continuous catalytic process which has been employed for hydrogen production over a period of several decades. The major reaction is the formation of carbon monoxide and hydrogen from methane and steam:

$$CH_4 + H_2O \rightarrow CO + 3H_2$$

while heavier hydrocarbons may also yield hydrogen:

$$C_3H_8 + 3H_2O \rightarrow 3CO + 7H_2$$

i.e.,

$$C_nH_m + nH_2O \rightarrow nCO + (0.5\,m + n)H_2$$

In the actual process (Figure 22) feedstock is first desulfurized by passage through activated carbon which may be preceded by caustic and water washes. The desulfurized material is then mixed with steam and passed over a nickel-based catalyst (730–845°C; 1350–1550°F; 400 psi). Effluent gases are cooled by the addition of steam or condensate to about 370°C (700°F) at which point carbon monoxide reacts with steam in the presence of iron oxide in a shift converter to produce carbon dioxide and hydrogen. The carbon dioxide is removed by amine washing; the hydrogen is usually high purity (>99%) material.

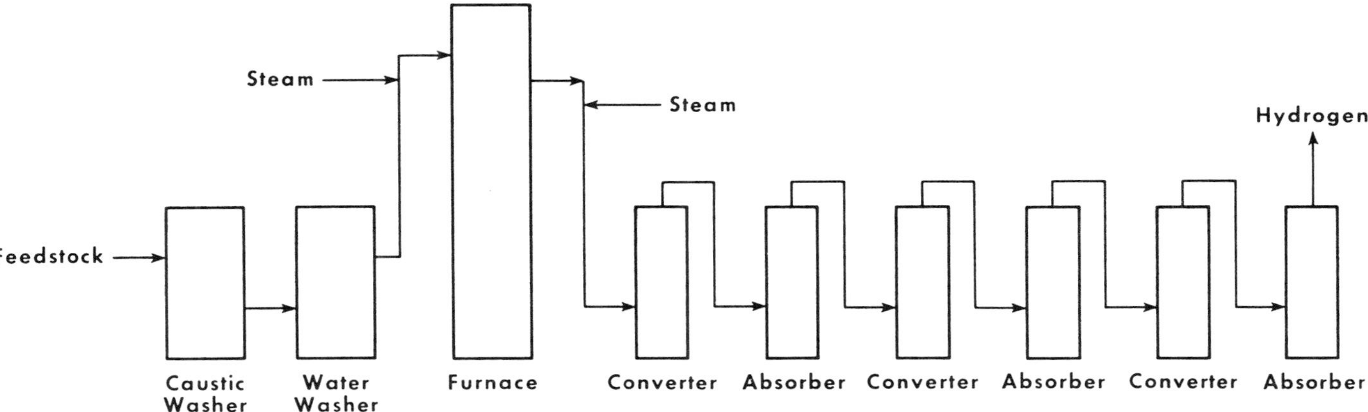

Figure 22 Hydrogen production by steam/methane reforming.

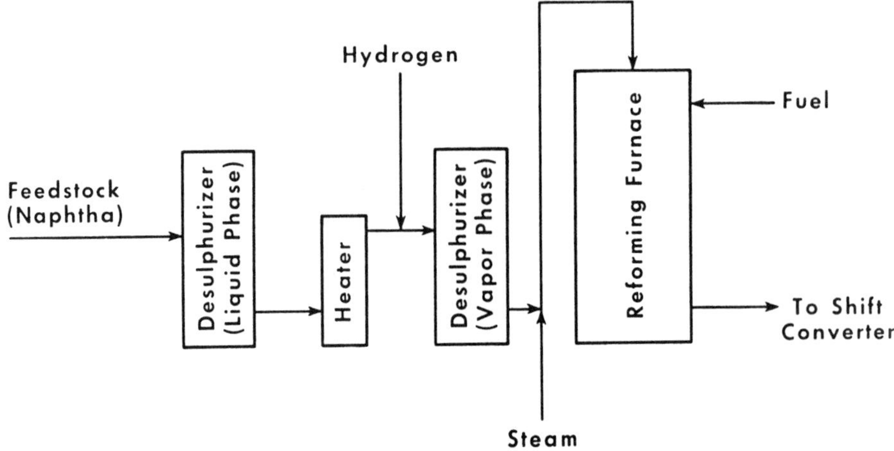

Figure 23 Hydrogen production by steam/naphtha reforming.

2. Steam-Naphtha Reforming

This is a continuous catalytic process (Figure 23) for the production of hydrogen from liquid hydrocarbons and is, in fact, similar to steam-methane reforming. A variety of naphthas in the gasoline boiling range may be employed, including feeds containing up to 35% aromatics. Thus, following pretreatment to remove sulfur compounds, the feedstock is mixed with steam and taken to the reforming furnace (675–815°C; 1250–1500°F; 300 psi) where hydrogen is produced.

3. Synthesis Gas Generation

This is a continuous, noncatalytic process which produces hydrogen by partial oxidation of gaseous or liquid hydrocarbons. A controlled mixture of preheated feedstock and oxygen is fed to the top of the reactor (Figure 24) where carbon dioxide and steam are the primary products. A secondary reaction between the feedstock and the gases forms carbon monoxide and hydrogen.

The effluent is then led to a shift conveyor with high-pressure steam where carbon monoxide is converted to carbon dioxide with the concurrent production of hydrogen at the rate of 1 mole of hydrogen for every mole of carbon dioxide. Reactor temperatures vary from 1095 to 1480°C (2000–2700°F) and pressures from atmospheric to

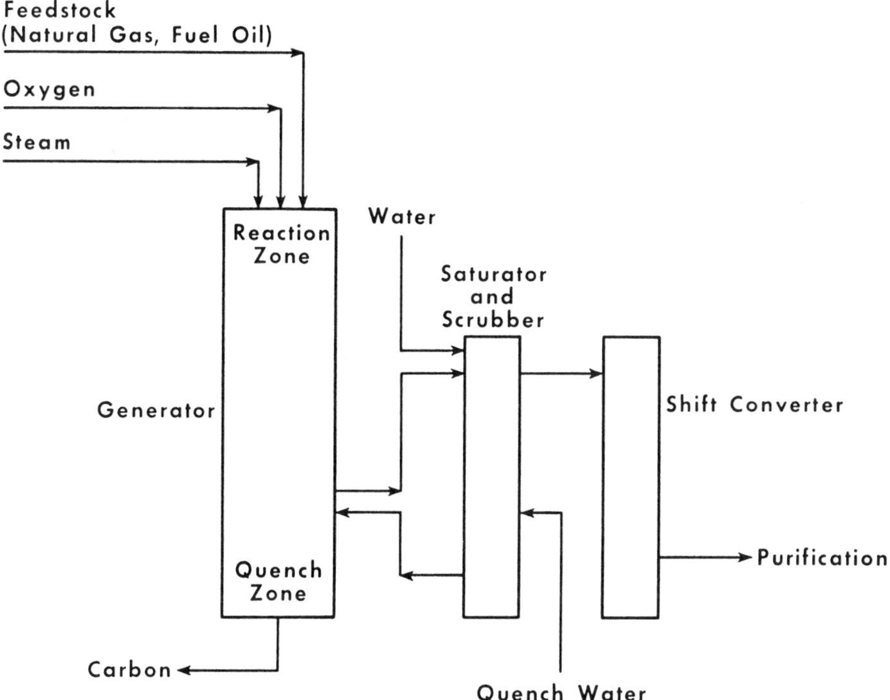

Figure 24 Synthesis gas generation.

more than 1500 psi. Gas purification depends on the use to which the hydrogen is to be put. For example, carbon dioxide is removed by scrubbing with various alkaline solutions while carbon monoxide is removed by washing with liquid nitrogen or various copper amine solutions if nitrogen is not desired in the product.

4. Hydrocarbon Gasification

The gasification of hydrocarbons to produce hydrogen is a continuous, noncatalytic process (Figure 25) which involves partial oxidation of the hydrocarbon. Air or oxygen (with steam or carbon dioxide) is used as the oxidant at 1095–1480°C (2000–2700°F). Any carbon produced (2–3% wt of the feedstock) during the process is removed as a slurry in a carbon separator and pelleted for use either as a fuel or as a raw material for carbon-based products.

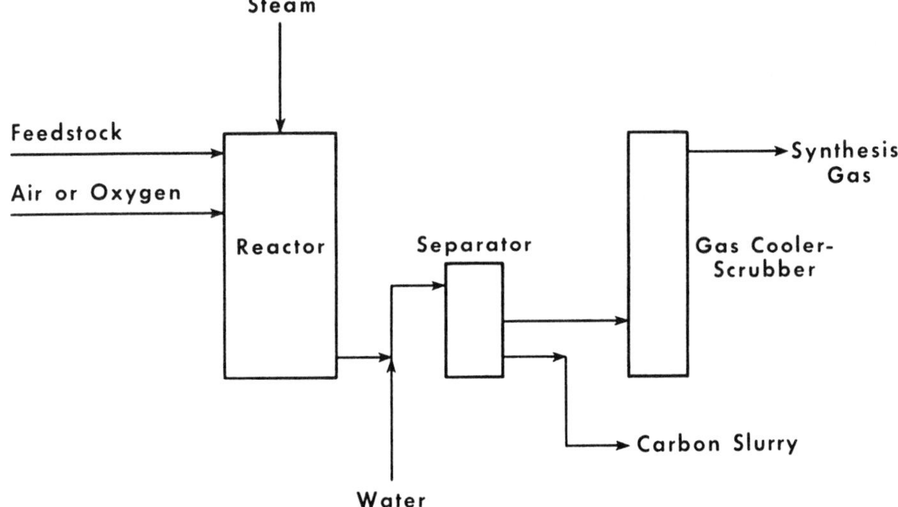

Figure 25 Hydrocarbon gasification.

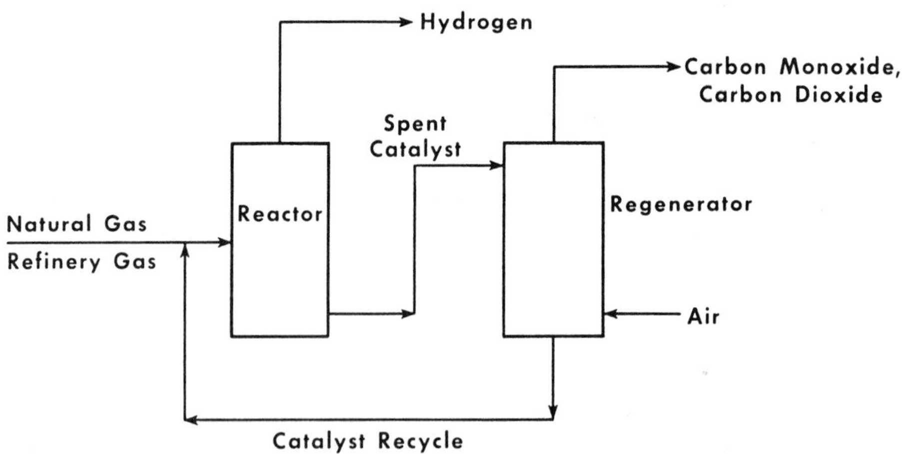

Figure 26 The Hypro process.

5. Hypro Process

The Hypro process is a continuous, catalytic method (Figure 26) for hydrogen manufacture from natural gas or from refinery effluent gases. The process is designed to convert natural gas:

$$CH_4 \rightarrow C + 2H_2$$

and recover hydrogen by phase separation to yield hydrogen of about 93% purity; the principal contaminant is methane.

7

Refining: Treatment Methods

I Introduction 203

II Commercial Processes 205

A. Caustic (Lye) Treatment 205

B. Acid Treatment 207

C. Clay Treatment 209

D. Oxidative Treatment 211

E. Solvent Treatment 215

F. Gas Cleaning 221

I. INTRODUCTION

The undesirable components in a petroleum fraction are referred to as impurities but they are almost invariably normal components of petroleum. The most common impurities are sulfur compounds and the naturally occurring acidic and nitrogenous compounds occur in smaller amounts. Sometimes olefins must be eliminated from a feedstock or aromatics removed from a solvent, and in these cases the olefins and aromatics are considered to be impurities. Similarly, polymerized material, asphaltic material, or resins may be impurities depending on whether or not their presence in a finished product is harmful.

Processes that are primarily concerned with the removal of hydrogen sulfide and mercaptans are known as sweetening processes.

Petroleum fractions containing these sulfur compounds are readily recognized by their odor and are called "sour." Fractions that are free of obnoxious sulfur compounds, either naturally or because of treatment, are called "sweet" but a sweet fraction may contain sulfur compounds that have no odor.

The more severe desulfurization of petroleum fractions is not classed as a treating process and, as such, has been included as a hydrotreating process (Chap. 6, Sec. III). Nevertheless, desulfurization is a serious problem worthy of brief mention here. Indeed, considerable importance has been attached to the lowering of the sulfur level in distillates and residual stocks. For example, the stability of sulfur compounds is greatly reduced when they are heated in the presence of adsorptive-type catalysts, and this fact is employed in a number of desulfurization processes. The noncyclic sulfur compounds (mercaptans, sulfides, and disulfides) in straight-run distillates (e.g., naphthas) are readily converted to hydrogen sulfide and olefins by contacting the vapors with clays, with aluminum oxide, or with alumina-silica cracking catalysts. These processes generally operate at about 345–425°C (650–800°F) and at about 50 psi pressure. When hydrogen is added and a dehydrogenation catalyst, such as cobalt and molybdenum sulfides on alumina, is employed, extensive desulfurization of a wider spectrum of compounds is brought about and even the refractory cyclic nitrogen and sulfur compounds are decomposed, e.g.:

Thiophene $\longrightarrow$ $CH_3.CH_2.CH_2.CH_3$ (n-Butane) + $(CH_3)_3.CH$ (Isobutane)

Methylthiophene $\longrightarrow$ $CH_3.CH_2CH_2.CH_2.CH_2.CH_3$ (n-Pentane) + $(CH_3)_3\ CH.CH_2.CH_3$ (Isopentane)

Pyrolle $\longrightarrow$ $CH_3.CH_2.CH_2.CH_3$ (n-Butane)

Quinoline $\longrightarrow$ $C_6H_5\text{-}CH_2.CH_2.CH_3$ (n-Propylbenzene)

Treatment processes for the removal of sulfur compounds are much less severe than the desulfurization techniques. When there are more than trace amounts (i.e., >0.1%) of sulfur present, it is often more convenient and economical to resort to methods such as thermal processes (e.g., hydrodesulfurization) that bring about a decrease in all types of sulfur compounds.

II. COMMERCIAL PROCESSES

Treating petroleum products by washing with solutions of alkali (caustic, lye) is almost as old as the petroleum industry itself. The industry discovered early that product odor and color could be improved by removing organic acids (naphthenic acids, phenols) and sulfur compounds (mercaptans, hydrogen sulfide) through the use of a caustic wash. The lye reacts with any hydrogen sulfide present to form water-soluble sodium sulfide:

$$H_2S + 2NaOH \longrightarrow Na_2S + 2H_2O$$

A. Caustic (Lye) Treatment

1. Lye Treatment

The lye treatment process involves the use of continuous treaters (Figure 1) each of which consists of a pipe containing baffles or other mixing devices into which the oil and lye solution are both pumped. The pipe discharges into a horizontal tank where the lye solution and oil separate. Treated oil is withdrawn from near the top of the tank while lye solution is withdrawn from the bottom and recirculated to mix with incoming untreated oil. A lye treatment unit may be incorporated as part of a processing unit. For example, the overhead from a bubble tower may be condensed and cooled and passed immediately through a lye treatment unit. Such a unit is often referred to as a worm-end treater since the treater is attached to the particular unit at a point beyond the cooling coil or cooling worm.

Caustic solutions ranging from 5020% w/w are used at 20–45°C (70–110°F) and 5–40 psi. High temperatures and strong caustic are usually avoided because of the risk of color body formation and stability loss. Caustic-to-product treating ratios vary from 1:1 to 1:10.

Spent lye is the term given to a lye solution in which about 65% of the sodium hydroxide content has been used by reaction with hydrogen sulfide, light mercaptans, organic acids, or mineral acids. A

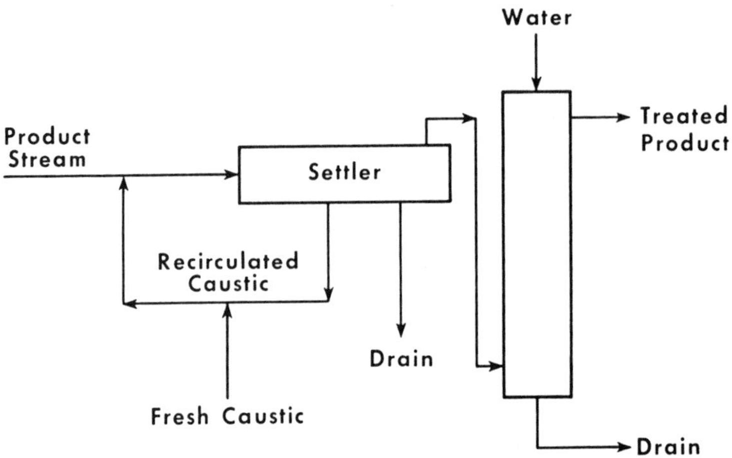

Figure 1 A caustic treatment unit.

lye that is spent as far as hydrogen sulfide is concerned may still be used to remove mineral or organic acids from petroleum fractions. Lye spent by hydrogen sulfide is not regenerated whereas lye spent by mercaptans can be regenerated by blowing the spent lye with steam, which reforms sodium hydroxide and mercaptans from the spent lye.

Spent lye can also be regenerated in a stripper tower with steam; the overhead consists of steam, mercaptans, and the small amount of oil picked up by the lye solution during treating. Condensing the overhead allows the mercaptans to separate from the water.

Nonregenerative caustic treating is generally economically applied when the contaminating materials are low in concentration and waste disposal is not a problem. However, the use of nonregenerative systems is on the decline because of the frequently occurring waste disposal problems which arise from environmental considerations and because of the availability of numerous other processes which can effect more complete removal of contaminating materials.

Steam regenerative caustic treating (Figure 2) is essentially directed toward removal of mercaptans from such products as gasoline. The caustic is regenerated by steam-blowing in a stripping tower. The nature and concentration of the mercaptans to be removed dictate the quantity and temperature of the process. However, the caustic does gradually deteriorate because of the accumulation of material that cannot be removed by stripping and the caustic quality has to be maintained either by continuous or intermittent discard, and replacement, of a minimum amount of the operating solution.

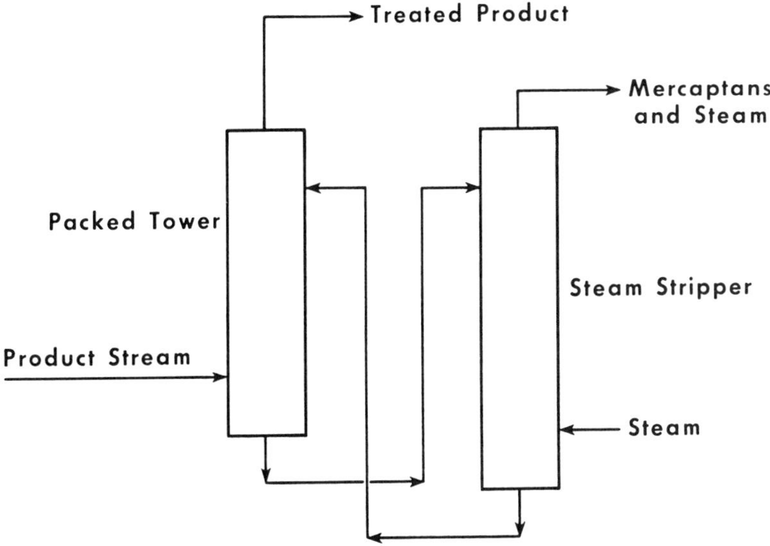

Figure 2 Steam regenerative caustic treatment.

B. Acid Treatment

Treating petroleum products with acids is, like caustic treatment, a procedure that has been in use for a considerable time in the petroleum industry. Various acids such as hydrofluoric acid, hydrochloric acid, nitric acid, and phosphoric acid have been used in addition to the more commonly used sulfuric acid, but in most instances there is little advantage in using any acid other than sulfuric.

The reactions of sulfur acid with petroleum fractions are complex. The undesirable components to be removed are generally present in small amounts, and large excesses of acid are required for efficient removal, which may cause marked changes in the remainder of the hydrocarbon mixture.

Paraffinic and naphthenic hydrocarbons in their pure forms are not attacked by concentrated sulfuric acid at low temperatures and during the short time of conventional refining treatment, but solution of light paraffins and naphthenes in the acid sludge can occur. Fuming sulfuric acid absorbs small amounts of paraffins when contact is induced by long agitation; the amount of absorption increases with time, temperature, concentration of the acid, and complexity of structure of the hydrocarbons. With naphthenes, fuming sulfuric acid causes sulfonation as well as rupture of the ring.

Aromatics are not attacked by sulfur acid to any great extent under ordinary refining conditions unless they are present in high concentrations. However, if fuming acid is used or if the temperature is allowed to rise above normal, sulfonation may occur:

$$C_6H_6 \xrightarrow{H_2SO_4} C_6H_5SO_3H + H_2O$$

where both aromatics and olefins are present, as in cracked distillates, alkylation can occur due to the action of the acid, e.g.:

$$C_6H_6 + CH_2{=}CH_2 \xrightarrow{[H+]} C_6H_5CH_2CH_3$$

The action of sulfuric acid on olefin hydrocarbons is very complex. The main reactions involve ester formation and polymerization:

$$R.CH{=}CH_2 + H_2SO_4 \longrightarrow R.CH_2.CH_2.SO_4H$$

$$R.CH{=}CH_2 \longrightarrow \left[\begin{array}{c}R\\ |\\ CH.CH_2\end{array}\right]_n$$

These esters are soluble in the acid phase but are also soluble, to some extent, in hydrocarbons, especially as the molecular weight of the olefin increases. The esters are usually difficult to hydrolyze with a view to removal by alkali washing. They are, however, unstable on standing for a long time and products containing them (acid-treated cracked gasolines) may evolve sulfur dioxide and deposit tarry condensates. The esters are quite unstable on heating, so that a redistilled, acid-treated cracked distillate usually requires alkali washing after the customary distillation.

Acid treatment of residua presents different problems. Most of these contain asphaltic substances and almost all the acid comes out as a sludge (acid tar). The separation of the sludge is aided by the addition of water or alkali solution, but there may be chemical changes and products may contain combined sulfur derived from the treating acid.

The disposal of the sludge is difficult because it contains unused free acid which must be removed by dilution and settling. The disposal is a comparatively simple process for the sludges resulting from treating gasolines and kerosene—the so-called light oils. The insoluble oil phase separates out as a mobile tar, which can be mixed and burned without too much difficulty. Sludges from heavier oils, however, separate out granular semisolids which offer considerable difficulty in handling.

1. Sulfuric Acid Treatment

Sulfur acid treatment is a continuous, or batch, method that is used to remove sulfur and even remove asphaltic materials from various refinery stocks. The acid strength varies from fuming (>100%) to 80% with 93% acid finding the most common use. The weakest suitable acid is used for each particular situation to reduce sludge formation from the aromatic and olefinic hydrocarbons.

The use of strong acid dictates the use of a fairly low temperature (-4 to 10°C; 25 to 50°F) but higher temperatures (20–55°C; 70–130°F) are possible if the product is to be redistilled.

C. Clay Treatment

The refining of petroleum distillates and residua by passing them through materials possessing decolorizing power has been in operation for many years. For example, various clays and similar materials are used to treat petroleum fractions to remove diolefins, asphaltic materials, resins, acids, and colored bodies. Cracked naphthas were frequently clay-treated to remove diolefins which formed gums in gasolines. This use of clay treatment has now been largely superseded by other processes and, in particular, by the use of inhibitors which, added in small amounts to gasoline, prevent gums from forming. Nevertheless, clay treatment is still used as a finished step in the manufacture of lubricating oils and waxes. The clay removes traces of asphaltic materials and other compounds that give oils and waxes unwanted odors and colors.

The original method of clay treatment was to percolate a petroleum fraction through a tower containing coarse clay pellets. As the clay absorbed impurities from the petroleum fraction, the clay became less effective. The activity of the clay was periodically restored by removing it from the tower and burning the absorbed material under carefully controlled conditions so as not to sinter the clay. The percolation method of clay treatment was widely used for lubricating oils but has been largely replaced by clay contacting.

1. Continuous Contact Filtration

This is a continuous clay treatment process (Figure 3) in which finely divided adsorbent is mixed with the charge stock and heated to 95–175°C (200–350°F). The slurry is then conveyed to a steam-stripping tower after which it is cooled, vacuum-filtered, and then vacuum-stripped for further product specification control.

2. Bauxite Treatment

Bauxite treatment involves passage of a vaporized petroleum fraction through beds of bauxite. The bauxite acts catalytically to convert many different sulfur compounds, in particular mercaptans, to hydrogen sulfide which is subsequently removed by a lye treatment.

A typical bauxite treatment unit consists of a fire-heated coil, two bauxite treatment towers, a bubble tower, a superheater for steam and air, and the usual exchangers, coolers, and pumps. Naphtha, raw kerosene, or other stock to be treated is preheated in heat exchangers and passed through the heating coil where it is heated to

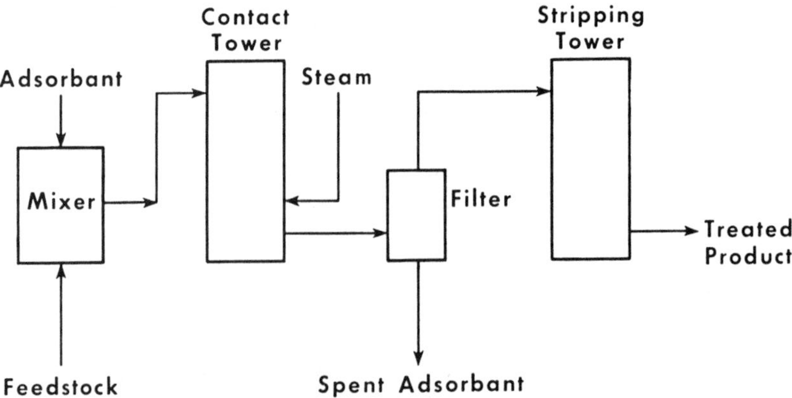

Figure 3 Continuous contact filtration.

415°C (780°F). At this temperature the stock is superheated. The vaporized feed is then passed downward through one of two bauxite towers under a pressure of about 40 psi. Three beds of catalyst in the tower convert mercaptans to hydrogen sulfide to enter a continuous lye treatment unit where hydrogen sulfide is removed.

After a time the bauxite towers are switched, since the bauxite progressively loses its catalytic activity. The spent catalyst is restored to its original activity by regeneration. This is done by passing superheated steam and air, carefully scheduled in proportions and rate, downward through the catalyst beds. The carbonaceous material that has accumulated on the catalyst is burned off. Combustion progresses downward through the beds; care is taken to prevent temperatures exceeding 595°C (1100°F) which would harm the bauxite. Air alone is finally used to burn away the last traces of carbonaceous material after which the bauxite is ready for use again.

D. Oxidative Treatment

Oxidative treatment processes are those processes which have been developed to convert the foul-smelling mercaptans to the less foul disulfides by oxidation:

$$4\text{R-SH} + O_2 \longrightarrow 2\text{R-S.S-R} + 2H_2O$$

Mercaptan (Thiol) — Disulfide

However, disulfides do tend to reduce the tetraethyl lead susceptibility of gasolines and recent trends are to processes which are capable of completely removing the mercaptans.

1. Doctor Method

The method of treating sour distillates (Figure 4) consists in agitating the distillate with alkaline sodium plumbite (Doctor solution) in the presence of a small amount of free sulfur. A black precipitate of lead sulfide is formed and the material is of improved odor, having been rendered sweet. The essential reactions of the Doctor process are:

$$2\,\text{RSH} + Na_2PbO_2 \longrightarrow Pb(SR)_2 + 2NaOH$$

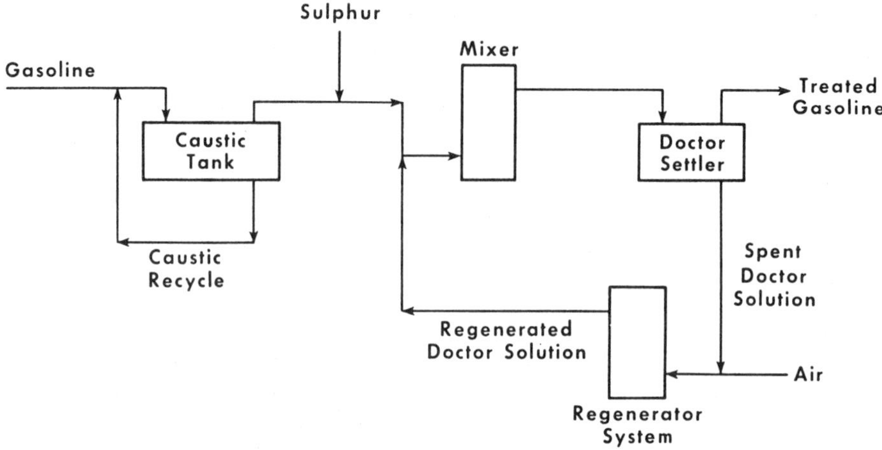

Figure 4 The Doctor (sodium plumbite) treatment process.

$$Pb(SR)_2 + S \longrightarrow PbS + R_2S_2$$

In practice, sour distillates are usually given an alkali wash before the Doctor treatment to remove traces of hydrogen sulfide and some of the lower molecular weight mercaptans, and which has a marked effect in reducing the plumbite treatment. Slightly more sulfur than the theoretical is required, owing to the formation of complex lead intermediates. In the presence of lead mercaptides the extra sulfur acts to form alkyl polysulfides, which are chemically analogous to peroxides:

$$Pb(SR)_2 + 2S \longrightarrow PbS + R_2S_3$$

$$Pb(SR)_2 + 3S \longrightarrow PbS + R_2S_4, \text{ etc.}$$

The precipitating effect is evidently a result of the presence of these polysulfides or possibly a sodium sulfide formed between mercaptans, sulfur, and the alkaline solution. The Doctor solution leaving the reactor consists essentially of a mixture of lead sulfide in free alkali, containing emulsified hydrocarbons, and this spent solution is pumped to steam-heated vessels, where it is air-blown for regeneration:

$$PbS + 4NaOH + 2O_2 \longrightarrow Na_2PbO_2 + Na_2SO_4 + 2HO_2$$

Considerable amounts of sodium thiosulfate are also formed:

$$2PbS + 2O_2 + 6NaOH \longrightarrow 2Na_2PbO_2 + Na_2S_2O_3 + 3H_2O$$

The thiosulfate, in turn, may react with the alkali present to form sodium sulfite (Na_2SO_3) and sodium sulfide (Na_2S). The loss of lead is very low and the main items of consumption are alkali and sulfur.

2. Copper Sweetening

The oxidizing power of cupric salts is also utilized to convert mercaptans directly to disulfides. Free sulfur is not employed, and polysulfides are not obtained. The process employs cupric chloride in the presence of strong salt solutions which are generally made up by dissolving copper sulfate in an aqueous solution of sodium chloride.

$$4RSH + 2CuCl_2 \longrightarrow R_2S_2 + 2CuSR + 4HCl$$

$$2CuSR + 2CuCl_2 \longrightarrow R_2S_2 + 4CuCl$$

$$4CuCl\ 4HCl + O_2 \longrightarrow 4CuCl_2 + H_2O$$

The cuprous chloride (CuCl) is soluble in the salt solution, and there is no precipitation. Under operating conditions, a certain amount of copper is retained by the sweetened petroleum fraction but can be removed by washing the material with aqueous sodium sulfide. Air-blowing the cuprous chloride solution after or during the sweetening operation regenerates the cupric chloride for further use.

Three methods of mechanical application of the copper chloride are used (Figure 5). If air will not cause the petroleum fraction to change color or form gum, a fixed-bed process may be used in which the sour material is passed through beds of an adsorbant which have been impregnated with cupric chloride. Air added with the sour fraction continuously regenerates the cupric chloride almost simultaneously with the sweetening reaction.

In the solution process, a solution of cupric chloride is continuously mixed in a centrifugal pump with the sour fraction. The mixture then enters a settling tank where the spent treating solution separates from the petroleum liquid and the treating solution is withdrawn to a tank where blowing with air regenerates cupric chloride.

The slurry process makes use of clay or similar material impregnated with cupric chloride. The clay is mixed with a small amount of, say, naphtha to form a slurry which is pumped into the sour naphtha

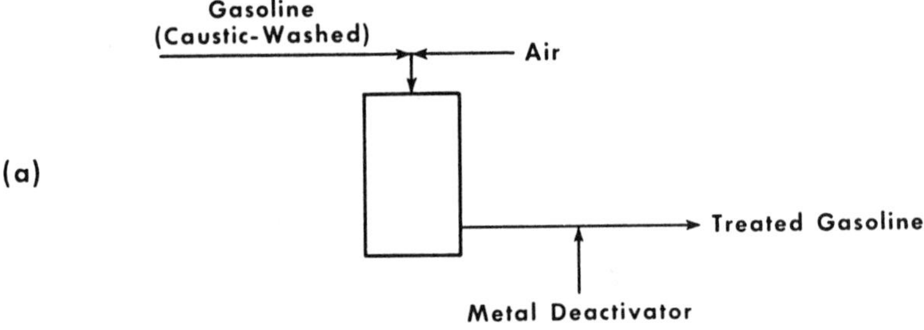

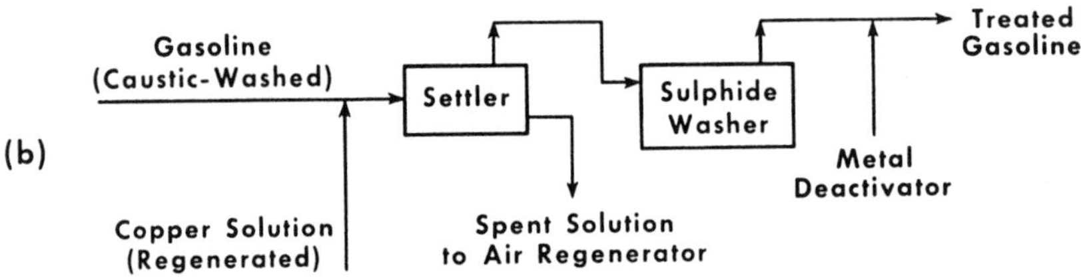

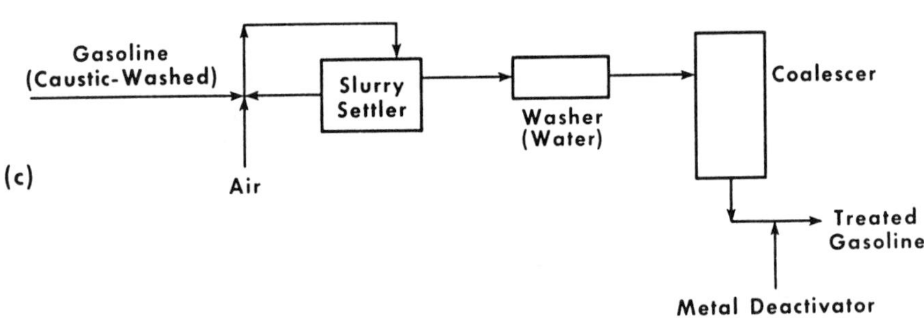

Figure 5 Variations of the copper treatment process. (a) The fixed-bed (solid) process, (b) the solution process, and (c) the slurry process.

stream. Air or oxygen gas is added with the sour stream and regenerates the cupric chloride continuously. The treated material and clay slurry flow into a settling tank, separate, and the clay slurry is recycled.

E. Solvent Treatment

Solvent refining processes are of a physical nature only and the desirable, as well as undesirable, constituents of the mixture can be recovered unchanged and in their original state. In addition, the processes which use solvents as a means of refining are extremely versatile insofar as both low- and high-boiling fractions can be used as feedstocks. In general, the solvent processes can be classified as deasphalting, refining, and dewaxing.

Deasphalting is usually applied to materials such as gas oils, lubricating stocks, residua, and even the heavier virgin crudes prior to further treating. The solvent acts as a precipitate and divides the feedstock into constituents which are usually based on molecular size.

Solvent refining is usually applied to materials boiling in the lubricating oil range and it is a means of selectively removing (by extraction) aromatics, naphthenes, or other constituents that adversely affect physical parameters such as viscosity (Chap. 3, Sec. III). The feedstock to these solvent refining processes has usually been through a solvent deasphalting treatment and, if desired, the effluent may be subjected to a dewaxing operation.

Dewaxing operations are applied to treated feedstocks prior to final treatment with clay. Wax in a finished lubricating oil adversely affects the properties of the oil such as by increasing the pour point and reducing the fluidity at low temperatures because of crystallization of the wax from the lubricating oil. Dewaxing operations are often combined with wax production when low-pour-point lubricating oil and high-melting-point wax can be produced simultaneously.

1. Deasphalting

The coke-forming tendencies of middle distillates, gas oils, lubricating oils, and residua can be reduced significantly through the precipitating, or extracting, action of solvents on the asphaltic and resinous materials which are present either in solution or in colloidal form. The solvents capable of being used in deasphalting operations may be divided into two major groups. The first group contains the low molecular weight liquid or liquefied hydrocarbons of which propane is the best and most widely used. The second group of solvents contains liquids such as the alcohols and ethers but, in general, these materials are not used to anywhere near the same extent as the hydrocarbons.

Solvent deasphalting is, in fact, a later addition to the petroleum refinery, although prior to its use many processes capable of removing asphaltic materials from feedstocks were employed in the form of distillation (atmospheric and vacuum) as well as clay treating and sulfur acid treating. In its present form, solvent deasphalting (when employed for feedstock preparation to catalytic cracking units) may be considered as complementary to vacuum distillation, visbreaking, and coking. Deasphalting is often referred to as decarbonizing.

Propane deasphalting: Liquefied petroleum gases have the ability to precipitate asphaltic and resinous materials from crude residues while the lubricating oil components remain in solution. While all liquefied hydrocarbon gases have this property to a marked extent, propane is used to deasphalt residual lubricating oils because of its relatively low cost and its ease of separation from lubricating oils.

In propane deasphalting (Figure 6), the crude residue and 3–10 times its volume of liquefied propane are pumped together through a mixing device and then into a settling tank. The temperature is maintained between 27°C (80°F) and 71°C (160°F)—the higher the temperature, the greater the tendency of asphaltic materials to separate. The propane is maintained in the liquid state by a pressure of about 200

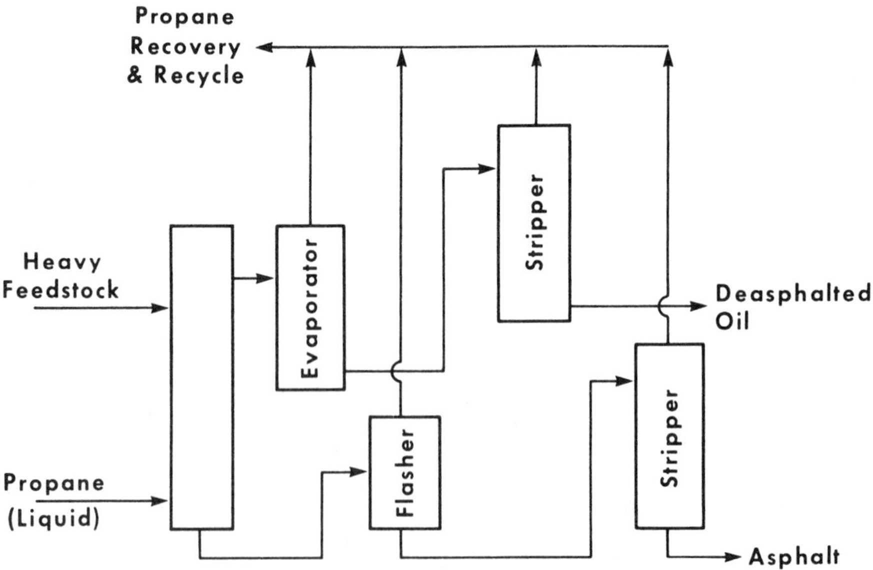

Figure 6 Schematic representation of the propane deasphalting process.

psi. The asphalt settles in the settling tank and is pumped to an asphalt recovery unit where propane is separated from the asphalt. The upper layer in the settling tank consists of lubricating oil dissolved in a large amount of propane which, while dissolved in the propane, may have more impurities removed by sulfuric acid or may even be dewaxed. In the latter case, evaporation of some of the propane cools the mixture to a sufficiently low temperature (-40°C; -40°F) where wax crystals form. Filtration at the low temperature separates wax from the liquid and propane is separated from the oil-propane mixture in evaporators heated by steam. The last trace of propane is removed from the oil by steam in a stripper tower and the propane is recondensed and reused.

In place of a mixer and settling tank, most modern deasphalting plants use a countercurrent tower. Liquefied propane is pumped into the bottom of the tower to form a continuous phase and lubricating stock or reduced crude (crude residuum) is pumped into the tower near the top. As the reduced crude descends, the oil components are dissolved and carried with the propane out the top of the tower and the asphaltic components are pumped from the bottom of the tower.

Duosol process: The Duosol process uses two solvents: one propane and the other a mixture of cresylic acid and a phenol called "Selecto." A series of treatment compartments is used where propane is pumped into one end of the series and Selecto into the other (Figure 7), and the solvents pass one another countercurrently. Raw lubricating oil is pumped into the third compartment from the propane end. The propane dissolves the paraffinic oil and wax components and the Selecto dissolves naphthenic, unsaturated, asphaltic, and resinous

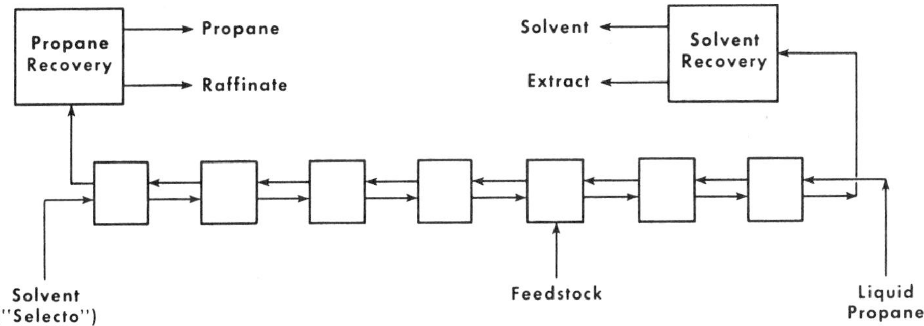

Figure 7 The Duosol process.

components. The raw lubricating oil is thus divided into two parts that move in opposite directions through the treatment system, each carried by its particular solvent.

After leaving the treatment system, each solvent stream passes through separate solvent recovery systems. Propane is separated from the treated oil in evaporators and Selecto is separated from the naphthenic and asphaltic extract by fractional distillation.

2. Dewaxing

Dewaxing processes are designed to remove wax from lubricating oils in order for the product to have good fluidity characteristics at low temperatures (i.e., low pour point). The mechanism of solvent dewaxing can either be the separation of wax as a solid which has been crystallized from the oil solution at low temperature or the separation of wax as a liquid which has been extracted at temperatures above the melting point of the wax through preferential selectivity of the solvent. However, the former mechanism is the usual basis for commercial dewaxing processes.

In the first solvent dewaxing process (developed in 1924), the waxy oil was mixed with naphtha and filter aid (fuller's earth or diatomaceous earth). The mixture was chilled and filtered and the filter aid assisted in building a wax cake on the filter cloth. This process is now obsolete and most of the modern dewaxing processes use a mixture of methyl ethyl ketone and benzene. Other ketones may be substituted for dewaxing, but regardless of what ketone is used, the process is generally known as ketone dewaxing.

The process is carried out by mixing waxy oil with one to four times its volume of ketone and heating the mixture until the oil is in solution (Figure 8). The solution is then chilled at a slow, controlled rate in double-pipe, scraped-surface exchangers. Cold solvent, such as filtrate from the filters, passes through the 2-in. annular space between the inner and outer pipes and chills the waxy oil solution flowing through the inner 6-in. pipe. To prevent wax from depositing on the walls of the inner pipe, blades or scrapers extending the length of the pipe and fastened to a central rotating shaft scrape off the wax. Slow chilling reduces the temperature of the waxy oil solution to 2°C (35°F) and then faster chilling reduces the temperature to the approximate pour point required in the dewaxed oil. The waxy mixture is pumped to a filter case into which the bottom half of the drum of a rotary vacuum filter dips. The drum (8 ft in diameter, 14 ft long), covered with filter cloth, rotates continuously in the filter case. Vacuum within the drum sucks the solvent and the oil dissolved in the solvent through the filter cloth and into the drum. Wax

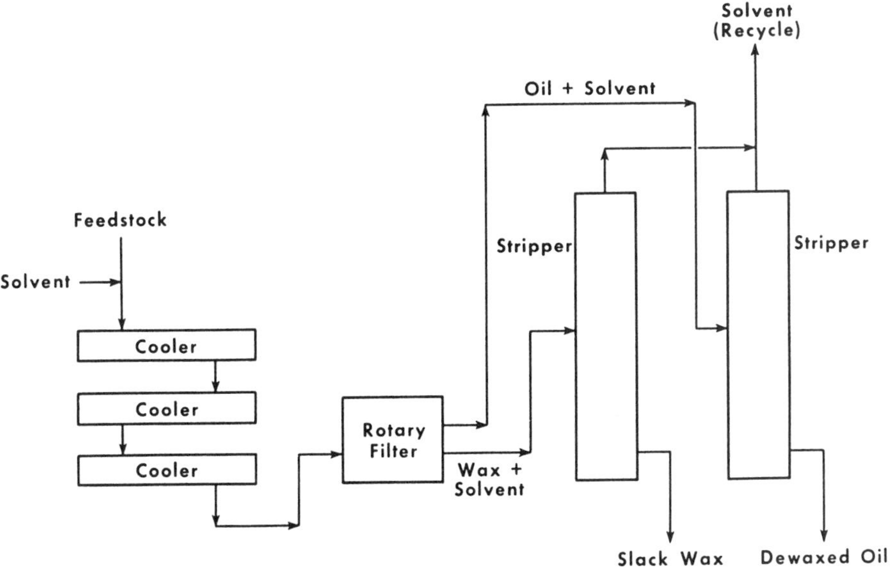

Figure 8 The solvent dewaxing process.

crystals collect on the outside of the drum to form a wax cake and, as the drum rotates, the cake is brought above the surface of the liquid in the filter case and under sprays of ketone that washes oil out of the coke and into the drum. A knife edge scrapes off the wax and the cake falls into the conveyor and is moved from the filter by the rotating scroll.

There are several processes in use for solvent dewaxing but all have the same general steps: (1) contacting the feedstock with the solvent; (2) precipitation of wax from the mixture by chilling; (3) solvent recovery from the wax and dewaxed oil for recycling. The processes use benzene-acetone (solvent dewaxing), propane (propane dewaxing), trichloroethylene (separator-Nobel dewaxing), ethylene dichloride–benzene (Bari-Sol dewaxing), urea (urea dewaxing), as well as liquid sulfur dioxide–benzene mixtures.

There are also later generation dewaxing processes that are being brought on-stream in various refineries. For example, BP has developed a hydrocatalytic process that is reputed to overcome some of the disadvantages of the solvent dewaxing processes. The operating costs for the solvent dewaxing processes are reputed to be high because of the solvent cooling that is necessary, and the pour point that can be achieved is limited by the high cost of refrigerating to very low temperatures.

Catalytic dewaxing is a hydrocracking process and is therefore operated at elevated temperatures (280--400°C; 550--750°F) and pressures (301--1500 psi) (see Chap. 6, Sec. III). However, the conditions for a particular dewaxing operation depend on the nature of the feedstock and the product pour point that is required. The catalyst employed for the process (Figure 9) is a mordenite-type catalyst (Chap. 9, Sec. I.A) that has the correct pore structure to be selective for normal paraffin cracking. Platinum on the catalyst serves to hydrogenate the reactive intermediates so that further paraffin degradation is limited to the initial thermal reactions. The process has been employed to successfully dewax a wide range of naphthenic feedstocks (Hargrove et al., 1979) but it may not be suitable to replace solvent dewaxing in all cases. The process has the flexibility to fit into normal refinery operations and can be adapted for prolonged periods on-stream.

Another catalytic dewaxing process has been developed by Mobil and also involves selective cracking of normal paraffins and those paraffins that might have minor branching in the chain (Smith et al., 1980). In the process (Figure 10), the proprietary catalyst can be reactivated to fresh activity by relatively mild nonoxidative treatment. Of course, the time allowed between reactivations is a function

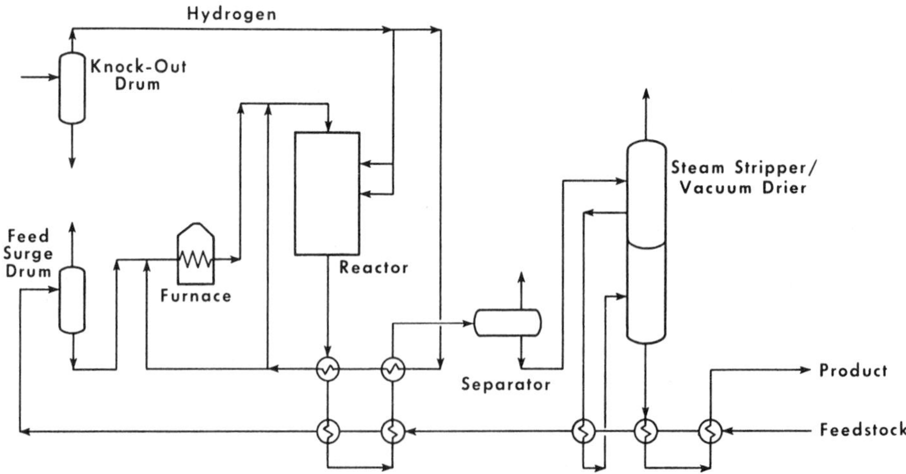

Figure 9 Schematic representation of the BP catalytic dewaxing process.

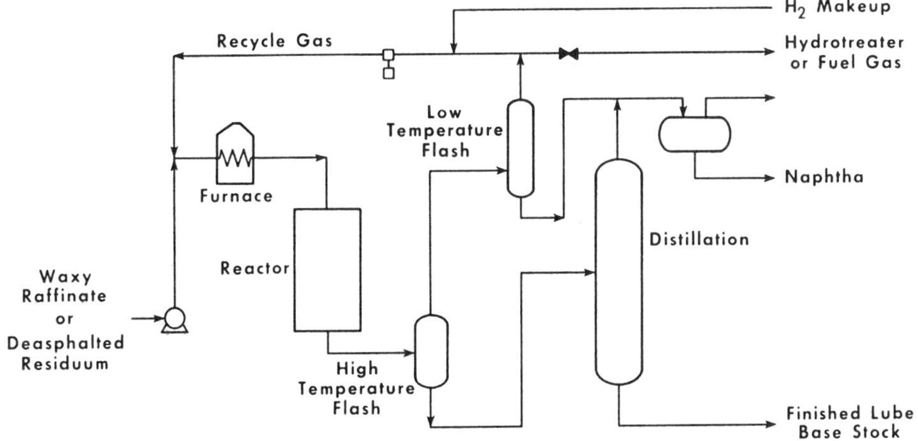

Figure 10 Schematic representation of the Mobil catalytic dewaxing process.

of the feedstock but after numerous reactivations it is possible that there will be coke buildup on the catalyst.

The process can be used to dewax a full range of lube base stocks and as such has the potential to completely replace solvent dewaxing (Figure 11) or can even be used in combination with solvent dewaxing (Figure 12). This latter option, of course, serves to debottleneck existing solvent dewaxing facilities.

Both catalytic dewaxing processes have the potential to change conventional ideas about dewaxing insofar as they are not solvent processes and may be looked on (more correctly) as thermal processes rather than treatment processes. However, both provide viable options to the solvent processes and represent an addition to the science and technology of refinery operations.

F. Gas Cleaning

Refinery and natural gas streams may contain substantial amounts of contaminant acidic gases such as hydrogen sulfide and carbon dioxide. Acidic gases corrode refining equipment, harm catalysts, pollute the atmosphere, and can prevent the use of hydrocarbon components in petrochemical manufacture. Where the amount of hydrogen sulfide is large, it may be removed from a gas stream and converted

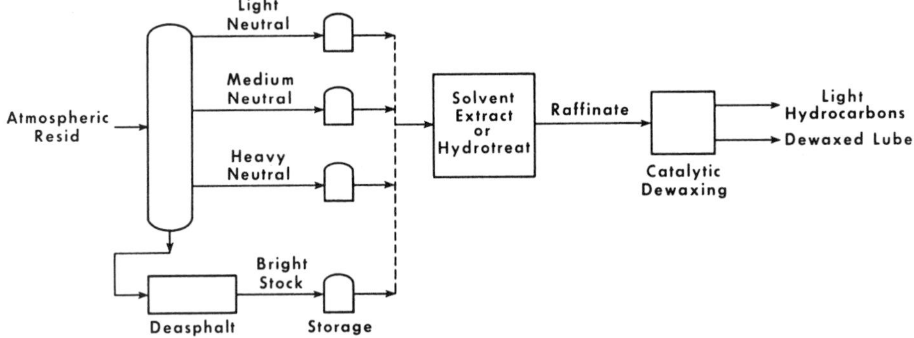

Figure 11 Schematic representation of the Mobil catalytic dewaxing process as a complete replacement for solvent dewaxing.

to sulfur or sulfuric acid. Some natural gases contain sufficient carbon dioxide to warrant recovery as "dry ice."

The usual contaminant of a hydrocarbon gas is hydrogen sulfide. Many chemicals may be used to react with hydrogen sulfide (Table 1). Among them lye, lime, iron oxide, sodium carbonate, sodium phenolate, potassium phosphate, and ethanolamine. Processes based on these chemicals have been used. Some of the treating chemicals are regenerated (by blowing with air or steam) and reused. Processes employing caustic are largely used for the removal of small, even trace, amounts of acid gases.

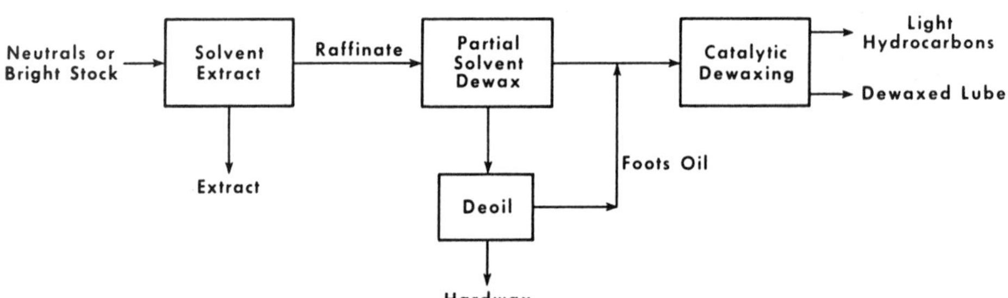

Figure 12 Schematic representation of the Mobil catalytic dewaxing process being used in conjunction with solvent dewaxing.

Table 1 Chemical Reactions for Hydrogen Sulfide Removal

Name	Reaction	Regeneration
Caustic Soda	$2NaOH + H_2S \longrightarrow Na_2S + 2H_2O$	None
Lime	$Ca(OH)_2 + H_2S \longrightarrow CaS + 2H_2O$	None
Iron Oxide	$FeO + H_2S \longrightarrow FeS + H_2O$	Partly by Air
Seaboard	$Na_2CO_3 + H_2S \rightleftarrows NaHCO_3 + NaHS$	Air Blowing
Thylox	$Na_4As_2S_5O_2 + H_2S \longrightarrow Na_4As_2S_6O + H_2O$	Air Blowing
	$Na_4As_2S_6O + \frac{1}{2}O_2 \longrightarrow Na_4As_2S_5O_2 + S$	
Girbotol	$2RNH_2 + H_2S \rightleftarrows (RNH_3)_2S$	Steaming
Phosphate	$K_3PO_4 + H_2S \rightleftarrows KHS + K_2HPO_4$	Steaming
Phenolate	$NaOC_6H_5 + H_2S \rightleftarrows NaHS + C_6H_5OH$	Steaming
Carbonate	$Na_2CO_3 + H_2S \rightleftarrows NaHCO_3 + NaHS$	Steaming

Other processes for hydrogen sulfide removal from refinery streams include oxidation to free sulfur:

$$H_2S \xrightarrow{O} H_2O + S$$

The sulfur is precipitated as a finely divided solid and subsequently recovered by settling or filtration. The liquid product may be recovered where it is possible to heat the suspension above the melting point of sulfur. This latter method of recovery is very difficult to apply as a good many petroleum liquids interact with sulfur at temperatures above 120°C (250°F).

The processes using ethanolamine and potassium phosphate are now widely used (Figure 13). The ethanolamine process, known as the Girbotol process, removes acidic gases (hydrogen sulfide and carbon dioxide) from liquid hydrocarbons as well as natural and refinery gases. The Girbotol treating solution is an aqueous solution of ethanolamines which are organic alkylamines that have the reversible property of reacting with hydrogen sulfide under cool conditions

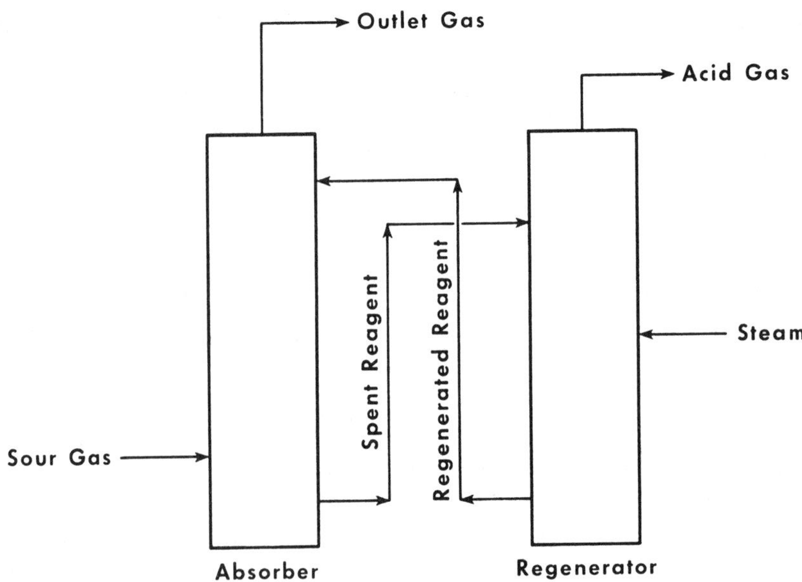

Figure 13 Schematic representation of a typical process for the removal of acidic gases from refinery gas streams.

and releasing hydrogen sulfide at high temperatures. The ethanolamine solution fills a tower called an absorber through which the sour gas is bubbled. Purified gas leaves the top of the tower while the ethanolamine solution leaves the bottom of the tower with the absorbed acidic gases. The ethanolamine solution enters a reactivator tower where heat drives the acidic gases from the solution. Ethanolamine solution, restored to its original condition, leaves the bottom of the reactivator tower to go to the top of the absorber tower, and acidic gases are released from the top of the reactivator.

The process using potassium phosphate is known as the phosphate desulfurization process and is used in the same way as the Girbotol process to remove acidic gases from liquid hydrocarbons as well as gas streams. The treatment solution is a water solution of tripotassium phosphate (K_3PO_4) which is circulated through an absorber tower and a reactivator tower in much the same way as the thanolamine is circulated in the Girbotol process; the treatment solution is regenerated thermally.

Moisture may be removed from hydrocarbon gases at the same time hydrogen sulfide is removed. Moisture removal is necessary to prevent harm to anhydrous catalysts and to prevent the formation of hydrocarbon hydrates (e.g., $C_3H_8.18H_2O$) at low temperatures. A widely used dehydration and desulfurization process is the glycol-amine process in which the treating solution is a mixture of ethanolamine and a large amount of glycol. The mixture is circulated through an absorber and a reactivator in the same way as ethanolamine is circulated in the Girbotol process. The glycol absorbs moisture from the hydrocarbon gas passing up the absorber while the ethanolamine absorbs hydrogen sulfide and carbon dioxide. The treated gas leaves the top of the absorber; the spent ethanolamine-glycol mixture enters the reactivator tower where heat drives off the absorbed acidic gases and water.

Other processes include the Alkazid process which removes hydrogen sulfide and carbon dioxide using concentrated aqueous solutions of amino acids. The hot potassium carbonate process decreases the acid content of natural and refinery gas from as much as 50% to as low as 0.5% and operates in a similar unit to that used for amine treating. The Giammarco–Vetrocoke process is used for hydrogen sulfide and/or carbon dioxide removal (Figure 14). In the hydrogen sulfide removal section, the reagent consists of sodium or potassium carbonates

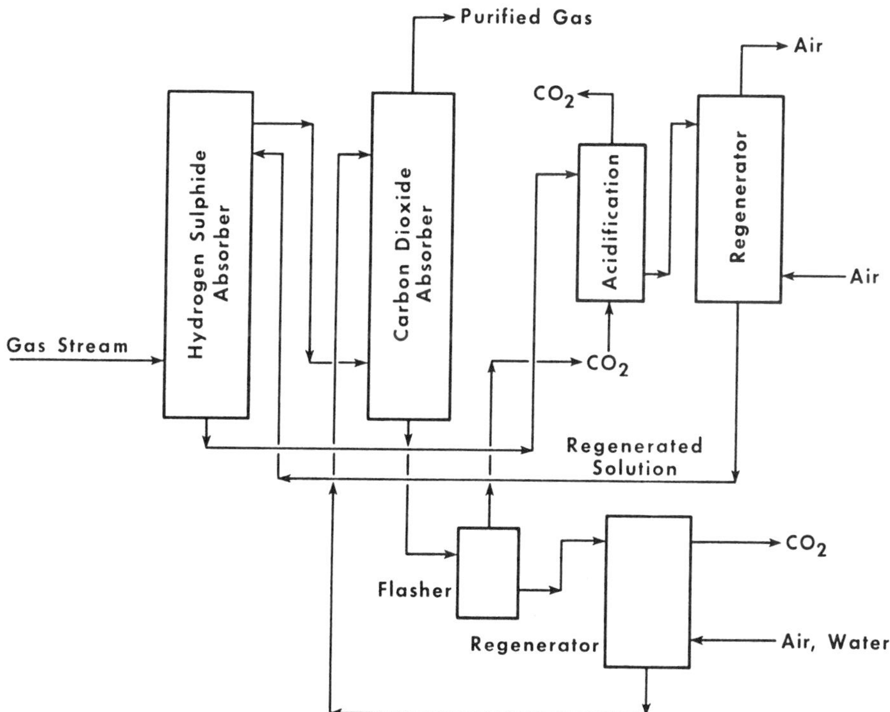

Figure 14 The Giammarco–Vetrocoke process for hydrogen sulfide and/or carbon dioxide removal from gas streams.

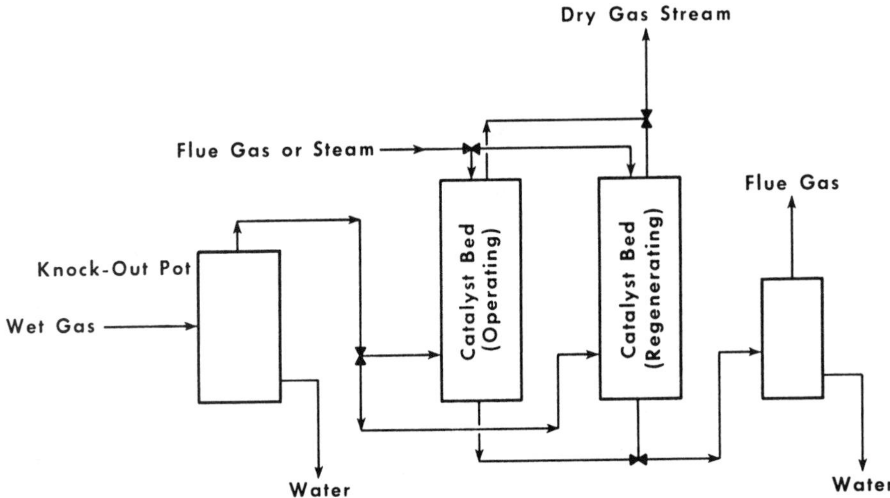

Figure 15 Process for drying liquids and gases.

containing a mixture of arsenites and arsenates while the carbon dioxide removal section utilizes hot aqueous alkali carbonate solution activated by arsenic trioxide, or selenous or tellurous acid.

The presence of water in refinery gases and refinery liquids causes corrosion and may even cause losses of water-soluble additives (in refinery liquids). The removal of water vapor from refinery gases is achieved by compression, cooling, and refrigeration, or by contact with such hygroscopic chemicals as solid caustic, calcium chloride, and so on (Figure 15). Water may even be removed from gases by adsorption into, say, diethylene glycol as for hydrogen sulfide removal. On the other hand, water removal from refinery liquids may be accomplished by contact with chemicals of the rock salt, calcium sulfate, etc., type. More recently, dehydrated aluminum silicate crystals (molecular sieves) have been used for drying gas streams and show a superior dehydration capacity to the conventional desiccants.

The purification of hydrocarbon gases by any of the above-mentioned processes is an important part of refinery operations, especially in regard to the production of liquefied petroleum gas (LPG or LP gas). This is actually a mixture of propane and butane that is an important domestic fuel as well as being an intermediate material in the manufacture of petrochemicals. The presence of ethane in liquefied petroleum gas must be avoided because of the inability of this lighter hydrocarbon to liquefy under pressure at ambient temperatures and its tendency to register abnormally high pressures in the liquefied petroleum gas containers. On the other hand, the presence of pentane in liquefied petroleum gas must also be avoided since this particular hydrocarbon (a liquid at ambient temperatures and pressures) may separate in a liquid state in the gas lines.

8

Refining Heavy Feedstocks

I	General	229
II	Effect of Asphaltenes	230
III	Effect of Porphyrins	233
	A. Overview	233
	B. Hydroprocessing	233
	C. Catalytic Cracking	236
	D. Demetallization	238
IV	Refinery Options	242

I. GENERAL

The increasing use of heavy feedstocks (such as the heavy crude oils and the tar sand bitumens) as refinery feedstocks is now widely recognized and it is further recognized that heavy crude oil and tar sand resources will be available in much larger quantities as the supplies of light crude oil are gradually diminished (Epstein and Barnes, 1985; *BP Stat. Rev. World Energy*, June 1987). Refiners are also going through internal revamping of their operations to handle these feedstocks with this expectation of increased availability of heavy crude oils in the future. Indeed, it has been estimated (Epstein and Barnes, 1985) that heavy crude oil production in the United States may be in excess of 1.5 million bbl/day in the very near future.

The problems of refining heavy feedstocks (in comparison to light crude oil) may be summarized very succinctly as being due to

Increased aromaticity
Higher molecular weights of the constituents
Higher hetero atom content (nitrogen, oxygen, sulfur, and the metals nickel, vanadium, and iron)

These issues can be conveniently (and generally) categorized as being due to an increased asphaltene content (Figure 1). This leads to increased catalyst deactivation, higher coke yields, and poorer quality products.

II. EFFECT OF ASPHALTENES

Thus, is is not surprising that the character and behavior of the asphaltene fraction of petroleum has been the subject of much attention (Bunger and Li, 1981; Speight, 1984c). Asphaltenes have been observed to be equatable to the yield of thermal coke and also to the rate of catalyst deactivation (Speight, 1984b). In summary, the asphaltene fraction of crude oils appears to be the embodiment of all that can adversely affect refinery processes.

As already noted, the definition of petroleum asphaltenes is an operational aid and it must never be forgotten that the asphaltene fraction is actually a *solubility class* (Chap. 4, Sec. II.A). However, the delineation of asphaltene structures has been the subject of many investigations (Speight, 1984a) but considering an asphaltene as a structural entity is difficult because asphaltenes are more clearly defined as a range of molecular types. This, however, has not deterred investigators from postulating "average" structures for asphaltenes using a spectroscopic method such as proton magnetic resonance ('H NMR) as the mainstay. The outcome of these investigations is that asphaltenes have been considered to contain a large central polynuclear aromatic system (Figure 2) which carries hetero atoms (nitrogen, oxygen, and sulfur), alkyl chains, and hydroaromatic ring systems. It now appears that the size of the polynuclear aromatic systems as deduced by these two methods has been grossly overestimated. In addition, pyrolysis (350–900°C; 660–1650°F) of asphaltenes produces low molecular weight nonaromatic and aromatic fragments; the predominant aromatics are one-, two-, and three-ring species (Speight and Pancirov, 1983) showing that any pendant systems attached to the core by thermally labile bonds must also include aromatic (as well as paraffinic) species. It is now apparent that there is no single analytical technique

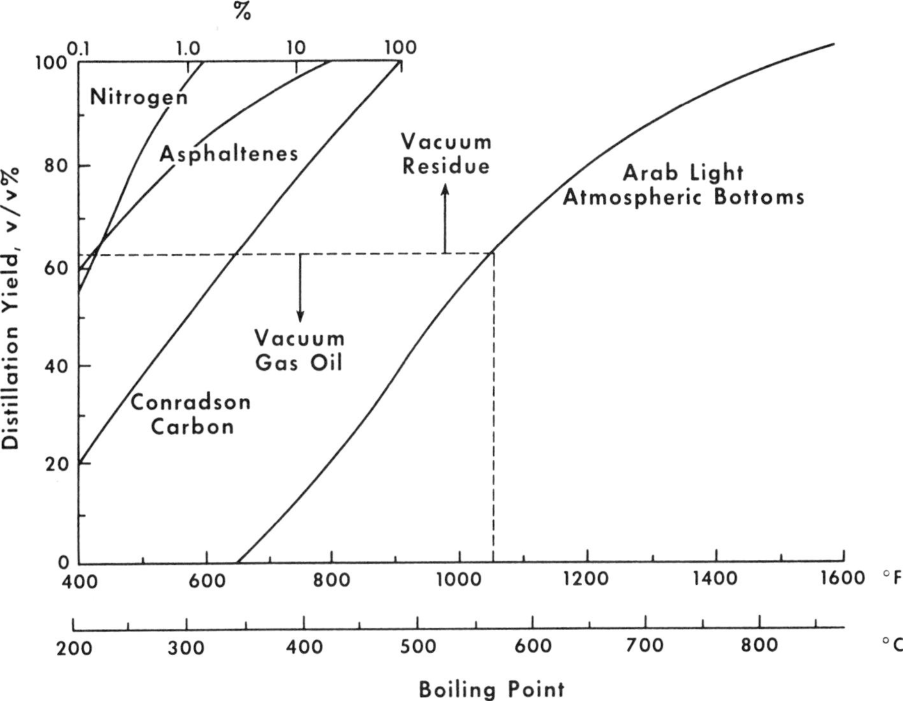

Figure 1 Problem materials in heavy oils and residua.

that can be used to define the structure of asphaltenes. A multidimensional approach (Speight, 1987) points to structures that are different from the highly condensed aromatic core systems. Such structures with the attendant alkyl, hydroaromatic, and hetero atom systems offers a unique alternate to the highly condensed polynuclear aromatic systems usually employed for petroleum asphaltenes. Furthermore, such a system has aromatic nuclei that are more in keeping with (1) vacuum gas oil constituents of petroleum; (2) the physicochemical behavior of the asphaltenes; and (3) the natural product or origins of petroleum.

Indeed, acceptance of the fact that the petroleum asphaltenes are a solubility class and are a melange of different molecular types without the presence of a large, central, condensed polynuclear aromatic system sheds a new light on the chemistry of coking.

It has been surmised throughout the literature that the formation of coke from petroleum asphaltenes during a thermal operation involves

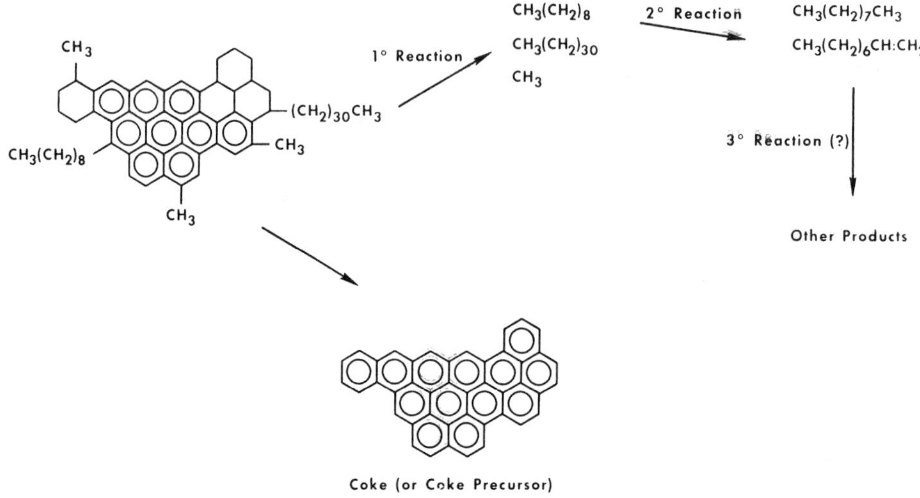

Figure 2 Asphaltenes are often represented as large polynuclear aromatic systems, but such systems cannot adequately explain the chemistry of coke formation.

the denuding of a central polynuclear aromatic core of alkyl (and other) attachments with the core material forming the coke (Figure 2). Such a concept is not only in complete contradiction to the actual chemistry (and products) of the reaction but is also in complete disagreement with current thinking and knowledge of petroleum asphaltenes (Speight, 1986, 1987).

Asphaltenes must be considered as a natural product, and on this basis the occurrence of such large polynuclear aromatic systems in petroleum is difficult to accept. It is more likely that molecular species within the asphaltene fraction which contain nitrogen and other hetero atoms (and have lower volatility than the pure hydrocarbons) are the prime movers in the production of coke. Such species, containing three- to six-ring polynuclear aromatic systems, would be denuded of the attendant hydrocarbon moieties and would undoubtedly (on the basis of solubility theory) be insoluble in the surrounding hydrocarbon medium. The next step would be gradual carbonization of such entities to form coke (Magaril et al., 1970).

It is apparent that the conversion of heavy oils and residua requires new lines of thought to develop suitable processing scenarios. Indeed the use of thermal and hydrothermal processes which were inherent in the refineries designed to process lighter feedstocks have been a particular cause for concern and has brought about the evolution of processing schemes that will accommodate the heavier feedstocks.

III. EFFECT OF PORPHYRINS

A. Overview

The deleterious effects of metal chelates in petroleum have been known for some time (Farrar, 1952). In addition to contaminating products, metal chelates cause catalyst poisoning and fouling, and corrosion of equipment. Metal chelates do not pose as great a problem for refiners of lighter crudes as they do for processors of heavy, high-sulfur feedstocks. However, since the energy crises of the 1970s, more refiners have found it necessary to process heavier feedstocks. This is a trend which may be expected to continue as supplies of lighter crudes dwindle or access to them is denied.

Processing studies have focused on the metal moiety of metalloporphyrins, which is either vanadium or nickel for petroleum. These metals represent about 10% of the mass of the average geoporphyrin and occupy the center of a fairly large molecule. The rest of the molecule contains four nitrogen atoms and varying amounts of aliphatic and aromatic carbon and hydrogen atoms. Some of the porphyrin aliphatic side chains are fairly long. Smaller homologs have measurable volatilities at temperatures experienced during some processing conditions. Removing metals from porphyrins does not eliminate all their undesirable properties. Mitchell and Scott (1986) report that free base porphyrins are worse catalyst poisons than metalloporphyrins. Nonporphyrin metal chelates, if they are not classical porphyrins, probably resemble them in having hetero atoms involved in metal coordination and in having a high degree of aromaticity. Any molecule having these structural features would not be easily processed even if it contained no metal. The chemical properties of metal chelates found in petroleum are influenced by their organic moieties as well as the metals involved in chelation. Thus chemical properties vary with ligand structure.

Asaoka et al. (1987) maintain that metal chelates are essential elements in asphaltene association and that metal removal is critical to the reduction of asphaltene molecular weights.

B. Hydroprocessing

Heavy crudes contain substantial quantities of sulfur-, nitrogen-, and oxygen-containing molecules, a minor proportion of which are metal chelates. Coals and shale oils usually contain much more nitrogen and oxygen than petroleum. Coals and some shale oils also are much more aromatic than crude oils. In order to convert these materials to salable products, hetero atoms and metals must be reduced to low levels. One way to accomplish this objective is to treat these materials or fractions derived from them with hydrogen.

Speight (1981) reviewed the desulfurization of heavy crudes and residua in detail. Hydrodesulfurization (HDS) is one of the most widely used processes for removal of sulfur from a variety of substrates, ranging from lower boiling distillates to heavy residua. Lower boiling fractions are relatively hydrogen-rich, contain little or no metals, and are composed of molecules of moderate size. Selective removal of sulfur and other hetero atoms is the objective of HDS of these materials. Heavy residua are composed of relatively large molecules and molecules which interact to form associations. They are more aromatic than distillate fractions and usually contain substantial amounts of metals. In HDS of residua, metals removal, reduction of molecular size, and lowering boiling range may be objectives in addition to removal of sulfur, nitrogen, and oxygen.

In HDS operations, a mixture of feedstock and hydrogen is contacted with a catalyst at high temperatures. The catalyst most commonly used is cobalt-molybdenum (Co-Mo) on an alumina support. Other catalysts that have been used in HDS are nickel-tungsten and nickel-molybdenum. The metal oxides are sulfided prior to use. It is generally accepted that molybdenum is the active metal in the Co-Mo catalyst, with cobalt acting as a promoter. This view is disputed by Vissers et al. (1987), who claim that cobalt is the active metal in HDS.

Reactions other than sulfur removal catalyzed by HDS catalysts are hydrogenation of some unsaturated molecules, hydrocracking, demetallization of metal chelates, and coking. These other reactions are not always desirable. Hydrogenation may result in excessive use of expensive hydrogen, and hydrocracking may yield substantial quantities of light gases. Coking results in laydown of carbonaceous deposits on HDS catalysts, and hydrodemetallization (HDM) forms metal sulfide deposits. The combination of the carbonaceous and metal sulfide deposits are referred to as coke. All the above reactions may take place at different sites on HDS catalysts.

During the first hours of operation of a HDS catalyst, substantial amounts of carbonaceous material are deposited on the catalyst. This results in considerable catalyst deactivation. After this initial phase, sulfides of nickel and vanadium deposit on the catalyst (Rankel and Rollman, 1983). This requires an increase in the HDS reaction temperature to maintain activity. After metal sulfide buildup reaches a critical level (about 25% of catalyst weight), activity decreases rapidly and themselves have some HDS activity (Rankel, 1987), but not enough to efficiently sustain HDS of refinery feedstocks.

The behavior of metalloporphyrins during HDS has been studied by several research groups. Morales and Galliasso (1982) determined that vanadyl porphyrins isolated from Boscan crude adsorb on acidic sites of HDS catalysts through the vanadyl oxygen atom. In the case of vanadyl tetraphenyl-porphins TPP, which is often used as a substrate in laboratory studies, adsorption on catalysts may involve the organic moiety of the molecule. Chen and Massoth (1987) report that both vanadyl and nickel TPP are thermally stable under HDS conditions (150–3000 psi, 550–850°F) in the absence of a catalyst.

After adsorption of a metalloporphyrin on a HDS catalyst, reaction of the porphyrin with hydrogen takes place to form a metallochlorin. Direct reductive demetallization is not a major pathway of metalloporphyrin decomposition during HDS. Metallochlorins have been isolated as intermediates in laboratory studies of HDS (Ware and Wei, 1985; Chen and Massoth, 1987; Rankel, 1987). The initially formed metallochlorins may add hydrogen to saturate another double bond or may demetallate to form a free base chlorin. After either second step, the chlorin molecule eventually undergoes hydrogenolysis and is broken up into smaller units.

Under refinery conditions, hydrogen sulfide is always present in HDS operations. This compound reacts with metalloporphyrins to form metal sulfides and pyrrole derivatives (Rankel, 1987). The reaction does not require a catalyst.

The mechanism of the decomposition of nonporphyrin metal chelates is not clear. These compounds are part of strong asphaltene associations. These asphaltene associations are too large to enter pores of standards HDS catalysts, so the asphaltene associations must be broken up before hydrogenation of metal chelates can take place. The asphaltene associations may be penetrated by hydrogen sulfide, which may react with metal chelates in the asphaltenes. Rankel and Rollman (1983) suggested that during HDS, metalloporphyrins are preferentially hydrogenated to chlorins while nonporphyrin metal chelates react with hydrogen sulfide. It may be mentioned parenthetically that reaction with hydrogen sulfide will be the pathway by which nonporphyrin metal chelates decompose during HDS regardless of their nature. It may be that these so-called nonporphyrin metal chelates really are classical porphyrins. However, the large asphaltene associations of which they are a part prevent catalytic hydrogenation of any metal chelate. These asphaltene associations persist for some time at surprisingly high temperatures, up to 480°F (Rao and Serrano, 1986). Even when unassociated, number-average molecular weights of asphaltene molecules are high (about 2000 D).

Reynolds et al. (1987) subjected a heavy resid to catalytic HDS and a thermal hydro treatment and analyzed the products by means of size exclusion chromatography (SEC). SEC of the unaltered resid results in the separation of a high molecular weight fraction containing nonporphyrin metal chelates from a low molecular weight fraction containing metalloporphyrins. Both the products from catalytic and noncatalytic hydro treatment were separated into the same fractions by SEC and metals analysis was performed on each fraction. Metal levels in both fractions were reduced by the noncatalytic treatment, while the catalytic treatment preferentially removed porphyrins from the resid. It is evident that access to the catalyst pores is required for demetallization during HDS and that demetallization of molecules of appropriate size by HDS catalysts is very efficient.

Laboratory studies of metalloporphyrins on HDS catalysts are complicated by the observation that crude oil porphyrins are less thermally stable than model compounds such as vanadyl TPP (Rankel, 1987; Reynolds et al., 1987). On the other hand, some other model porphyrins may be even less thermally stable than crude oil porphyrins.

The metal sulfides deposited during HDS may do more than foul catalyst surfaces and block pores. Aldag (1987) suggests that effective promoter levels also may be reduced. The alumina support of HDS catalysts has some hydrodemetallization (HDM) activity, and the effects of metal sulfide deposition on the support are not known with certainty. It is known that the interactions of the support with the catalytic metals influence HDS activity more than hydrogenation. If metal sulfide deposition on the support interferes with these interactions, HDS may be suppressed relative to hydrogenation. Catalytic metal–support interactions can be altered by impregnating the support with a variety of substances (Jiratova and Kraus, 1986).

The foregoing discussion is concerned with the fate of nickel and vanadyl chelates in HDS operations and the effect of the sulfides of these metals on HDS catalysts. It has been reported that organic iron compounds are particularly harmful to HDS catalysts even if the iron compounds are present in feedstocks at low levels (Howell et al., 1985). Some crudes may contain appreciable quantities of organic iron, so in processing these crudes, nickel and vanadium are not the only metals of concern.

C. Catalytic Cracking

Ideally, catalytic cracker feedstocks should contain no more than 20 ppm metals (5% conradson carbon residue, and 0.2 nitrogen (Howell et al., 1985. Most catalysts used in catalytic cracking operations were developed to process feedstocks with the above specifications. Typically, these are gas oils and coker gas oils. In recent years, some refiners have begun mixing heavy oil resids with gas oils for catalytic cracker feedstocks. This results in the deposition of metals on catalyst particles, in addition to the coke that invariably accompanies catalytic cracking. The term "coke" here refers to carbon deposits, and not the mixture of metal sulfides and carbonaceous deposits laid down on HDS catalysts and referred to as coke.

Zeolite catalysts used by refiners in catalytic cracker units are adversely affected by both nickel and vanadium. Nickel deposition does not appear to reduce overall catalyst activity but does cause excessive coking and hydrogen yields. The coke must be burned off in the regenerator, causing excessively high regenerator temperatures, which may result in spalling of catalyst particles. Vanadium attacks the zeolite structures embedded in the catalyst particles, destroying crystal structures and catalytic activity. This destructive action of vanadium probably is due to the low (690°C; 1274°F) melting point of vanadium pentoxide, the chemical form of vanadium in catalyst particles after coke burnoff in the regenerator. Vanadium pentoxide is mobile under conditions at which catalytic cracking units operate. Vanadium and its oxides also attack catalyst matrices. Compared with equivalent amounts of nickel, vanadium deposits do not cause as much coke or hydrogen production. Because most heavy crudes contain considerably more vanadium than nickel, vanadium will be the more troublesome of the two elements when heavy crudes or reside are used as catalytic cracker feedstocks.

Several methods to alleviate the adverse effects of metals on catalytic cracking operations are in use. One approach to the problem is to design catalysts that can tolerate heavy loading of metal deposits. Catalysts are manufactured with special trapping areas which serve as sinks for vanadium, thus sparing zeolite structures. In other catalysts, the matrix is altered. As is true for HDS catalysts, most asphaltene molecules are too large to enter the pores of conventional zeolite catalysts. Asphaltenes decompose on the silica-alumina matrix of the catalysts, depositing coke and whatever metals are contained in the asphaltenes. Some authorities claim that if the catalyst matrix has large pores and a low surface area, dispersion of metals throughout catalyst particles is inhibited (Ritter et al., 1981). Other authorities contend that a catalytically active, high-surface-area matrix is desirable. It is believed that such a catalyst matrix induces even metal deposition throughout catalyst particles rather than have vanadium concentrate in zeolite structures (Masselli and Peters, 1984). Desirable properties for catalysts used to crack residual fractions with large amounts of metals have been recommended by Nillson et al. (1986). Among these properties are a low zeolite content and a matrix with large pores and a high surface area.

The above methods involve alterations of the physical structure of catalysts to improve metal tolerance. The chemical composition of the catalyst also may be altered. Occelli et al. (1986) report that catalyst matrices that are silica-rich are more resistant to metal attack than conventional catalysts. Doping catalysts with zirconium oxide also is claimed to reduce damage done by metal deposition. In some circumstances, it may even be feasible to use catalysts of a different nature than conventional expensive zeolite catalysts. One such alternative would be to employ short-lived but cheap catalysts. Economics of processing some feedstocks might make this approach competitive with conventional practice.

Feedstocks for catalytic cracking units also may be treated to minimize metal damage to catalysts. The most obvious approach is to reduce metal levels to low values, but this is not always practicable. It is possible to passivate either vanadium or nickel by means of certain additives to catalytic cracker feedstocks. Antimony compounds passivate nickel (Dale and McKay, 1977). Antimony and nickel form an alloy which does not have the dehydrogenation activity of nickel. Some tin compounds are known to passivate vanadium.

When amorphous catalysts were in general use in catalytic cracking operations, a process was developed for removal of nickel from used catalysts by chemical treatment. Equivalent processes that remove metals from used zeolite catalysts with some restoration of catalyst activity are under development.

D. Demetallization

In demetallization processes, heavy crudes often behave unpredictably and do not all behave alike. Dolbear et al. (1987) report that for a number of heavy crude oils, hydrodemetallization (HDM) behavior varied significantly. Properties such as metal content, API gravity, asphaltene content, and Conradson carbon residue could not be used to accurately predict HDM behavior. It is unlikely that a more detailed characterization of the metal chelates in heavy crudes will be necessary for accurate prediction of their behavior in some demetallization and other processes. Each heavy crude has a unique metalloporphyrin suite. These compounds are distributed differently among resin and asphaltene fractions in different crudes. Relative amounts of porphyrins and nonporphyrins vary from one crude to another. Nonporphyrin metal chelates are largely associated with asphaltene "hard-cord" fractions, although there are exceptions. Efficient processing strategies should develop from good characterization studies. In the absence of good characterization studies, processing strategies may be empirical at best. It is difficult to accurately predict the course of chemical reactions of species whose structures are only vaguely known.

Dautzenberg and De DeKen (1987) analyzed the various options for demetallization of heavy crudes for grassroots refiners and retrofits. Carbon rejection strategies, usually flexicoking, were compared with hydrogen addition technologies. These authors recommend thermal hydrogenation as the most viable upgrading step for most operations. This conclusion is supported by Farcasiu and LaPierre (1987). These workers report that thermal treatment (425°C; 797°F) of a heavy crude mixed with appropriate acceptor species for 30 min at an initial hydrogen pressure of 100 psi resulted in a 70% reduction in the metal content of the charge. Aromatic fractions from crudes can serve as acceptors.

In other situations catalytic hydrogenation may be preferable to thermal hydrogenation. If the catalytic approach is taken, it appears best to separate HDS and HDM operations. One way in which this can be done is to use guard reactors which accomplish HDM using special large-pore catalysts which tolerate high-metal loadings. The HDM treatment is followed by HDS using a smaller pore catalyst. The HDM catalyst may be of different composition than the HDS catalyst. Iron or nickel may be used with molybdenum in the guard catalyst. Guard catalysts should have large pores so that asphaltenes have access to active sites and cracking of these molecules does not take place entirely on the catalyst matrix. If metals in a feedstock are largely porphyrinic, HDM with small-pore catalysts will be more efficient.

Carbon rejection strategies may be viable for crudes which contain relatively small amounts of vacuum resids and have low metal contents, but will not likely be satisfactory for high-sulfur crudes having large amounts of asphaltenes and metals. However, it may be possible to employ some combination of carbon rejection and hydrogen addition steps in upgrading heavy crudes. Howell et al. (1985) report that residuum hydrotreating followed by delayed coking yields smaller amounts of higher quality coke than would be obtained without the hydrotreating step. Liquid products obtained when coking is preceded by hydrotreatment also are of higher quality. This has a positive impact on subsequent catalytic cracking operations. Beaton et al. (1986) also report that catalytic hydrotreating of vacuum resids from poor-quality crudes followed by coking of unconverted resid provides higher yields of better quality products than coking the vacuum resids without preliminary hydrotreating. In this case, the vacuum resids are treated in three successive expanded bed reactors. The first reactor is a HDM reactor, and the other two effect HDS and Conradson carbon conversion.

Suchanek and Moore (1986) claim that coking of heavy resids using standard processes is not an efficient way to convert these materials to usable products, and that hydrogenation of heavy resids usually consumes too much hydrogen. These authors recommend a strategy of carbon rejection followed by hydrogen addition, but not employing the conventional coking process. In the process they recommend, resids are contacted with an additive with an affinity for metals and the mixture is heated for a short time. This results in less solid material being produced than in coking operations. The products from this operation can be efficiently hydrogenated using a nickel-molybdenum catalyst. Hydrogen consumption is less than is required for hydrogenation of products from conventional coking operations.

Solvent deasphaltening may be a method of choice for upgrading feeds in which metals, sulfur, and Conradson carbon residue are highly concentrated in asphaltenes. Mohammed et al. (1987) describe a process in which a heavy crude is deasphaltened with a fivefold excess of *n*-heptane. This treatment results in a small yield of asphaltenes. The resulting oil is lower in metal concentration than the parent crude and is more easily desulfurized and demetallized in a subsequent catalytic hydrogenation step using a nickel-molybdenum catalyst.

Asaoka et al. (1987) describe a process in which asphaltene structures in heavy feedstocks are reduced in number/average molecular weight, metals are removed, and hydrogen uptake is minimal. The process may be used alone or in combination with a hydrovisbreaking step. The catalyst used in the process is of the Co-Mo type and is able to tolerate large metal loadings without unacceptable loss of activity.

The removal of metal-containing species from heavy crudes by chemical treatments using oil-soluble reagents have been investigated by Kukes et al. (1987). Most of the reagents tested by these workers caused extensive degradation of feedstock quality. However, some alkyl phosphites were found to remove some vanadium from a heavy crude both in solution and on a catalyst surface. The vanadium removed homogeneously from solution deposits on catalyst surfaces as vanadium phosphorus compounds which themselves catalyze vanadyl porphyrin decomposition under processing conditions. The phosphorus compounds react mostly with lower molecular weight vanadyl chelates and are unreactive with nickel compounds.

It is clear that there is no routine procedure for dealing with the problems caused by metal chelates in all heavy crudes. Each one of the approaches to the solution of these problems is favored by one or more energy companies. For the foreseeable future, it is likely that dealing with crudes with high metal concentrations will involve a number of processes. These processes will depend on the nature of the crude and the situation of the refiner. Solutions to these problems will be helpful for future refiners of coal- and oil shale-rerived materials. For these substances, elements other than vanadium and nickel must be dealth with. Arsenic, iron, and titanium are often associated with coals and oil shales, and contaminate feedstocks derived from those sources. All three elements are catalyst poisons. However, catalyst lifetimes in processing coals and coal-derived materials (e.g., coal liquefaction) are much shorter than in petroleum refining due to much greater coke production. The use of cheap catalysts in processing coal liquids may be an approach of choice, and damage to cheap catalysts by metal deposition may not result in the inefficiencies that result when metal deposits are laid down on expensive, long-lived catalysts used in petroleum refining.

E. Summary

The processing of heavy crudes and tar sand bitumens poses serious problems for refiners. Among these is the problem of dealing with so-called metal contaminants. Up to now, solutions to the problem have been empirical. Full characterizations of metal chelate suites in large numbers of heavy crudes have not been performed, nor has their chemistry during refining operations been fully elucidated. These materials have been dealt with by processors as well as possible. Some refineries that process heavy crudes were not desitned to do so.

It appear that more thorough characterizations of metal chelates in a number of heavy crudes will be necessary if an understanding of the chemistry of these compounds in refinery operations is ever to be obtained. Since these compounds are concentrated in asphaltene fractions of petroleum, the role of metal chelates in asphaltene structures needs to be better known. Breaking down of asphaltene structures may be the critical step in processing many heavy crudes.

Most energy companies support substantial research efforts aimed at solving problems posed by heavy crude processing, among them the difficulties caused by metal chelates. The solution to this particular problem has been approached in a variety of ways. Some energy companies emphasize the design of catalysts that are relatively tolerant to metals deposition. Others have developed methods to reduce metal levels in feedstocks for operations employing expensive catalysts or to ameliorate the effects of these compounds. Various combinations of techniques are advocated to process crudes with high concentrations of metals. The lack of a routine, standard procedure is partly caused by the nature of heavy crudes. These materials are diverse in chemical composition and respond differently to a given method of processing. Methods for predicting how a particular heavy crude will behave in certain processes are not well developed.

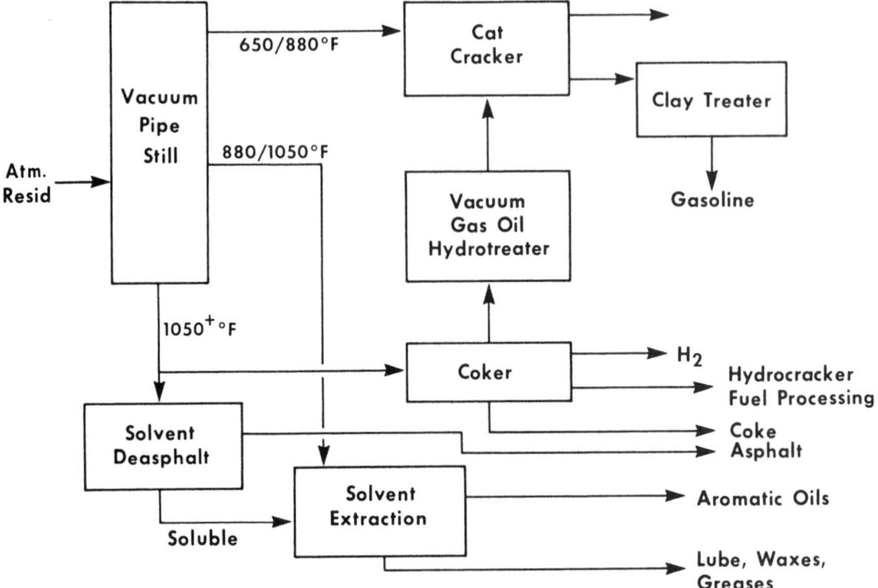

Figure 3 Schematic representation of the various options for processing heavy oils and residua. (See Figure 1 in Chap. 5 to put this in the perspective of the whole refinery.)

IV. REFINERY OPTIONS

In the early days of heavy ends processing, the refinery options were somewhat more straightforward than at present. For example, a heavy feedstock (such as an atmospheric residuum) would be distilled in a vacuum tower (Figure 3) to retrieve more overhead liquid products, which would then be sent to a catalytic cracker. The gas oil would only produce a low yield (less than 5 wt %) of coke in this operation and there was also the option of sending the heavier gas oil (880–1050°F; 470–565°C material) to the catalytic cracker. The vacuum residuum would then be used as a coking feedstock to produce more overhead products as well as a source of asphalt and lubricating stocks.

The yields of coke were not sufficient to put the refinery in the uneconomical position of having to stockpile the coke since the feedstock was low in residuum and the options were such that a balance could be reached between coke output (refinery fuel or specialty product) and liquids output.

Obviously, the situation has changed in the last two decades whereby refiners are now having to accept crude oils that contain higher proportions of residual material and suffer the consequence of high coke output and lower liquids output.

It is this dilemma that is stimulating the search for other methods to convert the heavier crude oils and residua to a more economical slate of products.

There are two general routes to heavy-feedstock upgrading: (1) The first is to remove carbon (coke) as a product having a low atomic hydrogen/carbon ratio and, at the same time, to produce overhead material (distillate) having a high atomic hydrogen/carbon ratio; (2) the second method involves the concept of hydrogen addition by a hydrocracking/hydrogenolysis mechanism by which the yield of coke is reduced in favor of enhanced yields of liquid products (Figure 4). An example of carbon rejection is the delayed coking process (Chap. 6, Sec. I.B) in which the feedstock is converted to overhead with the concurrent deposition of coke; such a process is used by Suncor at their oil sands plant (Berkowitz and Speight, 1975). The fluid coking and flexicoking

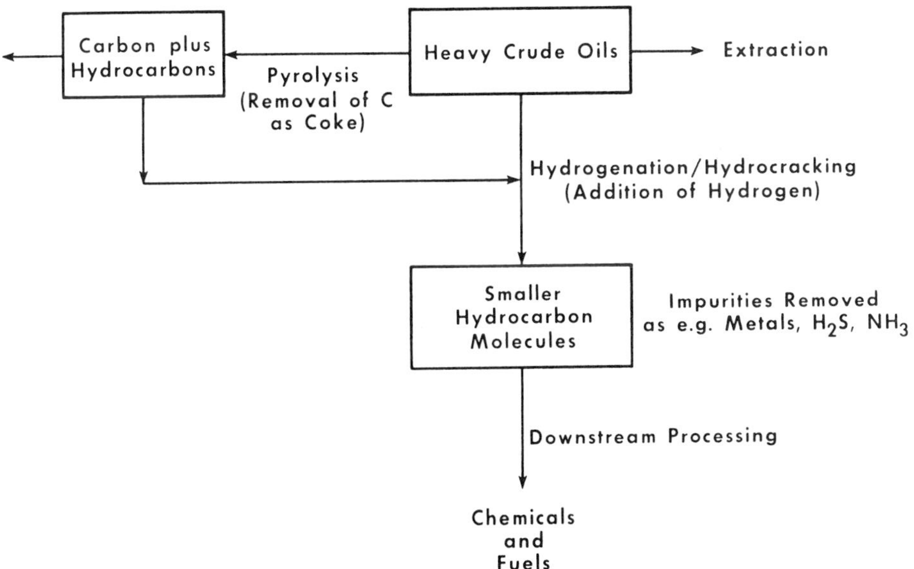

Figure 4 Schematic representation of the carbon rejection/hydrogen addition concept.

options are more sophisticated processes for carbon rejection that allow higher yields of overhead products through more intimate contact of the feedstock with hot coke particles (fluid coking) and through an additional step in which excess coke is gasified (flexicoking; Figure 5).

Residuum fluid catalytic cracking is an extension of the fluid catalytic cracking operation which produces a low-value byproduct (coke on the catalyst) that must be burned off in the regenerator. The ART process (Figure 6) is a comparatively recent innovation (Logwinuk and Caldwell, 1983) in which the feedstock is contacted with a fluidizable solid material and is essentially a combination of selective vaporization and fluid decarbonization/demetallization.

The hydrogen concept can be illustrated by the hydrocracking process (Chap. 6, Sec. III) in which hydrogen is used in an attempt to "stabilize" the reactive fragments produced during the cracking;

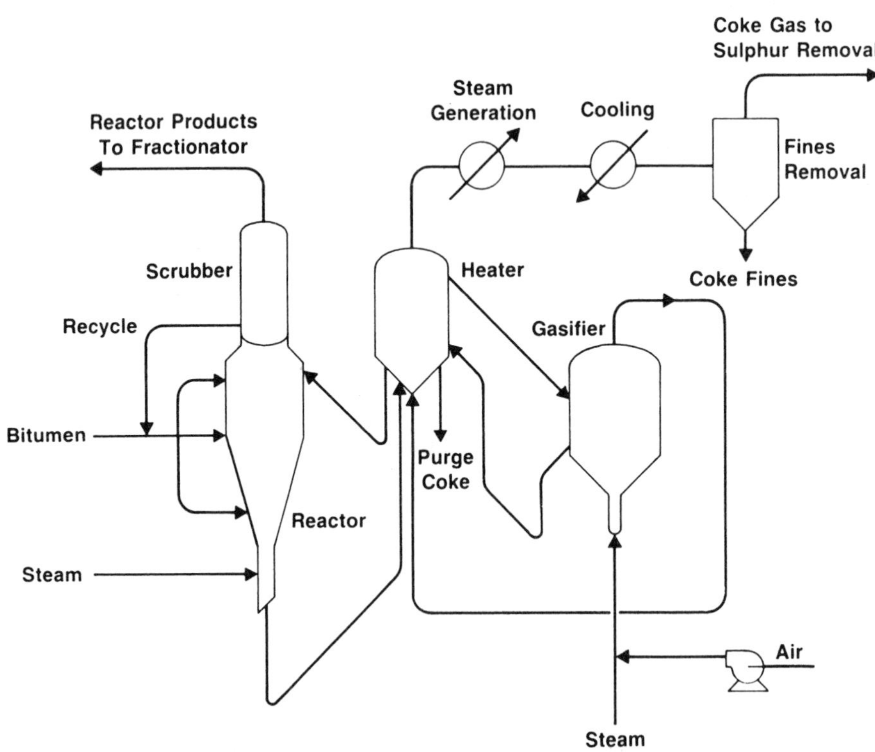

Figure 5 Schematic representation of the flexicoking process.

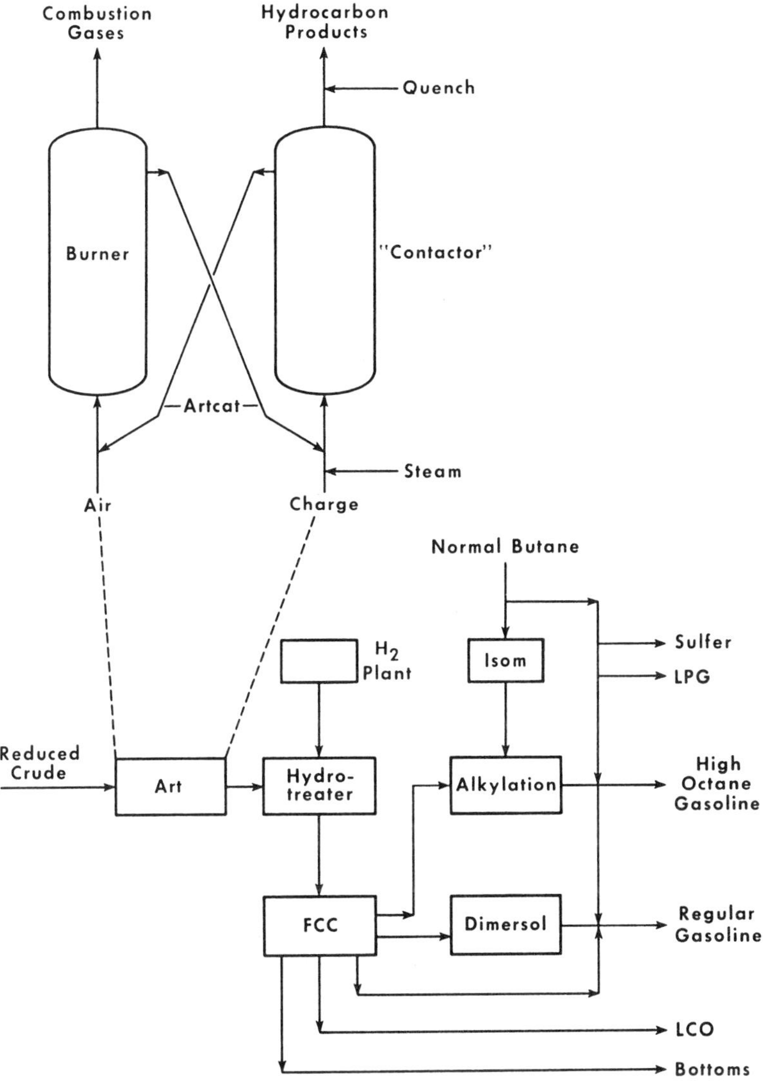

Figure 6 Simplified representation of the ART process and its possible integration into a refinery operation.

thereby decreasing their potential for recombination to heavier products and, ultimately, to coke. Residuum hydroprocessing effectively removes contaminants while achieving moderate to high levels of conversion without producing the low-value byproducts. The choice of a primary conversion step such as a carbon rejection option or a hydrogen addition option may, however, depend on the individual requirements of each refinery. Obviously, such a choice will also have to take into account the addition of hydrogen as a primary or downstream step.

A feedstock such as Athabasca bitumen which has 17% w/w pentane-asphaltenes capable of producing approx. 50% w/w carbon residue (Moschopedis et al., 1978) requires processing conditions that are not conducive to the formation of thermal coke or catalytic coke. Removal of the asphaltenes by a nonthermal technique (i.e., deasphalting) provides a deasphalted oil that is easier to process downstream but the asphaltenes still remain for disposal. It has been suggested that use of the asphaltenes as a fuel in a partial oxidation process is one answer to the disposal "problem" but the asphaltenes are better considered as a carbon source from which gas/liquid overhead products (approx. 50% w/w or more yield) can be produced.

Catalysts for use with heavy feedstocks must be selected on the basis of activity and stream life. Both are dependent on several factors not the least of which is the deposition of impurities such as nitrogen-containing species which eventually form a layer of coke on the surface and, within the pore system, of the catalyst. These feedstocks are so complex that any catalysts selected for the process must be extremely resistant to such depositional effects.

In addition, the deposition of inorganic species can produce rapid poisoning of the catalyst through the deactivation of the "active" sites. Hydrodemetallization of the feedstock occurs on the catalyst and metals such as the "porphyrinic" metals (nickel and vanadium) as well as iron (either natural to the feedstock or picked up during flow through pipes). The precise site of deposition of the metals depends, to a large extent, on the catalyst morphology; hence individual systems (either the feedstock, the catalyst, or a combination of both) may behave very differently.

In summary, heavy feedstocks are deleterious to catalyst activity and to on-stream life. The chemistry is complex and much remains to be resolved (Laine and Trimm, 1982).

Therefore, it is obvious that efficient conversion of heavy feedstocks (Table 1) also requires serious efforts to develop adequate catalysts as well as the modification of existing, or the development of new, processes to respond to market demands (Ternan, 1983). In the hydrogen addition options, particular attention must be given to hydrogen management thereby promoting asphaltene fragmentation to lighter products rather than coke formation. In this latter respect it is worth noting the reemergence of donor solvent processing of heavy oils, which has its roots in the 30-year-old hydrogen donor diluent visbreaking (Langer et al., 1962; Bland and Davidson, 1967; Fischer

Table 1 Summary of the Current Processes for Heavy Feedstock Refining

---

Current Processes
- Visbreaking
  - Low conversion
  - Slight quality improvement
- Coking
  - Good conversion
  - Product quality low
- Hydroprocesses
  - Not intended as major conversion process, used to reduce sulfur
  - Limited by susceptibility of catalyst to asphaltenes
- Catalytic cracking
  - Median-quality product
  - Catalyst intolerable to metals and coke deposition

Future Needs
- "Selective" thermal degradation
- Need to reduce coke formation
  - Prevent recombination of "radicals"
  - Saturation of aromatics
  - Catalyst development
- Need to prevent catalyst deterioration
  - Control affinity of asphaltenes for catalyst
- Hydroprocesses are the key
  - Reduce radical recombination
  - Saturate aromatics?
  - Hydrogen economics?

---

et al., 1982). Other options include a low-temperature primary conversion process (Speight and Moschopedis, 1979) as well as the potential as a medium to be used for coprocessing with coal (Speight and Moschopedis, 1986). Several other processes are also available at the commercial, or near-commercial, scale which show promise for refining heavy feedstocks (*Hydrocarbon Processing*, September 1984) and include Residfining, Canmet Hydrocracking, H-Oil, LC-Fining, Flexicracking,

Heavy Oil Cracking, and Eureka Cracking. Each has its own particular novel aspect and is certainly worthy of consideration for further use and development.

There are studies which take the approach that methods are already available for refining heavy crude oils (Scheutze and Hofmann, 1984; Chapel et al., 1987) which do indeed show promise. The process examples vary from modifications of distillate processes development to newer concepts for conversion to liquid products (Figures 7-9) as well as estimation of product slates for a heavy vacuum residuum (Scheutze and Hofmann, 1984). It may also be obvious from this study that as yet there is no one process that can be generally applied to heavy crude upgrading. It appears that each heavy feedstock will have to be individually evaluated for process application.

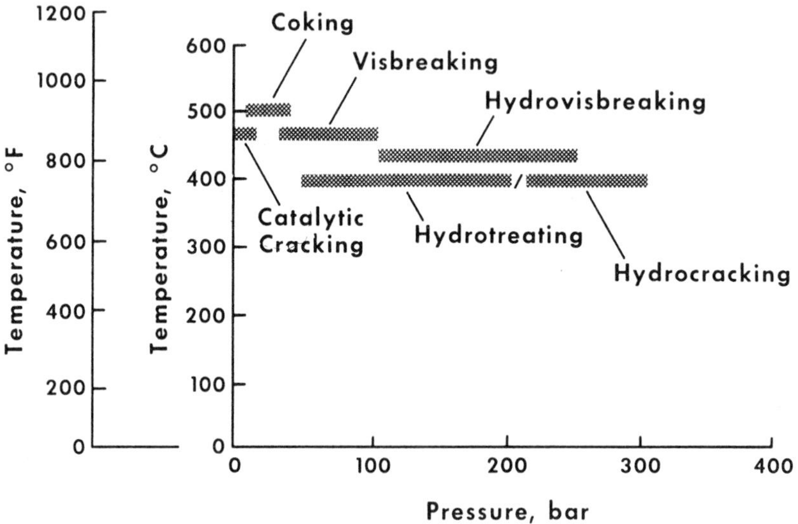

Figure 7 Summary of processing conditions currently used in conversion processes.

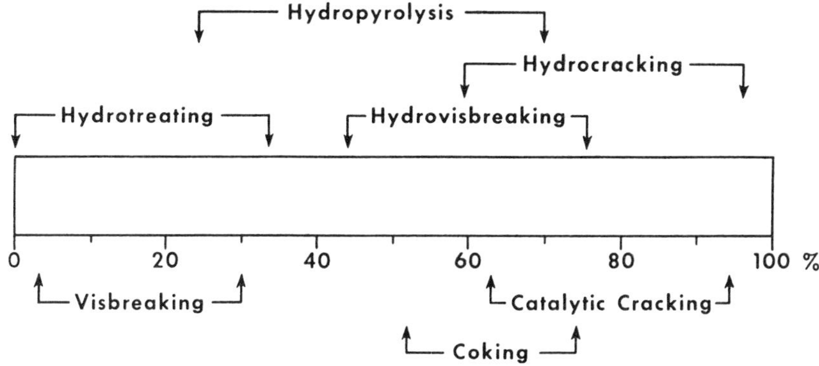

Figure 8 Summary of the degree of conversion achieved by various upgrading processes.

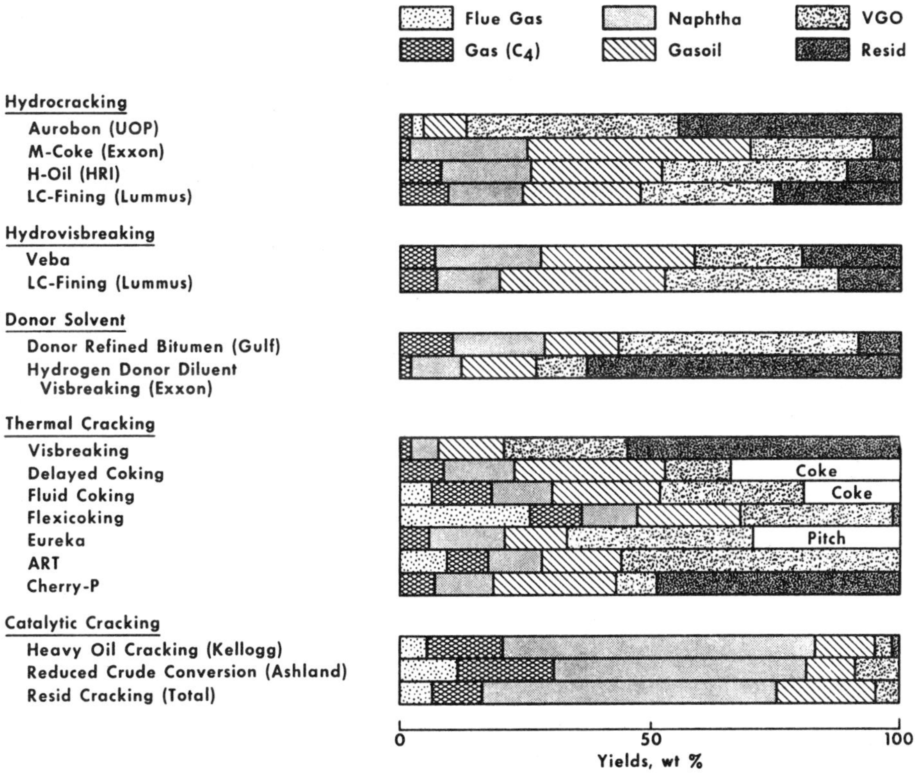

Flue Gas
Gas (C_4)
Naphtha
Gasoil
VGO
Resid

Hydrocracking
Aurobon (UOP)
M-Coke (Exxon)
H-Oil (HRI)
LC-Fining (Lummus)

Hydrovisbreaking
Veba
LC-Fining (Lummus)

Donor Solvent
Donor Refined Bitumen (Gulf)
Hydrogen Donor Diluent Visbreaking (Exxon)

Thermal Cracking
Visbreaking
Delayed Coking
Fluid Coking
Flexicoking
Eureka
ART
Cherry-P

Catalytic Cracking
Heavy Oil Cracking (Kellogg)
Reduced Crude Conversion (Ashland)
Resid Cracking (Total)

Coke
Coke
Pitch

0
50
100
Yields, wt %

Figure 9 Projected yields of products from the processing of a heavy feedstock by various process options.

9

Catalysts

I	Processes	251
	A. Cracking Processes	251
	B. Hydroprocesses	253
	C. Reforming Processes	255
	D. Isomerization Processes	256
	E. Alkylation Processes	257
	F. Polymerization Processes	257
II	Catalyst Treating	257
	A. Demet	258
	B. Met-X	259

I. PROCESSES

A. Cracking Processes

The cracking of crude oil fractions occurs over many types of catalytic materials but high yields of desirable products are obtained with hydrated aluminum silicates. These may be either activated (acid-treated natural clays of the bentonite type or synthesized silica-alumina or silica-magnesia preparations). The activity that yields essentially the same products may be enhanced to some extent by the incorporation of small amounts of other materials, such as the oxides of zirconium, boron (which has a tendency to volatilize away on use), and thorium. Both the natural and synthetic catalysts can be used as pellets or beads, and also as powder; in either case replacements are necessary because of attrition and gradual loss of efficiency (DeCroocq, 1984; Thakur, 1985; Le Page et al., 1987).

The catalysts must be stable to physical impact loading and thermal shocks, and withstand the action of carbon dioxide, air, nitrogen compounds, and steam. They should also be resistant to sulfur compounds; the synthetic catalysts and certain selected clays appear to be better in this regard than average untreated natural catalysts. The silica-alumina catalysts are reported to give the highest octane gasolines and silica-magnesia the largest yields, with the natural clays falling in between.

Neither silica nor alumina alone is effective in promoting catalytic cracking reactions. In fact, they (and also activated carbon) promote hydrocarbon decompositions of the thermal type. A mixture of anhydrous silica and alumina, or anhydrous silica with hydrated alumina, is also essentially noneffective. A catalyst having appreciable cracking activity is obtained only when prepared from hydrous oxides followed by partial dehydration (calcining). The small amount of water remaining is necessary for proper functioning.

The catalysts are porous and highly adsorptive and their performance is effected markedly by the method of preparation. The catalysts chemically identical, but having pores of different size and distribution, may have different activities, selectivities, temperature coefficients of reaction rates, and responses to poisons. While the intrinsic chemistry and catalytic action of a surface may be independent of pore size, small pores appear to produce different effects because of the manner and time in which hydrocarbon vapors are transported into and out of the interstices.

Commercial synthetic catalysts are amorphous and contain more silica than is called for by the above formulae; they are generally composed of 10–15% Al_2O_3 and 85–90% SiO_2. The corresponding natural materials, montmorillonite, a nonswelling bentonite, and halloysite, are hydrosilicates of aluminum, with a well-defined crystal structure and approximate composition of $Al_2O_3 \cdot 4SiO_2 \cdot XH_2O$. Some of the newer catalysts contain up to 25% of alumina and are reputed to have a longer active life.

Commercially used cracking catalysts are "insulator catalysts" possessing strong protonic acid properties. They function as catalysts by altering the cracking process mechanisms through an alternate mechanism involving chemisorption by proton donation and desorption resulting in cracked oil and theoretically restored catalyst. Thus it is not surprising that all cracking catalysts are poisoned by proton-accepting vanadium.

The catalyst/oil volume ratios range from 5:1 to 30:1 for the different processes although most processes are operated to 10:1. However, for moving bed processes the catalyst/oil volume ratios may be substantially lower than 10:1.

B. Hydroprocesses

The character of the hydrotreating processes are chemically very simple since it essentially involves removal of sulfur and nitrogen as hydrogen sulfide and ammonia, respectively:

$$-S- + H_2 \longrightarrow H_2S$$

$$2N\vdots + 3H_2 \longrightarrow 2NH_3$$

However, nitrogen is the most difficult contaminant to remove from feedstocks and processing conditions are usually dictated by the requirements for nitrogen removal.

In general, any catalyst capable of participating in hydrogenation reactions may be used for hydrodesulfurization. The sulfides of hydrogenating metals are particularly used for hydrodesulfurization and catalysts containing cobalt, molybdenum, nickel, and tungsten are widely used on a commercial basis.

Cobalt-molybdenum catalysts are widely used for hydrotreating and, under the conditions whereby nitrogen removal is accomplished, desulfurization usually occurs as well as oxygen removal. Indeed, it is generally recognized that fullest activity of the hydrotreating catalyst is not reached until some interaction with the sulfur (from the feedstock) has occurred with part of the catalyst metals being converted to the sulfides. Too much interaction may of course lead to catalyst deactivation.

The reactions of hydrocracking require a dual-function catalyst with high cracking and hydrogenation activities. The cracking function is usually supplied by the catalyst base, such as acid-treated clay, alumina, or silica-alumina, which is used to support the hydrogenation function supplied by metals such as nickel, tungsten, platinum, palladium, etc. These highly acidic catalysts are very sensitive to nitrogen compounds in the feed, which break down under the conditions of reaction to give ammonia and neutralize the acid sites. As many heavy gas oils contain substantial amounts of nitrogen (up to approx. 2500 ppm), a purification stage is frequently required. Denitrogenation and desulfurization can be carried out using cobalt/molybdenum or nickel/cobalt/molybdenum on alumina or silica-alumina.

Hydrocracking catalysts such as nickel (5% wt) on silica-alumina work best on feedstocks which have been hydrofined to low nitrogen and sulfur levels. The nickel catalyst will then operate well at 350--370°C (660--700°F), and a pressure of about 1500 psi to give good conversion of feed to lower boiling liquid fractions with minimum saturation of single-ring aromatics and giving a high iso/normal ratio in the lower paraffins. The poisoning effect of nitrogen can be offset to a certain degree by operation at a higher temperature, but this tends to increase the production of material in the C_1--C_4 range and decrease the operating stability of the catalyst so that it requires more frequent regeneration. Catalysts containing platinum or palladium (approx. 0.5% wt) on a zeolite base appear to be somewhat less sensitive to nitrogen than are nickel catalysts, and successful operation has been achieved with feedstocks containing 40 ppm nitrogen. This catalyst is also more tolerant of sulfur in the feed which acts as a temporary poison, the catalyst recovering its activity when the sulfur content of the feed is reduced.

On catalysts such as nickel or tungsten sulfide on silica-alumina, isomerization does not appear to play any part in the reaction, as uncracked normal paraffins from the feedstock tend to retain their normal structure. Extensive splitting produces large amounts of low molecular weight (C_3--C_6) paraffins, and it appears that a primary reaction of paraffins is catalytic cracking followed by hydrogenation to form isoparaffins. With catalysts of higher hydrogenation activity, e.g., platinum on silica-alumina, direct isomerization occurs. The product distribution is also different, and the ratio of low to intermediate molecular weight paraffins in the breakdown product is reduced.

Zeolite catalysts have also found use in the refining industry during the last two decades. Like the silica-alumina catalysts, zeolites also consist of a framework of tetrahedra, usually with a silicin atom or an aluminum atom at the center. The geometrical characteristics of the zeolites are responsible for their special properties which are particularly attractive to the refining industry (DeCroocq, 1984). Specific zeolite catalysts have shown up to 10,000 times more activity than the so-called conventional catalysts in specific cracking tests. The mordenite-type catalysts are particularly worthy of mention since they have shown up to 200 times greater activity for hexane cracking in the temperature range 360–400°C (680–750°F).

Other zeolite catalysts have also shown remarkable adaptability to the refining industry. For example, the resistance to deactivation of the type Y zeolite catalysts containing either noble or nonnoble metals is remarkable and catalyst life of up to 7 years has been obtained commercially in processing heavy gas oils in the Unicracking-JHC processes. Operating life depends on the nature of the feedstock, the severity of the operation, and the nature and extent of operational upsets. Gradual catalyst deactivation in commercial use is counteracted by incrementally raising the operating temperature to maintain the required conversion per pass. The more active a catalyst, the lower is the temperature required. When processing for gasoline, lower operating temperatures have the additional advantage that less of the feedstock is converted to isobutane.

Basic nitrogen-containing compounds in a feed diminish the cracking activity of hydrocracking catalysts. However, zeolite catalysts can operate in the presence of substantial concentrations of ammonia, in marked contrast to silica-alumina catalysts, which are strongly poisoned by ammonia. Similarly, sulfur-containing compounds in a feedstock adversely affect the noble metal hydrogenation component of hydrocracking catalysts. These compounds are hydrocracked to hydrogen sulfide, which will convert the noble metal to the sulfide form. The extent of this conversion will be a function of the hydrogen and hydrogen sulfide partial pressures. Removal of sulfur from the feed results in a gradual increase in catalyst activity returning almost to the original activity level. As with ammonia, the concentration of the hydrogen sulfide can be used to precisely control the activity of the catalyst. Nonnoble metal-loaded zeolite catalysts have an inherently different response toward sulfur impurities since a minimum level of hydrogen sulfide is required to maintain the nickel-molybdenum and nickel-tungsten in the sulfide state.

C. Reforming Processes

The composition of a reforming catalyst is dictated by the compositions of the reformer charge and the desired reformate. Reforming consists of two types of chemical reactions which are catalyzed by two different types of catalysts: (1) isomerization of straight-chain paraffins and isomerization (simultaneous with hydrogenation) of olefins to produce branched chain paraffins, and (2) dehydrogenation/hydrogenation of paraffins to product aromatics and olefins to produce paraffins.

The catalysts used are principally molybdena/alumina, or chromia/alumina, or platinum on a silica/alumina or alumina base. The non-platinum catalysts are widely used in regenerative processes for feeds containing, for example, sulfur which poison platinum catalysts, although pretreatment processes (e.g., hydrodesulfurization) may permit platinum catalysts to be employed.

The purpose of platinum on the catalyst is to promote dehydrogenation and hydrogenation reactions, i.e., the production of aromatics, participation in hydrocracking, and rapid hydrogenation of carbon-forming precursors. For the catalyst to have an activity for isomerization of both paraffins and naphthenes, the initial cracking step of hydrocracking, and participate in paraffin dehydrocyclization, it must have an acid activity. The balance between these two activities is most important in a reforming catalyst. In fact, in the production of aromatics from naphthenes it is important that hydrocracking be minimized to avoid loss of yield, and thus the acid activity should be lower than in the case of gasoline production from a paraffinic feed, where dehydrocyclization and hydrocracking play an important part. The acid activity can be obtained by means of halogens (usually fluorine or chlorine up to about 1% wt in catalyst) or silica incorporated in the alumina base. Platinum content of the catalyst is normally in the range 0.3–0.8% wt. At higher levels there is some tendency to effect demethylation and naphthene ring opening, which is undesirable, while at lower levels the catalysts tend to be less resistant to poisons.

Most processes have a means of regenerating the catalyst as needed. The time between regenerations, which varies with the process, the severity of the reforming reactions, and the impurities in the feedstock, ranges from a few hours to several months. Several processes use a nonregenerative catalyst which can be used for a year or more after which it is returned to the catalyst manufacturer for reprocessing. The processes that have moving beds of catalysts utilize continuous regeneration of the catalyst in separate regenerators.

The processes using bauxite (Cycloversion) and clay (Isoforming) differ from other catalytic reforming processes in that hydrogen is not formed; hence none is recycled through the reactors. Since hydrogen is not involved in the reforming reactions, there is no limit to the amount of olefins that may be present in the feedstock. The Cycloversion process is also used as a catalytic cracking process and as a desulfurization process. The Isoforming process causes only a moderate increase in octane number.

D. Isomerization Processes

During World War II aluminum chloride was the catalyst used to isomerize butane, pentane, and hexane. Since then, supported metal catalysts have been developed for use in high-temperature processes which operate in the range 370–480°C (700–900°F) and 300–750 psi, while aluminum chloride plus hydrogen chloride is universally used for the low-temperature processes. However, aluminum chloride is volatile at commercial reaction temperatures and is somewhat soluble in hydrocarbons and techniques must be employed to prevent its migration from the reactor. This catalyst is nonregenerable and is utilized in either a fixed-bed or liquid contactor.

E. Alkylation Processes

Sulfuric acid, hydrogen fluoride, and aluminum chloride are the only catalysts used commercially. Sulfuric acid is used with propylene and higher boiling feeds, but not with ethylene because it reacts to form ethyl hydrogen sulfate and a suitable catalyst contains a minimum of 85% titratable acidity. The acid is pumped through the reactor forming an air emulsion with reactants and the emulsion is maintained at 50% acid. The rate of deactivation varies with the feed and isobutane charge rate. Butene feeds cause less acid consumption than the propylene feeds.

Aluminum chloride is not widely used as an alkylation catalyst; however, when employed hydrogen chloride is used as a promoter and water is injected to activate the catalyst. The form of catalyst is an aluminum chloride hydrocarbon complex and the aluminum chloride concentration is 63--84%.

Hydrogen fluoride is used for alkylation of higher boiling olefins, and the advantage of hydrogen fluoride is that it is more readily separated and recovered from the resulting product. The usual concentration is 85–92% titratable acid, with about 1.5% water.

F. Polymerization Processes

Phosphates are the principal catalyst for polymerization; the commercially used catalysts are liquid phosphoric acid, phosphoric acid on kieselguhr, copper pyrophosphate pellets, and phosphoric acid film on quartz. The latter is the least active but the most used and easiest to regenerate simply by washing and recoating; the serious disadvantage is that tar must occasionally be burned off the support. The process using liquid phosphoric acid catalyst is far more responsive to attempts to raise production by increasing temperature than the other processes.

II. CATALYST TREATING

The latest technique developed by the refining industry to increase gasoline yield and quality is to treat the catalysts from the cracking units to remove metal poisons which accumulate on the catalyst. Nickel, vanadium, iron, and copper compounds contained in catalytic cracking feedstocks are deposited on the catalyst during the cracking operation thereby adversely affecting both catalyst activity and selectivity. Increased catalyst metal contents effect catalytic cracking yields by increasing coke formation, decreasing gasoline and butane-butylene production, and increasing hydrogen production.

The recent commercial development and adoption of cracking catalyst treating processes definitely improve the overall catalytic cracking process economics.

A. Demet

A cracking catalyst is subjected to two pretreatment steps (Figure 1). The first step effects vanadium removal and the second nickel removal--to prepare the metals on the catalyst for chemical conversion to compounds (chemical treatment step) which can readily be removed through water washing (catalyst wash step). The treating steps include use of a sulfurous compound followed by chlorination with an anhydrous chlorinating agent (e.g., chlorine gas) and washing with an aqueous solution of a chelating agent (e.g., citric acid). The catalyst is then dried and further treated before returning to the cracking unit.

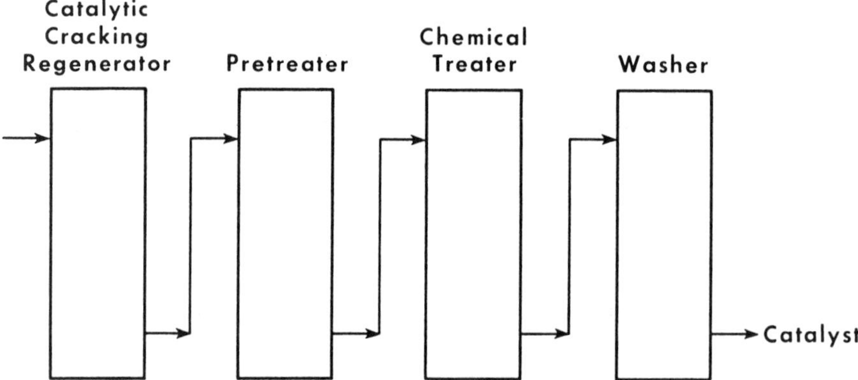

Figure 1 The Demet process for treating cracking catalysts.

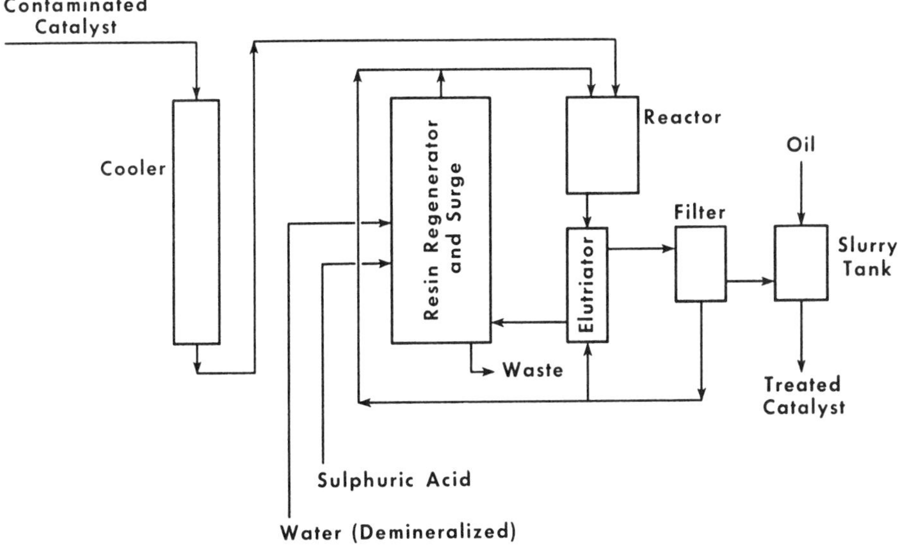

Figure 2 The Met-X process for treating cracking catalysts.

B. Met-X

The Met-X process consists of cooling, mixing, and ion exchange separation, filtration, and resin regeneration. Moist catalyst from the filter is dispersed in oil and returned to the cracking reactor in a slurry (Figure 2). On a continuous basis, the catalyst from a cracking unit is cooled and then transported to a stirred reactor and mixed with an ion exchange resin (introduced as slurry). The catalyst-resin slurry then flows to an elutriator for separation. The catalyst slurry is taken overhead to a filter; the wet filter cake is slurried with oil and pumped into the catalytic cracked feed system. The resin leaves the bottom of the elutriator and is regenerated before returning to the reactor.

10

Petroleum Products

I	Gasoline	263
II	Solvents	265
III	Kerosene	266
IV	Fuel Oils	266
V	Lubricating Oils	267
VI	Waxes	269
VII	Asphalt	271
VIII	Coke	273
IX	Acid Sludge	274

The constant demand for products such as liquid fuels is the main driving force behind the petroleum industry (Guthrie, 1960). Other products, such as lubricating oils, waxes, and asphalt, have also added to the popularity of petroleum as a national resource. There have, however, been many changes in emphasis on product demand since petroleum first came into use some five to six millennia ago (Table 7 in Chap. 1). These changes in product demand have been largely responsible for the evolution of the industry—from the asphalt used in ancient times to the gasoline and other liquid fuels of today (Figure 1).

A major group of products from petroleum (petrochemicals) are the basis of a major industry. They are, in the strictest sense, different from petroleum products insofar as the petrochemicals are the basic building blocks of the chemical industry (Chap. 11).

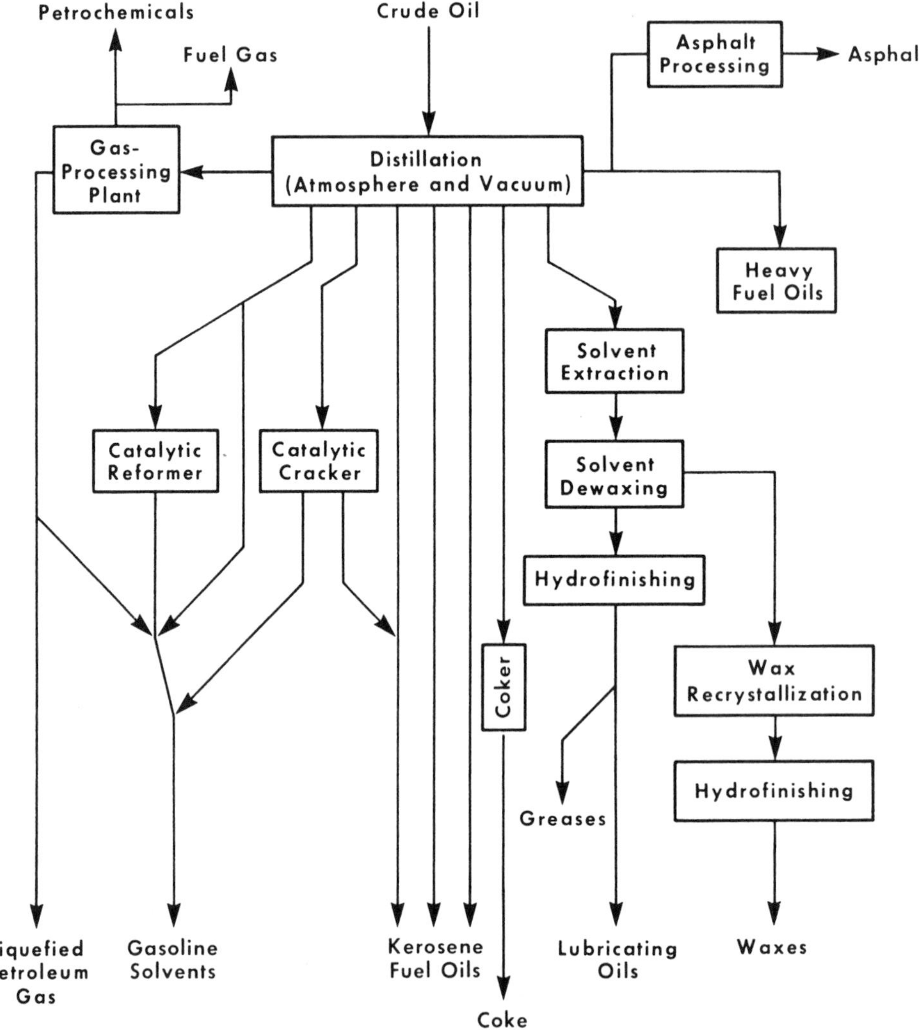

Figure 1 General overview of products from the various refinery processes.

Unlike processes, products are more difficult to place on an individual evolutionary scale. Processes changed and evolved to accommodate the demand for, say, higher octane fuels or longer lasting asphalts or lower sulfur coke. In this section, a general overview of some petroleum products is presented in order to show the *raison d'etre* of the industry.

I. GASOLINE

Gasoline is a complex mixture of hydrocarbons that boils below 200°C (390°F). The hydrocarbon constituents in this boiling range are those that have 4–12 carbon atoms in their molecular structure. Gasolines can vary widely in composition--even those with the same octane number may be quite different. For example, low-boiling distillates with high aromatics content (above 20%) can be obtained from some crude oils. The variation in aromatics content as well as the variation in the content of normal paraffins, branched paraffins, cyclopentanes, and cyclohexanes all involve characteristics of any one individual crude oil and influence the octane number of the gasoline.

Up to and during the first decade of the twentieth century, the gasoline produced was that which was originally present in crude oil or which could be condensed from natural gas. However, it was soon discovered that if the heavier portions of petroleum (such as the fraction which boiled higher than kerosene, e.g., gas oil) were heated to more severe temperatures, thermal degradation (or cracking) occurred to produce smaller molecules which were within the range suitable for gasoline (Chap. 6, Sec. I). Therefore, gasoline that was not originally in the crude petroleum could be manufactured.

At first cracked gasoline was regarded as an inferior product because of its comparative instability on storage, but as more gasoline was required the petroleum industry revolved around processes by which this material could be produced (e.g.,catalytic cracking, thermal and catalytic reforming, hydrocracking, alkylation, and polymerization) and the problem of storage instability was addressed and resolved.

Because of the differences in composition of various gasolines, gasoline blending is necessary. The physical process of blending the components is simple but determination of how much of each component to include in a blend is much more difficult. The operation is carried out by simultaneously pumping all the components of a gasoline blend into a pipeline that leads to the gasoline storage, but the pumps must be set to automatically deliver the proper portion of each component. Baffles in the pipeline are often used to mix the components as they travel to the storage tank.

Selection of the components and their proportions in a blend are the most complex problems in a refinery. Blending of up to a dozen different hydrocarbon streams may be needed to produce quality gasoline (Table 1). Each property of each stream is a variable and the effect of the product gasoline can be considerable. For example, the low octane number of straight-run naphtha limits its use as a gasoline component although its other properties may make it desirable. The problem is further complicated by changes in the properties of the component streams due to processing changes. For example, an increase in cracking temperature will produce a smaller volume of a higher octane cracked naphtha, but before this cracked naphtha can be included in a blend, adjustments must be made in the proportions of the other hydrocarbon components. Similarly, the introduction of

Table 1 Component Streams for Gasoline

Stream	Producing process	Boiling range (°C)	Boiling range (°F)
Paraffinic			
Butane	Distillation Conversion	0	32
Isopentane	Distillation Conversion Isomerization	27	81
Alkylate	Alkylation	40–150	105–300
Isomerate	Isomerization	40–70	105–160
Straight-run naphtha	Distillation	30–100	85–212
hydrocrackate	Hydrocracking	40–200	105–390
Olefinic			
Catalytic naphtha	Catalytic cracking	40–200	105–390
Steam-cracked naphtha	Steam-cracking	40–200	105–390
Polymer	Polymerization	60–200	140–390
Aromatic			
Catalytic reformate	Catalytic reforming	40–200	105–390

new processes and changes in the specifications of the finished gasoline dictate reevaluation of the components that make up the gasoline.

Aviation gasolines—now usually found in use in light aircraft and older civil aircraft—have narrower boiling ranges than conventional (automobile) gasoline, i.e., 38–170°C (100–340°F), compared to -1 to 200°C (30 to 390°F) for automobile gasolines. The narrower boiling range ensures better distribution of the vaporized fuel through the more complicated induction systems of aircraft engines. Since aircraft operate at altitudes where the prevailing pressure is less than the pressure at the surface of the earth [pressure at 17,500 ft is 7.5 psi (0.5 atmosphere) compared to 14.8 psi (1.0 atmosphere) at the surface of the earth], the vapor pressure of aviation gasolines must be limited to reduce boiling in the tanks, fuel lines, and carburetors.

II. SOLVENTS

Petroleum naphthas have been available since the early days of the petroleum industry. They are valuable as solvents because of their nonpoisonous character and good dissolving power. The wide range of naphthas available and the varying degree of volatility possible offer products suitable for many uses.

Petroleum naphtha is a generic term which is applied to refined, partly refined, or unrefined petroleum products. Naphthas are prepared by any one of several methods, including:

1. Fractionation of distillates or even crude petroleum
2. Solvent extraction
3. Hydrogenation of distillates
4. Polymerization of unsaturated (olefinic) compounds
5. Alkylation processes

The naphtha may also be a combination of product streams from more than one of these processes.

The main uses of petroleum naphthas fall into the general areas of (1) solvents (diluents) for paints, etc., (2) drycleaning solvents, (3) solvents for cutback asphalts, (4) solvents for the rubber industry, and (5) solvents for industrial extraction processes. Turpentine, the older, more conventional solvent for paints, has now been almost completely replaced by the cheaper and more abundant petroleum naphthas.

III. KEROSENE

Kerosene was the major refinery product before the onset of the "automobile age," but now kerosene might be termed as one of several other petroleum products after gasoline. Kerosene originated as a straight-run (distilled) petroleum fraction which boiled between approximately 205 and 260°C (400--500°F). In the early days of petroleum refining some crude oils contained kerosene fractions of very high quality, but other crudes, such as those having a high proportion of asphaltic materials, must be thoroughly refined to remove aromatics and sulfur compounds before a satisfactory kerosene fraction can be obtained.

The kerosene fraction is essentially a distillation fraction of petroleum. The quantity and quality of the kerosene vary with the type of crude oil; some crude oils yield excellent kerosene but others produce kerosene that requires substantial refining. Kerosene is a very stable product, and additives are not required to improve its quality. Apart from the removal of excessive quantities of aromatics, kerosene fractions may need only a lye (alkali) wash if hydrogen sulfide is present.

IV. FUEL OILS

Fuel oils are classified in several ways, but generally they may be divided into two main types: distillate fuel oils and residual fuel oils. Distillate fuel oils are vaporized and condensed during a distillation process; they have a definite boiling range and do not contain high-boiling oils or asphaltic components. A fuel oil that contains any amount of the residue from crude distillation or thermal cracking is a residual fuel oil. The terms distillate fuel oil and residual fuel oil are losing their significance, since fuel oils are now made for specific uses and may be either distillates, residuals, or mixtures of the two. The terms domestic fuel oils, diesel fuel oils, and heavy fuel oils are more indicative of the uses of fuel oils.

Domestic fuels are those used primarily in the home and include kerosene, stove oil, and furnace fuel oil. Diesel fuel oils are also distillate fuel oils, but residual oils have been successfully used to power marine diesel engines, and mixtures of distillates and residuals have been used on locomotive diesels. Heavy fuel oils include a variety of oils ranging from distillates to residual oils that must be heated to 260°C (500°F) or higher before they can be used. In general, heavy fuel oils consist of residual oils blended with distillates to suit specific needs. Included among heavy fuel oils are various industrial oils; when used to fuel ships, heavy fuel oil is called bunker oil.

Stove oil is a straight-run (distilled) fraction from crude oil whereas other fuel oils are usually blends of two or more fractions. The straight-run fractions available for blending into fuel oils are heavy naphtha, light and heavy gas oils, and residua. Cracked fractions such as light and heavy gas oils from catalytic cracking, cracking coal tar, and fractionator bottoms from catalytic cracking may also be used as blends to meet the specifications of the different fuel oils.

Heavy fuel oils usually contain residuum which is mixed (cut back) to a specified viscosity with gas oils and fractionator bottoms. For some industrial purposes where flames or flue gases contact the product (ceramics, glass, heat treating, open hearth furnaces), fuel oils must be blended to contain minimum sulfur contents; low-sulfur residues are preferable for these fuels.

The manufacture of fuel oils at one time largely involved using what was left after removing desired products from crude petroleum. Now fuel oil manufacture is a complex matter of selecting and blending various petroleum fractions to meet definite specifications.

V. LUBRICATING OILS

After kerosene, the early petroleum refiners wanted paraffin wax for the manufacture of candles, and at first lubricating oils were byproducts of paraffin wax manufacture. The preferred lubricants in the 1860s were lard oil, sperm oil, and tallow, and the demand that existed for kerosene did not develop for petroleum-derived lubricating oils. However, as the trend to heavier industry increased, the demand for lubricating oils increased, and after the 1890s petroleum largely replaced animal oils and vegetable oils as the source of lubricants.

Lubricating oils are distinguished from other fractions of crude oil by their unusually high (>400°C, >750°F) boiling point, as well as their high viscosity. Materials suitable for the production of lubricating oils are comprised principally of hydrocarbons containing from 25 to 35 carbon atoms per molecule, whereas residual stocks may contain hydrocarbons with 50–80 carbon atoms per molecule.

Lubricating oil manufacture was well established by 1880 and the method depended on whether the crude petroleum was processed primarily for kerosene or for lubricating oils. Usually the crude oil was processed for kerosene and primary distillation separated the crude into three fractions: naphtha, kerosene, and a residuum. To increase the production of kerosene, the cracking distillation technique was used, and this converted a large part of the gas oils and lubricating oils into kerosene.

The development of vacuum distillation provided the means of separating more suitable lubricating oil fractions with predetermined viscosity ranges and removed the limit on the maximum viscosity that might be obtained in a distillate oil. Vacuum distillation prevented residual asphaltic material from contaminating lubricating oils but did not remove other undesirable materials such as acidic components or components that caused the oil to thicken excessively when cold and to become very thin when hot.

Lubricating oils may be divided into many categories according to the types of service they are intended to perform. However, there are two main groups: (1) oils used in intermittent service, such as motor and aviation oils, and (2) oils designed for continuous service such as turbine oils.

Oils used in intermittent service must show the least possible change in viscosity with temperature and these oils must be changed at frequent intervals to remove the foreign matter that is collected during service. The stability of such oils is therefore of less importance than the stability of oils used in continuous service for prolonged periods without renewal. Oils used in continuous service must be extremely stable because the engines in which they are used operate at fairly constant temperature without frequent shutdown.

Grease is a lubricating oil to which a thickening agent has been added for the purpose of holding the oil to surfaces that must be lubricated. The most widely used thickening agents are soaps of various kinds and grease manufacture is essentially the mixing of soaps with lubricating oils.

The soaps used in grease making are usually made in the grease plant and usually in a grease-making kettle. Soap is made by chemically combining a metal hydroxide with a fat or fatty acid:

$$\underset{\text{Fatty Acid}}{R{\cdot}CO_2H} + NaOH \longrightarrow \underset{\text{Soap}}{R{\cdot}CO_2^{-}Na^{+}} + H_2O$$

The most common metal hydroxides used for this purpose are calcium hydroxide (lye), lithium hydroxide, and barium hydroxide. Fats are chemical combinations of fatty acids and glycerine:

$$\underset{\text{Fatty Acid}}{3\ R{\cdot}CO_2H} + \underset{\text{Glycerine}}{\begin{array}{c} CH_2{\cdot}OH \\ | \\ CH{\cdot}OH \\ | \\ CH_2{\cdot}OH \end{array}} \longrightarrow \underset{\text{Fat}}{\begin{array}{c} CH_2{\cdot}O{\cdot}COR \\ | \\ CH{\cdot}O{\cdot}COR \\ | \\ CH_2{\cdot}O{\cdot}COR \end{array}}$$

If a metal hydroxide is reacted with a fat, a soap containing glycerine is formed. Frequently, a fat is separated into its fatty acid and glycerine components and only the fatty acid portion used to make soap. Commonly used fats for grease-making soaps are cottonseed oil, tallow, lard, and degras. Among the fatty acids used are stearic acid (from tallow), oleic acid (from cottonseed oil), and animal fatty acids (from lard).

To make grease, the soap is dispersed in the oil as fibers of such a size that it may only be possible to detect them by microscopy. The fibers form a matrix for the oil and the consistency, texture, bleeding characteristics, and the other properties of grease are dictated by the type, amount, size, shape, and distribution of the soap fibers. Greases may contain from 50 to 30% soap and, although the fatty acid influences the properties of a grease, the metal in the soap has the most important effect. For example, calcium soaps form smooth, buttery greases that are resistant to water but are limited in use to temperatures under about 95°C (200°F). Soda (sodium) soaps form fibrous greases that disperse in water but which can be used at temperatures well over 95°C (200°F). Barium and lithium soaps form greases similar to those from calcium soaps but they can be used at both high and very low temperatures; hence barium and lithium soap greases are known as multipurpose greases.

The soaps may be combined with any lubricating oil from a light distillate to a heavy residual oil. The lubricating value of the grease is chiefly dependent on the quality and viscosity of the oil. In addition to soap and oil, greases may also contain various additives which are used to improve the ability of the grease to stand up under extreme bearing pressures, to act as a rust preventive, and to reduce the tendency of oil to seep or bleed from a grease. Graphite, mical, talc, or fibrous material may be added to greases that are used to lubricate rough machinery to absorb the shock of impact while other chemicals can make a grease more resistant to oxidation or modify the structure of the grease.

The older, more common method of grease making is a batch method, but grease is also made by a continuous method which involves soap manufacture in a series (usually three) of retorts. Soap-making ingredients are changed into one retort while soap is made in the second retort. The third retort contains finished soap which is pumped through a mixing device where the soap and the oil are brought together and blended. The mixer continuously discharges finished grease into suitable containers.

VI. WAXES

Petroleum waxes come in two general types: the paraffin waxes in petroleum distillates and the microcrystalline waxes in petroleum residua.

The melting point of wax is not directly related to its boiling point because waxes contain hydrocarbons of different chemical structures. Nevertheless, waxes are graded according to their melting point and oil content.

Paraffin wax is a solid crystalline mixture of straight-chain (normal) hydrocarbons ranging from C_{20} to C_{30} and higher. Wax constituents are solid at ordinary temperatures (25°C; 77°F) whereas petrolatum (petroleum jelly) does contain both solid and liquid hydrocarbons.

Wax production by "wax sweating" was originally used in Scotland to separate wax fractions with various melting points from the wax obtained from shale oils. Wax sweating is still used to some extent but is being replaced by the more convenient wax recrystallization process. In wax sweating, a cake of slack wax is slowly warmed to a temperature at which the oil in the wax and the lower melting waxes become fluid and drip (or sweat) from the bottom of the cake, leaving a residue of higher melting wax.

The amount of oil separated by sweating is now much smaller than it used to be due to the development of highly efficient solvent dewaxing techniques. Wax sweating is now more concerned with the separation of slack wax into fractions with different melting points. A wax sweater consists of a series of about nine shallow pans arranged one above the other in a sweater house or oven, and each pan is divided horizontally by a wire screen. The pan is filled to the level of the screen with cold water. Molten wax is then introduced, allowed to solidify, and the water is then drained from the pan leaving the wax cake supported on the screen. A single sweater oven may contain more than 600 bbl of wax and steam coils arranged on the walls of the oven slowly heat the wax cakes, allowing oil and the lower melting waxes to sweat from the cakes and drip into the pans. The first liquid removed from the pans is called "foots" oil, which melts at 38°C (100°F) or lower, and is followed by "interfoots" oil, which melts in the range of 38–44°C (100–112°F). Crude scale wax next drips from the wax cake and consists of wax fractions with melting points over 44°C (112°F).

Wax recrystallization, like wax sweating, separates wax into fractions but, instead of relying on differences in melting points, the process makes use of the different solubilities of the wax fractions in a solvent such as a ketone. When a mixture of ketone and wax is heated, the wax usually dissolves completely, and if the solution is cooled slowly, a temperature is reached at which a crop of wax crystals is formed. These crystals will all be of the same melting point, and if they are removed by filtration, a wax fraction with a specific melting point will be obtained. If the clear filtrate is cooled further, a second batch of wax crystals with a lower melting point will be obtained. Thus, by alternate cooling and filtration, the wax can be subdivided into a large number of wax fractions, each with different melting points.

This method of producing wax fractions is much faster and more convenient than sweating and results in a much more complete separation of the various fractions. Furthermore, recrystallization can also be applied to the microcrystalline waxes obtained from intermediate and heavy paraffin distillates, which cannot be sweated. Indeed, the microcrystalline waxes have higher melting points and differ in their properties from the paraffin waxes obtained from light paraffin distillates and, thus, wax recrystallization has made new kinds of waxes available.

VII. ASPHALT

Asphalt is a product of many petroleum refineries (Barth, 1962). It may be residual asphalt made up of the nonvolatile hydrocarbons in the feedstock, along with similar materials produced by thermal alteration during the distillation sequences, or it may be produced by air-blowing an asphaltic residuum (Figure 2).

Asphalt is a residuum and cannot be distilled even under the highest vacuum because the temperatures required to volatilize the residuum promote the formation of coke. Asphalts have complex chemical and physical compositions which usually vary with the source of the crude oil.

Asphalt manufacture is in essence a matter of distilling everything possible from crude petroleum until a residue with the desired properties is obtained. This is usually done by stages; crude distillation at atmospheric pressure removes the lower boiling fractions and yields a reduced crude which may contain higher boiling (lubricating) oils, asphalt, and even wax. Distillation of the reduced crude under vacuum removes the oils (and wax) as volatile overhead products and the asphalt remains as a bottom (or residual) product. At this stage the asphalt is frequently (and incorrectly) referred to as pitch.

There are wide variations in refinery operations and in the types of crude oils, so that different asphalts will be produced. Asphalts of intermediate softening points may be made by blending with higher and lower softening point asphalts. If lubricating oils are not required, the reduced crude may be distilled in a flash drum which is similar to a distillation tower (Chap. 5, Sec. III) but has few, if any, trays. Asphalt descends to the base of the flasher as the volatile components pass out of the top. Asphalt is also produced by propane deasphalting (Figure 6 in Chap. 7). Asphalt can be made softer by blending

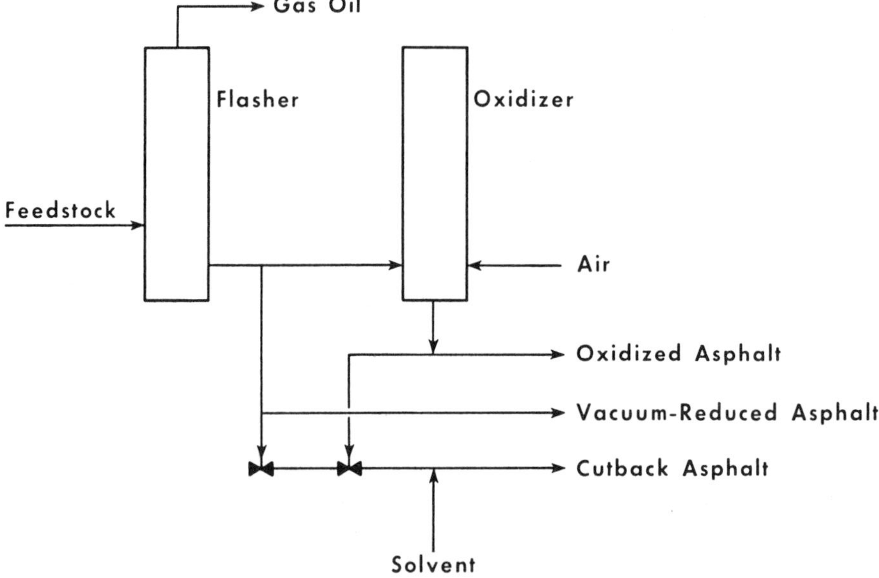

Figure 2 Asphalts are produced by several methods from the nonvolatile asphaltic residua obtained from crude oils.

the hard asphalt with the extract obtained in the solvent treatment of lubricating oils. On the other hand, soft asphalts can be converted into harder asphalts by oxidation (air-blowing).

Road oils are liquid asphalt materials intended for easy application to earth roads. They provide a strong base or a hard surface and will maintain a satisfactory passage for light traffic. Liquid road oils, cutbacks, and emulsions are of recent date, but use of asphaltic solids for paving goes back to the European practices of the early 1800s.

Cutback asphalts are mixtures in which hard asphalt has been diluted with a lighter oil to permit application as a liquid without drastic heating. They are classified as rapid, medium, and slow curing, depending on the volatility of the diluent, which governs the rate of evaporation and consequent hardening.

An asphaltic material may be emulsified with water to permit application without heating. Such emulsions are normally of the oil-in-water type. They reverse or break on application to a stone or earth surface, so that the oil clings to the stone and the water disappears. In addition to their usefulness in road and soil stabilization, they are useful for paper impregnation and waterproofing. The emulsions are chiefly

(1) the soap or alkaline type and (2) the neutral or clay type. The former break readily on contact, but the latter are more stable and probably lose water mainly by evaporation. Good emulsions must be stable during storage or freezing, suitably fluid, and amenable to control for speed of breaking.

As already pointed out (Table 4 in Chap. 1), asphalt has been known and used for six millennia or so. Nevertheless, it is only in the twentieth century that asphalt has grown to be a valuable refinery product. In the post-1980 period, a shortage of good-quality asphalts has developed in the United States. This is due in no short measure to the tendency of refineries in the post-1973 era to produce as much liquid fuel (e.g., gasoline) as possible. Thus, residua that would have once been used for asphalt manufacture are now being used to produce liquid fuels (and coke).

VIII. COKE

Petroleum coke is the residue left by the destructive distillation of petroleum residua. The coke formed in catalytic cracking operations is usually nonrecoverable because of adherence to the catalyst, as it is often employed as fuel for the process.

The composition of petroleum coke varies with the source of the crude oil but, in general, large amounts of high molecular weight complex hydrocarbons (rich in carbon but correspondingly poor in hydrogen) make up a high proportion. The solubility of petroleum "coke" in carbon disulfide has been reported to be as high as 50–80%, but this is in fact a misnomer since the coke is an insoluble, honeycomb-type material that is the end product of thermal processes.

Petroleum coke is employed for a number of purposes, but its major use is in the manufacture of carbon electrodes for aluminum refining, which requires a high-purity carbon--low in ash and sulfur-free. In addition, petroleum coke is employed in the manufacture of carbon brushes, silicon carbide abrasives, and structural carbon (pipes, Rashig rings, etc.), as well as calcium carbide manufacture from which acetylene is produced.

In the early days of refining (pre-1920s), coke was an unwanted refinery byproduct and was usually discarded. As more coking operations became integral parts of refinery operations in the post-1920 era, coke was produced in significant amounts by many refiners. The demand for petroleum coke as an industrial fuel and for graphite/carbon electrode manufacture increased after World War II.

The use of coke as a fuel must proceed with some caution with the acceptance by refiners of the heavier crude oils as refinery feedstocks.

The higher contents of sulfur and nitrogen in these oils means a product coke that contains substantial amounts of sulfur and nitrogen. Both of these elements will produce unacceptable pollutants—sulfur oxides and nitrogen oxides during combustion. These elements must also be regarded with caution in any coke that is scheduled for electrode manufacture, and removal procedures for these elements are continually being developed.

IX. ACID SLUDGE

The sludges produced during the use of sulfuric acid as a treating agent are mainly of two types: (1) those from light oils (gasoline and kerosene) and (2) those from lubricating stocks, medicinal oils, and the like. In the treatment of the latter oils, it appears that the action of the acid causes precipitation of asphaltenes and resins as well as the solution of color-bearing and sulfur compounds. Sulfonation and oxidation-reduction reactions also occur but to a lesser extent since much of the acid can be recovered. In the desulfurization of cracked distillates, however, chemical interaction is more important and polymerization, ester formation, aromatic olefin condensation, sulfonation, etc., also occur. Nitrogen bases are neutralized, and naphthenic acids are dissolved by the acid; thus, the composition of the sludge is complex and depends largely on the oil treated, the acid strength, and the temperature.

The action of sulfuric acid on hydrocarbons is indeed quite complex but it is obvious that reaction will occur readily with compound types such as aromatics and those tertiary carbon atoms in naphthenic rings, which are both present in the lubricating fractions of petroleum. Ordinarily, a charge stock for sulfuric acid treatment will have already been refined by solvent extraction with, say, furfural to remove those more highly aromatic constituents (Speight, 1980). Thus, the remaining hydrocarbons, which give higher yields of better sulfonates, are those in which aromatic rings are entirely absent or are low in proportion relative to the naphthene rings and paraffinic chains and, hence, the preferred sulfonic acids of commerce are probably naphthene sulfonic acids:

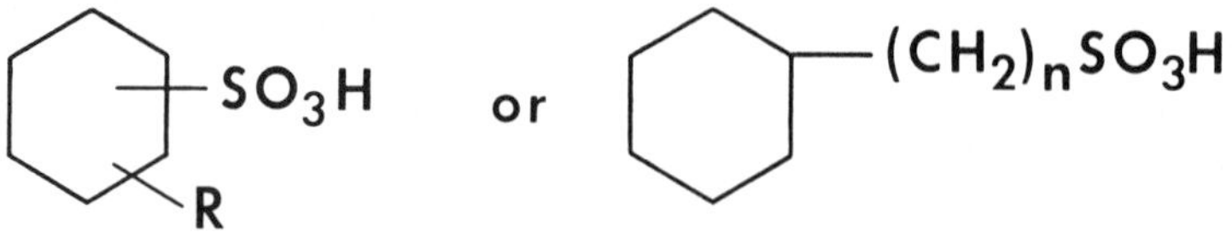

SO_3H (R) or $(CH_2)_nSO_3H$

where R is an alkyl group.

Sulfonic acids are also used as detergents which are made by the sulfonation of alkylated benzenes. The number, size, and structure of the alkyl side chains are important in determining the performance of the finished detergent.

Two general methods are applied for the recovery of sulfonic acids from sulfonated oils and their sludges. In one case (1) the acids are selectively removed by adsorbents or by solvents (generally low molecular weight alcohols) and in the other (2) the acids are obtained by salting out with organic salts or bases.

Petroleum sulfonic acids may be roughly divided into those soluble in hydrocarbons and those soluble in water. Because of their color, hydrocarbon-soluble acids are referred to as "mahogany" acids and the water-soluble acids as "green" acids. The composition of each type varies with the nature of the oil sulfonated and the concentration of the acids produced. In general, those formed during light acid treatment are water-soluble, while oil-soluble acids result from more drastic sulfonation.

The salts of mixed petroleum sulfonic acids have many commercial applications. They find use as anticorrosion agents, leather softeners, and flotation agents and have been used in place of red oil (sulfonated castor oil) in the textile industry. Lead salts of the acids have been employed in greases as extreme pressure agents and alkyl esters have been used as alkylating agents. The alkaline earth metal (Mg, Ca, Ba) salts are used in detergent compositions for motor oils and the alkali metal (K, Na) salts as detergents in aqueous systems.

11

Chemicals from Petroleum

I	Historical Background	277
II	Process Sources of Petrochemicals	278
III	Hydrocarbons as Chemical Intermediates	287
	A. Feedstocks for Alkylations	287
	B. Feedstocks for Oxygenates	288
	C. Feedstocks for Nitrogenates	297
	D. Polyolefins	302
	E. Polyaromatics	302
	F. Acrylates	303
	G. Others	303
	H. Engineering Plastics	304

I. HISTORICAL BACKGROUND

Almost two centuries ago, Thomas Jefferson wrote to a fellow Virginian, Dr. Thomas Ewell: "Of the importance of turning a knowledge of chemistry to household purposes, I have been long satisfied. The common herd of philosophers seem to write only for one another. The chemists have filled volumes on the composition of a thousand substances of no sort of importance to the purposes of life; while the arts of making bread, butter, cheese, vinegar, soap, beer, cider, etc., remain unexplained." The third president of the United

States was, indeed, sophisticated in his appreciation of chemistry and chemicals. In addition to his plea for clarity in terms of the teaching and appreciation of organic chemistry, Jefferson was one of the earliest consumers of young E. I. DuPont's high-quality gunpowder made in the early nineteenth century on the banks of the Brandywine across the road from the site of the world-famous DuPont Experimental Station.

While the petroleum era was ushered in by the 1859 finding at Titusville, Pennsylvania, the flourishing of chemicals from petroleum has been only since the early twentieth century. Natural gas and petroleum are, in fact, our chief sources of hydrocarbons. Natural gas is quite variable in composition but the major constituent (>60%) is methane. Other components are the homologous alkanes, ethane, propane, and higher hydrocarbons. In terms of volume, most of the natural gas produced is used for fuel, although a substantial amount is used as raw material for the synthesis of various types of chemicals.

Petroleum is a complex liquid mixture of organic compounds. It is generally defined (Webster, 1983) as "an oily flammable bituminous liquid that may vary from almost colorless to black, occurs in many places in the upper strata of the earth, is a complex mixture of hydrocarbons with small amounts of other substances, and is prepared for use as gasoline, naphtha, or other products by various refining processes." There are substantial differences between the various types of petroleum (Chap. 3, Secs. I and II), which vary from "sweet" to "sour" to "heavy" to "light." Each, however, is a very complex nature of hydrocarbons which may be aliphatic, alicyclic, or aromatic in varying proportions. In addition to carbon and hydrogen, petroleum typically contains 1–6% sulfur and nitrogen and lesser amounts of oxygen. There is considerable evidence for the presence of some metals, chiefly nickel and vanadium, as chelating metals in porphyrin or related structures.

The chemical industry depends very heavily on petroleum and natural gas and natural gas liquids as sources of raw materials. It is likely that in excess of 90% of the literally thousands of different basic organic chemicals employed today are derived from these sources. The petrochemical industry has grown with the petroleum industry (Goldstein, 1949; Steiner, 1961; Hahn, 1970) and is considered by some to be a mature industry. However, as is the case with the petroleum industry itself, the petrochemical industry must also keep pace with the latest trends in changing crude oil types and must also evolve to meet changing technological needs.

II. PROCESS SOURCES OF PETROCHEMICALS

The starting materials for the petrochemical industry (Table 1) are obtained from crude petroleum in one of two general ways. They may be

Table 1 Hydrocarbon Intermediates Used in the Petrochemical Industry

	Hydrocarbon type		
Carbon number	Saturated	Unsaturated	Aromatic
1	Methane		
2	Ethane	Ethylene Acetylene	
3	Propane	Propylene	
4	Butane	*n*-Butene Isobutene Butadiene	
5	Pentane	Isopentene (iso-amylene) Isoprene	
6	Hexane Cyclohexane	Methylpentene Cyclohexenes	Benzene
7		Mixed heptenes	Toluene
8		Diisobutylene	Xylene Ethylbenzene Styrene
9			Cumene
12		Propylene tetramer Triisobutylene	
18			Dodecylbenzene
6–18		*n*-Olefins	
11–18	*n*-Paraffins		

present in the virgin petroleum and as such are isolated by physical methods such as distillation (Chap. 5) or extraction (Chap. 7). On the other hand, they may be present in trace amounts, if at all, and are synthesized during the refining operations. In fact, unsaturated (olefinic) hydrocarbons which are not usually present in virgin petroleum are nearly always manufactured as intermediates during the various refining sequences (Table 2).

Table 2 Sources of Petrochemical Intermediates

Hydrocarbon	Source
Methane	Natural gas
Ethane	Natural gas
Ethylene	Cracking Processes
Propane	Natural gas, catalytic reforming, cracking processes
Propylene	Cracking processes
Butane	Natural gas, reforming and cracking processes
Butene(s)	Cracking processes
Cyclohexane	Distillation
Benzene	Catalytic reforming
Toluene	Catalytic reforming
Xylene(s)	Catalytic reforming
Ethylbenzene	Catalytic reforming
Alkylbenzenes	Alkylation
$>C_9$	Polymerization

The manufacture of chemicals from petroleum is based on the ready response of the various compound types to basic chemical reactions such as oxidation, halogenation, nitration, dehydrogenation, addition, polymerization, alkylation, etc. The low molecular weight paraffins and olefins, as found in natural gas and refinery gases, and the simple aromatic hydrocarbons have so far been of the most interest because it is these individual species which can be readily isolated.

When petroleum and natural gas (liquids) are converted to chemicals or chemical intermediates, inexpensive abundant raw materials are converted to more valuable products. The marriage between the petroleum business and the petrochemical business is easily seen in two facts: (1) chemical raw materials costs parallel movements in the cost of crude oil and (2) (petrochemical) installations are very frequently located as close as possible to a refinery site, if not on the very same site.

As far as can be determined, the first large-scale petrochemical process was the sulfuric acid absorption of propylene ($CH_3CH{=}CH_2$) from refinery cracked gases to produce isopropyl alcohol:

$$CH_3CH{=}CH_2 \xrightarrow{H_2SO_4} CH_3CHOHCH_3$$

This process was operated at Exxon's Bayway plant (Linden, NJ) as early as 1918. During the next 20 years, processes for the production of such simple molecules as ethylene glycol, acetone, and ethanol (all two- or three-carbon oxygen-containing molecules based ultimately on ethylene or propylene) as well as ammonia (NH_3) were implemented.

Although they are relatively unreactive organic molecules, paraffin hydrocarbons are known to undergo thermolysis when treated under high-temperature, low-pressure vapor phase conditions. The cracking chemistry of petroleum constituents has been extensively studied (Albright and Crynes, 1976; Oblad et al., 1979). Cracking is the major process for generating ethylene and the other olefins which are the reactive building blocks of the petrochemical industry.

In addition to thermal cracking, other very important processes which generate sources of hydrocarbon raw materials for the (petro)-chemical industry include catalytic reforming, alkylation, dealkylation, isomerization, and polymerization. Together these (and other smaller) processes accounted for the production of 180 *billion* pounds of organic chemicals in the United States in 1984 (*Chem. Eng. News*, 1985a). This growth is truly remarkable for an industry which had its modest beginnings at the end of the First World War.

Olefins (C_nH_{2n}) are the basic building blocks for a host of chemical syntheses. These unsaturated materials enter into polymers, plastics, rubbers, and with other reagents react to form a wide variety of useful compounds including alcohols, epoxides, amines, and halides.

Cracking reactions involve the cleavage of carbon-carbon bonds with the resulting redistribution of hydrogen to produce smaller molecules. Thus, cracking of petroleum or petroleum fractions is a process by which larger molecules are converted into smaller, lower boiling molecules. In addition, cracking generates two molecules from one, with one of the product molecules being paraffinic (saturated) and the other olefinic (unsaturated) (Chapter 6).

At the high temperatures of refinery crackers [usually >500°C (950°F)], there is a thermodynamic driving force for the generation of more molecules from fewer molecules, i.e., cracking is favored. Unfortunately, in the cracking process, certain of the products interact with one another to produce products of increased molecular weight from that in the original feedstock. Thus, some products are taken off from the cracker as useful light products (olefins, gasoline, etc.), while collisions between various intermediate species can and do lead to combination products ultimately generating fuel oil, tar, and even coke. Bond scission reactions occur at these high temperatures leading to the formation of free radicals (species where one free electron resides on a carbon atom). These free radicals are very reactive and it is their subsequent reactions which ultimately determine product mixes. An illustrative example of the manifold reaction schemes possible is shown below for the thermolysis of butane.

Initiation

$$CH_3CH_2CH_2CH_3 \rightarrow CH_3\cdot + CH_3CH_2CH_2\cdot$$

Propagation

$$CH_3\cdot + CH_3CH_2CH_2CH_3 \rightarrow CH_4 + \underset{\underline{1}}{CH_3CH_2CH\cdot CH_3}$$

$$\underline{1} \longrightarrow H\cdot + CH_3CH{=}CHCH_3$$

$$\underline{1} \longrightarrow CH_3\cdot + CH_2{=}CHCH_3$$

$$\underline{1} + CH_2{=}CHCH_3 \longrightarrow C_7H_{15}\cdot \rightarrow \text{etc.}$$

Termination

$$R\cdot + R'\cdot \rightarrow R{-}R'$$

Thus, there are two general kinds of chemical reactions that occur during the cracking process:

1. The decomposition of larger molecules into smaller ones. These are primary reactions typified by the conversion (via a number of individual steps) of butane into, e.g., methane and propylene, or into ethane and ethylene, and
2. Reactions in which some of the initially formed radicals undergo combination reactions to yield higher molecular weight materials via a secondary reaction pathway. The example of the butyl radical interacting with propylene to produce a heptyl (C_7) radical shown above is an example of these "growth" reactions.

While paraffinic thermal chemistry pathways are complex and while the feed to crackers is also complex, the basic features of paraffin-cracking mechanisms have been known since the pioneering work of Rice published in 1933. The specific product distribution will, of course, be a function of the cracking conditions as well as the feedstock to the cracker.

A great deal of work has been carried out to evaluate the second order effects in paraffin pyrolysis. One significant fact is that hydrocarbon free radicals (as opposed to carbocations) are very resistant to carbon skeleton rearrangements. This translates into a result that thermal cracking does not produce any appreciable increase in the amount of branching. Thus, linear paraffins will produce (mostly) linear olefins and linear smaller paraffins. Thus normal paraffins (straight-chain paraffins) typically decompose to form an α-olefin and a shorter linear paraffin:

$$CH_3(CH_2)_xCH_3 \rightarrow CH_3(CH_2)y + CH_3(CH_2)_zCH{=}CH_2$$

feedstock *α-olefin*

where $x > y + z$.

Branched paraffins have a greater tendency to lose methane than do straight-chain molecules, thus providing substantial yields of olefins having one carbon fewer than the parent hydrocarbon.

$$RCH_2CH(CH_3)CH_2CH_3 \rightarrow CH_4 + RCH_2CH{=}CHCH_3$$

Thus, additional linearity is achieved in the feed after thermal cracking according to this mechanism.

Cyclic paraffins (also called "naphthenes") undergo different kinds of reactions upon pyrolysis. According to Virk, Korosi and Woebcke (Oblad et al., 1979), hydrocarbon pyrolysis involves free radical and pericyclic types of reactions, reactions whose principles are well enough understood (Linstead, 1942; Kochi, 1973) that likely pyrolysis pathways for any particular molecule can be described. Cyclohexane, decahydronaphthalene (decalin), and perhydrophenanthrene simulate unsubstituted cyclic paraffins found in the naphtha, kerosene, and atmospheric gas-oil boiling ranges of petroleum.

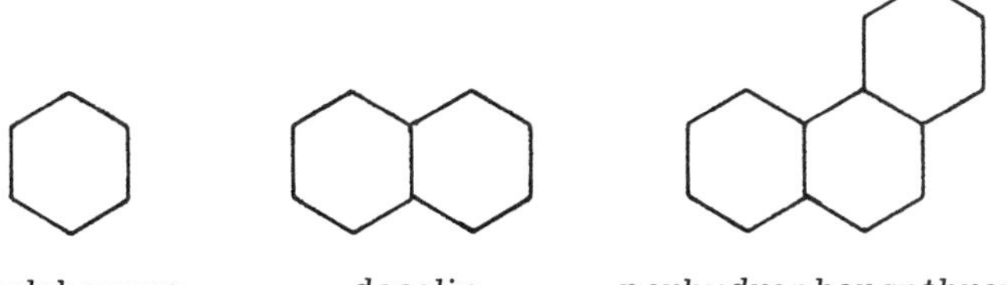

cyclohexane *decalin* *perhydrophenanthrene*

Pyrolysis of each of these three naphthenes at about 850°C (1560°F) yields ethylene as the major product in each case. Other products are also produced but only in relatively low yield. For cyclohexane the next largest products were 1,3-butadiene, benzene, propylene, and methane. On the other hand, for the two- and three-ring systems, benzene was the next largest product after ethylene, with toluene, methane, and propylene following in order. In addition, Virk showed that hydroaromatic molecules tend to revert to their fully aromatic form upon pyrolysis, thereby contributing mainly to the relatively undesirable fuel oil fraction. He showed that the incremental yield of the desired olefinic products increases most strongly as the feedstock aromatics content approaches zero.

Aromatic compounds are quite stable under moderate cracking temperatures (in fact, toluene is often used as an inert diluent for thermal chemistry experiments). Alkylated aromatics, in a manner analogous to alkylated naphthenes, are prone to dealkylation reactions rather than to aromatic ring destruction.

The chemical engineering community has expended enormous effort, with excellent results, on the optimization of the cracking process. Extensive correlations of product distribution with cracking conditions are available. The fundamental cracking variables again appear to be the nature of the feedstock, coil outlet temperature, residence time in the coil, hydrocarbon partial pressure, and temperature. The effect of feedstock composition has also been studied and correlations made.

As mentioned earlier, the major products from the cracker are olefins, the basic building blocks of the petrochemical industry. Ethylene, which is the largest volume organic chemical (projected worldwide production in 1987 to exceed 33 billion pounds) manufactured in the United States, finds major end uses in polyethylene, ethanolamine, ethylene oxide (which is used to make ethylene glycol, acrylate esters, surfactants, etc.), ethyl alcohol (which is converted to acetaldehyde, acetic acid, and anhydride), ethylene dichloride (polyvinylchloride and other vinyl resins), trichloroethylene (used as a solvent), ethyl chloride which finds use (although diminishing) in tetraethyl lead, fluoroethanes used as refrigerants and aerosols, vinylidene chloride (sold as Saran), converted to styrene, and many many other uses.

The interest, then, in thermal reactions of hydrocarbons has been high since the 1920s when alcohols were produced from the ethylene and propylene formed during petroleum cracking. The range of products formed from petroleum pyrolysis has widened over the past six decades to include the main chemical building blocks. These include ethane, ethylene, propane, propylene, the butanes, butadiene, and aromatics. Additionally, other commercial products from thermal reactions of petroleum include cokes and carbons, pitches and asphalts.

Changes in the petroleum world have brought about a variety of technical improvements in cracking furnace design over the past 10

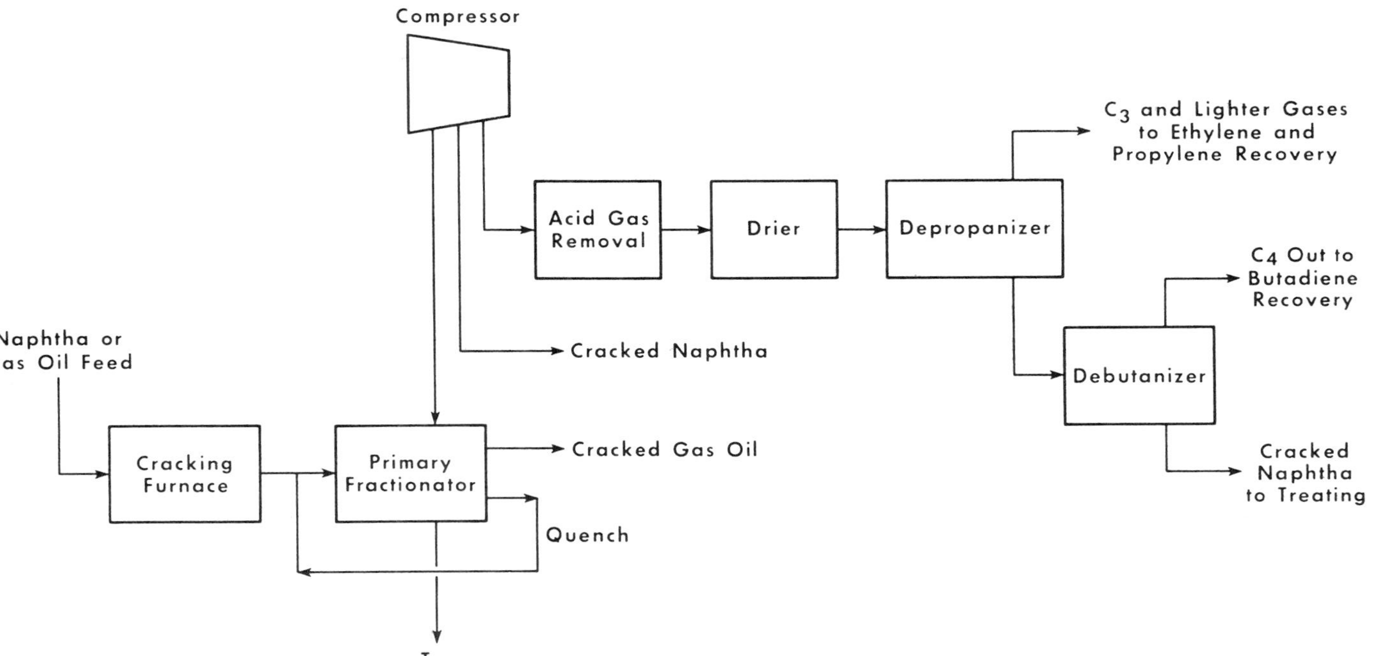

Figure 1 Schematic representation of a steam-cracking operation.

or 15 years (Figure 1). According to Lohr and Schwabe of Linde Ag (Oblad et al., 1979), chief among the critical changes have been the following:

1. Increases in costs of feedstocks requiring improved selectivity to ethylene
2. Increases in energy costs requiring higher thermal efficiencies
3. More stringent environmental regulations requiring lower noise emissions and lower NO_x emissions
4. A more diverse and demanding feedstock selection, requiring greater refinery flexibility and furnaces capable of processing a variety of feedstocks for olefin generation

Although petroleum prices have moderated as this is written from the >$25/bbl costs seen in 1986, costs are still six times higher than they were in the early 1970s and the incentives for improvements in cracking technologies remain. One major response to this challenge has been improvements in the process control of pyrolysis systems. A pyrolysis system usually consists of a cracking furnace where hydrocarbon feed is thermally converted to a mixture containing mostly olefins and aromatics. The product is sent to a quenching device, either a transfer line exchanger (TLE) or an oil quench where the olefin/aromatic stream is rapidly cooled below the thermal cracking temperature. Good control of such a system offers maximum production at minimum cost. Today's systems are sufficiently complex to justify—in fact, to demand—an on-line computer control system. Optimal pyrolysis system control requires the solution of a mathematical model and the use of an on-line computer (Albright and Crynes, 1976). For a given pyrolysis system, the selectivity is affected by TLE fouling, radiant coil fouling, total flow of hydrocarbon and dilution steam, and the ratio of hydrocarbon to dilution steam.

Optimal control strategy requires operation near system constraints, with the most important constraint in a pyrolysis system being the tube metal temperature of the radiant coil. When the radiant coil and the TLE are fouled, the inside surfaces of the radiant coil and the tube of the TLE have coke laid down. This coke acts as an insulator and retards heat transfer, and also reduces the available cross-section of flow area. As coke is deposited during a thermal cracking run, tube metal temperature of the radiant coil increases at constant levels of cracking severity and feed rates of both hydrocarbon and dilution steam. When the maximum allowable tube metal temperature is reached, the pyrolysis system is taken off-line and decoked. In practice, what happens is the feed rate is increased at a given severity, thus firing more fuel and maintaining the tube wall temperature near the maximum permitted at all times.

Thus, by increasing hydrocarbon and steam flow when the system is clean, production of olefins and aromatics can be maximized. As the system is fouled, adjustment of the total flow of dilution steam and hydrocarbon as well as adjustment of the dilution/steam ratio will allow maintenance of the selectivity for maximizing profit and/or minimizing production costs.

III. HYDROCARBONS AS CHEMICAL INTERMEDIATES

A. Feedstocks for Alkylations

Alkylation chemistry contributes to the efficient utilization of C_4 olefins generated in the cracking operations. Isobutane has been added to butenes (and other light olefins) to give a mixture of highly branched octanes (heptanes, etc.) by a process called alkylation. The reaction is thermodynamically favored at low temperatures ($<20°C$; $<68°F$) and thus very powerful acid catalysts are employed. Typically, sulfuric acid (85–100% or anhydrous hydrogen fluoride or a solid sulfonic acid are employed as the catalyst in these processes. The first step in the process is the formation of a carbocation by combination of an olefin with an acid proton:

$$(CH_3)_2C{=}CH_2 + H^+ \rightarrow (CH_3)_3C^+$$

Step 2 is the addition of the carbocation to a second molecule of olefin to form a dimer carbocation:

$$(CH_3)_3C^+ + (CH_3)_2C{=}CH_2 \rightarrow (CH_3)_3CH_2C^+(CH_3)_2$$

The next step is the key one. In the presence of isobutane [$(CH_3)_3CH$], the dimer carbocation does not deprotonate to form a C_8 olefin (an octene), nor does it undergo further growth reactions (to C_{12}, C_{16}, etc.). Rather, the isobutane rapidly transfers its tertiary hydrogen as a hydride (H^-) to the dimer carbocation to produce the saturated addition product along with a new tertiary carbocation:

$$(CH_3)_3CCH_2C^+(CH_3)_2 + H{:}C(CH_3) \rightarrow$$
$$(CH_3)_3CCH_2CH(CH_3)_2 + {}^+C(CH_3)_3$$

2,2,4-trimethylpentane

The extensive branching of the saturated hydrocarbon results in high octane. In practice, mixed butenes are employed (isobutylene, 1-butene, and 2-butene), and the product is a mixture of isomeric octanes that has an octane number of 94–94. With the phase-out of leaded additives in our motor gasoline pools, octane improvement is a major challenge for the refining industry. Alkylation is one answer. Only isobutane, and not normal butane, undergoes the hydride shift. The reason for this is that the resulting carbocation from isobutane is a tertiary carbocation, one which is stabilized. Thus the C--H bond involved in the hydride shift is a tertiary bond and is somewhat weaker than any C–H bond in normal butane, and this bond can undergo cleavage under the process conditions.

B. Feedstocks for Oxygenates

1. Methanol and Formaldehyde

Methanol is the largest volume oxygenated hydrocarbon produced in this country, with production in excess of 8 billion lb/year. Typically it is manufactured from reformed methane (or heavier hydrocarbons) reacted with carbon monoxide. The largest use, conversion to formaldehyde, is losing market share to such newer uses as methyl-*t*-butyl ether (MTBE), an octane value additive in the motor gasoline (mogas) pool, and the use of methanol itself as a fuel. Other major derivatives produced from methanol include acetic acid and chloromethanes. In addition to utilization in fuels (methanol and MTBE), major end uses in the United States for methanol derivatives include polymers used as adhesives, fibers, and plastics. Major producers of methanol are ARCO Chemical, Borden, Celanese, and Texaco (*Chem. Eng. News*, 1985b).

Formaldehyde (HCHO) is a product with substantial overcapacity, with major American producers including Borden, Celanese, DuPont, Georgia-Pacific, and Reichhold Chemicals. Formaldehyde is prepared commercially by the catalytic vapor phase oxidation of methanol using air as the oxidant and heated silver, copper, alumina, or coke as catalyst.

$$CH_3OH \xrightarrow{\text{oxidant}} CH_2{=}O$$

Reichhold uses a molybdenum iron oxide catalyst (U.S. Patent 2,849,492). Major derivatives of formaldehyde in the United States include urea-formaldehyde resins, phenol-formaldehyde resins, polyacetal resins, and butanediol. Major end uses in this country are for adhesives and plastics.

Formaldehyde is expensive to ship, thus accounting for the existence of negligible levels of foreign trade in the commodity. Furthermore, small plants are built to serve specific areas, e.g., formaldehyde may be used in the preparation of adhesive resins for particle board and plywood and the like. Formaldehyde plants then are likely to be located near mills, making such products as particle board and plywood, and these mills are typically located near appropriate timber lands. Because of the cyclical nature of wood-product production in such interest-sensitive industries as housing and nonresidential construction, it is no wonder that formaldehyde capacity and production goes through significant variations.

Formaldehyde has been listed as a carcinogen by the Environmental Protection Agency (1981). Studies are continuing on the toxicity of formaldehyde and decisions on whether to regulate the molecule in various situations will soon be made. The peak production year for formaldehyde was 1979 when a total of 6.5 billion pounds was produced.

2. Ethylene-Based Oxygenates

Ethylene oxide is the simplest molecule of a family of cyclic ethers. About 6 billion pounds of ethylene oxide was produced in 1984 (*Chem. Eng. News*, 1985b). In one of the oldest processes still accounting for a significant fraction of total production, chlorine and ethylene are passed into an absorption column countercurrent to a stream of water. The product, ethylene chlorohydrin, is removed from the bottom of the vessel as a 5% solution. Conversion to the epoxide (cyclic ether) is via heating the ethylene chlorohydrin solution to 100°C (212°F) with a slight excess of CaO slurry (lime). The product ethylene oxide is separated from the byproducts ethylene chloride, ethylene glycol, and α-chloroethyl ether by distillation.

$$CH_2{=}CH_2 \xrightarrow{Cl_2,\ H_2O} HOCH_2CH_2Cl \xrightarrow{Ca(OH)_2} \underset{\diagdown\ \diagup \atop O}{CH_2{-}CH_2}$$

More recent processes involve direct oxidation of ethylene with air or oxygen in the presence of a silver catalyst (U.S. Patent 2,960,511, 1960).

$$CH_2{=}CH_2 + 0.5\,O_2 \rightarrow \underset{\diagdown\ O\ \diagup}{CH_2{-}CH_2}$$

Ethylene oxide is used as a fumigant for foodstuffs and textiles, to sterilize surgical instruments, as an agricultural fungicide, and, most importantly, in organic syntheses. The major use is in the production of ethylene glycol. It also has major application in the manufacture of acrylonitrile and nonionic surfactants. Ethylene oxide can be highly irritating to eyes and mucous membranes. High concentrations of this material can cause pulmonary edema. Ethylene oxide is considered an industrial substance suspect of carcinogenic potential for man (American Conference of Governmental Industrial Hygienists, "Threshold Limit Values and Biological Exposure Indices for 1986–1987"). Worker exposure limits are set at a level of 1 ppm for a time weighted average (TWA). This substance is suspected of inducing cancer based on a demonstration of carcinogenesis in one or more animal species by appropriate methods.

Ethylene glycol. As mentioned above, the major end use for ethylene oxide (accounting for approximately two-thirds of all the ethylene oxide produced) is as a precursor for ethylene glycol. About 5 billion pounds of ethylene glycol was produced in 1984 (*Chem. Eng. News*, 1985b). In the 1920s, there was a program at the Mellon Institute aimed at the development of a cheaper process for acetylene manufacture. A process was developed for the thermal cracking of petroleum or natural gas to acetylene, but a large amount of ethylene was obtained as a byproduct. The conversion of ethylene to ethylene glycol was developed in order to make use of the ethylene byproduct. Although there were no important uses for ethylene glycol in the 1920s, today it is used in vast amounts (>3 billion lb/year) as a nonvolatile antifreeze for automobile radiators and as a coolant for airplane motors. It is also used as a solvent in the paint and plastics industries and in the formulation of printing inks, etc. Its derivatives, including the monoethyl ether, the monomethyl and dimethyl ethers, dioxane, and higher homologs and alkylaryl ethers of poly(ethylene glycol) are important products as well (see below).

Today much of the ethylene glycol is made via hydrolysis of ethylene oxide with either dilute sulfuric acid at around 60°C or with water at 200°C (390°F).

$$(CH_2)_2O + H_2O \xrightarrow[(390^\circ F)]{200^\circ C} HOCH_2CH_2OH$$

Ethylene glycol boils at 197°C (387°F) and can be separated from water and the byproducts diethylene and triethylene glycols by distillation. Ethylene glycol has some level of toxicity; large doses depress the central nervous system and can lead to cyanosis and respiratory failure.

When ethylene oxide is reacted with an alcohol or a phenol, a monoalkyl or monoaryl ether of ethylene glycol is generated.

$$(CH_2)_2O + ROH \rightarrow HOCH_2CH_2OR$$

$$(CH_2)_2O + ArOH \rightarrow HOCH_2CH_2OAr$$

Ethyl alcohol can be produced by a direct catalytic hydration of ethylene or via absorption of olefins into sulfuric acid to form esters, followed by dilution and hydrolysis, generally in the presence of steam. For example, ethylene is absorbed in concentrated (96–100%) sulfuric acid at a temperature of around 80°C (175°F), forming both the mono- and diethyl sulfates. Hydrolysis of the esters takes place upon dilution with water and heating.

$$CH_2{=}CH_2 \xrightarrow[H_2O]{H_2SO_4} CH_3CH_2OH$$

Conversion of ethylene to ethyl alcohol over phosphoric acid, diatomaceous earth, or promoted tungstic oxide occurs under about 100 psi pressure and at about 300°C (570°F). More than 1 billion pounds of ethanol per year is manufactured.

Ethanolamine (2-aminoethanol) is produced in large amounts by the ammonolysis of ethylene oxide. It is in widespread refinery use as an acid gas scrubber (chiefly for removal of CO_2 and H_2S).

Ethyl ether. This widely used solvent is produced on a large scale by the hydration of ethylene or by dehydration of ethyl alcohol. Both processes are carried out in the presence of concentrated sulfuric acid.

3. Oxygenates from Propylene

Products derived from propylene are analogous in many ways to those from ethylene. Thus important oxygenates resulting from propylene include isopropyl alcohol, acetone, diisopropyl ether, propylene oxide, propylene glycol, glycerine, etc.

Propylene oxide. Around 2 billion pounds of propylene oxide are produced each year. Analogous with ethylene oxide, propylene oxide is the product of base treatment of propylene chlorohydrin (2-chloro-1-propanol), which in turn is prepared from propylene in the presence of chlorine and water.

$$CH_3CH{=}CH_2 \xrightarrow{Cl_2,\ H_2O} CH_3CHOHCH_2Cl\ (90\%) + CH_3CH(Cl)CH_2OH\ (10\%)$$

Important manufacturers of propylene oxide include Dow, Union Carbide, and Arco.

Isopropyl alcohol. The absorption by sulfuric acid of propylene (derived from refinery cracked gases) to produce isopropyl alcohol at Exxon's Bayway refinery in 1918 may well have been the first large-scale petrochemical process. While sulfuric acid absorbs propylene more readily than it does ethylene, care must be taken to avoid oligomerization of the olefin. This is controlled by using relatively cool temperatures and relatively weak acid solutions.

$$CH_3CH{=}CH_2 \xrightarrow{H_2SO_4} CH_3CH(OSO_3H)CH_3 \xrightarrow{H_2O} CH_3CHOHCH_3$$

About 1.2 billion pounds of isopropanol was produced in 1984. Two major uses for isopropyl alcohol are as a solvent and as a precursor to acetone.

Acetone. Acetone is made not only by dehydrogenation of isopropyl alcohol (typically using a copper-based catalyst), but also as a side product in the conversion of cumene to phenol. In fact, the latter process is a more important source of acetone.

$$\text{Benzene} + \text{propylene} \xrightarrow{H^+} \text{Isopropylbenzene (cumene)}$$

$$\underset{\text{cumene}}{C_6H_5CH(CH_3)_2} + O_2 \longrightarrow \underset{\text{cumene hydroperoxide}}{C_6H_5C[O\text{-}OH](CH_3)_2}$$

$$C_6H_5C[O\text{-}OH](CH_3)_2 \xrightarrow{H^+} C_6H_5OH + CH_3COCH_3$$

Acetone is also produced by the high-temperature air oxidation of isopropyl alcohol, generating hydrogen peroxide as a side product.

$$(CH_3)_2CHOH + O_2 \rightarrow (CH_3)_2C{=}O + H_2O_2$$

Much of the acetone produced in the United States is consumed in the synthesis of other chemicals, including methyl isobutyl ketone and methyl methacrylate. Dimerization of acetone produces methyl isobutenyl ketone (mesityl oxide), which upon hydrogenation yields methyl isobutyl ketone. Methyl methacrylate is derived from acetone via the following scheme:

$$(CH_3)_2C{=}O + HCN \longrightarrow (CH_3)_2COHCN \xrightarrow{H^+} CH_2{=}C(CH_3)CONH_2{\cdot}H_2SO_4$$

$$CH_2{=}C(CH_3)CONH_2{\cdot}H_2SO_4 \xrightarrow[H_2SO_4]{CH_3OH} CH_2{=}C(CH_3)COOCH_3$$

Polymerization of methyl methacrylate using peroxide initiators such as *t*-butyl hydroperoxide or cumyl hydroperoxide produces poly(methyl methacrylate), a strong, transparent thermoplastic solid sold under such trade names as Lucite and Plexiglas. These kinds of polymers are also found in thermosetting enamels where the polymer is cross-linked with melamine, a cyclic triamine.

Propylene glycol is made via the hydrolysis of propylene oxide:

$$\underset{\diagdown\,O\,\diagup}{CH_3CH{-}CH_2} + H_2O \xrightarrow{H^+} CH_3CHOHCH_2OH$$

Propylene glycol has properties similar to those of ethylene glycol. Unlike ethylene glycol, however, propylene glycol is nontoxic and thus can be employed in food products and cosmetics. The likely reason for the nontoxicity is the fact that oxidation (chemical or metabolic) yields pyruvic and acetic acids, both of which are produced in the body. The oral LD_{50} in rats is 25 ml/kg (Bartsch, 1976).

An important use of propylene glycol, built on its very low toxicity, is as an antifreeze in breweries, dairy establishments, and other food-handling and packaging plants. It also finds use as a solvent for pharmaceuticals. It is used in the manufacture of synthetic resins, acts as an inhibitor of fermentation and mold growth, and acts to prevent cellophane from drying out and becoming brittle. One of the more interesting applications of propylene glycol is in aerosols used in hospitals and schools to reduce the incidence of airborne infections. The rationale is that the mist droplets take up moisture from the air and condense on bacteria, which are carried down to the floor where they are no longer so readily breathed in.

Glycerin may be derived from propylene by high-termperature chlorination to produce allyl chloride, followed by hydrolysis to allyl alcohol, then conversion with aqueous chlorine to glycerol chlorohydrin, a product that can be easily hydrolyzed to glycerol (glycerine).

$$CH_3CH{=}CH_2 + Cl_2 \rightarrow ClCH_2CH{=}CH_2$$

$$ClCH_2CH{=}CH_2 \xrightarrow{Cl_2,\ H_2O} ClCH_2CHOHCH_2OH \xrightarrow{base} HOCH_2CHOHCH_2OH$$

Glycerine has found many uses over the years; important among these are its use as solvent, emollient, sweetener, cosmetics, and precursor to nitroglycerine and other explosives.

Secondary butyl alcohol and methyl ethyl ketone (MEK). This alcohol is formed in a manner similar to that for the production of isopropyl alcohol from propylene. *sec*-Butyl alcohol is formed on adsorption of 1- or 2-butene in fairly concentrated sulfuric acid (typically in the area of 65–80%), followed by dilution and hydrolysis. Subsequent conversion of secondary butyl alcohol to methyl ethyl ketone is carried out by catalytic oxidation or by dehydrogenation.

$$\begin{matrix} CH_3CH_2CH{=}CH_2 \\ + \\ CH_3CH{=}CHCH_2 \end{matrix} \xrightarrow[H^+]{H_2O} 2CH_3CH_2CHOHCH_3$$

$$CH_3CH_2CHOHCH_3 \xrightarrow[-H_2]{} CH_3CH_2COCH_3$$

sec-Butyl alcohol is employed in the synthesis of flavors and fragrances, is used in industrial cleaners and paint removers, and is used as a solvent for many natural resins. The major use for methylethyl ketone is as a solvent in the surface-coating industry. Union Carbide, Exxon, and Shell are the major producers of MEK.

4. Higher Alcohols

There are a number of methods for preparing the higher alcohols. Commercially, the major method involves the so-called oxo reaction. The oxo or oxonation reaction involves the addition of carbon monoxide and hydrogen (synthesis gas) to an olefinic double bond in the presence of a cobalt catalyst. The active catalyst is believed to be one of the many forms of cobalt hydrocarbonyl [$HCo(CO)_4$].

A brief history of the oxo process is illustrative. The process was patented by Otto Roelen in Germany in 1938 and the first commercial plant was briefly operated by Ruhrchemie in Germany. The first plant startup in the United States occurred in Baton Rouge in 1948. The first oxoalcohol produced was isooctyl alcohol. In the 1950s Exxon Chemical commercialized isodecyl and tridecyl alcohols, and a number of oxo plants were built around the world. Most of the new capacity was for 2-ethyl hexyl alcohol production. In 1960, Exxon commercialized hexyl alcohol and in 1965 Shell (U.S.) commercialized linear detergent range oxoalcohols. In the late 1960s, Exxon commercialized hexadecyl and isononyl alcohols and in 1969 Monsanto commercialized their linear plasticizer range oxoalcohols.

The major technical advance in the 1970s occurred at the start of the decade when Union Carbide obtained a patent on rhodium oxo technology, with a plant on-stream (Puerto Rico) to produce *n*-butyl alcohol and 2-ethyl hexanol in 1976.

In the 1980s, Exxon continued to commercialize new oxoalcohols; isoheptyl in 1982 and dodecyl in 1985. Finally, in 1987 Idemitsu announced plans to commercialize detergent range linear oxoalcohols prepared via rhodium technology.

As the historical development above indicates, a wide range of olefins can undergo this reaction, with those containing terminal double bonds being the most active. Whereas straight-chain olefins add the hydrogen and carbon monoxide across the double bond in a nonstereospecific manner (propylene gives a mixture of 60% *n*- and 40% isobutyraldehyde), things are more selective in the case of branched structures. Thus isobutylene forms 95% isovaleraldehyde and only 5% trimethylacetaldehyde.

$$(CH_3)_2C{=}CH_2 \rightarrow (CH_3)_2CHCH_2CHO\ (95\%) + (CH_3)_3CCHO$$

Commercial application of the synthesis has been most successful in the manufacture of products like isooctyl alcohol from a refinery C_3 + C_4 dimer, decyl alcohol from propylene trimer, and tridecyl alcohol from the tetramer of propylene. Important outlets for the higher oxo alcohols include conversion with dibasic acids or anhydrides such as phthalic anhydride to make plasticizers and synthetic lubricants, and in sulfonation to make detergents. Plasticizers are compounds which, when mixed with a resin, make the resin pliable. For plasticizer applications branchiness is acceptable, whereas for detergents the preferred structure is straight chain. The issue in detergents is biodegradability: with an α-methyl group [$-CH(CH_3)CH_2OH$], the molecule is considerably slower to undergo biodegradation. In aqueous systems such as rivers, this factor led to visual problems of foaming, etc.

As a result of this issue, much work has been directed at the generation of more linear alcohol streams. One avenue of work has been the homologation of ethylene to give C_8, C_{10}, C_{12}, etc., linear olefins using a Ziegler type of catalyst. Upon oxonation, these olefins provide linear alcohols. Since the technology is ethylene-based, not all carbon numbered alcohols are available; in fact only every other carbon numbered alcohol can be made. Chevron, for example, using a triethyl aluminum-type catalyst oligomerizes ethylene. The oligomer is oxidized and hydrolyzed to produce an alcohol product which can contain about 99% normal alcohol. Empirically, as one goes up in carbon number, the percentage of normal alcohol product is reduced (Figures 2–5).

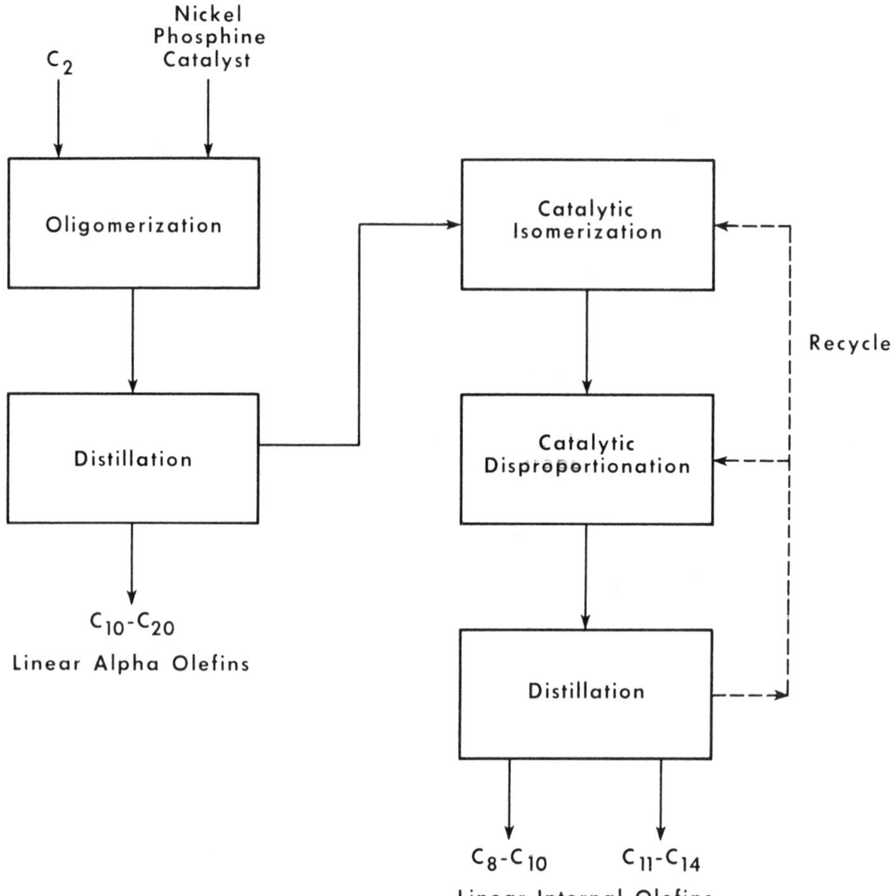

Figure 2 Simplified representation of the Shell process for olefin manufacture.

In 1986 oxoalcohol capacity in Europe and the United States totaled 3600 ktons (kt), with in excess of 2000 kt manufactured in Europe. With an additional capacity in the Pacific Basin and South Africa totaling >1000 kt, the total world capacity is almost 5000 kt.

BASF is the leading producer with a total plant capacity of 620 kt, with Exxon (500 kt) and Shell (460 kt) the next largest producers. Higher alcohol demand can be broken down into three segments; 2-EH and isooctyl alcohol (end use as plasticizers, lube additives, solvents, and acrylates), and linear C_{11}- and linear C_{12+}, largely employed

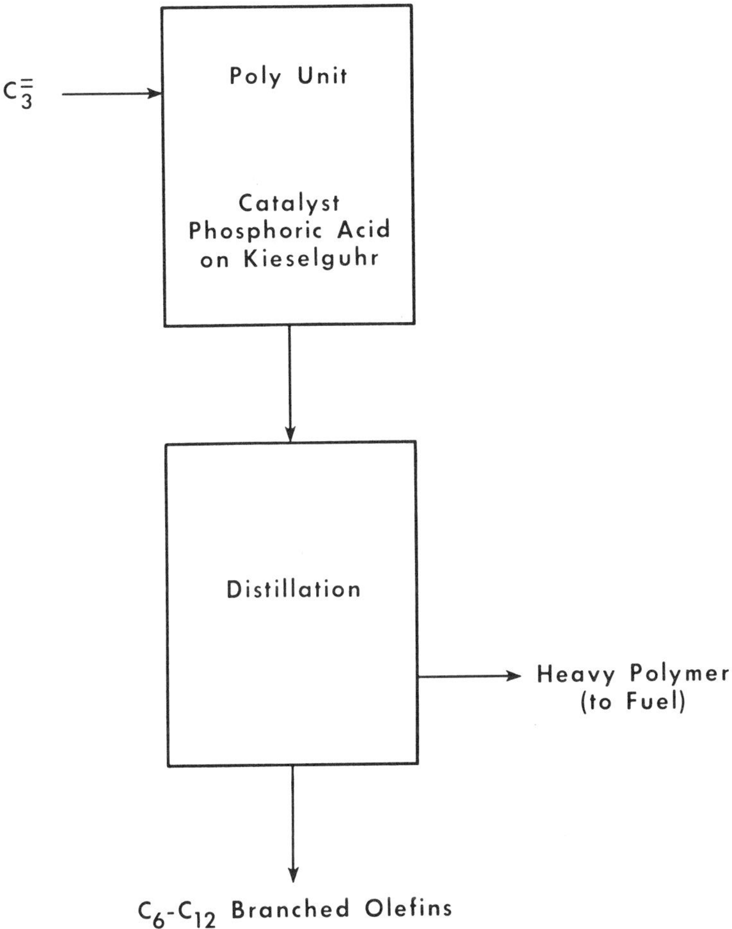

Figure 3 Simplified representation of the Exxon process for olefin manufacture.

as feedstocks for ethoxylates and as plasticizers and feedstocks for detergents and surfactants.

C. Feedstocks for Nitrogenates

The list of very high-volume nitrogen-containing organics is much shorter than the corresponding list for oxygenates. The only organic

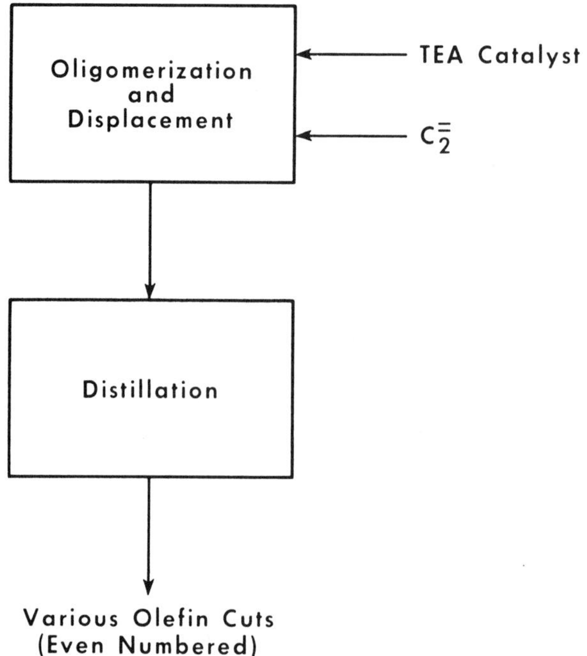

Figure 4 Simplified representation of the Chevron process for olefin manufacture.

nitrogenates on the list of the top 50 chemicals (volume) in 1986 (*Chem. Eng. News*, 1987) are urea (number 14 at 12.06 billion pounds) and acrylonitrile (2.31 billion pounds). In contrast, 12 of the 50 top chemicals are oxygenated organics (methanol, ethylene oxide, formaldehyde, ethylene glycol, acetic acid, phenol, propylene oxide, vinyl acetate, methyl-*t*-butyl ether, acetone, adipic acid, and isopropyl alcohol). Other commercially important nitrogen-containing organics include ethanolamine, hexamethylene, aniline and derivatives, nitrobenzene, amines, nitroparaffins, etc.

1. Nitroalkanes

The lower nitroalkanes are commercially prepared by the vapor phase reaction of propane with nitric acid at 420°C (790°F). Not only 1- and 2-nitropropane are formed, but nitromethane and nitroethane are obtained

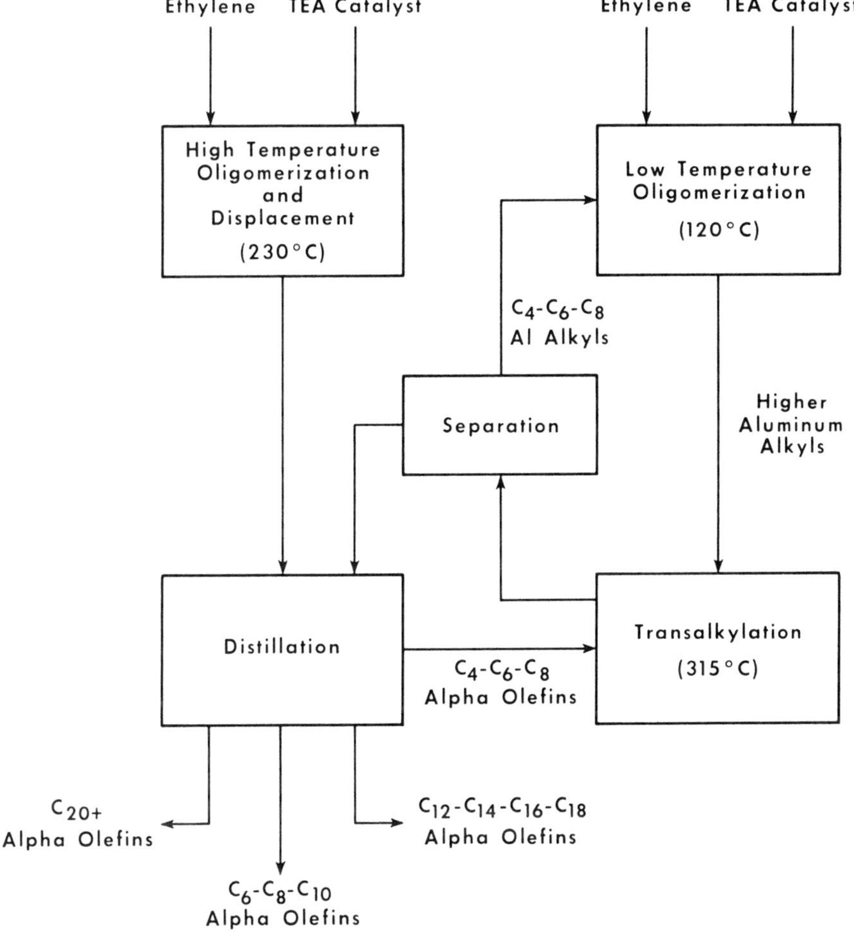

Figure 5 Simplified representation of the Ethyl Corporation process for olefin manufacture.

$$RH + HONO_2 \rightarrow RNO_2 + H_2O$$

by the cleavage of carbon-carbon bonds. The lower products are separated from the nitropropanes by distillation. The nitropropanes find use as solvents for cellulose acetate, vinyl resins, lacquers, synthetic rubbers, fats, dyes, etc. They also are used as high-energy propellants.

2. Nitroaromatics

Nitrobenzene is the most important of the nitroaromatic hydrocarbons. Almost all of it is converted by reduction to aniline. Production in the United States runs to several hundred million pounds/year. 2,4,6 Trinitrotoluene (TNT) is an important military explosive. It is used either alone or with other explosives (trimethylenetrinitramine) for firing of bombs, shells, and hand grenades. Since it melts at 81°C (178°F) and does not explode until ∿280°C (535°F), it can be poured into shells in a liquid state and allowed to solidify. TNT is relatively insensitive to shock and therefore must be exploded by a detonator. Production in the United States during World War II was at an annual rate of about 1 million tons/year

3. Aliphatic Amines

Dimethylamine is the most important amine of those formed when a mixture of methyl alcohol and ammonia is passed over heated zinc chloride.

$$CH_3OH + NH_3 \rightarrow CH_3NH_2 \rightarrow (CH_3)_2NH \rightarrow (CH_3)_3N$$

Some of the uses for dimethylamine include preparation of dimethylformamide (DMF), dimethylacetamide, and dimethyllaurylamine oxide used as a foam stabilizer for liquid detergents.

4. Aminoalcohols

1-Hydroxy-2-amino compounds are typically prepared commercially by the reaction of ethylene oxide with ammonia or with primary or secondary amines. For example, reaction of ethylene oxide with ammonia gives 2-aminoethanol or ethanolamine. Ethanolamine finds its largest use as an acid gas (H_2S and CO_2) scrubber from natural gas and from

$$\underset{\diagdown\,O\,\diagup}{CH_2-CH_2} \xrightarrow{NH_3} HOCH_2CH_2NH_2$$

refinery operations (Chapter 7). It is also used in the synthesis of surface-active agents, etc. Ethanolamine and its trimethylammonium salts, the cholines, constitute a portion of an important class of biological substances known as the phospholipids or phosphatides. The lecithins are mixed esters of glycerol and choline with fatty acids and phosphoric acid.

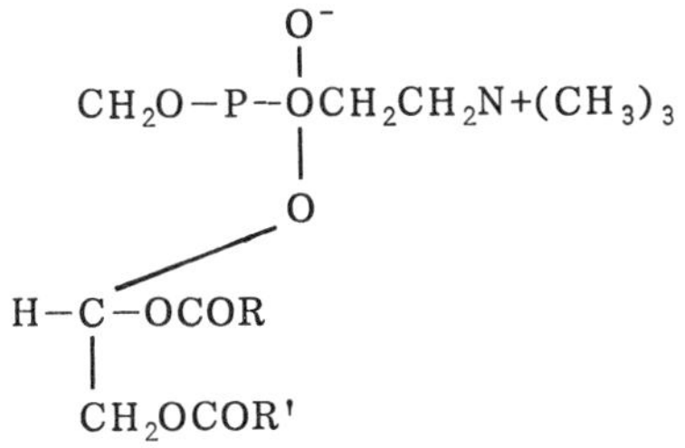

$$
\begin{array}{l}
\qquad\quad O^- \\
\qquad\quad | \\
CH_2O-P-OCH_2CH_2N^+(CH_3)_3 \\
\qquad\quad | \\
\qquad\quad O \\
\quad / \\
H-C-OCOR \\
\quad | \\
\quad CH_2OCOR'
\end{array}
$$

lecithins

About 550 million pounds of ethanolamine was produced in the United States in 1986.

Acrylonitrile. Along with methyl acrylate and methyl methacrylate, acrylonitrile is one of the most important α,β-unsaturated compounds from an industrial point of view. The output in 1986 of acrylonitrile by the U.S. chemical industry was in excess of 2.3 billion pounds. Acrylonitrile can be prepared via a number of synthetic routes starting from either ethylene, propylene, or acetylene. Probably the least costly route is one which yields acrylonitrile directly by means of air oxidation of a mixture of propylene and ammonia at 450°C (840°F). A typical catalyst system is a fluidized bed of molybdenum oxide promoted with bismuth oxide.

$$CH_2{=}CHCH_3 + O_2 \xrightarrow[450^\circ C]{MoO_3\text{-}Bi_2O_3} CH_2{=}CHCHO$$

$$CH_2{=}CHCHO \xrightarrow{NH_3} CH_2{=}CHCH{=}NH$$

$$CH_2{=}CHCH{=}NH \xrightarrow{O_2,\ MoO_3\text{-}Bi_2O_3} CH_2CHCN$$

When acrylonitrile is polymerized in an aqueous medium containing persulfate along with sodium bisulfite as an activator, a solid resin is obtained that is insoluble in common organic solvents. The discovery of solvents in which polyacrylonitrile could be dissolved and from which

$$xCH_2{=}CHCN \xrightarrow{persulfate} [-CH_2CH(CN)-]_x$$

it could be spun into fibers led to the production of so-called acrylic fibers. Solvents widely used for spinning are *N*,*N*-dimethylformamide

[$HCON(CH_3)_2$] and *N,N*-dimethylacetamide [$CH_3CON(CH_3)_2$], although polyacrylonitrile can be wet-spun from concentrated solutions of inorganic salts such as zinc chloride or calcium thiocyanate.

The acrylic fibers such as orlon, acrilan, etc., are copolymers or graft polymers with a small percentage of one or more other polymerizable monomers, such as methyl acrylate, vinyl acetate, vinylpyridine, etc., that increase the affinity of the fiber for either acidic or basic dyes. Other, less important end uses of acrylonitrile include its use as a chemical intermediate in the synthesis of antioxidants, pharmaceuticals, etc., and as a chemical modifier for natural polymers.

Adiponitrile. Adiponitrile is made from butadiene or acetylene, and is a necessary imtermediate in the preparation of the important product nylon 6-6.

Caprolactam. This lactam is prepared from cyclohexanone by conversion to the oxime with hydroxylamine, followed by an acid-catalyzed Beckmann rearrangement. Polymerization/ring opening occurs by heating the caprolactam with a trace of water. Over 1 billion pounds of caprolactam was produced in the United States in each of the last three years.

D. Polyolefins

Just as there are enormous roles for ethylene and propylene in the petrochemical industry, there are also important roles for diolefins, which are compounds containing two double bonds. Conjugated dienes (diolefins, including 1,3-butadiene, isoprene, cyclopentadiene, and methyl cyclopentadiene) are used in the manufacture of synthetic rubbers and as starting materials in insecticide synthesis. United States production in 1986 was 2.6 billion pounds of *butadiene*. Major uses for butadiene include incorporation into buna rubbers and as a starting material in numerous syntheses, including the preparation of adiponitrile used to make nylon 6-6.

Isoprene (2-methyl-1,3-butadiene) is used in the manufacture of "synthetic" natural rubber and of butyl rubber, and is a copolymer in the production of synthetic elastomers.

Cyclopentadiene is used in the manufacture of resins. It is also famous as the diene in a Diels–Alder reaction where it is used in the synthetic production of such products as sesquiterpenes, camphors, and alkaloids.

E. Polyaromatics

Polyaromatics are compounds containing more than one aromatic ring. Important polyaromatic compounds include naphthalene, phenanthrene, anthracene, quinoline, acridine, etc. These compounds were initially isolated and characterized from coal tars and will be discussed in the section on chemicals from coal.

F. Acrylates

Derivatives of acrylic acid include butyl acrylate, ethyl acrylate, 2-ethylhexyl acrylate, and methyl acrylate. These esters are homopolymerized using peroxide initiators or copolymerized with other monomers to generate acrylic or acryloid resins.

$$[CH_2CH(COOR)]_n$$

acrylate resin

Methyl α-methylacrylate, commonly called methyl methacrylate, can be made via esterification of α-methylacrylic acid or from acetone.

$$(CH_3)_2CO \xrightarrow{HCN} (CH_3)_2COHCN \xrightarrow{\text{Conc } H_2SO_4} CH_2{=}C(CH_3)CONH_2$$

$$CH_2{=}C(CH_3)CONH_2 \xrightarrow{CH_3OH,\ H^+} CH_2{=}C(CH_3)COOCH_3$$

This monomer is usually stabilized by the addition of an inhibitor. When methyl methacrylate is polymerized using a peroxide initiator such as *tert*-butyl benzoyl peroxide, poly(methyl methacrylate) is formed. This strong, thermoplastic solid is highly transparent and has a high refractive index. Trade names for poly(methyl methacrylate) include Lucite, Plexiglas, and Crystallite.

Starting in the early 1960s thermosetting enamels consisting of poly(methyl methacrylate) and melamine were being used for automobile finishes and household appliances. Baking of the acrylate resin with melamine produces crosslinking by reactions of the ester side chains of the acrylate polymer with the amino groups of the melamine. The amide bonds so produced results in a permanent finish requiring no polishing.

G. Others

The huge number of other chemicals derived from petroleum makes the selection of a few for inclusion here a very subjective exercise. Fluorocarbons, chlorofluorocarbons, acetic anhydride, ethylene dichloride, vinylidene dichloride, hydrogen sulfide, carbon black, and so on, are the choices I have made. They are large volume products with important end-use applications.

Petroleum-based chemicals and their derivatives are ubiquitous and have a profound impact on virtually every aspect of our lives, from the foods we eat, to the cars we drive, the clothes we wear, and the medicines we depend on. The number of petroleum-derived products of commercial importance continues to increase and the future will surely see this explosion of consumer products from petrochemicals continue.

H. Engineering Plastics

Engineering plastics is a growing $2 billion business in the United States. The definition of engineering plastics is a difficult one to generate in any but an operational mode. In terms of properties, engineering plastics usually demonstrate a good balance of high-tensile properties, compressive and shear strength, stiffness, and impact resistance. They are easily moldable. In terms of market/pricing criteria, engineering plastics fall between the high-volume/low-price commodity plastics and the low-volume/high-price specialty plastics. Five resin families are typically included in the engineering plastics group: nylon, acetal, thermoplastic polyester molding compounds, poly(phenylene oxide)-based resin (PPO), and polycarbonate. The selling price for all members of this family of products was in the $1.40–$2.00 per pound range in 1985. Major players in the area of engineering plastics are DuPont, General Electric, Celanese, Mobay, and Allied/Signal. Given the rapid expansion of the technology and its projected growth, many new entries to this field are expected.

1. Nylon

Nylon represents a family of polyamides, formed from various combinations of diacids, diamines, and amino acids. Nylon 66 is a polyamide formed when the hexamethylenediamine salt of adipic acid is heated. Hexamethylenediamine is the reduction product of adiponitrile (see above). The numbering system, 66, indicates that this particular nylon has two six-carbon components. The molecular weight of nylon 66 is about 10,000, it is spun from a melt (melting point = 260°C; 500°F), and the filaments are drawn to about four times their original length to orient the molecules along the length of the fiber. Of all the nylons, nylon 66 is best suited for textile fiber application due to its high melting point.

2. Acetals

Acetals are used almost exclusively in injection molding applications. Acetal homopolymers are formed by the polymerization of formaldehyde; the high molecular weight polymers are stabilized by conversion of the end group to esters. Acetal copolymers are made by the copolymerization of trioxane, a cyclic trimer of formaldehyde, with a small amount of comonomer. Celanese is the major supplier of acetal copolymer (Celcon) while DuPont dominates the acetal homopolymer market (Delrin).

3. Thermoplastic Polyester Molding Compounds

The main players in this family are poly(ethylene terephthalate) (PET) and poly(butylene terephthalate) (PBT). In their engineering plastics applications, the major end uses are in automotive and electronic or electric areas.

4. Poly(phenylene Oxide)-Based Resin

The major poly(phenylene oxide) (PPO)-based engineering thermoplastic currently available domestically is a blend of poly(phenylene oxide) and impact polystyrene marketed as Noryl by General Electric. Noryl has the lowest water absorption rate of the thermoplastic engineering resins, and it is virtually completely resistant to hydrolysis. It also has a high degree of flame recordings. Flame retardance is imparted to Noryl by the incorporation of a proprietary flame retardant into the compounding procedure. Special-purpose formulations can be made by varying the relative content of PPO and polystyrene.

5. Polycarbonate

Polycarbonate is an amorphous polyester of carbonic acid produced from dihydric or polyhydric phenols via a condensation reaction with a carbonate precursor.

References to Part I

Aalund, L. R. 1975. *Oil Gas J.* 77(35):339.

Abraham, H. 1945. *Asphalts and Allied Substances*, Van Nostrand, New York.

Albright, L. F., and B. L. Crynes (ed.). 1976. *Industrial and Laboratory Pyrolyses*, Sumposium Series No. 32, Am. Chem. Soc., Washington, D.C.

Aldag, A. W. 1987. Accelerated Aging Tests with a Resid Hydrotreating Catalyst. *Preprints, Div. Petrol. Chem., Am. Chem. Soc., 32*(2): 443–449.

Al-Kashab, K., and J. J. Neumann. 1976. *Strassen Tieflau. 30*(3): 44.

American Society for Testing and Materials. 1980. Standard No. D893-80, Philadelphia.

American Society for Testing and Materials. 1970. Standard No. D2006-70, Philadelphia.

American Society for Testing and Materials. 1980. Standard No. D2007-80, Philadelphia.

American Society for Testing and Materials. 1984. Standard No. 04124-84, Philadelphia.

Asaoka, S., S. Nakata, Y. Shiroto, and C. Takevchi. 1987. Characteristics of Vanadium Complexes in Petroleum Before and After Hydrotreating. In *Metal Complexes in Fossil Fuels* (R. H. Filby and J. F. Branthaver, eds.), Am. Chem. Soc., Washington, D.C., p. 275–289.

Atkinson, G., and J. J. Zuckerman. 1981. *Origin and Chemistry of Petroleum*, Pergamon Press, New York.

Baker, E. W. 1969. Porphyrins. In *Organic Geochemistry* (G. Eglinton and M. Murphy, eds.), Springer-Verlag, New York, pp. 464–497.

Baker, E. W., and S. E. Palmer. 1978. Geochemistry of Porphyrins. In *The Porphyrins, Vol. 1, Structure and Synthesis, Part A* (D. Dolphin, ed.), Academic Press, New York, pp. 485–551.

Baker, E. W., and J. W. Louda. 1986. Porphyrins in the Geological Record. In *Biological Markers in the Sedimentary Record* (R. B. Johns, ed.), Elsevier, Amsterdam, pp. 125–225.

Barth, E. J. 1962. *Asphalt Science and Technology*, Gordon and Breach, New York.

Bartsch, W. 1976. *Arzneimittel-Forschung. 26*:1581.

Bell, H. S. 1945. *American Petroleum Refining*, Van Nostrand, New York.

Berkowitz, N., and J. G. Speight. 1975. *Fuel 54*:318.

Bland, W. F., and R. L. Davidson. 1967. *Petroleum Processing Handbook*, McGraw-Hill, New York.

Bonnett, R. 1978. Nomenclature. In *The Porphyrins, Vol. I, Structure and Synthesis, Part A* (D. Dolphin, ed.), Academic Press, New York, pp. 1–27.

Bonnett, R., P. J. Burke, and F. Czechowski. 1987. Metalloporphyrins in Lignite, Coal, and Calcite. In *Metal Complexes in Fossil Fuels* (R. H. Filby and J. F. Branthaver, eds.), Am. Chem. Soc., Washington, D.C., pp. 173–185.

Boyd, M. L., and D. S. Montgomery. 1963. *J. Inst. Petroleum 49*: 345.

BP Statistical Review of World Energy. 1987. British Petroleum Company, June.

Brooks, B. T., B. S. Kurtz, O. E. Board, and L. Schmerling. 1954. *The Chemistry of Petroleum Hydrocarbons*, Reinhold, New York.

Brooks, J., and D. H. Welte. 1984. *Advances in Petroleum Geochemistry, Vol. 1*, Academic Press, New York.

Bunger, J. W. 1977. *Preprints. Div. Petrol. Chem., Am. Chem. Soc. 22*(2):716.

Chapel, D. G., R. E. Brown, D. D. Cobb, and D. L. Heaven. 1987. *Oil Gas J.*, November 16, p. 41.

Chem. Eng. News. 1985a. June 10.

Chem. Eng. News. 1985b. February 4.

Chem. Eng. News. 1987. June 8.

Chem. Week. 1980. January 16.

Chen, H. J., and F. E. Massoth. 1987. Intrinsic Kinetics of the Hydrodemetallization of V- and Ni-Porphyrin over Sulfided CoMo/Al_2O_3 Catalyst. *Preprints, Div. Petrol. Chem., Am. Chem. Soc., 32*(2): 437–442.

Corbett, L. W., and U. Petrossi. 1978. *Ind. Eng. Chem., Prod. Res. Dev. 17*:342.

Crouch, F. W., C. S. Sommer, J. F. Galobardes, S. Kraus, E. M. Schmauch, M. Galobardes, A. Fatmi, K. Pearsall, and L. B. Rogers. 1983. Fractionations of Nonporphyrin Metal Complexes of Vanadium and Nickel from Boscan Crude. *Oil. Sep. Sci. Technol. 18*:603–634.
Crynes, B. L., and W. H. Corcoran. 1983. *Pyrolysis: Theory and Industrial Practice*, Academic Press, New York.
Cummins, J. J., R. E. Poulson, and W. E. Robinson. 1975. *Preprints, Div. Fuel Chem., Am. Chem. Soc. 20*(2):154.
Dale, G. H., and D. H. McKay. 1977. Passivate Metals in FCC Feeds. *Hydrocarbon Proc. Sept. 1977*:97–102.
Dautzenberg, E. M., and J. C. De DeKen. 1987. Modes of Operation in Hydrodemetallization. In *Metal Complexes in Fossil Fuels* (R. H. Filby and J. F. Branthaver, eds.), Am. Chem. Soc., Washington, D.C., pp. 233–256.
Decroocq, D. 1984. *Catalytic Cracking of Heavy Petroleum Hydrocarbons*, Editions Technip, Paris.
Dickson, F. E., C. J. Kunesh, E. L. McGinnis, and L. Petrakis. 1972. Use of Electron Paramagnetic Resonance to Characterize the Vanadium-IV-Sulfur Species in Petroleum. *Anal. Chem. 42*:978–981.
Dolbear, G. E., A. Tang, and E. L. Moorehead. 1987. Upgrading Studies with Californian, Mexican, and Middle Eastern Heavy Oils. In *Metal Complexes in Fossil Fuels* (R. H. Filby and J. F. Branthaver, eds.), Am. Chem. Soc., Washington, D.C., pp. 220–232.
Eglington, G., and M. J. T. Murphy. 1969. *Organic Geochemistry*, Springer-Verlag, New York.
Environmental Protection Agency. 1981. *Second Annual Report on Carcinogens*, 184.
Epstein, W., and J. Barnes. 1965. *Wall Street Journal*, January 18.
Erdman, J. G., and P. H. Harju. 1963. Capacity of Petroleum Asphaltenes to Complex Heavy Metals. *J. Chem. Eng. Data 8*:252–258.
Falk, J. E. 1964. *Porphyrins and Metalloporphyrins*, Elsevier, Amsterdam.
Farcasiu, M., and R. B. LaPierre. 1987. Thermal Reactions of Whole Crude Oils and Related Model Compounds. *Fuel Sci. Technol. Int. 5*: 697–711.
Farrar, G. L. 1952. Metals in Petroleum. *Oil Gas J., Apr. 5, 1952*: 79.
Filby, R. H., and G. J. Van Berkel. 1987. Geochemistry of Metal Complexes in Petroleum, Source Rocks, and Coals: An Overview. In *Metal Complexes in Fossil Fuels* (R. H. Filby and J. F. Branthaver, eds.), Am. Chem. Soc., Washington, D.C., pp. 2–39.
Fish, R. H., J. J. Konlenic, and B. K. Wines. 1984. Characterization and Comparison of Vanadyl and Nickel Compounds in Heavy Crude Petroleums and Asphaltenes by Reverse-Phase and Size-Exclusion Liquid Chromatography/Graphite Furnace Atomic Absorption Spectrometry. *Anal. Chem. 56*:2452–2460.

Fish, R. H., J. G. Reynolds, and E. J. Gallegos. 1987a. Molecular Characterization of Nickel and Vanadium Nonporphyrin Compounds Found in Heavy Crude Petroleums and Bitumens. In *Metal Complexes in Fossil Fuels* (R. H. Filby and J. F. Branthaver, eds.), Am. Chem. Soc., Washington, D.C., pp. 332–349.

Fish, R. H., W. Walker, and R. S. Tannous. 1987b. Organometallic Geochemistry 2: The Molecular Characterization of Trace Organometallic and Inorganic Compounds of Arsenic Found in Green River Formation Oil Shale and Its Pyrolysis Product. *Energy Fuels 1*:243–247.

Fisher, J. P., F. Souhrada, and H. J. Woods. 1982. *Oil Gas J. 80*(47):111.

Forbes, R. J. 1958. *A History of Technology*, Oxford University Press, Oxford, England.

Franceskin, P. J., M. G. Gonzalez-Jiminez, F. Da Rosa, O. Adams, and L. Katan. 1986. First Observation of an Iron Porphyrin in Heavy Crude Oil. *Hyperfine Interactions 28*:825–828.

Francisco, M. A., and J. G. Speight. 1984. *Preprints. Div. Fuel Chem., Am. Chem. Soc., 29*(1):36.

Gary, J. H., and G. E. Handwerk. 1984. *Petroleum Refining: Technology and Economics*, Marcel Dekker, New York.

Germain, G. E. 1969. *Catalytic Conversion of Hydrocarbons*, Academic Press, New York.

Girdler, R. B. 1965. *Proc. Assoc. Asphalt Paving Technologists 34*: 45.

Glebovskaya, F. A., and M. V. Volkenshtein. 1948. Spectra of Porphyrins in Petroleum and Bitumen. *J. Gen. Chem. USSR 18*:1440–1451.

Goldstein, R. F. 1949. *The Petroleum Chemicals Industry*, E & F. N. Spon Ltd., London.

Goulon, J., A. Retournard, P. Frient, C. Goulon-Ginet, C. Berthe, J. F. Muller, J. L. Poncet, R. Guilard, J. C. Escalier, and B. Neff. 1984. Structural Characterization by X-Ray Absorption Spectroscopy (EXAFS/XANES) of the Vanadium Chemical Environment in Boscan Asphaltenes. *J. Chem. Soc. Dalton Trans. 1984*:1095–1103.

Gruse, W. A., and D. R. Stevens. 1960. *Chemical Technology of Petroleum*, McGraw-Hill, New York.

Guthrie, V. B. 1960. *Petroleum Products Handbook*, McGraw-Hill, New York.

Hahn, A. V. 1970. *The Petrochemical Industry: Market and Economics*, McGraw-Hill, New York.

Hargrove, J. D., G. J. Elkes, and A. H. Richardson. 1979. *Oil Gas J. January 15*:103.

Hirsch, D. E., R. L. Hopkins, H. J. Coleman, F. O. Cotton, and D. J. Thompson. 1972. *Anal. Chem. 44*:915.

Hobson, G. D., and W. Pohl. 1978. *Modern Petroleum Technology*, Applied Science, Barking, England.

Hobson, G. D., and E. N. Tiratsoo. 1975. *Introduction to Petroleum Geology*, Scientific Press, Beaconsfield, England.

Hodgson, G. W., B. L. Baker, and E. Peake. 1967. Geochemistry of Porphyrins. In *Fundamental Aspects of Petroleum Geochemistry* (B. Nagy and U. Columbo, eds.), Elsevier, Amsterdam, pp. 177–259.

Holberg, A. J. 1960. *Bituminous Materials: Asphalts, Tars and Pitches*, Interscience, New York.

Howell, R. L., C. Hung, K. R. Gibson, and H. C. Chen. 1985. Catalyst Selection Important for Residuum Hydroprocessing. *Oil Gas J., July 29, 1985*:121–128.

Hydrocarbon Processing. 1984. Refining update, September.

Institute of Petroleum. 1982. Standard No. IP 148/82, London.

Jewell, D. M., J. H. Weber, J. W. Bunger, H. Plancher, and D. R. Latham. 1972a. *Anal. Chem. 44*:1391.

Jewell, D. M., R. G. Ruberto, and B. E. Davis. 1972b. *Anal. Chem. 44*:2318.

Jewell, D. M., E. W. Albaugh, B. E. Davis, and R. G. Ruberto. 1974. *Ind. Eng. Chem. Fund. 13*:278.

Jiratova, K., and M. Kraus. 1986. Effect of Support Properties of the Catalytic Activity of HDS Catalysts. *Appl. Catal. 27*:21–29.

Kobe, K. A., and J. J. McKetta. 1958. *Advances in Petroleum Chemistry and Refining*, Interscience, New York.

Kochi, J. K. 1973. *Free Radicals*, John Wiley and Sons, New York.

Koots, J. A., and J. G. Speight. 1975. *Fuel 54*:179.

Kowanko, N., J. F. Branthaver, and J. M. Sugihara. 1978. Direct Liquid Phase Fluorination of Petroleum. *Fuel 57*:769–775.

Kukes, S. G., A. W. Aldag, and S. L. Parrott. 1987. Hydrodemetallization with Phosphorus Compounds Over Aluminas in a Trickle-Bed Reactor. In *Metal Compexes in Fossil Fuels* (R. H. Filby and J. F. Branthaver, eds.), Am. Chem. Soc., Washington, D.C., pp. 265–274.

Laine, J., and D. L. Trimm. 1982. *J. Chem. Tech. Biotechnol. 82*:818.

Landes, K. K. 1959. *Petroleum Geology*, John Wiley and Sons, New York.

Langer, A. W., J. Stewart, C. E. Thompson, H. T. White, and R. M. Hill. 1962. *Ind. Eng. Chem., Process Design and Devel. 1*:309.

Larkins, T. H. 1986. *Abstracts*, Division of Petroleum Chemistry, American Chemical Society, April.

Le Page, J. F., J. Cosyns, P. Courty, E. Freund, J. P. Franck, Y. Jacquin, B. Juguin, C. Marcilly, G. Martino, J. Miquel, R. Montarnal, A. Sugier, and H. van Landeghem. 1987. *Applied Heterogenous Catalysis*, Editions Technip, Paris.

Linstead, R. P. 1985. *J. Am. Chem. Soc. 64*:1985.

Logwinuk, A. K., and D. L. Caldwell. 1983. *Chem. Econ. Eng. Rev. 15*(3):31.

Long, R. B. 1981. In *The Chemistry of Asphaltenes* (J. W. Bunger and N. Li, eds.), Advances in Chemistry Series No. 195, Am. Chem. Soc., Washington, D.C., p. 17.
Magaril, R. L., L. F. Ramazeeva, and E. I. Aksenova. 1970. *Khim. Tekhnol. Topl. Masel.* *15*(3):15.
Masselli, J. M., and A. W. Peters. 1984. Preparation and Properties of Fluid Cracking Catalysts for Residual Oil Conversion. *Catal. Rev. Sci. Eng.* *26*:525–554.
McKay, J. F., J. H. Weber, and D. R. Latham. 1975. *Fuel* *54*:50.
McKay, J. F., J. H. Weber, and D. R. Latham. 1976. *Anal. Chem.* *48*:891.
McKay, J. F., P. J. Amend, T. E. Cogswell, P. M. Harnsberger, R. B. Erickson, and D. R. Latham. 1977. *Preprints, Div. Petrol. Chem., Am. Chem. Soc.* *82*(2):708.
Meyer, R. F. (ed.). 1977. *The Future Supply of Nature-Made Petroleum and Gas*, Pergamon Press, New York.
Mitchell, D. L., and J. G. Speight. 1973. *Fuel* *52*:149.
Mitchell, P. C. H., and C. E. Scott. 1986. *Polyhedron* 5:237.
Mohammed, A. M. A. K., A. A. Abbas, A. B. Ahmed, and A. S. K. A. Mayah. 1987. Catalytic Hydrotreatment of Petroleum Residue. *Fuel Sci. Technol. Int.* 5:655–675.
Morales, A., and R. Galiaso. 1982. Adsorption Mechanism of Boscan Porphyrins on MoO_3, Co_3O_4 and $CoMo/Al_2O_3$. *Fuel* *61*:13–17.
Morandi, J. R., and H. B. Jensen. 1966. Comparison of Porphyrins from Shale Oil, OIl Shale, and Petroleum. *J. Chem. Eng. Data* *11*:81–88.
Moschopedis, S. E., S. Parkash, and J. G. Speight. 1978. *Fuel* *57*:431.
Nagy, B., and V. Colombo. 1967. *Fundamental Aspects of Petroleum Geochemistry*, Elsevier, Amsterdam.
Narain, N. K., R. E. Tischer, G. J. Stiegel, D. L. Cillo, and M. Krishnamurthy. 1986. Demetallization of a Coal Liquid Residuum, Topical Report. Pittsburgh, PA. DOE Report DOE/PETC/TR-86/6.
Nelson, W. L. 1958. *Petroleum Refinery Engineering*, McGraw-Hill, New York.
Oblad, A. G., H. B. Davis, and R. T. Eddinger (eds.). 1979. *Thermal Hydrocarbon Chemistry*, Advances in Chemistry Series No. 183, Am. Chem. Soc., Washington, D.C.
Occelli, M. L., D. C. Kowalczyk, and C. L. Kibby. 1986. Fluid Cracking Catalyst with Carbon Selectivity. *Appl. Catal.* *16*:227–236.
Oudin, J. L. 1970. *Rev. Inst. Francais du Petrole*. *25*:470.
Pfeiffer, J. Ph. (ed.). 1950. *The Properties of Asphaltic Bitumen*, Elsevier, Amsterdam.
Purdy, G. A. 1957. *Petroleum: Prehistoric to Petrochemicals*, Copp Clark, Toronto.

Quirke, J. M. E., G. J. Shaw, P. D. Soper, and J. R. Maxwell. 1980. Petroporphyrins II. The presence of Porphyrins with Extended Alkyl Substituents. *Tetrahedron 36*:3261–3267.

Quirke, J. M. E. 1987a. Techniques for Isolation and Characterization of the Geoporphyrins and Chlorins. In *Metal Complexes in Fossil Fuels* (R. H. Filby and J. F. Branthaver, eds.), Am. Chem. Soc., Washington, D.C., pp. 308–331.

Quirke, J. M. E. 1987b. Rationalization for the Predominance of Nickel and Vanadium Porphyrins in the Geosphere. In *Metal Complexes in Fossil Fuels* (R. H. Filby and J. F. Branthaver, eds.), Am. Chem. Soc., Washington, D.C., pp. 74–83.

Rankel, L. A., and L. D. Rollman. 1983. Catalytic Activity of Metals in Petroleum and Their Removal. *Fuel 62*:44–46.

Rankel, L. A. 1987. Degradation of Metalloporphyrins in Heavy Oils Before and During Processing. In *Metal Complexes in Fossil Fuels* (R. H. Filby and J. F. Branthaver, eds.), Am. Chem. Soc., Washington, D.C., pp. 257–264.

Rao, B. M. L., and J. E. Serrano. 1986. Viscometric Study of Aggregation Interactions in Heavy Oil. *Fuel Sci. Technol. Int. 4*:483–500.

Reynolds, J. G., W. E. Biggs, and S. A. Bezman. 1987. Reaction Sequence of Metalloporphyrins During Heavy Residuum Upgrading. In *Metal Complexes in Fossil Fuels* (R. H. Filby and J. F. Branthaver, eds.), Am. Chem. Soc., Washington, D.C., pp. 205–219.

Rice, F. O. 1988. *J. Am. Chem. Soc. 55*:3035.

Ritter, R. E., L. Rheume, W. A. Walsh, and J. S. Magee. 1981. A New Look at FCC Catalysts for Resid. *Oil Gas J., July 6, 1981*:103–109.

Rossini, F. D., B. J. Mair, and A. J. Streiff. 1953. *Hydrocarbons from Petroleum*, Reinhold, New York.

Rostler, F. S. 1965. In *Bituminous Materials: Asphalts, Tars and Pitches, Vol. II, Part I*, (J. Holberg, ed.), Interscience, New York, p. 151.

Sanders, J. M., J. C. Waterton, and I. S. Dennis. 1978. Spin-Lattice Relaxation, Nuclear Overhauser Enhancements, and Long Range Coupling in Chlorophylls and Metalloporphyrins. *J. Chem. Soc. Perkin Trans. I 1978*:1150–1157.

Scheutze, B., and H. Hofmann. 1984. *Hydrocarbon Proc. February*: 75.

Scott, J. W., and A. G. Bridge. 1971. *Origin and Refining of Petroleum*, Am. Chem. Soc., Washington, D.C.

Seifert, W. K. 1975. *Advances in Organic Geochemistry*, Madrid.

Selucky, M. L., Y. Chu, T. O. S. Ruo, and O. P. Strausz. 1977. *Fuel 56*:369.

Smith, K. M. 1975. General Features of the Structure and Chemistry of Porphyrin Compounds. In *Porphyrins and Metalloporphyrins* (K. M. Smith, ed.), Elsevier, Amsterdam, pp. 3–28.

Smith, K. W., W. C. Starr, and N. Y. Chen. 1980. *Oil Gas J.* May 26, 1975.
Speight, J. G. 1979. Information Series No. 84, Alberta Research Council, Edmonton, Alberta, Canada.
Speight, J. G. 1980. *The Chemistry and Technology of Petroleum*, Marcel Dekker, New York.
Speight, J. G. 1981. *The Desulfurization of Heavy Oils and Residua*, Marcel Dekker, New York.
Speight, J. G. 1983. *The Chemistry and Technology of Coal*, Marcel Dekker, New York.
Speight, J. G. 1984a. In *Caracterization of Heavy Crude Oils and Petroleum Residues* (B. P. Tissot, ed.), Editions Technip, Paris, p. 32.
Speight, J. G. 1984b. In *Catalysis on the Energy Scene* (S. Kaliaguine and A. Mahay, eds.), Elsevier, Amsterdam.
Speight, J. G. (ed.). 1984c. *Liquid Fuels Technol.* *2*:211 et seq.
Speight, J. G. 1986. *Preprints. Div. Petrol. Chem., Am. Chem. Soc.* *31*(3/4):818.
Speight, J. G. 1987. *Preprints. Div. Petrol. Chem., Am. Chem. Soc.* *32*(2):413.
Speight, J. G., R. B. Long, and T. D. Trowbridge. 1984. *Fuel 63*: 616.
Speight, J. G., and S. E. Moschopedis. 1979. *Fuel Proc. Technol.* *2*:295.
Speight, J. G., and S. E. Moschopedis. 1986. *Fuel Proc. Technol.* *13*:215.
Speight, J. G., and R. J. Pancirov. 1988. *Preprints. Div. Petrol. Chem., Am. Chem. Soc.* *25*(3):155.
Steiner, H. 1961. *Introduction to Petroleum Chemicals*., Pergamon Press, New York.
Stephens, M. M., and O. F. Spencer. 1956. *Petroleum Refining Processes*, Pennsylvania State University.
Sugihara, J. M., and R. M. Bean. 1962. Direct Determination of Metalloporphyrins in Boscan Crude Oil. *J. Chem. Eng. Data* *7*:269--271.
Sugihara, J. M., T. Okada, and J. F. Branthaver. 1965. Reductive Desulfurization on Vanadium and Metalloporphyrin Contents of Fractions from Boscan Asphaltenes. *J. Chem. Eng. Data* *16*:190--194.
Ternan, M. 1983. *Can. J. Chem. Eng.* *61*:689.
Thakur, D. S. 1985. *Appl. Catalysis* *15*:197.
Thomas, K. P., J. Oberle, P. M. Harnsberger, D. A. Netzel, and E. B. Smith. 1987. The Effect of Recovery Methods on Bitumen Composition. Report to the U.S. Dept. of Energy, Contract No. DE-A020-RELD11071: December.
Tissot, B. P. (ed.). 1984. *Characterization of Heavy Crude Oils and Petroleum Residues*, Editions Technip, Paris.

Tissot, B. P., and D. H. Welte. 1978. *Petroleum Formation and Occurrence*, Springer-Verlag, New York.
Traxler, R. N. 1961. *Asphalt: Its Composition, Properties and Uses*, Reinhold, New York.
Treibs, A. 1934. The Occurrence of Chlorophyll Derivatives in an Oil Shale of the Upper Triassic. *Justus Liebigs Anal. Chem. 509*: 103–114.
Van Berkel, G. J., and R. H. Filby. 1987. Generation of Nickel and Vanadyl Porphyrins from Kerogen During Simulated Catagenesis. In *Metal Complexes in Fossil Fuels* (R. H. Filby and J. F. Branthaver, eds.), Am. Chem. Soc., Washington, D.C., p. 110–134.
Vandegrift, G. F., R. E. Winans, R. G. Scott, and E. P. Horwitz. 1980. Quantitative Study of the Carboxylic Acids in Green River Oil Shale. *Fuel 59*:627–633.
vanNes, K., and H. A. Westen. 1951. *Aspects of the Constitution of Mineral Oils*, Elsevier, Amsterdam.
Vissers, J. P. R., V. H. J. DeBoer, and R. Prins. 1987. The Role of the Cobalt Promoter in Hydrodesulfurization. *Preprints, Div. Petrol. Chem., Am. Chem. Soc. 32*(2):347–350.
Ware, R. A., and J. Wei. 1985. Hydrometallation Reaction Selectivity on Modified Co-Mo Catalysts. *Preprints, Div. Petrol. Chem., Am. Chem. Soc. 30*(1):62–69.
Webster's Ninth New Collegiate Dictionary. 1983. G. & C. Merriam Company, Springfield, Massachusetts.
Yen, T. F. 1975. Chemical Aspects of Metals in Native Petroleum. In *The Role of Trace Metals in Petroleum* (T. F. Yen, ed.), Ann Arbor Science Publishers, Ann Arbor, pp. 1–30.

Part II

Tar Sand

12

Origin, Occurrence, and Recovery

I	Definitions and Terminology	326
II	Deposits of Canada and the United States	327
	A. Athabasca	327
	B. Melville Island	329
	C. Utah	330
	D. California	331
	E. Texas	331
	F. Kentucky	331
	G. New Mexico	332
	H. Missouri	332
III	Other Deposits	332
	A. Venezuela	332
	B. Madagascar	333
	C. Albania	333
	D. Trinidad	333
	E. Rumania	334
	F. Soviet Union	334
	G. Colombia	334
	H. Miscellaneous Deposits	334
IV	Recovery	335
	A. Mining Operations	338
	B. Bitumen Recovery	343
	C. Direct Heating	348
	D. Nonmining Operations	351
	E. Modified In Situ Operations	362
	F. Product Quality	363

V Transportation 364
VI Environmental Aspects 367

Tar sand deposits have been recognized for a long time and were recognized in pre-Christian times. In the present context, the tar sands in North America (specifically those in northern Alberta) first received recognition by Peter Pond, an American fur trader who, in 1778, recorded that the local Indians used the bitumen (leached by the river from outcrops) to caulk their canoes.

On the other hand, until very recently there was very little reported data relating to the U.S. deposits. Out of necessity the material composing this section relies heavily on data from the Alberta oil sands.* It *must* be remembered, however, that the properties reported herein cannot be universally applied to the remaining oil sands of the world.

Because of the diversity of available information and the continuing attempts to delineate the various world oil sand deposits, it is virtually impossible to present "accurate" numbers which reflect the extent of the reserves in terms of the barred unit. Indeed, investigations of the deposits are continuing at such unknown rates that the in-place reserves may vary from one year to the next. Accordingly, the data contained herein must be recognized as being approximate with the potential of being higher than quoted by the time of publication.

Commercialization (Spragins, 1978) has dictated that the Alberta (Canada) oil sand reserves are the most widely known resource of this type. However, contrary to what may be the general opinion, oil sand deposits are widely distributed (Phizackerley and Scott, 1967; Demaison, 1977; Meyer and Dietzman, 1981) throughout the world (Figure 1 and Table 1).

Exact definition of any particular resource may not always be possible and is, in fact, in direct relation to the publicity that the resource has received as an alternate energy source. It is for this reason that the Alberta deposits are most widely known, as investigations have been ongoing since the latter days of the nineteenth century with commercialization being realized in 1967 (Berkowitz and Speight, 1975; Spragins, 1978; Hyndman, 1981).

---

*Where possible, data relating to other oil sands are included but such data are fragmentary.

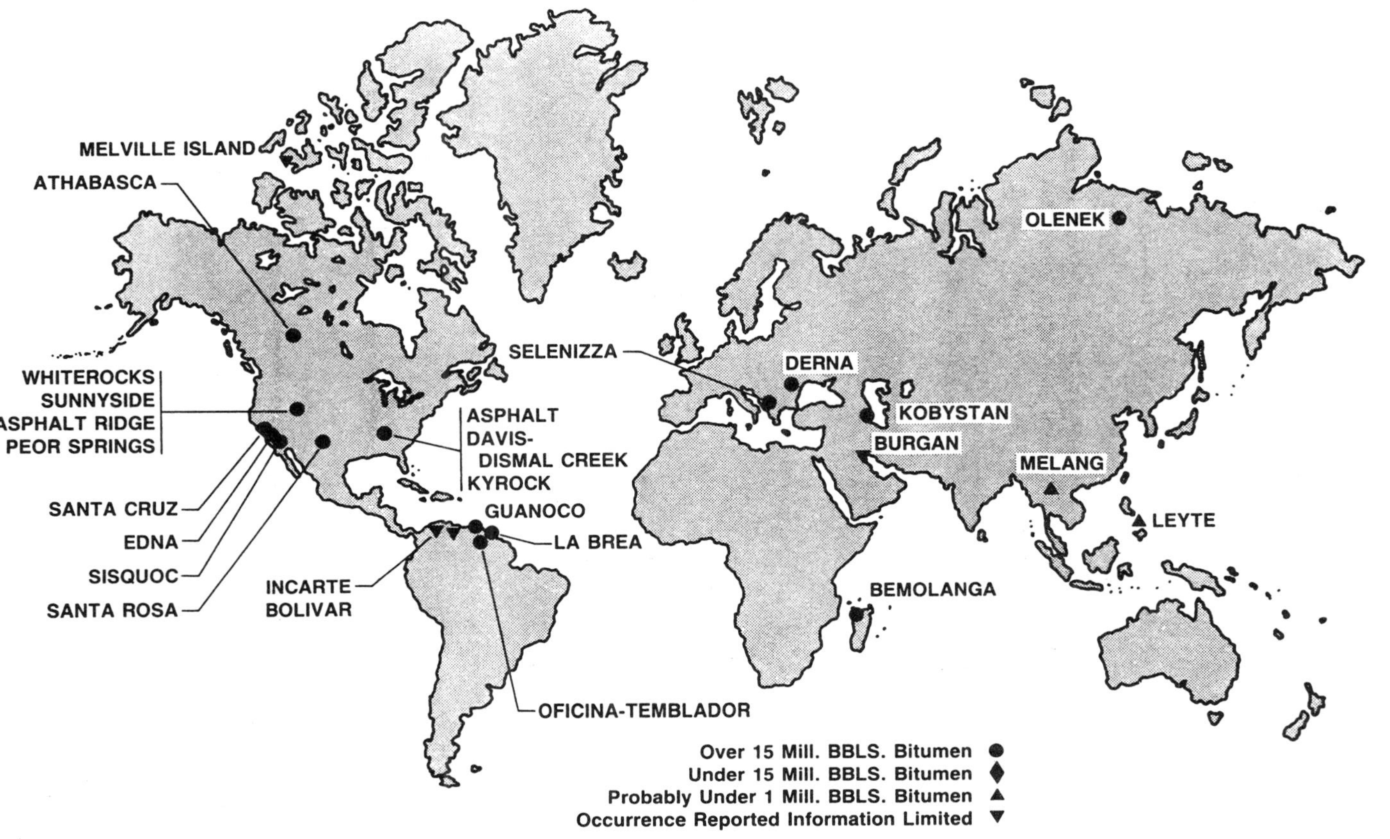

Figure 1 Major tar sand deposits of the world (see also Table 1).

Table 1 Characteristics of the Major Tar Sand Deposits of the World

Country	Name of Deposit	Age of reservoir rock	Area Acres
Canada	McMurray–Wabiskaw, Alberta		5,750,000
	Bluesky–Gething, Alberta		1,200,000
	Grand Rapids, Alberta		1,100,000
	Total "Athabasca" tar sands	L. cretaceous	8,000,000
	Melville Island, N.W.T.	Triassic	?
Venezuela	Oficina–Temblador tar belt	Oligocene	5,750,000
	Guanoco	Recent	1,000
Madagascar	Bemolanga	Triassic	96,000
USA	Asphalt Ridge, Utah	U. cretaceous Oligocene	11,000
	Whiterocks, Utah	Jurassic	1,900
	Edna, California	Mio-Pliocene	6,595
	Poor Springs, Utah	U. Eocene	1,735
	Santa Rosa, New Mexico	Triassic	4,630+
	Sisquoc, California	U. Pliocene	175
	Asphalt, Kentucky	Pennsylvanian	7,000
	Davis-Dismal Creek, Kentucky	Pennsylvanian	1,900
	Santa Cruz, California	Miocene	1,200
	Kyrock, Kentucky	Pennsylvanian	900*
	Sunnyside, Utah	U. Eocene	34,300*
Albania	Selenizza	Mio-Pliocene	5,306
Trinidad	La Brea	U. Miocene	126
Rumania	Derna	Pliocene	459
USSR	Cheildag, Kazakhstan	M. Miocene	82*

Source: Frazer et al., 1976; Phizackerley and Scott, 1978.

		Bitumen character			Overburden thickness (ft)
sq. mi.	Thickness (ft)	Content (wt%)	°API	% sulfur	
9,000	0–300	–	–	–	0–1900
1,875	0–400	–	–	–	700–2600
1,625	400	–	–	–	300–1400
12,500		2–18	10.5	4.5	
?	60–80	7–16	10	0.9–2.2	0–2000*
9,000	3–10		10		0–3000*
2	2–9	64	8	5.9	Nil
150	80–300	10		0.7	0–100
–	11–254	11	8.6–12	0.5	0–2000
–	24–200				
3	900–1000	10	12	0.5	Nil
10	0–1200*	9–16	13	4.2	0–600*
3	1–250	9*			Shallow
7	0–100	4–8			0–40
	0–185	14–18	4–8		15–70
11	5–36	8–10			6–30
3	10–50	5			15–30
2	5–50	10–12			0–100
–	15–40	6–8			15
–	10–350	9	10–12	0.5	0–150
8	33–330	8–14	4.6–13.2	6.1	Shallow
–	0–270	54	1–2	6.0–8.0	Nil
–	6–25	15–22		0.7	Shallow
–		5–13			Shallow

There are many definitions of petroleum-type materials based on API gravity. More recently, a system of classification has been developed (analogous to that for coal) which attempts to define in detail all heavy oil/bitumens (Figure 10 in Chap. 2). It should be noted that heavy oils, being more mobile than bitumens, are often noted as occurring in "fields" whereas bitumens are often noted as occurring in "deposits" (Meyer and Dietzman, 1981). This simple interchange of words may also indicate the means of recovery—thermal or mining, respectively.

Tar sand bitumen is typically found as a solid or near-solid material in consolidated or unconsolidated sandstone at depths that are usually less than 5000 ft where a portion of the formation may also appear as an outcrop to the surface. The bitumen is differentiated from heavy oil by viscosity and API gravity (Chap. 3, Sec. III). Tar sand bitumen will usually have an API gravity less than 10 and a viscosity in excess of 100,000 poise at formation temperature (about 4°C, 37°F; note that conventional petroleum has an API gravity in excess of 20° and a viscosity less than 100 poises; Table 2).

A major aspect of oil sand character that is particularly appealing is the amount of in-place bitumen (Table 3) relative to the other major fuel sources (i.e., conventional petroleum) and a source such as oil shale. It must be remembered, however, that figures given for in-place bitumen and recoverable bitumen (or oil equivalent) are often quite diverse. Therefore, all such estimates must be treated with some degree of caution as they are subject to different interpretation.

The concentration of the bitumen in the tar sand may be as high as 2000 bbl/acre foot which is equivalent to approximately 12% by weight or up to 40 gal/ton of tar sand.

A recently published study (Lewin Associates, 1984) indicated that 53.7 billion (53.7×10^9) barrels of bitumen exists in-place in the United States (Table 4). In addition over 400 small, unmeasured deposits are known to exist throughout the United States (Figures 2 and 3) but the major resource is found in the state of Utah and represents approximately 35–40% of the total tar sand bitumen in the United States. The U.S. resources are not well defined (in comparison to the Canadian resources) and only the tar sand resources in Utah, California, Kentucky, Texas, and New Mexico have received any attention, with the Utah resource receiving by far the most effort.

While commercialization has just begun to realize the potential of the Canadian deposits (McMillan, 1981), the prognosis for the deposits in the United States is not as straightforward. The deposits are not as concentrated (geographically), there are variations in overburden thickness that could influence choice of recovery options, and the deposits are not yet adequately defined.

Table 2 Comparison of Bitumen Properties With Those of Crude Oil

Property		Bitumen	Conventional
Gravity, °API		8.6	25–37
Distillation	Vol %	°F	
	IBP	–	
	5	430	
	10	560	–
	30	820	
	50	1010	
Viscosity, sus. @ 100°F (38°C)		35,000	<30
sus. @ 210°F (99°C)		513	
Pour point, °F		+50	≤0
Elemental analysis, wt %			
Carbon		83.1	86
Hydrogen		10.6	13.5
Sulfur		4.8	0.1–2.0
Nitrogen		0.4	0.2
Oxygen		1.1	
Hydrocarbon type, wt %			
Asphaltenes		19	≤5
Resins		32	
Oils		49	
Metals, PPM			
Vanadium		250	≤100 (Vanadium, Nickel, Iron, Copper)
Nickel		100	
Iron		75	
Copper		5	
Ash, wt %		0.75	0
Conradson carbon, wt %		13.5	1–2
Net heating value, btu/lb		17,500	About 19,500

Source: Speight, 1978a.

The oil sand deposits of the world (Figure 1 and Table 1) have been described as belonging to two types: in situ deposits resulting from breaching and exposure of an existing oil trap, and migrated deposits resulting from accumulation of migrating oil at outcrop, but

Table 3 Estimates of Oil Available in Tar Sand, Petroleum, and Oil Shale Resources

Commodity	Africa	Canada	Europe	Venezuela	USSR	USA	Totals
Tar sand							
Resource 10^9 bbl	1.75	2,000.00	0.4	1,000.00	168.00	36.2	3,206.35
Reserve 10^9 bbl[a]	.175	333.00	.06	100.00	16.00	2.90	452.135
Crude oil							
Resource 10^9 bbl	475.8[b]	50	198.3[b]	140.0	558.3[b]	260.0	1,542.4
Reserve 10^9 bbl	57.1	6.8	23.8	17.9	67.0	26.5	199.1
Oil shale							
Resource 10^9 bbl	356[c]	0	1,043[c]	0	—	2,800.00	4,199
Reserve 10^9 bbl	1.4		4.1			11.0	16.5

[a]Country estimates.
[b]Resource calculated on assumption that reserves are 12% of original oil in-place.
[c]Estimated on basis of U.S. recovery.

Table 4 Estimation of Bitumen In Place (bbl × 10^9) in the United States[a]

	Measured	Speculative	Total
(A) Major deposits (greater than 100 million barrels)			
Alabama	1.8	4.6	6.4
Alaska	–	10.0	10.0
California	1.9	2.6	4.5
Kentucky	1.7	1.7	3.4
New Mexico	0.1	0.2	0.3
S. Oklahoma	–	0.8	0.8
Texas	3.9	0.9	4.8
Tristate (KS, MO, OK)	0.2	2.7	2.9
Utah	11.9	7.5	19.4
Wyoming	0.1	0.1	0.2
Subtotal	21.6	31.1	52.7
(B) Minor deposits (between 10 and 100 million barrels)			
Alabama	–	0.1	0.1
California	–	0.2	0.2
Utah	–	0.7	0.7
Subtotal	--	1.0	1.0

[a]See also Figure 2.
Source: Lewin, 1984.

there are inevitably gradations and combinations of these two types of deposits. The deposits have been layed down over a variety of geologic periods and in different entrapments (Figure 4, Table 5) and a broad pattern of deposit entrapment is believed to exist since all deposits occur along the rim of major sedimentary basins and near the edge of pre-Cambrian shields. The deposits either transgress an ancient relief at the edge of the shield (e.g., those in Canada) or lie directly on the ancient basement (e.g., as in Venezuela, West Africa, and Madagascar).

A feature of major significance in at least five of the major areas is the presence of a regional cap (usually a widespread transgressive marine shale such as occurs in the Colorado Group in Western Canada,

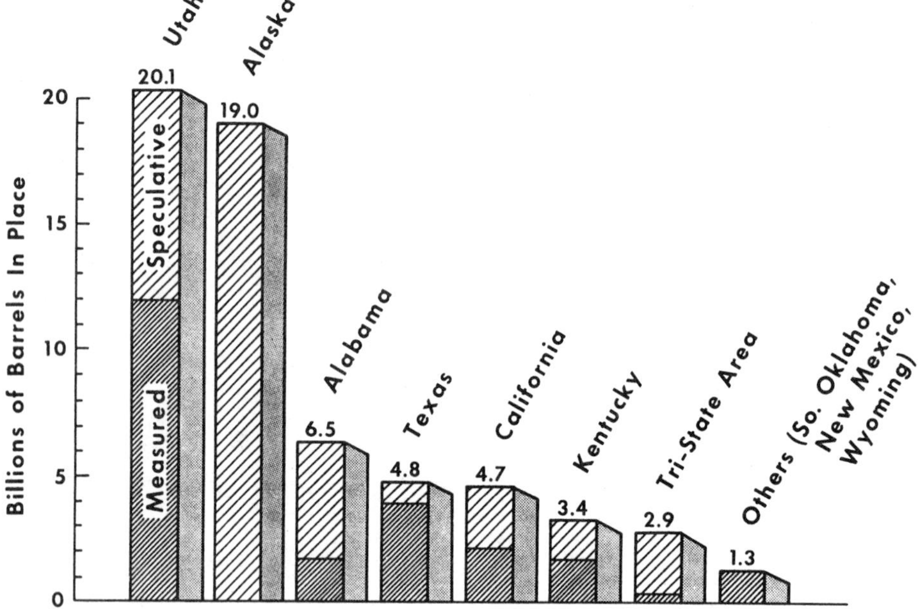

Figure 2 Distribution of the tar sand resources of the United States.

in the Freites formation in eastern Venezuela, or in the Jurassic formation in Melville Island) overlying the formation. The cap plays an essential role in restraining vertical fluid escape from the basin thereby forcing any fluids laterally into the paleodelta itself. Thus, the subsurface fluids were channeled into narrow outlets at the edge of the basin.

Entrapment characteristics for the very large tar sands all involve a combination of stratigraphic-structural traps; there are no very large ($>4 \times 10^9$ bbl) oil sand accumulations either in purely structural or in purely stratigraphic traps. In a regional sense, structure is an important aspect since all of the very large deposits occur on gently sloping homoclines.

I. DEFINITIONS AND TERMINOLOGY

The expression "tar sands" is commonly used in the petroleum industry to describe sandstone reservoirs that are impregnated with a heavy viscous black crude oil which cannot be produced through a well

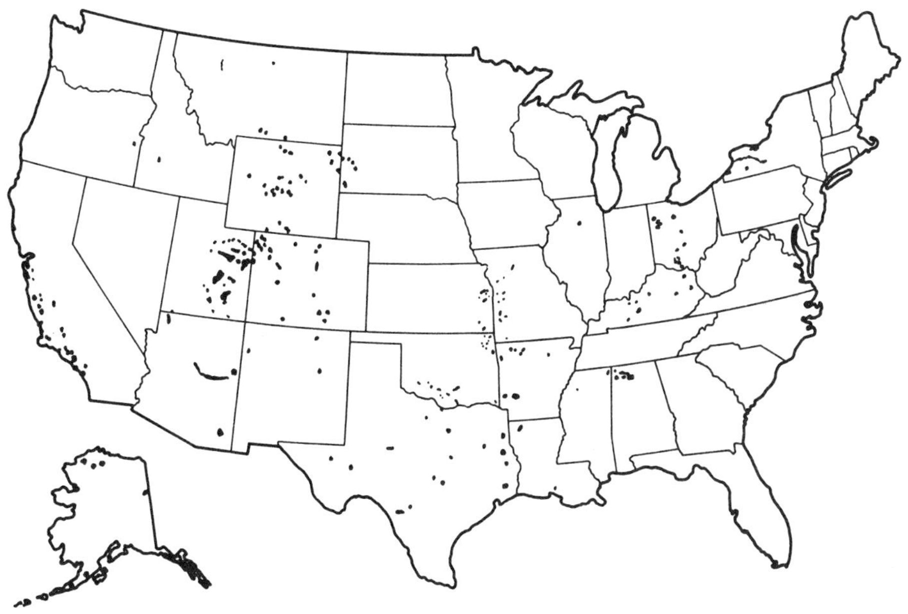

Figure 3 Occurrences of tar sand deposits throughout the United States (Ball Associates, 1965).

by conventional production techniques. However, the term tar sand is actually a misnomer; more correctly, "tar" is usually applied to the heavy product remaining after the destructive distillation of coal or other organic matter (Speight, 1980). Thus, alternate names such as bituminous sands or oil sands are gradually finding usage with the former name (bituminous sands) being more technically correct (Chap. 2, Sec. I). The term "oil sands" is used in the same way as the term "oil shale," i.e., in reference to the final product obtained. Nevertheless, the terms are used interchangeably and this will be continued throughout the present volume.

II. DEPOSITS OF CANADA AND THE UNITED STATES

A. Athabasca (Berkowitz and Speight, 1975; Strom and Dunbar, 1981)

The town of McMurray, about 240 miles north-northeast of Edmonton, Alberta, Canada, lies at the eastern margin of the largest accumulation

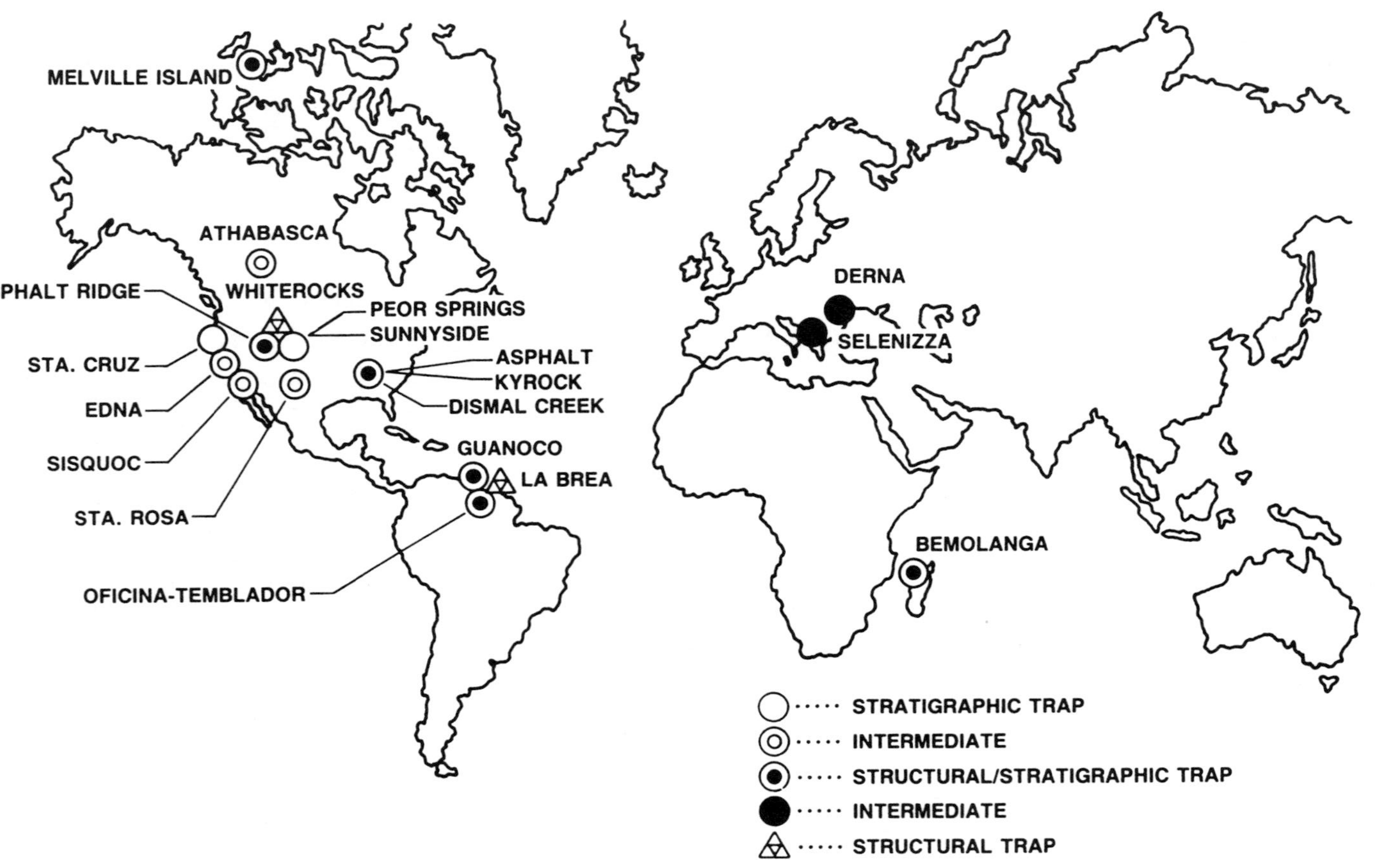

Figure 4 Types of traps of the major oil sand deposits (Walters, 1974).

Table 5 Mode of Entrapment of Various Tar Sand Deposits

1. Stratigraphic trap deposits. Structure is of little importance; short distance of migration is assumed.	Sunnyside, Peor Springs, Santa Cruz
2. Structural-stratigraphic trap deposits. Folding-faulting and unconformity are equally important.	Oficina–Temblador tar, Bemolanga, Asphalt Ridge, Melville Island, Guanoco, and the Kentucky deposits
3. Structural trap deposits. Structure is very important; long-distance migration is assumed; unconformity may be absent.	Whiterock, La Brea
4. Intermediate between 1 and 2.	Athabasca, Edna, Sisquoc, Santa Rosa
5. Intermediate between 2 and 3.	Selenizza, Derna

Source: Walters, 1974.

in the world, which is, in effect, three major accumulations within the lower cretaceous deposits.

The McMurray–Wabasca reservoirs are found toward the base of the formation and are characteristically cross-bedded coarse grits and gritty sandstones that are unconsolidated or cemented by tar; fine-to-medium-grained sandstones and silts occur higher in the sequence. Bluesky–Gething and Grand Rapids reservoirs are composed of subangular quartz and well-rounded chert grains. The sandstones of both of these deposits are frequently glauconitic and have a calcareous matrix. In addition, the McMurray–Wabasca tar sand deposit dips at between 5 and 25 ft/mile to the southwest. The Bluesky–Gething sands overlie several unconformities between the Mississippian and Jurassic deposits.

B. Melville Island (Walters, 1974; Phizackerley and Scott, 1978)

On the north shore of Marie Bay, Melville Island, some 1450 miles north of Edmonton, Triassic sandstones of the Bjorne Formation are impregnated with a bituminous material. This deposit was discovered in 1962 and the sands occur at intervals along a 60-mile outcrop. The richer sands tend to be associated with structurally high areas or are closely related with faults.

C. Utah (Walters, 1974; Glassett and Glassett, 1976; Phizackerley and Scott, 1978)

The major oil sand deposits of the United States occur within and around the periphery of the Uinta Basin. These include the Asphalt Ridge, Sunnyside, Tar Sand Triangle, and Peor Springs deposits.

Asphalt Ridge lies on the northeastern margin of the central part of the Uinta Basin at the contact of the tertiary beds with the underlying cretaceous Mesaverde Group. The Mesaverde Group is divided into three formations, two of which, the Asphalt Ridge and Rim Rock sandstones, are beach deposits containing the viscous oil. The Rim Rock sandstone is thick and uniform with good reservoir characteristics and may even be suitable for thermal recovery methods. The Duchesne River formation (lower Oligocene) also contains bituminous material but the sands tend to be discontinuous.

The Sunnyside deposits are of a greater areal extent than Asphalt Ridge and are located on the southwest flank of the Uinta Basin. The oil sand accumulations occur in sandstones of the Wasatch and lower Green River formations (Eocene). The Wasatch sandstones contain oil impregnations but are lenticular and occupy broad channels cut into the underlying shales and limestones; the Green River beds are more uniform and laterally continuous. The source of oil in the Asphalt Ridge and Sunnyside accumulation is considered to be the Eocene Green River shales.

The Peor Springs accumulation is about 60 miles (96.5 km) east of the Sunnyside deposit and occurs as lenticular sandstones (Eocene Wasatch formation). There are two main beds from 30 to 85 ft (9 to 26 m) thick with an estimated overburden thickness of 0–250 ft (0–76 m). The tilt of the southern flank of the Uinta Basin has left this deposit relatively undisturbed except for erosion, which has stripped it of its cover allowing the more volatile constituents to escape.

In the central southeast area of Utah, some deposits of oil-impregnated sandstone occur in Jurassic rock, but the great volume of in-place oil occurs in rocks of Triassic and Permian age. The Tar Sand Triangle is considered to be a single, giant stratigraphic trap containing the oil.

D. California (Hallmark, 1981)

The deposits are concentrated in the coastal region west of the San Andreas fault. The largest deposit is the Edna deposit, which is located midway between Los Angeles and San Francisco. It is fossiliferous and consists of conglomerate, sandstone, diatomaceous sandstone, and siliceous shale. The deposit occurs as a stratigraphic trap and outcrops in scattered areas on both flanks of a narrow syncline. The deposit extends over an area of about 7000 acres (28,327,600 m^2) and occurs from outcrop to 100-ft (30-m) depth. The accumulations are considered to have been derived from the underlying organic and petroliferous Monterey shale.

The Sisquoc deposit (upper Pliocene) is the second largest in California and occurs in a sandstone in which there are as many as eight individual oil sand units. The total thickness of the deposit is about 185 ft (56 m) occurring over an area of about 175 acres with an overburden thickness between 15 and 70 ft (4.6 and 21 m). The reservoir sands lie above the Monterey shale, which has been suggested to be the source of the bitumen.

The third California deposit at Santa Cruz is located approximately 56 miles (90 km) from San Francisco. The material occurs in sandstones of the Monterey and Vaqueros formations, which are older than both the Edna and Sisquoc reservoir rocks. The Santa Cruz oil sands are discontinuous and overlie precretaceous basement.

E. Texas (Whiting, 1981)

South Texas holds the largest reserves in the state of heavy crude and oil sand deposits. The oil sand deposits occur in the San Miguel "tar belt" (upper cretaceous) mostly in Maverick and Zavala counties as well as in the Anacadro limestone (upper cretaceous) of the Uvalde district.

F. Kentucky (Walters, 1974; Phizackerley and Scott, 1978; Whiting, 1981)

The oil sand deposits are located at Asphalt, Davis-Dismal Creek, and Kyrock; they all occur in nonmarine Pennsylvanian or Mississippian sediments. The three deposits appear as stratigraphic traps and are thought to have received their oil from the Devonian Chattanooga shale.

G. New Mexico (Walters, 1974; Phizackerley and Scott, 1978; Whiting, 1981)

The oil sand occurs in the Triassic Santa Rosa sandstone, which is an irregularly bedded, fine- to medium-grained, micaceous sandstone. The oil is thought to have migrated upward from the underlying Permian San Andreas limestone through sinkholes in the karst surface of this formation.

H. Missouri (Phizackerley and Scott, 1978; Whiting, 1981)

Oil sands occur over an area estimated at 2000 square miles (5180 km^2) in Barton, Vernon, and Cass Counties and the sandstone bodies which contain the oil are middle Pennsylvanian in age. The individual oil-bearing sands are approximately 50 ft (15 m) in thickness except where they occur in channels which may actually be as much as 250 ft (76 m) thick. The two major reservoirs are the Warner and Bluejacket sandstones, which at one time were regarded as blanket sands covering large areas. However, recent investigations suggest that these sands can abruptly grade into barren shale or siltstones.

III. OTHER DEPOSITS

A. Venezuela (Walters, 1974; Phizackerley and Scott, 1978)

The Officina/Tremblador tar belt is believed to contain bitumen-impregnated sands of a similar extent to those of Alberta. The Officina formation overlaps the Tremblador (cretaceous) formation and the organic material is a "typical" bitumen having an API gravity <10°.

The Guanoco asphalt lake occurs in deposits which rest on a formation of mio-Pliocene age. This formation, the Las Piedras, is principally brackish to freshwater sandstones with associated lignites. The Las Piedras formation overlies a marine upper cretaceous group; the Guanoco Lake asphalt is closely associated with the Guanoco oil field, which produces heavy oil from shales and fractured argillites of the upper cretaceous group.

B. Madagascar (Walters, 1974; Phizackerley and Scott, 1978)

The Bemolanga deposit is the third largest oil sand occurrence presently known and extends over some 150 square miles (388 km^2) in western Madagascar with a recorded overburden of 0–100 ft (0–30 m). The average pay thickness is 100 ft (30 m) with a total oil in-place quoted at 1.75×10^9 bbl (286,387,500 m^3).

The deposit is of Triassic age and the sands are cross-bedded continental sediments; the coarser, porous sands are more richly impregnated. The origin of the deposit is not clear; the most preferred source is the underlying shale or in down-dip formations implying small migration.

C. Albania (Walters, 1974; Phizackerley and Scott, 1978)

The largest oil sand deposit in Europe is that at Selenizza, Albania. This region also contains the Patos oil field throughout which there occur extensive asphalt impregnations. This deposit occurs in middle-upper Miocene lenticular sands, characterized by a brackish water fauna. Succeeding Pliocene conglomeratic beds, which are more generally marine, are also locally impregnated with heavy oil. The Selenizza and Patos fields occupy the crestal portions of a north-south trending anticline. The vertical distribution of the accumulation is also controlled by faulting. The Miocene rests on Eocene limestones and it is these which are thought by some to be the source of the tar.

D. Trinidad (Walters, 1974; Phizackerley and Scott, 1978)

The Trinidad Asphalt Lake [situated on the Gulf of Paria, 12 miles (19 km) west-southwest of San Fernando and 138 ft above sea level] occupies a depression in the Miocene sheet sandstones. It overlies an eroded anticline of upper cretaceous age with remnants of an early tertiary formation still preserved on the flanks. The deposit is thought to be derived from the argillite of the upper cretaceous formation which is known to contain heavy oil. After loss of the light ends during erosion and folding, the source beds were covered by Miocene clays after which tensional faulting allowed asphaltic material to rise and impregnate the overlying upper Miocene freshwater sands.

E. Rumania (Walters, 1974; Phizackerley and Scott, 1978)

The Derna deposits are located (along with Tataros and other deposits) in a triangular section east and northeast of Oradia between the Sebos Koros and Berrettyo rivers. The oil sand occurs in the upper part of the Pliocene formation and the asphalt is characterized by its penetrating odor. The reservoir rock is nonmarine, representing freshwater deposition during a period of regression.

F. Soviet Union (Walters, 1974; Phizackerley and Scott, 1978)

Oil sands occur at Cheildag, Kobystan and outcrop in the south flank of the Cheildag anticline; there are approximately 24 million barrels of oil in place. Other deposits in the USSR occur in the Olenek anticline (northeast of Siberia) and it has been claimed that the extent of asphalt impregnation in the Permian sandstones is of the same order of magnitude (areally and volumetrically) as that of the Athabasca deposits. Oil sands have also been reported from sands at Subovka and the Notanebi deposit (Miocene sandstone) is reputed to contain 20% bitumen by weight. On the other hand, the Kazakhstan occurrence, near the Shubar-Kuduk oil field, is a bituminous lake with an "oil" content that has been estimated to be of the order of 95%.

G. Columbia (Walters, 1974)

Oil sand occurrences also occur in the Southern Llanos of Colombia where drilling has presented indications of deposits generally described as asphalt, heavy oil, bitumen, and tar. Most of these occurrences are recorded below 1500 ft (457 m).

H. Miscellaneous Deposits (Walters, 1974; Phizackerley and Scott, 1978)

The tar sands at Burgan in Kuwait and at the Inciarte and Bolivar coastal fields of the Marcaibo Basin are of unknown dimensions. Those at Inciarte have been exploited and all are directly or closely associated with large oil fields. The oil sands of the Bolivar coastal fields are above the oil zones in Miocene beds and are in a lithological environment similar to that of the Officina–Tremblador tar belt.

The small Miocene asphalt deposits in the Leyte Islands (Philippines) are extreme samples of stratigraphic entrapment and resemble some of the Californian deposits. Those of the Mefang Basin in Thailand are in Pliocene beds which overlie Triassic deposits and their distribution is stratigraphically controlled.

Finally, there is a small accumulation at Chumpi, near Lima (Peru), which occurs in tuffaceous sands and it is believed to be derived from strongly deformed cretaceous limestones from which oil was distilled as a result of volcanic activity.

Oil sand deposits have also been recorded in Spain, Portugal, Cuba, Argentina, Thailand, and Senegal but most are poorly defined and are considered to contain (in-place) less than 1 million barrels ($<1 \times 10^6$ bbl) of oil.

IV. RECOVERY

Bitumen recovery processes can be conveniently divided into two categories: (1) above ground and (2) in situ. In the former type of process, the oil sand must first be removed from the formation by a mining technique and then transported to a bitumen recovery center. In the latter type of process, the bitumen (or a portion of the bitumen in-place) is recovered from the formation by a suitable thermal method, leaving the formation somewhat less disturbed than when the mining method is employed.

However, it must also be recognized that the successful recovery technique that is applied to one deposit/resource is not necessarily the technique that will guarantee success for another deposit. There are sufficient differences (Table 6) even between the U.S. and Canadian tar sand deposits that general applicability is not guaranteed. Hence, caution is advised when applying the knowledge gained from one resource to the issues of another resource. Although the principles may at first sight appear to be the same, the technology must be adaptable.

The mining technique has received considerable attention since it was chosen as the technique of preference for the only two commercial bitumen recovery plants in operation in North America. In situ processes have been tested many times in the United States, Canada, and other parts of the world and are ready for commercialization. There are also conceptual schemes that are a combination of both mining (above-ground recovery) and in situ (nonmining recovery) methods.

The mining recovery method consists of four individual operations:

1. Oil sands mining
2. Bitumen recovery

Table 6 Comparisons Between the U.S. and Canadian Tar Sand Resources

Canada	United States
1. The sand is water-wet. Thus the disengagement of the bitumen is very efficient using the hot water process. The caustic is sodium hydroxide. Bitumen recovery is $>98\%$.	1. The sand is oil-wet. The efficient disengagement of the bitumen requires a somewhat different process than that used in Canada. This process uses sodium carbonate as the caustic and high-shear rates. Bitumen recovery is about 95%.
2. The formations are usually unconsolidated.	2. The formations are usually consolidated to semi-consolidated. Mineral cementation is usually the cause of consolidation.
3. Only a few deposits have been identified. However, they are very large and thick. Alberta contains almost 2.5 trillion barrels of bitumen.	3. Numerous deposits have been identified. There are 33 major deposits containing more than 100 million barrels of bitumen each and 20 minor deposits containing 10–100 million barrels of bitumen each. The total resource is 54 billion barrels of bitumen, 22 billion measured and 32 billion speculative. The deposits are frequently thin and interbedded with shale.

4. The clays present in the tar sand deposits and in-process streams have been well studied. However, there are still problems encountered in the settling and removal of clay from the process streams.	4. Very little is known about the nature and effect on processing of the clays present in U.S. deposits.
5. Bitumen properties are fairly uniform from deposit to deposit. For example, sulfur varies from 4.5 to 5.5 wt % and nitrogen from 0.1 to 0.5 wt %. In addition, the H/C ratio is about 1.5 and the API gravity varies only from 6 to 12°.	5. Bitumen properties are quite diverse from deposit to deposit. For example, sulfur varies from 0.5 to 10 wt % and nitrogen from 0.1 to 1.3 wt %. In addition, the H/C ratio varies from 1.3 to 1.6 and the API gravity from -2 to 14.
6. The deposits are large with bitumen that is fairly uniform in quality. Recovery and upgrading plants have been on-stream for two decades.	6. The deposits are smaller than those in Canada and the bitumens are not of a uniform quality. The recovery and upgrading methods will need to be "site specific."

Source: Adapted from the *Heavy Oiler*, 1986.

3. Bitumen upgrading, and
4. Nonproduct disposal

On the other hand, the nonmining methods consist of two individual operations:

1. Bitumen recovery, and
2. Bitumen upgrading

There is also a final operation that is common to both methods and that is environmental cleanup. This can vary from cleanup of the surface as will be necessary for the mining operations to cleanup of underground systems (particularly aquifers) as will be required for any nonmining operation.

There have been many recovery processes enumerated for both the mining and nonmining methods (Figure 5) but the focus here will be on those methods that have passed beyond the conceptual and laboratory stages and that have the greatest potential for application to the Canadian and U.S. tar sand deposits.

A. Mining Operations

The bitumen occurring in oil sand deposits poses a major recovery problem. The material is notoriously immobile at formation temperatures and must therefore require some stimulation (usually by thermal means) in order to ensure recovery. Alternately, proposals have been noted which advocate bitumen recovery by solvent flooding or by the use of emulsifiers (Figure 5). There is no doubt that with time one or more of these functions may come to fruition, but for the present the two commercial operations rely on the mining technique (Dick and Wimpfen, 1980; Houlihan, 1984).

Even though estimates of the recoverable oil from the Athabasca deposits are only of the order of 27×10^9 bbl of synthetic crude oil (representing <10% of the total in-place material), this is, for the Canadian scenario, approximately six times the estimated volume of recoverable conventional crude oil. In addition, the comparative infancy of the development of the alternative options (Figure 5) almost ensured the adoption of the mining option for the first two (and even later) commercial ventures.

Underground mining options have also been proposed (Frazier et al., 1976) but for the moment have not been developed because of the fear of collapse of the formation onto any operation/equipment.

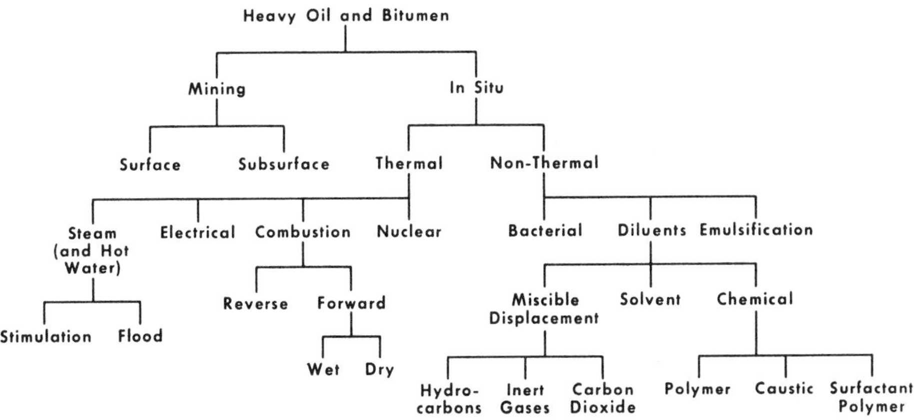

Figure 5 General overview of bitumen recovery process.

This particular option should not, however, be rejected out-of-hand because a novel aspect or the requirements of the developer (which remove the accompanying dangers) may make such an option acceptable.

The equipment employed at an oil sands mine is a combination of mining equipment and an on-site transportation system which may (currently) consist of conveyor belts and/or large trucks (Table 7). The mining operation is currently carried out using 8100 ton/hr bucket-wheel excavators (Suncor) and 60 m^3 (approx. 79 yd^3) capacity draglines (Syncrude) as the primary mining equipment. The

Table 7 Equipment Usually Considered for Tar Sand Mining Operations

Mining	Conveying
1. Draglines	1. Belt conveyors
2. Bucket-wheel excavators	2. 150- to 299-ton trucks
3. Power shovels	3. Trains
4. Scrapers	4. Scrapers
5. Bulldozer with front-end loaders	5. Hydraulic

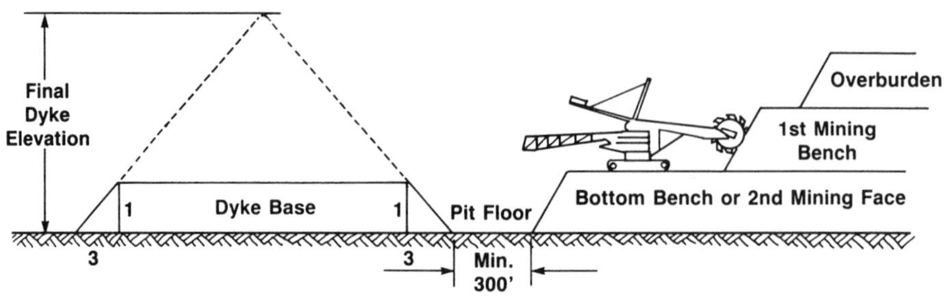

Figure 6 Schematic representation of an oil sands mine using bucket-wheel excavators.

mining operation itself differs in detail depending on the equipment; bucket-wheel excavators "sit" on benches (Figure 6) while the draglines sit on top of the surface (Figure 7).

The choice of a bucket-wheel excavator/conveyor belt system offered a means of continuously feeding ore to the processing plant but the technique was not without problems. Excessive bucket and belt wear and low machine availability all contributed to interruptions in production in the early years.

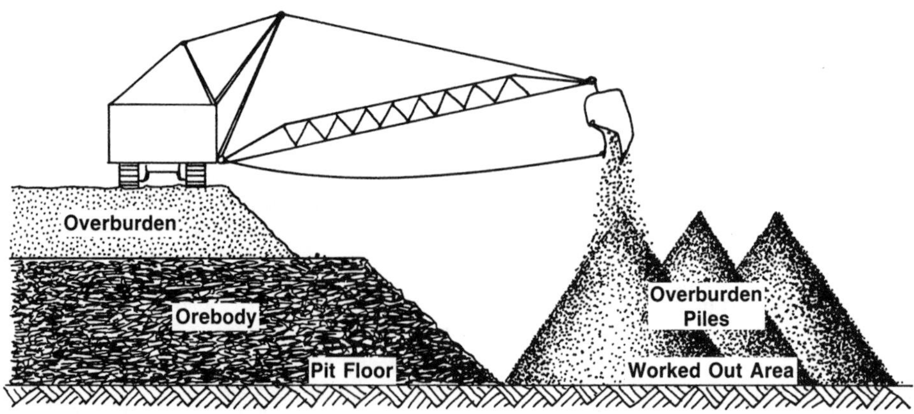

Figure 7 Schematic representation of an oil sands mine using a dragline.

The remoteness of the U.S. tar sands (Glassett and Glassett, 1976; Frazier et al., 1976; Kuuskraa et al., 1978) is often cited as a deterrent to development. However, there is more to tar sand development than a remote or nonremote location. Site topography, overburden-to-ore-body ratio, and richness of the ore body are equally important. Thus, location becomes only one of several factors.

In the present context of mining the U.S. tar sands, the Utah deposits (Tar Sand Triangle, P.R. Spring, Sunnyside, and Hill Creek deposits) generally have an overburden-to-net-pay-zone ratio above the 0.4–1.0 range with a lean oil content (Kuuskraa et al., 1978). On the other hand, the Asphalt Ridge deposit is loosely consolidated and could be mined using a ripper/front-end loader (without drilling and blasting) at the near-surface location of the deposit. Estimates (Kuuskraa et al., 1978) put oil recovery from mining at $100-200 \times 10^6$ bbl.

Application of bucket-wheel excavators to the U.S. oil sand deposits does not appear feasible because of the relatively small size of these deposits and the lack of sufficient oil sand to (economically) justify erection of a bucket-wheel/conveyor system (Reed, 1979). The terrain found in the largest U.S. deposits (i.e., pitching beds in mountainous regions) are just the opposite of the terrain (i.e., horizontal beds in a "prairie region") desirable for bucket-wheel excavators (Frazier et al, 1976; Kuuskraa et al., 1978).

Syncrude Canada Ltd. chose a dragline for their operation. The dragline affords a method for overburden stripping as well as oil sand mining but, like the bucket-wheel excavator, is best suited for the more level terrain containing relatively horizontal beds of the ore body. In addition, the dragline is usually limited to depths not exceeding 100 ft. Again, like the bucket-wheel excavators, draglines do not appear feasible for the U.S. deposits (Reed, 1979).

The U.S. deposits may be more suited to a mining technique involving power shovels and trucks. Such a technique provides a flexible mining operation which can be adapted to the many changes in terrain and ore body character. The venture would require a pit with a successive series of benches but such an operation would require stockpiling large quantities of overburden and waste solids until the time when backfilling can occur.

One other aspect of a mining operation that needs to be addressed is the abrasiveness of the sand to the cutting edges of the mining equipment. In-place tar sands are extremely hard and can cause severe "wear-and-tear" damage to the mining equipment. In addition, the northern deposits suffer from prolonged temperatures during the winter from -10° to -50°C (14° to -58°F). This problem may be circumvented to a degree by "loosening" the area to be mined in the autumn season by a series of explosive detonations.

Table 8 Brief Histories of Suncor and Syncrude

(a) Suncor Oil Sands Division

Started construction 1963 as Great Canadian Oil Sands (GCOS)

On-stream, September 1967

45,000 bbl/day synthetic crude (38° API) increased to 66,000 bbl/day in 1981

Schema: bucket-wheel excavators; belt conveyors; hot water extraction; froth dilution and centrifuging; delayed coking; naphtha, kerosene, gas oil hydrotreating; coke burned in boilers

Suncor since 1979: 72.8% owned by Sun; 25% Ontario; 2.2% public

(b) Syncrude Canada Ltd.

Started construction 1973

On-stream 1978

130,000 bbl/day

Schema: draglines; belt conveyors; hot water extraction; froth dilution and centrifuging; fluid coking; naphtha and gas oil hydrotreating; natural gas fuel

Owned by consortium; Esso 25%; Cities 17.6%; Gulf 13.4%; Petrocan 17%; AEC 10%; Alberta 8%; Pancanadian 4%; HBOG 5%

Frequent reference has been made to the two operational oil sand plants which are operated by Suncor and Syncrude (Table 8). Both have now been in operation for several years but are by no means adequate to fully realize the potential of the oil sand deposits. Both plants have, in their individual ways, acted as pioneering operations in a more remote and hostile part of the environment, which no doubt contributes to the phenomenal cost ($\$13 \times 10^9$) of an oil sands plant. Of course, it might now be conjectured that a nominal sized plant (about 50,000 bbl/day) might be more in order than the Syncrude plant (130,000 bbl/day) or the proposed Alsands (137,000 bbl/day) and Canstar (PetroCanada and Nova; 100,000--150,000 bbl/day) plants (Speight, 1978b; *Oilweek*, 1982a, 1982b).

B. Bitumen Recovery

The hot water process is to date the only successful commercial process to be applied to bitumen recovery from mined tar sands in North America (Clark, 1944; Carrigy, 1963a; Fear and Innes, 1967; Speight and Moschopedis, 1978). Many process options have been tested (Table 9) with varying degrees of success and one of these options may even supersede the hot water process. In view of the success of the hot water concept, a description with some degree of detail is warranted.

The process concept utilizes the linear and nonlinear variation of bitumen density and water density, respectively, with temperature (Chap. 14, Sec. III.A) so that the bitumen which is heavier than water at room temperature becomes lighter than water at about 80°C (180°F). Surface-active materials in the tar sand also contribute to the process (Moschopedis et al., 1977; Kessick, 1979; Sanford, 1983).

Table 9 Process Types Proposed for Bitumen Recovery

Hot water processes	
"Clark" process	Commercial, 1967
Guardian Chemical	Bench Scale, 1976
University of Utah	Bench Scale, 1976
Cold water processes	
Imperial Sand Reduction	Bench Scale, 1950s
NRC Spherical Agglomeration	Bench Scale, 1963
Hot water/solvent processes	
U.S. Bureau of Mines	Bench Scale, 1945
Union Oil	Pilot Scale, 1958
Major Oil/Arizona Fuels	Bench Scale, 1971
Cold water/solvent processes	
Canadian Mines Branch	Pilot Scale, 1940s
Fairbrim	Pilot Scale, 1975
Solvent processes	
Cities Service	Bench Scale, 1950s
Thermal processes	
NRC Direct Coking	Bench Scale, 1950s
Lurgi-Ruhrgas	Pilot Scale, 1970s

In the hot water extraction process (Figure 8), the oil sand feed is introduced into a "conditioning" drum. In this step the oil sand is heated, mixed with the slurrying water which is added, and agglomeration of the oil particles begins. The conditioning is carried out in a slowly rotating drum which contains a steam-sparging system for temperature control as well as mixing devices to assist in lump size reduction and a size ejector at the outlet end. The oil sand lumps are reduced in size by ablation and mixing action. The conditioned "pulp" has the following characteristics: (1) solids 60–85%; (2) pH 7.5–8.5; and (3) temperature 185 ± 15°F.

The conditioned pulp is screened through a double-layer vibrating screen. Water is then added to the screened material (to achieve more beneficial pumping conditions) and the pulp enters the separation cell through a central feed well and distributor. The bulk of the sand settles in the cell and is removed from the bottom as tailing, but the majority of the bitumen floats to the surface and is removed as froth. A middlings stream (mostly of water with suspended fines and some bitumen) is withdrawn from approximately midway up the side of the cell wall. Part of the middlings is recycled to dilute the conditioning drum effluent for pumping. Clays do not settle readily and generally accumulate in the middlings layer. High concentrations of clays increase the viscosity and can prevent normal operation in the separation cell. Thus, it is necessary to withdraw a drag stream to act as a purge; this is usually done at high clay concentrations but may not be as essential with a low-clay oil sand charge.

Under certain operating conditions, it may be necessary to withdraw a middlings stream to the scavenger cells (air flotation cells to recover bitumen from the drag stream). The froth from the scavenger unit(s) usually has a high mineral and water content which can be removed by gravity settling in froth settlers after which the froth is combined with the froth from the main separation cell from centrifuge plant for dewatering and demineralizing. Before the centrifuging operation, the froth is deaerated and naphtha added to lower the viscosity for a more efficient water and mineral removal operation.

The combined froth from the separation cell and scavenging operation contains an average of about 10% mineral material and up to 40% w/w water. The dewatering and demineralizing is accomplished in two stages of centrifuging; in the first stage the coarser mineral material is removed but much of the water remains. The feed then passes through a filter to remove any additional large-size mineral matter which would plug up the nozzles of the second stage centrifuges.

One of the major problems that comes from the hot water process is the disposal and control of the tailings. The fact is that each ton of oil sand in place has a volume of about 16 ft^3, which will generate about 22-ft^3 tailings giving a volume gain on the order of 40%. If

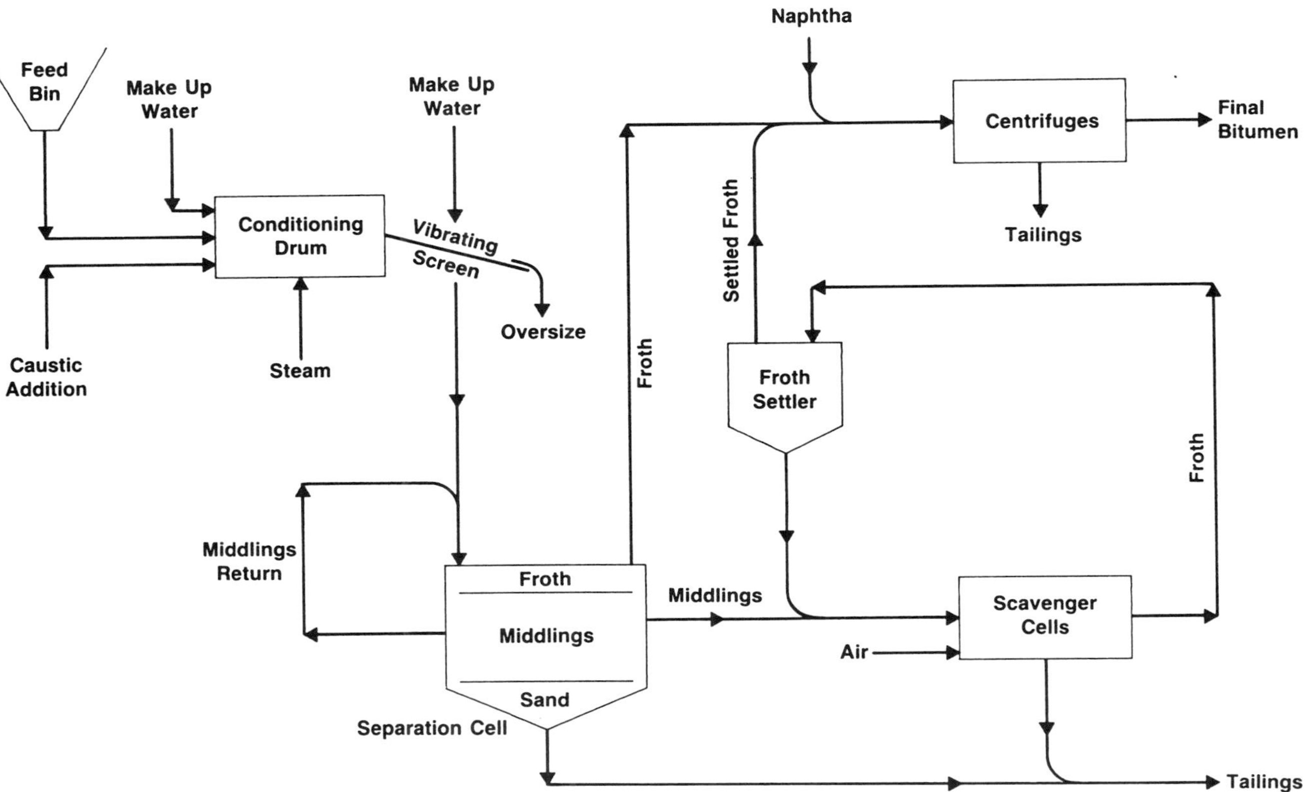

Figure 8 Schematic representation of the hot water recovery process.

the mine produces about 200,000 tons of oil sand per day, the volume expansion represents a considerable solids disposal problem.

Tailings from the process consist of about 49–50 wt % of sand, 1% bitumen, and about 50% water. The average particle size of the sand is about 200 μm and it is a suitable material for dyke building. Accordingly, Suncor used this material to build the sand dyke, but for fine sand, the sand must be well compacted.

The structure of the dyke may be stabilized on the upstream side by beaching (Figure 9). This gives a shallow slope but consumes sand during the season when it is impossible to build the dyke.

In remote areas such as the Fort McMurray (Alberta) site, the dyke can only be built in above-freezing weather because (1) frozen water in the pores of the dyke will create an unstable layer and (2) the vapor emanating from the water creates a fog, which can create a work hazard. The slope of the tailings dyke is about 2.5:1 depending on the amount of fines in the material. It may be possible to build with 2:1 slopes with coarser material, but steeper slopes must be stabilized quickly by bleaching.

The tailings sludge appears as a thick paste and is quite intractable with the following approximate composition:

Fines (-325 mesh; <50 μm)	25–30% w/w
Sand	5–10% w/w
Bitumen	2–5% w/w
Water	60–65% w/w

After discharge from the hot water separation system, it is preferable that attempts be made to separate the sand, sludge, and water; hence, the tailings pond. The sand is used to build dykes and the runoff which contains the silt, clay, and water collects in the pond. Silt and

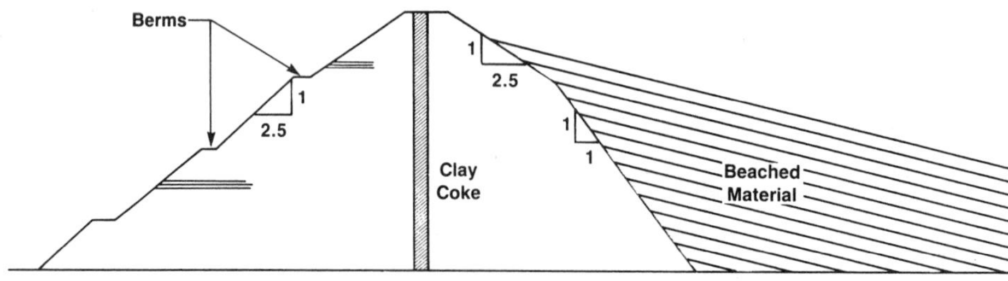

Figure 9 Schematic representation of the Tar Island dyke (Suncor Ltd).

some clay settle out to form sludge and some of the water is recycled to the plant (Figure 10).

One of the more promising concepts for reducing tailings pond size has been in the use of an oleophilic sieve which allows a reduction in the water requirements (Kruyer, 1983, 1984). The process is based on the concept that when a mixture of an oil phase and an aqueous phase is put through a sieve made from the oleophilic materials, the aqueous phase and any hydrophilic solids pass through the sieve but the oil adheres to the sieve surface on contact. The sieve is in the form of a moving conveyor; the oil is captured in a recovery zone and recovery efficiency is high.

The U.S. tar sands have received considerably less attention than the Canadian deposits. Nevertheless, approaches to recover the bitumen from U.S. tar sands have been made. In the present context an attempt has been made to develop the hot water process for the Utah sands (Misra et al., 1981; Miller and Misra, 1982). The process differs significantly from that used for the Canadian sands due to the contrast between the oil-wet Utah sands and the water-wet Canadian sands. This necessitates disengagement by hot water digestion in a high-shear force field under appropriate conditions of pulp density and alkalinity. The dispersed bitumen droplets can also be recovered by aeration and froth flotation. This concept has been taken (Hatfield and Oblad, 1982) to the pilot plant development stage

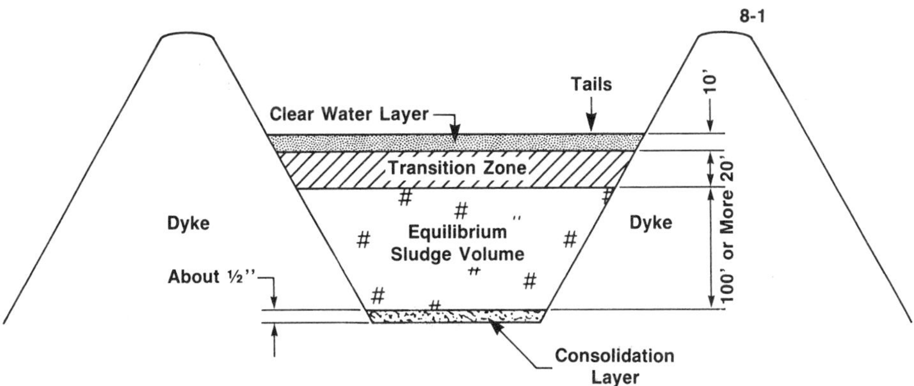

Figure 10 Schematic representation of the different aqueous zones in a tailings pond.

(125 tons/day tar sand feed) to produce 50--100 bbl/day of bitumen product. The final stage of the concept (Figure 11) uses a solvent for bitumen clean-up. The concept has been translated into a 5000 bbl/day process demonstration module with the option of expansion by addition of other 5000 bbl/day modules.

C. Direct Heating

The other above-ground method of separating bitumen from tar sands after the mining operation involves direct heating of the tar sand without any prior separation of the bitumen. Thus the bitumen is not recovered as such but as an upgraded overhead product. Although several processes have been proposed to accomplish this (Gishler, 1949; Steinmetz, 1969; Rammler, 1970; Donnelly et al., 1978), the

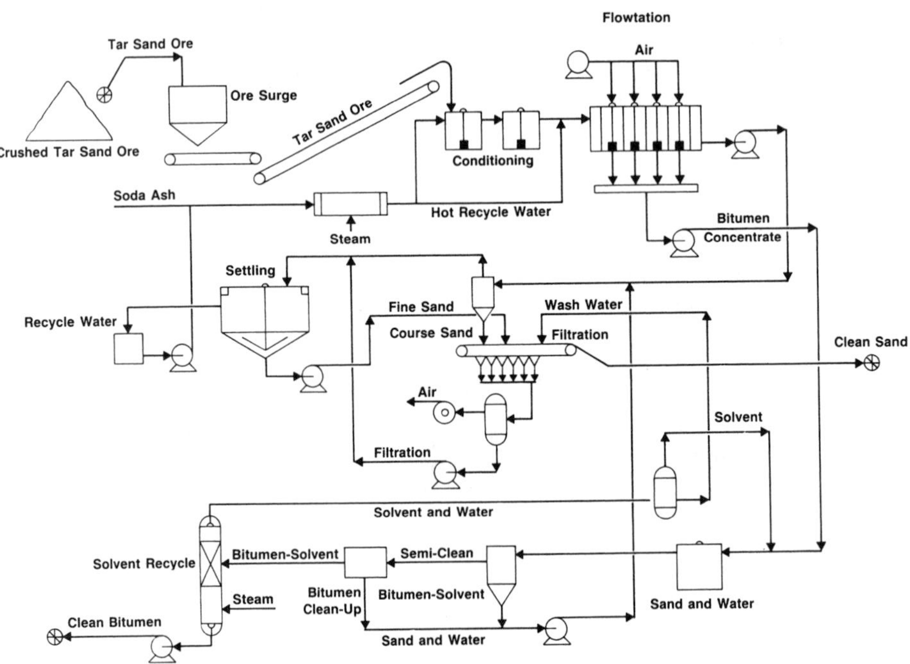

Figure 11 Schematic representation of a recovery scheme for the Utah tar sands (Miller and Misra, 1982).

common theme is to heat the tar sand to separate the bitumen as a volatile product. At this time, however, it must be recognized that the volatility of the bitumen is extremely low (Chap. 14, Sec. III.C) and what actually separates from the sand is a cracked product with the coke remaining on the sand.

In these processes, the sand is usually prepared (i.e., crushed) in a similar manner to that required for hot water extraction. The prepared sand is then introduced into a vessel where it is contacted with either hot (spent) sand or with hot product gases which furnish part of the heat required for cracking and volatilization. The volatile products are passed out of the vessel and are separated into gases and (condensed) liquids (Figures 12 and 13).

The coke that is formed as a result of the thermal decomposition of the bitumen remains on the sand, which is then transferred to a vessel for coke removal by burning in air. The hot flue gases can be used either to heat incoming tar sand or as refinery fuel.

As expected, processes of this type yield an upgraded product but require various arrangements of pneumatic and mechanical equipment for solids movement around the refinery.

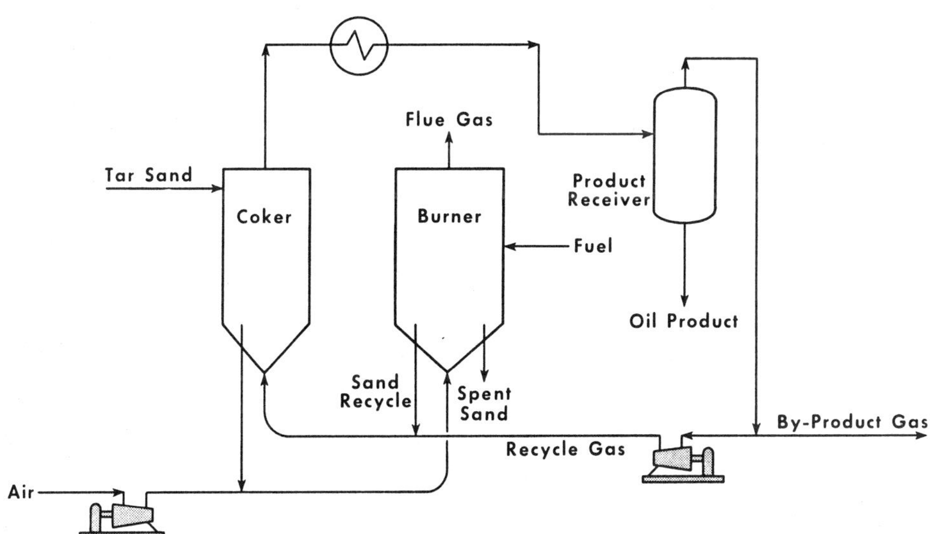

Figure 12 Schematic representation of a process for the direct heating of tar sand (Gishler, 1949).

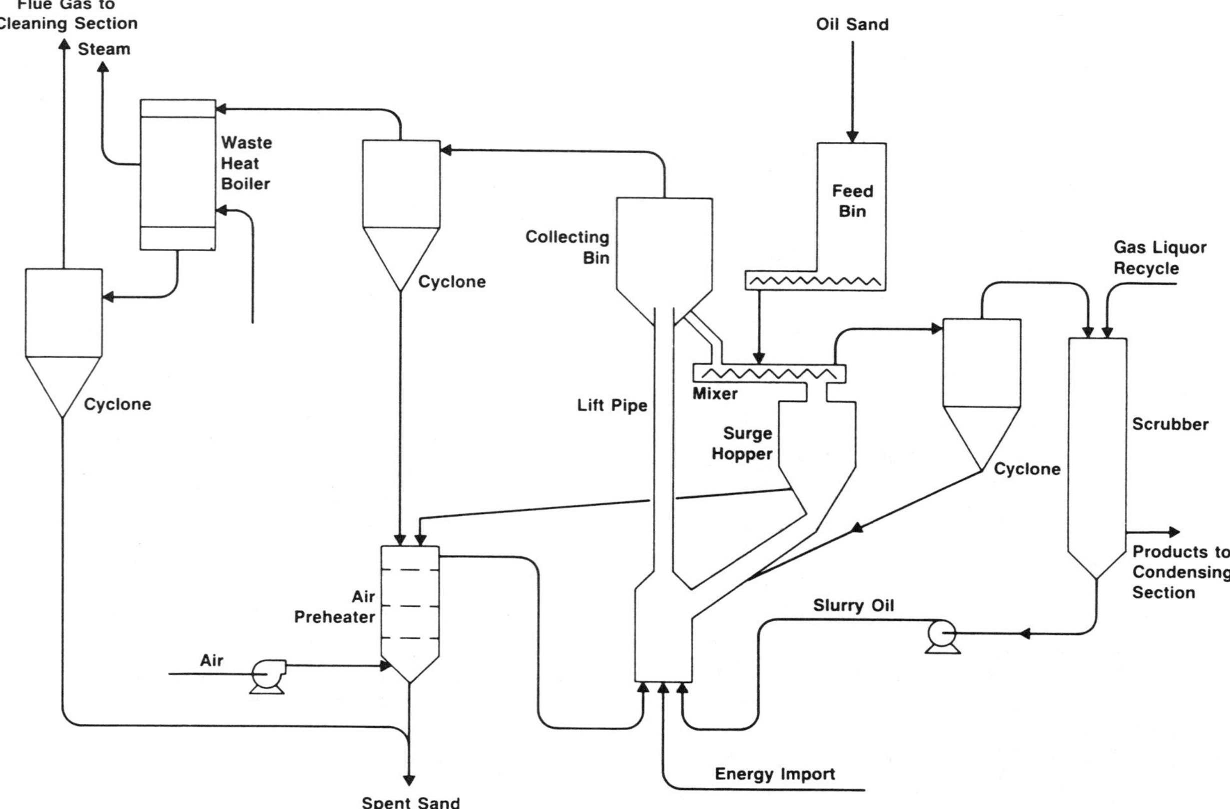

Figure 13 Schematic representation of the Lurgi-Ruhrgas process for the direct heating of tar sand (Remmler, 1970).

D. Nonmining Operations

One of the major deficiencies in applying mining techniques to bitumen recovery from tar sand deposits is (next to the immediate capital costs) the associated environmental problems. Moreover, in most of the known deposits, the vast majority of the bitumen lies in formations in which the overburden/pay zone ratio is too high (Govier, 1975; Burger, 1978). Therefore, it is not surprising that over the last two decades a considerable number of pilot plants have been applied to the recovery of bitumen by nonmining techniques from tar sand deposits—especially for the U.S. deposits (Table 10), where the local terrain and character of the tar sand may not always favor a mining option.

In principle, the nonmining recovery of bitumen from tar sand deposits requires the injection of a fluid into the formation through an injection well, the in situ displacement of the bitumen from the reservoir, and bitumen production at the surface through an egress (production well). There are, of course, variants around this theme but the underlying principle remains the same.

There are, however, several serious constraints that are particularly important and relate to bulk properties of the tar sand and the bitumen. In fact, both must be considered in toto in the context of bitumen recovery by nonmining techniques.

For example, the Canadian deposits are unconsolidated sands with a porosity ranging up to around 45% and have good intrinsic permeability. However, the U.S. deposits in Utah range from predominantly low-porosity, low-permeability consolidated sand to, in a few instances, unconsolidated sands (Table 6). In addition, the bitumen properties are not conducive to fluid flow under reservoir conditions in either the Canadian or U.S. deposits. Nevertheless, where the general nature of the deposits prohibits the application of a mining technique (as in many of the U.S. deposits), a nonmining technique may be the only feasible bitumen recovery option (Kuuskraa et al., 1978; Marchant and Koch, 1982).

The API gravity of bitumen varies upward from 5 to approx. 12 depending on the deposit and viscosities being very high (Chap. 3). Whereas conventional crude oils may have viscosities of several poise (at 40°C), the tar sand bitumens have viscosities on the order of 50,000–1 million poises at formation temperatures. This offers a formidable but not insurmountable obstacle to bitumen recovery.

Another general constraint to recovery by nonmining methods is the relatively low injectivity of the tar sand deposits. Thus, it is usually necessary to inject displacement/recovery finds at a pressure such that fracturing (parting) is achieved. Such a technique therefore changes the reservoir profile and introduces a series of channels

Table 10 Summary of Selected Nonmining Projects for the Recovery of Bitumen from U.S. Tar Sand Deposits

Project number	State location	Operator	Project status	Recovery method	Oil properties	
					Gravity (°API)	Viscosity (cp)
	California					
1	Bradley Canyon F.	Chevron	Cu. pilot	Steam drive	11	17,600
2	Cat Canyon F.	Conoco	Cu. comm.	Steam drive	8–12	1,600–10,000
3	Cat Canyon F.	Getty	Cu. comm.	Cyclic steam and steam drive	8–10	14,000–200,000
4	Cat Canyon F.	Husky	Co. pilot	Steam drive	8–10	4,000–20,000
5	E. Cat Canyon F.	Texaco	Cu. comm.	Steam drive	10.4	13,300–15,500
6	Cymric F.	Chevron (Gulf)	Cu. comm.	Steam drive	11	28,000
7	Cymric F.	Sun	Cu. pilot	Steam drive	–	>10,000
8	Kern Front F.	Chevron	Cu. comm.	Steam drive	13	23,000
9	McCool Ranch F.	Phillips	Cu. comm.	Cyclic steam	10.5	12,000
10	McKittrick	D. D. Feldman	Cu. pilot	Steam drive	11.5	10,000
11	McKittrick	Union	Cu. comm.	Steam drive	11.4	>10,000
12	Marport Area	Ogle	Co. pilot	Steam drive	2–5	1,000,000
13	Midway-Sunset F.	Arco	Cu. comm.	Steam drive	11.0	24,500
14	Midway-Sunset F.	Shell	Cu. comm.	Steam drive	10.5	4,000–30,000
15	Midway-Sunset F.	Tenneco	Cu. comm.	Steam drive	12	20,000
16	Midway-Sunset F.	Union	Cu. comm.	Steam drive	11.3	10,000
17	Oxnard F.	Chase	Cu. pilot	Steam drive	6	1,000,000
18	Oxnard F.	Sun	Cu. comm.	Steam drive	5	>500,000
19	Paris Valley F.	Husky	Cu. pilot	Wet comb	9–11	50,000–400,000
20	Yorba Linda F.	Tenneco	Cu. comm.	Cyclic steam	12–14	22,500

	Kentucky					
21	Edmonson C.	Gulf	Co. pilot	Comb. w/frac.	10.6	150,000
22	Edmonson C.	Kenoco	Pl. comm.	Steam w/comb.	12–14	>15,000
	Missouri					
23	Vernon C.	Mapco	Co. pilot	Steam with CO_2	14	11,000
24	Vernon C.	Phillips	Co. pilot	Rev. comb.	10	500,000
	Texas					
25	Little Tom F.	Electro Thermic	Co. pilot	Electric heater	—	>10,000
26	Saner Ranch F.	Conoco	Co. pilot	Steam drive	-2	2,000,000
27	Saner Ranch F.	Enpex	Co. pilot	Steam drive	-2	2,000,000
28	Saner Ranch F.	Exxon	Co. pilot	Steam drive	-2	>1,000,000
29	Saner Ranch F.	Mobil	Co. pilot	Combustion	7	>10,000
	Utah					
30	NW Asphalt Ridge D.	DOE-11TRI	Co. pilot	RF heating	14	1,000,000
31	NW Asphalt Ridge D.	DOE-LETC	Co. pilot	Comb. and steam drive	14	1,000,000
32	Sunnyside D.	Shell	Co. pilot	Steam soak and drive	9	100,000
33	Synnyside D.	Signal	Co. pilot	Steam soak	9–10	100,000
	Wyoming					
34	Burnt Hollow D.	Kirkwood	Co. pilot	Steam drive w/ caustic	9	1,000,000

Source: Marchant and Koch, 1984; Marchant, 1985.

through which fluids can flow from the injection well to the production well. On the other hand, the technique may be disadvantageous insofar as the fracture occurs along the path of least resistance, giving undesirable (i.e., inefficient) flow characteristics within the reservoir between the injection and production wells which have a large part of the reservoir relatively untouched by the displacement/recovery fluids.

From time to time, concepts arise for the recovery of bitumen by the injection of solvents into the formation. As already noted, solvent approach has had some success when applied to mined tar sand but when applied to nonmined material phenomenal losses of solvent and bitumen are always a major obstacle. This approach should not be rejected out of hand since a novel concept may arise which guarantees minimal (acceptable) losses of bitumen and solvent.

There are several methods that have the potential for application to the recovery of a bitumen product from tar sand deposits. They are developments of the enhanced recovery processes (Chap. 1, Sec. VI) and are discussed here in order to put them in the perspective of the current issues.

1. Cyclic Steam Injection

This technique involves the injection of steam at greater than fracturing pressure (1500–1600 psi for Athabasca sands) followed by a "soak" period after which production is commenced (Harmsen, 1971; Burger, 1978). The technique is also known (because of the alternate injection-production format as the "huff-and-puff" process (Figure 14).

The technique has been applied to the Cold Lake sands (Winestock, 1974; Mungen and Nicholls, 1975) in the Clearwater formation at a depth of about 1500 ft where the oil gravity was estimated at 10–12° API with a viscosity of about 100,000 cp and the porosity was 37%. Three pilot tests were reported: the Ethel and May pilots and the Leming pilot. The Ethel pilot acted as the preliminary venture; the May pilot produced up to 1500 bbl/day and the Leming pilot was quoted as producing up to 5000 bbl/day of oil (*Daily Oil Bull.*, 1975).

The technique has also been applied to the California tar sand deposits (Bott, 1967) and in some heavy oil fields north of the Orinoco deposits (Franco, 1976; Ballard et al., 1976).

The steam-flooding technique has been applied with some degree of success to the Utah tar sands (Watts et al., 1982) and has been proposed for the San Miguel (Texas) tar sands (Hertzberg et al., 1983).

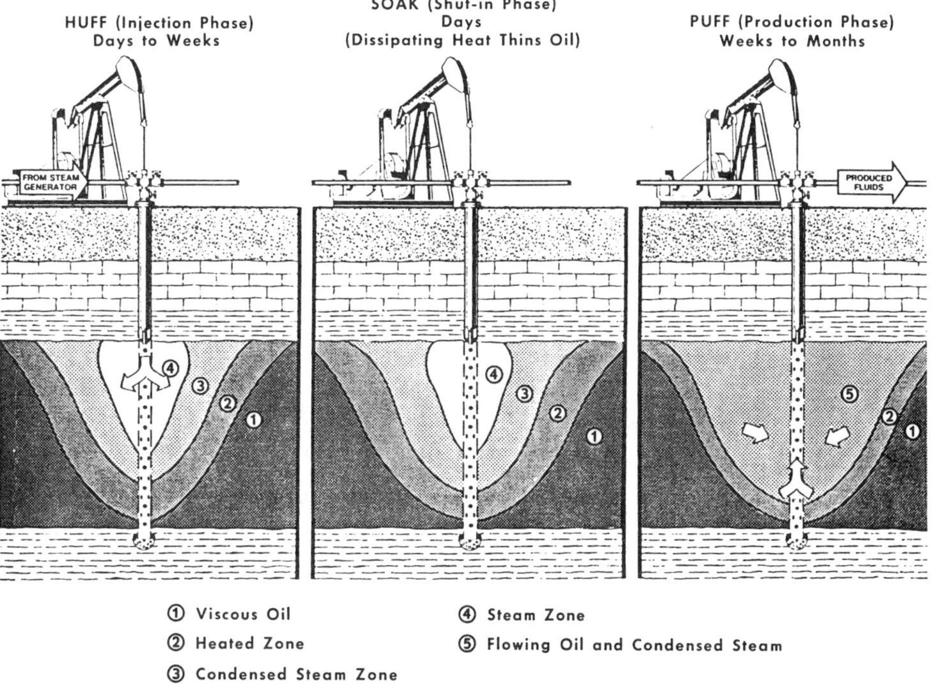

Figure 14 Simplified representation of the cyclic steam injection ("huff-and-puff") process.

Steam drive (Figure 15) involves the injection of steam through an injection well into a reservoir and the production of the mobilized bitumen and steam condensate from a production well. Steam drive is usually a logical follow-up to cyclic steam injection. Steam drive requires sufficient effective permeability (with the immobile bitumen in place) to allow injection of the steam at rates sufficient to raise the reservoir temperature to mobilize the bitumen.

Two expected problems inherent in the steam drive process are steam override and reservoir plugging. Any in situ thermal process tends to override (migrate to the top of the effected interval) because of differential density of the hot and cold fluids. These problems can be partially mitigated by rapid injection of steam at the bottom or below the target interval through a high-permeability water zone or fracture. The former option has been successfully utilized by Shell Canada

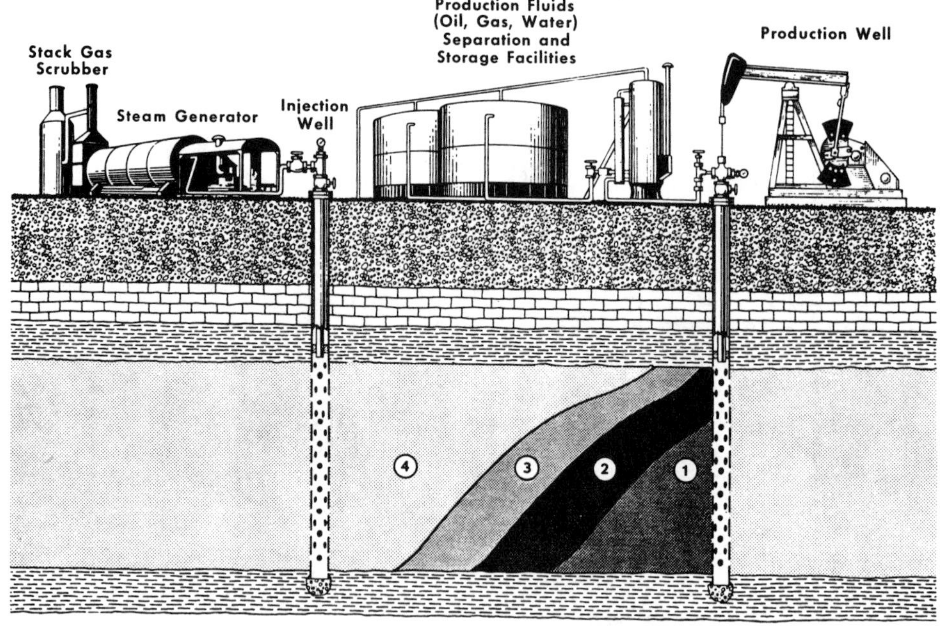

Figure 15 Simplified representation of the steam drive process.

in Alberta's Peace River deposit (Fraser et al., 1982). The latter option is the subject of a U.S. patent (Britton et al., 1981) and has been tested by Conoco in the San Miguel tar sand in southwest Texas (Britton et al., 1983). Each of these options will raise the temperature of the entire reservoir by conduction and, to a lesser degree, by convection. The bitumen will be at least partially mobilized and the effectiveness of the following injection of steam into the target interval will be enhanced.

For a successful steam drive project the porosity of the reservoir rock should be at least 20%; the permeability should be at least 100 md; and the bitumen saturation should be at least 40% (Mathews, 1983). The reservoir oil content should be at least 800 bbl/acre-ft (NPC, 1984). The depth of the reservoir should be less than 3000 ft (NPC, 1984) and the thickness should be at least 30 ft (Mathews, 1983); other preferential parameter have also been noted on the basis of success with several heavy-oil reservoirs (Table 11).

Table 11 Summary of Successful Steamflood Operations Applied to Heavy Oil Reservoirs

Field	Depth (ft)	Reservoir pressure (psig)	h Net pay (ft)	μ Oil viscosity (cp)	k Permeability (md)	Oil content (bbl/acre-ft)
Kern River, CA	900	35	60	4000	4000	1360
Inglewood, CA	1000	120	43	1200	6000	1580
Coalinga, CA	1500	300	35	100	5000	1250
Yorba Linda, CA	2100	200	32	600	500	1070
San Ardo Auginac, CA	2350	250	150	2000	3000	1690
Mount Poso, CA	1800	100	60	280	15000	1480
Yorba Linda, CA	650	—	325	6400+	—	—
South Beldridge, CA	1100	180	91	1600	3000	1820
Midway-Sunset, CA	1600	50	350	4000	4000	—
Slocum, TX	535	110	40	1300	3500	1400
Winkleman, Dome, WY	1200	210	73	900	600	1450

Source: Interstate Oil Compact Commission, 1984.

Economical steam drives applied to heavy oils commonly require injection of less than 4 bbl of (water-equivalent) steam for each barrel of oil produced (Mathews, 1983). For tar sand this ratio may approach or exceed 10 (Britton, 1983).

J. R. Bergeson and Associates (1978) describe the application of selected criteria to the design of a U.S. DOE steam flood in a Utah tar sand.

2. Steam Heating and Emulsification

This process involved fracturing using dilute aqueous alkaline solutions followed by hot caustic emulsification and displacement of the emulsified bitumen after which an emulsion was produced at the production well head by steam injection (Doscher, 1967). Over a 9-month period application of this technique to the Athabasca sands (McMurray formation) resulted in the production of about 2500 bbl of oil but it was estimated that some 7600 bbl of oil had leaked away from the pattern. The overall effect at the end of the test was that about 32% of the oil in-place had been displaced beyond the pattern or recovered.

Variations on this theme include the use of steam and the means of reducing interfacial tension by the use of various solvents (Ali, 1974; Ali and Abad, 1975; *Oil Gas J.*, 1976; Raplee et al., 1985).

The cost of steam generation is the most significant expense in steam injection projects. Several manufacturers package steam generators or boilers specifically for oil field use. These units are designed to burn a variety of fuels, including lease crude, diesel fuel, and natural gas. The feedwater that is converted to steam must be of relatively high quality. Contaminants in a typical feedwater must not exceed the following maximums (Asano et al., 1983):

TDS	1500–8000 ppm
Oil	1 ppm
Hardness	0.5 ppm
Silica	50 ppm

Typical oil field steam generators have capacities in the range of 10 million to 50 million btu/hr. In recent years several oil field steam generation plants have been designed and operated for cogeneration of electricity for lease use and for sale to power companies.

3. Forward and Reverse Combustion

The use of combustion to stimulate oil production is regarded as attractive for deep reservoirs (Finken and Meldau, 1972; Terwilliger,

1975); in contrast to steam injection, it usually involves no loss of heat. The duration of the combustion may be short (<30 days) or prolonged (about 90 days) depending on requirements. In addition, backflow of the oil through the hot zone must be prevented or coking will occur.

Forward combustion involves movement of the hot front in the same direction as the injected air while reverse combustion involves movement of the hot front opposite to the direction of the injected air.

Forward combustion (Figure 16) is particularly applicable to reservoirs containing a somewhat mobile bitumen and/or with a high effective permeability. Even though a lower effective reservoir permeability is required for air injection compared with steam injection, the reservoir ahead of the combustion front is subject to plugging as the vaporized fluids cool and condense. Consequently, a relatively high permeability (400–1000 md) and relatively low bitumen saturation (45–65% of pore volume) are most favorable for this process. The combustion process yields a partially upgraded product because the temperature gradient ahead of the combustion front mobilizes the lighter hydrocarbon components that move toward the cooler portion of the reservoir and mix with unheated bitumen. This mixture is eventually produced through a production well. The heavier components (e.g., coke) are left on the sand grains and are consumed as fuel for the combustion. Under certain operating conditions a significant cost saving is attained by injecting oxygen or oxygen-enriched air rather than atmospheric air because of reduced compression costs and a lower produced gas/oil ratio (Hvizdos et al., 1983).

Both methods have been used with some degree of success, with the forward combustion in the Orinoco deposits (Terwilliger et al., 1975) and in the Kentucky sands (Terwilliger, 1975).

Reverse combustion (Figure 16) is particularly applicable to reservoirs with lower effective permeability (in contrast with forward combustion). It is more effective because the lower permeability would cause the reservoir to be plugged by the mobilized fluids ahead of a forward combustion front. In the reverse combustion process, the vaporized and mobilized fluids move through the heated portion of the reservoir behind the combustion front. The reverse combustion partially cracks the bitumen, consumes a portion of the bitumen as fuel, and deposits residual coke on the sand grains (Reed et al., 1960). The maximum temperature of the process is over 900°F. Production of 50% of the bitumen in place is possible. Up to about 30% of the bitumen will be consumed as fuel and about 20% will be deposited on the sand grains as coke. This coke deposition serves as a cementing material, reducing movement and production of sand.

The reverse combustion technique has been applied to the Orinoco deposit (Burger, 1976) and Athabasca (Mungen and Nicholls, 1975), but the tests were largely unsuccessful owing to the difficulty of

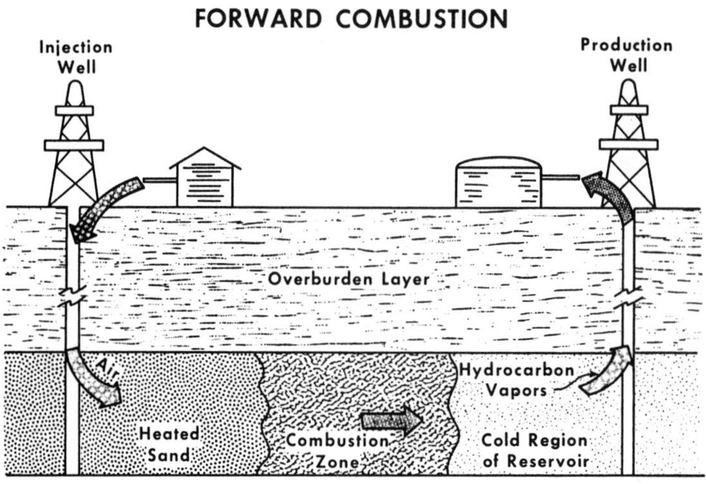

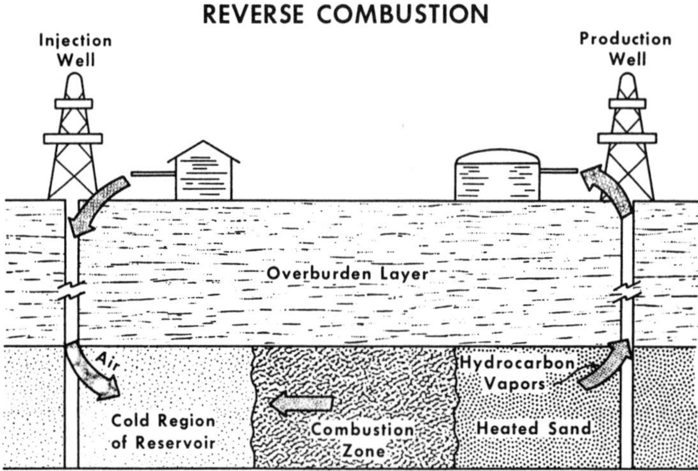

Figure 16 Simplified representation of the forward combustion process and the reverse combustion process.

controlling the air flow (Wilson et al., 1963) and the risk of spontaneous ignition (Wilson et al., 1963; Dietz and Weijdema, 1968; Burger, 1976). However, there has been some success in the application of the reverse combustion technique to the Missouri tar sands (Trantham and Marx, 1966).

A modified combustion approach has been applied to the Athabasca deposit (Mungen and Nicholls, 1975). The technique involved a heat-up phase, production (or blow-down phase), followed by a displacement phase using a fire-water flood (COFCAW process). In this manner over a total 18-month period (heat-up: 8 months; blow-down: 4 months; displacement: 6 months), 29,000 bbl of upgraded oil was produced from an estimated 90,000 bbl of oil in-place.

The addition of water or steam to an in situ combustion process can result in a significant increase in the overall efficiency of that process. Two major benefits may be derived. Heat transfer in the reservoir is improved because the steam and condensate have greater heat-carrying capacity than combustion gases and gaseous hydrocarbons. Sweep efficiency may also be improved because of the more favorable mobility ratio of steam-bitumen compared with gas-bitumen.

Modes of application include injection of alternate slugs of air (oxygen) and water or coinjection of air (oxygen) and steam. Again, the combination of air (oxygen) injection and steam or water injection increases injectivity costs, which may be justified by increased bitumen recovery.

The industry has far less experience with in situ combustion than with steam processes. Consequently, the knowledge base related to the combustion processes is much less. Process control for combustion is more difficult than for steam.

Sufficient bitumen must be consumed during the combustion process to raise the temperature of the reservoir sufficiently to mobilize and produce the balance of the bitumen. Sufficient oxygen must be supplied to support that combustion.

Process efficiency is affected by reservoir heterogeneities that will reduce areal sweep. The underburden and overburden must provide effective seals to avoid loss of injected air and produced bitumen.

Process efficiency is enhanced by the presence of some interstitial water saturation. The water is vaporized by the combustion and enhances the heat transfer by convection. The combustion processes are subject to override because of differences in the densities of injected and reservoir fluids.

Production wells should be monitored for, and equipped to cool, excessively high temperatures (over 2000°F) that may damage downhole production tools and tubulars.

The steam or combustion extraction processes may be significantly enhanced by applying a preheating phase before the bitumen recovery phase. Preheating can be particularly beneficial if the saturation of highly viscous bitumen is sufficiently great as to lower the effective permeability to the point of production being precluded by reservoir plugging. Preheating partially mobilizes the bitumen by raising its temperature and lowering its viscosity. The result is a lower required pressure to inject steam or air and move the bitumen.

Preheating may be accomplished by several methods. Conducting a reverse combustion phase in a zone of relatively high effective permeability and low bitumen saturation is one method. Steam or hot gases may be rapidly injected into a high-permeability zone in the lower portion of the reservoir. In Conoco's patented fracture-assisted steam technology (FAST) (Britton et al., 1981), steam is injected rapidly into an induced horizontal fracture near the bottom of the reservoir to preheat the reservoir. This process has been applied successfully in three pilot projects in southwest Texas (Britton et al., 1983). Shell has accomplished the same preheating goal by injecting steam into a high-permeability bottom water zone in the Peace River, Alberta, deposit (Fraser et al., 1982). Electrical heating of the reservoir by radiofrequency waves (Bridges and Sresty, 1985) may also be an effective method.

Preheating is, in effect, the application of a duplicate or second extraction process that obviously will significantly increase the capital and operating costs. Such increased costs may be justified by increased efficiency of the recovery phase.

E. Modified In Situ Operations

In modified in situ extraction processes (Figure 17), combinations of in situ and mining techniques are used to access the reservoir. A portion of the reservoir rock must be removed to enable application of the in situ extraction technology. The most common method is to enter the reservoir through a large-diameter vertical shaft, excavate horizontal drifts from the bottom of the shaft, and drill injection and production wells horizontally from the drifts. Thermal extraction processes then are applied through the wells. When the horizontal wells are drilled at or near the base of the tar sand reservoir, the injected heat rises from the injection wells through the reservoir and drainage of produced fluids to the production wells is assisted by gravity.

Through this method the effective well bore surface area is significantly increased with a minimum footage drilled. For example, in a 500-ft-thick reservoir at a total depth of 500 ft, 500 ft of drilling is required to obtain 117.8 ft^2 of effective well bore surface area for one 9-in.-diameter vertical well. An equivalent 500-ft-long, 9-in. diameter horizontal well drilled within the reservoir yields 1178 ft^2 of effective surface area. This result may be duplicated by drilling denser vertical well patterns or drilling horizontal wells from the surface. Neither of these alternatives is presently economically competitive with the in situ–mining combination.

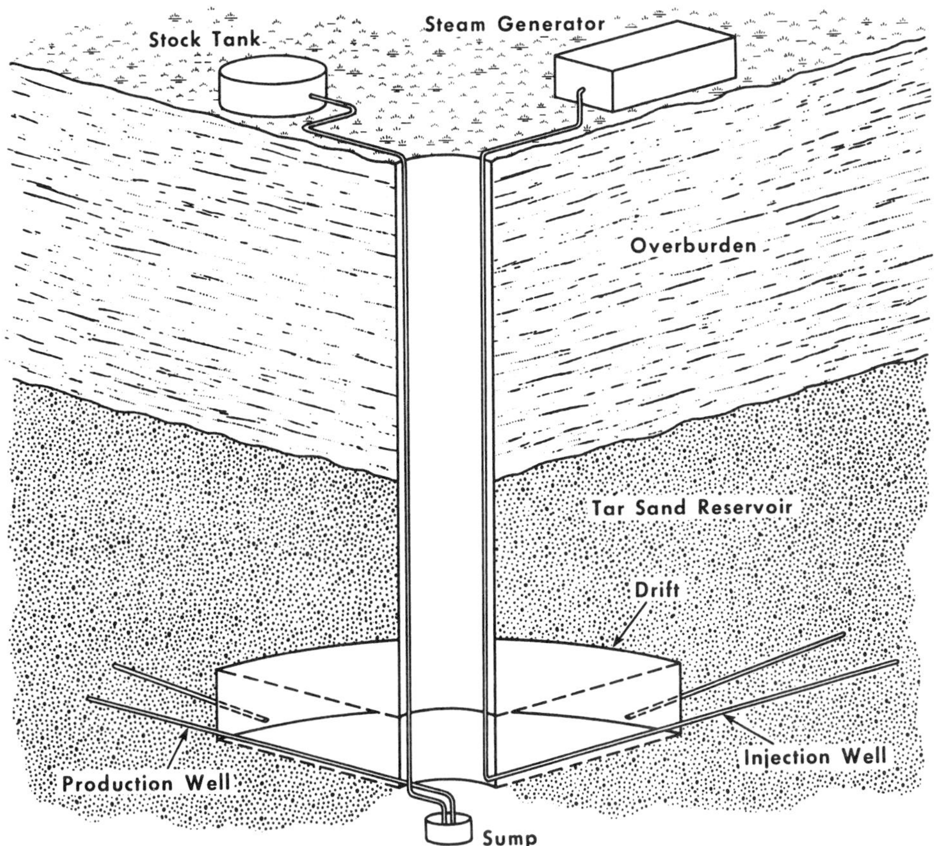

Figure 17 Simplified representation of a modified in situ operation.

Modified in situ technology has been applied in the USSR (Muslimov, 1979) for heavy oil extraction; it has been pilot tested in California for heavy oil extraction (Bleakly, 1980); it is currently the subject of research and development in the Alberta tar sand (Carrigy and Stephenson, 1985); and it has been pilot tested in Utah for shale oil extraction (Lekas, 1981).

F. Product Quality

In all senses of the word, the quality of the bitumen from tar sand deposits is poor as a refinery feedstock. Just as in any field where primary recovery operations are followed by secondary or enhanced recovery operations and there is a change in product quality (Chap. 1, Sec. IV.D), such is also the case for tar sand recovery operations. Thus, product oils recovered by the thermal stimulation of tar sand deposits show some improvement in properties over those of the bitumen in-place (Table 12).

Although this improvement in properties may not appear too drastic, it is usually sufficient to have major advantages for refinery operators. Any incremental increase in the units of hydrogen/carbon ratio can save amounts of costly hydrogen during upgrading. The same principles are also operative for reductions in the nitrogen, sulfur, and oxygen content. This latter occurrence will also improve catalyst life and activity as well as reduction in the metals content.

In short, in situ recovery processes (although less efficient in terms of bitumen recovery relative to mining operations) may have the added benefit of "leaving" some of the more obnoxious constituents (from the processing objective) in the ground.

V. TRANSPORTATION

Transportation is a major aspect of oil sands exploitation. There are four major aspects of liquid fuel production from an oil sand resource (Figure 18).

- Ore recovery
- Bitumen separation
- Bitumen conversion to synthetic crude oil, and
- Refining the synthetic crude oil to usable liquid fuels

Currently, the two commercial plants carry out the first three stages on site and in that respect the plants are completely self-contained. The synthetic crude oil is then shipped by pipeline to a more conventional refinery site (e.g., Edmonton) for fuel upgrading to liquid fuels.

There are, however, constraints on the character of liquids that may be shipped by pipeline. The synthetic crude oil conveniently meets the specifications for pipeline shipment but should an alternate means of bitumen upgrading be established, i.e., visbreaking* or even removal to another site, then problems of shipping will no doubt arise.

---

*Partial (limited) conversion as opposed to the more complete conversion found in coking operations (see Chap. 6, Sec. I.B., and Chap. 8).

Table 12 Comparison of the Properties of Oils Produced from Tar Sand Bitumens by Thermal Stimulation Methods[a]

Properties	Asphalt Ridge				Tar Sand Triangle		
		Product oils				Product oils	
	Bitumen	Reverse combustion	HGI-327	HGI-560	Bitumen	Forward combustion	HGI-493
Analysis, wt %							
Carbon	85.6	83.3	86.5	85.8	83.1	84.1	83.9
Hydrogen	11.5	11.9	11.6	12.2	10.0	11.1	10.7
Nitrogen	1.1	0.6	0.9	0.8	0.5	0.3	0.6
Sulfur	0.5	0.3	0.4	0.4	3.8	2.8	3.7
Oxygen	1.3	3.9	0.6	0.9	2.6	1.7	1.1
H/C atomic ratio	1.6	1.7	1.6	1.7	1.4	1.6	1.5
API gravity (15.6°C), °API	10.7	21.7	13.1	16.8	4.1	16.5	10.1
Viscosity 140°F, cp	2×10^4	16	4×10^3	173	7×10^{10}	20	490
Molecular weight	689	316	585	422	730	320	475
Distillation data, vol %							
<316°C (<600°F)	5	33	11	13	2	25	6
316 to 538°C	36	53	36	61	32	56	63
>538°C (>1000°F)	59	14	52	25	66	19	30

[a]See also Table 9 in Chap. 1.

Source: Thomas et al., 1987.

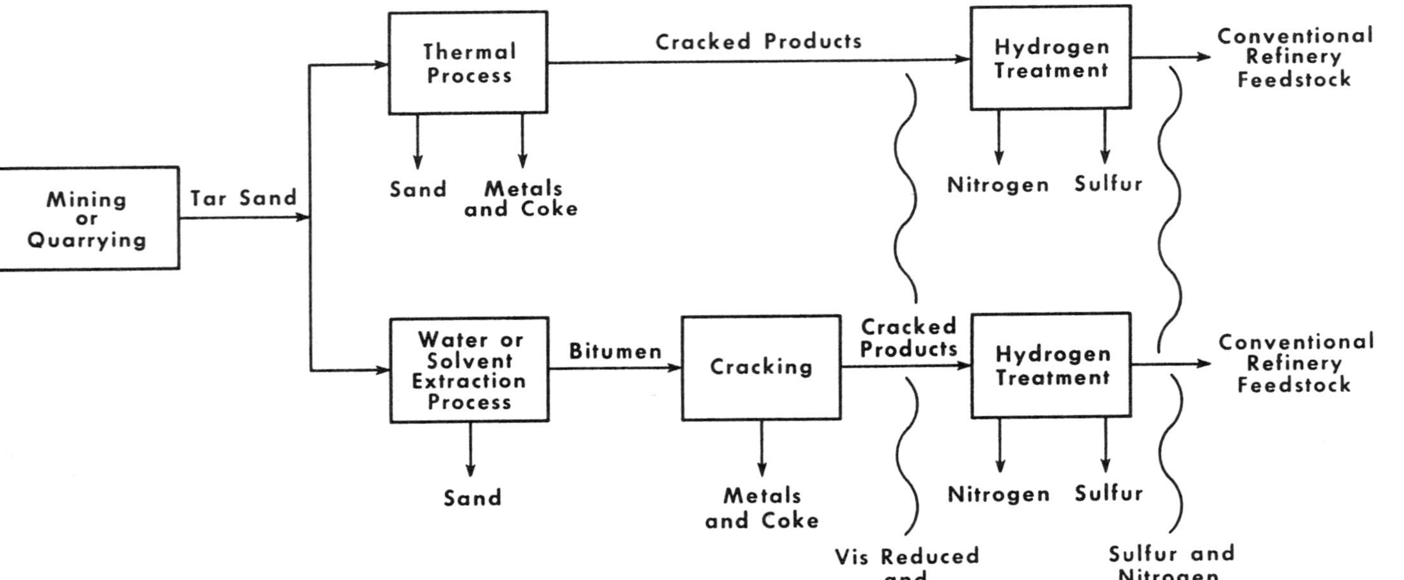

Figure 18 General overview of a tar sand processing operation showing the products available for transportation.

For example, it may be desirable to ship the whole bitumen feed by pipeline. Should this be the case, dilution with naphtha will be necessary. The naphtha may actually be produced at the recovery site by construction of a nominal conversion operation but it is to be anticipated that a feed having a viscosity in excess of 15,000 cs at 38°C (100°F) will require in excess of 0.5 bbl naphtha per barrel of bitumen. At higher temperatures, the amount of naphtha required will be reduced but light ends and even water may have to be removed to prevent undue pressure buildup in the pipeline if shipping temperatures of approximately 95°C (200°F) are considered.

One other aspect of transportation is the shipment of bitumen (or the whole oil sand or even bitumen-enriched oil sand) in trucks or trains. Currently, economic constraints related to the amount of material that would have to be moved to enable even a nominal conversion/upgrading operation to run continuously (hazards of weather and mechanical constraints not withstanding) have caused these types of operation to be downgraded in priority.

Finally, it is also possible for bitumen to be emulsified and shipped (by pipeline) as an emulsion. This particular idea has received some attention, especially in regard to bitumen recovery by aqueous flooding methods. The idea would be to produce the bitumen from the formation as an oil-in-water emulsion at the remote site followed by shipping of the emulsion to an oil recovery/upgrading site.

VI. ENVIRONMENTAL ASPECTS

A major aspect of oil sands resource development is the influence of the various integral parts of the development on the surrounding environment (Table 13). This aspect takes into account not only the influence of process emissions (gases, effluent liquids, and the like) but also mechanical incursions into the environment and the influence on indigent flora and fauna.

Of course, environmental constraints are all subject to federal and/or state/provincial regulations and development of an all-encompassing environmental policy for oil sands development is of utmost importance and urgency. Subjection of each potential developer to differing standards may not only cause confusion but also some degree of dissent.

The industrial activities that are capable of causing environmental concern can be classified as "conventional technology" and "possible future technology." In the former case, the technology is that which is already being employed and thus it is possible to design environmental programs that are capable of mitigating any adverse effects of

Table 13 Major Contaminants Occurring in Water and Air Emissions at a Tar Sands Plant

Sources	Potential contaminants
Process waste water	Suspended solids, dissolved solids, phenols, ammonia, oils, organics, sulfides, metals
Sanitary waste water	Suspended solids, dissolved solids, biochemical oxygen demand, organics nitrates, phosphate, residual chlorine, coliform organisms, metals
Runoff from upgrader area: Coke storage pile, sulfur storage pile, solid waste landfills	Suspended solids, oils, organics, inorganics, sulfur, metals
Power plant stacks	Particulates, SO_2, NO_x, CO
Sulfur plant stacks	Sulfides, SO_2, H_S particulates
Upgrader heaters	SO_2, NO_x, CO, hydrocarbons, particulates
Runoff from upgrader area, coke storage pile, sulfur storage pile, solid waste landfills	Suspended solids, oils, organics, inorganics, sulfur, metals

the technology. In the latter case, the design of environmental technologies is a higher risk since, at the time of site development, the technology to be applied to resource development is still an unknown. Obviously, the development of new recovery technologies must be accompanied by the development of adequate environmental technologies.

Obviously, the mitigation of environmental effects is a major aspect of tar sand development (in fact, a major aspect of any resource development) and, while a basic effort (Table 14) is essential, the final environmental plans may be much more complex (Table 15).

In the initial stages of resource development, plant construction can be a major trauma to any environment or to the developer. A major undertaking such as the construction of an oil sands plant cannot proceed without some disturbance of the environment. Therefore, government(s) and developer(s) should work closely to determine the extent of the disturbance to the environment by clearly delineating the means by which the development can proceed.

Thus, regulations could clearly state conditions for resource development in terms of:

- Foliage removal—to what extent and how or to where.
- Flora—endangered species should be transplanted.
- Fauna*—removal to another (similar) location.
- Till removal.

The mining opration poses a potential threat to the environment because

- Large "holes" have to be dug to retrieve the ore, and
- The overburden must first be removed from the immediate mine site

The first constraint is an inevitable part of the mining operation but regulations may require that the site be returned to at least the original (unspoiled) condition. This is being achieved in Alberta by the techniques in which the mineral matter from the hot water separation process is returned to the mine site as fill material thereby bringing the ground level back to the premine level**.

In a similar manner, overburden handling can also be achieved with minimal stockpiling of the earth.

Revegetation programs will complete the return of the site to the "predevelopment" state.

One of the greatest problems that has emerged from oil sands resource development is the disposal (and control) of the tailings streams that arise from the hot water separation. The extent of this problem has not been conceived until the commercialization of the oil sands commenced.

There are two major aspects of tailings disposal that need to be addressed:

- Each tone of oil sand in-place has a volume of $\sim$16 ft^3; the tailings has a volume of $\sim$22 ft^3—a volume gain of $\sim$40%.

---

*This classification may include not only animal life but also people.

**In fact, the postmine level may, because of expansion effects, be somewhat higher than the premine level.

Table 14 Potential Environmental Impact of the Surface Mining of Tar

Operation or source of impact	Surface changes: Increased landslide risk	Surface changes: Destruction of existing vegetation
Site preparation		X
Surface cleaning (cleared area)		(X)
Stripping (stripped area)		(X)
Tar sand extracting (mined area)		
Haul road transportation (construction)		(X)
Tailings disposal		
Bitumen in tailings or low grade tar sand waste		X
Fines in tailings		
Stripped waste		
Solubles or water-transportation particles in overburden		
New surface		
Increases in surface slope from waste disposal	X	
Rehandling of materials: Backfilling, grading, and recontouring		

Source: Frazier et al., 1976.

- The clay discharged as part of the tailings stream does not settle and, therefore, limits the amount of the stream that can be used as recycle water.
- The tailings stream contains bitumen.

Thus, tailings disposal requires a larger "hole" than that from which the oil sand originated or the level of the land will/must be raised at site abandonment.

Until a satisfactory means of clay settling can be achieved that will allow reuse of the water, the tailings ponds will grow in size with the

Sand Deposits

Alteration of habitats	Topographic changes	Drainage diversion	Increased noise	Changes in ground water regime	
				Physical	Chemical
X	X	X	X		
(X)	(X)	(X)	X		
X	X			(X)	(X)
	(X)		X	(X)	(X)
(X)	(X)	(X)	X		
			(X)		
			X		
					(X)
X				X	X
					X
			X		

life of the plant. Currently, some 40–60% of the daily water requirements of the oil sands plants are derived by a recycle stream from the tailings ponds; the remainder is makeup water from the nearby Athabasca river.

Environmental regulations in Canada or the United States will not allow the discharge of tailings streams (1) into the river; (2) onto the surface; or (3) onto any area where contamination of groundwater domains or the river may be contaminated. The tailings streams is essentially high in clays and contains some bitumen; hence the current need for tailings ponds.

Table 15 Matrix for Mitigation of Environmental Issues During Tar Sand Development (Originally Designed for the Athabasca Tar Sands)

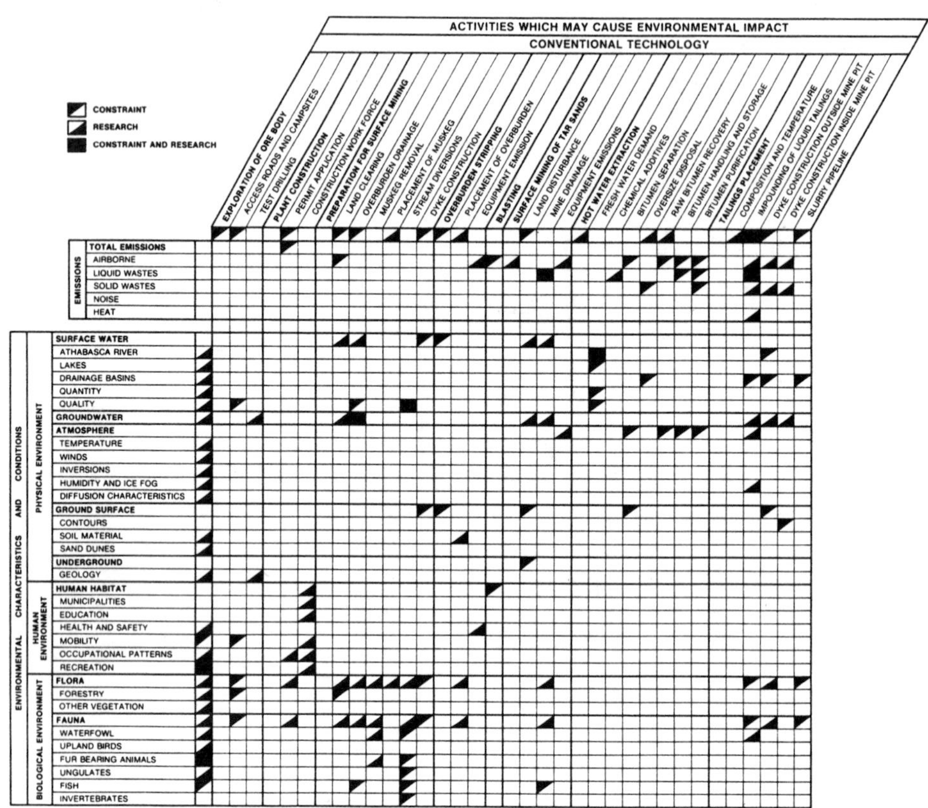

(continued)

Removal of the tailings ponds will have to be accommodated at the time of site abandonment. It is conceivable that the problems may be somewhat alleviated by the development of process options that require considerably less water in the sand/bitumen separation step. Such an option would allow a more gradual removal of the tailings ponds.

Table 15 (cont.)

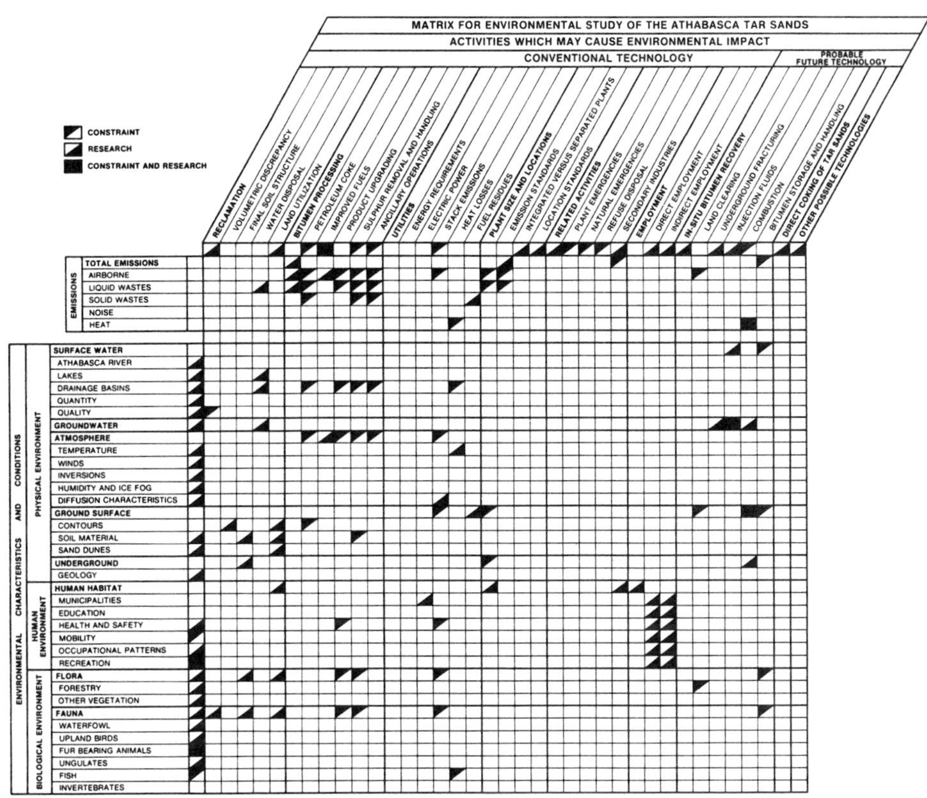

Source: Page, 1982.

All of the aspects must be given serious consideration since each adds to the capital cost of the oil sands plant. Some alleviation of the problem may occur as a result of the general remoteness of most (but not all) of the oil sand deposits *but* this should not be used as an opportunity for governments/developers to loosely interpret the regulations in terms of detriment to the environment.

13

Tar Sand Composition and Properties

I Physical Structure 375
II Minerals 378
III Bulk Properties 380
IV Bitumen Content 384

I. PHYSICAL STRUCTURE

Tar sand is a complex arrangement of sand, oil, water, and clay minerals (Berkowitz and Speight, 1975; Takamura, 1982). However, in many cases the two latter constituents may be difficult to define in an oil sand. The various oil sand deposits throughout the world have been described as being geologically similar in terms of the general structure/stratigraphy of the formations (Table 4 in Chap. 12) but physically these sands may have marked differences from locale to locale (Table 1). At present, in the Alberta deposits the sand is water-wet while in many other locales the sand is oil-wet. This is believed to be the major reason for the facile adaptation of the Alberta oil sands to the hot water process (Chap. 12, Sec. IV.B). The known properties do indicate that differences exist between the tar sands of Canada and those of the United States.

Table 1 Inspection Data for the Physical Properties of Selected U.S.

	Circle Cliffs	Edna	Santa Rosa
Deposit size, bbl	1.3×10^9	270×10^6	90×10^6
Deposit depth, ft	0–500+	0–250	0–1300
Deposit thickness, ft	260–310	0–1200	216–360
Deposit homogeneity	Poor	Variable	Variable
Bitumen, wt %	6–7	9–16 Av. 11	4.3–9.1
Water, vol %	2.7 (pore space)	1–3	0.4–2.5
Viscosity, cp	–	183-s Saybolt (210°F)	243
Fm permeability, md	228	–	240
Fm porosity, vol %	9.7–15 Av. 12.9	10–36	8–19
Overburden	Siltstone	None	Limestone, shale
Fm matrix	Conglomerate, sandstone, and siltstone	Sandy conglomerate	Sandstone
Fm strength, psi	–	–	–
Dip	10°–20° E&W	Variable 10°–40° E&W	1°–2° SE
Faults/fractures	Few	Few	Few
Sulfur, %	2.37–4.9 Av. 3.6	3.0–3.5	1.44–2.30

Tar Sand Deposits[a]

Uvalde	Tar Sand Triangle	PR Spring	Asphalt Ridge	Sunnyside
141×10^6	16×10^9	4.5×10^9	1.2×10^9	3.5×10^9
0--1500	0–2000	0–370	0–700+	42–400
50--124	5–300+	14–212 Av. 98	10 to 135	40–1175 Av. 615
Variable	Variable	Variable	Variable	Variable
6.5–15.5	0.2–12.9	6–7.5	5–12	8–11
–	0.9–31.4	3–4	1.8--6.5	0.7–10.5
2×10^6	1.3×10^6	32.5×10^6	2.95×10^6	280–2980 (100°F)
–	0.03–2560 Av. 268.25	121 to 1510	497	729
–	9.7–31.7	8.4–31	19.6–26.7 Av. 20.9	16–30
Limestone	Dolomite, quartz, and sandstone	Shale, limestone, and sandstone	Shale, limestone, and sandstone	Shale, dolomite, limestone, and sandstone
Limestone	Quartz Sandstone	Sandstone	Sandstone	Sandstone and limestone
–	–	–	–	–
SSE	1°–3° NW	2°–4° NW	9°–35° SW	1°–10° NE
Numerous high but minor	Few	Few	Few	Minor
10	2.67–6.27 Av. 3.8	0.22--0.75 Av. 0.33	0.19–0.77 Av. 0.52	0.44–0.80 Av. 0.55

(continued)

Table 1 (Cont.)

	Circle Cliffs	Edna	Santa Rosa
Thermocharacter, btu/lb	–	17 900	18 600
Hydrogen, %	–	–	11.64
Underburden	Dolomite	Shale	Sandstone

Summary and comparison to Canadian tar sand deposits:

	United States	Athabasca
Porosity, %	26–39	17–46
Permeability, air, md	10–3800	0–600
Saturation, % pore volume		
Oil	13–33	40–98
Water	23–83	1–39
Saturation, oil, wt %	4–22	0–18
Viscosity, cp at formation temperature	$1.8 \times 10^3 - 500 \times 10^3$	$3 \times 10^6 - 600 \times 10^6$
Gravity, °API	3.7–15.0	6–10
Sulfur, %	0.5–4.2	3.7–5.0

[a]See also Table 1 in Chap. 14.

II. MINERALS

Tar sand is a mixture of sand, water, and bitumen with the sand component occurring predominently as quartz. The arrangement of the sand, water, and bitumen has been assumed to be an arrangement whereby each particle of the sand is water-wet and a film of bitumen surrounds the water-wetted grains (Figure 1). The balance of the void volume is filled with bitumen, connate water, or gas; fine material, such as clay, occurs within the water envelope.

The majority (<99%) of the tar sand mineral is accounted for by the quartz sand and clays, although a wide variety of other minerals exist in the tar sand solids (Table 2 in Chap. 14). The quartz sand ranges from the largest grains commonly found in the tar sand (<99.5% is <1000 μm down to 44 μm (325 mesh). Minerals falling in

Uvalde	Tar Sand Triangle	PR Spring	Asphalt Ridge	Sunnyside
–	17 900	18 100	18 800	18 400
–	10.1	11.1	11.7	11.6
–	–	–	Shale	–

the size range 2–44 μm are commonly referred to as silt while minerals having an equivalent spherical diameter of <2 μm are clay. Clays are actually aluminosilicate minerals but are usually classified by virtue of their small size. Use of the terminology "fines" generally refers to the silt and clay fractions of the tar sands.

One additional aspect of the character of Athabasca oil sands which plays a role in research and practice is that the sand grains are not uniform in character (Figure 2). Grain-to-grain contact is variable and such a phenomenon influences attempts to repack mined sand, as may be the case in studies involving oil removal from the sand in laboratory-type in situ studies. This phenomenon also plays a major role in the "expansion" of the sand during processing where the sand to be returned to the mine site might occupy 120–150% of the volume of the original as-mined material.

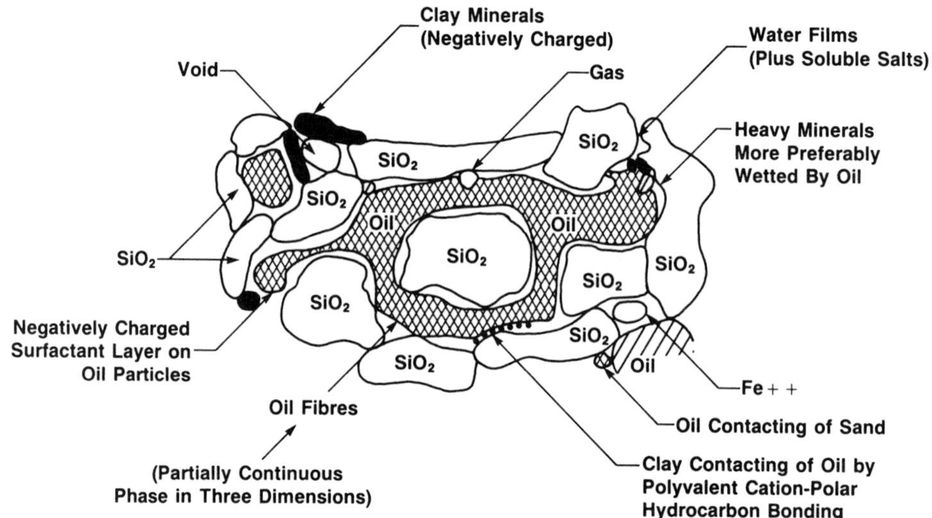

Figure 1 Schematic representation of the structure of (Canadian) tar sand (adapted from Bichard, 1982).

III. Bulk Properties

The tar sand mass can be considered a four-phase system composed of solid phase (siltstone and clay); liquid phase (from fresh to more saline water); gaseous phase (natural gases); viscous phase (black and dense bitumen, about 8° API).

In normal sandstone, sand grains are in grain-to-grain contact but tar sand is thought to have no grain contact due to the surrounding of individual grains by fines with a water envelope and/or a bitumen film, where some remaining void space might be filled with water, bitumen, and gas in various proportions. The sand material in the formation is represented by quartz and clays (99%), where fines content is approximately 30% by weight; the clay content and clay size are important factors which affect the bitumen content.

Tar sand properties that are of general interest are bulk density, porosity, and permeability (Table 3 in Chap. 14). Porosity is, by definition, the ratio of the aggregate volume of the interstices between the particles to the total volume and is expressed as a percentage. High-grade tar sand usually has a porosity in the range 30–35%

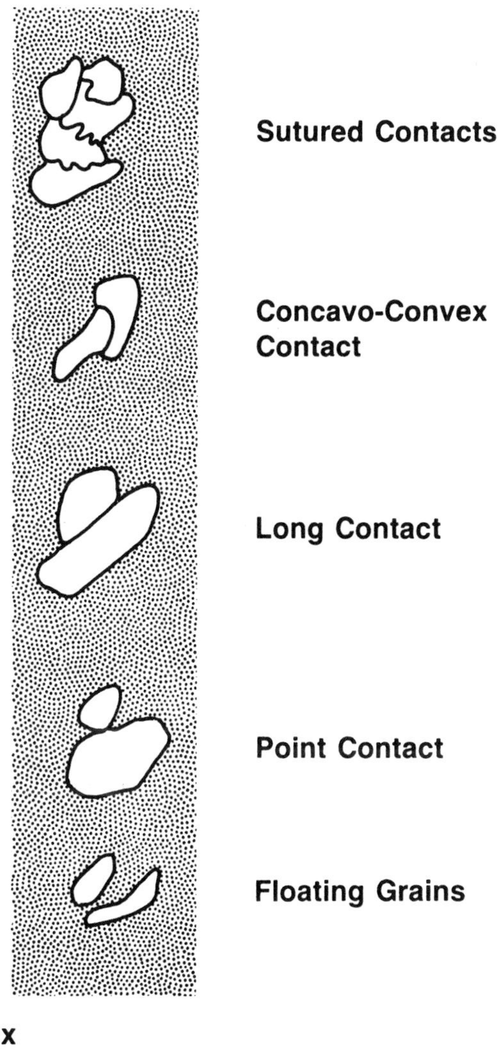

Figure 2 Schematic representation of the potential nonuniform grain-to-grain contacts (Harris and Sobkowicz, 1977).

which is somewhat higher than the porosity (5–25%) of most reservoir sandstones. The higher porosity of the tar sand has been attributed to the relative lack* of mineral cement (chemically precipitated material that binds adjacent particles together and gives strength to the sand, which in most sandstones occupies a considerable amount of what was void space in the original sediment.

Permeability is a measure of the sediment's or rock's ability to transmit fluids and is, to a major extent, controlled by the size and shape of the pores as well as the channels (throats) between the pores; the smaller the channel, the more difficult it is to transmit the fluid (water, oil, or gas). Fine-grained sediments invariably have a lower permeability than coarse-grained sediments, even if the porosities are equivalent. It is not surprising that the permeability of the bitumen-free sand from the Alberta deposits is quite high. On the other hand, the bitumen in the deposits, essentially immobile at formation temperatures (∿4°C; 40°F) and pressures, actually precludes any significant movement of fluids through the sands under unaltered formation conditions.

Subsurface properties of the tar sand are inconvenient for conventional mining systems due to instability of the ground for any safe mine structure (Jeremic, 1975). With the length of time that the structure remains open, one can expect an increase of deformation of the opening, an increase of temperature, and, respectively, an increase in bitumen flow or, at the time of immediate opening, a fast gas release and collapsing structure.

Bitumen (or oil) saturation (So) is expressed in the percentage of pore volume or percentage by weight of the reservoir rock. A high percentage of bitumen saturation is desirable, but when it approaches 100%, the available pore space (and consequently the effective permeability) for injection of fluids approaches zero. An optimum bitumen saturation is likely to be in the range of 50–70% of the pore volume for application of in situ recovery processes. At 20% porosity and 50% bitumen saturation, the reservoir would contain less than 800 bbl of bitumen per acre/ft, a marginal amount for in situ thermal recovery (NPC, 1984). At 20% porosity and 70% saturation the bitumen content would approach 1100 bbl/acre-ft above the theoretical practical minimum (NPC, 1984). An overall energy balance analysis of the resource segment and proposed extraction technology is necessary for

*The relative lack of mineral cement is the reason why the Alberta deposits are referred to as "sands" and not "sandstones."

determination of a project-specific minimum bitumen saturation as it relates to porosity (Figure 3).

The interstitial water (Sw) in a tar sand reservoir may be a help or a hindrance to an in situ recovery process.

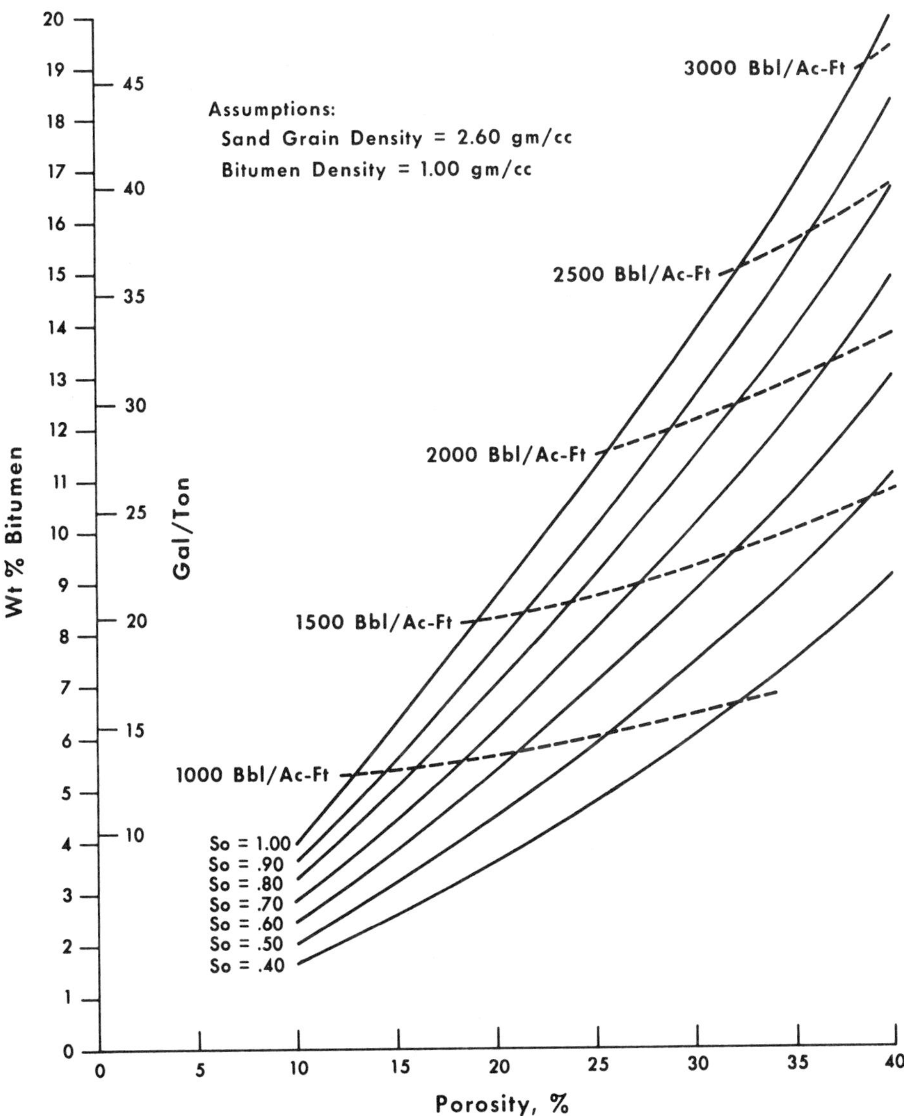

Figure 3 Relationship of porosity to bitumen content of tar sand.

Some interstitial water saturation is desirable for in situ combustion because it results in generation of steam in the reservoir and resultant improvement of heat transfer. But interstitial water saturation that is too high will either reduce process efficiency by soaking up too much heat or will quench combustion completely.

A minimum mobile interstitial water saturation is also desirable for steam processes because a continuous water phase aids in heat transfer. On the other hand, a water saturation (at the expense of bitumen saturation) in excess of 30% of the pore volume can reduce process efficiency by acting as a heat sink.

IV. BITUMEN CONTENT

The bitumen content of the U.S. and Canadian tar sands varies from zero to as much as 22% by weight (Table 1). There are, however,

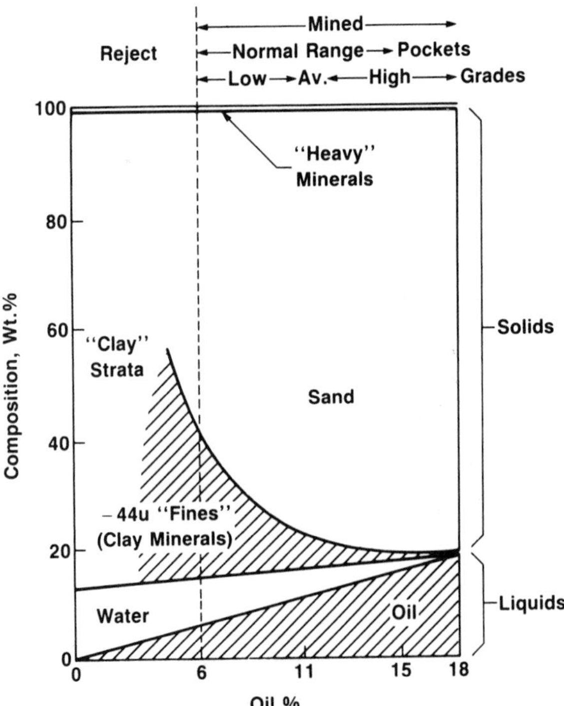

Figure 4 Interrelationships of bitumen, water, fines, and mineral matter contents of Athabasca oil sands (Camp, 1976; Bichard, 1982).

noted relationships between the bitumen, water, fines and mineral contents for the Canadian tar sands (Figure 4). Similar relationships may also exist for the U.S. tar sands but an overall lack of study has prevented the uncovering of such data.

For the Canadian tar sands, bitumen contents of 8–14 wt % may be considered as normal (or average). Bitumen contents above or below this range have been ascribed to factors which influence impregnation of the sand with the bitumen (or the bitumen precursor). There are also instances where bitumen contents in excess of 12 wt % have been ascribed to gravity settling (Figure 5).

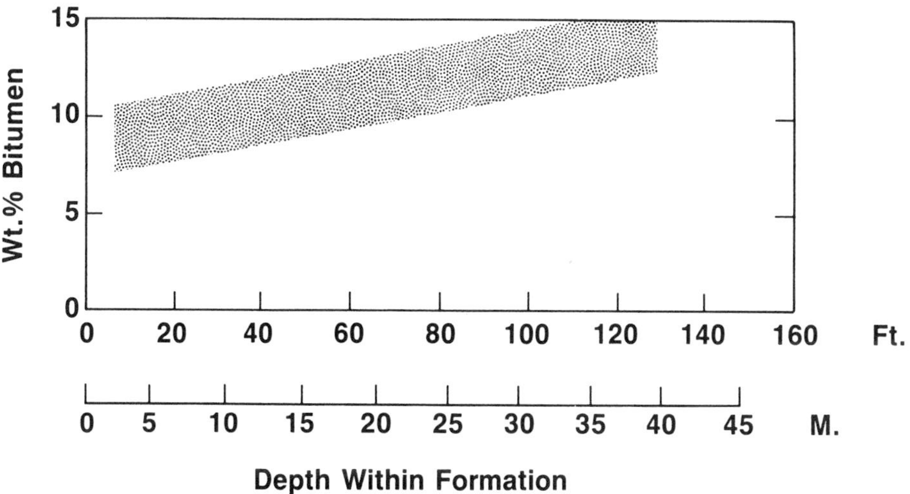

Figure 5 Variation of bitumen content of the oil sand with depth within the formation for the Athabasca deposit (Schutte, 1974; Speight, 1979).

14

Bitumen Composition and Properties

I	Ultimate Composition	387
II	Fractional Composition	388
III	Physical Properties	389
	A. Specific Gravity	389
	B. Viscosity	396
	C. Volatility	397
	D. Thermal Sensitivity	400
	E. Solubility	403
	F. Miscellaneous Properties	408

I. ULTIMATE COMPOSITION

The elemental analysis of oil sand bitumen has been widely reported (Camp, 1976; Meyer and Steele, 1981). However, the data suffer from the disadvantage that identification of the source is very general (i.e., Athabasca bitumen) or analysis is quoted for separated bitumen which may have been obtained by, say, the hot water separation or solvent extraction and may therefore not represent the total bitumen on the sand. The elemental composition of oil sand bitumen is generally constant and falls into the same narrow range as for petroleum (Chap. 3, Sec. I). In addition, the ultimate composition of the Alberta bitumen does not appear to be influenced by the proportion of bitumen in the oil sand or by the particle size of the oil sand minerals (Speight and Moschopedis, 1981) (Tables 1–3).

Bitumens from U.S. tar sands have a similar ultimate composition to the Athabasca bitumen (Table 4) but to note anything other than the H/C atomic ratio (which is an indicator of the relative amount of hydrogen needed for upgrading) or the amount of nitrogen is beyond the scope of general studies. When the many localized or regional variations in ordination conditions are assessed, it is perhaps surprising that the ultimate compositions are so similar.

The hydrogen/carbon ratio of a bitumen affects its viscosity (and, hence, the required supplementary heat energy for a thermal extraction process). It also affects the bitumen's distillation curve or thermodynamic characteristics, its gravity, and its pour point. Ratios of hydrogen to carbon as low as 1.3 have been observed for U.S. tar sand bitumens. However, a ratio of 1.5 is more typical. The higher the hydrogen/carbon ratio of a bitumen, the higher is its value as refinery feedstock. This is due to relatively lesser upgrading and refining requirements. Elements related to the hydrogen/carbon ratio are distillation curve, bitumen gravity, pour point, bitumen viscosity, and hydrogen/carbon ratio.

Even though sulfur is a valuable commodity, its inclusion in bitumen as organic or elemental sulfur or in produced gas as compounds of oxygen and hydrogen is an expensive nuisance. It must be removed from the bitumen at some point in the upgrading-refining process. Sulfur contents of U.S. tar sand bitumens can exceed 10% by weight. Elements related to sulfur content are nitrogen and hydrogen contents, hydrogen/carbon ratio, distillation curve, and viscosity.

Nitrogen contents of tar sand bitumens as high as 1.3% by weight have been observed. The nitrogen complicates the refining process by poisoning the catalysts employed in the refining process. Elements related to nitrogen content are sulfur and hydrogen content, hydrogen/carbon ratio, bitumen viscosity, and distillation curve.

II. FRACTIONAL COMPOSITION

An important property of the bitumen is the fractional composition of the material. For general purposes, attempts are usually made to determine the amounts of these fractions in petroleums and bitumens, namely, the asphaltene, resin, and oil fractions (Chap. 4, Sec. III). Compositional studies add detailed knowledge about the behavior of the material and, in particular, the processability since it is evident

(Speight, 1981a) that high-asphaltene feedstocks are difficult to process and produce high yields of coke or have a phenomenal deterimental effect on catalyst life (Chap. 8, Sec. II).

As with ultimate analysis, there are many reports of attempts to fractionate oil sand bitumen but the descriptions of either the sample history or the exact source of the sample have been too vague for specific identification. Some data are available (Table 5) which do indicate variations in the physical composition of the bitumen:

Asphaltenes	15.0–25.0%	about 50%
Resins	25.0–38.0%	
Oils	44.0–55.0%	

On a regional basis, there is fragmentary evidence (Table 6) that bitumen obtained from the northern locales of the Athabasca deposit (Bitumount, Mildred-Ruth Lakes) has a lower proportion of asphaltenes (16.7–20.1%) than the bitumen obtained from southern deposits (Abadand, Hangingstone River; 22.3–23.4% asphaltenes). In addition, other data (Figure 1) indicate a marked variation of asphaltene content in the tar sand bitumen with depth in the deposit.

III. PHYSICAL PROPERTIES

A. Specific Gravity

The specific gravity of bitumen shows a fairly wide range of variation (Table 7). The largest degree of variation is usually due to local conditions which affect material lying close to the faces, or exposures, occurring in surface oil sand beds. There are also variations in the specific gravity of the bitumen found in beds that have not been exposed to weathering or other external factors.

A very important property of the Athabasca bitumen (which also accounts for the success of the hot water separation process) is the variation of bitumen density (specific gravity) of the bitumen with temperature (Figure 2). Over the temperature range 30–130°C (85–265°F) the bitumen is lighter than water; hence (with aeration) floating of the bitumen on the water is facilitated and the logic of the hot water process is applied.

Bitumen gravity primarily affects the upgrading requirements needed because of the low hydrogen content of the produced oil. If not upgraded, the product oil has a lower selling price than conventional crudes. The API gravities of known U.S. tar sand bitumens range downward from about 14° API (0.973 specific gravity) to -2° API (1.093 specific gravity). Although only a vague relationship

Table 1 Detailed Comparative Data for U.S. and Canadian Tar Sand Deposits Showing Similarities and Differences

Deposit				
State or province	Alberta	Utah	Kentucky	Alaska
Deposit or area	Athabasca	Asphalt Ridge	Western Kentucky[2]	Kuparuk River
Reservoir	McMurray Fm.	Mesaverde Fm.	Big Clifty S.S.	Ugnu
Mineral ownership	Crown (Province)	Federal, state	Private	State
Geological properties				
Depth (overburden), ft	0–2500	0–2000	0–600	2000–4000
Thickness, ft	160	110	5–50	80
Clay (dispersed)	Significant	None	Minor	None
Consolidation	Unconsolidated	Well consolidated[1]	Well consolidated	Unconsolidated
Mineable, %	<10	>50	<40	0
Wettability	Water-wet	Oil-wet	Oil-wet	Mixed

Reservoir properties				
Oil-in-place, 10^9 bbl	826	1.1–2.0	1.2–2.1	10–50
Oil saturation, % pore space	70	75	50	75
Oil saturation, % wt	11	>10	<5	13
Water saturation, % pore space	30	<5	40	25
Porosity, %	31	30	16.4	35
Permeability, md	2000	2000	400	>1000
Bitumen properties				
Gravity, °API	9.1	11.6	8.7	8–12
Viscosity, 10^3 cp	640	1000	520	5–500
Nitrogen, % wt	0.44	0.96	0.64	Unknown
Sulfur, % wt	4.86	0.49	1.55	1.5
Hydrogen/carbon ratio	1.44	1.56	1.56	Unknown
Pour point, °F	75	126	84	5–50
Comments		[1]Some poorly to unconsolidated	[2]Several deposits scattered over six counties	

Table 2 Minerals Occurring in Oil Sand

Category	Mineral type	Specific gravity	Relative occurrence
Light minerals	Quartz	2.65	
	Clay (Kaolinite, illite)	2.6–2.69	High
	Feldspar	2.54–2.76	
	Dolomite	2.85	Low
	Calcite	2.72	
	Micaceous minerals	2.6–3.5	
Heavy minerals	Tourmaline	3.0–3.25	
	Iron minerals:		
	Pyrite, pyrrhotite	4.58–5.01	
	Siderite	3.83–3.88	
	Limonite	3.6–4.0	High
	Magnetite	5.18	
	Hematite	5.26	
	Titaniferous:		
	Ileminite	4.7	
	Rutile	4.18–4.25	
	Staurolite	3.65–3.75	
	Epidote	3.35–3.45	
	Corundum	4.02	
	Garnet	3.5–4.3	
	Kyanite	3.56–3.66	
	Apatite	3.15–3.20	
	Magnesite	3.0–3.2	
	Andalusite	3.16–3.20	
	Zoisite	3.3	
	Sillimanite	3.23	Low
	Pyroxenes	3.15–3.55	
	Amphiboles	2.85–3.2	
	Spinel	3.6–4.0	
	Rhodochrosite	3.45–3.6	
	Chalcopyrite	4.1–4.3	
	Bornite	5.06–5.08	
	Cassiterite	6.8–7.1	

Source: Carrigy, 1963a, 1966; Hamilton and Mellon, 1973.

Table 3 Bulk Properties of Alberta and U.S. Oil Sands

Property	Alberta	Asphalt Ridge	P.R. Springs	Sunny-side	Tar Sand Triangle	Texas	Alabama
Bulk density, g/cm^3	1.75–2.19		1.83–2.50				
Porosity, vol %	27–56	16–27	6–33	16–28	9–32	32	6–25
Permeability, md	10–600	497–603	56–1510	570–750	207–788	320	1–640
Specific heat, cal/g/°C	0.35–0.50						
Thermal conductivity, cal/sec/cm^2/°C/cm	0.0017–0.0035						

Source: Moftah, 1973; Camp, 1976; Glassett and Glassett, 1976; Frazier et al., 1976; Kuuskraa et al., 1978; Karim and Hanafi, 1981; Smith-Magowan et al., 1982.

Table 4 Elemental Composition of Tar Sand Bitumens

	Athabasca (Wabasca McMurray)	Asphalt Ridge	P.R. Springs	Tar Sand Triangle	San Miguel	Arroyo Grande	Sunnyside	Santa Rosa	Bemolanga
C, wt %	83.1	85.3	84.4	84.0		81.9	86.3	84.6	87.1
H, wt%	10.6	11.7	11.1	10.1		10.2	11.7	10.2	11.2
N, wt%	0.40	1.02	1.00	0.46	0.40	1.2	0.9	0.5	0.7
O, wt%	1.1	1.1	2.2	1.1		3.2	0.6	1.7	1.2
S, wt%	4.8	0.59	0.75	4.4	9.5	3.5	0.5	2.4	0.5
Molecular weight	590	668	820	578	—	820	680	600	—
Carbon residue, wt									
Conradson	18.5	—	—	—	24.5				
Ramsbottom	14.9	3.5	12.5	21.6					
Nickel, ppm	100	120	98	53	24				
Vanadium, ppm	250	25	25	108	85				
API gravity	7.5	14.4	10.3	11.1					
H/C (atomic)	1.531	1.646	1.578	1.443	—	1.495	1.627	1.447	1.543

Source: Bunger et al., 1979; Speight, 1981b; Hertzberg et al., 1983.

Table 5 Fractional Composition of Alberta and U.S. Tar Sand Bitumens

	Athabasca Wabasca/ McMurray	Cold Lake	Asphalt Ridge	P.R. Springs	Tar Sand Triangle	San Miguel
Asphaltenes[a]	17.0	15.0	6.3	16.0	26.0	37.4
Resins	39.0	23.0 }				{ ca 27.0
			65.9	58.3	48.3	
Aromatics	23.0	29.0 }			}	
						ca 35.0
Saturates	21.0	33.0	27.8	25.7	25.7 }	

[a]Pentane-insoluble.
Source: Bunger et al., 1979; Speight, 1981a; Hertzberg et al., 1983).

exists between density (gravity) and viscosity, very-low-gravity bitumens generally have very high viscosities. For instance, southwest Texas bitumen with a gravity of -2° API has a viscosity in the range of 2 million cp, while some California bitumens with gravities of 5 or 6° have viscosities in the range of 20,000–5000,000 cp. Some Utah bitumens with gravities ranging from 10 to 14° API have viscosities up to 1 million cp. Elements related to API gravity are viscosity, thermal characteristics, pour point, hydrogen content, and hydrogen/carbon ratio.

Table 6 Fractional Composition of Athabasca Tar Sand Bitumen

	Mildred–Ruth Lakes			Abasand			Bitumount			Hangingstone River
Asphaltenes	16.7	19.0	18.6	22.5	23.4	22.3	20.1	18.3	17.9	22.5
Resins	39.3	32.0	36.2	28.1	29.0	29.3	24.8	35.6	37.1	26.0
Oils	44.0	49.0	45.2	49.4	47.6	48.4	55.1	46.1	45.0	51.5

Source: Speight and Moschopedis, 1981.

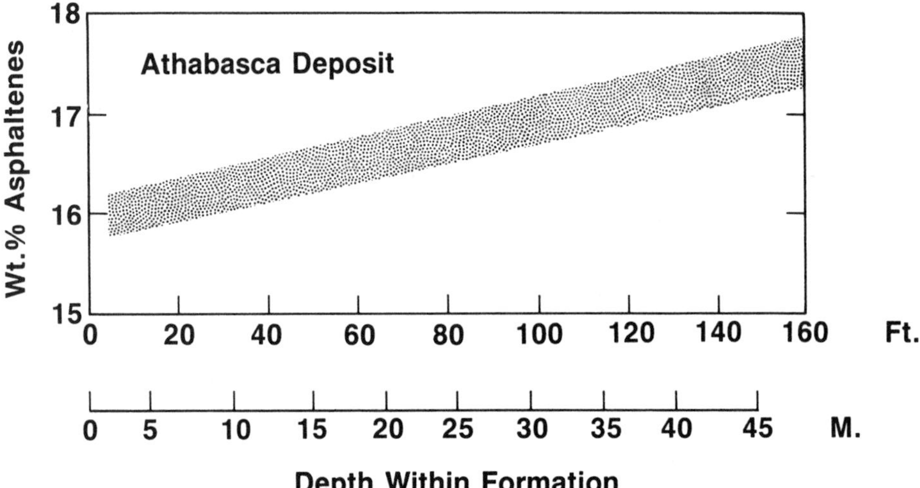

Figure 1 Variation of asphaltene content of the bitumen with depth within the formation (Schutte, 1974; Speight, 1979).

B. Viscosity

The viscosities of crude oils, heavy oils, and bitumens vary markedly over a wide range from less than 10 cp at room temperature to many thousands of centipoises at the same temperature (Chap. 3, Sec. III.E, as well as Figure 3 and Table 7). In the present context, oil sand bitumen occurs at the higher end of this scale where a relationship between viscosity and density between various crude oils has been noted (Figures 4 and 5).

It is also evident that not only are there variations in bitumen viscosity between the four Alberta deposits (Figure 6), but there is also considerable variation of bitumen viscosity within the Athabasca deposit and even within one location (Ward and Clark, 1950). There are relatively high proportions of asphaltenes in the denser, highly viscous samples, a trait which appears to vary not only horizontally but also vertically within a deposit (Schutte, 1974; Speight and Moschopedis, 1981).

The viscosity of the bitumen at original reservoir temperature and the viscosity-temperature relationship (viscosity reduction is proportional to temperature increase) dictate the amount of heat energy that must be added to the reservoir to mobilize the bitumen. The viscosity-temperature relationship is highly dependent on the chemical composition of the bitumen.

A bitumen's mobility is the ratio of its relative permeability to its viscosity (k/μ). A favorable mobility ratio is 1 or greater. Obviously, this ratio is very low (0.1) for a bitumen with a minimal viscosity of 10,000 cp in a reservoir with a reasonable permeability of 1000 md. In a reservoir with 1000-md permeability containing bitumen with 1 million cp viscosity the ratio becomes 10^{-3}—a mobility that is far from practical for commercial production. Elements related to mobility are bitumen viscosity, rock permeability, reservoir temperature, and reservoir pressure.

C. Volatility

One of the main properties of petroleum that serves to indicate the comparative ease with which the material can be refined is volatility, and investigating the volatility of petroleum is usually carried out under standard conditions thereby allowing comparisons to be made between data obtained from various laboratories. Thus, nondestructive distillation data (U.S. Bureau of Mines method; Table 8) show that tar sand bitumens are high-boiling materials in comparison to a more conventional crude oil (Figure 7). There is usually little or no gasoline (naphtha) fraction in bitumen and the majority of the distillate falls in the gas oil—lubrication distillate range (>260°C; >500°F). In excess of 50% of each bitumen is nondistillable under the conditions of the test; this amount of nonvolatile material responds very closely to the asphaltics (asphaltenes plus resins) content of each feedstock.

Early work on the Athabasca material (Table 9) shows that variations in distillate and residua yields occur with variations in bitumen composition and properties; data also illustrate the relationship of distillate yield to depth in the deposit and, hence, to asphaltene content, which also shows a relationship to depth of the bitumen sample in the deposit (Figure 8).

The distillation curve is a plot of the boiling fractions of a bitumen versus temperature and is a function of the bitumen's chemical composition. The distillation curve of the product of a thermal recovery process may show the influence of that process. The high temperatures of a combustion process will crack at least a portion of the bitumen, resulting in some upgrading. In a stream process, a majority of the product oil is very similar to the original bitumen. However, steam stripping of the lighter components can occur late in the life of the process to produce an upgraded oil. If a relatively high percentage of a tar sand bitumen boils above 425°C (>800°F), refining can be much more costly (Chap. 8, Secs. II and IV). A

Table 7 Specific Gravity and Viscosity[a] Data for Various Samples of Athabasca Bitumen (Clark and Blair, 1927; Ward and Clark, 1950; Pasternack and Clark, 1951; Bowman, 1967; Speight and Moschopedis, 1981) and Bitumens from Other Deposits (Various Literature Sources)

	Specific gravity 60/60°F	API gravity	SUS/100°F	SUS/210°F	CS/100°F (38°C)	CS/210°F (99°C)	SFS/210°F (99°C)
Athabasca							
Mildred–Ruth Lakes	1.025	6.5	35,100	513			
	1.025	6.5			35,000	302	
	1.012	8.3					
	1.010	8.6					
Abasand	1.027	6.3					
	1.031	5.7					
	1.027	6.3			500,000		
	1.022	6.9					
	1.029	6.0					
	1.030	5.9					775
	1.032	5.6					373
	1.034	5.4			570,000	1650	
Bitumount	1.007	9.0					98
	1.005	9.3					

Ells River	1.008	8.9	25,000
Horse River	1.020	7.2	
Hangingstone River	1.020	7.2	
Other Sources			
Asphalt Ridge	0.997	10.4	6.1×10^{4} (140°F)
Tar Sand Triangle	1.044	4.1	7.6×10^{10} (140°F)
Sunnyside	0.997	10.4	1.65×10^{6}
Arroyo Grande	1.055	2.6	1.3×10^{6} (220°F)
P.R. Springs	0.998	10.3	
Bemolanga	0.990	10.1	
Trinidad	1.092		

[a] $\frac{\text{Centipoise (cp)}}{\text{Specific gravity}}$ = centistokes (cs).

Centistokes (cs) × 4.632 = Saybolt Universal Seconds (SUS) at 200°F (38°C).
Centistokes (cs) × 4.664 = Saybolt Universal Seconds (SUS) at 210°F (99°C).
Centistokes (cs) × 0.4710 = Saybolt Furol Seconds (SFS) at 100°F (38°C).
Centistokes (cs) × 0.4792 = Saybolt Furol Seconds (SFS) at 210°F (99°C).

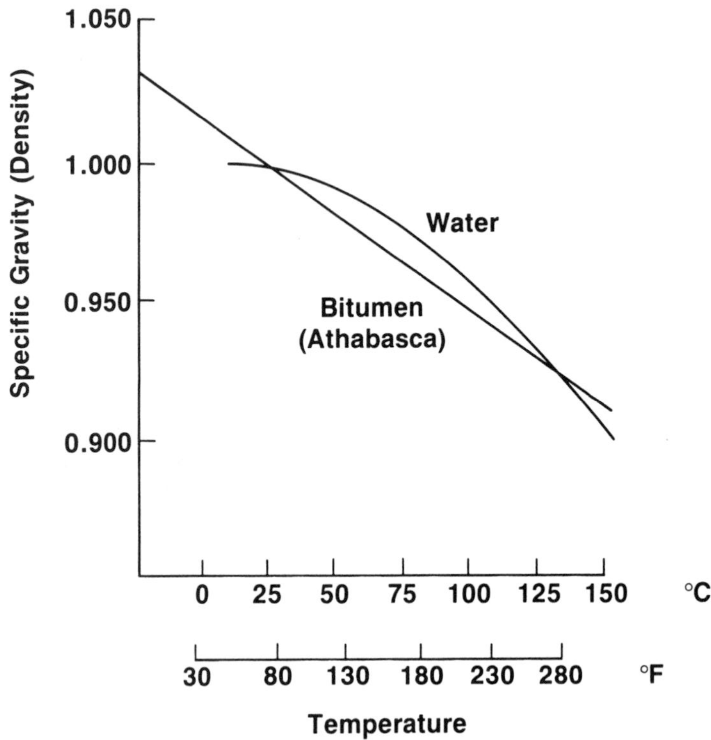

Figure 2 Variation of the specific gravity (density) of Athabasca bitumen and water with temperature (Camp, 1976).

separate upgrading process may be required to increase the hydrogen/carbon ratio of the produced oil. Elements related to the distillation curve include bitumen viscosity, hydrogen content, pour point, and hydrogen/carbon ratio.

D. Thermal Sensitivity

When separated from associated sand, bitumen is so sensitive to direct thermal treatment (Speight, 1978a) that even the asphaltene fraction commences decomposition at temperatures above 150°C (>300°F) and can be effectively cracked to a distillate at 460–470°C (860–880°F).

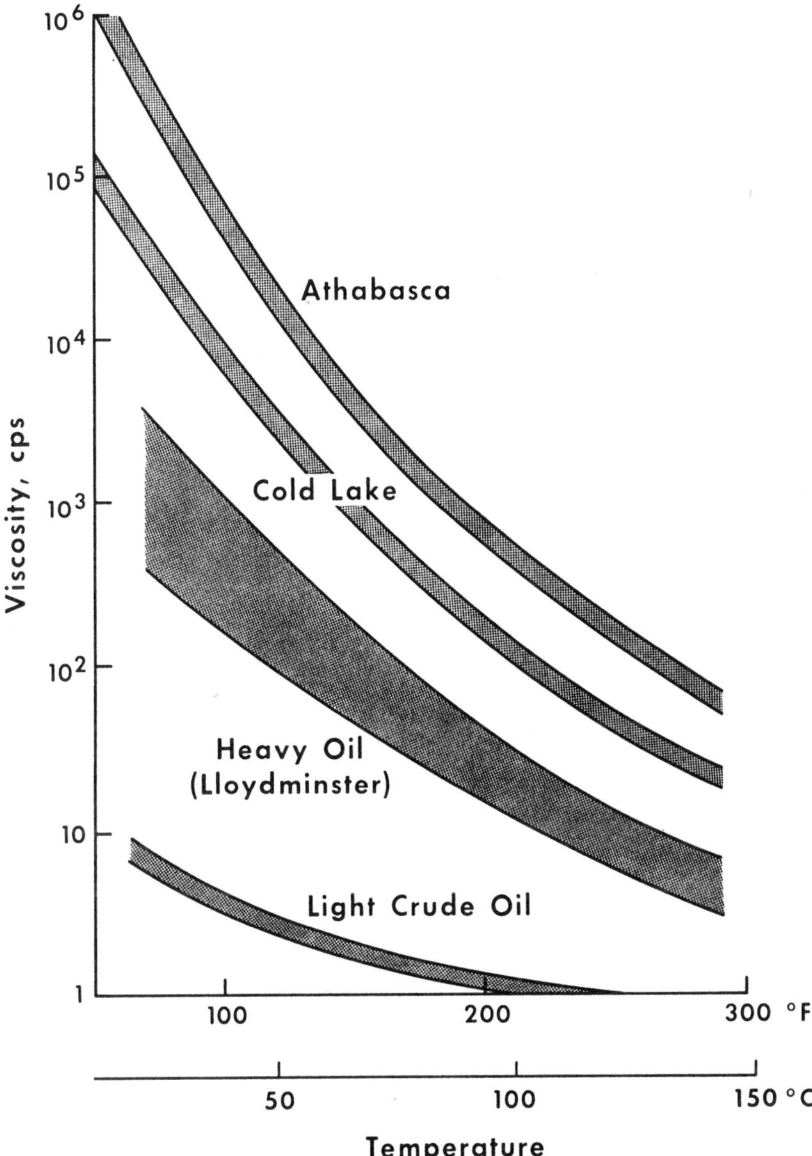

Figure 3 Simplified representation of the viscosity data for crude oils from Alberta.

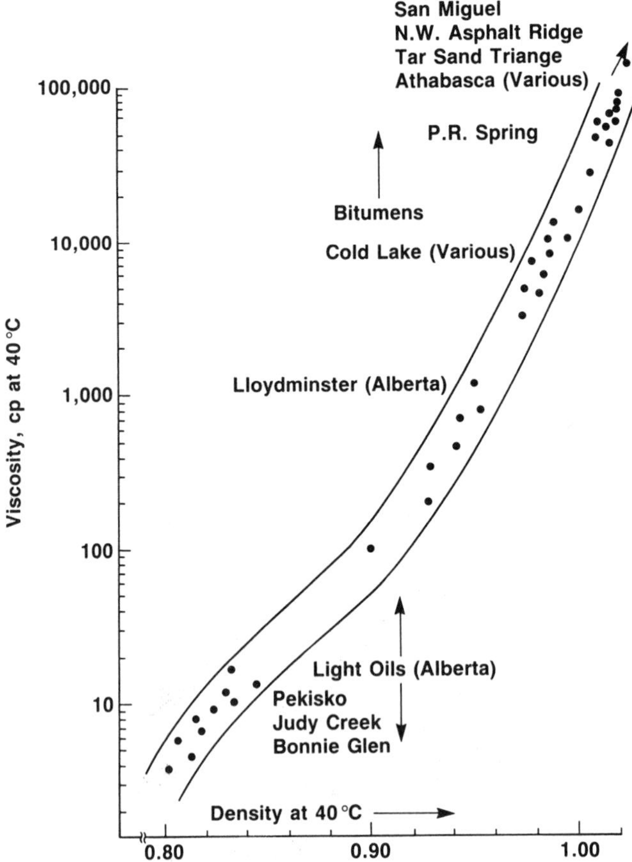

Figure 4 Schematic relationship of density and viscosity.

This thermal sensitivity is reflected in the low activation energy of thermal cracking, which has been determined as 49 kcal/mole for bitumen, relative to gas oil fractions from conventional petroleums, which have an activation energy of about 55 kcal/mole* (Henderson and Weber, 1965.

---

*It must also be noted that a gas oil has already been subjected to the distillation temperature, which may have removed any thermosensitive constituents; such comparisons must be treated with caution.

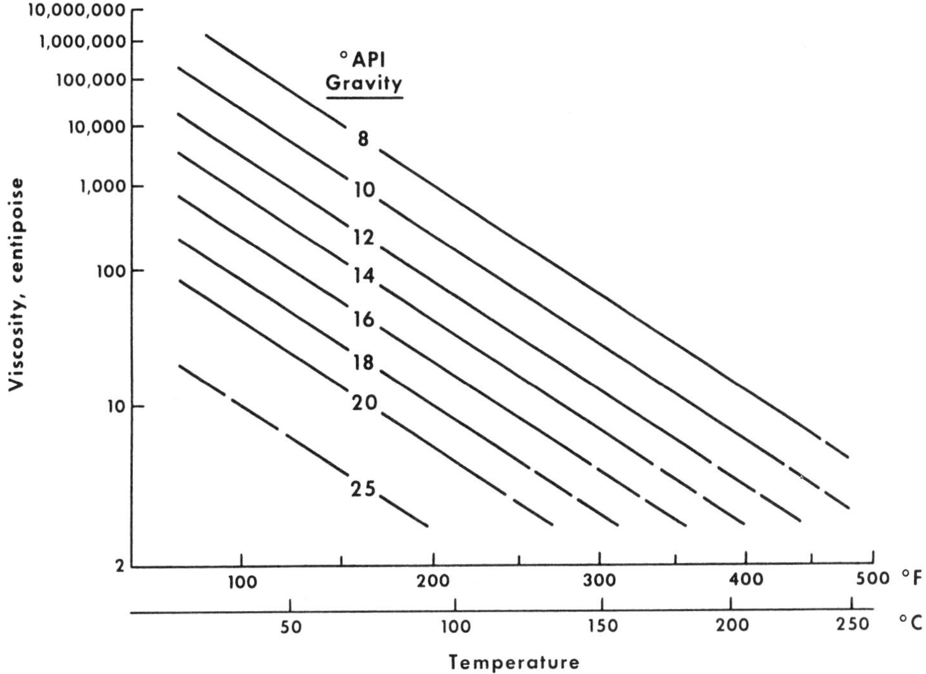

Figure 5 Oil viscosity as a function of temperature and gravity (adapted from Interstate Oil Compact Commission, 1983).

E. Solubility

For Athabasca bitumen, the data (Figure 9) show that *n*-paraffins, branched paraffins, or terminal olefins cause some variation (up to ±15%) in asphaltene yield. The solvent power (solubility parameter*) increases from 2-methyl paraffin to *n*-paraffin to terminal olefin. Cycloparaffins have a remarkable effect on asphaltene yield and give results totally unrelated to those from any of the other nonaromatic

*See Hildebrand et al. (1970) for derivation of the solubility parameter.

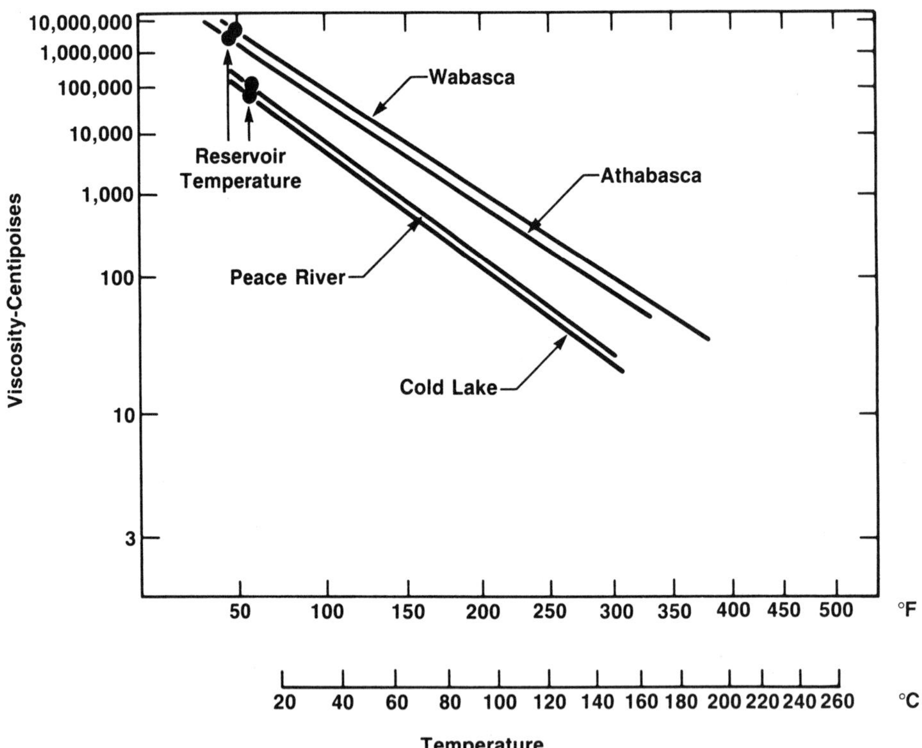

Figure 6 Relationship of viscosity to temperature for the various Alberta deposits (Bowman, 1967).

hydrocarbon solvents studied here. For example, when cyclopentane, cyclohexane, or their methyl derivatives are employed as the deasphalting media, only about 1% of the bitumen remains insoluble, and thus the solvent power of these solvents falls within the area of the solvent power of the aromatics.

Bitumen solubility can be correlated (Mitchell and Speight, 1973) with physical properties as well as with the chemical structure of the solvent. In addition, the physical characteristics of two different solvent types, in this case benzene and *n*-pentane, are additive on a mole-fraction basis and also explain the variation of solubility with temperature. The data also show the effects of blending a solvent with the bitumen itself and allowing the resulting solvent–heavy oil

Table 8 Distillation Data for Tar Sand Bitumens and a Conventional Crude Oil

Cut point						
°C	°F (±2%)	Athabasca wt % distilled[a]	N.W. Asphalt Ridge wt % distilled[a]	P.R. Spring wt % distilled[a]	Tar Sand Triangle wt % distilled[a]	Leduc.
200	390	3.0	2.3	0.7	1.7	35.1
225	435	4.6	3.3	1.4	2.9	40.1
250	480	6.5	4.4	2.4	4.4	45.2
275	525	8.9	5.8	3.8	5.9	51.3
300	570	14.0	7.5	4.9	8.4	
325	615	25.9	8.8	6.8	12.4	
350	660	18.1	11.7	8.0	15.2	
375	705	22.4	13.8	10.1	18.6	
400	750	26.2	16.8	12.5	22.4	
425	795	29.1	19.5	16.0	26.9	
450	840	33.1	23.7	20.0	28.9	
475	885	37.0	28.4	22.5	32.3	
500	930	40.0	34.0	25.0	35.1	
525	975	42.9	40.0	27.3	38.5	
538	1000	44.6	44.2	28.0	40.0	
538+	1000+	55.4	55.8	72.0	60.90	<5

[a]Cumulative.

Source: Speight, 1981a; Bunger et al., 1979.

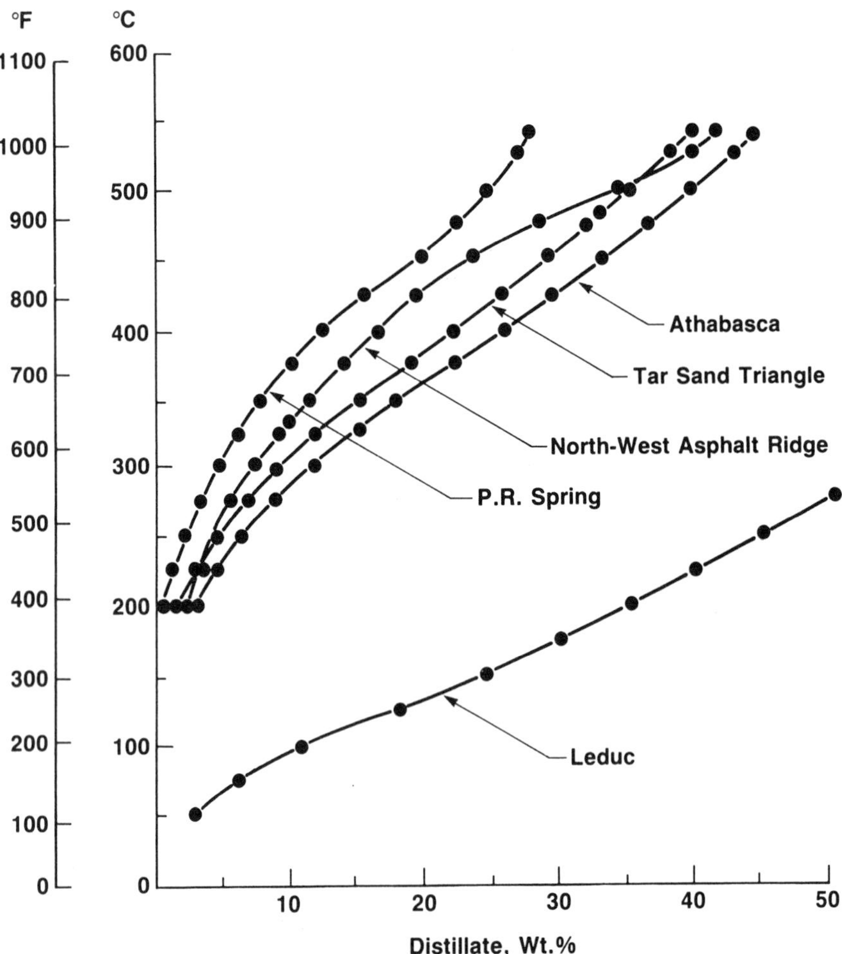

Figure 7 Simulated distillation data for Athabasca bitumen, Utah bitumens, and a conventional (Canadian) crude oil (Bunger et al., 1979; Speight, 1981a).

blend to control the degree of bitumen solubility. Varying the proportions of the hydrocarbon alters the physical characteristics of the oil to such an extent that the amount of precipitate (asphaltenes) can be varied accordingly within a certain range.

Table 9 Distillation Data (U.S. Bureau of Mines Method) for Samples of Athabasca Bitumen

Source of sample	Ellis River	Athabasca River	Athabasca River	Athabasca River	Athabasca River	Athabasca River
Sp. gr. of bitumen distilled 25°C/25°C	1.008	1.013	1.013	1.020	1.022	1.029
Distillation products:						
Gas oil	23%	15%	11%	8%	9%	7%
Nonviscous lubricating distillate	6%	5%	5%	5%	7%	5%
Medium lubricating distillate	5%	5%	9%	5%	11%	8%
Viscous lubricating distillate	—	6%	—	—	—	5%
Residuum	66%	69%	75%	82%	73%	75%
Properties of residuum:						
Specific gravity 25°C/25°C	1.035	1.040	1.038	1.049	1.046	1.053
Penetration,						
25°C, 5 s, 50 g	too	105	100	—	—	—
25°C, 5 s 100 g	soft	—	—	100	135	53

Definitions:

Gas oil—viscosity S.U. at 100°F, less than 50 sec, specific gravity at 60°F greater than 0.825.

Nonviscous lubricating distillate--viscosity between 50 and 100 sec.

Medium lubricating distillate—viscosity between 100 and 200 sec.

Viscous lubricating distillate—viscosity greater than 200 sec.

Source: Clark and Blair, 1927; Pasternack and Clark, 1951.

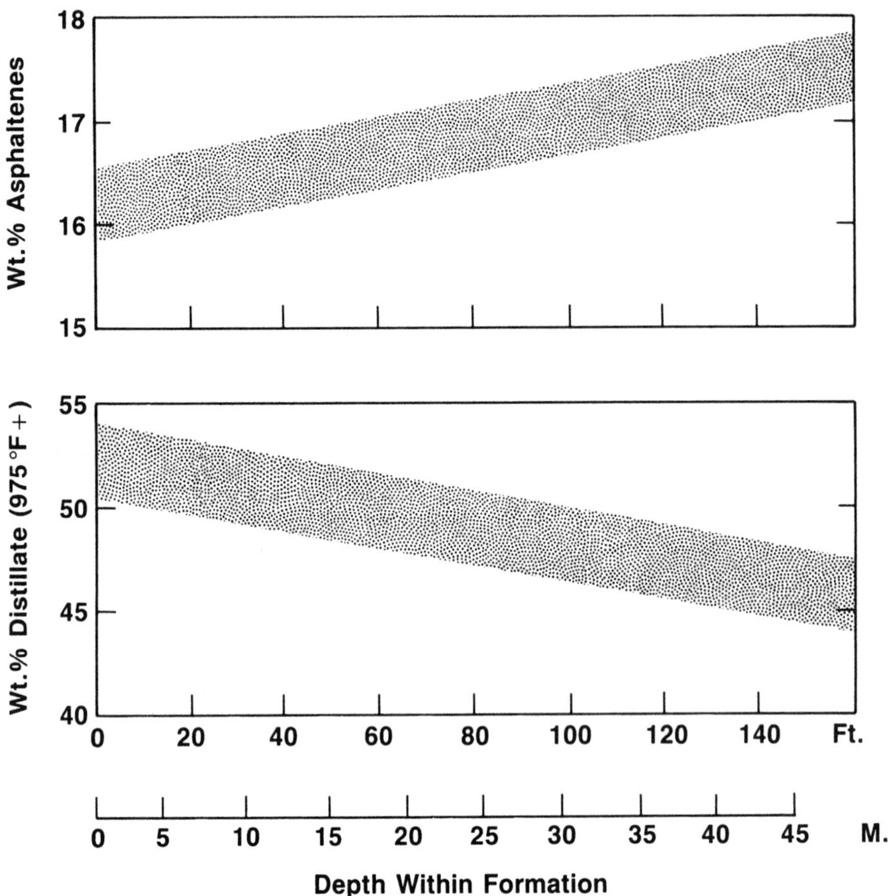

Figure 8 Variation of asphaltics content and distillate yield from Athabasca bitumen from different depths within the formation (Schutte, 1974).

F. Miscellaneous Properties

The pour point is the lowest temperature at which the bitumen will flow. Pour points for tar sand bitumens can exceed 300°F—far greater than the natural temperature of tar sand reservoirs. The pour point is a direct function of the bitumen's viscosity-temperature relationship. It is important to consider because for efficient production supplementary heat energy must be supplied by a thermal

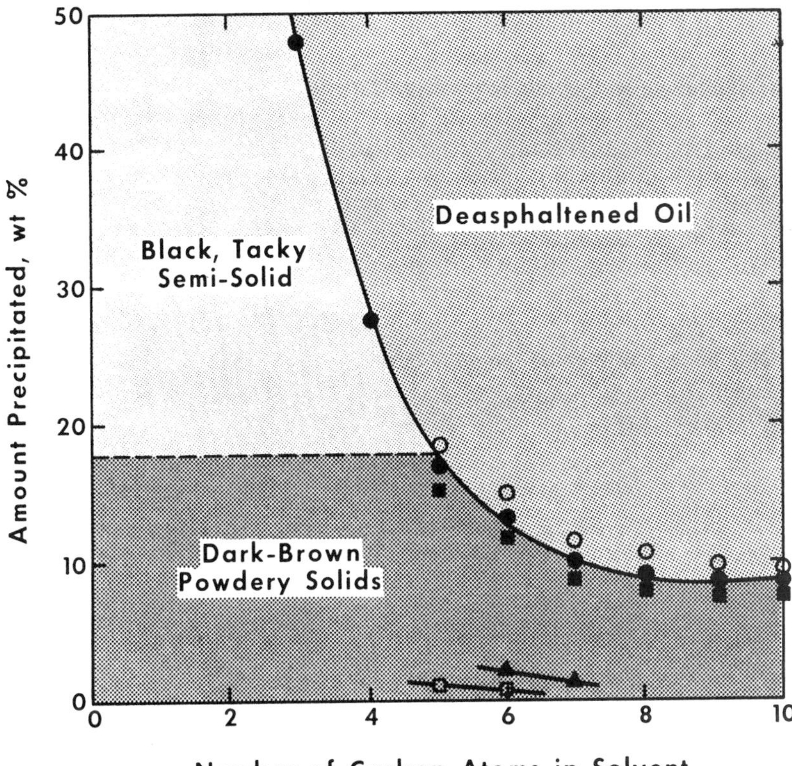

Figure 9 Relationship of precipitate yield to carbon atoms in nonaromatic solvents (Mitchell and Speight, 1973).

extraction process to increase the reservoir temperature to beyond the pour point. Elements related to pour point are depth, bitumen viscosity, original reservoir temperature, and hydrogen/carbon ratio.

Other properties such as calorific value, carbon residue, specific heat, softening point, flash point, molecular weight, and thermal conductivity (Table 10) are also used on occasion to determine the suitability of the bitumen for conversion options.

Table 10 Miscellaneous Properties of Tar Sand Bitumens

Property	Athabasca	Utah
Calorific value, Btu/lb	17,690–17,910	17,900–18,800
Carbon residue, wt %		
Conradson	13–19	3–24
Ramsbottom	10–13	3–22
Specific heat,[a] (cal/g/°C)	0.35–0.51	0.42–0.93
Softening point	60–90°F 16–32°C	
Flash point	250–350°F 120–175°C	
Molecular weight	500–800	660–820
Thermal conductivity[b]		
17.1% bitumen	0.0027–0.0032	
11.1% bitumen	0.0021	
8.6% bitumen	0.0024	
3.0% bitumen	0.0017	

[a]Specific heat of tar sand (17.1% bitumen, 0.9% water) is 0.218 cal/g/°C.
[b]Of tar sand, remolded samples; thermal conductivity of an undisturbed tar sand sample is 0.0035 cal/(sec)(cm^2)(cm/°C).

15

Bitumen Upgrading

I	Primary Conversion	412
	A. Delayed Coking	415
	B. Fluid Coking	416
	C. Other Options	417
II	Secondary Conversion	417

In general terms, the quality of oil sand bitumen is low compared to that of the "lighter" crude oils (Figure 1) and requires some degree of refining to approach the quality of a conventional crude oil (Speight, 1978a). The low volatility of the bitumen precludes refining by distillation and it is recognized that refining by thermal means is necessary to produce liquid fuel streams. A number of factors have influenced the development of facilities which are capable of converting bitumen to a synthetic crude oil. A visbreaking product would be a hydrocarbon liquid that was still high in sulfur and nitrogen with some degree of unsaturation; this latter property enhances gum formation with the accompanying risk of pipeline fouling and similar disposition problems in storage facilities and fuel oil burners. A high sulfur content in finished products is environmentally unacceptable. In addition, high levels of nitrogen cause problems in the downstream processes, such as in catalytic cracking where nitrogen levels in excess of 3000 ppm will cause rapid catalyst deactivation; metals (nickel and vanadium) cause similar problems (Speight, 1981a).

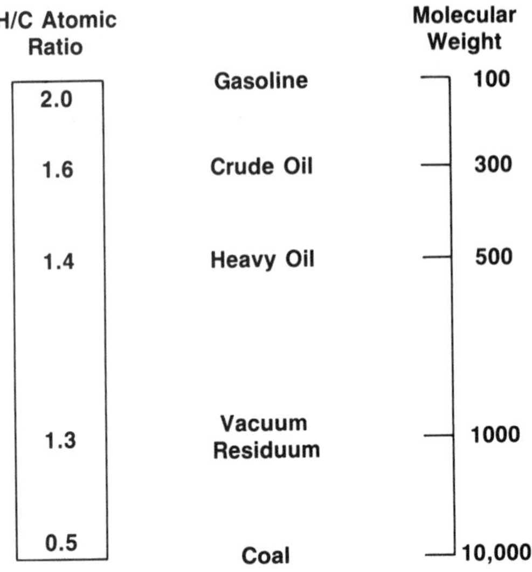

Figure 1 Schematic representation of the atomic hydrogen/carbon ratios of various feedstocks.

Examination of the distillation data for oil sand bitumen (Table 8 in Chap. 14) shows that about 40% of the sample boils above 540°C (1000°F), which approximates the amount of resins and asphaltenes normally found in whole bitumen. Thus, a product of acceptable quality could be obtained by distillation to an appropriate cut point but in excess of 50% of the bitumen would remain behind to be refined by whichever means would be appropriate—remembering, of course, the need to balance fuel requirements and coke production (Figure 2). It is therefore essential that any bitumen-upgrading program convert* the nonvolatile asphaltenes and resins to a lower-boiling, low-viscosity, low-molecular-weight, high hydrocarbon/carbon ratio oil (Figure 3).

I. PRIMARY CONVERSION

The two oil sand plants currently in operation employ a primary conversion process in which the whole bitumen feed is converted to a

---

*Conversion may be defined as weight percentage of original oil, boiling above a specified temperature usually between 525°C (975°F) and 565°C (1050°F), which disappears in the conversion process.

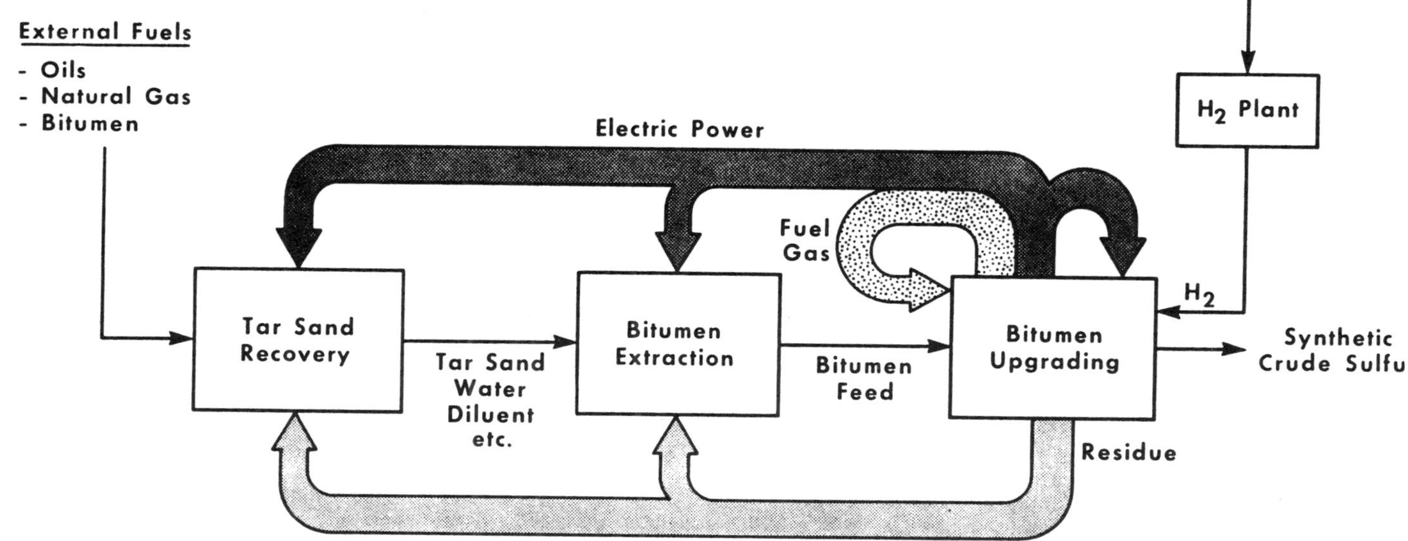

Figure 2 Simplified representation of fuel and energy input for a tar sands (hot water recovery method) plant.

low-carbon/high-hydrogen product but which still contains high concentrations of sulfur and nitrogen (Figure 4). The "primary" liquid product is then hydrotreated (secondary conversion/upgrading) to remove sulfur and nitrogen (as hydrogen sulfide and ammonia, respectively) and to hydrogenate the unsaturated sites exposed by the conversion process. It may be necessary to employ separate hydrotreaters for light distillates and medium-to-heavy fractions; for example, the heavier fractions require higher hydrogen partial pressures and higher operating temperatures to achieve the desired degree of sulfur and nitrogen removal. Commercial applications have therefore been based on the separate treatment of two or three distillate fractions at the appropriate severity to achieve the required product quality and process efficiency (Speight, 1981a).

Hydrotreating is generally carried out in downflow reactors containing a fixed bed of cobalt-molybdate catalysts. The reactor effluents are stripped of the produced hydrogen sulfide and ammonia.

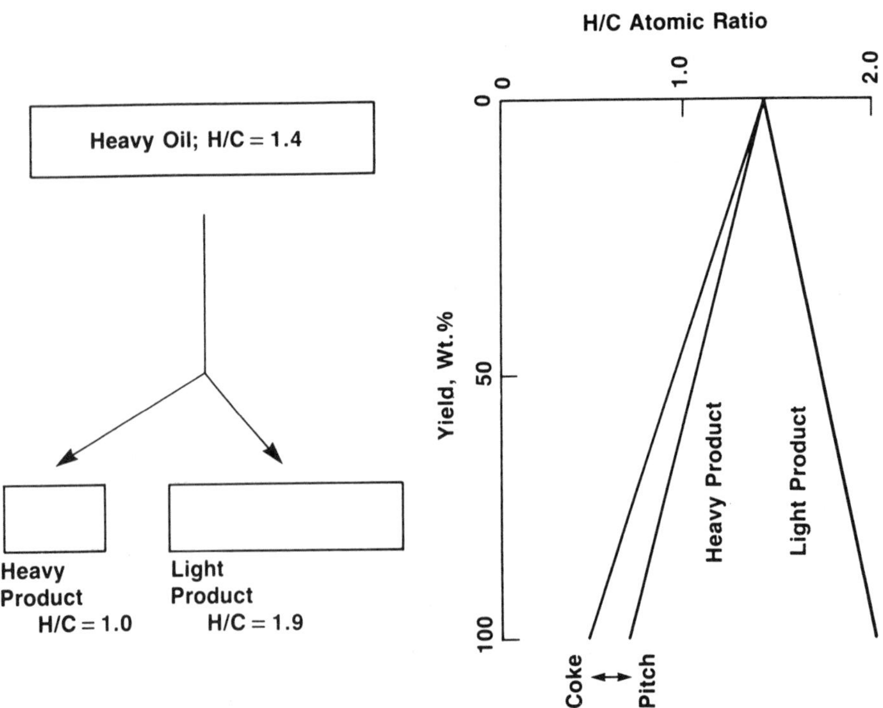

Figure 3 Simplified representation of the requirements of a bitumen-upgrading process.

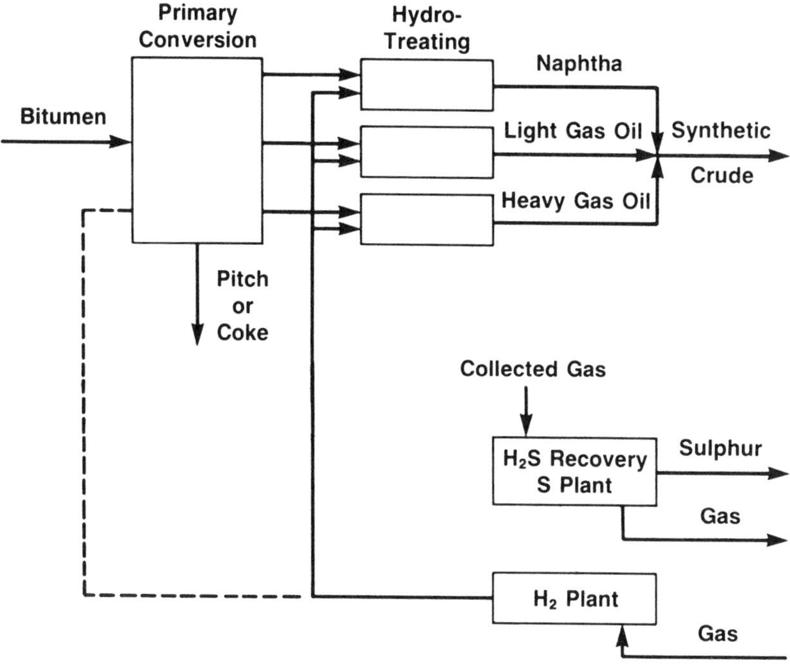

Figure 4 Schematic representation of bitumen conversion.

Any light ends are sent to the fuel gas system and the liquid products are recombined to form synthetic crude oil.

There are two processes that have been applied to the production of liquids from Athabasca bitumen. In this respect, these processes are "proven" but are not necessarily the best or ultimate processes (Chap. 8). Caution must therefore be exercised in the use of the word "proven." Delayed coking is practiced at the Suncor (formerly Great Canadian Oil Sands) plant, whereas Syncrude employs a fluid coking process which produces less coke than the delayed coking in exchange for more liquids and gases.

A. Delayed Coking

The Suncor plant (in operation since 1967) involves a delayed coking technique followed by hydrogen treating of the distillates to produce

Table 1 Properties of Suncor Synthetic Crude Oil

Property	Value
Gravity, °API	32
Sulfur, wt %	0.15
Nitrogen, wt %	0.06–0.10
Viscosity, cs @ 100°F	<10
Components, v/v %	
C4's	4
C5/430°F (220°C)	24
430/650°F (346°C)	32
650/1020°F (550°C)	40
1020°F+ (550°C+)	0

the synthetic crude oil (Table 1). The selection of delayed coking over less severe thermal processes, such as visbreaking, was based (at the time of planning, ca. 1960–1964) on the high yields of residuum produced in these alternate processes (which would exceed the plant fuel requirements, especially if the distillates had to be shipped elsewhere for hydrogen treating) as well as a more favorable product distribution and properties.

B. Fluid Coking

In the fluid coking process, whole bitumen (or topped bitumen) is preheated and sprayed into the reactor where it is thermally cracked in the fluidized coke bed at temperatures typically between 510 and 540°C (950 and 1000°F) to produce light products and coke.

The coke is deposited on the fluidized coke particles while the light products pass overhead to a scrubbing section in which any high-boiling products are condensed and recombined with the reactor fresh feed. The uncondensed scrubber overhead passes into a fractionator in which liquid products of suitable boiling ranges for downstream hydrotreating are withdrawn. Cracked reactor gases containing $<C_4$ material pass overhead to a gas recovery section. The $<C_3$ material ultimately flows to the refinery gas system and the condensed butanes/butenes may (subject to vapor pressure limitations) be combined with

the synthetic crude. The heat necessary to vaporize the feed and to supply the heat of reaction is supplied by hot coke which is circulated back to the reactor from the coke heater. Excess coke, which has formed from the fresh feed and deposited on hot circulating coke in the fluidized reactor bed, is withdrawn (after steam stripping) from the bottom of the reactor.

C. Other Options

Other process options for bitumen conversion have been investigated and, indeed, current work continues. The types of processes that are being given consideration include variations of the different cracking and catalytic cracking processes as well as the use of hydrogen in the processes. These options are discussed in more detail in Chap. 8. Concepts in the development stage include a pyrolysis/hydropyrolysis technique (Bunger and Cogswell, 1981; Dorius et al., 1984) that shows extremely promising yields of liquid products and has been planted to the pilot plant stage (Hatfield and Oblad, 1984). The concepts in the development stage also include hydrogen donor solvent cracking (Belinko et al., 1984), which is a revitalized version of the hydrogen donor diluent visbreaking process (Chap. 8, Sec, IV), and the Canmet hydrocracking process (Kriz et al., 1984).

II. SECONDARY CONVERSION

The term "secondary conversion" used in the present context refers to the removal of contaminant materials (nitrogen, sulfur, and others) from the liquid overhead products that are produced by the coking technologies currently applied to bitumen conversion. The liquid overheads still contain contaminant materials as well as olefinic species that confer undesirable properties on the products.

Thus, secondary conversion involves the application of hydrotreatment to the overhead products before they are blended into a synthetic crude oil. Each overhead fraction is treated separately because of the different process conditions needed to maximize the treatment process (Speight, 1981a; see also Chap. 9, Sec. III).

16

Current and Future Development

I Present 420
II Future 423

There have been numerous forecasts of western world producibility and demand for conventional crude, all covering varying periods of time, and even after considering the impact of the conservation ethic, the development of renewable resources, and the possibility of slower economic growth, nonconventional crude oil sources could well be needed to make up for the future anticipated shortfalls in conventional supplies.

This certainly applies to North America, which has additional compelling reasons to develop viable alternative fossil fuel technologies. Those reasons include, of course, the security of supply and the need to quickly reduce the impact of energy costs on the balance of payments.

There has been the hope that the developing technology in North America will eventually succeed in applying the new areas of nuclear and solar energy to the energy demands of the population. However, the optimism of the 1970s has been succeeded by the reality of the 1980s and it is now obvious that these energy sources will not be the answer to energy shortfalls for the remainder of the present century. Energy demands will most probably need to be met by the production of more liquid fuels from fossil fuel sources.

There are those who suggest that we are indeed faced with the inevitable decline of our liquid fuel culture. This may be so, but the potential for greater energy availability from alternative fossil fuel technologies is high. North America is rich in coal, oil shale, and oil sands—so rich that with the development of appropriate technologies, it could be self sufficient well into the 21st century.

I. PRESENT

There is very little doubt that unlocking energy from the oil sands is a complex and expensive proposition. With conventional production, the gamble is taken in the search and the expenses can be high with no guarantee of a commercial find. With oil sands, the oil is known to be there, but getting it out has been the problem and has required gambling on the massive use of untried technology.

There is no real market for the bitumen extracted from the oil sands and the oil sand itself is too bulky to be shipped elsewhere with the prospect of any degree of economic return. It is therefore necessary that the extraction and upgrading plants be constructed in the immediate vicinity of the mining operation.

To develop the present concept of oil from the oil sands, it is necessary to combine three operations, each of which contributes significantly to the cost of the venture:

1. A mining operation capable of handling 2 million tons, or more, of oil sand per day
2. An extraction process to release the heavy oil from the sand, and
3. An upgrading plant to convert the heavy oil to a synthetic crude oil

For Suncor (formerly Great Canadian Oil Sands Ltd.), being the first of the potential oil sands developers carried with it a variety of disadvantages. The technical problems were complex and numerous with the result that Suncor (on stream: 1967) had accumulated a deficit of $67 million by the end of 1976, despite having reported a $12 million profit for that year. Since that time, Suncor has reported steady profits and has even realized the opportunity to expand operations to 60,000 bbl/day of synthetic crude oil.

However, with hindsight it appears that such a situation is not without some advantages. The early start in the oil sands gave Suncor a relatively low capital cost per daily barrel for a nonconventional synthetic crude oil operation. Total capital costs were about $300 million which, at a production rate of 50,000 bbl/day, places the capital cost at about $6000 per daily barrel.

It is perhaps worthy of mention here that a conventional refinery of the Imperial Oil Strathcona-type (150–300 × 10^3 bbl/day) may have cost at that time $100–400 million and have an energy balance (i.e., energy output/energy input) of 90%+ while an oil sands refinery of the Suncor–Syncrude type may have an energy balance of the order of 70 ± 3%.

The second oil sands plant erected by the Syncrude Canada Ltd. faced much stiffer capital costs. In fact, it was the rapidly increasing capital costs which nearly killed the Syncrude project. Originally estimated at less than $1 billion (i.e., $1 × 10^9), capital needs began to escalate rapidly in the early 1970s. The cost was more than one of the four partners wanted to pay and the number of participants dropped to three. Since the company dropping out held one of the largest interests, the loss was keenly felt and for a while the project was in jeopardy. It was finally kept alive through the participation of the Canadian government and the governments of the provinces of Ontario and Alberta. The Canadian government took a 15% interest in the project while the Province of Alberta took a 10% interest and the Province of Ontario a 5% interest. The balance remained with three of the original participants: Imperial Oil Ltd., Gulf Oil Canada Ltd., and Canada-Cities Service Ltd.

After that initial setback, progress became rapid and the project (located a few miles north of the Suncor plant) was brought to completion (on-stream: 1978). The latest estimate of the cost of the plant is in the neighborhood of $2.5 billion. At a design level of 120,000–130,000 bbl/days, the capital cost is in excess of $20,000 per daily barrel (Figure 1).

For both the Suncor and Syncrude plants, the investment is broken down to four broad areas: mining (28–34%), bitumen recovery (12%), bitumen upgrading (28–30%), and offsites, including the power plant (16–24%) (Table 1).

In an economic treatment of, in this case, oil production, there are invariably attempts made to derive the costs on the basis of mathematical formulas. The effects of inflation on the capital outlay for the construction of an oil sands plant prohibit such mathematical optimism. For this reason, there are no attempts made here to "standardize" plant construction costs except to note that, say, the percentage of the outlay required for specific parts of a plant must be anticipated to be approximately the same whether the plant cost $300 × 10^6 or $15 × 10^9.

The economics of an in situ project will be somewhat different because of the ongoing nature of the project and the offset of costs by revenue.

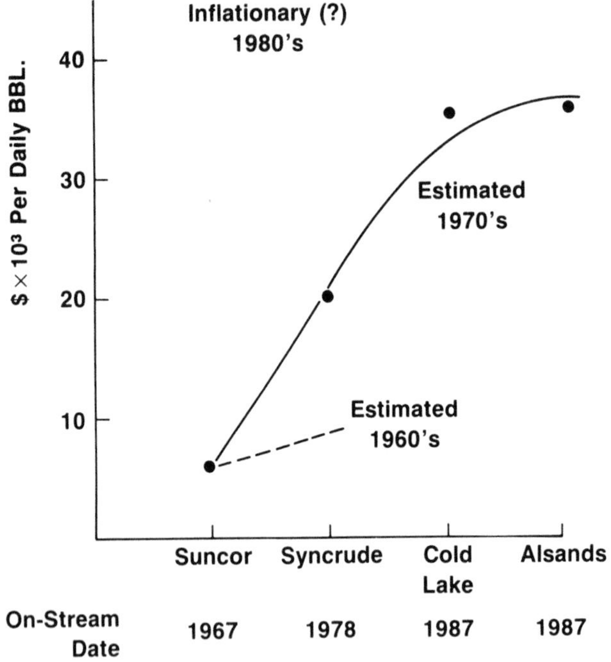

Figure 1 Schematic representation of the trends in the capital costs of oil sand plant construction.

Table 1 Breakdown of Construction Costs for the Suncor (GCOS) and Syncrude Oil Sand Processing Plants

Factor	% Capital cost
Mine	28–34
Bitumen recovery	10–14
Upgrading	26–34
Offsites (incl. power plant)	16–24

Source: Speight, 1978b.

Table 2 Projected Capital Costs for Various Canadian Oil Sand Plants[a]

			Proposed plants			
			Preproduction estimates		Final estimates[b]	
Factor	Suncor	Syn-crude	Cold Lake	Alsands	Cold Lake	Alsands
Bbl/day × 10^3	50	130	140	140	140	140
Initial capital $ × 10^9	0.3	2.5	5.0	5.02	12	13
K$/(bbl/day)	6.0	19.2	33.6	35.7	86	93
Completion year	1967	1979	1987	1987		

[a]See also Figure 1.
[b]These figures do not include income obtained in the production (post-1986) period.

In the United States, oil sand economics is still very much a matter for conjecture. The estimates (Table 2) published for current and proposed Canadian operations are, in a sense, not applicable to operations in the United States because of differences in the production techniques that may be required. As an example, one estimate in particular (Table 3; Reed, 1979) showed a construction cost (in 1978 dollars of $145 × 10^6 for a 10,000 bbl/day extraction plant and it was conjectured that such an extraction plant would have to operate and be maintained between $2.00 and 5.00 per barrel.

II. FUTURE

Any degree of North American self-sufficiency in liquid fuels will require* the development of heavy oil fields and deposits. Recent inflationary aspects of plant construction costs have brought to a slowdown

*In addition to projects which produce liquid fuels from coal and oil shale.

Table 3 Cost Estimates for a 10,000 Bbl/day Extraction Plant for U.S. Oil Sands

Factor	Cost ($ 1978)	
	$/ton	$/bbl
Equipment operation and ownership costs		
Strip and stockpile overburden	0.33–0.52	1.10–1.73
Mining ore	0.17–0.34	0.57–1.13
Hauling ore (1 mile)	0.16–0.18	0.53–0.60
Waste haul (1 mile)	0.16–0.18	0.53–0.60
Site reclamation	0.07–0.14	0.23–0.47
Mining equipment operation and ownership costs	0.89–1.36	2.96–4.53
Mining overhead and supervision		0.44–0.68
Mining profit and federal taxes		0.68–1.05
Total mining operation cost ranges		4.08–6.26
Exploration and development costs		0.50–1.00
Capital cost; 10,000 bbl/day extraction plant per barrel	= $145,000,000	
Charges to recover capital, pay taxes, and receive 9.5% equity return over 20 years	= $5.74/bbl	
Total mining operation and ownership and profit plus extraction plant capital recovery and profit	= $10.32–13.00	
Expected selling price	= $15.00/bbl	
Remainder available for extraction plant operation	= $4.68–2.00	

what was considered, for example, to be a natural evolution of a succession of oil sands plants in Canada.

The capital requirements ($10–15 × 10^9) for the construction of large recovery/upgrading plants will not decrease but the suggestion that commercial planning be directed toward smaller scale nominal size (i.e., Suncor-type plant, 45,000–60,000 bbl/day) (Speight, 1978b; *Oilweek*, 1982a) may be seeing fruition. On the other hand, oil sand plants could be developed on the minimodular concept (*Oilweek*, 1982c), thereby relieving some of the high capital outlay required before the production of 1 bbl of oil. Along these lines, there has also been a similar suggestion that several mining and recovery units produce feedstock for a larger upgrading facility (Speight, 1978b). Either of these modular concepts could be developed throughout the life of the lease by the sole owner or by a consortium.

There are plans for Esso Resources to develop the Cold Lake heavy oil area in phases consisting of six 9500 bbl/day of bitumen with maximum production being realized in the early 1990s.

Currently, Syncrude Canada Ltd. is giving serious consideration to the installation of a bitumen hydrocracker as part of a 5-year capital expansion program that will increase the capacity of the plant by some 30,000 bbl/day of synthetic crude oil at a cost of $\$500 \times 10^6$ (Canadian). In addition, Esso Canada Resources Ltd. is spearheading a drive to build a plant some 12 miles east of the Syncrude plant that will produce approximately 75,000 bbl/day of synthetic crude oil. This OSLO (Other Oil Sands Leases) project would also use surface mining for ore recovery but there is the possibility of as yet unpublicized new technology to cut production costs and make the price of the synthetic crude oil more competitive.

Activites related to the development of U.S. tar sand resources have declined substantially since the early 1980s in line with the decline in the price of crude oil. In 1981, when the price of crude oil was approximately $35/bbl, some 35 U.S. tar sand field projects were either operating or in the late planning stages. By mid-1985, these numbers had declined seriously (Marchant, 1985).

The projects in operation in the early 1980s included 34 in situ projects and 9 mining/extraction projects, and production totaled in excess of 10,000 bbl/day. Almost all of this production occurred in California by means of in situ steam operations but this was from reservoirs where the oil in-place had viscosities in the range of 10,000–25,000 cp (somewhat lower than the viscosities of "true" tar sand bitumen (Chap. 14, Sec. III.B).

Nevertheless, the projects did cover a wide range of reservoir conditions: porosities ranged from 15 to 37% and permeabilities up to 6000 md. Oil saturations were up to 90% of the pore space (up to 22% by weight of the tar sand) and the oil gravities ranged from 2 to 14° API and viscosities from 1–2 million cp.

Finally, the fact that most of the U.S. tar sand resource is too deep for economic development is reflected in the ratio of the numbers of in situ projects to mining/extraction projects (almost 4:1).

Obviously, there are many features (Table 4) to consider when development of tar sand resources is planned. It is more important to recognize that what are important features for one resource might be less important in the development of a second resource. Recognition of this facet of tar sand development is a major benefit that will aid in the production of liquid fuels in an economic and effective manner.

Table 4 General Summation of the Properties of Heavy Crudes and Tar Sand Bitumens (With Conventional Crude Oil Included for Reference) That Might Influence Development

Property	Heavy oil	Bitumen	Conventional crude
Physical			
Viscosity (at reservoir conditions)	<10,000 mPa/sec	>10,000 mPa/sec	Usually about 50 mPa/sec
Density (atm. conditions)			
kg/m^3	≥934	>1000	Usually <934
°API	≤20	<10	>20
Sulfur content	1.6–5.3%	1.6–5.3%	Usually ±2% except in sour type crudes
Presence of metals (most common)			
Vanadium	150–1100 ppm	150–1100 ppm	Usually 0 to traces
Nickel	40–150 ppm	40–150 ppm	Usually 0 to traces
Depth of deposit	Shallow to ±10,000 ft	Surface to ±3000 ft	Depends on the oil provinces, actual range 1000 to ±20,000 ft
Rock type	Unconsolidated sands, usually associated with swelling clays	Unconsolidated sands, often associated with swelling clays Limestones, dolomites	Consolidated sands Limestones, dolomites
Reservoir energy	Low	Lacking	High to very high

Exploration	Usually found when exploring for conventional oil. An exception has been the exploration appraisal of the Orinoco Oil Belt. There are now specific exploration programs (i.e., California)		Specifically designed programs for exploration.
Recovery	Short primary phase followed by secondary recovery. Before the secondary recovery method is selected, a pilot project is often established to test the commerciability of the recovery process. If successful, it is extended to the whole reservoir or field. Usually steam or in situ combustion methods are applied	Mining: surface, underground, or through horizontal tunneling	Long primary phase (natural flow and gas lift)
		In situ recovery methods may include steam or combustion	Secondary recovery: Pressure maintenance (water, gas) + enhanced
Well completion	Slotted liners Wire-wrapped screens Gravel pack completions Outside liner Inside liner	Same type of completion for similar cases	Open hole completion Single-tubing string Double-tubing string Ditto with gaslift mandrels Macaroni type Slim-hole completion Others

(continued)

Table 4 (Cont.)

Property	Heavy oil	Bitumen	Conventional crude
Well spacing (Care should be exercised to ensure that the greatest well spacing possible is also the most economical)	Reduced well spacing (i.e., 1 acre/well or less) increases drainage area and therefore recovery	Same applies	Ranges from ±200 acre/well to ±10 acre/well
	Well density for final recovery is a function of the viscosity of the crude	Same applies	Reduced well spacing may cause well interference and not an increase in the drainage area
	Smaller well spacing reduces land requirements		
Production Methods:			
1. Methods	Sucker-rod pumping (wells could sustain flow temporarily after the initial steam injection phase)	Mining Sucker-rod pumping (if in situ combustion or steam injection applies)	Natural flow Gaslift Hydraulic pumping Submergible electric pumping
2. Use of electric power	Intensive	Very intensive in Athabasca mining	Intensive
3. Use of water	Intensive	Intensive	Very intensive if water injection is used for pressure maintenance

Licensee has, generally through permission, the right to use the water in the area for operational purposes.

4. Use of gas	Intensive if gas is used as fuel to fire boilers and cogenerate steam and electricity	Intensive if gas is used as fuel to fire boilers and cogenerate steam and electricity	Very intensive if gas injection is used for pressure maintenance and/or gas for gaslift production
Licensee has, generally through permission, the right to use the gas in the area for operational purposes.			
5. Use of oil	Oil is sometimes used as fuel to fire burners. It takes about 1 bbl of oil to generate the energy to produce 3–5 bbl from the reservoir		
Transportation:			
1. To refineries or to loading piers	Through heated pipelines Mixed with diluents Mixed with water Through annular water core flow	Generally upgraded on site before transport Mixed with diluents	No special requirements except in the case of waxy crudes
2. In tankers	Tankers transport all crudes at a temperature of minimum 60°C (140°F). Water content is required to be not more than 1%		
Refining	Upgraded at refinery before processing Processes are generally catalyst-intensive Catalyst is poisoned by vanadium and nickel Needs to be desulfurized Main products are gasoline and lubes	The same applies except for the upgraded (30° API) portion which behaves like conventional crude	May need to be desulfurized if sour May need to be dewaxed if paraffinic Main products are gasoline, nafta, other light products and some lubricants from the resids

Source: Adapted from *The Heavy Oiler*, 1986.

References to Part II

Ali, S. M. F. 1974. In *Oil Sands: Fuel of the Future* (L. V. Hills, ed.), Can. Soc. Petrol. Geol., Calgary, p. 199.

Ali, S. M. F., and B. Abad. 1975. *Proc. 26th Ann. Meeting, Petrol. Soc., Can. Inst. Mining*, Banff, Canada.

Asano, B. H., and J. Kus. 1983. *4th Ann. Adv. in Petroleum Technology*, AOSTRA, Calgary, Canada, May.

Ball Associates, 1965. *U.S. Dept. of Interior (Bureau of Mines) Monograph No. 12.*

Ballard, J. R., E. E. Lanfranchi, and P. A. Vanags. 1976. *Proc. 27th Ann. Meeting, Petrol. Soc., Can. Inst. Mining*, Calgary

Belinko, K., L. Y. Cheung, T. E. Hogan, and B. B. Pruden. 1984. *Proc. 2nd International Conference on the Future of Heavy Crude and Tar Sands* (R. F. Meyer, J. C. Wynn, and J. C. Olson, eds.), McGraw-Hill, New York, p. 1268.

Bergeson and Associates. 1978. Golden, Colorado, September.

Berkowitz, N., and J. G. Speight. 1975. *Fuel 54*:138.

Bichard, J. A. 1982. Presented at the Petroleum Recovery Inst., Calgary, Canada, August.

Bleakey, W. B. 1980. *Petrol. Eng. Int.* June, *52*:132.

Bott, R. C. 1967. *J. Petrol. Technol. 19*:585.

Bowman, C. W. 1967. *Proc. 7th World Petrol. Cong. 3*:583.

Bridges, J. E., and G. Stresty. 1985. *Proc. 3rd Conf. on the Future of Heavy Crude and Tar Sands*, Long Beach, California.

Britton, M. W., W. L. Martin, J. D. McDaniel, and H. A. Wahl. 1981. U.S. Patent 4,265,310.

Britton, M. W., W. L. Martin, R. J. Leibrecht, and R. A. Harmon. 1983. *J. Petrol. Technol.* 35:511.
Bunger, J. W., and D. E. Cogswell. 1981. In *The Chemistry of Asphaltenes* (J. W. Bunger and N. C. Li, eds), Advances in Chemistry Series No. 195, Am. Chem. Soc., Washington, D.C., p. 219.
Bunger, J. W., K. P. Thomas, and S. M. Dorrence. 1979. *Fuel* 58: 183.
Burger, J. 1978. In *Bitumens, Asphalts and Tar Sands* (G. V. Chilingarian and T. F. Yen, eds.), Elsevier, New York, p. 191.
Camp, F. W. 1976. *The Tar Sands of Alberta*, Cameron Engineers, Denver.
Carrigy, M. A. 1963a. Bulletin No. 14, Alberta Research Council, Edmonton, Canada.
Carrigy, M. A. 1963b. *The Oil Sands of Alberta*, Information Series No. 45, Alberta Research Council, Edmonton, Canada.
Carrigy, M. A. 1966. Bulletin No. 18, Alberta Research Council, Edmonton, Canada.
Carrigy, M. A., and H. G. Stephenson. 1985. *Proc. 3rd Conf. on the Future of Heavy Crude and Tar Sands*, Long Beach, California.
Clark, K. A. 1944. *Trans. Can. Inst. Min. Met.* 47:257.
Clark, K. A., and B. M. Blair. 1987. Report No. 18, Alberta Research Council, Edmonton, Canada.
Daily Oil Bulletin. 1975. Calgary, Canada, August 14.
Demaison, G. J. 1977. In *The Oil Sands of Canada-Venezuela* (D. A. Redford and A. G. Winestock, eds.), Can. Inst. Mining and Metallurgy, Special Volume No. 17, 9.
Dick, R. A., and S. P. Wimpfen. 1980. *Si. Am.* 243(4):182.
Dietz, D. N., and J. Weijdema. 1968. *Prod. Monthly* 32(5):10.
Donnelly, J. K., R. G. Moore, D. W. Bennion, and A. E. Trenkwalder. 1978. *Proc. AIChE Meeting*, Florida.
Dorius, J. C., F. V. Hanson, and A. G. Oblad. 1984. *WRI-DOE Symposium on Tar Sands* Vail, Colorado, June.
Doscher, T. M. 1967. *Proc. 7th World Petrol. Cong.* 3:625.
Fear, J. V. D., and E. D. Innes. 1967. *Proc. 7th World Petrol. Cong.* 3:549.
Finken, R. E., and R. F. Meldau. 1972. *Oil and Gas J.* 70(29):108.
Franco, A. 1976. *Oil Gas J.* 74(14):132.
Fraser, J. E., I. G. Henderson, P. Kitzan, R. V. Schmitz, and N. A. Myhill. 1982. *Proc. 2nd Conf. on the Future of Heavy Crude and Tar Sands*, McGraw-Hill, New York, p. 788.
Frazier, N. A., D. W. Hissong, W. E. Ballentyne, and E. J. Mazey. 1976. Report for U.S. EPA Contract No. 68-02-1323.
Gishler, P. E. 1949. *Can. J. Res.* 27:104.
Glassett, J. M., and J. A. Glassett. 1976. Report for U.S. Dept. of Interior (Bureau of Mines) Contract No. 80241129.
Govier, G. W. 1975. *Proc. AIME 104th Ann. Meeting*, New York.

Hallmark, F. O. 1981. In *The Future of Heavy Crude and Tar Sands* (R. F. Meyer and C. T. Steele, eds.), McGraw-Hill, New York, p. 69.
Hamilton, W. N., and G. S. Mellon. 1978. In *Guide to the Athabasca Oil Sands Area* (M. A. Carrigy and J. W. Kramers, eds.), Information Series No. 65, Alberta Research Council, Edmonton, Canada.
Harmsen, G. J. 1971. *Proc. 8th World. Petrol. Cong. 3*:243.
Harris, M. C., and J. C. Sobkowicz. 1977. In *The Oil Sands of Canada-Venezuela* (D. A. Redford and A. G. Winestock, eds.), Special Volume No. 17, Can. Inst. Mining and Metallurgy, p. 270.
Hatfield, K. E., and A. G. Oblad. 1984. *Proc. 2nd International Conf. on Heavy Crude and Tar Sands* (R. F. Meyer, J. C. Wynn, and J. C. Olson, eds.), McGraw-Hill, New York, p. 1175.
Heavy Oiler. 1986. January.
Henderson, J. H., and L. Weber. 1965. *J. Can. Petrol. Technol. 4*:206.
Hertzberg, R., F. Hojabri, and L. Ellefson. 1983. Preprint No. 35e, AIChE Summer Meeting, Denver.
Hildebrand, J. H., J. M. Prausnitz, and R. L. Scott. 1970. *Regular Solutions*, Van Nostrand Reinhold, New York.
Houlihan, R. 1984. In *Proc. 2nd International Conference on the Future of Heavy Crude and Tar Sands* (R. F. Meyer, J. O. Wynnand, and J. C. Olson, eds.), McGraw-Hill, New York, p. 1076.
Hvizdos, L. J., J. V. Howard, and G. W. Roberts. 1983. *J. Petrol. Technol. 35*:1061.
Hyndman, A. W. 1981. In *The Future of Heavy Crude and Tar Sands* (R. F. Meyer and C. T. Steele, eds.), McGraw-Hill, New York, p. 645.
Interstate Oil Compact Commission. 1983. *Improved Oil Recovery*, Oklahoma City.
Jeremic, M. L. 1975. *Western Miner*, September, p. 25.
Karim, G. A., and A. Hanafi. 1981. *Can. J. Chem. Eng. 59*:461.
Kessick, M. A. 1979. *Clays and Clay Minerals 27*(4):301.
Kriz, J. F., M. Ternan, and J. M. Denis. 1984. *Proc. 2nd International Conference on the Future of Heavy Crude and Tar Sands* (R. F. Meyer, J. C. Wynn, and J. C. Olson, eds.), McGraw-Hill, New York, p. 1211.
Kruyer, J. 1983. Preprint No. 3d, AIChE Summer Meeting, Denver.
Kruyer, J. 1984. *Proc. 2nd International Conf. on Heavy Crude and Tar Sands* (R. F. Meyer, J. C. Wynn, and J. C. Olson, eds.), McGraw-Hill, New York, p. 1087.
Kuuskraa, V. A., S. Chalton, and T. M. Doscher. 1978. U.S. DOE Contract No. 9014-018-021-22004, Report No. HCP/T90141.
Lekas, M. A. 1981. *Proc. Oil Shale Symposium*, Golden, Colorado, April.
Lewin Associates. 1984. *Major Tar Sand and Heavy Oil Deposits of the United States*, Interstate Oil Compact Commission, Oklahoma City.

Marchant, L. C. 1984. *Proc. 3rd Conf. on the Future of Heavy Crude and Tar Sands*, Long Beach, California.
Marchant, L. C., and C. A. Koch. 1982. *Proc. National Tar Sands (Heavy Oil) Symposium*, Lexington, Kentucky.
Marchant, L. C., and C. A. Koch. 1984. *Proc. 2nd International Conference on the Future of Heavy Crude and Tar Sands* (R. F. Meyer, J. C. Wynn, and J. C. Olson, eds.), McGraw-Hill, New York, p. 1029.
Mathews, C. S. 1983. *J. Petrol. Technol. 35*:465.
McMillan, J. G. 1981. In *The Future of Heavy Crude and Tar Sands* (R. F. Meyer, and C. T. Steele, eds.), McGraw-Hill, New York, p. 775.
Meyer, R. F., and W. D. Dietzman. 1981. In *The Future of Heavy Crude and Tar Sands* (R. F. Meyer and C. T. Steele, eds.), McGraw-Hill, New York, p. 16.
Meyer, R. F., and C. T. Steele (Eds.). 1981. *The Future of Heavy Crude and Tar Sands*, McGraw-Hill, New York.
Miller, J. C., and M. Misra. 1982. *Fuel Processing Technol. 6*:27.
Misra, M., R. Aguilar, and J. D. Miller. 1981. *Separation Sci. Technol. 16*:1523.
Mitchell, D. L., and J. G. Speight. 1973. *Fuel 52*:149.
Moftah, I. Circular No. 89, Alabama Geological Survey, University of Alabama.
Moschopedis, S. E., J. F. Fryer, and J. G. Speight. 1977. *Fuel 56*:109.
Mungen, R., and J. H. Nichols. 1975. *Proc. 9th World Petrol. Cong. 5*:29.
Muslimov, R. K. 1981. In *The Future of Heavy Crude and Tar Sands* (R. F. Meyer and C. T. Steele, eds.), McGraw-Hill, New York, p. 586.
NPC. 1984. National Petroleum Council, *Enhanced Oil Recovery*, U.S. Department of Energy, Washington, D.C.
Oil Gas J. 1976. *74*(14):128.
Oilweek. 1982a. *33*(21):18.
Oilweek. 1982b. *33*(21):25.
Oilweek. 1982c. *33*(21):21.
Page, H. V. 1982. In *Tar Sands* (D. Ball, L. C. Marchant, and A. Goldburg, eds.), Interstate Oil Compact Commission, Oklahoma City, p. 223.
Pasternack, D. S., and K. A. Clark. 1951. Report No. 58, Alberta Research Council, Edmonton, Canada.
Phizackerley, P. H., and L. O. Scott. 1967. *Proc. 7th World Petrol. Cong. 3*:551.
Phizackerley, P. H., and L. O. Scott. 1978. In *Bitumens, Asphalts, and Tar Sands* (G. V. Chilingarian and T. F. Yen, eds.), Elsevier, New York, p. 57.

Rammler, R. W. 1970. *Can. J. Chem. Eng. 48*:552.
Raplee, B. G., F. Cottrell, S. Cotrell, and J. Raab. U.S. Dept. of Interior (Bureau of Mines), Report No. DOE/LC/10929-1921.
Reed, W. F. 1979. U.S. DOE Contract No. EF-77-C-01-2468, Report No. FE-2468-42.
Sandford, E. C. 1983. *Can. J. Chem. Eng. 61*:554.
Schutte, R. 1974. *Proc. 25th Ann. Meeting, Petrol. Soc., Can. Inst. Mining and Metallurgy*, Calgary, Canada.
Smith-Magowan, D., A. Skauge, and L. G. Hepler. 1982. *Can. J. Petrol. Technol. 21*(3):28.
Speight, J. G. 1978a. In *Bitumens, Asphalts and Tar Sands* (G. V. Chilingarian and T. F. Yen, eds.), Elsevier, New York, p. 123.
Speight, J. G. 1978b. Economics of Oil Recovery from the Alberta Oil Sands. Presented at the Economics Society of Alberta, Edmonton Branch, January 10.
Speight, J. G. 1979. Presented at the Div. Geochem., Am. Chem. Soc., Washington, D.C., September.
Speight, J. G. 1980. *The Chemistry and Technology of Petroleum*, Marcel Dekker, New York.
Speight, J. G. 1981a. *The Desulfurization of Heavy Oils and Residua*, Marcel Dekker, New York.
Speight, J. G. 1981b. Presented at the Div. Geochem., Am. Chem. Soc., New York, August.
Speight, J. G., and B. E. Moschopedis. 1978. *Fuel Proc. Technol. 1*:261.
Speight, J. G., and B. E. Moschopedis. 1981. In *The Fugure of Heavy Crude and Tar Sands* (R. F. Meyer and C. T. Steele, eds.), McGraw-Hill, New York, p. 603.
Spragins, F. K. 1978. In *Bitumens, Asphalts and Tar Sands* (G. V. Chilingarian and T. F. Yen, eds.), Elsevier, New York, p. 92.
Steinmetz, I. 1969. U.S. Patent 3,466,240.
Strom, N. A., and R. B. Dunbar. 1981. In *The Future of Heavy Crude and Tar Sands* (R. F. Meyer and C. T. Steele, eds.), McGraw-Hill, New York, p. 47.
Takamura, K. 1982. *Can. J. Chem. Eng. 60*:538.
Terwilliger, P. L. 1975. Paper 5568, *Proc. 50th Ann. Fall Meeting*, Soc. Pet. Eng., AIME.
Terwilliger, P. L., R. R. Clay, L. A. Wilson, and E. Gonzalez-Gerth. 1975. *J. Petrol. Technol. 27*:9.
Thomas, K. P., J. Oberle, P. M. Harnsberger, D. A. Netzel, and E. B. Smith. 1987. The Effect of Recovery Methods on Bitumen Composition, Report to the U.S. Dept. of Energy, Contract No. DE-AC20-85LC11071, December.
Trantham, J. S., and J. W. Marx. 1966. *J. Petrol. Technol. 18*:109.
Walters, E. J. 1974. In *Oil Sands: Fuel of the Future* (L. V. Hills, ed.), Can. Soc. Petrol. Geo., Calgary, p. 240.

Ward, S. H., and K. A. Clark. 1950. Report No. 57, Alberta Research Council, Edmonton, Canada.

Watts, K. G., H. L. Hutchinson, L. A. Johnson, R. V. Barbour, and K. P. Thomas. 1982. *Proc. 57th Ann. Fall Meeting, Soc. Pet. Eng., AIME*, New Orleans.

Whiting, R. L. 1981. In *The Future of Heavy Crude and Tar Sands* (R. F. Meyer and C. T. Steele, eds.), McGraw-Hill, New York, p. 90.

Wilson, L. A., R. L. Reed, D. W. Reed, R. R. Clay, and N. H. Harrison. 1963. *Soc. Pet. Eng. J.* 3:127.

Winestock, A. G. 1974. In *Oil Sands: Fuel of the Future* (L. V. Hills, ed.), Can. Soc. Petrol. Geol., Calgary, p. 190.

Part III

Coal

17

Origin, Occurrence, and Recovery

I	Introduction	438
II	Origin	439
	A. Formation of Peat	447
	B. Conversion of Peat into Coal	448
	C. Coalification and Coal Rank	448
	D. Coalification Models	451
	E. Mineral Matter in Coal	453
	F. Moisture in Coal	467
III	Occurrence	471
	A. U.S. Resources	473
	B. World Resources	497
IV	Recovery	506
V	Beneficiation	520
	A. Physical Methods	521
	B. Chemical Methods	533
VI	Transportation	540

I. INTRODUCTION

At the most fundamental level, the earth's crust is described as being composed of rocks.

Rock material can be differentiated further according to the physical and/or chemical processes by which it was formed, or according to its characteristics as mineral matter.

Sedimentary rocks—those that are formed physically by weathering or erosion, or chemically by precipitation or other reactions producing residua—cover about 75% of the earth's surface. Sandstone, shale, limestone, and chalk are four well-known examples of inorganic sedimentary rocks. Conglomerate, arkose, graywacke, and halite are less commonly encountered forms.

Minerals are generally defined as inorganic substances having a consistent and distinctive set of physical properties—among which crystalline structure and unique values for hardness, color, luster, specific gravity, cleavage, and fracture are included.

Organic sedimentary rocks are those which consist of the remains of animals or vegetable organisms. Such organic and sedimentary rocks are generally considered to be clastic or detrital, i.e., solidified sediment that in all likelihood was moved from its original location. Organic sedimentary rock is represented by one example: coal. Formed from consolidated peat, many of the original plant and animal features are often found essentially intact in coal.

In spite of its primarily organic composition, coal is nonetheless identified universally as a mineral. This can be understood insofar as that term is also used to include naturally formed material having an identifiable range of chemical composition and only usually a characteristic crystal form. Along with petroleum, natural gas, oil shale, and uranium ores, coal remains a very important mineral fuel worldwide.

The American Geological Institute defines coal as "a readily combustible rock containing more than 50% by weight and more than 70% by volume of carbonaceous material including inherent moisture, formed from compaction and induration of variously altered plant remains similar to those in peat. Differences in the kinds of plant materials (coal type), in degree of metamorphism (coal rank), and in the range of impurity (coal grade) are characteristic of coal and are used in classification" (*Dictionary of Geological Terms*, 1976).

Merely establishing that coal is an organic sedimentary rock is insufficient as soon as it becomes necessary or desirable to utilize coal in a manner more elaborate or sophisticated than burning it in a stove for cooking or in a furnace for domestic heat, i.e., using it without giving any consideration to its composition or structure and reactivity as one would in order to maximize its utility as a fuel and as a chemical feedstock.

Two scientific disciplines devoted to the study and analysis of coal are coal petrology and coal petrography. "Coal petrology is the science concerned with the nature, origin, evolution and significance of coal as a rock material. Coal petrography is the sub-science concerned with the description of coal materials and the practical use of compositional descriptions" (Larsen, 1975). In addition to the detailed study of the final coal product, which is the domain of coal petrologers and petrographers, coal paleobotanists provide indispensible information about the manner in which plant remains became coalified. By studying fossilized plant materials, paleobotany informs coal scientists about the relationships which existed among plants and the broader physical and geological environment during the initial stages of coal formation. Besides these specialized disciplines, however, investigations and inquiries about modern uses for and alteration of this most ancient of earth's rocks have attracted the interest of professionals in business, engineering, chemistry, biology, and physics as well as in geology.

Once a decision is made to investigate coal structure and reactivity, however, the task becomes, from any of these perspectives, almost formidable. First of all, none of the organic sedimentary materials related to coal, or coal itself, has a unique, easily described origin. The best that can be done is to identify the primary constituent materials and then make a distinction between a coal-containing group as one category and other naturally occurring hydrocarbons, exclusive of coal, in another (Bouska, 1981). The coal-containing group was derived primarily from plant organisms. The common belief is that these organisms decomposed to form humic material which became coal. This group has a significantly high oxygen content. By comparison, the natural hydrocarbon group was derived primarily from animal organisms whose decomposition led to products with a high lipid and protein content but very little, if any, oxygen (Bouska, 1981).

II. ORIGIN

Obviously then, the origin of coal can be tied closely with the origin and development of plant life on earth. The geologic history of the earth has been divided into periods on the basis of life forms identified by fossil remains found in successive layers of rocks (Figure 1). By the end of the pre-Cambrian (Table 1), a wide variety of algae, fungi, and small marine animals had appeared. Extensive geologic investigations have led to the conclusion that at various times in the past large areas of modern land surface were covered by seas, parts of which became isolated when land masses rose around them. Limestone

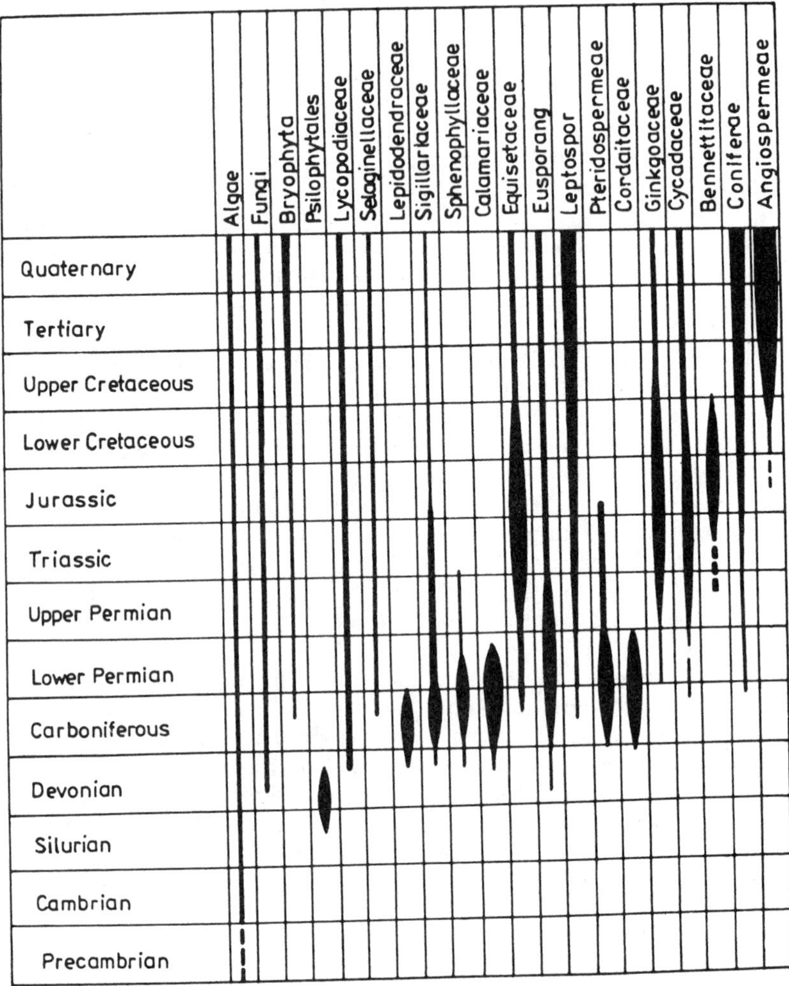

Figure 1 Development of vegetation during the geologic periods (K. Magdefrau, 1953. Reprinted with permission of the publisher.

and dolomite deposits up to hundreds of feet thick were formed when the remains of calcium-rich marine plants and animals accumulated on the sea floor. Similarly, evaporation led to the deposition of such salts as gypsum ($CaSO_4 \cdot 2H_2O$) and anhydrite ($CaSO_4$) (Speight, 1983).

Coal formation has similarly been correlated with these geologic time periods. The diagram of Bouska (1981) (Figure 2) shows that substantial coal did not appear before the formation of the carboniferous strata; a second large amount of coal is associated with the span of time corresponding to the Jurassic–Tertiary periods (Table 1).

In the eastern U.S. and Canada most coal mining has involved seams of the upper carboniferous, or Pennsylvanian, age. Although coals from all geologic periods–Silurian through Quaternary–are observed in places throughout the world, most commercially important coal is found in strata of lower carboniferous (Mississippean) age or younger (Speight, 1983). The thickest coal seam known (130 m, 390 ft) is a Permian seam located in the People's Republic of China (DOE/EIA, 1985). Pre-Cambrian coal is known in Kazakhstan and North America. The oldest mineable coal seams are upper Silurian coals in Tashkent and Kokand (Bouska, 1981). Devonian coals are mined in Europe and have been found on the Siberian and Canadian Shields; some on the Metvezhi Islands and in the Kuznetsk basin near Tomsk are also mineable (Bouska, 1981).

Permian coals are more important in the USSR and Australia, although some from this period is also mined in the United States. Triassic age coals are not significant; coals from the Jurassic are important in Siberia and western Canada. Likewise, western Canada and western United States, portions of Central and South America, and parts of Asia contain coal of cretaceous and Tertiary origin (Table 2).

Just as the material which was deposited and was converted to coal varies widely in age, type, and abundance (primarily according to prevailing climatic conditions), so does the final thickness of an individual coal seam which resulted in each time and location. Except for the seam in China mentioned above (130 m; 390 ft), a thickness of about 20–30 m (62–92 ft) is common, but seams smaller than 2.5 cm (1 in.) also exist.

Mapping coal seams has generally been limited to depths of less than 1621 m (5000 ft). In the United States most coal is found at depths up to 973 m (3000 ft) and little, if any, mining is being done on coal deeper than 324 m (1000 ft) (Speight, 1983). In some locations where the demand for coal warrants it, mining is done at much deeper levels. One report identified a mine in Belgium as being 1265 m (3900 ft) deep (Moore, 1940).

The first large coal-producing period is associated with tropical or subtropical climatic conditions which lasted until, at the end of the Permian, land masses began to shift away from equitorial locations and glaciation occurred. The plant material which accumulated during this era of damp, mild climate underwent incomplete decay and alteration by chemical and physical forces which resulted in the formation of peat.

Table 1 Description of Developmental Events in Geologic History

Era	Period	Time since period began (millions of yr)	Main developments of life	Important minerals
Cenozoic	Quaternary	1.2	Rise of civilized human beings; rise of primitive human beings and many species of mammals	Soil, peat, clay, nitrates, lignite
	Tertiary	55	Primitive forms of mammals, birds, plants; extinction of dinosaurs	Coal in western United States, Europe, Asia; oil and natural gas throughout the world; oil shale in western United States; phosphates, gold, silver
Mesozoic	Cretaceous	120	Dinosaurs and giant marine reptiles; rise of flowering trees and precursors of modern plants	Coal, oil, gas; copper in western United States, Mexico, and Chile; lead, zinc, tungsten, molybdenum, vanadium, uranium, radium
	Jurassic	150	Rise of flying lizards and primitive birds; rise of palmlike trees	Lithographic stone; salt; gypsum; coal in Asia and perhaps Alaska; oil and gas
	Triassic	180	Rise of giant lizards (dinosaurs) and primitive mammals	Some coal, small oil and gas deposits

Paleozoic	Permian	205	Large amphibians; primitive conifers; insects in modern form	Coal, salt, potash, oil, gas; gypsum; Rocky Mountain phosphate regions
	Pennsylvanian	240	Rise of reptiles, great spread of fernlike trees, plants, insects	Coal in eastern United States, Europe, Asia, Australia; oil, gas, salt, potash
	Mississippian	290	Rise of sharks; expansion of land flora	Limestone, coal, oil, gas; oil shale in eastern United States
	Devonian	340	Rise of amphibians and fernlike plants and trees	Oil, gas, glass sand
	Silurian	380	Rise of lung fishes and air-breathing insects	Iron ore in southeastern United States
	Ordovician	470	Development of fish, land plants, corals	Oil, gas, lead, zinc
	Cambrian	540	Algae and seaweeds abundant; large numbers of crustaceans and other marine invertebrates	Phosphates; marble
Proterozoic	Algonkian	740	Primitive aquatic plants and invertebrates	Iron ore in Great Lakes region; copper in Michigan; marble, nickel, cobalt, gold
	Archean	1740	Primordial forms of life originate	

Source: W. T. Thom, 1929. Reprinted with permission of the publisher.

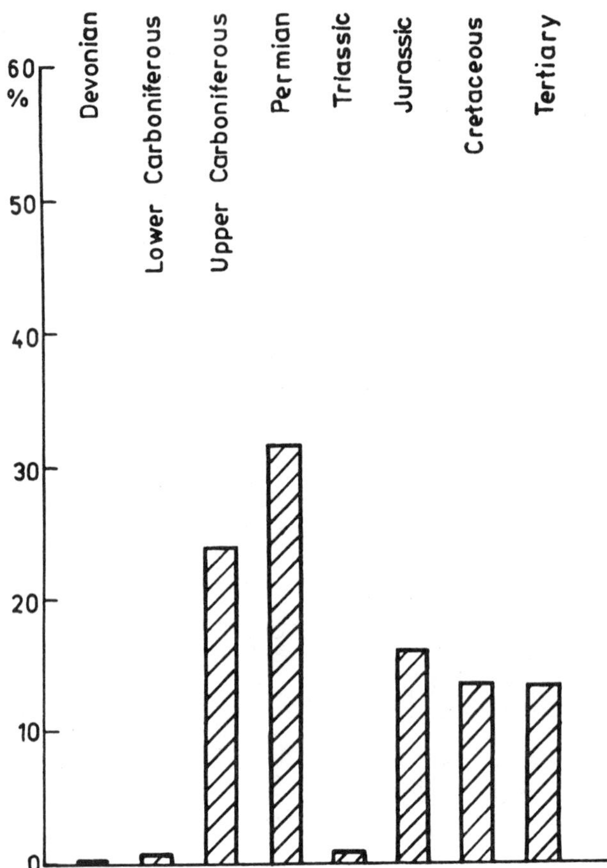

Figure 2 Relative distribution of coal in the geologic systems as related to total world reserves (V. Bouska, 1981. Reprinted with permission of the publisher).

Modern depositions of peat are located primarily in the USSR and Canada (Taylor and Smith, 1980). Recent paleobotanical studies resulted in the identification of 10 depositional microenvironments which are believed to model ancient peat-forming bogs (Table 3). In addition to identifying such archetypes, paleobotanists have identified as many as 3000 species of plant which grew during the carboniferous (Cargo and Mallory, 1974).

Table 2 Correlation of Geologic Age and Worldwide Deposition of Coal

Period	North America	South America	Europe	Asia	Africa	Australia
Quaternary	I	I	I	I	I	I
Tertiary	a,lvb,SB,L	B,SB,L	B,L	SB,B,L	L	L
Cretaceous	a,LVB,SB,L	SB,B	B,L	b	SB	l
Jurassic	B	—	sb,b,l	a,LVB,b,sb,L	—	b
Triassic	b	—	a,B,sb,l	—	—	—
Permian	b	b	A,B	B	B,c	B
Pennsylvanian	A,LVB,B,c	b	A,LVB,B,sb,l,c	A,LVB,B	A,B,l	LVB,b
Mississippian	a,lvb,B	—	B,sb,l	A,LVB,b	—	—
Devonian	—	—	b	—	—	—

L, lignite; SB, subbituminous; B, bituminous, high-, medium-volatile; LVB, low-volatile bituminous; A, anthracite; C, cannel. Capital letters indicate major deposits and lowercase letters indicate smaller, less significant deposits.

Source: E. S. Moore, 1940. Reprinted with permission of the publisher.

Table 3 Summary of Modern Microenvironments Typical of Coal-Forming Areas

Environment	Description
Kettle holes	Deglaciated areas located in northern regions of the Dakotas, Wisconsin, and Maine
Karst lakes	Limestone sinkholes in areas of artesian water in places such as South Dakota, Missouri, Iowa, and north-central Florida
Beach ridges and swale swamps	Sand ridges near shorelines; uncommon
Impoundments	Inland depressions caused by severe tectonic motion that block natural drainage channels. Reelfoot Lake in northwest Tennessee and areas near the Black Hills
Inland river swamps	Freshwater areas at or near sea level; usually extensive as in the Mississippi River delta and the Powder River Basin in Wyoming
Oxbow swamps	River edges where river meanders abandon a prior path and leave a low, wet region; Horseshoe Lake, in Illinois
Glacial lake basins	Flat former glacial basins; Glacial Lake, in Wisconsin
Back barrier	Barrier islands such as the Carolina coastal plain and the Okeefenokee swamp; rare but significantly large and subject to rapid burial
Deltas	Extensive marsh regions subject to marine flooding; variable size and frequent; Mississippi and Amazon deltas
Muskegs	Bog and marsh regions, very damp and frequently poor in nutrients, extensive in Canada and Ireland, some in Alaska

Source: Hessley, Reasoner and Riley, 1986. Reprinted with permission of the publisher.

A. Formation of Peat

Peat is a loosely compacted material composed of cellulose and with a high moisture content (as much as 90% in some places). The salinity and acidity of the water in which peat was formed and the nature of the rock substrate all affect the final character of peat. Highmoor peat is considered oligotrophic (poor in nutrients, especially calcium and potassium) and is most often found deposited on a clay substrate which is frequently mildly acidic. Lowmoor peat contains a greater diversity of plant material; it is eutrophic (rich in mineral nutrients) and rests on a slightly alkaline substrate (Bouska, 1981). It has generally been believed that the formation of peat, also called the biochemical stage of coal formation, or biogenesis, included an initial aerobic decay process which decomposed most of the cellulosic material in preference to more decay-resistant lignins, waxes, and resinous compounds. The resulting material is most commonly called humus or humic acids. Humic acids are very rich in oxygen. As subsidence occurred and anaerobic processes became predominant, protein decomposition was also extensive; as a result, the presence of accumulated nitrogenous byproducts is also observed in humus and peat.

Metal ions were incorporated into peat (and subsequently into coal) both during the aerobic, moldering processes, most probably as humic salts, and during the anaerobic, fermentation or putrefaction, processes. Alkali metals as well as nitrate, sulfate, and chloride species are commonly identified with the first group, while transition metals, principally iron (as FeS_2), are more commonly identified as having precipitated during the second phase of biogenesis (Bouska, 1981).

Some source beds of peat obviously contained very little mineral matter and were protected in some way from an influx of mineral-bearing species, as evidenced by the subsequent low quantities of mineral matter present in some coals. Other source beds had correspondingly high mineral content, at least by the time coalification was complete.

In addition to marked differences in mineral content, differences in the spatial orientation, distribution, and types of fossils, the presence or absence of clay substrata, and the overall size of the seams make it clear that in some regions an entire peat swamp was self-contained (in situ or autochthonous), i.e., it developed without any intrusion of material from some distant source. In other cases there is evidence that marine intrusions or freshwater flooding did transport "foreign" material to the final burial site (drift or allochthonous). It is likely that most coal seams contain at least small quantities of allochthonous material.

The end of all biogenic processes probably occurred when anaerobic species depleted their sources of nutrients and the extent of subsidence or the thickness of the overlying sediment could no longer admit sufficient percolating water to support microbial activity.

B. Conversion of Peat into Coal

Both physical and chemical changes occurred in peat to convert it to coal (the dynamochemical stage of coal formation). Temperature and/or pressure as well as the actual chemical constituents of the peat interacted in complex and not fully understood reactions to bring about the metamorphosis of peat into coals. The complexity and the multivariable nature of the processes are revealed in the diversity of coals which exhibit varying degrees of dehydration, decarboxylation, and devolatilization. It is clear that the conversion of peat to coal is brought about by the precise interaction of heat, pressure, and time regardless of when in geologic history a peat swamp was established; thus, coalification in either a carboniferous or in a Tertiary deposit may produce a fully mature coal, or may produce, ultimately, any of the less mature forms.

Several fundamental alterations occur during metamorphosis: (1) some carbon is released as CO_2 or methane gas, but most is retained. The overall carbon content increases from about 50% (daf) in peat to more than 92% (daf) in anthracite coal; (2) oxygen and hydrogen are gradually reduced; (3) moisture is lost and porosity decreases; (4) total volume decreases as volatile gases are expelled; (5) vitrification increases as gelification decreases; (6) polymeric structures are formed as constituent compounds condense; and (7) inherent heat content increases.

C. Coalification and Coal Rank

The extent to which these processes have proceeded in a particular location is described by the rank of that coal. Figures 3 and 4 illustrate the way in which time and depth of burial are believed to affect rank as it is reflected in the volatile matter content of coal. Coalification also brings about elemental changes in the organic matter (Table 4). While carbon content can be closely associated with coal rank, oxygen cannot. The amount of oxygen in coal decreases sharply as coalification begins, then changes more slowly. Peat does not contain large amounts of hydrogen, and the hydrogen content is affected only slightly during coalification until anthracite is formed. The nitrogen and sulfur content of coal are virtually independent of rank. Although hydrogen and oxygen content cannot be related directly to coal rank, Patteisky and Teichmuller (1960) developed an informative diagram (Figure 5) which illustrates how H/C and O/C ratios change with coalification. The heat content, or the calorific value, of coal is the property which reflects the fact that large quantities of chemical, thermal,

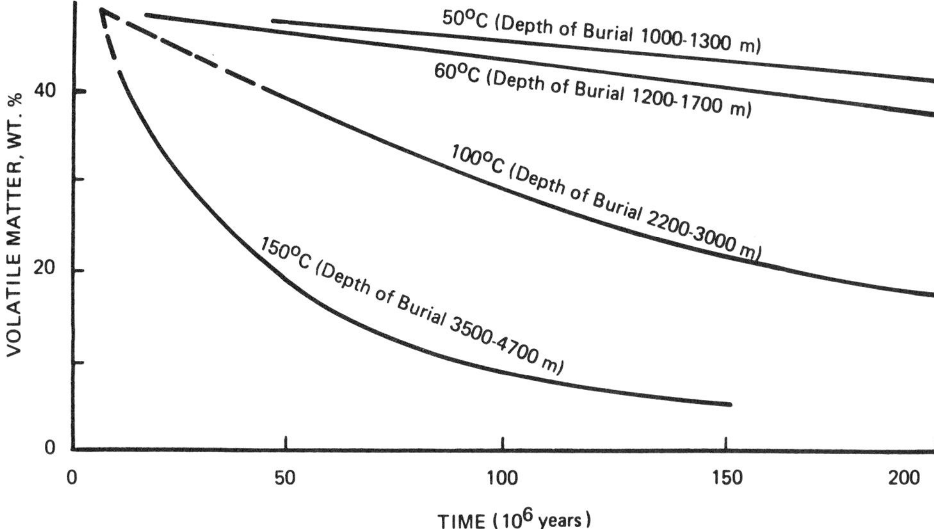

Figure 3 Variation in coal rank (volatile matter) with time (D. Murchison and T. S. Westoll, 1968. Reprinted with permission of the authors).

and light energy are stored in coal. The energy values determined for coals are, like virtually all properties of coal, the results of complex interactions of the elemental composition and overall structure of coal. Indeed, the empirically obtained value for this energy content is called the gross calorific value in order to reflect some additional intrinsic, but variable, features of coal, particularly moisture content. Data for heat content are reported according to the gross calorific values unless noted otherwise.

In actual practice, rank is determined after a prescribed analysis is carried out. It is based on the fixed carbon content or the heating value of the samples (see Section III below) and the coalification series has several of the parameters which are rank-dependent (Figure 6; Averitt, 1975). Lignite has been subjected to the least extensive metamorphosis. Increasing severity and/or duration of prevailing physiochemical conditions produced subbituminous, high-, medium-, and low-volatile bituminous, and, finally, anthracite coal.

In summary, then, coal is a variable material formed from decayed vegetation. Coal developed as the vegetal remains were subjected to elevated temperature and pressure (relative to the earth's surface)

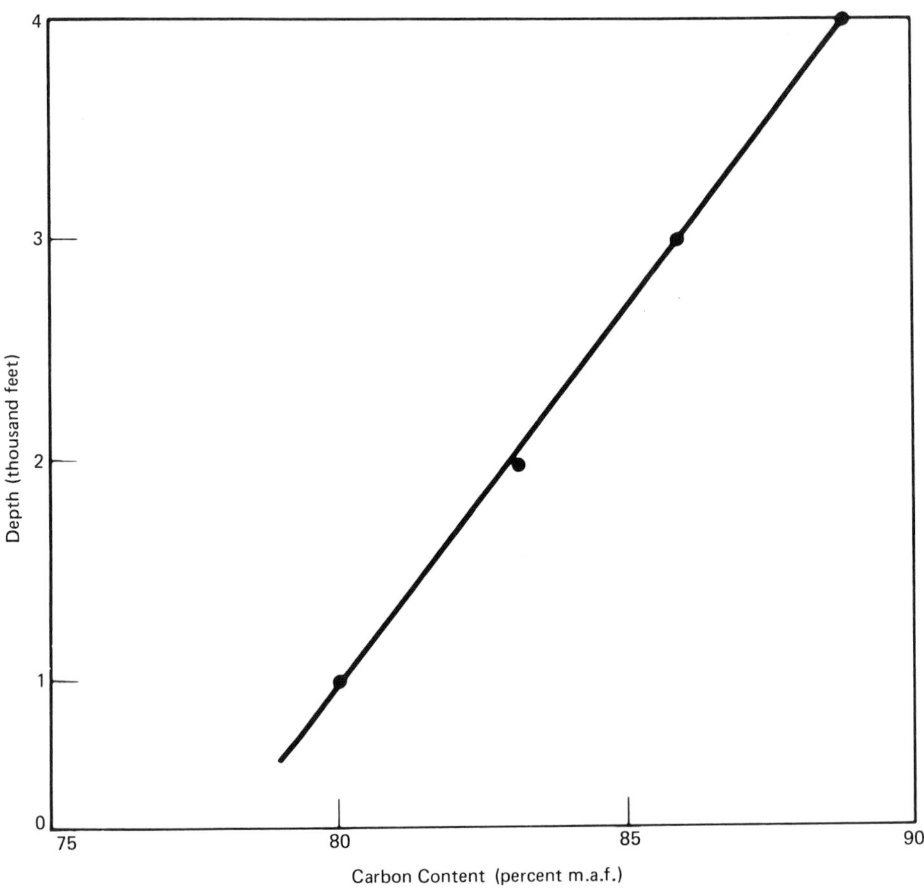

Figure 4 Variation in coal rank with depth (Hilt's rule) (D. W. van Krevelen, 1961. Reprinted with permission of the publisher).

associated primarily with depth of burial. Coalification proceeds from the soft brown coal to increasingly hard lignite, subbituminous, bituminous, and anthracite. If coalification occurs without the development of any unusual disturbances in the strata, bituminous coal will be formed eventually but the second stage of plant alteration, the dynamochemical phase, is virtually complete when low-volatile bituminous coal is formed (Figure 7). It is only when coal-bearing strata are subjected to substantially increased pressure and temperature associated with severe folding that additional devolatilization occurs

Table 4 Typical Elemental Composition of Peat and Some Representative Coals of Different Rank

Rank	Percent C	Percent H	Percent O	Percent N	Percent S
Peat	55	6	30	1	1.3[a]
Lignite	72.7	4.2	21.3	1.2	0.6
Subbituminous	77.7	5.2	15.0	1.6	0.5
High-volatile B bituminous	80.3	5.5	11.1	1.9	1.2
High-volatile A bituminous	84.5	5.6	7.0	1.6	1.3
Medium-volatile bituminous	88.4	5.0	4.1	1.7	0.8
Low-volatile bituminous	91.4	4.6	2.1	1.2	0.7
Anthracite	93.7	2.4	2.4	0.9	0.6

[a]Ash and moisture content constitute remaining weight percent.
Source: From H. Tschamler and E. deRuiter, 1966. Reprinted with permission of the publisher.

and anthracite is formed. The Appalachian Mountain region in eastern United States and the Atlas Mountain region in Morocco are the primary examples of this phenomenon (World Survey of Energy Resources, 1974). Along with the elemental content of coal, the volatile matter which remains trapped in the porous coal cavities, the calorific value and the quantity and type of a wide variety of mineral impurities determine the end-use and overall value of coal.

D. Coalification Models

Over the years many modeling and correlation studies have been made in an effort to gain insight into the physical demands required to achieve coal formation. In general, there seems to be a consensus that relatively mild temperatures (and, therefore, moderate depth)

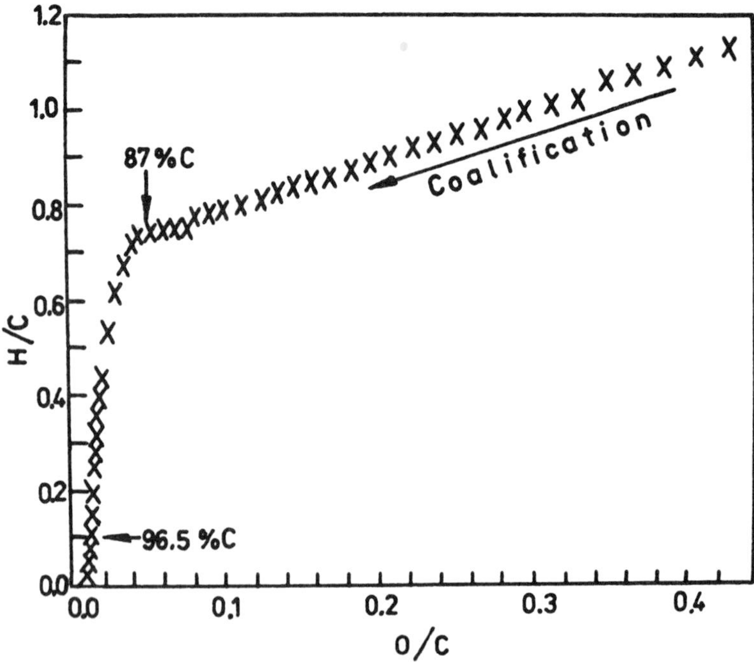

Figure 5 Changes in hydrogen and oxygen contents during coalification (Patteisky and Teichmuller, 1960. Reprinted with permission of the publisher).

would have been sufficient to bring about the formation of the various ranks of coal now found everywhere around the world in a span of several million years. However, in investigating catalytic aspects of coal formation, recent laboratory research has demonstrated dramatically that when lignin is held in contact with a clay material at only 150°C in the absence of air and for periods of time ranging from only 2 to 8 *months*, the lignin was converted directly to a form which is virtually indistinguishable from a low-rank coal (Hayatsu et al., 1984). Not only does this work apparently show conclusively that it is the lignin and not cellulosic components of woody plants which survives biogenesis and is coalified, but it strongly suggests that coalification proceeds *directly* from lignin without the formation of humates. Furthermore, it opens up the possibility that clay substrates, which are most often found in contact with coal seams, may act catalytically in coalification, and may be very important if not altogether necessary for coalification to proceed. Finally, this work is significant for the

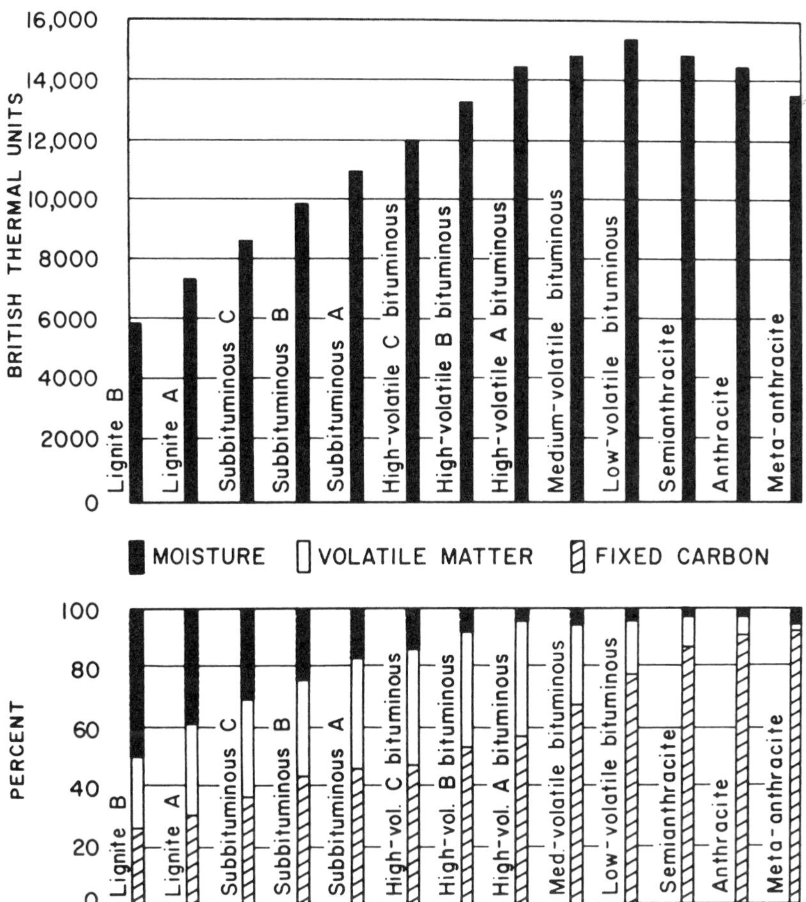

Figure 6 Variations in heat content, fixed carbon, volatile matter, and moisture with rank (P. Averitt, 1975).

fact that a coallike product was formed much more rapidly than was heretofore thought possible (Larsen, 1985).

E. Mineral Matter in Coal

With reference to coal, the mineral matter, ash and inorganic matter are often used interchangeably, but each term should be carefully

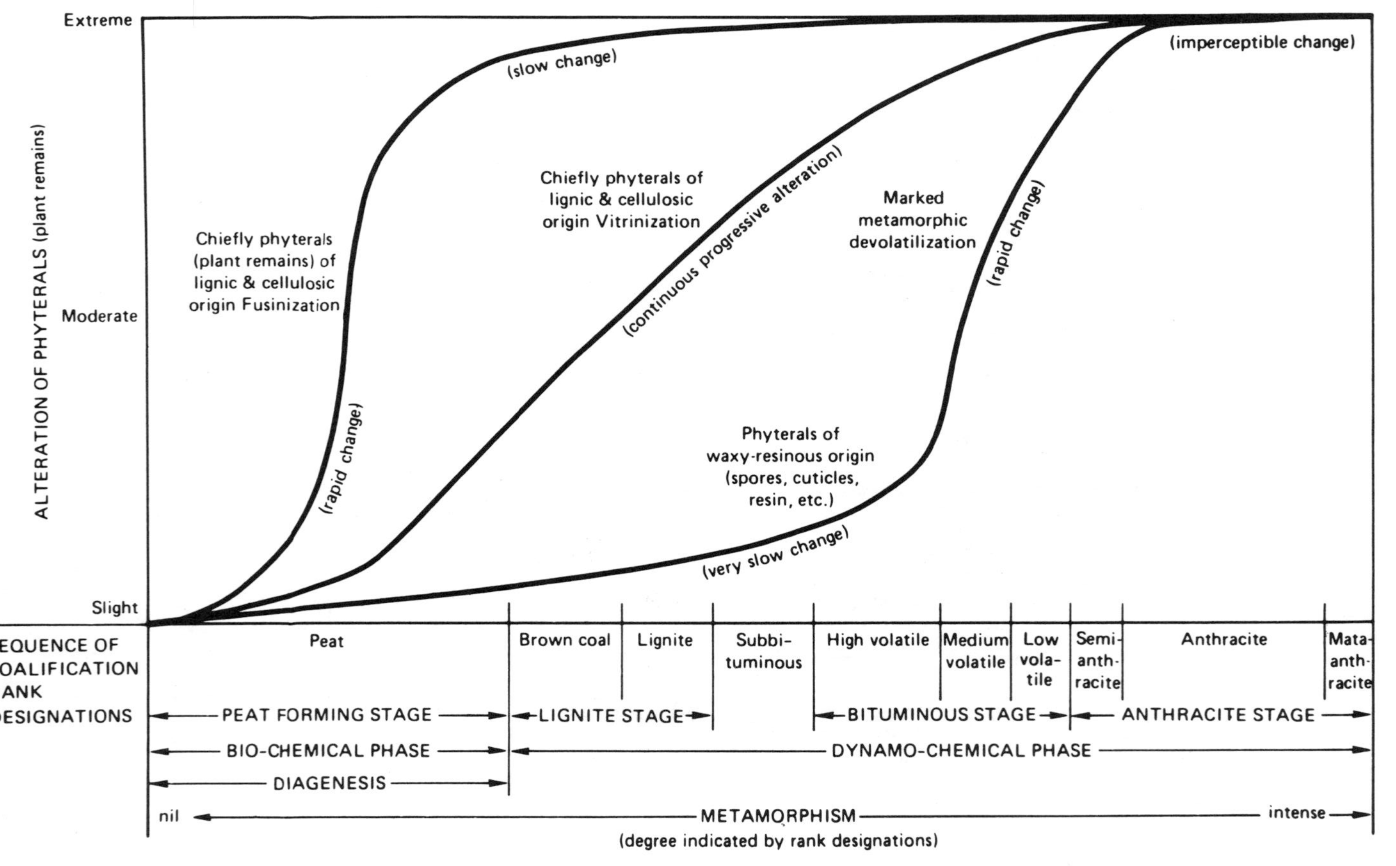

Figure 7 Metamorphic alteration of phyteral constituents (J. M. Schopf, 1948. Reprinted with permission of

distinguished. Mineral matter was defined in the first section. Ash is the residue after the coal has been completely combusted. The original mineral matter (the inorganic content) will be altered during combustion by oxidation and dehydration, so that the chemical form of ash is not precisely the same as the original mineral matter. In addition to terminology, there is also some variation (and often confusion) about how to describe or classify the sources of the mineral content in coal.

Some authors (Bouska, 1981) believe that the chemistry of the mineral matter in plants is particularly relevant for understanding the source(s) of minerals in coal. One comparison of the average composition of elements appearing in modern vegetation and in a bituminous coal has been compiled (Table 5). Fourteen elements were determined; the list shows which ones may have been enriched in coal from nonplant sources (Ca, K, S, Si) and which were seemingly depleted during coalification (Cl, N). Caution must be exercised against drawing too strong a conclusion from such data, however, because there is little justification for supposing that the elemental composition of primitive vegetation in any location was exactly identical to values determined on modern plant material. Nevertheless, this type of analysis does give some insight into the source(s) of minerals and elements in coal. On the other hand, other authors (Speight, 1983) suggest that plants were not the primary sources of mineral elements in coal; rather, it is proposed that most mineral matter washed or was otherwise transported to the peat swamps, and that these minerals are compositionally quite different from those of plant residue.

Four terms are most often used to describe the mineral matter in coal (Francis, 1961; Braunstein, 1981):

1. Primary or internal—minerals present in original vegetation
2. Secondary or external—minerals from sources other than the original vegetation
    (a) Syngenetic minerals—those transmitted to the peat and incorporated into the coal matrix during coalification (Table 6)
    (b) Epigenetic minerals—those transmitted, largely by percolation of water into fissures, after coalification has occurred (Table 6)

Some literature uses the term *authentic* to identify mineral matter which was part of the immediate vicinity of the peat swamp regardless of whether it was contained within the vegetation, and *allogenic* to identify the mineral matter that was transported to the bog from some distance (Speight, 1983).

Worldwide, all coal contains mineral matter, but the amount and the type of minerals varies considerably and may be as high as 35% by

Table 5 Average Analysis of Typical Plant Material and Bituminous Coals Containing 7.3% Ash

Element	Original plant material (%)	Average bituminous coal (%)
Carbon	45.0	80.0
Oxygen	43.0	5.0
Hydrogen	5.5	5.0
Nitrogen	2.1	1.5
Sulfur	0.1	1.0
Phosphorus	0.2	0.2
Subtotal	95.9	92.7
Potassium	1.80	3.20
Chlorine	0.60	0.30
Silicon	0.55	1.00
Calcium	0.38	0.67
Magnesium	0.18	0.20
Sodium	0.14	0.25
Iron	0.03	0.05
Aluminum	0.02	0.03
Other	0.40	1.50
Subtotal	4.10	7.30

Source: World Energy Survey, 1974. Reprinted with permission of the publisher.

weight of the coal (Speight, 1983). Generally, the mineral content is analyzed in the coal ash but only 10 elements are most often observed as the major constituents of ash (Table 7). Elements present as less than 1000 ppm in the earth's crust are considered to be trace elements in coal (Table 8).

That a wide diversity of minerals have been found in coal (Table 9) reflects the variations in the prior geologic and biological history of the coal. Clay minerals, hydrated aluminosilicates, are the most predominant minerals. Kaolinites are distinguished by their crystal structure, and montmorillinites have ions such as Ca^{2+} and Mg^{2+} substituted for aluminum at places throughout the crystal.

The occurrence of carbonates in coals suggests that marine intrusions took place during the early stages of coal formation. The carbonate ion can associate with many metal ions and calcite ($CaCO_3$), siderite ($FeCO_3$), dolomite ($CaCO_3 \cdot MgCO_3$), and ankerite ($2CaCO_3 \cdot MgCO_3 \cdot FeCO_3$) are common in coal (Speight, 1983).

Sulfide minerals appear in small amounts in most coals and are particularly important in subsequent coal utilization. Pyrite and marcasite are the two predominant iron sulfides (FeS_2) and differ only in their crystal structure (cubic and orthorhombic, respectively). Other sulfides include braoite [$(Ni \cdot Fe)S_2$] and laurite (RuS_2) (Speight, 1983).

Phosphates originate from protein material. Although present in only small amounts in coal, they also impact industrial uses of coal because they are particularly undesirable in metallurgical coke (Table 6).

The exact way(s) in which nitrogen exists in coal remains uncertain. A body of data has accumulated which does suggest that virtually all nitrogen is present in heterocyclic systems.

Other metals, especially heavy metals, are known to interfere with coal conversion processes by poisoning catalysts and are of special concern because of their toxicity if they are released into the environment as fly ash or in waste process water. Of course, this concern is not limited to heavy metals, such as mercury and lead, but includes a number of lighter elements (and minerals) such as arsenic, cadmium, beryllium, and asbestos (Braunstein, 1981). Zubovic (1975) compiled a list of the concentrations of the known toxic elements found in U.S. coal (Table 10). Lead, mercury, and zinc occur more frequently in Eastern coals, while antimony, arsenic, cadmium, and selenium occur in higher concentrations in Western samples. Of the toxic metals, mercury has been the most studied because there is considerable concern about its unusually high vapor pressure. Most U.S. coals contain less than 0.2 ppm mercury (Table 11).

Magee (1975) asserted that accumulation of 40 lb/day of trace elements in ash produced in industrial processing is not an unreasonable estimate. Thus, utilization or disposal of ash itself must also take into account the potential for introducing toxic materials into the environment. In this respect lead is of particular concern as an ash leachate. Lead was observed on average as less than 100 ppm in coal ash. For comparison, the average values for chromium in the same samples was 304 ppm in anthracite and 54 ppm in lignites and subbituminous coal; copper, 405 ppm (anthracites) and 655 ppm (lignites and subbituminous); strontium, 177 and 4660 ppm; and zirconium, 668 and 245 ppm (O'Gorman and Walker, 1972).

Table 6 Different Types of Minerals in Coal and Their Origin

Mineral group	Syngenetic formation (intimately intergrown)		Epigenetic formation	
	Transported by water or wind	Newly formed	Deposited in fissures and cavities (coarsely intergrown)	Transformation of syngenetic minerals (intimately intergrown)
Clay minerals	Kaolinite, illite, sericite, clay minerals with mixed-layer structure, "Tonstein"			Illite, chlorite
Carbonates		Siderite-ankerite concretions, dolomite, calcite, ankerite	Ankerite calcite dolomite	
		Siderite, calcite, ankerite in fusite		
Sulfide ores		Pyrite concretions melnikowite-pyrite coarse pyrite (marcasite) Concretions of FeS_2-$CuFeS_2$-ZnS	Pyrite marcasite zinc-sulfide (sphalerite) lead sulfide (galena) copper sulfide (chalcopyrite)	Pyrite from the transformation of syngenetic concretions of $FeCO_3$
		Pyrite in fusite		

Oxide ores		Hernatite	Goethite, lepidocrocite ("needle iron ore")	
Quartz	Quartz grains	Chalcedony and quartz from the weathering of feldspar and mica	Quartz	
Phosphates	Apatite	Phosphorite		
Heavy minerals and accessory minerals	Zircon, rutile, tourmaline, orthoclase, biotite		Chlorides, sulfates, and nitrates	

Source: Murchison and Westoll, 1968.

Table 7 Major Inorganic Constituents of Coal Ash

Constituent	% Range
SiO_2	40–90
Al_2O_3	20–60
Fe_2O_3	5–25
CaO	1–15
MgO	0.5–4
Na_2O	0.5–3
K_2O	0.5–3
SO_3	0.5–10
P_2O_5	0–1
TiO_2	0–2

Source: J. Speight, 1983. Reprinted with permission of the publisher.

Table 8 Major Trace Elements in Coal

Constituent	Range
Arsenic, ppm	0.50–93.00
Boron, ppm	5.00–224.00
Beryllium, ppm	0.20–4.00
Bromine, ppm	4.00–52.00
Cadmium, ppm	0.10–65.00
Cobalt, ppm	1.00–43.00
Chromium, ppm	4.00–54.00
Copper, ppm	5.00–61.00
Fluorine, ppm	25.00–143.00
Gallium, ppm	1.10–7.50

(continued)

Table 8 (Cont.)

Constituent	Range
Germanium, ppm	1.00–43.00
Mercury, ppm	0.02–1.60
Manganese, ppm	6.00–181.00
Molybdenum, ppm	1.00–30.00
Nickel, ppm	3.00–80.00
Phosphorus, ppm	5.00–400.00
Lead, ppm	4.00–218.00
Antimony, ppm	0.20–8.90
Selenium, ppm	0.45–7.70
Tin, ppm	1.00–51.00
Vanadium, ppm	11.00–78.00
Zinc, ppm	6.00–5350.00
Zirconium, ppm	8.00–133.00
Aluminum, %	0.43–3.04
Calcium, %	0.05–2.67
Chlorine, %	0.01–0.54
Iron, %	0.34–4.32
Potassium, %	0.02–0.43
Magnesium, %	0.01–0.25
Sodium, %	0.00–0.20
Silicon, %	0.58–6.09
Titanium, %	0.02–0.15
Organic sulfur, %	0.31–3.09
Pyritic sulfur, %	0.06–3.78
Sulfate sulfur, %	0.01–1.06
Total sulfur, %	0.42–6.47
Sulfur by X-ray fluorescence, %	0.54–5.40

Source: R. R. Ruch, H. J. Gluskoter, and N. F. Shimp, Environmental Geology Note No. 72, Illinois State Geological Survey, 1974, p. 18.

Table 9 Minerals Commonly Associated With Coal

Group	Species	Formula
Shale	Muscovite	$(K,Na,H_3O,Ca)_2(Al,Mg,Fe,Ti)_4$
	Hydromuscovite	$(Al,Si)_8O_{20}(OH,F)_4$ (general formula)
	Illite	$(HO)_4K_2(Si_6 \cdot Al_2)\ Al_4O_{20}$
	Montmorillonite	$Na_2(Al\ Mg)Si_4O_{10}(OH)_2$
Kaolin	Kaolinite	$Al_2(Si_2O_5)(OH)_4$
	Livesite	$Al_2(Si_2O_5)(OH)_4$
	Metahalloysite	$Al_2(Si_2O_5)(OH)_4$
Sulfide	Pyrite	FeS_2
	Marcasite	FeS_2
Carbonate	Ankerite	$CaCO_3 \cdot (Mg,Fe,Mn)CO_3$
	Calcite	$CaCO_3$
	Dolomite	$CaCO_3 \cdot MgCO_3$
	Siderite	$FeCO_3$
Chloride	Sylvite	KCl
	Halite	NaCl
Accessory minerals	Quartz	SiO_2
	Feldspar	$(K,Na)_2O \cdot Al_2O_3 \cdot 6SiO_2$
	Garnet	$3CaO \cdot Al_2O_3 \cdot 3SiO_2$
	Hornblende	$CaO \cdot 3FeO \cdot 4SiO_2$
	Gypsum	$CaSO_4 \cdot 2H_2O$
	Apatite	$9CaO \cdot 3P_2O_5 \cdot CaF_2$
	Zircon	$ZrSiO_4$
	Epidote	$4CaO \cdot 3Al_2O_3 \cdot 6SiO_2 \cdot H_2O$
	Biotite	$K_2O \cdot MgO \cdot Al_2O_3 \cdot 3SiO_2 \cdot H_2O$
	Augite	$CaO \cdot MgO \cdot 2SiO_2$
	Prochlorite	$2FeO \cdot 2MgO \cdot Al_2O_3 \cdot 2SiO_2 \cdot 2H_2O$
	Diaspore	$Al_2O_3 \cdot H_2O$
	Lepidocrocite	$Fe_2O_3 \cdot H_2O$

(continued)

Table 9 (Cont.)

Group	Species	Formula
	Magnetite	Fe_3O_4
	Kyanite	$Al_2O_3 \cdot SiO_2$
	Staurolite	$2FeO \cdot 5Al_2O_3 \cdot 4SiO_2 \cdot H_2O$
	Topaz	$2AlFO \cdot SiO_2$
	Tourmaline	$3Al_2O_3 \cdot 4BO(OH) \cdot 8SiO_2 \cdot 9H_2O$
	Hematite	Fe_2O_3
	Penninite	$5MgO \cdot Al_2O_3 \cdot 3SiO_2 \cdot 2H_2O$
	Sphalerite	Zns
	Chlorite	$10(Mg,Fe)O \cdot 2Al_2O_3 \cdot 6SiO_2 \cdot 8H_2O$
	Barite	$BaSO_4$
	Pyrophillite	$Al_2O_3 \cdot 4SiO_2 \cdot H_2O$

Source: J. Speight, 1983. Reprinted with permission of the publisher.

Table 10 Distribution of Environmentally Hazardous Trace Elements (ppm)

	Region			
Element	Powder River Basin	Western interior	Eastern interior	Appalachian
Antimony	0.67	3.5	1.3	1.2
Arsenic	3	16	14	18
Beryllium	0.7	2	1.8	2.0
Cadmium	2.1	20	2.3	0.2
Mercury	0.1	0.13	0.19	0.16
Lead	7.2		34	12
Selenium	0.73	5.7	2.5	5.1
Zinc	33		250	13

Source: Zubovic 1975, Table 3, p. 12A.

Table 11 Geographic Distribution of Mercury, 1971–1972 (ppm in coal)

Region and state	Analysis by			Averaged[d]	Total number of samples
	Neutron activation[a]	Neutron activation + atomic absorption[b]	Flameless atomic absorption[c]		
Appalachian					
Pennsylvania	0.16, 0.28	0.15		0.20(2.0)	3
Ohio	0.10, 0.13, 0.15	0.14, 0.28, 0.49		0.21	6
West Virginia		0.07, 0.18		0.12(6.6)	2
East Kentucky				(0.25)	
Eastern interior					
Illinois	0.04, 0.49, 0.60, 1.15			0.18(0.19)	53
Indiana		0.08		0.08(0.31)	
Western interior					
Missouri		0.19		0.19	

Northern Great Plains					
Montana	0.06	0.07, 0.09		0.07(33.0)	
Western United States					
Utah	0.04		0.03–0.08	0.05	15
Colorado	0.02, 0.02	0.05	0.03–0.06	0.04(0.22)	3
Wyoming			0.03–0.06	0.05(18.6)	6
Arizona	0.02	0.06	0.04–0.08	0.05	6
Nevada			0.04–0.05	0.05	7
New Mexico			0.05–0.29	0.15	37

[a]Illinois State Geological Survey Bulletin EGN-43, 1971 (as cited in Hall 1974).
[b]National Bureau of Standards, 1972 (as cited in Hall 1974).
[c]Southwest Energy Study, App. J. draft, January 1972 (as cited in Hall 1974).
[d]Values from Joensuu (1971) (as cited in Hall 1974) shown in parentheses for comparison, including lithotypes; extremes show no relationship to more representative average samples.
Source: Hall 1974, Table 3, p. 46.

The occurrence and identification of the minerals and elements in coals is also of interest insofar as it may prove economically feasible to recover these elements from coal. Nunn et al. (1953) investigated trace elements in anthracites. Zubovic (1975) and Ode (1963) collected similar data from a large number of samples for coals from several sites across the United States (Table 12) and worldwide (Table 13).

It is also of considerable interest to study the level of appearance of elements in coals relative to the level at which these same elements are found in the earth's crust. Ruch et al. (1974) conducted such a study. They cited and used the Clarke average percentage of an element in the crust and determined an enrichment factor by comparing the mean value of their data to the Clarke value (Table 14). Five elements exist in coal at a level an order of magnitude higher than they are found in the crust; eight elements occur an order of magnitude more rarely in coal. How this (apparent) selective concentration and selective exclusion of elements occurs is of considerable interest to coal scientists.

The most notorious, if not the most significant, element in coal is sulfur. Concentrations of 6% by weight or higher are not uncommon. Pyritic sulfur (or marcasite), organically bound sulfur, sulfate salts, and free elemental sulfur are all found in coal. Elemental sulfur is rare, but Ode (1963) reported concentrations as high as 15% by weight.

Saline water is believed to be the source of most iron sulfide, formed by reduction of sulfate by marine organisms. Marine conditions are associated with Appalachian and interior province coals (in the United States), which contain more than 80% of the "high" sulfur (>3% S by weight) coal in the country. Other regions were apparently formed in freshwater environments and show markedly lower concentrations of iron sulfide; thus distribution of the total sulfur content varies (Given 1973; Figures 8–10) for the different coal regions of the United States. These data show that coals from the interior region of the United States have the highest pyritic sulfur content. In the West, the total sulfur content is substantially lower. Sulfate sulfur itself is most often 0.1% or less by weight. Although exceptions to this are encountered, high sulfate values often indicate that the sample was oxidized prior to analysis.

Organic sulfur has not been as thoroughly studied. Attar and Dupuis (1979) published data which show that rank rather than depositional environment affects this moiety (Table 15). They observed that lignites and high-volatile bituminous coals have higher thiol (RSH) content, while higher rank coals investigated contained more thiophenic sulfur (an unsaturated cyclic structure). Interestingly, aliphatic sulfide (R-S-R) did not correlate with rank. However, it was suggested (Speight, 1983) that as condensation reactions occurred during coalification, thiolic groups are likely converted to thiophenic structures via the R-S-R species.

Table 12 Trace Element Content of American Coals (ppm in coal)

Element	Region			
	Northern Great Plains	Western interior	Eastern interior	Appalachian
Beryllium	1.5	1.1	2.5	2.5
Boron	116	33	96	25
Titanium	591	250	450	340
Vanadium	16	18	35	21
Chromium	7	13	20	13
Cobalt	2.7	4.6	3.8	5.1
Nickel	7.2	14	15	14
Copper	15	11	11	15
Zinc	59	108	44	7.6
Gallium	5.5	2.0	4.1	4.9
Germanium	1.6	5.9	13	5.8
Molybdenum	1.7	3.1	4.3	3.5
Tin	0.9	1.3	1.5	0.4
Yttrium	13	7.4	7.7	14
Lanthanum	9.5	6.5	5.1	9.4

Source: Zubovic 1975, Table 2, p. 11A

F. Moisture in Coal

There always tends to be some uncertainty about the nature of water (moisture) in coal; especially because moisture in coal is often described in two ways: (1) geologically and (2) analytically.

In the geologic sense, moisture occurs in coal either as pore moisture or as surface moisture. The former type of moisture is often referred to as inherent moisture but it is also known as natural moisture, bed moisture, seam moisture, capacity moisture, and even physically bound moisture. On the other hand, the surface moisture may also be referred to as extraneous moisture, visible moisture, adherent moisture, free moisture, or excess moisture.

Table 13 Minor Elements in Canadian, British, and German Coals and Ashes

	In ash (ppm)					In coal (ppm)				
	Sydney, Nova Scotia		Barnsley, Vitrain, Great Britain	Germany	Ruhr, Germany	Sydney, Nova Scotia		Germany		Ruhr, Germany
Element	Av	Range	Range	Max	Max	Av	Range	Av	Max	Max
Antimony			100–200	>1,000	3,000			10–30	≥10	17
Arsenic	900	280–2,300		10,000		100	33–270	100	500	
Barium	300	18–2,200		>1,000	a	35	2–257	a	100	a
Beryllium	14		50–100	4,000	1,000	2	1–2	13	40	20
Bismuth				2,000					100	
Boron	148	52–220	200–3,000	>1,000		17	6–25		≥10	
Chromium	45	18–79	100–1,000	a	5,000	5	2–9	a	a	50
Cobalt	87	26–196	100–300	2,000	2,000	10	3–34	14	30	12
Copper			800–1,000	>10,000	4,000			≥25	10,000	50

Gallium			80--300	>3,000	1,000			30	100	20
Germanium	44	9–70	300–1,000	5,000	1,000	5	1–8	19	50	20
Lead	572		200--800	31,000	3,000	66	25–120	140	3,000	30
Manganese	1,200	165–2,200	100--2,000	≥10,000	22,000	140	9--254	a	≥5,000	700
Molybdenum	60	18--105	80–200	1,000	6,000	7	2–12	21	200	50
Nickel	131	52–645	500--3,000	≥3,000	16,000	15	6–74	24	≥60	30
Silver				60	a			0.3	3	a
Strontium	560	225--750		>1,000	a	65	76–87	a	100	a
Tin	9	4–18	50--200	1,000	6,000	1		3	300	120
Titanium			3,000--8,000	12,000	30,000			700	1,000	1,500
Vanadium	120	61–244	400–5,000	>1,000	11,000	14	7–28	18	>100	20
Zinc	218	115–550	500–700	21,000	8,000	25	13–64	170	2,000	100
Zirconium			<100--500	a	7,000			a	a	140

[a]Not significant.

Source: Ode 1963, Table 6, p. 228. Reprinted by permission of the publisher.

Table 14 Enrichment Factors of Chemical Elements in Coal[a]

Region	Element	Enrichment factor	Mean value in coal (ppm): Author	Mean value in coal (ppm): Different source[b]
Illinois basin (81 samples)	Beryllium	11.38	113.79	10.0
	Cadmium	14.4	2.89	0.2
	Fluorine	0.09	59.30	625.0
	Manganese	0.06	53.16	950.0
	Plutonium	0.06	62.77	1050.0
	Selenium	39.80	1.99	0.05
Eastern United States (9 samples)	Fluorine	0.10	62.5	625.0
	Manganese	0.03	28.5	950.0
	Plutonium	0.09	94.5	1050.0
	Selenium	67.33	3.37	0.05
Western United States (8 samples)	Chromium	0.09	9.0	100.0
	Manganese	0.04	38.0	950.0
	Selenium	31.31	1.57	0.05

[a]Only those enriched or depleted by one order of magnitude or more are listed.
[b]Clarke and Washington, 1974 (as cited in Ruch, Gluskoter, and Shimp, 1974).
Source: Ruch, Gluskoter, and Shimp, 1974, Table 9, p. 32.

In the analytical sense, moisture in coal is measured by a two-step process which involves (1) drying under mild conditions (air-drying loss) and (2) drying under more severe conditions (residual moisture). Although the two components of the analytical moisture determination will equal the total (or geologic) moisture in the coal, the separate moisture components of each category are not equivalent. Thus, the inherent moisture and the surface moisture (as defined geologically) are both very different from the air dry loss moisture and the residual moisture. It is also worth a cautionary note that the terminology used here may also be different for different countries. Thus, confusion may exist.

For example, the ASTM and the ISO both recommend use of the term "surface moisture" but define it very differently; the ASTM defines the surface moisture as the difference between the inherent and

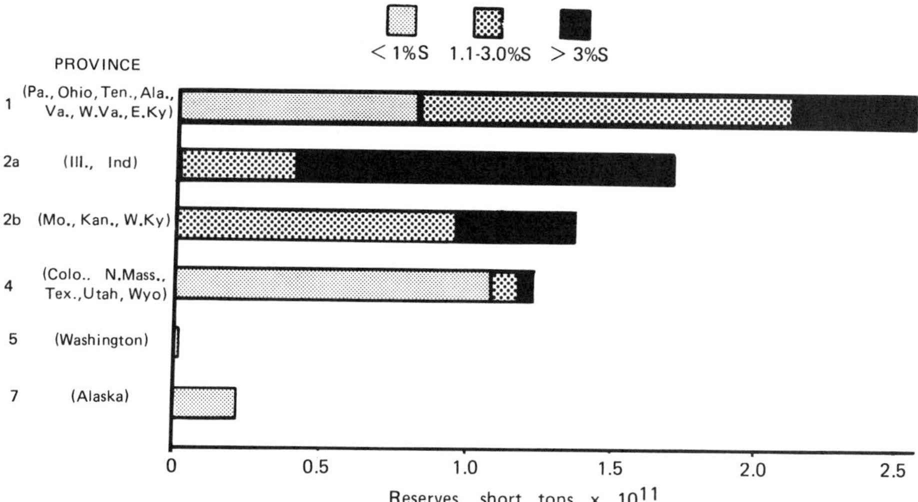

Figure 8 Distribution of U.S. bituminous coals by geological province and sulfur content (P. Given, 1973. Reprinted with permission of the publisher).

total moisture content, but the ISO defines it as the weight loss of the coal during air-drying.

An attempt was recently made (*Coal Quality Newsletter*, 1988) to relate the various definitions of moisture in coal and thus clear up the confusion that currently exists (Figures 11 and 12). Hopefully, this will be the start of moving in the direction of meaningful terminology that can be interchanged within one standards organization and even between two or more standards organizations.

III. OCCURRENCE

Because the geology of coal has been thoroughly defined and because prospecting and assessing coal resources has been somewhat less difficult than for other mineable resources, the determination of total coal resources is well established and the data have changed very little over time. That is not to say that exploration for new coal reserves is not being carried out; rather, it is intended to establish the foundation on which a discussion of resources can be based. Also, it should be stressed that not all the coal resources which have been

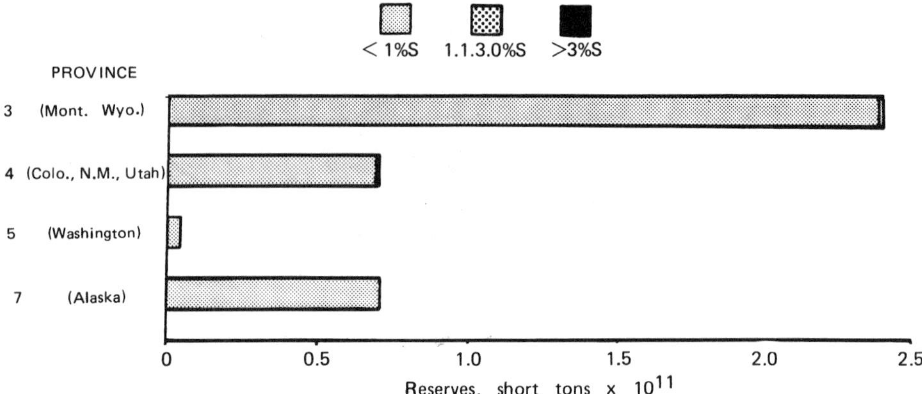

Figure 9 Distribution of U.S. subbituminous coals by geologic province and sulfur content (P. Given, 1973. Reprinted with permission of the publisher).

identified were, or are now, economically worth exploiting. The amount of coal within a reserve which is considered economically feasible to mine may actually change considerably over time if the economics of the fuels market changes abruptly or if environmental controls are altered significantly. Somewhat more gradual changes in

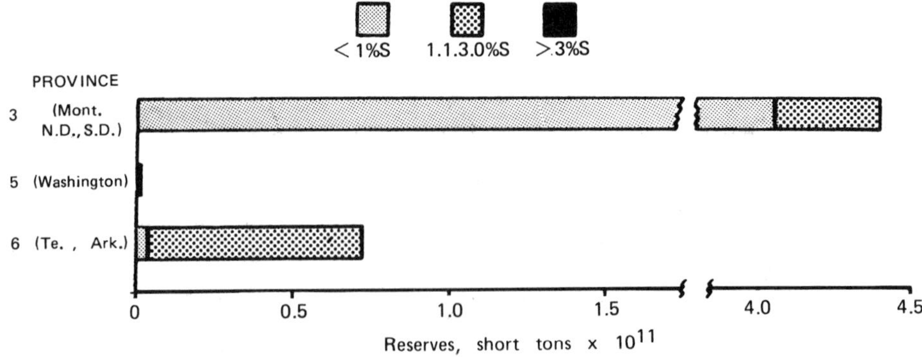

Figure 10 Distribution of lignites by geologic province and sulfur content (P. Given, 1973. Reprinted with permission of the publisher).

Table 15 Distribution of Organic Sulfur Functions in Various Coals

Coal	Organic S (wt %)	% Organic S accounted	Thiolic	Thiophenolic	Aliphatic sulfide	Aryl sulfide	Thiophenes
Illinois	3.2	44	7	15	18	2	58
Kentucky	1.43	46.5	18	6	17	4	55
Martinka	0.60	81	10	25	25	8.5	21.5
Westland	1.48	97.5	30	30	25.5	--	14.5
Texas lignite	0.80	99.7	6.5	21	17	24	31.5

Source: A. Attar and F. Dupuis, *Preprints*, Am. Chem. Soc., Div. Fuel Chem., 1979, *24*(1), 166.

the assessment of exploitable coal resources will occur as advancements are made in mining technology, in transportation systems, and as population and market centers change over time.

A. U.S. Resources

Coal is found in 35 states (Table 16) and is mined in 26 states. The U.S. Geological Survey (USGS), which is responsible for monitoring coal resources, uses the following terms to distinguish the accessibility of buried coal (Averitt, 1975):

Identified resources: coal-bearing rock whose location and existence is known.

Hypothetical resources: estimated tonnage of coal in the ground in unmapped and unexplored parts of known coal basins to an overburden of 6000 ft (2000 m); determined by extrapolation from the nearest area of identified resources and includes coal believed to be present in the continental shelves.

Resources: total quantity of coal in the ground within specified limits of bed thickness and overburden thickness; comprises identified and hypothetical resources.

Measured resources: tonnage of coal in the ground based on assured coal bed correlations and on closely spaced observations

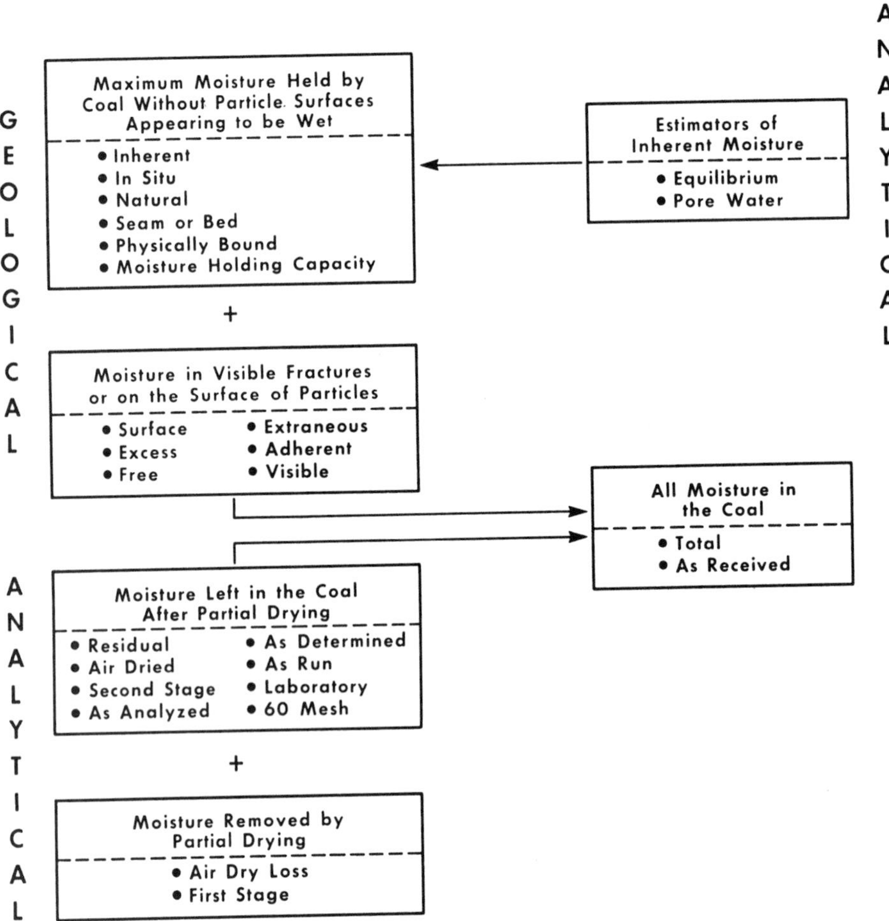

Figure 11 Simplified representation of the nomenclature of the different types of moisture in coal and the interrelationships of these types.

about one-half mile apart; computed tonnage judged to be accurate within 20% of the true tonnage.

Indicated resources: tonnage of coal in the ground based partly on specific observation, partly on reasonable geologic observations, and partly on reasonable geologic projections; points of observation and measurement are about 1 mile from beds of known continuity.

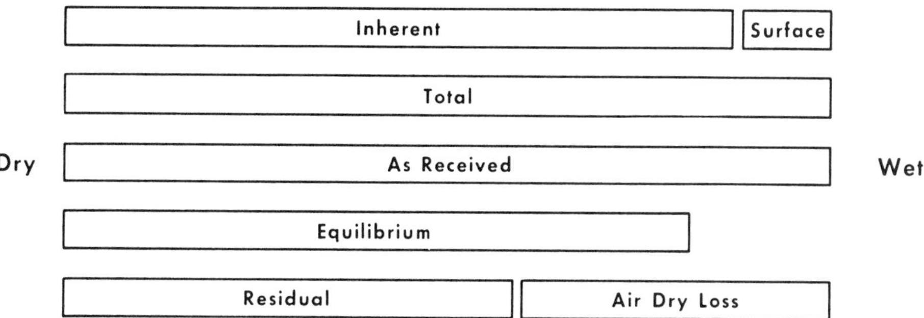

Figure 12 Representation of the moisture distribution for a lower rank coal.

Inferred resources: tonnage of coal in the ground based on assumed continuity of coal beds adjacent to areas containing measured and indicated resources.

Demonstrated reserve base: selected portion of coal in the ground in the measured and indicated category; restricted to coal in thick and intermediate beds less than 1000 ft (300 m) below the surface and deemed economically and legally available for mining at the time of the determination.

Reserve: tonnage that can be recovered from the reserve base by application of the recoverability factor.

The most recent USGS report (Bulletin 1412, 1974) lists identified resources at 1731 billion short tons at depths of less than 3000 ft (900 m) and total estimated hypothetical resources (to 6000 ft; 2000 m) as 2237 billion short tons for a grand total of 3968 billion short tons in coal resources. The U.S. Energy Information Administration (USEIA) estimated that as of January 1, 1984, 488.3 billion short tons of coal was in the demonstrated reserve base (DRB) (Table 17). Assuming a 50% production rate, the USEIA estimated that the coal in the DRB will last about 275 years (*EIA Review of Coal Resources*, 1985).

Ten states hold almost 90% of the DRB:

State	Reserve (billion short tons)
Montana	120.3
Illinois	79.0
Wyoming	64.4

Table 16 Size and Percentage Distribution of Coal-Bearing Areas in the United States

State	Total area of state (square miles)*	Area underlain by coal-bearing rocks	
		Square miles	%
Alabama	51,609	9,700	19
Alaska	586,412	35,000	6
Arizona	113,909	3,040	3
Arkansas	53,104	1,700	3
California	158,693	230	0.1
Colorado	104,247	29,600	28
Georgia	58,876	170	0.2
Idaho	83,557	500	0.6
Illinois	56,400	37,700	67
Indiana	36,291	6,500	18
Iowa	56,290	20,000	36
Kansas	82,264	18,800	23
Kentucky	40,395	14,600	36
Louisiana	48,523	1,360	3
Maryland	10,577	440	4
Michigan	58,216	11,600	20
Mississippi	47,716	1,000	2
Missouri	69,686	24,700	35
Montana	147,138	51,300	35
Nebraska	77,227	300	0.4
Nevada	110,540	50	—
New Mexico	121,666	14,650	12
New York	49,576	10	—
North Carolina	52,586	155	0.3
North Dakota	70,665	32,000	45
Ohio	41,222	10,000	24
Oklahoma	68,919	14,550	21
Oregon	96,981	600	0.6
Pennsylvania	45,333	15,000	33
South Dakota	77,047	7,700	10
Tennessee	42,244	4,600	11
Texas	267,338	16,100	6
Utah	84,916	15,000	18
Virginia	40,817	1,940	5
Washington	68,192	1,150	2

(continued)

Table 16 (Cont.)

State	Total area of state (square miles)[a]	Area underlain by coal-bearing rocks	
		Square miles	%
West Virginia	24,181	16,800	69
Wyoming	97,914	40,055	41
Other states	312,855	0	0
Total	3,615,122	458,600	13

[a]U.S. Bureau of the Census, 1973, Statistical Abstracts of the United States.
Source: P. Averitt, 1975.

Kentucky	39.6
W. Virginia	38.9
Pennsylvania	29.9
Ohio	18.8
Colorado	17.2
Texas	13.8
Indiana	10.4
Total	437.3
% of DRB	89.6

The distribution of the DRB across the United States and in Alaska is shown in Figure 13. These data show that 46.5% of the DRB still lies east of the Mississippi River and that 68.1% of the nation's coal is expected to be mined underground. In states east of the Mississippi River only 17.7% of the coal (only 8.2% of the U.S. total) is expected to be obtained by surface mining, while west of the Mississippi 44.2% of the DRB (23.6% of the U.S. total) will be surface-mined.

Across the United States, seam thickness ranges from less than 2 ft to 50 ft (Figure 14). In the East, where most of the mining activity is underground, the seams are not thick. Most average about 4 ft. In the West, where coal lies closer to the surface, seam thickness is much larger. As Figure 21a shows, the average thickness of all coal seams mined in 1979 was 10.7 ft.

The Bituminous Coal Act of 1937 classified the coal fields in the 48 contiguous states into districts, and subsequently these districts were grouped into three regions: Appalachian (districts 1–4, 6–8,

Table 17 Demonstrated Reserve Base of Coal,[a] January 1, 1984 (Billion Short Tons)

Region and state	Anthracite	Bituminous coal[b]		Lignite	Total		
	Underground and surface[a]	Underground	Surface	Surface[d]	Underground	Surface	Total
Appalachian							
Alabama	0	1.7	3.4	1.1	1.7	3.4	5.1
Kentucky, Eastern	0	17.1	2.1	0	17.1	2.1	19.2
Ohio	0	13.0	5.9	0	13.0	5.9	18.8
Pennsylvania	7.1	21.4	1.6	0	28.4	1.6	29.9
Virginia	0.1	2.3	0.8	0	2.4	0.8	3.2
West Virginia	0	33.8	5.1	0	33.8	5.1	38.9
Other[e]	0	1.3	0.4	0	1.3	0.4	1.7
Total	7.2	90.6	41.4	1.1	97.7	19.2	116.9
Interior							
Illinois	0	63.4	15.6	0	63.4	15.6	79.0
Indiana	0	8.9	1.6	0	8.9	1.5	10.4
Iowa	0	1.7	0.5	0	1.7	0.5	2.2
Kentucky, Western	0	16.9	4.0	0	16.9	4.0	20.8
Missouri	0	1.5	4.6	0	1.5	4.6	6.0
Oklahoma	0	1.2	0.4	0	1.2	0.4	1.6
Texas	0	0	0	13.8	0	13.8	13.8
Other[f]	0.1	0.3	1.1	(g)	0.4	1.1	1.5
Total	0.1	93.9	27.6	13.8	94.0	41.4	135.4
Western							
Alaska	0	5.4	0.7	(g)	5.4	0.7	6.2
Colorado	(g)	12.2	0.7	4.2	12.2	4.9	17.2

Montana	0	71.0	33.6	15.8	71.0	49.3	120.3
New Mexico	(g)	2.1	2.5	0	2.1	2.5	4.7
North Dakota	0	0	0	9.9	0	9.9	9.9
Utah	0	6.1	0.3	0	6.1	0.3	6.4
Washington	0	1.3	0.1	(g)	1.3	0.1	1.5
Wyoming	0	42.6	26.8	0	42.6	26.8	69.4
Other[h]	0	0.1	0.3	0.4	0.1	0.6	0.7
Total	(g)	140.8	65.0	30.2	140.9	95.2	236.0
U.S. Total	7.3	325.3	110.6	45.1	332.5	155.8	488.3
States East of the Mississippi River	7.2	179.8	39.1	1.1	186.9	40.3	227.2
States West of the Mississippi River	0.1	145.5	71.5	44.0	145.6	115.5	261.1

[a]Includes measured and indicated resource categories representing 100% of the coal in place. Recoverability varies from less than 40% to more than 90% for individual deposits. About one half of the demonstrated reserve base of coal in the United States is estimated to be recoverable.
[b]Includes subbituminous coal.
[c]Includes 130.3 million short tons of surface mine reserves of which 114.9 million tons are in Pennsylvania and 15.5 million tons are in Arkansas.
[d]There are no underground demonstrated coal reserves of lignite.
[e]Includes Georgia, Maryland, North Carolina, and Tennessee.
[f]Includes Arkansas, Kansas, and Michigan.
[g]Less than 0.05 billion short tons.
[h]Includes Arizona, Idaho, Oregon, and South Dakota.
Note: Sum of components may not equal total due to independent rounding.
Source: Energy Information Administration, *Coal Production 1981*.

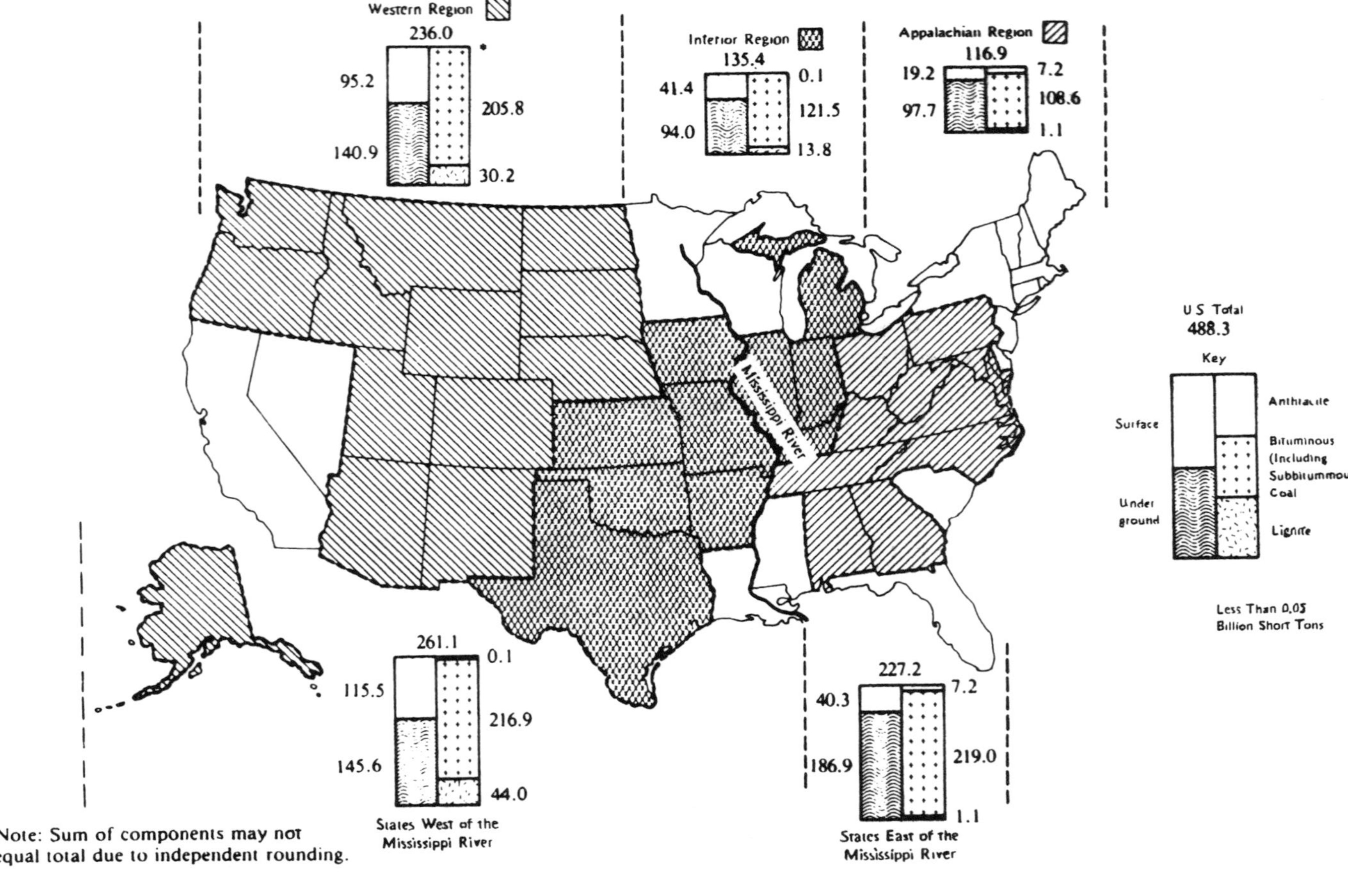

Figure 13 Demonstrated reserve base of coal, U.S., January 1, 1979 (billion short tons) (Energy Information Administration, Annaul Energy Review, 1985).

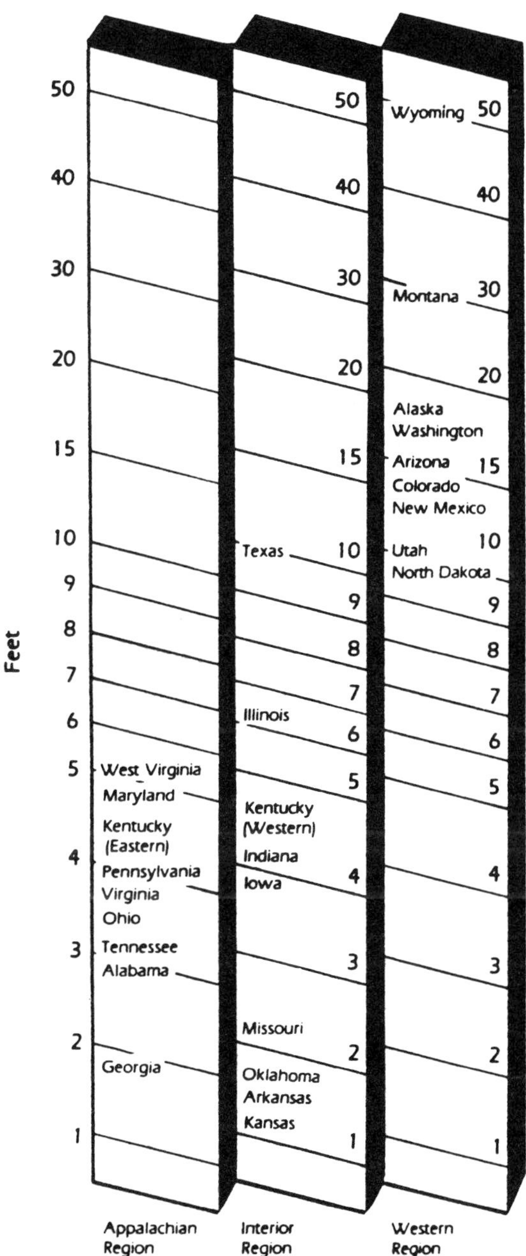

Figure 14 Average thickness of coal beds mined, U.S., 1979 (Energy Information Administration: Coal Data: A Reference, 1986).

13, 24); interior (districts 5, 9–12, 14, 15); and western (districts 16–23, and Alaska, added later) (Figure 15). The USGS groups the coal producing states into six provinces (Campbell, 1922):

Province	States
Eastern	Rhode Island Massachusetts Pennsylvania Ohio W. Virginia Maryland Virginia E. Kentucky Tennessee N. Carolina Northern Alabama
Interior	Michigan Illinois Indiana W. Kentucky Missouri Iowa Kansas Northwest Arkansas Oklahoma North central Texas
Gulf	S. Carolina Louisiana Southeast Arkansas Southern Alabama Mississippi Eastern Texas
Northern Great Plains	N. & S. Dakota Montana Wyoming Idaho Utah Colorado
Rocky Mountain	Arizona New Mexico

Coal-Producing Districts

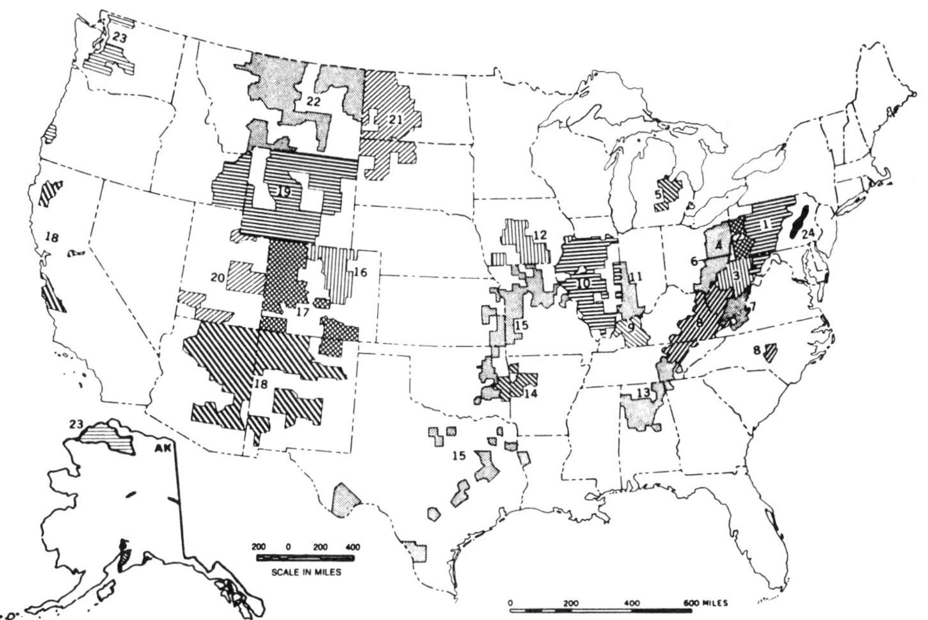

Bituminous Coal Producing Districts as Defined in the Bituminous Coal Act of 1937 and Amendments
The districts were originally established to aid in formulating minimum prices of bituminous coal and lignite. Because much statistical information was compiled in terms of these districts, their use for statistical purposes has continued since the abandonment of that legislation in 1943.
District 24 is the anthracite producing district in Pennsylvania.

Figure 15 U.S. coal-producing districts (Energy Information Administration: Coal Distribution January--December, 1985).

Pacific Coast	Alaska
	Washington
	Oregon
	California

This designation is used in most of the older literature, but all Department of Energy (DOE) statistics now follow the region/district format as described above.

The production and consumption of coal in the United States has waxed and waned throughout history as the economic and technological atmospheres have changed. The development of diesel engines, the increased availability of less expensive domestic and imported crude oil, the abundant domestic resources of natural gas, and increasingly rigid environmental restrictions on process emissions all sharply decreased the use for coal in the years after the Second

World War. International political and economic tensions coupled with a rapidly expanding demand for energy and the development of new coal conversion technologies for cleaner utilization of coal did, however, offset the negative trends to some extent. This optimism must nevertheless still be under caution. The development of nuclear power facilities in the United States has stalled, but the continued relatively low price for crude oil on the international market, even amid ongoing political tensions, has at the same time forestalled government and industry impetus for continuing to develop cost-effective processes for producing liquid and solid fuels from coal. As has been the case in recent decades, the primary application of coal in the next few years will be in combustion for the generation of electric power. Thus, the primary concern of the coal industry will be directed toward obtaining clean coal (coal producers) and developing technologies that can effectively use coal which, as it now comes from the mine, does not meet standards for pollution abatement (coal consumers).

Figure 16 shows that although there was a period from 1950 to 1961 in which production and domestic consumption of coal declined, after 1961 statistics in both categories increased markedly to about 900 and 800 million short tons, respectively. Table 18 shows the relative quantity of coal produced in each district since 1981. In spite of the greater expense of underground mining, the figures show that the Appalachian region produced almost 50% of the total in 1985, a decrease of only 2% from 1981.

Distribution of coal to domestic consumers (Figure 17) confirms that electric utilities remain far and away the largest coal user. Western coal is sold almost exclusively to utilities. Western coal has a lower sulfur content, which makes it more environmentally acceptable, and in the West hydroelectric power is not as available as it is in the East. Twenty-one percent of all Appalachian coal is exported. This value is 25–35 times higher than the exports from the other two coal regions.

Sulfur content in coal has become a critical issue that in some cases has been the overriding factor in determining the market value of a particular coal. Table 19 shows the relationship between the sulfur content and the sale of coal to various consumers. The most dramatic comparison can be made for eastern and western Kentucky coal. Kentucky has been the nation's leading coal-producing state since 1984 and coal in both sections of the state is bituminous rank. Yet, as the table shows, less than 30% of the state's production comes from the western district where coal contains more than three times more total sulfur (average of 3.6% by weight) than the coal in eastern Kentucky (only 1.1% by weight, average). Similarly, Indiana and Illinois, two other traditionally strong coal-producing states, can no longer compete as favorably on the market because of the sulfur content in the coal. Clearly, all interior region states hold high sulfur

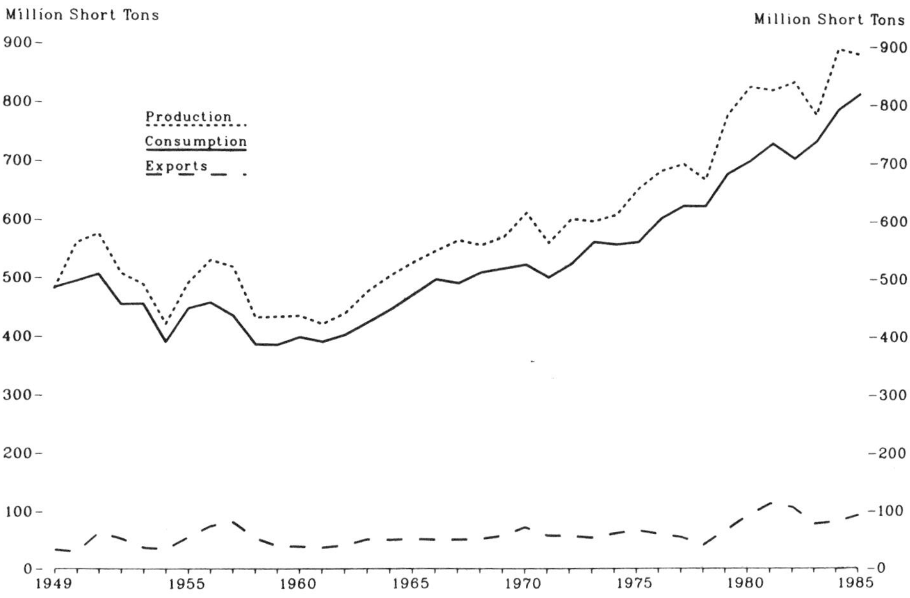

Figure 16 U.S. coal supply and disposition, 1949--1985.

coal. Of all the other states, only Ohio, which is in the Appalachian region, has such high sulfur content in the coal.

Coal exports have increased only slightly throughout the 25-year period 1960–1985. Table 20 shows that Canada and Japan are the two largest markets for U.S. coal. From 1960 to 1985 Canada imported an average of 16.6 million short tons and Japan accounted for the purchase of an average of 16.0 million short tons of coal from the United States annually. Since 1970, Japan has actually imported about 50% more coal from the United States than Canada has, but since 1981, when Japan imported 25.9 million short tons of U.S. coal, their imports have steadily declined.

Most coal exported from the United States is bituminous rank. Its major end uses are the production of steel and other metallurgical products (metallurgical coal) or the generation of power (steam coal). Metallurgical coal must produce a strong coke and be low in mineral impurities, especially phosphorus and sulfur. Usually midrank bituminous coals with 20--30% volatile matter by weight are the best coking coals. Steam coals require relatively low ash and at least a consistent moisture content, but overall, lower quality coals can be used to make coke to generate steam or for other boiler fuels.

Table 18 Distribution of U.S. Coal by Origin: January–December 1981–1985

Coal producing region and district of origin	January–December				
	1985	1984	1983	1982	1981[a]
Appalachian total	423,683	438,799	385,828	414,268	430,277
District 1	49,276	56,328	51,858	54,575	60,342
District 2	25,722	25,525	22,033	26,590	28,028
District 3	40,545	46,583	39,548	41,027	35,522
District 4	35,956	40,049	34,923	35,754	37,539
District 6	5,929	6,890	6,941	8,218	6,360
District 7	23,150	22,338	19,599	24,095	27,737
District 8	209,147	208,110	179,739	192,942	202,565
District 13	29,925	29,261	29,966	27,261	26,734
District 24[b]	4,032	3,716	4,220	3,807	5,451
Interior total	186,189	196,285	174,736	175,407	165,997
District 9	39,199	41,441	36,049	38,868	39,126
District 10	59,171	63,706	57,717	60,122	52,419
District 11	33,049	37,366	31,877	30,245	29,300
District 12	578	495	399	495	660
District 14	732	972	681	1,007	1,269
District 15	63,461	52,305	48,013	44,670	43,224
Western total	270,581	254,031	224,009	224,001	224,509
District 16	559	589	342	200	507
District 17	18,456	18,620	16,443	19,259	21,114
District 18	30,572	32,244	31,308	30,827	28,976
District 19	140,905	131,183	111,624	107,532	102,833
District 20	14,378	12,082	12,178	15,464	14,601
District 21	26,607	21,641	18,719	17,828	17,604
District 22	33,303	32,955	28,723	27,897	33,437
District 23	5,801	4,716	4,674	4,994	5,436
Districts 1–23[b] total	876,420	885,399	780,352	809,869	815,331
U.S. total	880,453	889,114	784,573	813,676	820,783

[a] UMWA Strike from March 27 to June 7.

[b] District 24 is the anthracite-producing district in Pennsylvania. Districts 1–23 represent the total U.S. production of bituminous, subbituminous coal, and lignite.

Note: Total may not equal sum of components because of independent rounding.

Source: Energy Information Administration, Form EIA-6, Coal Distribution Report.

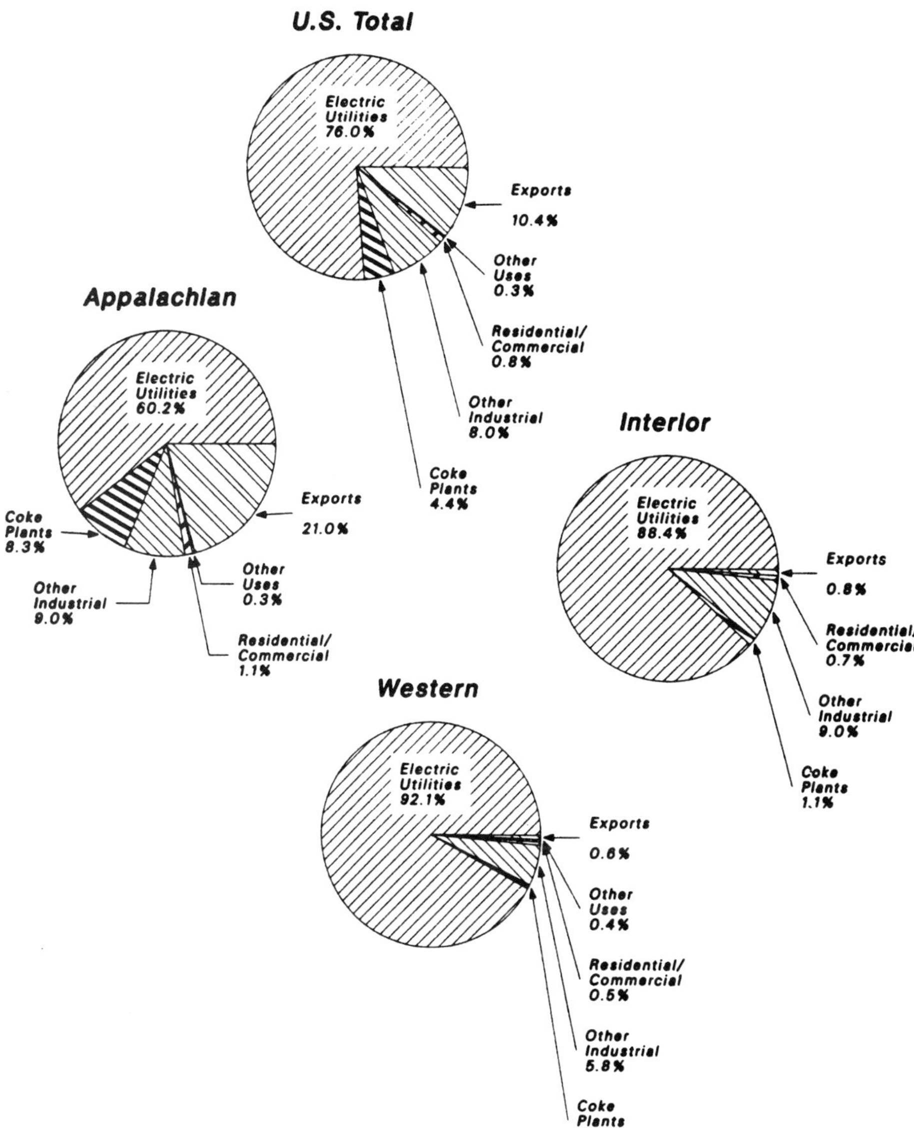

Figure 17 Distribution of U.S. coal by origin and consumer, 1985 (Energy Information Administration, Form EIA-6, Coal Distribution Report).

Table 19 Average Sulfur Content of Coal Shipments by State and Use,

State	Quantity shipped (thousand short				
	Electric utilities	Coke plants	Other industrial uses and retail dealers	All other uses[a]	Exports (include Canada & Mexico)
Alabama	12,324	3,630	1,168	1,539	1,893
Alaska	717	—	8	6	—
Arizona	9,054	—	—	—	—
Arkansas	—	304	89	90	35
Colorado	9,718	2,188	1,476	398	34
Georgia	—	3	20	—	90
Illinois	40,941	2,165	4,916	554	24
Indiana	21,749	—	2,280	153	—
Iowa	431	—	19	—	—
Kansas	1,014	—	188	24	—
Kentucky:					
Eastern	69,557	12,727	9,067	3,263	1,620
Western	36,856	77	1,130	1,394	—
Total Kentucky[b]	106,412	12,804	10,196	4,657	1,620
Maryland	2,105	16	399	419	60
Missouri	5,329	—	190	147	—
Montana	26,068	—	506	27	—
New Mexico	10,589	617	792	635	—
North Dakota	12,732	—	886	410	—
Ohio	34,553	28	4,855	1,793	7
Oklahoma	3,191	413	1,792	491	182
Pennsylvania	43,195	16,784	10,929	5,732	4,838
Tennessee	7,460	2	1,058	1,511	—
Texas	19,748	—	271	—	—
Utah	5,460	919	2,250	512	—
Virginia	11,905	7,763	4,105	3,861	4,313
Washington	4,694	—	14	—	—
West Virginia	41,457	22,876	5,629	2,118	13,234
Wyoming	54,157	214	3,935	22	—
U.S. total[b]	485,002	70,725	57,973	25,099	26,329

[a]Includes railroad fuel, shipments to Great Lakes and tidewater commercial docks (excluding Canada), mine fuel, sales to mine employees, and increases in mine inventory.

1978

tons)	Average sulfur content (percent)					
Total[b]	Electric utilities	Coke plants	Other industrial uses and retail dealers	All other uses[a]	Exports (include Canada & Mexico)	Weighted total
20,553	1.4	1.0	1.3	1.3	1.0	1.3
731	.2	—	.2	.2	—	.2
9,054	.6	—	—	—	—	.6
519	—	1.0	2.9	1.6	4.2	1.7
13,814	.6	.6	.7	.5	.7	.6
113	—	.6	1.0	—	.6	.6
48,600	3.2	1.0	2.7	2.1	4.1	3.1
24,182	3.2	—	2.8	2.1	—	3.2
450	4.0	—	3.3	—	—	4.0
1,226	3.5	—	3.1	4.1	—	3.4
96,233	1.1	.8	1.1	1.0	.9	1.1
39,456	3.7	3.4	3.0	3.2	—	3.6
135,689	2.0	.8	1.3	1.7	.9	1.8
2,998	1.7	1.5	2.0	1.9	1.4	1.8
5,665	4.4	—	3.5	3.5	—	4.4
26,600	.7	—	.9	.6	—	.7
12,632	.7	.5	.5	.6	—	.7
14,028	.8	—	.7	.6	—	.8
41,237	3.5	3.1	3.2	3.0	3.7	3.4
6,070	2.8	1.0	2.0	1.8	.6	2.3
81,477	2.1	1.3	2.1	2.0	2.0	1.9
10,032	1.6	.8	1.5	.9	—	1.5
20,020	1.2	—	.8	—	—	1.2
9,141	.6	1.0	.6	.6	—	.6
31,946	1.0	.8	.9	1.2	.8	.9
4,708	.6	—	.6	—	—	.6
85,314	2.1	.8	1.5	1.3	1.1	1.6
58,328	.6	1.0	.7	1.1	—	.6
665,127	1.9	.9	1.7	1.6	1.2	1.7

[b]Data may not add to totals shown due to independent rounding.
Source: Energy Information Administration, U.S. Department of Energy, *Sulfur Content in Coal Shipments, 1978.*

Table 20 Coal Exports[a] by Country of Destination, 1960–1965 (Million

			Europe				
Year	Canada	Brazil	Belgium / Luxem- bourg	Denmark	France	West Germany	Italy
1960	12.8	1.1	1.1	0.1	0.8	4.6	4.9
1961	12.1	1.0	1.0	0.1	0.7	4.3	4.8
1962	12.3	1.3	1.3	(b)	0.9	5.1	6.0
1963	14.6	1.2	2.7	(b)	2.7	5.6	7.9
1964	14.8	1.1	2.3	(b)	2.2	5.2	8.1
1965	16.3	1.2	2.2	(b)	2.1	4.7	9.0
1966	16.5	1.7	1.8	(b)	1.6	4.9	7.8
1967	15.8	1.7	1.4	0	2.1	4.7	5.9
1968	17.1	1.8	1.1	(b)	1.5	3.8	4.3
1969	17.3	1.8	0.9	0	2.3	3.5	3.7
1970	19.1	2.0	1.9	(b)	3.6	5.0	4.3
1971	18.0	1.9	0.8	0	3.2	2.9	2.7
1972	18.7	1.9	1.1	(b)	1.7	2.4	3.7
1973	16.7	1.6	1.2	0	2.0	1.6	3.3
1974	14.2	1.3	1.1	0	2.7	1.5	3.9
1975	17.3	2.0	0.6	0	3.6	2.0	4.5
1976	16.9	2.2	2.2	(b)	3.5	1.0	4.2
1977	17.7	2.3	1.5	0.1	2.1	0.9	4.1
1978	15.7	1.5	1.1	0	1.7	0.6	3.2
1979	19.5	2.8	3.2	0.2	3.9	2.6	5.0
1980	17.5	3.3	4.6	1.6	7.8	2.5	7.1
1981	18.2	2.7	4.3	3.9	9.7	4.3	10.5
1982	18.6	3.1	4.8	2.8	9.0	2.3	11.3
1983	17.2	3.6	2.5	1.7	4.2	1.5	8.1
1984	20.4	4.7	3.9	0.6	3.8	0.9	7.6
1985	16.4	5.9	4.4	2.2	4.5	1.1	10.3

[a]Excludes overseas shipments of anthracite to U.S. Armed Forces.
[b]Less than 50,000 tons.
Note: Sum of components may not equal total due to independent rounding.
Source: Bureau of the Census, *U.S. Exports by Schedule B Commodities*, EM 522.

Short Tons)

Nether-lands	Spain	United Kingdom	Other	Total	Japan	Other	Total
2.8	0.3	0	2.4	17.1	5.6	1.3	38.0
2.6	0.2	0	2.0	15.7	6.6	1.0	36.4
3.3	0.8	(b)	1.8	19.1	6.5	1.0	40.2
5.0	1.5	(b)	2.4	27.7	6.1	0.9	50.4
4.2	1.4	(b)	2.6	26.0	6.5	1.1	49.5
3.4	1.4	(b)	2.3	25.1	7.5	0.9	51.0
3.2	1.2	(b)	2.5	23.1	7.8	1.0	50.1
2.2	1.0	0	2.1	19.4	12.2	1.0	50.1
1.5	1.5	(b)	1.9	15.5	15.8	0.9	51.2
1.6	1.8	(b)	1.3	15.2	21.4	1.2	56.9
2.1	3.2	(b)	1.8	21.8	27.6	1.2	71.7
1.6	2.6	1.7	1.1	16.6	19.7	1.1	57.3
2.3	2.1	2.4	1.1	16.9	18.0	1.2	56.7
1.8	2.2	0.9	1.3	14.4	19.2	1.6	53.6
2.6	2.0	1.4	0.9	16.1	27.3	1.8	60.7
2.1	2.7	1.9	1.6	19.0	25.4	2.6	66.3
3.5	2.5	0.8	2.1	19.9	18.8	2.1	60.0
2.0	1.6	0.6	2.1	15.0	15.9	3.5	54.3
1.1	0.8	0.4	2.2	11.0	10.1	2.5	40.7
2.0	1.4	1.4	4.4	23.9	15.7	4.1	66.0
4.7	3.4	4.1	6.0	41.9	23.1	6.0	91.7
6.8	6.4	2.3	8.8	57.0	25.9	8.7	112.5
5.9	5.6	2.0	7.6	51.3	25.8	7.5	106.3
4.2	3.3	1.2	6.4	33.1	17.9	6.1	77.8
5.5	2.3	2.9	5.3	32.8	16.3	7.2	81.5
6.3	3.5	2.7	10.2	45.1	15.4	9.9	92.7

Table 21 Bituminous Coal Exports by Grade of Coal, 1973–1981 (Thousand Short Tons)

Year	Metallurgical (% of total)		Steam	Total
1973	42,607	80.6	10,263	52,870
1974	51,594	86.1	8,332	59,926
1975	51,597	78.6	14,072	65,669
1976	47,804	80.5	11,602	59,406
1977	41,891	78.0	11,796	53,687
1978	29,848	74.9	9,977	39,825
1979	50,698	78.2	14,085	64,782
1980	63,103	70.2	26,779	89,883
1981	65,234	59.1	45,010	110,243

Note: Total may not equal sum of components due to independent rounding.
Source: Bureau of the Census, U.S. Department of Commerce, Reports EM 522, Energy Information Administration, United States Coal: An Overview, 1985.

Exports of metallurgical quality bituminous coals have increased steadily over the past 20 years, but its percentage of the total exports has decreased relative to steam coal (Table 21).

This reflects the decreased demand worldwide for steel products in the face of the increasing production of synthetic construction and automotive materials, and lower labor market costs abroad. The rise in the demand for steam coal is in response to the sharp increase in the demand for electric power.

Table 22 shows that imports of coal reached the highest level of 3 million short tons in 1978 but have declined to values which average 1.25 million short tons since 1980.

The USEIA projects an annual growth of 2.3% in the decade 1985–1995 for U.S. Coal (*DOE/EIA Annual Energy Outlook*, 1985). If this occurs, production will have doubled from 1974, the benchmark year corresponding to the beginning of the first international crude oil embargo. It is expected that utilities will continue to be the primary domestic consumer and that they will also demonstrate the greatest rate of increase in coal consumption as declines in nuclear capability are expected to continue (Table 23).

Table 22 Coal Supply and Disposition 1949–1985 (Million Short Tons)

Year	Production	Imports
1949	480.6	0.3
1950	560.4	0.4
1951	576.3	0.3
1952	507.4	0.3
1953	488.2	0.3
1954	420.8	0.2
1955	490.8	0.3
1956	529.8	0.4
1957	518.0	0.4
1958	431.6	0.3
1959	432.7	0.4
1960	434.3	0.3
1961	420.4	0.2
1962	439.0	0.2
1963	477.2	0.3
1964	504.2	0.3
1965	527.0	0.2
1966	546.8	0.2
1967	564.9	0.2
1968	556.7	0.2
1969	571.0	0.1
1970	612.7	
1971	560.9	0.1
1972	602.5	
1973	598.6	0.1
1974	610.0	2.1
1975	654.6	0.9
1976	684.9	1.2
1977	697.2	1.6
1978	670.2	3.0
1979	781.1	2.1
1980	829.7	1.2
1981	823.8	1.0
1982	838.1	0.7
1983	782.1	1.3
1984	895.8	1.3
1985	886.1	2.0

Source: Energy Information Administration, Annual Energy Review 1985.

Table 23 Coal Supply, Disposition, and Prices

Supply, disposition, and price	Base case										
	1974	1979	1983	1984	1985	1986	1987	1988	1989	1990	1995
Production[a]											
East of the Mississippi	518	560	507	588	570	582	590	594	600	613	567
West of the Mississippi	92	221	275	308	316	327	339	347	358	372	449
Total	610	781	782	896	886	909	930	940	958	985	1116
Imports[b]	2	2	1	1	2	2	2	2	2	2	2
Exports[c]	61	66	78	81	85	85	85	86	87	89	104
Net imports	-59	-64	-77	-80	-83	-83	-83	-84	-85	-87	-102
Net storage withdrawals[d]	8	-36	27	-29	25	5	-5	-4	-5	-7	-6
Total supply	559	681	732	787	828	831	841	853	867	891	1000
Consumption by sector											
Residential and commercial	11	8	8	9	8	8	7	7	7	7	7
Industrial	65	68	66	74	77	75	76	78	81	83	87
Coking plants[f]	90	77	37	44	40	40	40	39	38	37	32
Electric utilities	392	527	625	664	693	711	719	728	742	764	882
Total consumption	558	681	737	791	818	833	842	853	867	891	1000
Discrepancy[g]	1	0	-5	-4	10	-3	•	•	•	•	•
Average minemouth price[h]	31.82	33.66	27.95	26.55	26.83	26.88	28.13	28.17	28.25	24.42	28.93

Delivered prices by sector											
Residential and commercial	63.82	55.44	43.38	44.45	44.44	44.78	48.69	48.87	49.07	49.33	51.06
Industrial	50.70	49.78	42.27	40.73	40.96	41.41	42.98	43.46	43.98	44.55	47.16
Coking plants[i]	73.32	71.80	63.78	58.62	60.24	60.71	62.07	62.42	62.86	63.31	65.09
Electric utilities[j]	31.11	37.05	37.62	36.39	35.65	35.86	36.33	36.44	36.64	36.92	38.24
Average to all sectors[k]	40.84	42.48	39.41	38.12	37.45	37.63	38.25	38.38	38.57	38.82	39.96

[a]Historical coal production includes anthracite, bituminous, and lignite. Projected coal production includes bituminous and lignite with anthracite included in bituminous.

[b]Coal imports are not projected beyond 1985, but are held constant at 2 million short tons per year.

[c]Excludes small quantities of anthracite shipped overseas to U.S. Armed Forces and coke exports.

[d]From stocks held by end-use sectors (secondary stocks held at industrial plants, coke plants, and electric utility plants). Net stock withdrawals are computed as the end-of-year stock levels from the current period subtracted from the end-of-year stock levels from the preceding period. A minus is treated as a deletion from total supply and a plus is treated as an addition to total supply.

[e]Total supply is equivalent to production plus net imports plus net storage withdrawals.

[f]Coke plants consume metallurgical coal which is a mixture of anthracite and bituminous coal. Historically, coking plant coal price is a weighted average of anthracite and bituminous coal types. In the projections, anthracite is included in bituminous coal.

[g]Historically, discrepancy represents revisions in producers (primary) stock levels, plus losses and unaccounted-for coal. In the projected period, discrepancy represents errors due to conversion factors.

[h]In historical years, the average production price of coal produced at the mine. Projected prices (1985–1995) are estimated and do not reflect market conditions.

[i]Projected residential and commercial prices (1983–1995) do not include dealer markup.

[j]Historically, electric utility price includes anthracite, bituminous, and lignite coal purchased under long-term contracts and on the spot market. In the projections, anthracite is included in bituminous coal with the bituminous coal price being used for anthracite coal price.

(continued)

Table 23 (Cont.)

[k]Weighted average price and the weights are the sectoral consumption values.
[•]Greater than zero but less than 0.5.
Note: The prices have been converted from nominal to real dollars by using the implicit Gross National Product deflator rebased to 1985 equals 1.00. Projected coal prices are based on cost estimates and do not reflect market conditions. Also, totals may not equal sum of components because of independent rounding.
Data sources: Historical prices through 1982 from the Energy Information Administration, *State Energy Price and Expenditure Report*, DOE/EIA-0376(82) (Washington, D.C., 1985), pp. 4–21. Historical quantities through 1982 are from the Energy Information Administration, *Annual Energy Review, 1984*, DOE/EIA-0384(84) (Washington, D.C., 1985), pp. 145–153, Tables 65, 66, and 67. Historical 1983 and 1984 quantities and prices (excluding residential and commercial) are from the Energy Information Administration, *Quarterly Coal Report*, DOE/EIA-0125(85/2C) (Washington, D.C., October, 1985). Historical quantities are through 1984. Projected values are outputs from the intermediate Future Forecasting System.
Input data file: Historical = D1230851; Projected = IFGMMM.D1118851. Tables printed on January 31, 1986.
Note: Million short tons per year; 1985 dollars per short ton.
Source: Energy Information Administration, Annual Energy Outlook 1985.

Industrial (nonutility) consumption of steam coal is expected to grow only slightly, and metallurgical coal use is expected to decline to about only one-third of its 1974 value of 91 million short tons.

Because the Great Plains Gasification facility in North Dakota was closed in 1986, little if any growth in the consumption of lignite is foreseen. In contrast to that projection, however, a strong average rate of growth (3.3%) is expected for all mines west of the Mississippi River while only a 1.6% increase is forecast for eastern mines for 1986–1990 (Table 23). Based on projections for new utility facilities coming on line and for some increases in the cost of imported crude oil, total production of coal is projected to rise more than 13.3% (8.9% in the East and 20.7% in the West between 1990 and 1995) (Energy Information Administration, 1986). The larger expected growth in the West will be due to lower surface-mining costs, lower sulfur content, and growing demand (population/industry) (Energy Information Administration, 1986). The distribution of this increased output is shown in Figure 18 for each region of the country.

Trends in demand for coal exports (Table 24) have followed the same trend as the domestic use of coal: between 1974 and 1985 the demand for steam coal grew from 14.8% to 31.7% of the export market and is expected to become 40% of the export market (104 million short tons, total exports) by 1995 (Energy Information Administration, 1986). Australia, Canada, Poland, and South Africa are the biggest competitors of the United States for world markets. Australia and Canada are expected to gain some former U.S. metallurgical markets in Asia by 1995, but U.S. losses may be offset by increased South American markets (Energy Information Administration, 1986).

Imports of coal (Table 23) may be decreased by pending legislation which seeks to impose a tariff on international suppliers who are not bound by comparable environmental, health, safety, and welfare standards which critically impact the cost of domestic coal. Restrictions are also being sought on companies within the United States that import coal.

B. World Resources

According to a 1974 survey (World Survey of Energy Resources, 1974), nearly 90% of the world's total solid fossil fuel resources was then found in the People's Republic of China, the United States, and the USSR. About 50% of all solid fossil fuel resources are believed to be located in the USSR, 25% in the United States, and 13% in the People's Republic of China.

In 1985, the total estimated recoverable reserve of coal worldwide was estimated to be nearly 987 billion short tons. Although as much as 50% of all coal resources may be located in the USSR (World Survey

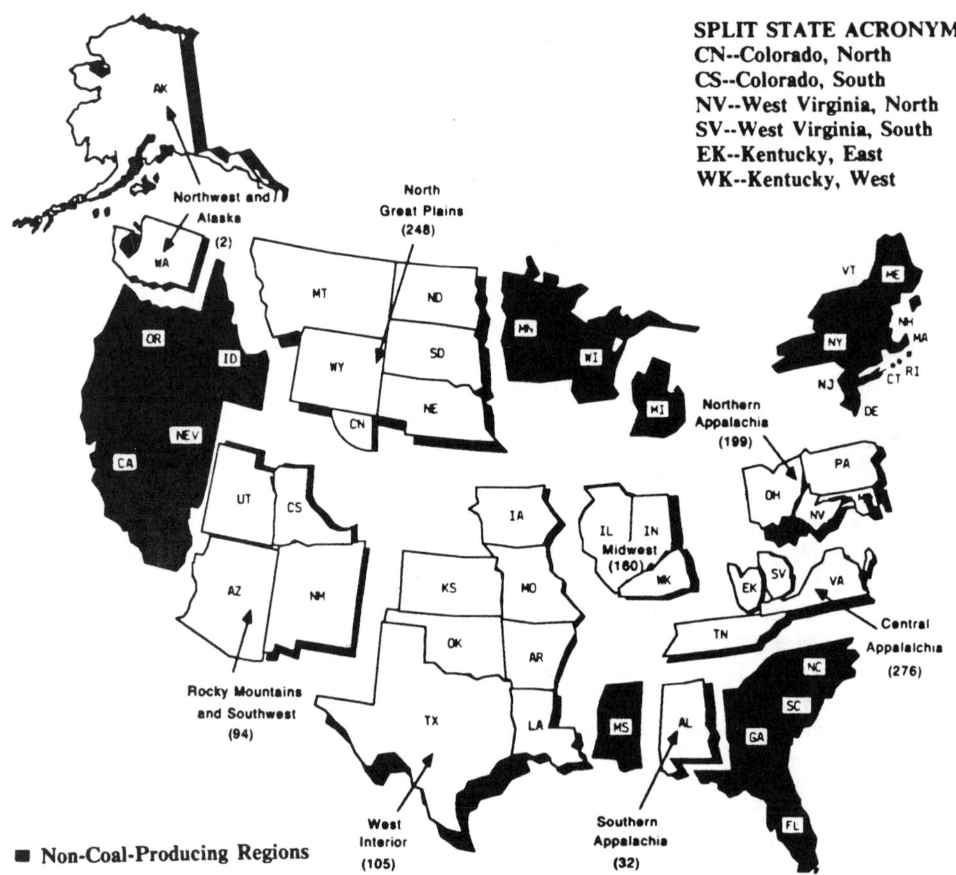

Figure 18 Projected U.S. coal production by region, 1995 (Energy Information Administration, *Annual Energy Outlook 1985*, DOE/EIA 0385(85), Washington, D.C., February 1986).

of Energy Resources, 1974), based on estimates of recoverable coal reserves, the USSR—with estimated recoverable reserves of 265 billion tons—is second to the United States which has 283 billion tons of estimated recoverable resources (*DOE/EIA International Energy Annual*, 1985). The third largest deposit of recoverable coal is located in the People's Republic of China. Seven countries—the United States, the USSR, People's Republic of China, Australia, Federal Republic of Germany (West Germany), South Africa, and Poland—together hold 91% of the world's total estimated recoverable coal resources (Table 24). It is also estimated that 63% of the recoverable coal in Russia is high rank: 12% anthracite, 22% coking coal, 25% steam coal, and 5% subbituminous. Russia is the world's largest producer of peat so it is not surprising that 36% of the recoverable reserves of coal are brown coal and lignite. However, Melnikov (1972) reported that much of the coal in the USSR is situated in Arctic regions, decreasing its accessibility considerably. A study carried out by the U.S. government showed that the extent of the vast Asiatic deposits in Siberia far exceeded known seams throughout the remainder of the country (Figure 19).

The People's Republic of China is advantaged by having much of its approximately 450 billion short tons of total coal resources located in the central or eastern part of the country, near population centers (World Survey of Energy Resources, 1974). Much of this coal is coking quality and it is estimated that 109 billion short tons of this reserve is recoverable (*DOE/EIA International Energy Annual*, 1985).

Although reliable current data are not available, a number of countries not identified specifically in Table 25 do possess measurable recoverable coal reserves and should also be listed for the sake of completeness (World Survey of Energy Resources, 1974):

Central and South America
- Argentina
- Mexico
- Peru

Western and Northern Europe
- Belgium
- Finland
- France
- Italy
- The Netherlands
- Portugal
- Spain

Africa
- Algeria
- Botswana
- Malawi
- Morocco
- Mozambique
- Nigeria
- Zimbabwe
- Tanzania
- Zambia

Far East and Oceania
- Bangladesh
- Indonesia
- Pakistan
- Turkey

Table 24 World Estimated Recoverable Reserves of Coal

Country	Anthracite and bituminous coal[a]			Lignite[b]		Total recov-erable coal
	Recov-erable	Portion surface-mineable	Portion of coking quality	Recov-erable	Portion surface-mineable	
North America						
Canada	4.18	0.00	1.38	2.33	2.33	6.51
Mexico	1.97	.31	1.37	.00	.00	1.97
United States	248.16	89.28	NA	35.25	35.25	283.41
Other	.00	.00	.00	.00	.00	.00
Total	254.31	89.59	2.75	37.58	37.58	291.90
Central and South America						
Brazil	14.33	1.43	1.15	.00	.00	14.33
Chile	1.30	NA	NA	.00	.00	1.30
Colombia	1.14	NA	.23	(•)	(•)	1.14
Other	.51	.04	.00	.02	.00	.53
Total	17.28	1.48	1.37	.02	(•)	17.30
Western Europe						
Germany, West	32.98	NA	19.78	38.66	38.66	71.64
Greece	.00	.00	.00	1.66	.00	1.66
Turkey	.20	.00	.00	1.90	.00	2.11
United Kingdom	5.06	.00	2.23	.00	.00	5.06
Yugoslavia	1.73	.06	.00	16.50	.00	18.23
Other	3.29	.66	1.03	.58	.07	3.87
Total	43.26	.72	23.04	59.31	38.74	102.58
Eastern Europe and USSR						
Bulgaria	.03	NA	.02	4.00	2.65	4.03
Czechoslovakia	3.00	NA	NA	3.15	.00	6.15
Germany, East	.00	.00	.00	NA	NA	NA
Hungary	.23	.00	.00	4.40	.00	4.63
Poland	30.00	.00	6.00	13.20	13.20	43.20
Romania	.00	.00	.00	.00	.00	.00
USSR	166.67	34.88	60.00	98.21	97.23	264.88
Total	199.93	34.88	66.02	122.96	133.08	322.89

(continued)

Table 24 (Cont.)

Country	Anthracite and bituminous coal[a]			Lignite		
	Recov-erable	Portion surface-mineable	Portion of coking quality	Recov-erable	Portion surface-mineable	Total recov-erable coal
Middle East	.00	.00	.00	.00	.00	.00
Africa						
Botswana	3.80	.00	.00	.00	.00	3.80
South Africa	57.03	NA	NA	.00	.00	57.03
Swaziland	2.00	.00	.00	.00	.00	2.00
Other	2.43	.55	.15	.11	(•)	2.54
Total	65.27	.55	.15	.11	(•)	65.38
Far East and Oceania						
Australia	32.52	7.00	13.01	39.90	39.90	72.42
China	108.90	10.91	40.29	.00	.00	108.90
India	NA	NA	NA	1.74	1.65	1.74
Japan	1.10	NA	.57	.02	.00	1.12
Other	1.92	.07	.12	.57	.51	2.49
Total	144.44	17.98	53.99	42.23	42.06	186.67
World total	724.50	145.21[c]	147.33[c]	262.22[c]	231.47[c]	986.72

[a]Includes subanthracite and subbituminous.
[b]Includes brown coal.
[c]Sum of reported totals only.
(•) Denotes less than 5 million short tons.
NA = not available.
Note: Sum of components may not equal total due to independent rounding.
Data source: The World Energy Conference has compiled these data from questionnaires submitted to National Committees of member countries or corresponding institutions of nonmember countries and from estimates based on examination of existing published sources. The reference year for most reserve data is 1981.
Source: Energy Information Administration, International Energy Annual, 1985.

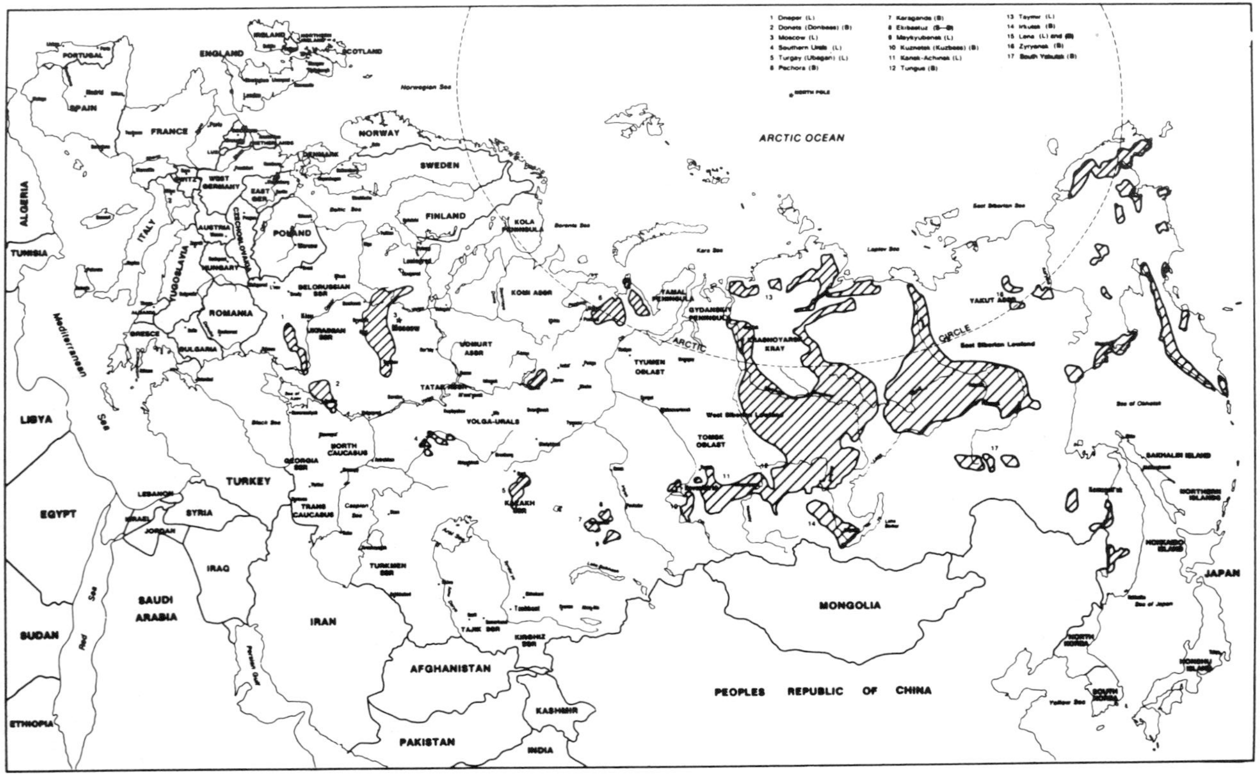

Figure 19 Soviet coal basins. S-B, subbituminous; B, bituminous; L, lignite (Congress of the United States, Office of Technology Assessment. *Technology and Soviet Energy Availability*. November 1981, p. 85).

Table 25 World Coal Production 1975–1985

Country	1975	1976	1977	1978	1979	1980	1981	1982	1983	1984	1985
North America											
Canada	28	28	32	34	37	40	44	47	50	63	67
Mexico	6	6	7	7	8	8	8	8	10	10	10
United States	655	685	697	670	781	830	824	838	782	896	886
Total	688	719	736	711	826	878	876	894	841	969	963
Central and South America											
Brazil	3	4	4	4	8	9	6	7	7	8	8
Chile	1	1	1	1	1	1	1	1	1	1	1
Colombia	4	4	4	4	5	6	6	7	6	8	8
Other	1	1	1	1	1	1	1	1	1	1	1
Total	9	10	10	10	16	16	14	16	16	18	18
Western Europe											
Austria	4	4	3	3	3	3	3	4	3	3	3
Belgium	8	8	8	7	7	7	7	7	7	7	7
France	28	28	27	26	23	23	24	22	22	21	19
Germany, West	238	247	229	228	239	239	241	247	236	233	231
Greece	20	25	26	25	26	26	30	30	33	35	38
Italy	2	2	2	2	2	2	2	2	2	2	1
Norway	(•)	1	1	(•)	(•)	(•)	(•)	(•)	1	(•)	1
Spain	15	16	19	22	24	32	38	43	44	44	44
Turkey	12	11	13	15	22	18	19	24	32	38	35
United Kingdom	142	137	135	136	135	141	138	137	127	55	100

(continued)

Table 25 (Cont.)

Country	1975	1976	1977	1978	1979	1980	1981	1982	1983	1984	1985
[Western Europe]											
Yugoslavia	39	41	43	44	46	52	58	60	65	72	72
Other	(•)	(•)	(•)	(•)	(•)	(•)	(•)	1	(•)	(•)	(•)
Total	509	518	506	509	529	543	561	577	571	510	552
Eastern Europe and USSR											
Albania	1	1	1	1	1	2	2	2	2	2	2
Bulgaria	31	28	28	28	31	33	32	35	36	36	36
Czechoslovakia	127	130	134	136	137	136	137	139	140	143	142
Germany, East	272	273	280	279	282	285	294	304	309	327	343
Hungary	27	28	28	28	28	28	29	29	28	28	27
Poland	233	241	250	258	264	254	219	250	258	267	275
Romania	30	28	30	32	36	39	41	42	39	49	49
USSR	773	784	796	798	792	790	776	792	789	785	798
Total	1494	1513	1546	1560	1571	1566	1529	1593	1601	1635	1671
Middle East											
Iran	1	1	1	1	1	1	1	1	1	1	1
Total	1	1	1	1	1	1	1	1	1	1	1
Africa											
Morocco	1	1	1	1	1	1	1	1	1	1	1
Mozambique	1	1	(•)	(•)	(•)	(•)	1	1	1	(•)	(•)
South Africa	77	85	94	100	114	127	144	151	161	179	180
Zambia	1	1	1	1	1	1	(•)	(•)	(•)	1	1
Zimbabwe	3	3	3	3	4	4	3	3	4	3	3

Other	(•)	(•)	1	1	1	(•)	1	1	1	1	1
Total	82	91	100	106	121	133	149	157	167	184	185
Far East and Oceania											
Australia	98	109	111	114	119	116	130	140	146	153	170
China	570	586	606	681	698	684	683	734	788	870	937
India	109	116	115	116	118	125	142	148	158	168	171
Indonesia	(•)	(•)	(•)	(•)	(•)	(•)	(•)	(•)	1	1	1
Japan	26	20	20	21	19	20	20	19	19	18	18
Korea, North	44	45	45	45	48	50	50	52	50	51	51
Korea, South	19	18	19	20	20	21	22	20	21	23	25
Mongolia	3	3	4	4	4	5	5	6	6	6	6
New Zealand	3	3	3	2	2	2	2	2	2	3	3
Pakistan	1	1	1	1	1	2	2	2	2	2	2
Philippines	(•)	(•)	(•)	(•)	(•)	(•)	(•)	1	1	1	1
Thailand	1	1	(•)	(•)	1	2	2	2	2	3	5
Vietnam	4	6	6	6	6	6	7	6	7	6	6
Other	4	4	3	3	3	3	3	4	3	3	3
Total	881	911	934	1014	1042	1036	1068	1136	1205	1307	1400
World total	3665	3763	3833	3911	4105	4173	4198	4375	4402	4623	4790

(•) Denotes less than one-half the unit of measure.
Note: Coal includes anthracite, subanthracite, bituminous, subbituminous, lignite, and brown coal. Also, sum of components may not equal total due to independent rounding.
Source: Energy Information Administration, International Energy Annual, 1985.

From 1975 to 1985 worldwide coal production rose nearly 3%/year to 4.8 billion short tons in 1985 (*DOE/EIA International Energy Annual*, 1985). China, the United States, and the USSR were the top three coal producers in 1985 (Table 25). Worldwide, anthracite and bituminous coal production accounted for 73% of the total; in the United States 92% of the total output was bituminous coal, while in China only 76.6% was bituminous (19.5% anthracite and 4% lignite), and in the USSR 78% of the coal production was anthracite plus bituminous (Table 26). Dramatic increases in coal production were realized in South Africa (134%) in 1985 compared with 1975, primarily in response to the commitment made by that country to develop coal conversion processes. Similarly, Australia increased production by 73%, China by 64%, India by 57%, and Canada by 139%. Japan has decreased production by 31% since 1975.

In 1985, the USSR accounted for 48% of all coal produced in Eastern Europe and 35% of the world's coal production came from the Eastern European countries, primarily the German Democratic Republic (East Germany) and Poland, in addition to the USSR. While Western Europe imports the largest quantity of coal, Japan alone imports about 25% of all exported coal (Table 27). The United States, Australia, and Poland are the three leading coal-exporting countries.

The United States and the People's Republic of China together consume about 41% of the world supply of coal. The USSR is the third largest consumer, accounting for about 15% of the total coal consumption (Table 28). The international flow of coal is depicted in Figure 20. This diagram shows that South Africa and the United States are the largest suppliers of coal to Europe and that Australia is the largest single supplier to Japan, providing about 50% of Japan's imports in 1984 (DOE/EIA Coal Data: A Reference, 1982).

IV. RECOVERY

Coal-mining operations are principally determined by geologic features or geologic occurrences that took place during peat formation and subsequent coalification. Speight (1983) summarized eight features associated with coal seams which are taken into account in planning and carrying out mining operations:

1. Clastic dikes and clay veins—clay, silt, or sand that interrupts the horizontal extension of a seam. Dikes may range from inches to hundreds of feet wide and may cut into overlying strata, altering the strength of the roof over a seam and affecting the release of mine gas from the seam. This material is additional waste which affects the cost of mining and the overall value of a seam.

Table 26 World Coal Production 1984

Country	Primary				Secondary[a]	
	Anthracite	Bituminous	Lignite	Metallurgical coke	Anthracite and bituminous briquets	Lignite briquets
North America						
Canada	–	52,342	10,932	5,342	–	–
Mexico	–	9,920	–	2,535	–	–
United States	4,162	828,689	63,070	30,561	–	–
Total	4,162	890,951	74,002	38,438	–	–
Central and South America						
Brazil	–	7,885	–	6,959	–	–
Chile	–	1,305	45	330	–	–
Colombia	–	7,564	–	330	–	–
Other	140	617	–	524	–	–
Total	140	17,371	45	8,143	–	–
Western Europe						
Austria	–	–	3,228	2,043	–	–
Belgium	112	6,830	–	6,532	48	–
France	2,597	15,695	2,674	9,920	1,596	–
Germany, West	6,390	87,160	139,706	23,663	1,584	7,024
Greece	–	–	34,806	17	–	203
Italy	–	–	1,929	7,863	–	–
Norway	–	497	–	347	–	–

(continued)

Table 26 (Cont.)

Country	Primary[a]				Secondary[a]	
	Anthracite	Bituminous coal	Lignite	Metallurgical coke	Anthracite and bituminous briquets	Lignite briquets
[Western Europe]						
Spain	6,100	10,742	26,814	3,618	18	—
Turkey	248	7,829	29,580	2,760	—	25
United Kingdom	1,320	53,245	—	7,696	188	—
Yugoslavia	—	429	71,302	3,875	—	1,045
Other	210	92	—	4,682	—	440
Total	16,977	182,519	310,039	73,016	3,434	8,737
Eastern Europe and USSR						
Albania	—	—	1,956	18	—	—
Bulgaria	93	153	35,420	1,307	—	1,335
Czechoslovakia	—	29,124	113,380	11,355	—	1,178
Germany, East	—	—	326,660	1,300	—	61,795
Hungary	—	2,836	24,773	775	—	1,710
Poland	—	211,194	55,532	18,866	627	165
Romania	—	9,323	39,485	4,960	825	1,040
USSR	—	613,436	171,740	94,800	225	6,095
Total	93	866,066	768,946	133,381	1,677	73,318
Middle East						
Iran	—	1,000	—	385	—	—
Total	—	1,000	—	385	—	—

Africa						
Morocco	920	–	–	–	–	–
South Africa	4,260	174,352	–	2,040	–	–
Zimbabwe	–	2,535	–	254	–	–
Other	138	1,780	–	1,050	–	–
Total	5,318	178,667	–	3,344	–	–
Far East and Oceania						
Australia	–	114,866[b]	38,028	3,927	–	902
China	170,000	666,785	33,190	39,548	–	–
India	–	159,692	8,432	12,091	–	159
Indonesia	–	1,196	–	–	–	–
Japan	25	17,040	1,282	56,520	337	–
Korea, North	40,000	10,000	500	3,750	19,672	–
Korea, South	22,750	–	–	5,730	–	–
Mongolia	–	450	6,000	–	–	–
New Zealand	–	2,525	259	18	6	–
Pakistan	–	2,060	–	–	–	–
Philippines	–	1,257	–	–	–	–
Thailand	–	–	2,576	–	–	–
Vietnam	5,500	–	–	–	–	–
Other	–	2,430	38,274	–	40	–
Total	238,275	978,301	128,541	121,884	20,055	1,061
World Total	264,965	3,114,875	1,281,573	378,591	25,166	83,116

[a]Primary coal includes all coal mines and when necessary washed and sorted. Secondary coal (e.g., coke, briquets) is derived from primary coal.
[b]Includes anthracite.
– Indicates data not applicable.
Note: Sum of components may not equal total due to independent rounding.
Source: Energy Information Administration, International Energy Annual, 1985.

Table 27 World Coal Supply and Disposition 1984 (Trillion Btu)

Country	Production	Imports	Exports	Apparent consump-tion
North America				
Canada	1,336	552	683	1,205
Mexico	179	13	0	191
United States	19,723	47	2,177	17,074
Total	21,237	612	2,860	18,470
Central and South America				
Brazil	163	228	0	391
Chile	33	14	0	47
Colombia	168	0	10	158
Other	16	27	1	42
Total	381	269	11	639
Western Europe				
Austria	41	117	0	158
Belgium and Luxembourg	160	399	41	517
Denmark	0	223	1	222
France	482	642	45	1,079
Germany, West	3,802	345	505	3,642
Greece	167	36	1	201
Italy	17	565	15	567
Netherlands	0	363	61	303
Norway	12	33	5	40
Spain	708	199	0	906
Turkey	494	65	0	558
United Kingdom	1,168	271	82	1,357
Yugoslavia	739	105	13	831
Other	7	366	7	366
Total	7,796	3,727	775	10,747
Eastern Europe and USSR				
Albania	25	8	0	32
Bulgaria	453	189	9	633
Czechoslovakia	1,905	134	140	1,899
Germany, East	2,496	148	77	2,566
Hungary	306	83	0	389
Poland	5,102	26	1,280	3,848
Romania	614	210	0	824

(continued)

Table 27 (Cont.)

Country	Production	Imports	Exports	Apparent consumption
[Eastern Europe and USSR]				
USSR	13,242	375	756	12,861
Total	24,143	1,172	2,262	23,053
Middle East				
Iran	23	11	0	34
Other	0	76	0	76
Total	23	87	0	110
Africa				
Morocco	22	6	(•)	27
Mozambique	10	5	4	11
South Africa	3,822	0	910	2,913
Zambia	12	0	0	12
Zimbabwe	56	3	6	53
Other	17	70	1	87
Total	3,939	85	921	3,103
Far East and Oceania				
Australia	3,058	1	2,070	989
China	16,965	67	200	16,832
India	2,959	12	2	2,969
Indonesia	28	1	22	6
Japan	398	2,399	70	2,726
Korea, North	1,161	23	2	1,183
Korea, South	432	371	0	803
Mongolia	60	(•)	0	60
New Zealand	57	(•)	9	47
Pakistan	36	16	0	52
Philippines	21	20	0	41
Thailand	21	8	1	28
Vietnam	139	2	27	113
Other	57	364	0	422
Total	25,393	3,282	2,404	26,271
World total	82,912	9,234	9,234	83,188

[a]Includes stock changes.
(o) Denotes less than one-half the unit of measure.
Note: Sum of components may not equal total due to independent rounding.
Source: Energy Information Administration, International Energy Annual, 1985.

Table 28 Overburden Ratios at Surface Mines, 1978

State	Overburden ratio (cubic yards excavated per ton of coal mined)	Percentage of total surface-mined coal for which data were reported
Alabama	32.5	52.4
Alaska	1.3	100.0
Arizona	7.9	79.8
Arkansas	37.1	51.9
Colorado	7.6	95.6
Illinois	14.7	83.7
Indiana	17.7	56.8
Iowa	9.6	31.6
Kansas	20.7	58.0
Kentucky	13.8	49.0
Maryland	16.4	21.0
Missouri	20.4	69.8
Montana	3.5	74.5
New Mexico	6.8	99.2
North Dakota	3.9	74.0
Ohio	19.1	55.8
Oklahoma	29.9	76.8
Pennsylvania	21.8	26.5
Tennessee	23.0	22.5
Texas	5.8	100.0
Virginia	13.8	19.6
Washington	6.0	100.0
West Virginia	11.8	46.5
Wyoming	3.0	75.2
U.S. total	11.3	60.9

Source: Energy Information Administration, U.S. Department of Energy, *Bituminous Coal and Lignite Production and Mine Operations, 1978.*

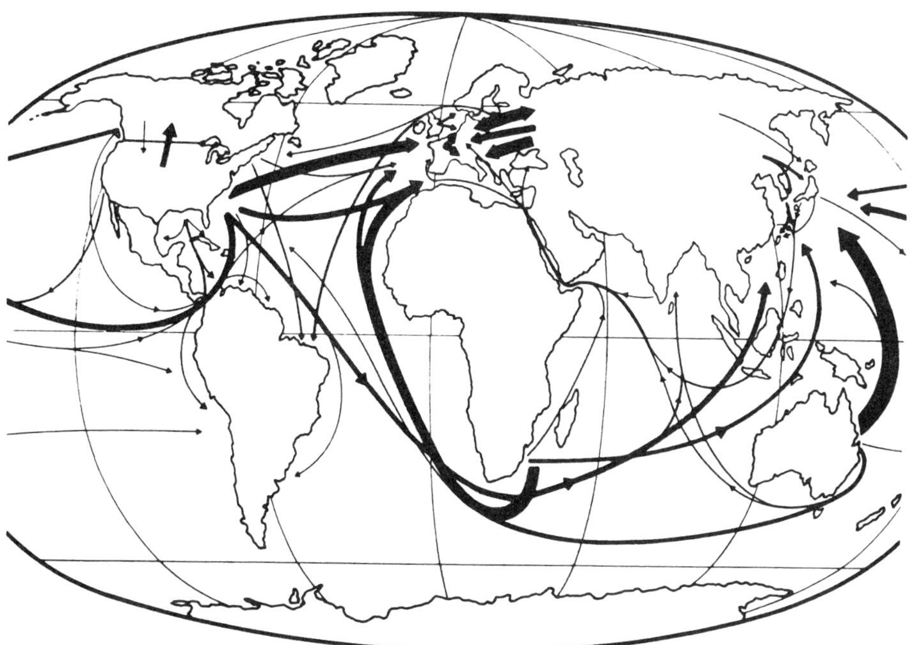

Figure 20 International coal flow, 1984. (Energy Information Administration, International Energy Annual, 1985.)

2. Cleats—breaks or fractures in a coal seam. Cleats most often lie parallel to the face of a seam (the face cleat) but may tend to be perpendicular to it (butt cleat). Identifying the trend of the cleat facilitates breaking of the coal seam during mining.
3. Concretions—commonly nodular concentrations of minerals such as calcite ($CaCO_3$), dolomite ($(CaMgCO_3)_2$), and pyrite (FeS_2). Coal balls are concretions containing petrified vegetation. Concretions vary in size and may be several feet in diameter. If concretions are deposited in an otherwise weak roof, they may collapse into a mine when the seam supporting the upper strata is removed.
4. Dipping and folded strata—distortions in the level of the strata caused by subsidence or horizontal compaction. In extreme instances a seam may be almost vertical to the surface, folded back on itself, or even overturned.
5. Faulting—fractures in a rock sequence such that strata adjacent to the fracture become offset vertically or horizontally (Figure 21). The extent of a fault may vary widely from a few inches to miles

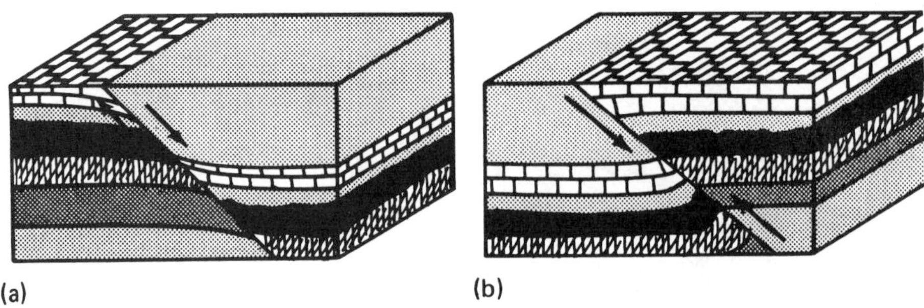

Figure 21 Diagram of (a) a normal fault and (b) a reverse fault (J. Speight, 1983. Reprinted with permission of the publisher).

(uncommon). Outcropping of coal may occur when a reverse fault brings a portion of a seam to the surface.

6. Igneous intrusions and sills—a special type of dike consisting of previously molten rocks. A sill forms when the instrusion seeps into a space at the intersection of bedding planes (Figure 22). Localized thermal effects on the adjacent coal may increase its carbon content (rank), forming coke in extreme cases. Because this effect is usually irregular in a seam and because the igneous material is waste, its presence may adversely affect the value of the entire seam.
7. Partings or splits—mud or silt washed into a swamp and later transformed into shale or limestone. Partings may separate the same main seam into several fingerlike sprays.
8. Washout or cutout—a discontinuity in the lateral extension of a seam caused by glacial ice or water erosion which was later filled in with other sediments. Washouts are distinguished from partings by their sharpness and by their frequent extension through one or more layers of strata above and/or below the coal (Figure 23).

Historically, most of the world's coal has been mined underground. In the United States, however, production from surface mining surpassed underground mining in 1971. The methods and technology used to mine coal varies according to the depth and the thickness of the seam, its inclination, the type of overburden, labor costs, land protection policies, and the cultural need to recover a particular coal.

When coal lies deeper than about 70 m (200 ft) below the surface underground methods are required. Typically, a seam is considered accessible up to 700 m (2,000 ft) below the surface; few mines in the

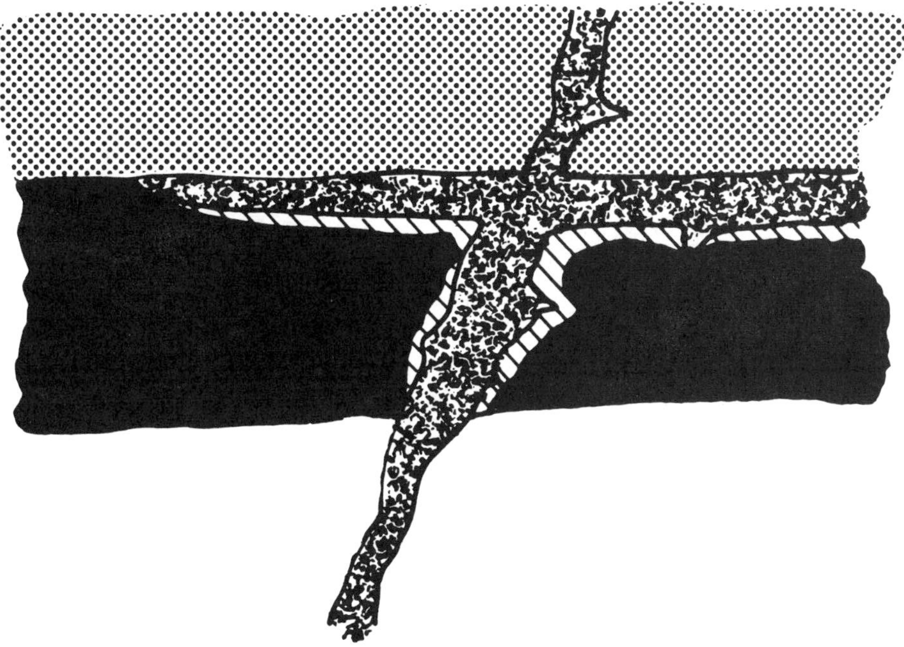

Figure 22 Diagram of an igneous dike through a coal seam spreading out into a sill at the top of the coal (J. Speight, 1983. Reprinted with the permission of the publisher).

United States are located deeper than 300 m (1,000 ft). Underground mining can be characterized by three forms of mine access and three patterns of mine architecture (Figures 24 and 25).

Shaft access is used for deep mines which lie relatively level. The shaft is a vertical opening equipped with elevators to transport miners, equipment, and mined coal.

A slope access is an entry inclined 15–20° into a shallow seam and uses conveyor belts, trolleys, or electric hoists to remove the coal.

Drift access provides a nearly horizontal tunnel into seams where excavation of rocks is not required, e.g., where an outcrop occurs. Conveyor belts are used most commonly to carry mined coal away from the seam.

The layout of an underground mine is often a variation of the room and pillar format. The room and pillar system is derived by mining in such a way that sets of perpendicular pillars are left standing amid large, mined-out areas called rooms. These large pillars of coal are

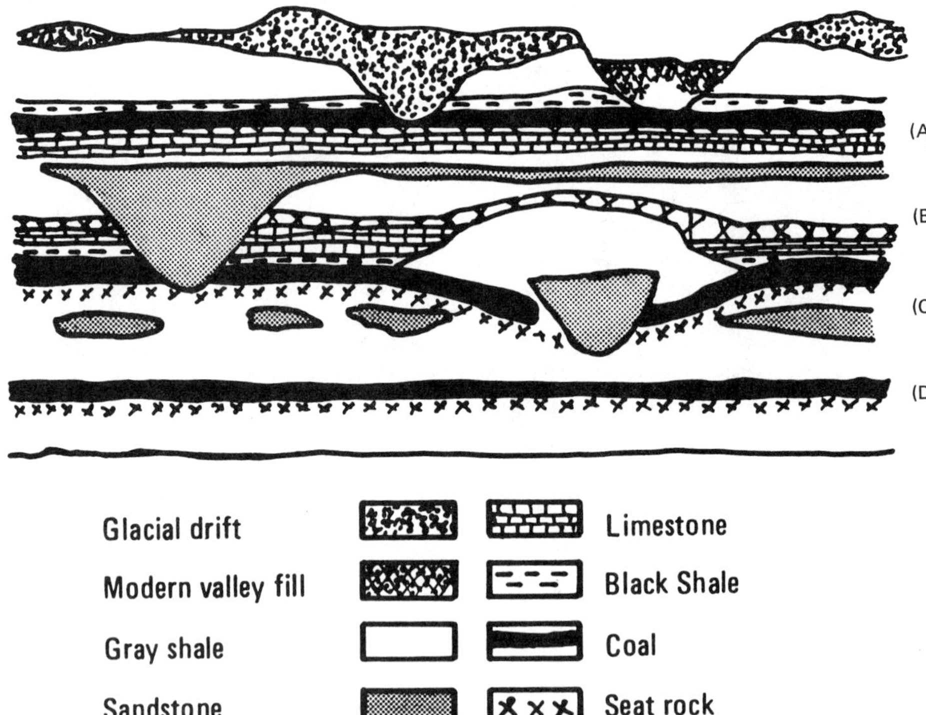

Figure 23 Diagram of coal seam discontinuities: (A) recent erosion; (B) preglacial erosion; (C, D) stream beds at the time of debris accumulation (J. Speight, 1983. Reprinted with permission of the publisher).

left to support the roof. This means that a significant portion of otherwise mineable coal (sometimes as much as 50%) is not recovered (Figure 25). Conventional methods for mining by room and pillar consist of first cutting into the seam in order to blast it with explosives, compressed air, or liquid CO_2 prior to mining out the pattern. Another option is called continuous mining. A single continuous mining machine digs and loads the coal onto conveyors without requiring prior blasting. Longwall mining (Figure 26) is a system which removes nearly all of the available coal. In this instance, a main tunnel and two parallel tunnels placed at right angles to the main one are cut into the seam. Self-advancing roof supports move at the face of the

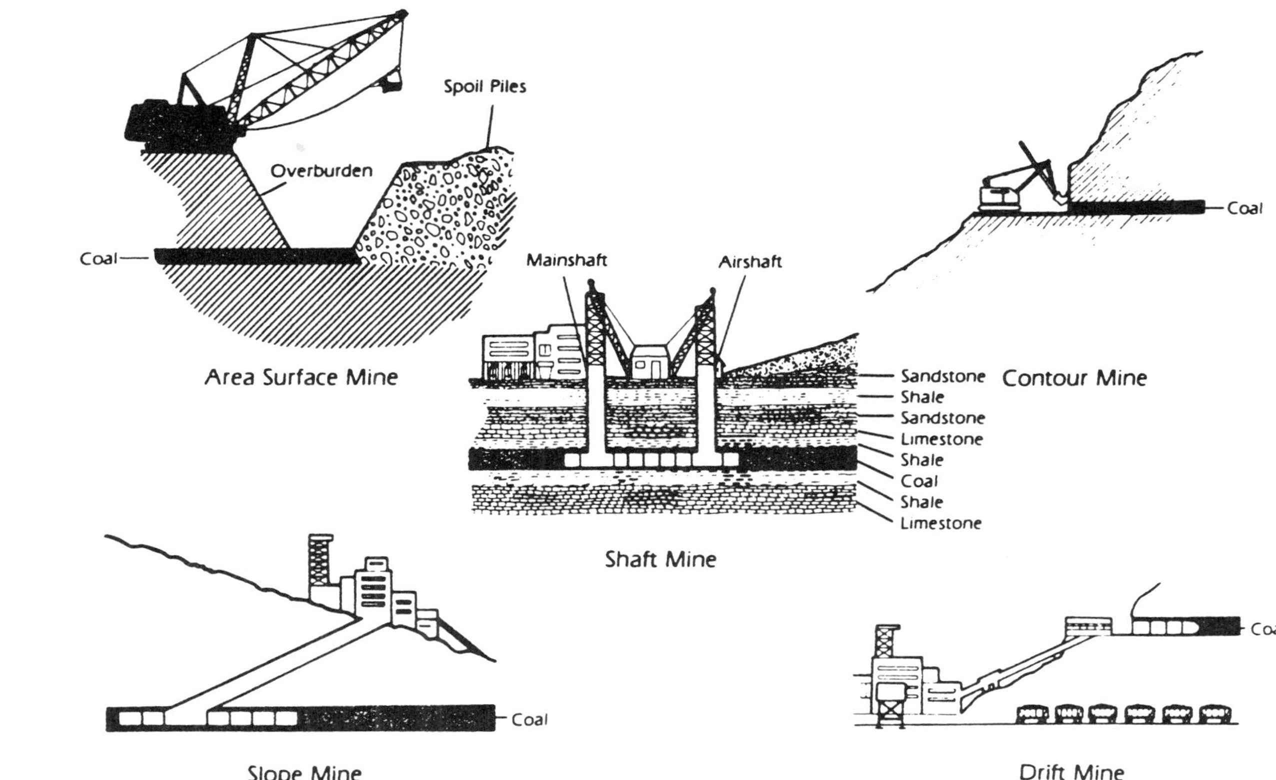

The method of mining a coal deposit depends on the depth of the coalbed and the character of the land.

Figure 24 Methods of coal mine access (Energy Information Administration, Coal Data: A Reference, 1986).

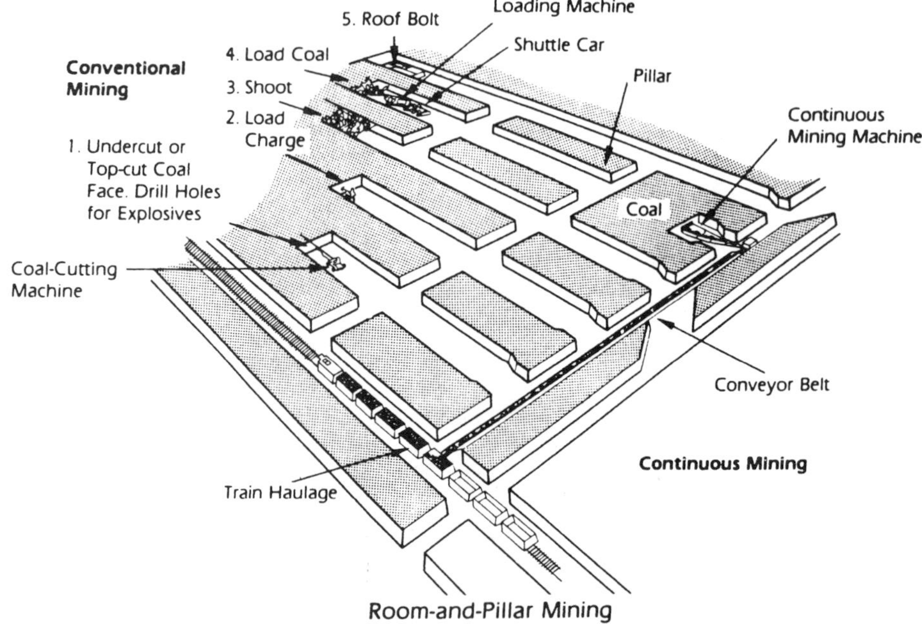

Figure 25 Schematic of room and pillar mining operation (Energy Information Administration, Coal Data: A Reference, 1986).

seam as the wall (the "long wall"—100—200 m; 300—600 ft) of coal between the two parallel tunnels is removed. A variation on this, called short-wall mining, uses a continuous mining machine to shear coal away from a shorter span (50 m; 150 ft) of coal (Speight, 1983).

The horizontal method is common in Europe where many mines are steeply inclined or severely faulted. Drift access is used but then shafts are constructed *up* into the seams where coal is removed by long-wall equipment (DOE Energy Information Administration, 1986).

Another mining method, common in the USSR and China, uses low-pressure streams of water to wash blasted coal out of the mine. In some instances, blasting is eliminated by the use of high-pressure streams of water (DOE Energy Information Administration, 1986).

Surface mines are used to remove coal buried less than 100 m (300 ft) deep. The economic feasibility of removing extensive overburden depends on the quality of the coal and the need. Generally the acceptable ratio of overburden to coal thickness is 30:1 for mid- and high-rank coals, 20:1 for lignites (Survey of Energy Resources, 1974). Table 28 shows the overburden ratio for U.S. surface mines

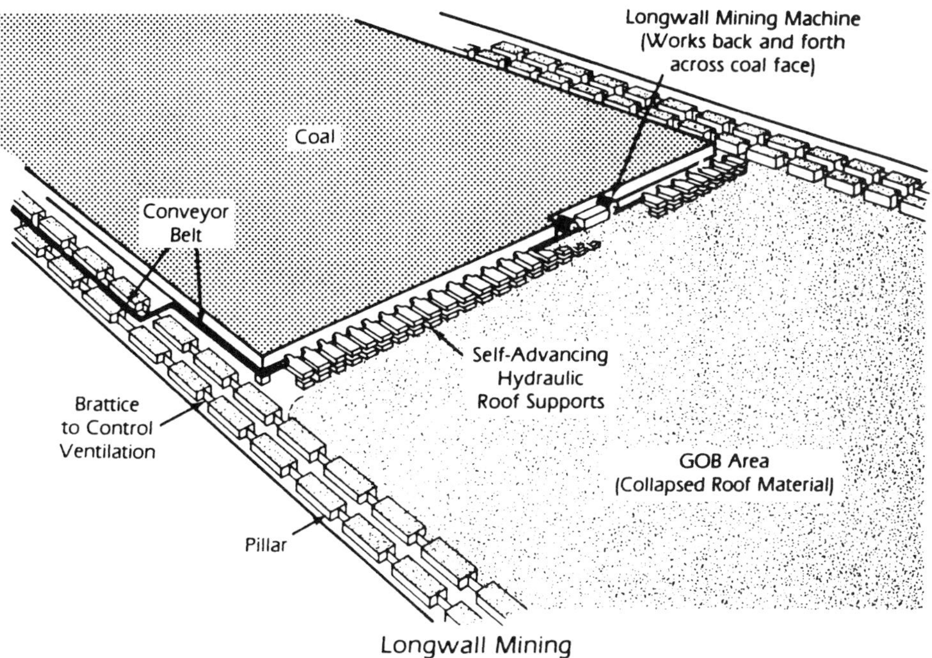

Figure 26 Schematic of longwall mining operations (Energy Information Administration, Coal Data: A Reference, 1986).

in 1978. Only two states, Alabama and Arkansas, mine coals with an overburden ratio >30. Eleven of the 24 states listed access coal with an overburden ratio of 10 or less.

Three methods of surface mining are area (or strip), contour, and open pit. In area mining the overburden is loosened by blasting and is removed by power shovels or draglines (Figure 24). Modern shovels can hold as much as 200 tons per scoop. Contour surface mining is used in hilly regions where power shovels can move around a hillside while remaining at about the same elevation (Figure 24). Open pit mining is only used in areas where the seams are very thick. In one variation of this method, called benching, a number of terraces or stages are established on which coal is mined simultaneously (Figure 27).

One other method, auger mining, finds application in both underground and surface mining. It involves boring into the seam and removing the coal along the auger spiral where it is collected. Recovery of as much as 90% of the coal is possible, labor and capital costs are

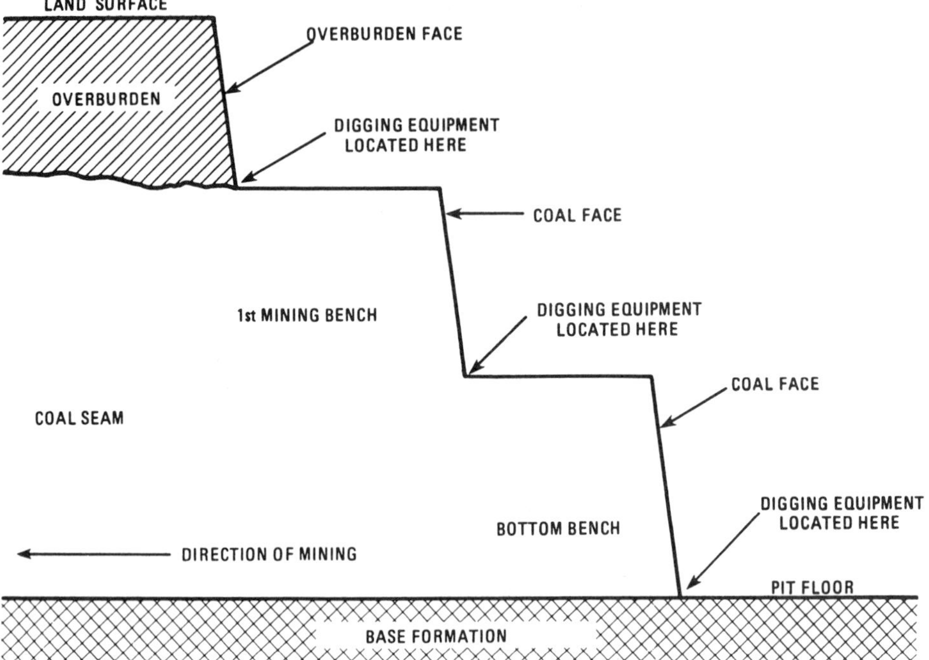

Figure 27 Schematic of mining by the benching method (J. Speight, 1983. Reprinted with permission of the publisher).

are low, and neither blasting, roof support, nor removal of overburden is necessary.

In the United States, little western coal is produced from underground mines. Although some seams are buried sufficiently deep to consider underground methods, the overburden is generally too weak to support a roof. It is therefore loose enough to be removed economically. In the East, on the other hand, seams are often deeper, usually thinner than in the West and covered with thick, rocky overburden which is both costly to remove and costly to recover.

V. BENEFICIATION

Beneficiation and clean coal technology is an area of major concern to coal producers and coal consumers worldwide. As environmental constraints on effluents grow more rigid, and as studies on the mobility and transport of fugitive materials becomes more sophisticated, it is imperative that technology and methodology for cleaning coal keep pace.

The main objective of coal cleaning is to give a uniform quality to freshly mined coal, most especially with regard to its sulfur content. Of the three chemical forms of sulfur in coal--pyrite, sulfate, and organic sulfur--some estimates place the organically bound sulfur as high as 70% in some coals, and 30–70% in most coals (Loftness, 1984). Because it is chemically bound to the coal matrix as thiols (R-SH), thioethers (R-S-R), and/or disulfides (R-S-S-R), physical methods are ineffective for removing it. Sulfate content, on the other hand, is primarily an artificial ingredient, i.e., it arises from the oxidation of coal subsequent to mining. Sulfate can be isolated in aqueous media and may be relatively easily washed out of coal. Inorganic forms of sulfur are believed to be most frequently associated with iron, and although pyrite and marcasite (FeS_2) are the most often encountered mineral forms of iron sulfide, at least eight other iron–sulfur minerals have been identified (Liu and Lin, 1976):

Troilite	FeS
Hexagonal pyrrhotite	Fe_7S_{11}
Monoclinic pyrrhotite	Fe_7S_8
Smythite	Fe_9S_{11}
Greyite	Fe_3S_4
Gamma iron sulfide	Fe_2S_3
–	Fe_9S_{10}
–	$Fe_{11}S_{12}$

Pyritic sulfur occurs either as discrete particles or as an intimately ingrained entity within the bulk of the coal structure. Pyrite has a specific gravity of nearly 5, more than three times the specific gravity of coal (Table 29). Most other impurities (not necessarily bound to the coal matrix) are also heavier than coal. This makes it possible to use physical washing methods to produce a cleaned coal (Figure 28).

A. Physical Methods

The first step in any washing process–crushing–is important to the overall efficiency of the process. Crushing separates gross impurities on the basis of grindability. Screens are used to control fines formation, and the crushing devices are controlled to deliver a desired top size of coal. With or without subsequent washing, sized coal may be

Table 29 Approximate Weights of Unbroken[a] (Solid) Coal in the Ground

Coal rank	Lb/ft^3	G/cm^3	Tons/acre-ft^2	Tons/mi^2-ft^2
Anthracite	91.7	1.47	2000	1,280,000
Bituminous	82.4	1.32	1800	1,150,000
Subbituminous	81.1	1.32	1770	1,130,000
Lignite	80.5	1.39	1750	1,120,000

[a]The weight of broken coal varies with the size of the coal. In general, a ft^3 of broken bituminous coal weighs 47--52 lb and a ft^3 of anthracite weighs 52–56 lb. A ton of broken coal occupies approximately 40 ft^3.

Source: Geological Survey, U.S. Department of the Interior, Bulletin 1412, Coal Resources of the United States, January 1, 1974, modified.

blended to obtain a particular size distribution or other average character required by a particular consumer. In the United States, plants process from 200 to 20,000 tons of coal per day, removing from 60 to 6000 tons of refuse per day (Speight, 1983).

Both recovery and control of coal fines is critical for three reasons: (1) coal mining itself generates a substantial quantity of fines which represent too much lost revenue if they are discarded as waste; (2) during cleaning, in spite of screening and control of crushing devices, as much as 50% of a batch fed to a crusher can end up as pieces of less than 1/4 in. for a 4-in. top size. In fact, formation of fines is most acute with prime coking coals; high-volatile coals are less susceptible to extensive fines formation; (3) although a large accumulation of fines may not be desirable from the point of view of handling, efficient physical coal cleaning depends directly on the liberation of impurities from the bulk coal. This requires, therefore, that the coal be crushed into the smallest pieces that can be tolerated (Tsai, 1982).

The washing step is based on differences in specific gravity between the coal and the impurities. When it is suspended in a medium whose specific gravity is adjusted to a desired value (somewhat greater than that of the coal), heavier impurities sink whereas the coal floats on the surface. There are several varieties of gravity devices in use for physical cleaning (Tsai, 1982; Loftness, 1984):

1. Static and agitated dense medium
2. Cyclones

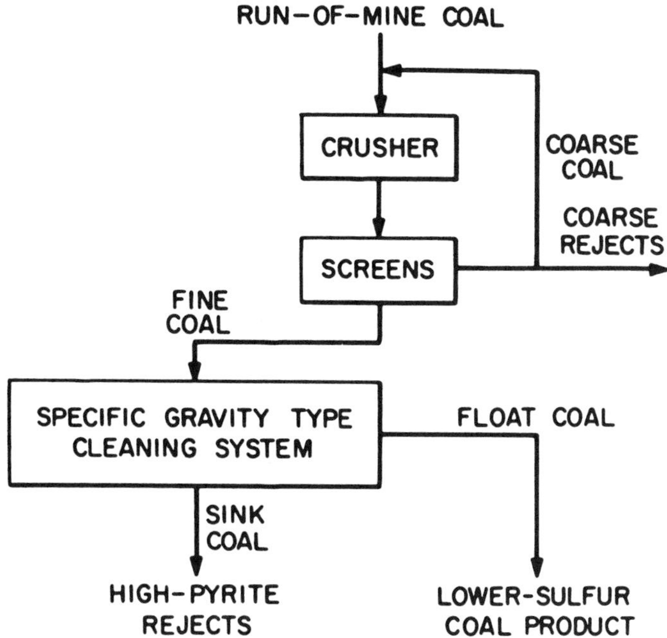

Figure 28 Simplified flow chart of physical coal-cleaning processes (U.S. Atomic Energy Commission, 1974).

3. Hydroclones
4. Froth flotation and oil agglomeration
5. Jigs
6. Wet concentration tables
7. Concentric spiral devices
8. Electrokinetic separators

A complete washing sequence may include more than one washing step or more than one washing process, depending on the amount of fines, the quantity of impurities, the efficiency of a process, and the final specified quality (Tsai, 1982).

Because the differences in specific gravity between coal and most impurities are the critical parameter, the amount of impurities of a given specific gravity value is particularly significant. Only material (refuse) with a specific gravity higher than that of the medium will sink. Thus, coals with a small amount of near-gravity material can often be cleaned satisfactorily with one washing, whereas repetitive washings or even several different processes may be necessary to clean a coal containing a high percentage of near-gravity refuse (Tsai, 1982). Near-gravity values are considered to be those within 0.10 unit of the specific gravity of the separation (Tsai, 1982).

The efficiency of recovery of washed coal is determined by comparing the yield of washed coal obtained from a commercial washing device to the yield of coal (of the same ash content) "floated" in a controlled laboratory test (Tsai, 1982).

Two other terms used to compare coal-washing devices are the misplaced material value and the ash error. Misplaced material occurs because of the near-gravity content of the feed. It includes both heavy material that did not sink and light material that did:

$$\text{Total misplaced material} = W_h \times \%R + (r_1 \times \%r)$$

where

W_h = fraction of washed coal which is actually heavier than the medium
R = recovery
r_1 = fraction of refuse which is actually lighter than the medium
r = rejection

The greater the content of near-gravity material in the sample, the more difficult the separation will be (and the higher the misplaced material value). Less than 10% near-gravity material presents only a slightly difficult separation; 15–20% is considered very difficult to separate and more than 25% is considered formidable (Tsai, 1982).

Ash error is the difference between the theoretical ash content of washed coal and the actual ash content realized after cleaning (Tsai, 1982).

Dense- or heavy-medium separation devices may use neat organic liquids, aqueous salt solutions, or solids dispersed in water. The Belknap chloride process and the Otisca process use homogeneous liquids; the Frases and Yancy and the Chance processes use sand suspended in water by agitation and aeration. The agitation and rate of air flow are used to adjust the specific gravity (Figure 29). Suspensions of magnetite (magnetic iron ore) in water are also used commercially. The high specific gravity (5.0) of magnetite permits media with a range of density values from about 1.3 to 2.0 to be used. This versatility is very useful for coals with large amounts of near-gravity impurities (Tsai, 1982). Figure 30 shows a magnetite dense-medium coarse coal flow diagram. For fine coal (<1/4 in.), a cyclone rather than a static device is used. Fine-coal washing also requires

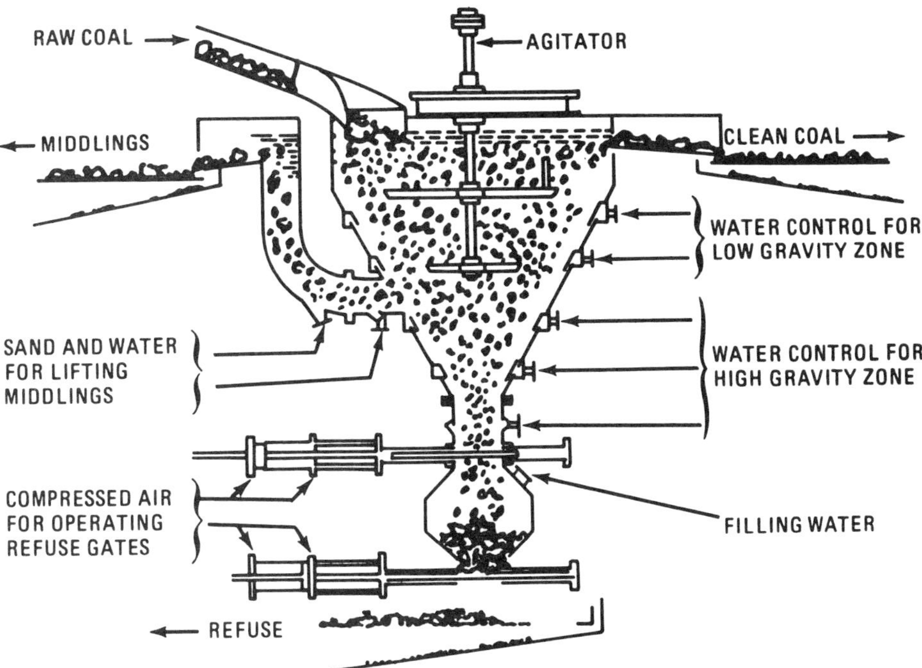

Figure 29 Schematic representation of the Chance sand–flotation coal-cleaning process (J. Speight, 1983. Reprinted with permission of the publisher).

a desliming screen positioned prior to the separator (Deurbrouk and Hudy, 1972).

Cyclone washers (Figure 31) rely on the rotational force of the medium moving through the conical unit to cause the heavy material to move to the walls and down to the apex with the descending vortex. The lighter coal rises with the central air flow throughout the vortex finder and is carried out at the top of the unit. Gravitational forces are enhanced in a cyclone compared to "static" float-sink baths by a factor of 20 at the inlet and by a factor of as much as 200 at the apex of the unit (Tsai, 1982). This increases the capacity of cyclones relative to other dense media separators and facilitates the cleaning of fines (Leonard and Mitchell, 1968). Also, because the centifugal force tends to push the medium particles outward and downward as well, the specific gravity of separation near the apex is higher than

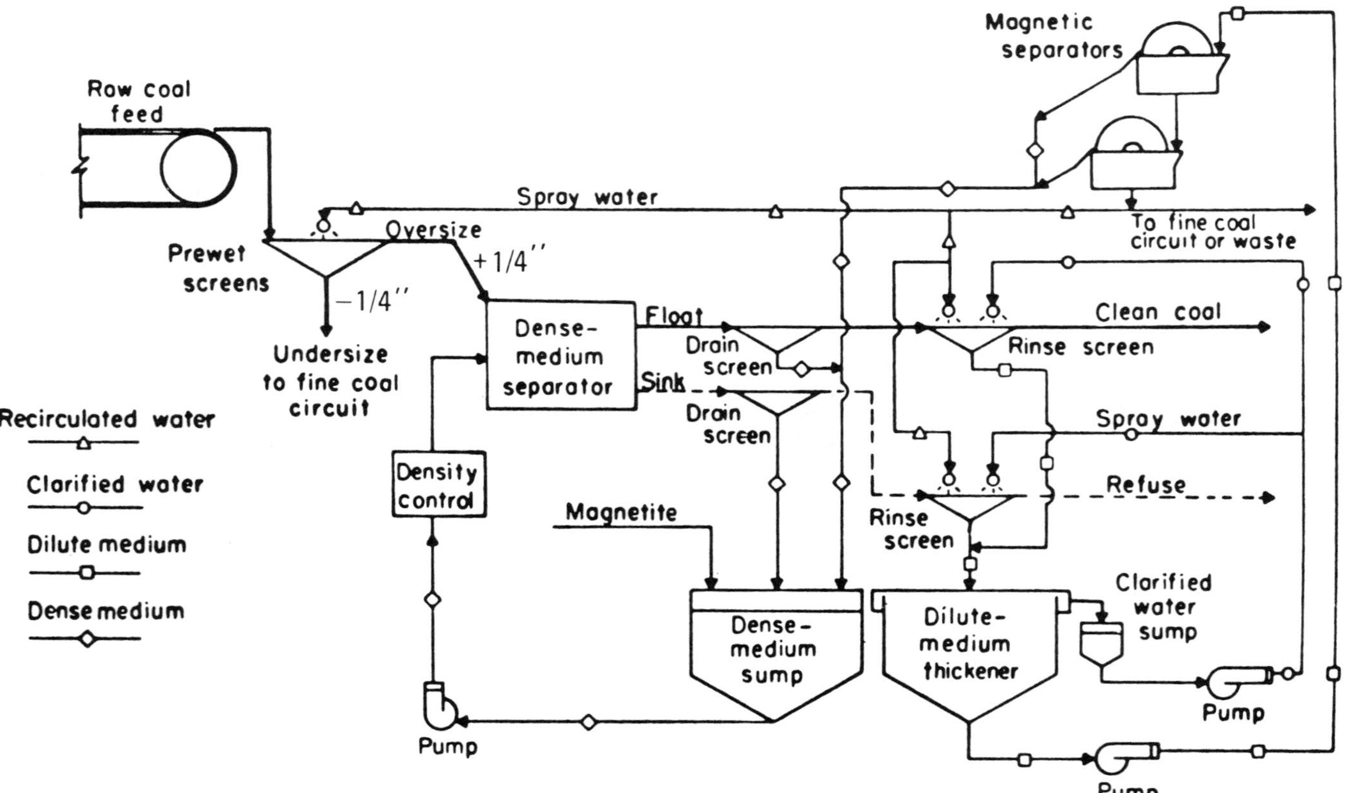

Figure 30 Generalized schematic for a dense-medium coarse coal washer (J. Hudy, Jr., 1968. Reprinted with permission of the publisher).

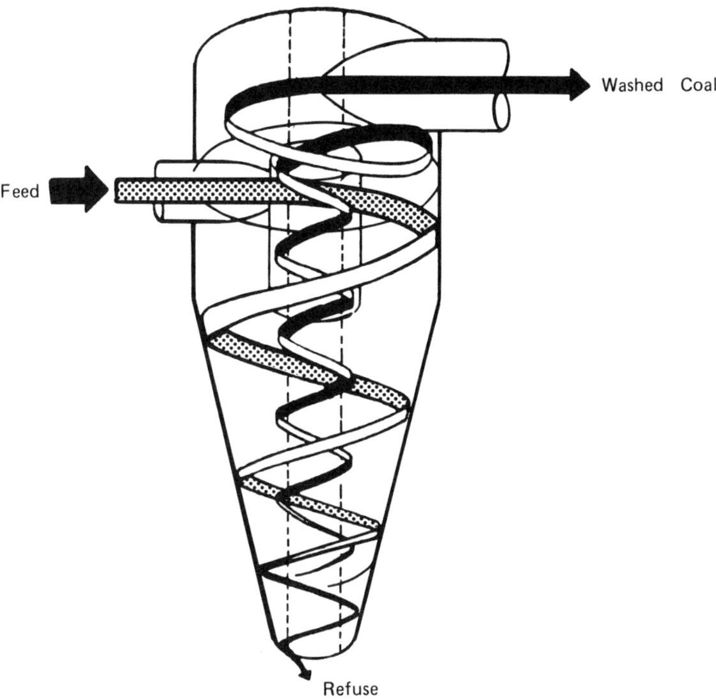

Figure 31 Stream flow pattern in a dense-medium cyclone coal-cleaning device (J. W. Leonard and D. R. Mitchell, 1968. Reprinted with permission of the publisher).

that of the bulk of the media (Leonard and Mitchell, 1968). The recovery efficiency of a cyclone unit has been reported to be greater than 97% (Tsai, 1982).

Figure 32 shows a Dynawhirlpool separator, which, like the conventional cyclone, uses centrifugal force to bring about separation. The cylindrical unit is inclined at an angle of 20–25° for operation. The coal is fed into the top end of the tube; the medium enters tangentially near the opposite, float discharge end. In contrast to the cyclone, the medium rises in this unit and creates an open vortex. This causes the coal to move downward while the heavy refuse moves to the outer wall and rises to the discharge outlet. Recovery and sharpness of separation are as high as with a cyclone (96–99%) (Tsai, 1982).

Froth flotation and oil agglomeration are often used to clean very dirty raw coal. The coal is crushed <0.5 mm (<0.02 in.) and fed into the separator, which contains an oil-water emulsion. For froth

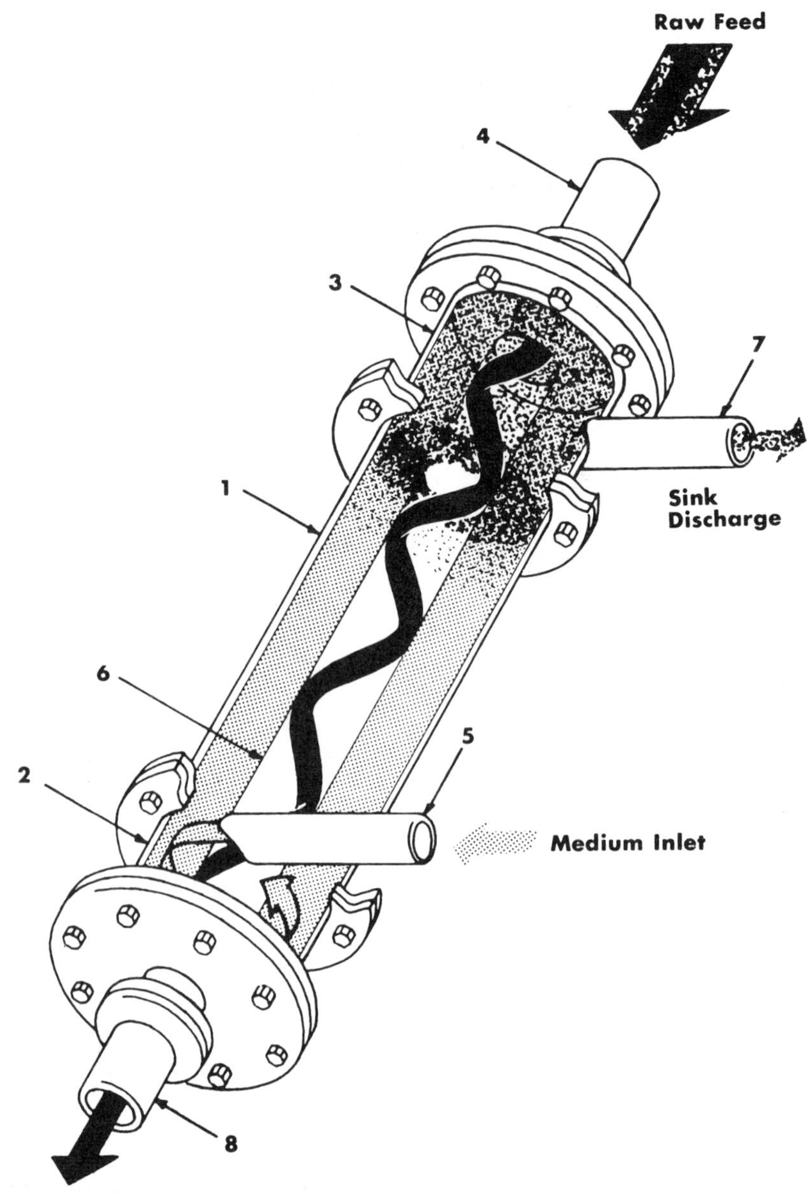

Figure 32 Schematic of the Dynawhirlpool coal-cleaning device (Bulletin, Mountain State Engineers, Tucson, Arizona).

flotation air is bubbled through the slurry to facilitate separation. The coal particles are preferentially coated with the organic oil and, being lighter, are carried to the surface by the air. The noncoal refuse is wetted by the water and sinks. The oil-coal foam or froth is scraped off and the clean coal is washed free of solvent (Speight, 1983). In the oil agglomeration method, no froth is created. Rather, the oil-coated fraction is separated from the remainder with screens (Tsai, 1982).

The flotation properties of coal have been studied by Sun (1954) among others. He determined that a floatability factor for coal can be calculated using the expression:

$$F = \frac{xH}{2.08} + \frac{yC}{12} - \frac{H}{2.08} + \frac{0.4S}{32.06} - \frac{zM}{18} - \frac{3.40}{16} - \frac{N}{14}$$

F is the calculated floatability. The letters H, C, S, O, and N represent the weight percentage of hydrogen, carbon, sulfur, oxygen, and nitrogen in the sample. The letter M is the as-received moisture value. The factors x, y, and z reflect the effects of ash (x), advanced coalification or graphitized carbon content (y), and the action of water on an already wet sample (smaller effect) or on a quite dry sample (larger effect) (z):

Ash (%) as received)	x	%H (maf)	y	%M	z
0–8.9	3.5	0.08–0.28	1.2	0–14.1	4
9–13.0	3.0	0.29–1	0.8	>14.1	3
>13.0	2.5	>1	0.6		

The expression H/2.08 is derived from an average hydrocarbon unit, $CH_{2.08}$. This same term is subtracted from yC/12 because graphitization is accompanied by loss of hydrogen. Sulfur, moisture, oxygen, and nitrogen are used on a number-atom basis (analyzed percent divided by mass number). A floatability index value compares F_c for a coal sample to the floatability of a Ceylon graphite (nonflotable) standard, F_{co}:FI = $F_c/F_{co} \times 100$. A small index value indicates large amounts of nonfloatable material. Coal floatability increases with rank from lignite to low-medium volatile bituminous, then decreases to anthracite. High oxygen and high moisture content effect the floatability of the low-rank coals, while increasing graphitic structure (nonhydrocarbon) decreases the floatability of high-rank coals (Tsai, 1982). Figure 33 shows the relationship between the amount of oil required to float various coals and coal-related material and the floatability index. Medium and low-volatile bituminous coals

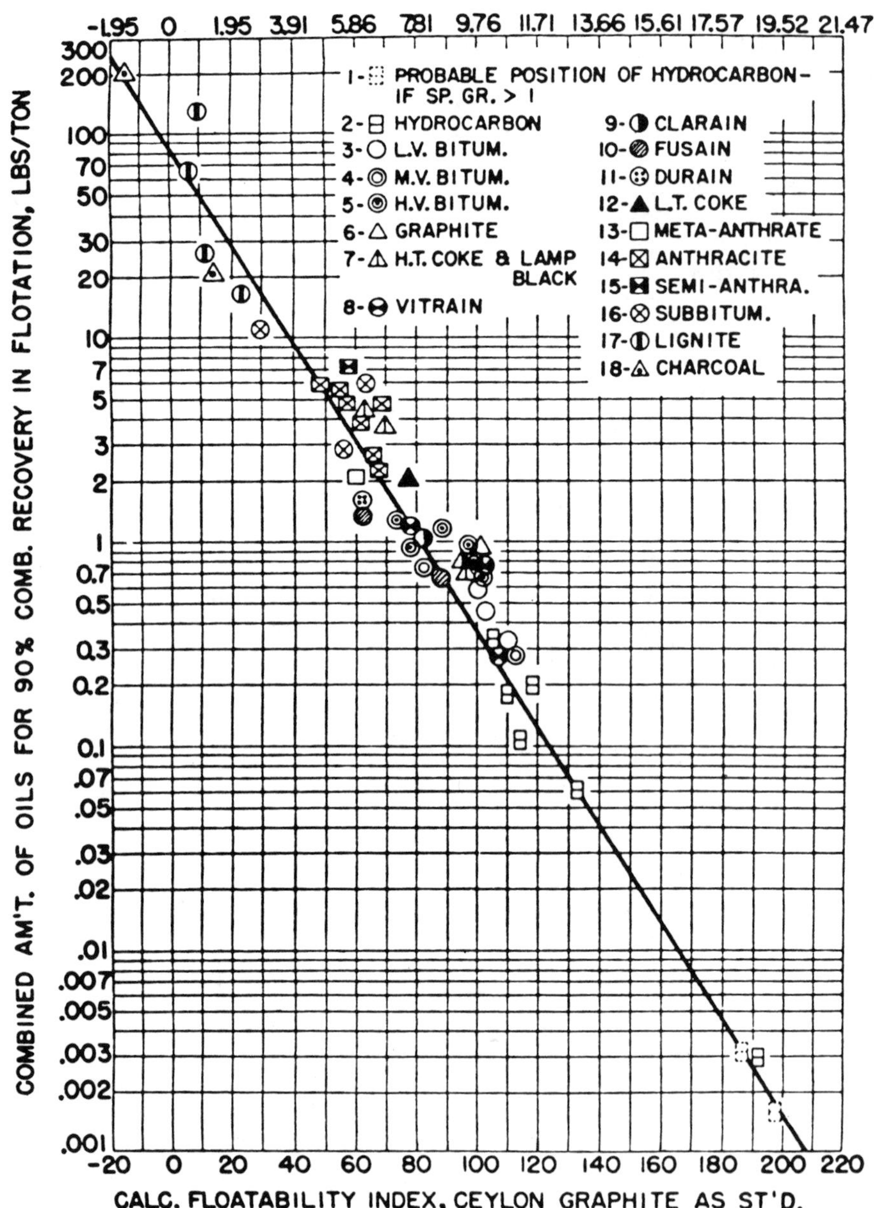

Figure 33 Correlation between the actual floatability and the floatability index for coals, cokes, and hydrocarbon minerals (J. W. Leonard and D. R. Mitchell, 1968. Reprinted with permission of the publisher).

have high floatability index values and require correspondingly less oil to achieve flotation.

A hydrocyclone differs from other dense-medium separators in that it does not use a suspension to achieve a desired specific gravity. Rather it achieves separation according to the dimensions of the discharge orifices. The specific gravity medium is derived from the medium-to-high specific gravity impurities in the raw coal feed. This is termed an autogenous dense medium (Tsai, 1982). The cone angle is 120° in a hydrocyclone (versus 20° in a dense-medium cyclone) and it has a larger vortex finder. Although there has been some controversy about precisely how the separation is achieved (Visman, 1960; Fontein, 1962), it has been reported that increasing the diameter of the vortex finder increases the yield of washed coal but it also results in increased ash content in the washed fraction; conversely, increasing the diameter of the underflow orifice decreases both the yield and the ash content of the recovered coal (Tsai, 1982).

Jigs and wet concentration tables are examples of hydraulic separators. A jig (Figure 34) supports a bed of coal on a perforated base. Water flows across the platform in alternating directions.

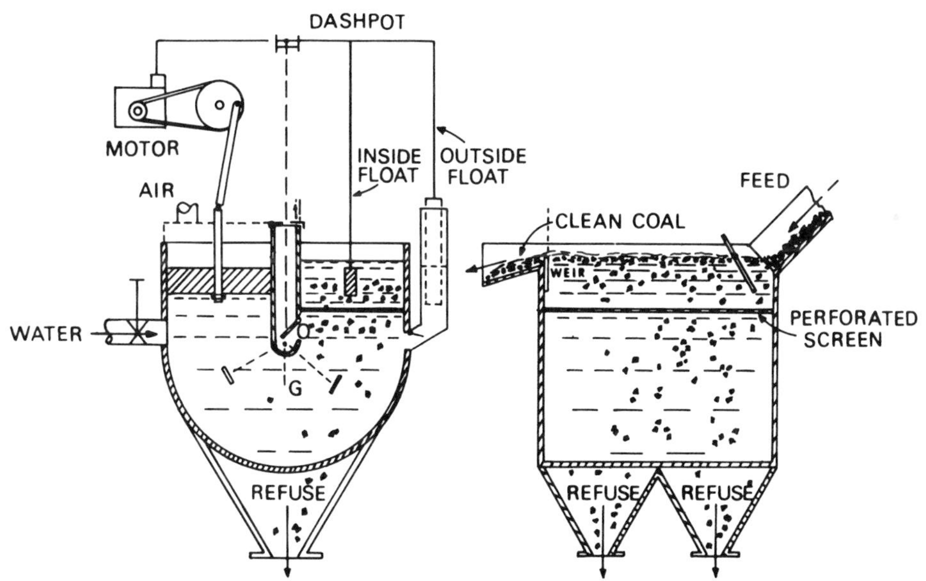

Figure 34 Schematic view of a jig coal-cleaning device (G. A. Vissac, 1955. Reprinted with permission of the publisher).

The water flow causes the particles to segregate as the bed expands and contracts under the pressure of the water. Higher specific gravity particles settle below lower specific gravity, cleaner coal. The actual separation is achieved by removing as much of the stratified material as desired, leaving the refuse behind (Tsai, 1982). The ability to separate material sharply according to specific gravity is much lower using jigs than it is for heavy-medium devices. The gravity of separation is usually in the range 1.94--1.30. However, a wide range of sizes can be cleaned (from 3 mesh to 8 in.) (Agarwal et al., 1976; Tsai, 1982). The capacity of jigs is also very high, about 2 tons/hr/ft^2 of bed. The rate at which coal is fed onto the jigs has been reported to be as high as 700 tons/hr (Tsai, 1982).

Concentration tables can accommodate small coal (48--200 mesh), but most often they are used with 3/8-in. to 0 coal (Deubrouk and Palowitch, 1963). A concentration table washes coal by allowing a stream of coal in water to flow over a series of riffles which are shaken rapidly. Thus, particle size and shape as well as specific gravity are involved in this process. The table is flat but rhomboid in shape, and is positioned with a slight slope to the short axis. Clean coal collects along the long side and refuse moves diagonally to the shorter, refuse side (Tsai, 1982). In one study (Deubrouk and Palowitch, 1963) involving several coals and a wide range of particle sizes, recovery ranged from 97.3 to 99.3%. However, it did decrease to about 91% for 100 × 200 mesh coal. Tsai (1982) tabulated data to compare four fine-coal washers (Table 30). Although the magnetite heavy-medium cyclone cannot use coal smaller than 28 mesh, it showed the best performance overall: 97–99% recovery and an ash error of only 0.3% on 28-mesh coal. The concentration table had a 98% recovery efficiency and could handle coal as small as 200 mesh. The jig processed coal to about 48 mesh with a high recovery efficiency. Jigs also offered the highest capacity (as much as 10 times that of a concentration table). The hydrocyclone showed the lowest recovery efficiency and the lowest ability to yield a sharp separation at the desired specific gravity values (Hudy, 1968).

A Humphrey's spiral concentrator (Figure 35) consists of a semicircular conduit containing three 120° sections per spiral turn. As coal moves down the spirals, centrifugal forces and channeling bring about separation. Heavier particles move toward the inner edge of the spirals and are allowed to fall through refuse ports. Lighter particles move outward and are removed at the bottom of the stream. A splitter may be inserted near the bottom to separate the clean coal into two fractions, outer and inner (heavier) (Figure 36). Recovery efficiency is about 95% (85% for the outer clean fraction). Actual recovery is higher with higher feed rates (Zeilinger and Deubrouk, 1976).

Electrokinetic cleaning of coal is based on the differences in dielectric or conductivity properties of coal and mineral matter. In one device of this type, coal is fed in a stream onto an electrically grounded rotating disk. As the rotor turns, coal passes under an active electrode which transfers a charge to the mineral matter but not to the coal. Attaining a potential equal to that of the rotor, the minerals (especially graphite) are repelled and fall away under the force of gravity. Coal is dielectric and adheres to the rotor until it is scraped off by a blade. In one study (Abel et al., 1973), a 52% reduction in pyrite was attained using this device. This is comparable to about 75% pyrite removal using the float-sink method with media of specific gravity of 1.6.

Maceral separation can also be carried out by physical float-sink methods. Macerals possess characteristic grindability and therefore can be separated by sizing techniques. Exinite contains substantial amounts of waxy components and is the hardest maceral; vitrinite is intermediate and fusinite is the softest maceral. Overall, grindability increases with rank from lignite to midrank coal (low-volatile bituminous), then decreases to anthracite (Tsai, 1982). The specific gravity of macerals also varies with rank. The specific gravity of exinite and micrinite increases throughout the rank series, exinite from 1.0 to 1.28 and micrinite from 1.35 to 1.45. Vitrinite specific gravity decreases from 1.43 to 1.27 as fixed carbon content (maf) increases from 70 to 87%; it then increases to 1.35 as the fixed carbon content increases from 88 to 91% (maf). Fusinite is the heaviest maceral. Its specific gravity is 1.5 or higher. Most minerals have specific gravity values greater than 2.77 (Tsai, 1982).

B. Chemical Methods

Several methods of chemical cleaning have also been developed. One method is the Meyer process (TRW) (Figure 37). In this system the crushed coal is contacted with a warm (90–130°C; 200–270°F) solution of iron(III) sulfate. Air or oxygen is introduced into the reactor for 4–6 hr and promotes reactions between pyrite (FeS_2) and iron(III) sulfate. This results in the formation of elemental sulfur, iron(II) sulfate, and sulfuric acid. The solution of iron(II) sulfate and sulfuric acid is recycled; elemental sulfur is removed by extracting it into a petroleum distillate solvent. The overall reaction is

$$FeS_2 + 2.4O = 0.6FeSO_4 + 0.2Fe_2(SO_4)_3 + 0.8S$$

It has been reported that 83–98% pyritic sulfur removal is achieved with no coal loss (Tsai, 1982).

Table 30 Performance Data of Fine Coal Washers

Process	Heavy and medium cyclone		Hydrocyclone	Concentrating table		Feldspar jig
Coal particle size	1/2 in. × 28 mesh		1.14 in. × 200 mesh	3/8 in. × 200 mesh		1/4 in. × 48 mesh
Ash, %						
Feed	19.2	25.5	17.5	10.8	10.8	23.3
Clean coal	6.0	3.8	7.0	4.7	3.7	11.0
Refuse	55.0	68.6	50.3	50.6	61.8	66.4
Actual recovery, %	73.0	66.5	75.8	86.8	87.8	77.7
Theoretical recovery, %	73.2	68.6	84.7	89.2	89.4	79.8
Efficiency, %	99.7	96.9	89.5	97.4	98.2	97.4
Ash error	0.1	0.3	—	0.6	0.4	—
Float in refuse, % of prod.	4.1	8.0	—	20.4	14.8	—
Sink in clean coal, % of prod.	1.6	1.4	—	2.1	1.1	—
Total misplaced material, % of feed	2.2	3.6	—	4.5	2.8	—

Near-gravity (±0.10) material	8.7	12.6	—	7.8	5.0	—
Specific gravity of separation	1.53	1.45	1.54	1.52	1.52	1.55
Probable error, Sp. gr.	0.030	0.029	0.12	0.08	0.074	—
Imperfection	0.020	0.020	—	0.154	0.142	—
Error area	18	25	78	63	52	77
Distribution, % to washed coal						
sp. gr. Fraction						
Under 1.30	100.0		93.1	99.1	99.5	99.3
1.30–1.35	99.8		86.0	96.4	98.2	98.4
1.35–1.40	99.4			90.2	95.1	
1.40–1.45	98.7		68.4	80.4	85.9	87.6
1.45–1.50	92.5			63.9	66.2	
1.50–1.60	13.9		47.4	41.5	39.5	51.1
1.60–1.70	2.8		25/1	22.3	16.2	28.2
1.70–1.80	1.0		13.7	17.4	6.3	23.1
1.80–Sink	0.7		5.2	8.7	3.0	2.9

Source: S. Tsai, 1982. Reprinted with permission of the publisher.

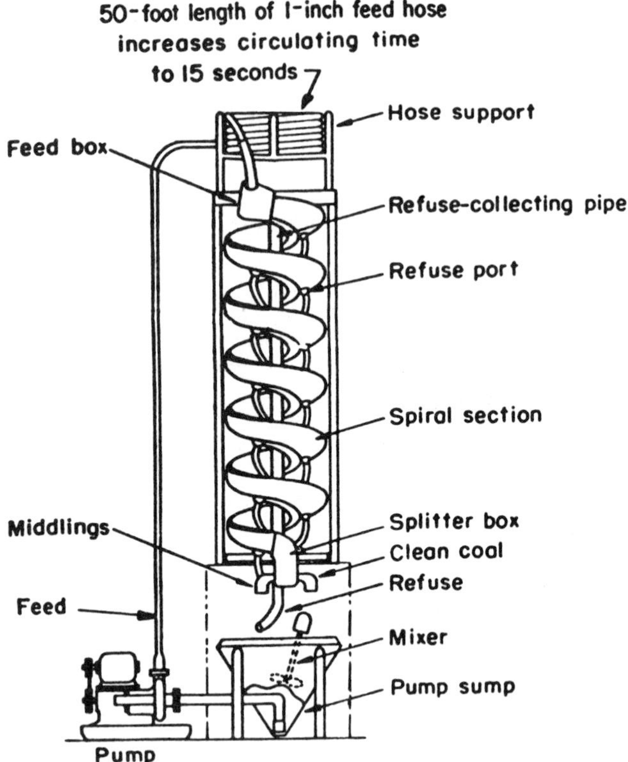

Figure 35 Humphrey's spiral concentrator (A. M. Gaudin, 1939. Reprinted with permission of the publisher).

Other successful processes for chemical cleaning involve alkali leaching and oxydesulfurization at elevated temperatures and pressures. Alkali leaching has been implemented in several forms. Masciantonia (1965) described the use of molten alkali for 30 min at 250°C (480°F). Total sulfur was reduced by 37%. A higher removal was realized by using higher temperatures, but this caused the coals to become fluid (Tsai, 1982). The use of aqueous alkali has been developed separately by the U.S. Bureau of Mines (USBM) and by Battelle Laboratory. The USBM system operates at 225°C (430°F) where no effects on the caking properties of the coal are observed. Most pyrite is removed and an acid wash is used to remove other minerals. Some organic sulfur is reported to be removed if the temperature is raised to 300°C (570°F), but at this temperature the caking properties of the coal are lost (Tsai, 1982).

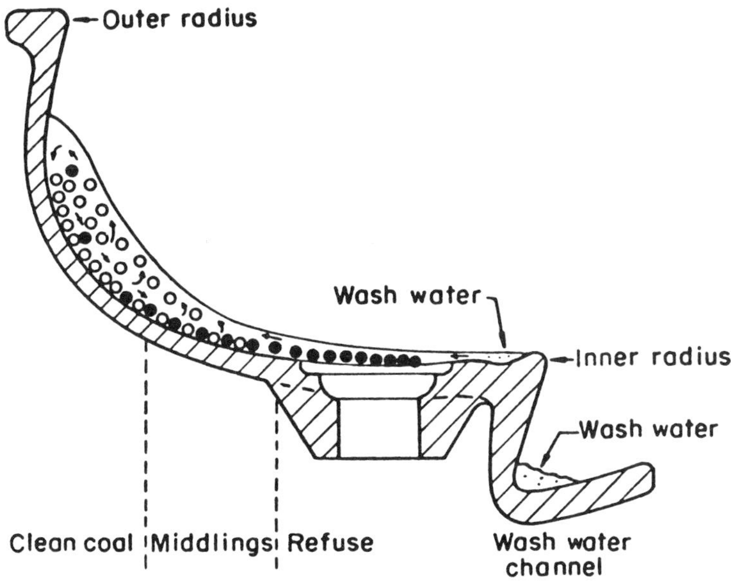

Figure 36 Cross-section of a spiral cleaning stream (A. M. Gaudin, 1939. Reprinted with the permission of the publisher).

The Battelle hydrothermal process (Figure 38) does not use any acid to wash the alkali-leached coal. The leachant is a mixture of NaOH and $Ca(OH)_2$ and the reaction is carried out at 225–270°C (430–525°F) under 350–2500 psi (2.41–77.2 MPa) pressure. The coal is rendered completely noncaking and nonswelling under these conditions, and about 5% of it is lost by dissolution in the leachant. Total sulfur removal is 50–84%, 24–70% of which is reported to be organic sulfur. Loss of caking and swelling properties, sulfur removal, and removal of some toxic metals as well makes this an attractive process for preparing coals for gasification (Tsai, 1982).

Oxydesulfurization is the name given to several similar processes: the Ledgement oxyleaching process (LOL), the promoted Ledgement process (PLOL), the Pittsburgh Energy Technology Center (PETC) process and the Ames process. Three reactions for sulfur removal are common to each of these processes:

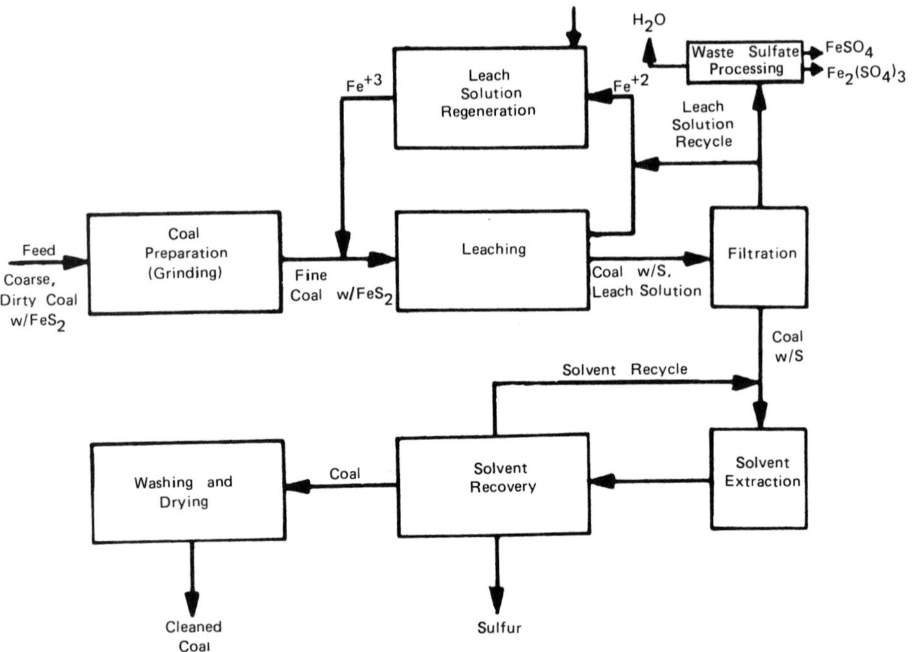

Figure 37 Flow diagram for the Meyers Chemical coal-cleaning process (L. Lorenzi, Jr., 1975).

$$2FeS_2 + 7O_2 + 2H_2O = 2FeSO_4 + 2H_2SO_4$$

$$4FeSO_4 + O_2 + 2H_2SO_4 = 2Fe_2(SO_4)_3 + 2H_2O$$

$$Fe_2(SO_4)_3 + 3H_2O = Fe_2O_3 + 3H_2SO_4$$

Some organic sulfur may also be solubilized, and some carbon loss has been observed. Table 31 shows the similarities and the reported sulfur removal values in these four processes. Because this aspect of coal cleaning is so critical to the successful development of coal utilization, continued efforts in this area are warranted and are of great interest.

Two nonconventional methods for chemically cleaning coal are the chloronolysis method developed at the Jet Propulsion Laboratory (JPL) and the use of microwave radiation. In the JPL method, chlorine gas is allowed to flow through a bed of moist powdered coal at 50--100°C (130–230°F), at atmospheric pressure for 1–2 hr. Pyrite is converted to iron(II) or iron(III) chloride plus some hydrochloric acid and some sulfuric acid:

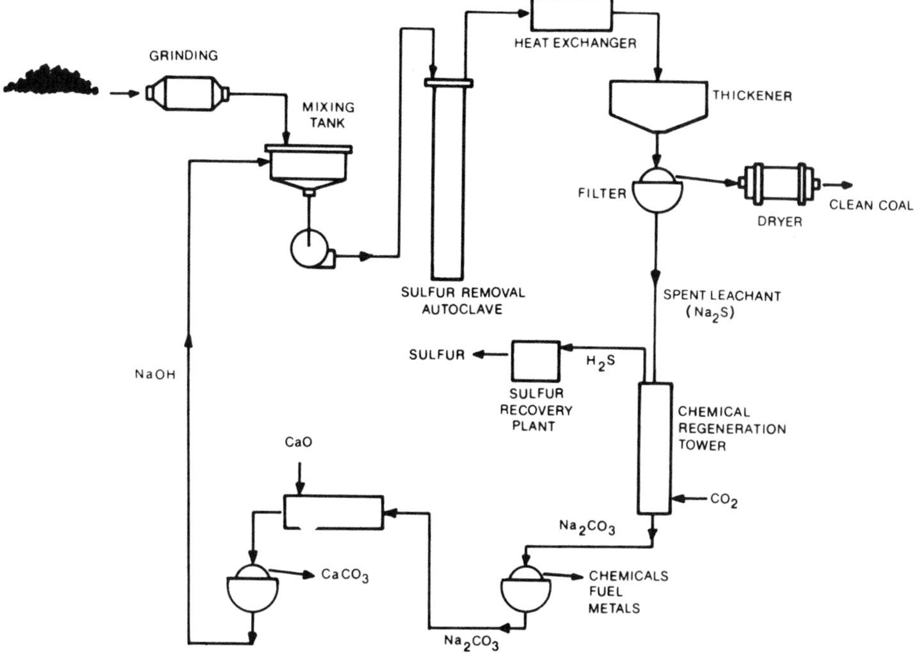

Figure 38 Batelle hydrothermal coal process (E. P. Stainbaugh, 1975. Reprinted with permission of the publisher).

$$FeS_2 + Cl_2 = FeCl_2(FeCl_3) + S_2Cl_2$$

$$S_2Cl_2 + 8H_2O + 5Cl_2 + 12HCl + 2H_2SO_4$$

Organic thioether and disulfide groups are cleaved to chlorosulfides:

$$R\text{-}S\text{-}R' + Cl_2 = RSCl + R'Cl$$

$$R\text{-}S\text{-}S\text{-}R' + Cl_2 = RSCl + R'SCl$$

$$RSCl(R'SCl) + 2Cl_2 + 3H_2O = RSO_3H + 5HCl$$

or

$$RSCl + 3Cl_2 + 4H_2O = RCl + H_2SO_4 + 6HCl$$

Chlorinated coals (RCl) can be reacted with water to liberate HCl and form hydroxy (ROH) species in the coal. The JPL has reported that only 0.06% chlorine is retained in the coal (Tsai, 1982).

Table 31 Summary of Technologies Being Developed to Chemically Remove Sulfur From Coal

Process	Operating pressure (atm)	Temp (°C)	Time (hr)	Condition	Remarks
LOL	10–20, O_2	130	1–2	Acid or alkali	80–90% pyrite removal; no organic S removed in acid, 30–40% in alkali; some C loss
PLOL	20, O_2	120	1	Acid + promotor	100% pyrite removal; up to 35% organic S at 350°C
PETC	30–70, air	180–200	1	–	100% pyrite removal; 45% organic, 20% ash reduction, loss of caking and 10% C loss
AMES	14, O_2	150	1	Alkali ($NaCO_3$)	95% pyrite removal; up to 50% organic S removal

Source: Data compiled from Tsai, 1982, Chap. 8.

Microwave radiation (2.4–8.3 GHz) applied for 1–3 min to an aqueous alkali coal slurry has been reported to remove as much as 95% of the pyritic sulfur and 60% of the organic sulfur in coal. The special feature of this technique is its speed. Also, although the coal may be heated to 250–300°C (480–570°F), no loss of heating value is observed, probably because of the rapid reaction time (Tsai, 1982).

While these methods are only in the preliminary developmental stages, there are no strong mitigating reasons why they may not prove to be widely applicable in the near future.

VI. TRANSPORTATION

Railroads have traditionally been the principal transporter of coal in all countries, although in Europe and the United States inland waterways are used extensively as well. In the United States more than 60% of the coal shipped to domestic markets and to ports in 1985 was carried by rail. In the West, more coal (68%) was shipped by rail than in the eastern (56%) or interior (48.5%) regions (Figure 39).

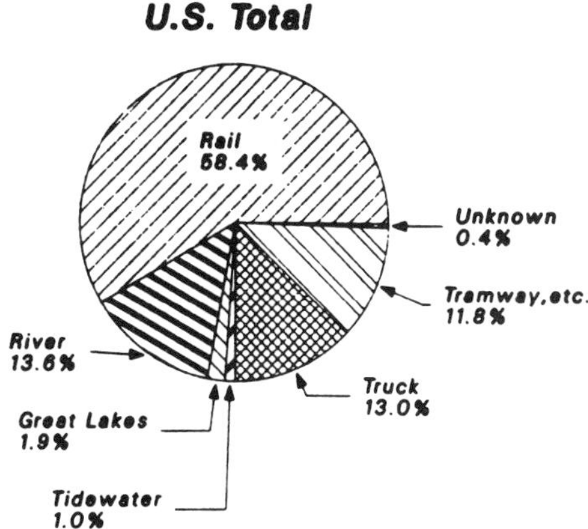

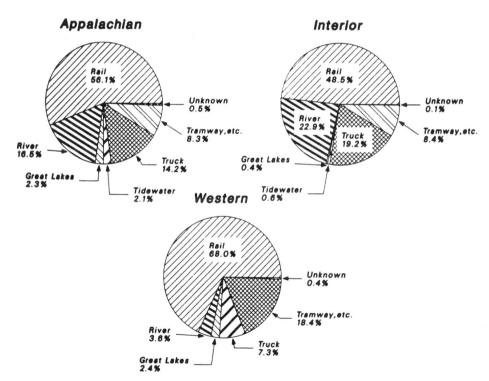

Figure 39 Domestic distribution of U.S. coal by origin and method of transportation: January–December 1985 (Energy Information Administration, Coal Distribution January–December 1985).

A generation ago coal was consumed very close to where it was mined, and it appears that a return to a similar situation is occurring in an effort to reduce costs incurred by transportation over large distances. Power companies and chemical industries alike are finding it more economical to transport their finished product to distant markets than to move the raw coal over the same distances.

Another development aimed at reducing the cost of transportation of raw coal is the coal slurry (oil or water) pipeline system. This may become an extremely attractive method as coal utilization technologies develop which can use the slurry directly.

18

Terminology and Classification

I	Nomenclature and Terminology	543
	A. Macerals	545
II	Macroscopic Classification	555
	A. Theissen–Bureau of Mines System	557
	B. International System	558
	C. U.S. Geological Survey System	559
III	Classification Based on Chemical Analysis	561
	A. American Society for Testing and Materials	561
	B. National Coal Board	561
	C. International System	563
IV	Correlation of Classification Systems	563
V	Classification of Brown Coal	568

I. NOMENCLATURE AND TERMINOLOGY

Establishing the rank to which a particular coal belongs is of very limited practical value. This is so primarily because rank does not correlate with basic structural features in coal. Rank alone does not explain structurally how (or if) the several bituminous coals differ, or how structurally each rank of coal differs from the others. On the other hand, empirical evidence has taught us that even two coals of the same rank may not behave similarly under identical reaction conditions. The reasons for this variability must surely lie in the fundamental structural differences existing among various coals.

Even casual visual inspection and comparison of coals reveals that a number of physical features can be used to separate coals into groups. This is in fact how classification of coals originated. Because people were making these observations simultaneously in different places, with different coals, more than one scheme for classifying coal came into use.

Unfortunately, even the basic geologic terminology used in conjunction with organic sedimentary materials is often ambiguous (Hessley et al., 1986). For a discussion of coal, two related but essentially different materials, bitumen and kerogen, should be distinguished clearly. Bitumen is often used generically for the organic constituents of rocks. Krauskopf (1979) asserts, however, that, strictly speaking, bitumen is hydrocarbon material (solid or liquid, and generally a mixture of compounds) that is "largely soluble" in carbon disulfide (CS_2). Bouska (1981) is less rigid in defining this term and does not identify a specific organic solvent in which bitumen must be soluble. Bitumen, then, includes petroleum, asphalt (relatively low-boiling mixtures of cyclic paraffins), and ozokerite (higher molecular weight paraffins) (Krauskopf, 1979). Solubility in organic solvent(s) should of course exclude insoluble organic matter found in sedimentary rock. While Bouska (1981) separates such insoluble organic matter, called kerogen, from bitumen, Krauskopf (1979) identifies kerogen as a solid bitumen. This contradiction can be clarified by noting that the American Geological Institute (AGI) defines bitumen as the organic material soluble in CS_2 and kerogen as the material in oil shales which yields oils upon destructive distillation. This distinction also permits coal to be defined as a largely organic material, not appreciably soluble in organic solvents, containing little or no "free" hydrocarbon material, but releasing both aliphatic and aromatic, saturated and olefinic hydrocarbons upon distillation (Krauskopf, 1979).

It follows from this that the rock mass of coal is a (variable) combination of carbonaceous and noncarbonaceous material. The carbonaceous, combustible fraction is greater than 50% (by weight; 70% by volume) of the rock mass and is called the maceral material in coal. The remainder—the noncarbonaceous, noncombustible material—is the mineral matter portion of coal.

This maceral-mineral characteristic of coal is one of six features which Spackman (1975) has identified as important "organizational levels" in coal:

1. Elemental
2. Molecular

3. Phyteral
4. Maceral-mineral
5. Lithotype
6. Lithobody

The elemental level is revealed at least in part by the rank of a particular coal as discussed earlier. Determination of the precise elemental composition is prescribed in the section addressing the analysis of coal. Characterizing coal at the molecular level is the task that still largely eludes coal scientists in spite of having been the focus of intense effort for more than 30 years (Speight, 1983). From paleobotanical studies of the phyteral aspects of coal, from the results of countless modeling studies, and from an increasing body of data from sophisticated analytical techniques, coal science still has only a speculative concept about the molecular structure of coal.

A. Macerals

Coal petrology and coal petrography have made perhaps the greatest strides in contributing to our understanding of the maceral-mineral character of coal. The term maceral was introduced in 1935 by Marie Stopes, a British paleobotanist. Different macerals were formed from different types of plants, or perhaps became differentiated because coalification processes occurred in different physical environments. Macerals are identified by oil immersion microscopic analysis according to their characteristic reflectance (see below). Some macerals also reveal characteristic differences when a thin section is exposed to transmitted light.

Three macerals were identified originally, but subsequent analyses and more sophisticated techniques have made it possible to refine the categories further. As a result, the three main macerals are termed maceral groups: vitrinite, exinite, and inertinite. The subgroups are now called macerals: collinite, telinite (vitrinites); alginite, exinite, resinite(exinites); macrinite, micrinite, sclerotinite, semifusinite and fusinite (inertinites) (Table 1). Spackman (1975) includes a maceral group called pseudovitrinite. It can be distinguished from vitrinite in that it exhibits distinct cell structure, visible relief, and higher reflectance than vitrinite. Sporinite and cutinite are often considered to be members of the exinite group. The photomicrograph (Figure 1) shows how macerals may appear when light is reflected from the sample surface and is observed through a microscope. It also illustrates that they can be distinguished by a trained specialist.

Table 1 Subdivision of the Three Maceral Groups and General Appearance of the Macerals

Group	Subdivision	Qualities
Vitrinite[a]	Collinite Telinite	Originates from humification and subsequent metamorphosis of cell wall materials from wood or cortex tissue. Transluscent dark or light orange in transmitted light, dark to light gray in reflected light.
Inertinite	Macrinite	A totally structureless material probably evolved from humic mud and particles of diverse origin; opaque to transmitted light. Reflective and white in incident light. Particles 10−>100 μm in diameter.
	Micrinite	Derived from plant material which was macerated before coalification. Micrinite from lignite appear transluscent yellowish brown to brown in transmitted light and dark gray in incident light but micrinite from higher rank coals is opaque to transmitted light and white in incident light. Particles 1−6 μm in diameter.
	Semifusinite	Intermediate between fusinite and vitrinite with some of the characteristics of both.
	Fusinite[b]	Fossil charcoal; exhibits cell structure; highly friable and hard. Formed by rapid alternation and charring of cell wall material before or soon after sedimentation. Opaque; polished faces highly reflective; white in vertically incident light.
	Sclerotinite	Fossil bodies of fungal sclerotia; opaque, highly reflective.
Exinite (or liptinite)	Resinite	High hydrogen content; formed from resinous material secreted by the plants; transluscent dark or light orange in

(continued)

Table 1 (Cont.)

Group	Subdivision	Qualities
[Exinite (or liptinite)]		transmitted light; dark to light gray in reflected light.
	Exinite	Hydrogen-rich; made up of fossil spores, pollens, cuticles (leaf surface materials), and their excretions. Translucent yellow with low reflectivity.
	Alginite	Fossil algal bodies making up boghead coal. Light yellow in transmitted light, dark in reflected light.

[a]Pseudovitrinite has the characteristics of vitrinite but has a higher reflectance.
[b]Although there is some evidence for the formation of fusain during ancient forest fires, microscopic evidence (such as the preservation of cellular forms) does not logically support a forest fire origin. It is therefore quite conceivable that fusain is actually of dual origin which would indeed be in keeping with the apparently conflicting evidence.
Source: Spackman, W.: *Proceedings of the NSF Workshop of the Fundamental Organic Chemistry of Coal*, University of Tennessee, Knoxville, July 1975, p. 12.

It is believed that this classification may help clarify technological behavior of coal during processing. For instance, there is evidence that exinite macerals are important in coke and tar formation. Resinite, an exinite maceral, yields as much as 80--90% tar by weight. This is substantially higher than the amount of tar produced with other maceral groups. Vitrinites give variable yields of tar that show some correlation to their volatile content, and inertinites show no reaction during coking. However, inertinite content is actually very critical for coke formation. It seems clear that it is the inertinite which forms a core or nucleus around which the plastic coal mass condenses to form coke (Speight, 1983).

An additional refinement of the maceral group scheme has been proposed by Spackman (1975). He suggests the use of the term microlithotype to designate three categories of combined macerals. In this arrangement eight subcategories are distributed in three groups using a 5% content as the threshold content, and it is required that

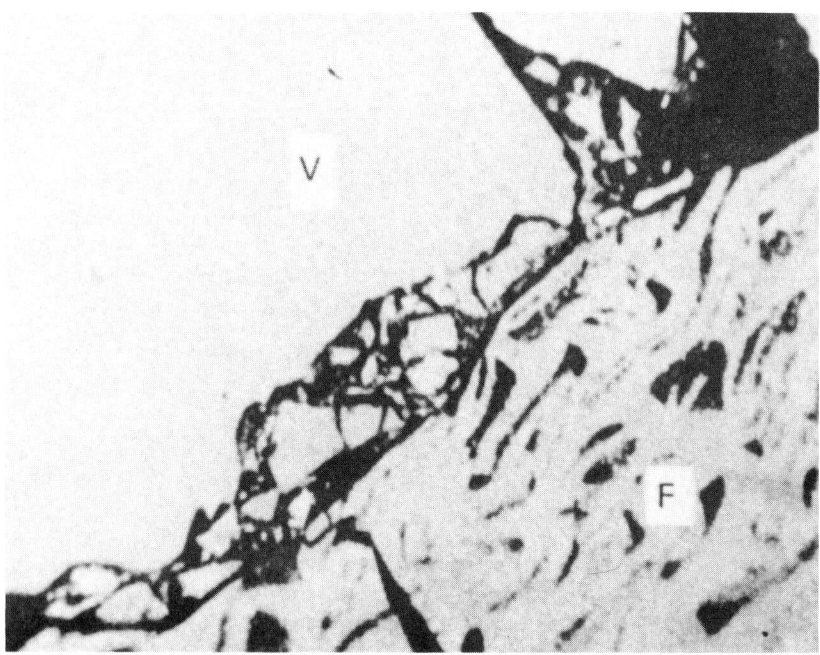

Figure 1 Photomicrograph of macerals from western Canadian coals: V, vitrinite; F, fusinite; E, exinite; SF, semifusinite; and MM, massive micrinite or macrinite. Coal, low-volatile bituminous (J. Speight, 1983. Reprinted with permission of the publisher).

the groups be associated within a 50-μm (5×10^{-3} mm) band (Spackman, 1975; Speight, 1983):

I.	Monomaceralic	
	Vitrite	>95% vitrinite (V)
	Fusite	>95% inertinite (I)
	Liptite	>95% exinite (E)
II.	Bimaceralic	
	Vitrinerite	>95% V + I
	Clarite	>95% V + E
	Durite	>95% I + E
III.	Trimaceralic	
	Duroclarite	V, I, and E all >5%; V > I
	Clarodurite	V, I, and E all >5%; I > V

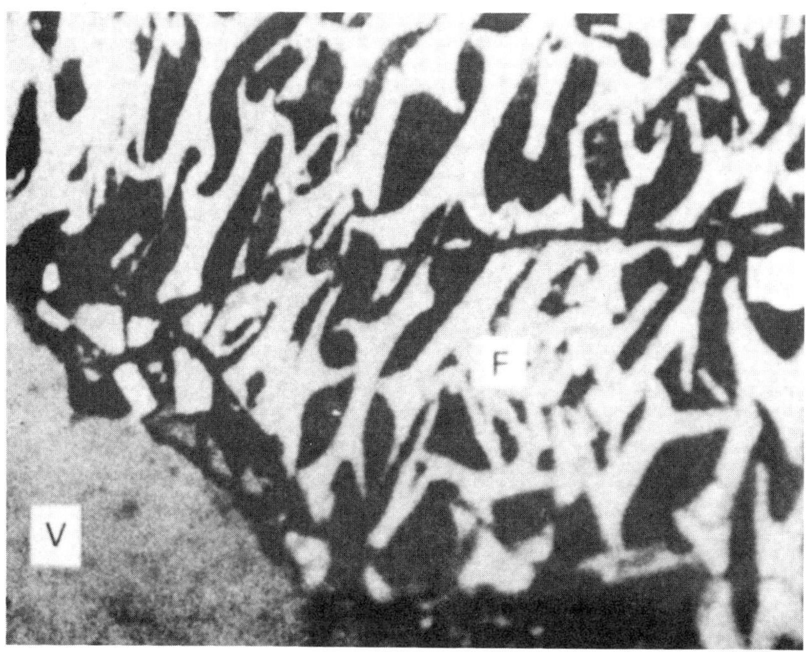

These distinctions are somewhat more cumbersome than the more straightforward maceral groupings and have not enjoyed widespread use.

Individual macerals can be isolated from bulk samples of coal and have been analyzed not only with regard to their elemental and volatile matter content, but also according to a variety of other parameters which seem to be closely related to specific structural features and which have been found to be related to the reactivity of coals. Extensive petrographic analysis of bituminous coal macerals has made it possible to summarize the average amount of carbon, the H/C atomic ratio, and the quantity of volatile matter for coals of this rank (Table 2). Exinites contain the most volatile matter and the most hydrogen; inertinite the least. Both the atomic C content and the H/C ratio of vitrinite overlap those of exinite at the low end of the range. In addition, the differences in atomic carbon content of all three maceral groups decreases as the weight percent of carbon (rank) increases (Figure 2). It is also found that exinite is richer in hydrogen, inertinite is rich in carbon, and vitrinite contains higher amounts of oxygen (Tsai, 1982). Data reported by Tschmaler and deRuiter

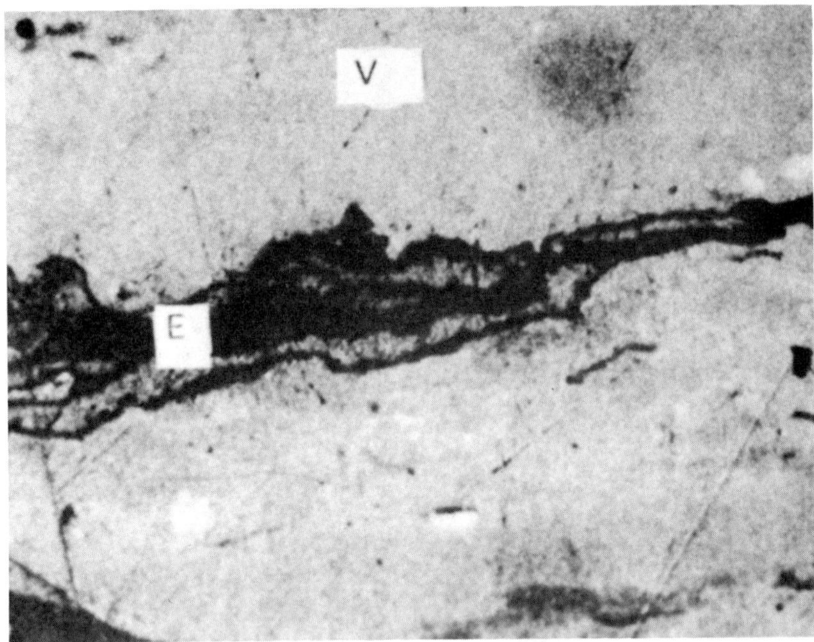

Figure 1 (Continued)

(1966) showed that the nitrogen and sulfur contents did not vary among the maceral groups analyzed. They reported 1.2–1.4% by weight) nitrogen and sulfur for all three groups (Tsai, 1982). However, van Krevelen and Schuyer (1957) had earlier reported values for nitrogen content of 1.15–1.45% (by weight) for vitrinite, exinite, and micrinite from coals with C content ranging from 81.5 to 92.2% (daf), but they found much lower sulfur content in the macerals (0.4–0.6% by weight).

The scarcity of hydrogen in coal presents a major obstacle for converting it to synthetic fuels. It has been determined that hydrogen is chemically present in three forms, aliphatic and hydroaromatic (Al), aromatic (Ar), and hydroxyl, and that each form reacts differently (Table 3). The data show that exinite contains a much larger amount of aliphatic/alicyclic hydrogen and that all three have few ($\leq$0.3%) hydroxyl moieties.

Among the physical properties of coal, specific gravity plays a very important role in coal cleaning (see below). In general, the density of bituminous coal increases with the ash content but as a function of rank density decreases and passes through a minimum at about 85% C

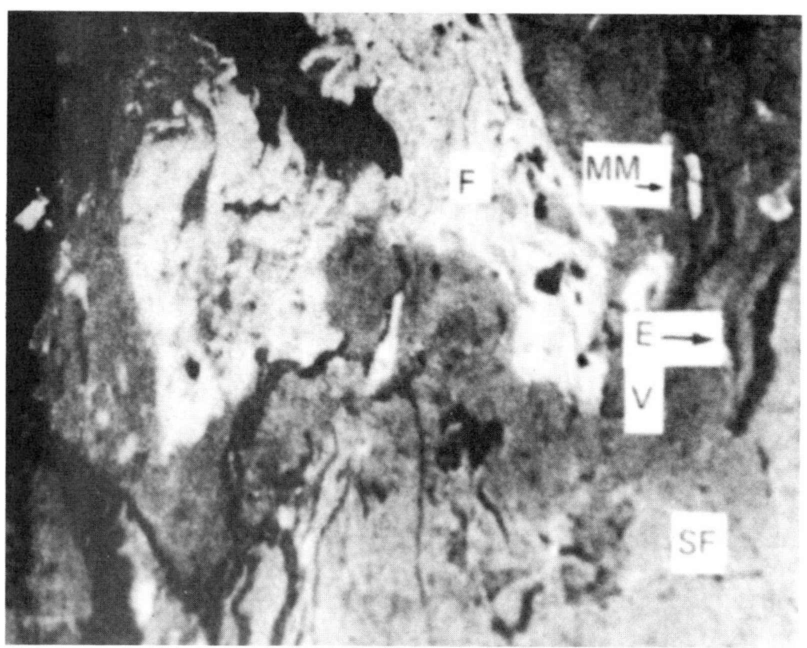

(daf) (Figure 3). It should be noted that the density of a coal reserve is often expressed as tons per acre-foot. Averett (1975) showed that there is a linear relationship, which increases with rank, between tons per acre foot and apparent density. Van Krevelen and Schuyer (1957) reported the density for individual maceral groups. As the percent carbon content (daf, rank) increased, the density of the vitrinite was found to decrease from 1.259 at 81.5% C to 1.240 at 85.0% C, but then to increase to 1.314 in coals with 87.0–91.2% C. The density of exinites was found to increase throughout the entire range of carbon values (81.5–91.2%, daf) from 1.120 to 1.320. Micrinite (an inertinite maceral) had the highest density values but showed the same trend to a minimum, 1.352, at 87.0% C (daf). The density then increased for this maceral to 1.416 at 91.2% C.

Vitrinite can also be used to determine coal rank. The intensity of light reflected from the vitrinite surface has been found to vary directly with rank (Hoffman and Jenker, 1932). Samples are prepared for analysis by pelletizing a finely ground sample. A microscope is used to measure the reflectance from the highly polished coal surface which is exposed to a beam of monochromatic (5460 Å) light.

Table 2 Average C, H, and Volatile Content of Maceral Groups

Maceral group	Atomic C (%)	H/C ratio	Volatile matter (%)
Exinite	43–62	1.18–0.59	79–18
Vitrinite	51–62	0.80–0.60	40–18
Inertinite	59–67	0.64–0.47	31–11

Source: Tsai, 1982. Reproduced with permission of the publisher.

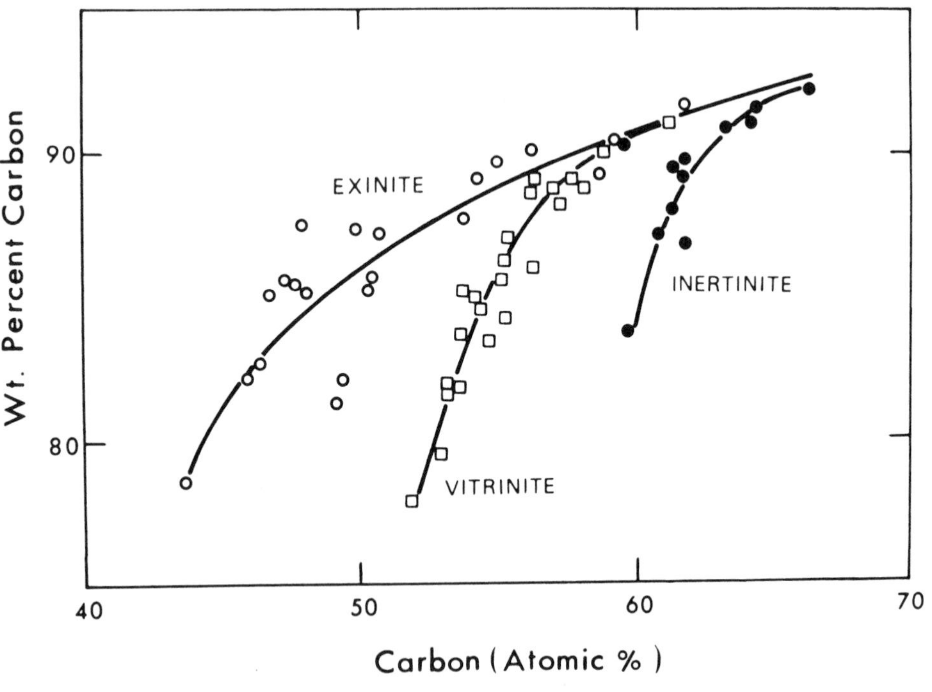

Figure 2 Percent atomic carbon versus wet % carbon of bituminous coal maceral groups. (S. Tsai, 1982. Reprinted with permission of the publisher).

Table 3 Distribution of Hydrogen in Maceral Groups

Maceral group	% H (wt)	H_{Al}/H	H_{Ar}/H
Exinite	7.0	0.77	0.20
Vitrinite	5.5	0.60	0.35
Micrinite	3.9	0.44	0.51

Source: van Krevelen and Schuyer, 1957. Modified with permission of the publisher.

The reflectance is reported as a percentage of the incident intensity and it has been shown (Davis, 1978) that the reflectance of vitrinite increases regularly with rank (Table 4). Other maceral groups also reflect light and these data have found application in sizing coal and in process control, particularly when several coals are to be blended (Stach, et al., 1982). Exinite macerals have the lowest reflectance (and are the least dense); inertinite macerals have the highest reflectance values (and are the most dense).

With regard to coal maceral composition and coal reactivity, van Krevelen (1963) introduced the concepts of aromaticity (f_a), condensation index (R_i), and average dimensions of aromatic groups (C_a). The value C_a is the number of aromatic atoms present in one condensed ring system. This value is compared to the total fixed carbon and is used to calculate f_a, that is, $f_a = C_a/C$. The condensation or ring index is the total number of carbon atoms in a ring structure (Table 5). Tschamler and deRuiter (1966) analyzed and tabulated similar data for coal macerals (Table 6). Together these data show that aromatic character is lowest in exinite. However, exinites contain the highest percent hydrogen and the most aliphatic hydrogen. Micrinites, which consistently have more total carbon (excluding fusinite) (Table 2), are more aromatic and are the most hydrogen-deficient of the three maceral groups.

The reactivity of maceral groups when coal is heated is particularly important with regard to the volatile matter content as it affects coal plasticity. Certain coals when heated reach a temperature at which they soften and become fluid. Continued heating accompanied by loss of additional volatile matter results in the loss of this plastic state as condensation reactions take place and coke is formed. In addition, for all three maceral groups the percentage of volatile matter decreases as the atomic carbon content increases (Table 2). Exinite is observed to devolatilize and soften to a greater extent

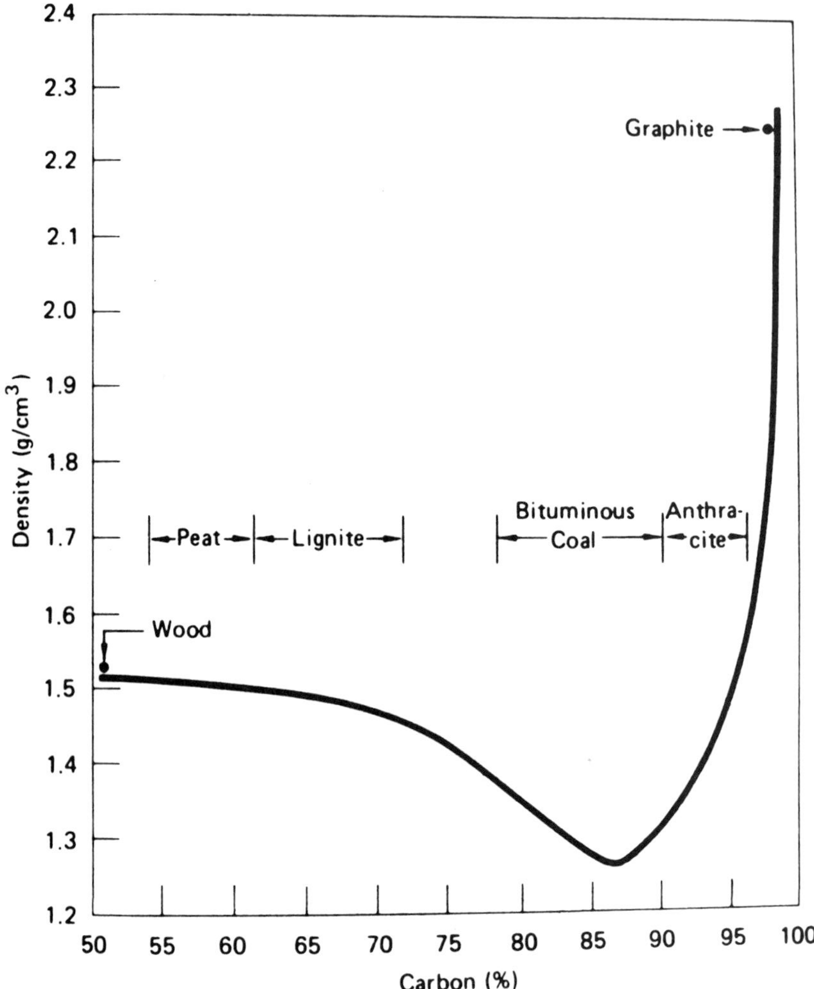

Figure 3 Relationship between apparent density and coal rank (% C content) (I. A. Williamson, 1967. Reprinted with permission of the author).

than vitrinite (Tsai, 1982). However, for midrank coals with volatile matter content between 19 and 33% by weight, it has been shown that it is the vitrinite content that governs the properties of the final coke product. On the other hand, neither anthracite nor low-rank, high-volatile coals will soften or form coke regardless of the vitrinite content (van Krevelen, 1961).

Table 4 Oil Reflectance of Coal by ASTM Rank

Rank	Max. reflectance (%)	Random reflectance (%)
Subbituminous	-0.47	--
High-vol. bit.	C: 0.47–0.57 B: 0.57–0.71 A: 0.71–1.10	0.50–1.2
Med-vol. bit.	1.10–1.50	1.12–1.51
Low-vol. bit.	1.50–2.05	1.51–1.92
Semianthracite	2.05–3.00 (approx.)	1.92–2.50
Anthracite	>3.00	>2.50

Source: Reproduced from Davis, 1978.

Very plastic and rapidly expanding coals are called euplastic coals. They contain abundant vitrinite and, generally, abundant clarain as well. Interestingly, it is the exinite/inertinite ratio that determines the final volume of expansion. Coals very rich in exinite often exhibit normal expansion which is not retained. Rather, these coals are observed to undergo subsequent collapse of the dilated mass and a final volume less than the volume of the original coal results. These are called fluidoplastic coals. A similar type of plastic behavior, but one that is less severe, is exhibited by perplastic coals. Intermediate between the behavior of euplastic and fluidoplastic coals, the final volume of the coke mass formed by perplastic coals is greater than the initial mass but less than that achieved at maximum expansion (Braunstein et al., 1979). In low-rank coals, or others having unusually high inertinite content, behavior termed subplastic is observed. This behavior is characterized by contraction of the mass on heating and no subsequent expansion (van Krevelen, 1961).

II. MACROSCOPIC CLASSIFICATION

Even a casual visual inspection of coals from a variety of places reveals that they possess a variable but identifiable macroscopic structure which also lends itself to a descriptive classification scheme.

Table 5 Typical Data of Maceral Compositions

% C in the vitrinite	Maceral	% C (daf)	F_A	2(r - 1)/C	C_A	R_{IM} (max)	% VM (daf)
81.5	V	81.5	0.83	0.42	18	0.67	39
	E	82.2	0.62	0.31	9	0.13	79
	M	83.6	0.90	0.54	21	1.27	30
85.5	V	85.0	0.84	0.40	23	0.92	34
	E	85.7	0.75	0.35	10	0.24	55
	M	87.2	0.92	0.51	27	1.50	24
87.0	V	87.0	0.86	0.41	26	1.07	30
	E	87.7	0.84	0.37	13	0.44	42
	M	89.1	0.93	0.51	31	1.66	20
89.0	V	89.0	0.89	0.43	31	1.26	26
	E	89.6	0.89	0.42	18	0.82	29
	M	90.8	0.94	0.51	37	1.90	16
90.0	V	90.0	0.91	0.43	34	1.39	23
	E	90.4	0.91	0.44	22	1.12	23
	M	91.5	0.95	0.54	40	2.08	14
91.2	V	91.2	0.93	0.48	39	1.64	18
	E	91.5	0.93	0.48	30	1.64	18
	M	92.2	0.96	0.57	43	2.44	11
85–91	F	94.2	1.0	0.62	—	—	10

V, vitrinites; E, exinites; M, micrinites; F, fusinites.
Source: D. W. van Krevelen and J. Schuyer, 1957. Modified with permission of the publisher.

First, coal often appears to consist of a number of more or less distinct laminar sections. Such coal is called banded coal. The bands, like macerals, were formed from a variety of original organic components, from the kinds of mineral inclusions, and from the fact that different tissues or compounds were affected differently by biogenesis and by dynamogenesis. The bands consist of woody tissue, finely pulverized debris, and charcoal-like granular particles. Banded and nonbanded are the two most commonly recognized types of coal. Some literature refers to banded coal as humic coal (Spackman, 1975) and

Table 6 Percentage of Hydrogen in Aliphatic and Aromatic Structures

Microliths	% H	% H_{aliph}	% H_{arom}	C_{arom}/C	R_{arom}
Exinite	7.0	5.4	1.4	0.62	2.2–3.5
Vitrinite	5.5	3.3	1.9	0.77	2.75–4.1
Micrinite	3.9	1.7	2.0	0.89	3.6–5.2

Source: H. Tschamler and E. deRuiter, 1966. Reprinted with permission of the publisher.

to nonbanded coal as sapropelic (Bouska, 1981) or liptobiolithic coal (Spackman, 1975). These latter terms are intended to distinguish banded coals as having been formed from land vegetation, while nonbanded coals were formed from aquatic (marine or freshwater) plants and microorganisms. The American Society for Testing and Materials (ASTM) also includes a category called impure coal. One example, bone coal, is rich in clay minerals (25–50% by weight ash, dry basis). Another coal, mineralized coal, is ambiguously described as coal containing large amounts of mineral matter which may be dispersed throughout or localized near fissures or cleats. Any related material which contains in excess of 50% by weight ash is called carbonaceous shale or siltstone (Speight, 1983).

A. Theissen—Bureau of Mines System

Separate European and U.S. systems for classifying banded species in coals by microscopic analysis still exist, although the U.S. system is of much more limited utility. Using an extremely thin section of coal in order to observe transmitted light, Robert Theissen and David White developed the American terminology adopted subsequently by the U.S. Bureau of Mines (Theissen, 1920):

Anthraxylon—translucent, bright red, lustrous woody bands remaining from stems and branches of plants
Attritus—dull, gray-to-black material; translucent or opaque; the remains of fats, oils, resins, leaves, pollen grains, and spores
Fusain—the remainder of the organic coal matrix; fibrous and friable

Banded coals must by definition exhibit at least 5% anthraxylon. Modification of the Theissen–Bureau of Mines classification led to the addition of subdivisions for banded coal (Spackman, 1975):

Bright–<20% opaque matter
Semisplint–20-30% opaque matter (called block coal)
Splint–>30% opaque matter

Nonbanded coal has little (<5%) or no anthraxylon. Only two examples have been identified: boghead coal, which is composed of attritus of algal origin, and cannel coal, composed of attritus of spore origin (Table 7). Spore remains appear quite yellow and are very easily recognized by their typical ellipsoidal or compressed ellipse shape with a visible central line. Remains of algae are generally orange and usually fluoresce (spore remains do not) (Spackman, 1975). Both of these nonbanded coals, lacking woody tissue, are dull in appearance. Boghead coal is frequently more brown than black.

B. International System

The International System, as it is now often called, or the Stopes–Heerlen system, was proposed initially by Marie Stopes in 1919 and

Table 7 Ultimate Analysis of Boghead and Cannel Coals

	Composition (wt %)						
Sample	C	H	N	S	O	Moisture	Ash
Boghead coal	74.5	7.8	0.7	0.9	8.5	1.2	6.2
Boghead coal (dry basis)	75.7	7.9	0.7	0.9	8.8	–	6.2
Boghead coal (daf basis)	80.7	8.5	0.7	0.9	9.2	–	–
Cannel coal (daf basis)	78.4	6.9	0.5	0.9	13.4	–	–

Source: A. Drath, *Bull. Inst. Geol. de Pologne*, 1932, *12*, 1.

was modified and adopted officially for international use in 1953 in Geleen (Heerlen), Netherlands. Stopes applied the term lithotype to four macroscopically visible ingredients in banded coal:

Vitrain—bright, uniformly glossy, distinct, and universally narrow bands; derived from woody tissue. Vitrain likely was formed slowly where swamps rested on a high water table. This permitted dead vegetation to sink quickly and the plant material was not subjected to extensive aerobic decomposition.

Clarain—smooth but having a variable silky sheen; variable thickness, often lenticular, and usually horizontal to the bedding plane. It is postulated that clarain formed in an environment of intermediate wetness. Under such conditions, the cell walls are largely preserved and contribute to the high vitrinite content which characterizes clarain. The presence of spores dispersed throughout the vitrinite in clarain indicates that substantial plant decomposition occurred before burial.

Durain—dull, black or gray; hard but may appear granular and produces a sooty residue when handled; has a variable thickness and is often mixed with clarain. Durain consists of the plant fragments most resistant to decay. This indicates that extensive decomposition associated with thoroughly wet but well aerated swamp conditions prevailed and formed the durain residue.

Fusain—patches of powdery, fibrous charcoal-like matter; porous and friable. Because of its similarity to charcoal, some scientists believe that fusain is the residue from prehistoric forest fires. Others postulate that it is the residue from microbial activity (Braunstein et al., 1979).

The Theissen—Bureau of Mines and the International systems have been correlated for a high-volatile bituminous coal (Petrakis and Grandy, 1980) (Table 8). The correlation is only approximate. The lithotype designations and the listed properties refer primarily to bituminous coals, while macroscopic descriptions do not take rank into account nor are they concerned with any particular chemical properties.

C. U.S. Geological Survey System

The U.S. Geological Society (USGS) developed the most basic purely macroscopic system for classifying banded coals. The system is based on the thickness of the vitrain and/or fusain and the luster of the remaining, nonbanded, attrital material. Using this system, a coal is

Table 8 Approximate Correlation of Nomenclature in the Theissen–Bureau of Mines System with the Stopes–Heerlen System

Common name	Coal type in Theissen–Bureau of Mines	Lithotype in Stopes–Heerlen	Properties of typical high-volatile bituminous lithotypes
	Anthraxylon	Vitrain	Relative hardness: 2 ‡$\rho \sim 1.3$ Lowest in ash 30–35% volatile matter Uniform, shiny black bands
Bright coals (U.S. & G.B.)	Fusain	Fusain	Relative hardness: 1 $\rho \sim 1.35$–1.45 (soft) $\rho \sim 1.6$ (hard) Highest in ash 10% volatile matter Charcoallike, forms dust
Semisplint (U.S.) Dull (G.B.)	Translucent Attritus	Clarain	Relative hardness: 3 $\rho \sim 1.3$ Moderately low in ash 40% volatile matter Laminated shiny and dull bands
Splint (U.S.) Dull (G.B.)	Opaque Attritus	Durain	Relative hardness: 7 $\rho \sim 1.25$–1.45 Moderately high in ash 50% volatile matter Dull, nonreflecting, poorly laminated

‡ρ = specific gravity.
Source: L. Petrakis and D. W. Grandy, 1980. Reprinted with permission of the publisher.

classified by assigning it a descriptive statement consisting of the appropriate characteristics, one from each of four categories (Table 9).

This is, of course, only a semiquantitative method useful primarily for rapid classification of samples in the field. However, because the

Table 9 Parameters of the USGS Classification of Banded Coal

Thickness (mm)	Litho-type	Abundance (%)	Attrital character
Thin: 0.5–2		Sparse: <15	Bright
Medium: 2–5	Vitrain or	Moderate: 15–30	Moderately bright
Thick: 5–50	fusain	Abundant: 30–60	Midlustrous
Very thick: >50		Dominant: >60	Moderately dull; dull

ultimate utility of a coal depends on its elemental (ultimate) and maceral composition quite irrespective of the banded nature, any system of classification which relies solely on a macroscopic physical, visual analysis is of very little value.

III. CLASSIFICATION BASED ON CHEMICAL ANALYSIS

A. American Society for Testing and Materials

As already noted, the proportion of carbon, hydrogen, and oxygen vary in a regular fashion as a function of coal maturation (rank), but since elemental composition and coal reactivity cannot be correlated in a straightforward manner, classification using only the elemental composition is not useful. However, the ASTM has established a classification of coal by rank based on fixed carbon content and several other parameters (Table 10). Fixed carbon is the residue, excluding mineral ash, which remains after the volatile material has been evolved from a sample using a specified method for carrying out the analysis. Coals which have more than 31% volatile matter content (daf) are then grouped according to their moist calorific value. This value represents the heat content (btu/lb) determined when the moisture content is evaluated according to a rigid set of conditions.

B. National Coal Board

In England, the National Coal Board uses another classification system which relates a code number to different categories of coal. The basis

Table 10 Coal Classification by Rank

Class and group	Fixed carbon[a] (%)	Volatile matter[a] (%)	Heating value[b] (btu/lb)
Anthracitic			
1. Meta-anthracite	>98	<2	–
2. Anthracite	92–98	2–8	–
3. Semianthracite	86–92	8–14	–
Bituminous			
1. Low-volatile bituminous coal	78–86	14–22	–
2. Medium-volatile bituminous coal	69–78	22–31	–
3. High-volatile A bituminous coal	<69	>31	>14,000
4. High-volatile B bituminous coal	–	–	13,000–14,000
5. High-volatile C bituminous coal	–	–	10,500–13,000[c]
Subbituminous			
1. Subbituminous A coal	–	–	10,500–11,500[c]
2. Subbituminous B coal	–	–	9,500–10,500
3. Subbituminous C coal	–	–	8,300–9,500
Lignitic			
1. Lignite A	–	–	6,300–8,300
2. Lignite B	–	–	<6,300

[a]Calculated on dry, mineral-matter-free coal.
[b]Calculated on mineral-matter-free coal containing natural inherent moisture.
[c]Coals with a heating value of 10,500–11,500 btu/lb are classified as high-volatile C bituminous coal if they have agglomerating properties and as subbituminous A coal if they are nonagglomerating.
Source: American Society for Testing and Materials, Standard Specifications for Classification of Coals by Rank (ASTM Designation D388-66).

for identifying the separate categories is the coking characteristics of the coal samples. The nature of the coke produced is determined according to the Gray–King assay, plus the determination of the amount of volatile matter produced. A three-digit code is assigned to each coal; low code numbers are indicative of higher rank (Table 11). The most serious limitations of this system are the time required to carry out the Gray–King assay for coke formation and the potential associated with that method for causing changes in the coal by oxidation, which unavoidably biases the results (Speight, 1983).

C. International System

The International System for Classification of Hard Coals uses the dry, ash-free volatile matter content, the ash-free calorific value, and the coking and caking properties of coal as the basis for its classification. Like the National Coal Board system, a three-digit code number is derived for each coal which reflects the interrelationship of the four experimentally determined parameters (Table 12). The first is the class number, 1–9. Up to class 5, this number reflects volatile matter content (class 5 is 33%, daf, the maximum for the system). Classes 6–9 are then assigned by the calorific value (maf) for samples with greater than 33% volatile matter content. The second number of the code is the group number, 1–3. This number relates to the caking properties determined by rapidly heating the sample according to either the free-swelling or the Roga index methods. Finally, the coking behavior, determined by the Gray–King assay or by the Audibert–Arnu test–both of which involve slow heating of the sample–is determined. The result of that test is translated into a number from 1 to 5, which is the third number in that code.

IV. CORRELATION OF CLASSIFICATION SYSTEMS

It is helpful to make comparisons and correlations among the various systems. Speight (1983) drew a correlation between the ASTM system and the International System which makes adjustments for the fact that mineral-matter free calorific values (ASTM) and ash-free calorific values (International System) are not the same (Table 13). Hodgkins (1961) constructed a chart showing the Russian classification of coals using a format similar to that used by the ASTM (Table 14). It should be noted in this case that in the Russian system, the total carbon content is used, and calorific values are determined on both moisture and ash-free basis rather than on a moist *but* ash-free basis, as the ASTM specifies. The Russian standards for both anthracite and bituminous coal are taken from the Donets basin. This chart also includes the typical range of sulfur and ash content values observed in Russian

Table 11 Coal Classification System Employed by the National Coal Board (UK)

Group ( )	Class	Volatile matter, dry, mineral matter free (%)	Gray–King coke type[a]	General description
100		Under 9.1	A	Anthracites
	101[b]	Under 6.1	A	
	102[b]	6.1–9.0		
200		9.1–19.5	A–G8	Low-volatile steam coals
	201	9.1–13.5	A–G	Dry steam coals
	201a	9.1–11.5	A–B	
	201b	11.6–13.5	B–C	
	202	13.6–15.0	B–G	Coking steam coals
	203	15.1–17.0	B–G4	
	204	17.1–19.5	G1–G8	
	206	9.1–19.5	A–B for V.M. 9.1–15.0 A–D for V.M. 15.1–19.5	Heat-altered low-volatile steam coals
300		19.6–32.0	A–G9 and over	Medium-volatile coals
	301	19.6–32.0	G4 and over	Prime coking coals
	301a	19.6–27.5	G4 and over	
	301b	27.6–32.0	G4 and over	
	305	Over 32.0	A–G3	(Mainly) heat-altered medium-volatile coals
	306	19.6–32.0	A–B	
400–900		Over 32.0	A–G9 and over	High-volatile coals
400		Over 32.0	G9 and over	Very strong caking coals
	401	32.1–36.0	G9 and over	
	402	Over 36.0		
500		Over 32.0	G5–G8	Strongly caking coals
	501	32.1–36.0		
	502	Over 36.0	G5–G8	
600		Over 32.0	G5–G4	Medium caking coals
	601	32.1–36.0		
	602	Over 36.0	G1–G4	

(continued)

Table 11 (Cont.)

Group ( )	Class	Volatile matter, dry, mineral matter free (%)	Gray–King coke type[a]	General description
700		Over 32.0	B–G	
	701	32.1–36.0		Weakly caking coals
	702	Over 36.0	B–G	
800		Over 32.0	C–D	
	801	32.1–36.0		Very weak caking coals
	802	Over 36.0	C–D	
900		Over 32.0	A–B	
	901	32.1–36.0		Noncaking coals
	902	Over 36.0	A–B	

[a]Coals of groups 100 and 200 are classified by using the parameter of volatile matter alone. The Gray–King coke types quoted for these coals indicate the ranges found in practice and are not criteria for classification.
[b]To divide anthracites into two classes, it is sometimes convenient to use a hydrogen content of 3.35% (dmmf) instead of a volatile matter of 6.0% as the limiting criterion. In the original Coal Survey rank coding system, the anthracites were divided into four classes then designated 101, 102, 103, and 104. Although the present division into two classes satisfies most requirements, it may sometimes be necessary to recognize four or five classes.
Source: Francis, 1961.

coals (Survey of Energy Resources, 1974). Table 30 in Chap. 17 compares terminology from a number of coal-producing and coal-consuming countries.

As should be expected, many if not all countries have devised a system by which they classify coal (Table 15), and it is important to recognize the variations in terminology each uses. While it is tedious to have to rely on the various correlation charts to move from one system to another, it is also unfortunate that a truly international system for classifying coal has not been developed to expedite international dialogue and commerce.

Table 12 International System for Classification of Hard Coals

Groups (determined by caking properties)			Code numbers										Subgroups (determined by coking properties)		
	Alternative group parameters		The first figure of the code number indicates the class of the coal, determined by volatile-matter content up to 33% volatile matter and by calorific parameters above 33% volatile matter. The second figure indicates the group of coal, determined by caking properties. The third figure indicates the subgroup, determined by coking properties.											Alternative subgroup parameters	
Group number	Free-swelling index (swelling number)	Roga index											Sub-group number	Audibert–Arnu dilatometer	Gray–King
3	>4	>45					435	535	635				3	>140	>C_0
						334	434	534	634	734			4	>50–140	C_0–C_0
						333	433	533	633	733			3	>0–50	C_0–C_0
						132a 332b	432	532	632	732	832		2	≤0	E–G
2	2-1/2–4	>20–45				323	423	523	623	723	823		3	>0–50	C_0–C_0
						322	422	522	622	722	822		2	≤0	E–G
						321	421	521	621	721	821		1	Contraction only	B–D
1	1–2	>5–20			212	312	412	512	612	712	812		2	≤0	E–G
					211	311	411	511	611	711	811		1	Contraction only	B–D
0	0–1/4	0–5	000	100 A, B	200	300	400	500	600	700	800	900	0	Nonsoftening	A
Class number →			0	1	2	3	4	5	6	7	8	9	As an indication, the following classes have an approximate volatile-matter content of: Class 6: 33–41% volatile matter Class 7: 33–44% volatile matter Class 8: 35–50% volatile matter Class 9: 42–50% volatile matter		
Class parameters	Volatile matter (dry, ash-free) →		0–3	>3–10 >3–6.5, >6.5–10	>10–14	>14–20	>20–28	>25–33	>33	>33	>33	>33			
	Calorific parameters →		–	–	–	–	–	–	>13,950	>12,960–13,950	>10,980–12,960	>10,260–10,980			

Classes: Determined by volatile matter up to 33% volatile matter and by calorific parameter above 33% volatile matter.

Gross calorific value on moist ash-free basis (86°F ∿6% relative humidity) bhu per pound.

Notes:
1. Where the ash content of coal is too high to allow classification according to the present system, it must be reduced by Laboratory Float-and-sink material (or any other appropriate means). The specific gravity selected for flotation should allow a maximum yield of coal with 5–10% of ash.
2. 332a > 14–16% volatile matter. 332b > 16–20% volatile matter.

Source: *Energy Technology Handbook*, Douglas M. Considine, ed., McGraw-Hill, New York, 1977, pp. 1–24.

Table 13 Comparison of Class Numbers of the International System of Coal Classification with Coal Rank from the American Society for Testing and Materials System of Coal Classification

International classification, class number	0	1	2	3	4	5	6	7	8	9
		5	10 Volatile-matter parameter[a]	15	20 25	30		14,000	13,000 Calorific value parameter[b] 12,000	11,000 10,000
ASTM classification group name	Meta an-thra-cite	Anthracite	Semianthracite	Low-volatile bituminous coal	Medium-volatile bituminous coal	High-volatile A bituminous coal		High-volatile B bituminous coal	High-volatile C bituminous coal and sub-bituminous A coal	Sub-bituminous B coal

[a]Parameters in International system are on ash-free basis; in ASTM system, they are on mineral-matter-free basis.
[b]No upper limits of calorific value for class B and high-volatile A bituminous coals.
Source: J. Speight, 1982. Reprinted with permission of the publisher.

Table 14 Typical Characteristics of Russian Solid Fossil Fuels,

Class	Group Name	Symbol	Total carbon (%)	Volatile matter (%)	Hydrogen (%)
Anthracite	Anthracite	A	89.4–96.4	2–7	1.2–3.0
	Semianthracite	T	88.0–92.4	8–15	3.8–4.6
Bituminous	Dry steam	PS	87.1–91.2	13–18	3.9–5.1
	Coking	K	86.7–90.7	18–26	4.0–5.4
	Fatty steam	Pzh	82.4–87.0	24–35	4.5–5.5
	Gassy	G	78.4–82.9	35–44	5.0–5.8
	Dry long flame	D	74.0–79.2	40–46	5.1–5.7
	Docpr basin	B	57.3–69.0	41–60	5.2–6.6
Subbituminous and lignite	Kansk–Achinsk	B	67.0–75.0	45–50	1.6–9.6
	Moscow basin	B	69.2–79.4	48–86	5.5–10.2
	Irkutsk basin	B	69.0–77.0	50–55	6.5–7.0

Source: S. A. Hodgkins, *Soviet Power: Energy Resources, Production and Potential*, Prentice-Hall, 1961.

V. CLASSIFICATION OF BROWN COAL

No mention has been made thus far about brown coal. Brown coal is a low-rank coal which is not always distinguished from lignite. It has a high moisture content, is friable, and crumbles easily after it is dried. The ASTM considers all low-rank coal as lignite (Table 10), but internationally, brown coal is regarded as that which has a calorific value of less than 10,260 btu/lb (5200 kcal/kg) (Speight, 1983). The International System categorizes brown coals on the basis of the total moisture content (ash-free) and the yield of tar produced from a dry ash-free sample (Table 16). The moisture content is indicated by the class number (10–15); the tar yield is described by the group number. A four-digit code number is finally derived from the combination of the class number and the group number categories (Speight, 1983).

Excluding Peat

Heating values		Moisture content		Ash content (%)	Sulfur content (%)
kcal/kg	MG/kg	As mined	Dry		
7950–8350	33.3–35.0	3.5–9.0	2	7.0–19.0	1.0–5.0
8300–8650	34.7–36.2	3.0–6.5	0.4–1.8	4.3–25.0	0.9–6.0
8300–8700	34.7–36.4	3.0–6.5	0.8	5.0–20.0	0.8–3.6
8450–8750	35.4–36.6	3.0–12.0	1.0	5.0–21.0	0.9–4.9
8250–8600	34.5–36.0	3.0–8.0	0.4–2.1	2.0–31.0	1.0–6.0
7650–8400	32.0–35.2	3.0–12.0	1.0–4.8	3.0–32.0	1.3–7.5
7400–7900	31.0–33.1	9.5–21.0	3.1–7.5	6.0–27.0	1.4–6.0
4380–6921	18.3–29.0	50.5–55.6		14.5–27.9	1.0–5.6
6500–6800	27.2–28.5	32.0–42.0	12.0–25.0	7.0–20.0	0.8
>4060	>17.0	7.9–31.8		5.7–40.9	1.0–4.5
7000–7200	29.3–30.1	25.0–28.0	12.0	12.0–45.0	1.0–2.0

Table 15 Schematic Comparison of the Various Systems for the Classification of Coal

Classes of the International system			Classes of national systems							
	Parameters									
Class no.	Volatile matter content	Calorific value (calculated to standard moisture content)	Belgium	Germany	France	Italy	Netherlands	Poland	United Kingdom	United States
0	0–3					Antraciti speciali		Meta-antracyt		Meta-anthracite
1A	3–6,5		Maigre	Anthrazit	Anthracite	Antraciti comuni	Anthraciet	Antracyt	Anthracite	Anthracite
1B	6,5–10							Polantracyt		
2	10–14		1/4 gras	Magerkohle	Maigre	Carboni magri	Mager	Chudy	Dry steam	Semi-anthracite
3	14–20		1/2 gras 3/4 gras	Esskohle	Demi gras	Carboni semi-grassi	Esskool	Polkoksowy Metakoksowy	Coking steam	Low volatile bituminous

4	20–28		Gras	Fett-kohle	Gras a courte flamme	Carboni grassi corta fiamma	Vetkool	Ortokoksowy	Medium volatile coking	Medium volatile bituminous
5	28–33			Gaskohle	Gras proprement dit	Carboni grassi media fiamma		Gazowo koksowy	High volatile	High volatile bituminous A
6	>33 (33–40)	8450–7750				Carboni da gas	Gaskool			
7	>33 (32–44)	7750–7200		Gas-Flamm-kohle	Flam-bant gras	Carboni grassi da vapore	Gas-vlam kool	Gasowy		High volatile bituminous B
8	>33 (34–46)	7200–6100			Flam-bant sec	Carboni secchi	Vlam-kool	Gazowo-plomienny		High volatile bituminous C
9	>33 (33–48)	<6100						Plomienny		Sub-bituminous

Source: *Coal Science: Aspects of Coal Constitution*, D. W. van Krevelen and J. Schuyer, Elsevier, New York, 1957.

Table 16 International System for the Classification of Brown Coals

Group no.	Group parameter tar yield (dry, ash-free) (%)[a]	Code number					
40	>25	1040	1140	1240	1340	1440	1540
30	20–25	1030	1130	1230	1330	1430	1530
20	15–20	1020	1120	1220	1320	1420	1520
10	10–15	1010	1110	1210	1310	1410	1510
00	10 and less	1000	1100	1200	1300	1400	1500
Class no.		10	11	12	13	14	15
Class parameeter, i.e., total moisture, ash-free, %[b]		20 and less	20–30	30–40	40–50	50–60	60–70

[a]Gross calorific value below 10,260 btu/lb. Moist ash-free basis (86°F/ 96% relative humidity).
[b]The total moisture content refers to freshly mined coal. For internal purposes, coals with a gross calorific value over 10,260 btu/lb (moist ash-free basis), considered in the country of origin as brown coals, but classified under this system to ascertain, in particular, their suitability for processing. When the total moisture content is over 30%, the gross calorific value is always below 10,260 btu/lb.
Source: *Energy Technology Handbook* (Douglas M. Considine, ed.), McGraw-Hill, New York, 1977, pp. 1–24.

19

Composition

I	Introduction	574
II	Analysis	575
	A. Sampling	575
	B. Proximate Analysis and Mineral Matter Content	575
	C. Ultimate Analysis and Forms of Sulfur	578
	D. Ash Analysis	581
	E. Minor and Trace Elements	583
	F. Mineral Matter Constitution	585
III	Organic Structure of Coal	585
	A. Introduction	585
	B. Oxidation	586
	C. Depolymerization and Alkylation Procedures	588
	D. Pyrolytic Methods	590
	E. NMR Spectroscopy	592
	F. Infrared Spectroscopy	597
	G. Mass Spectroscopy	598
	H X-ray Photoelectron Spectroscopy	601
	I. Hypothetical Structures	603
IV	Coal Liquids	605
	A. Solvent Fractionation	605
	B. Chromatographic Separation and Analysis	607
	C. Titration Methods	621
	D. Molecular Weight Determination	624
	E. Mass Spectrometry	625
	F. Ultraviolet and Luminescence Spectroscopy	627

G. Voltammetric Identification of Aromatic Groups 628
H. NMR Spectroscopy 631
I. Structural Analysis and Integration of Analytical Methods 639

I. INTRODUCTION

Standard test methods have long been used for coal classification but the past 20 years has seen a remarkable growth in the number of analytical techniques which are in widespread use for the chemical characterization of coals and coal products. This growth has resulted from both the major developments in analytical chemistry and the upsurge in research during the 1970s to develop more effective coal liquefaction processes for the production of chemical feedstocks and transport fuels. Analytical methods for the characterization of coals and coal liquids fit broadly into the following categories:

1. Standard test methods such as proximate and ultimate analyses
2. Spectroscopic and chromatographic methods which have found numerous applications but, in general, are not standardized, and
3. Other advanced techniques which, either due to high cost or limited applicability, have only been used occasionally (typically less than 10 literature references).

This section covers all three categories but the emphasis is placed on the more widely used chromatographic and spectroscopic methods, particularly nuclear magnetic resonance (NMR), mass spectrometry (MS), and gas and liquid chromatography (GC and LC). The aim is to provide concise coverage with many references included for specialist articles and texts. To provide a common thread between the techniques used for coal liquids, many of the examples given are from the authors' own work. With the exception of proximate analysis, all the analytical techniques described provide chemical information on coals and coal products. For information on the measurement of physical properties and the microscopic examination of coals and cokes, the other chapters should be consulted, together with texts by van Krevelen (1961), Speight (1983), and Berkowitz (1985).

II. ANALYSIS

A. Sampling

Coal samples can be taken from in situ sources, such as seams or boreholes, and from sources after extraction, e.g., conveyors and stockpiles. For analytical results to have any meaning, it is extremely important that the sample analyzed represent the coal from which it was obtained. Therefore, appropriate standard methods for sampling should always be used (BS, 1973a; ASTM, 1987a). British and American Standards for proximate and ultimate analysis require 50-g samples not having a top size greater than 200–250 μm to be prepared.

B. Proximate Analysis and Mineral Matter Content

This covers the determination of the concentrations of inherent moisture, noncombustable minerals (ash), volatile matter, and fixed carbon, volatile matter contents corrected for ash and moisture being widely used in coal classification (Chap. 18).

Inherent moisture (that held within the pore structure) is defined as the weight loss when air-dried coal (<200–250 μm) is heated at 105–110°C (220–230°F) with minimal oxidation occurring (BS, 1973b; ASTM, 1987b). It should not be confused with the total moisture content, which includes surface moisture in the coal as sampled. Methods of determination include drying (1) in a minimum space oven with a flow of nitrogen, (2) in a vacuum, and (3) in nitrogen, and collecting the water released in an adsorption train (BS, 1973b; ASTM, 1987b). It should be realized that the air-dried moisture content excludes that present in the structure of clays and other minerals. Temperatures in excess of 500°C (930°F) are required to decompose such moisture which, in principle, can be determined from the weight loss of low temperature ash prepared in a plasma furnace at ∿150°C (300°F).

Ash contents are determined by heating coal in a ventilated furnace to a specified temperature, the British standard (BS, 1973b) and the U.S. standard (ASTM, 1987c) differing in the temperature used (815 and 700–750°C, respectively; 1500°F and 1290–1380°F). The heating rate is slow and a two-stage heating cycle is used (temperature held at 500°C at end of first stage) to prevent sulfur fixation in the ash. Pyrite oxidizes to sulfur oxides and iron oxides at temperatures close to 500°C (930°F) while carbonates decompose to metal oxides and carbon dioxide at around 700°C (1290°F). A number of formulas have been used to relate the ash content to the concentration of mineral matter in the parent coal (Table 1). Alternatively, mineral

Table 1 Formulas for Calculating Coal Analyses on Different Bases

Given results	As analyzed (air-dry)	Dry	Dry, ash-free	Dry, mineral matter-free
As analyzed (air-dry)	—	$\frac{100}{100 - M_{ad}}$	$\frac{100}{100 - (M_{ad} + A_{ad})}$	$\frac{100}{100 - (M_{ad} + MM_{ad})}$
Dry	$\frac{100 - M_{ad}}{100}$	—	$\frac{100}{100 - A_d}$	$\frac{100}{100 - MM_d}$
Dry, ash-free	$\frac{100 - (M_{ad} + A_{ad})}{100}$	$\frac{100 - Ad}{100}$		$\frac{100 - Ad}{100 - MM_d}$
Dry, mineral matter-free	$\frac{100 - (M_{ad} + MM_{ad})}{100}$	$\frac{100 - MM_d}{100}$	$\frac{100 - MM_d}{100 - A_d}$	

Note: ad = air-dry; M = moisture; d = dry, MM = mineral matter; A = ash.
Source: BS 1016, part 1b, 1973.

matter contents can be determined by acid demineralization with hydrofluoric and hydrochloric acids (extract coal successively with 5 N HCl, 42% HF, and 5% HCl), but this is time consuming and corrections have to be made for dissolved pyrite and chlorine that cannot be washed out (Radmacher and Mohrauer, 1953, 1955; Bishop and Ward, 1958; Brown et al., 1949). Probably the most satisfactory procedure for estimating mineral matter contents is the use of plasma ashing at relatively low temperatures (∿150°C; 300°F) (Gluskoter, 1965; O'Gorman and Walker, 1971; Fraser and Belcher, 1972). Oxygen atoms, excited by a radiofrequency discharge in an evacuated chamber, oxidize the organic matter over a period of typically at least 2 days. Some pyrite can be oxidized but otherwise changes in mineral composition are relatively minor.

Volatile matter is defined as the material released when coal is heated in the absence of air to a prescribed temperature and is measured from the weight loss. The British standard specifies heating coal in a silica crucible to 900°C (1650°F) for 7 min (BS, 1973b) while, in contrast, the American standard uses a temperature of 950°C (1740°F) and a platinum crucible (ASTM, 1987d). Since results are likely to differ, the method used should always be stated. Also, values should ideally be corrected for contributions from minerals, in particular, carbon dioxide from carbonates, sulfur dioxide from sulfates, and water from clays (Table 2).

For purposes of coal classification, volatile matter (VM) is expressed on a dry ash free basis; fixed carbon is not actually determined since it is defined by difference as

$$\% \text{ fixed carbon} = 100 - \% (\text{moisture} + \text{ash} + \text{VM})$$

$$\% \text{ fixed carbon, daf basis} = 100 - \% \text{ VM}$$

The procedures outlined above for proximate analysis are relatively time consuming for laboratories with a high throughput of samples and more than one manufacturer has designed an instrument for the sequential determination of moisture, volatile matter, and ash on approximately 20 samples simultaneously (Instrument Manuals, 1980, 1983). These systems comprise furnaces, a control console, and associated microprocessor and electronics and proceed automatically through the analysis cycle while continuously weighing the samples. However, appropriate corrections should be included to give results in close agreement with those from standard procedures (BS, 1973b; ASTM 1987b–d). Formulas in Table 1 are used for transposing results between, as analyzed, dry, dry ash free, and dry mineral matter free bases.

Table 2 Some Formulas Used to Predict Mineral Matter Contents and to Correct Volatile Matter, Carbon and Hydrogen Contents for Contributions from Minerals

Parr, mineral matter	$= 1.03\text{Ash} + 0.55\text{S}$
Modified Parr	$= 1.13\text{Ash} + 0.47\text{Pyr} + 0.3\text{Cl}$
BCURA	$= 1.10\text{Ash} + 0.53\text{S} + 0.74\text{CO}_2 - 0.36$
King–Maries–Crossley	$= 1.13\text{Ash} + 0.80\text{CO}_2 + 0.5\text{S pyr} + 2.85\text{SO}_4 - 2.85\text{S ash} + 0.5\ \text{Cl}$
Demineralization, mineral matter	$=$ wt. loss + residual ash + 1/3 FeS_2 + HCl
Volatile matter (VM)[a]	$= \text{VM}_{\text{DET}} - (0.13\text{Ash} + 0.2\text{S pyr} + 0.7\text{CO}_2 + 0.7\text{Cl} - 0.2)$
Carbon (C)[a]	$= \text{C}_{\text{DET}} - 12/44\text{CO}_2$
Hydrogen (H)[a]	$= \text{H}_{\text{DET}} - 0.014\text{Ash} - 0.018\text{S pyr} - 0.019\text{CO}_2 - 0.014\text{SO}_3\ (\text{Ash})$

[a]BS 1016 part 16 (1971).
det = determined value; S pyr = pyritic sulfur; Pyr = pyrite.

C. Ultimate Analysis and Forms of Sulfur

Ultimate analysis refers to the determination of carbon, hydrogen, oxygen, nitrogen, and sulfur, the major elements in coal. Some European coals in particular contain over 0.5% chlorine and therefore for completeness the determination of chlorine is included here. Sulfur is present in organic structures, as sulfides (principally in pyrite) and sulfates (usually in low concentrations), and possibly in an elemental form.

Carbon and hydrogen have traditionally been determined by heating coal in a stream of dry oxygen and using absorption tubes to collect the liberated carbon dioxide and water with provision being made to prevent interference by oxides of nitrogen and sulfur (BS, 1977;

ASTM, 1987e). Low-temperature (Liebig, 800°C; 1470°F) and high-temperature (1300°C; 2370°F) combustion methods are both covered in the British Standard (1977). The results are corrected for contributions principally from inherent moisture and carbonate-derived CO_2 to give concentrations of organically bound carbon and hydrogen; Table 2 gives formulas used in the British Standard. Nitrogen does not occur in coal minerals and, hence, its determination poses fewer problems. The Kjeldahl method, which uses sulfuric acid digestion to convert the nitrogen in coals to ammonium sulfate, has been the most widely used procedure.

Recently, a number of instrument manufacturers have marketed "CHN" analyzers for the simultaneous determination of carbon, hydrogen, and nitrogen. In these systems, combustion gases are analyzed using thermal conductivity detectors (TCDs), infrared analyzers, and chromatographic columns equipped with TCD detection after reduction of nitrogen oxides to nitrogen and the removal of solids, sulfur oxides, and oxygen. TCD detectors demand small sample sizes (2–3 mg) but IR detectors can handle much larger sizes (50–100 mg), which is a big advantage in terms of obtaining a representative coal sample.

The total sulfur content of coals can be determined using a number of procedures (ASTM, 1987f, g; BS, 1977). The Eschka method uses magnesium oxide and sodium carbonate to oxidize coal at 800°C (1470°F), barium chloride then being added to form barium sulfate which is determined gravimetrically (ASTM, 1987f). In the high-temperature method (ASTM, 1987f; BS, 1977), coal is burned in oxygen at 1350°C (2460°F) to yield sulfur dioxide which is absorbed in hydrogen peroxide solution and the resultant sulfuric acid can be determined titrimetrically (a correction is needed for hydrochloric acid formed from chlorine in coals). Alternatively, as in commercial S analyzers, sulfur dioxide in the combustion gases can be determined using an IR absorption cell after moisture and particulate matter have been removed. The bomb-washing method determines sulfur after calorific value has been measured in an oxygen bomb calorimeter (ASTM, 1987f). After titration with base to determine the acid correction for the calorific value, the solution is heated and ammonium hydroxide added to precipitate ferric hydroxide. Barium chloride is then added to the filtered solution to precipitate the sulfate.

Pyritic and sulfatic sulfur are determined using the standard procedures (ASTM, 1987h; Figure 1). Sulfate concentrations are generally low except in weathered coals. Organic sulfur has traditionally been determined by difference but recently a number of groups used transmission and scanning electron microscopy (TEM and SEM) for direct measurements. A thin section of coal (0.5 μm) is used in TEM so that an electron beam is largely undispersed in passage through the sample, the weight of organic sulfur being proportional to the intensity

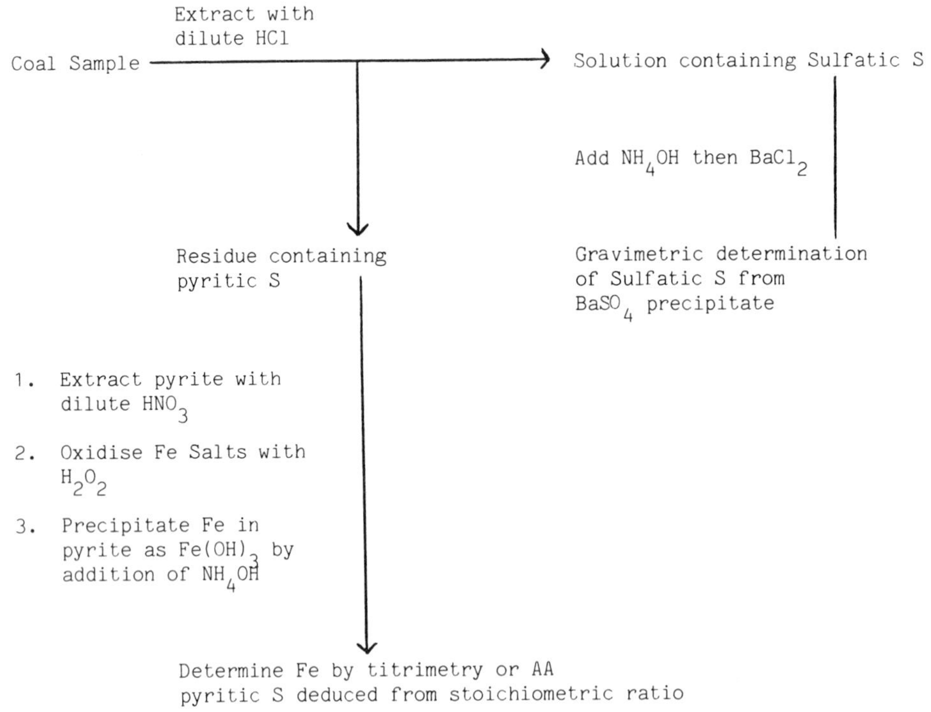

Figure 1 Schematic of methods for determining pyritic and sulfatic sulfur in coals (ASTM, 1987h).

of the sulfur Kα X ray (Hsieh and Wert, 1985; Tseng et al., 1986). The SEM method employs a high-energy beam (15–30 keV) focused on a small area of sample; the X-ray emissions resulting from exposed sites are analyzed by an energy-dispersive analyzer. The Kα X ray is monitored and the intensity is compared with that of a suitable standard in a similar matrix (Timmer and van der Burgh, 1984; Straszheim et al., 1983; Maisgren et al., 1983). The number of points sampled (mineral matter is avoided) is largely dependent on the instrumentation used.

The oxygen contents of coal are usually determined by difference, i.e., 100 - % (C + H + N + S), ash or mineral matter free basis (e.g., by ASTM, 1987i), and cumulative errors may be considerable. However, this state of affairs arises not from a lack of available methods but the corrections that have to be used to deduce concentrations of organically bound oxygen. Even if coals are dried to remove inherent moisture, corrections still have to be made for oxygen derived from clays, water of crystallization, and carbonates. The most widely used procedures are based on the Schutze–Unterzaucher method in which coal is pyrolyzed at ∿900°C (1650°F) in a stream of dry nitrogen (Unterzaucher, 1952; Oita and Conway, 1954). The resultant volatiles are passed over carbon or a platinum-carbon catalyst which converts the oxygen present to carbon monoxide. This is oxidized to carbon dioxide using iodine pentoxide. The released iodine can be determined titrimetrically or an absorption train can be used to measure the carbon dioxide. Neutron activation analysis has also been used for measuring oxygen contents of coal (Hamrin et al., 1975). This method determines total oxygen irrespective of the forms present. Therefore, organic oxygen can be determined by difference between the value for a whole coal and its low-temperature ash.

The foregoing discussion has indicated that to determine both carbon and oxygen, corrections are needed for CO_2 from carbonates. This can be determined by dissolving the carbonates in hydrochloric acid and measuring the amount of CO_2 released either gravimetrically with an absorption train or titrimetrically. As for proximate analysis, formulas (Table 1) can be used to transpose results from one basis to another.

D. Ash Analysis

The composition of ash obviously gives a clear indication of the types of minerals present in coal. For a detailed account of the origin and mode of occurrence of aluminosilicates, quartz, carbonates, sulfides, sulfates, and other minerals present in coal, the reader is referred to the work of Watt (1968).

The British Standard method for ash analysis (BS, 1963) (Figure 2) is broadly similar to the ASTM method (1987i), the 11 constituents determined typically accounting for over 99% of most ashes. Three solutions are prepared and analyzed as follows. The first by fusing the ash with sodium hydroxide and dissolving the melt in dilute hydrochloric acid is used for the colorimetric determinations of (1) silica by adding molybdate tartaric acid and reducing solution (molybdenum blue method) and (2) alumina by preparing calcium-aluminum alizarin red-S complex. The second solution is prepared by digesting ash with hydrofluoric and sulfuric acids and is used to determine:

1. Sodium and potassium by flame photometry

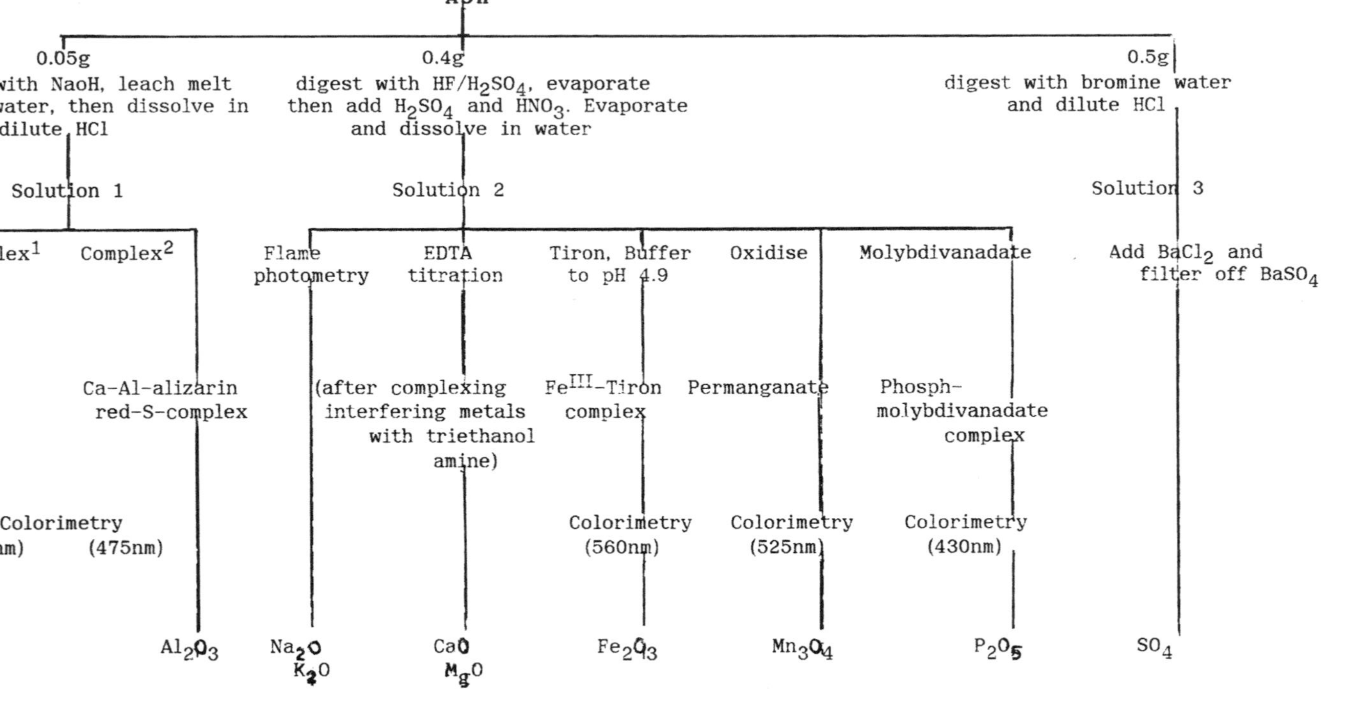

Key 1 = Add molybdate tartaric acid and reducing solution.

2 = Add thioglycollic acid to reduce Fe interference, then add $CaCl_2$ and alizarin-red-S.

Figure 2 Schematic of ash analysis by the British Standard method (BS, 1963).

2. Calcium and magnesium by titration with EDTA after triethanolamine has been added to complex interfering elements, and
3. Iron and manganese by atomic absorption (AA) spectrophotometry

The final solution is prepared by digesting the ash with bromine water and dilute hydrochloric acid, sulfate being determined gravimetrically as barium sulfate.

This procedure is time consuming and, not surprisingly, X-ray fluorescence (XRF) and inductively coupled plasma emission spectroscopy (ICP-ES), which can determine a number of elements simultaneously, are now being used in many laboratories for both major and minor elements (see following). The determination of all the major constituents in ash except phosphorus and sulfur can be accomplished by AA (ASTM, 1987j) but each element has to be determined individually.

E. Minor and Trace Elements

Coal contains a large number of elements present in minor and trace amounts (Table 3). AA, XRF, ICP-ES, neutron activation, and mass spectrometric methods have all been used for the determination of these elements, each procedure having its advantages and limitations.

AA is an established technique in many coal laboratories. Spectral interferences are virtually absent but flames are generally preferable to graphite furnaces for atomization because they are less prone to chemical interferences. However, the sensitivity with graphite furnaces is significantly higher and these are needed for trace metal analysis. Savings in analysis time for both AA and ICP methods can be achieved by direct injection of a coal slurry (Fuller et al., 1981; Sober et al., 1981) rather than use conventional ash preparations.

XRF is a relatively simple method in that X-ray spectra generally contain fewer signals than optical spectra but the method is dependent on the availability of suitable standards for calibration. Elements present in levels greater than a few ppm are readily determined (Kuhn et al., 1975) but XRF is a surface technique and the analysis of fluxed ashes is more straightforward than that of coals. ICP-ES can carry out multielement determinations (including some nonmetals such as S, P, and B) over a wide concentration range, the detection limits generally falling between those of flame and furnace AA. Although slurries can be analyzed (Fuller et al., 1981; Sober et al., 1981), normal practice has been to ash coals prior to analysis and fuse (with lithium metaborate/tetraborate) or acid digest (hydrofluoric acid) the resultant ash for introduction to the plasma.

Table 3 Typical Concentration Ranges for Trace Elements in Coals

Constituent	Range (ppm)
Arsenic	0.50– 93.00
Boron	5.00– 224.00
Beryllium	0.20– 4.00
Bromine	4.00– 52.00
Cadmium	0.10– 65.00
Cobalt	1.00– 43.00
Chromium	4.00– 54.00
Copper	5.00– 61.00
Fluorine	25.00– 143.00
Gallium	1.10– 7.50
Germanium	1.00– 43.00
Mercury	0.02– 1.60
Manganese	6.00– 181.00
Molybdenum	1.00– 30.00
Nickel	3.00– 80.00
Phosphorus	5.00– 400.00
Lead	4.00– 218.00
Antimony	0.20– 8.90
Selenium	0.45– 7.70
Tin	1.00– 51.00
Vanadium	11.00– 78.00
Zinc	6.00–5350.00
Zirconium	8.00– 133.00

Source: Speight, 1983.

The principal advantage of NAA over other methods is that it is nondestructive and involves the minimum of sample preparation. The use of NAA for trace elements analysis and other purposes in coal utilization has been reviewed (Tripathi, 1979) but the need for a neutron source will probably limit its general applicability although on-line measurements have been reported (Boyce, 1983). Spark source mass spectrometry can determine trace elements in whole coals as well as ashes (Sharkey et al., 1975) but relative standard deviations for transition metals have been found to be 6–15% compared to 2–3% for AA (Guidoboni, 1976). It is anticipated that cyclotron Fourier transform mass spectrometry with its extremely low detection limits will make a significant contribution to trace element analysis in the future.

F. Mineral Matter Constitution

A number of instrumental techniques have assisted the identification of minerals present in coal. X-ray diffraction has been used to measure concentrations of clay minerals, pyrite, dolomite, calcite, and quartz in low-temperature ash (O'Gorman and Walker, 1971; Renton, 1978). However, samples need to be spun to remove orientation effects and calibration mixtures containing similar proportions of each crystalline material as the sample are required. SEM and scanning TEM with energy-dispersive X-ray analysis have been used to determine the composition of minerals in both coals and ashes (Moza et al., 1980; Finkelman and Stanton, 1978). Mineral grains as small as 1 μm can be analyzed and the proportions of Al, Si, S, Ca, Fe, Ti and other elements present determined. Micropetrographic analysis can give an indication of the major minerals present, but it is difficult to identify clays and other species from their optical properties and minerals dispersed in organic matter can be overlooked.

IR spectroscopy provided little information on mineral composition before the advent of low-temperature ashing because of the severe overlap between Si-O, Al-O, and other mineral bands with bands due to organic material. Addition and subtraction routines available with Fourier transform instruments has considerably expanded the use of IR in coal analysis (Painter et al., 1978) (Sec. III.F). For example, subtraction of the spectrum of kaolinite from that of low-temperature ash has enabled less strongly absorbing minerals, such as mortmorillonite, to be identified. ^{29}Si and ^{27}Al NMR have been used to differentiate silicon in coal minerals according to the number of adjacent aluminum sites and to resolve tetrahedronally and octahedronally bound aluminum (Barnes et al., 1986).

III. ORGANIC STRUCTURE OF COAL

A. Introduction

The goal of coal chemists since the early 1950s has been to elucidate the nature of the organic structure in coals. This has mainly involved estimating the concentrations of aromatic, aliphatic, and heteroatomic groups present. However, in relation to coal liquids (Sec. IV), the fact that coals are largely insoluble in common organic solvents poses considerable problems. More recently, attention has focused on the macromolecular nature of coal (Table 4); many degradative, pyrolytic, spectroscopic, and other physical methods have been applied to coal (Given, 1984; Berkowitz, 1985) but space permits only a brief discussion here.

B. Oxidation

Coals oxidize readily even at ambient temperatures and when powerful oxidizing agents are used, such as potassium permanganate, nitric acid, and hydrogen peroxide, an array of low molecular weight carboxylic acids is obtained (Wender et al., 1981). Clearly, reagents that selectively oxidize either aromatic or aliphatic substituents are the most informative concerning coal structure.

Aqueous sodium hypochlorite was thought to oxidize only adamantyl-type structures at high pH and initial results suggested that over 50% of carbon in bituminous coals is sp^3-hybridized (Chakrabartty and Berkowitz, 1974), which obviously contradicts the generally accepted view that coal is largely aromatic. However, further studies have indicated that, as the pH drops, a range of aromatic and aliphatic structures can be oxidized to a significant extent (Mayo, 1975; Landolt, 1975). More recently, aqueous sodium dichromate at ∿250°C has been found to be a selective oxidant for aliphatic groups (Hayatsu et al.,

Table 4 Summary of Chemical Methods for Investigating Organic Coal Structure

Method/reaction	Measurement/product
Oxidation—NaOCl	Adamantyl together with various aromatic and aliphatic groups are oxidized
—$Na_2Cr_2O_7$	Aliphatic groups are oxidized selectively leaving aromatic carboxylic acids
—RuO_4	Aromatic groups are oxidized selectively leaving aliphatic carboxylic acids

(continued)

Table 4 (Cont.)

Method/reaction	Measurement/product
Depolymerization/alkylation/acylation	
–BF_3/phenol –reductive alkylation	Both reactions give high conversions to soluble products but side reactions complicate the results
–transalkylation	Toluene acts as receptor for alkyl groups
–methylation/acetylation	Widely used to estimate OH concentrations
Dehydrogenation	Used to estimate concentrations of hydroaromatic groups
Nonaqueous potentiometric and enthalpimetric titrations	Estimation of acidic and basic groups
Ion exchange	Estimation of carboxylic acids in low-rank coals
Pyrolytic methods–TPR	Specify S functionality
–pyrolysis	Identify long alkyls and other aliphatic groups
–MS	Characteristic fingerprints of coals
Spectroscopic methods	
–high-resolution ^{13}C NMR	Aromaticity and other C-type concentrations
–low-resolution ^{1}H NMR	Proportions of mobile and rigid components in structure
–FTIR	Proportion of OH, aromatic and aliphatic H
–MS (soft ionization, FAB)	Spectra of decomposition products
–XPS	Distribution of C, O, and N groups

1978), a number of benzene, polycyclic, and heterocyclic ring systems in lignites and bituminous coals being identified by GC/MS analysis of the methylated carboxylic acids (Table 4). The proportion of three-ring aromatic nuclei in the oxidized products was found to increase with increasing rank.

Trifluoroperoxyacetic acid (Deno et al., 1980, 1985) and ruthenium tetroxide (Stock and Tse, 1983) are selective oxidizing reagents that leave aliphatic carbons intact; concentrations of alkyl substituents on aromatic nuclei can be estimated from the yields of the corresponding aliphatic monocarboxylic acids (i.e., arylmethyl gives acetic acid). However, the origin of aliphatic polycarboxylic acids is less clear as these can arise from both hydroaromatic and bridging alkylene groups.

C. Depolymerization and Alkylation Procedures

Transalkylation with phenol and boron trifluoride (Heredy and Neuworth, 1962; Heredy et al., 1965) is a well-established procedure for breaking the macromolecular structure of coals into soluble molecular units. The yield of soluble product and the amount of incorporated phenol in the product decreases markedly with increasing coal rank, which probably reflects the lower concentrations of aliphatic-linked aromatic structures that can exchange with phenol in the higher rank coals. More recently, transalkylation catalyzed by trifluoromethanesulfonic acid with toluene as the receptor for alkyl groups has been used to identify and measure concentrations of alkyl chains up to six carbons long in coals (Benjamin et al., 1985).

Reductive alkylation has been widely used to break down the macromolecular structure of coals (Sternberg et al., 1971; Sternberg and Delle Donne, 1974; Wender et al., 1981). In the procedure developed by Sternberg and coworkers (1971, 1974), an alkali metal with naphthalene in tetrahydrofuran (THF) is used to form coal anions which then can be readily alkylated (see below). Naphthalene acts as an electron transfer reagent since neither coal nor alkali metals are soluble in THF. Electron transfer to the coal results in the cleavage of

$$Na + C_{10}H_8 \rightarrow Na^+ + C_{10}H_8^{\dot{-}}$$

$$Coal + nC_{10}H_8^{\dot{-}} \rightarrow Coal^{n-} + nC_{10}H_8$$

$$Coal^{n-} + nR\text{-}I \rightarrow Coal - (R)n + nI^-$$

$$ArOR + C_{10}H_8^{\dot{-}} \rightarrow Ar\text{-}\dot{O}^- + \dot{R} + C_{10}H_8$$

$$ArCH_2Ar + C_{10}H_8^{\dot{-}} \rightarrow Ar\text{-}\dot{C}H_2 + Ar^{\dot{-}} + C_{10}H_8$$

methylene bridges between two aromatic groups and of ethers (see above). The naphthalene anion can abstract protons from phenols to give the corresponding phenoxide ions. It has been demonstrated that the increases in coal solubility arise mainly from the alkylation step, protonation of coal anions giving only marginal improvements in solubility (Wender et al., 1981; Stock, 1982). Significant quantities of naphthalene are incorporated into the coal but this can be avoided by using alkali metals in liquid ammonia. Coals containing 75–90% daf carbon have yielded over 90% pyridine solubles and up to 90% benzene solubles for low-volatile bituminous coals (Sternberg and Delle Donne, 1974). Typically, 10 alkyl groups per 100 C atoms in bituminous coals are added via reductive alkylation. The number of average molecular masses ($\bar{m}_n$) of benzene-soluble materials is significantly higher for high-rank bituminous coals (1300–2000, 88% daf C versus 500–800, 78% dafC) which is probably indicative of the fewer easily reducible cross-links (arylethers and diarylmethanes) in these coals.

Significant improvements in the solubility of high-rank coals (87–90% daf C) can be achieved by alkylation or acylation with acid catalysts, such as aluminum chloride in the Friedel–Crafts reaction (Wender et al., 1981; Hodek and Kolling, 1973). The introduction of five acyl groups (C_{15}) per 100 C atoms has rendered a steam coal 85% soluble in pyridine. These findings suggest that higher rank coals have the largest number of easily accessible aromatic C-H groups and that associative interactions involving aromatic groups must limit yields of low-temperature solvent-extractable material.

Acetylation has been the most widely used procedure to estimate phenolic oxygen contents of coals, the phenolic oxygen typically accounting for over half of the total oxygen (Yarzeb et al., 1979; Given, 1984) in coals containing 76–86% daf coal. Nonaqueous potentiometric titration and other derivatization procedures which can be used to estimate total acidity and phenolic group concentrations are covered elsewhere (Sec. IV.C). It is generally accepted that only lignites and brown coals contain carboxyl and carbonyl groups in significant concentrations. Appreciable amounts of inorganic ions are present as salts of carboxylic acids and the ion exchange properties of low-rank coals form the basis of methods for measuring acidity (Schafer, 1970), such as exchange with barium acetate and titration of the released acetic acid. The extent of cation exchange is a function of pH with a value of ∿8 and ∿13 being required for carboxyls and phenols (total acidity measured), respectively. However, total acidities measured by ion exchange are generally lower than those determined by nonaqueous photentiometric titration in ethylenediamine (Maher and Schafer, 1976), presumably because some phenols are not exchanged by barium salts in aqueous solution.

A number of reductive procedures have been used to estimate carbonyl concentrations in low-rank coals. However, problems arise in the specificity of reagents since both ketones and quinones may be present. The highest apparent concentrations for brown coals have been determined by sodium borohydride reduction followed by acetylation of the resultant OH groups (Chaikin and Brown, 1949; Kroeger et al., 1965).

Relatively few attempts have been made to measure the concentrations of ether groups because of their low reactivity in relation to other oxygen functional groups and concentrations are often estimated by difference between the total and phenolic oxygen for bituminous coals. Protonation of coal anions formed by treatment with potassium and naphthalene in THF has increased the number of hydroxyl groups which can be acetylated (Ignasiak and Gawlak, 1977). Similarly, reduction with sodium in liquid ammonia increases the intensity of OH-stretching vibrations (3300–3500 cm^{-1}) in the IR spectra of the products compared to the initial coal. Along with reducing systems in basic media, ether bond cleavage can be achieved by refluxing coal in hydrogen iodide (Wender et al., 1981). However, heterocyclic ethers (e.g., dibenzofuran) are extremely stable and are not affected by the reactions outlined above.

Dehydrogenation has been used extensively to estimate concentrations of hydroaromatic groups in coals. The studies by Reggel and coworkers (1968, 1973) using $Pd/CaCO_3$ in boiling phenanthridine indicated that the concentration of hydroaromatic groups falls with increasing rank and, for U.S. coals containing ∿82% daf C, the hydroaromatic groups accounted for 30–35 H (∿15 C) atoms per 100 C atoms. However, secondary reactions involving groups other than hydroaromatic/naphthenic rings can affect dehydrogenation results and the yields of hydrogen are dependent on the metal used, palladium giving the highest yields. Indeed, NMR analysis of coal extracts and tars (Sec. III.E) has suggested that hydroaromatic/naphthenic groups account for no more than half of the aliphatic carbon (10 H per 100 C atoms) in bituminous coals containing ∿82% daf C.

D. Pyrolytic Methods

Pyrolytic methods can be broadly classified as requiring sufficiently high temperatures (>350°C; >660°F) to cleave C-C, O, N, and S bonds (information on the use of pyrolysis MS to fingerprint the organic matter in coals can be found in Sec. III.G).

Temperature-programmed reduction (TPR), which involves the reduction of coal in a stream of hydrogen and with suitable reducing agents has been used to determine the distribution of organic sulfur groups in coals (Attar and Dupuis, 1981), different groups having characteristic reduction temperatures. Thiols and aliphatic thioethers reduce at much lower temperatures than thiophenes. Indeed, much of the sulfur in bituminous coals is not reduced at temperatures below 600°C (1110°F) in TPR at atmospheric pressure, suggesting that it is present as stable benzo- and naphthothiophenes. Results for U.S. coals (Attar and Dupuis, 1981) have indicated that lignites contain a smaller proportion of thiophenic sulfur than bituminous coals but the distribution of sulfur groups does not vary markedly with increasing organic sulfur content for coals of similar rank. Two alternative procedures to TPR that have been used to probe the distribution of organic sulfur groups are:

1. Temperature-programmed oxidation where the sulfur is oxidized to sulfur dioxide at characteristic temperatures for different groups (Lacount et al., 1986), and
2. Fluidized bed pyrolysis where the yields of hydrogen sulfide, carbon disulfide, and carbonyl sulfide (COS) at particular temperatures can be related to the sulfur group concentrations in the parent coals (Calkins, 1984).

High-yield extracts obtained via mild liquefaction and tars from low-temperature (<600°C; <1100°F) pyrolysis have been particularly useful in providing information on the chemical nature of the parent coals, although thermolysis is invariably responsible for some structural modification. NMR has proved to be the preeminent spectroscopic technique for evaluating the bulk composition of high molecular mass extract and tar fractions, such as asphaltenes (Sec. IV.A). Indeed, it is much less problematical deriving reliable structural data by solution state ^{1}H and ^{13}C measurements than by solid state ^{13}C techniques (Sec. III.E). Extract characterization has confirmed that lignites contain less condensed aromatic structures than bituminous coals (Ladner et al., 1980) (one to three rings for ∿82% daf C) and indicated that alkyl substituents account for at least half of the aliphatic carbon in low-rank coals (Snape et al., 1985). Long alkyl chains account for up to 10% of the carbon in lignites but only, typically, 1–2% in bituminous coals (Calkins, 1984; Snape et al., 1985). Although it is recognized universally that coal is extremely heterogeneous in character, in both low-temperature hydropyrolysis (Bolton et al., 1988) and base-mediated liquefaction in carbon monoxide and steam (Ross et al., 1986), where catalysis rather than thermolysis governs the level of conversion, there is little variation in the structure of the liquid products as conversion increases up to ∿80% daf coal. These findings suggest that there could be a degree of homogeneity in the chemical nature of the vitrinite mineral group which, for most coals, accounts for the bulk of primary liquefaction products.

E. NMR Spectroscopy

During the 1950s, bulk composition data, including the proportion of aromatic carbon of the total carbon (aromaticity) and the number of carbons per aromatic group, were derived from XRD and statistical constitution analysis (van Krevelen, 1961). The latter uses structural correlations with physical properties, such as density (d) and refractive index (n), n-d-m (m is molecular mass) and is still being used in the petroleum industry to determine proportions of aromatic, paraffinic, and naphthenic carbon. However, for coal tars and extracts, these approaches were superseded during the 1960s and early 1970s by solution state ^{1}H and ^{13}C NMR (Sec. IV.H). More recently, the combination of dipolar decoupling and magic-angle rotation (MAR) to remove broadening in the ^{13}C spectra of solids from dipolar interactions and chemical shift anisotropy (CSA), respectively, has enabled high-resolution ^{13}C spectra of coals to be obtained in which aromatic and aliphatic carbon bands are well resolved (Axelson, 1985; Davidson, 1986) (Figure 3). The use of cross-polarization in which magnetization is transferred from abundant ^{1}H to dilute ^{13}C spins obviates the need for long relaxation delays between successive scans. It has been found that that signal-to-noise levels in ^{13}C spectra of coals can be further improved by irradiating at the Larmor frequency of the free electron (dynamic nuclear polarization), so that magnetization is transferred from the free unpaired electrons in coals ($\sim 10^{19}$ spins/g) (Wind, 1986).

At relatively low magnetic field strengths (<2.3 tesla equivalent to an observation frequency of 25 MHz for ^{13}C), MAR speeds of about 3 kHz are usually adequate to remove CSA and obtain spectra free of sidebands (Figure 3). Unfortunately, much greater speeds are required as the field strength increases (CSA roughly proportional to field strength) and, although spinning speeds in excess of 5 kHz are feasible, it has been necessary to resort to sideband suppression techniques, such as TOSS (Axelson, 1985), to obtain sideband free spectra of coals at high fields (^{13}C frequencies >50 MHz). However, relative peak intensities may be affected by differences in spin-spin relaxation (characterized by time constant, T_2), since even at MAR speeds of 3 kHz, the 0.3-msec time scale of the TOSS pulse sequence is considerable in relation to the T_2 values for ^{13}C spins in coals (Axelson, 1985).

The aromatic and aliphatic carbon bands in the ^{13}C spectra of coals are usually broad and featureless but relaxation techniques, in particular dipolar dephasing, have resolved peaks for different carbon types. In the dipolar dephasing technique, a short delay (~ 50 μsec) before data acquisition is introduced into the cross-polarization pulse sequence (Figure 4) so that ^{13}C spins relax via dipolar mechanisms. Quaternary aromatic and mobile CH_3 carbons have the weakest ^{1}H-^{13}C dipolar interactions and, consequently, these carbons relax much more

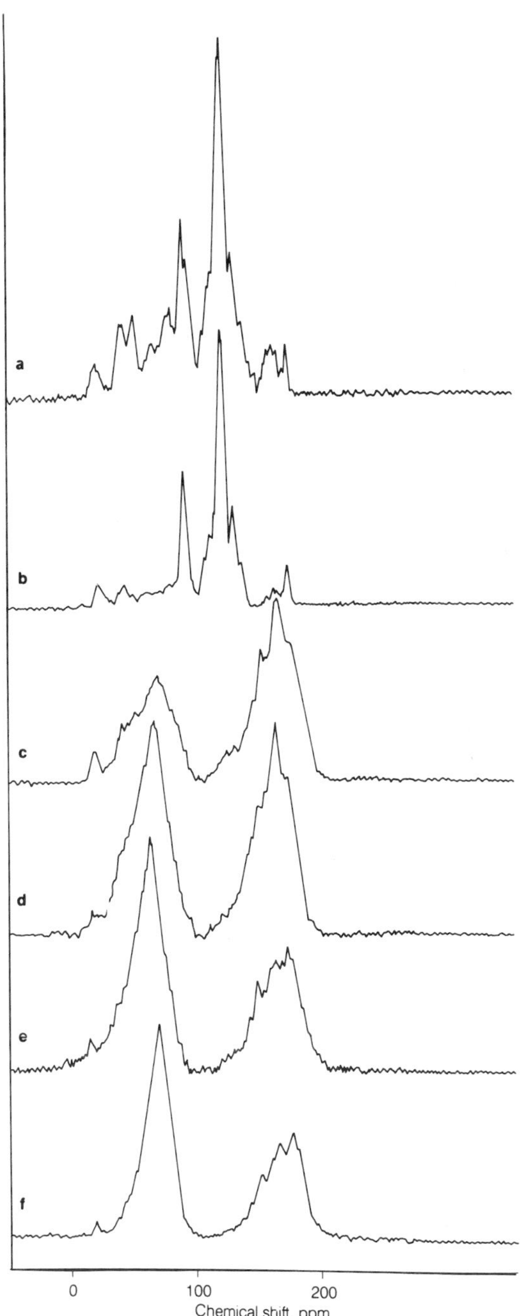

Figure 3 Solid state CP/MAR ^{13}C NMR spectra of peats (a, b), lignites (c, d, e), and bituminous coal (Dereppe et al., 1983). Reproduced with permission of Butterworth Scientific Ltd.

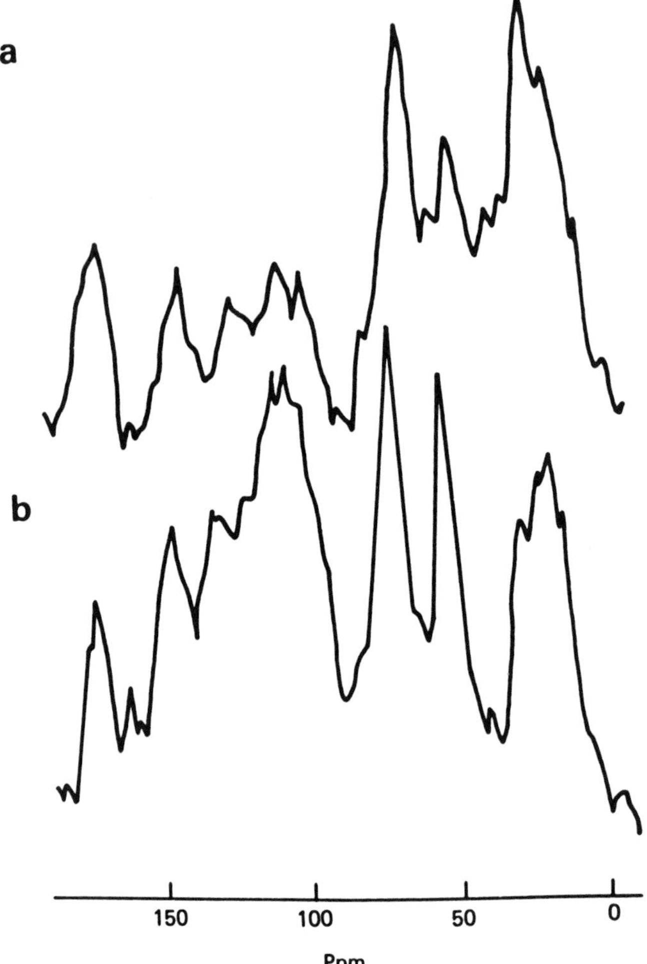

Figure 4 50-MHz CP/MAR ^{13}C NMR spectra of a peat. (a) Normal spectrum; (b) TOSS spectrum (Axelson, 1987). Reproduced with permission of Butterworth Scientific Ltd.

slowly than tertiary aromatic and rigid aliphatic carbons. Indeed, after about a 50-μsec delay, peaks due to quaternary aromatic carbon and mobile CH_3 groups are observed selectively by dipolar dephasing (Figure 4). Variable cross-polarization time experiments (Davidson, 1984) and the analysis of aromatic sideband patterns (Burgar et al.,

1985) have also been used to resolve different carbon types in the ^{13}C spectra of coals.

The derivation of aromaticity and other parameters relating to the distribution of carbon types by high-resolution ^{13}C NMR has undoubtedly been a major advance in coal science. However, there has been considerable debate over whether CP/MAR ^{13}C spectra of coals can be considered quantitative. Although the rate of magnetization transfer during CP for different carbon types varies considerably (Axelson, 1985) with quaternary aromatic carbons being polarized more slowly than tertiary ones, the major problem for coals is the presence of organic free radical centers. These are responsible for many of the ^{1}H spins relaxing too quickly during CP, i.e., before magnetization can effectively be transferred to ^{13}C spins. Moreover, carbons in the vicinity of free radicals are not likely to be observed because of shielding effects and short relaxation times. Internal standard and carbon-counting procedures have suggested that, typically, about half of the carbon in bituminous coal is observed by CP (Botto et al., 1985; Hagaman et al., 1986). Plots of aromaticity versus H/C ratio for a range of coals show considerable scatter, which is considered to arise because not all of the carbon is observed. More of the carbon is generally observed when normal accumulation (single-pulse) methods are used instead of CP for spectra accumulation (Botto et al., 1985), but because of the relatively long delays required between successive scans, long accumulation times are generally needed to achieve adequate signal-to-noise ratios.

Despite the uncertainties concerning quantification, there are numerous important applications of solid state ^{13}C NMR in coal research (Davidson, 1986). The spectra of peats, brown coals, lignites, bituminous coals, and anthracites (Figure 3) have shown unequivocably that OCH_2 (cellulose-derived material), carboxyl, and aliphatic carbon groups are lost as coalification progresses. Wide variations in aromaticity have been found between exinite, vitrinite, and inertinite group macerals. Dipolar dephasing experiments have indicated that anthracites are composed of highly condensed aromatic structures and that hydrogen aromaticity (fraction of aromatic hydrogen from total hydrogen) increases with increasing coal rank (Davidson, 1986). Concentrations of carboxyl and phenolic groups have been estimated from the spectra of coals methylated with ^{13}C-enriched reagents (see Sec. IV.H).

Solid state ^{1}H NMR spectra of coals in which aromatic and aliphatic hydrogen bands are partially resolved have been obtained by combining high-power multipulse methods (E.g., MREV-8, BR24; Axelson, 1985) and MAR to remove peak broadening principally from homonuclear dipolar interactions. Combined rotation and multipulse spectroscopy (CRAMPS) is particularly informative for low-rank coals (Figure 5, Rosenberger et al., 1983) with methoxyl and carboxyl-phenolic peaks being resolved in their spectra.

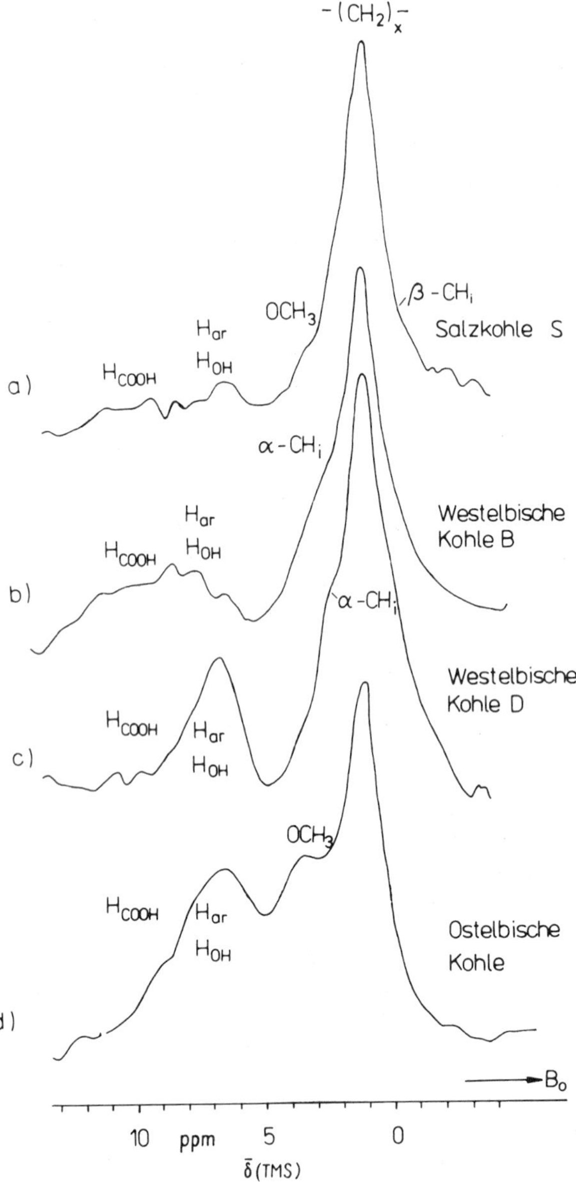

Figure 5 High-resolution 270-MHz ^{1}H NMR spectra of German brown coals (Rosenberger et al., 1983).

Broadline pulsed ^{1}H NMR measurements have been useful in probing the macromolecular structure of coal and mechanisms involved in coal pyrolysis (Davidson, 1986). The transverse magnetization free induction decays for coals have been resolved into faster and slower decaying components (time constants, T_{2S}, being ∿10 and 35 μsec for bituminous coals). It is generally accepted that the two components arise from differences in mobility (Given et al., 1986) but it is probably an oversimplification to assume that the faster relaxing, less mobile component can be attributed to molecules held within the macromolecular structure. Pyridine extraction of coals removes most of the slower relaxing hydrogen but, on the other hand, swelling coals in pyridine significantly increases the proportion of the slower relaxing hydrogen presumably due to greater mobility in the macromolecular structure.

F. Infrared Spectroscopy

The first IR spectra of coals were published during the 1950s using ground coal dispersed in KBr disks and variations in band intensities were correlated with rank and carbonization treatments (Brown, 1955). However, the broad nature of the principal bands in coal spectra due to C-H, C-C, C-O, O-H, and C=O vibrations (see Table 5 for assignments) and differences in extinction coefficients meant that, at best, IR was only semiquantitative. Since the introduction of Fourier transform spectrometers, the use of IR to characterize the organic structure of coals has expanded considerably. Much improved signal-to-noise levels can be achieved by accumulation of a number of scans, and wavenumber definition is superior compared to that achieved with traditional grating instruments. Spectra of neat powdered coal can be obtained by diffuse reflectance without the need to prepare KBr disks.

Data manipulation techniques which have been applied to coals include spectral subtraction (Sec. II.F), Fourier self-deconvolution, and curve fitting (Painter and Coleman, 1980). Fourier self-deconvolution can enhance resolution considerably (Wang and Griffiths, 1985) (Figure 6) if signal-to-noise levels are high and the shapes of absorption bands are known. Curve fitting requires particular care because many combinations of bands can be used to fit a given multiplet. Although IR spectra of most solids and liquids are assumed to consist of Lorentzian bands, the assumption of a Gaussian band shape has been found to give the more acceptable results when curve-fitting routines are applied to coals (Solomon, 1979). However, individual Lorentzian profiles in a broad unresolved band could well be fitted in this way (Wang and Griffiths, 1985).

Table 5 Summary of IR Band Assignments for Coals and Coal Products

Frequency (cm^{-1})	Assignment
750–810	Substituted aromatic rings
850–870	Unsubstituted aromatic rings
1250	Ethers, phenols
1390	Methyl
1450	Aliphatic, alicyclic C-C
1600	Aromatic C-C, intensity increased by polar substituents
1700–1780	Carbonyl, carboxyl
2800–2960	Aliphatic C-H
2869, 2956	$-CH_3$
2852, 2924	$-CH_2$
2890	$-CH$
3030–3050	Aromatic CH
3300	OH

Despite the uncertainties, curve-resolving techniques have helped in the measurement of OH concentrations for coals using either the broad O-H band (3200–3600 cm^{-1}) directly (Solomon and Cavangelo, 1982) or the carboxyl band (1740–1770 cm^{-1}) resulting from acetylation (Given, 1984). Fourier transform infrared (FTIR) spectroscopy has been used to estimate proportions of aromatic and aliphatic hydrogen in coals but few attempts have been made to calibrate measurements using soluble extract fractions whose hydrogen aromaticity can be determined by ^1H NMR (Painter et al., 1983).

G. Mass Spectroscopy

The application of mass spectrometry to the analysis of coals is severely limited by the requirement that the molecules be in the gas phase during ionization; vapor pressures of 10^{-6}–10^{-5} mmHg are

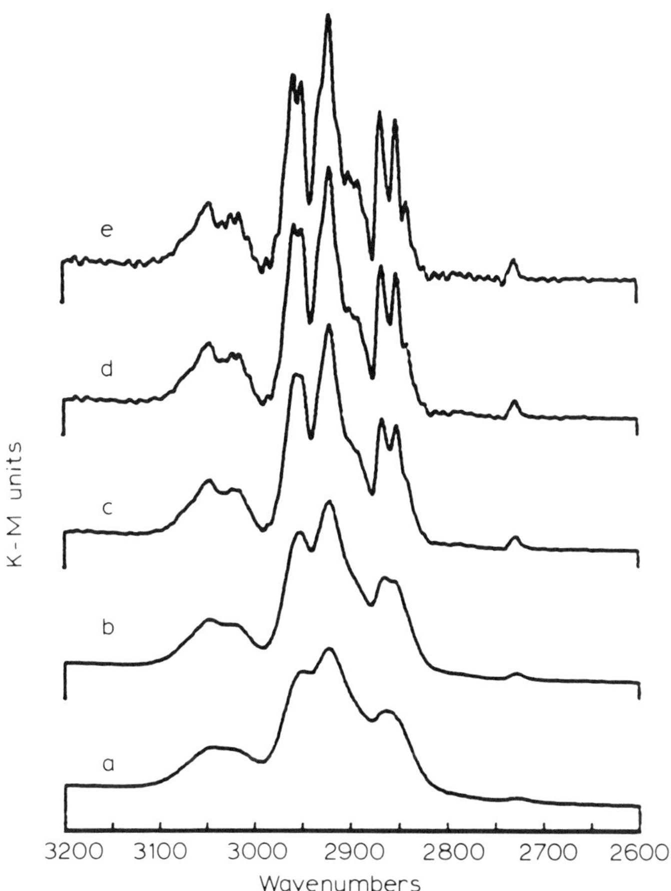

Figure 6 Effect of Fourier self-deconvolution on the resolution of an FTIR spectrum of coal (Wang and Griffiths, 1985). Reproduced with permission of Butterworth Scientific Ltd.

generally necessary. The larger molecules present in coal are likely to decompose before significant conversion to the gas phase can occur, and methods are required by which solids may be ionized without volatilization. This may be brought about by a number of "soft," or desorption, ionization processes which have the further advantage that molecular ions rather than fragment ions are produced in contrast with electron impact methods.

Energy is transferred to the surface to disrupt the solid coal and produce fragments, some of which are ionized. Energy input via a laser beam to an area of 10^{-7}–10^{-2} cm^2 at low-beam energies yields water, methane, ethene, and carbon oxides as molecular ions (Gaines and Page, 1983; Vastola and McGahan, 1987). At higher beam energies (up to 80 W/cm^2), substantial decomposition of the coal macromolecule occurs and $C_nH_2^+$ ions characteristic of polyacetylenes are observed, in addition to the molecular ions of alkylbenzenes and naphthalenes, dihydroxyphenols, and benzofurans. Bombardment with high-velocity particles, such as noble gas ions with up to 10 keV energy, is known as secondary ion mass spectrometry (SIMS). This method gives significant peaks with mass numbers up to 250 at low-beam energies but spectra contain mainly low-mass number peaks (i.e., more fragmentation) at higher energies (Wolf, 1983; Gaines and Page, 1983). Molecular beams of fast noble gas atoms (fast atom bombardment, FAB) are also thought to produce mass spectra with features similar to those of laser bombardment or SIMS. In all cases the decomposition of the coal occurs rather than evaporation of individual molecules.

Combination of MS with pyrolysis has proved to be a very useful method of characterizing macromolecules, including coals. Curie point pyrolysis, in which the coal is coated on a ferromagnetic wire and heated very rapidly, may precede mass spectrometry (PyMS), gas chromatography (PyGC), or GC in combination with MS (PyGCMS). PyMS of a wide range of coals produced ions with mass numbers up to 260 and a gradual shift in the composition of the pyrolysis products with rank. The relative intensities of hydrocarbon ions (alkylbenzenes, -naphthalenes, -phenanthrenes, and -biphenyls) all increase with increasing rank, whereas the intensities of alkylphenols decrease, as do the intensities of ions from dihydroxybenzenes and methoxyphenols for lignites. Correlations of numerous coal properties with PyMS data have been reported (Meuzelaar et al., 1984a, b, 1987).

More detailed information on the composition of Curie point pyrolysis products was obtained by PyGCMS (Tromp et al., 1988). With increasing coal rank, the amount and number of pyrolysis products decrease significantly. Thus for the anthracite and semianthracite, hardly any products with functional groups are formed, with polynuclear aromatic compounds, extending beyond the retention of benzopyrene and benzofluoranthene isomers (some with sulfur in the ring system), being the major compounds. Aliphatic products, especially *n*-alkanes up to C_{31} and, for low-rank coals, *n*-alkenes, have also been detected by PyGCMS.

Soft ionization MS in combination with pyrolysis of coals has extended the recorded mass range up to at least ∿1000, with almost exclusive formation of molecular ions. For example, Schulten (1982) used pyrolysis field desorption MS (FD-MS) in which coal deposited on a wire emitter was heated to temperatures up to 2000°C (3630°F).

The wire was coated with graphitic microneedles and subjected to a very high electric field. Ions with m/z from 200 to 800 were desorbed up to 300°C (570°F); at higher temperatures, intense ions with m/z from 1000 to 3000, believed to originate in large organic fragments attached to Na^+ and K^+, were observed.

More recently, the related technique of field ionization MS (FI-MS) was applied to coal (Shulten and Marzec, 1987). Samples were heated to 500°C with time and temperature resolution in an aluminum crucible held near an emitter wire in the ion source. The integrated spectrum contained a continuous range of molecular ions with mass up to m/z 800 (Figure 7).

H. X-ray Photoelectron Spectroscopy

The technique of X-ray photoelectron spectroscopy (XPS), also known as electron electroscopy for chemical analysis (ESCA), determines the

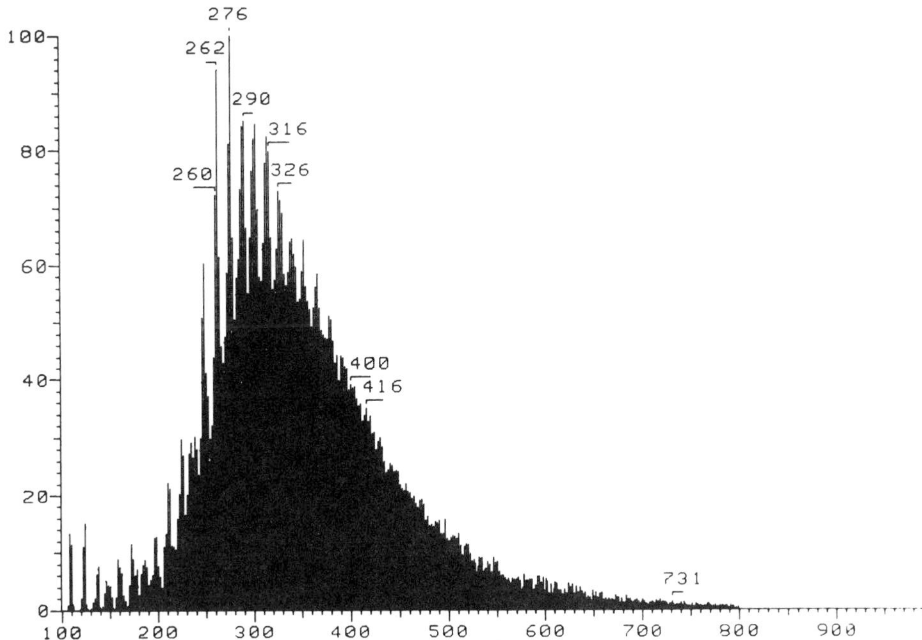

Figure 7 Integrated (40-spectra) field ionization mass spectrum of a coal heated from 50 to 500°C (Shulten and Marzec, 1987). Reproduced with permission of Elsevier Publishing Company.

binding energies of both core and valence electrons by measurement of the kinetic energies of electrons photoemitted from a solid when irradiated in vacuo with near-monochromatic X rays. Since the core electron binding energies are characteristic of the element, XPS provides elemental analysis for all elements except hydrogen and helium. Small shifts in the characteristic binding energy can be measured and it is therefore possible to use XPS to identify the functional group in which the element is present. XPS has the advantage of constant sensitivity to an element irrespective of the functional group to which it may contribute. Functional group analysis obtained is therefore quantitative and may be calibrated with known sensitivity factors for a given element.

XPS is an excellent probe of the chemistry of coal surfaces (Perry and Grint, 1983; Clark and Wilson, 1983). The surface composition of coal determined by XPS has been compared with the results of bulk analysis. For coals ground under heptane a good correlation was observed for oxygen, but the surface-to-bulk correlation was less good for sulfur, nitrogen, and chlorine, and poor for silicon and aluminum (enriched at the surface) and iron (depleted) (Perry and Grint, 1983). For coals ground in air and for a series of float-sink fractions of coking coal, the oxygen correlation was less good (Frost et al., 1974).

The chemical shift to different binding energies of XPS signals from elements in different oxidation states and functional groups has been exploited in the analysis of coal and to determine chemical changes brought about by oxidation and carbonization. For example, sulfur is observed in most coals in two binding states: one corresponds to organic sulfur and pyrite, and the other to oxidized sulfur species, both organic and inorganic. Jones et al. (1981) exploited this difference by using oxidation with performic acid to produce a new peak attributed to sulfones from which they concluded that the organic sulfur in the original coal is present as thiophenic or sulfidic groups.

Both carbon and oxygen in coal give rise to single XPS peaks, but the C(Is) signal may be curve-resolved or peak-synthesized to reveal different C-O functional groups (Perry and Grint, 1983), i.e., ethers, hydroxyls, carbonyls, or carboxyls. With the aid of these components it is possible to fit the carbon peak envelope of coal to a synthesized spectrum. Using this approach, oxidation and carbonization of a bituminous coal has been investigated (Perry and Grint, 1983). Oxidation was seen to occur initially via the exterior surface, producing a distribution of carbon-oxygen groups. Singly bonded C-O species predominate at all temperatures, stable carboxyl groups being formed in significant proportions only at temperatures above 250°C (480°F). Carbonization resulted in the formation of ether linkages by condensatio of hydroxyl groups.

Nitrogen gives a single peak for all coals, which can be resolved into two components (Jones et al., 1981; Perry and Grint, 1983). The attribution of these to pyrole and pyridine structures was recently confirmed by comparing the N(Is) spectrum of pure compounds with those of coal liquids and their nitrogen-rich fractions (Bartle et al., 1987). The variation of nitrogen functionality with coal rank and in macerals has been investigated by XPS (Burchill, 1987). Pyrrolic nitrogen predominates throughout the bituminous range, but the proportion of pyridinic nitrogen increases with rank. Coal nitrogen is concentrated in the vitrinite group of macerals, which also contain the highest proportion of pyridinic groups.

I. Hypothetical Structures

It has been customary to present bulk compositional data derived using a combination of elemental analyses and the spectroscopic and physical methods discussed here by either presenting the results on the basis of 100 C atoms (see Table 6 as an example) or constructing the hypothetical average structures whose compositions closely fit the experimental data (see Figure 8; Given, 1960). While the latter give an indication of the types of structure present, they cannot be used to provide information on how the various groups are

Table 6 Summary of ^{13}C NMR Chemical Shift Assignments for Coal Liquids

Chemical shift (ppm)	Carbon type
170–210	Carboxyl, carbonyl
148–170	Mainly aromatic C-O
129.5–148	Mainly aromatic C-C
105–129.5	Mainly aromatic C-H
22.5–60	Aliphatic CH_2, CH
32, 29.7, and 23	Characteristic alkyl CH_2 groups
18–22.5	Mainly arylmethyl
10–18.5	CH_3 not substituted in aromatic and naphthenic rings
14	Terminal CH_3 in $>C_3$ *n*-alkyl chains

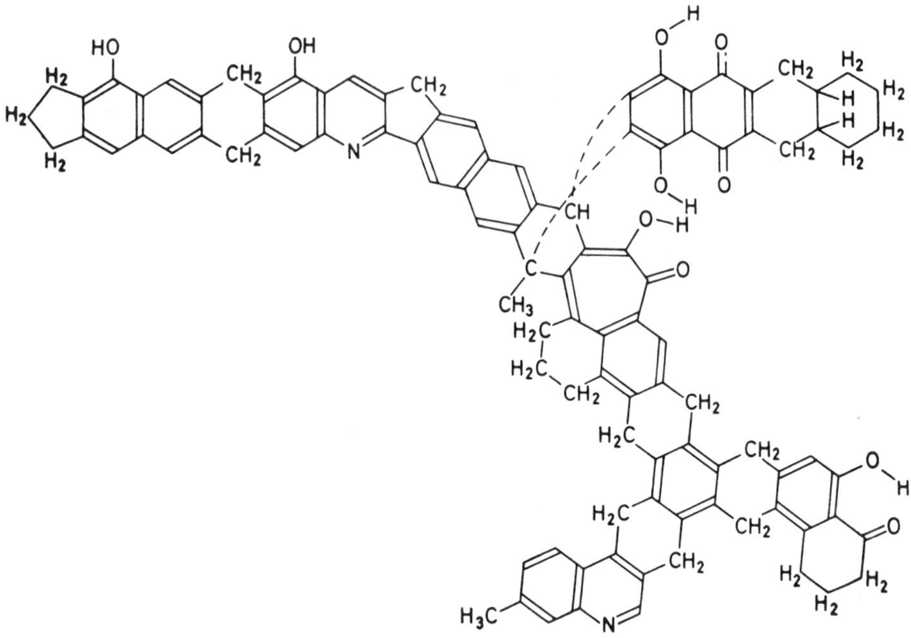

Figure 8 Hypothetical structure for a bituminous coal (Given, 1960). Reproduced with permission of Butterworth Scientific Ltd.

bound together. Therefore, it is extremely misleading to draw conclusions on the two- and three-dimensional arrangement of the macromolecular structure of coals based solely on bulk compositional data. For example, swelling and extraction results suggest that hydrogen bonding is extensive within the macromolecular framework and, hence, phenolic OH, basic N, and other polar groups are likely to be in close proximity. The authors are not alone in questioning the usefulness of hypothetical structures (Given, 1984) and it is recommended that the 100 C atom basis is used to compare coals of differing rank.

It will be some time before coal scientists have a reliable tool for the precise determination of aromaticity despite the developments in solid state ^{13}C NMR (Sec. III.E). In early studies, values derived by statistical constitution analysis (van Krevelen, 1961) were higher than those estimated from IR and other spectroscopic measurements (Dryden, 1964). An important consequence of underestimating aromaticities has been to foster the idea that hydroaromatic groups account for the bulk of aliphatic carbon in bituminous coals (low aliphatic H/C ratios are deduced if aromaticities are underestimated).

Although dehydrogenation results have supported this viewpoint, extraction studies have suggested that alkyl groups account for about half of the aliphatic carbon (Snape et al., 1985) and this area clearly needs further investigation. In contrast, it is generally accepted that the aromatic groups in bituminous coals containing ∿82% cmmf C (vitrinite, exinite, maceral groups) are not highly condensed and consist mainly of one to three ring structures. Indeed, oxidation and extract characterization studies (Hayatsu et al., 1978; Ladner et al., 1980) have indicated that the aromatic structure is probably considerably more open chain than that based on early spectroscopic data (Figure 8).

One additional aspect of structural analysis that is very important and which is related to swelling/molecular mass studies is the description of the porous nature of coal. In this regard coal is similar to a molecular sieve, i.e., its thoroughly porous structure makes it possible for some molecules to move in and out of the cavities easily while others cannot. From the time of its formation, some material has remained trapped inside the pores (either because of its size or because of its bonding interactions) but may be released when the coal is heated or is subjected to swelling and solvent extraction. Micropores range in size from less than 4 to 12 Å; mesopores are defined as being from 12 to 300 Å and macropores are those from 300 to 29,600 Å in diameter (Tsai, 1982). Low-rank coal is extremely porous and the pores are large. Midrank coals have lower porosity and as much as 80% of the pores are micro- or mesopores. Anthracite has large pores and is composed primarily of macropores. The internal surface area of coal (measured by CO_2 diffusion at -78°C) is related to the pore size distribution; the internal surface area increases with decreasing rank and is a minimum for the midrank prime coking coals (Figure 9). The internal surface area then increases with rank to the hvC bituminous coals. Lignites have an internal surface area slightly higher than the mvB bituminous coals primarily because the pores are in the macropore range (Tsai, 1982). From a correlation of the internal surface area with maceral composition and moisture content of coals it has been noted that vitrinite has the largest internal surface area at any rank and that it exhibits greater surface area values than whole coal.

IV. COAL LIQUIDS

A. Solvent Fractionation

Solvent fractionation is widely used in the analysis of coal-derived materials. A decreasing sequence of solvent power is observed: pyridine > tetrahydrofuran > benzene > cyclohexane > *n*-pentane. The most

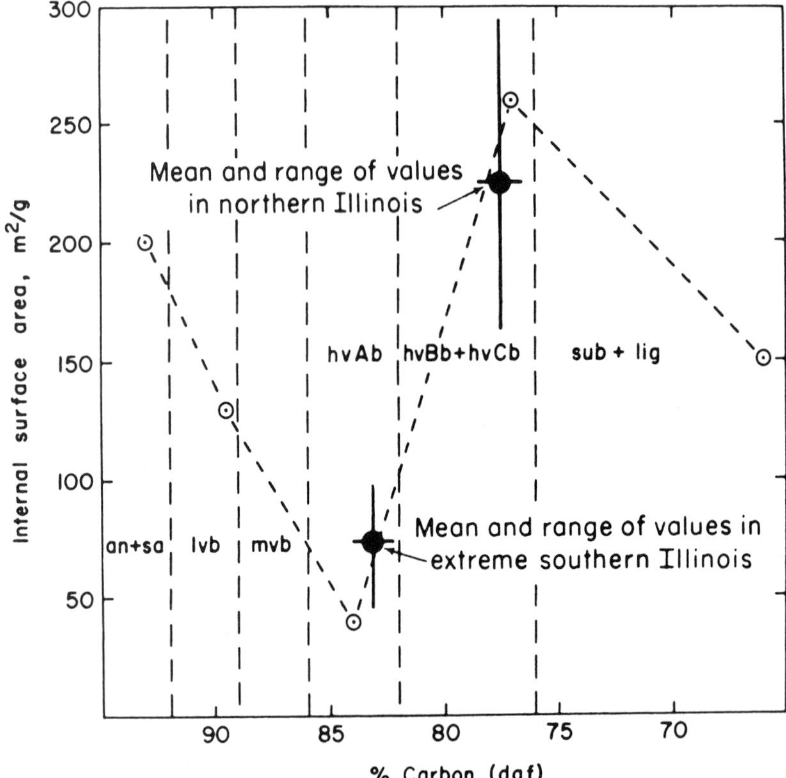

Figure 9 Effect of coal rank on internal surface area of coal (S. Tsai, 1982. Reprinted with permission of the publisher).

usual classification is in terms of solubility in benzene or toluene and an aliphatic hydrocarbon (*n*-pentane, *n*-hexane, *n*-heptane, or cyclohexane), typically defined as the proportion of sample extractable in an approximate 10% w/v slurry refluxed close to the boiling point of the solvent. Solubility in hot quinoline is an important parameter in characterizing coal tar pitches.

Preasphaltenes and asphaltenes are operationally defined as material derived from coal which is, respectively, insoluble in benzene or toluene and soluble in benzene or toluene but insoluble in a low-boiling alkane solvent (Mima et al., 1978). Precipitation of asphaltenes is carried out (Bartle et al., 1979) by addition of large (at least 20) volumes of *n*-pentane or *n*-heptane (to minimize adsorption and/or occlusion of the solubles within the precipitate) to hot concentrated benzene or toluene solutions containing approximately 1 g solubles per 5 cm^3 of solution (cf. recommendations of Speight et al., 1984 for petroleum asphaltenes; Chap. 4, Sec. II). Size exclusion chromatography (Snape et al., 1987) has shown that low molecular mass material is present in coal-derived asphaltenes prepared by standard procedures, but most of this is removed by a single reprecipitation.

The solubility of various fractions derived from coal products can be rationalized empirically (Snape and Bartle, 1984) in terms of a solubility parameter SP:

$$SP = 0.75 \log_{10} (\bar{M}_n/200 + 0.1 (\% \text{ acidic OH}) + 1.5C_{int}/C$$

which takes into account contributions from number average molecular mass $\bar{M}_n$, acidic hydroxyl as a measure of polarity, and C_{int}/C (proportion of internal, or bridgehead aromatic carbon of total carbon), a measure of the degree of condensation of aromatic structures and believed to represent propensity for π-π bonding interactions.

B. Chromatographic Separation and Analysis

Coal-derived materials are extremely complex, and separation into simpler fractions based on polarity, functionality, or molecular size is generally a prerequisite to more detailed characterization. A variety of chromatographic methods is therefore employed to separate coal derivatives before further spectroscopic analysis. High-resolution chromatographic methods also allow the separation and identification of individual constituents. Virtually the entire range of chromatographic procedures, with a gas, or liquid, or a supercritical fluid as mobile phase and often coupled to a spectroscopic method, has been used in the separation and analysis of coal derivatives.

Open-column chromatography, ranging from simple chemical-type separation into gross fractions to very detailed separations using more than one method, is used extensively in the fractionation of coal liquids. Coal derivatives are commonly separated on the basis of molecular size by size exclusion chromatography (SEC).

High-resolution chromatographic analysis is usually carried out by capillary column gas chromatography (GC) for fairly volatile compounds, but for nonvolatile or thermally unstable molecules, high-performance liquid chromatography (HPLC) or supercritical fluid chromatography (SFC) are more useful.

1. Column Adsorption Chromatography

The simplest chromatographic separation of coal derivatives is on an open column of silica gel to yield aliphatic hydrocarbons (eluted by *n*-pentane or *n*-hexane), neutral aromatics (benzene/toluene), and a polar fraction (methanol or tetrahydrofuran). The method is applicable to fairly high molecular mass (MM) fractions such as asphaltenes. Sequential elution from silica gel with specific solvents (SESC) yields further subfractions based on polarity (Farcasiu, 1977). Aromatic fractions may be further fractionated into mono- (single ring), di- (two ring), and polyaromatics (three or more rings) on an alumina column by elution with benzene/pentane or dichloromethane/pentane mixtures. Comprehensive open-column LC methods for the separation of coal liquids into discrete polar compound classes have been developed based on some of the above procedures (Later et al., 1983; Later, 1985). The use of small quantities of neutral alumina and silicic acid allows separation into aliphatic hydrocarbons, polycyclic aromatic compounds (PAC) plus sulfur heterocycles, nitrogen polycyclic aromatic compounds, and hydroxylated polycyclic aromatic hydrocarbons (Figure 10). The nitrogen-containing aromatic compounds can be further separated into secondary nitrogen polycyclic heterocycles, amino-substituted aromatics, and tertiary nitrogen polycyclic aromatics (azarenes). Derivatization of the amino PAH-rich fraction with pentafluoropropionic anhydride enables the complete separation of this class of compounds, which is then recovered by hydrolysis on alumina (Later et al., 1983).

In subsequent work the above scheme was modified to permit separation of hydroaromatic compounds and of polycyclic aromatic nitriles from synthetic fuels. The former are obtained from a dual column of picric acid-modified alumina over alumina; aliphatics, hydroaromatics, and PAC are eluted with increasing concentration of dichloromethane in hexane (Wozniak and Hites, 1983). Chromatography on picric acid-doped alumina also permits separation of the aromatic nitriles from the other neutral PAC by elution with chloroform (Later, 1985).

The SARA separation originally devised for the fractionation of petroleum products (Jewell et al., 1974) has been applied to coal liquids. Acidic and basic components are removed on anion and cation exchange resin columns, and neutral nitrogen-containing compounds are separated by complexation with $FeCl_3$ deposited on Attapulgus clay. Aliphatics and aromatics are then separated on silica gel and alumina.

2. High-Performance Liquid Chromatography

HPLC is very widely used in the separation of PAC. The small particle sizes, which necessitate the pumping of the liquid mobile phase

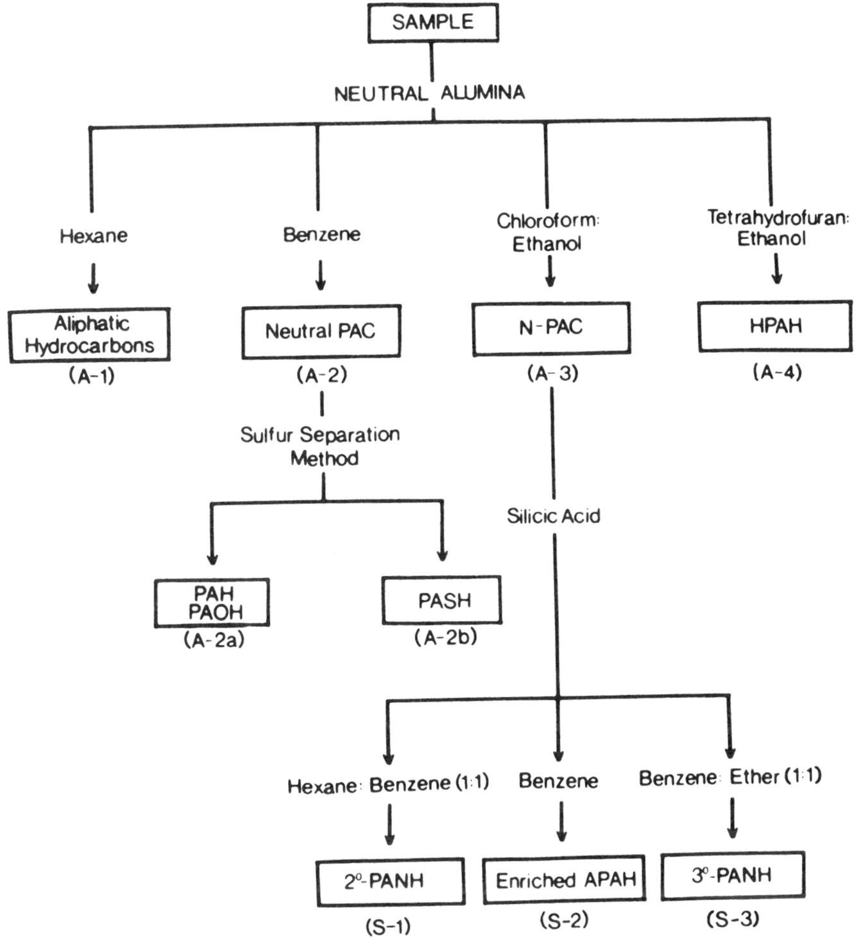

Figure 10 Fractionation of coal liquids by open-column chromatography using neutral alumina and silicic acid (Later, 1985). Reproduced with permission of the American Chemical Society.

through the column, result in much more efficient separation than can be achieved on open colunns. Unless microcolumns with very small internal diameters are used, the high separation efficiency of capillary gas chromatography is not achieved in HPLC. However, HPLC does offer a variety of stationary phases capable of providing unique selectivity both for functional group types and for isomers which are difficult to separate because of interactions of the solute with both stationary and mobile phases. HPLC thus provides both a useful fractionation technique and the means for high-resolution analysis of compounds with MM up to 600 (Bartle et al., 1981).

Elution of a silica or amino column by an alkane gives saturated hydrocarbons followed by aromatic compounds, polar compounds being eluted by backflushing. Class fractionation of coal liquids by functional group has also been investigated for a wide variety of other normal phase columns (e.g., chemically bonded NO_2, CN, diol, sulfonic acid) but NH_2 and NO_2 phases have been found to be the most selective for fractions containing hetero functions (Bartle et al., 1979). Excellent resolution of PAC from azarenes has been reported on a nitrophenyl phase, and about 70% of a coal tar could be analyzed with this phase (Blumer et al., 1978). Normal phase HPLC is commonly used to separate PAC on the basis of ring number (Figure 11) with alkyl derivatives eluting with the corresponding parent compound. Such separations usually precede analysis by another chromatographic method (see below) or by a spectroscopic procedure. The elution profiles of coal liquid distillates and residues during normal phase (*n*-heptane mobile phase) semipreparative HPLC, and analysis of fractions by mass spectrometry showed that separation had been achieved on the basis of the number of double bonds (d.b.): saturates (0 d.b.); monoaromatics (3 d.b.); dicyclic PAC (5 d.b.); dicyclic PAC (6 d.b.); tricyclic PAC (7 d.b.); tetracyclic PAC (8 d.b.); tetracyclic PAC (9 d.b.); pentacyclic PAC (10–11 d.b.); and hexacyclic and greater PAC (12 + d.b.) (Boduszynski et al., 1983).

High-resolution HPLC analysis is generally achieved on reverse phase columns, such as octadecylsilane (ODS) with a mobile phase of a polar solvent (methanol, acetonitrite, etc., in water), often with its composition changed continuously during the run (gradient programming); retention is decreased by increasing the proportion of organic solvent. Methyl and other alkyl derivatives are usually separated from parent compounds. Numerous separations of PAC and polycyclic aromatic sulfur heterocycle (PASH) fractions from coal which show the excellent selectivity for the separation of PAC isomers and alkyl derivatives have been reported. The great advantage of reverse phase packings is their compatibility with a variety of mobile phases and with gradient elution. Selective separations are achieved on the basis of the length-to-breadth ratio of solutes, the more nearly linear molecules being retained longer (Wise, 1983).

A major advantage of HPLC in the analysis of coal liquids is the availability of sensitive and selective detectors. The UV detector is universal for PAC, and the sensitivity and selectivity may be increased by monitoring at specific wavelengths for given compounds, e.g., benzo(*a*)pyrene exhibits nearly maximum absorbance at 290 nm with very little interference from perylene. A scanning UV detector or photodiode array detector allows the possibility of identifying

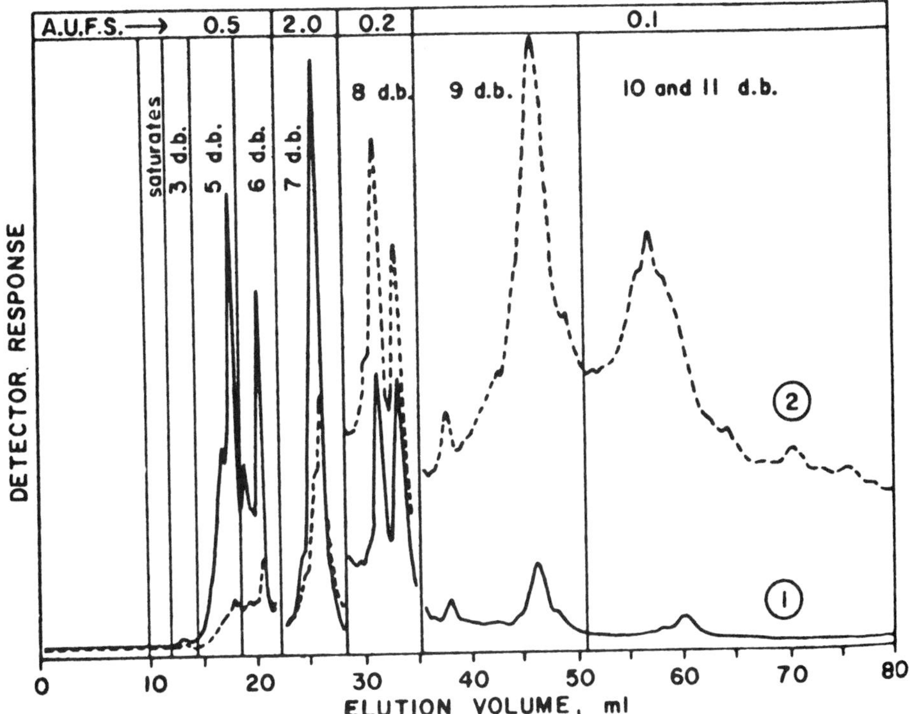

Figure 11 HPLC separation of PAC in coal liquids by ring size (Boduszynski et al., 1983). Reproduced with permission of the American Chemical Society.

chromatographic peaks from complete spectra or absorbance ratios at several wavelengths. Compositional changes in coal liquids as a function of process conditions have been demonstrated by this approach (Klatt, 1979).

Fluorescence detection provides unique selectivity to the identification of individual compounds separated by HPLC (Wise, 1983). For example, six isomeric PAC of molecular mass 252 are generally found in coal-derived mixtures and are not completely resolved in HPLC with ODS columns, but all may be determined by varying the emission and excitation wavelengths. Close-eluting compounds such as phenanthrene and anthracene may also be analyzed in this way. Distinction between alternant and nonalternant systems is also possible by HPLC with fluorescence detection (Bartle et al., 1979).

Novel developments in detection in HPLC include videofluorometry, in which fluorescence spectra are rapidly recorded to yield an emission/

excitation matrix as components are eluted from the column (Fogarty et al., 1981). Combination of HPLC with mass spectrometry is also proving fruitful. Depositing the HPLC column effluent onto a moving belt from which the mobile phase is evaporated allows introduction of sample into the mass spectrometer. A very detailed analysis of coal tar from the mass spectra with both electron impact and chemical ionization of PAC classes separated by normal phase HPLC allowed identification of a range of aromatic types with alkyl substituents containing up to 51 carbon atoms (Herod et al., 1987a).

Microcolumn HPLC has been proposed as a method of improving the efficiency of HPLC separations and has proved capable of resolving many constituents of coal-derived materials containing between five and nine rings (Novotny et al., 1984). The columns are typically 200 μm internal diameter and are packed with 3-μm ODS particles to generate over 200,000 theoretical plates, although extremely long analysis times are necessary. Fractions containing nitrogen compounds from coal liquids have also been separated on similar columns: over 170 peaks were resolved and 600 nitrogen compounds characterized by mass spectrometric and fluorescence analysis of trapped peaks (Borra et al., 1987).

Microbore HPLC columns with internal diameters of about 1 mm reduce considerably the volume of solvent containing the various separated fractions (to ∿100 μl) compared to open-column LC and normal HPLC. The fraction solutions may then be transferred directly to an analytical GC capillary column from an automatically and pneumatically operated 10-port valve (Davies et al., 1987). On-line HPLC/GC is a rapid and highly reproducible method for the analysis of complex samples such as coal liquids (Davies et al., 1988).

3. Size Exclusion Chromatography

The distribution of molecular mass (MM) is the factor of greatest importance after chemical composition in considering the upgrading of coal extracts and derivatives to transport fuels and chemical feedstocks. The method of choice for determining MM distributions is HPSEC, in which molecules of different sizes are separated according to their degree of penetration into the pores of a gel packed in a column as small-diameter spheres. The gel is typically a three-dimensional network of crosslinked polymeric chains of controlled porosity such as crosslinked polystyrene, and a number of mobile phases, most usually tetrahydrofuran (THF), have been investigated. Eluted components are detected by monitoring a physical property such as UV absorption or refractive index. Typical size exclusion chromatograms showing the MM range of coal liquids between approximately 200 and

5000 is shown in Figure 12. The recent advances in the technology of gas chromatography and supercritical fluid chromatography have extended the limits of these techniques, but of availabile chromatographic techniques only SEC is capable of providing information on MM distribution above 1000 (Bartle et al., 1979).

The choice of eluting solvent is vital in SEC. Unless a solvent minimizing solute self-association and adsorption on the column is chosen, large errors may result in MM distribution (Bartle et al., 1986). When

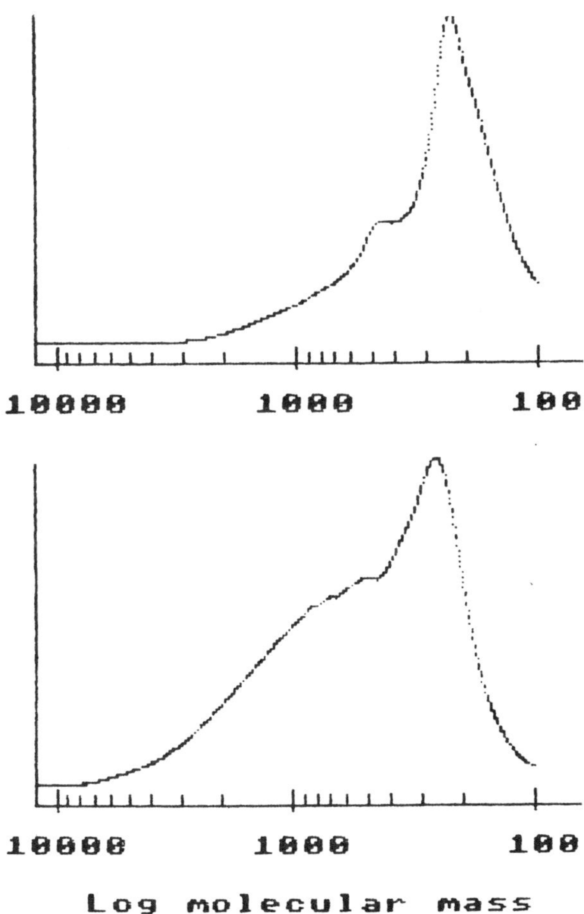

Figure 12 Size exclusion chromatogram of two coal liquids.

a nonpolar mobile phase such as toluene is used, much less of the high-MM coal-derived material is eluted than for THF. Similar interactions of asphaltenes from petroleum residues with crosslinked polystyrene have been reported for toluene elution. The absence of significant self-association of coal-derived asphaltenes during elution with THF at the concentrations employed in SEC has been proved by measurements of $\bar{M}n$ by vapor pressure osmometry and comparisions with $\bar{M}_n$ (number average MMs) for silylated asphaltenes (Bartle et al., 1986).

The gel-solvent combination of polystyrene and THF is the most commonly used in SEC of coal liquids and extracts although some work with hydrophilic gels has been reported. THF is not appropriate for coal-derived materials that contain high concentrations of alkanes, which are eluted early, or highly condensed aromatic molecules which are retained by adsorption. Low-MM polar compounds such as phenols, other oxygen-containing compounds, and nitrogen-containing compounds are solvated by THF through hydrogen bonding and therefore also elute early because of increased apparent molecular size (Evans et al., 1986). This effect results in apparently bimodal MM distributions of coal tars in THF, but not in toluene or chloroform, which do not hydrogen-bond. SEC must be applied with caution in determining MM distributions of polar low-MM materials, but the effect is small for higher MM fractions such as asphaltenes.

For coal derivatives, such as coal-tar pitches, etc., containing material insoluble in THF, quinoline and trichlorobenzene have been recommended as solvents (Lewis and Petro, 1976). The use of a highly polar mobile phase dimethylformamide with a neutral macroporous gel modifies the separation process considerably from one based mainly on molecular size to include significant contributions from association, partition, and adsorption effects (Mulligan et al., 1987).

Number ($\bar{M}_n$) and weight ($\bar{M}_w$) average MMs and molecular mass distributions for heavy coal liquids are determined by high-performance SEC by interfacing a flow rate sensor and a detector to a microcomputer. Detector response and flow rate are sampled at intervals; from stored calibration data, a table of retention time, detector response, and MM is produced. Number and weight average MM and a normalized mass distribution curve are calculated with facility for taking account of the variation of detector response with MM (Bartle et al., 1986).

Reliable MM distributions of coal derivatives can only be derived if the detector responds equally to a given concentration of solute throughout the MM range (Bartle et al., 1983). For the differential refractive index detector, a reversal of polarity with increasing MM is observed. The distribution of groups which absorb in convenient regions of the infrared spectrum is not uniform. The UV detection is sensitive and linear, but the specific response varies with MM.

These effects originate in the variation of structure with MM, e.g., the trend to lower aromaticity with increasing MM for asphaltene subfractions is reflected in smaller UV absorbances for the same concentration. Quite erroneous MM distributions are therefore calculated for coal derivatives by SEC if the detector response is not calibrated. Different MM distributions are also calculated if different wavelengths are used (Bartle et al., 1986).

Narrow polystyrene fractions are often employed as calibration standards in SEC of coal derivatives via a graph of log MM against retention volume V_R. The derived constants A and B in the equation:

$$\log MM = A - BV_R$$

are used to calculate the distribution of MM. However, polystyrene is not a satisfactory model for the retention behavior of coal-derived molecules, especially above 1000. A variety of alternative procedures both rigorous and empirical have been investigated, but calibrations should be made (Bartle et al., 1984) with narrow preparative SEC subfractions of coal derivatives. Figure 13 shows chromatograms of typical asphaltene subfractions with $\bar{M}_n$ determined by vapor pressure osmometry. Alternatively, other, more suitable polymers should be sought. The SEC retention behavior of subfractions of asphaltenes and preasphaltenes from high-yield extracts of coals has been comared with that of polymer standards. Above MM > 1000, coal extract subfractions are approximated better by polyacenaphthylene (Bartle et al., 1984).

4. Thin-Layer Chromagoraphy

TLC has found fairly limited application in the separation and analysis of coal-derived materials mainly because of the susceptibility of many PAC to photooxidation on the TLC plate. However, pyrolysis TLC with flame ionization detection (FID) has been used in the analysis of coal-tar pitch volatiles (Boden and Roussel, 1983) and synthetic fuels (Poirier and George, 1983). The separation is carried out on fused-silica rods coated with activated silica gel. Separation is achieved by elution with solvents of increasing polarity which ascend the rod by capillarity; separated bands on the rod are scanned with an FID detector where the components are burned and detected. The analysis of coal-derived liquids is especially rapid by this method (Selucky, 1983).

5. Gas Chromatography

For the low-MM constituents of coal derivatives, GC is the method of choice (White, 1983). The extreme complexity of coal derivatives also

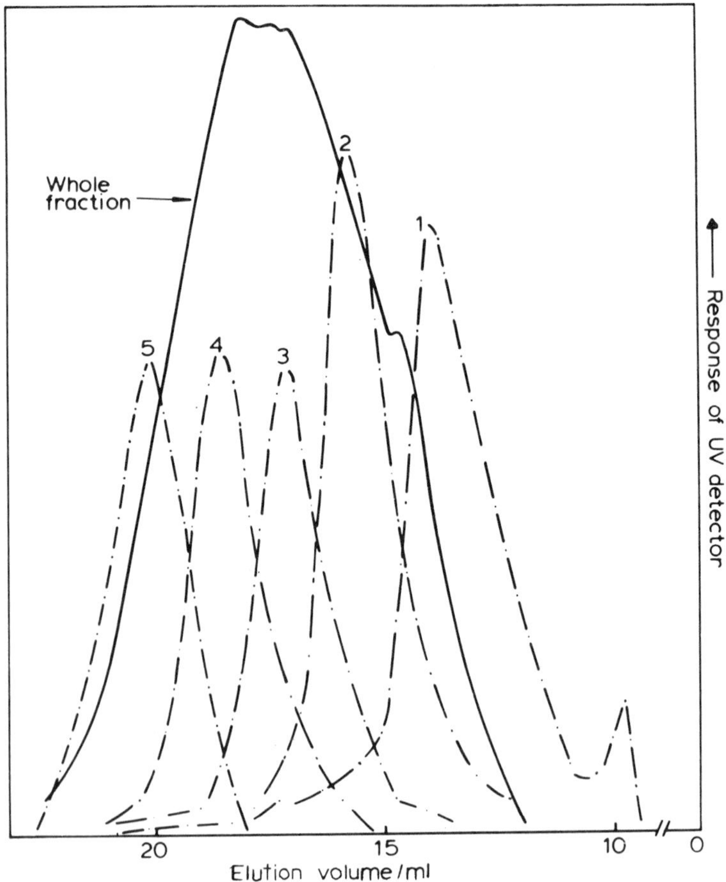

Figure 13 Size exclusion chromatograms of asphaltene subfractions.

demands the greatest possible resolution in their analysis and, in this respect, GC on packed columns even as long as 20 m has fallen short of that available on capillary columns. Fused-silica columns are now used universally, and if surface activity is controlled by silanization treatments, trace compounds are eluted as sharp peaks. For most analytical work, columns 10–25 m in length are suitable, with internal diameters of 0.2–0.3 mm. Column performance is readily judged from the separation of isomer pairs such as anthracene/phenanthrene and chrysene/benz[*a*]anthracene (Lee et al., 1984).

Capillary column GC has been widely used in the analysis (Lee et al., 1981, 1984; Bartle, 1985) of coal-derived products (Figure 14). The nonpolar or slightly polar stationary phases—methylsiloxanes (OV-1, OV-101) or "small-content" phenylsiloxanes (SE-52 and SE-54)—have generally been used especially after free radical crosslinking to improve thermal stability, although high-phenyl-content siloxanes and more polar phases such as Poly S-179 are also used. Liquid crystal blends have been observed to give remarkably selective separations, but use is limited by the low thermal stability. Poly(mesogenmethyl)-siloxanes are, however, gum phases which show high column efficiencies and stabilities but retain high selectivity for PAC isomers of coal origin and have a wide nematic temperature range (70–300°C). More polar stationary phases such as polyethylene glycols, Superox 20M, etc., are required for the capillary GC separation of coal-derived phenols.

As described elsewhere (Sec. IV.E) GC with mass spectrometric identification is often the final step of combined procedures in which open-column LC and HPLC are used to separate the coal liquid. Literally hundreds of compounds from coal have been separated and identified in this way (see, for example, Chang et al., 1988); Figure 15 shows the chromatogram of an HPLC subfraction from an aromatic ring size separation containing phenanthrenes.

More routinely, flame ionization is the most commonly used detector in the GC analysis of coal-derived oils since response is based on the amount of carbon present in eluting compounds. Capillary column flame ionization chromatograms may be used for many identifications

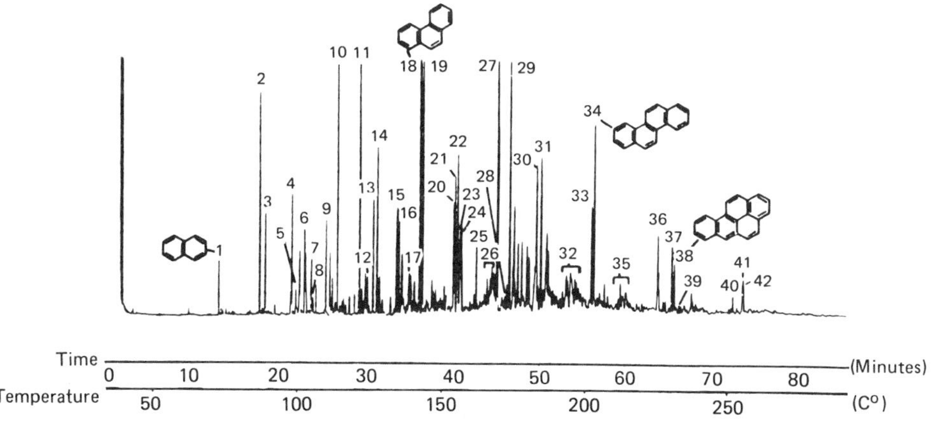

Figure 14 Typical capillary column gas chromatogram of a coal liquid.

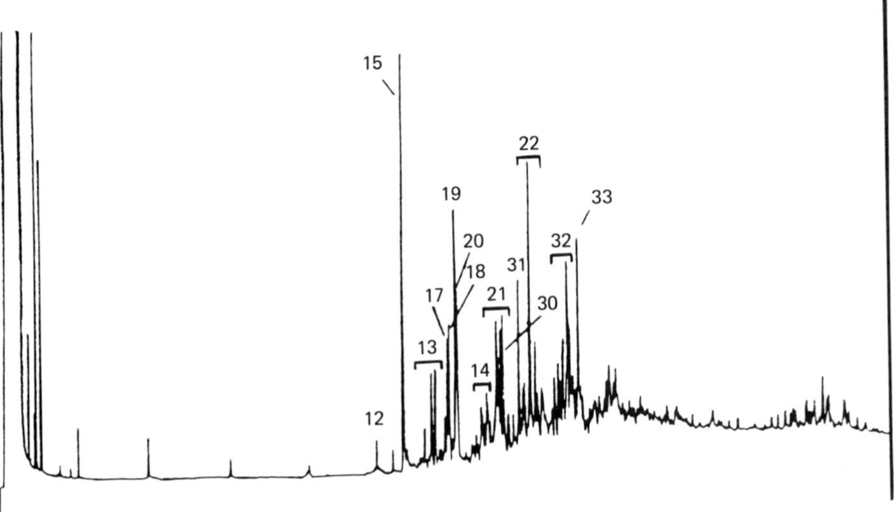

Figure 15 Capillary column gas chromatogram of an HPLC subfraction containing principally phenanthrenes from a coal liquid (Chang et al., 1988).

without having to resort to GC/MS by making use of retention indices—systems of reproducible retention parameters. Variations in chromatographic conditions and column film thicknesses makes the use of a retention index I based on a homologous series of retention standards most useful. For PAH, the Kovats system is less reliable than that proposed by Lee et al. (1979) which is based on the internal standards naphthalene, phenanthrene, chrysene, and picene. Values of I for a total of 310 PAC have been listed, with standard deviations (generally less than 0.10 unit) and 95% confidence limits. This system has the advantage that most of the reference standards are generally present in coal-derived oils (Vassilaros et al., 1982).

The high concentrations of nitrogen- and sulfur-containing compounds makes selective detection especially useful in GC analysis in coal liquids. Figures 16a and 16b show dual-trace nitrogen (NPD)- and sulfur-(FPD)-selective chromatograms of enriched fractions from solvent-refined coal. Limited use has also been made of infrared and UV detection (Lee et al., 1984; Bartle, 1985).

The wide volatility range of coal-derived mixtures means that care is necessary to exclude discrimination against higher MM compounds during injection by splitless injection. Only cold on-column injection completely avoids fractionation during sample vaporization and is also more reproducible, since the liquid sample is injected directly into the column (Lee et al., 1984; Bartle, 1985).

6. Supercritical Fluid Chromatography

Although chromatography with a supercritical fluid as mobile phase was reported more than 20 years ago, the advantages of supercritical fluid chromatography (SFC) have only recently been fully realized—in particular its rapidity, flexibility, and ability to allow the analysis of substances which cannot be analyzed by GC.

Above its critical point a substance has density and solvating power approaching that of a liquid but viscosity similar to that of a gas, and diffusivity intermediate between those of a gas and a liquid. Hence, above their critical temperatures, some substances are fluids with properties which make their use as chromatographic mobile phases very favorable: extraction and solvation effects allow the migration of materials of high molecular weight; the low viscosity means that the pressure drop across the column is greatly reduced for given flow rates, and high linear velocities can be achieved. The high diffusivity confers very useful mass transfer properties, so that higher efficiencies in shorter analysis times are achieved than in HPLC are possible. Furthermore, the density of the supercritical fluid and hence the solubility and chromatographic retention of different substances can easily be varied by changing the applied pressure. The analogue of temperature programming in GC and gradient elution in HPLC is SFC with pressure programming—slowly increasing the mobile phase density and decreasing solute retention. The much lower operating temperatures used in SFC compared with GC allow high-resolution chromatography to be applied to mixtures which would normally be separated by HPLC, but only after considerable investment of time to determine the appropriate mobile phase composition. In SFC, the separating power is midway between that of GC and HPLC (see Figure 17).

Standard HPLC columns are suitable for SFC, and many applications in the separation of polycyclic aromatic compounds have been reported. Efficient and rapid analysis of coal-derived oils is possible with CO_2 as mobile phase at low temperatures (Jackson et al., 1985). For example, Figure 18 is the SFC chromatogram of an anthracene oil on a conventional HPLC packed ODS column; the separation (within 9 min) of benzopyrenes and benzofluoranthenes is similar to that obtained by capillary GC of the same mixture (Bartle et al., 1987).

The greater permeability of capillary columns allows column lengths to be greatly increased, with subsequent high efficiency. The advent of new column technology has made small-diameter fused-silica columns for SFC widely available. The stationary phase, usually a polysiloxane,

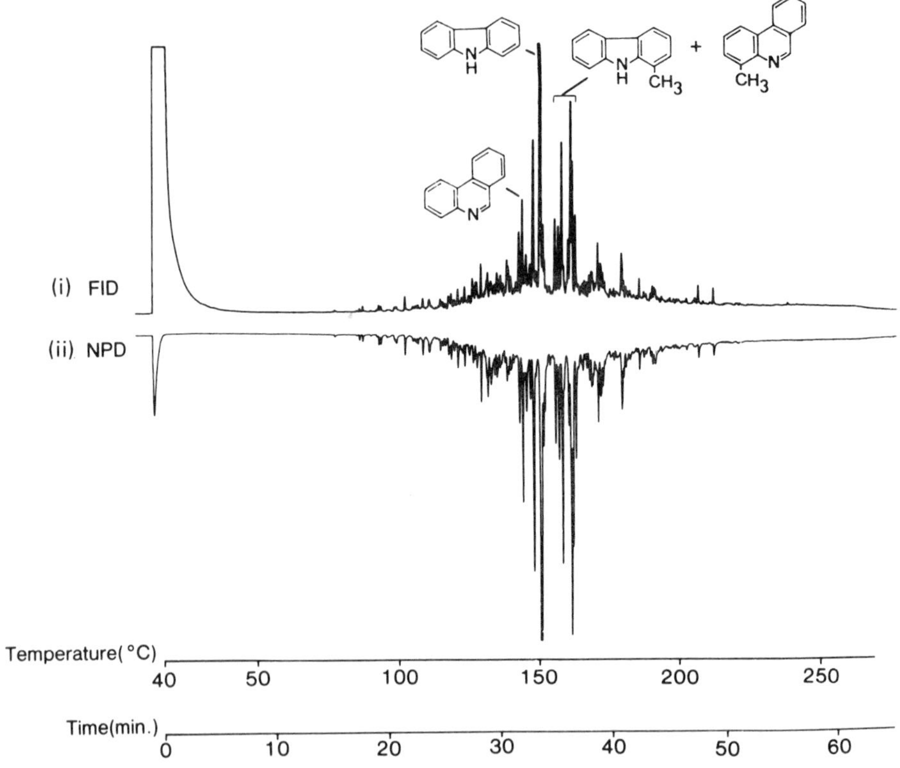

Figure 16a Dual-trace FID- (i) and nitrogen-NPD (ii) selective chromatograms of a nitrogen enriched fraction from solvent-refined coal.

must be subjected to free radical crosslinking to prevent extraction by mobile phase. Capillary columns offer a number of further advantages in SFC; high sensitivity, the maximum use of density programming, and compatibility with a variety of detectors including the universal flame ionization detector.

Fractionation of complex coal-derived mixtures is also possible by elution with supercritical CO_2 from columns packed with NH_2-modified stationary phase. Fractions are collected in pressurized vessels, according to the number of aromatic rings, with much "cleaner" separation than is possible in preparative HPLC (Campbell and Lee, 1986).

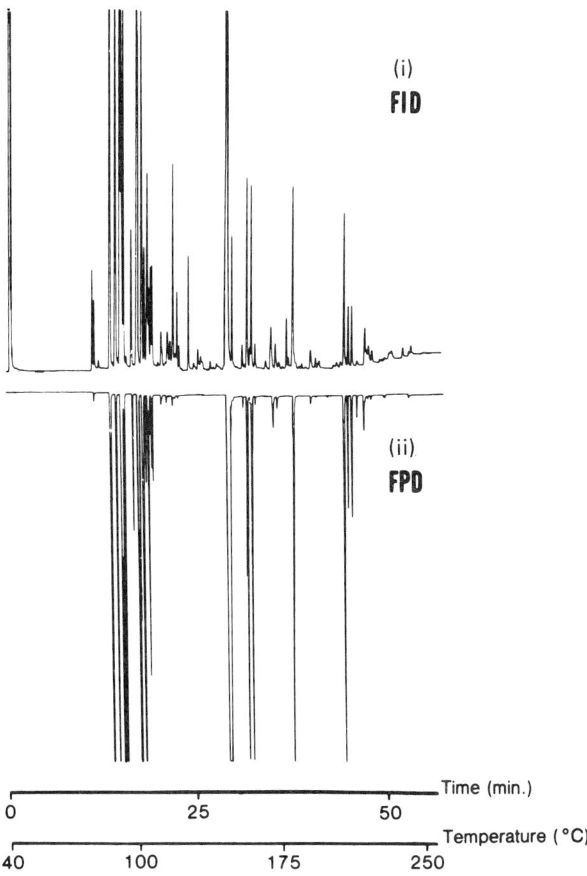

Figure 16b Dual-trace FID- (i) and sulfur-FPD (ii) chromatogram of a sulfur enriched fraction from solvent-refined coal.

C. Titration Methods

Acetylation has probably been the most widely used procedure for determining OH concentrations in coals and coal liquids. The uptake of acetyl groups can be monitored by (1) IR (Brown and Wyss, 1955), (2) NMR (Chapter 19, Section IV), (3) radiochemically if ^{14}C-labeled acetic anhydride is used (Yarzeb et al., 1979), and (4) hydrolysis and titration of the acetic acid released (Blom et al., 1957).

Acidic OH and basic N groups in coals and coal liquids can be determined by nonaqueous potentiometric and enthalpimetric titration.

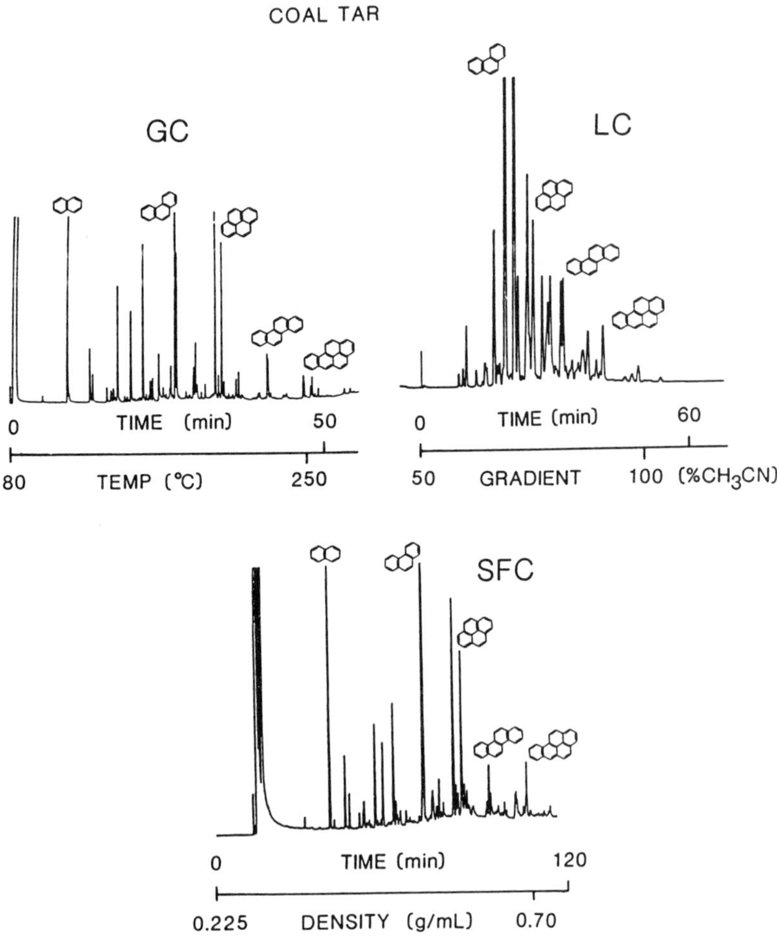

Figure 17 Comparison of SFC with capillary GC and HPLC analysis for a coal tar.

For acidic groups, the latter makes use of the dimerization of acetone to diacetone alcohol which is base-catalyzed and exothermic (Vaughan and Swithenbank, 1965, 1970). The dimerization reaction is sufficiently slow that the titration of acidic groups is completed before it is catalyzed by excess base, the exotherm at the end point being sensed with a thermistor. Results obtained for coal extracts have been found to be in close agreement with those from derivatization procedures (Snape,

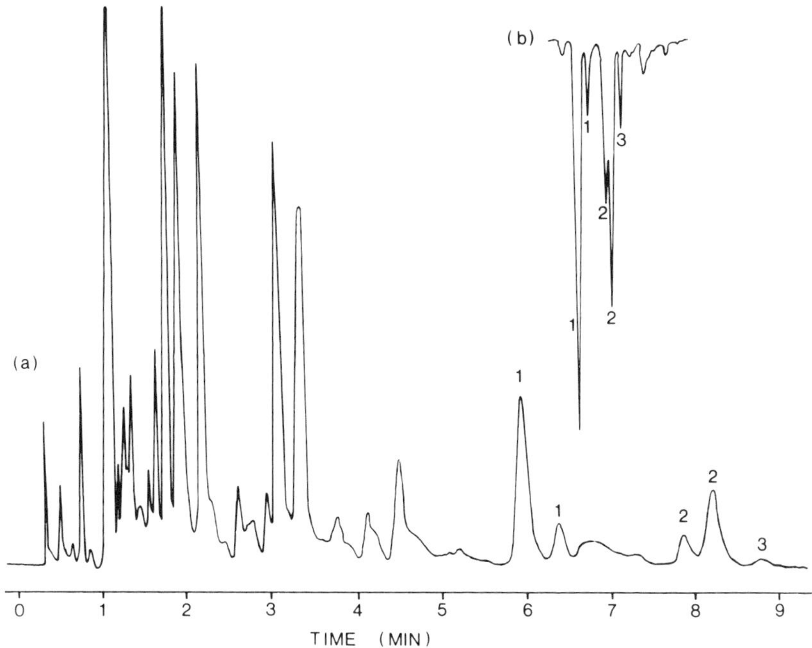

Figure 18 (a) SFC analysis of an anthracene oil on a conventional HPLC packed ODS column. (b) Capillary GC separation of (1) benzofluoranthrenes, (2) benzopyrenes, and (3) perylene.

et al., 1982). The method appears to be applicable to coals provided they are finely ground (Vaughan and Swithenbank, 1970).

Titration with sodium aminoethoxide in ethylenediamine has been used to determine the total acidity of coals. However, prior extraction of coal with a number of solvents increased the measured acidic contents considerably (Maher and O'Shea, 1967), suggesting that not all the phenolic groups are readily accessible in the original macromolecular framework.

Basic N concentrations in coals and extracts have been determined titrimetrically using perchloric acid in a glacial acetic acid medium with conventional glass and calomel reference electrodes (Darlage et al., 1978). Coal liquids are generally insoluble in glacial acetic acid and have to be predissolved in a solvent, such as nitrobenzene. Bases with $pK_B < 11$ are titrated (Moore et al., 1951) and these include primary amines and aza compounds, such as pyridines and quinolines but not aromatic secondary amines. Results for bituminous coals and their extracts indicate that basic N accounts for 40–50% of the total N (Snape et al., 1984).

D. Molecular Weight Determination

Vapor pressure osmometry (VPO) has been the most widely used colligative method for the determination of number average MMs ($\bar{M}_n$). The difference in temperature between drops of solvent and a dilute solution of the sample is measured. This temperature difference is proportional to the difference in vapor pressure at the two drops which, in turn, is proportional to the number of moles of solute present. $\bar{M}_n$ can be related to the temperature difference ΔT, by the general expression:

$$\frac{\Delta T}{C} = \frac{K}{\bar{M}_n}(1 + \Gamma_2 C + \Gamma_3 C^2 + \cdots)$$

where C = solution concentration. For ideal solutions, the virial coefficients Γ_2 and Γ_3 are zero, but for coal liquids and other nonideal systems it is necessary to take enough readings to extrapolate the plot of $\Delta T/C$ versus C to infinite dilution. Rectilinear plots have been reported for coal liquids (Chung et al., 1979; i.e., Γ_3 is zero) but, as pointed out by Larsen et al. (1981), these have no thermodynamic significance. The nonideality of coal liquids arises in part from intermolecular associative interactions in solution and these can be limited by using relatively polar solvents, such as THF. The fact that $\bar{M}_n$ values are affected by small concentrations of low-MM impurities has been well documented (Given, 1984). Moreover, extremely high MM species cannot be detected by VPO because they have essentially no vapor pressure.

To investigate the MM range of heavy products and their fractions, such as asphaltenes, and to help calibrate SEC measurements (Chapter 26), it has proved valuable to isolate narrow-MM-range subfractions by preparative scale SEC that have much lower polydispersities ($\bar{M}_w/\bar{M}_n < \sim 1.2$) than the parent fraction. It has been found that asphaltenes have MMs in the approximate range 200 to over 2000 (Richards et al., 1983). Once SEC columns have been calibrated with appropriate standards, good agreement with VPO-determined $\bar{M}_n$ values has been obtained (Sec. IV.B).

Weight average MMs ($\bar{M}_w$) of coal liquids have been determined by light scattering (Hombach, 1981; Olson et al., 1987). However, the strong absorption in the visible light range of most coal liquids poses difficulties for modern instruments which use a laser source and small incident beam angles. Olson et al. (1987) reduced lignite humic acids to give virtually colorless derivatives to eliminate this interference. $\bar{M}_W$ values of over 10^5 have been reported both by Hombach (1981) and Olson et al. (1987) and these cannot be reconciled with the $\bar{M}_n$ values determined by VPO and SEC. However, the extent of light scattering increases with molecular weight and it is exceedingly difficult to identify low-MM species (the opposite trend to VPO). Clearly, as Given (1984) discussed in much greater detail than here, the measurement of MMs for coal derivatives is an area that still requires more investigation.

E. Mass Spectrometry

"Conventional" electron impact (EI) mass spectrometry at ionization voltages of $\sim$70 eV has widespread application in the analysis of coal liquids by combined gas chromatography/mass spectrometry (GC-MS) (Lee et al., 1984; Sec. IV.B). The molecular mass of a compound giving rise to a GC peak is determined from the molecular ion, and the fragmentation pattern gives structural information. The compound may be identified by a prior interpretation of the spectrum or by comparison with literature spectra, often via computer search of libraries of spectra. EI-MS of PAC is generally insensitive to isomer structure, but differentiation of isomers is possible by use of mixed charge exchange chemical ionization (CI) reagent gas, by pulsed positive ion/negative ion CI, or by careful choice of CI reagent gas (Lee et al., 1981). However, isomers may be identified from GC retention data once their molecular formulas are known. Figure 19 is the mass spectrum of a dimethyl naphthalene in the capillary gas chromatogram in a coal liquid; the measured retention index (248.5) is identical with that of the 1,2 isomer.

Single-ion monitoring is particularly useful in GC-MS. The spectrometer is focused on one m/z value characteristic of the compound under consideration, so that only the compound and its isomers give peaks in the chromatogram. Such an approach is often rendered necessary by the complexity of coal-derived mixtures which cannot be resolved even by capillary column GC with the highest resolution possible; as many as six molecular ions representing three different aromatic types were detected in the mass spectrum of compounds from a hydropyrolysis tar contributing to a single GC peak (Herod et al., 1987a).

The most widely used method for MS analysis of coal liquids has been low-voltage EI-MS in which the ionizing voltage is reduced to eliminate fragment ions and produce spectra containing only molecular

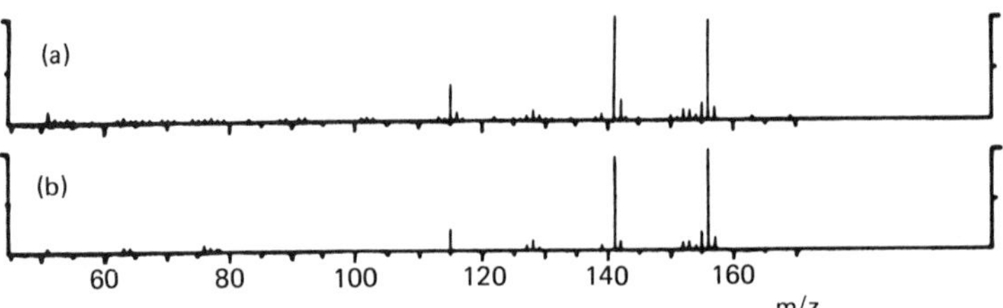

Figure 19 Identification of 1,2-dimethylnaphthalene in a coal liquid by GC/MS using an automated library search. (a) Mass spectrum of compound giving rise to a single peak in the chromatogram of a coal liquid. (b) Mass spectrum of 1,2-dimethylnaphthalene from spectral library.

ions (Bartle et al., 1981). Heating the direct-insertion probe up to 370°C yields peaks with m/z values up to 800 (Schultz et al., 1965). Mass sensitivy data must be employed; the sensitivity increases rapidly with size for parent compounds, but rises and then falls for alkyl derivatives (Lee et al., 1981). Such spectra, scanned over the full temperature range and obtained under conditions of low resolution, allow carbon number distributions to be determined and some structural types to be identified. Results can be presented by a matrix of carbon number versus Z number with molecular formulas being represented by C_n-H_{2n-z} (Herod et al., 1987b).

High-resolution low-ionizing voltage MS, which takes advantage of the small mass differences between hydrocarbon and heterocyclic types with the same mass number, has been profitably applied to coal derivatives. For example, at nominal mass 208, peaks are separated for molecular formulas $C_{15}H_{12}O$, $C_{14}H_8S$, and $C_{16}H_{16}$ (Kessler et al., 1969). A computer system is necessary for data acquisition in these experiments for the rapid calculation of formulas and intensities in a format suitable for further calculations and quantitative analysis. In favorable cases, mass spectra of coal-derived mixtures may be simple enough to allow accurate mass determinations of prominent peaks. For example, a base fraction from coal-tar pitch gave three mass series of highly condensed (6–11 rings) nitrogen-containing compounds (Wallace et al., 1987). Two of the series consisted of azarenes and their methyl derivatives, while the third series consisted of compounds containing both pyrrole- and pyridine-type functional groups.

Tandem mass spectrometry (MS-MS) allows direct analysis of individual components of coal-derived mixtures. Scanning of both parent and daughter ions has been used to analyze the alkylated aromatic compounds in coal liquids, e.g., naphthalenes, acenaphthenes, and cyclized acenaphthenes, etc. (Singleton et al., 1987).

For the higher MM fractions of coal liquids, the same problems of volatility are met as was discussed in the description of coal analysis by MS (Sec. III). Similar methods have been applied: integrated FD-MS (Yoshida et al., 1979; Herod et al., 1987a) and FI-MS (Bodzek and Marzec, 1981) of coal liquids yield broad peak envelopes extending beyond m/z 1000. A marked shift to lower MM is observed for FI-MS of a coal liquid in comparison with the original coal (Schulten and Marzec, 1987). FD-MS with chemical ionization (DCI) with isobutane as reagent gas gives spectra similar to 10-eV direct-probe MS; the main difference is the greater constancy of CI response with degree of alkylation (Herod et al., 1987a).

Ionization of the constituents of the asphaltene fraction of a coal tar has been achieved by fast atom bombardment--molecular beams of fast noble gas atoms—for samples introduced via a moving-belt interface (Herod et al., 1987b). Masses in excess of 1000 were observed.

Soft ionization spectra of coal liquids are often used as indicators of the molecular mass distribution, but caution is necessary in such interpretations: while SEC, which has no volatility restrictions, generally agrees with MS-determined distributions of MM for tars, SEC of extracts reveals an MM range extending far beyond the limits of MS (Ladner and Snape, 1985).

F. Ultraviolet and Luminescence Spectroscopy

UV spectroscopy is useful for the identification of polycyclic aromatic ring systems, and applications have been made in the analysis of coal tar fractions (Zander, 1966), but the spectra contain broad lines and have low specificity. However, more information may be derived by low-temperature or derivative adsorption spectroscopy (Bartle et al., 1979).

Luminescence offers advantages over UV absorption in greater sensitivity and selectivity, and also in the availability of four spectra (excitation and emission of fluorescence and phosphorescence) (Lee et al., 1981). Thus, benzo[*a*]pyrene and benzo[*ghi*]perylene have similar UV (and fluorescence excitation) spectra but are easily distinguished by fluorescence emission.

Fluorescence spectroscopy has therefore found more application than UV absorption in the analysis of coal-derived mixtures. Twelve polyaromatic ring systems were identified with up to six rings in hydrogenation products (Kershaw, 1978). Identifications were aided by the general similarity of fluorescence spectra of large aromatic hydrocarbons and their alkylated derivatives. Fluorescence spectra of oil, asphaltene, and preasphaltene fractions of coal extracts have also been used in analyses for constituent structural types (Aigbehinmua et al., 1987; Clark et al., 1987). Three-dimensional plots of total luminescence [the function I (λ_{ex}, λ_{em}) where I is the intensity and λ_{ex} and λ_{em} are all possible excitation and monitoring wavelengths] were obtained. The weak fluorescence emission of certain compounds (e.g., phenanthrene) make their identification difficult, and there are substantial problems in quantitative analysis because of quenching and self-absorption.

Shpol'skii fluorescence emission spectra in frozen solution glasses show much more fine structure than spectra recorded for solutions at room temperature and can be assigned to individual compounds (Lee et al., 1981); individual PAH containing between three and 10 rings were thus identified without prior isolation in coal extracts (Drake, 1978). The x-ray-excited optical luminescence spectra of coal extracts in frozen solution (Woo et al., 1980) also showed characteristic fine structure. Fluorimetric procedures have been developed which permit selective quenching of the luminescence of alternant hydrocarbons (by electron acceptors such as nitromethane) and nonalternants (by electron donors such as 1,2,4-trimethoxybenzene) (Bartle et al., 1981) and applied to coal derivatives.

Phosphorimetry is more selective than fluorimetry but is generally observed in solvent glasses at low temperature; it is most useful when a mixture contains strongly fluorescent but weakly phosphorescent interfering species (Zander, 1968; Bartle et al., 1979). For example, perylene interferes with the fluorescence of dibenz[*a,i*]pyrene but gives negligible interference in phosphorimetry. The longer lifetimes of phosphorescence compared with fluorescence allows this parameter to be used in analysis. The sensitivity and selectivity of phosphorescence may be improved in solvents containing heavy atoms (e.g., methyl iodide), and silver nitrate selectively enhances the phosphorescence of aza-aromatics relative to PAH (Lee et al., 1981). Numerous analyses of coal tar fractions by phosphorimetry, including the use of the above effects, have been reported (Zander, 1968).

G. Voltammetric Identification of Aromatic Groups

The characteristic reduction potential of many PAC have been determined and provide a basis for the detection of various functional groups in solvent extracts of coals. Modern microprocessor-based polarographic systems coupled with "dispense-type" mercury drop electrodes afford many advantages over those used in early work.

In particular, the contribution of charging current of the electrical double layer at current sampling is reduced, while a wide range of voltage waveforms in a variety of pulse modes can be employed; three-element polarographic cells also reduce potential drop effects in the electrolyte and improve the precision of the measured reduction potentials. The hanging mercury drop electrode (HMDE) allows fuller use of the technique than the dropping mercury electrode, which shows mechanical instability in DMF solutions (Bartle et al., 1982).

Reduction potentials for a wide range of PAC have been determined by differential pulse voltammetry at the HMDE (Pappin et al., 1987). Graphs of pulse current against concentration were linear. The presence and concentration of different PAC structures, e.g., anthracenes, fluoranthenes, pyrenes, and chrysenes, etc., could therefore be determined for coal-derived oils and compared with results from analysis by open-tubular-column gas chromatography. Voltammetric curves for asphaltenes and preasphaltenes from a variety of coal derivates show considerable electrochemical activity, with voltammetric peaks at similar potentials to those of the corresponding lower MM material (see Figure 20) (Tytko et al., 1987). DPV of specially synthesized model compounds shows (Tytko et al., 1987) how linkages through alkyl groups have only a small effect on $E_{1/2}$ values of common polycyclic aromatic groups, so that the peaks in voltammograms of high-MM fractions can be assigned to the reduction of specific aromatic species even though these are linked together.

On this basis, the asphaltenes and preasphaltenes of a supercritical gas (SCG) extract of lignite show little electrochemical activity (Figure 20) other than the reduction of aromatic clusters containing two aromatic rings (naphthalene, biphenyl, fluorene, etc.). The same fractions from an SCG extract of bituminous coal show similar voltammograms to those from the lignite extract, except that small signals from more condensed structures such as phenanthrenes, fluoranthenes, and pyrenes are also present (Figure 20). For hydrogen donor solvent (HDS) or anthracene oil extract fractions of bituminous coals, the reduction of the three- and four-ring aromatic structures are much more prominent (Figure 20). This evidence correlates well with the structural parameters determined by NMR. In particular, DPV confirms that the polycylic aromatic structures suggested as being consistent with NMR-derived average structures are actually present in the asphaltenes and preasphaltenes obtained from bituminous coals under relatively severe conditions of extraction and carbonization (Tytko et al., 1985). Experiments have been carried out to confirm that signals in the voltammograms of asphaltenes from which low molecular mass material has been removed originate from molecules containing linked aromatic structures rather than "independent" aromatic molecules (Tytko et al., 1987).

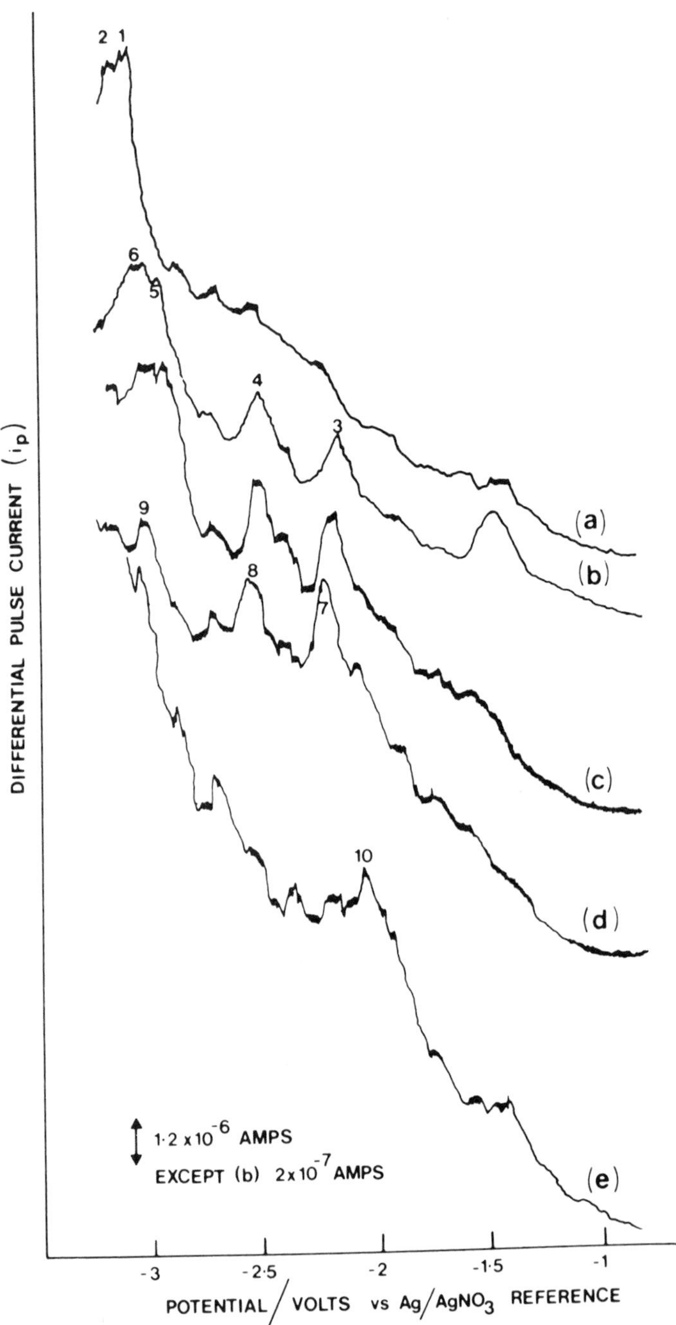

Figure 20 Differential pulse voltammograms of coal extract asphaltenes. (a) Supercritical gas extract of bituminous coal; (b, c) hydrogen donor solvent extracts of bituminous coal; (d) anthracene oil extract of bituminous coal; (e) supercritical gas extract of lignite.

H. Nuclear Magnetic Resonance Spectroscopy

1. Introduction

The current widespread use of NMR for the characterization of coal liquids stems from the fact that it is an inherently quantitative technique with both ^{1}H and ^{13}C NMR providing direct information on the distribution of aromatic and aliphatic groups. Also, the complexity and involatility of high-MM fractions, such as asphaltenes, makes it easier to compare different samples using bulk compositional parameters, such as aromaticity, rather than to identify particular molecular components. Before Fourier transform (FT) spectrometers were commercially available, high-resolution NMR analysis of coal liquids was limited to ^{1}H measurements, which date back to 1960 (Rao et al., 1960; Brown and Ladner, 1960). The continual introduction of improved instrumentation for multipulse FT experiments has expanded considerably the role of NMR in the characterization of coal liquids. Indeed, two-dimensional (2D), multinuclear and ^{13}C spectral editing have all found applications as covered in this section. The use of structural analysis schemes in which NMR data are combined with elemental compositions, molecular masses, and other analytical results to derive parameters describing the bulk composition of coal liquids is described in Sec. IV.I.

2. Experimental Considerations

Chloroform-d is generally considered the solvent of choice. It is a reasonably powerful solvent for coal liquefaction products (better than toluene) and interfering peaks in their spectra are avoided, the ^{13}C chemical shift at 77.1 ppm falling conveniently between the aromatic and aliphatic carbon bands. For primary liquefaction products containing chloroform insolubles, pyridine-d_5 can be used for ^{1}H spectra but suffers generally from lower isotropic purity than chloroform-d, which gives rise to interfering aromatic peaks. Morover, it is difficult to eliminate water completely and, if present, it may give an interfering peak overlapping the aromatic hydrogen bands.

Sym-triazine has been proposed as a more powerful alternative to chloroform-d for heavy coal liquids (Saito et al., 1981) but its ^{13}C chemical shift at 170 ppm lies close to the low field end of the aromatic carbon region. The chloroform extractability of primary coal liquids can be increased considerably by silylation of methylation (hydrogen bonding vastly reduced; Snape and Bartle, 1979) if concentrations of phenolic OH are high. For coal tar pitches which contain relatively low concentrations of phenolic OH, solubility has been improved by chlorination with a mixture of sulfuryl chloride (SO_2Cl_2) and sulfur monochloride (S_2Cl_2) (Grienke, 1984). As an alternative to chemical modification as a means of improving the solubility of heavy coal liquids for ^{13}C NMR analysis, spectra have been obtained by using melts of elevated temperatures (Dorn et al., 1979) but the resolution does not match that achieved for solutions.

Sensitivity is affected by a number of factors including field strength and homogeneity, probe design, and, for FT spectra, accumulation and processing parameters. As a rough guide, FT ^1H spectra of coal liquids can be obtained using milligram quantities of sample but considerably more is required for CW spectra ($\sim$25 mg) and ^{13}C spectra (at least 100 mg).

3. ^1H Spectra and Peak Assignments

Typical high-field spectra of a hydrogenated anthracene oil and an asphaltene fraction are shown in Figure 21 which also gives the assignments of the principal bands. The major separation is between aromatic and aliphatic hydrogen but distinct regions can be identified within these types, particularly for the oil. The vastly superior resolution in the oil spectrum arises from the relatively low molecular mass compounds present, such as phenanthrene ($H_{4,5}$ at $\sim$8.5 ppm), which give narrow peaks. The asphaltenes are much more complex (MM range $\sim$250–2000), the broad bands also arising from the protons in the high-MM species present being significantly shorter than those in the oil (resolution α $^1/T_2$, where T_2 is spin-spin relaxation time).

The chemical shift of phenolic hydrogen depends on solvent and sample concentration, which both influence the extent of hydrogen bonding. Indeed, the phenolic hydrogen peak in the asphaltene spectrum overlaps the aromatic band (Figure 21) and the difficulty in identifying and measuring phenolic hydrogen by ^1H NMR has led to the use of derivatization procedures (Sec. IV.H.4).

For aliphatic groups, the chemical shifts are governed principally by the proximity of the hydrogen to an aromatic ring. In the earliest ^1H spectra of coal extracts (Brown and Ladner, 1960), it was observed that a separation occurred at $\sim$2.0 ppm between hydrogen adjacent (positioned α) to an aromatic ring and other aliphatic hydrogen. Peaks in the range 3.4–4.5 ppm can be assigned to hydrogen adjacent to two aromatic rings (e.g., CH_2 in fluorene, $H_{\alpha,2}$ region in Figure 21). The most prominent peak in the 0–2 ppm range for oils containing significant concentrations of *n*-alkanes occurs at 1.25 ppm from long-chain methylene. A division at $\sim$1.0 ppm is found in the spectra of most samples between CH_2, CH not adjacent to an aromatic ring, plus β-CH_3 (1–2 ppm) and any other CH_3 present (paraffinic and that

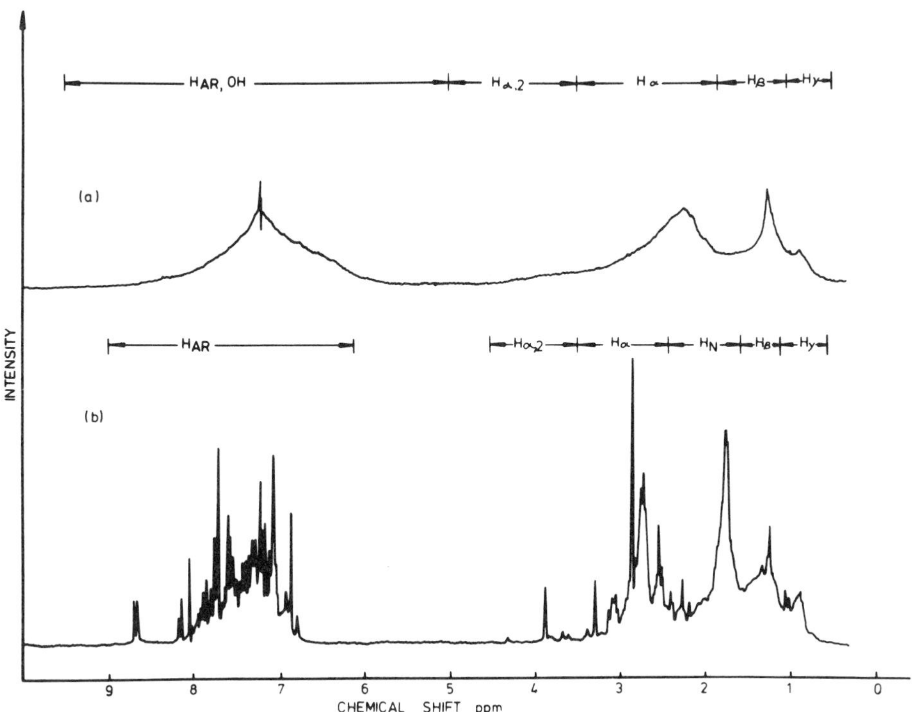

Figure 21 220-MHz ^{1}H NMR spectra of (a) an asphaltene fraction and (b) hydrogenated anthracene oil.

positioned γ or further from an aromatic ring) although some CH_2 in cycloalkanes and naphthenic substituents may also contribute to the H_γ band. When hydroaromatic compounds are present in significant concentrations, a prominent band between 1.5 and 2.0 ppm (H_n in Figure 21) attributable to βCH_2 and CH is observed in the spectra of liquefaction products. The assignments discussed above are based on literature data for model compounds but the recent application to coal liquids of two-dimensional methods in which spin-spin coupling between adjacent spins can be correlated has enabled many peaks to be assigned unequivocally (Sec. IV.H.3).

Provided that a suitably long relaxation delay (typically >5 sec) is used with a relatively small pulse angle (<40°) to acquire FT spectra and that correct phasing and flat baselines are obtained before integration, proportions of aromatic and aliphatic hydrogen can be determined with reasonable accuracy and precision (reproducibility of ∿1% or less of the total hydrogen). Errors involved in estimating concentrations of the aromatic and aliphatic hydrogen types discussed above (Figure 21) are probably greater because only partial separation is achieved between the different bands.

Although the combined use of ^1H and ^{13}C NMR is preferable for deriving detailed structural information (Sec. IV.I), ^1H NMR is ideally suited for the rapid monitoring of the compositions of chromatographic fractions both from aromatic ring size and from molecular mass (SEC) separations. Effluents from HPLC can be analyzed by ^1H NMR (Haw et al., 1980, 1981) but the necessity to use deuterated or nonprotonated solvents places limitations on the separations.

4. ^{13}C Spectra and Peak Assignments

The first ^{13}C spectra of coal liquids were obtained by continuous sweep methods (Friedel and Retcofsky, 1966) without proton decoupling to eliminate ^1H-^{13}C spin-spin coupling and reduce multiplet peaks to singlets. However, since the introduction of commercially available FT spectrometers, virtually all spectra have been acquired with proton decoupling because of the vast improvement in resolution, although initially there were problems concerning quantification.

Well-resolved peaks are generally obtained in both the aromatic and aliphatic regions of the proton-decoupled spectra of oils even at low field (Figure 22). Broader bands are usually found in the spectra of heavier fractions because of the greater structural complexity and shorter ^{13}C relaxation times, but even spectra of asphaltenes contain sharp alkyl peaks (32, 29.5, and 14 ppm; Figure 22). Depending on the nature of the sample, a number of divisions in both the aromatic and aliphatic regions of the spectra of coal liquids can usually be made and these are summarized in Table 6. These assignments were based initially on literature data for suitable model compounds (Snape et al., 1979) and were recently confirmed and extended through the use of spectral editing and two-dimensional methods.

Two methods—distortionless enhancement by polarization transfer (DEPT) and gated spin-echo ^{13}C NMR in which the intensities of CH, CH_2, and CH_3 peaks are modulated as a function of their ^{13}C-^1H coupling—have enabled the generation of C, CH, CH_2, and CH_3 subspectra for coal liquids (Cookson and Smith, 1983; Snape, 1982, 1983), including heavy fractions such as asphaltenes. DEPT requires more stringent pulse programming and data manipulation but has advantages over the gated spin-echo method in that CH and CH_3 subspectra can be obtained much more easily and it does not require separate experiments to edit aromatic and aliphatic carbon. However, it is often not necessary to separate aliphatic CH and CH_3 peaks because their chemical shift ranges do not usually overlap (see Table 6, p. 605).

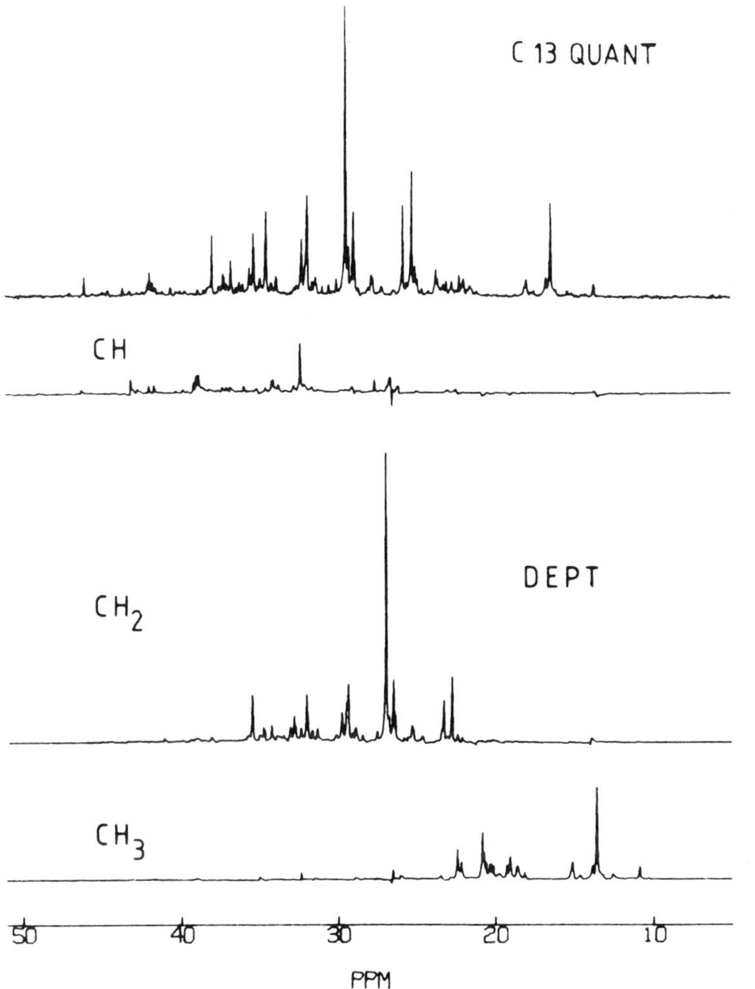

Figure 22 50-MHz ^{13}C subspectra for aliphatic groups in a coal liquid obtained by the DEPT method (Gerhards, 1983).

The use of spectral editing methods to resolve tertiary and quaternary aromatic carbon peaks has confirmed that a reasonable separation occurs at ∿129.5 ppm for mono- and diaromatic species. However, peaks for certain bridgehead and internal quaternary carbons in polyaromatic compounds—notably those for $C_{10b,c}$ in pyrenes at 124.5

ppm—occur in the range for tertiary aromatic carbon peaks (∿115–129.5 ppm). Spectral editing methods have also indicated that there is considerable variation in the relative proportions of aliphatic CH, CH_2, and CH_3 groups in coal liquids (Snape, 1983; Cookson and Smith, 1983) and that, in nonparaffinic fractions, quaternary aliphatic carbons are present in low concentrations (<2% of the aliphatic carbon).

The two-dimensional ^1H-^{13}C chemical shift correlation experiment which identifies spin-spin coupling between adjacent ^1H and ^{13}C has been particularly informative for aliphatic groups in coal liquids (Snape, 1986; Cookson and Smith, 1987). Figure 23 shows the aliphatic region from a high-field correlation spectrum for a monoaromatic fraction from a coal liquefaction solvent, the cross-peaks arising from coupled ^1H and ^{13}C spins. Numerous groups can be differentiated, e.g., CH_3 substituted in aromatic and hydroaromatic rings. All the ^{13}C peaks from these two types of CH_3 occur in the 17–23 ppm chemical shift range (Table 6) but ^1H peaks occur between 2.1 and 2.4 ppm for CH_3 in aromatic rings and between 1.0 and 1.5 ppm for CH_3 in hydroaromatic rings.

It is now well established that quantitative ^{13}C spectra of coal liquids can be obtained using gated decoupling (decoupler switched off during relaxation period following data acquisition) to suppress nuclear Overhauser enhancements (variable increase in peak intensity up to a maximum of ∿3 arising from removal of ^1H-^{13}C spin-spin coupling) in conjunction with doping sample solutions with low concentrations of a noninteracting paramagnetic compound, such as chromium acetylacetonate (Shoolery and Budde, 1976; Hajek et al., 1978). Chromium acetylacetonate concentrations of 0.02–0.05 considerably shorten thermal relaxation times (T_{1S}; a delay of $5.T_1$ is needed to ensure complete relaxation between successive scans). If samples have to be recovered for other analyses, chromium acetylacetonate cannot be readily separated from polar material and hence gated decoupling has to suffice for quantification. This increases spectral accumulation times considerably because relatively small pulse angles (<40°) with long relaxation delays (>10 sec) have to be used to ensure that complete thermal relaxation occurs. Quaternary aromatic carbons have considerably longer T_{1S} than aliphatic and tertiary aromatic carbons in fuels (Alger et al., 1979). Although T_{1S} decrease with increasing field strength (due to contribution from chemical shift anisotropy to the relaxation rate; Alger et al., 1979) and sample viscosity, these are still ∿1s for quaternary aromatic carbons in asphaltenes (Sklenar et al., 1980).

The reproducibility of aromaticity determinations should be typically 1% or less provided that appropriate conditions for spectral accumulation are used. Reasonably accurate data on the proportions of C, CH, CH_2, and CH_3 groups can be obtained both by the DEPT (Netzel, 1987) and gated spin-echo (Snape, 1983) methods although the precision is probably not as good as for aromaticity measurements.

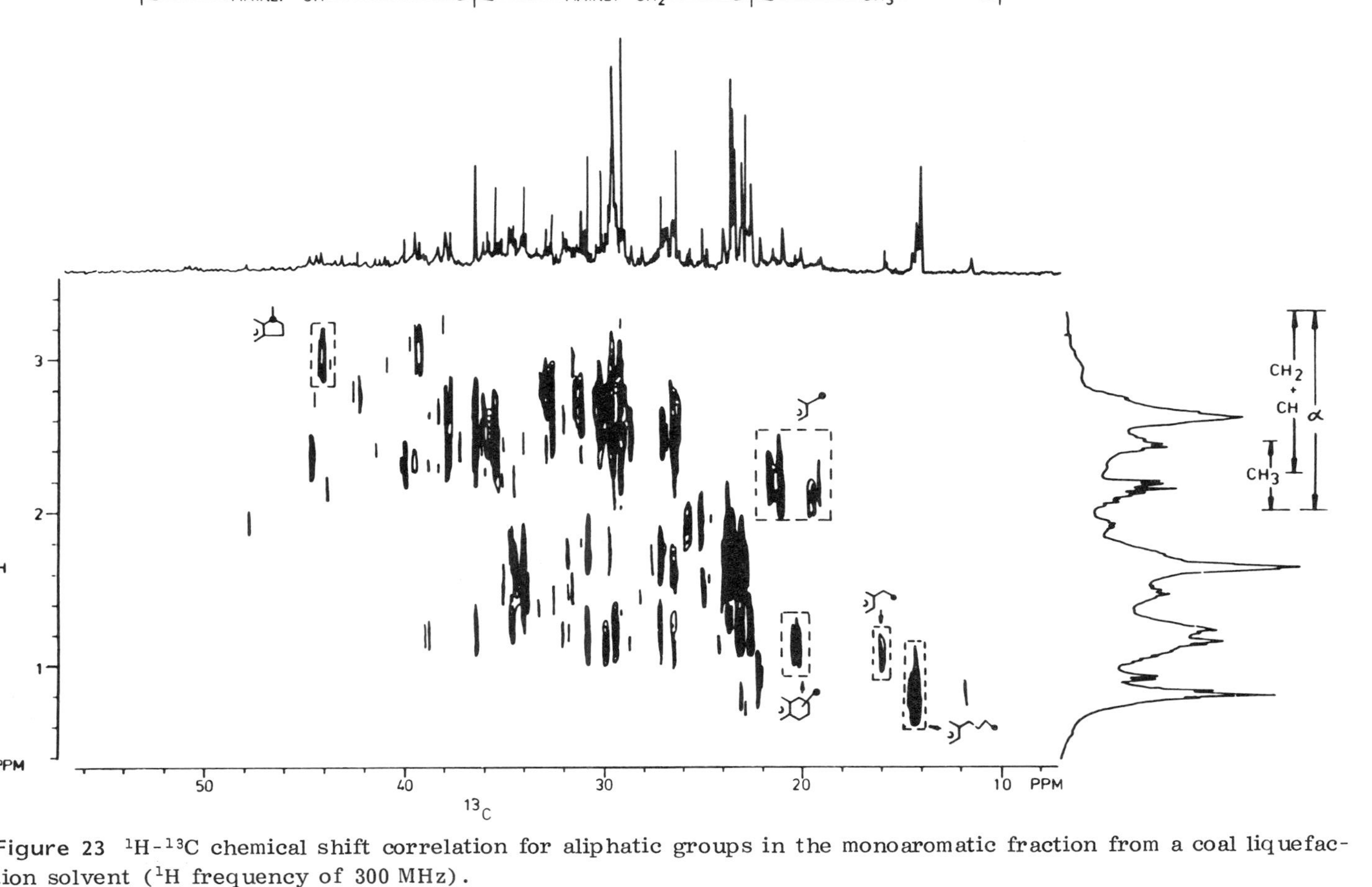

Figure 23 ^1H-^{13}C chemical shift correlation for aliphatic groups in the monoaromatic fraction from a coal liquefaction solvent (^1H frequency of 300 MHz).

5. Techniques for Heteroatomic Groups

Although titration methods can be used for the estimation of the concentrations of strongly acidic OH and basic N groups in coals and coal liquids (Sec. IV.C), there is a need to develop methods for the measurement of neutral and weakly acidic O, N, and S groups, such as ethers, furans, and secondary amines. ^{17}O, ^{15}N, and ^{33}S all suffer from low receptivity for fuels while ^{17}O, ^{14}N, and ^{33}S are quadrupolar nuclei and generally give broad signals. Not surprisingly, there have been few reports of the direct identification of heteroatomic groups in fuels by NMR. The application of ^{17}O NMR to coal liquids has been reported (Grandy et al., 1984) but poor signal-to-noise levels were obtained.

Indirect methods in which labile hydrogens are replaced with groups containing a convenient magnetic label (^{1}H, ^{13}C, ^{19}F, ^{29}Si, and ^{31}P) have been widely used to estimate OH (mainly phenols and carboxylic acids) concentrations but NH, NH_2, and SH groups have also been identified. Some of the procedures that have been applied to coals and coal liquids are summarized in Table 7. The obvious advantage of a multinuclear label (^{19}F, ^{29}Si, and ^{31}P) is that there are no overlapping sample peaks, although trimethylsilyl, acetyl, and methyl derivatives all give well-resolved peaks in either ^{1}H or ^{13}C spectra. Methylation in conjunction with ^{13}C NMR has arguably been the most successful of the various approaches (Table 7). OH and COOH groups can be differentiated because carboxylic acids give methyl ester peaks close to 50 ppm while peaks for methyl ethers of unhindered phenols occur at ∿55 ppm and those of hindered phenols plus alcohols at ∿60 ppm. The use of ^{13}C-enriched iodomethane enables the intensities of the peaks from the introduced methyl groups in the spectra of methylated coals to be measured readily (Liotta et al., 1981).

6. ^{2}D Studies on Coal Hydrogenation

Both deuterated tetralin and gaseous deuterium have been used in isotopic tracer experiments and the products analyzed by ^{2}D NMR (Franz, 1979; Cronauer et al., 1982). Figure 24 compares the ^{1}H and natural abundance ^{2}D spectra of a coal liquid, the bands being broader in the ^{2}D spectrum despite the use of proton decoupling (Farnum et al., 1984). Gate spin-echo and DEPT methods have been used to clarify the various tetralin isomers that arise after exchange with deuterium. The results of these studies have indicated that initial deuterium incorporation into coal-derived molecules is primarily in benzylic groups but, at long reaction times, considerable scrambling occurs with incorporation of deuterium into aromatic and nonbenzylic groups.

Table 7 Summary of Derivatization and NMR Methods Used to Investigate Heteroatomic Groups in Coal Liquids

Method	Derivatives from phenols/ alcohols	Other groups derivatized	Magnetic label	Refs.
Silylation	$-OSi(CH_3)_3$	$-COOH$, $-NH$	^1H	Snape et al. (1982)
			^{29}Si	Coleman and Boyd (1982)
Hexafluoro-acetone adduction	$-OC(CF_3)_2OH$	$-NH_2$	^{19}F	Bartle et al. (1980)
Trifluoro-acetylation	$-OCOCF_3$	$-NH$, $-NH_2$	^{19}F	Sleevi et al. (1979)
Acetylation	$-OCOCH_3$	$-NH$	^{13}C	Snape et al. (1982)
Methylation	$-OCH_3$	$-COOH$, $-NH$	^{13}C	Snape et al. (1982)
		$-NH_2$		Liotta et al. (1981)

I. Structural Analysis and Integration of Analytical Methods

Since the 1960s (Brown and Ladner, 1960), NMR data have been combined with elemental composition, average MMs, and concentrations of heteroatomic groups, usually with some assumptions, to derive more detailed information on the aromatic and aliphatic groups present in coal liquids. Structural analysis results have been presented in three forms:

1. On a 100 C atom basis, as for coals, and using structural parameters, such as the average size of aliphatic groups
2. Hypothetical average structures, the difference from coal molecules (Sec. IV.I) being the inclusion of average MMs into the calculations so that the structures have a finite size, and

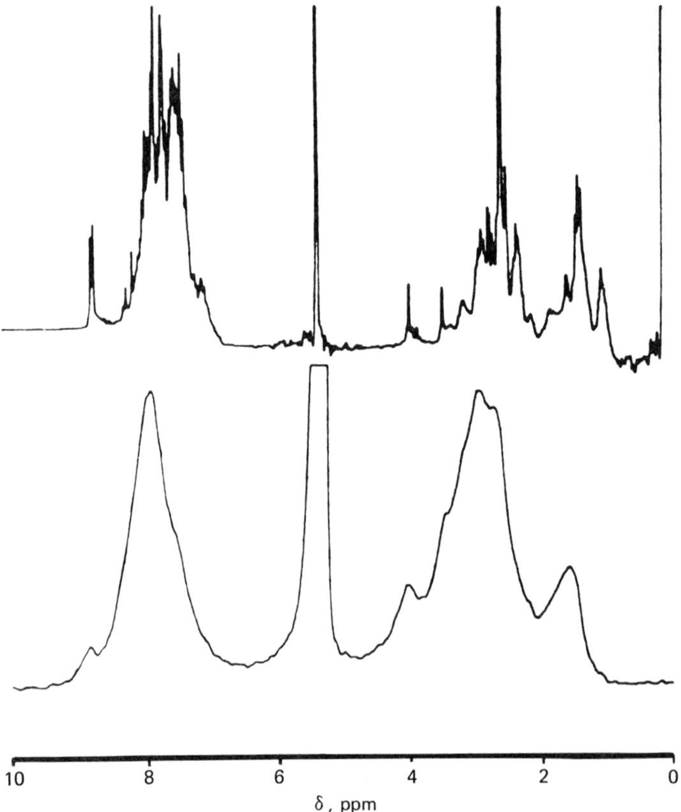

Figure 24 200-MHz ^1H (top) and 31.7 MHz ^2D (bottom) NMR spectra of an aromatic fraction from a lignite liquefaction product prepared with deuterium and carbon monoxide (Farnum et al., 1984). Reproduced with permission of the American Chemical Society.

3. Concentrations of particular groups (Table 8), a relatively small number being used to describe the complex molecular compositions of heavy coal liquids (Allen et al., 1984).

The last approach was devised to alleviate the main problem of average structures and certain structural parameters in that little information is derived on the distribution of structures present. The accuracy of structural calculations is obviously dependent on that of the NMR and other data used but also on the assumptions made, and both random and systematic errors can easily arise (Shenkin, 1983).

Table 8 Groups Used to Define Compositions of Heavy Coal Liquid Fractions

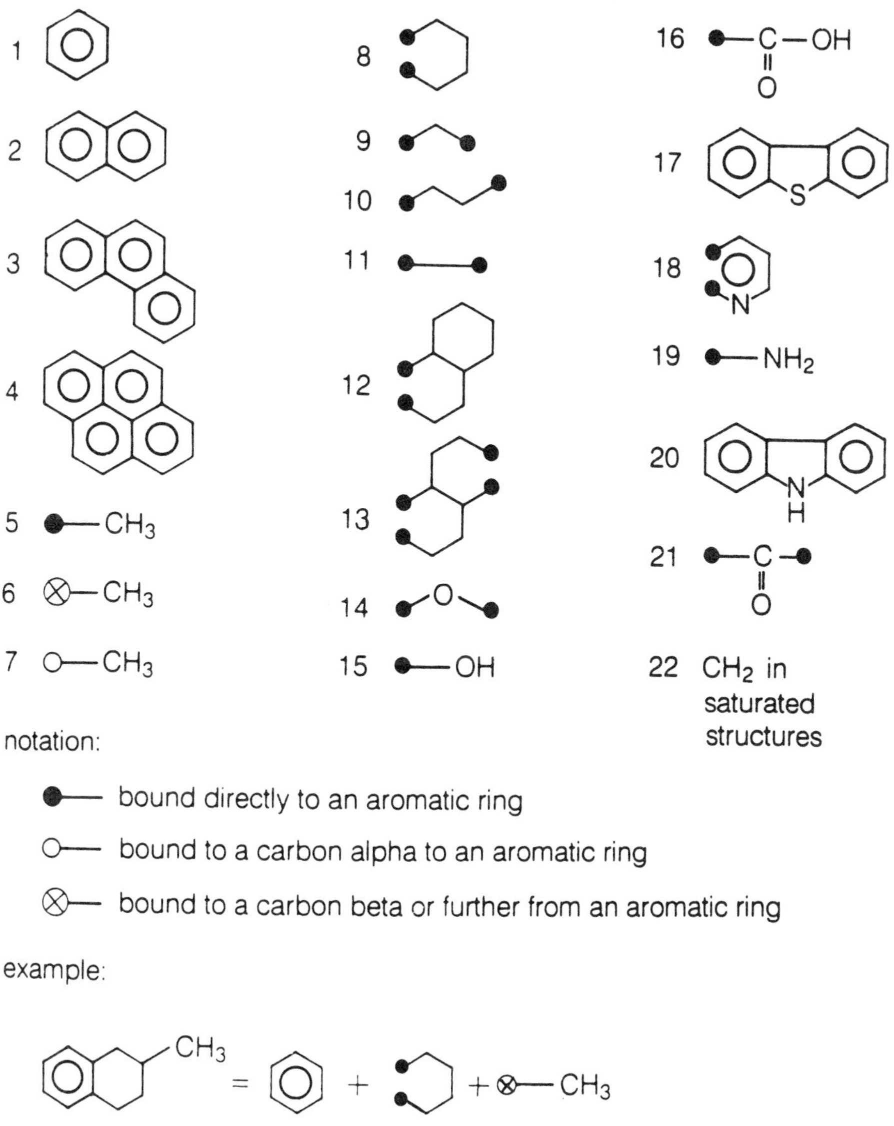

1
2
3
4
5 ●—CH_3
6 ⊗—CH_3
7 ○—CH_3
8
9
10
11
12
13
14
15 ●—OH
16 ●—C(=O)—OH
17
18
19 ●—NH_2
20
21 ●—C(=O)—●
22 CH_2 in saturated structures

notation:

●— bound directly to an aromatic ring

○— bound to a carbon alpha to an aromatic ring

⊗— bound to a carbon beta or further from an aromatic ring

example:

CH_3 = + + ⊗—CH_3

Source: Allen et al., 1984. Reproduced with permission of Butterworth Scientific Ltd.

There is also an urgent need for greater standardization in the terminology used in structural calculations. As already highlighted (Sec. II.B), it is desirable to separate lower MM coal liquids into discrete fractions (e.g., mono-, diaromatics, etc.) for detailed characterization in order to limit the number of structural possibilities.

Much of the early work on structural analysis was concerned with estimating aromaticities from ^1H NMR data and C/H ratios by the formula initially used by Brown and Ladner (1960):

$$fa = \frac{C/H - H\alpha/x - H\beta/y}{C/H}$$

where $H\alpha$ and $H\beta$ are the proportions of aliphatic hydrogen adjacent and not adjacent to an aromatic ring, respectively, and x, y are the assumed aliphatic H/C ratios for these environments. Systematic errors can arise from the choice of aliphatic H/C ratios but this indirect calculation procedure has been superseded because aromaticities are now routinely determined by ^{13}C NMR.

For aromatic groups in coal liquids, the general objective of structural analysis has been to deduce proportions of bridgehead or internal aromatic carbon so that the degree of condensation of aromatic structure can be gauged. Although signals from individual cata- and peri-condensed aromatics can be identified in the ^{13}C spectra of liquefaction oils and high-temperature tars, there is usually little resolution for heavier fractions (Sec. IV.H). Thus, concentrations or numbers of internal aromatic carbons (C_{int}) have been deduced from summing all the peripheral aromatic carbons and subtracting from the total concentrations or numbers of aromatic carbons (Snape et al., 1984). Knowledge of heteroatom environments, particularly oxygen, is essential for this calculation but often has not been considered (Shenkin, 1983) with the result that internal aromatic concentrations are overestimated.

Because of the uncertainties in calculating the concentrations of internal aromatic carbons using NMR-based schemes, confirmatory evidence from other techniques is highly desirable. Indeed, LC, UV spectrometry, selective oxidation, and differential pulse voltammetry all provide information on the distribution of aromatic groups present in heavy fractions not amenable to GC (Secs. III.B, IV.B, IV.F, and IV.6).

Aliphatic H/C ratios give a good indication of the nature of the aliphatic substituents in heavy coal liquids. These and numbers of hydroaromatic/naphthenic rings per molecule (R_N) have traditionally been calculated indirectly but, as already discussed for the former, these calculations are prone to large errors (Shenkin, 1983). Highly condensed naphthenic groups have been proposed as being the major aliphatic substituents in primary coal liquefaction products (Farcasiu,

1979) on the basis of grossly underestimated aliphatic H/C ratios (1.3–1.7). The critical assessment of structural analysis by Shenkin (1983) highlighted the lack of precision in calculating R_N. As discussed earlier (Sec. IV.D), spectral editing ^{13}C NMR methods offer the best approach for providing detailed information on aliphatic substituents. Also, the use of dehydrogenation in conjunction with ^{13}C NMR should prove useful for characterizing naphthenic substituents in heavy fractions, such as asphaltenes. Selective oxidation and transalkylation procedures provide information on the distribution of alkyl chains on aromatic substituents (Sec. III.C), and it has been shown that for an asphaltene fraction, the concentrations of arylmethyl derived by ruthenium tetroxide oxidation and ^{13}C spectral editing are in good agreement (Snape et al., 1985).

Thus, there are ways in which NMR can be integrated with chemical and other spectroscopic techniques to provide more reliable and detailed information. In order to gain some insight into the distribution of structural parameters about their statistical average values in heavy coal liquids, NMR has been used to characterize subfractions separated by SEC (e.g., Richards et al., 1983); wide variations in aromaticity and the average size of aliphatic substituents with increasing MM were found for asphaltenes and preasphaltenes.

NMR-based structural analysis has been particularly useful for assessing the chemical nature of primary liquefaction products (Snape et al., 1984; Kershaw, 1985) although some of the structural variations reported in the literature could well reflect inherent deficiencies in the procedures used (Shenkin, 1983). A number of studies indicated that, with increasing process severity, tars and extracts become more aromatic in character and contain more highly condensed nuclei (e.g., Snape et al., 1985; Kershaw, 1985). The use of the group composition approach (Allen et al., 1984) has enabled thermodynamic properties such as heat capacity to be predicted for coal liquids (Le and Allen, 1985). The predicted variation with increasing temperature has been found to be in close agreement with experimental results.

20

Coking and Carbonization

I	Introduction	645
II	Coking	648
III	Carbonization	651

I. INTRODUCTION

The primary areas of research and development with regards to coal utilization are:

1. Carbonization, pyrolysis, and combustion
2. Gasification
3. Liquefaction
4. Pollution abatement and clean coal

In the near term it appears that efforts to commercialize gasification and liquefaction technologies will be minimal, but efforts to provide clean coal for combustion uses will receive increased attention.

The heat content, or calorific value, of coal was referred to earlier. However, since the utilization of coal necessarily involves its being heated and burned, a few remarks about the thermal properties and the thermal behavior of coal are in order here.

Heat capacity C_p (and specific heat) and the thermal conductivity of coal are two parameters used to describe the thermal properties of coal (in addition to calorific value). The heat capacity of any material is the thermal energy required to raise the temperature of a unit weight of a sample 1° Celsius (from 14.5 to 15.5°C, by definition). The units are either cal/g/°K or btu/lb/°F. When the heat capacity of a sample is compared to the heat capacity of water, taken as 1/cal/g/°C, the value is called the specific heat (unitless). Values for the specific heat of coal may range from 0.19 to 0.33 or higher. The value is observed to increase with increasing moisture and volatile matter content, but to decrease with increasing percentage of fixed carbon (rank) (Table 1). An estimate of the specific heat of coal can be made from

Table 1 Specific Heat Data for Air-Dried Coals

	Proximate analysis				Mean specific heat at:			
Source	Moisture (wt %)	Volatile matter (wt %)	Fixed carbon (wt %)	Ash (wt %)	28–65°C	25–130°C	25–177°C	25–227°C
West Virginia	1.8	20.4	72.4	5.4	0.261	0.288	0.301	0.314
Pennsylvania (bituminous)	1.2	34.5	58.4	5.9	0.286	0.308	0.320	0.323
Illinois	8.4	35.0	48.2	8.4	0.334			
Wyoming	11.0	38.6	40.2	10.2	0.350			
Pennsylvania (anthracite)	0.0	16.0	79.3	4.7	0.269			

Source: G. L. Baughmann, 1981. Reprinted with permission of the publisher.

Kopp's law, which uses the elemental (wt %) composition of the coal (Speight, 1983):

$$C_p = 0.89C + 0.874H + 0.491N + 0.360O + 0.215S$$

Thermal conductivity measures the heat flow per unit time through a material of unit volume for a unit temperature difference. The expression used to determine this parameter is dQ/dt = kdT where Q = heat flow, dt = unit time (sec), dT = difference in temperature (°K). Thermal conductivity k has units of cal/sec/cm/°K or btu/hr/ft/°F. There is not a unique value of thermal conductivity determined for all coals. As with other empirical parameters used to describe coal, the thermal conductivity varies with rank; it has also been observed to vary in samples from the same seam as well as with particle size, with apparent density, moisture content, volatile matter content, ash content, and temperature range of interest. Typically observed values range from 2.5×10^{-4} to 9×10^{-4} cal/sec/cm/°K (Speight, 1983).

Pyrolysis of coal is a thermal decomposition process which results in the formation of a variety of gases, tars, and a solid, carbon-rich char. Although the term pyrolysis is often used synonymously with the term carbonization, pyrolysis describes more pervasive thermal decomposition processes which are carried out under a variety of conditions and over a wide range of temperatures. While a solid residue is formed during coal pyrolysis, strictly speaking, pyrolysis reactions are not designed primarily to produce a useful coke.

Reactions occurring during general pyrolysis of coal can be described in summary fashion as follows:

Temperature		
(°C)	(°F)	Observed reaction or products
60–100	140–210	Moisture evolved; trapped gases released: CO, H_2, CH_4
100–200	210–390	CO_2 evolved from carboxylic acid groups; H_2S and NH_3 evolved
200–370	390–700	Low molecular weight organics released (aliphatic and aromatic)
>370	>700	High molecular weight polycyclic aromatics; phenols, nitrogen heterocyclic compounds; methane and hydrogen gas from structural decomposition

Source: Speight, 1983. Reproduced with permission of the publisher.

Coals of different rank will produce different quantities of these compounds, and different heating rates, and the rate of product removal will also affect the overall product formation. Speight (1983) cites the use of steam, zinc or aluminum chloride, sodium hydroxide, iron(III) oxide, phosphoric acid, clays, and activated carbon as having very pronounced effects on product formation during coal pyrolysis.

Very rapid thermolysis of coal—10^2–10^6°C/sec—is called flash pyrolysis and results in a significantly higher yield of gaseous products, especially acetylene, hydrogen, and methane (Speight, 1983). Thus, it has been observed that when a coal was heated at 38°C/sec, volatile materials accounted for about 47% of the products formed, but at a heating rate of 5540°C/sec, volatile production increased to about 55%, substantially above the analytically determined volatile matter content for the coal. A wide variety of techniques have been used to study flash pyrolysis, but there is no current commercial application of this technique.

II. COKING

Prime coking coals are those having 20–31% (dmmf) volatile content. It is observed that as coals with this range of volatile matter content are heated, they become soft, then plastic. The plastic material undergoes decomposition and coke forms when the decomposing material resolidifies into a hard and porous but very strong-walled solid by about 1000°C (1830°F) (Speight, 1983). The principal chemical reactions occurring during this process are hydrogen outgassing accompanied by ring condensation. These account in part for the puffing up of the coal and the formation of the porous product. Initial decomposition of coal begins at temperatures at or even below 200°C (390°F), but softening of coking coals usually occurs at 350–450°C (660–840°F) depending largely on the volatile matter content (Figure 1). True coke is generally formed in the temperature range 900–1100°C (1650–2000°F). In noncoking coals, the residue formed by heating and devolatilization has a highly porous, but weak, thin-walled structure associated with more extensive decomposition during the plastic transition period.

Analysis has supported the hypothesis that various maceral groups exhibit pronounced effects on the overall thermal behavior of coal. Maceral groups, like whole coal, can be characterized according to their volatile matter content (Table 2 in Chap. 18), and weight loss with temperature has been correlated with macerals from the same rank of coal (Figure 2). Prime coking coal (medium-volatile bituminous)

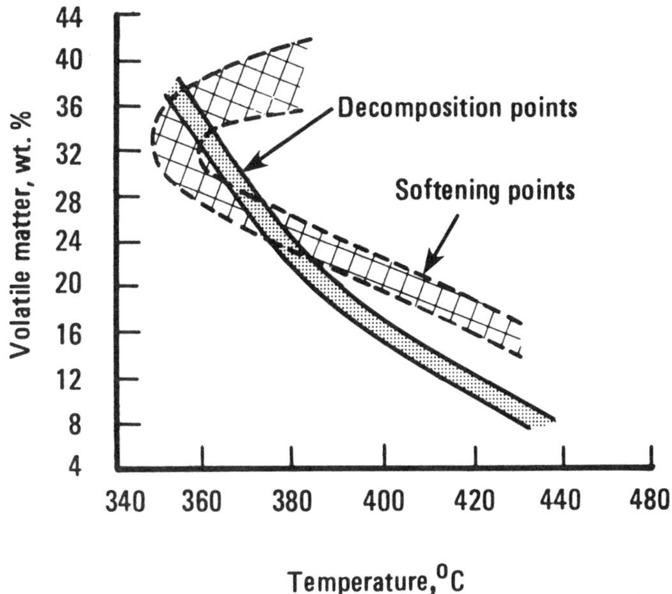

Figure 1 Relationship of volatile matter content to softening and decomposition points of coal (G. J. Pitt and G. R. Millward, 1979. Reprinted with permission of the publisher).

is rich in vitrinite, which swells and becomes fluid upon heating. Also, it has been determined that although inertinites release volatile matter upon being heated, they do not soften; rather inertinite appears to be the "core" material in coke to which vitrinite and exinite bind (Table 2). The amount of inertinite may also play a role in determining the extent of both swelling on the one hand and the degree of contraction on the other.

Over the years, experience has shown that blending less desirable coals with better coking coals was advantageous for optimizing coke formation. In this way, not only can the physical and chemical properties of the coke be improved, but value is added to the otherwise less useful coals, and damage to coking ovens has been reduced because proper blending can serve to diminish the extremely high pressures that developed during devolitilization (Speight, 1983). Besides maceral constitution, five other considerations which are important for blending coals are moisture content, particle size, hardness, fusability, and qualitative and quantitative mineral matter content.

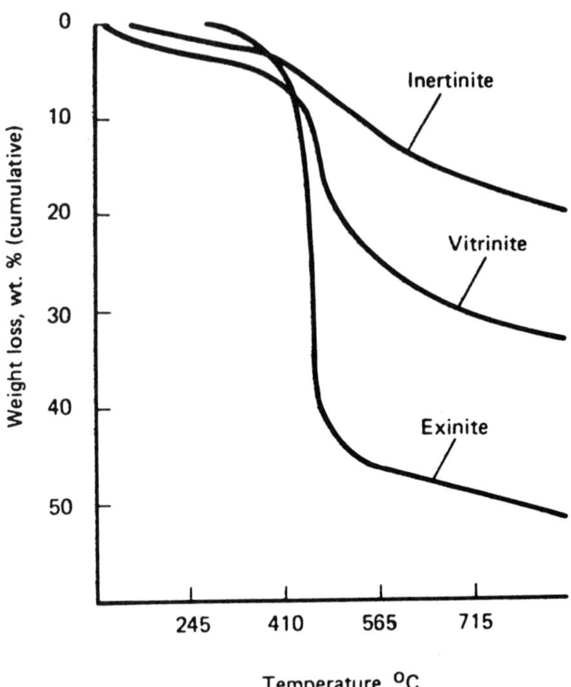

Figure 2 Correlation of maceral weight loss with heating temperature (C. Kroeger and A. Pohl, 1957. Reprinted with permission of the publisher).

Table 2 Reactivities of Coal Macerals During Coking

Relative reactivity	Maceral constituents
Reactive	Vitrinite with reflectance (r) = 0.5–2.0%; exinite
Partially reactive	Semifusinite (r < 2.0%)
Unreactive	Vitrinite (r > 2.0%); fusinite, micrinite

Source: Speight, 1983. Reproduced with permission of the publisher.

III. CARBONIZATION

Carbonization of coal is historically the most important utilization process for coal. In this process, coal is heated in the absence of air to temperatures of 750–1500°C (1380–2730°F). Coal gas, or town gas, as it was originally called, is evolved. It is a mixture of hydrogen, carbon monoxide, methane, ethylene, and several other gases in small amounts. Prior to the widespread use of natural gas, coal gas was a primary domestic heating and industrial fuel. In addition, coal gas was the principal source for the production of ammonia. While coal gas per se is no longer commercially important, the recovery of ammonia from the gas mixture is an attractive alternative source for this important industrial byproduct. Ammonia gas can be recovered from the product gas stream at high pH or by converting it, by addition of

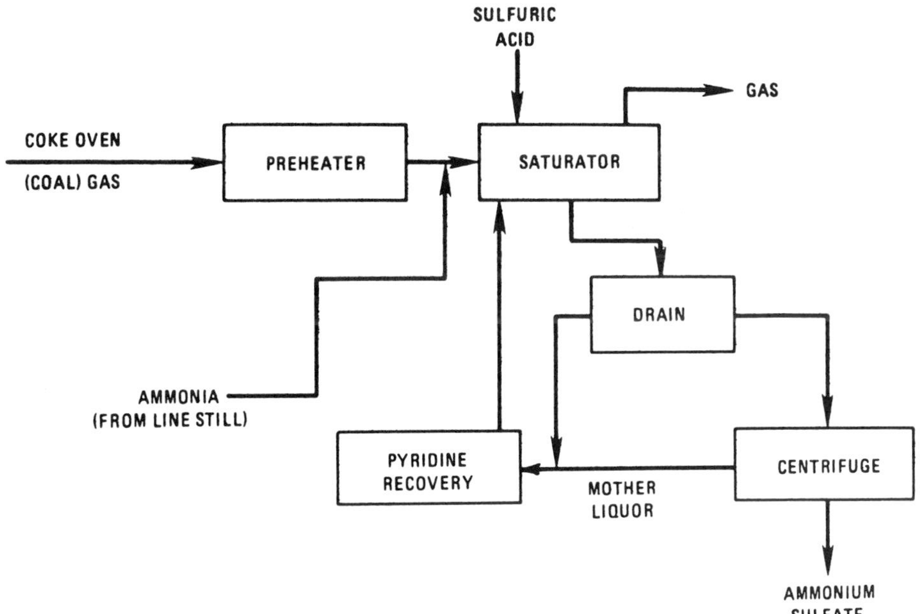

Figure 3 Generalized flow chart for the recovery of ammonium sulfate from coal gas (J. Speight, 1983. Reprinted with permission of the publisher).

acid, to an ammonium salt. Figure 3 shows the flow chart for such a process. The product gas stream plus tar is condensed. Lime (CaO) is added to convert ammonium ions to ammonia gas. After returning the concentrated ammonia to the main gas stream, it is adsorbed in 5–10% sulfuric acid at 50–60°C (120–140°F). Under these conditions ammonium sulfate precipitates from the solution. Figure 3 also shows a step for pyridine recovery. Pyridine, along with hydrogen cyanide and hydrogen sulfide, is a commercially valuable component of the coal gas which can be recovered from coal gas as nonvolatile salt (Speight, 1983).

Low temperature carbonization is carried out at 500–750°C (930–1380°F). Semicoke (coalite or char), a solid smokeless fuel, is the primary product (Crabbe and McBride, 1978). Some but not all of the volatile and tar components of coal are driven off at these temperatures. The semicoke has a heat content of about 12,270 btu/lb compared to an average value for bituminous coal of 12,500 btu/lb (Crabbe and McBride, 1978). While some semicoke, or low-temperature char, is not suitable for metallurgical processes, it is very reactive and therefore an energy-efficient fuel. Furthermore, because it is smokeless and contains little sulfur or other trace elements, it is environmentally acceptable. This is a particularly attractive end use for low-rank coals, especially lignites and brown coal. These low-heat, high-moisture coals will experience a greatly improved market value if they are converted to this type of reactive char. High-rank (high-volatile-content) coals which do not produce useful metallurgical coke may also be used in a low-temperature carbonization process if treatment to eliminate any caking tendencies is carried out (Speight, 1983).

Over the years a variety of retorts have been used for low-temperature carbonization. The most recent developments in carbonization technology have developed one scheme in which a continuous downward flow of coal is heated by an upward flow of hot combustion gas. The Lurgi–Spulgas retort (Figure 4) and the Koppers oven (Figure 5) are two variations of this technique.

Another alternative use for coals which are unsuitable for direct coking or even blending is the production of formed coke. Formed coke is made from finely crushed coal which is heated to only 600–800°C (1110–1470°F) to form a char but to minimize or prevent swelling or caking by extensive devolatilization of the coal. Noncoal binders are mixed with the char to agglomerate it into briquettes, after which additional heating (900–1000°C; 1650–1830°F) is carried out to consolidate the mass. Formed coke is stronger than low-temperature semicoke, and several types are used as metallurgical fuels as well as domestic fuels (Speight, 1983).

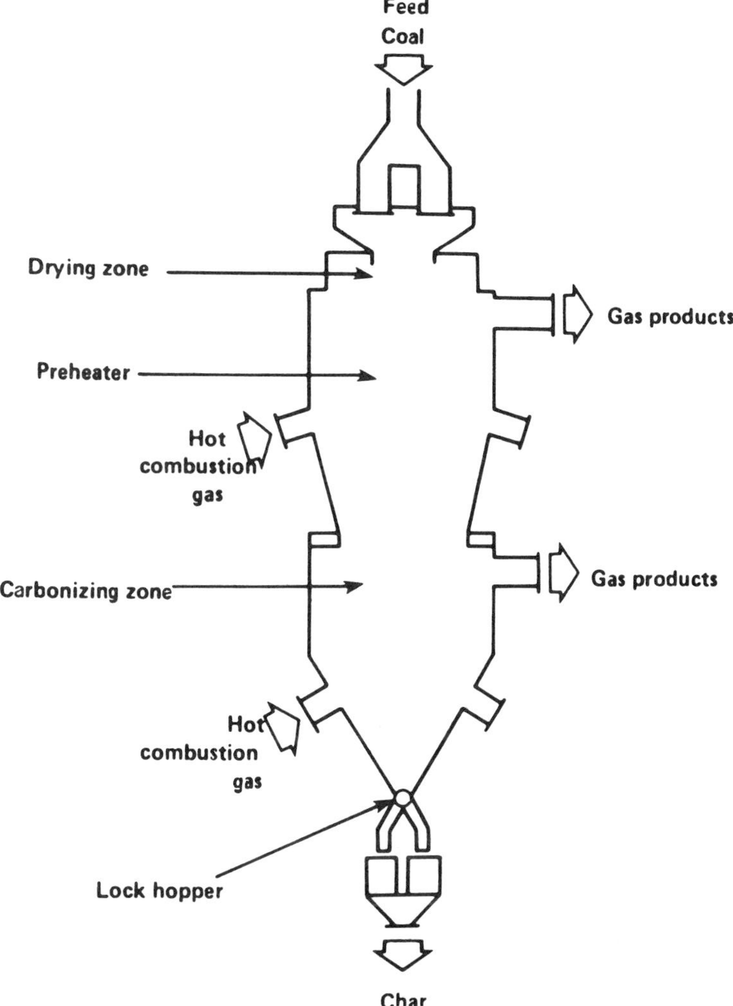

Figure 4 Schematic diagram of the Lurgi–Spulgas retort (N. Berkowitz, 1979. Reprinted with permission of the publisher).

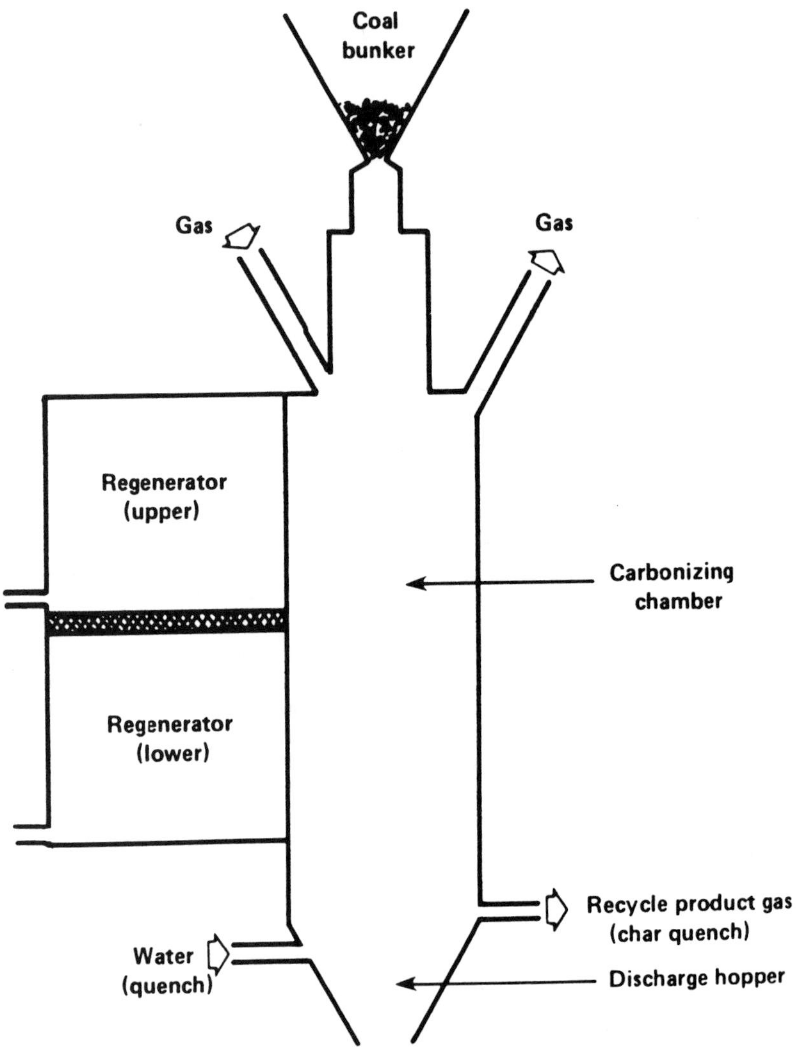

Figure 5 Schematic diagram of the Koppers coke oven (J. Speight, 1983. Reprinted with permission of the publisher).

21

Combustion

I	Effects of Coal Properties	655
II	Chemical Aspects	657
III	Combustion Systems	657

Combustion is the oldest use of coal and is now probably the single most important use worldwide. Because combustion technology has developed steadily, the primary focus in this era is to improve methods that can use low-rank coal efficiently and that can meet increasingly stringent environmental demands.

I. EFFECTS OF COAL PROPERTIES

For the generation of heat, whether it is to be used to produce steam for power production or to heat industrial boilers, the heat content (calorific value) of coal is of primary importance. This necessarily implies that the low-rank, high-moisture coals are less suitable for all combustion processes and that high-rank coals are always preferred. But ease of ignition is also critical for the economical implementation of coal combustion. Ignition of coal can occur at temperatures as low as 150°C (390°F) for lignite and can range as high as 600°C (1110°F) for anthracite. In spite of their lower btu values, low-rank coals are easier to ignite than high-rank coals, which often require additional oil firing and require longer time and larger furnaces to achieve complete combustion (Speight, 1983). On the other hand, low-rank coals have a higher moisture content which inhibits combustion by decreasing flame temperature and which consumes useful heat. An optimum moisture content is 6–8%.

Volatile matter is important in coal combustion for several reasons. First, it is important because it enhances the ease of ignition and the rate of combustion. Second, volatile matter accounts for at least some of the btu content of the coal. Third, it is the evolution of volatiles by outgassing that gives rise to the (undesirable) swelling and caking of coal. There is also a relationship between volatile matter content and the grindability of coal (hardness). Grindability affects the cost of preparing coal for combustion. This is especially true for fluidized bed systems which use very small particles (2–3 mm diameter). The ease of grinding coals (high grindability index value) varies with the volatile matter content (rank) and reaches a maximum at about 22% volatile content (Berkowitz, 1979).

Mineral matter (ash) content and ash fusion properties as well as trace elements are also important parameters in coal combustion. Ash disposal after combustion adds substantially to the cost of coal combustion for power generation. Also, high ash may contribute to abrasion, erosion, and breakdown of metal boilers. Ash deposition, or slagging, in the furnace, and fouling, the deposition of ash in other areas of the system, are related to the ash fusion properties of the coal. Fouling is most often related to iron, sodium, potassium, magnesium, and calcium oxides, but phosphorus and sulfur contribute to ash depositions as well. The ash fusion temperature is determined by standard procedures (described elsewhere in this volume). A low ash content and a high ash fusion temperature are undesirable. In addition, the viscosity of molten ash is also important in that it affects the rate of flow of ash. The type of equipment used to remove ash deposits will be determined by the ash viscosity.

Elemental composition is important because of the detrimental effects trace elements can have on combustion processes in addition to ash formation. Chlorine, for instance, is highly corrosive to boiler materials and may be evolved into the air with stack gases; sulfur is similarly corrosive if it forms sulfuric acid in the boilers and is definitely problematic when it is evolved (as SO_2 or sulfuric acid vapors) into the atmosphere.

II. CHEMICAL ASPECTS

The chemical aspects of coal combustion are far from simple and are not completely understood. Speight (1983) summarized the principal reactions and associated enthalpy values for combustion processes:

Process	Enthalpy	
	Kcal/mole	Btu/lb
$C_S + O_2 = CO_2(g)$	-94.4	-169,290
$2C_S + O_2 = 2CO(g)$	-52.8	-95,100
$C_S + CO_2(g) = 2CO(g)$	+41.2	+74,200
$2CO(g) + O_2 = 2CO_2(g)$	-135.3	-243,490
$2H_2(g) + O_2 = 2H_2O(g)$	-115.6	-208,070
$C_S + H_2O(g) = CO(g) + H_2(g)$	+31.4	+56,490
$C_S + 2H_2O(g) = CO_2(g) + 2H_2(g)$	+21.5	+38,780
$CO(g) + H_2O + CO_2 + H_2(g)$	-9.8	-17,710

Although the enthalpy values for these reactions are known, the actual overall heat balance for coal combustion reactions is very complex. It is not only affected by the heat(s) of combustion, as shown by the various equations listed above, but it also involves the sensible and latent heats of the air and of the carbonization products, the heat associated with exothermic reactions other than coal combustion, heat retained in uncombusted coal, endothermic reactions, and heat losses to the surroundings (Speight, 1983). Such factors as particle size, surface area, pore structure, volatile matter content, rate and extent of pyrolysis prior to and during combustion, additives, catalysts, and impurities are among a large number of physical and chemical parameters involved in coal combustion (Essenhigh, 1981).

III. COMBUSTION SYSTEMS

The systems in use for most commercial applications of coal combustion are fixed-bed, fluidized bed, and entrained flow combustors. Fixed and fluidized beds are able to use sized particles while entrained flow systems require finely pulverized coal.

Fixed-bed combustion is characterized as being either up-draught or down-draught combustion. In an up-draught configuration, the primary air source is at or slightly below the level of the fuel. The fuel is ignited at the bottom and the flame travels upward with the air flow. A secondary air inlet is positioned above the level of the bed to facilitate combustion of volatiles distilled away from the bed prior to being combusted. However, smoke containing incompletely combusted volatiles, particularly hydrocarbons and heteroatom-containing pollutants, can easily escape from this system. In the down-draught configuration, the air flows downward onto the fuel bed, i.e., counter to the combustion front; the flue is positioned below the grate. Ignited so that the flame front moves counter to the direction of the air, distilled volatiles are kept in the flame by the air stream. More complete combustion and reduction of pollution can be achieved if proper conditions of turbulence, temperature, and residence time are maintained. An underfeed stoker is one example of a down-draught combustor. A worm feeder stokes coal into the bottom of the furnace, continually forcing the coal upward into the air stream (Speight, 1983). In another configuration, the chain-grate system, the air flow is from below the grate onto which the coal is fed. However, the coal, being fed into the chamber on a continuously moving belt, is ignited at the top and the flame travels downward against the air flow. Coal fines and caking coals are not desirable for the chain-grate system because they tend to clog the grate and contribute to high loss of unburned carbon (Speight, 1983).

A fluidized bed process is one in which finely divided coal solids are caused to "float" in an upward stream of gas. The high heat transfer which characterizes these combustors permits the use of lower temperatures and smaller combustion chambers. Low operating temperatures are significant because they reduce the emission of nitrogen oxides (NO_X) (Crabbe and McBride, 1978) and a fluidized bed can operate efficiently under high pressure. In addition, this configuration is advantageous because ash sinks to the bottom even against the upward air flow. Not only does this facilitate ash removal, but lower quality coals are therefore amenable to use. The fact that efficient combustion can be achieved with as little at 1–5% coal feed has prompted the use of additives such as limestone ($CaCO_3$) or dolomite ($CaCO_3$-$MgCO_3$) in fluidized bed reactors. These highly reactive minerals effectively remove sulfur pollutants by forming calcium or magnesium sulfates (plus CO_2) from SO_2 released during combustion. Furthermore, recovery and recycle of the calcium or magnesium is achieved by treatment of the sulfates with H_2 of CO according to the following reactions:

$$CaSO_4(s) + H_2 = CaO(s) + H_2O + SO_2(g)$$

or

$$CaSO_4(s) + CO = CaO + CO_2(g) + SO_2(g)$$

The sulfur dioxide released in these reactions is not fugitive; it is a suitable feedstock for both sulfuric acid manufacture and further recovery of elemental sulfur (Speight, 1983). This form of pollution abatement does add to operating costs, especially if separation and recycle of the limestone is desired. Fluidized bed combustors in general require the added cost also of separating ash from uncombusted char in order to reclaim the useful char. Loss of fines also occurs during fluidized bed combustion and must be controlled. This is usually done by incorporating cyclones or electrostatic precipitators into the system. Finally, the energy cost of power to maintain the fluidized conditions must be taken into account.

The entrained flow combustor system uses a high-velocity (100 ft/sec) carrier to suspend finely dispersed coal particles. Expansion of the fuel stream into the combustor is achieved by forcing the stream through a jet. The most common carrier gas is air, but other gases or coal-water and coal-oil slurries have also been used. The operating temperatures are high (1400–1700°C; 2550–3090°F) compared to that of the fluidized bed (800–850°C; 1450–1560°F) and ash fusion or ash volatilization may be a drawback. Corrosion problems are encountered and nitrogen oxide emissions are also very high under these conditions. The most important example of the entrained flow system is the cyclone furnace. The tubular unit is water-cooled. Accumulated molten slag (at about 1700°C; 3090°F) is removed at the bottom and reduces heat losses. A secondary furnace is also used to recover thermal energy. A much greater release of heat compared to the fluidized bed system has been reported (Speight, 1983). Little ash is lost in a cyclone system and the use of crushed, rather than pulverized, coal helps offset increased costs incurred to maintain high air velocity and pressure.

The development of other experimental combustion systems is also ongoing. While these processes are not yet commercially feasible, they all possess the potential to impact future combustion systems (Speight, 1983).

1. Colloidal fuel system: Involves coal-oil slurries using pulverized coal plus surfactants; ash removal and control of particulates are the major problems.
2. Ignified system: A variation on the chain-grate combustor, which uses much higher velocity air to blow coal off the upper end of

an inclined chain stoker. Smaller particles burn rapidly, but even larger particles are held in suspension and are recirculated to maximize combustion.

3. Submerged systems: Crushed coal is suspended in a hot oxygen-saturated, pressurized aqueous slurry (Zimpro process). Oxidation occurs at 200–350°C (390–660°F). Alternatively, coal, air, and steam are reacted in a bath of molten iron (Atgas process). Coal is oxidized at about 1425°C (2600°F) to CO, and sulfur is effectively removed by reacting it with the molten iron and transferring it to a covering of CaO slag (Speight, 1983).
4. Superslagging system: Coal is burned in a molten ash/limestone slag at about 1540°C (2840°F), which should effectively remove sulfur by formation of calcium sulfide. Reported sulfur removal is low even under only moderately oxidizing conditions (Speight, 1983).
5. Magnet hydrodynamic systems: Combustion gases from coal are seeded with an alkali or alkaline earth metal. Flowing at a high velocity in a magnetic field, electricity is generated directly. Molten metal or metal vapors could be used instead of combustion gas (Speight, 1983).

22

Gasification

I	Pretreatment	666
II	Primary Gasification	667
III	Secondary Gasification	668
IV	Reactor Design	670
V	Cool Water Gasification	675
VI	Catalytic Gasification	678
	A. Introduction	678
	B. Catalyst Deactivation	684
	C. Catalyst Deactivation in Carbon-Oxygen Reactions	697
	D. Catalyst Deactivation in Carbon-Steam Reactions	699
	E. Catalyst Deactivation in Carbon-Carbon Dioxide Reactions	707
	F. Catalyst Deactivation in Carbon-Hydrogen Reactions	710
VII	Mild Gasification	716
VIII	Underground Gasification	726
IX	Future Development	734

Coal combustion to produce gaseous fuels was one of the earliest uses of coal. In fact, one of the leading examples of current gasifier technology, the Lurgi pressure gasifier, is an upgraded version of the century-old gas generator.

As expected, the reactions that are important in coal combustion are equally important in gasification processes (Chap. 21, Sec. II). While the goal of combustion is to produce a maximum amount of heat, the goal of gasification is to produce large quantities of combustible gases with the desired btu content. Depending on the end use, gas with a high-btu, medium-btu, or low-btu value may be most appropriate. As a substitute for or a supplement to natural gas supplies, a high-btu product (600–1000 btu/scf) is required.

The components of product gas from coal gasification reactions vary with conditions (Table 1). Low-btu gas is rendered "low" by the high nitrogen content (30–50% by volume), which is impossible to remove economically. Water and carbon dioxide are more easily removed. Hydrogen sulfide is a minor component and is removed by washing. No nitrogen is present in medium-btu gas and this, along with higher methane content, accounts for the higher heat content. High-btu gas is the product most closely resembling natural gas. To be considered synthetic natural gas (SNG), it must be 95% methane.

Chemicals production generally operates with a medium-btu (225–500 btu/scf) mixture of CO and H_2 containing some methane. Ammonia and methanol production are two chemicals obtained in high quantity from medium-btu synthesis gas mixtures. Medium-btu gas is also an acceptable boiler fuel, but implementation of its use may be limited by the cost of retrofitting oil or gas boilers.

Table 1 Products from Coal Gasification Systems

Product	Characteristics
Low-btu gas (150–300 btu/scf)	Around 50% nitrogen, with smaller quantities of combustible H_2 and CO, CO_2, and trace gases such as methane
Medium-btu gas (300–550 btu/scf)	Predominantly CO and H_2 with some incombustible gases and sometimes methane
High-btu gas (980–1080 btu/scf)	Almost pure methane

Source: J. Speight, 1983. Reprinted with permission of the publisher.

Electric power generation and some industrial consumers use low-btu (90–150 btu/scf) mixtures. The use of low-btu H_2-CO mixtures is being developed by several utility companies for the generation of electricity by integrated combined cycle (ICC) turbines.

Several different reactions which may occur in the main gasifier (Figure 1). This unit is not held at constant temperature; most contain several sections and the temperatures can vary by as much as 540°C (1000°F). Devolatilization of coal occurs in the temperature range of 600–820°C (1100–1500°F). Some methane is produced, but char and hydrogen are the main products of devolatilization. Direct reaction of char and hydrogen to form methane can, in principle, occur at temperatures of about 930°C (1700°F). This process, called hydrogasification or methanation, is exothermic but quite slow. Increasing the temperature further makes the equilibrium unfavorable for methane formation. Some increase in methane formation by hydrogasification has been achieved by increasing the gasifier pressure (Figure 2). Two other reactions, the steam-char and the water-gas shift reactions, can also occur in the main gasifier. In the former reaction, carbon monoxide and hydrogen gas are produced in an endothermic equilibrium reaction which is rapid above 930°C (1700°F). The water-gas shift reaction produces hydrogen and carbon dioxide when carbon monoxide reacts with steam. Thus, the raw product stream contains variable amounts of methane, hydrogen, water, and carbon oxides, but also sulfur and nitrogen contaminants.

Before catalytic methanation can be carried out, sulfur compounds, excess water, and CO_2 must be removed from the synthesis gas stream. Additional water-gas shift conversion is used to maintain the proper H_2/CO ratio (≥ 3) for catalytic methanation, which produces a pipeline quality gas.

Two critical features of these reactions which must be taken into account in dealing with these gasification reactions is the relative rates of reaction and the fact that production of methane from coal requires substantial addition of hydrogen to compensate for the low natural abundance of hydrogen in coal. Walker (1973) estimated relative rates for several critical reactions:

Reactions	Relative rate @ 800°C 0.1 atm
$C-CO_2$ (endothermic)	1
$C-H_2O$ (endothermic)	3
$C-O_2$ (exothermic)	1×10^5
$C-H_2$ (exothermic)	3×10^{-3}

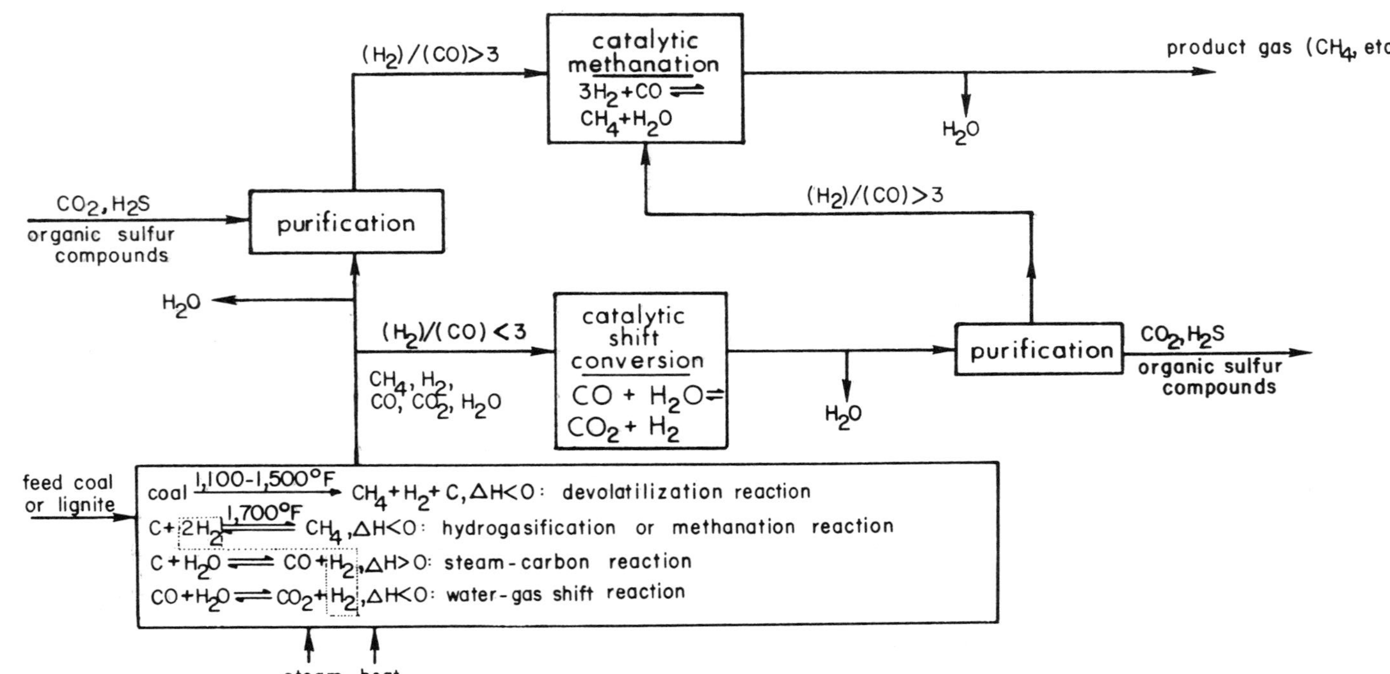

Figure 1 Schematic diagram for conversion of coal to methane.

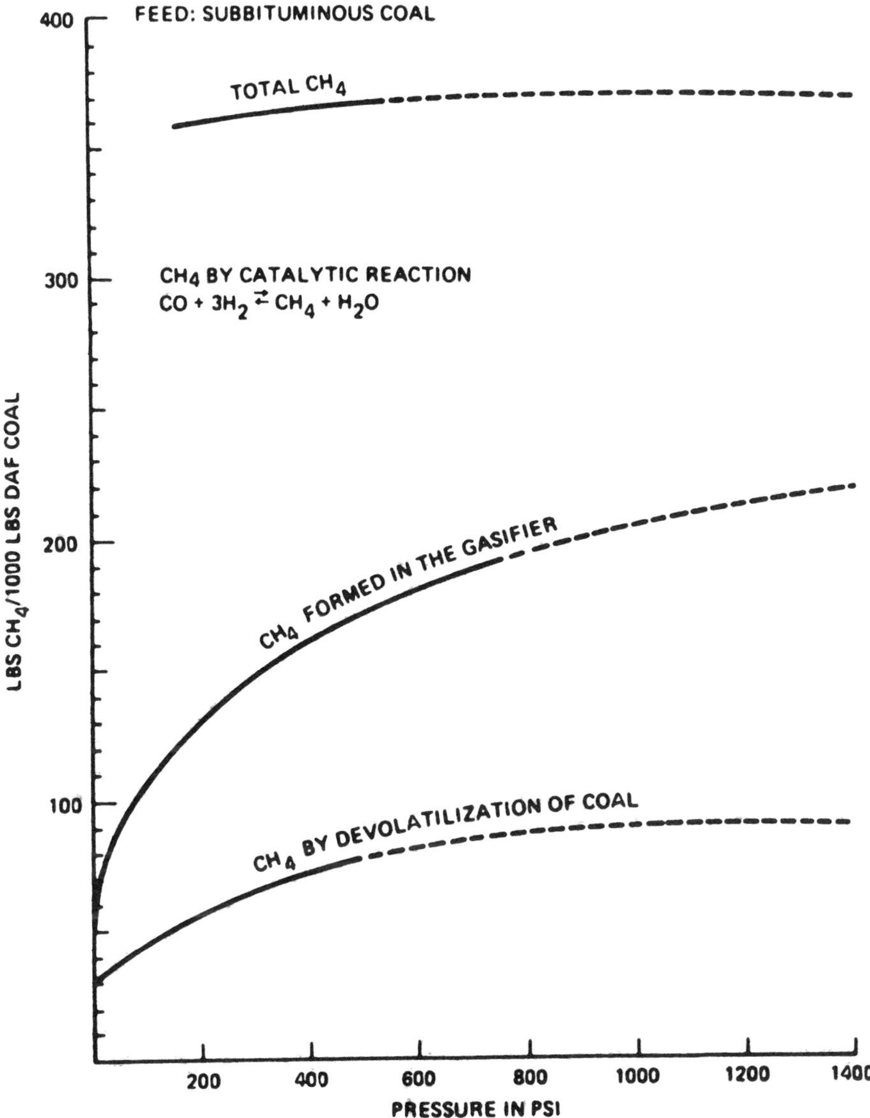

Figure 2 Formation of methane during coal gasification as a function of pressure (Synthetic Fuels from Coal Task Force Report, Project Independence Blueprint, Federal Energy Administration, November 1974).

It is important to notice that while combustion (oxidation) is quite exothermic and relatively very rapid, the more desirable reactions, i.e., those producing H_2 and CH_4, are both considerably slower and therefore require a catalyst. They are also much less exothermic: -18 kcal/mole for CH_4 oxidation compared to -94 kcal/mole for oxidation of C to CH_2. Thus, for the sequence (Hessley, 1986):

$$2C + 2H_2O = 2CO + 2H_2 \qquad +62.8 \text{ kcal/mole}$$

$$CO + H_2O = CO_2 + H_2 \qquad -9.8 \text{ kcal/mole}$$

$$CO + 3H_2 = CH_4 + H_2 \qquad -42.3 \text{ kcal/mole}$$

$$2C + 2H_2O = CH_4 + 2H_2 \qquad +3.7 \text{ kcal/mole}$$

The first step (gasification) requires very high heat input (≥927°C; 1700°F) which cannot be recovered from the exothermic methanation step because, to maintain favorable equilibrium, that step is carried out at much lower temperatures (370°C; 700°F) using a catalyst.

In addition to the thermodynamic considerations, sulfur and nitrogen removal, corrosion (especially from chlorides) and waste disposal are also extremely important and are as problematic and costly in coal gasification as they are in all coal conversion processes. The final composition of the synthesis gas depends on the interrelation of several factors regardless of what grade of product gas is ultimately prepared. There are (1) the air or oxygen flow rate; (2) the coal feed flow rate; (3) the operating pressure and temperature; and (4) the rate of flow of the product gas from the main gasifier. Beyond the primary gasification step, however, the unit operations are essentially the same (Figure 3); differences occur in actual handling operations at particular stages, such as pretreatment, mode of admission of the coal, off-gas cleanup, and the extent of shift conversion reactions that are required by the quality of the final product.

I. PRETREATMENT

Coal pretreatment is often necessary to destroy the swelling and caking tendencies of some coals. During gasification the accumulation of a plastic mass will plug the gasifier, and pressure buildup can also damage the unit. Most Eastern coals with 24–26% volatile matter (by weight) are strongly caking coals. Mild oxidation by heating in air or oxygen is usually sufficient to destroy the caking properties. Unfortunately, as much as 20% of the weight of the coal can be lost during pretreatment.

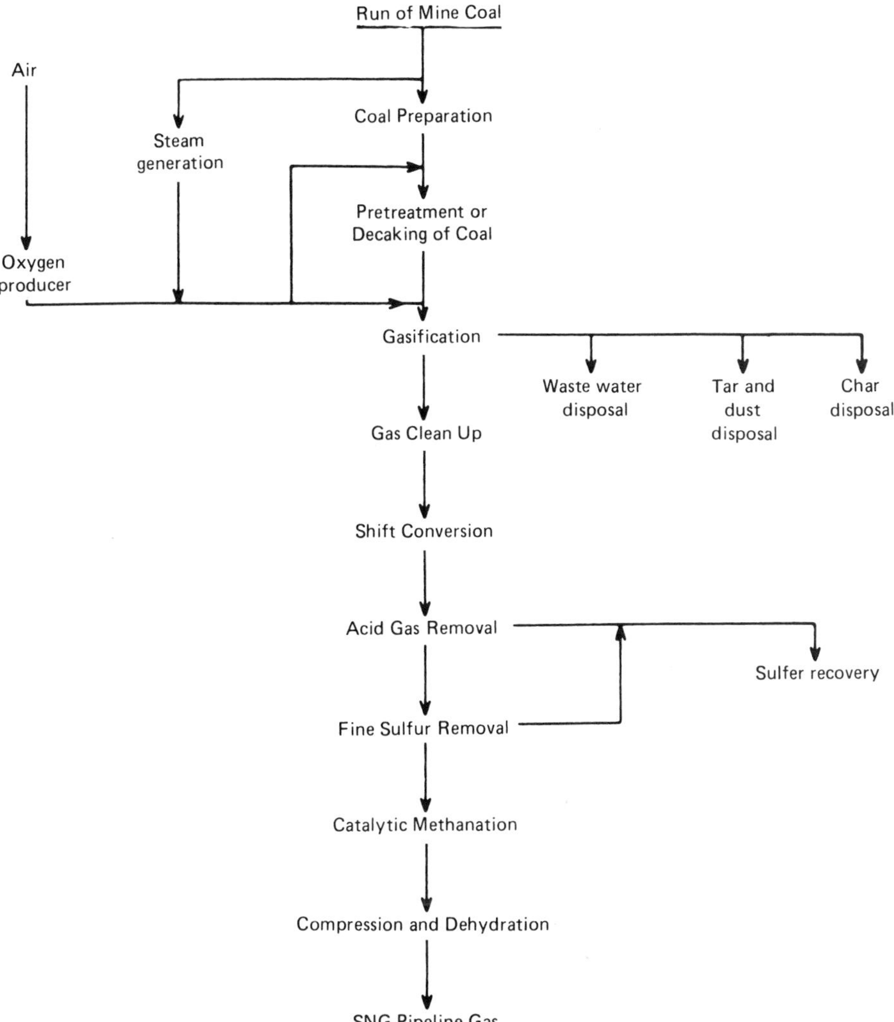

Figure 3 Schematic diagram for coal gasification processes (Haynes and Forney, 1975. Reprinted with permission of the publisher).

II. PRIMARY GASIFICATION

Primary gasification uses air to oxidize coal and to produce a low-btu (150–300 btu/scf) product gas, or oxygen if a medium-btu (300–400 btu/scf) gas is desired. In either case, the product gas is contaminated with varying amounts of carbon monoxide, carbon dioxide, hydrogen sulfide, nitrogen, water, and methane, plus tars, oils, and char.

Gas cleanup is a significant aspect of coal gasification. Before catalytic methanation can be carried out, all other components besides hydrogen and carbon monoxide must be removed. Table 2 lists the components of raw product gas which must be removed and Table 3 shows the specific methods used to treat the gas for H_2S and CO_2 removal. These methods are primarily neutralization and washing processes which are not very expensive. However, every added step in a process adds time and increases expense.

III. SECONDARY GASIFICATION

Secondary gasification converts char to carbon monoxide and hydrogen by reacting the char with steam. This is a necessary but very energy-intensive reaction. Some char is sacrificed to supply heat for secondary gasification by reacting it with air or oxygen. As

Table 2 Byproducts From Coal Gasification Processes and Their Means of Removal

Byproduct	Process type
CO_2	Acid gas scrubbing
H_2S	Acid gas scrubbing, Stretford process, amine treatment, Rectisol process
COS, CS_2	Removed with H_2S
NH_3	Scrubbing and ammonia stripping
HF, HCl, HCN	Scrubbing
Ash	Removed from gasifier for landfill (or minefill)
Suspended particles	Cyclone separators, electrostatic precipitators, scrubbing
Tar, oils	Scrubbing

Source: J. Speight, 1983. Reprinted with permission of the publisher.

Table 3 Methods for the Removal of Acid Gases

Sorbent or reactant	Gases removed	Process
A. Solvent absorption		
20–30% potassium carbonate in hot water solution + catalyst	H_2S, CO_2	Benfield
15% monoethanolamine in water	H_3S, CO_2	Amine
Cold methanol	H_2S, CO_2	Rectisol
B. Solid surface adsorption		
Carbon	H_2S	Activated carbon
Iron metal	H_2S	Iron sorption
C. Chemical reaction of acid gases		
Ferric oxide	H_2S	Iron sponge
Zinc oxide	H_2S	Zinc oxide

Source: J. Speight, 1983. Reprinted with permission of the publisher.

shown above, this is a highly exothermic reaction producing carbon dioxide. Once carbon monoxide and hydrogen are formed, the water-gas shift reaction, which is the reaction between CO and water (steam) to yield CO_2 and hydrogen, is carried out to increase the amount of hydrogen. Since coal contains only small amounts of hydrogen compared to carbon, the amount of hydrogen evolved during either primary or secondary gasification is insufficient to carry out methanation. The water-gas shift reaction is allowed to proceed until a 3:1 hydrogen/carbon monoxide ratio is reached, after which methanation can occur:

$$CO + 3H_2 = CH_4 + H_2O$$

Some hydrogen from the shift reaction may also be used for direct hydrogenation of char to methane:

$$C + 2H_2 = CH_4$$

Direct hydrogenation of char (hydrogasification) can be carried out within the main gasifier at the temperature of primary gasification

(930°C; 1700°F). However, the reaction is very slow (see above) and increasing the temperature does not favor the desired equilibrium formation of methane.

Methanation by reaction of hydrogen and carbon monoxide requires a separate reactor in order to operate at a lower temperature than that of the main gasifier, about 370°C (700°F). Catalysts are also used to carry out this reaction. Hydroxides and carbonates of alkali metals have been shown to be very effective catalysts for the steam-char reaction which produces CO and H_2 in the primary gasification step (Veraa and Bell, 1978), but for methane formation from synthesis gas, only five metals—cobalt, iron, molybdenum, nickel, and ruthenium—have been effective (Braunstein et al., 1979). Cost, selectivity for methane formation, and resistance to deactivation are the important characteristics for a suitable catalyst. Although it is very active, ruthenium is prohibitively expensive; nickel is usually the catalyst of choice because it is active, selective, and relatively inexpensive. Molybdenum is the least active of the five metals, but it is highly selective for methane formation and is very resistant to sulfur poisoning. It is often used in combination with nickel or cobalt (Braunstein et al., 1979).

IV. REACTOR DESIGN

The three commercial reactor designs for coal gasification are fixed-bed, fluidized bed, and entrained flow systems (Figure 4).

In the fixed-bed gasifier, coal enters at the top and is heated as it falls onto a rotating grate near the bottom. Air and steam enter below the grate and create the hottest combustion zone at the level of the grate. Gasification occurs throughout the chamber, however, and raw gas exits at the side of the reactor near the top. Relatively large-size coal (3–50 mm) can be used in this gasifier. The most common fixed-bed system, the Lurgi gasifier, operates at 400 psi. South Africa's SASOL fuels and chemicals production industry has the largest number of Lurgi gasifiers in the world (Spencer et al., 1982).

A variation of the fixed-bed configuration is the moving-bed ignified gasifier (Figure 5). In this system coal enters the bottom and is combusted upon entry by reaction with the oxidant which enters below the grate. A moving, continuous chain belt moves the layer of burning coal and removes char at the upper end. A cyclone and blower at the top of the chamber returns uncombusted fines to the combustion zone.

In contrast to the fixed- or moving-bed gasifiers, both the fluidized bed and entrained flow gasifiers use turbulent mixing to maintain

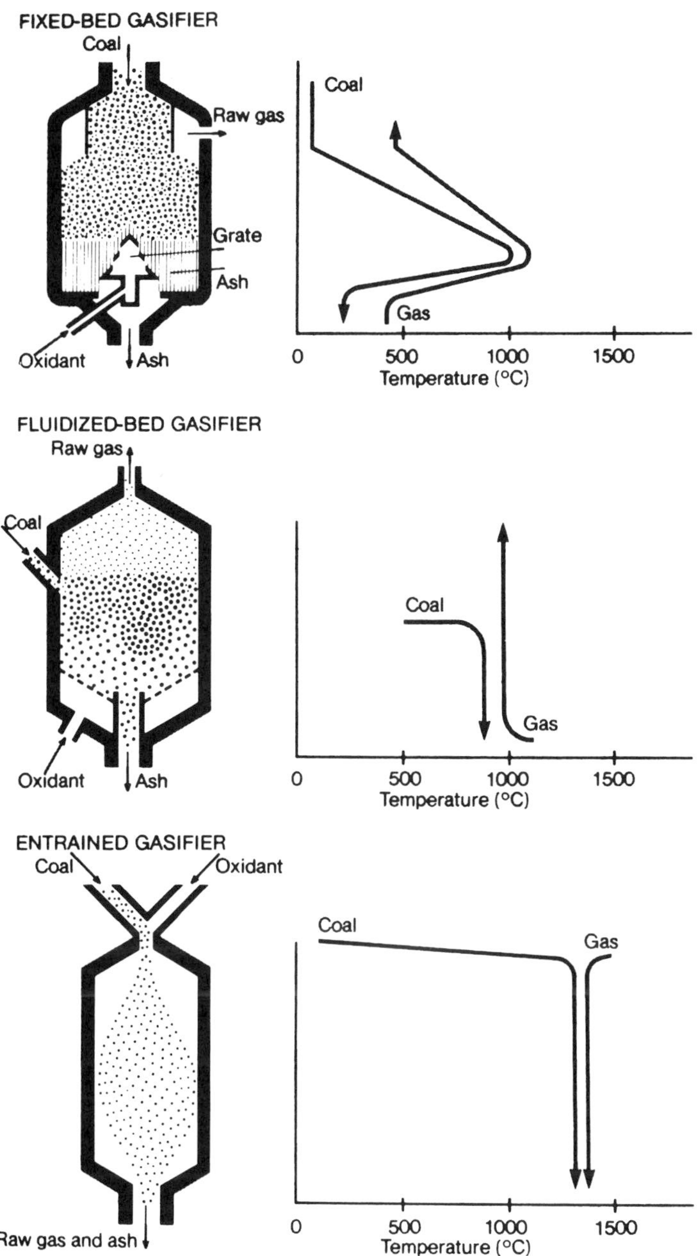

Figure 4 Schematic diagrams and temperature profiles for three principal types of coal gasifier units (*EPRI Journal*, Electric Power Research Institute, Palo Alto, CA, April 1979. Reprinted with permission of the publisher).

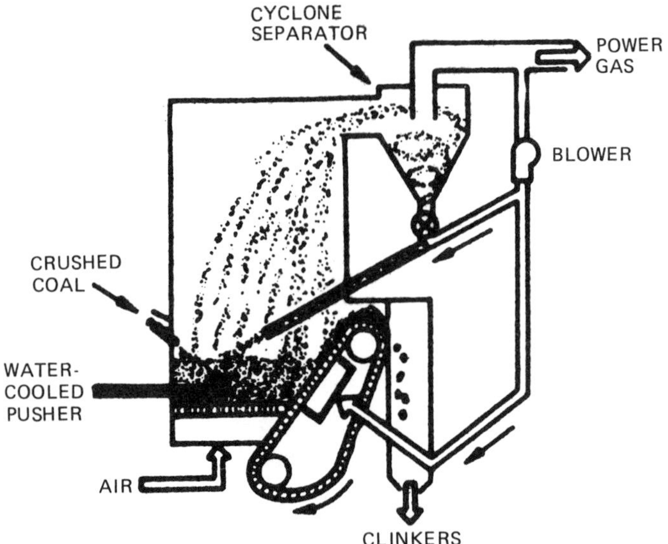

Figure 5 The Mark I ignified coal gasifier unit (Synthetic Fuels from Coal Task Force Report, Project Independence Blueprint, Federal Energy Administration, November 1974).

a constant combustion and gasification temperature throughout the reaction chamber. In the fluidized bed reactor, an upward flow of air (or oxygen) continuously stirs the bed of coal. Small-particle-size coal ($\leq$8 mm) is used in this system. The main disadvantages of the fluidized bed technology are its limited ability to consume all the coal fed into it and its inability to use caking coal without pretreatment (Spencer et al., 1982). The Winkler process is the best known commercial application of the fluidized bed design.

Pulverized coal (<0.1 mm) is required for the entrained flow system. The powdered coal is mixed with a carrier gas and enters the gasifier chamber under pressure. One configuration of the entrained flow design is the Koppers–Totzek gasifier. This unit operates at atmospheric pressure at about 1700°C (3000°F). Since high temperatures generally increase reaction rates, this system typically has the highest throughput capacity. Product gas is CO and H_2 with small amounts of CO_2, but no methane, tar, or oils are formed. The main application of this gasification process is for the production of ammonia.

Coal residence time in the gasifier affects the composition of the raw gas and the composition and character of the ash. The gas composition for four commercial systems is shown in Table 4. For SNG

Table 4 Typical Gas Composition[a]

Component	Moving bed, Lurgi oxygen-blown	Fluidized bed, Westinghouse	Entrained: Texaco oxygen-blown	Entrained: Combustion engineering air-blown
CH_4	4.2	7.2	0.3	1
C_2^+	0.5			
H_2	21	29	29.6	9
CO	8	43	41	16
CO_2	15	6	10	6
H_2S + COS	0.7	1	1.1	<1
N_2	0.2	1.5	0.8	62
NH_3	0.4	0.3	0.2	
H_2O	50	12	17	5
Tars and oils (weight fraction)	0.02	0.0	0.0	0.0
Temperature (°C)	540	985	1315	985

[a]Values are percentages by volume except for the last two rows.
Source: D. F. Spencer et al., 1982. Reprinted with permission of the publisher.

production maximum methane formation and high hydrogen content is desirable; for electric power production, a high concentration of carbon monoxide is preferred. This is generally achieved with high-temperature operation and is usually accompanied by low tar yield. Table 5 shows that based on these caracteristics, the fluidized bed gasifier developed by Westinghouse is superior to either the moving-bed or the entrained flow units described.

Entrained flow systems exhibit the overall best performance but operate at higher temperatures (Table 5). As a consequence, less chemical energy is recovered in the product gas relative to the chemical energy content of the feed coal (the "cold gas" efficiency) when compared to units which operate at lower temperatures. The Lurgi-type gasifiers produce more tar, consume more steam, and cannot consume coal fines adequately. A major advantage of the Koppers–Totzek

Table 5 Comparison of Coal Gasification Reactor Types

Function	Moving-bed		Fluidized bed	Entrained flow
	Dry ash	Slagging		
Capacity potential	Low	High	Intermediate	High
Ability to handle caking coals without pretreatment	Moderate	Shown at 300-ton-per-day scale	Shown on small scale	Excellent
Temperature of operation	1100–450°C	1550–450°C	870–1050°C	1650–950°C
Temperature control	Poor	Poor	Good	Moderate
Refractory problems	Moderate	Poor	Moderate	Poor
Byproduct tar formation	Yes	Yes	Possibly	Probably not
Ability to extract ash low in carbon	Moderate	Good	Moderate	Good
Ability to consume fine carbon particles	Poor	Good	Probably poor	Good

Source: D. F. Spencer et al., 1982. Reprinted with permission of the publisher.

entrained flow process is the ability to use a wide variety of both caking and noncaking coals (Spencer et al., 1982).

Many experimental processes using each major type of reactor design are being investigated in order to improve the overall efficiency and utility of coal gasification for both fuels and chemicals production. These projects are addressing one or more critical aspects of the gasification chemistry and technology:

1. Coal pretreatment
   a. Effective use of pretreatment off-gases
   b. Recovery and use of pretreatment coal fines

2. Reactor design
   a. Use of improved refractory materials
   b. Alternative methods for slag and ash removal
   c. Improved char recovery and high-temperature particulates removal
   d. Improved methanation with regard to temperature control and catalyst deactivation
3. Process control systems
   a. Improved construction materials for scrubbing corrosive gases at high temperature and pressure
   b. Improved methods for containment and clean-up of acid gases
4. End use
   a. Improved gas turbines for utilization of low-btu fuel
   b. Improved retrofit of gas, oil, and coal boilers

Although several experimental units are fully operational at greater than 100 tons of coal throughput per day (Table 6), most are still in early stages of development. By one estimate, the cost of bringing a new unit on-line commercially is between \$200 and \$500 million (Spencer et al., 1982). Such an exorbitant investment will prompt decisions to put many of these projects into abeyance.

V. COOL WATER GASIFICATION

Perhaps the most ambitious proposal for advanced coal gasification is the Cool Water project, joint between Southern California Edison and Texaco Oil Co. with additional funds committed from the Electric Power Research Institute (EPRI), Bechtel Power Corporation, the General Electric Co., a group of New York State utilities, and a consortium of four Japanese companies.

The project uses the Texaco gasifier process (Figure 6), which consists of an entrained bed reactor into which a coal-water or coal-oil slurry is pumped. The unit operates with air or oxygen at high temperature (1200–1500°C; 2200–2800°F). Medium-btu (280 btu/scf synthesis gas containing some methane but little oil or tar is produced. The gas and molten slag are passed into cool water in a spray chamber and a slag quench bath; the gas is treated for acid gas removal and sulfur is removed from the recovered hydrogen sulfide. The clean product gas is heated and is coupled to a gas and a steam turbine for generation of electricity. The unit has operated successfully to generate 120 MW of gross power (Siegart et al., 1986).

Table 6 Status of Second-Generation Coal Gasification Technologies (Capacity Greater than 100 Ton/Day)

Technology	Type	Plant capacity (ton/day)	Plant location	Product	Status
Texaco	Entrained	165	Oberhausen, West Germany	Gaseous fuel, synthesis gas	Operational
	Entrained	150	Muscle Shoals, Alabama	Synthesis gas, fertilizer	Startup
	Entrained	150	Plaquemine, Louisiana	Gaseous fuel, electricity	Operational
Shell	Entrained	150	Harburg, West Germany	Gaseous fuel, synthesis gas	Operational
Combustion Engineering	Entrained	120	Windsor, Connecticut	Gaseous fuel	Operational
British Gas/Lurgi	Moving-bed-slagging	350	Westfield, Scotland	Gaseous fuel	Operational
British Gas/Lurgi	Moving-bed-slagging	700–800	Westfield, Scotland	Gaseous fuel	Construction operation 1982
KilnGas	Rotating kiln	600	Wood River, Illinois	Gaseous fuel	Construction, operation 1982
Lurgi	Moving-bed (high-pressure)	175	Dorsten, West Germany	SNG, fuel gas	Operational
Saarberg-Otto	Entrained slag bath	250	Volkingen, West Germany	SNG, fuel gas	Operational

Source: D. F. Spencer et al., 1982. Reprinted with permission of the publisher.

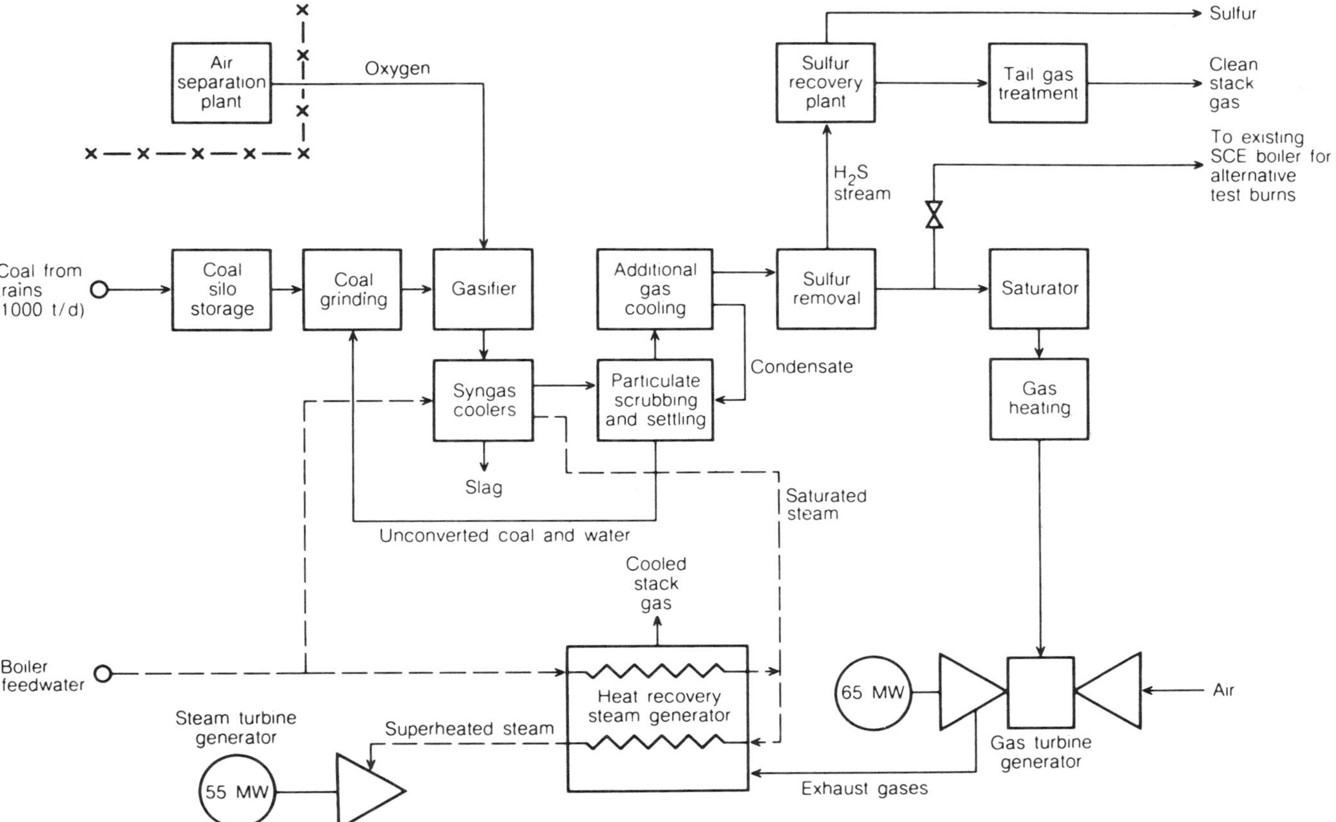

Figure 6 Schematic diagram for the Cool Water Coal Gasification Project (Electric Power Research Institute. Reprinted with permission of the publisher).

VI. CATALYTIC GASIFICATION

In recent years, interest in the gasification of coal has been stimulated by the concept of gasification by the presence of catalytic materials. The focus has been on the application of catalytic principles to the gasification process in order to minimize secondary reactions and move the process toward the ultimate goal: the production of a high-energy fuel gas or synthesis gas.

The work was initiated by the recognition that coal contains many different types of minerals (Chap. 1, Secs. II and V) which are often removed before use of the coal. However, it was also recognized that in certain instances the mineral matter may act as an enhancement (catalyst) for a particular reaction rather than as a detriment. Thus, the concept of catalytic gasification was born.

The major part of the research in this area has focused on the catalytic gasification of char since it is widely believed that, under gasification conditions, the coal rapidly chars and it is, in fact, the char that is actually gasified by the catalyst. Thus, fundamental studies of the gasification of char are extremely important for all gasification and combustion processes since char is, without doubt, the intermediate product which governs the kinetics controlling the gasification and combustion processes.

A. Introduction

Research activity directed at an understanding of the fundamental aspects of carbon gasification has undergone tremendous oscillations over the past 30 years. The direction of these studies has been dictated to a large degree by the particular process or application which was in vogue. In the case of CO_2-cooled graphite-moderated nuclear reactors and present day space applications of carbon-carbon composites, the emphasis is placed on developing methods of preventing carbon gasification. In contrast, in processes such as steam gasification of coal, catalytic combustion of diesel soot in monolith filters, and decoking (regeneration) of supported metal catalysts used in hydrocarbons conversion reactions, the goal is to enhance the rate of carbon gasification.

It has been known for many years that a small amount of inorganic impurity can have a profound effect on the rate of gasification of carbonaceous materials and comprehensive reviews have been written on this topic (Walker et al., 1959, 1968; Thomas, 1965; Lewis, 1970; McKee, 1981; Wood and Sancier, 1984).

Catalytic gasification of carbonaceous solids is a unique example of heterogeneous catalysis, where the catalyst support (i.e., carbonaceous material) is also a reactant and is consumed during the reaction. It is a fascinating reaction since one can directly observe, with the aid of a microscope, the catalyst in action. It is important to appreciate that because coal and coke deposits are complex and heterogeneous materials, it is difficult to identify the key steps which control the catalyzed and uncatalyzed reactions. In an attempt to circumvent this problem, structurally simpler and much purer forms of carbonaceous substrates are frequently used in basic studies, most commonly graphite.

The factors which control the mode of action of a catalyst in graphite-gas reactions can be divided into chemical and physical aspects. Several mechanisms have been proposed to account for the role of the catalyst in carbon oxidation reactions, of which the most widely considered are the oxygen transfer and electron transfer mechanisms. In the oxygen transfer mechanism, the surface of the catalyst is converted to the oxide state by dissociative adsorption of the gaseous reactant. The oxidized state is reduced by reaction with carbon resulting in the formation of carbon monoxide/dioxide. In this mechanism the catalyst operates by an oxidation-reduction cycle.

The electron transfer mechanism, first proposed by Long and Sykes (1952), is based on the notion that in the presence of a catalyst the electrons in the carbon structure undergo a rearrangement causing a change in the Fermi level of the carbon. This effect causes a weakening of the carbon-carbon bonds at edge sites and a consequent decrease in the activation energy of the steps in which these bonds are broken during gasification.

In the mechanism presented by Holstein and Boudart (1981, 1982) for carbon gasification in hydrogen, steam, and carbon dioxide, the catalyst is required to perform two roles. These include the ability to directly break carbon-carbon bonds without the participation of adsorbed gas species on the catalyst surface. The catalyst must also be capable of dissociatively adsorbing gas molecules, which react with the carbide species on the catalyst surface to form gaseous products. Thus the catalyst must be bifunctional, interacting with both the carbon and the gas.

Experiments performed with controlled atmosphere electron microscopy (CAEM) have revealed the physical factors which determine whether a particular particle will act as a catalyst for graphite gasification and the mode by which it will operate (Baker, 1982a). The most likely condition for a particle to become catalytically active is if it is located at a graphite edge or step site. Detailed examination of the particles in these regions shows that prior to catalytic attack they undergo distinct changes in morphology, resulting in a change in contact angle from obtuse to acute, i.e., a transformation from nonwetting to a wetting state.

A more fundamental understanding of the mechanism of adherence and the factors controlling the contact of metal particles on graphite can be obtained from a treatment of the surface forces affecting the

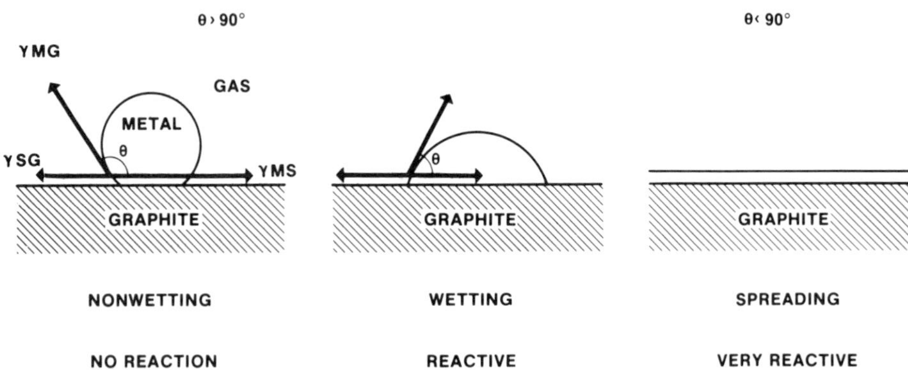

Figure 7 Changes in shape of particles as a function of the wettability of the catalyst on graphite (Baker, 1982).

metal melt–solid substrate systems. For a metal particle resting on the edge site of a graphite surface (Figure 7), the contact angle θ at equilibrium is determined by the surface energy of the support γ_{gs}, the surface energy of the metal γ_{mg}, and the metal-support interfacial energy γ_{ms}, and is expressed in terms of Young's equation:

$$\gamma_{gs} = \gamma_{ms} + \gamma_{mg} \mathrm{Cos}\theta \qquad (1)$$

Equation (1) can be written in the form:

$$\mathrm{Cos}\ \theta = \frac{\gamma_{gs} - \gamma_{ms}}{\gamma_{mg}} \qquad (2)$$

If γ_{ms} is larger than γ_{gs}, the contact angle is greater than 90° and the particle is in a nonwetting state; if the reverse relationship holds, then θ will be less than 90° and wetting will occur, and if $\gamma_{gs} > \gamma_{ms} + \gamma_{mg}$, then the particle will spread out over the support surface. The ability of the particle to undergo transformation from a nonwetting to a wetting state and, in certain cases, to a spread condition suggests that a significant degree of atomic mobility exists within the particles at temperatures well below that of the bulk melting points of the metals or metal oxides. Provided that the chemical requirements are satisfied, the mode of catalytic attack is governed by the degree of wettability of the graphite (Figure 8).

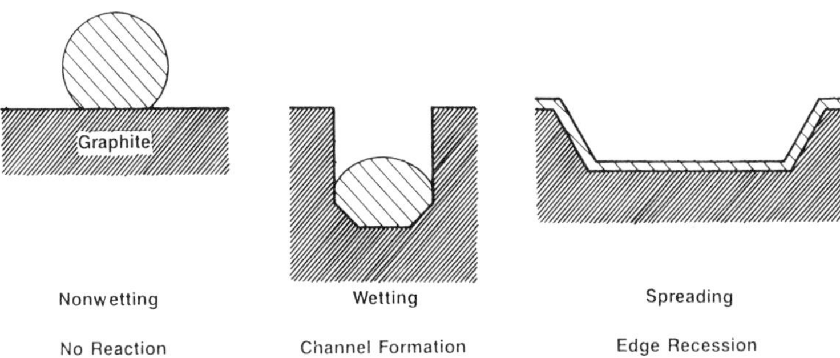

Figure 8 Influence of the metal-graphite interaction on the mode of catalytic attack (Baker, 1986).

For the intermediate wetted state where the catalyst is present as a discrete cap-shaped particle, then the mode of attack would be expected to proceed by a channeling action. Throughout the reaction, the active particles maintain contact with the graphite interface and, as a consequence, always remain at the leading face of the channel.

For the situation where the catalyst spreads along the graphite edges in the form of a thin film, then the subsequent mode of catalytic attack is by edge recession. Spreading of the catalyst in this manner results in the most efficient use of the additive in that the contact area between the catalyst and carbon atoms is maximized.

In addition to providing qualitative information on the various modes of catalytic action in carbon-gas reactions, the data from CAEM studies is also amenable to detailed kinetic analysis (Baker and Harris, 1972). From the measurements of many catalytic channeling sequences it has been possible to establish the following relationships:

(a) For carbon oxidation reactions at a given temperature, the rate of channel propagation decreases with increasing particle size provided that the channels are of similar depth (Figure 9). These data can be expressed in the form (Baker et al., 1980a):

$$\text{Channel propagation rate} \propto \frac{1}{(\text{particle diameter})^{1/2}}$$

(b) In hydrogenation of carbon at a given temperature, it is found that the larger the catalyst particle, the faster the rate of catalytic attack for channels of similar depth. This dependence (Figure 10) can be expressed according to the relationship (Baker et al., 1980b):

$$\text{Channel propagation rate} \propto (\text{particle size})^2$$

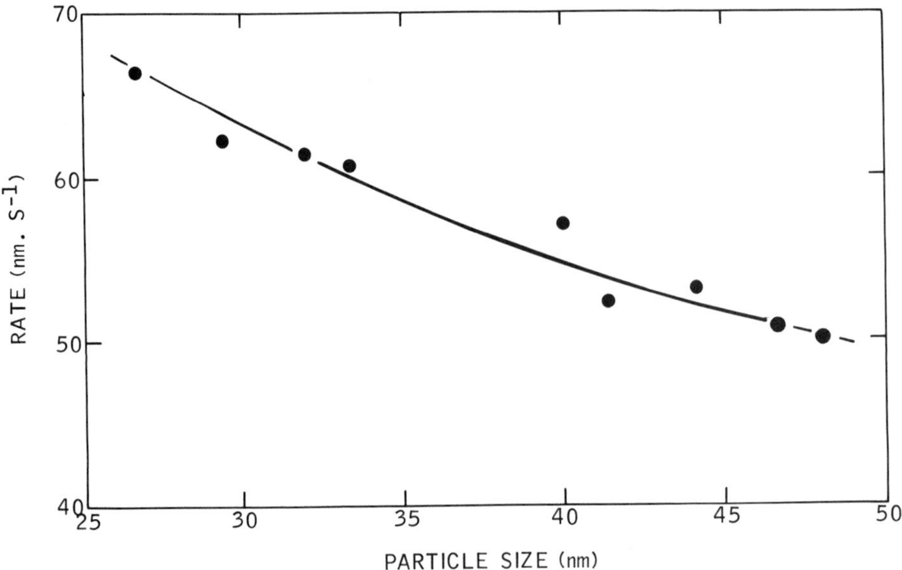

Figure 9 Relationship between channel propagation rate and catalyst particle width in the graphite-oxygen reaction (Baker et al., 1980a).

This variation clearly points to the existence of fundamental differences in the reaction mechanisms of catalyzed oxidation and catalyzed hydrogenation of carbon.

Baker and Sherwood (1981) suggested that in strong oxidizing conditions, carbon at the graphite-catalyst interface is dissolved in the particle and diffuses through a viscous outer layer of the particle and at the cooler face is converted to CO_2/CO by reaction with atomic or molecular oxygen.

Recently, Choi and coworkers (1987) set up a mathematical model of the channeling activity of metal particles and compared their theoretical results with experimental data obtained from CAEM studies (Baker et al., 1976). In the model, they assumed that the particle and the support are isothermal and the surface concentration of carbon atoms in the oxide layer of the particle is not uniform, inducing surface diffusion within the layer. Using this approach, they found an excellent fit between the experimental data and that predicted by the model, which supports the view that catalytic oxidation of carbon occurs via the following sequence: the weakening of carbon-carbon bonds at the active metal-carbon interface followed by dissolution of carbon atoms in the metal, steps which can be accounted

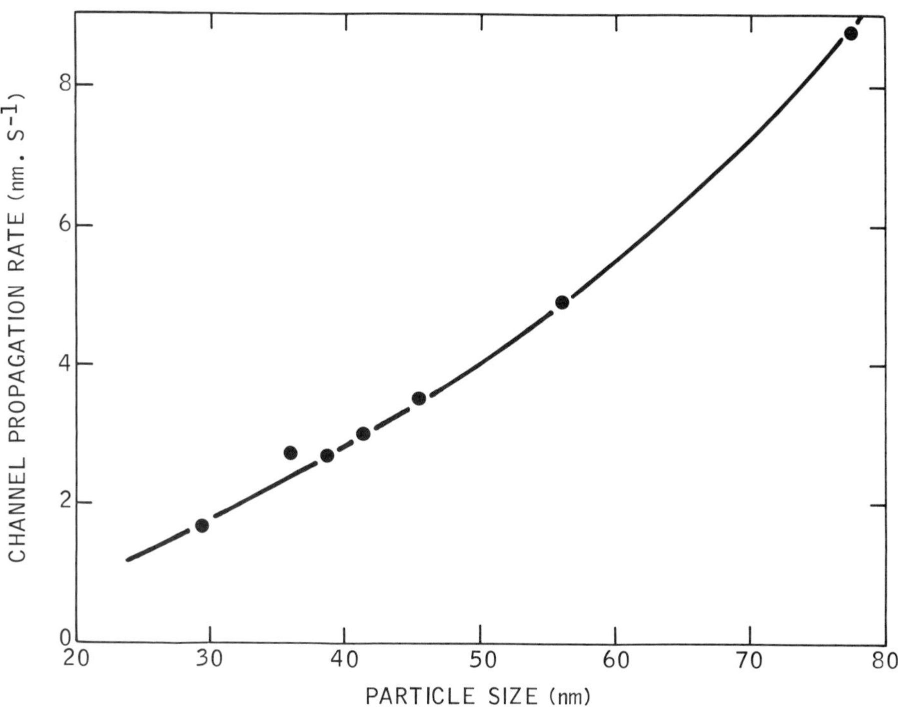

Figure 10 Relationship between channel propagation rate and catalyst particle width in the graphite-hydrogen reaction (Baker et al., 1980b).

for by the electron transfer mechanism, and reaction between the carbon atoms and the adsorbed oxygen in the thin oxide layer of the catalyst which undergoes a redox cycle as the reaction proceeds.

It is clear from the foregoing discussion that considerable progress has been made on the elucidation of the mechanisms relating to the activation and modes by which catalysts function in various carbon gas reactions.

The inhibitory effects of chlorine and its compounds and phosphorus oxides on the graphite-oxygen reaction was presented by McKee (1981). It is the purpose of this section to focus on the various factors which give rise to the decrease in catalytic activity in carbon gasification reactions. In order to treat the subject in a comprehensive manner, it is necessary to make some general remarks regarding deactivation of gasification catalysts before proceeding to deal with aspects related to a decrease or loss of catalytic activity in a specified carbon-gas system.

B. Catalyst Deactivation

In the study of the catalytic gasification of carbon one experiences the tremendous advantages offered by the use of the CAEM technique (Baker and Harris, 1972). It is possible to directly observe the "antics" of individual catalyst particles as they move across the surface of a graphite crystal. The catalytic action of some materials is particularly intriguing, being seen as the development of a channel behind each particle as it appears to eat into the underlying substrate. Attempts have been made to acquire this type of information using conventional transmission electron microscopy from examination of the appearance of a specimen before and after reaction, or over successive periods of time, with the result that the investigator is confronted with the problem of deciphering from a comparison of micrographs what has occurred during the reaction sequences. While this procedure can yield valuable information, some caution must be exercised in interpretation, since the specimen has been removed from the reaction environment and, as a consequence during cooling, could have undergone chemical and physical changes. Furthermore, during examination there is the possibility that specimens become contaminated and subsequently exhibit unpredictable behavior.

The following features observed by CAEM of the change in activity patterns of some metal/graphite/gas systems point to factors which have an influence on the behavior of a catalyst.

1. Modification of Catalyst Wetting Properties

Occasionally it is found that the wetting characteristics of a particular catalyst on graphite are modified as the reaction temperature is increased. In some cases this effect is manifested by the nucleation of discrete particles at edge regions where initially the material was present as a thin film. This transformation is accompanied by a change in mode of catalytic attack from edge recession (catalyst in the spread condition) to channeling (catalyst in a particulate form).

The following example is taken from a study of the barium-catalyzed gasification of graphite (Baker et al., 1984) to demonstrate how the overall catalytic activity is decreased from edge recession to channeling (Figure 11). At 825°C (1515°F), the catalyst exhibits a slow transformation in its wetting properties on graphite and as a consequence both forms of catalytic attack can be observed on the specimen at the same time. From the quantitative data it is possible to obtain an estimate of the relative contributions of the two modes of catalytic attack toward the overall steam gasification rate. The number of moles of carbon gasified per second, dn/dt, is given by

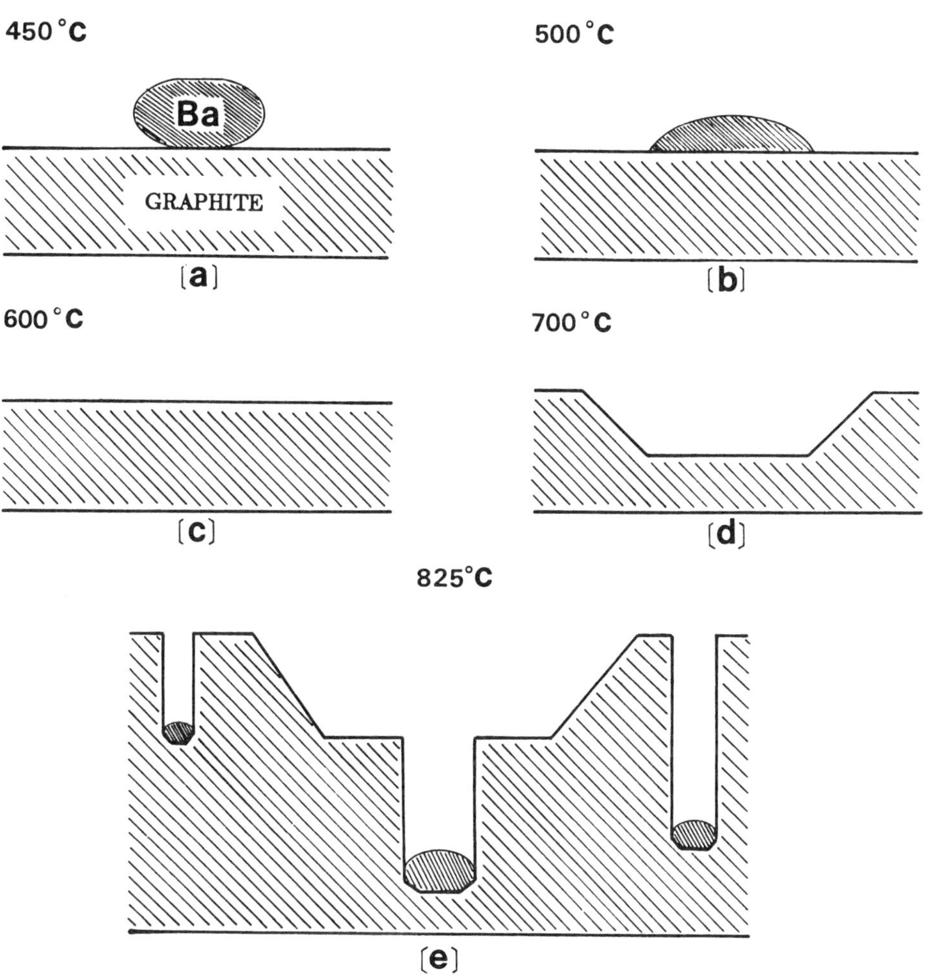

Figure 11 Change in wetting properties of a particle on graphite and the concomitant change in catalytic action (Baker et al., 1984).

$$\frac{dn}{dt} = \frac{2\ell D\delta\rho}{M}$$

where ℓ is the rate of edge recession or channel propagation (nm/sec), D is the width of the receding edge or channel (nm), δ is the depth of the edge or channel (nm), ρ is the density of graphite (2.25×10^{-21} g nm^3), and M is the atomic weight of carbon. At 825°C (1515°F), the amount of carbon gasified per second by a receding edge of total length 1500 nm, and thickness 10 nm, moving at a rate of 1.11 nm/sec, is 3.09 $\times 10^{-18}$ moles. As the reaction proceeded, the catalytic film covering this edge nucleated to form a single particle, 30 nm in diameter, which then created a channel at a linear propagation rate of 3.30 nm/sec. If one assumes that the depth of the channel is also 10 nm, then the number of moles of carbon gasified per second by this particle is 1.86 $\times 10^{-19}$. It is therefore apparent that the change in wetting characteristics of the catalyst on the graphite produces a dramatic decrease in catalytic activity.

2. Loss of Catalyst Channeling Activity

Loss of channeling activity can occur for a number of reasons, the most common one being the result of a loss of contact between the catalyst particle and the graphite step at the head of the channel. This situation is usually observed when a particle crosses a region of the surface where it has already been active and proceeds to become isolated from the graphite edge region (Figure 12). At low temperatures, the particle remains within the confines of the pit which it has created. However, as the temperature is raised, it may regain mobility and once again become activated if it regains contact with an edge or step.

3. Sintering of Catalyst Particles

Particle sintering is a major factor associated with the overall loss of catalytic activity in supported metal systems (Wynblatt and Gjostein, 1975). Overall the past few years this subject has provided one of the greatest sources of controversy in the field of heterogeneous catalysis. The argument concerns the mechanism by which small metal catalyst particles grow on a support. Two mechanisms have been proposed to account for sintering: (1) particle migration and coalescence, (2) transfer of atomic species from one stationary particle to another stationary particle.

The particle migration model was proposed by Ruckenstein and Pulvermacher (1973) based on the concept that in a supported metal system the interactions between adjacent metal atoms is stronger than the interactions between metal atoms and the support. As a consequence when the reaction temperature approaches the Tammann temperature of the metal, which is calculated from 0.5 times the bulk melting point (K), the metal particles will exhibit mobility on the surface. Under these conditions the probability that particle diffusion across the surface will lead to coalescence is extremely high.

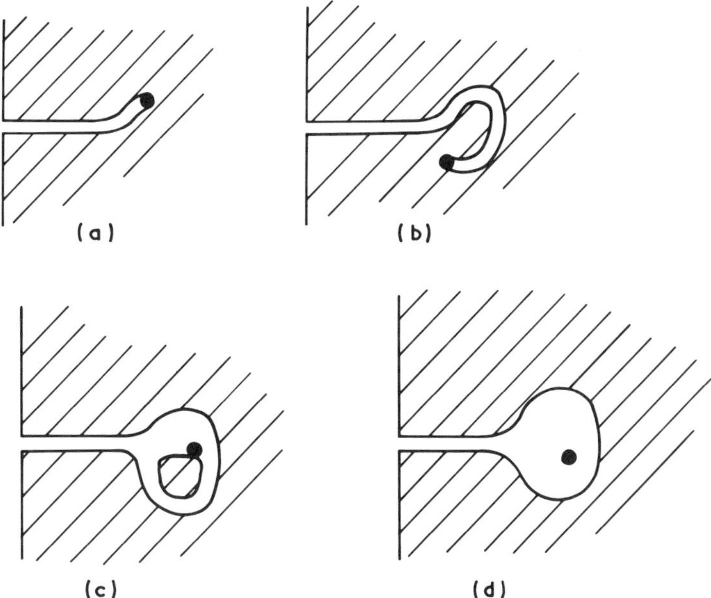

Figure 12 Deactivation of catalyst particles by loss of contact with graphite edges (Baker & Harris, 1973).

The second model, commonly referred to as the atomic or molecular migration model, was developed by Flynn and Wanke (1974) as an extension of classical Ostwald ripening mechanisms. They suggest that sintering occurs via the surface transport of dissociated atomic or molecular species which ultimately collide with and are captured by a second, larger stationary metal particle. The overall process results in the growth of larger particles at the expense of smaller particles. Such behavior is predicted by the Kelvin equation, which suggests that the rate of loss of atoms is lower than the rate of capture for large particles, whereas for small particles the rate of loss is higher than the rate of capture. The driving force for molecular transport is the minimization of surface free energy.

Sintering by either of these mechanisms is expected to be highly temperature-dependent and this aspect is particularly apparent in graphite-supported metal systems where it has been known for many years that particles will exhibit mobility at a characteristic temperature (Bassett, 1961; Thomas and Walker, 1964). Early studies with the CAEM showed that when iron particles supported on graphite were heated in various gas environments, the particles became mobile on the basal plane at 700°C (1292°F) (Baker et al., 1972). Subsequent studies with other metals dispersed on graphite showed that this was a general phenomenon and that there was a characteristic mobility temperature which was related to the bulk melting point of the particular metal or metal oxide. Figure 13 is a collection of these data, from which it can be seen that the experimentally observed temperature for the onset of mobility of 10-nm particles of various metals and oxides is approximately half their respective bulk melting point when the temperature is expressed in absolute units. These studies provide direct evidence that particle motion occurs at the Tammann temperature in systems where there is a weak interaction between a particle and support (Baker, 1982a).

The ramifications of this effect in terms of particle sintering and concomitant overall decrease in catalyst activity can be seen from the particle size distribution curves for palladium on graphite at 550°C (1020°F) and at 750°C (1380°F) (Figure 14). At the lower temperature,

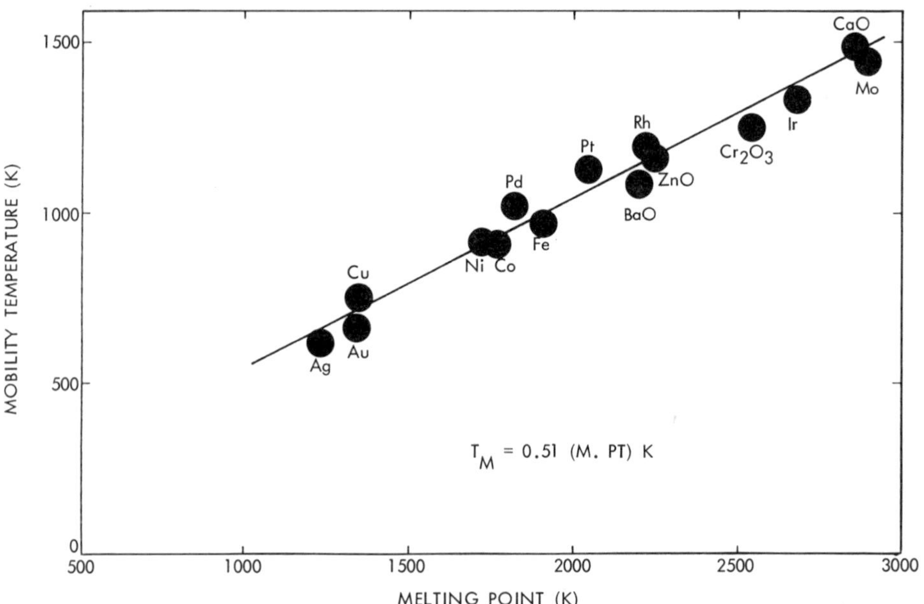

Figure 13 Relationship between the mobility temperature of particles supported on graphite and their bulk melting points (Baker, 1982).

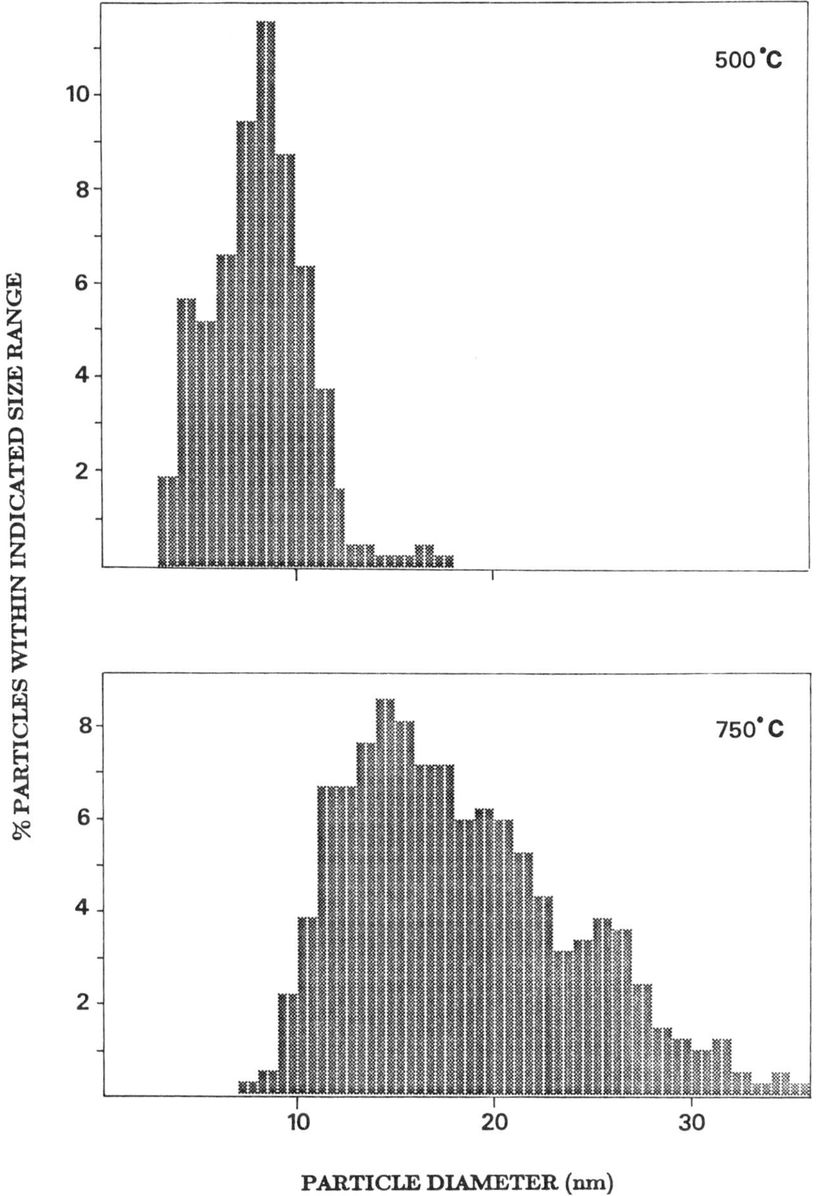

Figure 14 Particle size distribution curves for palladium on graphite at 550°C (1020°F) and at 750°C (1380°F).

particle growth occurs exclusively by the atomic migration mechanism and, as expected, a very symmetrical distribution is found. In contrast, at 750°C (1380°F), which is above the Tammann temperature for palladium, the overriding mode of growth is via particle migration and the distribution exhibits a Gaussian-shaped curve skewed toward the large particle diameter side. These curves have been produced from the measurement of over 800 particles at each temperature covering several different areas of specimens, but it should be appreciated that since the experiments were performed by CAEM there is a resolution limit of 2.5 nm and therefore there will be inaccuracies in the distribution curves at the lower end. Nevertheless, it is evident that the rate of catalyst particle growth increases dramatically at the Tammann temperature of the metal.

There are a number of reports in the literature where deactivation of catalytic gasification reactions has been attributed to particle sintering (Rewick et al., 1974; Nishiyama and Tamai, 1977; Kim et al., 1982; Colle et al., 1983; Radovic et al., 1983a, b).

Several approaches have been investigated in an attempt to overcome this phenomenon. One of these involved the modification of the surface state of the carbon using chemical treatment which resulted in an improvement in catalytic activity (Haga and Nishiyama, 1983). Another method which was developed for preventing sintering and maintaining high metal dispersions in alumina-supported systems (McVicker et al., 1978) was subsequently applied to carbon-supported metals (Nishiyama, 1986). The key feature of this treatment was the introduction of group IIA oxides onto the support, which functioned as trapping agents for mobile molecular species and thus prevented the capture process by large metal crystallites. Since the escape of molecular species from the metal crystallites was unaffected by the presence of added oxide, the net result was that all crystallites decreased in size.

4. Effect of Gas Phase Poisons

It is well known that metal catalysts are poisoned by compounds of group 5B and 6B elements (Maxted, 1951). However, the precise effect of a given poison may vary from system to system. In addition, catalyst deactivation may also result from the adsorption of a product of the reaction onto the active surface. The poisoning effect may be reversible and in such cases catalytic activity can be restored by eliminating the source of poison. On the other hand, when poisoning occurs by an irreversible process. regeneration of the catalyst may not be possible so that it has to be discarded.

Poisoning of metals by sulfur is a major problem encountered in many catalytic processes (Bartholomew et al., 1982). Various mechanisms have been proposed for the poisoning action including (1) an electronic interaction involving donor covalency of the lone pair of the sulfur to the d band of the metal (Dilke et al., 1948); (2) geometric blocking of a number of metal sites by an adsorbed sulfur atom (Demuth et al., 1974; Fitzharris et al., 1982); and (3) chemical interaction of the metal surface layers with sulfur to form a metal sulfide.

In addition, it has been found that adsorption of sulfur can induce crystallographic transformation in the metal particles which then results in modifications of the adsorption properties of reactant gases and the course of the reaction. Adsorption of sulfur has a strong influence on the wetting properties of metals and supports (Halden and Kingrey, 1955; Felsen and Regnier, 1977), and as a consequence can have a direct impact on the sintering process (Brill and Schaeffer, 1969).

A literature search shows that surprisingly little work has been reported on the influence of sulfur in the catalyzed gasification of carbon. Otto and coworkers studied the effect of sulfur on the steam gasification of graphite in the presence of alkaline earths and concluded that the deactivation was due to two effects. The first one was an elimination of active sites by chemisorption of sulfur species on carbon edge atoms. This mode, which was established quickly, was found to be completely reversible and consistent with earlier observations by Yang and Steinberg (1977) who investigated the influence of SO_2 on the oxidatio of petroleum coke. The second effect was believed to be a result of a chemical change in the nature of the catalytic species from the oxide to the sulfide state. This effect was also found to be largely reversible.

Matsumoto and Walker (1986) studied the effect of H_2S and COS on the steam gasification of char catalized by potassium, sodium, calcium, and iron. The degree to which the sulfur compounds inhibited the gasification reaction was found to depend on whether they were added in the presence of wet N_2 or wet H_2, the latter being the most effective mixture. They also found that the poisoning effect was reversible and that this recovery process was also slower in wet H_2 than in wet N_2. Other workers have demonstrated that addition of copper to another metal increases not only its catalytic activity for the oxidation of char, but also its resistance to poisoning by sulfur (Lopez-Peinado et al., 1986).

It is well known that gaseous halogens act as inhibitors for the gasification of carbons (Arthur and Bowring, 1951; Day et al., 1958; Asher and Kirstein, 1968). McKee and Spiro (1985) demonstrated that pretreatment of graphite with organohalogen compounds or chlorine at temperatures in excess of 700°C (1292°F) had an inhibiting effect on the subsequent reaction in air over the range 600–900°C (1112–1652°F). Moreover, the inhibiting effect persisted for long periods of time. Baker and coworkers (1983) used CAEM to study the influence of chlorine on the catalytic gasification of graphite. In these experiments platinum/ graphite specimens were reacted in either oxygen or hydrogen to the point where catalytic attack became widespread and then chlorine in the form of carbon tetrachloride was introduced into the gas stream. In both systems, after a short period of time, all catalytic channeling action and particle motion came to a complete halt. When the halogen was removed from the oxygen stream, the catalytic channeling was restored (Figure 15). In contrast, removal of chlorine from hydrogen did not result in regeneration of catalytic activity. However, if an intermediate oxygen treatment was used, then catalytic action could

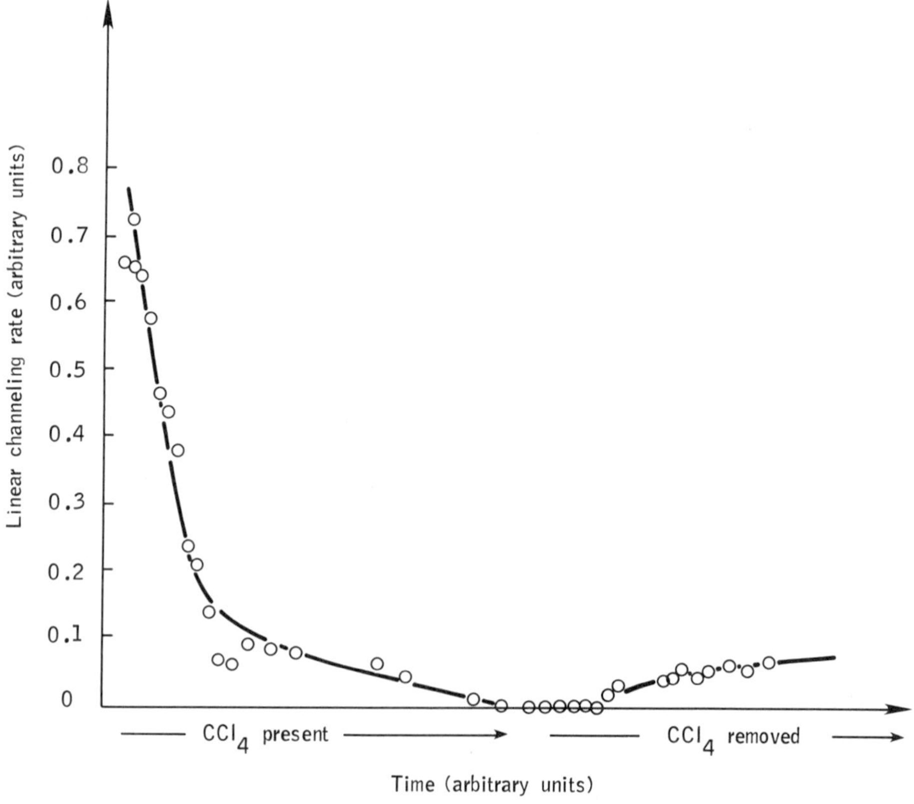

Figure 15 Averaged intrinsic channeling rate of platinum on graphite during an $O_2/CCl_4/O_2$ cycle (Baker et al., 1983).

be recovered during a subsequent hydrogen exposure. Based on these observations the authors concluded that oxygen was more effective in removing adsorbed chlorine species from both carbon and platinum surfaces than was hydrogen. They further suggested that under the conditions used, the platinum surface was only partially stripped of chlorine and hence did not exhibit its initial catalytic activity.

Other gas phase species are known to poison the catalytic activity of metals and these include compounds of nitrogen, phosphorus, and arsenic. As stated earlier, McKee (1981) summarized the literature up to 1981 relating to the inhibiting effect of gaseous phosphorus compounds on the graphite-oxygen reaction. On the other hand, there does not appear to be any reports pertaining to investigations of the influence of the above-mentioned poisons on the action of catalysts during carbon gasification reactions. In other related systems it has been shown that nitrogen in the form of either ammonia or gaseous aliphatic amines caused deactivation of copper, nickel, and cobalt catalysts during amination of alcohols and disproportionation of amines (Baiker et al., 1984).

5. Poisoning Effect of Solid Inclusions in Carbon

Much of the research which has been conducted on the influence of solid inclusions in carbons on the activity of added catalysts has been driven by the need to establish the role of indigenous impurities in coal on the gasification process. Huttinger (1983) discussed the effects of organic and inorganic bound sulfur on the iron-catalyzed gasification of a variety of carbons and coals. He pointed out that major poisoning of the iron catalyst was encountered with organic or heterocyclic bound sulfur, but the problem could be partially overcome by the use of high pressure or temperature and a hydrogen-rich gasification environment which favors the hydrodesulfurization reaction (Adler et al., 1985). Inorganic bound sulfur, usually present in coal as a constituent of mineral matter, e.g., pyrite, FeS_2, was also a poison, but presented only a minor problem, since it could be removed prior to gasification by one of the following desulfurization procedures:

1. Pretreatment with iron trichloride (Huttinger and Krauss, 1980, 1982).
2. Thermal treatment in a gasification environment (reducing or oxidizing) at 500°C (930°F) (Sinha and Walker, 1971; Huttinger and Krauss, 1981a).
3. Sulfur exchange using alkaline earth oxides or carbonates (Huttinger and Krauss, 1981b).

The detailed chemical steps occurring during thermal desulfurization (procedure 2) were studied by Jacobs and coworkers (1982). They used Mossbauer spectroscopy to examine the behavior of indigenous iron-bearing phases during coal gasification and found that in the presence of steam the dominant impurity, pyrite, was converted to a mixture of FeO, ferrous silicate, and metallic iron.

Tomita et al. (1983) and Higashiyama et al. (1985) studied the influence of mineral constituents on nickel catalysts during the gasification of coal in both steam and hydrogen environments. Nickel was found to exhibit an extremely high activity in the steam gasification of coals containing about 0.3% sulfur at temperatures as low as 475°C (885°F). However, with coals containing larger amounts of sulfur (2.2–2.9%) no rapid gasification was observed under these conditions. Furthermore, when mixtures of such coals were investigated, then gasification at the low-temperature region was suppressed. It was suggested that hydrogen sulfide which was evolved in the devolatilization stage was adsorbed by the nickel and this resulted in poisoning of the catalyst. Although similar events occurred during hydrogasification, the influence of sulfur on nickel was less severe. Sulfur initially incorporated in nickel at the devolatilization stage was removed by hydrogen at 1000°C (1830°F) and as a consequence catalytic gasification proceeded at an appreciable rate. Surprisingly, the major cause of deactivation under these conditions resulted from the interaction of nickel with iron-containing material to produce particles with low activity.

In contrast to its detrimental effect on the catalytic action of transition metals, sulfur was found to exert little or no influence on the steam gasification of carbon when potassium was used as the catalyst (Huttinger and Minges, 1985). However, in this case, a strong irreversible deactivation was observed which resulted from the interaction of active alkali species with silica inclusions in the carbonaceous material to form catalytically inactive silicates (McKee et al., 1982; Kuhn and Plogmann, 1983; Leenhardt et al., 1983). The extent of the deactivation by silicate formation is illustrated in Figure 16 (Huttinger and Minges, 1985). Similar conclusions were reached by Radovic and coworkers (1984) who suggested that the deactivation of potassium species in the oxidation of coal was due to an interaction of the alkali with clays in the mineral matter to form insoluble aluminosilicates.

In addition to undergoing reaction with catalytic materials to form silicates, incorporation of silicon species into metal particles produces a reduction in both the solubility of carbon and the rate at which it diffuses through the catalyst particle (ASTM, *Metals Handbook*, 1973). Both these effects promote the formation of an encapsulating layer of graphite at the metal surface, which results in a decrease in the rate of carbon gasification and, ultimately, complete deactivation of the catalyst. Indeed, it was this feature which was used in the successful

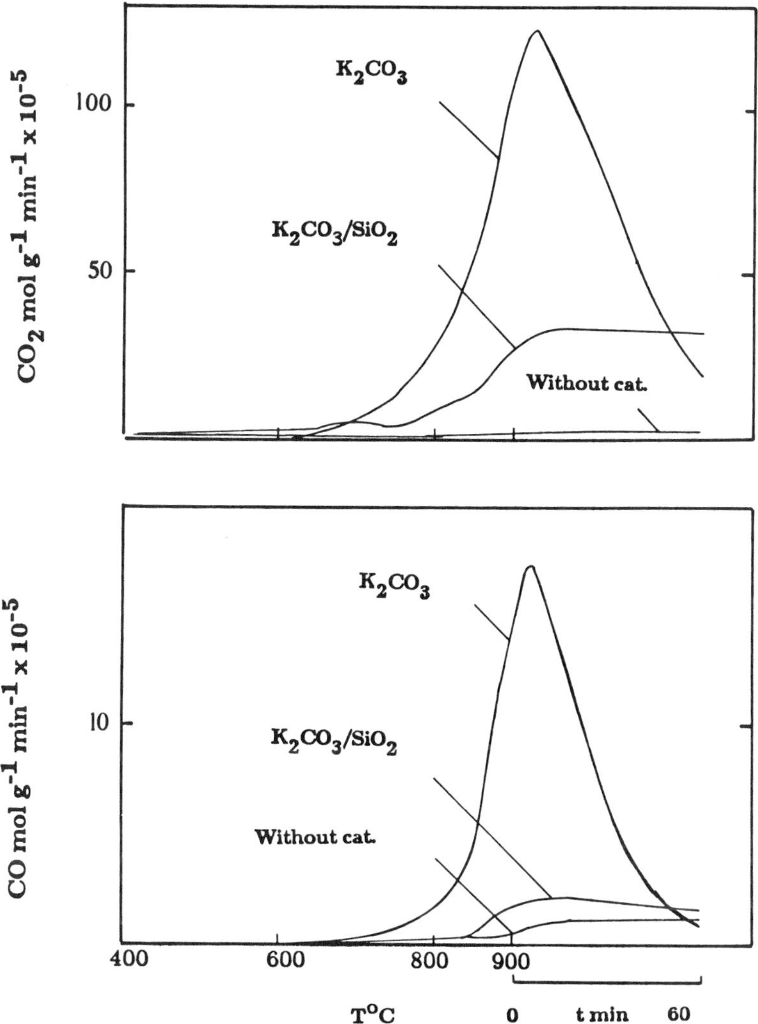

Figure 16 CO_2 and CO formation rates during gasification of graphite in an equimolar argon/steam environment (adapted from Huttinger and Minges, 1985).

development of a treatment to inhibit the growth of filamentous carbon on nickel-iron particles (Baker and Chludzinski, 1980). Introduction of silicon species into the catalyst particles reduced the rates of the crucial steps involved in the growth of this form of carbon, namely, carbon solubility and carbon diffusion through the metal particles, with a resultant buildup of excess carbon at the catalyst surface causing premature deactivation.

Several workers (Long and Sykes, 1952; Haynes et al., 1974; Otto et al., 1979; McKee, 1979; Coates et al., 1983) have reported that calcium is a very efficient catalyst for steam gasification of carbonaceous materials. Direct observation of graphite specimens undergoing reaction in steam revealed that the high catalytic activity of calcium was associated with its ability to wet and spread along graphite steps and edges, which subsequently underwent gasification by the edge recession mode (Baker and Chludzinski, 1985).

When nickel-calcium/graphite specimens were reacted under similar conditions, the mixed catalyst was found to exhibit a lower intrinsic activity than either of its pure components. This result was attributed to a fundamental difference in the mechanisms by which these two elements catalyze the graphite-steam reaction. It was argued that when the mixed catalyst system was reacted in steam, preferential segregation of the calcium to the catalyst/gas interface occurred and caused a corresponding enrichment of nickel at the graphite interface. As a consequence the carbon supply necessary for the efficient conversion of calcium peroxide species to calcium oxide was attenuated by the presence of a nickel diffusion barrier, resulting in an overall decrease in catalytic activity (Baker and Chludzinski, 1985).

6. Volatilization of Catalytic Material

The loss of catalytic activity due to volatilization is a situation encountered in certain gasification systems. Baker and coworkers (1975) used CAEM to study the catalytic oxidation of graphite by vanadium oxide. They reported that at 725°C (1335°F) catalytic material gradually disappeared from the specimen surface and this behavior coincided with a dramatic decrease in the overall rate of reaction. Although volatilization of catalytic material had been suspected in an earlier CAEM study of the oxidation of graphite in the presence of zinc oxide (Baker and Harris, 1973), the experiments with vanadium oxide represented the first time the phenomenon had been observed directly.

Volatilization of catalyst material from carbonaceous solids is most frequently found in systems where alkali metals are used as the gasifying agents. McKee and Chatterji (1978) observed the vaporization and subsequent condensation of sodium at 900°C from a graphite specimen containing an equal amount of sodium carbonate. Similar behavior was reported by Huhn and coworkers (1983) from a sample of carbon containing 45 wt % potassium when heated to 800°C. Other workers

(Wigmans et al., 1984) found that during pyrolysis of coal containing 10 wt % potassium carbonate, loss of active catalyst material took place between 725°C (1335°F) and 925°C (1695°F). Sama and coworkers (1985) carried out a detailed kinetic study of catalyst loss during potassium-catalyzed gasification of carbon in carbon dioxide and established that the extent of the loss was dependent on the reaction start-up procedure. Temperature-programmed reaction experiments showed that under inert atmospheres both KOH and K_2CO_3 reacted with carbon to give a reduced form of potassium-carbon complex, which appeared to be a precursor for potassium vaporization. They demonstrated that under gasification conditions only a fraction of the catalyst was in the reduced form and as a consequence catalyst loss was not as severe as that under an inert atmosphere.

Finally, it is worthwhile to engender a note of caution with respect to the interpretation of electron micrographs which show the existence of channels without catalyst particles at their head. At first sight this condition might be construed as catalyst loss due to volatilization; however, it could be due to other causes such as change in the wetting characteristics between the catalyst and carbon. This might be manifested by spreading of the catalyst on the carbon resulting in the formation of a thin film of material along the channel walls and the consequent disappearance of the original catalyst particle.

C. Catalyst Deactivation in Carbon-Oxygen Reactions

Early work by Amariglio and Duval (1966) showed that only metals which were able to oscillate between two states of oxidation under a given set of conditions could function as active catalysts for the graphite-oxygen reaction. McKee (1970b) later extended these arguments to cover the situation where the catalyst was initially present in the form of a metal oxide. He concluded that only those oxides which could be reduced by graphite to a lower oxide or to the metallic state appeared to behave as active catalysts. Based on this rationale one can readily understand why a number of workers (Thomas and Walker, 1965; Amariglio and Duval, 1966; Heintz and Parker, 1966; McKee, 1970b; Baker and Sherwood; 1981; Baker et al., 1985) found that the metals, iron, nickel, aluminum, tin, titanium, and tantalum, which all form stable oxides under these conditions, gradually lose their catalytic gasification activity when exposed to oxygen at about 750°C (1380°F).

In a recent study, Baker (1986) correlated the mode by which a catalyst operates in the graphite-oxygen reaction with the wetting characteristics of the catalyst material on the graphite edge regions. It was shown that metals which adsorb oxygen dissociatively and readily form oxides exhibit a strong interaction with the oxygenated carbon edge atoms. The strength of the interaction is sufficient to induce spreading of the catalyst material along graphite edges with subsequent attack by edge recession. In contrast, metals which adsorb oxygen nondissociatively tend to remain in the noble state and exhibit a somewhat weaker interaction with the oxygenated carbon surface. As a consequence these materials remain in the energetically preferred configuration of a discrete particle and promote gasification of the graphite by the channeling mode.

It has also been found that some metals—ruthenium, rhodium, and iridium—exhibit both forms of attack: edge recession at temperatures below 1000°C (1830°F) and channeling at higher temperatures. These two activity regions have been found to correlate with the existence of oxides as the stable solid phases at the lower temperatures and metals at temperatures greater than 1000°C (1830°F) (Baker and Sherwood, 1980; Baker and Chludzinski, 1986). It is clear from the previous discussion (Sec. VI.B.1) that such a transformation in the mode of the catalytic attack is also accompanied by an overall decrease in the rate of gasification.

Alloying of one metal with another was found to have unpredictable effects on the subsequent catalytic activity toward gasification of carbon. A 20% addition of either rhodium or iridium to platinum produced catalyst particles with a reduced activity compared to that found with platinum alone (Baker et al., 1980a). Addition of zinc, aluminum, or tin to copper was found to bring about a significant decrease in the rate of the copper-catalyzed oxidation of graphite (McKee, 1970a). It was suggested that the inhibitive effect of these metals was due to their ability to form stable oxide coatings on copper and effectively interrupt the redox cycle with carbon, the latter being one of the critical steps in the mechanism of catalytic oxidation of graphite. In sharp contrast, bimetallic particles consisting of equal amounts of tungsten and rhenium exhibited a reactivity which was more than an order of magnitude higher than either of the single components (Baker et al., 1983).

In a recent publication, McKee (1986) described the influence of various refractory metal compounds as carbon oxidation catalysts. Of particular interest was the behavior of the borides of vanadium, molybdenum, zirconium, chromium, silicon, and aluminum. In the presence of oxygen these materials all underwent thermal decomposition to form the respective metal oxides and boron oxide. Wide variations in the catalytic action of the borides was observed; large increases in the gasification rate was found in graphites containing added VB_2 and MoB, but significant decreases in the presence of the other borides. This difference in behavior was accounted for in terms of a competitive effect resulting from the relative catalytic gasification activity of the metal oxide versus the known inhibiting action of boron oxide (McKee and Spiro, 1984). It was claimed that since vanadium and molybdenum oxides were both extremely active catalysts for carbon oxidation (McKee, 1981), they could overcome the influence of boron oxide. On the other hand, the stable oxides of aluminum, silicon, and zirconium were not active catalysts and the inhibiting effect of boron oxide was dominant in these systems. Chromium oxide was a moderately active catalyst; however, its effect was suppressed by boron oxide.

Finally, mention should be made of a very interesting result obtained by Harris and coworkers (1974) from their CAEM investigation of the influence of silver on the graphite-oxygen reaction. They found that at temperatures between 365°C (690°F) and 875°C (1605°F) silver was a powerful catalyst of this reaction, operating mainly by the channeling mode. However, at temperatures greater than 875°C the metal particles were observed to lose their activity and become inhibitors. In the inhibited state the recession of edges was faster than the recession of occupied edges and the result was the formation of promontories (Figure 17). This feature was rationalized according to the "compensation effect," whereby the catalyzed reaction with a lower activation energy than that of the uncatalyzed reaction would be expected to be faster than the latter at temperatures below the isokinetic temperature and slower than the uncatalyzed reaction at higher temperatures.

D. Catalyst Deactivation in Carbon-Steam Reactions

In the steam gasification of carbon the initial activity of nickel is considerably higher than that of the other materials. Unfortunately, this activity is not maintained as the metal tends to undergo rapid deactivation (McKee, 1974; Wigmans and Moulijn, 1981; Wigmans et al., 1981; Colle et al., 1983; Lund, 1985). This effect is clearly evident in Figure 18.

Magnetic susceptibility studies performed by Colle and coworkers (1983) indicated that the deactivation of nickel in this reaction was associated to some degree with sintering of the metal particles. This conclusion was supported by X-ray and electron microscopy determinations of particle sizes. However, they found that the calculated loss of metal surface area did not entirely account for the observed decrease in catalytic activity. Other workers (Wigmans and Moulijn, 1981; Wigmans et al., 1981) argued that particle sintering was of minor importance and instead attributed the deactivation of the catalyst to the formation of a layer of inactive, amorphous carbon, which blocked the rate-determining step in the gasification process—the diffusion of carbon species through the nickel particle.

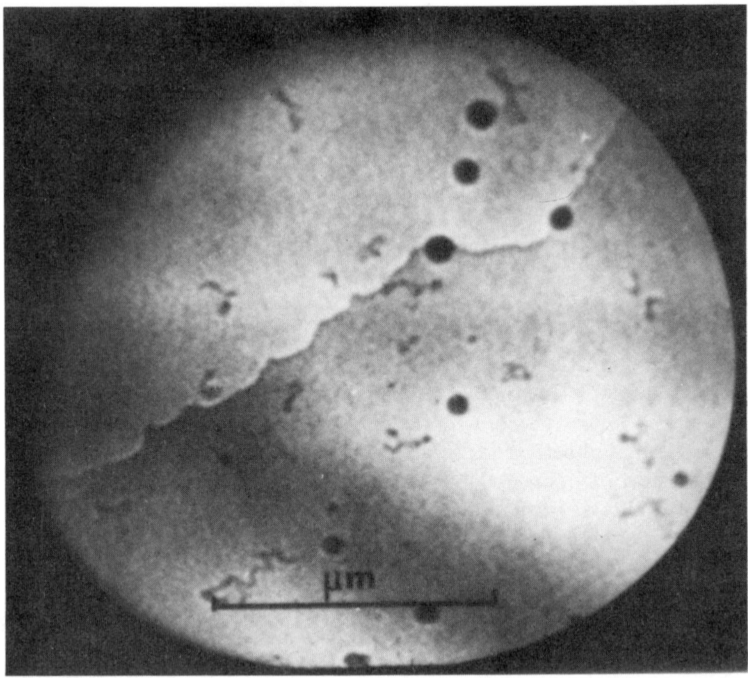

Figure 17 Inhibition of edge recession by silver particles on graphite in oxygen at 885°C (1625°F) (Harris et al., 1974).

In the most comprehensive investigation of the nickel deactivation phenomenon, Lund (1985, 1987) studied the steam gasification activity of nickel as a function of the chemical nature of the carbonaceous material. He found that for graphite catalytic deactivation was only moderate. In contrast, with less ordered carbons, like Spherocarb, deactivation was extremely severe and very rapid, which was believed to arise from encapsulation of the metal particles by an epitaxial carbon layer. A dual-bed reaction configuration was used to probe the causes for the differences in the behavior of nickel on the various carbonaceous supports. In this arrangement the products of the reaction in the first bed passed over the second bed, which was located below and separated by a quartz wool plug. In all cases the lower downstream bed was a nickel/graphite sample and the upstream bed a sample of various carbon without added nickel (Figure 19).

It was found that an upstream bed of graphite had no effect on the catalyzed reaction. Spherocarb and activated carbon caused an

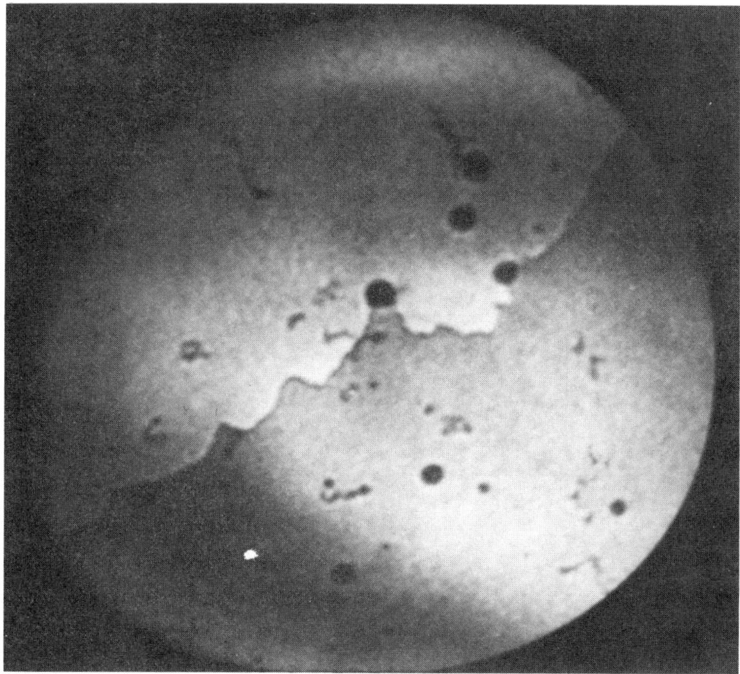

increase in the rate of deactivation, and coal char completely eliminated the catalytic action in the downstream bed. Based on these data it was concluded that the proposed carbon overlayer was produced from the decomposition of hydrocarbon species released during reaction of less ordered carbons. The information derived from this study was later used to develop a mathematical model, from which it was possible to predict the advantages of conducting the gasification process in two isothermal stages as opposed to a single step (Lund et al., 1987).

According to McKee (1974), the metals iron, cobalt, and nickel only exerted a catalytic effect on the graphite-steam reaction when they were maintained in the fully reduced state, the oxides being generally inactive. This feature was demonstrated very clearly for the case of iron/graphite samples treated in various gas environments (Figure 20). Initially, using wet nitrogen, no change in weight of graphite was detected and under these conditions iron would be present as oxide. On replacing wet nitrogen with wet hydrogen, where the H_2/H_2O ratio favored metal formation, a very rapid weight loss

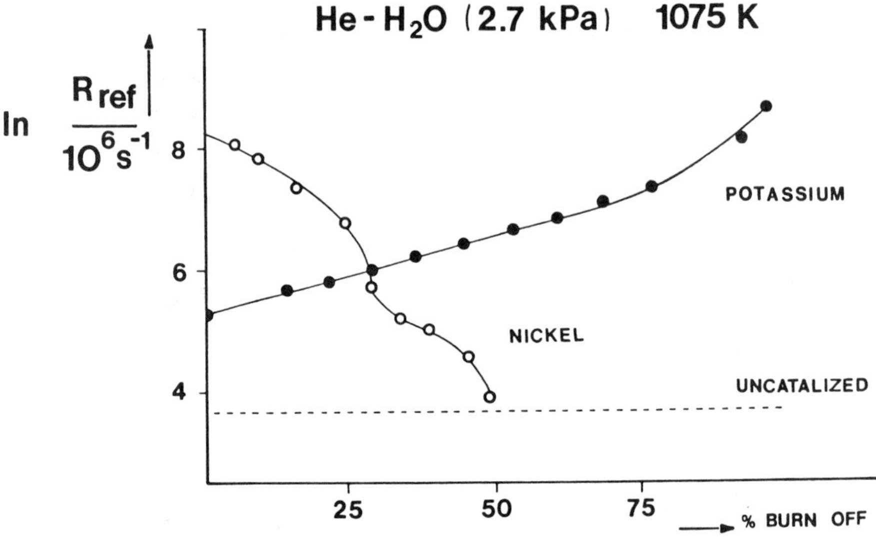

Figure 18 Comparison of the catalytic activity of nickel and potassium in steam gasification of carbon.

ensued. Finally, if water vapor was removed from the system, then the rate of gasification dropped to a lower value indicating that the hydrogasification reaction was slower than steam gasification at this temperature.

Some fascinating effects were observed when iron/graphite samples were reacted in an ethane/steam (40:1) environment (Baker and Sherwood, 1985; Lund et al., 1987). It was found that the gasification activity of the metal particles was extremely sensitive to the thermal history of the sample. Kinetic analysis of the channeling action of iron particles was performed from CAEM experiments. From the experimental data (Figure 21), it can be seen that during the initial reaction cycle and subsequent heating cycles where the upper temperature had not exceeded 965°C (1770°F) the rate followed pathway A. If specimens were heated to temperatures above 965°C (1770°F), the gasification rate exhibited a dramatic decrease to a new, lower activity pathway C. During subsequent reheating of these samples, the channeling rate always remained on pathway C and never regained the higher initial values. Mossbauer spectroscopy studies of similarly treated samples showed that particles which exhibited high activity were α-iron, whereas those which followed the lower activity path were γ-iron. This variation in reactivity patterns

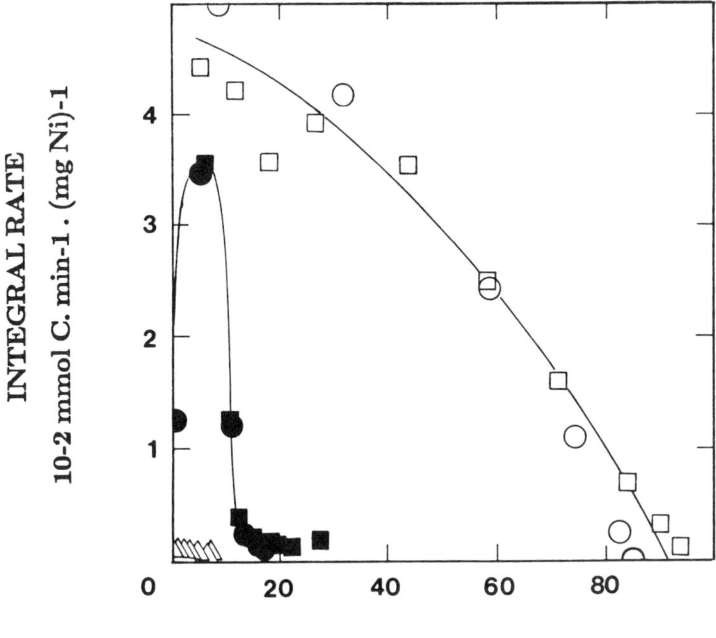

Figure 19 Dual-bed steam gasification burnoff curves. In all cases the downstream bed is 5 wt % nickel/graphite. Upstream beds are o, none; □, graphite; •, spherocarb; ■, activated charcoal; and △, demineralized Illinois No. 6 coal (Lund, 1985).

was ascribed to differences in the properties of the two phases with respect to carbon solubility and diffusion (Mims et al., 1984).

In a separate series of experiments with the same system, attention was focused on the modifications in catalytic action of iron particles as the temperature was cycled between 925°C (1695°F) and 1025°C (1875°F) (Baker and Sherwood, 1985). As the temperature was gradually raised to 1025°C, the rate of channeling by large particles (>50 nm width) slowed appreciably; channeling action by particles in the size range 30 to 50 nm ceased, and the smallest particles (<30 nm width) started to move in a reverse direction and deposited carbon within the tracks they had created at lower temperatures (Figure 22). If the temperature was reduced to below 975°C, then gasification activity of all particles was restored. The reversible nature of these processes can be seen from the kinetic data (Figure 23), which

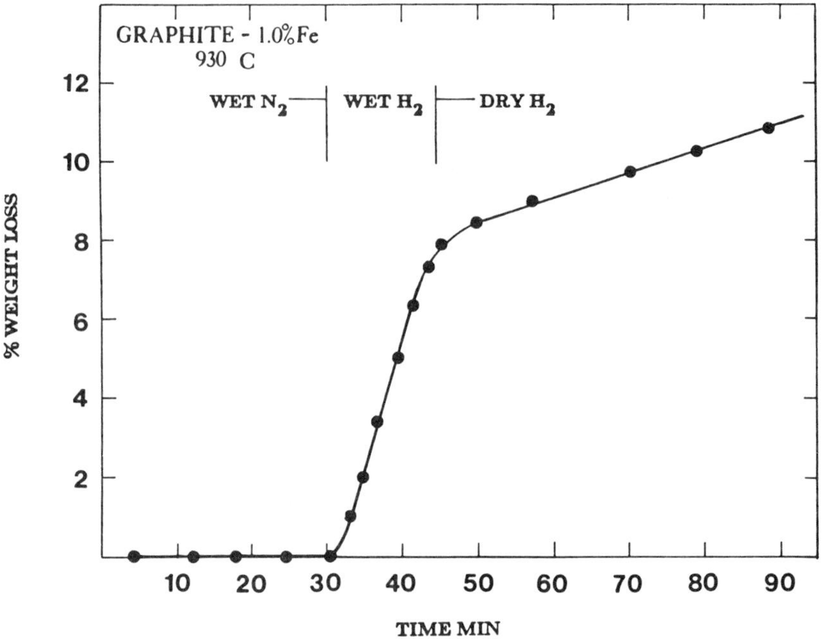

Figure 20 Reaction of 1.0 wt % iron/graphite in wet nitrogen, wet and dry hydrogen (McKee, 1974).

show the effect of multiple cycling between 850 and 1050°C (1560 and 1920°F). It was suggested that this pattern of behavior resulted from changes in the composition of the gas phase, which resulted in the competition between two catalytic processes, hydrogenation of graphite and carbon deposition. This situation was created by the thermal decomposition of ethane, which produced a mixture of ethylene and hydrogen. As a consequence, sufficient amounts of these products were formed so that the gasification reaction was controlled by catalytic hydrogenation and a potentially carbon-depositing environment was created with respect to small iron particles.

Mims and coworkers (1984) used CAEM to follow the manner by which potassium hydroxide catalyzed the gasification of graphite in steam. They found that the potassium species exhibited a strong interaction with the graphite and that gasification proceeded by recession of the edges of layer planes and occasionally portions of actively receding edges suddenly stopped. It was significant that areas immediately adjacent to deactivated edges continued to react. A plot of

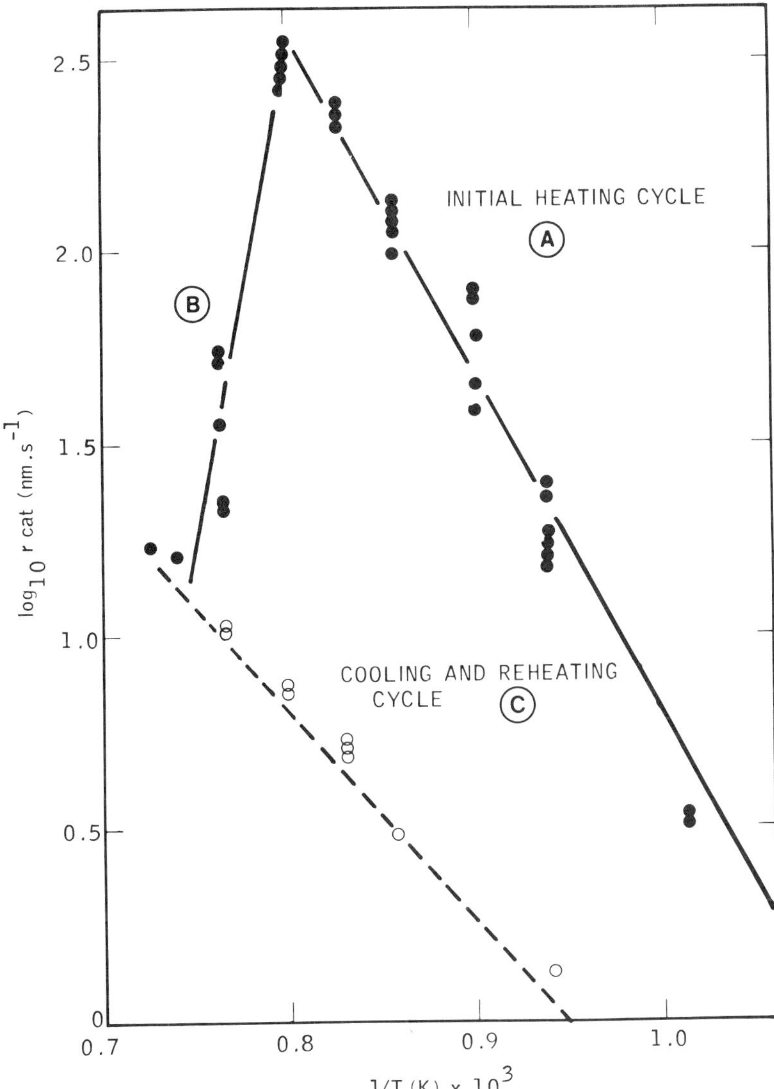

Figure 21 Arrhenius plots of iron-catalyzed gasification of graphite (A) during initial heating cycle in ethane/steam (40:1); (B) when the upper temperature exceeded 965°C; and (C) relationship which prevailed after the specimens were heated at temperatures in excess of 965°C and cooled to lower temperatures. ●, Data points from particles initially present as α-Fe°, o, data points from particles which were initially present as γ-Fe° (Lund et al., 1987).

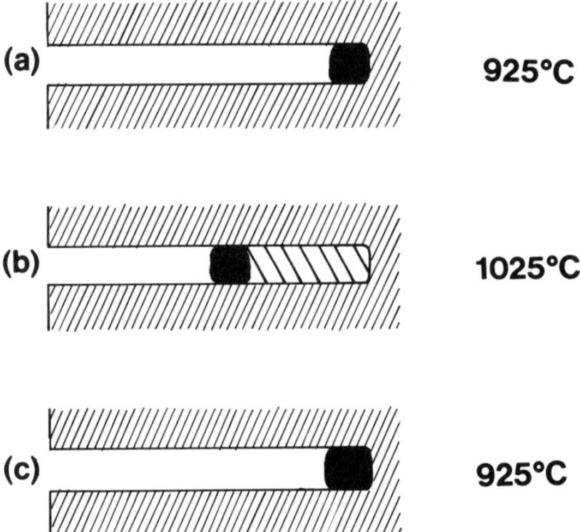

Figure 22 Schematic representation of the catalytic reversal action of iron on graphite in ethane/steam (40:1) (Baker and Sherwood, 1985).

the distance moved by a receding edge as a function of time (Figure 24) shows that cessation of edge movement occurred suddenly without a gradual falloff in activity. Although no definite explanation was presented to account for this phenomenon, the subsequent reactivity in oxygen of these "dead" areas indicated that catalyst was still present after deactivation.

Wong and Yang (1984) used the etch decoration/transmission electron microscropy technique to study the influence of potassium salts on the reaction of steam with the basal plane of graphite. One of the main findings from this investigation was that potassium, particularly in the form of carbonate, inhibited the gasification of monolayer pits which had been produced on the basal plane. This behavior was rationalized according to the notion that the electronic interaction between the catalyst and the graphite at multilayer edges resulted in a decreased reactivity on the single-layer edge sites.

A further effect which must be taken into consideration is the influence of gaseous reaction products, particularly hydrogen, on the rate of the catalyzed steam gasification reaction. It was found that hydrogen reduced the catalytic activity of potassium, sodium, and calcium for the carbon steam reaction (Matsumoto and Walker, 1986; Mims and Pabst, 1983). This inhibition was attributed to the reduction in the steady-state concentration of metal-oxygen intermediates

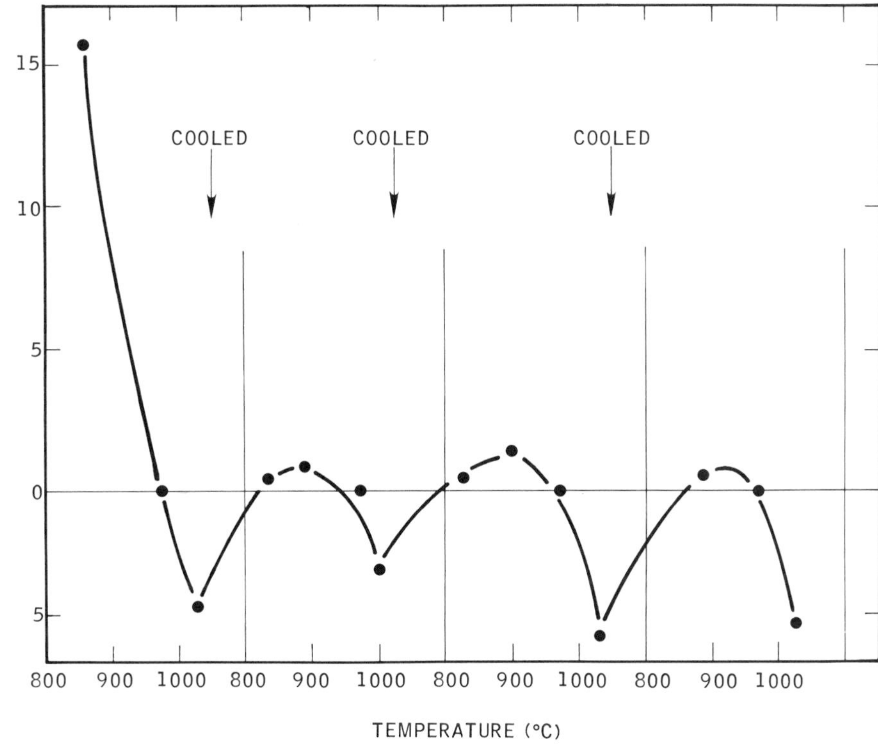

Figure 23 Example showing the rates of gasification and deposition of a 21-nm iron particle as a function of temperature in ethane/steam (40:1) (Baker and Sherwood, 1985).

formed in the presence of steam, which ultimately decomposed to form carbon monoxide and resulted in gasification of carbon.

In contrast to the inhibiting effect of hydrogen on the rate of gasification by steam when alkali and alkaline earth metals were employed as catalysts, it was mentioned earlier that hydrogen acted as a promoter when the reaction was catalyzed by iron (McKee, 1974).

E. Catalyst Deactivation in Carbon-Carbon Dioxide Reactions

A very comprehensive review of the catalysts of the carbon–carbon dioxide has been presented by Walker and coworkers (1968). In that

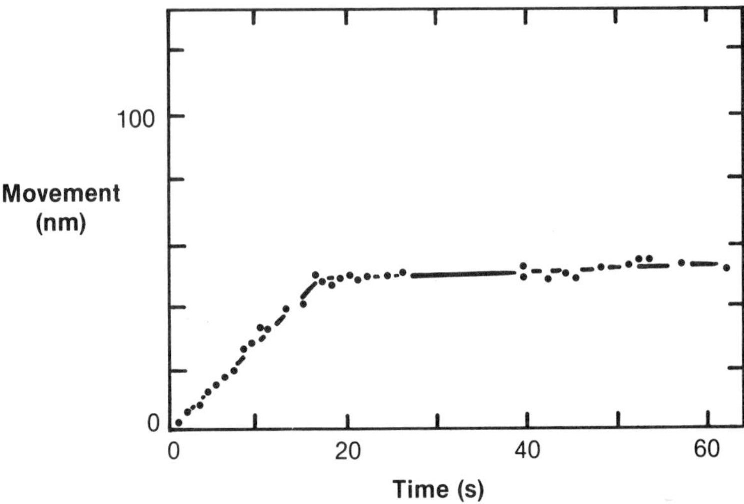

Figure 24 Relative rate of potassium-catalyzed graphite edge recession at short time intervals showing the suddenness of deactivation (Mims et al., 1984).

work particular emphasis was placed on the influence of iron on the reaction. They used a combination of magnetic susceptibility and bulk reactivity measurements to identify the active phases involved in the catalytic reaction. They concluded that metallic iron was the most active state and as the catalyst was converted to wustite, $Fe_{0.95}O$, its activity was reduced, and finally when it formed magnetite, Fe_3O_4, it became inactive. Regeneration of a spent iron catalyst could be achieved by treatments in hydrogen or carbon monoxide, or by heating the sample to a sufficiently high temperature where the carbon acted as a reducing agent. It was significant that the deactivation step was not as pronounced when either cobalt or nickel was used as catalyst.

Baker and coworkers (1972) used CAEM to follow the behavior of iron particles supported on graphite when heated in various gas mixtures containing carbon dioxide. It was found that as the temperature was progressively raised to 900°C (1650°F), a number of events were observed: the rate of catalytic gasification declined and particles which had exhibited mobility at lower temperature suddenly became stationary. This transition in particle behavior coincided with the conversion of the metal to the carbide phase. On continued reaction at this temperature platelets were observed to grow from the particles. The development of the crystalline platelet deposit on the edge of a large metal particle (Figure 25) continued until the iron particle was completely encapsulated. It was suggested that the deposit formed by a mechanism involving the decomposition of cementite, Fe_3C,

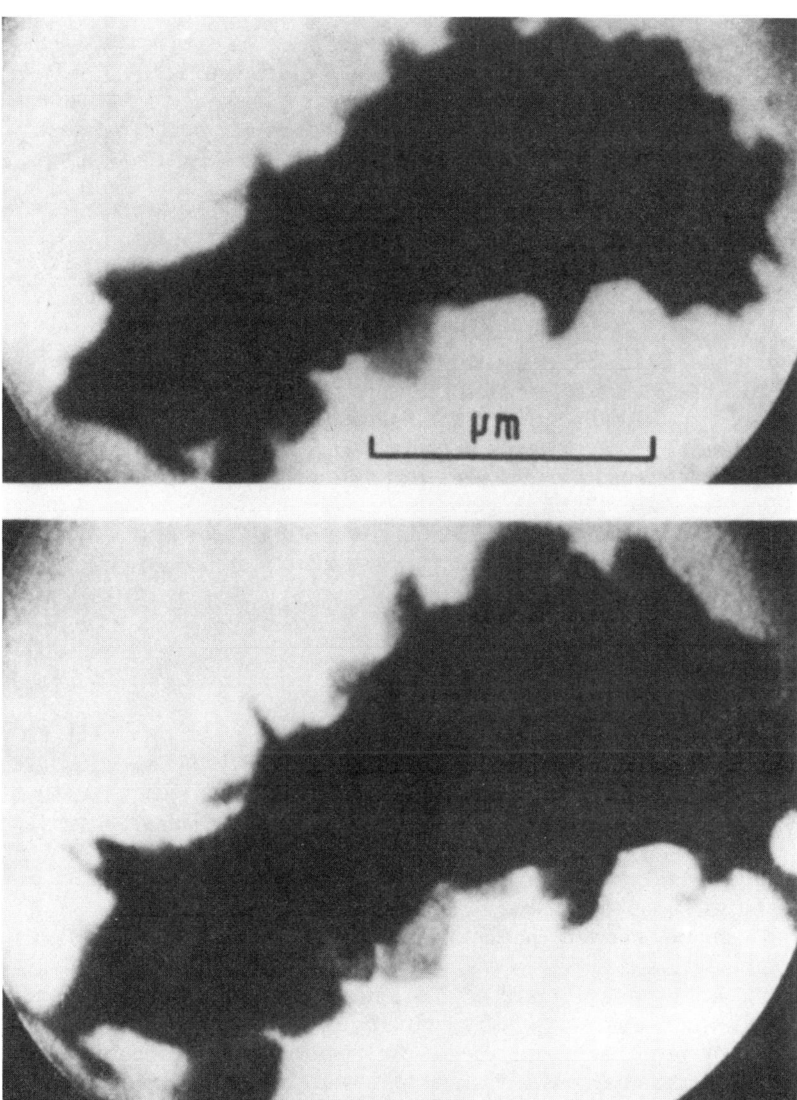

Figure 25 Formation of graphitic platelets on iron at 925°C in CO_2 (Baker et al., 1972).

to form metallic iron and precipitation of carbon as graphite platelets on the gas-cooled upper surfaces.

The influence of nickel on the graphite–carbon dioxide reaction was investigated in the CAEM by Keep and coworkers (1980b). They reported that the onset of catalytic attack occurred at 700°C (1290°F) by the channeling mode; however, this action ceased in most cases as the temperature was raised to 900°C (1650°F). The deactivation was associated with the gradual disappearance of particles from the leading face of the channels, which they suggested was due to diffusion of nickel species into the graphite structure. Electron diffraction examination of the remaining particles showed that at this stage they were in the form of nickel oxide.

Lund and coworkers (1985) used a similar experimental approach to study the platinum/graphite/carbon dioxide system. The first signs of catalytic attack, seen as the creation of channels, were observed at 940°C (1720°F). Continuous observations revealed that not only was platinum a relatively poor catalyst for this reaction but also very susceptible to deactivation. It was common to find a situation where particles had stopped at the end of a channel for no apparent reason. In adjacent regions particles were still active and the inactive particles did not appear to have lost contact with the carbon. Although the authors did not present any definitive conclusions regarding the nature of the deactivation phenomenon, it was believed to originate from carbon deposition or precipitation during the reaction which caused blocking of the exposed metal surface.

F. Catalyst Deactivation in Carbon-Hydrogen Reactions

In the hydrogenation of carbon there does appear to be a critical upper temperature for many systems where the activity of the catalyst starts to decline. Furthermore, if this temperature is exceeded, then on subsequent cooling and reheating in hydrogen, the catalyst particles exhibit a new, lower level of activity. This effect was first reported by Rewick (1974) and coworkers from their bulk studies of various platinum-doped carbons.

CAEM studies showed that catalytic attack of graphite by nickel particles in the presence of hydrogen commenced at 845°C (1555°F) and was seen as the development of fine channels (Baker and Sherwood, 1981). As the temperature was gradually raised to 980°C (1796°F), it became apparent that active nickel particles were disappearing from the heads of the channels. Figure 26 is a sequence showing the depletion in catalyst material as the channel increases in length. This behavior was also reported by other workers (Keep et al., 1980a) who considered the disappearance and concomitant loss of catalytic activity of nickel to be the result of diffusion of metal into the graphite structure. Continued heating in hydrogen up to 1250°C (2282°F) produced no further catalytic action or restoration of the original particles. A detailed quantitative analysis of these events combined with hydrogen chemisorption measurements indicated that the nickel was deposited on the walls of the channels as a near-monolayer film (Baker et al., 1982).

When hydrogen was replaced by oxygen and the sample reheated, then at 850°C (1562°F) small particles started to reform along the edges of the original channels and eventually at 1065°C (1949°F) these particles proceeded to create very fine channels (Figure 27a).

If steam was used as the oxidant, then "breakup" of the nickel film occurred at 830°C (1525°F) and on continued reaction fresh channels were propagated from the edges of the original ones at 935°C (1715°F) (Figure 27b).

In an attempt to learn more about the extraordinary behavior of nickel, in both the wetted and redispersed states, the nickel/graphite system was investigated by ferromagnetic resonance following treatment under various conditions (Simoens et al., 1982). From these studies it was concluded that the material which spread along the channel walls was carbidic material, which had a higher surface energy than the metal. Upon cooling the nickel carbide decomposed to form metallic nickel with a graphite monolayer coating. The existence of this "sandwich" model was used to explain the loss of catalytic activity and decrease in hydrogen chemisorption capacity.

When iron/graphite specimens were treated in hydrogen and the reaction followed by CAEM, it was found that many of the particles which were located at edges and steps on the graphite became activated toward channel propagation at 890°C (1635°F) (Baker et al., 1985). In contrast to the behavior observed with nickel, these channels remained parallel-sided and the integrity of the catalyst particles was preserved throughout the propagation process. In this case there was a tendency for active particles to slow down during prolonged reaction at a given temperature and occasionally become completely deactivated.

It was significant that when the hydrogenation reaction was carried out with a catalyst containing equal amounts of nickel and iron, the channeling action was maintained in an uninterrupted fashion up to 1250°C (2280°F) (Baker et al., 1985). The active particles did not exhibit either the wetting and spreading action characteristic of nickel or the deactivation phenomenon seen with iron. In discussing the differences between these systems, the authors suggested that the deactivation step was linked to the ability of the catalyst to interact with graphite to form a carbide phase—a reaction which occurred readily with nickel and iron, but not with the alloy.

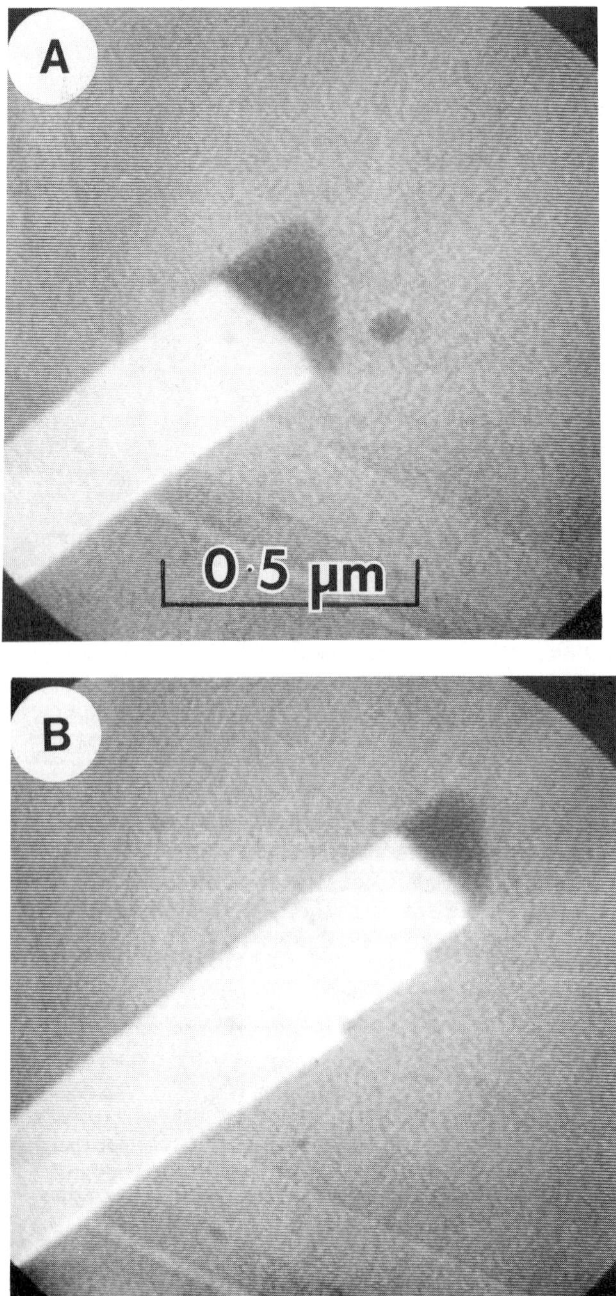

Figure 26 Sequence showing the gradual loss of activity of a nickel catalyst particle caused by deposition of metal along the edges of the channel (Baker et al., 1982).

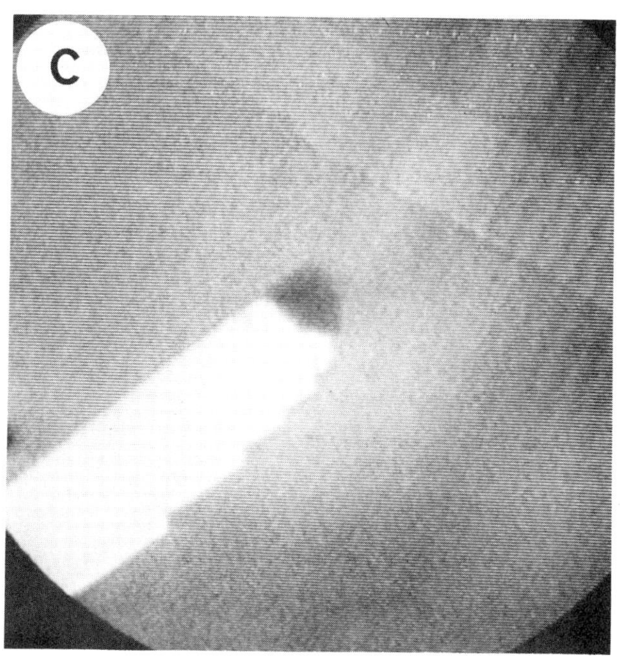
C

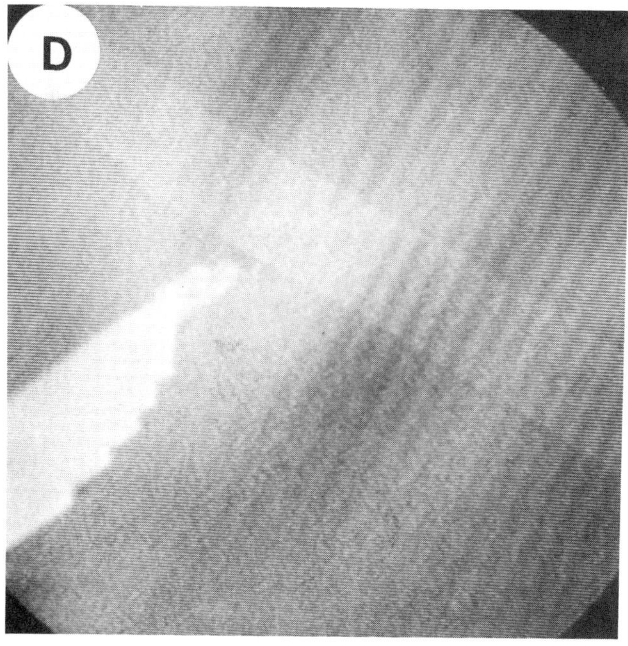
D

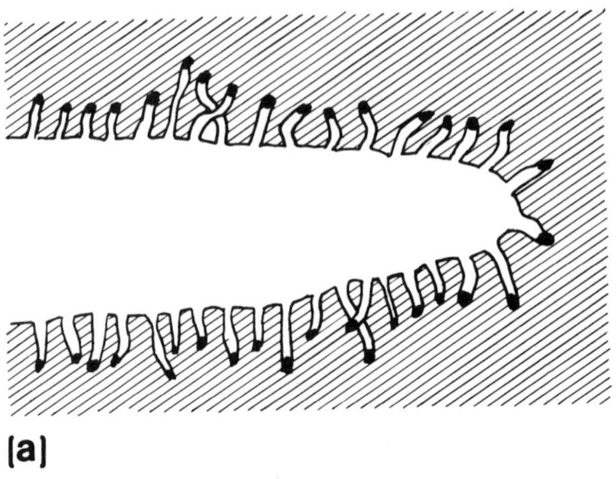

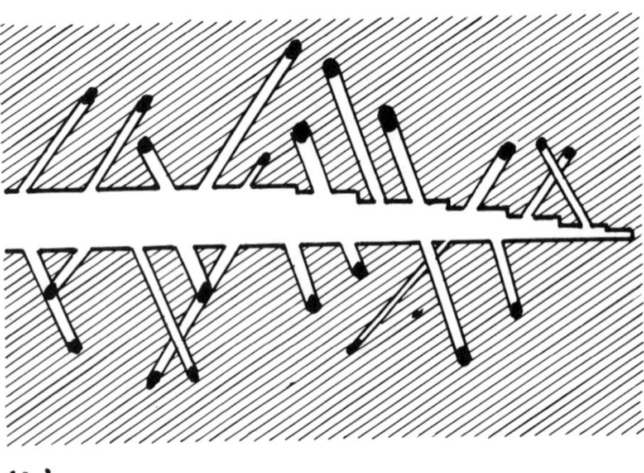

Figure 27 (a) Schematic representation of reactivation of nickel in oxygen; (b) reactivation of the metal in steam.

Recently, Baker and coworkers (1988) extended their hydrogasification studies to cover the influence of other nickel alloys. With copper-nickel they found that gasification of graphite took place over the temperature range 415–675°C (780–1245°F) by the channeling mode. At higher temperatures active particles underwent a spreading action; however, in contrast to nickel, the dispersed material retained its catalytic activity and continued to attack the graphite by the edge recession mode. By analogy with the previous studies one might speculate that addition of certain amounts of copper to nickel inhibits carbide formation and as such prolongs the active life of the catalyst toward hydrogasification of graphite.

The differences in deactivation behavior of various metals and alloys observed during the carbon-hydrogen reaction at temperatures in excess of 950°C (1740°F) is probably related to differences in the metal-carbon chemistry of these systems. The formation of metal-carbon bonds with subsequent reaction of carbidic carbon with atomic hydrogen to form methane are the key steps in the proposed mechanism of catalytic hydrogenation of carbon (Holstein and Boudart, 1981). Surface carbides are readily formed when metals are heated in the presence of a carbon source. If the carbide species are not removed fast enough to form gaseous products, then a decrease in catalytic activity will be observed, and under extreme circumstances where total blocking of the particle surface occurs, then the catalyst will completely lose its activity. For the case where the catalyst particle becomes partially blocked by carbon (graphitic) overlayers, then a uniform drop in the rate of channeling of a given sized particle during a subsequent cooling cycle would be expected. A partially deactivated large particle would exhibit the same activity as that displayed by a "fresh" smaller particle. Consequently, one would not expect to find large differences in the values of the activation energies during the heating and cooling cycles, which is consistent with the reported data (Rewick et al., 1974; Baker and Chludzinski, 1986).

In general, the loss of catalyst activity can be attributed to many causes. However, in gasification reactions two modes appear to predominate: changes in the wetting characteristics of the catalyst material on the carbon substrate and encapsulation of catalyst particles by graphitic overlayers.

The former situation is unique to catalytic gasification reactions and is most frequently encountered when the carbonaceous material is in the form of graphite. Although it is well known that the wetting characteristics of a metal—metal oxide particle on a carbonaceous support are governed by a number of factors including the chemical state of the particle and that of the carbon surface, and the nature of the gas environment, it is a difficult task to control the subsequent catalytic action of the particle. This dilemma might influence the catalyst-carbon interaction. In this regard there is a need to understand how changes in the composition of the gas phase and the presence of impurity atoms in both the catalyst and the carbon modify the particle geometry. Only when these questions have been answered will it be possible to control the entire course of the catalytic reaction.

The problem of catalyst particle encapsulation by graphite is another area where there is a need for more basic studies. At present it is not clear which is the rate-determining step in the process; carbon diffusion through the particle, precipitation of carbon at the exposed surfaces of the particle, or reaction of deposited carbon with adsorbed species on the catalyst surface. There is also a question of whether the process involves the participation of an intermediate metal carbide phase. It has been found that in some cases encapsulation can be prevented by addition of a second metal to the catalyst. This is an aspect which obviously requires a more detailed investigation. Finally, it is imperative to establish the influence of typical impurities, such as sulfur, nitrogen, and chlorine, on the development of the graphite deposit. These elements may well have an inhibitory effect on the encapsulation process.

VII. MILD GASIFICATION

One other aspect of coal conversion that should not be ignored is the concept known as mild gasification. For the purpose of definition, mild gasification is essentially a pyrolysis technique (Chap. 20, Sec. I) to bring about conversion of the coal to gaseous, liquid, and solid products at temperatures below 700°C (1290°F; Table 7). In contrast, conventional gasification of coal (this chapter, Sec. II) is usually carried out at temperatures in excess of 900°C (1650°F).

The U.S. Department of Energy is investigating the mild gasification concept as a means of opening up new markets for coal (Kahn and Kurata, 1985; Bajura and Ghate, 1986). This includes use of all of the products—gases, liquids, and solids—with the solids (char) product being an added-value product to improve the overall process economics.

Coal pyrolysis (Chap. 20, Sec. I) is an established technique that has been widely applied to the production of liquid products from coal (Tables 8–10). In comparison with other processes, pyrolysis of coal offers several potential advantages. For example, conventional gasification and liquefaction options may involve complex combinations of unit processes such as coal preparation, gas processing, gas/liquid upgrading, as well as complex reactor configurations. On the other hand, pyrolysis is somewhat less complex and can be conveniently divided into two general groups: (1) modern processes and (2) early processes. In essence, the early processes were quite simple in terms of reactor configuration and were aimed at the production of coke rather than liquids.

On the other hand, the modern processes are more concerned with maximizing the production of liquid products. To achieve this goal, there is the added expense of reactor design and careful control of

Table 7 Generalized Description of the Various Coal Pyrolysis Operations

Pyrolysis process	Final temp.	Products	Processes
	Slow Heatup		
Low-temperature	500–700°C (930–1290°F)	Relatively more reactive coke and high-tar yield	Rexco (700°C) made in cylindrical vertical retorts. Coalite (650°C) made in vertical tubes
Medium-temperature	700–900°C (1290–1650°F)	Relatively more reactive coke with high-gas yield, or domestic briquettes	Town gas and gas coke (obsolete). Phurnacite, low-volatile steam coal, pitch-bound briquettes carbonized at 800°C
High-temperature	900–1050°C (1650–1920°F)	Hard, relatively unreactive coke for metallurgical use	Foundry coke (900°C). Blast furnace coke (950–1050°C)
	Rapid Heatup		
Low-temperature	500–700°C (930–1290°F)	Relatively higher tar yield (20–35 wt %) compared to gas (5–15 wt %)	COED (multiple fluidized bed); occidental flash pyrolysis (entrained flow)
Medium-temperature	700–900°C (1290–1650°F)	Lower tar yield compared to that in low-temperature operation	Clean coke process (oxidized clean coal in fluid bed)

reactor process parameters to minimize the occurrence of secondary reactions; it is these latter reactions that are often instrumental in the formation of coke.

Table 8 Summary of the Low-Temperature Batch Pyrolysis Process for

Process	Krupp--Lurgi	Brennstoff-Technik	Otto
Nation/yr	Germany/1930s	Germany/1940s	Germany/1940s
Objective	Char production	Lumpy char from weakly coking coals	Char and gas production
Plants (yr)	Wanne-Eickel (1943)	Berlin-Neukölln (1944)	–
Yields			
Char	2.04×10^8 kg/yr	0.75 kg/kg dry coal	–
Tar	1.2×10^6 kg fuel oil/year 1.9×10^6 kg motor fuel/yr	0.1 l/kg dry coal	–
Gas	105 l/kg coal	142 l/kg dry coal	
Gas HV	–[a]	2.5×10^4-3×10^4 kJ/m^3	–
Reactor	Vertical fixed-bed retort	Vertical fixed-bed retort	Vertical fixed-bed retort
Capacity	270 kg/day	10^5 briquettes/day	3600 kg/day
Heating	Indirect hot gas recycling	Indirect hot gas recycling	Indirect heating by gas burned in flues
Temperature range	560–620°C	∿650°C	–
Residence time	2–3 hr	2–4 hr	–
Current state	–	Semi-commercial plant still may be operating in Germany	Abandoned

[a]Information not available.

Coal Conversion

Weber	Phurnacite	Parker report	Rexco
Germany/1940s	United Kingdom/ 1940s	United Kingdom/ 1920s	United Kingdom/ 1930s
Char production	Smokeless domestic fuel	Smokeless domestic fuel	Smokeless domestic fuel
–	South Wales (1942)	Barnsley (1927) Bolsover (1936)	Nottingham, England (1936)
–	–	0.7 kg/kg coal	–
–	–	0.06 l/kg coal	0.08 l/kg coal
		–	–
–	–	125 l/kg coal	Unknown
–	–	26,000 kJ/m^3	5,200 kJ/m^3
Vertical fixed-bed retort	Vertical fixed-bed retort	Vertical fixed-bed retort	Vertical fixed-bed retort
Unknown	11,340 kg/day	0.3 kg/kg charge	50,000 kg/day
Unknown	Indirect heating by recycling hot combustion gas	Indirect radiant heat by combustion gas	Direct heat by combustion gas
–	–	∿650°C	∿650°C
–	4 hr	4-1/2 hr	13-1/2 hr
–	–	–	–

Table 9 Summary of the Low-Temperature Continuous Processes for Coal Conversion

Process	Coalite and Chemical Products, Ltd.	Lurgi–Spullgas	Disco Process	Koppers Continuous Vertical Ovens
Nation/year	United Kingdom/ 1950s	Germany/1930s	USA/1930s	Germany/1940s
Objective	Char, liquid fuels	Char/motor fuels	Lump char	Char
Plants (yr)	Bolsover, England (1952)	Offleben, Germany, Japan (1941), New Zealand (1931), Lehigh, North Dakota (1940)	Pittsburgh, PA	Kattowitz (1945)
Yields				
Char	0.74 kg/kg coal	0.45 kg/kg briquette	6.8×10^5 kg/day	1.5×10^4 kg/day-retort
Tar	0.06 l/kg coal	0.125 kg/kg briquette	–	2.1×10^3 kg/day-retort
Gas	111 l/kg coal	∿144 l/kg briquette	–	0.32 l/kg coal

Gas HV	26,000 kJ/m^3	~8,380 kJ/m^3	—	1.8×10^4 kJ/m^3
Reactor	Parker vertical retort	Continuous vertical retort	Continuous rotating horizontal retort	Continuous vertical retort
Capacity	0.3 kg/kg coal charge	2.7×10^5 kg briquettes/day	1.4×10^5 kg/day	25,000 kg/day
Heating	Radiant heat supplied by combustion gases	Direct contact with combustion gases	Indirect heating by circulating hot combustion gas through flues built in hearths	Indirect/direct heating by passing hot gases alternately upward and downward through flues, regenerators were employed
Temperature range	Combustion chamber held between 600 and 700°C	600–700°C	~550°C	~700°C
Residence time	Coal carbonized for 4 hr	~20 hr	1-1/2 hr	21 hr/retort
Current status	In operation	—	Abandoned	—

Table 10 Summary of the Operating Conditions for Pyrolysis,

Process	Status	Reactor type	Reactor temperature
COED, FMC Corp.	Developed	Multiple fluidized-bed	290–565°C (555–1050°F)
TOSCOAL	Developed	Rotating horizontal retort	520°C (970°F)
Lurgi–Ruhrgas	Commercial	Mechanical (twin-screw) mixer	595°C (1105°F)
Occidental flash pyrolysis	Developing	Entrained flow	610°C (1130°F)
Clean Coke, U.S. Steel	Developing	Fluidized bed	∿800°C (∿1470°F)
Rockwell/Cities Service flash hydropyrolysis	Developing	Entrained flow	845°C (1555°F)
Supercritical gas extraction, NCB[c]	Developing		400°C (750°F)

[a]In pyrolysis stages.
[b]Water plus tar/oil sum 39%, individual values estimated.
[c]NCB: National Coal Board, United Kingdom.
[d]Tar extract, softening point 70°C.

However, the early processes (although relatively simple in terms of process conditions and reactor design) may suffer from any one of a number of limitations that prevent them from being employed in a more modern mild gasification process. Such limitations vary from uncontrolled noxious emissions to being employed only for a specific type of coal. Other limitations, such as precise temperature control and stable product slate, were also evident in the early processes. Thus, in any modern process, the process parameter variable will need

Hydropyrolysis, and Extraction Processes as Applied to Coal Conversion

Reactor pressure (MPa)	Residence time (sec)	Coal	Yield, mass % dry coal Char	Tar/oil	Gas	Water	Size of facility (in 1000 kg/day coal)
0.2–0.19	1200[a]	Bituminous	62	21	14	3	32
0.1	300–600	Subbituminous	69	13	9	9	23
0.1	∿3	Subbituminous	50	∿32[b]	11	∿7[b]	1600
0.3	1.5	Bituminous	56	35	7	2	3
∿0.8	∿3000	–	66	14	15	–	–
3.5	∿0.1	Bituminous	46	38	16	–	22
10	1800	Bituminous	63	33[d]	2	2	–

detailed attention (Table 11) and it may well be that the desire for a particular product slate will be controlled by careful choice of the process parameters. In summary, the desired product slate will be available only by the choice of suitable process conditions (Howard, 1981; Gavalas, 1982).

Currently, the Department of Energy is funding several research programs to develop the fundamental concept of a coal pyrolysis process (Figure 28) and are actively assessing the possible uses of the

Table 11 Summary of the Parameters That Can Influence Coal Conversion by Means of Pyrolysis Processes

Heating Rate

- High heating rate increases liquid/gas yield and reduces char yield.
- The tar obtained at a high heatup rate is of poorer quality (i.e., lower H/C ratio) than that obtained at a slower heating rate.
- High heating rate increases open char structure and char reactivity (in reactive gases).
- Sophisticated (often expensive) systems needed to achieve high heating rate.
- High heating rate increases the thermoplastic (softening and swelling) behavior of coal.

Temperature

- Low-temperature operation (500–700°C) improves liquid yield.
- Affects heteroatom distribution among char, liquid, and gas.
- At elevated temperatures (>1300°C), inorganics are removed as slag.
- Longer residence time needed for reaction to be completed at lower temperatures.

Pressure

Inert gas atmosphere:

- Higher pressure operation reduces reactor size needed (i.e., increases throughput).
- Higher pressure reduces tar yield.
- Coal feeding, product separation are more difficult at high pressure.
- Better gas-solid heat transfer is achievable at higher pressure.

H_2 atmosphere:

- Cost of H_2 must be considered versus the quality of the product generated.
- Improves yields of liquid and light products.
- Requires sophisticated pressure systems.
- May increase the undesirable agglomerating properties of coal.

Other atmospheres (H_2O, CO_2, CO, CH_4, CH_2):

- Probably improve liquid/gas yield.
- Not much information available.

(continued)

Table 2 (Cont.)

[*Pressure*]

Vacuum:

- Plastic behavior of coal is reduced.
- Increases liquid/gas yield.
- Difficult to achieve gas-solid heat transfer (solid-solid heat transfer feasible).
- Nut much information available.

Particle size

- Smaller particle size improves product gas/liquid yield.
- Smaller particle size reduces secondary reactions.
- Grinding cost increases with the reduction in size.

Coal rank

- High-volatile A (HVA) bituminous coals produce largest quantities of tar.
- Lignites are rich in oxygen function groups which lead to overall reduction in the calorific value of the product.
- Types of sulfur (pyritic versus organic) present in various coals can influence its distribution in products.

---

end products. There is interest in the application of coal pyrolysis technology in conjunction with electric utility companies as a means of reducing the costs of power generation. One such study (Bechtel, 1982) examined the potential for using the char (produced from a high-volatile bituminous coal) as utility feedstock while the liquid products would be sold to offset the cost of the char. It was generally concluded that overall power generation costs could indeed be reduced by means of this type of technology. In fact, it was concluded (Karr, 1963) over two decades ago that the oldest and simplest method of pyrolysis would be the most economical means of producing liquid fuels from coal. Even though the economics of producing liquids by this method have varied (depending on the current world market price for crude oil), there is always the tendency to return to the pyrolytic method for liquids production. It will be very interesting to follow these current developments.

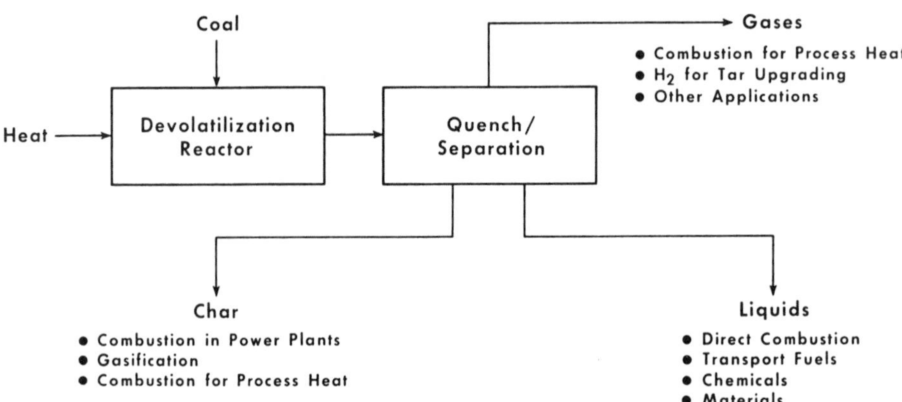

Figure 28 Simplified representation of pyrolysis for coal conversion including suggestions for product use.

VIII. UNDERGROUND GASIFICATION

The United States began developing underground gasification technology in 1973 although such technology had been in use elsewhere for more than 50 years. The fundamental principles of underground gasification are that air (or oxygen) and steam are injected into one of two wells drilled into a coal seam. The coal is partially combusted to a synthesis gas mixture plus tar which is recovered through the second well. Low-btu gas (50–280 btu/scf) is generated, the range of heat content varying with the oxidants used. Air, for instance, yields the lowest btu product (50–140 btu/scf) whereas an air-steam cycling sequence produces gas with 250–280 btu/scf) (Nadkarni et al., 1974).

The success of underground coal gasification depends in large part on effective contact between the coal and the injected gases. A major limitation, then, to efficient underground gasification is the low permeability of coal seams. In some instances large shafts are drilled into the seam and, after some mining, panels of coal are sectioned off by brickwork. This is called the chamber or warehouse method (Figure 29). It is costly in terms of the labor required to construct the chambers and to break into or bore into the panels to create passageways for the oxidants.

In a borehole gasification scheme, a set of parallel boreholes (galleries) are drilled into the seam and are connected by a series of horizontal channels. Air is forced into the galleries and remote electric ignition initiates combustion (Figure 30). The borehole method is

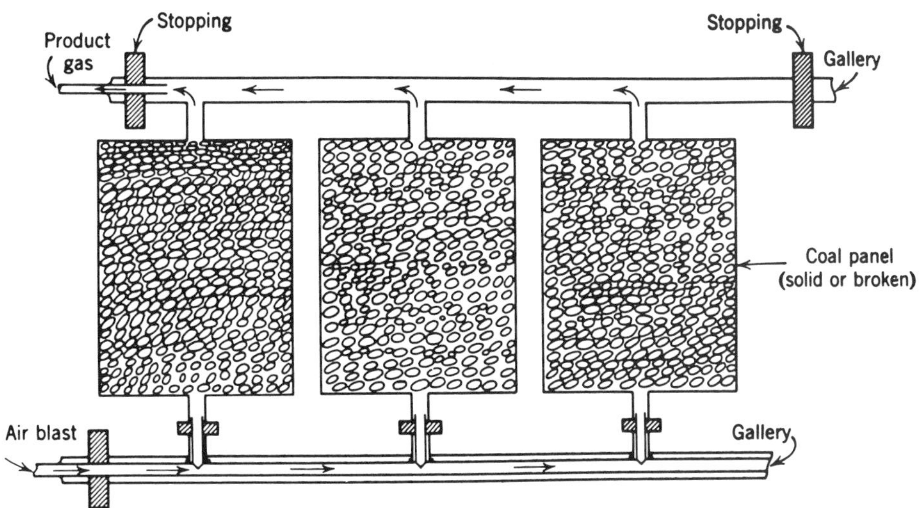

Figure 29 Schematic diagram for the chamber method of underground coal gasification.

most appropriate for level seams. For seams with a large dip, inclined galleries called streams are sunk into the seam and are connected at the bottom by a horizontal "fire drift" channel. The gasification is initiated in the fire drift channel and the vertical openings carry inlet and product gases. This method uses the process called reverse combustion, i.e., the combustion front moves toward the source of injected oxidant (Figure 31). In the chamber and borehole methods, combustion and inlet gas moved in the same direction (forward combustion). Shaftless methods for in situ gasification include the percolation method. After an array of paired boreholes is drilled, the combustion gases are forced into one hole of the pair and product gases are released through another, depending on the permeability of the coal. The combustion zone can be forward or reverse and the process is repeated sequentially throughout the array (Figure 32).

In a design intended to produce high-btu gas from underground seams up to 3000 ft (1000 m) below the surface, Lawrence Livermore Laboratory at the University of California devised a scheme that first fractures the coal using explosives. Boreholes can then be drilled and cased. Combustion with oxygen and steam is proposed in order to produce a gas rich in methane (Figure 33). Cleanup and additional methanation could be carried out at the surface.

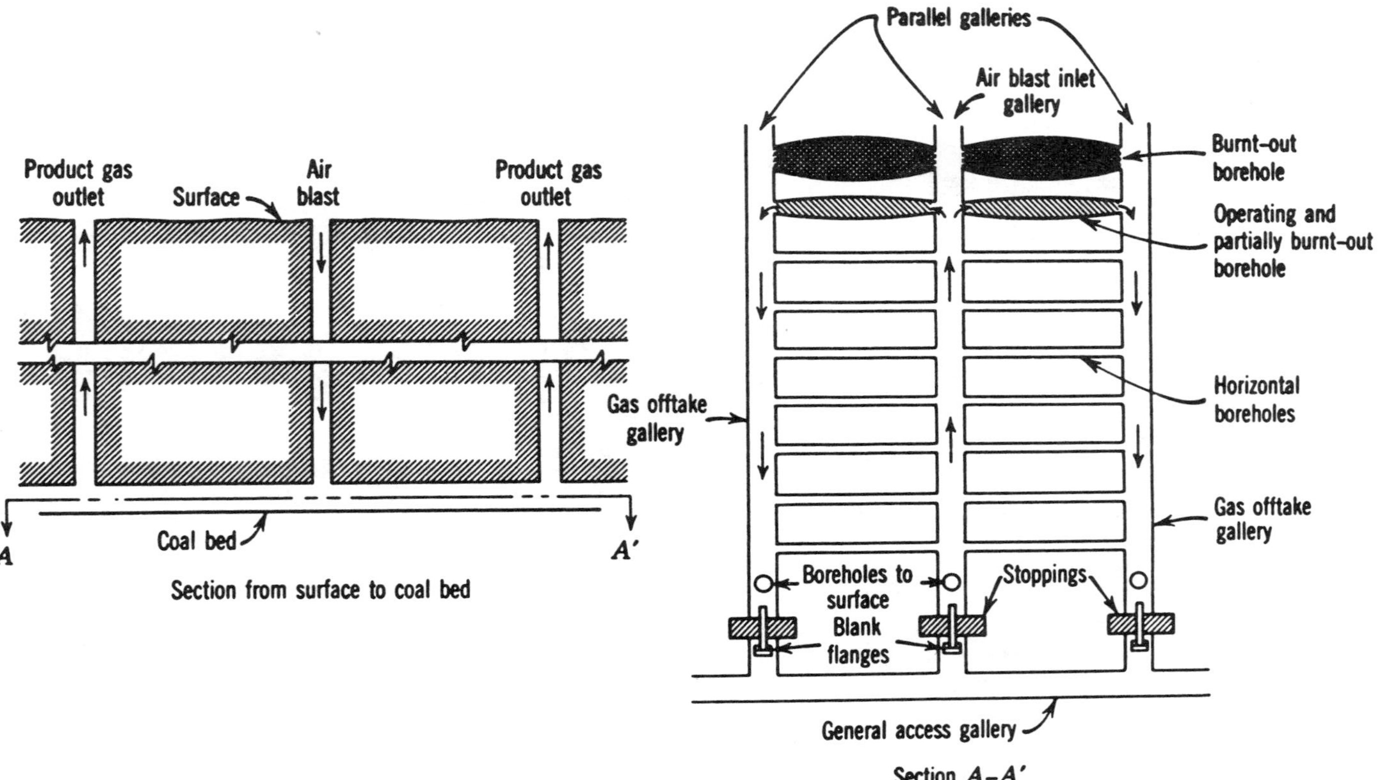

Figure 30 Schematic diagram for the borehole method of underground coal gasification.

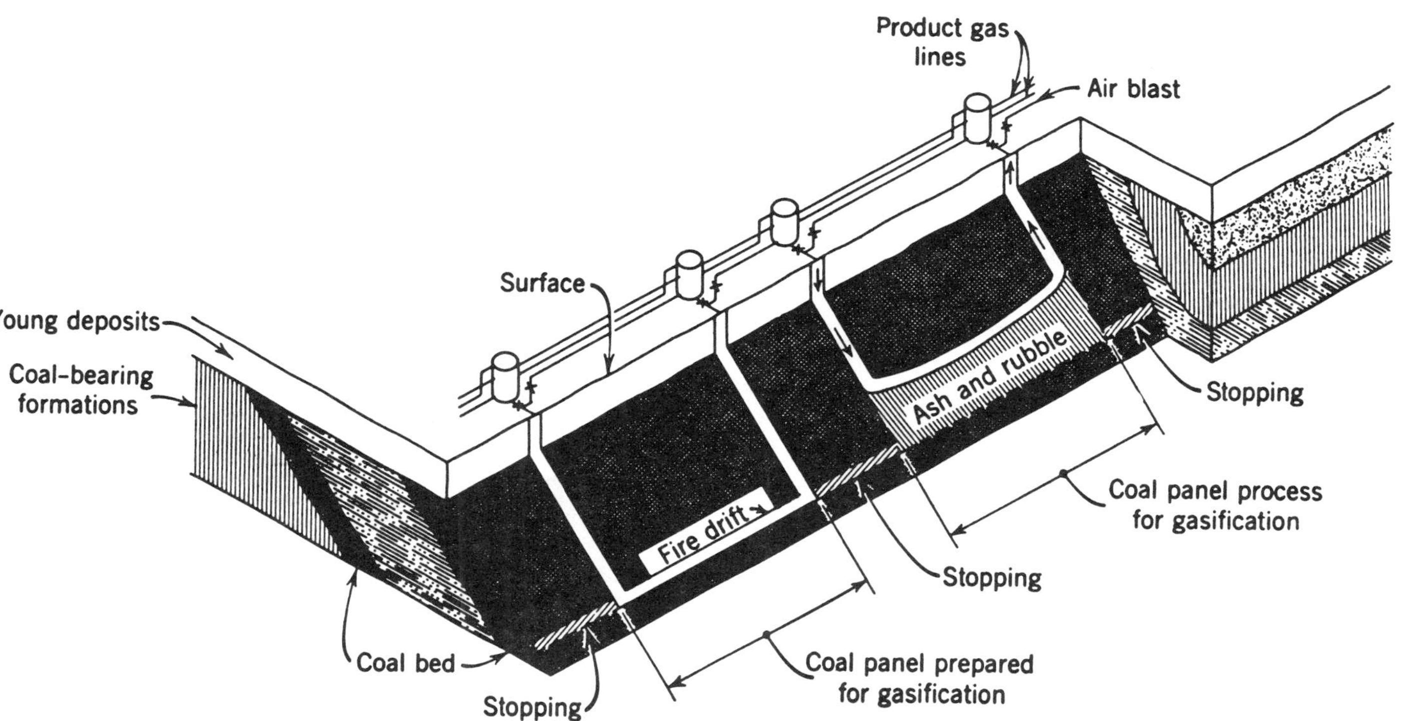

Figure 31 Schematic diagram for the stream method of underground coal gasification.

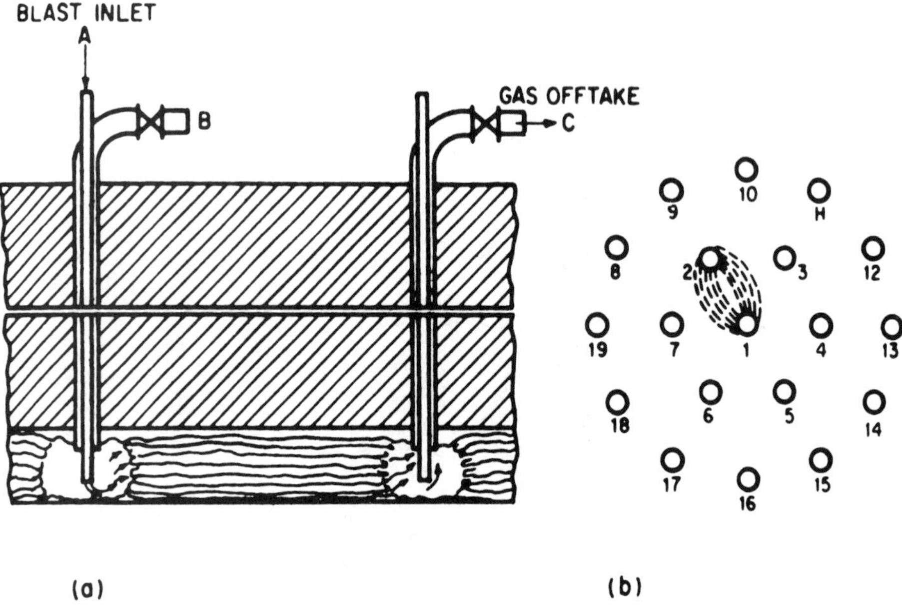

Figure 32 Schematic diagram for the percolation method of underground coal gasification. (a) Drilled boreholes. (b) Array.

Underground coal gasification technology has been tested in the United States, but in terms of commercial development the USSR has been the most active country (Table 12).

Currently, there is no commercial underground coal gasification facility operational in the United States. However, a test at Rawlins (Wyoming) has just been completed and was sponsored by the U.S. Department of Energy, the Gas Research Institute, Amoco, the Electric Power Research Institute, and Union Pacific Resources.

In this test, two commercial-sized underground gasification modules of different configurations were simultaneously operated to produce medium heating value gas. The extended linked well (ELW) module consists of a horizontal borehole connected to vertical injection wells. The product gases were produced out of the horizontal well.

The other module configuration tested was the CRIP (controlled retracting injection point) system (Figure 34). In UCG operations, gasification efficiency generally declines with time due to increased heat losses to the overburden. More overburden is exposed as the UCG

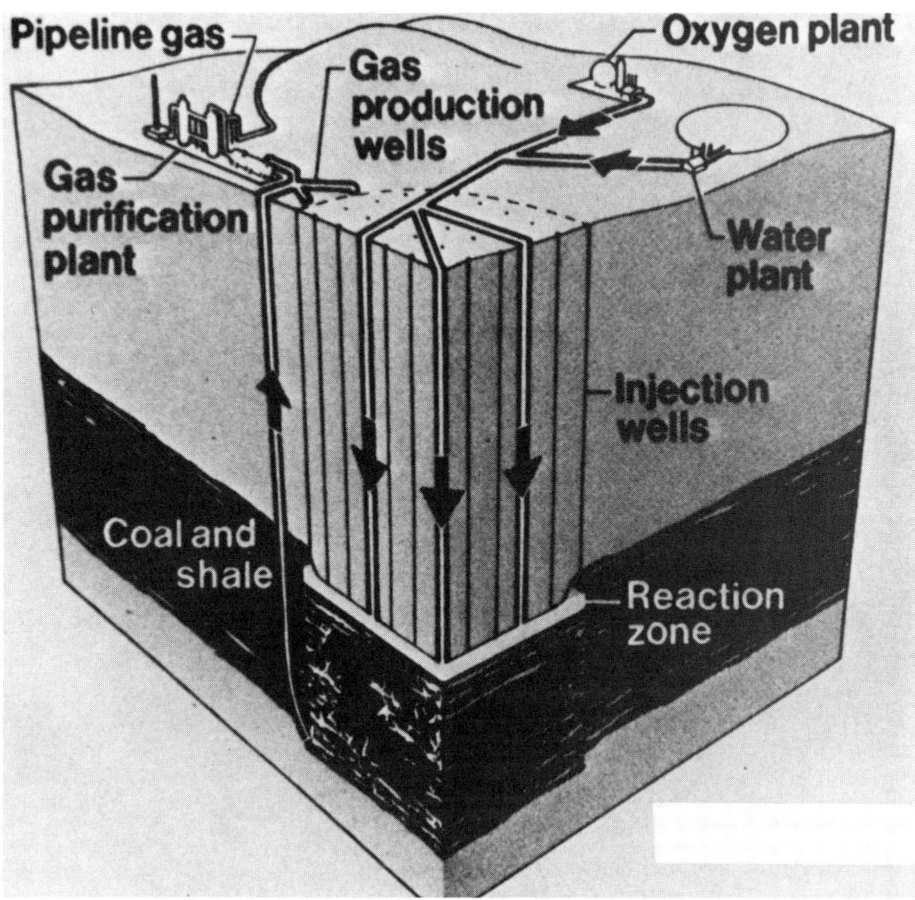

Figure 33 Sketch of the Lawrence Livermore Laboratory underground gasification concept (U.S. Atomic Energy Commission).

module matures. The CRIP system permits the movement of the injection point away from the existing cavities to expose fresh coal resource to the process and reduces heat losses to exposed overburden. In the CRIP module, a liner is inserted in a horizontal borehole in the coal seam. Steam and oxygen are injected through the liner to react with the coal and its products to fuel the gasification reactions. When

Table 12 Methods Used for Testing and Commercialization of Underground Coal Gasification Technology in Various Countries

Country	Location	Coal seam: Type	Thickness (in.)	Depth (ft)	Dip	CV (btu/lb)	Technique	Type operation	Linkage	Pattern	Blast	CV, product gas (btu/scf)	Remarks
Shaft (underground development)													
Russia	Lisichansk	Bituminous	30	100	40	11,600	Streaming	Commercial	Gallery	Panels	30% O_2	86	
Russia	Gorlovka	Bituminous	72	200	75	10,000	Streaming	Commercial	Gallery	Panels	30% O_2	130	
Poland			39				Producer	Experimental	Drill holes	Parallel	Air	80	60% heat loss
U.S.	Gorgas	Bituminous	35	25	2	14,000	Streaming	Experimental	Gallery	U-shape	Air	47	High gas loss
Belgium	Bois-la-Dame	Semianthracene	36	525	87	14,000	Streaming	Experimental	Gallery	37-ft panel	Air	56	35–45% thermal efficiency
Shaftless (boreholes)													
U.S.	Gorgas	Bituminous	37	180	2	15,000	Percolation	Experimental	Electro	Square	Air Oxygen Water gas	93 195 279	30–50% gas loss
U.S.	Gorgas	Bituminous	37	180	2	15,000	Percolation	Experimental	Hydraulic	X pattern	Air	84	44% thermal efficiency
U.S.	Gorgas	Bituminous	37	180	2	15,000	Percolation	Experimental	Hydraulic	Straight line	Oxygen	124	
U.S.	Gorgas	Bituminous	37	180	2	12,000	Percolation	Experimental	Hydraulic	30-ft circle	Air	90	40% gas loss
U.K.	Newman-Spinney	Bituminous	36	75	8	12,800	Streaming	Experimental	Boreholes	Straight line	Air	85	
U.K.	Newman-Spinney	Bituminous	36	100	8	12,800	Percolation	Experimental	Pneumatic	Rectangular	Oxygen	100	
Russia	Yushno Abinsk	Bituminous	276		75	12,000	Percolation	Commercial	Pneumatic	Inclined	Air	130	65% heat recovery
Russia	Lisichansk	Bituminous	30	100	40	11,600	Percolation	Commercial	Pneumatic	Inclined	Air	100	High pressure used
Russia	Moscow Field	Lignite	72	65	0	4,900	Percolation	Commercial	Pneumatic	75-ft square	Air	85	
Russia	Shatsky	Lignite	120	150	0	4,900	Percolation	Commercial	Pneumatic	25-m grid	Air		
Russia	Stalinsk	Bituminous	98	1500	80	14,000	Streaming	Commercial	Pneumatic	Inclined	Air		
Russia	Tula	Lignite	390	180	0	4,900	Percolation	Commercial	Pneumatic	25-m grid	Air	105	
Combination (shafts plus boreholes)													
U.S.	Gorgas	Bituminous	42	125	2	14,000	Streaming	Experimental	Gallery	Straignt line	Air	70–90	4–40% gas loss
U.K.	Newman Spinney	Bituminous	36	240	8	12,800	Streaming	Commercial	Boreholes	Parallel unit	Air	57	84% coal recovery
Morocco	Djerada			160	90		Streaming	Experimental	Gallery	Panels	Air		Combustible gas produced
Russia	Lisichansk	Bituminous	30	100	40	11,600	Streaming	Commercial	Boreholes	Panels	43% O_2	100	47% efficiency

[a]Proposed

Source: A. D. Little, 1972. Reprinted with permission of the publisher.

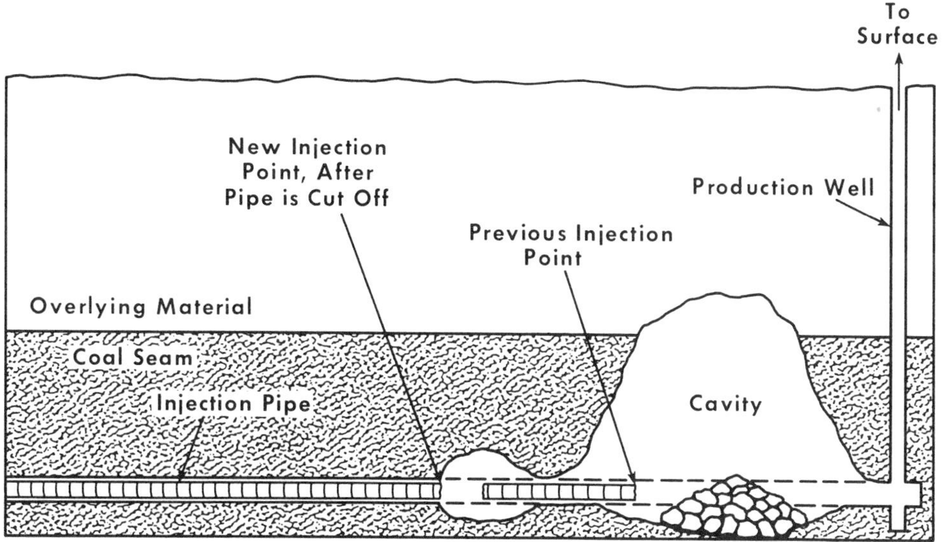

Figure 34 Schematic representation of the controlled retracting injection point (CRIP) system.

the gasification efficiency declines to some economically determined value, "CRIP maneuver" is performed. A burner in the liner is retracted to a new location where the liner is burned off. The coal is ignited at this point, initiating the development of a new gasification cavity.

Operations of both UCG modules were very successful; however, the CRIP module performed better. The CRIP module was operated for approximately 100 days, consumed 9000 tons of coal, and produced an average gas heating value of 295 btu/ft^3. Three CRIP maneuvers were accomplished which in each case improved the heating value of the product gas. The ELW module was operated for approximately 60 days, consumed 4000 tons, and produced an average gas heating value of 275 btu/ft^3.

One of the most important aspects of the RM1 UCG test is that it appears to have been conducted with minimal environmental impact. Previous UCG tests have raised concerns about the environmental acceptability of the UCG process, particularly with respect to the groundwater. Based on evaluation of past UCG tests and laboratory research, the Western Research Institute with the support of the Department of Energy and the Gas Research Institute developed a set of environmental control methods to minimize or eliminate groundwater contamination associated with UCG. These methods were demonstrated at the RM1 test site and preliminary results have shown that the methods were successful in reducing the generation, deposition, and transport of contaminants (Covell, 1988).

IX. FUTURE DEVELOPMENT

The potential for any product is determined by economics. UCG must be shown to be profitable before any commercial development will be attempted. With the current domestic energy oversupply situation, the risks of an unproved technology in UCG may inhibit commercialization. However, a return to higher natural gas and oil prices will definitely make UCG more attractive.

Many potential uses for UCG product gases are possible. There is very little difference between product gas composition from surface gasifiers and that from UCG reactors. Therefore, in many applications, gas produced by UCG can be substituted for gas produced by surface gasifiers. These uses include methanol and synthetic gasoline production, feed for combined cycle power plants, and ammonia production.

23

Liquefaction

I	Technical Objectives	735
II	Chemical Objectives	736
III	Liquefaction Processes	736
IV	Indirect Liquefaction	746
V	Environmental Aspects of Coal Utilization	758

I. TECHNICAL OBJECTIVES

The successful development and implementation of coal liquefaction technology is desirable for three reasons (Penner and Icerman, 1984):

1. To produce refinery-ready feedstocks as a substitute for petroleum crude oil
2. To provide heavy boiler fuels as a substitute for petroleum resid fuels, and
3. To provide a more acceptable solid fuel as a substitute for coal

II. CHEMICAL OBJECTIVES

Like the chemistry of coal itself, the chemistry of coal liquefaction reactions is very complex. Indeed, the inability to definitively resolve the structure of coal aggravates the difficulty of the task of efficiently harnessing its reactivity.

We do understand enough about both the structure and the reactivity of coal to have identified four basic chemical objectives for coal liquefaction reactions (Speight, 1983):

1. To isolate or at least separate segments of the macromolecular structure of coal by eliminating van der Waals and hydrogen bond effects
2. To cleave strategic bridging unit bonds to reduce the molecular weight of coal fragments
3. To remove noxious or otherwise troublesome mineral matter, oxygen, sulfur, and nitrogen contaminants, and
4. To increase the H/C ratio from about 0.8 (in a coal with 6% H, daf) to values more closely approximating crude oils (1.5–1.8; Chap. 3, Sec. I).

III. LIQUEFACTION PROCESSES

Historically, three basic methodologies for the conversion of coal to liquid (primarily) fuels have been explored:

1. Pyrolysis (carbonization) which produces coke (or char) as the principal product and 10–15% by weight tars and creosote-type oils
2. Direct catalytic hydrogenation of coal (often at extremely high pressures) to produce high yields of liquids and little char
3. Hydrogenation coupled with solvent extraction to produce a wide variety of liquids, and
4. Indirect catalytic synthesis of liquids from gasified coal

Pyrolysis is generally carried out at 450–750°C (840–1380°F). As a means of producing coal liquids, this method is of minimal value because most of the coal is converted to char rather than to gases or liquids. Although the char may be used as a fuel, it requires treatment for removal of mineral contaminants, particularly sulfur. The liquids derived from pyrolysis reactions must also be hydrogenated to be useful fuels. The pyrolyses used for coal conversion have been acknowledged elsewhere in a somewhat different sense (mild gasification; See Chap. 23, Sec. VII) but are also summarized here for convenient reference in the current context (Table 1).

The Lurgi-Ruhrgas and the Occidental processes use very rapid heating to achieve more extensive fragmentation of the coal (flash pyrolysis). The high gas yield from the Lurgi-Ruhrgas process is the result of secondary reactions which occur in the reactor (Speight, 1983).

Direct catalytic hydrogenation of coal has been the subject of study since 1912 and has been used to produce fuels on a large scale since 1916 (Donath, 1963) (Figure 1). Although particular processes may alter the sequence somewhat, most follow a similar series of steps: the coal is dried, pulverized, and slurried in a carrier solvent. The slurry is heated and contacted with a catalyst and hydrogen. The heating often takes place very rapidly, and the coal is reacted for very short periods of time. The effluent from the heater is quenched by cooling the stream; hydrogen and other gases are separated and low-boiling liquids are condensed. Additional clean-up steps remove mineral matter, sulfur components, and unreacted solids. As the process was originally designed and carried out by Bergius, coal was slurried in a heavy oil and was reacted using an iron oxide catalyst, with hydrogen gas at pressures as high as 10,000 psi (68.8 MPa). The liquids yield was high, but the operating cost was prohibitive.

Since its inception, a variety of modifications have been made to reduce the severe process demands of the Bergius system. Better catalysts and altered reactor design have been introduced and have made significant advances in direct hydrogenation technology (Table 2).

Using zinc chloride catalysis and a residence time of 10–12 sec, the University of Utah process converts about 60% of the coal feed into liquid products. Although the catalyst is initially impregnated into the coal, recovery of catalyst material is as high as 99% (Speight, 1983).

The Schroeder process uses a molybdenum catalyst and produces a wide range of light and heavy liquids, gases, and char (Speight, 1983). Like the Utah process, this method also entrains the slurry in a stream of hydrogen and is also characterized by the use of a very short residence time (<1 min) in the reactor.

The Dow process operates using a water-soluble salt catalyst which is mixed with the coal slurry (Moll and Quarderer, 1979; Nowacki, 1979). The slurry and hydrogen are mixed in a preheater before being pumped to an entrained flow reactor. After reacting, gases are removed and liquids are separated into light and heavy fractions. The catalyst is not recovered; it remains in the light-oil fraction which is then recycled to the slurry preparation stage.

Table 1 General Summary of Several Pyrolysis and Hydropyrolysis Processes

Process	Developer	Reactor type	Reaction temperature		Reaction pressure (psi)	Coal residence time	Yield (wt %)		
			°C	°F			Char	Oil	Gas
Lurgi-Ruhrgas	Lurgi-Ruhrgas	Mechanical mixer	460–600	840–1110	15	20 sec	55–45	15–25	30
COED	FMC Corp.	Multiple fluidized bed	290–815	550–1500	20–25	1–4 hr	60.7	20.1	15.1
Occidental coal pyrolysis	Occidental	Entrained flow	580	1075	15	2 sec	56.7	35.0	6.6
Toscoal	Tosco	Kiln-type retort vessel	425–540	795–1005	15	5 min	80–90	5–10	5–10
Clean coke	U.S. Steel Corp.	Fluidized bed	650–750	1200–1380	100–150	50 min	66.4	13.9	14.6
Union Carbide Corp.	Union Carbide Corp.	Fluidized	565	1050	1000	5–11 min	38.4	29.0	16.2

Source: J. Speight, 1983. Reprinted with permission of the publisher.

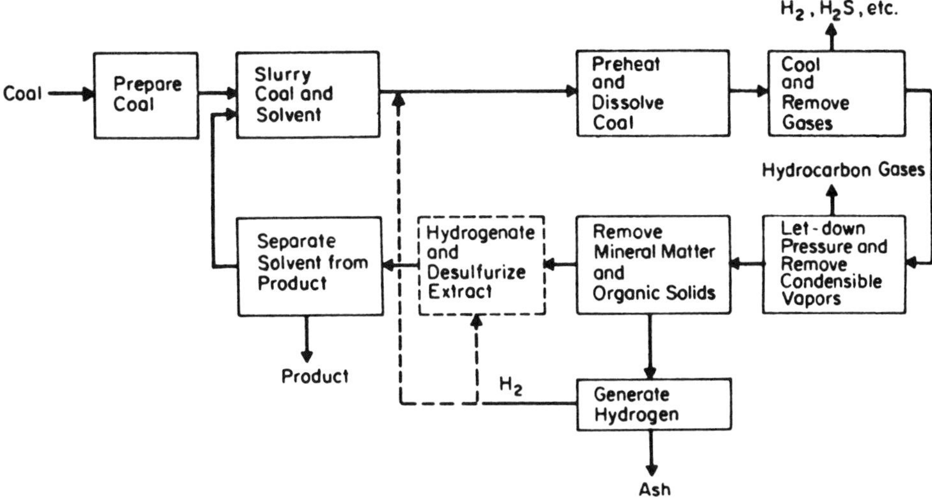

Figure 1 Flow chart for coal dissolution and liquefaction (Electric Power Research Institute, 1974).

There are several other technologies for catalytic liquefaction of coal which differ from the methods described above primarily in that they do not mix the catalyst into the coal-hydrogen slurry. Rather, the slurry-hydrogen mixture is preheated before being transferred to a bed reactor which contains the catalyst. One other process, developed by Conoco, is in some ways an intermediate form of these two general methods. The unique feature of this process is that it uses large amounts of catalyst (1:1 by weight with coal) which are kept in a molten state. The catalyst is zinc-chloride and gasoline-type fuels are produced by substantial hydrocracking of the feed coal slurry (Speight, 1983). The slurry and hydrogen are separately pumped into the reactor containing the molten zinc. Gases and liquids in the product stream are separated and liquids are fractionated into light, middle, and heavy (recycle). Solids from the reactor are oxidized in a fluidized bed combustor where zinc chloride is regenerated by separating it from the heteroatoms (sulfur, oxygen, and nitrogen) with which it reacts during the course of the process (Zielke et al, 1969).

The Synthoil, CCL, H-coal, and multistage processes (Table 2) are examples of bed-type technologies. The Synthoil and CCL processes use a fixed-bed reactor containing the catalyst. The force with which the slurry moves through the reactor is the only means for achieving contact between the catalyst and the slurry-hydrogen

Table 2 Summary of Several Catalytic Liquefaction Processes

Process	Developer	Reactor	Catalyst	Temperature		Pressure (psi)
				°C	°F	
Bergius	Bergius	Plug flow	Co-Mo/Al_2O_3	480	895	3000–10,000
University of Utah	University of Utah	Entrained	$ZnCl_2$, $SnCl_2$	500–550	930–1020	1500–2500
Schroeder	Schroeder	Entrained	$(NH_4)_2MoO_4$	500	930	2000
DOW	Dow Chemical	Entrained	Aqueous salts	450–460	840–860	2000
H-coal	Hydrocarbon Research Inc.	Ebullated bed	Co-Mo/Al_2O_3	450	840	2250–3000
Synthoil	ERDA	Fixed bed	Co-Mo/Al_2O_3	450	840	2000–4000
CCL	Gulf	Fixed bed	Co-Mo/Al_2O_3	400	750	2000
Multistage	Lummus	Expanded bed	Co-Mo/Al_2O_3	400–430	750–805	1000
Zinc chloride	Conoco	Liquid phase	$ZnCl_2$	360–440	680–825	1500–3500

Source: Modified from J. Speight, 1983.

mix. After exiting the reactor, the gas-liquid separation and product clean-up are carried out, followed by upgrading of the recovered oils.

The H-coal process is able to use low-rank as well as bituminous coals. The reactor is an ebullating bed which improves catalyst-slurry contact over that of the fixed-bed mode. It also permits continuous removal of product and addition of catalyst. Heavy fuel oil or a lighter, synthetic crude can be produced from the H-coal operation.

One major drawback to the use of bed reactors is the high potential for mineral deactivation of the catalyst. This can (and does) occur both when the mineral matter physically plugs the pores within the catalyst and when the various metals chemically deactivate the catalyst. The multistage process was developed in an attempt to minimize this problem. It uses an expanded-bed reactor which reduces plugging and maximizes catalyst-slurry contact.

Direct catalytic hydroliquefaction methods necessarily require that intimate contact of the coal, hydrogen, and catalyst occur. Even with the very short residence times which are typically used, catalyst deactivation and the thermodynamic and physical barriers involved in solid-gas reactions often result in inefficient reactions and low yields. These difficulties can be overcome in part by a modification of the process such that the slurry solvent itself acts as the source of hydrogen for the coal. These processes are a type of solvent extraction (Table 3). They generally require residence times of 1/2 to 2 hr and are carried out at high temperatures and pressures (500°C, 930°F; up to 4000 psi, 27.5 MPa). The product is most often a liquid fuel with a low heteroatom content.

The solvent-refined coal processes (SRC-I and SRC-II) use a process-derived solvent and hydrogen gas. In SRC-I, gases and light hydrocarbons are obtained from the process stream leaving a slurry of undissolved solids in the solvent. The solvent is recycled and the solid can be used as a fuel (Table 4).

In SRC-II, part of the solvent slurry is recycled and the remainder is fractionated to produce a wide range of solid and liquid products (Table 5).

The Exxon donor solvent (EDS) process also uses a process-derived donor solvent, but one which is catalytically hydroprocessed. The coal liquefaction reaction is carried out without any catalyst in an entrained flow reactor (Nowacki, 1979). Distillate vacuum bottoms are coked to produce additional liquids (Table 6). While SRC technology is advantaged by being able to accommodate high-sulfur, high-ash coals, the EDS system can use lignite, subbituminous and bituminous coals, and relies to a limited extent on the mineral matter in coal to act as catalyst for the process.

A technology designed to use lignite in the SRC process has been developed (Speight, 1983). Called solvent-refined lignite (SRL),

Table 3 General Summary of Several Solvent Extraction Processes

Process	Developer	Reactor	Temperature		Pressure (psi)	Residence time (hr)
			°C	°F		
Consol synthetic fuel (CSF)	Conoco	Stirred-tank	400	750	150–450	<1
Solvent-refined coal (SRC)	Pittsburgh and Midway Mining Co.	Plug flow	∼450	840	1000–1500	<1
Solvent-refined lignite (SRL)	University of North Dakota	Plug flow	370–480	700–895	1000–3000	∼1.4
Costeam	ERDA	Stirred-tank	375–450	705–840	2000–4000	1–2
Exxon donor solvent (EDS)	Exxon Research and Engineering Co.	Plug flow	425–480	795–895	1500–2000	0.25–2.0

Source: J. Speight, 1983. Reprinted with permission of the publisher.

Table 4 Product Inspections for the Solvent-Refined Coal Process

Component	Analysis (wt %)	
	Raw coal	Product[a]
Carbon	70.7	88.2
Hydrogen	4.7	5.2
Nitrogen	1.1	1.5
Sulfur	3.4	1.2
Oxygen	10.3	3.4
Ash	7.1	0.5
Moisture	2.7	0.0
	100.0	100.0
Volatile matter	38.7	36.5
Fixed carbon	51.5	63.0
Ash	7.1	0.5
Moisture	2.7	0.0
	100.0	100.0
	12,821 btu/lb	15,768 btu/lb

[a]Heavy semisolid bituminous material.
Source: J. Speight, 1983. Reprined with permission of the publisher.

Table 5 Product Inspections for the SRC II Process

	Solid fuel	Distillate fuel
Gravity (° API)	-18.3	5.0
Boiling range: °C	425+	400–800
°F	800+	
Fusion point: °C	175	
°F	350	
Flash point: °C	–	168
°F	–	
Viscosity: SUS at 100°F	–	50
Sulfur	0.8	0.3
Nitrogen	2.0	0.9
Heating value (btu/lb)	16,000	17,300

Source: G. L. Baughman, 1981.

Table 6 Yields for Liquefaction of Illinois No. 6 Bituminous Coal in Exxon Donor Solvent Process

	Liquefaction	Liquefaction plus coking
H_2O, CO_2, CO	10	10
H_2S, NH_3	4	4
C_1–C_3	6	9
C_4, C_5	3	4
Naphtha	15	16
Fuel oil	17	25
Liquefaction btms.	48	–
Coke and Ash	–	35
Total	100	100
Liquid yield wt % on dry coal	35	45
H_2 consumption scf/bbl liquid	5600	4100

Source: J. Speight, 1983. Reprinted with permission of the publisher.

the process can use hydrogen or a process-derived gas to yield a mixture of gases, liquids, and the primary, solid product, which must be deashed before use (Table 7). The attractive feature of SRL is the availability of substantial reserves of lignite. Although lignite has been considered less valuable and less useful because of its high moisture content, in the SRL process this is used to advantage for an in situ water-gas shift reaction which provides hydrogen for the system.

Solvent extraction of coal, even at the elevated temperatures and pressures used for liquefaction reactions, is not very efficient. The major drawback is the presence of mineral matter in coal, which is undesirable from virtually every perspective. However, the necessity of using high temperatures to thermally cleave covalent bonds in coal prior to hydrogenation and the necessity of operating at elevated pressure to maximize the rate and extent of reactions led to the development of technologies employing supercritical fluid extraction to bring about coal liquefaction.

Table 7 Yields from Solvent-Refined Lignite (SRL)

Liquid hourly space velocity	0.90	1.41
Gas hourly space velocity	164	321
Solvent/coal ratio	2.30	1.91
Coal charged (lb/hr/ft^3 reactor)	18.6	32.9
Gas charged (scf/ton coal)	17,700	19,500
H_2 equivalent consumed (wt % daf coal)	2.35	1.50
Yields (wt % daf coal)		
Gas	10.4	15.9
Total product yield	69.4	66.8
Light oil product	–	(9.5)
Heavy product	(69.4)	(57.3)
H_2O and ash	-6.5	-4.4
Unconverted daf coal	26.7	21.7
Solvent recycle (%)	85.9	89.2
Conditions		
Temperature: °C	370	400
°F		
Preheater outlet	395	410
Reactor exit temperature	310	314

Source: J. Speight, 1983. Reprinted with permission of the publisher.

A supercritical fluid, SCF (or supercritical gas, SCG) is one which, above the critical temperature and pressure possesses the density and the solvent powers of the liquid but the mobility and the compressibility of the gas. An added advantage of fluids under these conditions is that even small changes in either temperature or pressure can bring about large changes in the fluid's properties (Hoyer, 1985).

For coal liquefaction, supercritical fluids have enhanced ability to penetrate the pore structure of coal and can produce an essentially mineral-free product stream suitable for recovery of solid and liquid feedstocks.

In addition to using neat solvents, a wide variety of solvent mixtures have been investigated for use as SCF solvents. Some examples are water-CO, water-H_2, toluene-tetralin, toluene-tetrahydroquinoline, alcohol-NaOH, and even coal tar itself (Towne et al., 1985; Chen et al., 1985). Operating conditions vary widely, depending on the reaction fluid in use.

Supercritical fluid extracts of coal are often compared to products produced by flash pyrolysis or to liquids recovered from more conventional hydrogenation reactions. For an Australian bituminous coal, for instance (Kershaw, 1985), supercritical extraction was carried out at 450°C (840°F) for 1 hr using a toluene/coal ratio of 12:1. Flash pyrolysis was carried out at 685°C (1400°F) with the same coal, and hydrogenation, using stannous chloride catalyst, was carried out in tetralin (2:1 ratio) at 400 and 450°C (830 and 930°F) for 1 hr at an initial pressure of 1500 psi (70.3 MPa). The H/C ratio (daf) of the feed coal was 0.79 and the ash, sulfur, nitrogen, and oxygen contents were 6.8% (dry), 0.6% (daf), 2.0% (daf), and 8.3% (daf), respectively (Table 8). The higher H/C ratio, lower aromaticity, and lower heteroatom content all make the SCG extract more amenable to subsequent upgrading, especially as aviation or diesel fuel—based on the amount of unsubstituted alkyl chain compounds present. However, the SCG data were not as favorable when compared to those obtained from the hydrogenation reactions (Table 9): about twice as much oil was produced during hydrogenation, and the SCG oil had a higher heteroatom content. At the same time, the SCG product did have a higher H/C ratio, and it did contain more carbon in long alkyl chains. The aromaticity and the oil/asphaltene ratio for the SCG was intermediate between the 400°C (750°F) and 450°C (840°F) hydrogenation products.

It is evident, then, that obtaining liquids from coal with supercritical fluids has substantial merit. It remains to be seen whether overall improved product yields can be achieved and whether the necessary use of high pressure is offset by gains associated with the relatively facile separation of unwanted solid residue, quantitative product recovery, and recovery of a product with comparatively lower process demands for upgrading.

IV. INDIRECT LIQUEFACTION

Indirect liquefaction of coal follows the production of synthesis gas ($CO + H_2$) from coal after coal is reacted with oxygen and steam (595–765°C; 1105–1410°F; and 350–450 psi). Liquid fuels can then be produced when the synthesis gases are reacted over a catalyst (Table 10). Ultimately, it is the catalyst which governs the nature of the final product and its quality.

The BASF company in Germany pioneered catalyst development for liquefaction of synthesis gas; an early mixed catalyst which proved successful consisted of Co, Mo, and ThO_2 supported on diatomaceous earth (kieselguhr). Subsequently, many other metals were found to be active and suitable catalysts: Re, Os, Rh, Ru, Pd, and Pt. All are used in reactions taking place at temperatures ranging from about 200°C (390°F) (Ru) to as high as 600°C (1110°F) (Os) (Kirk-Othmer, 1978). With all of these metals, however, methane is the primary

Table 8 Comparison of a Supercritical Gas Extract and a Flash Pyrolysis Tar from Liddell Coal

	SCG extract	Flash pyrolysis tar
Oil		
Yield (wt % liquid product)	39	36
(wt % coal daf)	12.8	9.7
H/C atom, ratio	1.29	1.19
N, S, O (wt %)	4.4	8.8
Mol wt	270	290
f_a	0.55	0.57
Aromatic H[a] (% H)	20	26
% Carbon as unsubstituted alkyl chains (>C_3)	15	13
Number-average chain length of alkyl chains[b]	12	14
% Alkyl chains having a $CH=CH_2$ end group	5	25
Asphaltene		
Yield (wt % liquid product)	41	45
(wt % coal daf)	13.6	12.1
H/C atom, ratio	0.95	0.86
N, S, O (wt %)	9.7	13.2
Mol wt	504	538
f_a	0.71	0.75
Aromatic H[a] (% H)	36	39
Pre-asphaltene		
Yield (wt % liquid product)	20	15
(wt % coal daf)	6.5	4.1
H/C atom, ratio	0.88	0.75
N, S, O (wt %)	16.7	19.8
Mol wt[c]	925	924
f_a	0.66	0.71
Aromatic H[c] (% H)	33	36

[a]Includes phenolic OH

[b]Assuming a majority of alkyl chains are present as alkyl substitutents as opposed to *n*-alkanes.

[c]From silylated preasphaltene and corrected for $Si(CH_3)_3$ content.

Source: J. R. Kershaw, 1984. Reprinted with the permission of the publisher.

Table 9 Comparison of a Supercritical Gas Extract and Coal Hydrocarbon Liquids from Liddell Coal

		SCG extract	Hydrogenation liquid	
			400°C	450°C
Oil (wt % coal daf)		12.8	24	34
Asphaltene (wt % coal daf)		13.6	44	25
Preasphaltene (wt % coal daf)		6.5	N.D.[a]	N.D.[a]
Gas (wt % coal daf)		0	16[b]	38[b]
Oil/asphaltene ratio		0.94	0.55	1.36
H/C atom, ratio	oil	1.29	1.18	1.11
	asphaltene	0.95	0.97	0.88
N, S, O (wt %)	oil	4.4	2.4	3.3
	asphaltene	9.7	6.4	10.3
Mol wt	oil	270	382	282
	asphaltene	504	794	550
Aromatic H[c] (% H)	oil	20	17	26
	asphaltene	36	23	36
f_a	oil	0.55	0.50	0.57
	asphaltene	0.71	0.66	0.75
% Carbon as unsubstituted alkyl chains (>C_3 in oils)		15	13	8

[a]N.D. = not determined.
[b]Includes low-boiling liquid product.
[c]Includes phenolic OH.
Source: J. R. Kershaw, 1984. Reprinted with permission of the publisher.

product, while C_3 and C_4 hydrocarbons account for 20% (with Pt) or less, and as low as 3% of the product mixture. One significant exception is observed with Ru. At temperatures above 464°C (>870°F), $C_3 + C_4$ products are produced in low yield (<3%); however, as the reaction temperature is decreased, $C_3 + C_4$ yield increases and reaches values approaching 60% of the product stream at <185°C (<365°F) (Kirk-Othmer, 1978). Unfortunately, Ru is 8000 times more expensive than Co; 40,000 times more expensive than Ni (Dry and Erasmus, 1987).

Table 10 Product Composition from Fischer–Tropsch Process (SASOL)

	Composition (vol %)	
	Fixed-bed reactor	Entrained bed reactor
Liquefied petroleum gas (C_3–C_4)	5.6	7.7
Gasoline (C_5–C_{11})	33.4	72.3
Middle oils (diesel, furnace, oil, etc.)	16.6	3.4
Waxy oil	10.3	3.0
Medium wax	11.8	—
Hard wax	18.0	—
Alcohols and ketones	4.3	12.6
Organic acids	Traces	1.0

Source: W. W. Bodle and K. C. Vygas, 1974. Reprinted with permission of the publisher.

For liquid fuels, methane production is ideally minimized if not eliminated. The Fischer–Tropsch process for synthesis of liquid fuels was initiated in the 1920s. The Ni catalyst used first led to almost exclusive formation of methane (Kirk-Othmer, 1978):

$$CO + 3H_2 \rightarrow CH_4 + H_2O$$

Later, Fe (as iron oxide) or Co was used, and the product was a mixture of higher hydrocarbons, such as gasoline, diesel oil, kerosene, and waxes (Nowacki, 1979). With iron, operating conditions were 400°C (750°F) and 10–15 MPa (1500–2200 psi) pressure. The cobalt catalyst was used at lower temperatures (180–220°C; 360–430°F) and at greatly reduced pressure, 0.1–1 MPa (15–150 psi). Other early low-cost catalysts were molybdenum (Mo) and tin (Sn).

With a Co catalyst, the Fischer–Tropsch process yields a mixture of saturated, straight-chain compounds and some small olefins (unsaturated compounds). A small quantity (<1%) of aocohols as well as high molecular weight acids are also found. Operating pressure affects product distribution: normal pressure (2.4 MPa, 350 psi) leads to 60% gasoline and 30% gas oil, while slightly higher pressures produce

35% gasoline, 35% gas oil, and correspondingly more (30%) waxes (Kirk-Othmer, 1978).

After World War II little new synthetic fuel development occurred until 1951, when South Africa initiated a visionary program committing that country to energy independence. The South African Coal, Oil and Gas Corporation (SASOL) began construction at a site now called the township of Sasolburg, on the Vaal River. The original plant consisted of nine Lurgi gasifiers which are suitable for South Africa's low-rank, high-ash (25–27%) coal. Originally, Mark One gasifiers were installed but modifications have been made to improve gasifier performance. The first modification improved gas volume yield from 18,500 m^3/hr to 27,500 m^3/hr; the second improvement increased the yield to 34,000 m^3/hr. All these gasifiers were 3.66 m in diameter. The next level of improvement included increasing the volume to 3.8 m diameter. This Mark Four gasifier produces 46,200 m^3/hr. Most recently, the Mark Five unit (4.7 m diameter) has been developed and reportedly produces 95,000 m^3/hr of raw synthesis gas (Dry and Erasmus, 1987).

After gasification, synthesis gas is reacted on a fixed-bed alkaline iron oxide gel catalyst promoted with a potassium salt (K_2O generally) (Dry and Erasmus, 1987) and operated at 2 MPa pressure (Arge process). Catalysis results in the formation (after $C_3 + C_4$ oligomerization) of C_5–C_{18}, high-octane (93 RON; 85 R100) gasoline (22%), C_{13}–C_{18} diesel fuel (15%) heating oil (kerosenes), and waxes (C_{19}^+, 41%) (Kirk Othmer, 1978). Alternatively, synthesis gas is fed to an unpromoted entrained flow (circulating fluidized bed) magnetite (iron oxide) catalyst reactor from which gasoline (39%), heating oil, and low molecular weight hydrocarbons ($C_1 + C_2$, 20%; $C_3 + C_4$, 23%) are recovered (Kellogg Synthol process) (Kirk-Othmer, 1978). The latter are compressed and cooled to yield medium-btu LP fuels (Grainger and Gibson, 1981; Schobert, 1987). A number of chemical byproducts are also produced: sulfur, ammonia, phenol, creosote oil, and pitch (Schobert, 1987). Methane is recovered and sold, although the process can also recycle it.

Initial oligomerization of $C_3 + C_4$ gases produces gasoline, jet fuel, and a light diesel fuel, all of which are highly branched. To improve the quality of the diesel fuel, alternative catalysts are used. Amorphous silica-alumina and the crystalline zeolite ZSM-5 silica-alumina catalyst have been tested. The former yields a branched but high molecular weight diesel fuel. The zeolite catalyst produces a much less branched product, but at higher cost (Dry and Erasmus, 1987). The zeolite is also used to improve C_5–C_6 fractions and to dewax diesel fractions (Dry and Erasmus, 1987).

At SASOL 1, 10,000 tons/day (tpd) of coal is utilized (Johnson, 1981). The Arge units (five) produce hydrocarbons at the rate of about 550 bbl/day, and the Synthol units (three) produce 2000 bbl/day (Nowacki, 1979). In addition, ammonia is recovered and diverted to fertilizer production, and facilities exist to recover butadiene, styrene, and ethylene, major polymer feedstocks (Dry and Erasmus, 1987). The SASOL 1 operation was eventually enlarged to 13 Lurgi gasifiers. Based on the success of this investment, SASOL 2 was undertaken and began operation in 1980 at Secunda, South Africa.

SASOL 2 produces gasoline using only the Kellogg Synthol reactor. Compared to SASOL 1, SASOL 2 has about 3 times more gasifiers (36) (Considine and Considine, 1983), and each gasifier has 50% greater capacity (Grainger and Gibson, 1981). Designed to use 40,000 tpd of coal, SASOL 2 produces 50,000 bbl/day of fuel products (Johnson, 1981). All methane is recycled with steam to yield a reformed synthesis gas exiting the reformer at 800°C (1470°F) (Dry and Erasmus, 1987):

$$1.0CH_4 + 0.24H_2O(g) + 0.62O_2 \rightarrow 0.52CO + 0.48CO_2 + 2.24H_2$$

Methane reforming reduces the thermal efficiency (heat content of fuels/heat content of coal feed) to 40–44%; without reforming the value approaches 60% (Guccione, 1981). Because oxygen is used in the initial gasification process and in methane reforming, SASOL 2 has a very large oxygen demand. While much specific information about SASOL is not made public, it appears that the six air separation units constitute the largest single oxygen-producing operation in the world (Johnson, 1981).

SASOL 3 is a duplicate of SASOL 2 in terms of unit operations (40,000 tpd coal utilization) (Johnson, 1981). In full operation since 1985 at the Secunda location, this third plant is also power-independent. As at SASOL 2, only the Synthol process is used to produce primarily gasoline; however, creosote hydrotreating is also carried out to improve the quality and quantity of diesel fuel (Mako and Samuel, 1979; Schobert, 1987). SASOL 1 produced about 25% of South Africa's motor fuel; with SASOL 3 that value is close to 85% (Johnson, 1981).

Fischer–Tropsch reactors such as those employed at SASOL are very versatile. Because catalyst, temperature, pressure, residence time, gas purity, H_2/CO ratio, and reactor design can all be varied, the product mixtures can be customized; gasoline to diesel can be tailored from about 4:1 to about 1:1, alternatively, gasoline production can be reduced to about 20 or 25% while SNG is maximized to as much as 50% of the output (Johnson, 1981). Taking methane production from the Arge process as 1, overall selectivity of the SASOL 1 processes is (Dry and Erasmus, 1987):

	Arge	Synthol
CH_4	1	2.25
C_2–C_6 saturates	2.5	2
C_2–C_6 olefins	2	9.5
C_7+ to 160°C	2	4.25
160–350°C	4.75	3.5
>350°C	12	2
Small oxygenates	0.75	1.5

Chain length increases as temperature decreases, as H_2/CO ratio decreases, and as potassium content (promoter) increases (Dry and Erasmus, 1987). Further catalyst development will focus on total elimination of C_1–C_3 compounds (Dry and Erasmus, 1987).

As it is currently operated in the SASOL processes, the Synthol reactor receives raw and recycled (from methane) gas at 160°C (320°F) and 2.1 MPa (306 psi) pressure at the bottom of a loop unit where it is mixed with heated (335°C; 635°F) catalyst flowing downward through valves at the bottom of a standpipe. Thermal equilibrium is reached at about 315°C (600°F) as the gas-catalyst mixture moves through the bottom segment of the loop. Because the conversion reactions are all exothermic, the gas stream temperature increases. Excess heat is recovered in the upward segment of the reactor loop (the riser). As the gas flows through the top curve of the reactor, the catalyst is carried to cyclones and a settling hopper where it passes into the standpipe and begins recycle into the loop. The product gas stream is removed at the point where catalyst recovery occurs (Nowacki, 1979).

South Africa is the only country currently using Fischer–Tropsch indirect liquefaction on a commercial scale. The Fluor Co., the U.S. company that built SASOL 2 and 3, estimates that a similar facility, suitable for coal and oil shale use, could be built in the United States at a cost of about \$4 billion in 1979 dollars, increasing at a rate estimated at \$0.50 per year (Johnson, 1981, p. 45). Water supplies are a critical consideration in coal-rich but water-poor regions of western U.S. Sites for such a facility are Wyoming, Colorado, and New Mexico in the west, the Ohio River Valley in the east (Johnson, 1981, p. 45). As an alternative to Lurgi gasification, it is possible, but unproven, that the Texaco gasifier would be capable of meeting the operational demands (defined as 7300 hr operation per year and >50,000 m^3/hr raw gas at >3.0 MPa; 437 psi pressure) (Dry and Erasmus, 1987, p. 10). In addition to mere durability, ash fusion temperature and moisture content of the coal feed are significant. Lurgi gasifiers operate at low temperatures, well below the ash fusion temperature of South African coal (∿1400°C, ca 2550°F), and can accommodate the 10% moisture typical of South African coal. The Texaco unit is ash-slagging, and fluxing agents (limestone or blending) may be necessary to control ash fusion processes if such a gasifier is used (Dry and Erasmus, 1987, p. 10). The Lurgi configuration uses less oxygen than other gasifiers, but methane reforming increases the oxygen demand dramatically. This would not be the case with Texaco gasifiers which do not yield much CH_4 (Dry and Erasmus, 1987). Other gasifier considerations are steam demand, CO_2 production, H_2/CO ratio, and byproduct formation and recovery (e.g., H_2S, organic sulfur, ammonia, cyanides) (Dry and Erasmus, 1987).

The Texaco synthesis gas generation process uses a preheated (500°C; 930°F) coal-water slurry which is reacted with oxygen or air at up to (8.3 MPa; 1215 psi) and at temperatures variously reported as up to 1370°C (2500°F) and as high as 2500°C (4530°F) (Grainger and Gibson, 1981; Schobert, 1987). It is an entrained flow reactor (Meyers, 1984). Gas and slag are quenched at the bottom in a water spray. No methane is produced and CO_2 is 15% or less of the product gas stream. Pilot plant units operate in the United States and in Germany; two commercial units have been operated, one by Olin Matheson and one by Tennessee-Eastman (Schobert, 1987).

The Shell–Koppers system developed by Shell Oil is a high-pressure, entrained flow gasifier (3 MPa; 450 psi) that recycles product gas heat to generate steam. Such high pressure (the Koppers–Totzek gasifier is an atmospheric unit) avoids the need to compress the product gas and the senible heat is not lost as it is in the Kopper–Totzek system (LeBlanc et al., 1981; Schobert, 1987). Finely pulverized coal of any rank can be used because the operating temperature (to as high as 1500°C or 2730°F in the interior) is above coal ash fusion temperatures. For the Shell–Koppers process, the product gas is about 60% CO, 30% H_2, and 10% CO_2 (Schobert, 1987).

In the Shell process CO + H_2 formation exceeds 90% and has been reported to be as high as 98%. Pulverized coal is dried to ∿2% water. In a short residence time reaction (about 4 sec), no tars or phenols are formed and 99% carbon conversion is achieved. The Shell process provides a high-quality synthesis gas suitable for methanol formation, Fischer–Tropsch synthesis, or SNG production. Coal with ash up to 40 wt % can be used.

Besides the Texaco and the Shell–Koppers gasifiers, the Koppers–Totzech and the Winkler are also high-temperature gasifiers. Typical operating temperatures are 1315–1480°C (2400–2700°F). In addition to the Lurgi system, the BGC-Lurgi is the only other low-temperature gasifier. All the high-temperature gasifiers are slagging (remove ash in the molten state), except the Winkler (LeBlanc et al., 1981).

While Fischer–Tropsch technology has enabled South Africa to produce an array of motor fuel and salable byproducts, two other approaches to indirect coal liquefaction are also at advanced levels of operational development: methanol synthesis and methanol-to-gasoline. The reaction for methanol formation from coal (LeBlanc et al., 1981):

$$2.143CH_{0.8}O_{0.1} + 0.965O_2 + 1.143H_2O \rightarrow CH_3OH + 1.143CO_2$$

illustrates the inherent complexities of such a process. Not all the carbon in coal is converted to product, largely because coal contains insufficient hydrogen. What is not shown in the reaction as written is the release and/or formation of H_2S and other sulfur-containing compounds which poison catalysts and the environment (LeBlanc et al., 1981).

In 1959 Imperial Chemical Industries (ICI) developed a Cu-ZnO catalyst which functions optionally at 200–300°C (390–570°F) and 5–10 MPa (750–1500 psi) to produce methanol from coal synthesis gas (LeBlanc et al., 1981).

$$CO + 2H_2 \rightarrow CH_3OH$$

$$CO_2 + H_2 \rightarrow CO + H_2O$$

Early technology utilized a ZnO/chromic catalyst which required both hotter temperatures (340–455°C; 650–850°F) and higher pressures 24–34 MPa (3500–5000 psi). Also, zinc oxide catalysts leads to more undesirable byproducts than the ICI Cu-based catalyst (LeBlanc et al., 1981).

In general, methanol synthesis follows a sequence in which synthesis gas (plus any recycled gas) is fed into the catalytic converter. Hot product gas (the conversion reactions are exothermic) is necessarily cooled to recover process heat and to condense crude methanol. Uncondensed vapor is recycled, while the condensate is distilled to separate pure methanol from water. A methanol conversion unit fits into most any conventional coal gasification process (LeBlanc et al., 1981).

Methanol can be used as a general motor fuel but it is not widely acceptable because it boils at a low temperature and produces only about one-half as much energy as gasoline; it is toxic (as are some components of gasoline), absorbs moisture, and is corrosive. Additionally, the market value of gasoline from crude oil is still low enough that methanol fuel is not economically compelling in spite of the wisdom of having a long-range policy in place. Such vision would consider methanol seriously as an alternative motor fuel.

Table 11 Product Yield From the Methanol-to-Gasoline Process

Bed temperature, °C	410
°F	775
Pressure (psi)	25
Yields (wt %)	
Methanol + ether	0.2
Hydrocarbons	43.5
Water	56.0
CO, CO_2	0.1
Coke	0.2
	100.0
Hydrocarbon products (wt %)	
Light gas	5.6
Propane	5.9
Propylene	5.0
i-Butane	14.5
n-Butane	1.7
Butenes	7.3
Gasoline (C_5+)	60.0
	100.0

Source: G. L. Baughman, 1981.

At the same time, a far more attractive process is one which converts methanol to gasoline. By far, the Mobil MTG process is the most highly developed and the most successful (Table 11). Having developed a successful moving-bed catalytic cracker in the late 1940s, Mobil Oil was a leader in the development of the crystalline zeolite cracking catalysts (Meisel, 1988). Since their introduction in 1962, more than 40 zeolite catalysts have been invented and processes for petroleum refining, chemicals formulation, and fuel production have followed in steady succession. The most successful zeolite-type catalyst is called ZSM-5, a tetrahedral structure consisting of 96 silica and alumina moieties arranged in 10-member rings (Meisel, 1988). The acidic aluminum provides the catalytic properties, while the shape selectivity of the structure provides the major advantage of this particular catalyst over related zeolites (Meisel, 1988). Diffusion of molecules into the catalyst pore system is limited to straight-chain, monomethylated or small aromatic compounds (reactant shape selectivity). Likewise, the size and shape of product molecules are also limited (transition state selectivity). Finally, only molecules that are sufficiently small diffuse *out* of the catalyst at a useful rate (product shape selectivity) (Meisel, 1988). In actuality, transition state selectivity has not been proved, but it is a useful model for explaining why certain molecules are not observed to form (Meisel, 1988).

It has been shown that at 370°C (700°F), 20 psi, and 0.02 moles/cm^3/hr flow rate, methanol is converted to a mixture about 18 wt % $C_1 + C_3$, 17 wt % iso-C_4 and slightly less than 45% mononuclear, mostly methyl-substituted aromatics (Chang and Silvestri, 1987).

The MTG technology, even with the most favorable ZSM-5 catalyst, is not free of problems. A major difficulty is temperature control. The conversion reactions are highly exothermic (1740 kJ/kg; 400 cal/g methanol) and has the potential to cause a 600°C (1100°F) rise in temperature. A second dilemma that must be faced is a dual mode of ZSM-5 catalyst deactivation: irreversibly by steaming and reversibly by coke deposition. Third, unless the reaction is complete (no unreacted methanol), additional cost is incurred when it is recovered by distillation from water (Squires et al., 1985).

Mobil investigated both fixed and fluidized bed reactors with these factors (among others) in view. Reaction exotherms and catalyst deactivation were best addressed with a fluidized bed configuration because such a system can accommodate condensed methanol, which consumes excess heat for vaporization. The bed can be circulated to a cooler (a circulating fluidized bed, CFB); or cooling coils can more easily be inserted into the fluidized environment. Similarly, continuous (cycling) catalyst regeneration is much more feasible in the fluid system. Furthermore, a fluidized reactor is cheaper, the yield of product is higher (92% versus 85% for a fixed bed) as is the research octane number (96 versus 93) (Squires et al., 1985).

As implemented by Mobil, the MGT process converts methanol to an equilibrium mixture with dimethyl ether (DME) (Maiden, 1988).

$$n/2(2CH_3OH \longleftrightarrow CH_3OCH_3 + H_2O) \xrightarrow{-H_2O} C_nH_{2n} \longrightarrow n(CH_2)$$

where the final product is an aromatic paraffin mixture (Maiden, 1988). It is the methanol-DME-water mixture which is converted on ZSM-5 to gasoline (a mixture of paraffins, olefins, and aromatics) (Bem, 1988). In New Zealand, fixed-bed reactors were chosen in spite of the advantages associated with the fluidized bed reactor primarily because of the perceived urgency which favored the fixed-bed configuration (Squires et al., 1985). Production in New Zealand began in 1985 and produces a blended gasoline that has a research octane number (RO) of 93.7 and consists of about 38% evaporating at 70°C (160°F), 56%

evaporation at 100°C (210°F), 98% evaporating at 190°C (375°F) (Maiden, 1988). No hydrocarbon heavier than gasoline is produced in this MTG unit (Meisel, 1988). Projections forecast that the MTG operation will cut crude oil imports to New Zealand by more than 50% of the 1973–1974 volume and will provide about 33% of that nation's transportation fuel (Meisel, 1988). With regard to the highly exothermic reactions, the New Zealand operation removes about 20% of the excess heat in the first stage reactor where methanol-DME-H_2O equilibrium is established. The catalytic formation occurs in a second stage reactor. Further cooling is carried out prior to final product isolation. This heat is recycled to the methanol preheater unit (Chang and Silvestri, 1987). Catalyst deactivation, called band aging, is a serious drawback to fixed-bed reactors. To compensate for the resultant need to regenerate the catalyst, multiple reactors are used at different levels of deactivation, one being regenerated at any given time (Chang and Silvestri, 1987). One method for catalyst regeneration has been tested at Paulsboro, New Jersey. In this test case, nitrogen as 200 scf/lb catalyst-hours, initially at 340°C (650°F) and 2MPa (300 psi), was passed over the catalyst. Air sufficient to yield a 0.3 vol % oxygen is used to create a combustion environment. Combustion results in a temperature increase to about 480°C (900°F). After combustion ceases (as detected by cessation of oxygen consumption), the oxygen level is increased and the reactor cooled (Nowacki, 1979).

The MTG facility in Weiseling, Federal Republic of Germany (West Germany) has successfully demonstrated the fluidized bed technology on the pilot plant scale (Meisel, 1988). At 400°C (750°F) and 2.2 MPa methanol pressure, olefin production maximizes at about 40% C_2–C_5 compounds, then decreases with continued residence time. Olefin formation can be optimized by catalyst modification and by operating at higher temperature (500°C; 930°F), and has been shown to produce primarily propylene and butylene, plus high-octane gasoline (Meisel, 1988). Band aging is avoided and regeneration is obviated by the use of fluidized bed unit. However, catalyst deactivation still occurs, and evidence indicates that conversion levels are not as high as desired; the system may also require a water-methanol separation and recovery unit (Chang and Silvestri, 1987). Based on propane/propene ratios, catalyst breakthrough is reached at a ratio of 0.25 but is not observed until more than 2000 hr of operation (Keim et al., 1984). Total hydrocarbon yield for both reactor types is the same, but the fluidized bed produces more C_1–C_4 compounds (40% versus 20%) and correspondingly less C_{5+} gasoline (60% versus 80%) (Chang and Silvestri, 1987). Catalyst regeneration is carried out at a maximum temperature of 480–540°C (900–1000°F) at 0.5 vol % oxygen in nitrogen. After coke is combusted, the catalyst is recalcined at 590°C (1100°F) (Nowacki, 1979).

An early developmental process problem for MTG technology was plugging of distillation condensers with crystalline durene (1,2,4,5-tetramethylbenzene). Durene removal from the gasoline product is also required if it is to be suitable for engine use. The pore dimensions of ZSM-5 are 5.4 × 5.6 Å while durene measures 6.7 Å, so it is not readily accommodated by the catalyst. Both temperature and pressure affect its formation. Durene formation generally decreases with temperature, while overall aromatics formation decreases at lower pressure (Chang and Silvestri, 1987). In product road tests 4% (wt) durene has proved tolerable; 2% (wt) is the goal (Chang and Silvestri, 1987).

In a related Mobil process, olefin-gasoline-distillate (MOGD), low molecular weight olefinic oligomers are formulated over ZSM-5 into gasoline and lubricant products (Meisel, 1988). The MOGD process is ready for commercial operation in Germany, likely at the beginning of the next decade.

V. ENVIRONMENTAL ASPECTS OF COAL UTILIZATION

The environmental impact of coal utilization technology encompasses three main stages which all conversion technologies have in common (Braunstein et al., 1979):

1. Feedstock storage and preparation
2. Conversion operations
3. Waste stream handling

The type of potential pollutants associated with these stages and a generalized approach to treatment processes have been identified and are summarized in Table 12. It is not unexpected that these data show that all coal conversion processes have similar impact on the environment. Many of the pollutants listed in the table are treated routinely during the various processes, but several (radioactive materials, catalysts, and slag disposal) still present difficult handling problems.

For most pollutants associated with coal utilization technologies, environmental pollution standards are being continually evaluated and revised. Studies of the environmental impact of coal conversion processes have accompanied the development of each process. The main objective of control and treatment is to deal with pollutant emissions in such a manner that they can be converted to a more acceptable

Table 12 Potential Coal Conversion Pollutants

Potential pollutants	Process source	Treatment processes
Runoff water	Storage	Collection, neutralization, demineralization, ion exchange, settling
Coal-washing wastes	Processing	Collection, neutralization, demineralization, ion exchange, settling
Coal dusts/fines	Storage and processing	Cyclones, filters, grinders, scrubbers, precipitators
Mineral debris	Processing	Landfill, minefill
Water treatment	Processing	Landfill, minefill
N_2	Oxygen	Venting
Stack gas	Utility	Limestone scrubbing, precipitators
H_2S	Conversion	Acid gas scrubbing
NO_x	Conversion	Venting, catalytic absorption
CO_2	Conversion	Stack gas sulfur and combustibles removal, venting
COS, CS_2	Conversion	Recovered with some H_2S scrubbing; fugitive emission from sulfur recovery unit tail gas
NH_3	Conversion	Stripped for recovery; residual NH_3 treated in wastewater
HCN	Conversion	Follows ammonia to wastewater treatment
HF, HCl, CO	Conversion	Wet-scrubbed from product gas stream and transferred to wastewater treatment processes

(continued)

Table 12 (Cont.)

Potential pollutants	Process source	Treatment processes
Volatile trace metals	Conversion	Stack gas cleaning, limestone scrubbing
Trace metals	Conversion	Recovered in ash, tars, chars, and wastewater
Ash	Conversion	Precipitators, minefill, landfill, some resale
Particulates	Conversion	Separators, filters, grinders, scrubbers
Char	Conversion	Recycled as fuel; requires sulfur recovery; wastewater treating potential
Tar	Conversion	Hydrodesulfurized for fuel oil; gasified or combusted, residual treated in wastewater
Oils	Conversion	Hydrodesulfurized for fuel oil; residual treated in wastewater
Gaseous hydrocarbons	Conversion	Scrubbed into wastewater
Phenols	Conversion	Stripped for recovery and resale; residual removal by absorption on carbon, biological oxidation or evaporation
Process wastewater	Conversion	NH_3 and phenol stripping, neutralization, carbon absorption, biological oxidation ponding, tar and oil removal, evaporation, zeolite treatment
Spent catalysts	Conversion	Regeneration, disposal in landfill or minefill

(continued)

Table 12 (Cont.)

Potential pollutants	Process source	Treatment processes
Sulfur recovery tail gases	Control and treatment	Off-gas treating, venting, or combustion
Slag	Control	Granulated for landfill, minefill, or resale
Radioactive wastes	Conversion	Protected landfill or minefill
Thermal emissions	Conversion	Design improvement
Noise	Conversion	Design improvement
Odor	Conversion	Design improvement

Source: Braunstein et al., 1979. Modified with permission of the publisher.

form. Although it may be supposed that the production of pollutants will expand along with the research and development of utilization technologies, it should also be stressed that research and implementation of pollution controls will expand as well. As broad spectrum of methods for treating the pollutants has been identified (Table 12) as already in use (Braunstein et al., 1979).

24

Chemicals from Coal

I Introduction 763
II Chemical Products 765

I. INTRODUCTION

Coal is partially decomposed old plant matter. Typically when a rainforest is covered with silt and then subjected to temperature and pressure over a geologic age, coal ensues (Chap. 17, Sec. II). Coals vary from anthracites (high carbon, low hydrogen, little oxygen) to bituminous, subbituminous, and lignites (rich in oxygen and low in carbon). Coal has been used as a source of heat for several centuries and fueled the industrial revolution.

When a coal such as a bituminous coal is heated to temperatures above about 350°C (660°F) or higher, in the absence of air, the coal undergoes what is known as destructive distillation or carbonization or pyrolysis. Volatile products are formed along with a solid residue (coke). Metallurgical coke or met coke is extremely important in the reduction of ores (iron, etc.) in blast furnaces. A ton of bituminous coal yields about 1500 lb of coke under these conditions, 8 lb of coal tar (condensable volatiles) and 10,000 ft^3 of noncondensable coal gas (hydrogen, methane, carbon oxides, light hydrocarbons, water vapor, etc.).

In addition to coke (or char) and gases, tars and liquids are also formed during the thermal decomposition of coal, the nature of which is related to the maceral constituents of the coal (Table 1).

The nature of the liquid products derived from the thermal decomposition of coal is also related to the final, or ceiling, temperature to which the heating is carried out, and the rate of heating to the final

Table 1 Types of Condensible Products Formed from Reactive Macerals During Thermal Decomposition of Coal (wt %)

	Exinite	Vitrinite
Light oils	2.8	1.0
Heavy oils	29.8	2.3
Acids (nonphenolic)	0.3	2.0
Phenols	3.8	21.6
Bases	1.6	6.4
Neutrals	90.5	70.0

Source: Macrae, 1943. Reproduced with permission of the publisher.

temperature. For comparative purposes, two analytical methods are recognized as standard procedures for determining the yields of gaseous and liquid products. Both of these, the Fischer assay and the Gray–King assay, are described elsewhere in this volume. A wide variety of low molecular weight liquids are obtained from the gas stream of carbonization processes by fractional steam distillation of an oil wash of the gas stream or by stripping the compounds from a solid absorbant. This permits the recovery of several benzene derivatives (plus benzene itself), and some small, e.g., C_5–C_{10}, alkanes, cycloalkanes, and olefins. This low-boiling fraction is typically 50–60% benzene but may contain as much as 90% benzene. Carbon disulfide and thiophene, two primary sulfur-containing components, plus some substituted thiophenes are also condensed or are recovered by sulfuric acid extraction (Speight, 1983).

The tars obtained either by low-temperature or high-temperature carbonization of coal are complex mixtures of a myriad of aromatic compounds. Crude tars can be separated by steam or vacuum distillation to yield oils and refined pitch. Investigators have identified more than 100 condensed polynuclear aromatic hydrocarbon and heterocyclic compounds in coal tar, but it was estimated that as many as 5000 compounds may be present. Light oils are those boiling below 220°C (430°F) that can be used as gasoline, aviation fuel, solvent, or petrochemical feedstock. Middle distillates are recovered in the range 220–375°C (430–710°F) and are suitable for use in diesel fuel, kerosene, or creosote. The high-boiling fraction is recovered up to temperatures of about 550°C (1020°F), and yields many of three- and four-fused ring compounds. The principal commercial hydrocarbon compounds and heterocyclic aromatics derived from coal tar have been identified by Fieser and Fieser (1961) (Figure 1). Minor products appear as impurities. For instance, benzene is usually contaminated with thiophene, and anthracene contains the linear, four-ring analog naphthacene (which gives the liquid a yellow color) and carbazole, a three-ring nitrogen heterocyclic compound.

II. CHEMICAL PRODUCTS

The production of chemical materials from coal can be traced back to the seventeenth century when various types of coal tars were produced for a host of different uses (Khan and Kurata, 1985). Such uses for coal continued well into the present century (beyond World War II) but the industry lost some of the previous popularity with the emergence of the petrochemical industry (Speight, 1980). Nevertheless, coal has been the starting material for a whole host of chemical products (Figure 2). Some are used as a bulk fraction (Table 2) while other products are isolated in the pure (or near-pure) state as distinct chemical species (Table 3). One of the major sources of these chemical products is the low-temperature tar (as might be produced in the mild gasification process; Chap. 23, Sec. VII). In all, several hundred individual chemical compounds have been isolated from coal tar with the first being naphthalene (isolated for the first time in 1820).

More recently, efforts have been directed either at rendering coal into a petroleum substitute or converting coal to carbon monoxide and hydrogen, which in turn are recombined into a number of useful organic chemicals. The conversion of coal into a petroleum substitute is an area of research driven by geopolitical as well as economic reasons. With the advent of mechanized warfare, Germany recognized the need for sources of petroleum to fuel its war machine. Thus it reached out to the oil fields of Romania early in World War II. In a similar manner, the Japanese looked to Malaysia for a source of oil. It is not surprising that, in addition, Germany and Japan have been in the forefront of research on conversion of coal to oil. Both nations have indigenous supplies of coal, none of oil. The two techniques of most value are both German in origin. Bergius published his works on conversion of coal to oil starting in the years prior to 1920. Fundamentally, he reacted coal, which is a solid and which is hydrogen-poor relative to oil ($H/C < 0.9$ for coal versus $H/C > 1.2$ for oil), with molecular hyrogen under very high pressures (10,000 psi) in a slurry made from coal-derived oil at temperatures at which the coal starts to thermally decompose.

$$\text{Coal} \xrightarrow{\text{heat}} R^{\cdot}$$

$$R^{\cdot} + H_2 \longrightarrow R\text{-}H + H^{\cdot}$$

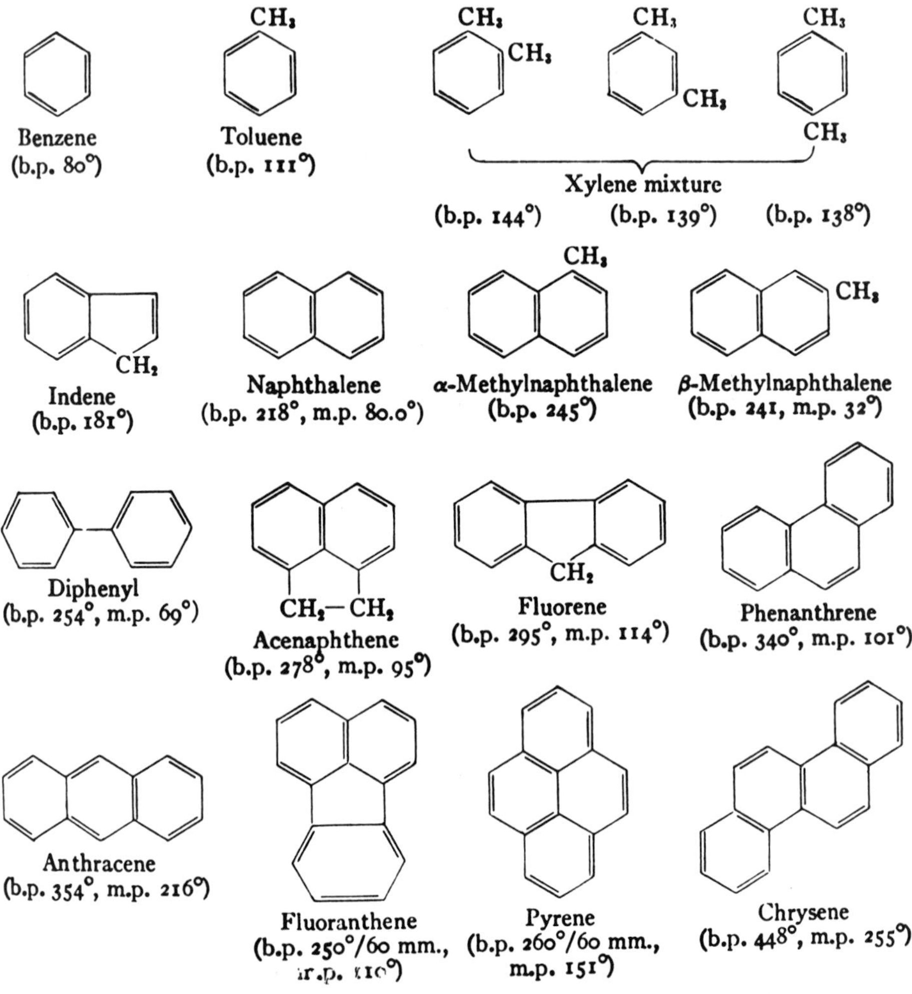

(a)

Figure 1 (a) Principal aromatic hydrocarbons from coal tar. (b) Principal nitrogen compounds from coal tar. (c) Principal oxygen-containing compounds from coal tar (Fieser and Fieser, 1961. Reprinted with permission of the publisher).

In this manner, coal is thermally degraded in a manner similar to the carbonization process. Once thermally generated radicals are formed, however, in the Bergius process they are capped with hydrogen. This generates additional moieties which contain carbon and hydrogen in

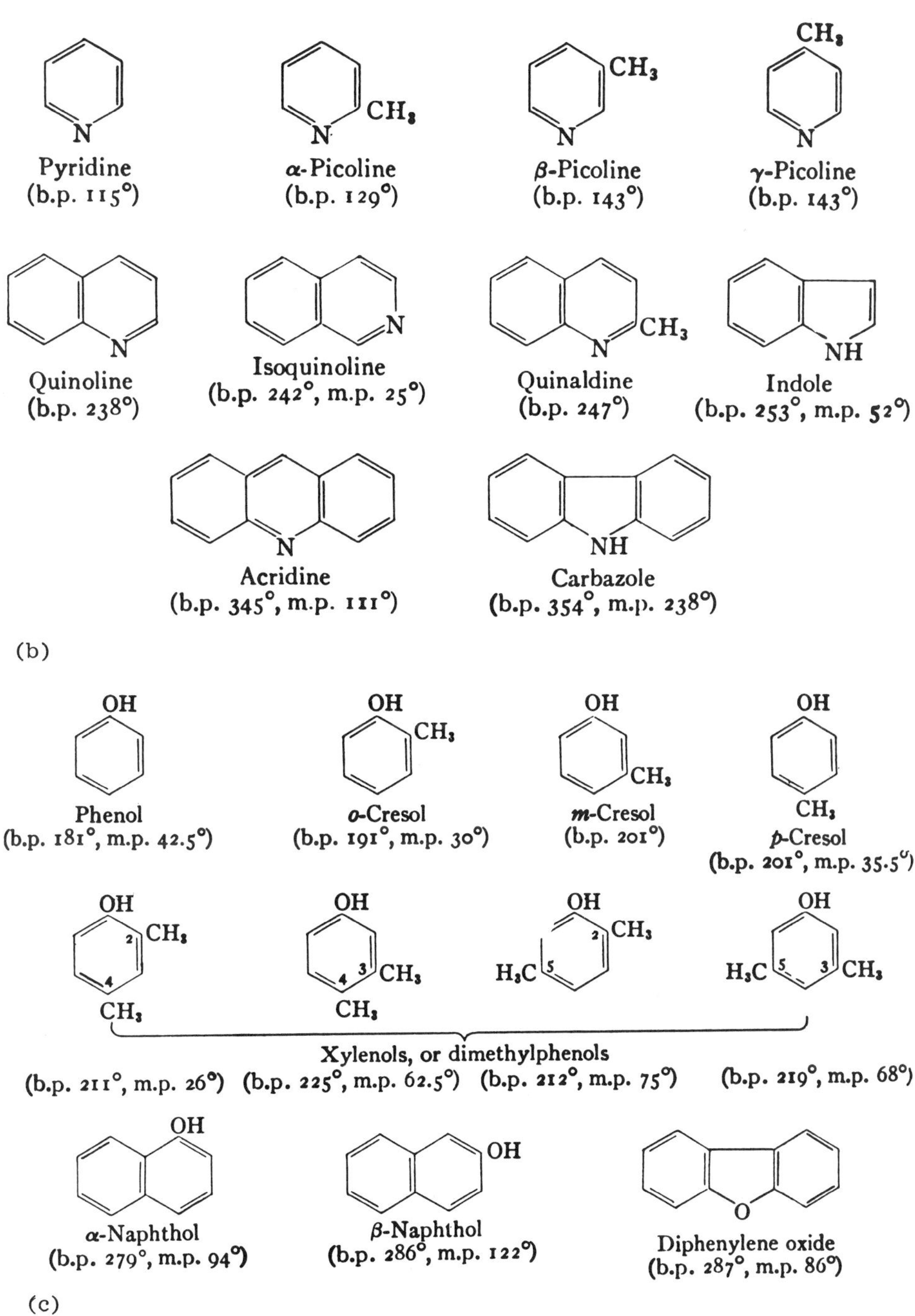

N
Pyridine
(b.p. 115°)
CH_3
N
α-Picoline
(b.p. 129°)
CH_3
N
β-Picoline
(b.p. 143°)
CH_3
N
γ-Picoline
(b.p. 143°)
N
Quinoline
(b.p. 238°)
N
Isoquinoline
(b.p. 242°, m.p. 25°)
N
CH_3
Quinaldine
(b.p. 247°)
NH
Indole
(b.p. 253°, m.p. 52°)
N
Acridine
(b.p. 345°, m.p. 111°)
NH
Carbazole
(b.p. 354°, m.p. 238°)
(b)
OH
Phenol
(b.p. 181°, m.p. 42.5°)
OH
CH_3
o-Cresol
(b.p. 191°, m.p. 30°)
OH
CH_3
m-Cresol
(b.p. 201°)
OH
CH_3
p-Cresol
(b.p. 201°, m.p. 35.5°)
OH
2
CH_3
4
CH_3
OH
3
4
CH_3
CH_3
OH
2
CH_3
H_3C
5
OH
H_3C
5
3
CH_3
Xylenols, or dimethylphenols
(b.p. 211°, m.p. 26°)
(b.p. 225°, m.p. 62.5°)
(b.p. 212°, m.p. 75°)
(b.p. 219°, m.p. 68°)
OH
α-Naphthol
(b.p. 279°, m.p. 94°)
OH
β-Naphthol
(b.p. 286°, m.p. 122°)
O
Diphenylene oxide
(b.p. 287°, m.p. 86°)
(c)

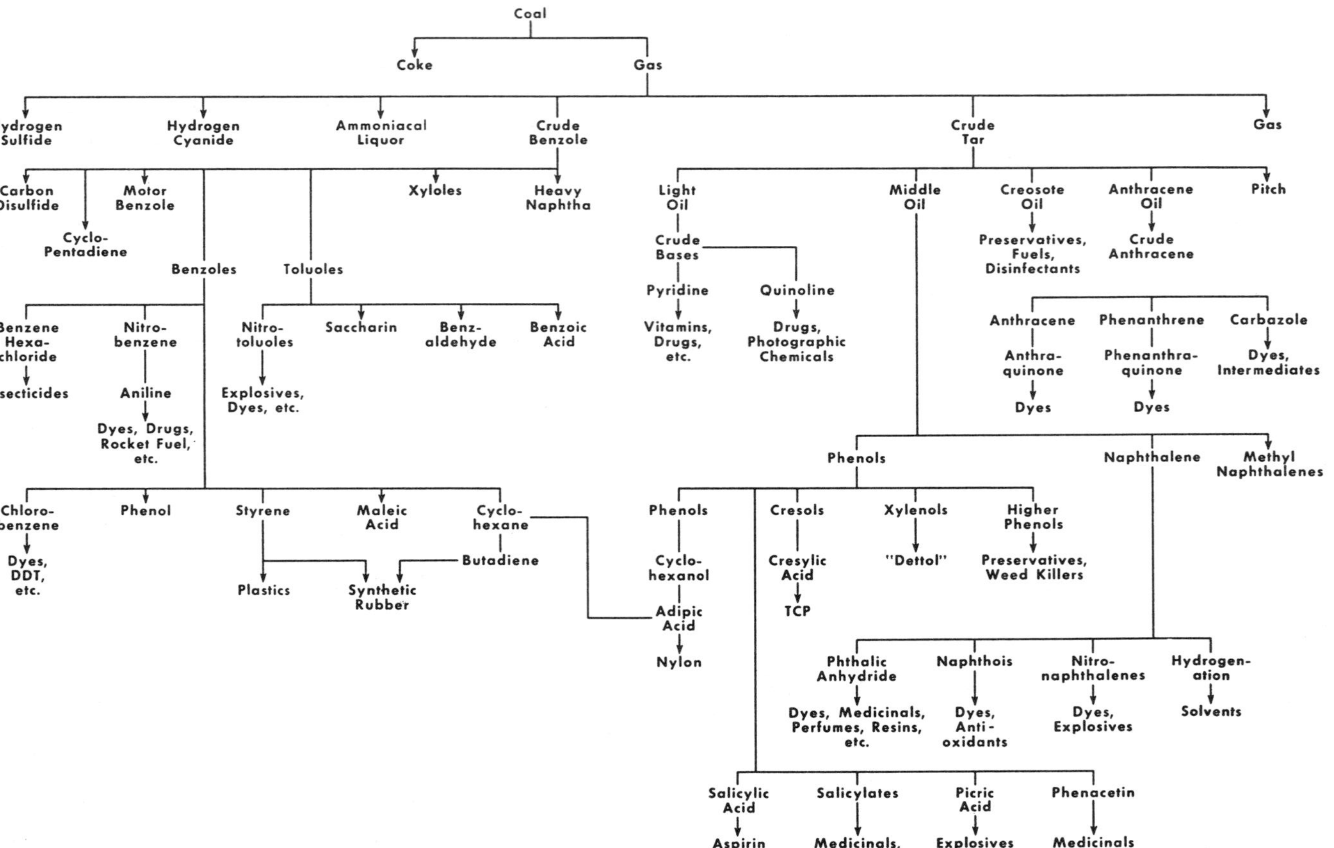

Figure 2 General pathways for the production of chemicals from coal.

amounts to put them into the liquid boiling range. Over the past seven decades great engineering advances have occurred which have led to substantial improvements in coal liquefaction technology. Oil-importing developed nations, such as the United States, West Germany, Britain, and Japan, have taken the leadership roles in this work.

The other major coal conversion technology also developed in Germany is the so-called Fischer--Tropsch chemistry. Here coal is treated under pressure with steam at high temperatures to generate carbon monoxide and hydrogen. The carbon monoxide and hydrogen are recombined in the presence of a catalyst to generate useful products. Use of additional hydrogen in the carbon monoxide–hydrogen reaction in the proper stoichiometry allows for the formation of a gasoline fraction material. The carbon monoxide plus hydrogen mixture is passed over a cobalt-thoria-type catalyst on a suitable support at temperatures of from 200 to 250°C at atmospheric pressure.

$$nCO + (2n + 1)H_2 \rightarrow C_n H_{2n+2} + nH_2O$$

To achieve desirable octane ratings the hydrocarbon product needs to be reformed.

When the ratio of carbon monoxide to hydrogen is diminished in hydrogen compared to the stoichiometry shown for gasoline production above, oxygenated molecules are produced. Perhaps the worldwide leader in technology to produce chemicals from coal is Eastman. At their plant at Kingsport, Tennessee, Eastman is producing industrial chemicals from coal. In a report at the Symposium on Chemicals from Syngas and Methanol presented before the Division of Petroleum Chemistry, Inc. of the American Chemical Society in New York (April 13–18, 1986), T. H. Larkins, Jr. indicated that in the fall of 1983, Eastman made "chemical history as a modern generation of industrial chemicals were produced from coal." The Eastman facility is not a synfuels plant, although it does produce syngas. Rather it uses the syngas to produce acetic anhydride.

As reported in the January 16, 1980 issue of *Chemical Week*, Eastman makes synthesis gas (carbon monoxide and hydrogen) from coal in a Texaco gasifier. The coal employed is approximately 900 tons/day of high-sulfur coal (about 4% sulfur) from southwest Virginia and eastern Kentucky. The Texaco gasifiers (there are two in operation) run at about 2500°F and 1000 psig. As a result of high operating temperature and the Texaco gasifier flow pattern, very rapid reaction rates are realized and byproduct tars are destroyed within the unit. In addition, the methane production is minimal. Thus the product gas is largely carbon monoxide and hydrogen, making it particularly suitable for chemical reactions.

A gas clean-up plant uses a cold methanol wash to remove hydrogen sulfide and carbon dioxide. The CO_2 is vented to the atmosphere, and

Table 2 Bulk Chemicals From Coal and Their Potential Uses

Primary products	Chemicals extracted	Principal derivatives	Commercial outlets
Aqueous	Catechol and homologs	Various mono-, di-, and tri-alkyl compounds	Dyeing, pharmacy, oxidation, inhibitors. Resins, adhesives, dyestuffs
	Phenol	98% phenol mono- di- and tri-chloro compounds	Resins, explosives, dyestuffs, insecticides, fungicides, plasticizers
Bulked crude tar acids	Cresols Xylenols	Mono-, di-, and trialkyl compounds	Various agricultural chemicals, antioxidants, inhibitors, antiseptics, and disinfectants, selective weedkillers, pharmaceuticals
	High-boiling tar acids	Various grades—some chlorinated	Disinfectants and high-grade antiseptics

Name	Use
Pitch	Binder in production of colored pitch mastic flooring
Coarse brown resinous powder	Rubber extender
Creosote	Wood preservative
Miscellaneous oils	Flotation agents in froth flotation plants for separation of coal from coal slurry
Phenolic materials	Production of various chlorinated by reacting different phenols with chlorine, e.g., 2,4-dichlorophenol. Leads to herbicides, weed killers, dyestuffs
Xylenols (chlorinated)	Antiseptic compounds

(continued)

Table 2 (Cont.)

Fuel	Sp. gr. at 60°F	Color	Boiling range (°C)	Sulfur content (%)	Use
Coal petrol (light oil)	0.766	Water white	60–170	0.25	Spark ignition engine
Diesel	0.900	Clear reddish brown	177–314	0.8	Compression ignition engine
Fuel oil	1.032	–	200–400	–	External combustion

Source: Pound, 1952; Pitt and Millward, 1979.

the H_2S is recovered and sent to the sulfur recovery plant. At the sulfur recovery plant, elemental sulfur (about 99.7% of the sulfur in the coal) is recovered and sold to sulfuric acid manufacturers as high-quality molten sulfur. The last traces of sulfur are removed from the vent gas via a Shell-Claus off-gas treatment (SCOT) unit. After cryogenic separation of the CO/H_2O stream, pure carbon monoxide is sent to the acetic anhydride plant while a mixed stream containing carbon monoxide, hydrogen, and carbon dioxide is sent to the methanol plant. Using an Eastman-developed proprietary process, methyl acetate is produced using the methanol along with acetic acid produced as a byproduct in Eastman's cellulose acetate operations.

Finally, the acetic anhydride plant produces this important intermediate from methyl acetate and carbon monoxide. Acetic acid is a byproduct of the acetic anhydride process. The plant capacity for acetic anhydride production is 500 million lb/year. This amounts to about one-half of the 1986 Eastman requirements for this intermediate. Acetic anhydride is used in aspirin, cigarette filters, photographic film, cellulose esters, acetate ester solvents, triacetin, and other products. In the *Chemical Week* article of January 16, 1980, T. F. Reid, executive vice-president of Eastman Kodak and general manager of Eastman Chemicals Division, said; "To our knowledge, Eastman will be the first company in the U.S. to commercially produce a new generation of chemicals from coal. We believe that this first step is just that--that there is a potential for growth and development to make other chemical products from coal."

With the downslide in world oil prices over the past several years, other chemical companies have not seen fit to invest in a chemicals-from-coal complex, but in this author's opinion it is a matter of when

Table 3 Monohydric and Dihydric Phenols Isolated from Coal and Their Potential Uses

Name	Use
Phenol	Dyestuffs, explosives, pharmaceuticals, perfumes, synthetic resins, etc.
o-Cresol	Dyestuffs, explosives, weed killers, perfumes, pharmaceuticals, plant washes, etc.
Xylenol	Synthetic resins, dyestuffs, inhibitors, manufacture of PCMX and synthetic resins
2,3,5-Trimethylphenol	Vitamine E synthesis
3-Methyl-5-ethylphenol	Manufacture of styrene-type resins and special phenol-formaldehyde resins
β-Naphthol	Intermediate in dyestuff industry
Catechol	Antioxidant azo dyes, tanning agents, resins, pharmaceuticals
3-Methyl catechol	Photographic developers, dyestuffs
4-Methyl catechol	Inhibitors, pharmaceuticals
Resorcinol	Dyestuffs, resins, adhesives, glues, antiseptics, and pharmaceutical preparations
2-Methyl resorcinol	Fur-dyeing dyestuffs, synthetic glues, drugs
4-Methyl resorcinol	Synthetic resins, glues, dyestuffs
5-Methyl resorcinol	Synthetic resins, glues, dyestuffs
2,4-Dimethyl resorcinol	Dyestuffs, synthetic glues
2-Methyl quinol	Photographic developer

Source: Pound, 1952.

(not if) other complexes will be built to take advantage of America's hugh coal resources. Judging from projected increases in methanol production from 5.6 million tons (1985) to 7.8 million tons in 1988 and in acetic acid production from 2.9 million tons (1985) to 3.15 million tons in 1988 (*Chem. Eng. News*, December 14, 1987, p. 28), it appears that Eastman is well positioned in these major chemicals areas (Figure 3).

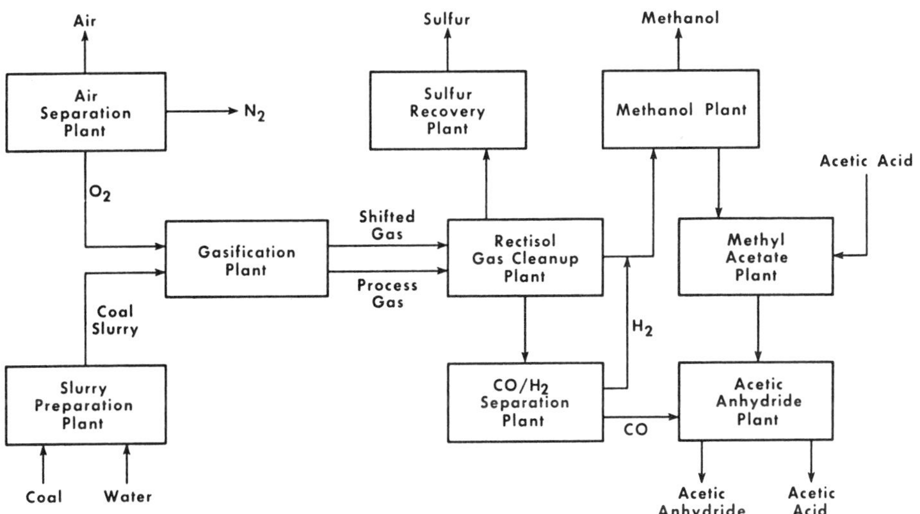

Figure 3 Schematic representation of the Eastman chemicals-from-coal operation.

References to Part III

Abel, W. T., M. Zulkoski, G. A. Brady, and J. W. Eckerd. 1973. Removing Pyrite from Coal by Dry-Separation Methods. United States Bureau of Mines report RI 7732.

Adler, J., K. J. Huttinger, and R. Minges. 1985. *Fuel 64*:1215.

Aigbehinmua, H. B., J. R. Darwent, and A. F. Gaines. 1987. *Energy and Fuels 1*(5):386.

Alger, T. D., R. J. Pugmire, W. D. Hamill, and D. M. Grant. 1979. *Prepr Am. Chem. Soc. Div. Fuel Chem. 24*(2):334.

Allen, D. T., L. Petrakis, D. W. Grandy, G. R. Gavalas, and B. C. Gates. 1984. *Fuel 63*:803.

Amariglio, H., and X. Duval. 1964. *Carbon 4*:323.

American Geological Institute. 1976. *Dictionary of Geologic Terms*, rev. ed. Doubleday, New York.

American Society for Testing and Materials. 1973. *Metals Handbook*, Vol. 8, Philadelphia.

American Society for Testing and Materials. 1986. Classification of Coal by Rank D-388. *Annual Book of Standards*, Part 26, Philadelphia.

Annual Book of ASTM Standards, Vol. 05.05, 1987a, D2013-72; 1987b, D3175-85; 1987c, D3174-82; 1987d, D3175-82; 1987e, D3178-84; 1987f, D3177-84; 1987g, D4239-85; 1987h, D2492-84; 1987i, D3180-84; 1987j, D3682-78; 1987k, D3683-78.

Arthur, J. R., and J. R. Bowring. 1951. *Ind. Eng. Chem. 43*:528

Asher, R. C., and T. B. A. Kirstein. 1968. *J. Nuclear Materials 25*:344.

Attar, A., and F. Dupuis. 1981. Data on the Distribution of Organic Sulfur Functional Groups in Coal. In *Coal Structure*, Adv. Chem. Ser. 192 (M. L. Gorbaty and K. Ouchi, eds.), Am. Chem. Soc., Washington, D.C., pp. 239–257.

Averitt, P. 1975. Coal Resources of the United States, January 1974. *United States Geological Survey Bulletin No. 1412, document No. I19. 3: 1412*, United States Geological Survey, Washington, D.C. pp. 7, 17.
Axelson, D. E. 1985. *Solid State Nuclear Magnetic Resonance of Fossil Fuels*, Multiscience, Canada.
Baiker, A., D. Monti, and Y. S. Fan. 1984. *J. Catalysis 88*:1.
Bajura, R. A., and M. R. Ghate. 1986. *Modern Power Systems 6*(10): 31.
Baker, R. T. K. 1982a. *Chem. Ind. 18*:698.
Baker, R. T. K. 1982b. *J. Catalysis 78*:473.
Baker, R. T. K. 1986. *Carbon 24*:715.
Baker, R. T. K., and J. J. Chludzinski. 1980. *J. Catalysis 64*:464.
Baker, R. T. K., and J. J. Chludzinski. 1985. *Carbon 23*:635.
Baker, R. T. K., and J. J. Chludzinski. 1986. *J. Phys. Chem. 90*: 4730.
Baker, R. T. K., and P. S. Harris. 1972. *J. Phys. Ed.* 5:792.
Baker, R. T. K., and P. S. Harris. 1973. *Carbon 11*:25.
Baker, R. T. K., and R. D. Sherwood. 1980. *J. Catalysis 61*:378.
Baker, R. T. K., and R. D. Sherwood. 1981. *J. Catalysis 70*:198.
Baker, R. T. K., and R. D. Sherwood. 1985. *J. Catalysis 95*:101.
Baker, R. T. K., J. J. Chludzinski, C. A. Bernardo, and J. L. Figueiredo. 1988. *Proc. 9th International Congress on Catalysis*, Calgary, Canada.
Baker, R. T. K., J. J. Chludzinski, N. C. Dispenzieri, and L. L. Murrel. 1983. *Carbon 21*:579.
Baker, R. T. K., J. J. Chludzinski, and R. D. Sherwood. 1985. *Carbon 23*:245.
Baker, R. T. K., F. S. Feates, and P. S. Harris. 1972. *Carbon 10*: 98.
Baker, R. T. K., J. A. France, L. Rouse, and R. J. Waite. 1976. *J. Catalysis 41*:22.
Baker, R. T. K., C. R. F. Lund, and J. J. Chludzinski. 1984. *J. Catalysis 77*:87.
Baker, R. T. K., C. R. F. Lund, and J. A. Dumesic. 1983. *Carbon 21*:469.
Baker, R. T. K., R. D. Sherwood, and E. G. Derouane. 1982. *J. Catalysis 75*:382.
Baker, R. T. K., R. D. Sherwood, and J. A. Dumesic. 1980a. *J. Catalysis 62*:221.
Baker, R. T. K., R. D. Sherwood, and J. A. Dumesic. 1980b. *J. Catalysis 66*:56.
Baker, R. T. K., R. B. Thomas, and M. Wells. 1975. *Carbon 13*: 141.
Barnes, J. R., A. D. H. Calgue, N. J. Clayden, C. M. Dobson, and R. B. Jones. 1986. *Fuel 65*:437.

Bartholomew, C. H., P. K. Agrawal, and J. R. Kalzer. 1982. *Adv. Catalysis 31*:135.
Bartle, K. D. 1972. *Rev. Pure Appl. Chem. 22*:79.
Bartle, K. D. 1985. In *Handbook of Polycyclic Aromatic Hydrocarbons*, Vol. 2 (A. Bjorseth and T. Ramdahl, eds.), Marcel Dekker, New York, Chap. 6.
Bartle, K. D. 1988. In *Supercritical Fluid Chromatography* (R. Smith, ed.), Royal Society of Chemistry.
Bartle, K. D., I. K. Barker, A. A. Clifford, J. P. Kithinji, M. W. Raynor, and G. F. Shilstone, 1987. *Anal. Proc.* 428.
Bartle, K. D., G. Collin, J. W. Stadelhofer, and M. Zander. 1979. *J. Chem. Tech. Biotechnol. 29*:531.
Bartle, K. D., C. Gibson, D. Mills, M. J. Mulligan, N. Taylor, T. G. Martin, and C. E. Snape. 1982. *Anal. Chem. 54*:1730.
Bartle, K. D., W. R. Ladner, T. G. Martin, C. E. Snape, and D. F. Williams. 1979. *Fuel 58*:413.
Bartle, K. D., M. L. Lee, and S. A. Wise. 1981. *Chem. Soc. Rev. 10*:113.
Bartle, K. D., R. S. Matthews, and J. W. Stadelhofer. 1980. *Appl. Spectrosc. 36*:615.
Bartle, K. D., D. G. Mills, M. J. Mulligan, I. O. Amaechina, and N. Taylor. 1986. *Anal. Chem. 58*:2403.
Bartle, K. D., M. J. Mulligan, N. Taylor, T. G. Martin, and C. E. Snape. 1984. *Fuel 63*:1556.
Bartle, K. D., D. L. Perry, and S. Wallace. 1987. *Fuel Proc. Technol. 15*:351.
Bartle, K. D., N. Taylor, M. J. Mulligan, D. G. Mills, and C. Gibson. 1983. *Fuel 62*:1181.
Bassett, G. A. 1961. *Proc. Eur. Reg. Conf. Electron Microsc. 1*:270.
Baughman, G. L. 1981. *Synthetic Fuels Data Handbook: U.S. Oil Shale, U.S. Coal, Oil Sands*, 2nd ed., Cameron Engineers, Denver, p. 172.
Bechtel Group Inc. 1982. Coal Pyrolysis Feasibility Study.
Bem, J. Z. 1988. From Concept to Concrete, *Chemtech*, January, p. 42–46.
Benjamin, B. M., E. C. Douglas, P. M. Herschberger, and J. W. Gohdes. 1985. *Fuel 64*:1340.
Berkowitz, N. 1979. *Introduction to Coal Technology*, Academic Press, New York.
Berkowitz, N. 1985. *The Chemistry of Coal*, Elsevier, New York.
Bishop, M., and D. W. Ward. 1958. *Fuel 37*:191.
Blom, L., L. Edelhausen, and D. W. van Krevelen. 1957. *Fuel 36*:135.
Blumer, G. P., R. Thoms, and M. Zander. 1978. *Erdol Kohle 31*:197.
Boden, H., and R. Roussel. 1983. *Light Metals* 35.

Bodle, W. W., and K. C. Vygas. 1974. Clean Fuels from Coal. *Oil Gas J.* *72*(34):73–88.

Boduszynski, M. M., R. J. Hurtibise, T. W. Allen, and H. F. Silver. 1983. *Anal. Chem.* *55*:225.

Bodzek, D., and A. Marzec. 1981. *Fuel 60*:47.

Bolton, C., C. Riemer, C. E. Snape, F. J. Derbyshire, and M. T. Terrer. 1988. *Fuel 87*, 901.

Borra, C., D. Wiesler, and M. Novotny. 1987. *Anal. Chem.* *59*:339.

Botto, R. E., R. Wilson, R. Hayatsu, R. L. McBeth, R. G. Scott, and R. E. Winans. 1985. *Prepr. Am. Chem. Soc. Div. Fuel Chem.* *30*(4):180.

Bouska, Vladimir. 1981. *Geochemistry of Coal*, Elsevier, Amsterdam, pp. 10, 13, 16–20, 73–74.

Boyce, I. S. 1983. *Int. J. Appl. Radioact. Isot.* *34*:45.

Braunstein, H. M., E. D. Copenhaver, and H. A. Pfuderer (eds.). 1979. *Environmental, Health and Control Aspects of Coal Conversion: An Information Overview*, Vol. 1, Ann Arbor Science, Ann Arbor, MI.

Brill, R., and H. Schaeffer. 1969. *Z. Physik. Chem.* *64*:333.

British Standard. 1963. BS1016, Part 14.

British Standard. 1971. BS1016, Part 16.

British Standard. 1973a. BS1017, Part 1, 1973b, BS1017, Part 3; 1973c, BS1016, Part 1.

British Standard. 1977. BS1016, Part 6.

Broom, H. R., R. A. Durie, and H. N. S. Shafer. 1959. *Fuel 38*: 295, 1960; *39*:59.

Brown, J. K. 1955. *J. Chem. Soc.* 744.

Brown, J. K., and W. R. Ladner. 1960. *Fuel 39*:87.

Brown, J. K., and W. F. Wyss. 1955. *Chem. Ind.* 1118.

Burchill, P. 1987. *1987 International Conference in Coal Science* (J. A. Moulijn, ed.), Elsevier, Amsterdam, p. 5.

Burgar, M. I., J. R. Kalman, and J. F. Stephens. 1985. *Proc. 1985 Int. Conf. on Coal Science*, Pergamon Press, Sydney, p. 784.

Calkins, W. H. 1984. *Fuel 63*:1125.

Calkins, W. H. 1987. *Energy and Fuels 1*(1):59.

Campbell, M. R. 1983. Geology of the Big Stone Gap Coal Field of Virginia and Kentucky. United States Geological Survey Bulletin No. 111-B. United States Geological Survey, Washington, D.C.

Campbell, R. M., and M. Lee. 1986. *Anal. Chem.* *58*:1748.

Cargo, D. N., and B. F. Mallory. 1974. *Man and His Geologic Environment*, Addison-Wesley, Reading, MA, p. 217.

Chaikin, S. W., and W. B. Brown. 1949. *J. Am. Chem. Soc.* *71*:122.

Chakrabartty, S. K., and N. Berkowitz. 1974. *Fuel 53*:240.

Chakrabartty, S. K., and N. Berkowitz. 1976. *Fuel 55*:362.

Chang, Clarence D., and A. J. Silvestri. 1987. MTG: Origin, Evolution, Operation, *Chemtech*, October, 624–631.

Chang, H. C. K., M. Nishioka, K. D. Bartle, S. A. Wise, J. M. Bayona, K. E. Markides, and M. L. Lee. 1988. *Fuel 87*, 45.

Chen, J. W., C. B. Muchmore, T. C. Lin, and K. E. Tempelmeyer. 1985. Supercritical Extraction and Desulfurization of Coal With Alcohols. *Fuel Proc. Tech. 11*:289–295.

Choi, A. S., A. L. Devera, and M. C. Hawley. 1987. *J. Catalysis 106*:313.

Chung, K. E., L. L. Anderson, and W. H. Wiser. 1979. *Fuel 58*:847.

Clark, D. T., and R. Wilson. 1983. *Fuel 62*:1034.

Clark, E. R., J. R. Darwent, B. Demirci, K. Flunder, A. F. Gaines, and A. C. Jones. 1987. *Energy and Fuels 1*(5):392.

Coal Quality Newsletter, 1988. A. P. Hoeft, Denver, Colorado, *3*(1):1.

Coates, D. J., J. W. Evans, and H. Heinemann. 1983. *Appl. Catal.* 7:233.

Coleman, W. M., and A. R. Boyd. 1982. *Anal. Chem. 54*:133.

Colle, K. S., K. Kim, and A. Wold. 1983. *Fuel 62*:155.

Considine, Douglas M. (ed.). 1977. International System for Classification of Hard Coals. *Energy Technology Handbook*, McGraw-Hill, New York, pp. 1–24.

Considine, Douglas M., ed., and Glenn D. Considine. *Scientific Encyclopedia*, 6th ed., Vol. 1, *Coal Conversion Processes*, Van Nostrand, New York.

Cookson, D. J., and B. E. Smith. 1987. *Energy and Fuels 1*:111.

Cookson, D. J., and B. E. Smith. 1983. *Fuel 62*:34.

Covell, J. R. 1988. Rocky Mountain One (RM1) UCG Test–Compliance and Support, Contract No. DE-FC21-86MC11076, Presented at the WRI-DOE (Western Research Institute–U.S. Department of Energy) Technical Progress Review, Laramie, Wyoming, March 30 and 31.

Crabbe, D., and R. McBride. 1978. *The World Energy Book*, Nichols, New York.

Cronauer, D. C., R. I. McNeil, D. C. Young, and R. G. Ruberto. 1982. *Fuel 61*:610.

Darlage, L. J., H. N. Finkbone, S. J. King, J. Ghosal, and M. E. Bailey. 1978. *Fuel 57*:479.

Davidson, R. M. 1986. Nuclear Magnetic Resonance of Coal ICTIS/TR32, IEA Coal Research, London.

Davies, I. D., M. W. Raynor, P. T. Williams, G. E. Andrews, and K. D. Bartle. 1987. *Anal. Chem. 59*:2579.

Davis, A. 1978. The Measurement of Reflectance of Coal Macerals: Its Automation and Significance. Department of Energy Report No. DOE-FE-2030-TR10, Washington, p. 68.

Day, R. J., P. L. Walker, and C. C. Wright. 1958. *Carbon and Graphite*, Soc. Chem. Ind., London, p. 348.

Demuth, J. E., D. W. Jepson, and P. M. Marcus. 1974. *Phys. Rev. Lett. 32*:1182.

Deno, N. C., B. A. Greigger, and S. G. Stroud. 1985. *Fuel 64*:1340.
Deno, N. C., B. A. Greigger, A. D. Jones, W. G. Rakitsky, K. A Smith, and R. D. Minard. 1980. *Fuel 59*:700.
Deurbrouck, A., and J. Hudy, Jr. 1976. *Dictionary of Geological Terms*, ref. ed., Anchor Press, Garden City, NY.
Deurbrouck, A. W., and J. Hudy, Jr. 1972. Performance Characteristics of Coal-Washing Equipment: Dense-Medium Cyclones. United States Bureau of Mines Bulletin Report RI 7673, Washington, D.C.
Deurbrouck, E. W., and E. R. Palowitch. 1963. Performance Characteristics of Coal Washing Equipment: Concentration Tables. United States Bureau of Mines Bulletin Report RI 6239, Washington, D. C.
Dilke, M. H., D. D. Eley, and E. B. Maxted. 1948. *Nature 161*:804.
Donath, E. E. 1963. Hydrogenation of Coal and Tar. In *Chemistry of Coal Utilization*, suppl. vol. (H. H. Lowry, ed.), John Wiley and Sons, New York, pp. 1041–1088.
Dorn, H. C., L. T. Taylor, and T. E. Glass. 1979. *Anal. Chem. 51*: 947.
Drake, J. A. G., D. W. Jones, B. S. Cowsey, and G. F. Kirkbright. 1978. *Fuel 57*:663.
Drath, A. 1932. Analysis of Boghead and Cannel Coal. *Bull. Inst. Geol. de Pologne 12*:1.
Dry, M. E., and H. B. deW. Erasmus. 1987. "Update of the SASOL Synfuels Process," *Annu. Rev. Energy 12*:1–21.
Dry, M. E., and L. C. Ferreira. 1967. Structural Promotion of Reduced Magnetite. *J. Catalysis* 7:352.
Dry, M. E., J. A. K. duPlessis, and G. M. Leuteritz. 1966. Influence of Structural Promoters on the Surface Properties of Reduced Magnetite Catalysis. *J. Catalysis* 6:194.
Dryden, I. G. C. 1964. In *Kirk-Othmer's Encyclopedia of Chemical Technology*, John Wiley and Sons, p. 640.
Elder, J. W. 1963. The Underground Gasification of Coal. In *Chemistry of Coal Utilization*, suppl. vol. (H. H. Lowry, ed.), John Wiley and Sons, New York, pp. 1023–1040.
Electric Power Research Institute. 1974. Evaluation of Coal Conversion Process to Provide Clean Fuels. Report No. EPRI 206-0-0, prepared by the University of Michigan, College of Engineering for EPRI.
Electric Power Research Institute. 1979. Schematic and Temperature Profile for Three Types of Coal Gasifiers. *EPRI J.*, Electric Power Research Institute, Palo Alto, CA, April.
Electric Power Research Institute. 1981. *EPRI J.*, April.
Essenhigh, R. H. 1981. Fundamentals of Coal Combustion. In *Chemistry of Coal Utilization*, suppl. vol. (H. H. Lowry, ed.), John Wiley and Sons, New York, Chap. 19, pp. 1153–1312.
Evans, N., T. M. Haley, M. J. Mulligan, and K. M. Thomas. 1986. *Fuel 65*:694.

Farcasiu, M. 1979. *Prepr. Am. Chem. Soc. Div. Fuel Chem.* *24*(1): 121.
Farnum, S. A., B. W. Farnum, J. R. Rindt, D. J. Miller, and A. C. Wolfson. 1984. *Prepr. Am. Chem. Soc. Div. Fuel Chem.* *29*(1):144.
Felsen, M. F., and P. Regnier. 1977. *Surf. Sci.* *68*:410.
Fieser, L. F., and M. Fieser. 1961. Aromatic Hydrocarbons. In *Advanced Organic Chemistry*, Reinhold, New York.
Finkelman, R. B., and R. W. Stanton. 1978. *Fuel* *57*:763.
Fitzharris, W. D., J. R. Katzer, and W. H. Manague. 1982. *J. Catalysis* *76*:369.
Flynn, P. C., and S. E. Wanke. 1974. *J. Catalysis* *34*:390.
Fogarty, M. P., D. C. Shelley, and I. M. Warner. 1981. *J. High Res. Chrom. Chromatogr. Commun.* *4*:561, 616.
Francis, W. 1961. *Coal: Its Formation and Composition*, Edward Arnold, London, pp. 630, 639.
Franz, J. A. 1979. *Fuel* *58*:405.
Fraser, F. W., and C. B. Belcher. 1972. *Fuel* *53*:41.
Friedel, R. A., and H. L. Retcofsky. 1966. *Chem. and Ind.* 455.
Frost, D. C., W. R. Leeder, and R. L. Tapping. 1974. *Fuel* *53*:206.
Fuller, F. W., R. C. Hurran, and B. Preston. 1981. *Analyst* *106*: 913.
Gaines, A. F., and F. M. Page. 1983. *Fuel* *62*:1041.
Gaudin, A. M. 1939. *Principles of Mineral Dressing*, McGraw-Hill, New York.
Gavalas, G. R. 1982. In *Coal Science and Technology*, Vol. 4, Elsevier, New York.
Gibson, D. L. 1985. Soviet Coal Basins. *Energy Graphics*, Prentice-Hall, Englewood Cliffs, NJ.
Given, P. J. 1960. *Fuel* *39*:147.
Given, P. H. 1973. How May Coals Be Characterized for Practical Use? Workshop Oct. 29–Nov. 2, 1973. Penn. State University, University Park, PA.
Given, P. H. 1984. In *Coal Science*, Vol. 3 (M. L. Gorbaty, J. W. Larsen, and I. Wender, eds.), Academic Press, Orlando, p. 63.
Gluskoter, H. 1965. *Fuel* *44*:285.
Grainger, L., and J. Gibson. 1981. *Coal Utilization*, Halsted Press, New York.
Grandy, D. W., L. Petrakis, D. C. Young, and B. C. Gates. 1984. *Nature* *308*:175.
Green, T. K., and T. A. West. 1986. Coal Swelling in Straight-Chain Amines: Evidence for Specific Site. *Fuel* *65*(2):298-299.
Grienke, R. A. 1984. *Fuel* *63*:1374.
Guccione, Eugene. 1981. The Fischer–Tropsch Synthesis: A Cornucopia of Hydrocarbons, *Coal Mining and Processing* *18*(2):43.
Guidoboni, R. J. 1976. *Anal. Chem.* *45*:1275.

Haga, T., and Y. Nishiyama. 1983. *Carbon 21*:219.
Hagaman, E. W., R. R. Chambers, Jr., and M. C. Woody. 1986. *Anal. Chem. 58*:387.
Hajak, M., V. Skeinar, G. Sebar, I. Lang, and O. Weisser. 1978. *Anal. Chem. 50*:773.
Halden, F. A., and W. D. Kingrey. 1955. *J. Phys. Chem. 59*: 557.
Hall, H. J., G. M. Vargo, and E. M. Magee. 1974. Trace Elements and Potential Pollutant Effects in Fossil Fuels. In Symposium Proceedings: Environmental Aspects of Fuel Conversion Technology, May 1974. Environmental Protection Agency Report No. EPA-R2-73-188 PB-220 376. Washington, D.C., pp. 35–47.
Hamrin, C. E., P. S. Mia, L. L. Chyi, and W. D. Fhman. 1975. *Fuel 54*:70.
Harris, P. S., F. S. Feates, and B. G. Reuben. 1974. *Carbon 12*: 189.
Haw, J. F., T. E. Glass, D. W. Hausler, E. Motell, and H. C. Dorn. 1980. *Anal. Chem. 52*:1135.
Haw, J. F., T. E. Glass, and H. C. Dorn. 1981. *Anal. Chem. 53*: 2332.
Hayatsu, R. R. E. Winans, R. G. Scott, L. P. Moore, and M. H. Studier. 1978. *Fuel 57*:541.
Hayatsu, Ryoichi, Robert L. McBeth, Robert G. Scott, Robert E. Botto, and Randall E. Winans. 1984. Artificial Coalification Study: Preparation and Characterization of Synthetic Macerals. *Org. Geochem. 6*:463–471.
Haynes, W. P., and A. J. Forney. 1975. The Synthane Process. In *Papers—Clean Fuels from Coal Symposium II*. June 23–27, Chicago, pp. 150–157.
Haynes, W. P., S. J. Gasior, and A. J. Farney. 1974. *Advances in Chemistry Series No. 131*, Am. Chem. Soc., Washington, D.C., p. 179.
Heintz, E. A., and W. E. Parker. 1966. *Carbon 4*:473.
Heredy, L. A., and M. B. Neuworth. 1962. *Fuel 41*:221.
Heredy, L. A., A. E. Kostyo, and M. B. Neuworth. 1965. *Fuel 44*: 125.
Herod, A. A., W. R. Ladner, B. J. Stokes, A. J. Berry, D. E. Games, and M. Hohn. 1987a. *Fuel 66*:935.
Herod,A. A., W. R. Ladner, and B. J. Stokes. 1987b. *1987 International Conference on Coal Science* (J. Moulijn, ed.), Elsevier, Amsterdam, p. 343.
Hessley, R. K., J. W. Reasoner, and J. T. Riley. 1966. *Coal Science: An Introduction to Chemistry, Technology, and Utilization*, John Wiley and Sons, New York, pp. 44–45, 166.
Higashiyama, K., A. Tomita, and Y. Tamai. 1985. *Fuel 64*:1525.
Hodek, W., and G. Kolling. 1973. *Fuel 52*:220.

Hodgkins, S. A. 1961. *Soviet Power: Energy Resources, Production and Potential*, Prentice-Hall, Englewood Cliffs, NJ.
Hoffman, E., and A. Jenkner. 1932. Die Inkohlung und ihre Erkennung im Mikrobild. *Gluckauf 68*:81–88.
Holstein, W. L., and M. Boudart. 1981. *J. Catalysis 72*:328.
Holstein, W. L., and M. Boudart. 1982. *J. Catalysis 75*:337.
Hombach, H. P. 1981. *Fuel 60*, 663 and 1982, *Fuel 61*, 215.
Howard, J. B. 1981. In *Chemistry of Coal Utilization*, 2nd suppl. vol. (M. A. Elliott, ed.), John Wiley and Sons, New York.
Hoyer, G. G. 1985. Extraction with Supercritical Fluids: Why, How, and So What? *Chemtech* July pp. 440--448.
Hsieh, K. C., and C. A. Wert. 1985. *Fuel 63*:255.
Hudy, J., Jr. 1968. Performance Characteristics of Coal-Washing Equipment. United States Bureau of Mines Report RI 7154. United States Bureau of Mines, Washington, D.C.
Huhn, T., J. Klein, and H. Juntgen. 1983. *Fuel 62*:196.
Huttinger, K. J. 1983. *Fuel 62*:166.
Huttinger, K. J., and W. Kraus. 1981a. *Proc. Int. Conf. Coal Sci.* Dusseldorf, p. 320.
Huttinger, K. J., and W. Kraus. 1981b. *Fuel 60*:93.
Huttinger, K. J., and W. Kraus. 1982. *Fuel 61*:291.
Huttinger, K. J., and R. Minges. 1985. *Fuel 65*:1122.
Ignasiak, B. S., and M. Gawlak. 1977. *Fuel 56*:216.
Instrument Manual. 1980. Fischer Model 490 Analyser, Fischer Scientific.
Instrument Manual. 1983. MAC-400 Proximate Analyser, LECO.
Jackson, W. P., R. C. Kong, and M. L. Lee. 1985. In *PAH Mechanisms, Methods and Metabolism* (M. Cooke and A. J. Dennis, eds.), Battelle Press, Columbus, Ohio, p. 609.
Jacobs, I. S., C. Federighi, D. W. McKee, and H. J. Porchen. 1982. *J. Appl. Phys. 53*:8326.
Johnson, R. W. 1981. South Africa's Sasol Project: How to Succeed in Synfuels. *Coal Mining and Processing 18*(2):42–45
Jones, R. B., C. B. McCourt, and P. Swift. 1981. In *Proceedings International Conference in Coal Science*, Verlag Gluckhauf, Dusseldorf, Essen, p. 657.
Kahn, M. R., and T. M. Kurata. 1985. The Feasibility of Mild Gasification of Coal: Research Needs. Morgantown Energy Technology Center, United States Department of Energy, Morgantown, West Virginia, DOE/METC-85/4019.
Karr, C. 1963. In *Chemistry of Coal Utilization*, suppl. vol. (H. H. Lowry, ed.), John Wiley and Sons, New York.
Keep, C. W., S. Terry, and M. Wells. 1980a. *J. Catalysis 66*:451.
Keep, C. W., Terry, S., and R. J. Waite. 1980b. In *Gas Chemistry in Nuclear Reactors and Large Industrial Plants* (A. Dyer, ed.), Heyden, New York, p. 196.

Keim, K. H., J. Mazluk, and A. Tonnesmann. 1984. The Methanol to Gasoline (MTG) Process. *Erdol und Kohle 37*(2):558–562.
Kershaw, J. R. 1978. *Fuel 57*:299.
Kershaw, J. R. 1984. The Chemical Nature of a Supercritical Gas Extract Including Comparison with Other Coal Liquids. *Fuel Proc. Tech. 9*:235–250.
Kershaw, J. R. 1985. *Liquid Fuels Technol. 3*(2):205.
Kessler, T., R. Raymond, and A. G. Sharkey. 1969. *Fuel 48*:179.
Kim, K., R. Kershaw, K. Dwight, A. Wold, and K. Colle. 1982. *Mater. Res. Bull. 17*:591.
Kirk-Othmer *Encyclopedia of Chemical Technology*. 1978. Vol. 5, *Catalysis*, and Vol. 11, *Fuels, Synthesis*, John Wiley and Sons, New York.
Klatt, L. N. 1979. *J. Chromatogr. Sci. 17*:225.
Krauskopf, K. B. 1979. *Introduction to Geochemistry*, 2nd ed. McGraw-Hill, New York.
Kroeger, C. A., and A. Pohl. 1957. Properties of Hard Coal Constituents. V: Calorific Effects During Thermal Decomposition. *Brennstoff Chemie 38*:102.
Kroeger, C., G. Darsow, and K. Furh. 1965. *Erdol Kohle Erdgas Petrochemie 18*:701.
Kuhn, L., and H. Plogmann. 1983. *Fuel 62*:205.
Kuhn, J. K., W. F. Harfst, and N. E. Shrimp. 1975. In *Trace Elements in Fuel* (S. P. Babu, ed.), Adv. Chem. Ser. No. 141, Am. Chem. Soc., p. 66.
Lacount, R. B., R. R. Anderson, S. Friedman, and B. D. Blaustein. 1986. *Prepr. Am. Chem. Soc. Div. Fuel Chem. 31*(1):70.
Ladner, W. R., T. G. Martin, C. E. Snape, and K. D. Bartle. 1980. Insights into the Chemical Structure of Coal from the Nature of Extracts. *Prepr. Am. Chem. Soc. Div. Fuel Chem. 25*(4):67.
Landolt, R. G. 1975. *Fuel 54*:299.
Larsen, J. W. 1985. From Lignin to Coal in a Year. *Nature 314*:316.
Larsen, J. W., L. Urgan, C. Lawson, and D. Lee. 1981. *Fuel 60*:267.
Later, D. W. 1985. In *Handbook of Polycyclic Aromatic Compounds*, Vol. 2 (A. Bjorseth, ed.), Marcel Dekker, New York, Chap. 9.
Later, D. W., T. G. Andros, and M. L. Lee. 1983. *Anal Chem. 55*:2126.
Le, T. T., and D. T. Allen. 1985. *Fuel 64*:1754.
LeBlanc, J. R., D. O. Moore, and A. E. Cover. 1981. Coal Can Be Gasoline," *Hydrocarbon Processing 60*(6):133–138.
Lee, M. L., M. Novotny, and K. D. Bartle. 1981. *Analytical Chemistry of Polycyclic Aromatic Compounds*, Academic Press, New York.
Lee, M. L., F. J. Yang, and K. D. Bartle. 1984. *Open Tubular Column Gas Chromatography*, John Wiley and Sons, New York.
M. L. Lee, D. L. Vassilaros, C. M. White, and M. Novotny. 1979. *Anal. Chem. 51*:768.

Leenhardt, P., A. Sulima, K. H. van Heek, and H. Juntgen. 1983. *Fuel 62*:200.
Leonard, J. W., and D. R. Mitchell. 1968. *Coal Preparation*, AIME Seeley W. Mudd Series.
Lewis, J. B. 1970. In *Modern Aspects of Graphite Technology* (L. C. F. Blackman, ed.), Academic Press, New York, p. 129.
Lewis, I. C., and B. A. Petro. 1976. *J. Polym. Sci. 14*:1975.
Liotta, R., K. Rose, and E. Hippo. 1981. *J. Org. Chem. 46*:227.
Little, A. D. 1972. *A Current Appraisal of Underground Coal Gasification*, Report PB-207 274, A. D. Little, Cambridge, MA, p. 38.
Liu, Y. A., and. C. J. Lin. 1976. Assessment of Sulfur and Ash Removal from Coals by Magnetic Separation. *IEEE Trans. on Magnetics*. MAG-12(No. 5). IEEE, pp. 538–550.
Long, F. J., and K. W. Sykes. 1952. *Proc. Roy. Soc. London, Ser. A215*:100.
Lopez-Peinado, A. J., J. Rivera-Utrilla, F. J. Lopez-Garzon, I. Fernandez-Morales, and C. Moreno-Castilla. 1986. *Fuel 65*:1419.
Lorenzi, L., Jr. 1975. Chemical Coal Cleaning. Economic Commission for Europe, Second Seminar on Desulfurization of Fuels and Combustion Gases, Report ENV/SEM 4/Aa.6, Washington, D.C.
Lund, C. R. F. 1986. *J. Catalysis 95*:71.
Lund, C. R. F. 1987. *Carbon 25*:337.
Lund, C. R. F., J. J. Chludzinski, and R. T. K. Baker. 1985. *Fuel 64*:789.
Lund, C. R. F., R. D. Sherwood, and R. T. K. Baker. 1987. *J. Catalysis 104*:233.
Macrae, J. C. 1943. Thermal Decomposition of Spore Exines from Bituminous Coal. *Fuel 22*:117–129.
Magdefrau, K. 1953. *Paleobiologie der Pflangen*. Jena, p. 438.
Magee, E. M. 1975. Environmental Impact and Research and Development Needs. EPA Symposium: Environmental Aspects of Fuel Conversion Technology II, Hollywood, FL, Dec. 15–18, 1975.
Maher, T. P., and J. M. O'Shea. 1967. *Fuel 26*:283.
Maiden, Colin J. 1988. A Project Overview, *Chemtech*, January, 38–41.
Maisgren, B., W. Hubner, K. Norgard, and S. B. Sundvall. 1983. *Fuel 62*:1076.
Mako, Peter F., and Wm. A. Samuel, P. E. 1978. A Fluor Perspective on Synthetic Liquids: Their Potential and Problems, July.
Masciantonia, P. X. 1965. The Effect of Molten Caustic on Pyritic Sulfur in Bituminous Coal. *Fuel 44*:209–275.
Matsumoto, S., and P. L. Walker. 1986. *Carbon 24*:277.
Maxted, E. B. 1951. In *Advances in Catalysis* (W. G. Frankenberg, V. I. Komarewsky, and E. K. Rideal, eds.), Vol. 3, Academic Press, New York, p. 129.

Mayo, F. 1975. *Fuel 54*:273.
McKee, D. W. 1970a. *Carbon 8*:131.
McKee, D. W. 1970b. *Carbon 8*:623.
McKee, D. W. 1974. *Carbon 12*:453.
McKee, D. W. 1979. *Carbon 17*:419.
McKee, D. W. 1981. In *The Chemistry and Physics of Carbon*, Vol. 16 (P. L. Walker, ed.), Marcel Dekker, New York, p. 1.
McKee, D. W. 1986. *Carbon 24*:331.
McKee, D. W., and D. Chatterji. 1978. *Carbon 16*:53.
McKee, D. W., and C. L. Spiro. 1984. *Carbon 22*:507.
McKee, D. W., and Spiro, C. L. 1985. *Carbon 23*:437.
McKee, D. W., C. L. Spiro, P. G. Kosky, and E. G. Lamby. 1982. *Preprints, Div. Fuel Chem., Am. Chem. Soc. 21*:74.
McVicker, G. B., R. L. Garten, and R. T. K. Baker. 1978. *J. Catalysis 54*:129.
Meisel, S. L. 1988. Catalysis Research Bears Fruit, *Chemtech*, January, 32–37.
Melnikov, N. V. 1972. Role of Coal in the Energy Fuel Resources in the USSR. *Can. Mining Met. Bull.* June *65*:77–82.
Meuzelaar, H. L. C., A. M. Harper, G. R. Hill, and P. H. Given. 1984. *Fuel 63*:640; 1984b; *63*:793.
Meuzelaar, H. L. C., B. L. Hoesterey, W. Windig, and G. R. Hill. 1987. *Fuel Proc. Technol. 15*:59.
Mima, M. J., H. Schulz, and W. E. McKinstney. 1978. In *Analytical Methods for Coal and Coal Products*, Vol. 1 (C. Karr, Jr., ed.), Academic Press, New York, Chap. 19.
Mims, C. A., and J. A. Pabst. 1983. *Fuel 62*:176.
Mims, C. A., J. J. Chludzinski, J. A. Pabst, and R. T. K. Baker. 1984. *J. Catalysis 88*:97.
Moll, N. G., and G. J. Quarderer. 1979. The Dow Coal Liquefaction Process. *Chem. Engr. Progr.* 75(11):46.
Moore, E. S. 1940. *Coal: Its Properties, Analysis, Classification, Geology, Extraction, Uses and Distribution*, 2nd ed., John Wiley and Sons, New York, p. 350.
Moore, R. T., P. McCutchan, and D. A. Young. 1951. *Anal. Chem. 23*:1639.
Moza, A. K., D. W. Strickler, and L. G Austin. 1980. *Scanning Electron Microsc.*, 91.
Mulligan, M. J., K. M. Thomas, and A. P. Tytko. 1987. *Fuel 66*:1050.
Muetterties, E. L., and J. Stein. 1979. Mechanistic Features of Catalytic Carbon Monoxide Hydrogenation Reactions. *Chem. Rev. 79*(6): 479.
Murchison, D., and T. S. Westoll (eds.). 1968. *Coal and Coal-Bearing Strata*, Elsevier, New York.

Nadkarni, R. M., C. Bliss, and W. I. Watson. 1974. Underground Gasification of Coal. *Chem. Tech. 1974*:230–237.
Nadkarni, R. M., C. Bliss, and W. I. Watson. 1975. Underground Gasification of Coal. *Papers—Clean Fuels from Coal Symposium II*, June 23–27, pp. 625–651.
Netzel, D. A. 1987. *Anal. Chem. 59*:1775.
Nishiyama, Y. 1986. *Fuel 65*:1404.
Nishiyama, Y., and Y. Tamai. 1977. *Preprints, Div. Fuel Chem. Am. Chem. Soc. 24*:219.
Novotny, M., A. Hirose, and D. Wiesler. 1984. *Anal. Chem. 56*:1243.
Nowacki, P. 1979. *Coal Liquefaction Processes*, Noyes Data, Park Ridge, NJ, pp. 104–114.
Nowacki, Perry. 1978. Indirect Liquefaction, *Coal Liquefaction Processes*, Noyes Data Corp., Park Ridge, NJ, pp, 161–182.
Nunn, R. C., H. L. Lovell, and C. C. Wright. 1953. Spectrographic Analysis of Trace Elements in Anthracite. In *Trans. Ann. Anthracite Conf. Lehigh Univ. 11*:51–65.
Ode, W. H. 1963. Coal Analysis and Mineral Matter. In *Chemistry of Coal Utilization*, suppl. vol. 1 (H. H. Lowry, ed.), John Wiley and Sons, New York, pp. 202–231.
O'Gorman, J. V., and P. L. Walker, Jr. 1971. *Fuel 50*:135.
O'Gorman, J. V., and P. L. Walker, Jr. 1972. Mineral Matter and Trace Elements in U.S. Coals. Report OCR-RDR-61-IR-2. U.S. Government Printing Office, Washington, D.C.
Oita, I. J., and H. S. Conway. 1954. *Anal. Chem. 26*:600.
Olson, E. S., J. W. Diehl, and M. L. Froelich. 1987. *Fuel Proc. Technol. 15*(1-3):319.
Otto, K., L. Bartosiewicz, and M. Shelef. 1979. *Carbon 17*:351.
Painter, P. C., and M. M. Coleman. 1980. *Int. Lab.* 17.
Painter, P. C., R. W. Snyder, M. Starsinic, M. M. Coleman, D. W. Kuehn, and A. Davis. 1981. *Appl. Spectrosc. 35*(5):475.
Painter, P. C., P. Y. Whang, and P. L. Walker. 1978. *Fuel 57*:337.
Pappin, A. J., A. P. Tytko, K. D. Bartle, N. Taylor, and D. G. Mills. 1987. *Fuel 66*:1050.
Patteisky, K., and M. Teichmuller. 1960. Inkohlungs-Verlauf, Inkohlungs-Ma stabe und Klassifikation der Kohlen auf Grund von Vitrit-Analysen, *Brennstoff-Chemie 41*:79–84, 97–104, 113–137. Essen.
Peaden, P. A., and M. L. Lee. 1982. *J. Liquid Chromatogr. 5* (Supp. 2):179.
Penner, S. S., and L. Icerman. 1984. *Non-Nuclear Energy Technologies*, Vol. 2, 2nd ed., Pergamon Press, New York.
Perry, D. L., and A. Grint. 1983. *Fuel 62*:1024.
Petrakis, L., and D. W. Grandy. 1980. Coal Analysis, Characterization and Petrography. *J. Chem. Ed. 57*:689–694.

Pitt, G. J., and G. R. Millward (eds.). 1979. *Coal and Modern Coal Processing: An Introduction*, Academic Press, New York, p. 57.
Poirier, M. A., and A. E. George. 1983. *J. Chromatogr. Sci. 21*: 144.
Pound, G. S. 1952. *Coke and Gas. 14*:355.
Radmacher, W., and P. Mahrauer. 1953. *Gluckauf 89*:503.
Radmacher, W., and P. Mahrauer. 1955. *Brenst Chem. 36*:236.
Radovic, L. R., P. L. Walker, and R. G. Jenkins. 1983a. *Fuel 62*: 209.
Radovic, L. R., P. L. Walker, and R. G. Jenkins. 1983b. *J. Catalysis 82*:382.
Radovic, L. R., P. L. Walker, and R. G. Jenkins. 1984. *Fuel 63*: 1028.
Rao, H. S., G. S. Murti, and A. Lahiri. 1960. *Fuel 39*:263.
Reggel, L., I. Wender, and R. Raymond. 1968. *Fuel 47*:373.
Reggel, L., I. Wender, and R. Raymond. 1973. *Fuel 52*:162.
Renton, J. J. 1978. *Energy Sources 4*:91.
Rewick, R. T., P. R. Wentrcek, and H. Wise. 1974. *Fuel 53*:274.
Rich, F. J. 1983. Modern Wetlands and Their Potential as Coal Forming Environments. *Bulletin—Corpus Christi Geological Society* (March), Corpus Christi, TX, p. 4.
Richards, D. G., C. E. Snape, K. D. Bartle, C. Gibson, M. J. Mulligan, and N. Taylor. 1983. *Fuel 62*:724.
Riesser, B., M. Starsinic, E. Squires, A. Davis, and P. C. Painter. 1984. *Fuel 63*:1253.
Ross, D. S., R. M. Laine, T. K. Green, A. S. Hirschon, and G. P. Hum. 1985. *Fuel 64*:1323.
Ruch, R. R., H. J. Gluskoter, and N. F. Shimp. 1974. Occurrence and distribution of Potentially Volatile Trace Elements in Coal: A Final Report. *Environ. Geol.* I11. St. Geol. Survey Notes 72.
Ruckenstein, E., and B. Pulvermacher. 1973. *Am. Inst. Chem. Eng. J. 19*:356.
Saito, K., R. J. Baltisberger, V. I. Stenburg, and N. F. Woolsey. 1981. *Fuel 60*:1039.
Sama, D. A., T. Talverdian, and T. Shadman. 1985. *Fuel 64*:1208.
Schafer, H. N. S. 1970. *Fuel 49*:197, 271.
Schobert, Harold H. 1989. *Coal*, American Chemical Society, Washington, D.C.
Schulten, H. R. 1982. *Fuel 61*:670.
Schulten, H. R., and A. Marzec. 1987. *Fuel Proc. Technol. 15*:307.
Schultz, J. L., R. A. Friedel, and A. G. Sharkey. 1965. *Fuel 44*: 55.
Selucky, M. L. 1983. *Anal. Chem. 55*:141.
Sharkey, A. G., Jr., T. Kessler, and R. A. Friedel. 1975. In *Trace Elements in Fuel* (S. P. Babu, ed.), Adv. in Chem. Series No. 141, Am. Chem. Soc., p. 1.

Shenkin, P. S. 1983. *Prepr. Am. Chem. Soc. Div. Pet. Chem.* *28*(5): 1367.
Shoolery, J. N., and W. L. Budde. 1976. *Anal. Chem.* *48*:1458.
Siegart, W. R., P. F. Curran, and S. B. Alpert. 1986. Texaco Coal Gasification Process: Commercial Plant Applications. *Prepr. Am. Chem. Soc., Div. Fuel Chem.* *31*(2):274–281.
Squires, Arthur M., Mooson Kwauk, and Amos A. Ardan. 1985. Fluid Beds: At Last, Challenging Two Entrenched Practices. *Science* *230*: 1329–1337.
Simoens, A. J., E. G. Derouane, and R. T. K. Baker. 1982. *J. Catalysis* *75*:175.
Singleton, K. E., R. G. Cooks, K. V. Wood, A. Rabinowich, and P. H. Given. 1987. *Fuel* *66*:74.
Sinha, R. K., and P. L. Walker. 1971. *Fuel* *50*:125.
Sklenar, V., M. Hajek, G. Sebor, I. Lang, M. Suckanek, and Z. Starcuk. 1980. *Anal. Chem.* *52*:1794.
Sleevi, P. S., T. E. Glass, and H. C. Dorn. 1979. *Anal. Chem.* *51*: 1931.
Snape, C. E. 1982. *Fuel* *61*:775.
Snape, C. E. 1983. *Fuel* *62*:621.
Snape, C. E., and K. D. Bartle. 1979. *Fuel* *58*:898.
Snape, C. E., and K. D. Bartle. 1984. *Fuel* *63* 883.
Snape, C. E., K. D. Bartle, I. O. Amaechina, and D. G. Mills. 1987. *Fuel Proc. Technol.* *16*:89.
Snape, C. E., W. R. Ladner, and K. D. Bartle. 1979. *Anal. Chem.* *51*:2189.
Snape, C. E., W. R. Ladner, and K. D. Bartle. 1985. *Fuel* *64*:1394.
Snape, C. E., W. R. Ladner, L. Petrakis, and B. C. Gates. 1984. *Fuel Proc. Technol.* *8*:155.
Snape, C. E., G. J. Ray, and C. T. Price. 1986. *Fuel* *65*:877.
Snape, C. E., C. A. Smith, K. D. Bartle, and R. S. Matthews. 1982. *Anal. Chem.* *54*:20.
Sober, C. S., W. E. Rhine, and K. K. Eisentraut. 1981. *Anal. Chem.* *53*:1099.
Solomon, P. R. 1979. *Prepr. Am. Chem. Soc. Div. Fuel Chem.* *24*(3): 184.
Solomon, P. R., and R. N. Cavangelo. 1982. *Fuel* *61*:663.
Spackman, W. 1975. The Nature of Coal and Coal Seams and A Synopsis of North American Coals. In *The Fundamental Organic Chemistry of Coal* (John W. Larsen, ed.), University of Tennessee, Knoxville, pp. 10–41.
Spackman, W., A. Davis, and G. D. Mitchell. 1976. The Fluorescence of Liptinite Macerals. Brigham Young University Geol. Studies 22 (Part 3):60–61.
Speight, J. G. 1978. *Analytical Methods for Coal and Coal Products*, Academic Press, New York, Chap. 22.

Speight, J. G. 1983. *The Chemistry and Technology of Coal*, Marcel Dekker, New York.
Speight, J. G., R. B. Long, and T. D. Trowbridge. 1984. *Fuel 63*: 616.
Spencer, D. F., M. J. Gluckman, and S. B. Alpert. 1982. Coal Gasification for Electric Power Generation. *Science 215*(4540):1572, 1573.
Stach, E., M.-Th. Mackowshy, M. Teichmuller, G. H. Taylor, D. Chandra, and R. Teichmuller. 1975. *Coal Petrology* (D. G. Murchison, G. H. Taylor, and F. Zierke, eds.), Gebruder Borntraeger, Berlin.
Stainbaugh, E. P. 1975. Batelle Hydrothermal Coal Process. *Coal Age* August, pp. 72–74.
Sternberg, H. W., and C. L. Delle Donne. 1974. *Fuel 53*:172.
Sternberg, H. W., C. L. Delle Donne, P. Pantages, E. C. Moroni, and R. E. Markby. 1971. *Fuel 50*:432–442
Stock, L. M. 1982. *Coal Science*, Vol. 1 (M. L. Gorbaty, J. W. Larsen, and I. Wender, eds.), Academic Press, New York, pp. 161–282.
Stock, L. M., and K-T. Tse. 1983. *Fuel 62*:974.
Straszheim, J. M., R. T. Green, and R. Markuszewski. 1983. *Fuel 62*:1070.
Sun, S. C. 1954. Hypothesis for Different Floatabilities of Coals, Carbons and Hydrocarbon Minerals. *Trans. AIME 199*:67–75.
Sun, S. C. 1974. *Survey of Energy Resources*. World Energy Conference, London, pp. 30, 31.
Sun, S. C. 1974. *Synthetic Fuels from Coal Task Force Report*, Project Independence Blueprint, Federal Energy Administration, November.
Taylor, J., and R. Smith. 1980. Power in the Peat Lands. *New Scientist 88*:644–646.
Thiessen, R. 1920. Occurrence and Origin of Finely Disseminated Sulfur Compounds in Coal. *Min. and Met.* No. 157 Sect. 12:40–44.
Thom, W. T. 1929. *Petroleum and Coal: The Keys to the Future*, Princeton University Press, Princeton, NJ, pp. 42–44.
Thomas, J. M. 1965. In *The Chemistry and Physics of Carbon*, Vol. 1 (P. L. Walker, ed.), Marcel Dekker, New York, p. 121.
Thomas, J. M., and P. L. Walker. 1964. *J. Chem. Phys. 41*:587.
Thomas, J. M., and P. L. Walker. 1965. *Carbon 2*:434.
Timmer, J. M., and N. van der Burgh. 1984. *Fuel 63*:1645.
Tomita, A., Y. Ohtsuka, and Y. Tamai. 1983. *Fuel 62*:150.
Towne, S. E., Y. T. Shah, G. D. Holder, G. D. Deshpande, and D. C. Cronauer. 1985. Liquefaction of Coal Using Supercritical Fluid Mixtures. *Fuel 64*:883–889.
Tripathi, P. S. M. 1979. *Erdol Kohle Erdgas Petrochemie* 6:256.

Tsai, S. 1982. *Fundamentals of Coal Beneficiation and Utilization*, Elsevier, New York.

Tschamler, H., and DeRuiter, E. 1966. A Comparative Study of Exinite, Vitrinite and Micrinite. In *Coal Science*, Adv. in Chem. Ser. 55 (R. F. Gould, ed.), Am. Chem. Soc., Washington, D.C., pp. 330–341.

Tseng, B-H., M. Buckentin, K. C. Hsieh, C. A. Wert, and G. Dyrkacz. 1986. *Fuel 65*:385.

Tytko, A. P., K. D. Bartle, N. Taylor, I. O. Amaechina, and A. Pomfret. 1987. *Fuel 66*:1060.

Tytko, A. P., K. D. Bartle, N. Taylor, M. A. Thomson, W. Kemp, and W. Steedman. 1985. *Fuel 64*:1024.

Unterzaucher, J. 1952. *Analyst 77*:584.

U.S. Department of Energy, *Annual Energy Outlook 1985* DOE/EIA-0383(85), Washington, D.C.

U.S. Department of Energy, *Annual Energy Outlook for U.S. Coal 1986* DOE/EIA-0484(86), Washington, D.C.

U.S. Department of Energy, *Annual Energy Review 1985* DOE/EIA-0384(85), Washington, D.C.

U.S. Department of Energy, *Coal Data: A Reference 1982* DOE/EIA-0064(82), Washington, D.C.

U.S. Department of Energy, *Coal Distribution January–December 1985* DOE/EIA-0125(85/4Q), Washington, D.C.

U.S. Department of Energy, *International Energy Annual 1985* DOE/EIA-0219(85), Washington, D.C.

van Krevelen, D. W. 1961. *Coal*, Elsevier, New York.

van Krevelen, D. W. 1963. Geochemistry of Coal. In *Organic Geochemistry* (I. A. Breger, ed.), Pergamon Press, New York, pp. 183–247.

van Krevelen, D. W., and J. Schuyer. 1957. *Coal Science*, Elsevier, Amsterdam, p. 352.

Van Ness, K., and H. A. Van Westen. 1951. *Aspects of the Constitution of Mineral Oils*, Elsevier, New York.

Vaughan, G. A., and J. J. Swithenbank. 1965. *Analyst 90*:594, 1970; *95*:890.

Vassilaros, D. L., R. C. Kong, D. W. Later, and M. L. Lee. 1982. *J. Chromatogr. 252*:1.

Vastola, F. J., and L. J. McGahan. 1987. *Fuel 66*:886.

Veraa, M. J., and A. T. Bell. 1978. Effects of Alkali Metal Catalysts on Gasification of Coal. *Fuel 57*:194.

Vissac, G. A. 1955. Coal Preparation with the Modern Feldspar Jig. *Trans. AIME 202*:649–655.

Vogt, Erich V., Paul J. Weller, and Maarten J. Vanderburgt. 1984. The Shell Coal Gasification Process, Chap. 3-2, 3-27–3-44. In *Handbook of Synfuels Technology* (R. A. Meyers, ed.), McGraw-Hill, New York.

Walker, P. L. 1973. Coal Characteristics and Gasification Behavior. Papr. Short Course on Coal Characteristics and Coal Conversion Processes, Oct. 29–Nov. 2, Penn. State University, University Park, PA.

Walker, P. L., F. Rusinko, and L. G. Austin. 1959. In *Advances in Catalysis*, Vol. 11 (D. D. Eley, P. W. Selwood, and P. B. Weisz, eds.), Academic Press, New York, p. 133.

Walker, P. L., M. Shelef, and R. A. Anderson. 1968. In *The Chemistry and Physics of Carbon*, Vol. 4 (P. L. Walker, ed.), Marcel Dekker, New York, p. 287.

Wallace, S., K. D. Bartle, D. L. Perry, M. G. Hodges, and N. Taylor. 1987. *1987 International Conference in Coal Science* (J. A. Moulijn, ed.), Elsevier, Amsterdam, p. 9.

Wang, S-H., and P. R. Griffiths. 1985. *Fuel 64*:229.

Watt, J. D. 1968. The Occurrence, origin, distribution and estimation of the mineral species in British Coals, BCURA Lit. Survey.

Weiss, G. 1976. K. prubehu zimen proukelneni s hloubkou v cs. casti hornoslezske panve. *Sb Geol. pruzkumu Ostrave* Ostrava *11*:7–34.

Wen, C. Y., ed., and E. Stanley Lee. 1979. *Coal Conversion Technology*, Addison-Wesley, Reading, MA.

Wender, I., L. A. Heredy, M. B. Neuworth, and I. G. C. Dryden. 1981. In Chemistry of Coal Utilisation, 2nd Suppl. Vol. (M. A. Elliott, ed.), John Wiley, New York, p. 425.

White, C. M. 1983. *Handbook of Polycyclic Aromatic Hydrocarbons*, Vol. 1 (H. Bjorseth, ed.), Marcel Dekker, New York, Chap. 13.

White, D., and R. Thiessen. 1913. The Origin of Coal. United States Bureau of Mines Bulletin No. 38, Pittsburgh, PA.

Wigmans, T., and J. A. Moulijn. 1981. In *New Horizons in Catalysis*, Vol. 7A (T. Seiyama and K. Tanabe, eds.), Elsevier, New York, p. 501.

Wigmans, T., H. Haringa, and J. A. Moulijn. 1984. *Fuel 63*:185.

Wigmans, T., F. van Dam, and J. A. Moulijn. 1981. *Carbon 19*:309.

Williamson, I. A. 1967. *Coal Mining Geology*, Oxford University Press, New York.

Winans, R. E. 1984. Artificial Coalification Study: Preparation and Characterization of Synthetic Macerals. *Org. Geochem. 6*:463–471.

Wind, R. A. 1986. *Prepr. Am. Chem. Soc. Div. Fuel Chem. 31*(1): 223.

Wolf, M. 1983. *Int. J. Mass Spec. Ion Phys. 46*:487.

Wong, C., and R. T. Yang. 1984. *Ind. Eng. Chem. Fund. 23*:298.

Woo, C. S., A. P. D'Silva, and V. E. Fassel. 1980. *Anal. Chem. 52*: 159.

Wood, B. J., and K. M. Sancier. 1984. *Catalysis Rev. Sci. Eng. 26*: 233.

Wozniak, T. J., and R. A. Hites. 1983. *Anal. Chem. 55*:1791.

Wynblatt, P., and N. A. Gjostein. 1975. *Prog. Solid State Chem.* *9*:21.
Yang, R. T., and M. Steinberg. 1977. *J. Phys. Chem.* *81*:1117.
Yarzeb, R. F., Z. Baset, and P. H. Given. 1979. *Geochim. Cosmochim. Acta 43*:281.
Yoshida, T., R. Yoshida, T. Maekawa, Y. Yoshida, and Y. Itagaki. 1979. *Fuel 58*:153.
Zander, M. 1966. *Erdol Kohle Erdgas Petrochemi 19*:279.
Zander, M. 1968. *Phosphorimetry*, Academic Press, New York.
Zeilinger, J. E., and A. W. Deurbrouck. 1976. Physical Desulfurization of Fine-Size Coals on a Spiral Concentrator. United States Bureau of Mines Report RI-8152, Washington, D.C.
Zielke, C. W., R. T. Struck, and E. Gorin. 1969. Fluidized Combustion Process for Regeneration of Spent Zinc Chloride Catalysts. *Ind. Eng. Chem. Process Des. Dev.* *8*(4):552.
Zubovic, P. 1975. Geochemistry of Trace Elements in Coal. United States Environmental Protection Agency, Off. Res. Dev. EPA Report No. EPA-600/2-76-149, U.S. Geological Survey, Reston, VA. p. 12A.

ADDED IN PROOFS

Axelson, D. E. 1987. *Fuel 66*:197.
Davies, I. L., M. W. Raynor, D. J. Unwin, and K. D. Bartle. 1988. *J. High Resoln. Chromatogr. 11*:792.
Dereppe, J. M., J. P. Boudou, C. Moreaux, and B. Durand. 1983. *Fuel 62*:575.
Gerhards, R. 1983. *Z. Anal. Chem. 316(2)*:231.
Jewell, D. M., E. W. Albaugh, B. E. Davis, and R. C. Ruberto. 1974. *I. & EC Fundamentals 13*:278.
Ladner, W. R., and C. E. Snape. 1985. *Proc. Int. Conf. on Coal Sci. 948*, Pergamon.
Rosenberger, H., G. Scheler, and K.-H. Rentrop. 1983. *Z. Chem. 23*:34.

Part IV

Oil Shale

25

Origin, Occurrence, and Recovery

I	Introduction	795
II	Definitions and Terminology	798
III	Origin	799
	A. Sedimentation and Mineralogy	801
	B. Production of Organic Matter	815
	C. Kerogen Types and Maturation	820
	D. Preservation of Organic Matter	823
	E. Marine Oil Shales	828
	F. Lacustrine Oil Shales	831
IV	Occurrence	832
V	Recovery	832
	A. Underground Mining	832
	B. Surface Mining	836
	C. Size Reduction	837

I. INTRODUCTION

Oil shales comprise a truly enormous and largely untapped fossil fuel resource. As readily accessible petroleum sources dwindle (Chap. 1, Sec. I), causing petroleum prices to rise, utilization of the oil shale resource will become economically attractive. Worldwide oil shale deposits are estimated to contain 30 trillion (22×10^{12}) barrels of shale oil, but only a small fraction (3.7×10^{12} bbl) of this amount is easily recoverable using current technology. Thus, the utilization of oil shale to replace petroleum will mean finding economically efficient and environmentally acceptable methods for recovering the energy-rich organic material locked inside the oil shale's rock matrix and for upgrading the recovered shale oil. This is a formidable challenge. Oil shales are complex, intimate mixtures of organic and inorganic materials and vary widely in their compositions and properties. Some, such as the "oil shale" of the Green River formation, are not even true shales. Today, some three centuries of research and development are behind us as we address the many scientific, technological, and environmental aspects of the oil shale challenge.

Oil shale technology has a long history. Oil shales were sources of oil as early as 800 A.D. and the British oil shale deposits were worked in Phoenician times (Russell, 1986). The use of oil shale was recorded in Austria in 1350. The first shale oil patent, British Crown Patent No. 330, was issued in 1694 to Martin Eele, Thomas Hancock, and William Portlock, who "*after much paines and expences hath certainely found out a way to extract and make great quantityes of pitch, tarr and oyle out of a sort of rock*" (Eele et al., 1694). Oil shale utilization on an industrial scale, however, did not follow immediately. Not until 1838 was the first industrial oil shale plant put into service at Autun, France (Duncan and Swanson, 1965). Soon plants in Scotland (1850), Australia (1865), and Brazil (1881) followed. In addition to being a source of refined shale oil products, it was soon discovered that torbanite, an especially organic-rich type of oil shale, was useful for increasing the luminosity of illuminating gas flames (Silliman, 1869). This provided an important market for the early Scottish oil shale industry and later, as the Scottish torbanite deposits were depleted, for the early Australian shale industry. By the 1870s, Australian torbanite was being exported not only to Great Britain, but also to the United States, Italy, France, and Holland (Cane, 1979). The invention of the Welsbach gas mantle and the advent of low-cost, high-quality kerosene from American petroleum spelled the end of this period. As the need for liquid transportation fuels increased, the Australian oil shale operations consolidated. Elsewhere, oil shale plants followed in New Zealand (1900), Switzerland (1915), Sweden (1921), Estonia (now USSR, 1921), Spain (1922), China (1929), and South Africa (1935) (Prien, 1976). The high point of this stage of oil shale development was reached during, or just after, World War II. However, the oil shale industry in Estonia and neighboring Leningrad Province still flourishes, with most of the mined shale being burned directly in electric power generating plants with the remaining 10% being retorted to provide chemical feedstock and smaller quantities of refined products (Schmitz and Tolle, 1987). The Japanese began in 1926 the commercial production of shale oil from the large Chinese oil shale deposits at Fushun in Manchuria.

Improved retorts were installed at this complex in 1941 to provide important supplies of liquid fuels for the Japanese forces during World War II. At Maoming, near Canton in southern China, a second oil shale project was developed. Shale oil production in the PRC peaked about 1975 and has since declined as emphasis has shifted to newly discovered petroleum supplies (Russell, 1986; Schmitz and Tolle, 1987).

Oil shale utilization will involve several different kinds of technologies. In any case, the major areas will include mining, size reduction, retorting or other means of recovering shale oil from the rock. disposal of the spent shale, and upgrading the shale oil into marketable products.

Estimates of required selling prices for synfuels have, through the years, been very badly off the mark. However, estimating the potential for lowering shale oil costs by improving the economics of one or another of the various technologies can be useful in assigning priorities for research and development efforts. Cost breakdowns have been estimated for shale oil production from Green River oil shale and Estonian kukersite (Tables 1 and 2).

It is interesting to note how similar these estimates are, even though their cost bases are obviously different, they differ by years in time (1980 versus 1987) and were made for shales half a world apart.

Table 1 Estimated Cost Breakdowns for Shale Oil Production from Green River Oil Shale

Operation	Capital expense (%)	Operating expense (%)	Scalability
Mining	8	35	Marginal
Shape preparation	10	15	Marginal
Preheat-retorting	7	5	Very good
Combustion	15	12	Very good
Power/heat systems	20	12	Medium
Product upgrading	25	16	Very good
Infrastructure	15	5	Good

Source: Voetter et al., 1987.

Table 2 Estimated Cost Breakdown for Production of Shale Oil and Chemical Byproducts from Estonian Kukersite

		Production costs (% of 1959 total)	
		1959	1977
Shale feed (incl. mining, prep.)		64.5	58.6
Utilities:	Electric	5.1	3.9
	Heat	12.0	2.2
	Water	0.6	0.9
Labor		6.3	2.3
Fixed costs		11.4	6.5
By-product credit		-19.6	-24.4
Net production cost		80.4	74.4

Source: Mashin et al., 1980.

Yet, in each case the costs of getting the shale out of the ground and ready to process represent fully one-half the total cost. It is also interesting to note that substantial reductions in the cost of the Estonian shale oil were achieved, even though this industry is about 50 years old and hence should be relatively mature.

The near-term future of oil shale is uncertain. Very clearly, this future will be influenced by international crude oil prices and supplies. Indeed, as pointed out (Hutton et al., 1987b), the rise of interest in oil shale during the late 1970s was due largely to the high prices and tight supply of crude oil. With the decline of crude prices and growth of a crude oil surplus, interest in oil shale and other synfuels waned. How long this will continue is difficult to predict.

II. DEFINITIONS AND TERMINOLOGY

There is no "easy" scientific definition of oil shale; the definition is strictly an economic one: According to Gavin (1924), "Oil shale is a compact, laminated rock of sedimentary origin, yielding over 33% of ash and containing organic matter that yields oil when distilled, but not appreciably when extracted with the ordinary solvents for petroleum." Materials containing <33% ash should be considered coals, however, this distinction will be of little importance in the following discussion. Thus, the term "oil shale" will be used to denote an organic-rich rock that contains little or no free oil.

Shale oil is defined as the oil produced from an oil shale on heating. Three other terms will be used extensively, hence their definitions are important: *Bitumen* is defined as the organic material which can be extracted by ordinary organic solvents, such as benzene, toluene, tetrahydrofuran (THF), and chloroform ($CHCl_3$), or mixtures (generally azeotropes) of solvents, such as benzene methanol (60:40) (Chap. 2, Sec. I.A.). *Kerogen*, which comprises the major part of the organic material, is *not* soluble in such solvents (Steuart, 1912). Of course, these are operational definitions and the relative proportions of bitumen and kerogen depend on the choice of solvent and extraction conditions. Nevertheless, these are useful definitions provided their limitations are kept in mind. The third term, *kerogen concentrate*, refers to the organic concentrate that is produced by beneficiation or chemical demineralization of an oil shale. Strictly speaking, this term should refer only to that part of the organic concentrate that is insoluble in organic solvents. However, in common usage, kerogen concentrate refers to the total organic material (kerogen + bitumen) obtained by removing minerals. The common usage will be employed in this discussion, except where specifically indicated.

III. ORIGIN

Oil shales were formed by the inclusion of materials containing organic carbon into sediments that eventually become sedimentary rocks. To understand their origin and formation, we begin by considering the distribution and cycling of carbon in the environment (Figure 1; see also Chap. 1, Sec. II).

Plants use CO_2 from the atmosphere to build their cells. Animals eat the plants and return CO_2 to the atmosphere as a product of metabolism. The decay of dead organisms returns additional CO_2 to the environment. The action of acids on carbonate rocks liberates CO_2 into the atmosphere, while rain and the precipitation of insoluble carbonates removes CO_2. Only a minute fraction of the carbon that cycles through this marvelously complex cycle is preserved in the sediments that yield fossil fuels.

Hunt estimated the amount of carbon present in sedimentary rocks as hydrocarbons (including N-, S-, and O-containing hydrocarbon derivatives) that contain reduced carbon and as oxidized carbon forms (Table 3).

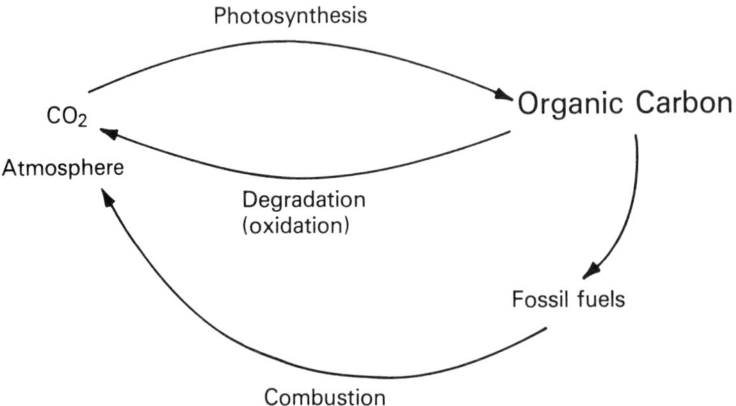

Figure 1 Simplified representation of the carbon cycle.

While these data are only approximate, they do indicate that about one-sixth of the global carbon is present in reduced forms. Unfortunately, much of this reduced carbon is widely dispersed and not easily recovered. This is especially true for the very large amount of reduced carbon held in clays and shales, where the organic matter is almost always a minor—often a trace—constituent. Consequently, only a small fraction of the carbon in clays and shales is likely to be useful as a fuel resource.

The current oil shale technology is well suited for processing the richest (organic-rich) oil shale deposits; those that yield over 10 bbl of shale oil per ton of rock. Using the available data (Tables 3 and 4), it can be calculated that present oil shale technology can efficiently recover only 0.06% of the potential oil in relatively rich shales (>10 gal/ton)—and only about one-billionth of the total carbon in clays and shales! The challenge is clear; even a marginal improvement in technology could greatly increase this minute fraction of recoverable shale oil.

Compared to known or even anticipated petroleum resources, the oil shale resource is enormous. This is not surprising, since shales are source rocks for many of the world's crude oils (Chap. 1, Sec. II). Moreover, geochemical studies indicate that only a small fraction of the carbon in a source rock is typically released as liquids into migration channels that lead to a petroleum reservoir. Indeed, many crude oils can be regarded as shale oils produced by natural heating and upgraded into petroleum during migration from the shale source rock to the reservoir (Tissot and Welte, 1984). Thus, a more useful comparison

Table 3 Global Carbon Inventory

	Organic carbon (reduced forms)	Carbonate carbon (oxidized form)
All sediments	Insoluble Organic Matter	
Clays and shales	8,900	9,300
Carbonates	1,800	51,100
Sands	1,300	3,900
Coal beds thicker than 15 ft	15	
Nonreservoir rocks	Spluble Organic Matter	
Asphalt	275	
Petroleum	265	
Reservoir rocks		
Asphalt	0.5	
Petroleum	1.1	
	∿13,000	∿64,000

[a]Carbon, 10^{18} g in sedimentary rocks. One barrel of oil contains about 10^5 g of carbon.
Source: Hunt, 1979.

is between the proven and anticipated reserves of crude oil, coal, and natural gas with the shale oil that could be recovered using current or improved technology (Table 5) (Thorne, 1964; Duncan and Swanson, 1965; Ward, 1984).

Thus, both the coal and oil shale resources dwarf the petroleum resource; the problem is economical recovery. This is difficult to achieve because of the widely varying properties of oil shales (Table 6).

A. Sedimentation and Mineralology

Oil shales are fine-grained sedimentary rocks in which is dispersed a minor fraction of organic matter as tiny particles. Formation of an oil shale requires simultaneous sources of both fine-grained minerals and organics, under conditions where the organics can be preserved. In addition to these clastic, detrital components, biogenic and authigenic

Table 4 Shale-Oil Resources of the World's Land Areas (10^9 bbl[a])

Range in grade, oil yield in gal/ton[c]	Recoverable known resources	Known resources			Total resources[b]		
	10–100	5–10	10–25	25–100	5–10	10–25	25–100
Africa	10	ne[d]	ne	ne	4,000	80,000	450,000
Asia	20	2	3,700	ne	5,500	110,000	590,000
Australia and New Zealand	24	ne	84	200	1,000	20,000	100,000
Europe	30	100	200	ne	1,400	26,000	140,000
North America	80	900	2,500	4,000	3,000	50,000	260,000
South America	50	ne	3,200	4,000	2,000	40,000	210,000
Total	214	1,000	9,600	8,000	17,000	325,000	1,750,000

[a]To convert bbl to m^3, divide by 6.29.
[b]Includes oil shale in known resources and anticipated extensions of known resources.
[c]To convert gal/ton to liters/t, divide by 0.2397.
Source: Duncan and Swanson, 1965.
[d]Not estimated.

Table 5 Estimated World Reserves of Oil Shale, Coal, Petroleum, and Natural Gas

	Total (estimated)		Recoverable (estimated)	
Resource	tons ($\times 10^9$)	bbl ($\times 10^9$)	tons ($\times 10^9$)	bbl ($\times 10^9$)
Shale oil	3,100	22,700	500	3,700
Coal	11,000	--	987	–
Petroleum[b]	400	3,000	95	700
Natural gas	200	–	20	–

[a]See also Table 5 in Chap. 1, Table 3 in Chap. 12, and Table 24 in Chap. 17.
[b]Tar sand resources are not included in these estimates.

minerals (see also Chap. 1, Sec. II.E) are also present in most shales. Detrital minerals typically include quartz, feldspar, and clays (often including illite, montmorillonite, and kaolinite), and sometimes volcanic ash. Biogenic minerals include amorphous silica and calcium carbonate, usually in very minor amounts. Authigenic minerals typically include pyrite and other metal sulfides, carbonates (calcite, dolomite, siderite), chert, and phosphates. Authigenic silica from clay diagenesis is also an important mineral in many shales, serving to cement together the larger detrital particles. The saline minerals trona, nacholite, dawsonite, and halite are often important in oil shales (e.g., Green River shales) formed in stratified lacustrine environments. Not surprisingly, halite (NaCl) is usually present in marine oil shales.

The mineralogy of some representative oil shales from around the world is summarized in Tables 7 and 8.

Mineral assemblages vary from shale to shale; from top to bottom within a given oil shale deposit; even from sample to sample within a given stratum. This complicates mineralogy studies. As a result, comprehensive mineralogy studies of oil shales are few; most studies have focused on the occurrence or genesis of one particular mineral or mineral type, while other studies have concentrated on minerals of economic interest. This makes it difficult to get a broad view of the mineral assemblages in even major deposits. The Green River formation has been extensively studied and a summary of its mineralogy, based largely on the work of Bradley (1929, 1931, 1970, 1973) and

Table 6 Properties of Various Oil Shales

	Australia (Glen Davis)	Brazil (Trem-embe-Taubate)	Canada (Nova Scotia)	France (Autun)	Israel (Um Barek)	Lebanon	Manchuria (Fushun)	New Zealand (Ore-puki)	Scotland (West-wood mine)	South Africa (Ermelo)	Spain (Puertol-leno)	Sweden (Kvarn-torp)	Thailand (Maesod)	United States (Colo-rado)
Modified Fischer assay														
Oil, liters/t[a]	414	156	257	129	78	307	38	331	111	228	234	70	357	122
Oil, wt %	30.0	11.5	18.8	9.7	6.4	24.8	3.0	24.8	8.2	17.6	17.6	5.7	26.1	9.3
Water, wt %	0.7	6.2	0.8	3.2	2.2	11.0	4.9	8.3	2.2	3.0	1.8	2.0	3.8	1.0
Spent shale, wt %	64.1	78.4	77.7	84.0	88.4	56.5	90.3	57.6	86.6	75.6	78.4	87.2	66.3	87.5
Gas and loss, wt %	4.3	3.9	2.7	3.1	3.0	7.7	1.8	9.3	3.0	3.8	2.2	5.1	3.8	1.6
Conversion of organic material to oil,[b] wt %	66	59	60	44	48		33	45	56[c]	34	57[c]	26	71	70
Rock characteristics														
sp. gr. (at 16°C)	1.60	1.70		2.03			2.29	1.46	2.22	1.58	1.80	2.09	1.61	2.23
Heating value, mJ/kg[d]	18.8	8.2	12.6	8.9			3.4	21.3	5.9	19.1	12.5	9.0	15.4	5.1
Ash, wt %	51.6	71.4	62.4	70.8	60.0	18.8	82.7	32.7	77.8	42.5	62.8	72.1	56.4	66.9
Organic carbon, wt %	39.8	16.5	26.3	18.8	10.6		7.9	45.7	12.3	43.8	26.0	18.8	30.8	11.3
Assay oil														
sp. gr. (at 16°C)	0.89	0.88	0.88	0.90	0.97	0.96	0.92	0.90	0.88	0.93	0.90	0.98	0.88	0.91
Carbon, wt %	85.4	84.3		84.9	79.6	83.2	85.7	83.4		84.8		85.0	84.4	84.6
Hydrogen, wt %	12.0	12.0		11.4	9.8	10.3	10.7	11.8		11.1		9.0	12.4	11.6
Nitrogen, wt %	0.5	1.1		0.8	1.4	0.6		0.6			0.9	0.7	1.1	1.8
Sulfur, wt %	0.4	0.2		0.3	6.2	1.5		0.6		0.6	0.3	1.7	0.4	0.5
Ash analysis, wt %														
SiO_2	81.5	55.8	61.1	55.1	∿26		62.3	44.2	55.7	61.3	56.6	62.4	60.8	43.6
Al_2O_3	10.1	26.7	30.1	27.6			26.7	28.1	25.1	30.5	27.6	17.6	19.9	11.1
Fe_2O	3.0	8.5	5.0	9.3			6.1	20.5	9.9	2.9	9.1	10.7	4.8	4.6
CaO	0.8	2.8	1.1	1.7	∿45		0.1	4.6	2.6	1.6	2.6	1.2	3.3	22.7
MgO	0.8	3.7	1.6	1.9			1.8	1.4	3.1	1.7	2.2	1.7	3.8	10.0
Other oxides	3.8	2.5	1.1	4.4			3.0	1.2	3.6	2.1	1.9	6.4	7.4	8.0

[a]To convert liters/t to gal/short ton, multiply by 0.2397.
[b]Based on recovery of carbon in oil from organic carbon in shale.
[c]Carbon content of oil is 84 wt %.
[d]To convert mJ/k to btu/lb, multiply by 430.4

Source: Thorne, 1964.

reviews by Milton (1971) and Smith (1983), serves to illustrate the range of minerals that can be present in a single oil shale deposit (Tables 9 and 10).

The Green River oil shales were formed in the waters of ancient lakes that covered much of what is now northwestern Colorado, southwestern Wyoming, and northeastern Utah (Figure 2). There were two, and possibly three, of these lakes: Lake Gosuite in Wyoming, Lake Uinta in Colorado, and an unnamed lake in Utah. These lakes existed for a very long time, some 4–6 million years through the early and middle Eocene. As climatic conditions varied the lakes expanded, sometimes (at least partially) merging, and contracted. Nevertheless, continued subsidence caused by downward warping of the lake beds enabled the deposition of sediments that are more than 2000 ft thick in some areas (Figure 3).

Green River oil shales have a very fine texture. In a composite sample from the U.S. Bureau of Mines site at Rifle, Colorado, Tisot and Murphy (1963) found that >99 wt % of the mineral particles were smaller than 44 μm, 75 wt % were in the range of 2–20 and 15 wt % were smaller than 2. The surface area of these particles was low and mainly external. Some mineral micropore structure was found; pore sizes ranged from 10 to 100 Å. Essentially all of the organic matter was in particles lodged between the mineral grains; it was estimated that less than 4% of the organic matter was contained within mineral particle pores.

The lithology of the Green River formation is complicated, in part at least because fluvial material was deposited by creeks and rivers at the edges of the lakes—edges that moved as the lake levels rose and fell. There are, however, some striking uniformities (McDonald, 1972). The composition of the organic matter in the oil shale is remarkably uniform, especially within the Piceance Creek Basin on which the following discussion will concentrate. The most prominent variation going from top to bottom of the deposit is a decrease in carboxylic acid groups (increased decarboxylation) in the more deeply buried shales. Lateral uniformity is also remarkable. Many fine details in core samples can be tracked across distances of 100 miles and major features can be tracked over twice this distance. This uniformity of organic matter is even more remarkable given the great diversity of the accompanying minerals and suggests that it may be possible to draw some general conclusions about the conditions of deposition. To explain the lithology and the chemistry that formed the organic-rich oil shales, it has been proposed that the ancient lakes were stratified through much of their existence (Bradley, 1925; Smith, 1980, 1983; Figure 4).

The lake history can be divided into three main periods. During an initial period of 1–2 million years, clay-rich oil shales (Garden Gulch and Douglas Creek Members in the Piceance Creek Basin,

Table 7 Minerology of Some Oil Shales

Shale	Ash (%)	Mineral matter (%)	Amorphous silica and quartz (%)	Feldspar (%)
Kukersite, Estonia	36.3	47.87	9.0	6.75
Kohat, N.W.F.P. India	68.7	88.83	12.40	2.47
Broxburn, Main	67.4	76.15	16.55	11.30
Kimmeridge, Dorset	37.8	40.89	38.97	5.74
Ermelo, Transvaal	44.9	47.85	50.13	5.14
Tasmanite, Tasmania	79.2	82.05	56.3	6.0
Amherst, Burma	43.9	46.78	34.33	5.63
Boghead, Autun	65.0	79.2	32.4	n.d.
Pumpherston I	75.0	83.45	24.6	n.d.
Pumpherston II	66.3	86.77	19.3	n.d.
Middle Dunnet	77.6	84.76	26.5	n.d.
Newnes, N.S.W.	20.1	20.79	74.0	n.d.
Cypris shale, Brazil	65.9	69.50	66.33	n.d.
Massive shale, Brazil	72.8	48.5	n.d.	37.2

[a]Apparently Himus did not consider the presence of dolomite (editorial comment).
n.d. = not determined.
Source: After Himus, 1951.

Tipton Shale in the Green River Basin) were deposited. Lake Uinta probably began as a normal, shallow lake and deposited typical lacustrine sediments for a considerable period as it grew, in both depth and extent. Thermal density differences abetted by low circulation in the isolated lake probably initiated stratification and hydrolysis of aluminosilicates began to build up chemical stratification. Using albite as an example (Garrels and Mackenzie, 1971);

$$\underset{\text{albite}}{2NaAlSi_3O_8} + 2CO_2 + 11H_2O \rightarrow \underset{\text{kaolinite}}{Al_2Si_2O_5(OH)_4} + 2Na^+ + 2HCO_3^- + 4H_4SiO_4$$

Clay minerals (%)	Gypsum ($CaSO_4$) *$2H_2O$ (%)	Pyrite (FeS_2) (%)	Calcite ($CaCO_3$) (%)	Magnesite[a] ($MgCO_3$) (%)	Siderite ($FeCO_3$) (%)
13.9	1.1	4.25	56.1	–	–
40.68	0.49	trace	22.68	8.34	3.76
45.85	0.43	1.76	2.91	2.63	11.24
20.68	8.56	4.64	3.51	–	–
29.45	0.24	2.03	1.73	0.24	–
23.75	1.45	1.64	–	–	–
27.45	5.49	0.19	trace	–	–
17.4	1.1	0.7	37.3	–	–
22.9	trace	2.35	5.8	4.15	2.15
22.9	0.3	1.35	26.7	12.1	5.1
54.65	0.3	0.55	4.25	3.65	–
17.9	0.3	0.4	–	3.3	–
17.13	0.78	1.24	5.25	0.62	–
37.2	0.4	1.4	2.6	1.1	–

In this reaction, CO_2 (from decomposition of organic matter) is consumed as albite is transformed into the clay, kaolinite, while sodium and bicarbonate ions are produced. Thus, as the lower, clay-rich sediments were deposited, the concentrations of dissolved sodium carbonates increased in the lower stratum of water just above. This would increase the density of the lower layer of the stratified lake. At the same time, CO_2 production depleted the oxygen content of the lower layer providing a reducing environment conducive to preservation of new organic sediment falling from the oxygenated and productive upper layer. As H_2S was produced by sulfate-reducing

Table 8 Mineralogy of Selected Australian and Overseas Oil Shale Samples

Sample	Specific gravity	Quartz	Clay minerals	Calcite	Siderite	Dolomite	Pyrite	Apatite	Others	% of Clays			
										Ka	Mo	Il	Ch
Condor (Carb.)	1.66	A	A				T			85	15	T	
Mt. Coolon	—	A	A				T			85	15		
Nagoorin (Carb.)	1.60	A	A		T		T			75	25		
Morwell	1.12	Not determined, less than 1% mineral matter											
Alpha	1.10	A	A				T	T		70	20	10	
Glen Davis	1.16	A	A				T		F		100		
Joadja	1.17	A	A				T			85	15		
Byfield	1.92	A	A		T		T	T	B	30	20	50	
Condor (brown)	2.41	A	A		T		T	T	B	70	20	10	
Duaringa (top)	1.52	A	A				T	T			95	5	
Duaringa (base)	1.60	A	A				T	T		30	55	15	
Lowmead	1.66	C	A				T	T		80	20		
Nagoorin (brown)	1.94	A	A		T		T			80	5	15	
Nagoorin South	1.65	A	A	T			T			25	75	T	
Rundle (MC member)	1.41	A	A			T	T			20	70	10	

Rundle (RC member)	1.62	A	A	T	T	T			25	65	10	
Stuart (HC member)	1.76	A	A	T		T			15	55	30	
Stuart (KC member)	1.52	A	A	C		T			5	90	5	
Yaamba	1.48	C	A			T			60	40		
Green River	2.01	C	T	A	T	T		A^c				
Camooweal	2.50	A	A	A	A	T			25	25	50	
Julia Creek	2.02	A	A	A		T			20	20	50	10
Irati*	2.15	A	A	T		T			35	65	5	
Kentucky (C member)*	2.24	A	A	T		T			5	5	80	10
Kentucky (S shale)*	2.12	A	A	T		T			5	5	80	10
Paris Basin*	2.26	A	A	C		T	T		25	10	55	10
Mersey River	2.04	A	A	T		T				5	75	20
Coorongite	1.01	Only surface sand										

Carb, carbonaceous; MC, Munduran Creek; RC, Ramsay Crossing, HC, Humpy Creek; KC, Kerosene Creek; Ka, kaolinite; Mo, montmorillonite; Il, illite; Ch, chlorite; A, abundant; C, common; T, trace (much less than 5%); B, buddingtonite (trace); F, feldspar (trace); A^c analcime (common); *Percentage of total clay minerals. No entry indicates not detected.

Table 9 Major Minerals in Green River Oil Shale

Ubiquitous	Common
Quartz, SiO_2	Illite, $KAl_2(AlSi_3)O_{10}(OH)_2$
Dolomite, $CaMg(CO_3)_2$	Calcite, $CaCO_3$
Pyrite, marcasite, FeS_2	Nacholite, $NaHCO_3$
Soda feldspar, $NaAlSi_3O_8$	Shortite, $Na_2Ca_2(CO_3)_2$
Potash feldspar, $KAlSi_3O_8$	Trona, $Na_2CO_3 \cdot NaHCO_3 \cdot 2H_2O$
	Dawsonite, $NaAl(OH)_2CO_3$

bacteria, the precipitation of pyrite and marcasite would be limited only by the availability of iron in solution. Eventually, precipitation of the sodium carbonates, such as nacholite ($NaHCO_3$), commenced. The resulting transition from normal lacustrine sediments to oil shale was not abrupt and there is no evidence for evaporite deposition at this point. Low circulation in the lake is indicated by even deposition of organic matter in tiny layers (varves). Under these conditions large clastic matter would be deposited near the lake shore and only airborne particles and the finer particles would be deposited in the deeper, stratified zone further from shore. Note that the anoxic lower layer (monimolimnion) in Figure 5 does not extend to the shoreline. The upper layer (mixolimnion) was oxygenated, and therefore capable of depositing normal lacustrine sediments, including the larger clastic particles, in the shallow water along the water's edge. Of course, as the lake expanded and contracted with variations in rainfall and runoff, the shore moved—and with it the region where lacustrine sediment was deposited.

The second stage in the evolution of the Green River formation was a relatively arid period. As the water level dropped, the large lakes separated into several smaller lakes in what we now know as the Piceance Creek (Colorado), Uinta (Utah), Green River, and Washakie (Wyoming) basins. These isolated lakes had individual characteristics that are reflected in a series of saline minerals, largely nacholite ($NaHCO_3$), halite (NaCl), dawsonite [$NaAl(OH)_3CO_3$], and some nordstrandite [$Al(OH)_3$] in the Piceance Creek Basin. Mostly trona ($Na_2CO_3 \cdot NaHCO_3 \cdot H_2O$) and shortite [$Na_2Ca_2(CO_3)_3$] were deposited during this period in the Green River and Washakie Creek Basins of Wyoming.

Table 10 Authigenic Silicates in Green River Formation

Name of mineral	Formula	Abundance
Clay Minerals		
Kaolinite	$H_4Al_2Si_xO_9$	Locally abundant
Stevensite	$(Al_{0.06}Fe_{0.04}Mg_{2.81}Li_{0.04})(Si_{3.98}Al_{0.02})O_{10}(OH)_2Na_{0.04}$	Locally abundant
Loughlinite	$H_{16}Na_2Mg_3Si_6O_{24}$	Locally abundant
Sepiolite	$H_4Mg_2Si_3O_{10}$	Rare
Talc	$Mg_3(OH)_2Si_4O_{10}$	Rare
Zeolites		
Analcite	$NaAlSi_2O_6 \cdot H_2O$	Widespread
Natrolite	$Na_2Al_2Si_3O_{10} \cdot 6H_2O$	Rare
Harmotome-wellsite	$(Ba,Ca,K_2)Al_2Si_6O_{10} \cdot 6H_2O$	Rare
Clinoptilolite-mordenite	$(Ca,Na_2K_2)(AlSi_5O_{12}) \cdot H_2O$	Rare
Borosilicates		
Searlesite	$NaBSi_2O_6 \cdot 6H_2O$	Locally abundant
Garrelsite	$(Ba,Ca,Mg)B_2SiO_6(OH)_3$	Rare
Leucosphenite	$CaBaNa_3BTi_3Si_9O_{29}$	Rare
Reedmergnerite	$NaBSi_3O_8$	Rare
Other Silicates		
Quartz	SiO_2	Ubiquitous
Orthoclase	$KAlSi_3O_8$	Widespread
Albite	$NaAlSi_3O_8$	Widespread
Acmite	$NaFeSi_2O_6$	Rare
Riebeckite	$Na(FeMg)_3Fe_2(OH,F)(Si_4O_{11})_2$	Rare
Labuntsovite	$(K,Ba,Na,Ca,Mn,Ti,Nb)(Si,Al_2(O,OH)_7H_2O$	Rare
Vinogradovite	$Na_5Ti_4AlSi_6O_{24} \cdot 3H_2O$	Very rare
Elpidite	$H_6Na_2ZrSi_6O_{18}$	Very rare
Natron-catapleiite	$H_4(Na_2Ca)ZrSi_3O_{11}$	Very rare
Biotite	$K(Fe,Mg)_3AlSi_3O_{10}(OH)_2$	Very rare
Hydrobiotite	$(K,H_2O)(Mg,Fe,Mn)_3AlSi_3O_{10}(OH,H_2O)_2$	Very rare

Source: Modified after Milton, 1971.

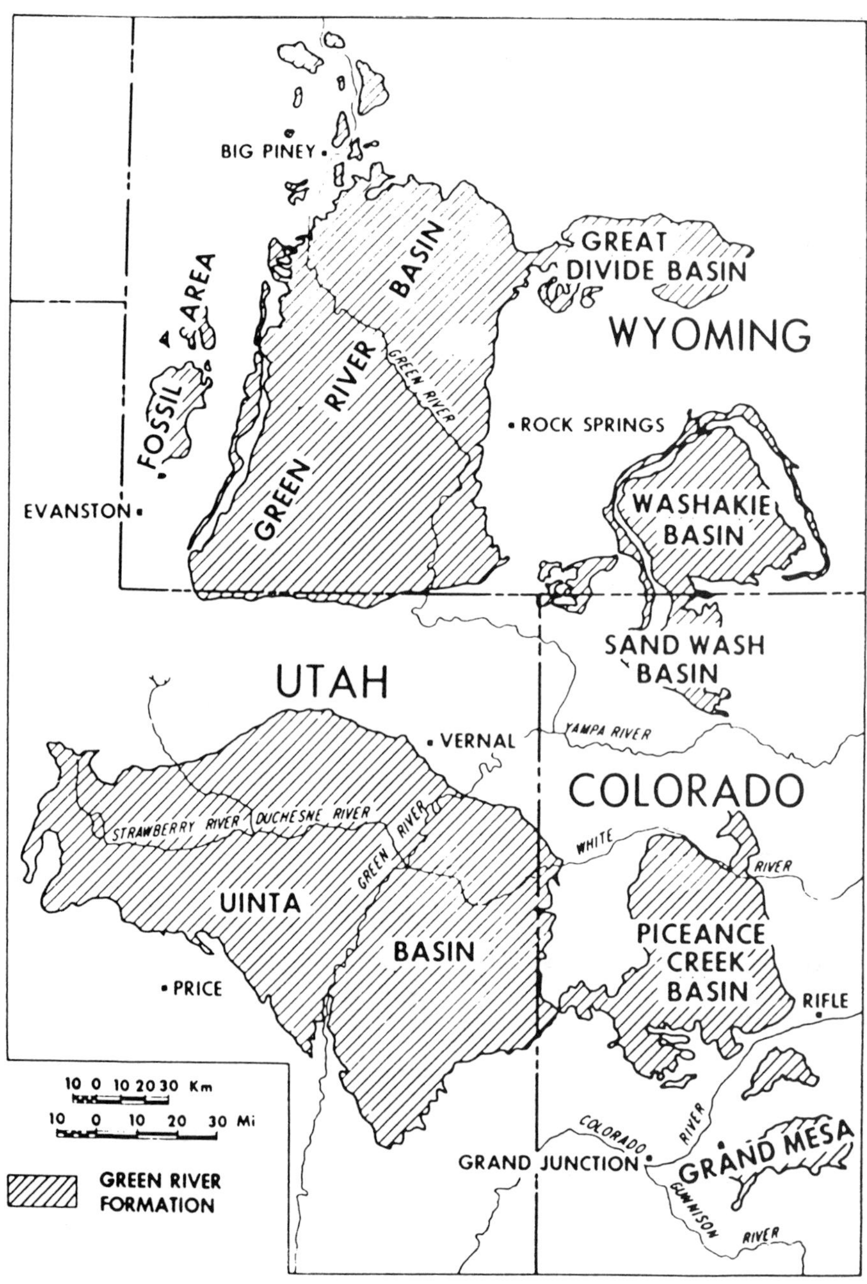

Figure 2 The Green River formation (Smith, 1980).

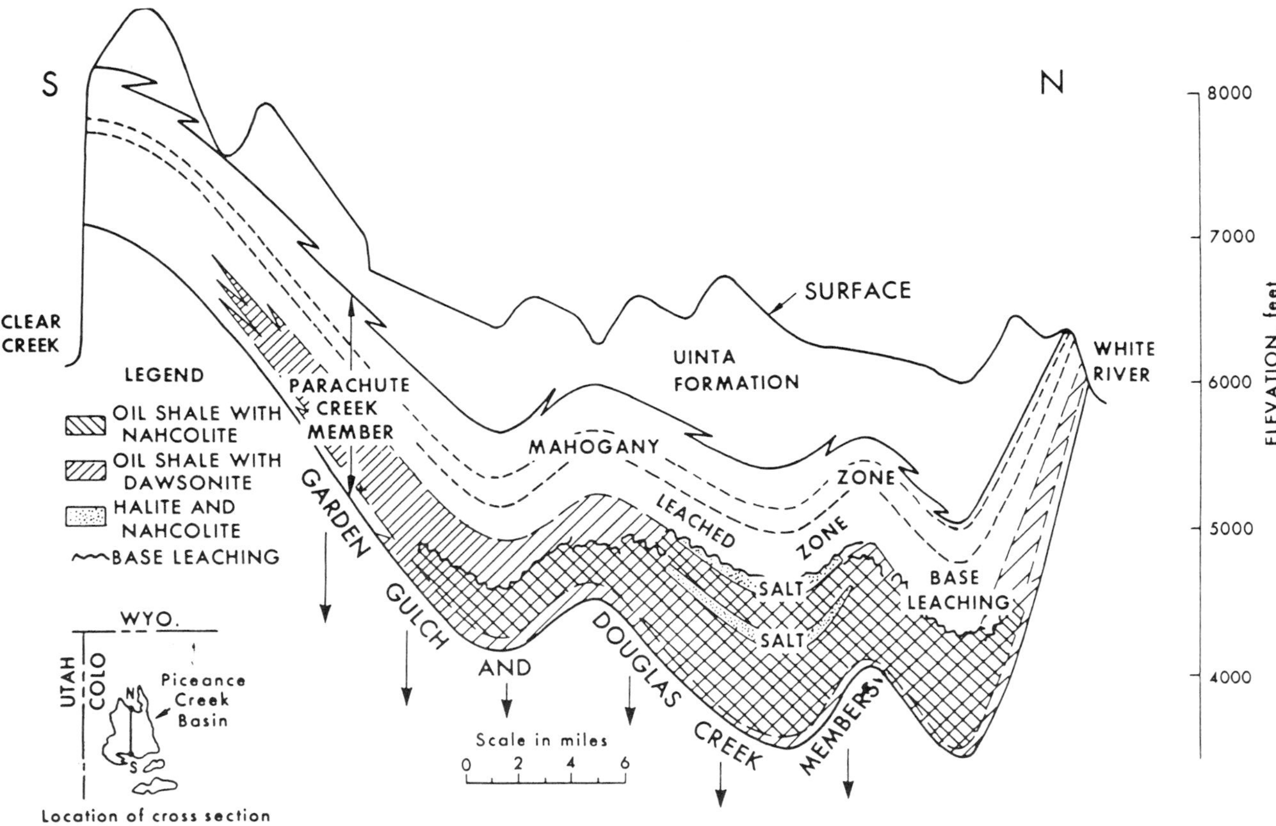

Figure 3 South-north cross-section of the Piceance Creek basin (Smith, 1980).

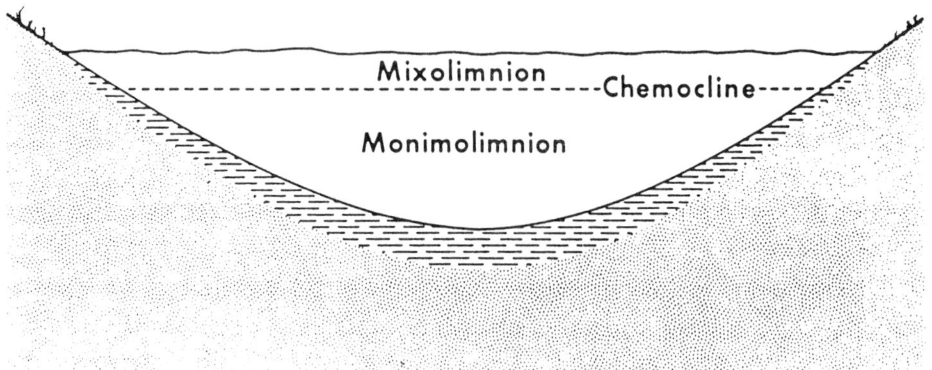

Figure 4 Schematic representation of a stratified lake.

In the Piceance Basin, the lake was relatively deep and remained stratified. As evaporation proceeded, the concentrations of sodium and aluminum ions in the lower layer increased to the point that precipitation of the nacholite, which is 3–4 times less soluble than Na_2CO_3, commenced. When both CO_2 and aluminum ions were available, dawsonite was precipitated, with occasional nordstrandite [$Al(OH)_3$] precipitation occurring under conditions of CO_2 depletion.

In contrast, the Wyoming lakes were shallow and did not remain stratified as evaporation proceeded to dryness, or nearly so. This exposed the lower layer of the lake to the atmosphere leading to increased oxidation of organic matter. As a result, sediments formed during this period contain little organic material. Also, exposure of the concentrated aqueous Na_2CO_3 solution of the lower layer to atmospheric CO_2 led to trona precipitation, accompanied by calcite and shortite limited by the availability of calcium. Intermittent rain falling in this playa lake may have contributed to trona evaporite formation by dissolving previously deposited Na_2CO_3, thereby exposing it to atmospheric CO_2. Under these conditions, trona deposition would no longer be limited by the rate of CO_2 production from organic decomposition. Deposition of today's thick trona beds was the result.

The presence and types of authigenic minerals is an important indication of the conditions present during sedimentation and subsequent burial. For example, the presence of authigenic sulfides and high levels of organic material together in an oil shale suggests deposition and burial in an anoxic, reducing environment of low Eh (Demaison and Moore, 1980). Since the typical surface waters (rainwater, flowing streams, ocean surface) are strongly oxidizing, this probably means an environment of low circulation, such as a lower level in a stratified lake or sea, where oxidation of the organic matter is minimal. The fine grain size of the detrital clays generally found in oil shales is also consistent with this kind of stagnant depositional environment.

In general, the deposition of nonclastic, nonevaporite minerals can be understood in terms of oxidation-reduction (redox, Eh*) and hydrogen ion (pH) potentials, subject of course to the availability of aqueous ionic species (Figures 5–7). Most oil shales of interest were formed within a very restricted region of Eh–pH space, namely, the region where pH > 6 and Eh is below the "sulfate-sulfide fence."

B. Production of Organic Matter

Kerogen (insoluble organics) and bitumen (soluble in organic solvents; see Chap. 2, Sec. I) are the biogenous materials that are the major reason for our interest in oil shales. Many oil shales also contain readily discernible skeletal fragments, fossilized fish bones, insect parts, diatoms, and even bacterial cell walls. These fossils are interesting and are important indicators of conditions through geologic time, but generally have little effect on the technological properties of oil shales. Oil shales deposited in near-shore areas generally contain significant amounts of woody terrestrial material that has an important adverse effect on oil-generating potential.

As shown (Figure 1), only a minute fraction of the carbon cycling through the biosphere becomes incorporated into sediments, and of that only a small part is preserved to become fossil fuels. It is important to understand how conditions affect this process because it helps to identify potential oil shale basins, to locate oil shale deposits within these basins, to predict the geometry of oil shale seams, and to pinpoint the most valuable areas that have thick deposits of high-yielding oil shale. Without trying to be too precise, three oil shale types can be identified by depositional environment: marine shales that were deposited in saltwater are widespread, lacustrine shales that were deposited in freshwater are extensive as discussed above, while the less common, coal-associated shales were formed in areas subjected to uplift which altered the environment from aqueous to swampy. In each case, characteristics of the oil shale organics were determined by two factors: production and preservation.

---

*In geology, the redox character of an environment is commonly expresed as Eh, which is the oxidation potential of the aqueous phase (volts) referred to the standard hydrogen half-cell.

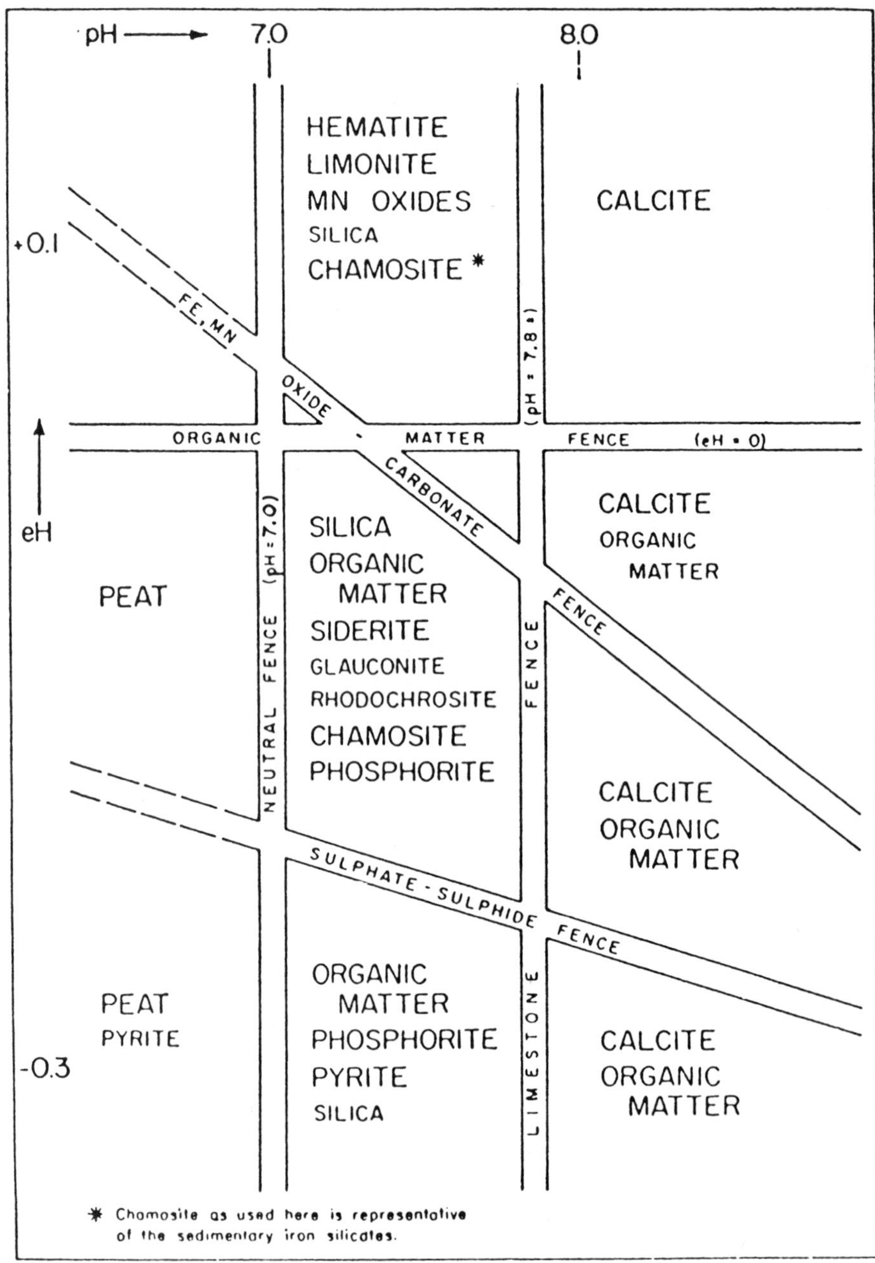

Figure 5 Relationship of Eh and pH to the (nonevaporite) minerals in sedimentary rocks (Pettijohn, 1957; Krumbein and Garrels, 1965).

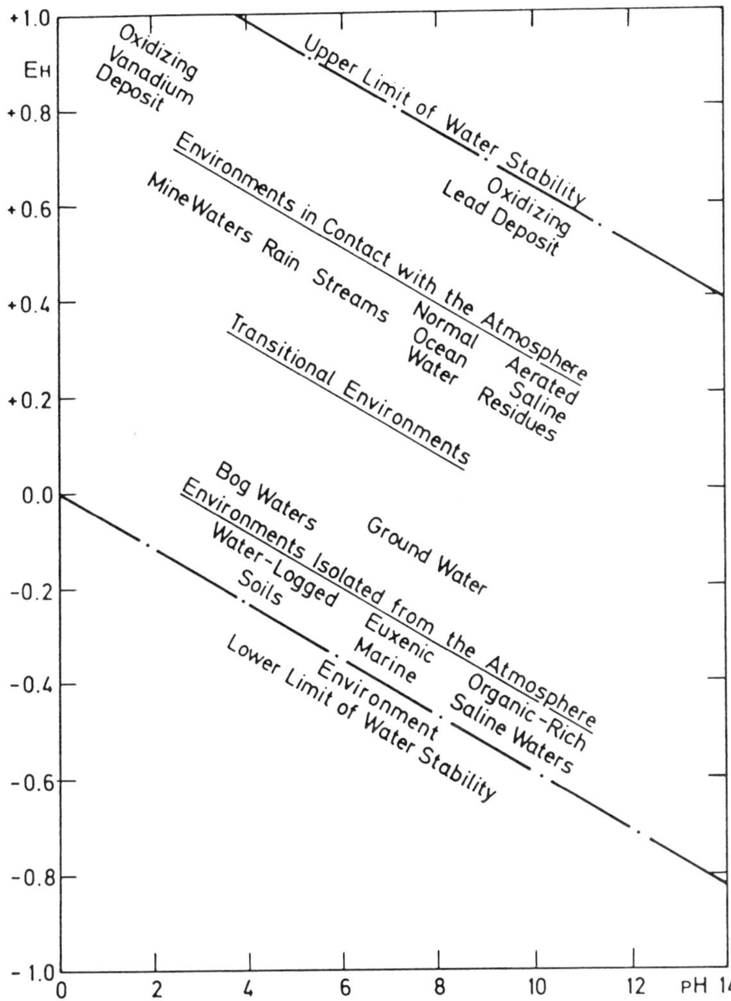

Figure 6 Eh and pH characteristics of some natural environments (Garrels and Christ, 1965).

Photosynthesis by phytoplankton, especially blue-green algae (Cyanophycae), diatoms, dinoflagellates, and in warmer waters phytobacteria, is ultimately responsible for producing nearly all of the organic matter in the oceans (Figure 8) (Tappin and Loeblich, 1970; Sorokin, 1971; Menzel, 1974).

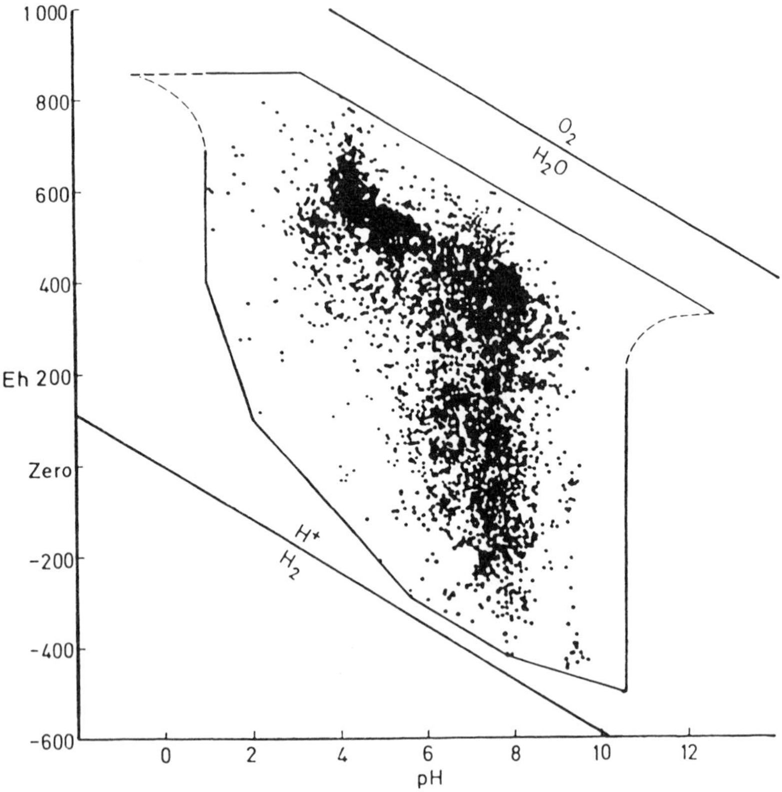

Figure 7 Relationship of Eh to pH for some natural environments (Baas-Becking et al., 1960).

A precise quantitative relationship between the fossil record and organic productivity is not available; many productive species have no hard body parts that are easily preserved, hence their productivity may not be reflected in the surviving fossils. Nevertheless, as pointed out by Moore (1969), the fossil record indicates significant differences between production in marine and lacustrine environments (Figures 9 and 10).

Photosynthetic productivity is mainly controlled by light, temperature, and nutrient availability. Light seems not to be the controlling factor, except in deep water, polar regions, or turbid coastal regions (Menzel, 1974). The availability of CO_2, which is essential for photosynthesis, is not limiting in the "eutrophic zone"—the top

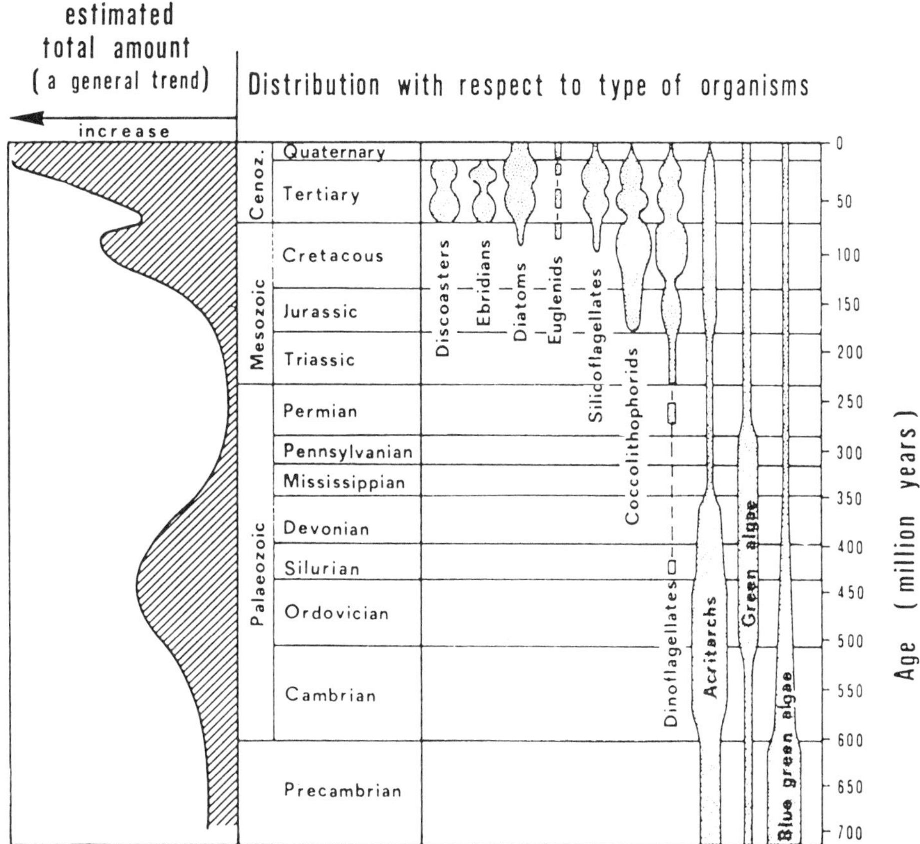

Figure 8 Importance of phytoplankton types to primary production through geologic time (after Tappan and Loeblich; reprinted with the permission of the Geological Society of America).

200 m or so, where productivity is highest. Rather, nutrient availability—often nitrogen (as ammonium or nitrate) in marine, or phosphate in lacustrine, environments—seems to be the most common limit on productivity (Strickland, 1965). The availability of vitamins, especially B_{12}, and other trace organics is important, and seems to be a limiting factor in some areas. Productivity in lakes follows the same general pattern as in oceans, but deposition in the confines of a lake is much more subject to local climatic variations and periodic changes in clastic input (e.g., volcanic ash) than is marine deposition.

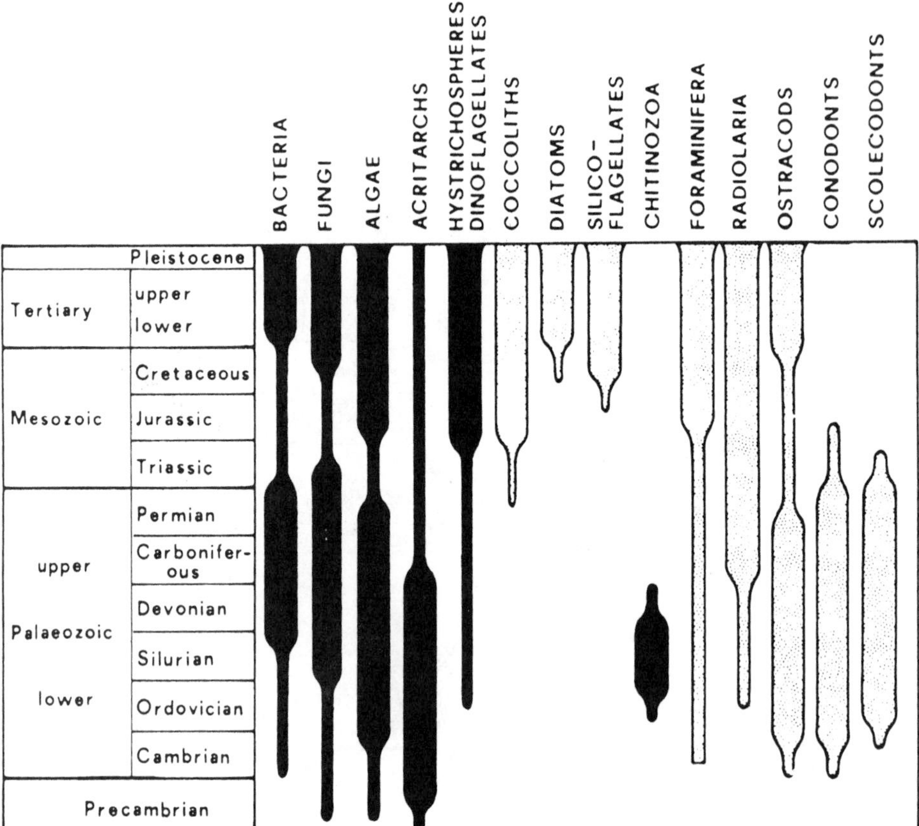

Figure 9 Main groups of microfossils in marine environments (Moore, 1969; reprinted with the permission of Springer-Verlag).

C. Kerogen Types and Maturation

While most of the organic matter in oil shales was produced in an aqueous environment, some terrestrial material is often present. Moreover, conditions of burial affect the nature of the organic matter. The concepts of kerogen type and kerogen maturation provide a useful framework for understanding and classifying these effects.

Kerogen types (I, II, and III) are determined by the kind of debris that is deposited in the sediment. As initially deposited in a "recent sediment," each type of sediment has a (reasonably) characteristic

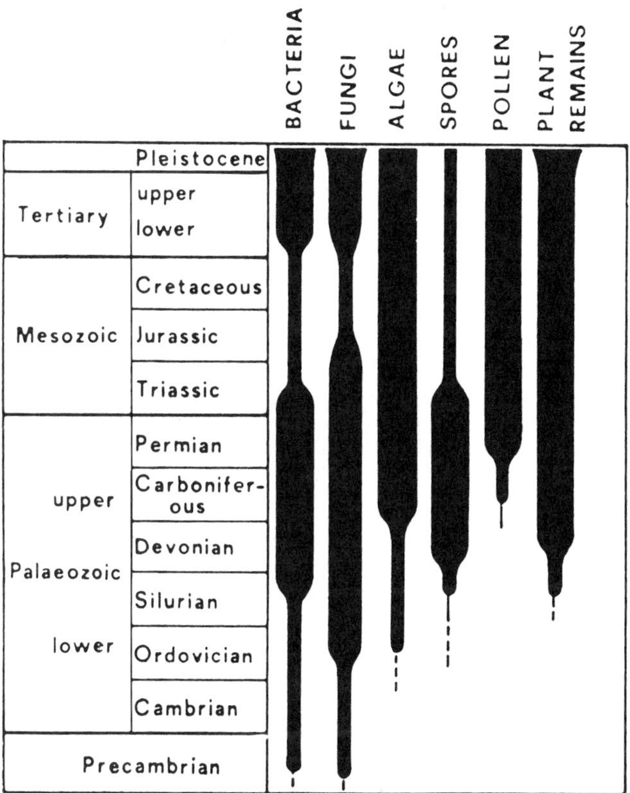

Figure 10 Main groups of microfossils in lacustrine environments (Moore, 1969; reprinted with the permission of Springer-Verlag).

range of composition. As the sediment becomes buried deeper and/or hotter and for a longer time, the organic material in the sediment undergoes "maturation" to give oil, gas, or a mixture of the two. The following discussion is based largely on information about kerogen type and maturation that has been developed in the context of petroleum exploration (Broughton, 1920; Hunt, 1979; Tissot and Welte, 1984).

Type I kerogen is rich in lipid-derived aliphatic chains and has relatively low polyaromatic and heteroatom contents. The initial atomic H/C ratio is high (1.5 or more) and the atomic O/C ratio is generally low (0.1 or less). Such kerogens are generally of lacustrine origin and have very high oil-generating potentials. Organic sources for the type I kerogens include the lipid-rich products of "algal blooms" and the finely divided and extensively reworked lipid-rich biomass in oil shales that, like those of Green River, were deposited in stable stratified lakes. The rubbery material, coorongite, that results from periodic *Botryococcus* blooms in the Coorong district of Australia and the sediments of Big Soda Lake (Nevada), a stratified, saline lake, provide contemporary examples of type I kerogens. Oil shales that contain type I kerogen include the Autun and Campine boghead shales, and torbanite (Scotland), as well as the Green River oil shales. Tasmanite is a marine sediment that contains type I kerogen.

Type II kerogens include most of the marine oil shales, of which many are important petroleum source rocks. Atomic H/C ratios are generally lower than for type I kerogens, while generally higher O/C ratios reflect more ketones, carboxylic acids, and esters. Organic sulfur levels are also generally higher, generally reflecting more thiophenes and in some cases sulfides as well. The oil-generating potentials of type II kerogens are generally lower than those of the type I kerogens, i.e., less of the organic material is liberated as oil upon heating a type II kerogen (at the same level of maturation; see below). The organic matter in these kerogens is usually derived from a mixture of zooplankton, phytoplankton, and bacterial remains that have been deposited in a reducing environment. The Devonian shales of the United States and Canada (Chattanooga, Sunbury, New Albany, Duvernay), the Jurassic shales of Europe (e.g., Toarcian, Paris basin), and the Paleozoic shales from North Africa containe type II kerogen.

Type III kerogen is found in coals and coaly shales. Easily identified fossilized plants and plant fragments are common, making it clear that this type of kerogen is derived from woody, terrestrial material. These materials have relatively low atomic H/C ratios (usually <1), relatively high atomic O/C ratios (0.2–0.3 or even higher). Aromatic and heteroaromatic contents are high and ether units (especially of the diaryl type) are important, as expected for a lignin-derived material. Oil-generating potentials are low, while gas-generating potentials are high. No oil shales contain predominantly type III kerogen, but many oil shales contain some clastic material of terrestrial origin. Because this material is easily identified under the microscope, it is often disproportionately emphasized in petrographic studies. Moreover, because of its high aromaticity, a small contribution of clastic type III material can greatly complicate attempts to elucidate the chemical structure and reactivity of type I or type II kerogens. Thus, in oil shale studies it is important to identify and understand the contribution of type III kerogen.

After the relatively rapid alterations that take place shortly after the initial deposition of organic matter in a sediment, the surviving (preserved) organic matter undergoes additional changes. This is the process of *kerogen maturation*, which is responsible for the generation of crude oil and natural gas. Oil shales contain relatively immature kerogen, i.e., the kerogen has not been extensively "cooked" except where exposed to an unusual geothermal gradient, such as an intrusion of volcanic magma (geologic sill or dike). Over geologic time, however, further alteration does occur. Thus to understand the chemical structure and reactivity of oil shale organics, it is necessary to know not only the origin but also the maturity of the kerogen.

The van Krevelen diagram (see also Chap. 18) provides a simple way of graphically representing the relationships between kerogen type, composition, and maturation (Figure 11).

As shown above, maturation involves the loss of hydrogen and oxygen from the kerogen. Oil shales are generally immature and only the earlier stages of maturation, termed *diagenesis*, need be considered. During diagenesis, hydrogen is lost primarily as methane and other light hydrocarbon gases, water, and hydrogen gas, while oxygen is lost primarily as water and carbon oxides. These losses are, however, significant in determining the processing characteristics of the oil shale kerogen in retorting for liquid fuel production.

As maturation proceeds into the "oil generation window," the atomic H/C ratio decreases due to hydrogen loss as hydrogen-rich liquids that are expelled from the rock (Figure 12; see Brooks, 1981 and 1984 for more detail).

Maturation increases with increased exposure of the organic matter to time, temperature, and pressure. The catalytic effects of minerals in the oil shale accelerate this process. Thus, maturation involves exposure of organic matter to a time-temperature-pressure-catalysis history. Similar effects are seen in "simulated maturation" experiments in the laboratory and in oil shale retorting to produce shale oils.

D. Preservation of Organic Matter

A fundamental problem in basin geochemistry is to understand how organic stratigraphy data from rock analyses relate to the occurrence and variability of oil shale beds. Efforts to understand how the preservation of sedimentary organic matter forms kerogen types I, II, and III led to the concept of organic facies* (Figure 13).

The relationship between kerogen type and the origin of the sedimentary organics was discussed above, as was the relationship between organic preservation and the redox potential of the benthic

*The geologic term *facies* refers to a stratigraphic rock unit, differentiated from adjacent or associated units by appearance or characteristics that usually reflect its origin.

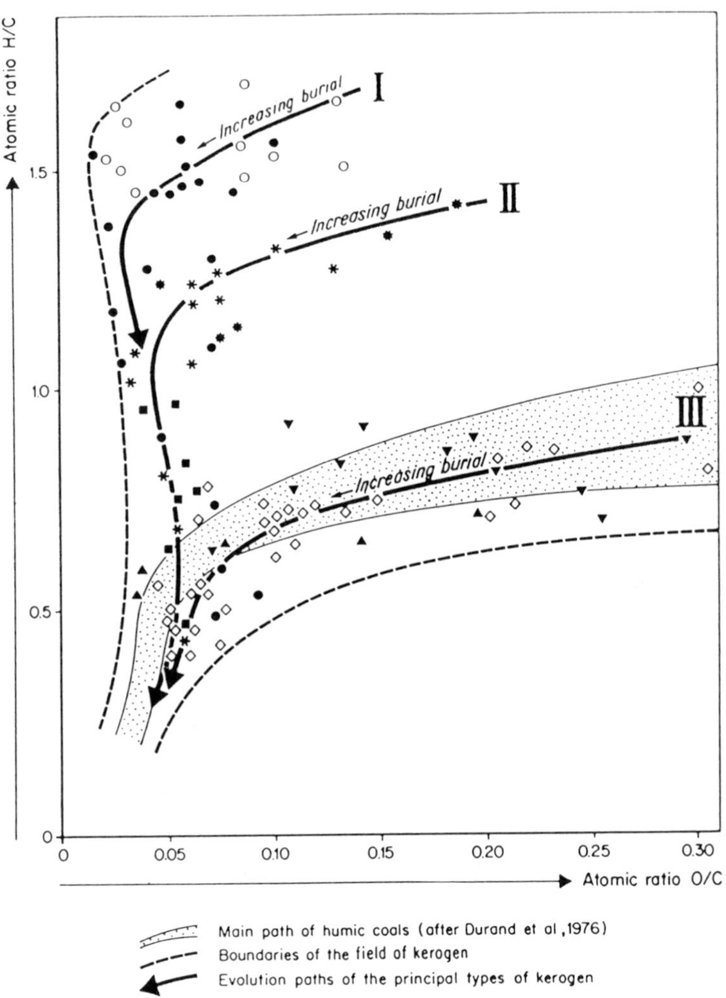

Type	Age and /or formation	Basin, country	
I	Green River shales (Paleocene - Eocene)	Uinta, Utah, U.S.A.	●
	Algal kerogens (Botryococcus, etc...). Various oil shales		○
II	Lower Toarcian shales	Paris, France; W. Germany	*
	Silurian shales	Sahara, Algeria and Libya	■
	Various oil shales		✱
III	Upper Cretaceous	Douala, Cameroon	◇
	Lower Mannville shales	Alberta, Canada	▲
	Lower Mannville shales (Mc Iver, 1967)	Alberta, Canada	▼

Figure 11 The van Krevelen diagram traces the evolution paths of types I, II, and III kerogen as a function of increasing burial (increased time-temperature exposure). Note that maturation eventually leads to a hydrogen- and oxygen-depleted, carbon-rich residue without regard for kerogen type (after Tissot and Welte, 1984; reprinted with the permission of Springer-Verlag).

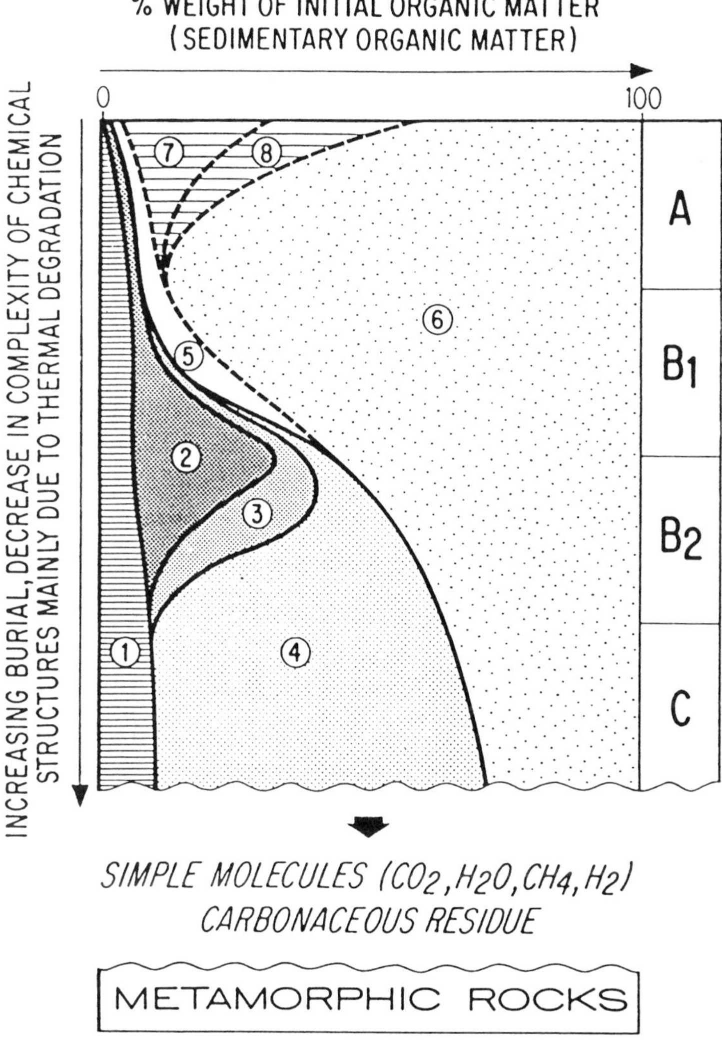

Figure 12 The atomic H/C ratio decreases due to hydrogen loss as maturation proceeds into the oil generation window.

(bottom) environment. Two additional facts are needed to round out the background for the organic facies concept: First, kerogen type strongly correlates with depositional redox potential, i.e., types I and II kerogens are usually found in situations where the depositional environment was strongly reducing. Under these conditions, organic preservation was high and reworking was done primarily by anaerobic

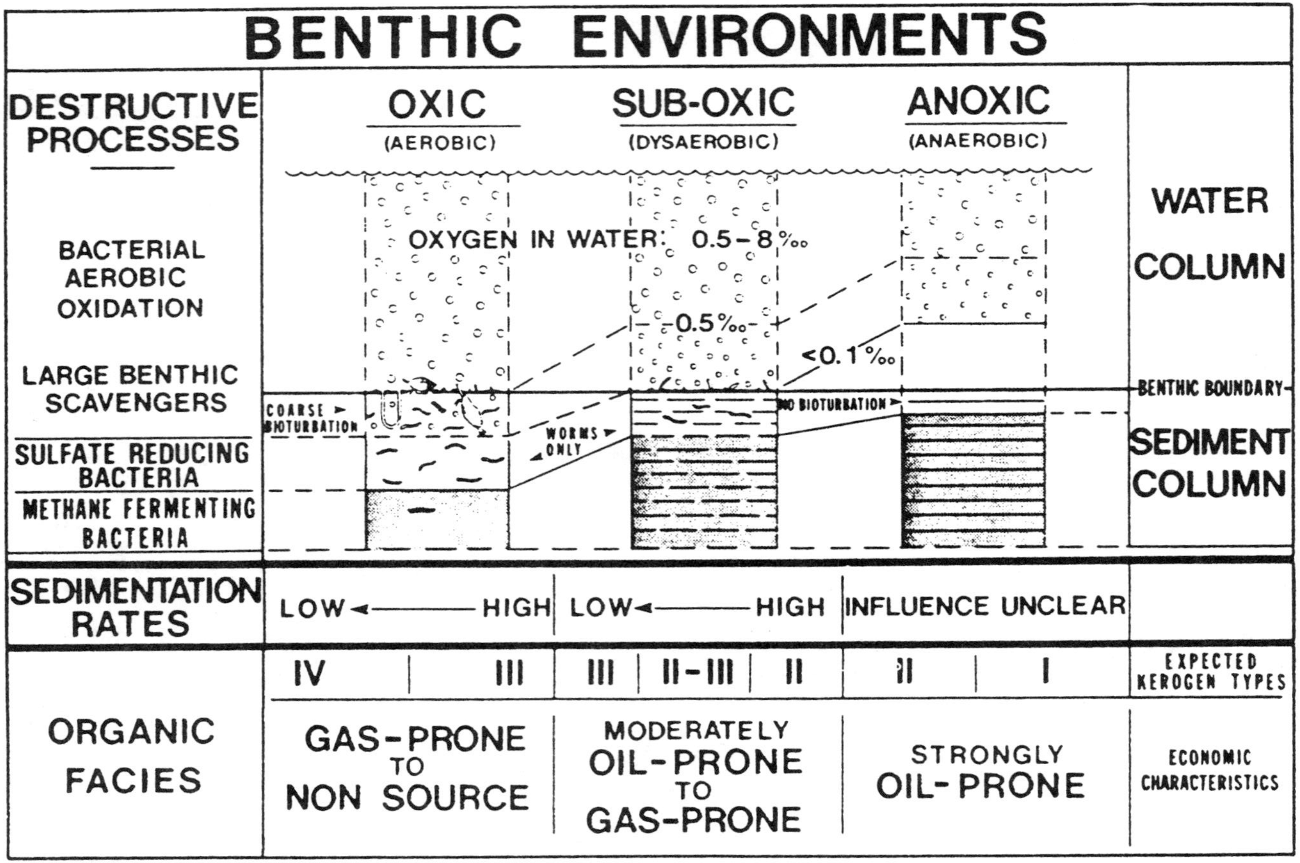

Figure 13 Organic facies mapping depends on a clear understanding of the relationships between early diagenetic factors and kerogen type (after Demaison et al., 1984; by permission of John Wiley and Sons).

bacteria which attack hydrocarbons, especially aliphatic chains (e.g., lipid-derived waxy material), much less aggressively than their aerobic counterparts. Further, anaerobic bacteria feeding on nonlipidic plankton remains are efficient producers of lipids. Thus anaerobic bacterial action can actually increase the content of lipid material that converts easily to shale oil upon heating. In contrast, organic degradation in the oxidizing surface waters of a shallow swamp tends to give coaly material containing highly aromatic type III kerogen. In part, at least, this is because aromatics and hydroaromatics are much less attractive substrates for bacterial growth than the aliphatic chains of lipids (i.e., aromatics are less biodegradable). Thus, not only is the initial aromatic content of terrestrial material generally higher than that of the algal and bacterial material, but the aromatics are selectively preserved.

Second, type II kerogens are often associated with situations where the environment was moderately oxic but the depositional rate was high. Under these conditions, the role of high depositional rate was to minimize the time the organic sediment was exposed to benthic reworking. This means that organic productivity also had to be high if the organic content of the resulting oil shale is high enough to be useful.

This leads to a description of the four organic facies types:

1. Organic facies I (strongly oil-prone) is typical of the strongly anoxic environments of stratified lacustrine and marine systems. Organic contents are often high. The strata are laminated, with the absence of bioturbation (mixing) indicating the absence of benthic worms in the depositional environment.
2. Organic facies II (oil-prone) is typical of anoxic environments or of the moderately oxic environment with high depositional rate. Carbon contents are generally lower than in facies type I, often in the range of 1–10% TOC (total organic carbon). The sediments may be partially laminated, showing evidence of benthic worm burrows and possibly evidence of bioturbation by other benthic organisms.
3. Organic facies III (gas-prone) is typical of the mildly oxic conditions in coal swamps or shallow marine environments. Any planktonic or algal material deposited in such an environment will usually be degraded quickly by benthic bacteria and/or worms. Preserved material of aquatic origin is usually thoroughly bioturbated.
4. Organic facies IV (nonsource) is typical of aquatic environments where organic matter spends a long time in the oxic zone where aerobic bacteria are active. Such environments occur even at great ocean depths with circulation, as well as in shallow, circulating seas with high energy input. Low organic contents in the rocks reflect efficient degradation, even where the initial organic

productivity was high. Extensive bioturbation usually precludes the observation of laminations (varves).

E. Marine Oil Shales

Marine oil shales are usually associated with one of two settings (Figure 14).

The anoxic silled basin shown (Figure 14a) can occur in the shallow water of a continental shelf. High phytoplankton growth rates near the surface will give a high deposition rate. The sill shields the trough from the circulation of oxygen-laden water. Under these conditions, the decomposition of organic sedimentary matter will rapidly deplete oxygen within the confines of the basin, thereby providing the strongly anoxic (reducing, low-Eh) environment that is needed for efficient preservation.

The anoxic zone in an upwelling area (Figure 14b) arises from circulation of an open-ocean current over a cold, oxygen-depleted bottom

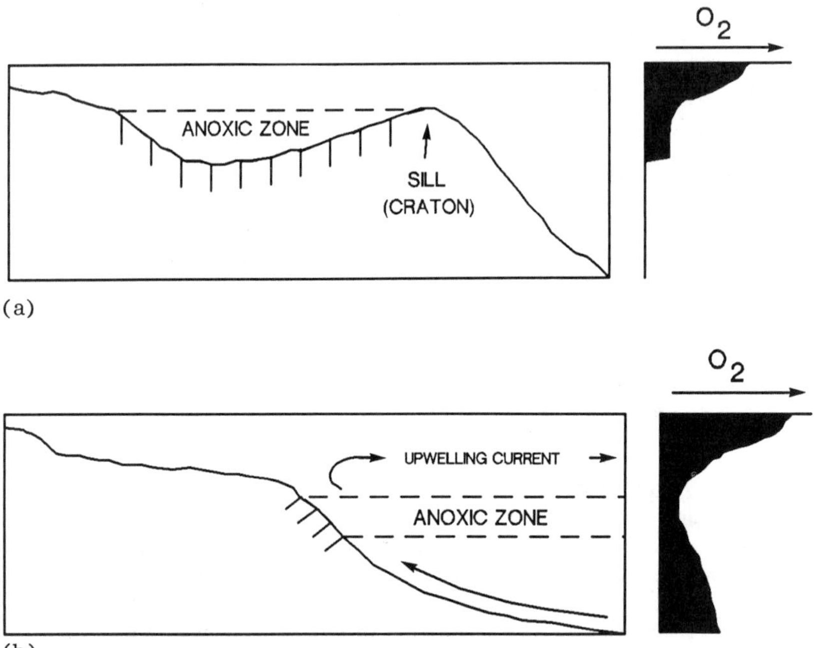

Figure 14 Schematic representation of (a) an anoxic silled basin and (b) an anoxic zone caused by upwelling.

layer. Mixing of nutrient-rich current, such as the Gulf Stream, into the CO_2- and light-rich eutrophic zone gives an environment capable of sustaining very high organic production. Such environments occur today along the west coasts of Africa and the Americas, where good fishing is found along with the potential for organic-rich sediments (Debyser and Deroo, 1969).

Information about the nature of organic matter in marine environments has resulted from studies of recent deposits and the contemporary oceans (Trask, 1932; Bader et al., 1960; Bordovsky, 1965). Only a small part of primary production in the oceans reaches the bottom. Of an estimated annual production of 9×10^{19} tons of dry matter, Trask (1932) estimated that about 2% reaches the floor in shallows and only about 0.02% in the open sea. The major part of marine primary production is consumed by predators; most of the rest by microbes. The principal marine microbial scavengers are bacteria that live free in the water or are attached to organic particles. In ocean water, organics occur in solution, in colloidal suspension, and as particulate matter comprising bodies and body fragments of living and dead organisms. Except in regions of a seaweed or plankton "bloom," the dissolved organics usually predominate. As a result, marine bacteria are most abundant only in the very upper part of the water column and in the organic detritus at the very bottom. Even in the oceans, the adsorption of organics onto inorganic detritus, such as the silica parts of diatoms, plays an important part in sedimentation.

After the organic sediment reaches the bottom, reworking begins. Bottom-dwelling (benthic) organisms feed on both the sediment and the dissolved organics and, in turn, are fed on by predators (e.g., crustaceans). In this sphere, the benthic bacteria are largely responsible for the decomposition of organics and the synthesis of new organics through enzymatic transformations. Moore estimated that 60–70% of the sedimentary organic carbon is typically liberated as CO_2 during this reworking, while most of the rest is converted into new compounds. The result is a very complex mixture.

The various organic compound classes in oil shales (see also Chap. 1, Sec. II) include carbohydrates, lignins, humates and humic acids, lipid-derived waxes and the saturated and polyene acids in algal lipids which can serve as precursors of these waxes, and biological pigments and their derivatives (e.g., carotenoids, porphyrins). Only the latter three were judged to have sufficient inertness to be major contributors to oil shale kerogens (Cane, 1976).

Degens and coworkers studied the role of proteins, carbohydrates, and humates in marine sediments (Degens, 1963; Degens et al., 1961, 1964, 1975). Protein-derived materials included both original and altered proteins and their decomposition products (amines, amino acids, amino complexes). Carbohydrates are rapidly hydrolyzed and generally not important in oil shales. Humates can be important, even in marine shales, when deposited near shore, though obviously humic material has been found in marine sediments that were deposited far from land. It is possible that some humic material may be derived from proteins and/or carbohydrates, perhaps when these materials are adsorbed on inorganic particulates (e.g., clays, volcanic ash) in a moderately oxidizing environment.

Lipids are produced by phytoplankton and also synthesized from carbohydrates by microbial activity (Bodorovsky, 1965). With respect to oil shales, the polyene fatty acids are especially interesting. It is well known that adverse conditions can lead to very high lipid production by algae. At low temperature and with limited oxygen, for example, *Chlorella* may produce lipids to >75% of their body weight. Much of these lipids is unsaturated. Abelson and coworkers (1964) found that polyene acids in lipid fats from *Chlorella* disappeared upon heating, while saturated acids remained unaltered. Evidently, the polyene acids polymerized. Cane (1976) studied extensively the role of unsaturated lipid acids in the products left by *Botryococcus braunii* blooms in the Coorong of southern Australia. Polymerization of the unsaturated lipid residue gives "coorongite," a tough, resilient, insoluble material that resembles kerogen in many respects. Iso- and anteiso-fatty acids have been found in some oil shales, but these are probably secondary products. Leo and Parker (1966) found that bacteria are active in transforming *n*-alkanoic acids into the branched iso- and anteiso-acids. Other bacterially induced transformations include the hydrogenation of oleic and linoleic acids, decarboxylation and polymerization of alkanoic acids, and lipid hydrolysis (Schonbrunner, 1940; ZoBell, 1944; Rosenfeld, 1948).

Carotenoid pigments have been found in many oil shales and in petroleums and coals as well. A recent article (Repeta and Gagosian, 1983) discussed the carotenoids and carotenoid transformation products in the waters and sediments of the Peruvian upwelling and provides leading references to much earlier work on carotenoids in marine sediments. Studies of the carotenoids isolated from DSDP (Deep Sea Drilling Project) cores from the Quaternary sediments in the Cariaco Trench shows that the chemistry of these materials is largely reductive and traceable over 50,000–350,000 years (Watts and Maxwell, 1977; Watts et al., 1977). This work gives useful insight into the diagenetic transformations of carotenoids which lead to the observance of partially and perhydrogenated carotenoids in marine oil shales (Kimble et al., 1964).

The black marine shales formed in shallow seas have been extensively studied, as they occur in many places. These shales were deposited on broad, nearly flat sea bottoms, and therefore usually occur in thin deposits (10–50 m thick), but may extend over thousands of square miles. The Irati shale (Permian) in Brazil extends over more than 1000 miles from north to south (Costa Neto, 1983). The

Jurassic marine shales of Western Europe [Toarcian, Paris Basin, also source rock for Paris Basin crudes (Tissot and Welte, 1984); Kimmeridge, England (Williams, 1987) also an important source rock for North Sea crude oil], Silurian shales of north Africa, and the Cambrian shales of northern Siberia and northern Europe are other examples of this kind of marine oil shale.

F. Lacustrine Oil Shales

The lacustrine oil shales of the Green River formation which were discussed above, are among the most extensively studied of sediments. However, their strongly basic depositional environment is certainly unusual, if not unique. Therefore, it is useful to discuss the characteristics of the organic material in other lacustrine shales.

A particularly thorough study (Hau and Douglas, 1983), was made of the lacustrine sequences from the Permian oil shales of Autun (France) and the Devonian bituminous flagstones of Ciathness (Scotland). Several series of biomarkers were prominent in extracts from these shales: hopanes, stearanes and carotenoids. Algal remains were abundant in both shales. Blue-green algae, similar to those that contributed largely to the Green River oil shale kerogen, were found in the Devonian shale, for which a stratified lake environment similar to Green River has been proposed (Donovan and Scott, 1980). In contrast, *Botryococcus* remains were found in the Permian Autun shale and are presumed to be the major source of organic matter, except for one sample. No *Botryococcus* remains were found in this sample and the oil produced by its retorting was nearly devoid of the straight-chain alkanes and 1-alkenes which are prominent in oils from *Botryococcus*-derived shales. Evidently, some as yet unidentified algae contributed to the organic matter in this stratum. Biodegradation cannot be ruled out but seems unlikely due to the lack of prominent iso- and anteisoalkanes. Straight-chain alkanes and 1-alkenes were also prominent in gas chromatograms of the retorted oils from the Devonian shale. However, in this case a pronounced hump, which usually indicates polycyclic derivatives, was also prominent. Both extracts and oil from the Devonian shale were found to be rich in steranes and tricyclic compounds. Di- and triterpenoids have been suggested as precursors for the di- and tricyclic compounds found in many oil shales. Rock–Eval pyrolysis results indicate that these shales have high hydrogen indices; the kerogens are all type I or type II, with one of the Devonian samples being clearly type I.

Other major lacustrine oil shale deposits include the Triassic shales of the Stanleyville Basin in Zaire and the Albert shales of New Brunswick, Canada (Mississipian).

IV. OCCURRENCE

Oil shale deposits are located throughout the world (Jaffe, 1962; Duncan and Swanson, 1965; Duncan, 1977) but it is the features of the major deposits (Table 11) that are of particular interest in the current context. Such deposits are of interest because of their potential to produce oil in addition to (hopefully) having minimal recovery problems.

V. RECOVERY

Very thick seams of well-consolidated oil shale are characteristic of the Green River formation. Around the basin rims (especially of the Piceance Basin), outcrops are numerous, while considerable overburden covers the rich seams near the basin center. As a result, both underground mining and surface (open-pit) techniques may have economic advantages in different locations (see also Chap. 1, Sec. IV). In some locations, in situ methods may have advantages, but even in these cases some mining will probably be required to provide a void into which rubble can be blasted. Several general reviews have been made of the mining technology applicable to oil shale (Jee et al., 1977; Farris, 1980; Kauppila, 1982a,b).

A. Underground Mining

Underground mining was used at the Anvil Points Oil Shale Research Facility (APF), which was operated from 1944 to 1956 by the U.S. Bureau of Mines (East and Gardner, 1964), then by others until 1984 when it was decomissioned (Virgona, 1986). Underground mining was also used successfully by Union Oil, Mobil Oil, and Colony Development, and was planned for Exxon's Colony Shale Oil Project (now terminated) (Willmon, 1987). In these cases, the mine was of a room-and-pillar design (Figure 15).

The mine at APF was begun as three levels, but in later operations only two levels were mined. Nevertheless, the mining at APF provided a wealth of information about the safe design, operation, and maintenance of an underground oil shale mine in the unique environment of the Green River formation. Variations on this basic design can be tailored to mine the rich strata while selectively leaving the lean ones.

Such designs, however, leave much of the shale in place, and hence are not suited for zones that are both thick and uniformly rich.

Bulk underground mining methods, such as sublevel stoping with full subsidence or block caving methods, can provide more complete recovery of rich oil shale from thick zones. However, spent shale cannot easily be disposed of in the mined void. A stoping method with spent-shale backfill may be a compromise that is both reasonably efficient and environmentally acceptable. Longwall mining, a method widely used in underground coal mining, is another alternative. Longwall methods have been used for oil shale mining in the USSR (Kurapei and Cheshko, 1977), and its use in the United States has been contemplated (Trent and Dunham, 1977). Given the high cost of mining, it is clear that even incremental improvements in oil shale mining technology could have a major impact on the economics of a proposed oil shale project. However, it must be realized that mine design is very much site-specific and an improvement applicable at one site may be inappropriate for another. Moreover, safety considerations dictate a conservative policy with respect to the amount of rock that must be left in place. Perhaps improvements in subsurface imaging can be made to enable more precise three-dimensional modeling of subsurface features using computer techniques. This will require a much better fundamental understanding of wave propagation in inhomogeneous media. However the benefits should be both lowered cost of mine design and more efficient designs that significantly increase the amount of rich oil shale that can be removed without compromising mine safety.

From 1981 to 1984, an extensive oil shale fragmentation research program was conducted at Anvil Points by a consortium composed of Cities Service, Getty Oil, Mobil Research and Development, Phillips Petroleum, Sohio Shale, and Sunoco Development, Science Applications Inc. (SAI) managed the program and provided technical direction. Los Alamos National Laboratory and, in later stages, Sandia National Laboratory participated with SAI and the consortium and shared the experimental work. This program included a wide range of tests, ranging from single-level/single-borehole tests to obtain basic fragmentation data, to multiple-level/multiple-borehole tests to explore fragmentation as a function of explosive type, charge placement, and detonation sequencing (Dick et al., 1984; Parrish, 1986). Some of the heavily instrumented tests carried out late in the program provided results that were used to verify and refine computer programs that simulate with remarkable precision the behavior of the Green River oil shale under explosive stress. These results should prove invaluable in designing new, more efficient mines and in formulating efficient operating procedures for these mines of the future.

Table 11 General Features of the Major Oil Shale Deposits of the World

Deposit	Type of material	Geologic age	Geologic setting	Average grade (gal/ton)	Estimated resource (10^6 bbl)
Green River	Oil shale	Tertiary Eocene	Stratified lake	25	4,300,000
Eastern U.S.					
Chattanooga, etc.	Oil shale	Devonian	Shallow sea	10	2,600,000
Phosphoria	Oil shale	Permian	Marine platform		
Alaska	Tasmanite	Jurassic	Marine	10	Large
	Oil shale	Mississippian	Marine platform	10	?
Austria					
Australia					
Rundle, Stuart	Oil shale	Tertiary			65,000
Toolebuc	Oil shale	Cretaceous	Sloping marine platform	18	365,000
Condor	Oil shale	Tertiary			107,000
Dauringa	Oil shale	Tertiary		20	63,000
New South Wales	Torbanite	Permian	Marine	33	40
Tasmania	Tasmanite	Permian	Lacustrine	42	55
Brazil					
Irati	Oil shale	Permian	Marine platform	21	800,000
Tremembe-Taubate	Oil shale	Tertiary	Lacustrine	15–18	2,000
Marahu	Marahuito	Tertiary	Boghead, coaly	42	Small
Canada (Albert)	Oil shale	Mississippian	Lacustrine	13–30	150,000
China (PRC), Fushun	Oil shale	Oligocene	Boghead	15	1,000

France					
Autun	Oil shale	Permian	Boghead, coaly	10–18	160
Severac-le-Chateau	Oil shale	Jurassic	Marine	10	250
Germany	Oil shale	Jurassic	Marine	12	2,000
Great Britain					
Kimmeridge	Oil shale	Jurassic	Marine	10–45	1,000
Scotland (Lothians)	Oil shale	Carboniferous	Lacustrine, estuary	16–40	400
Italy (Sicily)	Oil shale	Triassic	Marine	25	7,000
Israel, Jordan, Syria	Oil shale	Cretaceous	Marine	10–25	120,000
Morocco					
New Zealand					
Orepuki	Oil shale	Tertiary		45	8
Nevis, Otago Central	Oil shale	Tertiary	Boghead	11–15	300
Portugal	Oil shale	Carboniferous	Boghead	30	80
Spain	Oil shale	Tertiary	Lacustrine	25	280
Sweden	Oil shale	Cambrian-Ordovician	Marine	10–25	3,000
Union of South Africa, Ermelo	Torbanite	Permian-Carboniferous	Boghead, marine	25	130
USSR (Estonia)	Kukersite	Ordovician	Marine platform	50	6,500
Yugoslavia, Aleksinac	Oil shale	Tertiary	Lacustrine	45	210
Zaire (Congo)	Oil shale	Triassic	Lacustrine	25	100,000

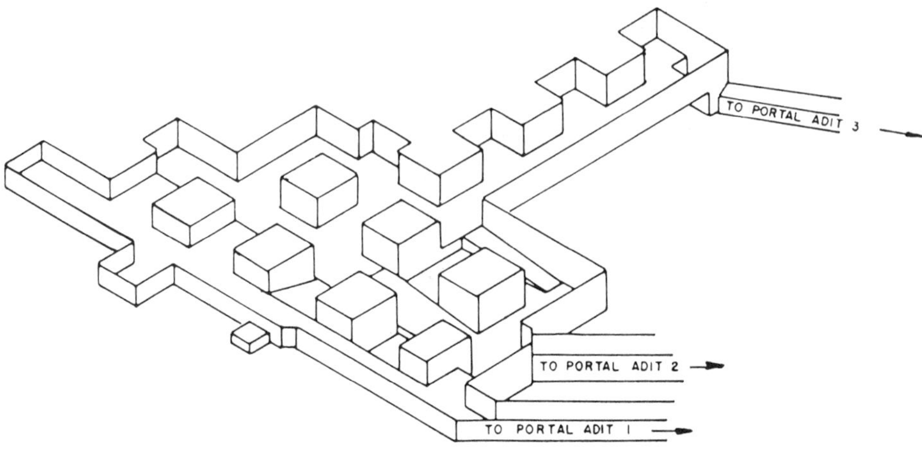

Figure 15 One level of a room-and-pillar oil shale mine (after East and Gardner, 1964).

B. Surface Mining

At Fushun, in Manchuria (People's Republic of China) there is a very large open-pit mine where 450 ft of low-grade oil shale (15 gal/ton) overlies one of the world's thickest coal deposits. Oil shale processing has accompanied coal production since the Japanese initiated large-scale operations at Fushun in 1929 (Whitworth, 1979a,b; Wyllie, 1979). Surface (open-pit) mining is also practiced on a large scale in the USSR, especially in Estonia (*Coal Miner*, 1978). Surface mining was planned for the (now deferred) Rundle Project (Queensland, Australia) by a consortium comprising Esso Exploration and Production Australia, Central Pacific Minerals, and Southern Pacific Petroleum (Siwinski, 1987). Surface (strip) mining has also been envisioned for the Toolebuc shale, Australia's largest oil shale resource, especially along the St. Elmo structure where the shale is 7–14 m thick below a nearly barren oxidized zone about 18 m thick (Hutton et al., 1987).

Surface mining of Green River oil shale has never been practiced on a large scale, though about 15% of the total reserve is potentially recoverable by surface mining methods. A detailed engineering economics study was carried out to explore the potential of open-pit mining of Green River oil shale (Adams et al., 1976a,b). The potential environmental impact of such a large-scale surface mining operation in the Piceance Basin was examined in a recent U.S. EPA study

(Lappi et al., 1982). Surface mining of eastern U.S. (Indiana, Kentucky) shales was evaluated as one part of an effort to develop plans for Devonian oil shale utilization (Gitel, 1985).

Surface mining is used extensively in ore recovery and for mining coals, especially low-rank coals. It is anticipated that technology developed for these applications will be adaptable to oil shale mining.

C. Size Reduction

In situ retorting will not require size reduction past that obtained by blasting to rubblize the shale. However, size reduction by crushing and grinding will be required for above-ground processing. Moreover, size reduction is expensive and becomes rapidly more so as the target particle size decreases below about 0.5 in. Nevertheless, closely controlled size reduction is a necessary preliminary to the use of heavy-medium cycloning and other of the newer, more efficient methods for oil shale beneficiation. In addition, most current aboveground retorts have severe particle size restrictions, especially with respect to fines.

A detailed study of size reduction of Green River oil shale was made by Salotti and Datta (1983). Three types of crushers (gyratory, impact, roll) were studied, using 200-ton samples collected from the R-6 (20.7 gal/ton) and R-5 (24.9 gal/ton) zones of the Rio Blanco mine (tract C-a). The authors caution that extrapolation of the results to leaner shales ($\sim$15 gal/ton) should be done with caution and that extrapolation to appreciably richer shales should be made only with trepidation. Major findings of this study are as follows:

1. Power consumption for a similarly sized feed and product is roughly the same for all three types of crushers.
2. Variations in feed grade (20–25 gal/ton) have little or no effect on the performance of the three crusher types for product having a top size of 3/8 in. or larger.
3. When a crusher is producing 3/8-in. or finer top size product, the richer oil shale tends to have a significantly finer size distribution.
4. Crusher throughput was about 70% of rated capacity.
5. Relative ability to process larger pieces (6 in. $\times$ 0) was in the order gyratory > impact > roll.
6. Relative percentage of the recirculating load present as -3/4 in. material is in the order roll > impact > gyratory.
7. Relative production of oversize material is in the order roll > impact > gyratory.

8. Grinding of these relatively rich shales was roughly equivalent to grinding a moderately hard limestone, and wear during oil shale grinding is anticipated to be similar to that suffered during grinding such a limestone.

A similar study of size reduction was made as part of a study of the beneficiation of an eastern U.S. Devonian oil shale (Datta and Salotti, 1982).

26

Characterization, Testing, and Classification of Oil Shales

I	Fischer Assay	840
	A. Modified Fischer Assay	840
	B. Shortcomings	842
II	Alternative Methods	843
III	Elemental Composition and Oil Yield	847
IV	Bitumen Content and Composition	847

Early Scottish workers estimated the quality of cannel coal (Chap. 18, Sec. I) by the curl of its shaving when cut with a knife. Similarly, workers at Fushun, Manchuria found that rich shale curled when chipped with a glass shard, while lean shale crumbled. There is a need for additional methods to evaluate oil shale quality, as such qualitative methods are still used, especially in the field. However, these methods have now been supplemented with a variety of quantitative methods employing the latest in laboratory instrumentation to determine chemical and physical properties. In this section, the more important methods currently used for characterizing, testing, and classifying oil shales will be discussed; the properties of 12 representative oil shales will be compared and the methods used to determine these properties will be discussed to illustrate the general approach.

I. FISCHER ASSAY

In the above discussion, the term "oil yield" was used repeatedly but without definition of how the values are measured. Related terms include "grade" (usually in gallons per ton), which is synonymous with oil yield, whereas "rich" and "lean" refer to shales with higher and lower oil yields, respectively. The Fischer assay—more specifically, the modified Fischer assay (ASTM D3904-80)—is the standard method for measuring oil yield. Because of its importance, it is appropriate to introduce the Fischer assay as the first topic in this section which deals with measuring the properties of oil shales.

The aluminum retort was originally designed for assaying coals by pyrolysis (Fischer and Schrader, 1920). In early practice, a sample of ∿20 g was heated rapidly to a maximum temperature of 520–550°C (970–1020°F) using a gas burner, and the experiment was concluded 15 min after the final drop of distillate was collected. Stanfield and Frost produced a version holding 100 g of sample and containing aluminum disks to minimize local overheating (Figure 1).

A. Modified Fischer Assay

Using the new retort, modifications to the procedure have been made to improve both accuracy and reproducibility. These include the use of retorts with standardized heat capacity, standardized heating programs, carefully controlled condenser and cooling bath temperatures, and continuation of the assay at a maximum temperature of 500°C (930°F; Figure 2) until no more oil is produced. In its present form, the modified Fischer assay (ASTM D3904-80) is a widely used and generally accepted standard method for evaluating oil shales (Figure 3).

The shale to be assayed is crushed to provide a 1-lb sample of 8-mesh material. An aliquot is removed for drying to determine total moisture. The retort is charged with 100 g of the crushed oil shale in layers, the layers being separated by the perforated aluminum disks. The retort is then heated to 500°C over 40 min, then held at that temperature for 20 min or until no more oil is collected (10–20 min longer for shales richer than 20 gal/ton). The heavier liquid products and some water are condensed in the centrifuge tube, which is maintained at 100°C (210°F). Lighter products and the balance of the water are liquefied in the condenser, which is maintained at 0°C (32°F) by a circulating coolant.

When the heating period is complete, the retort and centrifuge tube are cooled to room temperature and weighed. The tube is centrifuged to separate oil and water. The water is determined volumetrically and subtracted from the liquid product weight to obtain the oil yield. The

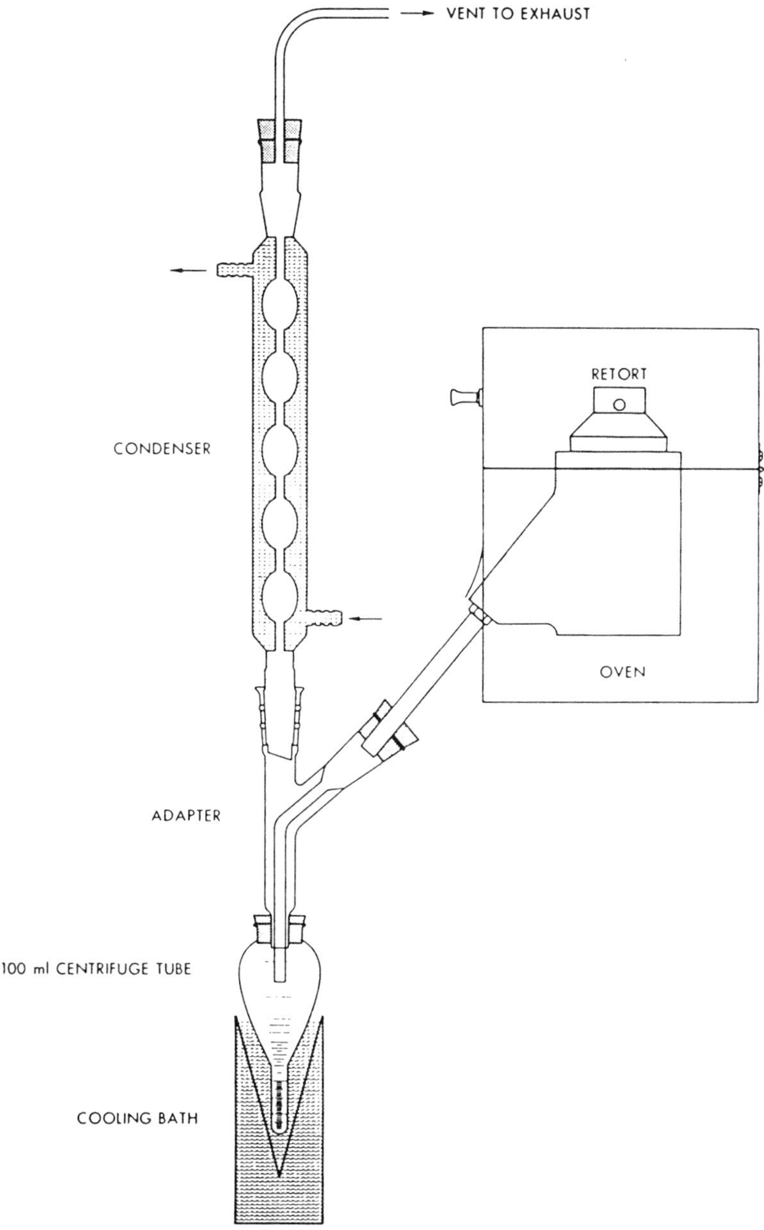

Figure 1 Schematic representation of the standardized Fischer assay retorting equipment (ASTM D3904; see also Hubbard, 1965; Heistand, 1976).

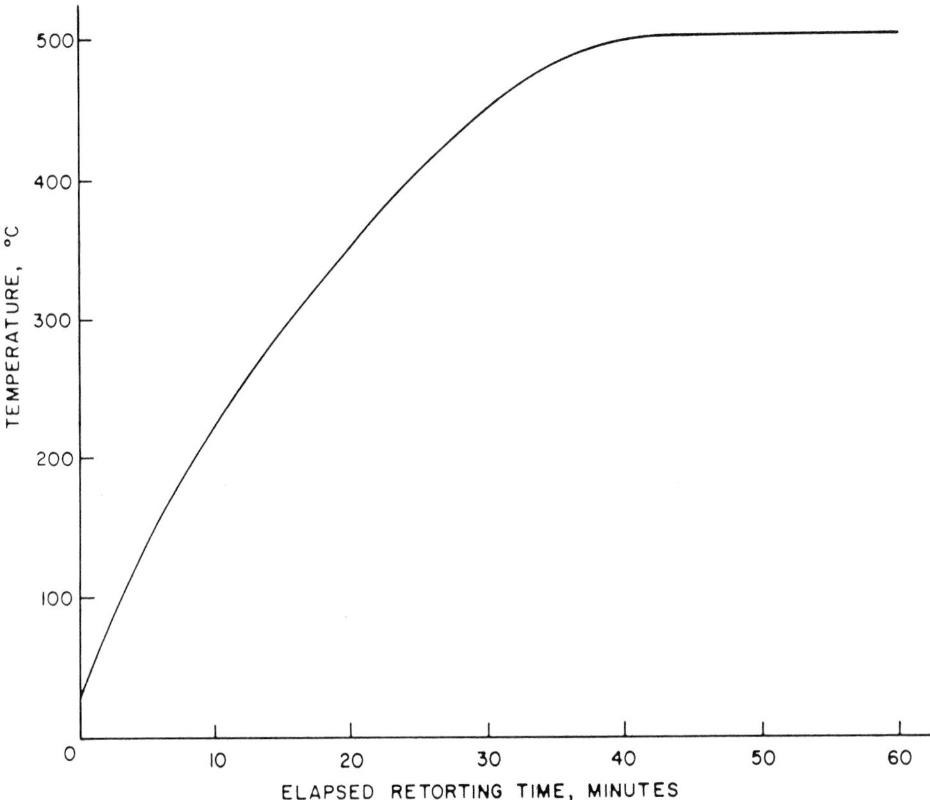

Figure 2 Temperature profile for the Fischer assay retorting procedure.

specific gravity of the oil is determined and used to convert the weight of oil produced to the oil yield in gallons per ton.

B. Shortcomings

Despite acceptance as a standard method, the Fischer assay has technical shortcomings. Gases are not determined directly but by difference as "gas and loss." Some liquid hydrocarbons are also lost as mist. Thus, the yields of these important parts of the product slate can only be inferred. To provide a more complete assay, the TOSCO

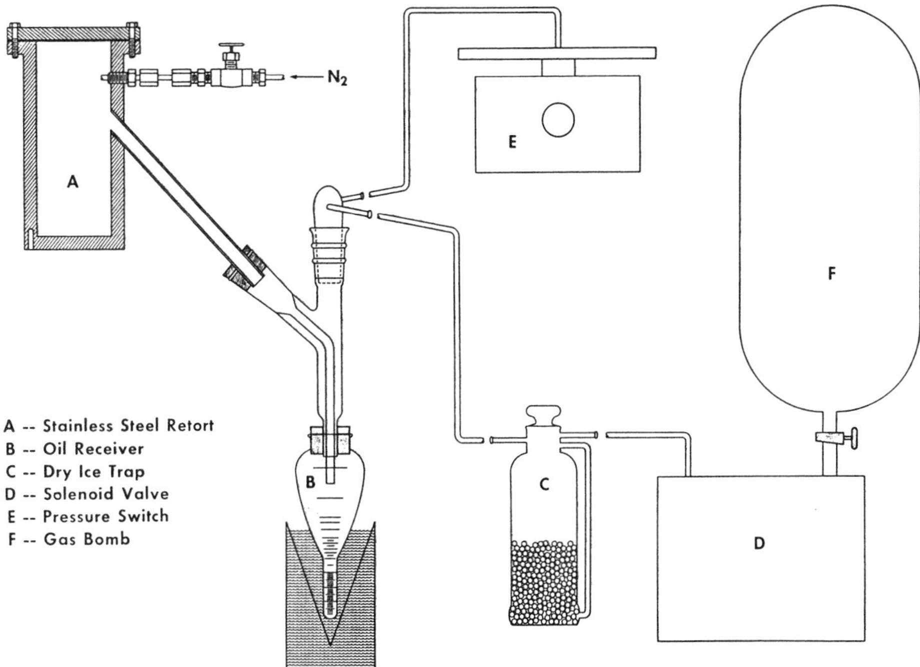

Figure 3 Schematic representation of the Western Research Institute modified (mass balance) Fischer assay retorting equipment (reproduced by permission of the Western Research Institute).

material balance assay was introduced in 1974 by Goodfellow and Atwood (Figure 4).

An extensive survey comparing the results of modified Fischer assay and several closely related alternatives for Green River oil shales was recently reported by workers at Lawrence Livermore National Laboratory (Singleton et al., 1986). These workers also conclude that lean shales may give product distributions that are significantly different from richer samples from the same area.

II. ALTERNATIVE METHODS

Oil yields depend not only on retorting temperature but on heating rate as well. High oil yields are generally obtained by rapid heating

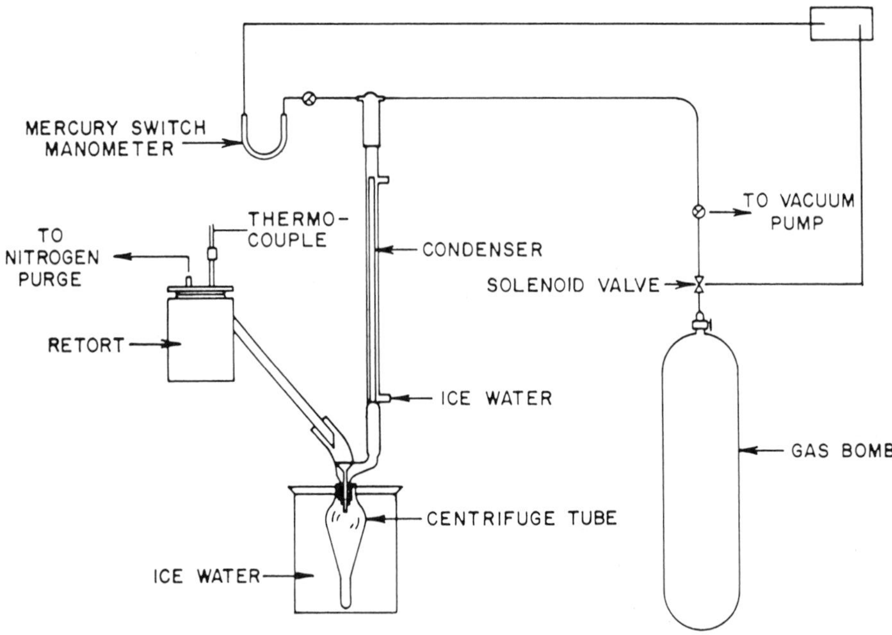

Figure 4 Schematic representation of the Tosco material balance assay equipment (Goodfellow and Atwood, 1974).

to an optimum temperature. The Fischer assay is not useful for evaluating oil yield under such conditions. This has led to the development of alternative thermal assay methods (of which the Rock-Eval method is notable) (Espitalie et al., 1977; see also Williams, 1983). The more important of these and some nonpyrolytic alternatives to the Fischer assay will be discussed below in the context of the various instrumental analysis methods.

Most oil shale studies have been limited in scope. One of the handful of studies that has been reported encompassed both a wide variety of shales and a wide variety of techniques (Robinson and Dineen, 1967). The geologically youngest of the shale samples was the Tertiary Pliocene Sao Paulo shale of Brazil (1–13 million years old), whereas the oldest is the Canadian shale of the Carboniferous period (280–350 million years) (Figure 5). The 12 samples are also representative of a wide range of depositional environments, kerogen types, and organic facies (Table 1).

The Colorado sample is a Green River shale from the Piceance Basin whose formation was discussed in an earlier section; it and the Brazilian

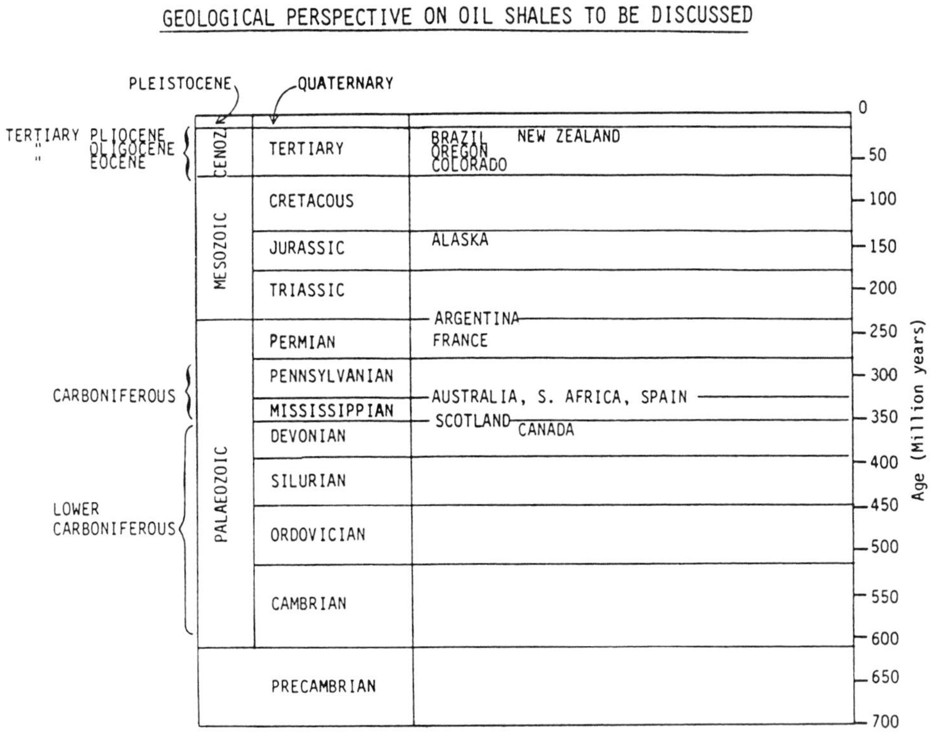

GEOLOGICAL PERSPECTIVE ON OIL SHALES TO BE DISCUSSED

Epoch / Sub-period	Era	Period	Oil shales
PLEISTOCENE		QUATERNARY	
TERTIARY PLIOCENE, " OLIGOCENE, " EOCENE	CENOZ.	TERTIARY	BRAZIL, OREGON, COLORADO, NEW ZEALAND
	MESOZOIC	CRETACOUS	
		JURASSIC	ALASKA
		TRIASSIC	
			ARGENTINA
	PALAEOZOIC	PERMIAN	FRANCE
CARBONIFEROUS		PENNSYLVANIAN	
			AUSTRALIA, S. AFRICA, SPAIN
		MISSISSIPPIAN	SCOTLAND, CANADA
LOWER CARBONIFEROUS		DEVONIAN	
		SILURIAN	
		ORDOVICIAN	
		CAMBRIAN	
		PRECAMBRIAN	

Age (Million years): 0, 50, 100, 150, 200, 250, 300, 350, 400, 450, 500, 550, 600, 650, 700

Figure 5 Schematic representation of the geologic ages of different oil shales used as comparisons of oil shale behavior (see text and Robinson and Dinneen, 1967).

(Sao Paulo) sample are lacustrine shales that are representative of organic facies type I and contain type I kerogen. The Alaskan, Argentine, Oregon, and Scottish shales are of marine origin, representing organic facies type II with type II kerogen. The remaining samples represent the coal-associated organic facies type III. However, as we shall see, not all contain primarily type III kerogen.

At this point, the casual reader may be wondering about the introduction of two apparently redundant concepts, that of kerogen types I, II, and III and organic facies types I, II, and III. It is tempting to try to simplify matters by eliminating one. The South African sample provides a compelling illustration of the need for both concepts.

Microscopic examination revealed that the organic material of the Alaskan, Australian, and South African shales contains mostly algal

Table 1 Geological Perspective on the 12 Oil Shales: Location, Depositional Environment, Geologic Period, Kerogen Type, and Organic Facies Type

Name	Location	Geologic period	Environment	Kerogen type	Organic facies type
Brazil	Sao Paulo	Tertiary (Pliocene)	Lacustrine	I	I
Oregon	Shale City	Tertiary (Oligocene)	Marine	II	I
Colorado	Piceance Creek	Tertiary (Eocene)	Lacustrine	I	I
New Zealand	Orepuki	Tertiary	Assoc. lignite	II	III
Alaska, Tasmanite	Kiligwa River	Jurassic	Marine	II	I
Argentina	San Juan	Triassic-Permian	Marine	II	II
France	St. Hilaire	Permian	Assoc. coal	II	III
Australia	Coolaway Mt.	Permian-Carboniferous	Assoc. coal	(Probably II,	probably III)
Spain	Puertollano	Permian-Carboniferous	Assoc. coal	(Probably II,	probably III)
South Africa	Ermelo	Permian-Carboniferous	Assoc. coal	II	III
Scotland	Dunnet	Lower Carboniferous	Lagoon-marine	II	II
Canada	New Glasgow	Carboniferous	Assoc. coal	II	III

Source: After Robinson and Dineen, 1967.

remains, "Yellow bodies" which are generally accepted as being the remains of algal colonies. Thus, these shales contain type II kerogen and are classified as *torbanites*. The Alaskan and Australian shales are apparently the result of algal blooms in the shallow waters of a marine platform or shelf. The Alaskan and Australian shales show little evidence of woody plant remains (terrestrial input). Thus, these fit cleanly into organic facies type II as well as having type II kerogen. However, the South African shale deposit is coal-associated and has appreciable terrestrial input. Presumably, this shale was formed from the remains of the algal blooms in a stagnant lagoon or stream near a coal-forming swamp. Thus, the South African sample is an example of a shale that has mostly type II kerogen yet fits into organic facies type III.

III. ELEMENTAL COMPOSITION AND OIL YIELD

A summary (Table 2) of elemental compositions and oil yields of the 12 oil shales and data for their corresponding kerogen concentrates shows that the total organic carbon (TOC) contents vary from 7.9% for the Canadian shale to 81.4% for the Australian sample. The oil shales contain hydrogen not only in organics, but also as water and in the minerals (e.g., as silanols). Hence, the hydrogen contents of the oil shales are not reported. The kerogen concentrates were freed of their hydrogen-containing minerals by HCl-HF treatment and carefully dried. Consequently, the hydrogen content of the kerogen concentrates reflects only the organic hydrogen present in the starting oil shales. Not surprisingly, the assay oil yields generally increase with the organic carbon content of the shale. However, the situation is not *that* simple! Note, for example, that the Colorado and Brazilian shales have nearly identical TOC contents, but the oil yield from assay of the former shale is much higher (27.7 versus 17.8 gal/ton). Evidently, factors other than TOC are also important in determining the quality of an oil shale.

IV. BITUMEN CONTENT AND COMPOSITION

It is interesting to note that there is no simple correlation between the organic carbon contents and the amounts of extractable bitumen (Table 3). The Australian shale with the highest organic carbon content gave one of the lowest bitumen yields, whereas the Colorado shale with a

Table 2 Elemental Compositions and Oil Yields of the 12 Oil Shales

	5	6	8	1	12	4
Geologic age	Alaska	Argentina	Australia	Brazil	Canada	Colorado
Raw oil shale						
Organic C, wt % TOC	53.86	8.88	81.44	12.79	7.92	12.43
Total N, wt %	0.30	0.26	0.83	0.41	0.54	0.41
Total S, wt %	1.30	0.48	0.49	0.84	0.70	0.63
Ash, wt %	34.1	82.6	4.4	75.0	84.0	65.7
Mineral CO_2, wt %	0.1	4.5	0.1	0.6	2.4	18.9
Assay oil, gal/ton	139.0	14.2	200.0	17.8	9.4	27.7
Kerogen concentrate						
Organic C, wt %	83.28	51.66	83.58	64.88	54.78	66.38
Organic H, wt %	11.48	6.60	10.69	8.52	5.57	8.76
Ash, wt %	1.5	34.9	1.7	12.5	33.4	12.9

	7	2	3	11	9	10
	France	New Zealand	Oregon	Scotland	South Africa	Spain
Raw oil shale						
Organic C, wt % TOC	22.25	45.69	25.83	12.33	52.24	26.01
Total N, wt %	0.54	0.78	0.51	0.46	0.84	0.55
Total S, wt %	2.32	4.79	2.20	0.73	0.74	1.68
Ash, wt%	66.3	32.7	48.3	77.8	33.6	62.8
Mineral CO_2, wt %	8.4	0.1	0.2	3.2	0.9	2.3
Assay oil, gal/ton	25.0	66.2	48.0	22.2	99.8	46.9
Kerogen concentrate						
Organic C, wt %	70.47	63.89	47.32	67.61	81.14	61.08
Organic H, wt %	6.45	7.48	6.00	7.62	9.12	7.04
Ash, wt %	17.7	9.9	36.2	18.6	1.3	26.0

Source: Data from Robinson and Dinneen, 1967.

Table 3 Bitumen Contents, Extraction Ratios, Organic Carbon Contents, and Contents of Carbonate Minerals for the Oil Shales

Location	Bitumen (wt %)	Extraction ratio[a] (wt %)	TOC (wt %)	Carbonates (wt %)
Oregon	4.0	15.5	25.83	0.2
New Zealand	4.3	9.3	45.69	0.1
Colorado	2.7	9.0	12.43	18.9
France	1.4	6.2	22.25	8.4
Scotland	0.7	5.4	12.33	3.2
Canada	0.4	5.3	7.92	2.4
Argentina	0.4	5.1	8.88	4.5
Spain	0.9	3.4	26.01	2.3
Brazil	0.4	3.3	12.79	0.6
Alaska	0.5	0.9	53.86	0.1
Australia	0.5	0.7	81.44	0.1
South Africa	0.3	0.7	52.24	0.9

[a]Extraction ratio = 100 × % bitumen/% organic C.
Note: The benzene extracts were fractionated into asphaltenes (pentane insolubles) and pentane solubles at 0°C (32°F). The pentane solubles were further fractionated by chromatography to obtain *n*-alkanes, branched- and cycloalkanes, aromatics, polars, and resins (Table 4).
Source: Data from Robinson and Dinneen, 1967.

relatively low organic carbon content gave one of the highest. Likewise, no simple correlation exists between age and bitumen content. The relatively young Tertiary shales of Colorado, Oregon, and New Zealand gave higher bitumen yields per weight of carbon (9–15.5%) than did the older shales (3.4–6.2%). However, the South African shale, which is also of Tertiary origin, gave the lowest bitumen yield, both in absolute terms and on a per-weight-of-carbon basis (0.3% and 0.7%, respectively). There does appear to be a relationship between carbonate mineral content and bitumen yield for this limited set of samples, but such a relationship would not be apparent were the data for a much larger sample set examined.

The bitumen from the Canadian shale (one of the oldest) contains 47% hydrocarbons and 53% nonhydrocarbons, while the New Zealand shale (one of the youngest) contains only 3% hydrocarbons, the other 97% being nonhydrocarbons. Evidently, oil shale bitumen, like petroleum, loses heteroatoms as it matures. What is not clear is whether heteroatom loss to produce hydrocarbons or reaction of heteroatom species to produce insolubles is responsible for this phenomenon. The observation that the older bitumens contain less of the heteroatom-rich pentane insolubles suggests that the latter may be more important (Table 4).

The *n*-alkane fractions were analyzed by GC and the results were used to calculate the carbon preferences indices (CPI*) for the C_{14}–C_{19} and C_{25}–C_{29} ranges (Table 5; Bray and Evans, 1961).

In the higher molecular weight range, alkanes with an odd preference often indicate terrestrial input; higher plants produce even-numbered carboxylic acid groups which give odd alkanes upon decarboxylation. No such preference is observed in material of marine origin (Koons et al., 1965). Because of the abundance of cuticular waxes in higher plants, the terrestrial contribution usually determines CPI in a mixture of marine and terrestrial material. Generally, CPI in the higher molecular weight range decreases with age, as cracking produces lower molecular weight species.

Five acyclic isoprenoid compounds expected in the oil shale bitumens were 2,6,10,14-tetramethylhexadecane (C_{20}, phytane); 2,6,10, 14-tetramethylpentadecane (C_{19}, printane); 2,6,10-trimethylpentadecane (C_{18}); 2,6,10-trimethyltridecane (C_{16}); and 2,6,10-trimethyldodecane (C_{15}). The iso- and cycloalkane fraction chromatograms were obtained using identical injection amounts and conditions (Figure 6).

The Brazilian and New Zealand bitumens contained only small amounts of the expected isoprenoids, while the Oregon bitumen contained the

---

*Several ways of expressing the odd preference have been proposed. The original definition by Bray and Evans (1961) used the C_{24}–C_{34} interval:

$$\text{CPI} = 0.5 \left| \frac{C_{25} + C_{27} + \cdots + C_{33}}{C_{24} + C_{26} + \cdots + C_{32}} + \frac{C_{25} + C_{27} + \cdots + C_{33}}{C_{26} + C_{28} + \cdots + C_{34}} \right|$$

Analogous expressions that average the preferences in overlapping even-carbon ranges were used to calculate CPIs in the ranges discussed above.

Table 4 Distribution of Components (wt %) in Bitumen Extracted from Oil Shales

	n-Alkanes	Branched + cyclic alkanes	Aromatic oil	Polars	Resins	Pentane insolubles	Total hydro-carbons
Canada	14.9	31.7	0.5	0.6	40.7	11.6	47.1
Scotland	11.4	32.6	2.1	0.1	44.0	9.8	46.1
Alaska	4.7	34.5	5.3	0.5	35.7	19.3	44.5
Spain	6.9	35.1	2.2	0.1	54.2	1.5	44.2
Australia	10.6	28.4	0.9	0.1	40.9	19.1	39.9
South Africa	16.5	20.0	0.9	0.1	49.4	13.1	37.4
Argentina	0.4	29.8	3.6	0.3	61.5	4.4	33.8
France	0.2	23.9	3.5	0.1	56.8	15.5	27.6
Colorado	1.1	13.7	0.9	0.1	66.2	18.0	15.7
Brazil	1.1	9.6	1.3	0.1	37.4	50.5	12.0
Oregon	0.2	7.0	0.2	0.2	43.2	49.2	7.4
New Zealand	1.1	1.4	0.2	0.1	9.5	87.7	2.7

Source: Data from Robinson and Dinneen, 1967.

Table 5 Bitumen CPI Values for the C_{14}–C_{19} and C_{25}–C_{29} *n*-Alkanes

	CPI values	
Location	C_{14}–C_{19}	C_{25}–C_{29}
New Zealand	1.04	4.76
Brazil	0.50	4.00
Colorado	1.18	3.17
Australia	0.60	2.03
South Africa	1.00	1.44
Oregon	0.72	1.33
Alaska	1.03	1.28
Scotland	1.04	1.21
Spain	1.03	1.18
Canada	1.04	1.16
France	0.41	0.99
Argentina	1.08	0.96

most. Phytane/pristane ratios varied widely. In general, the younger bitumens contained more phytane than pristane. The compounds eluting in the range of 250–325°C (480–615°F) include the steranes and hopanes. Gammacerane, a pentacyclic triterpane with five six-membered rings (not a hopane), was isolated from this sample of Colorado bitumen (Hills et al., 1966).

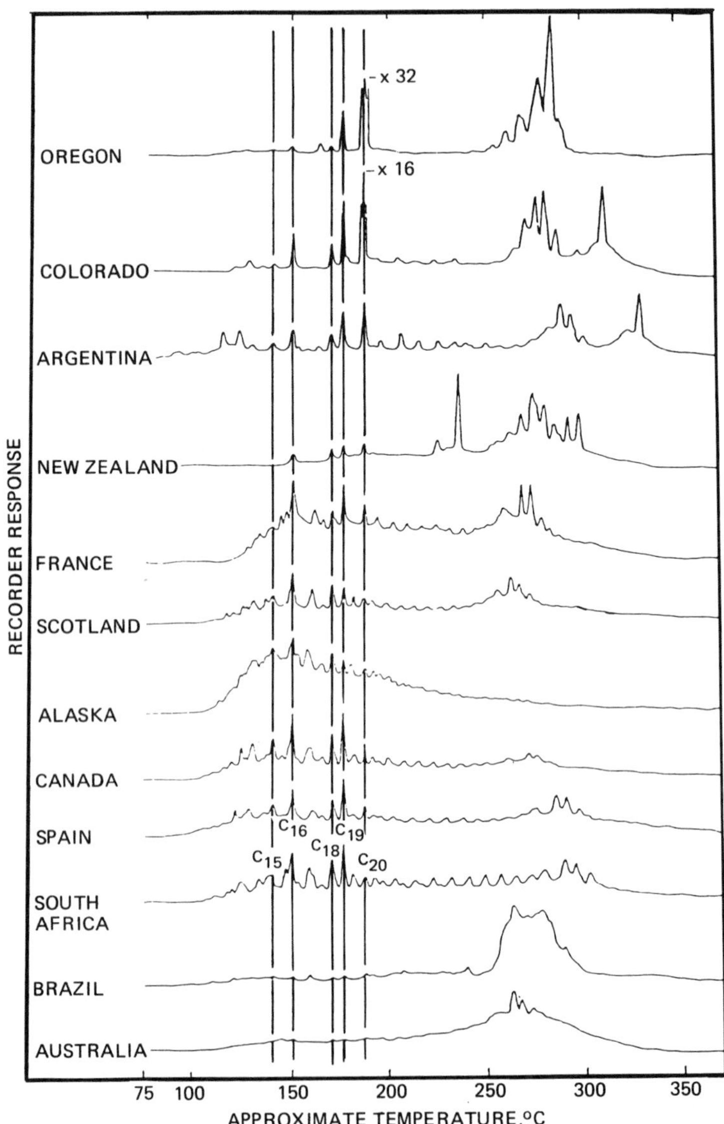

Figure 6 Isoprenoid compounds in the oil shale bitumens. Chromatograms are arranged in descending order of phytane/pristane ratios (from Robinson and Dinneen, 1967; reprinted with the permission of Elsevier Publishing Company).

27

Physical/Chemical Properties

I	H/C and N/C Ratios	856
II	Aliphatic and Aromatic Carbon	857
III	NMR Data	857
IV	NMR Techniques for Kerogen Characterization	859
V	ESR Data	861
VI	Density Methods	861
VII	Thermal Methods of Analyses	864
VIII	Heat Capacity and Heat of Retorting	870
IX	Mechanical Properties	873
X	Porosity and Permeability	875

The correlations of the oil yield and heating value of oil shales with their chemical and physical properties are based on many different kinds of measurements. These range from simple, qualitative tests that can be performed in the field to highly sophisticated quantitative measurements using the latest in laboratory instruments. This section describes some of the more useful correlations and the measurement methods on which they are based.

I. H/C AND N/C RATIOS

The relationships of the organic hydrogen and nitrogen contents, and assay oil yields, to organic carbon contents is shown in Table 1. Stoichiometry suggests that oil shales with higher organic H/C ratios can yield more oil per weight of carbon than those that are hydrogen-poor (Figure 1). However, H/C is not the only important factor. The South African shale with an H/C ratio of 1.35 has an oil yield/TOC of 0.72, while the Brazilian shale with an H/C ratio of 1.57 gives an oil yield/TOC of only 0.52. In general, the shales whose kerogen is converted efficiently to oil contain relatively low levels of nitrogen. However, the Brazilian and Colorado shales have very similar nitrogen contents yet give very different oil yields.

Table 1 Ratios of Hydrogen, Assay Oil Yield, and Nitrogen to Organic Carbon

Kerogen concentrate	H/C (atomic)	Oil shale	Oil/TOC	Oil shale	N/C × 10^2 (atomic)
Alaska	1.65	Alaska	0.95	Alaska	0.48
Colorado	1.58	Australia	0.92	Australia	0.87
Brazil	1.57	Colorado	0.85	South Africa	1.38
Australia	1.53	South Africa	0.72	New Zealand	1.46
Argentina	1.53	Oregon	0.70	Oregon	1.69
Oregon	1.52	Spain	0.68	Spain	1.81
New Zealand	1.41	Scotland	0.67	France	2.09
Spain	1.38	Argentina	0.61	Argentina	2.51
South Africa	1.35	New Zealand	0.54	Brazil	2.74
Scotland	1.35	Brazil	0.52	Colorado	2.82
Canada	1.22	Canada	0.44	Scotland	3.19
France	1.10	France	0.43	Canada	5.85

Source: Data of Robinson and Dineen, 1967.

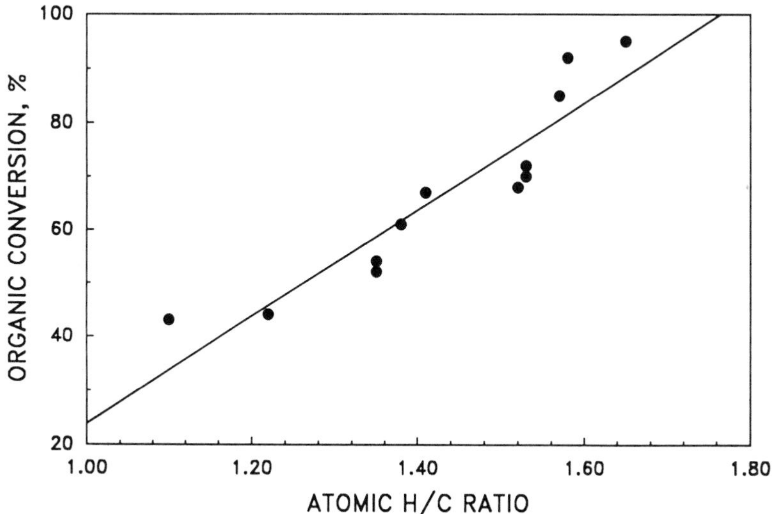

Figure 1 Conversion of oil shale organics as a function of atomic H/C ratio.

II. ALIPHATIC AND AROMATIC CARBON

A high atomic H/C ratio, as in the Alaskan and Colorado shales, indicates little unsaturation or aromatic, coallike structure. Conversely, a low atomic H/C ratio generally indicates high content of material with coallike aromatic structure.

III. NMR DATA

The advent of high-resolution, solid state NMR techniques (see also Chap. 26, Sec. III.D) has provided the geochemist a powerful tool for probing the structures of the organic material in solid oil shale kerogens and coals (Resing et al., 1978; Miknis et al., 1979; Maciel and Dennis, 1981; Hagaman et al., 1984) (Figure 2). ^{13}C NMR with cross-polarization and magic-angle spinning, the so-called CPMAS technique, has been used to determine the fractions of aliphatic and aromatic carbon in a variety of oil shales and the corresponding kerogen concentrates.

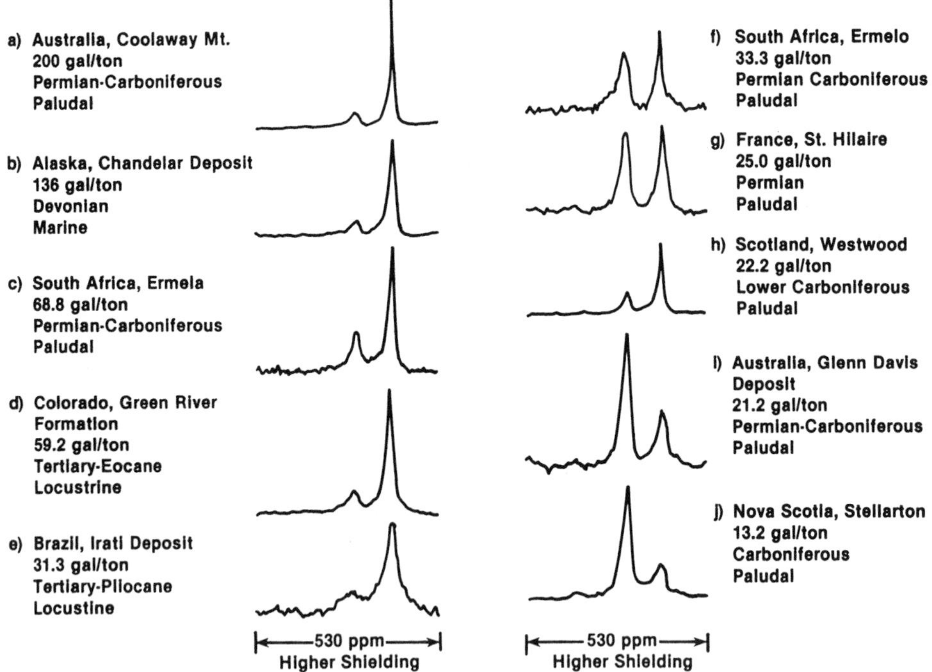

Figure 2 Solid state ^{13}C NMR spectra of oil shales obtained under CPMAS conditions. Note that the samples used in this work are not identical to the 12 shales discussed in the previous section, though several are very similar (Miknis et al., 1979; reprinted with the permission of Pergamon Press).

Paramagnetic and ferromagnetic species can affect the quantitative accuracy of NMR spectra obtained using the CPMAS technique. The problem of quantitative reliability in solid state ^{13}C NMR spectra of oil shales was first pointed out by Maciel and Dennis (1981) who found significant differences between the observed aromaticities of some oil shales and their kerogen concentrates. Paramagnetics were suggested as the cause, but the available data did not allow confirmation of this hypothesis. Subsequently, these effects have been studied extensively in the context of coals and products obtained by heating coals (Botto et al., 1985; Dereppe and Moreaux, 1987).

From TOC content and the fraction of aliphatic carbon determined by NMR, the wt % aliphatic carbon in the oil shale or kerogen concentrate can be calculated. A good correlation was obtained between oil yield and aliphatic carbon content (Figure 3). This technique has also been applied to the evaluation of petroleum source rocks (Miknis et al., 1982).

IV. NMR TECHNIQUES FOR KEROGEN CHARACTERIZATION

Three other solid state NMR techniques warrant mention, as they show great promise for further unraveling the structural features and thermal reactivities of the kerogens in oil shales and coals.

The "dipolar dephasing" technique introduced by Opella and Frye makes use of strong dipolar coupling between ^{13}C nuclei and directly bonded ^{1}H nuclei to selectively relax the ^{13}C NMR signals of the protonated carbons in relatively immobile rings and chains, but not in methyl groups which freely rotate. Thus, allowing dephasing of the ^{13}C polarization for a short time (30–60 μsec), then refocusing and adqiring the residual signal, enables the observation of a MAS ^{13}C NMR spectrum due almost entirely to nonprotonated and methyl carbons. However, it was recently pointed out that signals due to mobile methylene groups can also be observed in dipolar dephasing

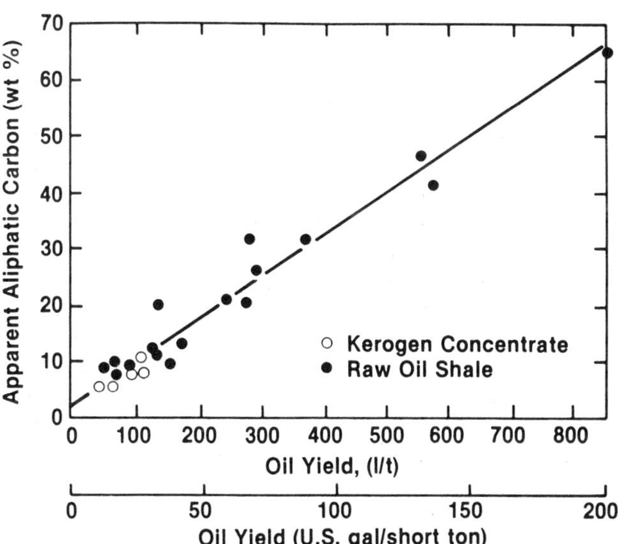

Figure 3 Aliphatic carbon content versus oil yields for 22 oil shales and kerogen concentrates (after Maciel et al., 1978, 1979).

experiments (Soderquist et al., 1987). Thus, dipolar dephasing results should be interpreted with some care and with considerable structural insight. Nevertheless, the dipolar-dephasing magic-angle spinning" (DDMAS) technique is a very powerful tool, especially for probing the aliphatic components of oil shales, kerogens, and coals.

The variable-angle sample spinning (VASS) technique recently introduced by Sethi et al. (1987) affords detailed structural information through evaluation of the tensor components of chemical shift anisotropy. In the CPMAS technique the sample is spun rapidly (about 3000 Hz, or $\sim$180,000 rpm!) at the "magic angle" of 58°44' to average the components of the chemical shift tensor and obtain an isotropic chemical shift similar to that observed in liquid phase spectra. In doing this, the structural information contained in the individual chemical shift tensor components is lost. The VASS technique enables recovery of this information by spinning the sample at angles other than the magic angle. The recovered information can then be used to evaluate the contributions of different types of aromatic carbons: protonated, substituted, and inner. Thus, DDMAS and VASS are complementary techniques, providing information about the structural details of the aliphatic and aromatic components of the kerogen, respectively. Used in conjunction with the CPMAS technique, DDMAS and VASS make solid state ^{13}C NMR one of the most useful tools available for characterizing the organic material in soild oil shales.

The third of the newer NMR methods is the dynamic, in situ ^{1}H NMR technique (Parks et al., 1987). The power of the method lies in the fact that the measurements can be carried out while the sample is being heated. In contrast, the currently available solid state ^{13}C NMR equipment is generally restricted to operations near or below room temperature. Samples can be heated, then observed, but not observed during heating. The ^{1}H signals observed by Lynch are sensitive to the molecular dynamics of the hydrogen-containing material. Using this technique it is possible to estimate the fractions of molecular structure that are "rigid" and "mobile" on a time scale of about 10^{-5} sec. Oil shale and coal samples typically exhibit dynamic ^{1}H NMR spectra that are composites containing both rigid and mobile components. Presumably, crystalline or glassy material contributes to the former, while the latter, mobile component is due to rubbery, amorphous material. As the sample is heated, an initial increase in the mobile fraction (ascribed to conversion of rigid material into mobile) is typically observed, followed by a decrease as volatile hydrocarbons are evolved. After the initial decrease in the rigid component, an increase indicates the formation of coke precursors and/or coke. The kinetics of both changes can be followed by monitoring the dynamic ^{1}H spectrum as a function of temperature and time. This technique has been used to probe the pyrolysis (i.e., retorting) behavior of a wide variety of oil shales and coals.

V. ESR DATA

Electron spin resonance (ESR), also known as electron paramagnetic resonance (EPR), methods have been used extensively to monitor the changes in free radicals that occur when coals are heated. ESR has also been used to study the process of kerogen maturation, both natural and simulated, by heating in the laboratory--mostly in the context of petroleum exploration, but also in the context of oil shale retorting. These studies were surveyed in a recent article by Silbernagel and coworkers (1987), who also reported the changes in free radical content and character that occurred upon heating oil shales from the Green River and Rundle (Australia) formations, and their kerogen concentrates, for varying times at relatively low temperatures of 350–375°C (600–705°F). Free radical concentrations generally increased with heating but these increases were shale-dependent, being much larger for the Rundle samples. Changes in *g* values and line widths were also observed and discussed.

Sousa and coworkers (1987) used ESR in conjunction with thermal alteration index (TAI) determinations to study the natural maturation of the Irati oil shale of Brazil. Qualitative agreement between TAI and ESR results were obtained for stratigraphic column CERI-1 for which previous workers had hypothesized an unusual exposure to paleotemperature at one particular point in the column. However, the ESR results proved much more sensitive to this local heating than TAI.

To summarize, the potential of ESR for oil shale studies is clearly great; the technique is very sensitive to heating. Qualitative ESR methods have been useful. However, quantitative ESR studies have generally raised more questions than they have answered. The reason for this may be that the free electron spins observed by ESR are not directly involved in the chemistry of interest, but merely reflect exposure to the environment (heat, pressure, structure, etc.) in which this chemistry takes place. If this is the case, then ESR can at best be expected to provide correlations, not a direct probe of the chemistry.

VI. DENSITY METHODS

It is evident that the difference in density between the organic and mineral components of an oil shale is important as the basis for dense medium beneficiation. However, density is also important as a characterization tool. The relationship between density and oil yield was pointed out by Frost and Stanfield (1950) who reported measurements on 32 samples of Green River oil shale spanning the ranges of 1.67–2.54 g/cm^3 in density and 10–77 gal/ton in oil yield. Later, Smith

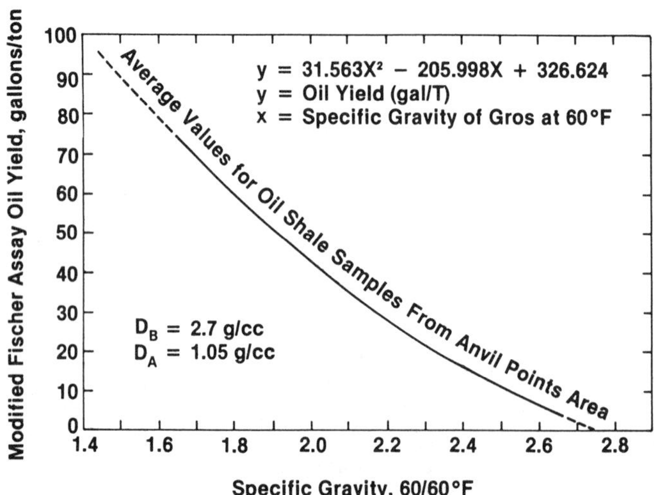

Figure 4 Specific gravity versus oil yield for Colorado oil shales. The curve is calculated from Eq. (2) (Smith, 1956, 1969; reprinted with the permission of the American Chemical Society).

derived the following equation to describe this relationship quantitatively (Figure 4):

$$D_T = \frac{D_A D_B}{A(D_B - D_A) + D_A}$$

where

D_T = density of the shale rock
A = weight fraction of organic matter
B = weight fraction of mineral matter
D_A = average density of the organic matter (g/cm^3)
D_B = average density of the mineral matter (g/cm^3)

Vadovic used density separations in another way to estimate the amounts of hydrogen and nitrogen associated with the mineral matter in oil shales (Vadovic, 1983). This approach begins with separation into several fractions of increasing density. A pulverized oil shale sample is separated into the fractions which sink and float in a medium of specific gravity 1.5. The sink fraction is freed of the dense medium and dried for analysis, while the float fraction is subjected to another sink-float separation in a more dense medium. In this way, fractions having specific gravities of <1.6, 1.6–1.7, 1.7–1.8, 1.8–1.9, 1.9–2.1, 2.1–2.2, and >2.2 were obtained. The sink-float fractions were then analyzed for carbon, hydrogen, and nitrogen. For Green River oil shale from the Colony mine, a plot of hydrogen results versus organic carbon results for the different fractions gave a straight line (Figure 5). The intercept of this line at 0% organic carbon (no organic matter) provides an estimate of the atomic ratio of mineral hydrogen to organic carbon content. From this ratio and wt % organic carbon, a mineral hydrogen content of 0.08 wt % (3% of the total H) was estimated for the Colony oil shale. Similar treatments yielded mineral

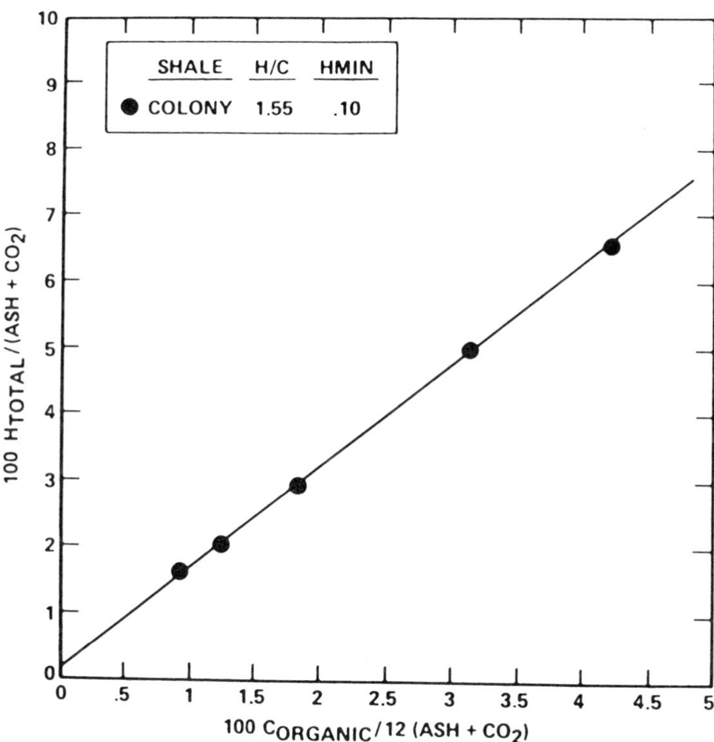

Figure 5 Hydrogen analysis for Green River oil shale (Colony mine, Colorado) (Vadovic, 1983; reprinted with the permission of the American Chemical Society).

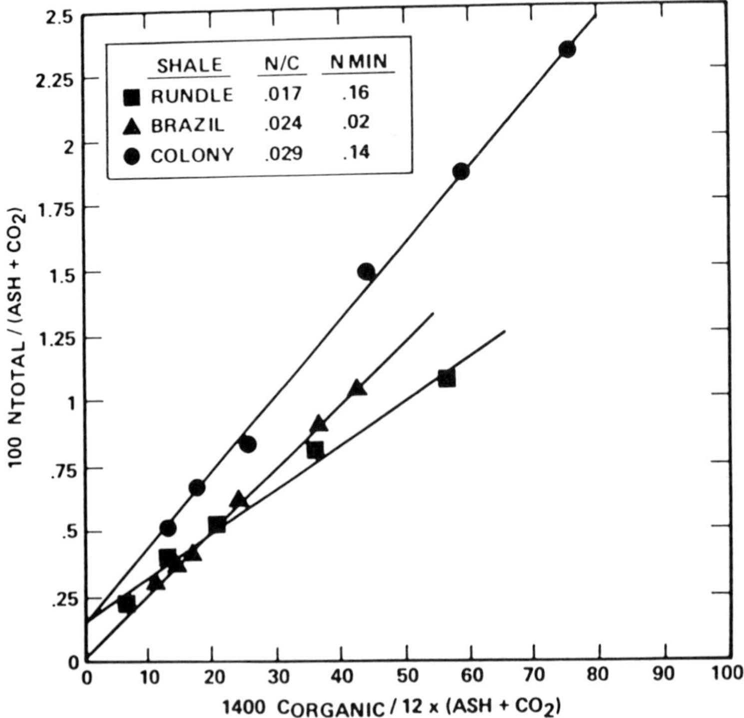

Figure 6 Nitrogen analyses for Green River, Rundle, and Brazilian oil shales (Vadovic, 1983; reprinted with the permission of the American Chemical Society).

hydrogen estimates of 0.25 and 0.28 wt % (11 and 15% of the total H) for Rundle and Brazilian oil shales, respectively. This approach is also applicable to other mineral elements; mineral nitrogen contents of 0.14, 0.13, and 0.02 wt % (18, 31, and 4% of total N) were estimated for the Colony, Rundle, and Brazilian shales, respectively (Figure 6).

VII. THERMAL METHODS OF ANALYSES

Heat is the objective of most interest in oil shale and large amounts of heat must be supplied during retorting to produce shale oil from the rock. Moreover, when the shale rock is heated, its weight decreases

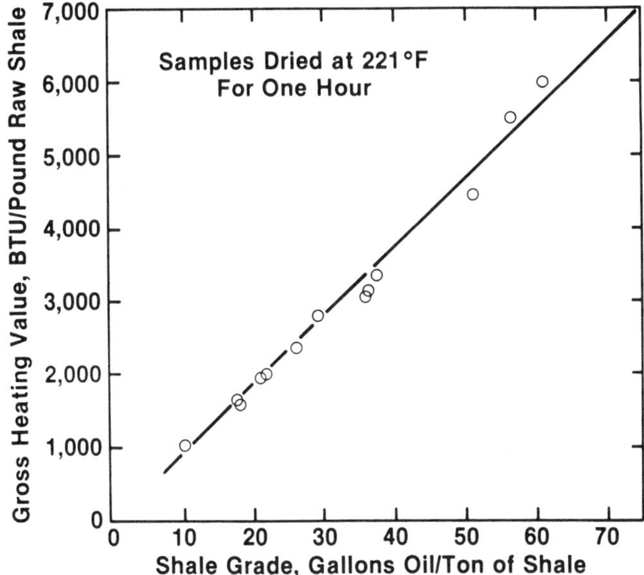

Figure 7 Heating values of dried Green River oil shale from the Mahogany Ledge, Rifle, Colorado.

as volatile species are evolved. Consequently, thermal methods of analysis have long been important in oil shale characterization.

Not surprisingly, there is a good linear correlation between the heating value of dry oil shale and its Fischer assay oil yield (Figure 7).

It is also possible to estimate the heating value of a dry oil shale from its elemental composition. In the following equations, Q is the heating value in btu/lb, while C, H, O, and S are contents of the respective elements in wt %.

Boie Equation

$$Q = 15,120C + 49,977H + 2700N + 4500S - 4770O$$

Dulong Equation

$$Q = 14,544C + 62,028(H - O/8) + 4.050S$$

The kerogen (and most of the bitumen) in oil shales is not volatile. As oil shale is heated during retorting, these nonvolatile organics

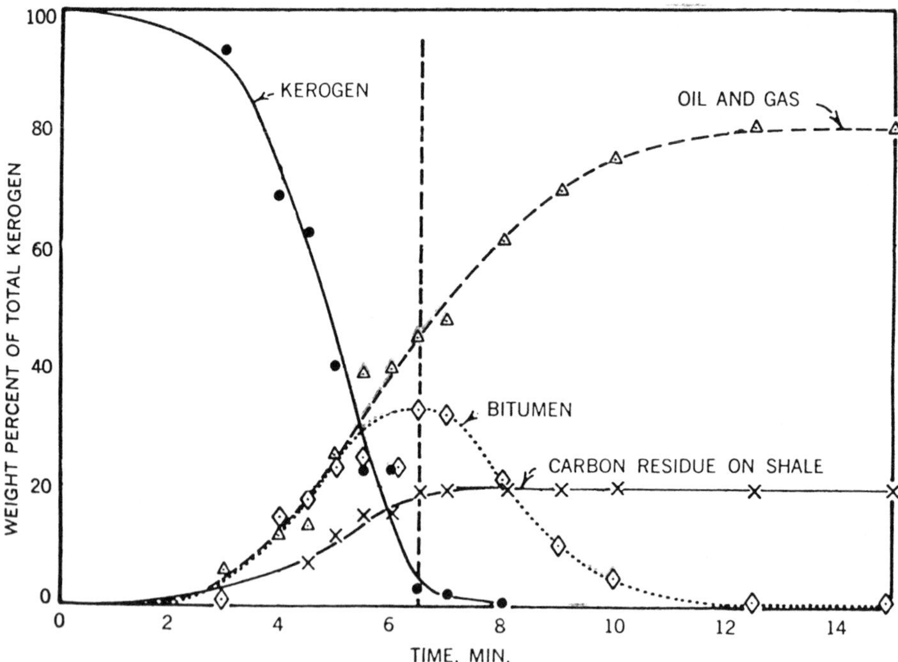

Figure 8 Rates of kerogen disappearance and product appearance during heating of Green River oil shale at 900°C, 1650°F (after Allred, 1966, with data from Hubbard and Robinson, 1950).

crack to give volatile products that are evolved from the rock. Thermogravimetry (thermogravimetric analysis, TGA), which measures the change in weight as a sample is heated, has been widely used to study oil shale retorting (Rajeshwar et al., 1979; Rajeshwar, 1983).

Many workers have used TGA to study the kinetics of oil shale retorting. Hubbard and Robinson (Figure 8) deduced first order kinetics in a detailed study of the thermal decomposition kinetics of Green River oil shales; they proposed a two-step decomposition pathway:

$$\text{Kerogen} \xrightarrow{k_1} \text{pyrobitumen} \xrightarrow{k_2} \text{oil + gas + char}$$

Further studies on Green River oil shale using both isothermal and nonisothermal TGA techniques (Allred and Nielsen, 1965; Allred, 1966) led to the conclusions that the thermal decomposition involved three first order steps:

$$\text{Kerogen} \xrightarrow{k_1} \begin{matrix}\text{gas}\\ \text{pyrobitumen}\\ \text{char}\end{matrix} \xrightarrow{k_2} (\text{oil} + \text{gas})_{\text{liq.}} \xrightarrow{k_3} (\text{oil} + \text{gas})_{\text{vap.}}$$

$$\ln\left[\frac{1 - X}{X}\right] = -kt + \ln I$$

where X = (weight of oil + gas)/(weight of starting kerogen).

Campbell and coworkers (1978) interpreted their nonisothermal TGA results on Green River oil shale in terms of a single first order step, as did Herrell and Arnold (1976) in their study of Chattanooga oil shale. Rejeshwar (1981), in a later nonisothermal TGA study, concluded that the data required two first order steps. First order kinetics were also found by Haddadin and Mizyed (1974) in a TGA study of Jordanian oil shales, while Bekri and coworkers (1983) and Thakur and Nuttall (1987) found first order kinetics and a two-step organic decomposition mechanism in studies of Timhadit shales from Morocco. Williams (1985) found similar kinetics in TGA decompositions of Kimmeridge oil shales from Great Britain and good agreement between the TGA results and those obtained in artificial maturation. Interpretations of the results of the foregoing studies were all based on assumed mechanisms of kerogen decomposition.

Braun and Burnham (1986) reported a new analysis of the kinetics of product evolution from Green River oil shale. The isothermal retorting data had been earlier obtained in a quartz fluidized bed apparatus with a flame ionization detector (from a GC) to detect evolved organics (Richardson et al., 1982). This apparatus does not respond to water (unlike the TGA which records water as weight loss, along with oil + gas), but does respond to hydrocarbon gases that do not contribute to oil yield. Therefore, the FID response was corrected for C_1-C_3 hydrocarbon evolution using other rate data (Richardson et al., 1982). It was discovered that the results could be fitted equally well with *either two parallel first order rates or one three-halves order rate* (Figures 9 and 10).

There is still some uncertainty regarding the maximum amount of oil that can be evolved from Green River oil shale under the rapid heat-up conditions. However, there is good evidence that this is at least 110% of the Fischer assay yield that can be obtained for shale with particles smaller than about 3 mm (Wallman et al., 1981; Braun and Burnham, 1986; Campbell et al., 1978).

Both pressure and atmospheric composition can affect the rates of product evolution during pyrolysis. It was concluded that very high heating rate (>1000°C/sec) pyrolysis did not enhance oil yield over

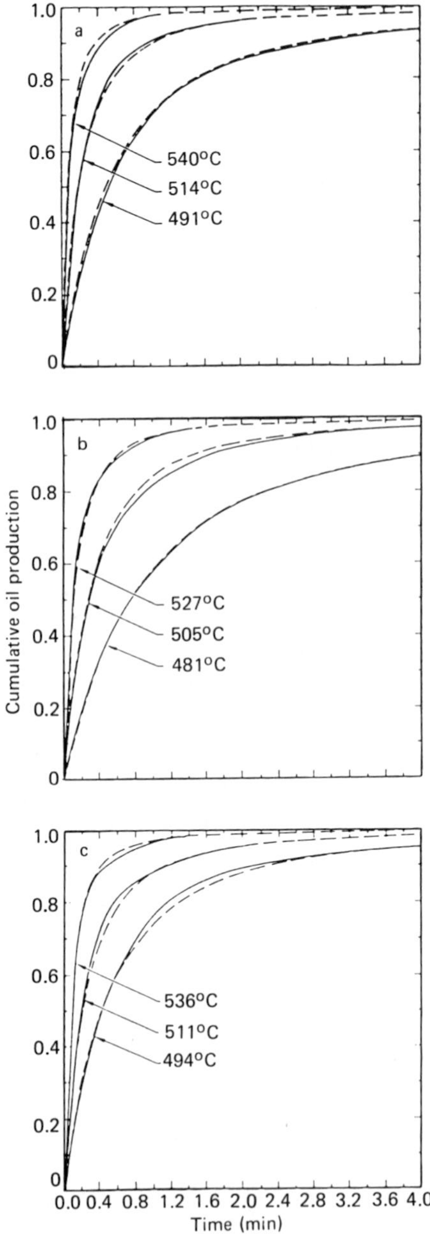

Figure 9 Comparison of calculated (----) and measured (——) oil production for one 1.51 order reaction. (a) Anvil Points sample; (b) Tract Ca, Sample RB-1; (c) Tract Ca, Sample RB-2 (Braun and Burnham, 1986; reprinted with the permission of Butterworth & Co.).

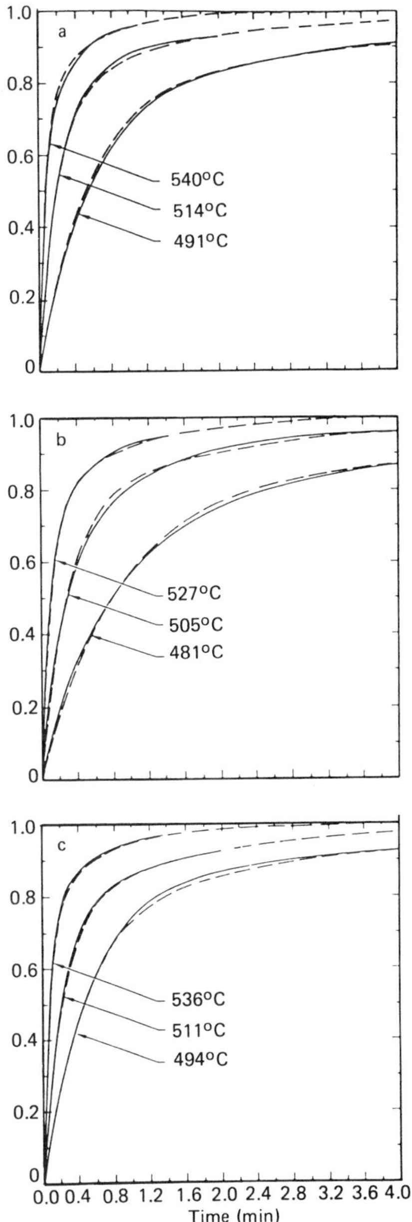

Figure 10 Comparison of calculated (----) and measured (——) oil production for two parallel first order reactions. (a) Anvil Points sample; (b) Tract Ca, Sample RB-1; (c) Tract Ca, Sample RB-2 (Braun and Burnham, 1986; reprinted with the permission of Butterworth & Co.).

Fischer assay for a rich (50 gal/ton) Green River shale (Suuberg et al., 1987). Little effect of pressure was observed near or below 1 atm, but higher pressures significantly retarded oil evolution and altered the product slate to favor lower molecular weight products. Similar observations were made under slow heating conditions by Burnham and Singleton (1983), who also observed that the pressure effect was less pronounced at lower heating rates. It was pointed out that the residence times of both liquid oil and oil vapor increase with increasing pressure. Moreover, the low porosity of the oil shale is likely to retard oil evolution, at least until some products depart generating appreciable porosity. Thus, the lower oil yield and lighter product slate are both ascribed to longer residence time of the liquid products, which allows more extensive cracking and char formation.

Because of the general acceptance of the Fischer assay, TGA has not been generally used for evaluation of oil yield. Moreover, TGA methods do not readily provide some of the information obtained from the Fischer assay, e.g., the water yield and the distribution of liquid and gaseous products. TGA is, however, well suited for comparative studies in the laboratory; it requires only small samples (1–25 mg) and automated equipment is commercially available. Smith et al. (1969), Johnson et al. (1978), and Rajeshwar et al. (1983) reported the utility of TGA for evaluating Green River oil shales, while Earnest (1982, 1983) obtained good correlations between TGA and oil yield for both Green River and Australian oil shales. Williams (1982) found reasonable correlation between TGA weight loss and oil yield, and good correlation between Rock-Eval peak area and oil yield, for Kimmeridge oil shales. However, in a more extensive study of Kimmeridge shales, both the water/oil and gas/oil ratios in retorting products varied with the richness of the shale (Williams, 1985). These findings led to the recommendation for caution in the interpretation of oil shale TGA results, especially when comparing results for shales of widely differing grade.

VIII. HEAT CAPACITY AND HEAT OF RETORTING

Knowledge of enthalpy changes is important to both modeling and design of oil shale processes. What is needed are predictive relationships that enable the calculation of enthalpy changes on arbitrary changes in the temperature and composition of the oil shale.

Most of the reported studies (Camp, 1987) deal with Green River oil shales. Several workers measured the enthalpies required to heat oil shale and most studied shales of varying oil yield (Shaw, 1947; Sohns et al., 1951; Cook, 1969; Wise et al., 1971; Mraw and Keweshan

1984). The data provided precise estimates of the heat requirements in retorting. However, the most useful correlations of heat capacity with temperature and shale grade were purely empirical and linear in temperature (Carley, 1975). These linear relationships have been criticized (Camp, 1987) with note of the pronounced nonlinearity at temperatures above 325°C (615°F) of the heat capacities of several major minerals in Green River oil shale.

As an alternative, the approach of summing the heat capacities of the organics (char in the case of spent shale), volatile products evolved from the oil shale, the individual minerals, plus bound and free water was taken (Camp, 1987). Using the resulting nonlinear model and published heat capacities (Robie et al., 1979) for a set of about 20 minerals, excellent fits were obtained to the reported heat capacities of spent and burned shale over the temperature range of 25–600°C (75–1110°F). Moreover, the equation fit the reported heat capacities of raw oil shales of different grades at temperatures up to 240°C (465°F), where decomposition of the Green River oil shale kerogen is negligible. The heat capacity of the organic fraction was modeled using published heat capacity data for graphite and 60 organic compounds and petroleum fractions, for which data were available over a wide range of temperature, to obtain nonlinear equations for the heat capacities of the kerogen and char, respectively. These equations fit the low-temperature data well, but predict lower, more reasonable heat capacities than the linear equation at the higher temperatures encountered in retorting:

$$C_{kerogen} = 0.2232 + 5.254 \times 10^{-3}T - 1.6536 \times 10^{-6}T^2$$

$$C_{char} = -0.1179 + 4.308 \times 10^{-3}T - 1.786 \times 10^{-6}T^2$$

where

heat capacity = C in kJ/°K kg
temperature = °K

The resulting model, incorporating nonlinear predictions for heat capacities of both organic and mineral fractions, was then used to fit the measured heat capacities of several Green River oil shale samples varying in both oil yield and water content (Figure 11).

Subtraction of the mineral heat capacity from the total heat requirement enabled estimation of the heat required for kerogen decomposition to volatile products and char. This heat was found to be substantially larger for shale samples heated slowly to 350°C (660°F) than for samples heated rapidly to 500°C (930°F) and held for only 2–3 min for pyrolysis (370 kJ/kg of kerogen versus 275 kJ/kg). Of this

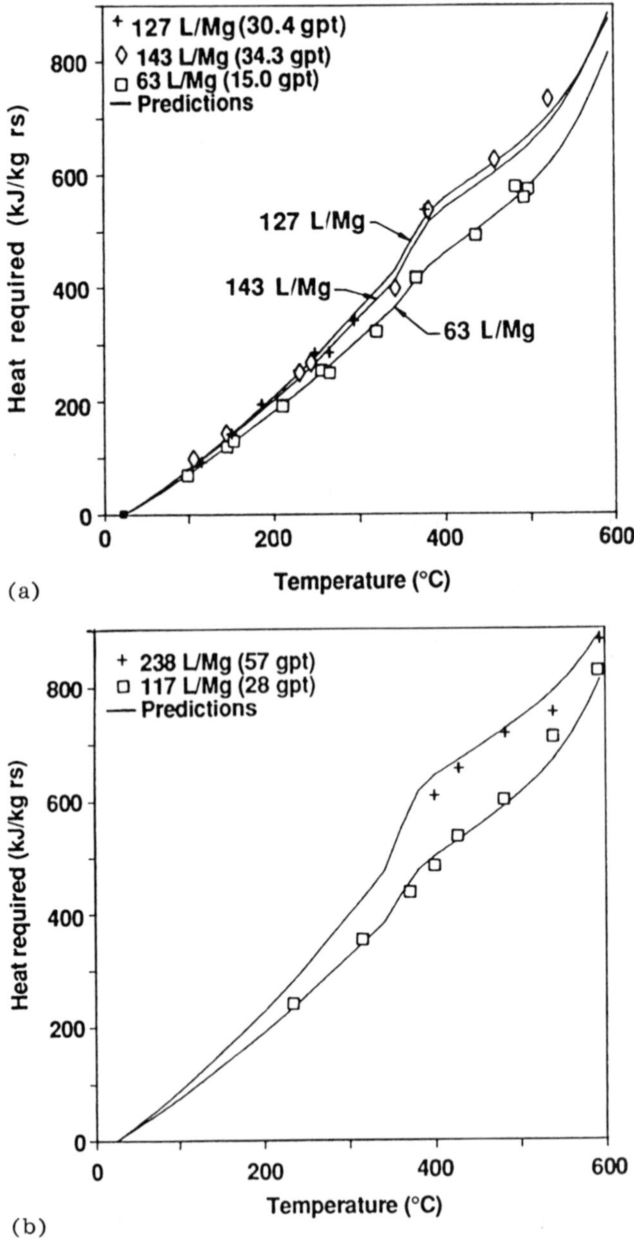

Figure 11 Measured and predicted heat requirements above 25°C for Green River oil shale samples (Camp, 1987; reprinted with the permission of the Colorado School of Mines). (a) Sample assaying 127 liters/mg (127 liters/10^6) required more heat than the 143 liters/mg sample due to its higher content of hydrated minerals (data of Wise et al., 1971). (b) Data of Sohns et al., 1951.

difference, 50 kJ/kg could be accounted for by the larger sensible heat of the kerogen relative to that of the evolved products. The remaining difference, 45 kJ/kg of kerogen (12% of the total heat), clearly reflects an endothermic reaction that occurred to a greater extent during the long heating at 350°C (660°F). It is suggested that either increased char formation (coking) or a mineral decomposition reaction not accounted for in the model may be responsible.

IX. MECHANICAL PROPERTIES

Minerals play a major part in determining the strength and hardness that are important in mining and processing (e.g., crushing, grinding) oil shales, especially those of the more prevalent leaner grades [see Figures 12 and 13, where both strength and hardness decrease markedly as the organic content of the oil shale increases (Sellers et al., 1972)].

The strength and hardness of oil shales decrease upon retorting and, not surprisingly, this decrease is greater for richer shales. Dinneen found that lean oil shale cores (<15 gal/ton) retained high compressive strength (∿90% of initial value both parallel and perpendicular to the bedding plane) after heating to 510°C (950°F) (Dineen, 1968). Richer samples lost progressively larger fractions of their

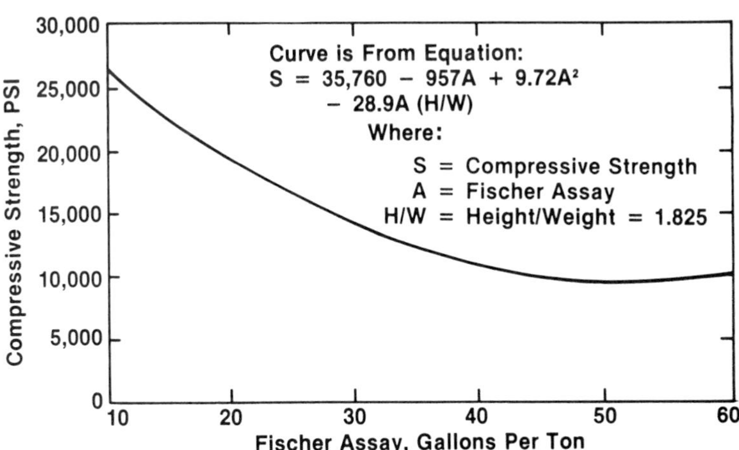

Figure 12 Relationship of Fischer assay oil yield to compressive strength (Sellers et al., 1972).

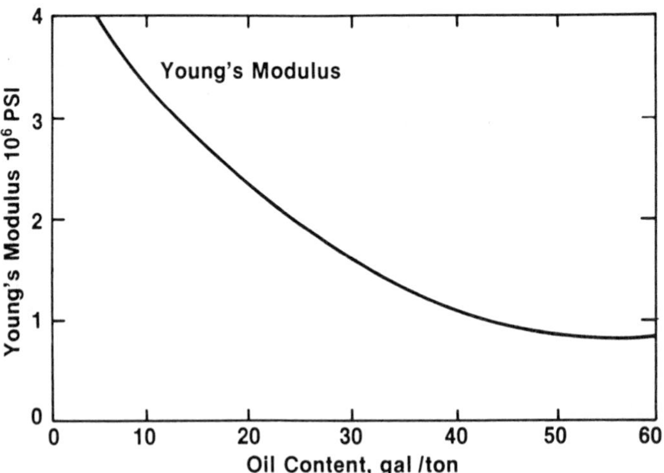

Figure 13 Relationship of Fischer assay oil yield to Young's modulus (Sellers et al., 1972).

strength on heating, a 15 gal/ton sample lost about 75% of its initial strength, while a 60 gal/ton sample lost >99% to give a very soft, almost chalklike ash. Further heating to 815°C (1500°F) to decompose the carbonate minerals caused little further loss in strength (see also Baughman, 1978).

Acoustic wave propagation experiments reflect several phenomena, but do provide a sensitive probe of the state of the oil shale matrix, which may prove useful in noncontact monitoring of in situ and aboveground retorting processes. Mraw and coworkers (1983) determined the acoustic wave propagation characteristics of some relatively rich Green River oil samples. Both cores and rubble samples were studied. The results were interpreted in terms of a complex set of thermal alterations, including the release of free and bound water, pyrolysis and release of the organic matter, and recementation of the shale matrix by pyrolysis products. The results were strongly dependent on the initial organic content, but only slightly upon temperature. Measurements on rubble (broken fragments) showed an expected dependence on the compaction pressure and revealed an unexpected anomaly that suggested the presence of small voids in the heat treated samples.

X. POROSITY AND PERMEABILITY

Porosity and permeability of shales are generally very low. Because of this, shales often comprise the cap rock or seal in petroleum and natural gas reservoirs. However, in retorting, permeability is critical to allow the escape of products from the rock. Dinneen (1968) found that appreciable porosity was developed by heating Green River oil shale. Tisot and Sohns (1971) monitored the permeability of columns of Green River oil shale particles as a function of temperature, heating rate, and compressive load/strain. Thermoplastic flow was observed with samples that contained >50 vol % organic matter (>48.5 gal/ton). The loss of strength observed in rich samples led these workers to conclude that permeability losses under the influence of heating and overburden pressure could significantly impair in situ retorting of rich zones (>27 gal/ton). In contrast, a column containing fragments of 27-gal/ton shale underwent significant compression upon heating but retained its high initial permeability. These results show that permeability changes upon heating must be considered in planning an in situ retort operation.

Porosity and permeability have been extensively studied in the context of petroleum exploration and production (Hunt, 1979; Ward, 1984). A recent article by Hall et al. (1983) discusses the methods available for measuring the porosity and permeability of very nonporous shaly rocks and describes a new technique that uses small-angle neutron scattering (SANS). This technique allows definition of the size distribution for pores in the range of about 2–75 nm and is capable of detecting anisotropy in pore orientation. Presently, this technique does not provide information about the connectivity of the pores (topology of the pore network). As a result, it is not clear that SANS, or the closely related small-angle X-ray scattering (SAXS) technique, is ready to displace the more traditional methods for measuring permeability but further study of these methods is clearly warranted.

28

Chemical Structure of Kerogen

I Kerogen Isolation 878
II Structural Inferences from Analyses 879
III Structural Analysis of Kerogen 881
A. Oxygen Functional Groups 881
B. Oxidation 883
C. Depolymerization 890
D. Micropyrolysis–Gas Chromatography–Mass Spectrometry 891
E. Hydrogenolysis 893
IV Structural Models for Kerogen 895

Typical oil shales are nonporous, impermeable rocks containing 80–95 wt % minerals, and only 5–20 wt % organics. Of these organics, only a minor part, the bitumen, is extractable into organic solvents. By far the major part of the organics in most oil shales is present as kerogen, an insoluble solid of variable composition which is usually finely dispersed throughout the mineral matrix. The difficulty of achieving quantitative and selective reactions of insoluble organic solids under mild conditions and the scarcity of good methods for probing the structures of such materials have discouraged attempts to characterize the organic material in solid oil shales. As a result many of the ideas regarding the structures of oil shale kerogens have been derived from the many reported studies of bitumens (oil shale extracts) and thermally generated shale oils (see also Chap. 19, Sec. III for comparison). Nevertheless, over the past 25 years several research groups have begun to probe the chemical structures of oil shale kerogens. In general, the first step in these studies has been isolation of the corresponding kerogen concentrate. A wide variety of chemical and physical techniques have been used to characterize the resulting kerogens.

I. KEROGEN ISOLATION

By far the most common technique for kerogen isolation involves acid demineralization of the oil shale to produce the corresponding kerogen concentrate. Recently, a rapid two-step base-acid treatment as well as a combined chemical-physical concentration method offering mild conditions and high kerogen recovery were applied to Green River oil shale. Physical separations were also used. All of these methods involve comminution as the initial step and it is important to select samples that have not been unduly altered by weathering (Forsman and Hunt, 1958; Forsman, 1963; Robinson, 1969; Saxby, 1976; Durand and Nicase, 1980).

Because of their low permeability, oil shales do not weather as rapidly as coal samples. Nevertheless, weathering can alter oil shale kerogens that have been exposed to the atmosphere. Whenever possible, core samples or samples taken more than 12 in. below an exposed surface should be used for kerogen isolation to avoid weathering effects. Storage under nitrogen has not been found necessary for oil shale samples that are larger than 1/2 in. or so.

Larger pieces of oil shale are crushed in a jaw crusher to pass 8-mesh, then ground to pass 100-mesh in a hammer mill or disk mill. Care should be taken to keep temperatures low during fine grinding, and grinding under an inert atmosphere (e.g., nitrogen) is recommended to avoid possible oxidation. Finely ground oil shale should be stored under nitrogen.

Physical methods to produce an organic-rich kerogen concentrate are of interest for two reasons: exposure of the kerogen to the strong acid and/or base is avoided, thereby lessening the chance of chemical alteration; and because such "beneficiation" methods are practiced on a large scale in the coal and ore-processing industries, they may be useful in a commercial oil shale operation. Such methods do generally involve the potential for contamination of the kerogen with materials used to effect the separation. However, in many cases the potential impact of such contaminants can be limited by using only one, or a small number, of known and easily identified chemical species. Among the more important physical methods for kerogen concentration are sink-float, oil agglomeration, and froth flotation methods (Vadovic, 1983; Luts, 1928).

Hubbard and coworkers (1952) concentrated Green River oil shale kerogen by centrifugation in benzene—carbon tetrachloride mixtures.

In this work, the finely ground oil shale was first extracted to remove benzene solubles, dried, then subjected to successive sink-float separations in solvent mixtures having densities of 1.40, 1.20, and 1.15 g/ml. The concentrate that floated in the 1.20 g/ml mixture represented 6% of the starting organics and contained 14 wt % ash. The concentrate from the last stage contained 9 wt % ash but represented only 1% of the starting organic material. In this case the constancy of atomic H/C ratios and assay oil/C_{org} ratios suggests that little fractionation of the kerogen occurred. Similar results were observed by Vadovic.

Thus, laboratory sink-float methods offer mild conditions to minimize chemical alteration and, in favorable cases, kerogen concentrates with low ash contents. Also, elemental analyses can be carried out at each step and the results extrapolated to zero ash to obtain an estimate of the mineral-free kerogen composition. Disadvantages include high rejection of organics, leading to low kerogen recoveries, and the possibility of kerogen fractionation along with mineral rejection.

The oil agglomeration method (Quass, 1938) relies on selective wetting of kerogen particles by an oily pasting material, such as hexadecane.

From the results of model compound-model mineral tests, interactions between acidic clay minerals and N-containing organics have been identified as being much stronger than other likely candidates for kerogen-mineral interactions in Green River oil shale (Siskin et al., 1987 a,b). Ammonium sulfate was selected to serve both as a source of ammonia, a base, and as a source of acid, bisulfate ion, to generate porosity by attacking carbonate minerals. These workers found that treatment of 80- to 100-mesh Green River oil shale with aqueous ammonium sulfate at 85°C for 72 hr effectively disrupted kerogen-mineral interactions and liberated the kerogen. However, efficient kerogen recovery was only obtained by adding an organic solvent that would wet and swell the kerogen, thereby aiding the physical sink-float separation. Toluene was used in the laboratory, but a shale-derived naphtha could also be used.

II. STRUCTURAL INFERENCES FROM ANALYSES

Inferences regarding kerogen structure drawn from the results of bitumen and shale oil analyses are generally based on the premise that (1) the bitumen is analogous to residual monomer in a polymer, i.e., the bitumen represents units of the precursor that did not become bound into the insoluble three-dimensional macromolecular network of the kerogen, or (2) the bitumen comprises units of the kerogen structure that have been cleaved more or less intact from the kerogen by thermal treatment. In either case, it is assumed that the kerogen is structurally similar to the extractable bitumen.

Some of the compound types present in oil shale bitumens and the geochemical inferences that have been drawn from these results were discussed in an earlier section. Much more information can be drawn from bitumen composition in cases where the analysis is comprehensive (Robinson, 1976).

There are many postulates of the structural types found in kerogen, but there is—perhaps not surprisingly—a similarity in the types of compounds found by these analyses (Table 1). Of course, there are differences in the details of bitumen composition and the reader is directed to the cited papers for discussions of how these differences are related to the origin, depositional environment, and thermal history of the particular oil shales.

Most oil shales contain only small amounts of bitumen; rarely does the bitumen content exceed 15% of the total organic matter. Moreover, bitumens are generally richer in hydrogen and poorer in aromatics and N, S, O heteroatoms than the corresponding kerogen. This limits the usefulness of structural inferences drawn from the bitumen composition. Shale oil, on the other hand, generally represents a much larger fraction of the organic matter, over one-half in most cases and in favorable cases even more. Methods for shale oil analysis are generally similar to those used for petroleum. Structural inferences drawn from the results of shale oil analyses typically parallel those from bitumen analyses. However, as a consequence of their thermal treatment, shale oils are usually richer in aromatics and olefins than the starting kerogen, and poorer in N and S heteroatoms since these become concentrated in the nonvolatile char. Thus shale oils reflect both the structure of the starting kerogen and the thermal treatment used to produce the oil. Consequently, the results of shale oil analyses cannot be relied on to give an accurate picture of kerogen structure.

While many useful inferences about kerogen structure have been drawn from bitumen and shale oil analyses, the limitations outlined above have led to the development of more direct methods for probing kerogen structure.

III. STRUCTURAL ANALYSIS OF KEROGEN

A. Oxygen Functional Groups

Only one attempt to characterize the oxygen functional groups in Green River oil shale kerogen is reported in the literature. Fester and Robinson used acid demineralization (successive treatments with HCl and HF) to prepare a Green River kerogen concentrate containing 14 wt% mineral matter, and wet chemical methods to determine the distribution

Table 1 Representatives of Major Compound Types Found in Oil Shale Bitumens and Suggested as Structural Units in Oil Shale Kerogen

n-Paraffins

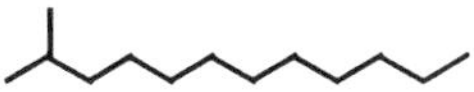

Isoparaffins (2-methylalkanes)

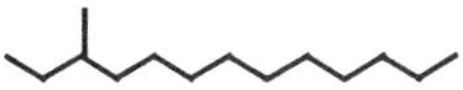

Anteisoparaffins (3-methylalkanes)

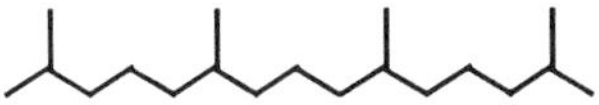

Pristane

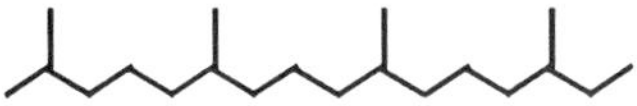

Phytane

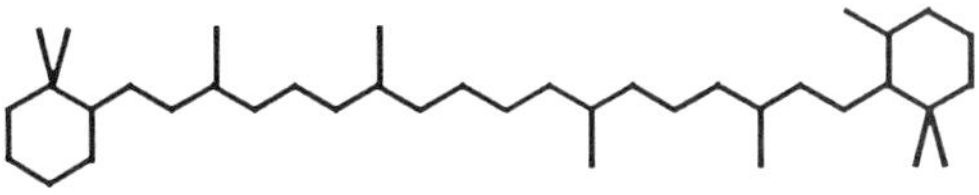

Carotane

Isoprenoid Paraffins and Isoprenoid Cycloalkanes

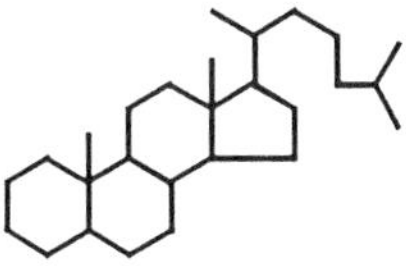

Sterane Series

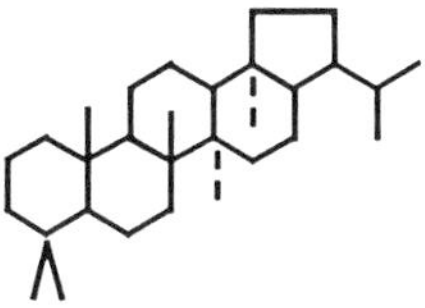

Hopane Series

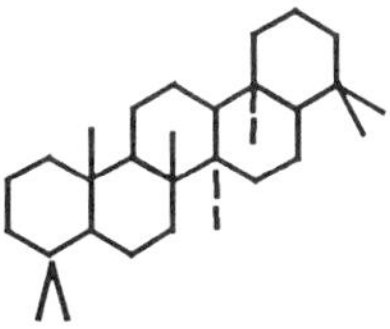

Gammacerane Series

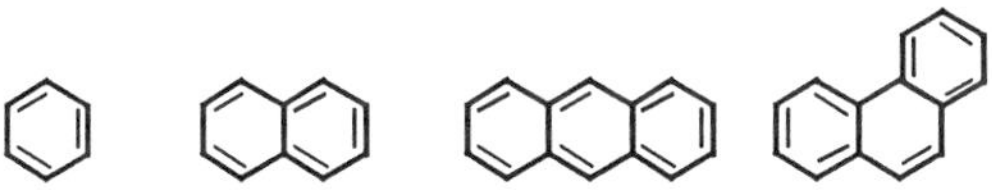

Aromatics

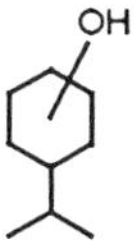

OH

Alcohols

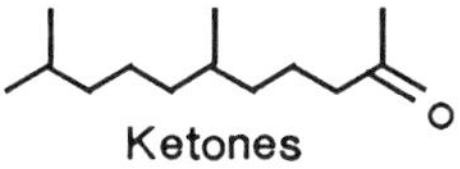

O

Ketones

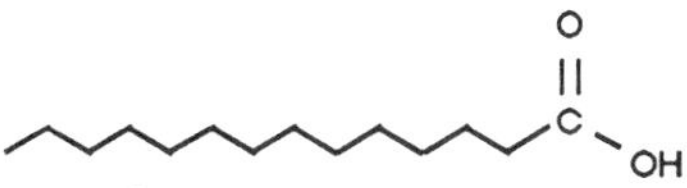

O

C

OH

Carboxylic Acids

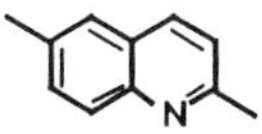

N

Pyridines

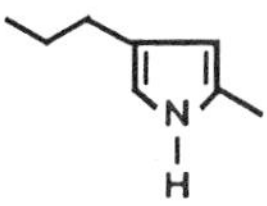

N

H

Pyrroles

Table 2 Distribution of Oxygen Functional Groups in Green River Oil Shale

Oxygen group	% of Total oxygen
Carboxyl	15.3
Ester	24.7
Amide	0.6
Carbonyl (R-CHO, R_2CO)	1.2
Hydroxyl (R-OH, Ar-OH)	4.7
Ether (by difference)	53.5

Source: Fester and Robinson, 1964, 1966.

of oxygen functional groups (Table 2) as illustrated by the following equations:

Carboxylic acids:

$$2R\text{-}CO_2H + Ca(OAc)_2 \rightarrow (R\text{-}CO_2)_2Ca + 2HOAc$$

- HOAc was steam-distilled and titrated with 0.02N NaOH
- Solid was filtered, washed with dilute NaOH (pH ∿ 8), dried, and analyzed for Ca

Esters:

$$R\text{-}CO_2R' \xrightarrow[\text{18–24 hr reflux}]{\text{0.2N NaOH}} \xrightarrow{\text{HCl}} \xrightarrow{Ca(OAc)_2} \text{as above}$$

Amides:

$$R\text{-}CONH_2 \xrightarrow[\text{3 hr reflux}]{\text{1 N NaOH}} R\text{-}CO_2Na + NH_3 \xrightarrow{\text{Trap in } H_3BO_3} \text{titrate}$$

$R\text{-}CO_2Na$ → 1. Acidify; 2. Extract with Et_2O → IR analysis for amines; N analysis

Aldehydes, ketones:

$$\text{R-CHO, R}_2\text{CO} \xrightarrow[\text{pH = 7.5–8, 8–24 hr, 20°C}]{5\%\ \text{H}_2\text{N-OH·HCl/H}_2\text{O}} \xrightarrow[\text{2. H}_2\text{O wash}]{\text{1. Filter}} \text{N analysis}$$

3. Dry

$$\text{R-CHO, R}_2\text{CO} \xrightarrow{\text{NaBH}_4 \text{ in 0.1 N NaOH}} \xrightarrow[\text{2. H}_2\text{O wash}]{\text{1. Filter}} \text{Quantitative IR analysis using 5.8 peak (KBr pellet)}$$

3. Dry

Alcohols, phenols:

$$\text{R-OH, Ar-OH} \xrightarrow[\text{2. Filter}]{\text{1. Ac}_2\text{O pyridine}} \xrightarrow[\text{Hydrolysis}]{\text{0.2 N NaOH}}$$

3. Wash

1. Neutralize
2. Steam-distill HOAc
3. Titrate HOAc

While directionally correct, the reported results (Table 3) did not take into account the organic structural changes which may have occurred during the preparation of the kerogen concentrate or during the derivatization reactions, nor did they consider the inabilities of aqueous reaction media to wet and swell the nonporous kerogen concentrates.

B. Oxidation

Oxidative degradation, one of the primary tools of classical natural product chemistry, has been widely used to probe kerogen structure. Oxidation methods for kerogen characterization have been reviewed by Vitorovic (1980). Alkaline permanganate (Ambles et al., 1983) and chromic acid (Simoneit and Burlingame, 1974) have been the two most widely used oxidants.

Alkaline permanganate oxidations of kerogens have been carried out in two very different ways. Older work was generally done using the "carbon balance" method that was developed for coal studies (Chap. 19, Sec. III.B). The products of this exhaustive oxidation are CO_2, oxalic acid (HO_2C-CO_2H, from aromatic rings), nonvolatile, nonoxalic acids (mostly benzene polycarboxylic acids), and unoxidized organic carbon (reported values generally include both handling losses and analytical errors). Because aliphatic material is oxidized mainly to CO_2, this method is not well suited to probe the structures of kerogens that are highly aliphatic. This led to the development of stepwise procedures to give products that retain more structural information about the starting kerogen.

Table 3 Carboxyl, Ester, and Hydroxyl Oxygen in Kerogen Concentrates from Around the World

Kerogen	Oxygen as COOH (mg/g C)	Kerogen	Oxygen as COOR (mg/g C)	Kerogen	Oxygen as OH (mg/g C)
New Zealand	64	Argentina	109	New Zealand	42
Brazil	39	Oregon	99	Brazil	33
Oregon	28	New Zealand	78	Oregon	29
Argentina	19	Canada	62	Colorado	22
Colorado	14	Brazil	61	Argentina	17
Canada	8	Spain	39	Canada	17
Spain	8	South Africa	38	South Africa	15
Scotland	7	Alaska	37	Australia	13
South Africa	5	Scotland	31	Scotland	12
France	3	Colorado	25	France	11
Alaska	2	France	20	Spain	10
Australia	1	Australia	1	Alaska	4

Source: After Dinneen and Robinson, 1967.

1. Alkaline Permanganate

The careful development of the stepwise alkaline permanganate method and its application to a wide variety of oil shale kerogens have been extensively documented (Vitorovic et al., 1987). In general, these workers attempt to minimize unwanted secondary oxidation of the first-formed product by adding the oxidant ($KMnO_4$ in 1% aqueous KOH) in small portions, heating until the violet (MnO_4^-) and green (MnO_4^{2-}) colors disappear, then separating the base-soluble oxidation products. The base-insoluble residue is then subjected to another oxidation step with a fresh portion of permanganate. At 75–80°C (165–175°F) the time required to discharge the oxidant color increases from a few minutes for the initial portions to several hours, until eventually the color is not discharged after an extended time (12–36 hr). Ether extraction yields the neutral and basic products, acidification with HCl yields a high molecular weight acid fraction and a second ether extraction yields the ether-soluble acids. The latter are then analyzed, either directly or as the methyl esters, using conventional methods.

In some cases, the acids obtained from stepwise oxidation proved to be of such high molecular weight that they precipitated upon acidification and were insoluble in ether. Such acids were difficult to characterize. In these cases, the precipitated acids were subjected to further stepwise oxidation to produce the desired ether-soluble acids of lower molecular weight.

The results of stepwise alkaline permanganate oxidations of kerogen concentrates from four representative oil shales show that, while the distributions vary widely, the *n*-monocarboxylic acids and α,ω-dicarboxylic acids together comprise the major part of the product. Tricarboxylic acids were much less abundant in the product from the Pumpherston shale, while isoprenoid acids were abundant (5.74% of total acids) only from the Irati shale. The Aleksinac and Irati shales yielded more aromatic acids than the Green River or Pumpherston shales (Tables 4–7).

Obviously, much effort has been expended in developing methodology for the stepwise oxidation procedure and in characterizing the oxidation product acids. Considerably less effort seems to have gone

Table 4 Compositions of the Kerogen Concentrates

	Green River (Rifle, CO)	Aleksinac (Yugoslavia)	Irati (Brazil)	Pumpherston (Scotland)
Organics, wt %				
C	77.39	71.87	78.83	75.70
H	10.26	8.73	9.47	10.37
N	3.10	3.21	3.92	4.18
O + S (diff)	9.25	16.19	7.78	9.75
Atomic H/C	1.59	1.46	1.44	1.64
Atomic O/C	0.09	0.17	0.07	0.10
% Organics in the kerogen concentrate	49.61	65.57	51.49	81.98

Source: After Vitorovic et al., 1987.

Table 5 Product Yields from Stepwise Alkaline Permanganate Oxidations

Oxidation of kerogen concentrate

	Product yield based on starting organics (wt %)				
	Neutrals + bases	Ether-soluble acids	Precipi-tated acids	Acids, aqueous solutions	Total yield
Green River (25)[a]	0.86	24.32	51.20	8.12	84.50
Aleksinac (23)	1.00	27.97	45.57	18.30	92.84
Irati (31)	1.69	47.85	31.38	8.34	89.26
Pumpherston (29)	3.73	35.94	22.14	3.33	65.14

Oxidation of precipitated acids

	Product yield based on precipitated acids (wt %)				
	Neutrals + bases	Ether-soluble acids	Precipi-tated acids, last step	Acids, aqueous solutions	Total yield
Green River (30)	11.22	95.56	5.26	10.13	123.17
Aleksinac (26)	3.25	72.86	13.35	14.01	103.47
Irati (25)	7.96	80.21	0.25	9.75	98.17
Pumpherston (31)	19.24	33.76	12.16	5.44	70.60

[a]The value in parentheses below each shale name is the number of steps required for complete oxidation of the kerogen or precipitated acids.
Source: Data of Vitorovic et al., 1987.

into defining the susceptibility of kerogen structural features to oxidation and the precise relationships between oxidation products and kerogen structure.

Table 6 Acids Obtained by Alkaline Permanganate Oxidation

	Green River	Aleksinac	Irati	Pumpherston
Aliphatic acids				
n-Monocarboxylic	C_{10}–C_{34}	C_{10}–C_{34}	C_{10}–C_{29}	C_9–C_{36}
Monocarboxylic, branched	–	–	C_{15}	C_{15}, C_{17}
α,ω-Dicarboxylic	C_6–C_{34}	C_5–C_{33}	C_6–C_{26}	C_6–C_{33}
Dicarboxylic, branched	C_8, C_9	C_8	–	–
Isoprenoid	C_{14}–C_{17}; C_{19}, C_{20}	C_{14}–C_{17}; C_{19}–C_{21}	C_{14}–C_{17}; C_{19}–C_{21}	–
Tricarboxylic	C_6–C_{14}	C_6–C_{15}	C_4–C_9	C_3–C_9
Tricarboxylic, branched	–	C_6–C_9	–	–
Tetracarboxylic	C_8–C_{14}	C_8–C_{18}	–	C_6, C_8–C_{13}
Tetracarboxylic, branched	–	C_9, C_{10}	–	C_{10}, C_{13}
Aromatic acids	Mono- to tetra-carboxylic	Mono- to tri-carboxylic	Mono- to tetra-carboxylic	Mono- to tetra-carboxylic

Source: Data of Vitorovic et al., 1987.

Generally speaking, alkaline permanganate will oxidize alkylbenzenes, alkylthiophenes, and alkylpyridines—but not alkylfurans—to the corresponding carboxylic acids. This is not true in cases where the aromatic ring bears an electron-donating group (e.g., -OH, -OR, $-NH_2$). In such cases degradation of the aromatic portion is usually rapid. Condensed aromatics are also attacked, and benzene polycarboxylic acid results:

$KMnO_4^-$, OH^- →

HO_2C, CO_2H, CO_2H, HO_2C, CO_2H, CO_2H

Table 7 Distributions of the Oxidation Product Acids

	Wt % of total acids			
	Green River	Aleksinac	Irati	Pumpherston
Aliphatic acids				
n-Monocarboxylic	27.94	18.80	56.81	54.32
Monocarboxylic branched	–	–	0.02	0.10
α,ω-Dicarboxylic	58.06	55.77	21.04	36.75
Dicarboxylic, branched	0.04	–	–	–
Isoprenoid	1.95	0.68	5.74	–
Tricarboxylic	5.26	6.05	5.26	1.37
Tricarboxylic, branched	–	0.10	–	–
Tetracarboxylic	4.08	10.77	–	3.67
Aromatic acids	2.87	7.93	11.13	3.79

Source: Data of Vitorovic et al., 1987.

However, the need for care can hardly be overemphasized; even benzene is slowly attacked by hot alkaline permanganate solutions.

Olefins are rapidly converted into the corresponding glycols, which are then cleaved to carboxylic acids. Cyclic olefins yield dicarboxylic acids. Enolizable ketones are also cleaved, presumably via the enol.

$$\text{R-CH=CH-R'} \xrightarrow{\text{KmnO}_4,\ \text{OH}^-} \text{R-}\underset{\text{HO}}{\text{CH}}\text{-}\underset{\text{OH}}{\text{CH}}\text{-R'} \longrightarrow \text{R-CO}_2\text{H} + \text{HO}_2\text{C-R'}$$

Tertiary and benzylic C-H groups are attacked by a free radical mechanism. Good yield of tertiary alcohols have been obtained in selected cases. In simple alkyl systems, the presence of an alcohol group markedly accelerates the rate of this reaction, probably by increasing water solubility of the unit that bears the tertiary C-H. A free radical mechanism was established by isolating the coupling products from dimerization of the intermediate benzylic radicals:

$$C_6H_5\text{-}CH_2CH_2CH_3 \xrightarrow{KMnO_4^-} \underset{\displaystyle C_6H_5}{\overset{}{CH_3CH_2CH}}\text{-}\overset{\displaystyle C_6H_5}{CHCH_2CH_3}$$

A very facile free radical attack has also been observed when a tertiary hydrogen is gamma to a carboxyl group, as in 4-methylhexanoic acid:

$$H_3C\text{-}\underset{\displaystyle H_3C}{\overset{\displaystyle CH_2\text{-}CH_2\text{-}CO_2H}{C}}\text{-}H \xrightarrow{KMnO_4^-,\ OH^-} (H_3C)_2\text{-lactone}$$

Primary and secondary alcohols are oxidized to the corresponding acids and ketones, with the rates depending on both pK_a of the alcohol and pH, and being roughly proportional to the concentration of the corresponding alkoxide ion. Evidently, permanganate attacks the alkoxide anion much more rapidly than the alcohol itself.

Alkaline permanganate oxidation degrades the porphyrin nucleus, giving pyrrole-2,4-dicarboxylic acid derivatives under mild conditions. Under these conditions, the porphyrin side chains -Me, -Et, $-CH_2CH_2CO_2H$, and $COCH_3$, and $-CH(OH)CH_3$ persist in the degradation products, but $-CH{=}CH_2$ and -CHO side chains are both oxidized to $-CO_2H$.

On the basis of this outline, it is clear from the data (Tables 4–7) that aliphatic chains are of major importance in all four oil shales. The high yield of α,ω-dicarboxylic acids and the low yield of aromatic acids obtained from Green River kerogen suggests that this structure includes a large number of alkylene bridges joining small, activated aromatic rings that would be destroyed by oxidation. The appreciable amounts of tri- and tetracarboxylic acids suggests that some cross-links in the macromolecular network involve aliphatic or cycloaliphatic (possibly cyclo-olefinic) units. In this basic shale, the *n*-monocarboxylic acids may be present salts, especially of divalent cations (e.g., Ca^{2+}), though some contribution from alkyl side chains cannot be ruled out.

The situation with the Aleksinac shale is similar, but the even higher yields of tri- and tetracarboxylic acids and the abundance of aromatic acids suggest that cyclic structures may be more important in this case.

The α,ω-dicarboxylic acids are much less abundant in the product from the Irati shale. Aromatic acids are relatively abundant and *n*-

monocarboxylic acids are by far the major component. These results suggest that relatively large aromatic clusters with *n*-alkyl side chains and joined by alkylene bridges are important in the Irati shale.

The Pumpherston shale stands out in giving a very low yield of oxidation products (65%) and in giving a relatively large amount of neutral and basic product (∿8 wt % based on the starting shale). Presumably, some structures in this shale were oxidized to CO_2; however, the high atomic H/C and the low yield of aromatic acids suggest that these were not activated aromatics of the type found in coals. The high nitrogen content suggests that pyrrolic compounds may be the source of the CO_2, but knowledge of the fate of nitrogen would be needed to make this more than mere conjecture.

The above discussion shows that oxidation, using mild selective techniques like the stepwise alkaline permanganate method, can provide a wealth of information about kerogen structure. However, it should also be clear that much additional research is needed to define the reactivity of different structural features and establish clear relationships between oxidation products and the structure of the starting kerogen.

2. Chromic Acid

Chromic acid oxidations of kerogens (Vitorovic, 1980) gives structural information obtained from kerogen oxidations with chromic acid and other chromium oxidants that is usually similar to that obtained with alkaline permanganate. However, the recovery of organic carbon in the chromic acid oxidation products is often low. For example, the total product obtained (Simoneit and Burlingame, 1974) in a stepwise chromic acid oxidation of Green River oil shale kerogen amounted to less than 10% of the total organic material, though essentially all of the organic material was oxidized. Consequently, the alkaline permanganate procedure appears to be superior for elucidating kerogen structure.

C. Depolymerization

While the oxidations discussed above can give reasonably high organic recoveries in favorable cases, extensive alteration of the structure and some features are obliterated (see also Chap. 19, Sec. III.B). This is especially a problem in the case of easily oxidized nitrogen functionalities. Therefore, while the Fischer assay remains the standard for oil recovery in retorting and most thermal treatments give oil recoveries comparable to that of the Fischer assay, the search for ways to improve thermal oil recovery continues. Indeed, thousands of patents and papers have been published on the subject. One particularly attractive approach to minimizing the formation of intractable residues has been to heat the shale at a moderate temperature for a long time, then extract the depolymerized kerogen (Hubbard and Robinson, 1950). Another approach is to heat the shale at a low temperature for a time, then increase the temperature to the conventional retorting temperature, cool, then extract (Freeman, 1928). In either case, the idea is to depolymerize the kerogen with minimal condensation to intractable material, then recover even nonvolatile oil by extraction.

Until recently, it was not realized that there is an optimum time-temperature window for thermal depolymerization (see also Chap. 19, Sec. III.C). For example, Hubbard and Robinson reported that a two-step process (two heating + extraction cycles) was not necessary for Green River oil shale, since the same conversions could be obtained by simply heating for a longer period before extraction. Bock and his co-workers (1984a,b) found that this was not the case. These workers heated oil shales under a nitrogen sweep to obtain volatile oils and gases, then extracted the residue to obtain depolymerized, but nonvolatile, products. For example, a sample of Green River oil shale was heated at 400°C (750°F) for 1 hr to obtain 40 wt % (based on total organic matter) volatile products, and then extracted to obtain 44 wt % extractable liquids. The results for other times and temperatures show that for each temperature in the range of 350-425°C (660-795°F) there is an optimum heating time, and that heating for 400°C for 1 hr gives an optimum conversion (Figure 1). Similar results were obtained for Ramsay Crossing oil shale from the Rundle deposit (Queensland, Australia, Figure 2). Similar results but somewhat lower overall conversions were obtained from a more highly aromatic eastern U.S. Devonian shale.

After heating at the optimum time, the unextractable organic residue is not yet converted into the intractable char obtained from conventional thermal methods. Heating the THF-extracted residue a second time at 400°C (750°F) for 1 hr and again extracting with THF enabled recovery of the balance of the organic material. Together, the two heat-soak/extraction cycles achieved conversion of 100 wt % (±2 wt %) of the total organic matter in Green River oil shale—over 90 wt % of which was recovered as liquid products. Moreover, because of the relatively mild conditions, absence of added reagents, and controlled time, alteration of the recovered organics should be minimal. Thus, the mild heat-soak/extraction method is an attractive alternative to oxidation for providing liquid products for structural studies.

D. Micropyrolysis—Gas Chromatography—Mass Spectrometry

Schmidt-Collerus and Prien (1974) used micropyrolysis coupled with GC-MS to obtain information about the structural units in Green River

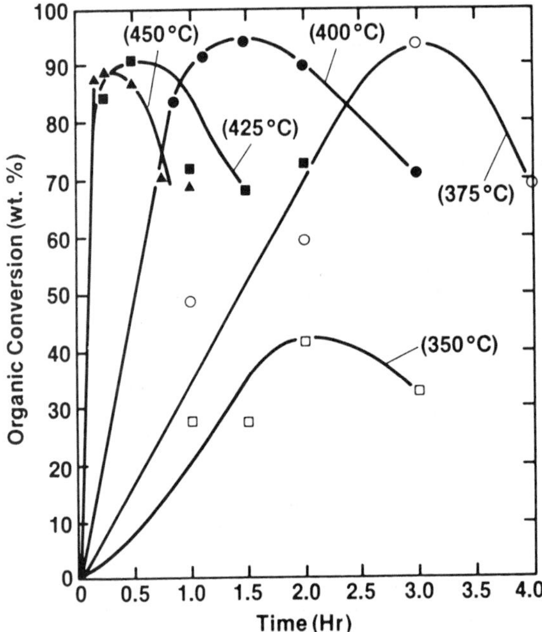

Figure 1 Plots of conversion versus time show that there is an optimum time-temperature window for thermal depolymerization of Green River oil shale kerogen (reaction under 1 atm N_2).

oil shale kerogen. In addition to the on-line micropyrolysis/GC-MS studies, larger samples of kerogen were pyrolyzed to obtain products that were fractionated by chromatography (ion exchange, $FeCl_3$ complexation, silica gel) into compound classes, then by GPC into fractions of increasing molecular weight. These samples were analyzed by conventional MS techniques. Additional information was obtained by microscopic and microspectrophotometric studies on the whole kerogen.

These studies led to the conclusion that Green River kerogen contains two distinct types of material: α-kerogen, an alginite-like material of low aromatic content, and β-kerogen with a much higher content of aromatic (probably polycondensed) material. The latter, representing about 5% of the total, was a reddish brown color.

In micropyrolysis, α-kerogen yielded several types of products: normal and branched alkanes, alkyl naphthalenes, and tetralins, alkyl-substituted tricyclic or phenanthrene derivatives (Table 8). These results led to the important conclusion that most of the cyclic units

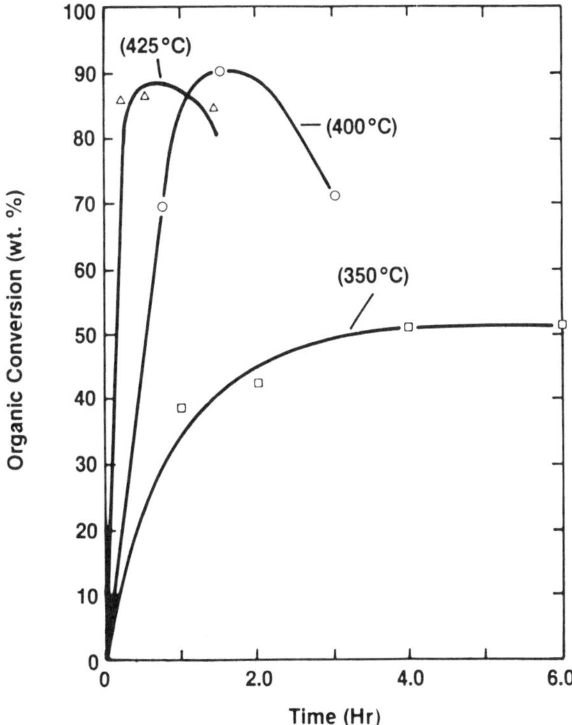

Figure 2 An optimum time-temperature window was also found for the thermal depolymerization of Ramsay Crossing oil shale kerogen.

(alicyclics, naphthenics, aromatics) in the Green River kerogen are small, containing one to three rings.

E. Hydrogenolysis

Hubbard and Fester (1958) used hydrogenolysis in the presence of tin dichloride to degrade Green River oil shale kerogen:

$$\begin{array}{c}\text{Green River kerogen}\\ \text{concentrate}\end{array} + SnCl_2 \cdot 2H_2O \xrightarrow[335°C,\ 4\ hr]{4200\ psig\ H_2} \begin{array}{c}9.3\%\ \text{gas}\\ 86.6\%\ \text{benzene}\\ \text{solubles}\end{array}$$

Although the conditions were severe, the gas make was low and a good yield of liquid products was obtained for characterization. Under

Table 8 Principal Fragmentation Products from Micropyrolysis of Green River Oil Shale Kerogen

Aliphatic hydrocarbons:	$n\text{-}C_{10}$ - $n\text{-}C_{34}$ Branched C_{10}-C_{36}
Alicyclic hydrocarbons:	
Cyclohexanes	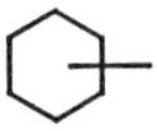 (cyclohexane)—$C_{10-13}H_{21-27}$
Decalins	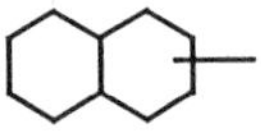 (decalin)—$C_{5-8}H_{17-25}$
Hydroaromatic hydrocarbons:	
Dialkyltetralins	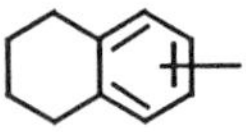 (tetralin, aromatic ring)—$C_{2-5}H_{5-11}$
	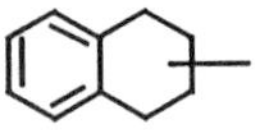 (tetralin, saturated ring)—$C_{8-12}H_{17-25}$
Hexahydrophenanthrenes	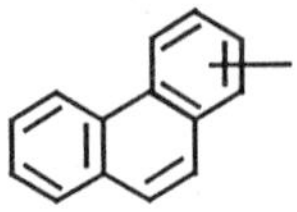 (phenanthrene)—$C_{1-3}H_{3-7}$ + 6 H
Dialkylbenzenes	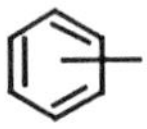 (benzene)—$C_{8-13}H_{17-27}$
Dialkylnaphthalenes	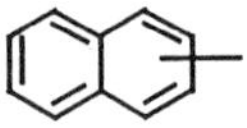 (naphthalene)—$C_{3-4}H_{7-9}$
Alkylphenanthrennes	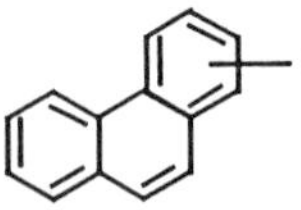 (phenanthrene)—$C_{1-3}H_{3-7}$

these conditions, most of the heteroatoms were removed: N, 90% as NH_3; S, 90% as H_2S, O, 84% as CO_2 (56%) and H_2O (28%). This led to the conclusion that N, S, and O functionalities comprise weak links in the kerogen structure. This may be so but seems at variance with other results. Given the high yields of liquids, this technique looks promising, but the extreme severity suggests that reinvestigation under milder conditions might be profitable.

IV. STRUCTURAL MODELS FOR KEROGEN

The need to put the very large mass of information about kerogen structure into a compact form that is useful for guiding research and development has led to models for kerogen structure. These models are not intended to depict *the* molecular structure of kerogen, at least not in the sense that the double helix describes the structure of DNA, or even in the sense that synthetic polymers are described in terms of monomers joined to form chains which then segregate into crystallites of well-defined structure and liquid-like amorphous regions. Kerogens are not so well ordered. Instead, the kerogen models attempt to depict a representative collection of skeletal fragments and functional groups connected into a three-dimensional network in a way that seems reasonable based on the available data.

No one analytical technique provides sufficient information to construct a usefully detailed model of a kerogen structure (see Chaps. 4 and 19). Thus, most workers now use a multiple technique approach. However, even within one laboratory, the results of such an approach are diverse—and often conflicting. This complicates kerogen modeling. Moreover, the collections of techniques used by different workers have different strengths, hence tend to emphasize different features of kerogen structure. Sample-to-sample variation further clouds the picture. Consequently, it is not surprising that several models have been proposed for the structure of Green River oil shale kerogen. These will be discussed to illustrate the state of kerogen modeling. For contrast, the section will end with a discussion of the characterization and structural modeling of the kerogen in Ramsay Crossing oil shale from the Rundle deposit in Australia.

As discussed above, Burlingame and his coworkers used chromic acid oxidation to degrade Green River oil shale kerogen and mass spectrometry (MS) as the key tool for characterizing the oxidation products (Burlingame and Simoneit, 1969). Additional information was provided by studies of the bitumen, again primarily by MS, and incorporated into a structural model (Figure 3).

Key features of this model include very large regions of undefined structure containing trapped organics of unknown nature and bearing side chains linked to the main structure by nonhydrolyzable C-C and hydrolyzable ester linkages. An ester linkage is shown connecting two such large regions. Also, the model includes a carbocyclic ring,

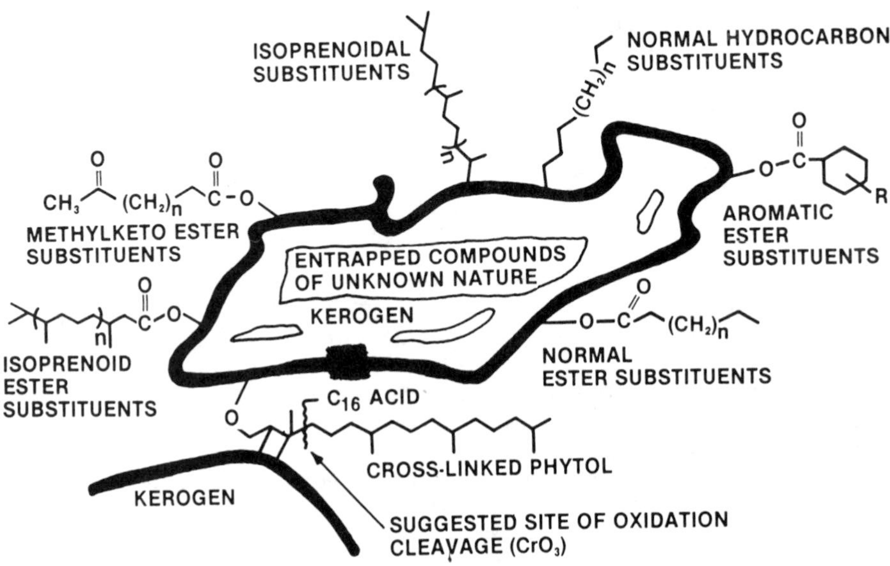

Figure 3 Burlingame model of Green River oil shale kerogen (Burlingame et al., 1968; Burlingame and Simoneit, 1969; reprinted with the permission of the publisher).

presumably derived from a Diels–Adler reaction involving the allyl C=C of phytol (3,7,11,15-tetramethyl-2-hexadecen-1-ol), and the *gem*-dialkyl site on this ring is identified as a possible location for oxidative cleavage to give an isoprenoid C_{16} carboxylic acid. To understand this model, it is important to recall that the oxidation products (acids, ketones) on which most of this structure is based represented only 0.7 wt % of the total organic carbon.

Other workers used stepwise alkaline permanganate oxidation to obtain carboxylic acids in high yield (70% of total organic carbon) from Green River oil shale kerogen (Djuricic et al., 1971). The more recent studies that were discussed in a previous section extended and confirmed this work. Based on their oxidation results, a crosslinked macromolecular network structure was proposed (Figure 4).

The most striking feature of this model is the predominance of straight-chain groups in the backbone of the network. The network bears both branched and unbranched side chains. The branching points (indicated in the model by open circles) must be of a type susceptible to oxidative attack, alkaline hydrolysis, or both in order to yield the observed mixture of mono-, di-, tri-, and tetracarboxylic

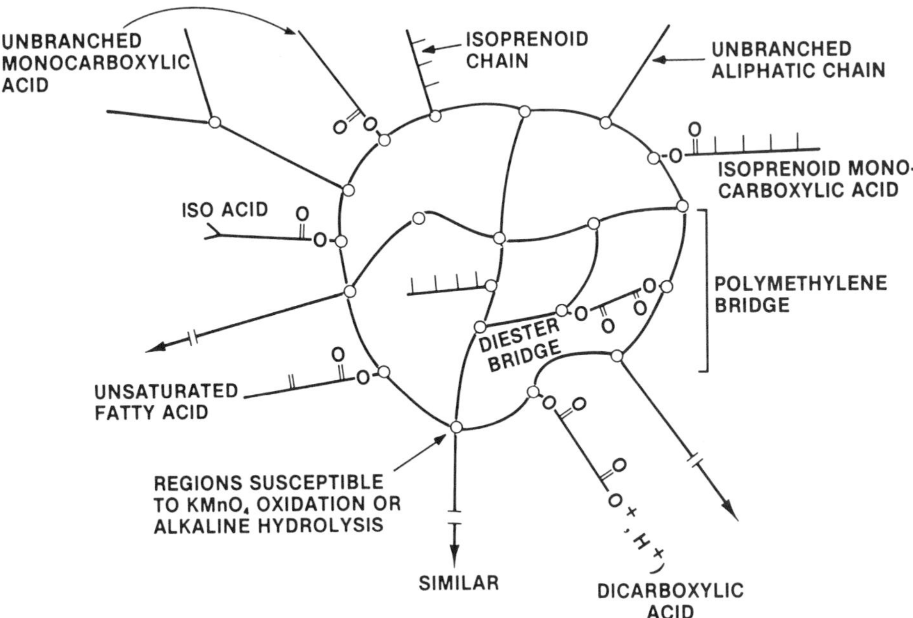

Figure 4 Representation of the Djuricic et al. (1971) model for Green River oil shale kerogen.

acids. This model accommodates many important experimental observations, including reversible swelling and gellike "rubbery" behavior in the swollen state, but does not satisfactorily account for the aromatic carbons observed by ^{13}C NMR or the appreciable N and S contents found by elemental analysis.

The kerogen model proposed by Schmidt-Collerus and Prien was assembled from the subunits identified by their micropyrolysis-MS studies. Key features of this model include formulation as a three-dimensional macromolecular network and a very uniform hydrocarbon portion comprising mostly small alicyclic and hydroaromatic subunits with few heterocyclic rings (Figure 5). Long-chain alkylene and isoprenoid units and ethers serve as interconnecting bridges in this structure. Entrapped species (bitumen) include long-chain alkanes and both *n*-alkyl and branched carboxylic acids. This model provides a useful view of the types and role of hydrocarbon units, but deemphasizes heteroatom functional groups and rings. This is not surprising. The concentration of N and S in residues from thermal treatments of Green River oil shale kerogen is well known. Hence,

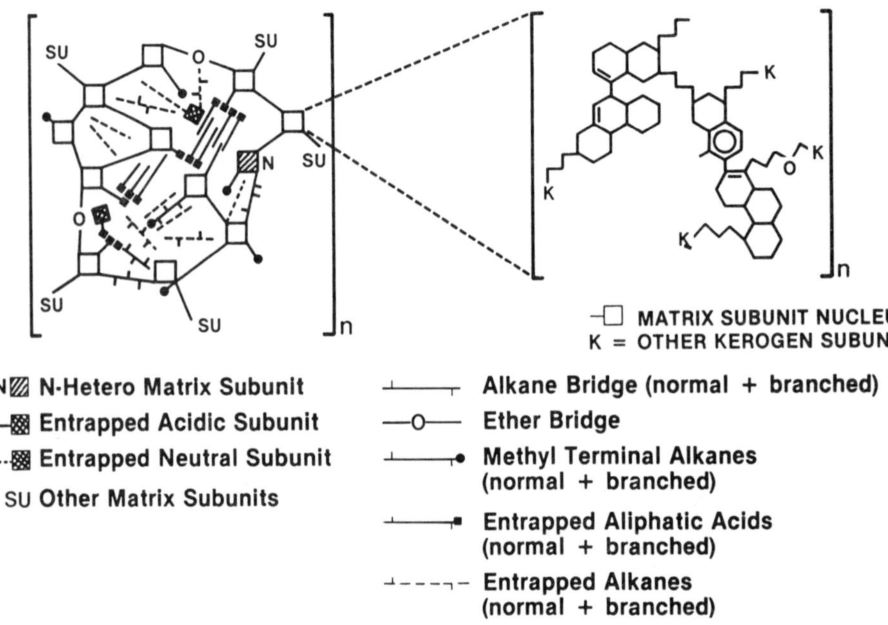

Figure 5 Small ring systems are highlighted in the Green River kerogen model proposed by Schmidt-Collerus and Prien (reprinted with the permission of the American Chemical Society).

groups containing these elements would not be seen with high efficiency by the micropyrolysis technique used by these workers.

The structure of Green River oil shale kerogen has been probed by Yen and his coworkers using a wide variety of techniques, including stepwise alkaline permanganate and dichromate/acetic acid oxidations, electrochemical oxidation and reduction (in nonaqueous ethylenediamine/LiCl), and X-ray diffraction techniques. Based on the results of these studies, Yen concluded that aromaticity was low ("approaches zero") but that isolated carbon-carbon double bonds were possible, that the structure was largely comprised of three- to four-ring naphthenes, that oxygen is present mostly as esters and ethers, that the kerogen structure comprises a three-dimensional network, and that ethers serve as crosslinks in this network (based on results obtained by Fester and Robinson) with additional linkages provided by disulfides, nitrogen heterocyclic groups, unsaturated isoprenoid chains, hydrogen bonding, and charge-transfer interactions (Figure 6) (Yen, 1976a,b). This model is characterized by a

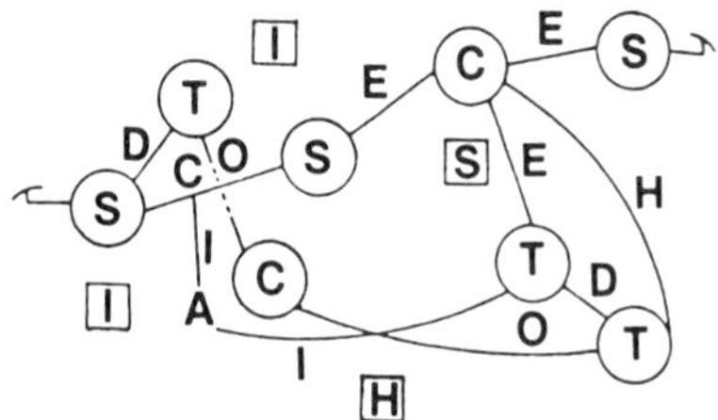

Circles: Essential Components of Kerogen
Squares: Molecules trapped in Kerogen network

I. isoprenoids. **S.** steroids. **T.** terpenoids. **C.** carotenoids.

Bridges:

D. disulfide. **O.** ether. **E.** ester. **H.** heterocyclic. **A.** alkadiene.

Figure 6 The Yen–Young–Shih multipolymer model of Green River oil shale kerogen (reprinted with the permission of the American Chemical Society).

very irregular hydrocarbon structure consistent with their X-ray results that indicated that the kerogen is amorphous–lacking any long-range order. This is in sharp contrast to the very uniform structure envisioned in the model proposed by Schmidt-Collerus and Prien (Figure 5). Yen points out that the extractable bitumen molecules could reside, more or less freely depending on their size, in cavities (molecular free volume) within this network.

To account for the observed variations in the products obtained from the individual steps of stepwise permanganate oxidation, Yen suggested an unusual "core plus shell" arrangement for the individual kerogen particles. The core in this arrangement is a rather loosely crosslinked region containing most of the alkyl and alkylene chains and the bulk of the kerogen as naphthenic ring structures. The shell, on the other hand, is more tightly crosslinked and contains most of the heteroatom functional groups and heterocyclic rings. This has interesting geochemical implications. Recall that the outside of a particle of Yen's kerogen concentrate is precisely the part of the kerogen that was in contact with the mineral matrix in the starting shale, and that the heteroatom groups tend to interact more strongly with minerals than do the hydrocarbon chains. Perhaps, soon after deposition and before the kerogen became rigidly crosslinked, the kerogen components organized themselves into a micelle-like arrangement with the polar heteroatom groups on the outside and the nonpolar hydrocarbon groups within. Crosslinking could then "lock in" this arrangement. Organic-mineral interactions in the resulting composite would then be ideally situated to hinder physical separation of minerals from kerogen. This picture is consistent with the results of Siskin et al. (1987a,b) on chemically assisted oil shale enrichment. Clearly, confirmation of this idea by more direct means would be desirable to provide a basis for the development of more efficient oil shale beneficiation and/or grinding technologies.

The organic material in oil shale from the Ramsay Crossing seam of the Rundle Deposit of Queensland, Australia (RXOS) was characterized and modeled by Scouten and coworkers (1987) using an integrated, multitechnique approach. Acid demineralization yielded the kerogen concentrate (RXOS-KC) and the chemistry that accompanied this demineralization was studied. Selective derivatizations under mild conditions with isotopically labeled reagents followed by solid state ^{13}C and ^{29}Si NMR analysis enabled a comprehensive study to chemically characterize the organic functionalities in the kerogen concentrate. Combining these data with in-depth MS and NMR analyses on shale oils produced under mild conditions from RXOS and variable-temperature X-ray diffraction studies on the kerogen concentrate led to the development of a detailed structural model of the organic material.

The model with a formula weight of 30,000 Da (and an empirical formula of $C_{100}H_{160}N_{2.25}S_{0.68}O_{9.22}$) was required to accommodate the large range of heteroatom functionalities and long side chains present in RXOS (Figure 7). It was not possible to accurately represent the range of compound types in the bitumen because of its small amount (8.5%) and the need to maintain a finite model size. Because carboxylic acids comprise the major part of the bitumen, all the bitumen in the model was represented as carboxylic acid.

A comparison with models for Green River oil shale (GROS) kerogen serves to illustrate some of the key features of the RXOS kerogen.

Aliphatic material is the most obvious feature of the RXOS model and RXOS is more aliphatic than typical samples of GROS. Aliphatics in RXOS are longer and more linear than those in GROS and are present both as alkylene bridges and as alkyl side chains. Isoprenoid chains are distinctly less abundant. Some of the polymethylene chains in RXOS are very long and proximately situated to give appreciable regions that have the ordering characteristic of paraffin wax crystals. Thus, secondary structure due to paraffin-paraffin interactions is important in the RXOS kerogen. No such waxy paraffin interactions are apparent.

The nitrogen content of RXOS is lower than that of GROS (2.3 N's/100 C's versus about 3 N's/100 C's). Some of the nitrogen in RXOS is present as primary amides and additional nitrogen is present as aliphatic amines. However, the major part of the nitrogen in RXOS is

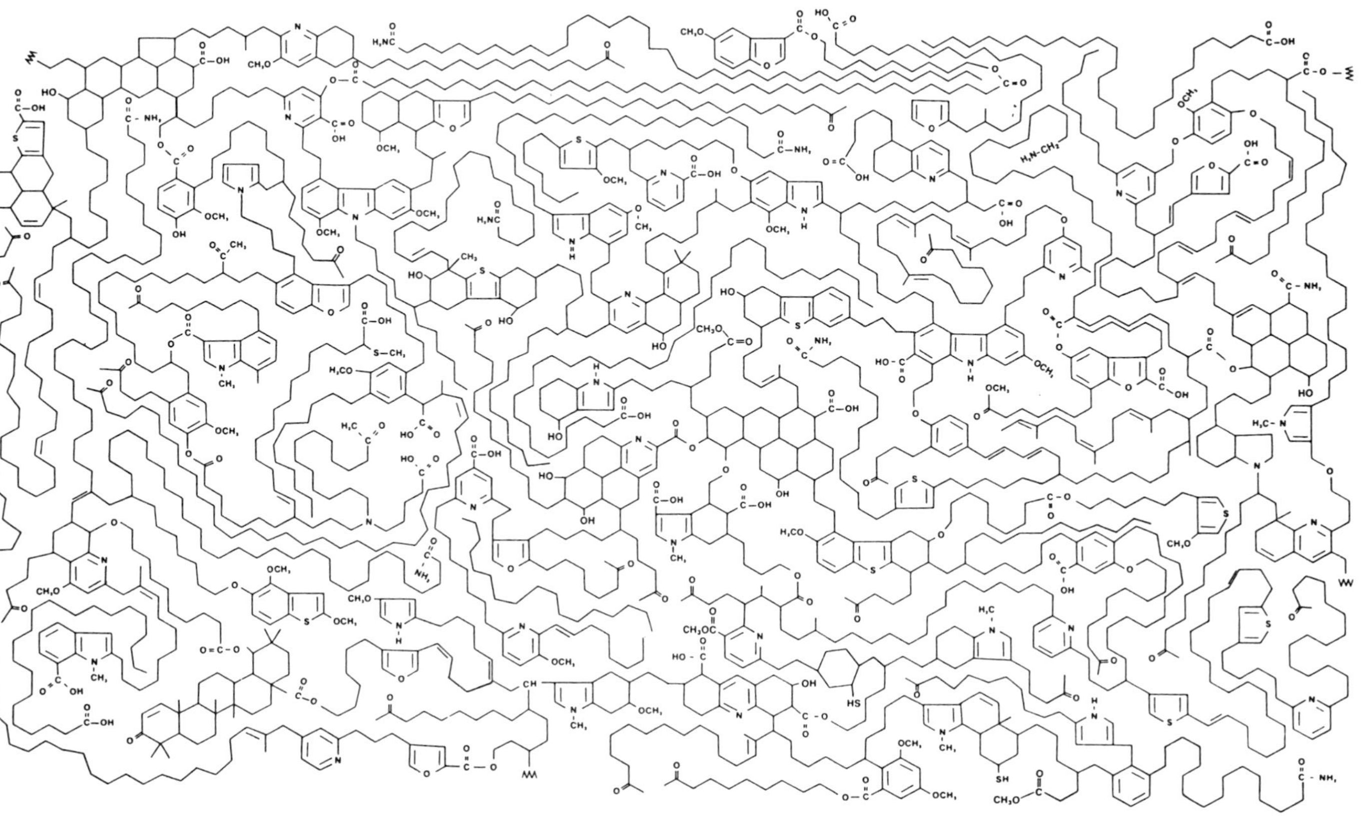

Figure 7 Structural model of representative organic material in Rundle Ramsay Crossing oil shale.

present in very stable aromatic nitrogen heterocycles, derivatives of pyridine and pyrrole.

The organic oxygen content in RXOS is much higher than that in GROS (9 O's/100 C's versus 3 O's/100 C's) and present in a variety of functional group types. Carboxylic acids are important and some of the carboxylic acids that were observed in the kerogen concentrate are initially present in the rock as amides. Some free carboxylic acids are evidently present in RXOS, as they are observed in the extractable bitumen. How the balance of the carboxylic acids are bound in the starting shale is not yet known.

This discussion serves to illustrate the power of the closely integrated characterization-modeling-reactivity approach to provide a model of molecular structure in sufficient detail to be useful as a tool for guiding oil shale research and interpreting experimental results.

Workers at the Institut Français du Pétrole (IFP) have taken a distinctly different approach to kerogen modeling. The studies discussed above were directed toward modeling the structure of kerogen in a particular oil shale. In contrast, the IFP workers have constructed generalized models representative of the three types of kerogen and of the asphaltenes from the corresponding oils as a function of maturity (Tissot and Espitalie, 1975). Emphasis in this work was placed on elucidating the chemistry of maturation for the three kerogen types—in the context of the IFP approach to the organic aspects of petroleum geochemistry.

The latest IFP models, those proposed by Behar and Vandenbroucke (1986) represent the kerogen at the beginning of diagenesis *sensu stricto* (excluding the early stages of diagenesis, which is probably dominated by microbial action), at the beginning of catagenesis (start of the oil generation window), and at the end of catagenesis where late gas begins to be generated. Type I kerogen was modeled only at the beginning of diagenesis and the end of catagenesis, while models of types II and III kerogen were constructed at each stage of maturation (Figures 8–10). The corresponding asphaltenes were modeled only at the end of diagenesis/beginning of catagenesis, where they become most abundant (Figure 11). For the asphaltenes, a molecular weight of 8000 amu was chosen.

These models are based on the results of elemental, infrared, and ^{13}C NMR analyses, pyrolysis (Rock-Eval, artificial maturation) and electron microscopy (fringe analysis) results. The functional group contents that were used in assembling these models are summarized in Table 9. The IFP models provide an interesting view of the structural relationships between the three types of kerogen.

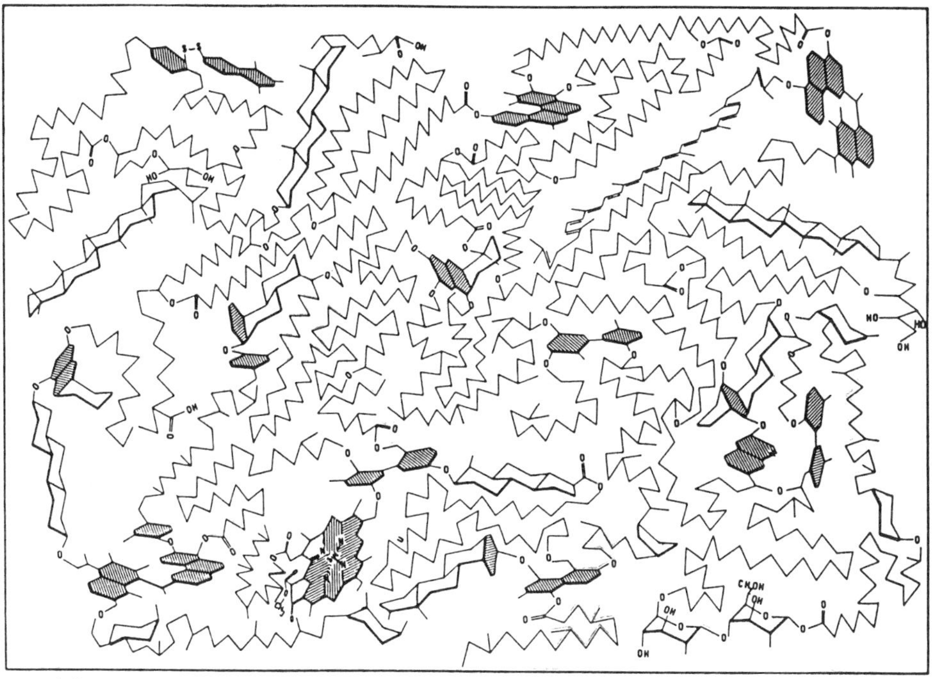

(a)

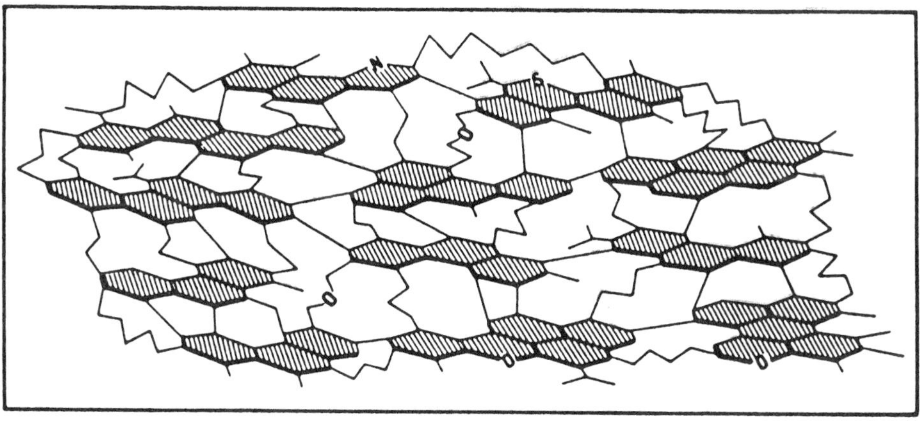

(b)

Figure 8 Generalized IFP model of type I kerogen. (a) At the beginning of diagenesis (start of oil generation window). (b) At the end of catagenesis (start of late gas generation).

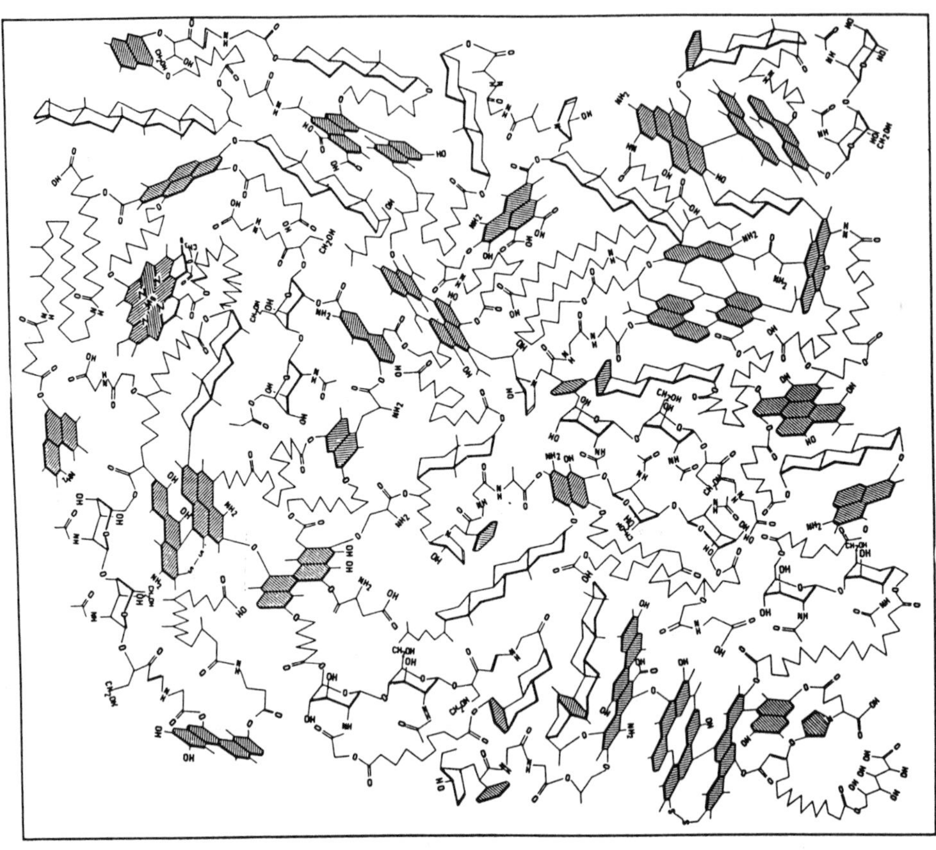

(a)

Figure 9 Generalized IFP model of type II kerogen. (a) At the beginning of diagenesis. (b) At the beginning of catagenesis (start of oil generation). (c) At the end of catagenesis (start of late gas generation).

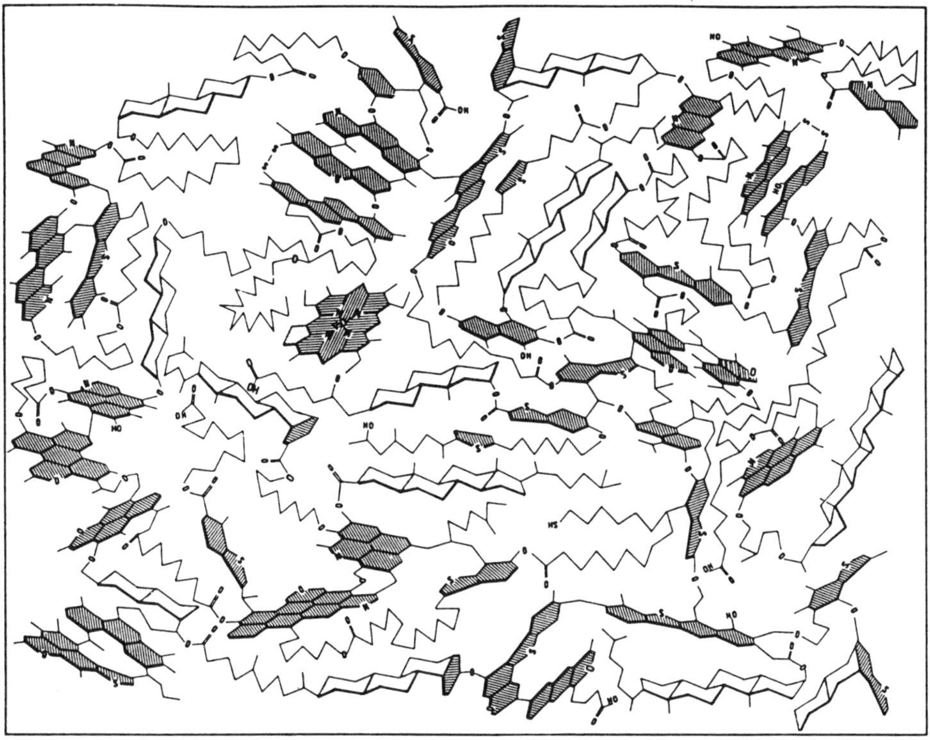

(b)

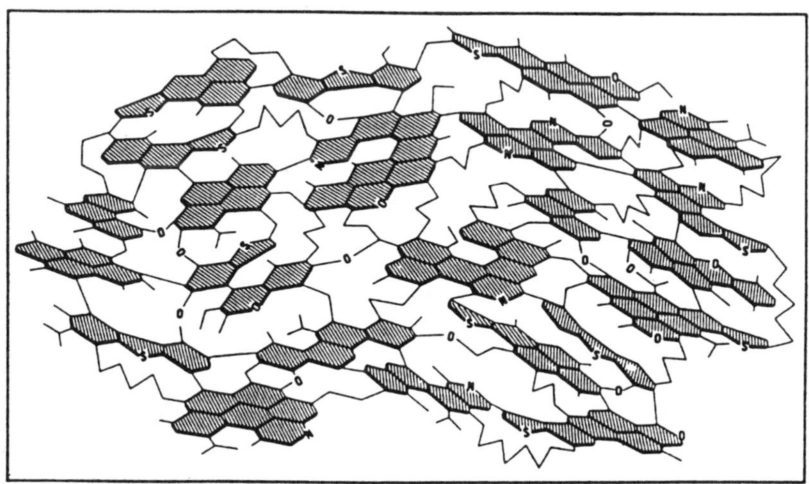

(c)

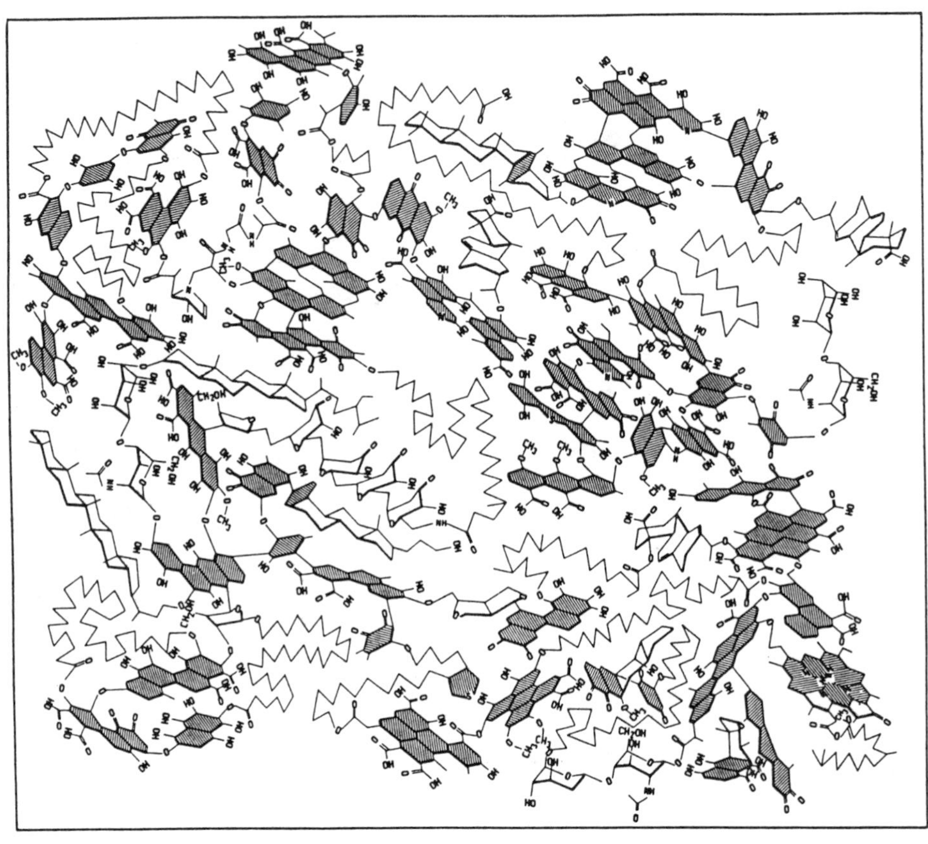

(a)

Figure 10 Generalized IFP model of type III kerogen. (a) At the beginning of diagenesis. (b) At the beginning of catagenesis (start of oil generation). (c) At the end of catagenesis (start of late gas generation).

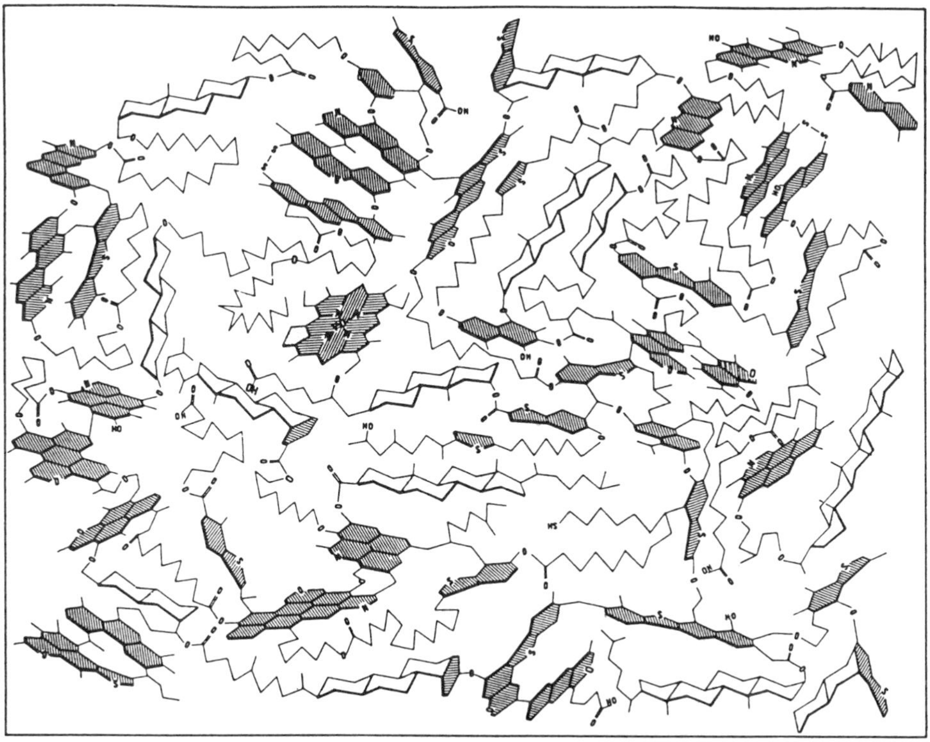

(b)

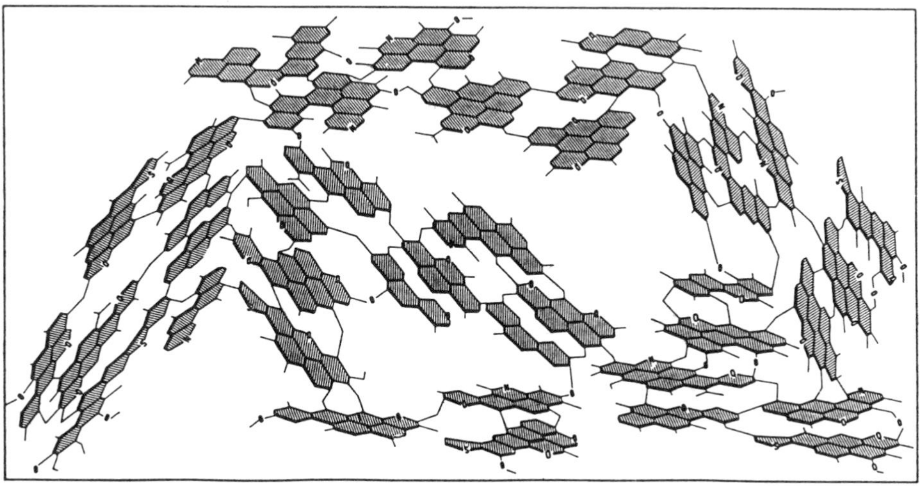

(c)

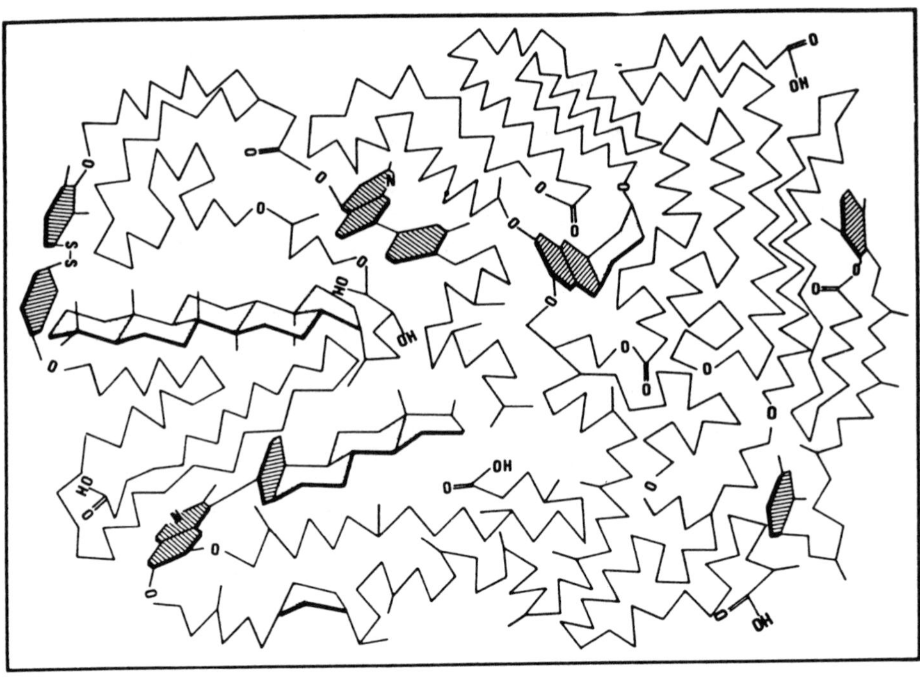

(a)

Figure 11 IFP models of generalized asphaltenes. (a) Type I. (b) Type II. (c) Type III.

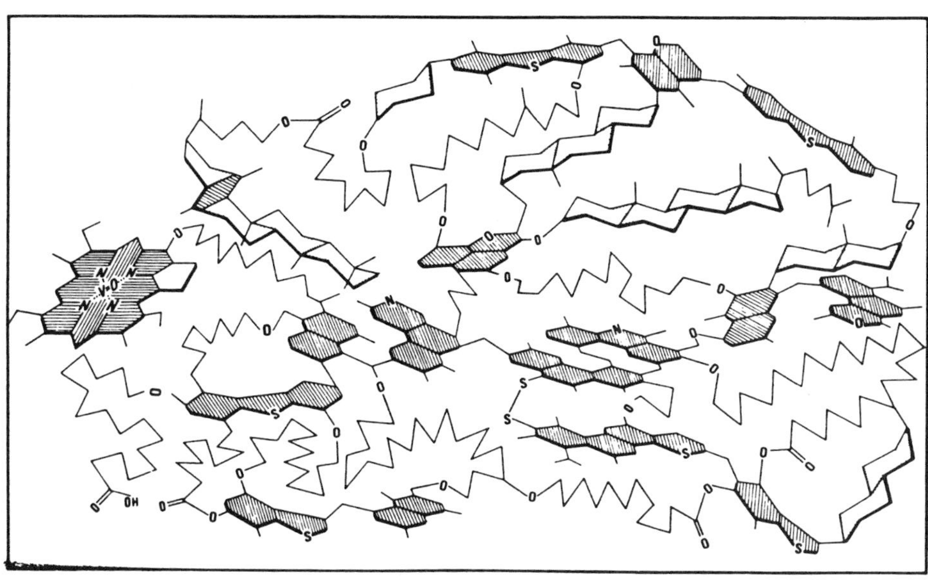

(b)

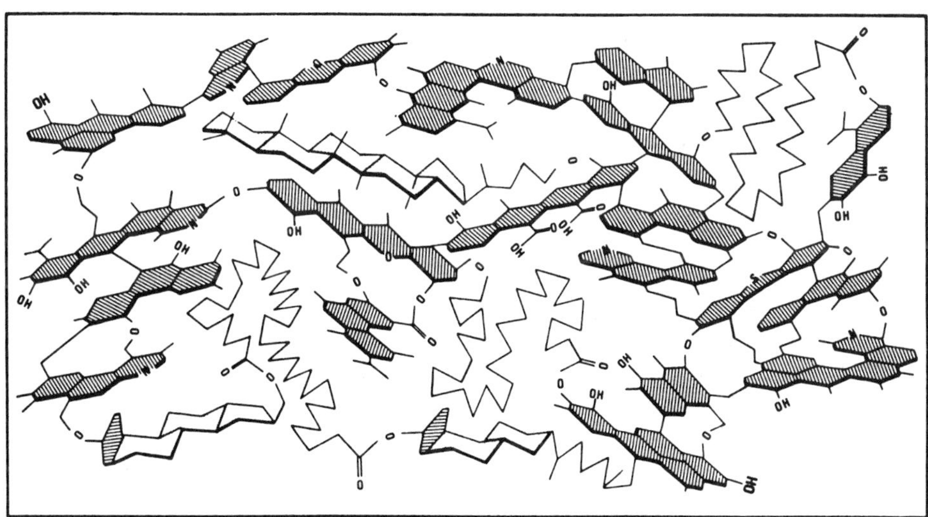

(c)

Table 9 Heteroatom Functional Group Contents of the IFP Models

Kerogen	Heteroatom functional group content of the models											
	Alcohol	Phenol	Aliphatic acid	Aromatic acid	Ester	Amide	Ketone + quinone	Aliphatic ether	Furan	Pyridine	Thiophene	Thiol + disulfide	Amine
Type I-a	9	—	2	—	14	—	—	44	1	5	—	2	—
-b	—	—	—	—	—	—	—	2	2	1	1	—	—
Type II-a	43	20	15	7	45	31	9	52	2	4	2	2	26
-b	1	5	3	1	27	—	1	33	12	16	22	1	—
-c	—	—	—	—	—	—	—	9	6	8	11	—	—
Type III-a	26	102	4	48	16	6	48	80	7	14	2	—	2
-b	1	54	—	6	14	—	10	46	17	18	5	—	1
-c	—	—	—	—	—	—	—	19	37	9	6	—	—
Asphaltene (at start of catagenesis)													
Type I	2	—	3	—	4	—	—	13	—	1	—	2	—
Type II	—	—	1	—	5	—	—	15	5	6	6	2	—
Type III	—	14	—	2	5	—	—	14	2	6	1	—	—

[a]Each kerogen type was modeled with an initial collection of 1400–1500 carbons, some of which were lost as the kerogen matures. Thus, the numbers shown *do not* represent the functional group contents based on a constant number of carbon atoms. The contents are given at each modeled stage of maturation: Stage a corresponds to the beginning of diagenesis, stage b to the beginning catagenesis, and stage c to the end of catagenesis. Asphaltenes were modeled only at stage b, the beginning of catagenesis, where they become most abundant.

Source: Data from Behar and Vandenbroucke, 1986.

29

Retorting

I	Surface	912
	A. N-T-U Process	912
	B. Gas Combustion Process	916
	C. Lurgi-Ruhrgas Process	921
	D. Tosco II Process	925
	E. Shell Pellet Heat Exchange Retorting (SPHER) Process	931
	F. Shell Shale Retorting Process (SSRP)	934
	G. Exxon Shale Retort (ESR) Process	937
	H. Chevron (STB) Process	938
	I. Petrosix Process	940
	J. Moving Grate Processes	943
	K. Paraho Process	948
	L. Unocal Processes	954
	M. Kiviter and Galoter Processes	966
	N. Hytort Process	971
II	In Situ	980
	A. U.S. Department of Energy Process	981
	B. Geokinetics Process	985
	C. Occidental Vertical Modified In Situ (VMIS) Process	985
	D. Rio Blanco Modified In Situ Process	994
	E. Equity Oil/Arco BX Process	998

Retorting is the process of heating oil shale in order to recover the organic material as shale oil plus gas (less commonly, just as gas (see also Chap. 13, Sec. IV.C). To get reasonable rates of product recovery, temperatures of 400–600°C (750–1100°F) are generally used. Therefore, to avoid unwanted combustion a *retort* in its simplest form is a vessel in which the shale can be heated without exposure to air and from which the product gases and vapors can escape to a collector. Retorts used in early shale oil processes were just that. Modern retorts are usually tailored to meet the needs of an integrated oil shale process and therefore are somwhat more complicated. Therefore, this section outlines the major features not only of the different retorts, but also of the corresponding integrated processes that have been developed/improved during the past two decades.

Shale oil recovery can be carried out above ground or underground (in situ). In situ processing (Chap. 13, Sec. IV.D for comparison) is attractive because the requirements for mining, hauling, crushing, and grinding the oil shale rock are eliminated or greatly reduced. Thus, in situ retorting offers the potential for corresponding savings in both capital and operating expenses. Above-ground retorting (Chap. 12, Sec. IV.D), on the other hand, generally affords better control of retorting conditions that can minimize heat loss due to carbonate decomposition and lead to a better yield of higher quality products. Novel processes using supercritical solvent extraction or bioleaching to recover shale oil are still in the early experimental stages, but appear to have the potential for use either above or underground. Most work to date has been on above-ground retorting.

Above-ground retorting processes fall into three broad classes depending on whether the process heat is generated internally (direct-heated retort), externally (indirect-heated retort), or have the potential for both. Currently, only the Kiviter retort used in the USSR for retorting of Estonian kukersite simultaneously derives major fractions of its process heat from both internal and external sources (Table 1).

I. SURFACE

Although interest in directly heated retorting has waned, it is useful to begin the discussion of retorting by examining two directly heated retorts of relatively simple design.

A. N-T-U Process

Invented in 1923 by Dundas and Howes, the N-T-U retort takes its name from the N-T-U Company (the letters stand for Nevada-Texas-

Table 1 Summary of Oil Shale Retorting Technologies

Above-ground retorting process	Underground retorting process
• Direct-heated	OXY modified in situ retort process (Occidental Oil Shale, Inc.)
N-T-U	
Gas combustion	Horizontal modified in situ (LETC) retorting process
• Indirect-heated	
Lurgi-Ruhrgas	Laramie true in situ retorting process
TOSCO-II	
Petrosix (Brazil)	Multi-Mineral in situ (MIS) process
Exxon shale retort (ESR) process	Geokinetics in situ project
Shell shale retorting process	Equity Oil BX in situ
HYTORT (IGT hydrogen atmosphere)	RISE (rubble in situ extraction)
Superior circular grate	
Allis–Chalmers roller grate	
Galoter (USSR)	
• Combination	
Kiviter (USSR)	
Union Oil	
Paraho	

Utah), which took the lead in its early development. The N-T-U batch retort is relatively simple and inexpensive to construct and proved to be durable in operation (Figure 1).

During the 1920s, the N-T-U Company constructed a 40-ton retort near Santa Maria, California and the U.S. Bureau of Mines (USBM) constructed a small N-T-U retort at its Anvil Points facility. This first phase ended about 1930, but interest was revived by the fuel demands of World War II. In Australia, N-T-U retorts were used by Lithgow Oil Pty. Ltd. at their facility at Mangaroo. Three 35-ton retorts were constructed but seldom operated simultaneously due to lack of feed shale. Nevertheless, nearly 2 million gallons of liquids were produced during 1944–1945. Interest was also revived in the United States. Upon passage of the Synthetic Liquid Fuels Act of April 5, 1944, the USBM constructed two identical 40-ton N-T-U retorts at Anvil Points to provide design data and quantities of shale

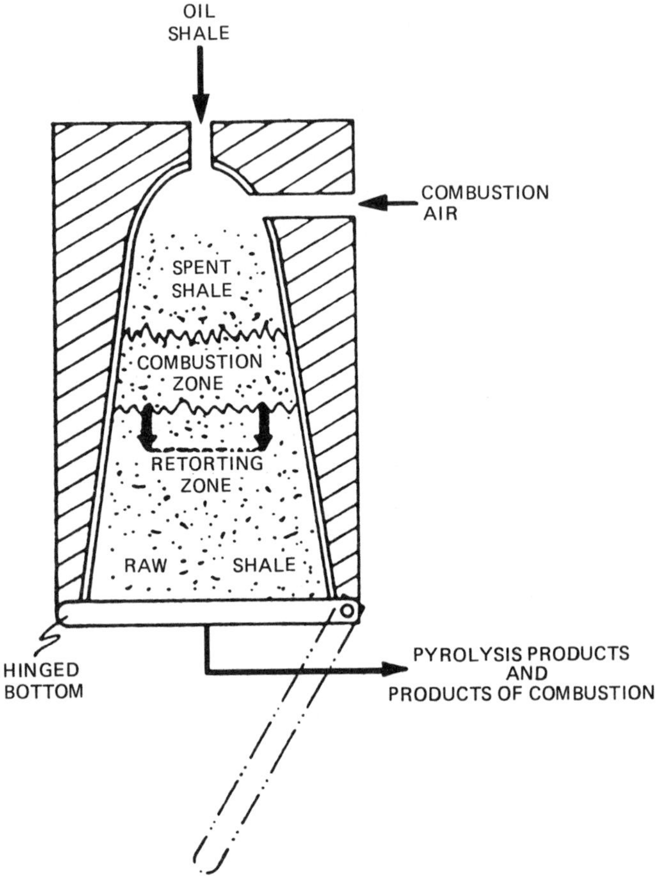

Figure 1 Schematic of the N-T-U retort. Heat from the downward-moving flame front drives the retorted oil before it. Gas produced by retorting can be recycled to provide additional heat.

oil sufficient for refining studies. By the time these retorts were dismantled in 1951, each retort had operated nearly 7000 hr to process about 18,000 tons of oil shale and produce 6000 bbl of shale oil. Inspection showed that the retorts were still in good condition (Cattell et al., 1951).

In the 1960s, a 10-ton N-T-U retort was constructed at USBM's Laramie Energy Technology Center (LETC), later the Laramie Energy Research Center (LERC) of the Energy Research and Development administration (ERDA) and now the Western Research Institute. This

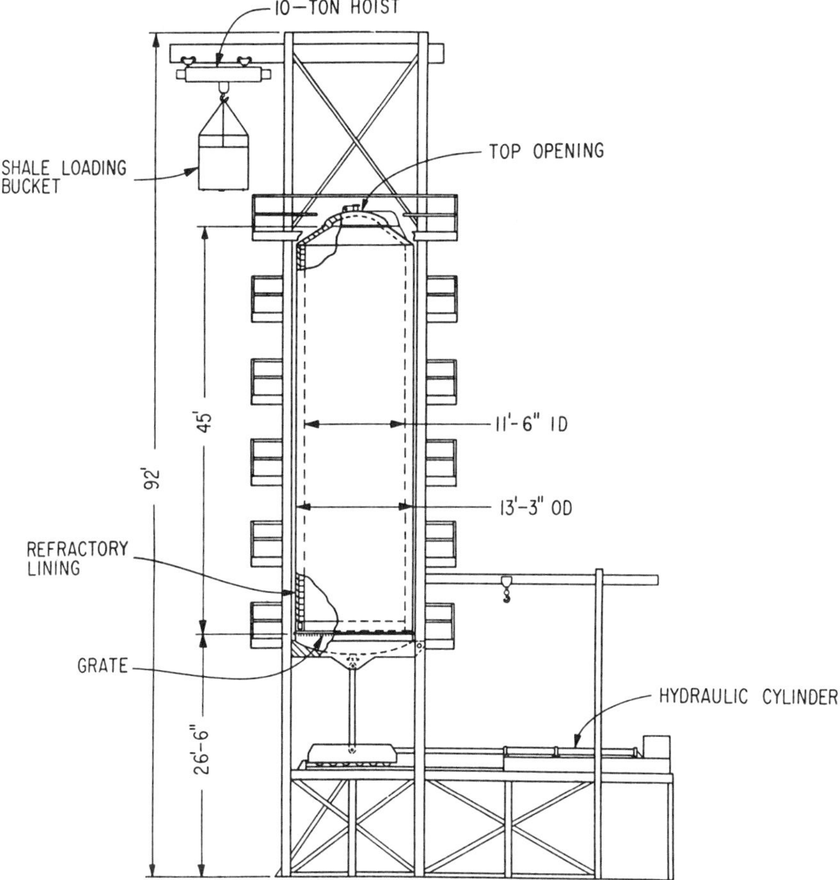

Figure 2 Schematic representation of the 150-ton oil shale retort at the Western Research Institute site, Laramie, Wyoming.

retort, and the 150-ton version added in 1968, have been used extensively to study many of the parameters important to in situ retorting. The 150-ton retort, located just north of Laramie, Wyoming, has an internal diameter of just over 6 ft and a height of 45 ft (Figure 2).

To simulate an in situ retort, the 150-ton unit is generally charged with upgraded shale selected to simulate the wide range of sizes obtained by blasting to produce a rubblized shale bed in a underground retort. Individual shale blocks weighing up to 10,000 lb have been retorted successfully (Ruark, 1956; Harak, 1971; Docktor, 1972). Oil recovery from oil shale in the range of 1/2 to 3-1/2 in. ranged from

80% of Fischer assay for 30 gal/ton grade to 87.5% of Fischer assay for 50 gal/ton grade. Even very large pieces of shale could be retorted. Harack (1971) described the results of one run where 20% of the feed was larger than 20 in., 10% was smaller than 1 in., and one very large piece of shale weighed 7500 lb (Table 2).

The N-T-U retorts served well as research tools, but as batch units they were not well suited for commercial use. This led to a search for a more efficient and continuous retorting process.

B. Gas Combustion Process

Design objectives for the gas combustion process grew out of the USBM work (see previous section). These included gravity flow to minimize mechanical complexity, heating of the raw shale by hot gas, and generation of this hot gas by burning the residual carbon on retorted shale within the retort vessel.

Work began in 1949 on the first version, known as the dual-flow retort. A second and much improved version, the countercurrent retort, soon followed. By flowing combustion gases up and shale down, oil recoveries above 90% of Fischer assay were obtained from 25–20 gal/ton shale at throughputs of 200 lb/hr/ft^2. However, having the combustion zone at the bottom of the retort led to the spent shale being being discharged at high temperature, thereby wasting valuable heat. Also, handling the hot shale was a problem. The gas combustion process was developed to alleviate these problems (Figure 3).

Table 2 Results Summary For a Run in the 150-Ton N-T-U Retort at the Western Research Institute[a]

Length of run	days	12.25
Operating conditions:		
Shale charge	tons	178.67
Retort pressure	psig	3.0
Air rate (dry)	scfm	135
Do	scf/ton shale	13,300
Avg. air temp. into retort	°F	28
Recycle gas rate (dry)	scfm	67
Do	scf/ton shale	6,600

(continued)

Table 7 (cont.)

[Operating conditions]		
Avg. recycle gas temp. into retort	°F	43
Oxygen content of retorting gas	pct	14.5
Space velocity	ft^3 gas/ft^2 bed/min	1.94
Stack gas rate (dry)	scfm	177
Do	scf/ton shale	17,400
Gas produced in retort (dry)	scfm	42
Do	scf/ton shale	4,100
Max. retort differential press	in. H_2O	0.6
Avg. ambient temp.	°F	26
Avg. retorting advance rate	in./hr	1.75
Max. bed temp.	°F	1,600
Bed compaction	pct of initial height	5.6
Oil shale properties:		
Fischer assay	gal/ton	25.4
Water content	gal/ton	2.9
Bulk density	lb/ft^3	80.0
Gross heating value	btu/lb	2,267
Recovery:		
Oil	gal	2,830
Spent shale	tons	125.86
Oil recovery	vol % of Fischer assay	62.2
Oil properties:		
Gravity	API	25.2
Pour point	°F	70
Viscosity	SUS at 100°F	79
Hydrogen	wt %	11.76
Nitrogen	wt %	1.77
Sulfur	wt %	0.76
Carbon	wt %	84.58
Ash	wt %	0.01
Gross heating value	btu/lb	18,660
Spent shale properties:		
Fischer assay	gal/ton	0
Gross heating value	btu/lb	117

[a]This run included a single block of shale that weighed 7500 lb.
Source: Data of Harack, 1981.

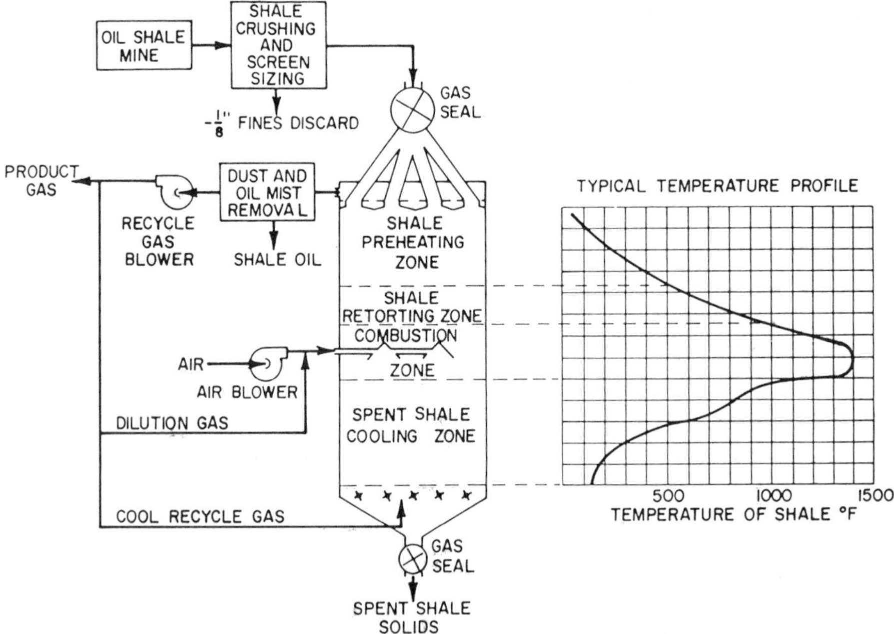

Figure 3 Simplified flow diagram of the Petrosix retort (reprinted with the permission of the Pace Company, Consultants and Engineers).

The gas combustion retort is relatively simple; no internal baffles divide the zones and the flow of shale from the retort is controlled by a movable grate at the bottom. Shale fed into the top of the retort through a rotating lock is heated by the hot product vapors. This serves to cool the oil vapors past the dew point into a mist that is swept out into the collector by the upflowing gases. The preheated shale then moves down into the retorting zone where, upon further heating, kerogen decomposition yields the oil and gas products and a carbonaceous residue that adheres to the spent shale. Next the shale moves into the hottest part of the system where air is introduced for combustion of the residue and of the hydrocarbons in the recycle gas. In the lowest part of the retort, heat from the combusted shale is transferred to the recycled gas as it flows upward. Finally, the cooled, combusted shale is discharged from the bottom of the retort.

Three gas combustion retorts were constructed by USBM at Anvil Points: 6 and 25 ton/day pilot units and a 150 ton/day unit designed to furnish engineering design data for a commercial unit (Tables 3 and 4).

Table 3 General Data Summary, USBM Evaluation Runs on the 150 tpd Gas Combustion Retort

	Test number					
	25(1–5)	26(1–2)	26(3–5)	27(1–3)	28(1–4)	28(5–6)
Length of test, hr	120	48	72	72	96	48
Rates and quantities:						
Shale size, in.	3/8-3	1-2	1-2	4-2	1-2	1-2
Bed height, ft and in.	9,11	9,11	9,11	9,11	7,2	7,2
Raw shale rate, lb/(hr) (ft^2)	299	222	299	350	299	300
Air rate, std ft^3/ton shale	3,940	4,230	3,910	3,840	4,010	4,290
Dilution gas rate, std ft^3/ton shale	2,860	3,800	2,950	3,140	3,100	4,040
Recycle gas rate, std ft^3/ton shale	13,340	12,400	12,650	12,660	12,500	10,260
Temperatures:						
Product outlet, °F	162	142	141	143	141	126
Retorted shale out, °F	376	348	356	345	378	447
Raw shale in, °F	40	34	32	30	28	25
Recycle gas, °F	241	247	250	246	224	213
Dilution gas, °F	83	90	92	92	79	86
Air, °F	128	129	131	144	110	98
Yields:						
Oil, vol %/Fischer assay	82.8	92.3	86.2	86.7	85.1	86.1
Gas, std ft^3/ton shale	6,040	6,440	6,000	6,020	6,400	6,090
Retorted shale, wt % of raw shale	81.8	82.9	82.3	82.1	83.3	83.0
Liquid water, lb/ton shale	0.2	5.0	0.9	1.1	4.9	11.2
Miscellaneous:						
Retort pressure drop, in H_2O/ft bed	0.90	0.37	0.73	1.02	0.58	0.45
Carbonate decomposition, wt %	24.9	24.1	26.9	25.6	23.3	23.6

Table 4 Material Balances, USBM Evaluation Runs on the 150-Ton Gas Combustion Retort

	Test number					
	25(1−5)	26(1−2)	26(3−5)	27(1−3)	28(1−4)	28(5−6)
Materials in:						
Raw shale, lb	2000	2000	2000	2000	2000	2000
Recycle gas, lb	986	933	952	946	927	773
Dilution gas, lb	211	286	222	235	230	304
Air, lb	302	324	299	294	307	329
Total materials in, lb	3499	3543	3473	3475	3564	3406
Materials out:						
Retorted shale, lb	1636	1658	1646	1642	1666	1660
Product oil, lb	185	185	178	183	189	184
Offgas, lb	1644	1705	1625	1631	1609	1536
Water in oil, lb	nil	5	1	1	5	11
Total materials out, lb	3405	3553	3450	3457	3469	3391
Recovery, %	99.0	100.3	99.3	99.5	97.3	99.6

The first stage in development of the gas combustion process ended in 1955 when work was halted by USBM. Stage II began in 1961 with the passage of Public Law 87-796 empowering the Secretary of the Interior to lease the facility, in order to encourage further development of oil shale technology. After evaluating several proposals, the Interior Department leased the Anvil Points facility to the Colorado School of Mines Research Foundation (CSMRF). Under the terms of the lease, CSMRF became lessor, provided administrative and logistic support, and made the facility available to a consortium that eventually included Mobil Oil (project manager), Humble Oil & Refining, Pan American Petroleum Corp., Sinclair Research, Inc., Continental Oil Co., and Phillips Petroleum Co. Under this agreement, the facility was reactivated in 1964 and was operated for about a year in 1966–1967. During this time, nearly 300 runs were made, mostly in the 25 and 150 ton/day units (Ruark, 1971a,b). No work was carried out on the Gas Combustion Process after 1967, but the Paraho Process in its directly heated mode is nearly identical, so Paraho results can be used as a guide to the potential of the gas combustion process.

C. Lurgi-Ruhrgas Process

The Lurgi-Ruhrgas oil shale process (LR process) grew out of an earlier Lurgi process for making high-btu gas from coal fines (see also Chap. 12, Sec. IV.C, and Chap. 20). Initially, a solid heat carrier (e.g., sand) was mixed with the coal fines, but later the hot coal char product was recycled to supply the needed heat. The process was tested on a variety of coals and was used commercially in Germany, England, Yugoslavia, Argentina, and Japan. Application to oil shale was a logical extension.

In the LR shale retort, hot solids recycled from a combustor and the raw shale are fed (ratio 6–8:1) to a screw-type mixer (Figure 4). Retorting takes place in the mixer and in the surge bin (accumulator) that follows. Gas-solid separation takes place in the surge bin and

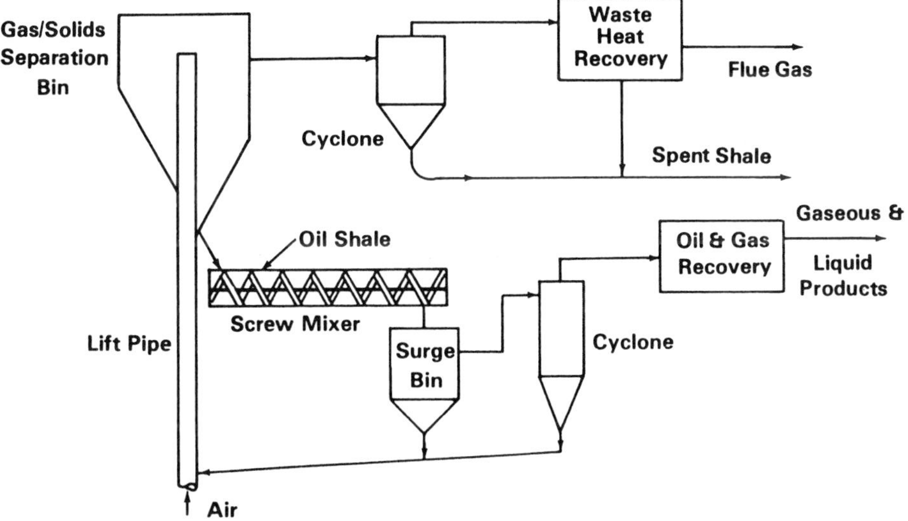

Figure 4 Simplified flow diagram of the Lurgi-Ruhrgas oil shale retorting process shows that the screw mixer section where retorting takes place is actually a very small part of the system (after Rammler, 1982).

the following cyclone. Most of the solids are diverted to the lift pipe combustor where burning of the residual carbon raises the temperature to about 650°C (1200°F). The hot solids are separated from the gases and returned to the retort, which is maintained at an optimum retorting temperature in the neighborhood of 540°C (1000°F).

The use of the mixer provides rapid heat transfer and allows very short residence times for the product vapors in the hot zone to minimize unwanted cracking. The forced mixing also reduces the tendency for rich shales to agglomerate upon heating, thereby enabling high throughputs. Significantly, this section, which must withstand both heat and abrasion, is not large. As a result, the critical materials problems are confined to a small, manageable section of the overall facility. Also, it is important that no part of the LR retorting system operates at high pressure; maximum pressures are a few inches of water.

The product collector section is designed in stages to provide two, and preferably three or more, product fractions. This is not to give fractions defined by boiling range but rather to avoid oil/water emulsions and for improved dust control. The first section is an underflow scrubber, where the product stream is concurrently cooled and scrubbed by injection of the aqueous product and heavy oil. Fine dust not collected in the cyclone is concentrated in the heavy oil and the subsequent fractions are essentially dust-free.

For commercial use, provision would be made to recover process heat from spent shale and heat exchangers would be used to recover additional heat. Also, two or three cyclones may be staged for more complete dust removal from the product vapors; and scrubbing would be used in all stages to give efficient cooling and condensation of the product vapors. Electrostatic precipitators would probably be added to reduce particulate emissions in the flue gas from the lift-pipe combustor.

One of the important features of the LR process is its ability to process shale fines; the entire oil shale resource can be utilized in a single type of reactor (Table 5). Good material balances were obtained in pilot plant operations and oil yields were high from a variety of oil shales; yields were commonly above 95% of Fischer assay (Table 6). In favorable cases oil yields went as high as 110% of Fischer assay. The product oil is 85–90% volatile, though nitrogen and sulfur levels are high, and pour points are desirably low (Table 7). Dust in the heavy oil is troublesome, but Lurgi has a patented process for oil dedusting. Alternatively, the dust-laden heavy oil can be recycled to the mixer. Thus, the LR retort solved two of the most persistent problems of the gas combustion retort, namely, low oil yields and an inability to process shale fines (i.e., the entire oil shale resource).

Gas products from the LR process are not diluted with combustion gas, so btu content is high and the gas should be suitable for reforming to produce the hydrogen needed in product upgrading.

The aqueous product is only slightly basic and should not require special materials of construction. It does contain phenols and ammonia that may be recovered by steam stripping; however, further treatment (e.g., biological oxidation) may be required before the process water is environmentally acceptable for moistening the spent shale.

After removal of the kerogen by retorting and decomposition of some of the carbonates during combustion, the spent shale is friable (Table 8). Shear applied in the mixer retort does reduce particle size. On the one hand, these characteristics make the combusted shale a very good SO_x acceptor, while on the other, the fine particle size means that special care must be exercised in above-ground disposal. This should not normally be a problem, since most of the spent shale would be returned to the mine. Moreover, when moistened, the spent shale acts like a cement and gives a hard, rocklike mass (Watson et al., 1982). This will reduce the tendency for dusting of the spent shale, and hence should simplify environmentally acceptable disposal.

In summary, the LR process uses relatively simple, inexpensive, and reliable hardware to achieve high oil shale throughput and high oil yields with the ability to process the entire oil shale resource. Major environmental concerns, save leachate, have been addressed (Gulf Oil/Standard Oil, 1976). On the other hand, the high nitrogen and sulfur contents of the LR shale oil product will make upgrading difficult and expensive.

Several suggestions were recently advanced by workers at Monash University to improve overall efficiency of the LR process when handling oil shale from the Rundle deposit in Australia (Potter et al., 1984). These suggestions included (1) using a fluidized bed solid-solid heat exchanger for improved heat recovery from spent shale, (2) staged drying of the raw shale with the added benefit of raising low-pressure steam, (3) the use of fluidized bed combustion to give better temperature control than the lift-pipe combustor and thereby lower heat requirements for carbonate decomposition, and (4) to co-pyrolysis of coal with shale to obtain higher liquid yields from a retort of given size.

As a historical note, it is interesting that Lurgi, in collaboration with Ruhrbach, developed a fluid bed retorting system in the 1950s (Schmalfeld, 1975). This retort was not designed to produce oil, but rather to produce heat for steam generation of electricity and a residue that is valuable in cement manufacture. Two plants with a combined capacity to retort 720 tons of shale a day and generate 6 MW of

Table 5 Properties of Lurgi-Ruhrgas Oil Shale Pilot Plant Feeds[a]

	Type of shale					
	A	B	C	D	E	F
Fischer assay						
Moisture, wt %	0.4	0.4	3.5	2.7	1.8	5.9
Gas liquor, wt %	0.8	1.4	2.1	3.3	1.7	2.3
Oil, wt %	11.6	9.4	5.6	4.2	3.4	8.0
Residue, wt %	83.4	84.9	84.4	87.5	91.0	79.7
Gas and loss, wt %	3.8	3.9	4.4	2.3	2.1	4.1
Ultimate analysis						
Carbonate CO_2, wt %	18.7	16.4	13.7	9.1	19.6	14.6
Organic carbon, wt %	15.7	12.9	10.2	7.3	6.7	14.3
Hydrogen, wt %	2.0	1.9		1.3	1.1	1.8
Nitrogen, wt %		0.5	0.3	0.2	0.2	0.6
Total sulfur, wt %		0.9	3.4	1.0	1.6	2.7
Gross calorific value, kJ/kg	7400	6150	4770	3520	2830	5690
Grindability index (Hardgrove)		30	76	49	90	69
Size analysis[b]						
+5 mm, wt %	2.8	0.0	0.0	0.0	0.1	0.1
4 -5 mm, wt %	25.2	0.0	2.7	0.0	2.1	1.4
3 -4 mm, wt %	25.2	0.1	12.2	0.2	7.6	10.0
2 -3 mm, wt %	17.1	16.7	17.5	14.0	13.8	58.2
1 -2 mm, wt %	18.3	26.8	23.2	25.8	19.5	29.1
0.8 -1 mm, wt %	14.0	11.4	10.2	9.7	7.7	0.2
0.5 -0.8 mm, wt %	14.0	9.2	7.8	8.5	7.3	0.1
0.315 -0.5 mm, wt %	10.4	13.9	8.4	10.0	9.2	0.1
0.2 -0.315 mm, wt %	10.4	9.0	5.5	7.6	7.8	0.1
0.1 -0.2 mm, wt %	5.2	6.0	5.1	8.4	8.6	0.1
0.063 -0.1 mm, wt %	7.0	3.1	3.1	5.2	6.0	0.2
-0.063 mm, wt %	7.0	3.8	4.3	10.6	10.4	0.4
Median grain size, mm	2.05	1.09	1.54	0.97	1.20	2.31
Bulk weight, kg/liter	1.15	1.25	1.09	1.23	1.18	1.00
Ash analysis						
CaO, wt %			24.0	11.5	31.3	25.9
SiO_2, wt %			38.4	46.6	29.5	39.1
Al_2O_3, wt %			13.8	15.5	10.0	11.1
Fe_2O_3, wt %			6.6	7.5	7.0	4.4

(continued)

Table 5 (Cont.)

	Type of shale					
	A	B	C	D	E	F
[Ash analysis]						
SO_3, wt %			9.7	2.6	4.0	4.6
MgO, wt %			2.9	3.4	2.6	5.5
K_2O, wt %			2.1	2.0	1.6	1.4
Na_2O, wt %			0.7	3.1	0.3	0.2
Others, wt %			1.8	7.8	13.7	7.8

[a]Ability to process fines means the entire oil shale resource can be processed.
[b]Crushing to 0–4 mm used in pilot plant work is not typical for larger units.
Source: Data of Rammler, 1982.

electricity were constructed at Dotternhausen, Wurttemberg (West Germany) in 1960. These plants are still in use (Speight and Vawter, 1988).

D. Tosco II Process

Development of this process by Tosco Corporation (formerly the Oil Shale Corporation) began in 1955 as an attempt to overcome the shortcomings of the USBM gas combustion retorting process. Exploratory work proceeded through a 25 ton/day pilot plant built in 1957 at Golden, Colorado and a 1000 ton/day semiworks plant constructed at Parachute Creek, Colorado in 1965 to detailed engineering design work completed in 1968 for a 66,000 ton/day commercial facility—the Colony project. To implement the recommendations of the 1968 study, a second phase of developmental work was carried out in the 1000 ton/day plant for acquisition of the data necessary for construction of the commercial facility (Whitcombe and Vawter, 1982).

Initially, the commercial facility was a joint venture of Tosco (40%) and Arco (60%), with Arco as the operator, but in 1980 Exxon acquired Arco's interest and assumed the role of operator. Construction of the facility and the associated town of Battlement Mesa began in 1980. However, faced with increasing project cost estimates, ultimately reaching about $7 billion (nearly twice the original estimate, due in

Table 6 Material Balances Were Good in LR Oil Shale Pilot Plant Operations[a]

	Type of shale		
	B	C	F
Feed			
Quantity, kg	1000	1000	1000
Moisture, wt %	0.4	3.5	5.9
Oil (Fischer assay), wt %	9.4	5.6	8.0
Output			
Circulated material discharged, kg	266.7	464.3	510.5
Dust in flue gas, kg	490.5	330.9	186.2
Dust in make gas, kg	9.7	13.5	55.8
Heavy in oil (dust-free), kg	62.9	28.8	50.6
Middle oil, kg	26.5	18.8	23.5
Gas naphtha, kg	10.7	5.4	5.5
Gas liquor, kg	24.5	32.9	85.9
Make gas (dry, free of C_{4+}), kg	29.4	27.1	32.8
CO_2 from carbonate decomposition in flue gas, kg	42.7	42.9	11.1
Fixed C burned, kg	36.4	35.4	38.1
Totals, kg	1000.0	1000.0	1000.0
Yield of oil (% of Fischer assay)	106.5	94.6	99.5

[a]Oil yields were high, generally over 95% of Fischer assay.
Source: Data of Rammler, 1982.

large part to high interest rates), and lower crude oil prices, the project was terminated early in 1982 (*Oil and Gas J.*, 1982a,b).

In the Tosco process, hot ceramic balls are mixed with smaller oil shale particles in a rotating drum retort. After retorting, the ceramic balls are separated from the spent shale and reheated in a separate ball heater using gas as the fuel (Figure 5).

The feedstreams to the Tosco retort are 1/2-in. ceramic balls heated to about 690°C (1270°F) and the oil shale crushed to pass a 1/2-in. screen and preheated by contact with hot flue gases from the ball heater. The streams are rapidly mixed under an intert atmosphere. Heat transfer is rapid and at the retort's exit the shale and ceramic balls are essentially the same temperature and the shale is fully retorted.

Table 7 Inspections of Dust-Free Total Oils from the LR Retort[a]

	Type of shale		
	E	F	H
Density (20°C), g/cm^3	0.985	0.971	0.882
Viscosity (20°C), 10^{-6} m^2/sec	21.5	11.5	6.3
(60°C), 10^{-6} m^2/sec	5.1	3.7	2.3
Pour point, °C	-20	-23	+16
Flash point, °C	103	n.d.	+23
Conradson C, wt %	5.3	5.6	2.1
Bromine number, g/100 g	56	n.d.	56.5
H/C atomic ratio	1.37	1.49	1.57
Nitrogen, wt %	0.89	1.47	0.84
Sulfur, wt %	3.25	6.47	0.70
Gross calorific value, kJ/kg	40,900	40,195	43,340
Initial boiling point, °C	130	78	69
5 vol %, °C	175	148	142
10 vol %, °C	200	167	170
20 vol %, °C	232	195	201
30 vol %, °C	260	227	228
40 vol %, °C	290	268	245
50 vol %, °C	318	308	276
60 vol %, °C	340	341	310
70 vol %, °C	370	377	347
80 vol %, °C	405	423	391
90 vol %, °C	450 (88.0%)	450 (85.4%)	440
Loss vol %, wt %	1.0	0.9	0.4

[a]Does not include gas naphtha. Pour points of the LR oils are low, but only 85–90% is volatile and the nitrogen and sulfur levels are high.
Source: Data of Rammler, 1982.

Retorting in the Tosco process is carried out in a rotating drum that is mechanically a simpler device than the screw mixer of the Lurgi process, but like the Lurgi retort is only a small part of the total facility (Figure 6). However, unlike the Lurgi process, heat for the Tosco retort is provided by gas, not combusted shale. The independence of gas flow and shale flow effectively decouples retorting from

Table 8 Inspections of Spent Shale From the LR Retort Show That Particle Size is Low and That the Spent Shale is Much More Friable Than the Starting Shale (i.e., Grindability is Increased)

	Type of shale		
	E	F	H
Organic C, wt %	0.3	1.3	2.0
Carbonate CO_2, wt %	15.8	19.3	12.3
Hydrogen, wt %	0	0.2	0.2
Total S, wt %	0.3	1.5	2.5
Size analysis			
>5 mm, wt %	0	0	0
4 –5 mm, wt %	0	1.1	0
3 –4 mm, wt %	0	2.9	2.9
2 –3 mm, wt %	3.1	12.6	7.1
1 –2 mm, wt %	7.7	2.1	8.8
0.8 –1 mm, wt %	3.3	5.2	2.0
0.5 –0.8 mm, wt %	3.0	5.7	4.7
0.315–0.5 mm, wt %	6.9	8.3	8.0
0.2 –0.315 mm, wt %	7.9	8.2	6.5
0.1 –0.2 mm, wt %	5.4	11.8	14.6
0.063–0.1 mm, wt %	4.4	10.7	11.1
0.045–0.063 mm, wt %	3.8	7.9	28.0
<0.045 mm, wt %	54.5	23.5	6.3
Median size, mm	0.32	0.68	0.55
Grindability of the circulated material	150	124	n.d.

Source: Data of Rammler, 1982.

heat generation and gives the Tosco retort unusual latitude for processing shales of widely varying grades. Oil yields from the Tosco retort generally exceed Fischer assay. An oil yield nearly 108% of the Fischer assay was obtained from one 7-day test in the semiworks plant.

Product vapors from the retort are passed through a separator to remove fines, then into a fractionator that yields heavy oil, distillate, naphtha, and product gas streams. Because combustion is separated from retorting in the Tosco process, the product gas is a high-btu

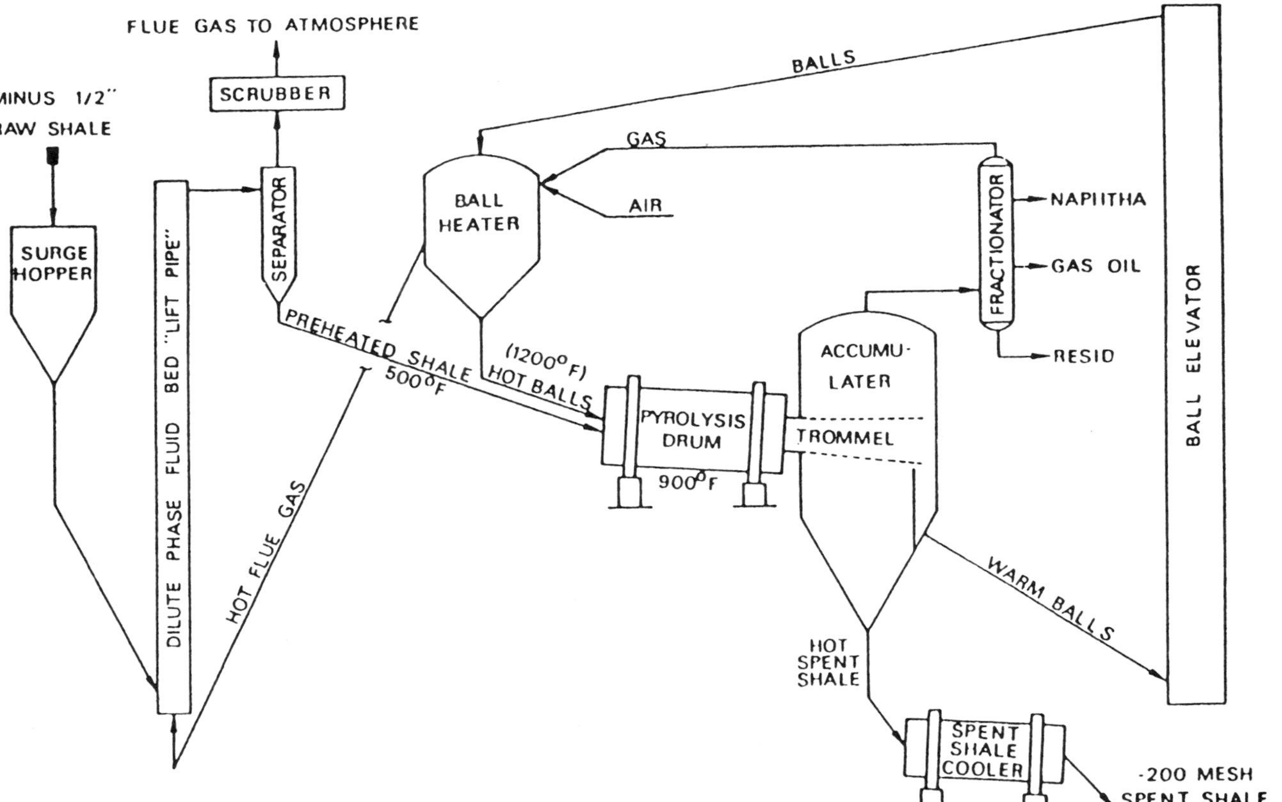

Figure 5 Tosco oil shale retorting unit decouples the heat generation and kerogen pyrolysis steps, thereby simplifying process control (reprinted with the permission of the Center for Professional Advancement, New Brunswick, NJ).

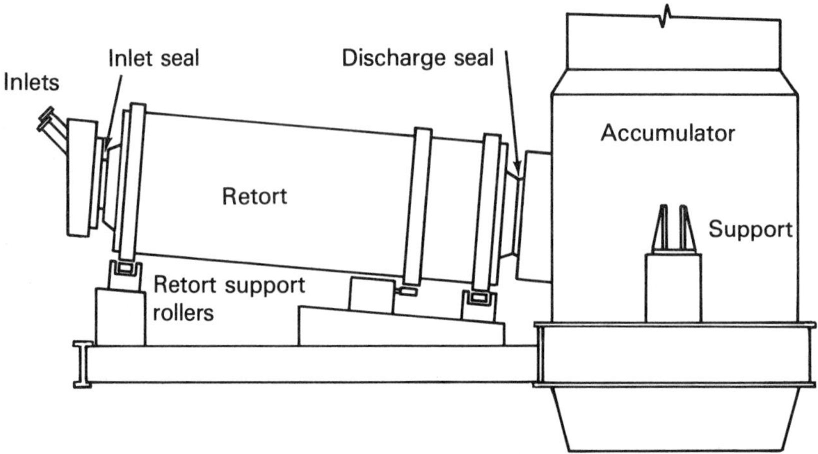

Figure 6 The Tosco retort is a rotating drum that constitutes only a small part of the total facility (reprinted with the permission of The Center for Professional Advancement, New Brunswick, NJ).

stream that may be used as plant fuel or reformed to give hydrogen for liquid product upgrading.

The ceramic balls and spent shale discharged from the retort are fed into a cylindrical screen (trommel) inside the spent shale accumulator housing. The ceramic balls, being larger than the holes in the screen, pass through the accumulator and are recycled to the ball heater by a bucket elevator. The spent shale falls into the accumulator at about 480°C (900°F) and passes into a rotating heat exchanger where its heat is used to raise steam in water-filled tubes. The spent shale is further cooled in the moisturizer where its water content is increased to 12–13% to reduce dusting before disposal.

One commercial size Tosco retort would process about 11,000 tons of oil shale per day to produce 4500 bbl of crude shale oil (Table 9). Properties of a typical oil produced from Mahogany Zone Green River oil shale are similar to those of a low-sulfur crude—with three very important exceptions: olefin and nitrogen contents are high, which makes the oil very unstable with respect to sludge and sediment formation (Table 10). Like other oils from Green River shales, the Tosco oil also has a high content of arsenic that is a powerful poison for the catalysts needed for efficient upgrading.

In summary, the Tosco II retorting process achieved several important technical goals: oil yields were high; consistently at or above

Table 9 Projected Products From a Single 11,000 Ton/Day Tosco II Retort

Shale feed, 20 gal/ton	11,000 tons/day
Pipelineable shale oil	4,500 bbl/day
Ammonia	150 tons/day
Sulfur	177 tons/day
Coke	836 tons/day

Source: Data of Whitcombe and Vawter, 1982.

Fischer assay. Shale fines could be processed, making it possible to utilize the entire shale resource. The gas byproduct was not diluted with combustion gases, and therefore had a high value for heating or hydrogen production. The process design afforded both good operability and an unusually high degree of flexibility in control and in an ability to handle shales of varying richness. However, operability and flexibility were obtained only at the cost of expensive hardware, making the Colony Project very capital-intensive. This last factor, coupled with the high interest rates of the early 1980s and uncertain crude oil prices, ultimately led to the demise of the Colony Project that would have used Tosco II retorting technology. Finally, even though flue gas heat was recovered in preheating the shale and spent shale heat was used to raise steam, the Tosco II process was relatively heat-inefficient because of the concurrent flows of heated balls and cooler shale.

E. Shell Pellet Heat Exchange Retorting (SPHER) Process

Workers at Shell Development addressed the Tosco drawbacks of high capital cost, mechanical complexity, and heat inefficiency in their design of the Shell pellet heat exchange retorting (SPHER) process. While never progressing past the experimental stage, a brief discussion of this process is worthwhile, since it was an important conceptual bridge in the evolution of the newer fluidized bed processes. The SPHER design grew out of Shell's experience with fludized beds in refinery processes, such as riser transport and catalytic cracking in dense beds, the former operating at relatively high and the latter at relatively low superficial gas velocities. The key features of the

Table 10 Inspection Data for a Typical Tosco II Crude Shale Oil

Gravity, °API	21.2
Pour point, °F	25[a]
Carbon, wt %	85.1
Hydrogen, wt %	12.6
Nitrogen, wt %	1.9
Sulfur, wt %	0.9
Arsenic, ppm	41
Nickel, ppm	6
Vanadium, ppm	3
Iron, ppm	100
Viscosity, SUS	
100°F	106
212°F	39
Distillation	
5 vol % at	200°F
10 vol % at	275°F
20 vol % at	410°F
30 vol % at	500°F
40 vol % at	620°F
60 vol % at	775°F
70 vol % at	850°F
80 vol % at	920°F

[a]The 25°F pour point is for oil that has been "heat-treated" under conditions described in U.S. patent 3,284,336.
Source: Data of Baughman, 1978.

SPHER process are countercurrent flows of shale and the heat exchange pellets for improved heat efficiency and the use of fluidized beds for low capital cost (Gwyn et al., 1981).

As conceived, the SPHER process uses two loops for circulation of the heat transfer pellets or balls (Figure 7). In the cool ball loop, balls

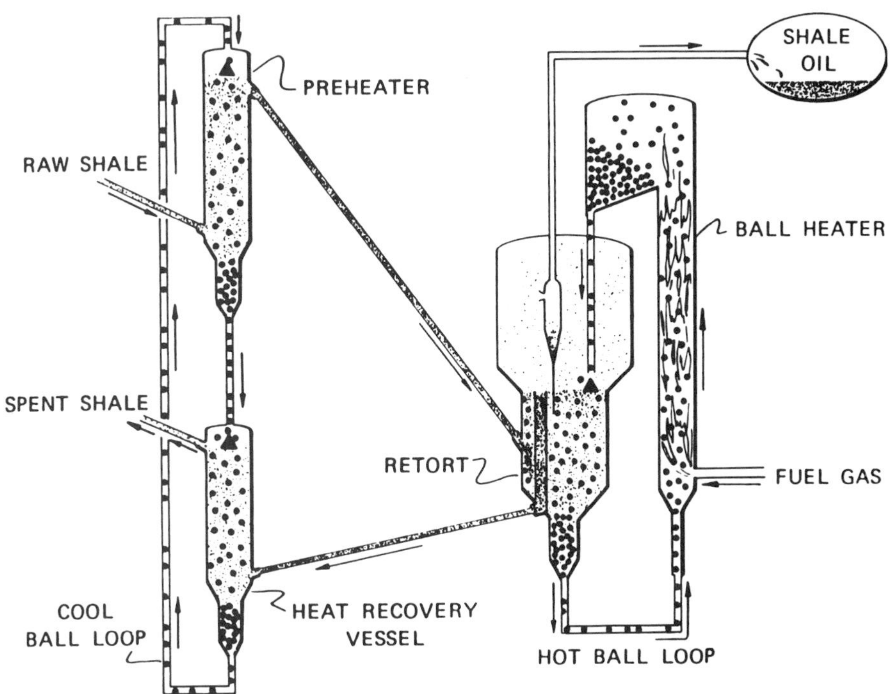

Figure 7 The SPHER process was designed with fluidized beds for low capital cost and mechanical simplicity, and countercurrent flows of shale and heat transfer pellets for high thermal efficiency.

fall from the preheater into a countercurrent fluidized bed for recovery of heat from the spent shale. A pneumatic riser transports the balls to the top of the cool ball loop, where they rain down through the up-flowing shale in the preheater. In the hot ball loop, after heating in a ball heater (riser/lift-pipe combustor or Tosco-type), the heated balls fall through a dense bed of shale fluidized by superheated steam in the retorting vessel. Segregation of the two ball loops allows the size of the different elements and the size and material of the balls to be tailored to each specific task, while high throughput and mechanical simplicity are maintained throughout. Projected thermal efficiency of the SPHER process was 67%, giving it a 4% advantage over Tosco II and a 9% advantage over the Paraho process.

While preliminary Shell economics indicated that shale oil from the SPHER process would be about 15% less expensive than Tosco shale oil, due primarily to lower capital cost, the retorting operation was still relatively complex and required heat transfer pellets. To obtain reliable fluidization in the SPHER beds was found to require grinding to -1/16 in. size. Such grinding would be expensive, even using staged grinding with oversize recycle, especially for hard, relatively rich shales, such as those from Green River and Jordan. Also, agglomeration of small particles of relatively rich shale would seriously impair the operability of the fluidized beds. Balanced against these factors are the benefits of fast heat-up and low vapor residence time in the retort, both of which contribute to increased yields of high-value liquids. Other workers at Shell and at Exxon have realized the capital cost advantages of fluidized bed retorting and have begun to study ways of further reducing costs and improving operability in fluidized bed retorting.

F. Shell Shale Retorting Process (SSRP)

A recent report outlines the current design basis for oil shale retorting at Shell (Voetter et al., 1987). Concentrating on the retort itself, van Wechem and his coworkers set out the following objectives for retort design:

Achieving low capital cost by minimizing solids recycle, which means the use of simple reactor designs and minimum reactor volumes throughout

Rapid particle heat-up to minimize coking that limits the conversion of kerogen to volatile products

Maximum oil recovery, which means vapor residence times of seconds to minimize cracking reactions that produce gas and thereby reduce the amount of liquid products that can be recovered

Scale-up capability to achieve economies of scale without undue recourse to multiple trains, while maintaining a solids residence time of 5–8 min to allow complete retorting

Flexibility with respect to shale grade—capability to process both rich and lean shales from a wide range of deposits with reliable operability and efficiency

Capability to process shale fines to enable use of the entire resource

Consideration of these factors led to the selection for SSRP of a staged, cross-flow fluidized bed, operating in the bubbling dense bed regime.

In a gas-solid system, rapid heat-up to a temperature where retorting proceeds rapidly will require the use of small particles due to the low volumetric heat capacity of the gas. Based on the results of Forgac, 2- to 3-mm particles would give acceptable average heating rates of 100–600°C/min. This requirement for small particle size is expected to lead to a substantial penalty in both capital and operating costs for size reduction in SSRP versus processes that can use larger particles.

Retorting kinetics vary with shale type due in large part to differences in heat and mass transport within the individual oil shale particles. Using the model of Wallman (Figure 8), the Shell workers concluded that staged solids flow would be required to obtain the desired kerogen conversion and selectivity toward oil.

Using this model (and data reported by Wilkins et al., 1981), calculations led to a total vapor residence time (extraparticle + bulk vapor) of 1–2 sec to maintain cracking to a level of about 10% of total product oil at normal retorting temperatures. To achieve such a short vapor residence time is not easy. A moving fixed bed would have to have a very low gas flow to avoid entrainment of the finer shale particles. In turn, this would require low bed depth and a very large bed area that could only be met with parallel trains. Increasing the particle size would allow increased gas flow but would also increase particle heat-up time, thereby increasing the bed area requirement, and lead to increased coking and lower product yields.

Because of the need for a relatively long solids residence time (8–12 min) to achieve complete retorting, selection among the various fluidized bed designs was driven by ability to maximize particle holdup

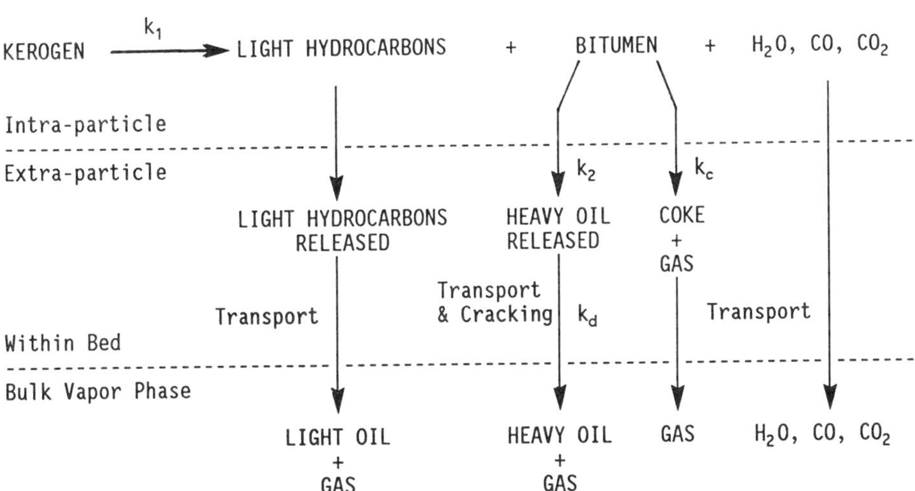

Figure 8 Model of the decomposition of oil shale to products in the Shell (SSRP) process (see Wallman et al., 1980).

(1 - ε). Whereas a dense bed offered particle holdup in the range of 0.45–0.55 and a turbulent bed holdup would be about 0.4, holdup in a riser bed design would only be in the range of 0.1–0.2. Moreover, it was found desirable to minimize turbulence to narrow the distribution of particle residence times and hence minimize the reactor volume required to achieve 99% retorting of the average particle. Thus, for both its higher holdup and its narrower distribution of residence times, the dense bed was favored. However, gas fluidized beds are inherently unstable and tend to have a large degree of backmixing associated with the circulation patterns around rising gas bubbles. Beds with large height/diameter ratios (L/D) restrict this circulation, and therefore tend to lower backmixing. To obtain desirable L/D while keeping the bed height (hence vapor residence time) at an acceptably low value would require multiple trains or horizontal staging. For its lower capital cost, the Shell workers chose the latter option with steam as the fluidizing gas (Figure 9).

In order to facilitate close-coupling the various stages within a single housing, the use of rectangular bed sections (instead of the usual circular bed cross-section) was suggested. This suggestion emphasizes the importance Shell attaches to a single-train/single-unit design in minimizing capital cost. The report indicates that a single unit capable of processing 100,000 tons of oil shale to produce 50,000 bbl of crude shale oil per day could be constructed using the SSRP design.

Other important features of the horizontally staged design include the following:

The fluidizing gas has the primary function of keeping the solids fluidized; its flow can be restricted to that duty.

Product vapors leave the reactor along a short path, thereby minimizing unwanted cracking.

A large bed area can be made available without incurring a severe penalty from backmixing due to turbulence.

Linear vapor velocity can be kept low enough to minimize fines carryover without sacrificing vapor residence time.

Different sections could be operated at different conditions selected for optimum product yield/quality. Presumably, some heavy product might be recycled to one or more sections for additional cracking.

Products from the different sections could be collected separately by partitioning the gas space without compromising solids staging.

Heat for SSRP would be provided by burning the carbon on the spent shale. Three types of fluid bed combustors (FBC) were indicated as candidates. A bubbling dense bed FBC, similar to those in use for coal combustion but with a gas velocity somewhat below the 2 m/sec typical for coal, was one option. It was pointed out that this

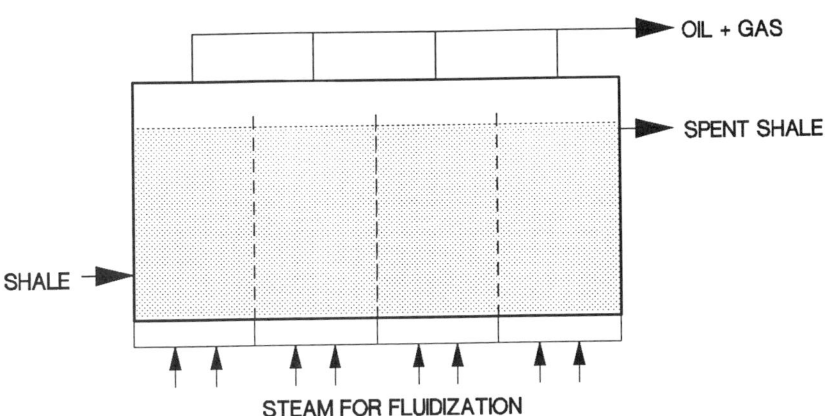

Figure 9 Principle of the horizontally staged fluidized bed selected for the Shell shale retorting process (after Voetter et al., 1987).

would require a large bed height and a large shale inventory due to the limited amount of air available for combustion. At the other extreme, a dilute phase riser FBC would offer high shale throughput and no shortage of air but would require a tall reactor to provide sufficient residence time for complete combustion. As a compromise, the use of a fast circulating fluidized bed combustor (FFBC), similar to those now being commissioned for coal combustion, was suggested. The FFBC combines good lifting power making gravity flow of the hot combusted shale possible with good contacting and burnout, while maintaining a moderate reactor size.

G. Exxon Shale Retort (ESR) Process

Recent work at Exxon also led to a fluidized bed retort. As reported by Bauman and coworkers, research on the Exxon shale retort process (ESR) began with exploratory work in 1980, and by 1982 had progressed to the point that a decision was made to proceed to the pilot plant stage for the development of a proprietary oil shale retorting technology (Bauman et al., 1987). A 5-ton/day pilot plant was constructed at Exxon's Baytown laboratory and operated from July 1984 until September 1986, processing both Colorado and Rundle (Australian) oil shales. Evidently, this facility, which included a fully redundant computerized data acquisition and control system, was constructed with an eye to the installation of a second, larger pilot plant at a later date. However, such plans were not discussed.

According to the report, fluidized bed designs were selected for both the retort and the combustor. It was claimed that during 26 months of operation, this pilot plant logged 3800 hr of operation and completed 73 yield periods, including a continuous operating period of 1040 hr on feed. It was also claimed that conditions affording oil yields over 100% of modified Fischer assay were identified. No product information or further details of the process were provided.

H. Chevron (STB) Process

The Chevron staged turbulent bed (STB) oil shale retort (Figure 10) represents a very different approach to fluidized bed retorting from that taken by the workers at Shell and Exxon. To approach plug flow performance in an open fluidized bed reactor requires minimizing the backmixing associated with circulation around rising bubbles. Also,

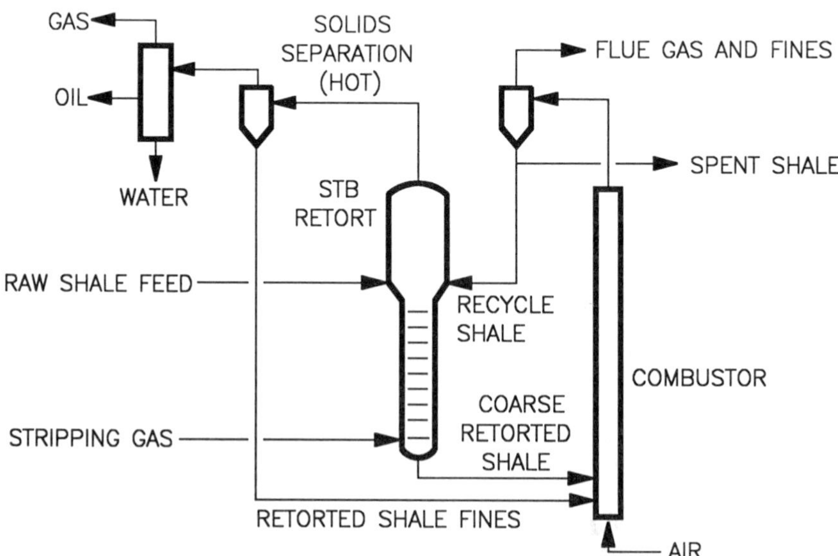

Figure 10 Process flow sheet for the Chevron STB retorting process. Preheat and heat recovery sections are not shown.

to avoid generating shale fines in the retort little turbulence is desirable. For these reasons the Shell and Exxon workers chose to avoid the turbulent bed regime. In contrast, the STB retort operates in the turbulent flow regime and staging by restricting the flow at intervals is used to approach plug-flow conditions for the solids (Piper and Ivo, 1986a). Like the Shell and Exxon retorts, the STB is a small-particle retort and hence can process the entire oil shale resource. However, top size of the shale feed to the STB retort is 1/4 in. (6.4 mm). Grinding to <1/4 in. is more expensive than the crushing required for lump shale processes but considerably less expensive than find grinding (to <2 mm) required by more conventional fluidized bed processes.

Calculations (Wallman et al., 1981) indicated that retorting of even the largest particles should be complete in about 4 min at 500°C (932°F). Experimental studies at this residence time (4 min) revealed a broad maximum in oil yield in the temperature range of 480–500°C (895–930°F). In this temperature range, yields of C_5+ oil were 100% (±3%) of the Fischer assay.

Locally, the bed of solids in the STB appears to be fluidized, but the superficial gas flow is well below that required for fluidization of the larger particles. However, rapid local mixing and good solid-solid heat transfer help avoid local overheating. This minimizes cracking and coking reactions that lower oil yields.

The STB retort is flexible with respect to fluidizing/stripping gas; it can be operated with steam, a recycle gas stream, or a mixture of the two. The superficial gas velocity is in the range of 0.3–1.5 m/sec (1–5 ft/sec) at the bottom of the retort but increases up the retort as product vapors add to the gas volume.

A combination of thermal shock as particles enter the bed, the removal of kerogen, and turbulence in the bed causes some breakage to generate fines within the retort. Particles smaller than about 200 mesh are elutriated with the product vapors, but most are recovered before the oil is condensed. These fines are rich in carbon and therefore are sent to the combustor for recovery of their fuel value.

Testing for development of the STB concept was carried out in a variety of units, including a 1 ton/day pilot plant and a 320 ton/day semiworks unit at Chevron's Salt Lake City refinery. The semiworks unit was commissioned in 1983, operated for about 2 years, and has now been demolished. The nominal operating conditions show that throughput of the STB retort is very high—in the range of 1–2.5 tons/hr/ft^2 (Table 11). Properties of oil produced from Anvil Points shale in the STB are similar to those of oils from other processes—nitrogen and arsenic contents are notably high (Table 12). It was reported that oil properties were essentially invariant with shale grade over the entire range of grades (14–38 gal/ton) studied in the STB pilot plant (Rezende, 1982).

Table 11 Nominal Operating Conditions for the STB Retort

Shale throughput	10–25 tonnes/hr/m^2 (2000–5000 lb/hr/ft^2)
Retort temperature	475–510°C (890–950°F)
Residence time in retort	2–8 min
Stripping gas velocity	0.3–1.5 m/sec (1–5 ft/sec)
Residence time in combustor	1–5 sec
Combustor outlet temperature	595–815°C (1100–1500°F)
Recycle shale:Raw Shale Ratio	2:1 to 5:1

Source: Rezende, 1982.

I. Petrosix Process

The Petrosix process was developed in Brazil, especially for processing oil shale from the Irati formation. A large demonstration plant (1600 tons/day) with a 18-ft diameter retort has logged over 82,000 hr of commercial operation and has processed over 4.6 million metric tons of Irati shale to produce about 2 million barrels of shale oil since 1972 (Piper and Ivo, 1986). Although the process was developed for the Brazilian Irati shales, the Petrosix technology has also been considered for use with the eastern U.S. Devonian shales.

The Petrosix process is indirectly heated. In the demonstration plant, a hot recycle gas stream is further heated by gas and injected into the retort to heat and retort the shale (Figure 11). Fluidized bed combustors would probably be used in the commercial Petrosix plant.

Recognizing the high capital and operating costs of fine grinding, the Brazilian workers developed Petrosix to handle large pieces of shale up to 6 in. in one dimension. However, to obtain the higher throughput possible with faster heat-up, secondary crushers and screens are used to provide shale in the size range of -2 in. to +1/4 in. Fines that pass the 1/4-in. screen are briquetted and added to the retort feed.

Crushed shale is fed into the retort at the top through a feeder designed to prevent horizontal segregation. The shale moves downward

Table 12 Properties of a Typical Oil From the STB Retort Pilot Plant[a]

Specific gravity	0.934	
Carbon, wt %	85	
Hydrogen, wt %	11	
Nitrogen, wt %	2.1	
Oxygen, wt %	1.2	
Sulfur, wt %	0.6	
Arsenic, ppm	20	
Viscosity, cs at 100°F	22	
cs at 130°F	12	
Pour point, °F	19	
Ramsbottom carbon, wt %	3.5	
Distillation, ASTM D1160		
Vol % distilled	*°C*	*°F*
IBP/5	146/196	295/385
10/20	220/273	428/523
30/40	322/371	612/700
50	411	772
60/70	452/489	846/912
80/90	529/–	984/–
95/EP	–/534	–/993

[a]Oil from 27 gal/ton (93 liter/tonne) shale from the Anvil Points Mine.
Source: Rezende, 1982.

through drying, heating, and retorting zones against an upflowing stream of heated recycle gas. The retorted shale then moves down to the lowest section of the retort, where it is cooled by an unheated recycle gas stream before being discharged through one of the hydraulically sealed spent shale hoppers.

Gases and product vapors are carried out of the top of the retort by the recycle gas stream and pass successively through a cyclone for fines removal, an electrostatic precipitator for collection of heavy oil mist, heat exchangers where the light oil and water are condensed, and into the gas treatment section for H_2S removal and recovery of light naphtha and LPG (liquefied petroleum gas). Noncondensible gases are compressed; part is used to cool the spent shale, part is heated and

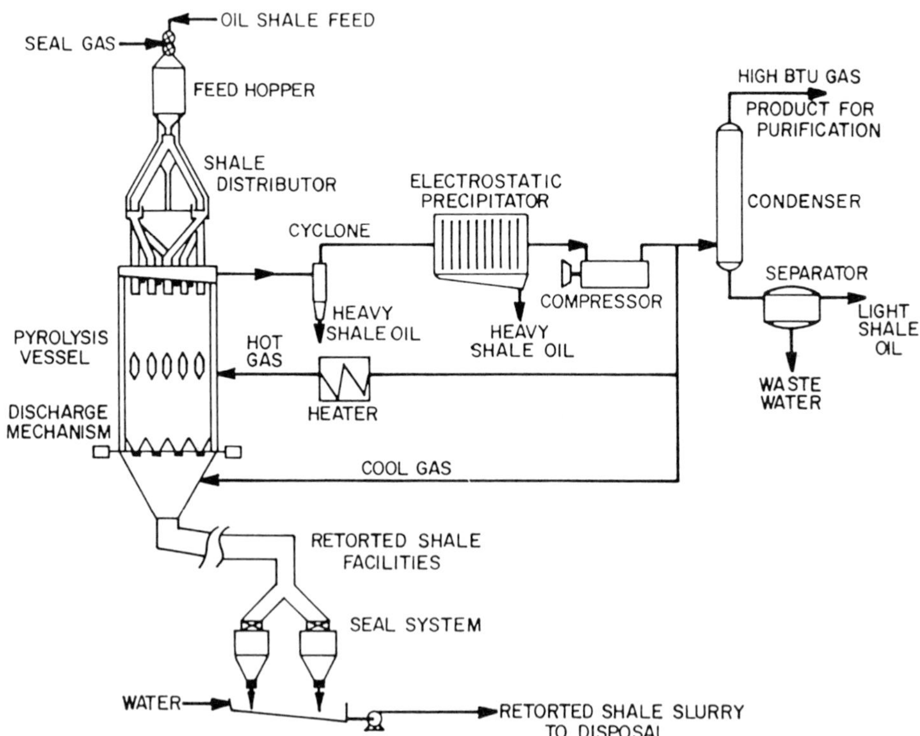

Figure 11 Schematic representation of the gas combustion retort for oil shale processing.

injected into the retorting zone, and the balance is used as fuel. Sulfur is recovered in a conventional Claus plant.

Atmospheric and vacuum distillation are used to recover distillate from a composite of the heavy and light oils. The vacuum bottoms will be used as fuel within the oil shale complex. Most oil shale facilities planned for the United States have included on-site hydrotreating to stabilize the crude shale oil before pipelining. In contrast, Petrosix plans to pipeline a composite of distillate and naphtha to a refinery for hydrotreating. The hydrotreated shale oil will then be refined along with petroleum crudes.

Relatively little information is available on the properties of Petrosix shale oil; however, some properties of oil produced from Irati shale in the demonstration plant are given in Table 13.

The 36-ft-diameter (11-m) commercial retort is designed to retort 260 tons of oil shale per hour to produce 2600 bbl of crude shale oil

Table 13 Some Properties of Petrosix Oil Retorted from Irati Shale

Density, °API	19.6
Pour point, °F	25
°C	-4
Aniline point, °F	86
°C	30
Viscosity at 100°F, cs	20.76
Sulfur, wt %	1.06
Nitrogen, wt %	0.85
Paraffin, wt %	0.02
Diolefins, wt %	15.0

Source: Data of Bruni, 1968.

and 52 tons of sulfur per day. A commercial facility, using 20 such retorts to produce 50,000 bbl/day of shale oil, was originally planned. Projected cost of this facility would have been $2.2 billion (1982 U.S. dollars). However, in 1982 the project was scaled back to a single retort in response to the world petroleum situation. During the mid 1980s, construction of the single retort was slowed to free Brazilian funds for offshore petroleum exploration and the unit is now scheduled to come on-line during 1990. Cost of the single unit will be about $57 million (1986 U.S.), and it is being constructed at Sao Mateus do Sol, in order to share existing facilities with the demonstration plant (Piper and Ivo, 1986).

J. Moving Grate Processes

In the earlier discussion of the Shell fluidized bed retort (SSRP), it was mentioned that both short vapor residence time (to minimize cracking) and long solids residence time (to obtain complete kerogen conversion) could be achieved in a moving bed, provided the bed height were small and the bed area very large. The Allis–Chalmers, Dravo, and Superior Oil moving grate retorts were designed to meet these criteria. In each case, a support grate moves through zones where the oil shale is loaded to form a dense bed, preheated, retorted, cooled, and discharged. There are, however, differences in grate design, degree of bed agitation, and the way gas flows are controlled to remove products and generate heat. Because these three processes are so closely related in concept, they will be discussed as a group.

1. Allis–Chalmers Roller Grate Process

The Allis–Chalmers process grew out of experience in design and construction of large iron ore and cement plants on the scale of 10,000 tons/day, about one-fifth the size needed for a 50,000 bbl/day shale oil retorting plant. In this process, the shale is conveyed along a straight-line path by a series of closely spaced slotted rollers (Figure 12).

Raw shale crushed to 1-3/4 × 1/4 in. is fed into the roller grate. Fines (-1/4 in.) are screened from the feed, agglomerated, and fed on top of the bed of crushed shale. Bed depth can be up to 3 ft. The rollers impart a mild tumbling motion to the shale particle. This has two important benefits in retorting: by exposing new surface to the

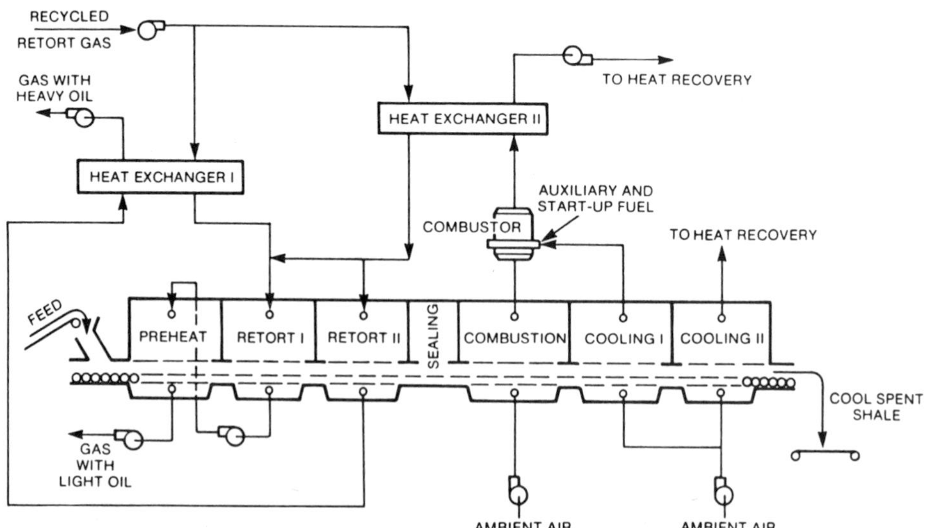

Figure 12 Flowsheet of the Allis–Chalmers oil shale retorting process (reprinted with the permission of Allis–Chalmers Corporation).

hot gas sweep it speeds heat-up and causes fines to quickly migrate down through and out of the bed, thereby preventing pockets of fines that would cause gas channeling.

In the preheater, the shale is heated by off-gas from the retorting zone. This dries the shale and liberates some light oil. Next the shale moves into the first retorting zone where it is further heated by recycled noncombustible retort gases in the temperature range of 480–540°C (900–1000°F). The heating gas flows down through the shale bed and slotted rollers, sweeping liberated oil into the preheat zone where it is partially condensed. In the second retorting zone, a 650°C (1200°F) gas stream completes the retorting process to give a heavy oil that is taken through a heat exchanger and condenser.

Sealing between the retorting and combustion zone is accomplished by the use of solid rollers (instead of slotted ones), drag plates at the entrance and exit of the sealing zone, and careful maintenance of equal overbed pressures.

In the combustion zone, air flow is upward through the grate and shale bed in order ot keep the grate temperature as low as possible. The two cooling zones are kept separate, since the hotter gases from the first cooling zone are combined with the hot off-gas from the combustion zone and passed through a heat exchanger where their heat is transferred to the noncondensible retort gas stream used in the retorting section.

The Allis–Chalmers process was tested in a process development unit (PDU) 18 in. wide and 8 ft long. In this unit, less than 0.2% of crushed Green River oil shale was carried out as dust, while about 1% was captured in the windbox. Size analysis showed that little size reduction occurred during retorting. However, the oil collection system of the PDU was not designed for efficient capture of light oils. As a result it is not possible to draw firm conclusions regarding the overall performance of this system (Faulkner et al., 1983).

2. Dravo Circular Traveling Grate Process

The Dravo retort also comes from a firm that is well established as a leader in materials-handling equipment and processes for the mining and mineral industries. The retort design is similar to traveling grate machines supplied by Dravo for the sintering or pelletizing of iron ore. The grate is a continuous chain of wheeled pallets that can accommodate a bed of shale up to 95 in. thick as it travels around a circular track. The shale would be crushed and screened to give a -1-in. by +1/4-in. shale feed. Unlike the Allis–Chalmers roller grate, the shale bed is not agitated on the Dravo grate. This minimizes the generation of fines during retorting. Some of the fines generated in crushing could be burned if needed to heat the recycle gas streams. The remainder and oil fines collected throughout the process would be agglomerated and fed as pellets or briquettes onto the top of the shale bed. This capability enables processing of the entire resource and avoids the expense of environmentally acceptable disposal of oily fins.

Retorting takes place in four zones (Figure 13) (Forbes et al., 1984). In the first zone, the shale is heated by oxygen-free gases produced by combustion of gas recycled from the heat recovery section (natural gas is used for startup and additional natural gas is added if needed to maintain temperature). The upper 20–30% of the bed is retorted in this zone.

A major part of the process heat is generated in the second zone, which is fed with a mixture of recycle gas and air to burn the carbon in the spent shale. As the combustion front moves down through the shale bed, the retorting front moves ahead of it to retort the middle portion of the bed. The amount of air is controlled to limit combustion

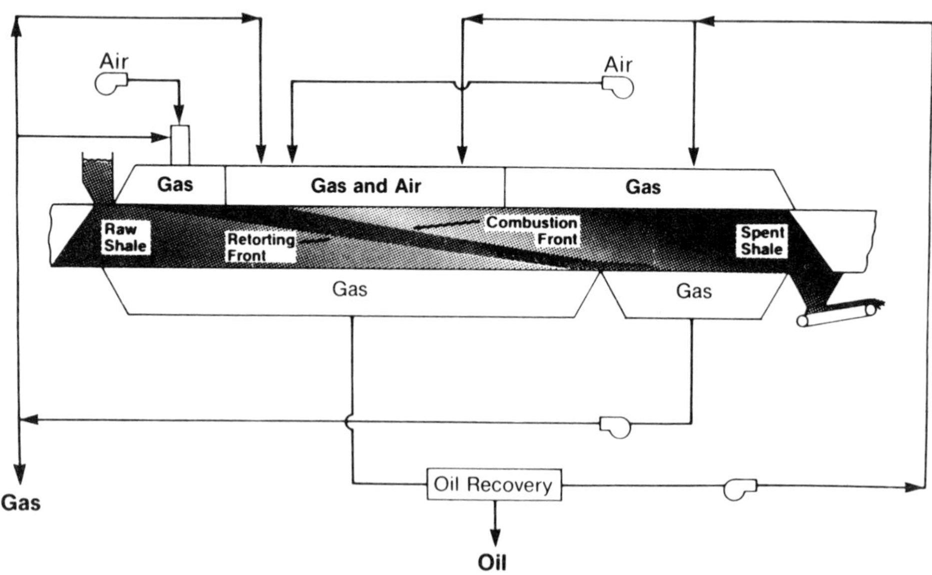

Figure 13 Simplified flow schematic of the Dravo process (reprinted with the permission of Dravo Engineering Companies, Inc.).

and prevent breakthrough of oxygen-rich gas into the product collection system. In zone 3, oxygen-free gases are used to transfer heat from the hot combusted shale in the upper part of the bed to the cooler shale at the bottom of the bed. Thus, the bottom of the bed is retorted adiabatically. Finally, oxygen-free gases cool the bed to below 120°C (250°F) before the grate is tipped to dump the spent shale into a water-sealed discharge hopper.

The Dravo process was tested in lab units and in an integrated 300 ton/day pilot plant (Figure 14). The use of separate windboxes in the pilot plant allowed demonstration of a key feature of the Dravo process, i.e., isolation of the process steps by varying gas flows to control local pressures above the various windboxes.

In the oil recovery section, oil and water are condensed from the retort off-gas, using direct contact quench towers or air heat exchangers.

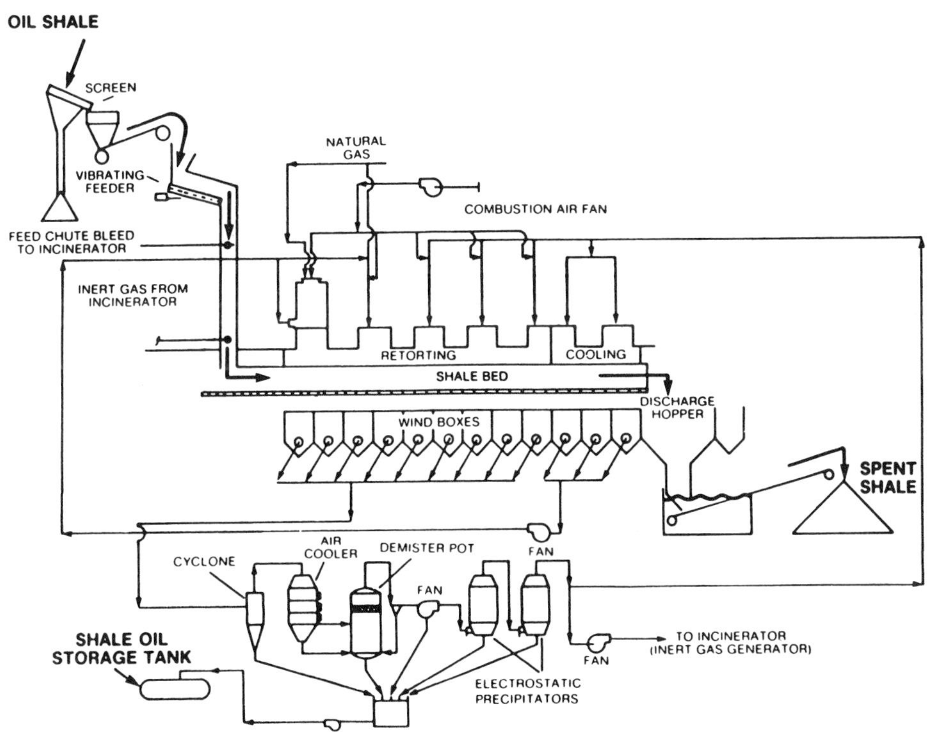

Figure 14 Flowsheet for the integrated 300 ton/day Dravo pilot plant (reprinted with the permission of Dravo Engineering Companies, Inc.).

Heavy oil mist is recovered by electrostatic precipitaion. Desalting and deashing of the raw shale oil is accomplished using (Petrolite) technology developed for petroleum applications. Overall oil recoveries in the range of 95–100% of Fischer assay are reported, depending on the grade and type of shale fed. Western (Colorado) and eastern (Kentucky Devonian Sunbury and Cleveland High Grade) U.S. shales and Australian shales have been tested in the Dravo 300 tons/day pilot plant. In each case, excellent operability was reported.

The quality of shale oil obtained from the Dravo retort is very similar to that obtained from the Paraho retort operating with direct heating. Pour point is high, as is the content of nitrogen, arsenic, iron, and nickel. In several cases, samples of raw shale oil were sent to Gulf Oil for upgrading via two-stage hydrotreating (Jones et al., 1984). Details of the Gulf results were not discussed by Dravo.

3. Superior Oil/Davy-McKee Circular Grate Retort

The Superior Oil/Davy-McKee retort was developed as a joint effort of Superior Oil (subsidiary of Mobil Corporation) and Davy-McKee, another firm well established in the materials-handling and mineral-processing field. The retort is not designed just to recover fuel; it is but one part of the Superior Multi-Mineral approach to efficient oil shale utilization.

The distinctive feature of the Superior Oil/Davy-McKee retorting process is the way crushed shale is fed onto the grate in three layers; the smallest particles (+1/4 in., the -1/4-in. material is not retorted but may be burned in a separate unit) are fed directly onto the slotted grate, intermediate-sized particles form the center layer, and the largest particles are placed on top of the bed (Figure 15). As a result, the largest particles that heat up slowest are exposed to the hottest recycle gases for the longest time. This minimizes the residence time needed to achieve complete retorting and has the added benefit of minimizing overheating, which would be detrimental to subsequent recovery of soda ash and aluminum trihydrate. Laboratory tests carried out in an adiabatic batch reactor were followed by testing of the retorting step in a 250 tons/day pilot plant. Product quality from Colorado oil shale retorted in the Superior retort seems somewhat inferior to that from the Paraho retort.

K. Paraho Process

Above were discussed several new approaches to oil shale retorting that resulted from attempts to overcome the shortcomings of the USBM gas

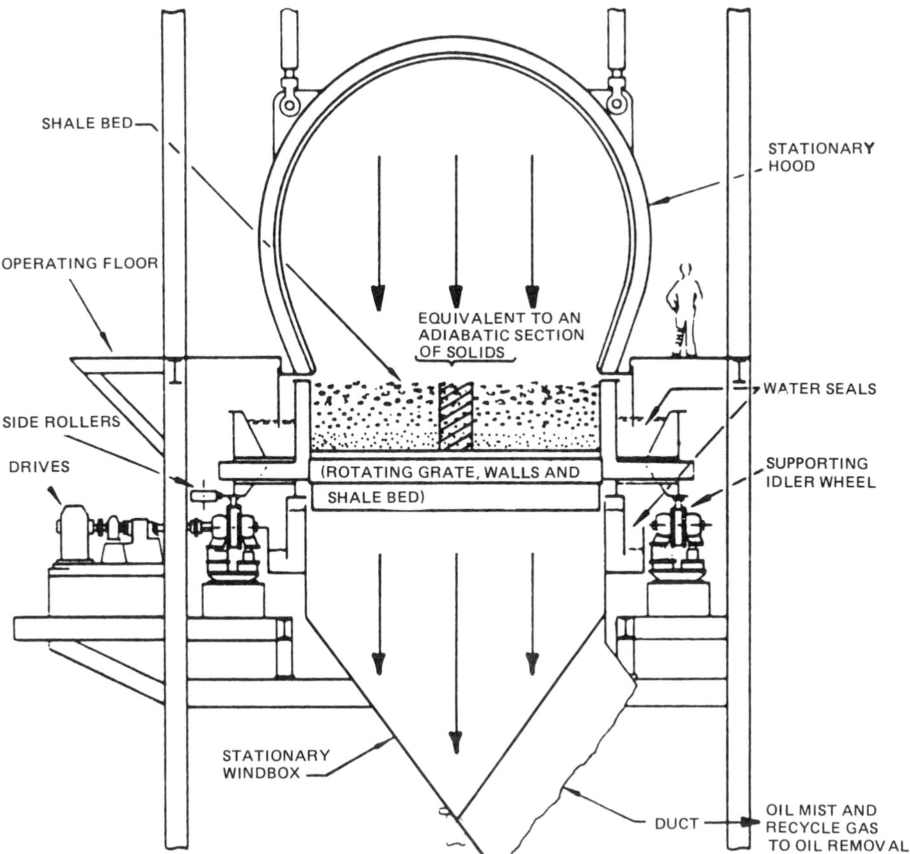

Figure 15 A cross-sectional view of the Superior Oil/Davy-McKee circular grate retort shows the distinctive layering that puts the largest shale particles on top of the bed where they are heated quickest. This view also shows how simple water troughs provide effective seals that are tolerant of both thermal distortion and dimensional variance in construction (Weichman, 1976; reprinted with the permission of the Colorado School of Mines).

combustion retort: inability to process fines and hence to process the entire resource, and oil yields substantially below Fischer assay. In contrast, the Paraho retort and retorting process seem best described as evolutionary. The design philosophy was outlined in 1981 by Pforzheimer, who listed his criteria for a good oil shale retort (Pforzheimer, 1981):

Countercurrent flow of raw shale and heat transfer medium.
Gravity feed of raw shale through the retort and out.
Lumps in–lumps out (minimal crushing and no fine grinding).
Direct heating–low btu product gas used as plant fuel for electric power generation.
Simple configuration.
It works.

In 1981 Paraho already had demonstrated ability to operate their retort for long times with yields approaching 100% of Fischer assay on a variety of oil shales. A comparison of these 1981 criteria with those recently set out by developers of the Shell shale retorting process (SSRP) (Voetter et al., 1987) provides some insight into the profound changes a few years have wrought in oil shale R&D.

Paraho was a small, privately held, publicly reporting corporation formed in 1971 to commercialize oil shale technology. Organization of the Paraho oil shale demonstration began in 1973 after Sohio and Cleveland Cliffs Iron Company had surveyed available technologies and concluded that Paraho's was the most promising. Eventually, 17 companies (Sohio, Cleveland Cliffs Iron, Southern California Edison, Kerr-McGee, Gulf, Sun Oil, Shell, Amoco, Exxon, Mobil, Webb Venture, Davy-McKee, Texaco, Marathon, ARCO, Phillips, Chevron) participated in this Paraho-managed project located at Anvil Points, Colorado. Based on the results, a 10,000 bbl/day commercial facility, the Paraho-Ute Project, was planned for a location on the White River near Vernal, Utah. Sohio led a group providing financial backing. However, the demise of the Synthetic Fuels Corporation removed the possibility of loan guarantees, thereby increasing financial risk. As a result, Sohio and its partners withdrew their support for the project and late in 1985 Paraho Development Corporation filed for protection under Chapter 11 (*Platt's Oilgram News*, 1985). The company has since emerged from bankruptcy as the New Paraho Development Corporation and is developing nonfuel uses for oil shale (*Platt's Oilgram News*, 1987). The Paraho operations are an important part of the oil shale story: the retort did work, Paraho addressed many of the pressing environmental issues in spent shale disposal, and much of the readily available information about oil shale upgrading and refining comes from work on 100,000 bbl of Paraho shale oil that was refined at Sohio's Toledo refinery under a U.S. Navy contract and from a variety of studies on comparable Paraho oil that were carried out elsewhere in smaller units.

The Paraho technology is a modification of a commercially available lime-burning kiln (Jones and Reeves, 1971) in which the key component in the design is the discharge grate which ensures even solids flow down the retort. An artist's rendition of the Paraho oil shale retort appeared in Sohio's 1973 *Annual Report* (Figure 16).

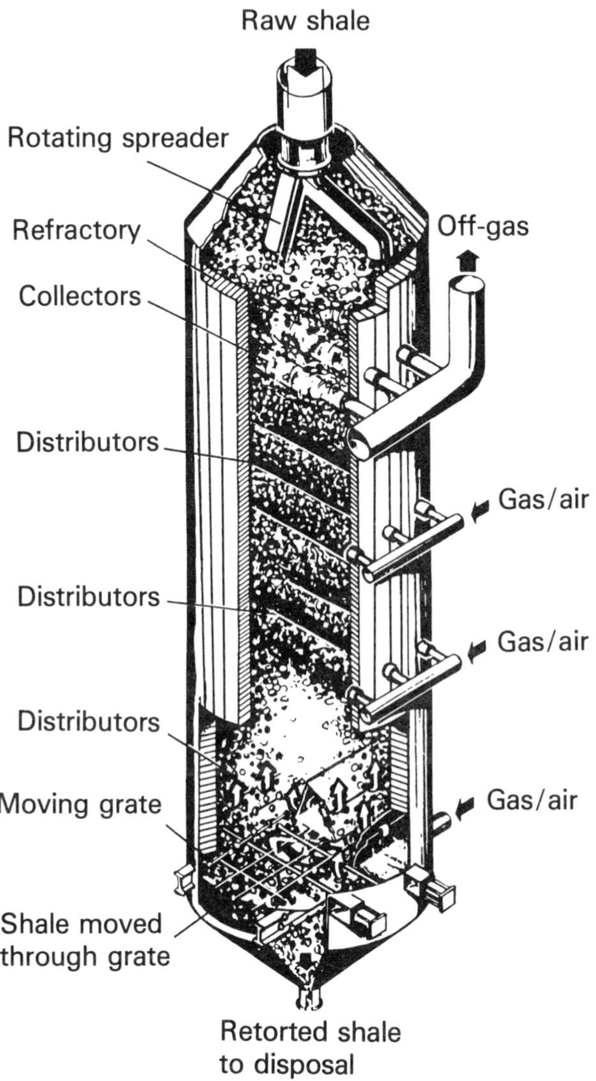

Figure 16 Cutaway view of the Paraho retort (1973 Annual Report, Standard Oil Company of Ohio; reprinted with permission).

The Paraho semiworks retort used in the demonstration project was a refractory-lined open tabular retort. Coarsely crushed shale is fed into the top of the retort by a rotating feeder (to prevent size segregation) and flows down against the gas flow. This retort was designed so it could be operated with either direct- (Figure 17) or by indirect-heating (Figure 18; Jones, 1976).

With indirect heating, the Paraho retort is similar to the gas combustion retort in concept, but oil yields are substantially better, approaching 100% of Fischer assay (Table 14). With indirect heating, gross oil yields are similar, but thermal efficiency is lower because more carbon is rejected with the spent shale. No facility for external combustion of the spent shale was provided at the Paraho semiworks stage.

Water and energy requirements for the Paraho process have been discussed extensively (McKee and Kunchal, 1976). Water requirements were of particular concern, since water is a precious commodity in northwest Colorado and northeast Utah (also in many other shale-

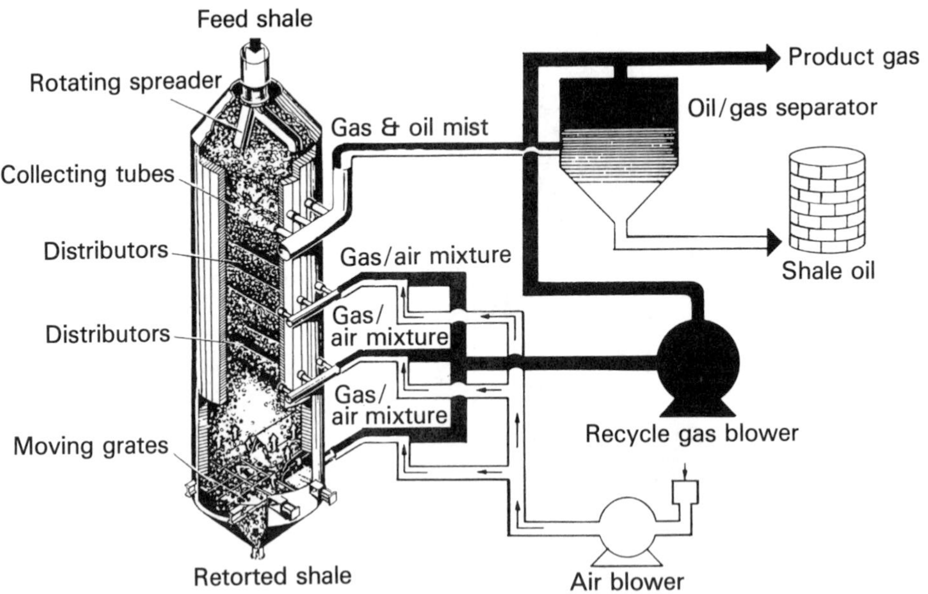

Figure 17 Paraho semiworks retort operating in the directly heated mode (Jones, 1976; reprinted with the permission of the Colorado School of Mines).

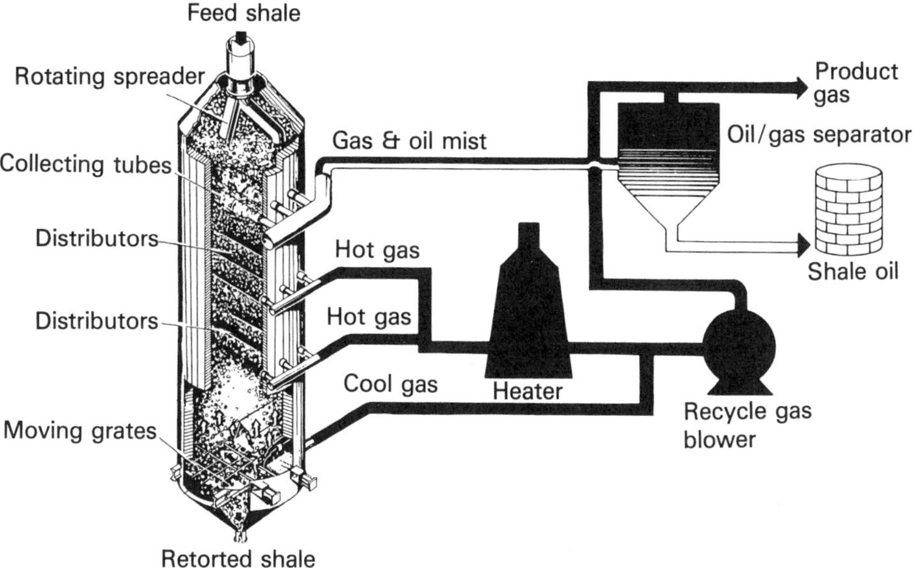

Figure 18 Paraho semiworks retort operating in the indirectly heated mode (Jones, 1976; reprinted with the permission of the Colorado School of Mines).

rich areas). The retort section of the Paraho process is a net producer of water, albeit sour water, in the direct mode. Water in the shale is recovered as is the water of combustion. With the retort in this mode, the overall process requires about 2.1 gal of water for each gallon of shale oil. The majority of this water goes to the cooling tower with smaller amounts required for refining, power generation, dust control, and vegetation uses.

Scale-up issues and the prospects for commercialization were discussed in several articles as the Paraho technology advanced (Pforzheimer, 1981). The commercial design consisted of 40 rectangular sections, each roughly equal in area to the Paraho semiworks unit, and with the rotating feeder of the semiworks unit replaced by 320 stationary rock hoppers fed by a moving conveyor (Figure 19). A more detailed cutaway view (Figure 20) of one of the 400 ft^2 Paraho retort emphasizes the overall nature of the retort as well as the internal complexity.

Table 14 Product Yields From the Paraho Retort in Directly and Indirectly Heated Modes

	Directly heated	Indirectly heated
Raw shale, Fischer assay, gal/ton	28.0	28.0
Yields: Shale oil, gal/ton	27.2	27.2
Vol % of Fischer assay	97	97
Oil quality		
Gravity, °API	21.4	21.7
Pour point, °F	85	65
°C	29	18
Viscosity, SUS at 130°F (54°C)	90	68
Ramsbottom carbon, wt %	1.7	1.3
Water content, wt %	1.5	1.4
Solids content, B.S. wt %	0.5	0.6
Gas properties, vol % (dry basis)		
H_2	2.5	24.8
N_2	65.7	0.7
O_2	0	0
CO	2.5	2.6
CH_4	2.2	28.7
CO_2	24.2	15.1
C_2H_4	0.7	9.0
C_2H_6	0.6	6.9
C_3's	0.7	5.5
C_4's	0.4	2.0
H_2S	2660 ppm	3.5
NH_3	2490 ppm	1.2
Heating value, HHV, btu/scf	102	885

Source: Data of Jones, 1976.

L. Unocal Processes

Unocal (nee Union Oil) has been active in oil shale research and development since its acquisition of Colorado oil shale properties in the 1920s. The three closely related Unocal processes are known as the retort A, retort B, and SGR (steam-gas recirculation) oil shale processes. All three processes move shale upward by means of a "rock pump," a reciprocating mechanical piston that alternately loads oil

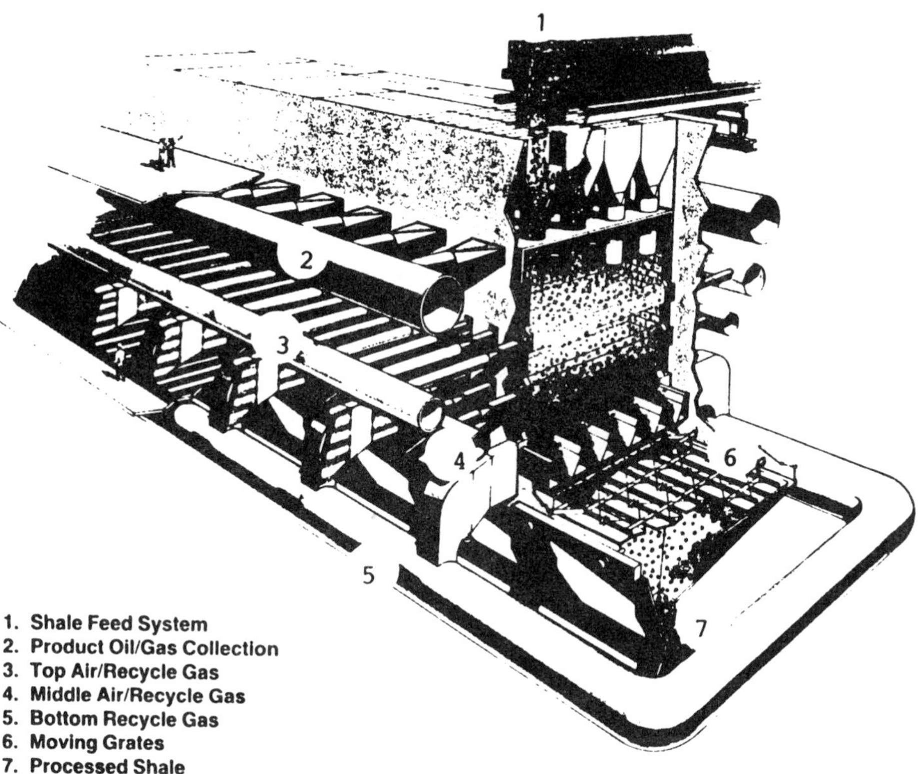

Figure 19 Cutaway view of the 10,000 bbl/day commercial retorting module proposed for the Paraho-Ute project (Pforzheimer, 1981; reprinted with permission of the Colorado School of Mines).

shale from a feedhopper and rams the loaded shale up into the bottom of the cone-shaped retort (Figure 21) (Berg, 1950). In all arrangements, the rock pump piston operates totally immersed in relatively cool product oil.

Technical feasibility of the rock pump concept was demonstrated during the 1940s in a 2 ton/day pilot retort. This work was followed by construction and operation of a 50 ton/day unit and scaled up to a semiworks unit at Unocal's Parachute Creek site in Colorado during the late 1950s (Hartley and Brinegar, 1959). This semiworks unit, termed retort A, had a feed piston diameter of 5.5 ft affording a shale pumping rate of 1200 tons/day (Figure 22). Activities through the 1960s

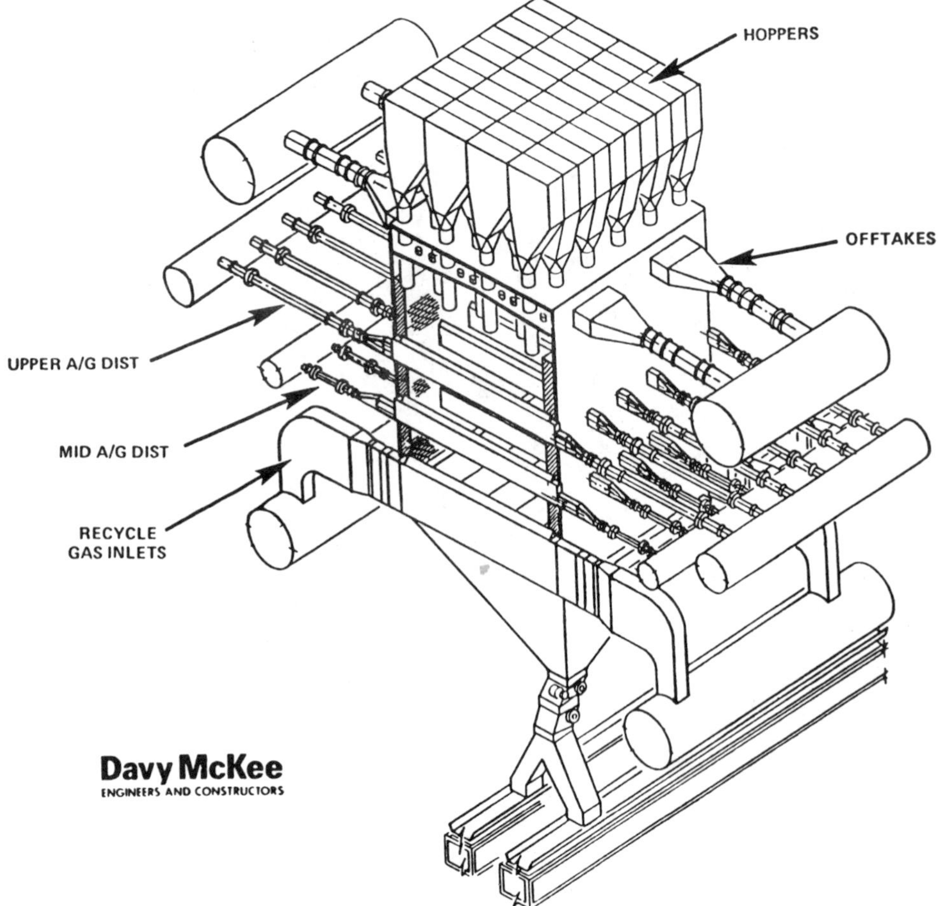

Figure 20 A cutaway view of one 400 ft^2 Paraho retort submodule shows its internal compexity. Eight such units are combined to form the commecial retort module (Greaves, 1981; reprinted with the permission of the Colorado School of Mines).

were at a low level due to unfavorable economics for shale oil, but were resumed in the early 1970s leading to retort B, an indirectly heated version that gave higher oil yield and a high-btu gas product. An attempt to improve thermal efficiency by burning the carbon rejected with spent shale from retort B led to the steam-gas recirculation (SGR) concept (Duir et al., 1983). In 1981, Unocal began construction of an

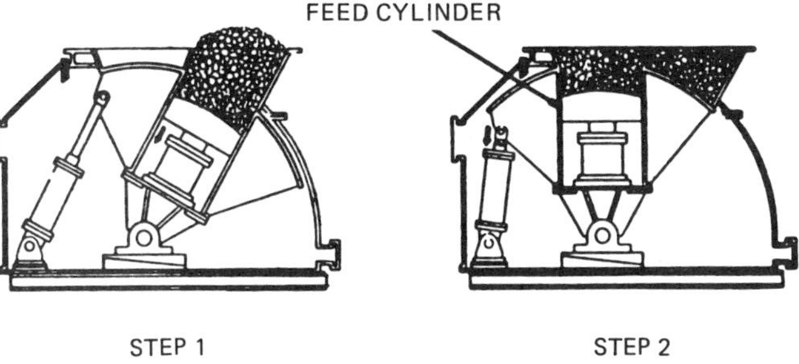

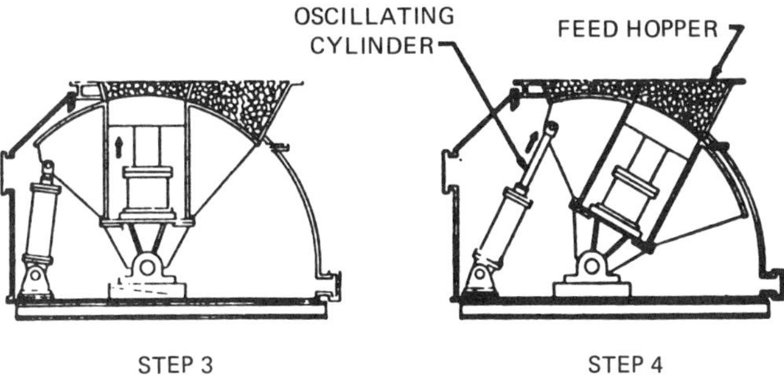

Figure 21 The rock-pump is a reciprocating piston that alternately loads oil shale from a feedhopper and rams the loaded shale up into the bottom of the retort (after Hartley and Brinegar, 1959).

integrated 10,000 bbl/day shale oil facility (mine, retort B, upgrading facility) at Parachute Creek. This facility was placed in service in 1983 and with continuing development is the only commercial shale oil facility now operating in the United States.

As solids are pumped upward through the conical retort A, air is blown in to burn the carbon on the spent shale. The amount of air is limited to confine combustion to the spent shale in the uppermost part of the moving shale bed but is sufficient to complete combustion within the retort. The hot combustion gases continue downward, retorting the shale in the central part of the cone and picking up product vapors. As the gases laden with product vapors contact the cooler shale in the lower part of the bed, most of the heavy oil condenses and trickles down through the shale bed; the balance forms a mist. The cooled stream of combustion gases, light product vapors, and mists exits the retort through the slots at the bottom and pass into the

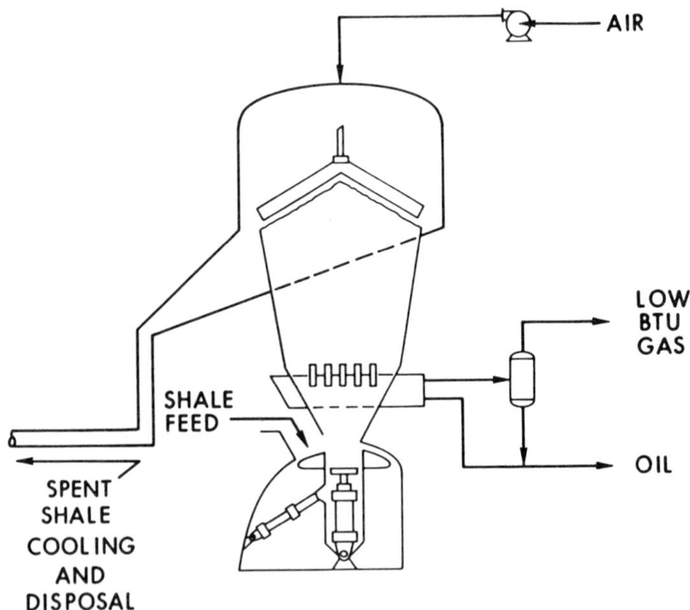

Figure 22 Retort A incorporated the "rock-pump" into an internally heated, solids upflow design. A limited amount of air is blown into the top of the retort to support combustion of the spent shale in the uppermost part of the moving shale bed (reprinted with the permission of Unocal, Inc.).

light product recovery section. A disengaging section surrounds this part of the retort cone and is sealed by product oil. As the retorted shale rises above the retort cone, it forms a freestanding pile. Spent shale falls off the pile by gravity and falls down discharge chutes through the dome wall into a vessel where it is cooled by a water spray. The steam generated during cooling helps to strip products from pores in the spent shale.

Condensation of the product oil within the retort is thermally efficient and significantly reduces the need for external heat exchangers and condensers, as well as the need for external cooling capacity for product recovery. Also, a filtering action that gives the product oil a relatively low solids content occurs as the condensed oil trickles down through the incoming shale.

Rich shales tend to become plastic and agglomerate during retorting. This agglomeration leads to uneven gas and solid flows, or even plugging in extreme cases, in retorts where solids flow is gravity-driven. Because there is always a positive mechanical force available to move the shale up through the retort, anything more than local agglomeration within the bed is avoided. A slowly revolving rake is provided to break up any agglomerates that do form. Also, conditions can be selected so that retorting takes place near the top of the bed where interparticle pressures are low; hence the tendency to agglomerate is lowest. Thus, agglomeration is not generally a problem, even with very rich shales.

In retort A, with once-through air as combustion gas, peak temperatures at the top of the retort reached 1095–1150°C (2000–2200°F). As a result, cracking limited the oil yields to about 75% of Fischer assay. Moreover, heating value of the gas product was low, about 120 btu/scf, because of dilution with nitrogen and CO_2 from combustion and carbonate decomposition.

Efforts to improve oil yield and heating value of the gas product led to the development of the second generation concept, retort B and the Unishale B process (Figure 23). In this case the heat required for retorting is provided indirectly using a recycle gas stream heated to 510–540°C (950–1000°F) in an external furnace. Rundown oil yields are high, essentially 100% of Fischer assay, and the C_4+ oil yields are significantly above Fischer assay. Moreover, the gas product has a high heating value of over 800 btu/scf.

In retort B, the space above the freestanding pile of spent shale is enclosed by a dome to exclude air. Spent shale falls off the retort by gravity and down discharge chutes through the dome wall into a vessel where it is cooled by a water spray. Steam generated during cooling helps to strip products from pores in the spent shale. The coolied shale is then moistened before being discharged.

As in retort A, the retort B shale oil and the gas stream containing product gas, vapors, and mist exit at the bottom of the retort through slots that lead into the disengaging section. Gases from the disengaging section are scrubbed and cooled. Oil and water collected as mist and condensate are separated. The oil is returned to the disengager, while the water is used to moisten spent shale. Part of the scrubbed gas is compressed and heated before being recycled to the retort. The balance is processed by compression and scrubbing to remove heavy ends and sweetened using the Unisulf process to remove H_2S. The sweetened gas is then suitable for use as plant fuel. However, hydrogen requirements for shale oil upgrading are substantial and the made gas is rich in hydrogen (Table 15). As a result, fractionation to recover this hydrogen may be desirable.

Retort B produces a very high-quality shale oil (Table 16). Pour point, a very important parameter if the oil is to be pipelined, is notably low and the Conradson carbon value also has a very desirable

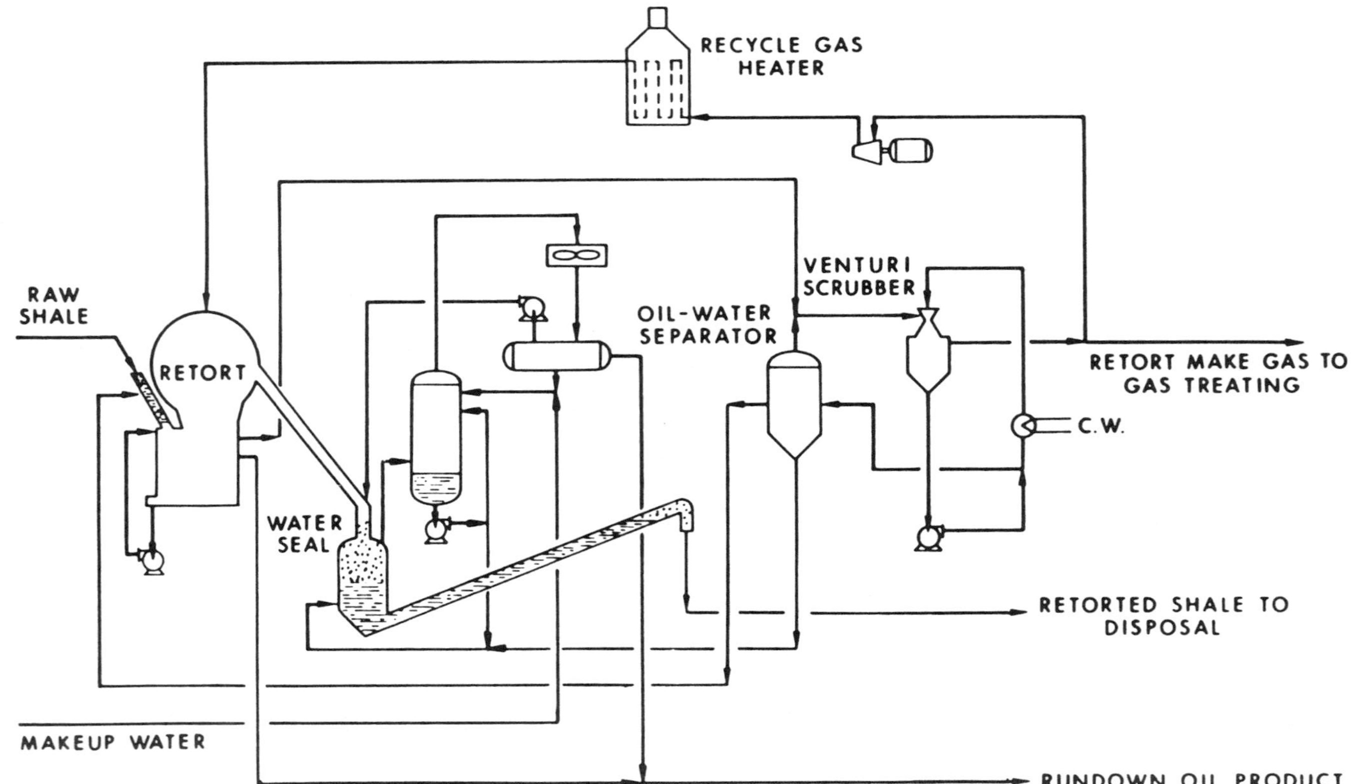

Figure 23 In the Unishale B process, retort B incorporates the rock-pump but is indirectly heated (reprinted with the permission of Unocal, Inc.).

Table 15 Unocal Retort B Make Gas Properties (Dry Basis)

	Mol%
H_2	25
Methane	24
C_2's	10
C_3's	8
C_4's	5
C_5's	2
C_6^+	1
CO	5
CO_2	16
H_2S	4
Total	100
Heating value, gross btu/scf	980 (37 MJ/m^3)

low value. Treatment of this crude shale oil involves water washing (two stages) to remove solids, removal of chemically bound arsenic to a level of 1 ppm (using a proprietary absorbent), and stripping of light ends to stabilize the oil. Unocal plans to upgrade the crude shale oil by hydrotreating to produce a syncrude that is a premium quality feedstock for a conventional refinery. Thus, the high quality of the retort B shale oil will eliminate the need for the coking step used to remove refractory material in upgrading retort A shale oil. Shale for the Unocal processes is crushed in two stages to <2 in. No fine grinding is required and little of the mined shale is discarded.

Water is a very valuable commodity in the area of the Green River deposit. Therefore, it is important that water requirements be low. Unocal estimates that retort B process water requirements will be only 1–2 bbl/bbl of shale oil. Water needs will go up somewhat if an SGR module (*vide infra*), or a shale oil upgrading facility is added, but should still be relatively low. By way of comparison, the water required for producing synthetic oil from coal would amount to 6–8 bbl/bbl oil, and the water requirements of an oil-fired steam generating plant are about 10 bbl/bbl oil. Thus, shale oil production has relatively low water requirements.

Table 16 Crude Shale Oil From the Unocal Retort B Has Both a Low Pour Point and a Low Conradson Carbon Value

Gravity, °API	22.2	
ASTM D-1160 distillation	°F	°C
IBP	150	66
10%	390	199
50%	770	410
90%	1010	543
Max	1095	590
Nitrogen, wt %	1.8	
Oxygen, wt %	0.2	
Sulfur, wt %	0.8	
Water (Karl Fischer), wt %	0.2	
Arsenic, ppm	50	
Conradson carbon, wt %	2.1	
Pour point, °F	60	
°C	15	
Heating value, gross M btu/gal	142	

From a nominal 34 gal/ton Green River oil shale, the Unishale B process recovers 87% of the available energy as oil and gas, but rejects 13% as coke. Some of the oil and gas produced must be burned to provide heat for retorting. If the retorted shale were burned to recover its energy, more oil and gas would be available for sale and thermal efficiency would rise. The steam-gas recirculation (SGR) concept is Unocal's way of achieving this goal of higher thermal efficiency. In the original SGR concept, carbon on the spent shale was gasified with steam to produce a hot synthesis gas stream that provided heat for retorting. Carbon conversion was excellent, but carbonate decomposition was extensive leading to a high CO_2 content (up to 60 mol %) in the recycle gas. Further work led in 1977 to SGR-3, an add-on that replaces the recycle gas heater for retort B to avoid these shortcomings (Duir et al., 1977). In the SGR-3 process, carbon on the spent shale is burned in a vessel isolated from the retort by a steam seal to prevent dilution of the made gas by flue gas (Figure 24). As a result, the product slate obtained using SGR-3 is identical to that given above for retort B. Heat from combustion is recovered from the hot flue gases to heat the recycle gas stream for retorting and to raise steam for other process heat needs, including most of the power plant requirements.

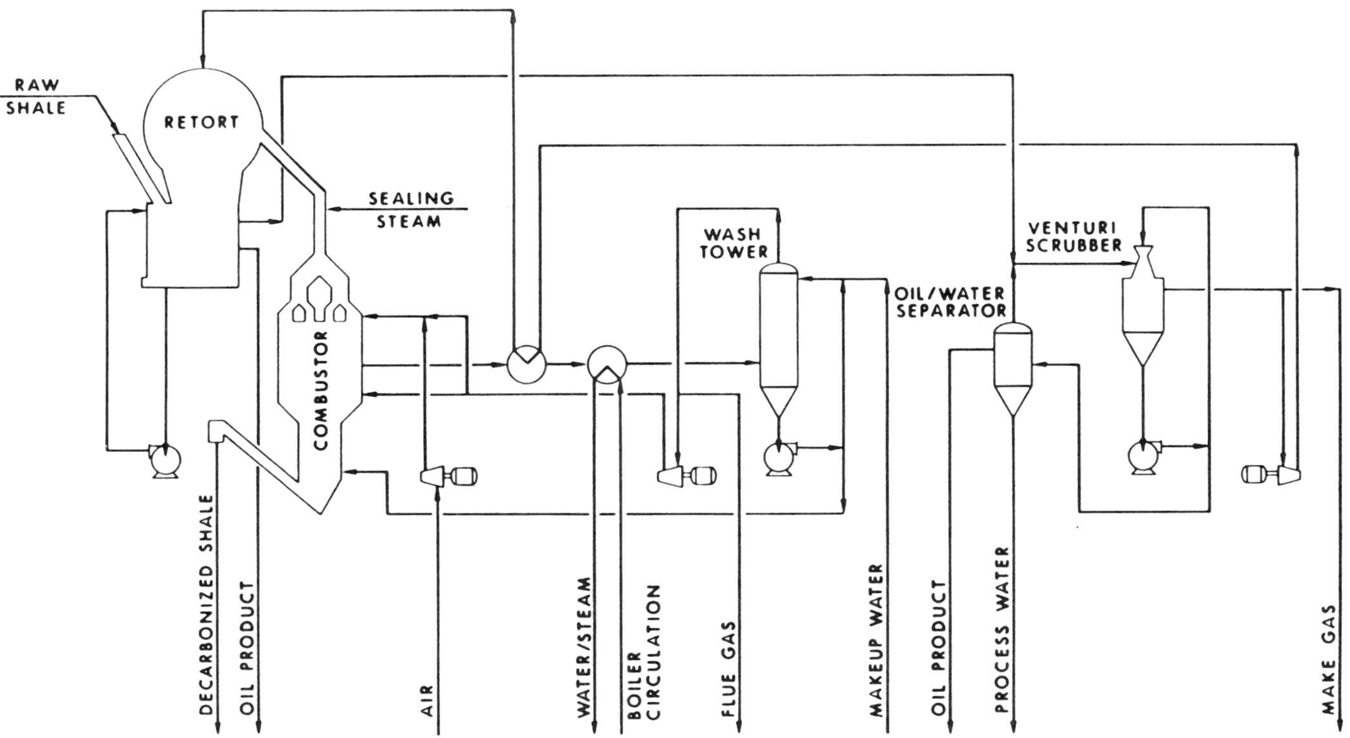

Figure 24 SGR-3 process flowsheet (after Duir et al., 1977).

Typical operating conditions for SGR-3 include a temperature of 1550°F at the combustor outlet (Table 17). At this temperature, endothermic carbonate decomposition will consume an appreciable fraction of the heat produced, but apparently such a high combustor temperature is necessary to heat the recycle gas stream to the 535°C (1000°F) required for efficient retorting. The combustor temperature is, however, maintained well below the fusion/clinkering temperature of 1800°F by varying the flows of air and flue gas recycle to the combustor. Flue gases from the combustor are very low in SO_x because of scrubbing action of the basic calcium-containing components, but treatment to lower the high NO_x levels will probably be required to meet emission standards (Table 18).

A comparison shows that the SGR-3 add-on substantially improves thermal efficiency over that achieved with retort B alone (Table 19). Not shown in Table 19 are the higher capital and operating costs, and the added complexity of operation, associated with the SGR-3 add-on. These factors will at least partially offset the higher thermal efficiency of SGR-3.

In summary, Unocal's shale upflow retorting technology has several strong points:

Oil vapors liberated from the shale are quickly carried down away from the retorting zone and cooled. This limits polymerization and cracking that lead to undesirable gas and heavy oil.

Table 17 Typical SGR-3 Pilot Plant Operating Conditions for 36 Gal/Ton Green River Oil Shale

Shale feed rate, tons/day	3.0
Shale size, in.	+1/4–1
Pressure—retort B top, psig	3.0
Recycle gas temperature at retort inlet, °F	1,000
Combustor outlet temperature, °F	1,550
Retort recycle gas flow, scf/ton	13,500
Air to burning zone of combustor, scf/ton	14,000
Flue gas to burning zone, scf/ton	12,000
Flue gas to combusted shale cooling zone, scf/ton	11,000

Source: Data of Duir et al., 1977.

Table 18 SGR-3 Flue Gas Composition (Dry Basis)

N_2	71	mol %
CO_2	28	mol %
O_2	0.5	mol %
SO_x	5	ppm
NO_x	300	ppm

Source: Data of Duir et al., 1977.

Table 19 Thermal Efficiency of SGR-3 is Substantially Higher Than That of Unishale B

Yield	Retort B		SGR-3	
Inputs				
Raw shale, 34 gal/ton	1.0 ton	6350	1.0 ton	6350
Purchased power	21.6 kWh	221[a]	5.9 kWh	60[a]
Diesel fuel[b]	0.3 gal	41	0.3 gal	41
Total input		6612		6451
Outputs[c]				
Shale oil product	32.86 gal	4650	34.04 gal	4816
Pipe line gas[d]	0	0	649 scf	519
Sulfur	0.0008 UK ton	7	0.0008 UK ton	7
Total output		4657		5342
Thermal efficiency (100 × energy output/input)	70%		83%	

[a]Power plant fuel.
[b]Fuel for mining and spent shale disposal.
[c]Includes all facilities to produce salable products.
[d]800 btu/scf gross heating value.
Source: Data of Duir et al., 1977.

Condensation of oil within the retort is thermally efficient and reduces the need for expensive external heat exchangers and condensers.

The high heat capacity of the retort B recycle gas and high gas-solid heat transfer rates enable very high shale throughput.

Positive mechanical force provided by the rock-pump coupled with process control that limits retorting to the region near the top of the retort where pressure between shale particles is least enables reliable retorting of rich shales that tend to agglomerate.

The Unocal technology has been proven over many years of operation at scales that include laboratory, pilot plant, and commercial units.

M. Kiviter and Galoter Processes

Oil shale provides a vital energy source for the Baltic region, Estonia, and Leningrad. About 75% of the oil shale mined is burned directly to raise steam for electric power generation, while the balance is retorted. Kukersite, the richest Estonian oil shale, is the preferred feed. The two processes are complementary: Kiviter handles coarse shale, Galoter the fines.

The older of the two Soviet processes, termed the Kiviter process, handles coarse oil shale lumps of 1- to 5-in. size in a single multifunction retort that is generically similar to the direct-fired Paraho and gas combustion retorts in the United States. Shale is gravity-fed downward through the Kiviter retort. However, relatively rich kukersite (like very rich Colorado shale) becomes plastic on slow heating and the resulting agglomeration can lead to uneven gas and solids flows and/or plugging of the retort. Gas flow patterns selected to avoid these problems, the retort design to accommodate these flows, and the water-sealed spent shale discharge with mechanical shovel unloading are distinctive features of the Kiviter retort, which is covered by a U.S. patent (Figure 25).

As shale is fed into the top of the Kiviter retort, it enters a relatively long and thin annular space (1–1.5 m thick × 10–12 m tall), the "semicoking zone," where it is rapidly heated by hot (700–900°C; 1290–1650°F) gases that flow rapidly outward through the annular shale bed from the heat carrier preparation chamber in the center of the retort. Because the bed is thin, the shale can be heated rapidly and the product vapors can be removed with very short gas residence times. The rapid heating and efficient removal of product vapors enable reliable retorting of all but the richest Estonian kukersites.

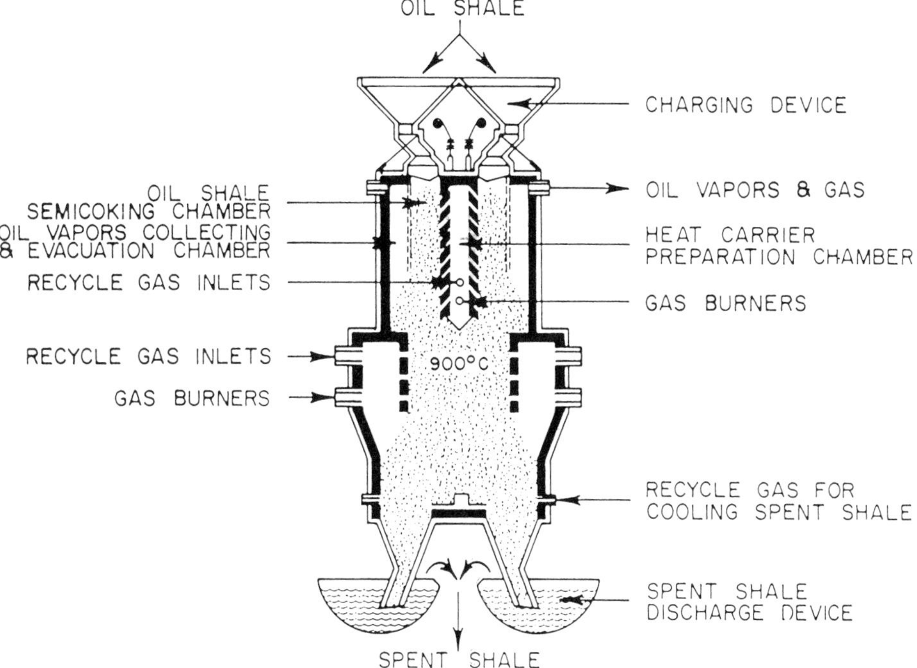

Figure 25 The Kiviter process uses a single, multifunction retort. Rapid, cross-current gas flow through the thin annular semicoking region gives very short residence times for the evolved product vapors (after Doilov et al., 1977; reprinted with the permission of Williams Brothers Engineering Company).

The retorted shale, which the Soviets term "semicoke," then moves down into the gasifying zone for gasification at 900°C (1650°F) using a steam-air mixture. This mixture is injected into the very bottom of the retort and helps to cool the spent shale as it travels upward. However, the steam-air mixture cannot provide enough heat to sustain gasification, so that additional heat must be supplied by gas burners built into the side of the retort.

Now the curious term "heat carrier preparation chamber" becomes more understandable. Part of the hot gas for retorting is the hot synthesis gas stream flowing up from the gasification zone, while the balance is generated by gas burners in the central chamber. Thus, the functions of the heat carrier preparation chamber include generation of hot flue gas, mixing this flue gas with the hot synthesis gas, and distributing the resulting hot gas mixture to give even heating of the annular shale bed and rapid product vapor removal.

As the spent shale moves further down toward discharge, it is cooled both by the upflowing steam-air mixture in the center of the retort and by cool recycle gas streams injected around the periphery. Obviously, there are large temperature gradients in this region; therefore the geometry of the discharge chutes and cooling gas injection points and flows must be carefully designed to achieve efficient heat recovery. Few details concerning these key aspects were presented.

Large amounts of gas for cooling the spent shale and for gasification are injected into the lower part of the Kiviter retort. This gas must travel a long path before passing through the annular retorting zone at the retort top. Even though care is taken to maintain bed permeability, the bottom of the retort operates under appreciable pressure. Also, reliably maintaining an even flow of shale through the retort was apparently as troublesome for the Soviet workers as for other retort developers. To provide sealing against pressures up to 1000 mm H_2O effective solids flow control, the Kiviter retort uses discharge chutes that project down into a water trough that is fitted with a reciprocating mechanical shovel (Figure 26). Spent shale flows down the discharge chutes until it piles up to the chute exit. The mechanical shovel is pivoted to sweep shale from the semicircular bottom of the trough onto a sloping drain tray. The shovel is lifted for its return past the discharge chute, then lowered for another sweep. The arrangement is claimed to work well. It does seem that a simpler, and probably more reliable, mechanical shovel would result if the discharge chutes were made vertical and the shovel were made double-acting, i.e., drain trays were provided on both sides of the seal trough and the shovel were made to sweep spent shale on both forward and reverse pivot strokes. However, other methods for controlling solids flow have proven better suited for continuous operations on a large scale (e.g., the nonmechanical steam seal used in Unocal's SGR-3 system).

Control of shale fines was necessary to obtain the even gas flows critical to even heating in the Kiviter retort. For this reason, the feed shale is screened to remove shale particles smaller than 1 in. (25 mm).

With a Baltic shale (presumably kukersite), the Kiviter oil yield is about 85% of Fischer assay. Gas production is also significant, but the Kiviter gas is diluted with nitrogen from the combustion air and CO_2 produced by combustion and carbonate decomposition (40–50% carbonate decomposition). As a result, it is a low-btu gas suitable for use in firing the retort or for providing heat for upgrading/refinery operations, but not for pipelining (unless blended with a high-btu gas). Overall thermal efficiency of 74% is claimed for operations with a 50

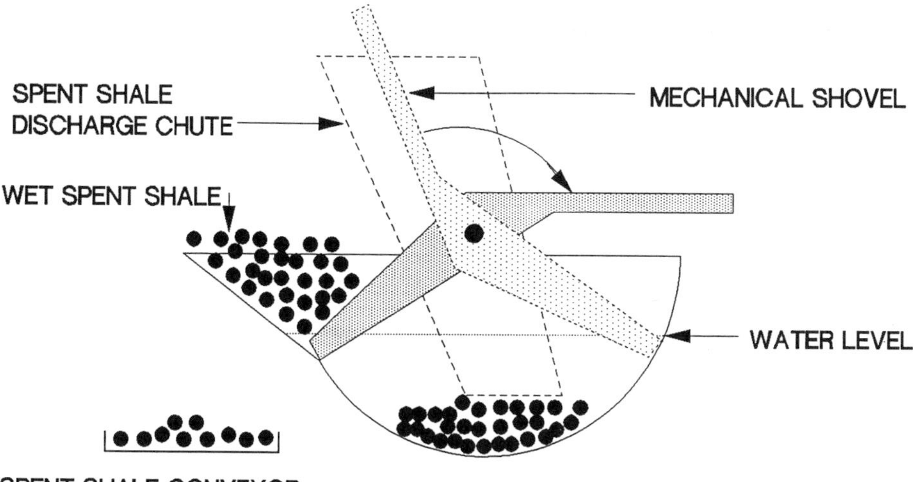

Figure 26 The Kiviter discharge section provides sealing against pressures up to 1000 mm H_2O and a pivoted mechanical shovel that both unloads the spent shale and controls solids flow through the retort (after Doilov et al., 1977).

gal/ton Baltic shale (enriched at the mine). Byproducts also generate significant credits; a high-quality cement is produced from the spent shale from selected retorts.

Kiviter retorts capable of processing 250–300 tons of shale per day are routinely operated, generally in banks of 10 retorts connected to a common product collection system and sharing a common upgrading facility. A prototype 1000 ton/day Kiviter retort was under construction in 1981. Presumably, this retort has been placed in service, but no operational details are at hand.

The Galoter process is similar to the Lurgi-Ruhrgas, Tosco II, and Shell (SPHER, SSRP) processes in using hot spent shale as the heat carrier. Dried oil shale (smaller than 1-in., 110°C; 230°F) is mixed with hot spent shale (800°C; 1470°F) in a screw mixer, then passed into a 500°C rotary kiln, similar to Tosco II (Figure 27).

From the rotary kiln, the retorted shale and product vapors pass into a gas-solid separator, from which the vapors are sent to the product recovery section and a part of the spent shale is discareded. The balance of the spent shale is fed into an air-blown riser combustor where burning of residual carbon raises the solid temperature to 800°C (1470°F). The hot shale stream is used to provide heat for retorting,

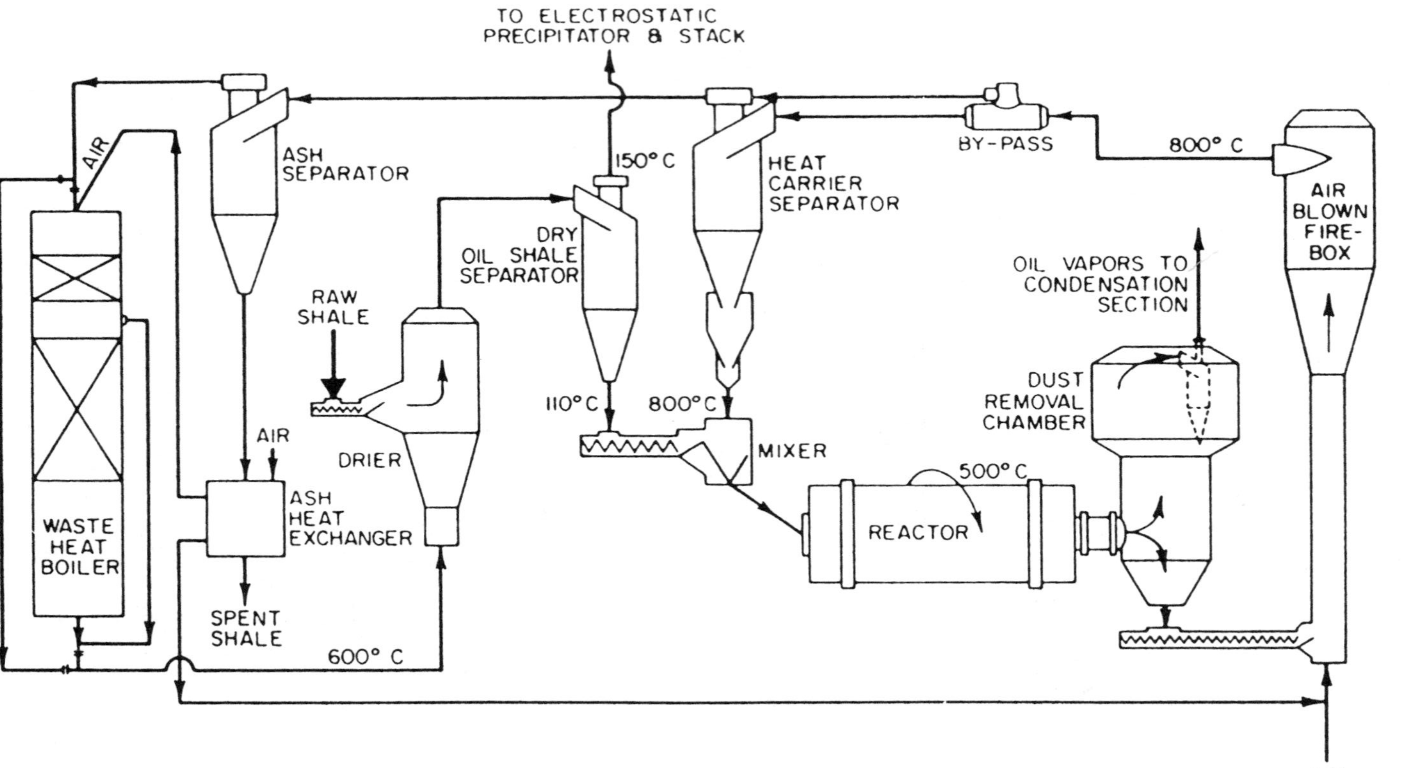

Figure 27 Flowsheet shows that the Galoter process shares many features with the Lurgi-Ruhrgas and Tosco II processes (after Resource Sciences Corporation, 1975; reprinted with the permission of Williams Brothers Engineering Company).

while the hot gases are used to raise steam and then to dry the wet incoming shale.

Oil yields 85–90% of Fischer assay and an 82% overall thermal efficiency are claimed for Galoter process operations with 40 gal/ton wet Baltic shale. It should, however, be realized that the comparison with Kiviter is unfavorable to Galoter for two reasons. First, shale for the Kiviter retort is processed at the mine to remove sand and rock, while shale for the Galoter retort is not cleaned. Second, the reported Galoter results were obtained with a shale that contained appreciably more water (12.4 wt % versus 9.0 wt %); removing this extra water will lower overall thermal efficiency. Also, the gas product from the indirectly heated Galoter retort is not diluted with combustion gases; therefore, it has a high heating value of 1170 btu/scf and is suitable for pipelining.

The Galoter process was tested extensively in a 500 ton/day plant. As of April 1975, this plant had operated for 68,000 hr, processed 1.4 million tons of shale, and produced 186,300 metric tons of oil and 2.5 billion scf of gas. Based on these results, a larger Galoter retort was designed to process shale at the rate of 3300 tons/day. A Galoter complex would include four of these larger retorts together with facilities for gas treatment and upgrading shale oil to produce finished products. The first unit of this complex was scheduled for completion in 1978. Eventually, Soviet workers envision shale-processing complexes that contain both Galoter and Kiviter retorts for more efficient processing of the entire mined resource. In such complexes, the low-btu gas from the Kiviter retorts would serve primarily as plant fuel, while the high-btu Galoter gas would be pipelined (perhaps as a blend containing a small amount of excess Kiviter gas).

The shale oils produced by Kiviter and Galoter retorts are reportedly of essentially the same composition (Table 20).

The above results show that oils from the Baltic shales are much richer in oxygenates, especially phenols, than a typical Green River shale oil, which is almost devoid of phenols. Also, the H/C ratio of the Baltic shale oil is much lower than the 1.55–1.60 that is typical of retorted Green River shale oils. In both respects, the Baltic shale oil is much more similar to shale oils from retorting of Devonian shales from the eastern United States. Peterson and Spall (1983) reported a detailed comparison of Estonian and Green River shale products.

N. Hytort Process

The Institute of Gas Technology (IGT) has been a leader in exploring the use of a hydrogen atmosphere in retorting (hydrotorting). Initially, the objective was gasification at high temperatures. However, most of the products obtained below the carbonate decomposition

Table 20 Typical Properties of Crude Shale Oil Produced from Baltic Shale in Either Kiviter or Galoter Retorts

Density at 68°F (20°C) g/cm^3			1.01
Viscosity at 167°F (75°C), Engler			4.5
Pour point, °F (°C)			5 (−15)
Coking value, wt %			8
Phenols, wt %			28
Calorific value, gross, btu/lb			17,010
Distillation:	*°C*	*°F*	*Vol %*
(IBP)	190	374	
	200	392	1
	250	482	6
	300	572	21
	360	680	45
Elemental composition (dry basis)			Wt %
Carbon			83.3
Hydrogen			10.0
Sulfur			0.7
Oxygen + nitrogen			6.0
Atomic H/C ratio			1.43

Source: Data presented by Resource Sciences Corporation, 1975).

temperature were liquid. Next, IGT devised and patented a process for gasifying the shale liquids (Linden, 1972). More recent hydrogen retorting studies at IGT have focused on obtaining high liquid yields and improved product quality, especially from eastern U.S. shales. In addition to IGT, Texaco was active in this area during the 1960s and Phillips Petroleum studied process response for hydrotorting Indiana New Albany shale in a recirculating loop reactor system (Ewart and Scinta, 1983). Phillips also participated with IGT, Bechtel, and Hycrude Corporation in process design studies aimed at the commercial use of the IGT technology with eastern U.S. shales. Work at Mobil Research and Development has resulted in the development of a rapid head-up (RHU) assay, especially designed to evaluate eastern shales (Audeh, 1984).

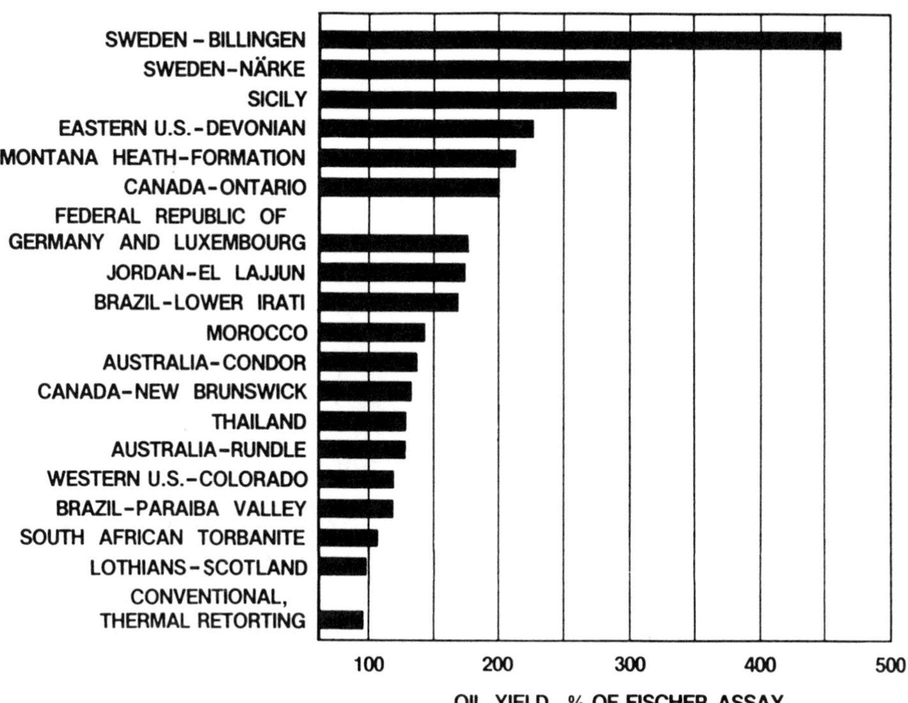

Figure 28 Hydrogen retorting dramatically increases oil yield over Fischer assay for many of the world's oil shales (Rex et al., 1983; reprinted with the permission of the Colorado School of Mines).

Most retorting studies have used the Fischer assay oil yield as a benchmark. Some of the more efficient fluidized bed processes have achieved oil yields of 110% or so of Fischer assay. However, Fischer assay recovers only a small part of the organic matter in many of the world's oil shales. Thus, potential oil yield is several times Fischer assay in these cases, notable among which are the Devonian shales of the eastern United States. Hydrogen retorting affords greatly enhanced oil yields in many such cases (Figure 28).

In comparison to Fischer assay oil, the quality of the oils obtained by hydrogen retorting depends markedly on the type of shale (Table 21). The oil produced from Green River oil shale by the HYTORT process is very similar to the Fischer assay oil; the main differences in elemental composition are a somewhat lower oxygen content. Nitrogen

Table 21 Elemental Analysis of Shale Oils and Hydrotreated Shale Oils

	Elemental analysis (wt %)			
	Fischer assay			IGT-HYTORT hydro-treated
	A	B	IGT-HYTORT	
Colorado shale oil				
Carbon	85.08	84.30	84.57	
Hydrogen	11.44	11.42	11.40	
Sulfur	0.73	0.63	0.62	
Nitrogen	1.80	1.90	2.13	
Oxygen	1.31	1.50	0.98	
H/C atomic ratio	1.61 (avg.)		1.61	
Kentucky New Albany Shale Oil				
Carbon	84.91	84.97	85.46	88.27
Hydrogen	9.83	10.08	9.42	11.70
Sulfur	1.02	1.61	1.52	0.05
Nitrogen	1.52	1.56	2.12	0.40
Oxygen	1.72	2.00	1.61	0.12
H/C atomic ratio	1.40 (avg.)		1.31	1.58
Kentucky Sunbury Shale Oil				
Carbon	84.36	84.59	85.45	87.30
Hydrogen	9.96	9.98	9.56	12.60
Sulfur	1.33	1.59	0.99	0.06
Nitrogen	1.33	1.48	2.12	0.17
Oxygen	2.12	2.36	1.22	0.28
H/C atomic ratio	1.41 (avg.)		1.33	1.72

Source: Data of D. Netzel and F. Muknis, 1982.

content was slightly higher and this seems to be a general phenomenon in hydrogen retorting. In sharp contrast, the oils from hydrogen retorting of the two eastern shales were quite different from their Fischer assay counterparts. Moreover, the two eastern shales were quite different in their response to hydrogen. In both cases, the HYTORT oils had appreciably lower atomic H/C ratios than the Fischer assay oil. However, the sulfur content was about the same and the oxygen content only slightly lower for the New Albany HYTORT oil, while both sulfur and oxygen contents were much lower for the Sunbury HYTORT oil. The reason for this behavior is not known.

Most hydrogen production involves the production of synthesis gas, which is further processed to produce hydrogen. In coal liquefaction, synthesis gas in the presence of liquid water has been shown superior to hydrogen alone for improving reaction rates and oil yields. Hydrogen retorting was compared to retorting in a mixture of synthesis gas and steam for three eastern U.S. shales. The oil yields and organic carbon conversions achieved with the synthesis gas mixture were approximately those that would have been obtained by hydrogen retorting at the hydrogen partial pressures used. However, the synthesis gas mixture yielded oils with appreciably higher H/C ratios and API gravities, and lower nitrogen contents (Table 22). Sulfur contents were, however, higher for the oils produced with synthesis gas.

Early work at IGT (1972–1979) focused on moving-bed hydroretorting, the HYTORT process. Tests were made with selected shales at rates up to 1 ton/hr. Additional progress was made during 1980–1983 by the joint Hycrude-IGT-Bechtel-Phillips Petroleum effort, especially in reducing hydrogen consumption. Based on these results, an engineering design study was made for a commercial HYTORT facility (Figure 29). The possible environmental impact of such a HYTORT facility versus other alternatives for retorting eastern shales has been examined for a central Tennessee site.

The HYTORT effort served to bring into focus the advantages of hydrogen retorting for eastern oil shales. However, the HYTORT moving-bed technology is mechanically complex and its large, high-pressure vessels are expensive to construct. Moreover, fines are not readily processed in the moving-bed system; hence they are usually discarded. Therefore, the emphasis of more recent work at IGT has shifted to a pressurized fluidized bed hydroretorting (PFBH) concept. The PFBH process will use a vertically staged fluidized bed (Figure 30). The PFBH is expected to afford higher throughput, and therefore lower reactor capital costs, than the moving-bed HYTORT process. Moreover, for eastern shales the use of smaller shale particles in the PFBH should boost oil yields while decreasing gas yields. and having little effect on overall carbon conversion (Figure 31). Thus, not only should the PFBH process more shale per capital dollar than the moving bed, it should also give about one-third more shale oil per unit weight of shale. As a result, the PFBH is expected to substantially reduce the cost of producing shale oil from eastern U.S. Devonian oil shales.

Other advantages of the PFBH versus the moving-bed HYTORT process stem from the characteristics of the fluidized bed. The efficient gas-solid contacting and good heat transfer characteristics of the fluidized bed should improve thermal efficiency and the need to briquette beneficiated shale (required for moving-bed hydroretorting) will be eliminated.

Table 22 Comparison of Oil Yields and Properties of Oils Produced by Hydroretorting and Synthesis Gas Retorting

	Syngas	Hydrogen[a]	Syngas	Hydrogen	Syngas	Hydrogen[b]
Maximum temperature, °C	649	654	568	569	502	487
°F	1201	1209	1055	1056	935	909
Total pressure, atm	69.4	28.4	68.9	28.3	68.9	28.4
psig	1006	402	998	401	998	403
Hydrogen partial pressure, atm	24.3	28.4	24.3	28.3	24.3	28.4
psig	350	402	350	401	350	403
Shale oil ultimate analysis, wt %						
Carbon	83.30	84.67	83.89	84.81	83.99	84.61
Hydrogen	9.86	9.62	10.16	9.63	10.02	9.97
Sulfur	1.79	1.28	1.52	1.23	1.50	1.39

Nitrogen	1.70	2.07	1.64	2.05	1.80	1.93
Ash	0.0	0.0	0.0	0.0	0.0	0.0
Oxygen (by difference)	3.35	2.36	2.79	2.28	2.69	2.10
Total	100.00	100.00	100.00	100.00	100.00	100.00
C/H ratio	8.45	8.80	8.26	8.81	8.39	8.49
Specific gravity [15.6/16.6°C (60/60°F)]	0.947	0.978	0.945	0.986	0.960	0.967
API gravity, °API	17.9	13.2	18.2	12.0	15.9	14.8
Oil yield						
Liters/tonne	89.2	100.5	90.06	103.4	92.1	97.6
Gal/ton	21.4	24.1	21.6	24.8	22.1	23.4
Organic carbon conversion, %[c]	69.4	78.6	72.9	75.6	68.1	72.9

[a]Average of three tests.
[b]Average of two tests.
[c]Elemental conversions based on solids and ash balance.
Source: Data of Punwani et al., 1986.

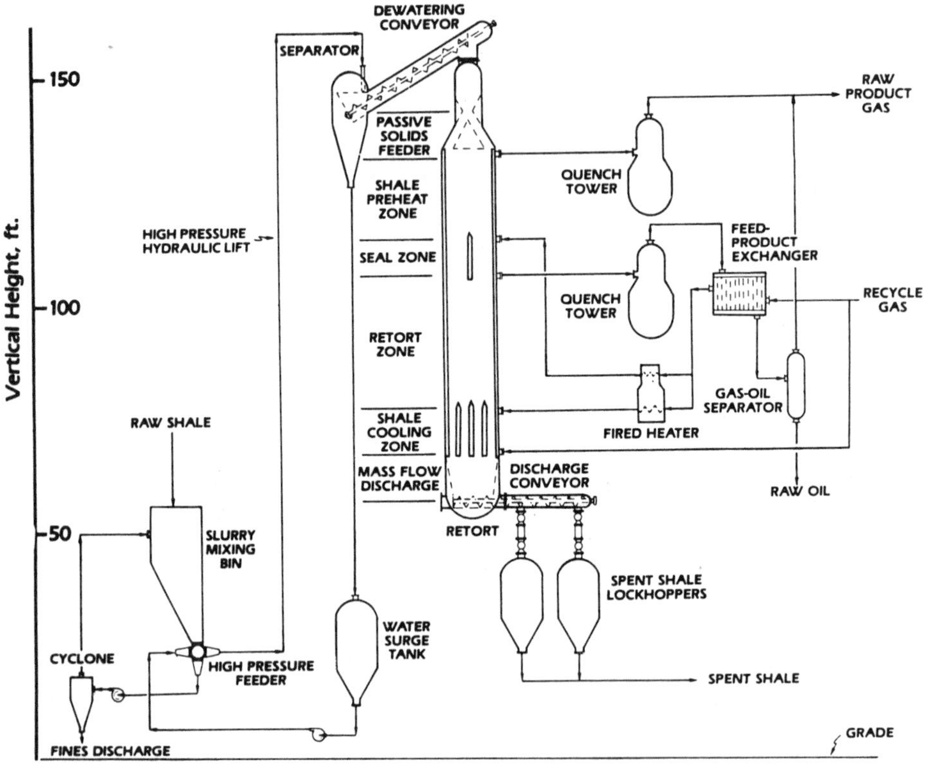

Figure 29 Retort area flowsheet for a commercial HYTORT facility (reprinted with the permission of the Institute of Gas Technology).

Additional work is under way to develop a fluidized bed gasification system that will utilize the carbon content of the fines to produce hydrogen (Lau et al., 1987). However, the shale-processing facility will require process heat and steam, and the carbon on spent shale is usually the energy source that meeds these needs. Therefore, even if technically successful, the impact of gasification as a hydrogen source may not be dramatic. However, if chemistry can be found that will enable the use of synthesis gas to increase the rate and/or lower the overall pressure requirement for hydrotorting, the impact of gasification may be very important.

Finally, it should be pointed out that the molecular chemistry responsible for the striking yield enhancements often obtained by hydrogen retorting are not yet well understood. Nor, for that matter, are the reasons why shales that are remarkably similar in elemental and mineral composition sometimes respond very differently to

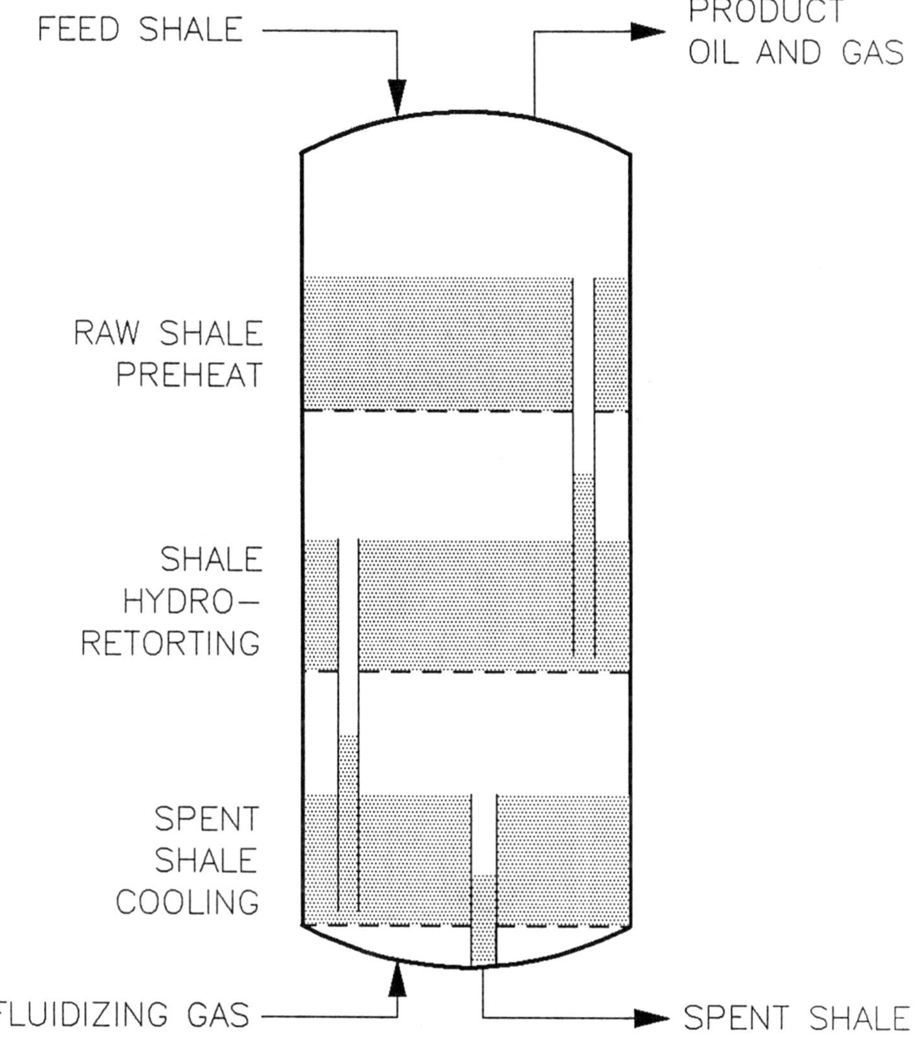

Figure 30 The PFBH retort is a vertically staged fluidized bed unit for retorting hydrogen-deficient oil shales, such as the eastern U.S. Devonian shales, in a hydrogen atmosphere at high pressures (reprinted with the permission of the Institute of Gas Technology).

hydrogen pressure. Perhaps a better understanding of the molecular structure of the kerogen, and of kerogen-mineral interactions, in the eastern shales might help to clear up some of these mysteries.

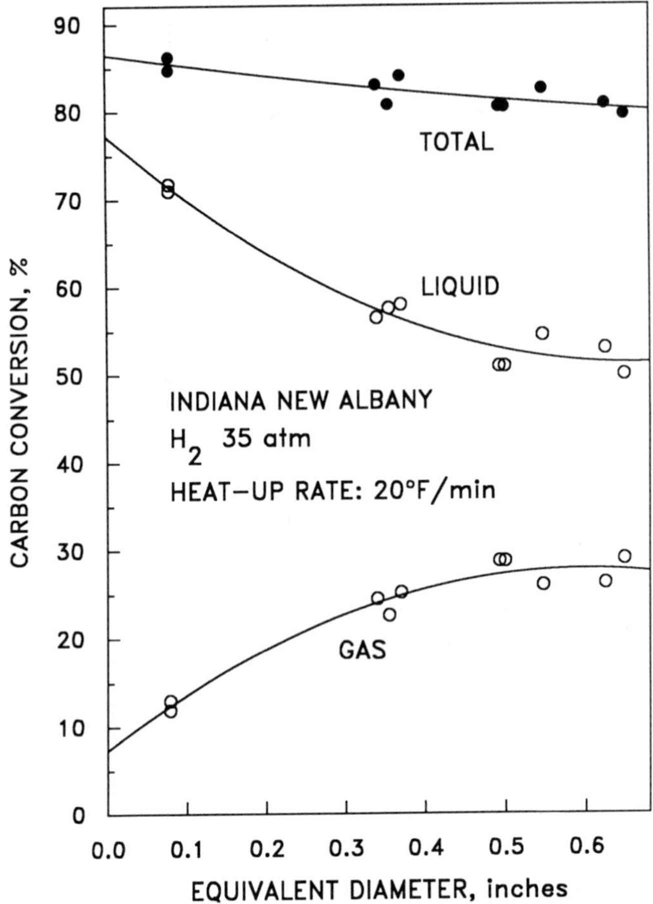

Figure 31 Decreasing particle size substantially increases oil yield and reduces gas yield, while having only a small effect on the overall carbon conversion, in hydrotorting of an eastern U.S. oil shale (reprinted with the permission of the Institute of Gas Technology).

II. IN SITU

In situ retorting (see also Chap. 12, Sec. IV.D) involves heating the shale in place (Chap. 22, Sec. VIII) to produce shale oil and gas. Because the shale remains underground, the requirements for mining, transporting, crushing, and grinding the shale rock are eliminated or greatly reduced. True in situ retorting also eliminates the costs of an above-ground retort. Thus, in situ retorting offers the potential for corresponding savings in both operating and capital costs. However, making an in situ retort work is generally difficult because most oil shales have very low porosity and almost no permeability. Without permeability, getting combustion air into the oil shale formation, or oil and gas products out of the formation, is not possible. Permeability can be induced, for example, by blasting or by injecting high-pressure air or water. However, maintaining such induced permeability is difficult because shale oil tends to fill the void space in the bed and because oil shale swells (exfoliates) upon heating. Attempts to induce and maintain permeability have led to the two main in situ retorting approaches: "true in situ" methods, which involve blasting and/or other fracturing techniques, but no mining; and "modified in situ" methods, which involve mining part of the shale to generate free underground space followed by blasting to generate a permeable zone of "rubblized" oil shale for retorting. Both true and modified in situ methods use a moving flame front to generate heat for retorting, and hence are similar in basic principle to the above-ground NTU retort described earlier. In a few areas of the Green River formation, leaching of water-soluble minerals affords a porous zone of high permeability. One such case is the "leached zone," located below the rich "Mahogany zone" and extending across much of the Piceance Creek Basin (see Figure 3 in Chap. 21). The BX in situ oil shale project was an attempt to take advantage of this natural permeability.

A. U.S. Department of Energy Process

In the early 1960s, Laramie workers began to lay the foundation for development of in situ retorting technology for Green River oil shale. Data obtained from the above-ground 10- and 150-ton NTU retorts at Laramie provided basic information about the ignition of rubblized shale beds and the use of a moving combustion front to provide heat for retorting. A series of nine field experiments to explore true in situ retorting were carried out in the Green River Basin near Rock Springs, Wyoming (Figures 32 and 33).

The major object of these experiments was to demonstrate that fracturing of the oil shale formation could induce sufficient permeability to support underground combustion. A variety of techniques—electrical, hydraulic, and explosive—were evaluated for inducing permeability (Table 23).

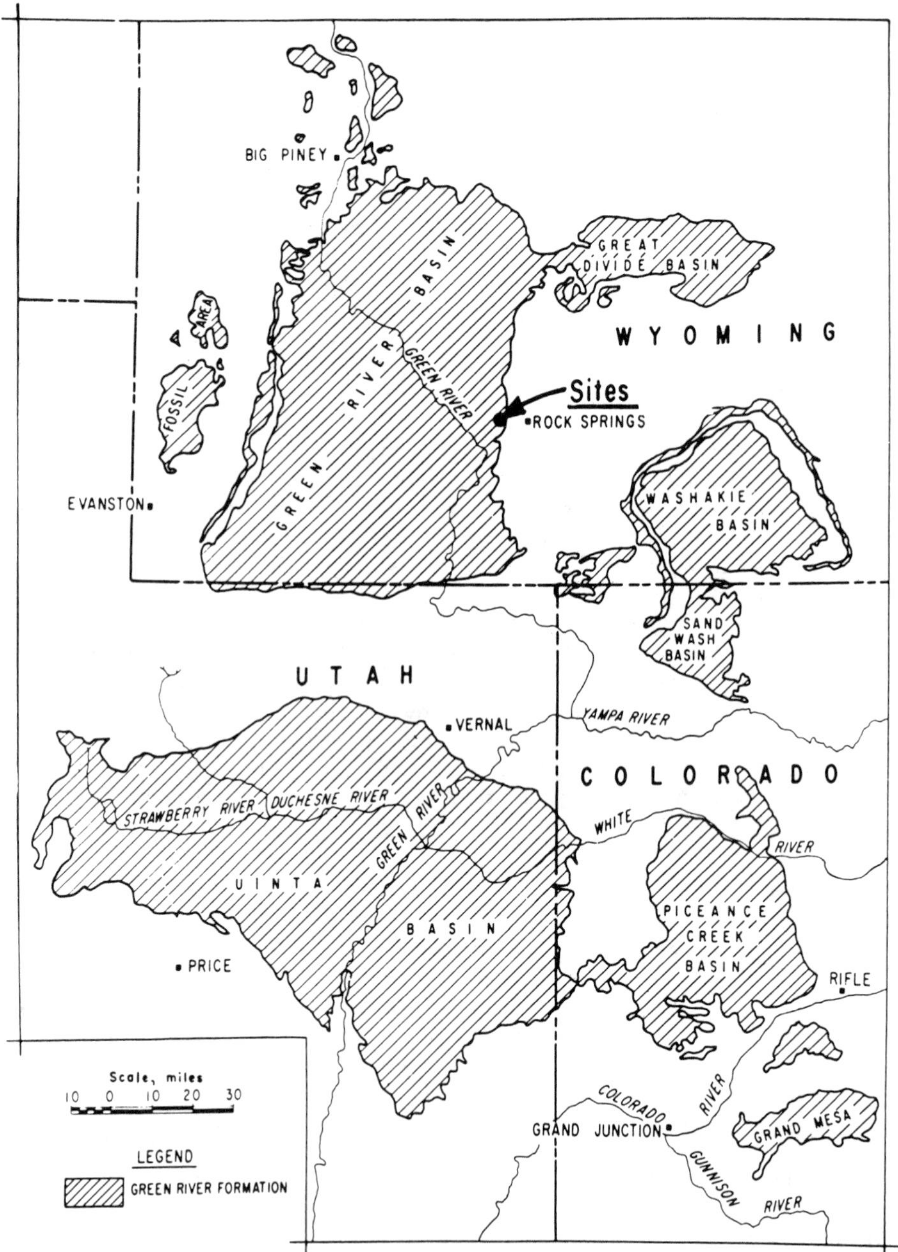

Figure 32 The Laramie in situ retorting experiments were carried out near Rock Springs, Wyoming in the Green River Basin, which lies north of the Piceance Creek Basin where most oil shale activity has been focused (Burwell et al., 1973).

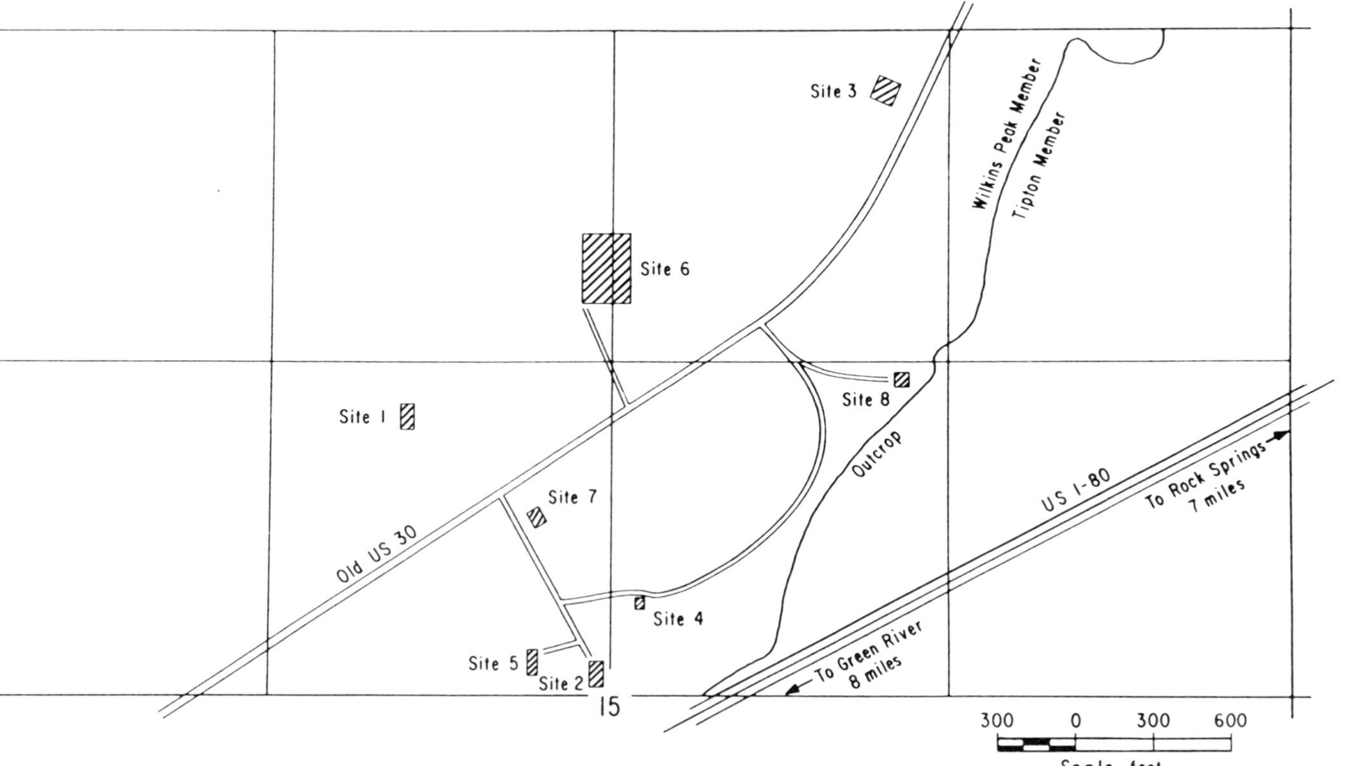

Figure 33 Layout of the Rock Springs exploratory in situ retorting sites (Burwell et al., 1973).

Table 23 Types of Research Carried Out at Rock Springs Sites 1–9

Site	Fracturing research			In situ combustion	Steam injection
	Electrolinking	Hydraulic	Explosive		
1	X	X	X	–	–
2	X	–	X	–	X
3	–	–	X	–	–
4	X	X	X	X	–
5	–	X	X	–	–
6	–	X	X	X	–
7	–	X	X	X	–
8	X	X	–	–	–
9	–	X	X	X	–

The major objective of the Laramie studies was met: sufficient permeability was induced for combustion to be initiated, and for the combustion front to be maintained and moved through the rubblized shale bed. Although postburn coring established that significant amounts of shale were retorted in the cases where in situ combustion was sustained, oil recoveries were poor. In part this was due to recurring problems with pumps that plugged with shale debris. However, nonuniform combustion and inadequate product containment also contributed to low oil recovery in these exploratory studies.

In parallel with, and following, the test burns, Laramie workers carried out an extensive program of environmental studies. These studies identified the disposition of the produced retort water as a key environmental concern. This water is produced during thermal retorting and is derived from dehydration of shale minerals as well as from combustion. The intrusion of groundwater may also contribute to the water recovered with the oil. Retort water is odoriferous, yellow to brown in color, and contains high levels of both organic and inorganic dissolved constituents. It is usually quite basic, with a pH in the range of 8–9.5. The amount of retort water is large; about equal to the volume of oil produced, in favorable cases. However, testing revealed that the retort water is not particularly toxic and indicated that standard purification techniques should be adequate for its environmentally acceptable disposal.

B. Geokinetics Process

The Geokinetics horizontal in situ retorting process (HISP) is a true in situ process, designed to retort shallow shale seams with no mining (Lekas, 1981). Because no mine construction is involved, a key feature of HISP is its very low initial capital cost. In 1974, HISP development began as a joint effort of Geokinetics Inc. and Aminoil USA. In 1976, ERDA (later DOE) joined the effort. In 1978, Aminoil withdrew, leaving Geokinetics and DOE as the remaining participants. In 1984, DOE withdrew and Geokinetics completed the project at its own expense.

In the process, a pattern of holes is drilled through the overburden and shale seam to be retorted (Figure 34). Explosive charges are then loaded and detonated sequentially to create the in situ retort. The blasting pattern is precisely designed so that the initial blast lifts its rubble some 10–20 ft at the surface (Britton, 1985). While this rubble is still aloft, about 0.5 sec after the initial blast, the subsequent rows of charges fire at approximately 0.1-sec intervals to laterally displace rich shale into the void. The result is a disruption of the surface (spalling) only at one end of the retort. Off-gas wells are then drilled at the disrupted end and air injection wells at the other. Oil production wells are associated with sumps to facilitate oil collection (Figure 35). Observation, thermocouple, and gas sample wells complete the required drilling.

Geokinetics field studies were carried out at the Seep Ridge site located south of Vernal, Utah. A total of 28 retorts were designed. Twenty were blasted and burned; six were blasted, but not burned; and one retort was abandoned prior to blasting. The final five, retorts 24–28, were of commercial size.

Oil recoveries from retorts 25 and 26 were 59 and 51% of Fischer assay, respectively, where superficial gas velocities were maintained at or above 0.8. Due to substantially lower air injection rates, oil recoveries and production rates from the other retorts were much lower.

Quality of the oil produced by HISP will vary from retort to retort, due to variations in shale composition, and also varies with time during production from a given retort. However, a composite analysis of oils from retorts 27 and 28 shows that oil quality is quite good (Table 24). In particular, the metals, asphaltenes, and residuum contents are desirably low.

C. Occidental Vertical Modified In Situ (VMIS) Process

The true in situ retorting processes discussed above involved drilling but no underground mining. In contrast, the modified in situ methods

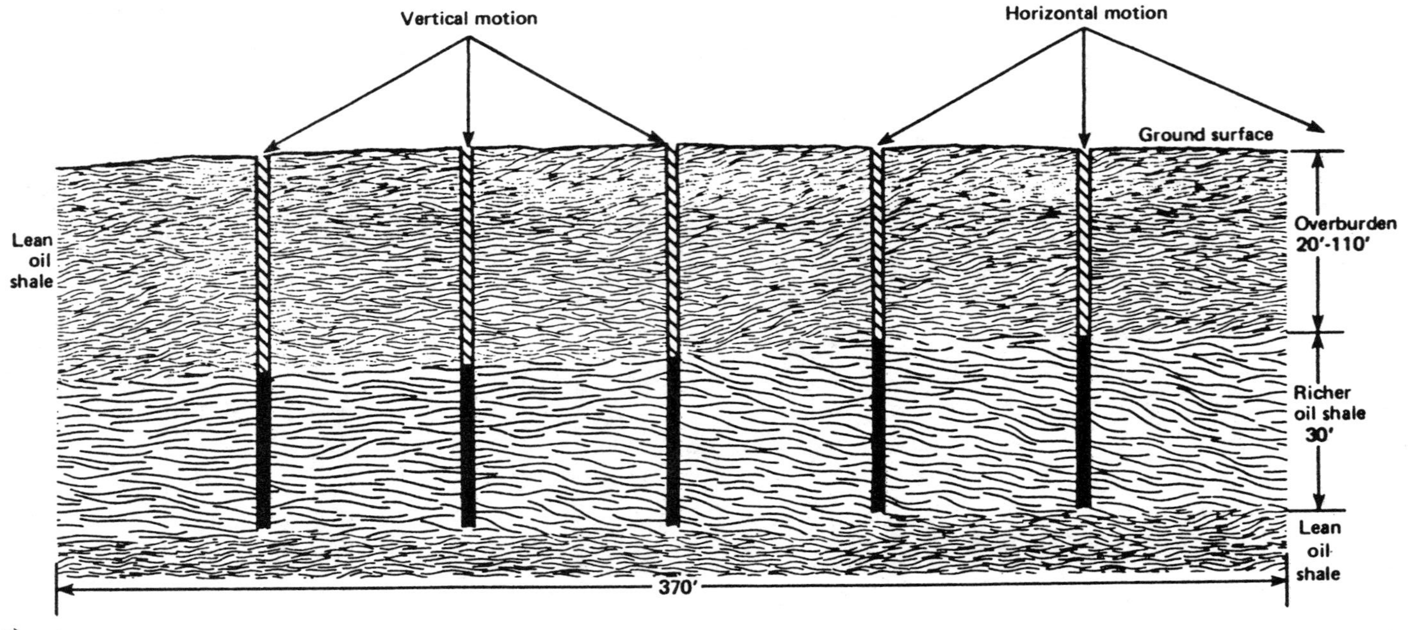
Vertical motion
Horizontal motion
Ground surface
Lean
oil
shale
Overburden
20'-110'
Richer
oil shale
30'
Lean
oil
shale
370'

(a)

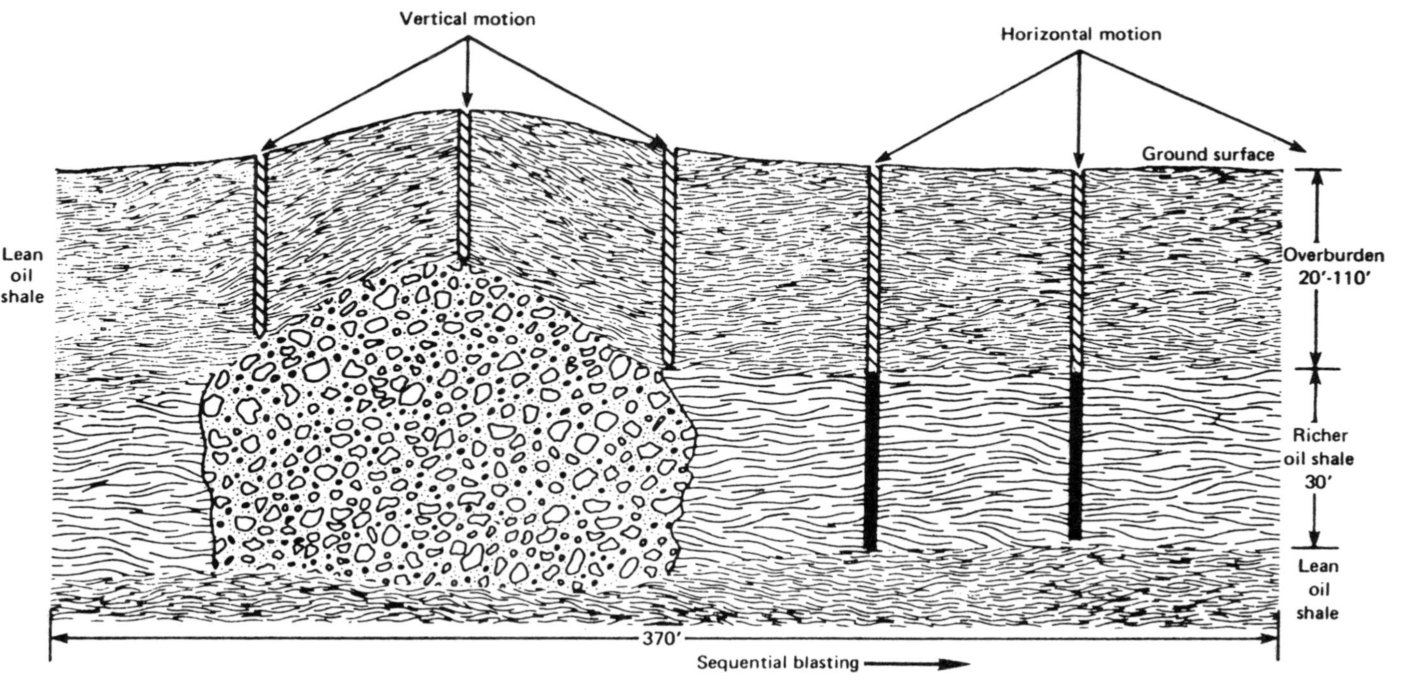

Figure 34 Geokinetics blasting sequence. (a) First blast initiates vertical uplift. (b) Subsequent charges displace rich shale laterally into created void space without disrupting overburden. (c) Retort at the end of the blasting sequence (after Lekas, 1985).

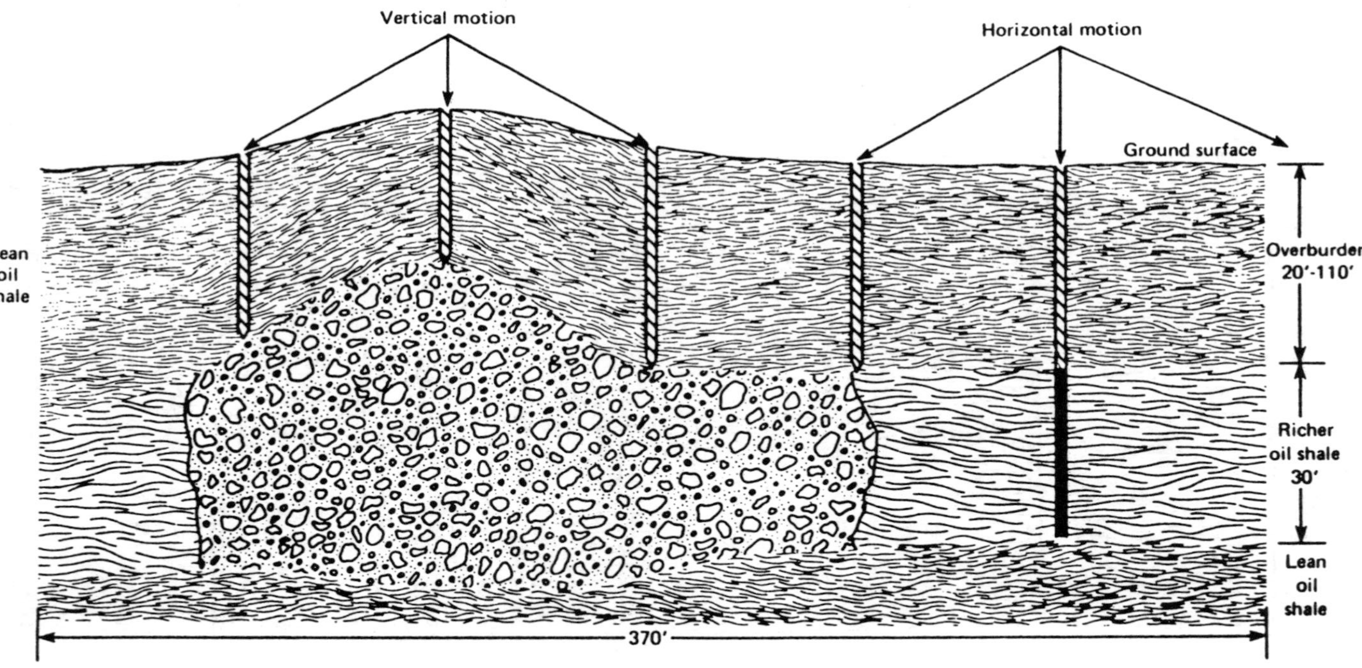

Fractured oil shale displaced into created void space.

(c)

Figure 34 Continued.

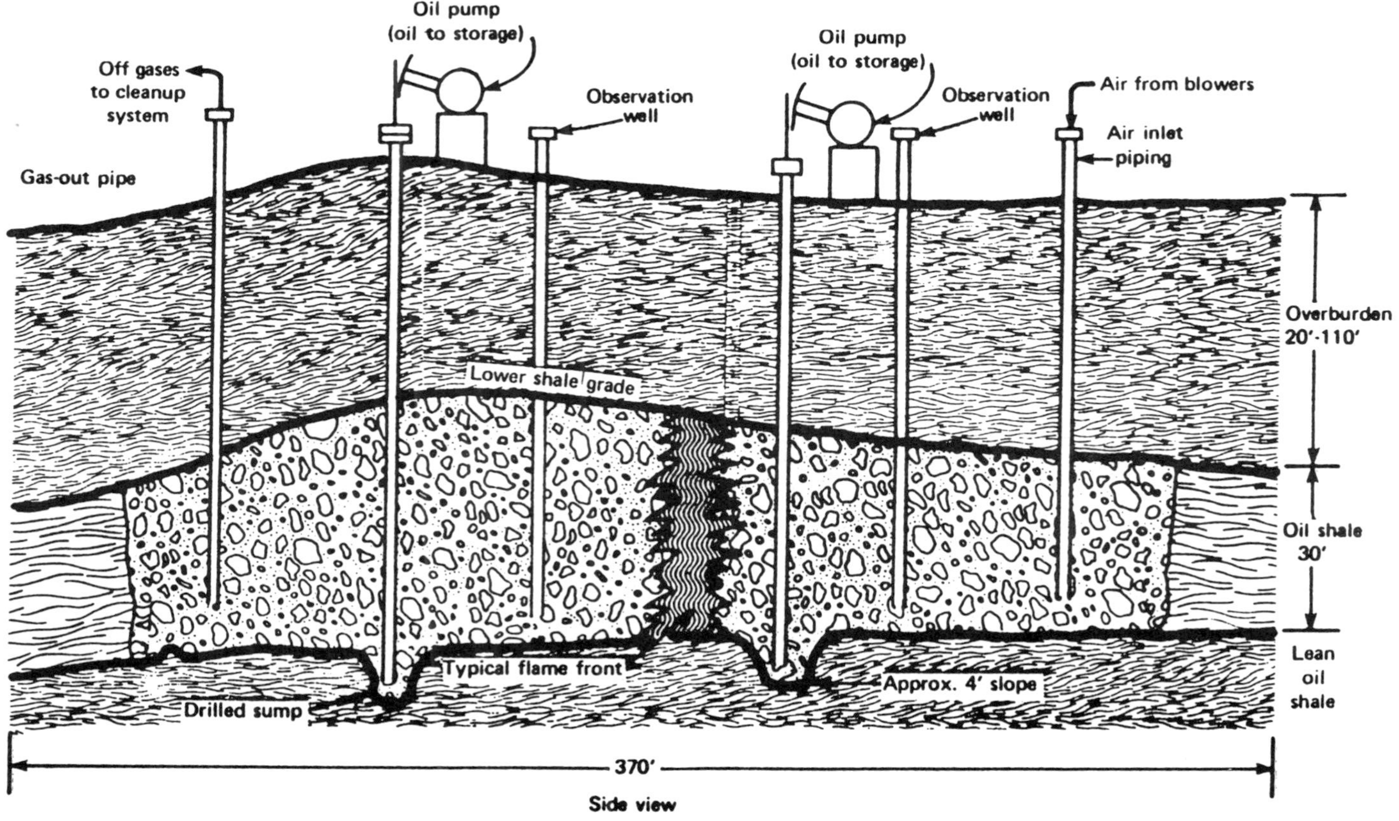

Figure 35 A Geokinetics horizontal in situ retort in operation (after Lekas, 1985).

Table 24 Properties of Crude Shale Oil From the Geokinetics Horizontal In Situ Retorting Process[a]

Gravity	25–26°API		
Viscosity: at 100°F (38°C)	12–16 cst		
at 140°F (60°C)	6–8 cst		
Flash point (ASTM D9323)	180–200°F (82–93°C)		
BS & W (maximum)	1.0 wt %		
Ash	0.015–0.030 wt %		
Pour point	70–80°F (21–27°C)		
Asphaltenes	0.5–1.5 wt %		
Elemental analysis			
Carbon	83.0–84.7 wt %		
Hydrogen	11.8–11.9		
Oxygen	0.9–1.6		
Nitrogen	1.5–1.6		
Sulfur	0.6–1.0		
Metals			
Iron	87–740 ppm		
Arsenic	8–11		
Vanadium	1–3		
Nickel	6–58		
Heat of combustion (gross)	19,000–19,500 btu/lb		
Distillation (ASTM D1160)	*Vol %*	*°F*	*°C*
	IBP	160–255	71–124
	10	420–470	216–243
	30	520–580	271–304
	50	600–675	316–357
	70	775–790	413–421
	90	900–920	482–493
	FBP	980–1150	527–621

[a]Data derived from analysis of raw shale oil samples produced from HISP retorts 27 and 28 during Geokinetics' development program.
Source: Data of Lekas, 1985.

discussed below involve mining 15–40% of the shale to create void space within the formation, then blasting to rubblize the remaining shale to fill the resulting retort. The retort is ignited at the top and burned with a downflow of air (Figure 36).

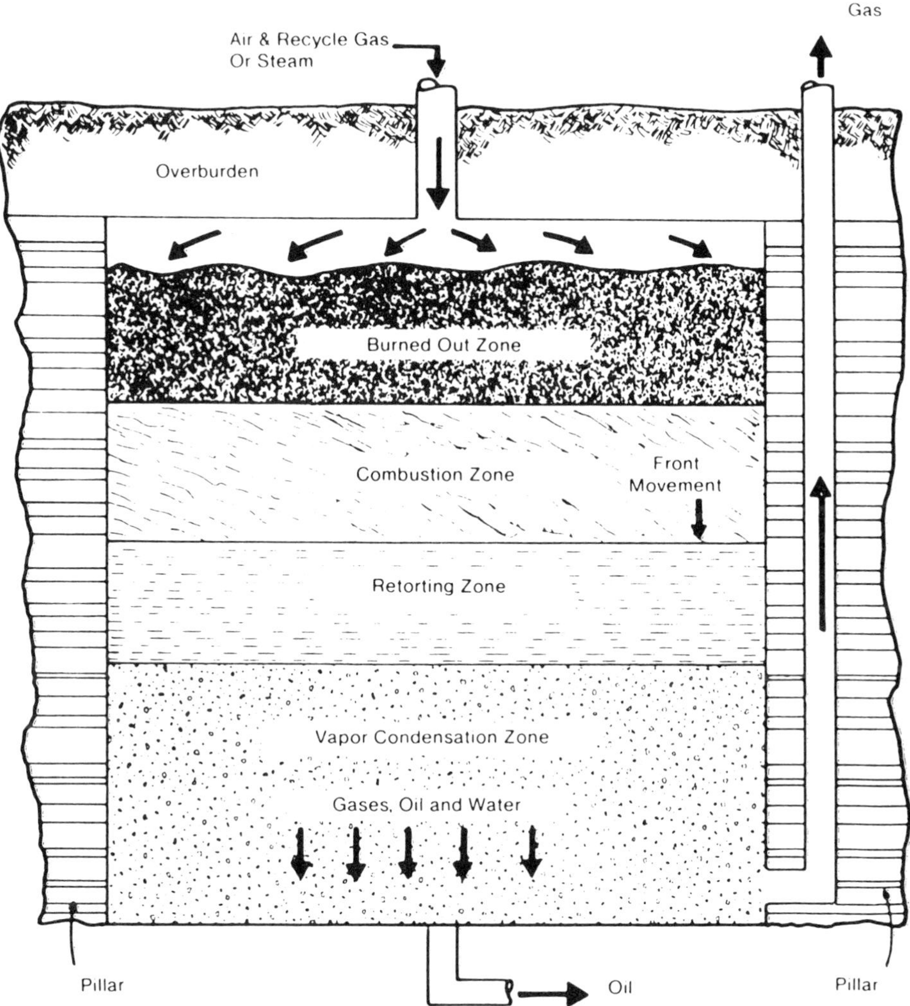

Figure 36 Occidental Oil's vertical modified in situ (VMIS) retort (after McCarthy and Cha, 1976; reprinted with the permission of the Colorado School of Mines).

In 1972, Occidental began field development of VMIS retorting at its Logan Wash oil shale mine north of DeBeque, Colorado. Early work at Logan Wash included the construction and processing of three small retorts and one commercial sized retort. The small retorts were about 30 ft^2 by 72–113 ft in height. Retorts 1E and 2E used a vertical raise in the center of each retort, with a single room at the bottom to provide void volume. Three horizontal rooms were spaced vertically to distribute void volume in retort 3E, which improved yield relative to 1E and 2E. Retort 4, with two vertical slots parallel to each other across the retort width, was an attempt to scale up the vertical void concept. The lower yields obtained from retort 4 were ascribed to limitations imposed by rock mechanics that led to nonuniform flows and bed stability problems.

Retorts 5 and 6 were commercial sized units that were developed under a 1976 cooperative agreement with U.S. DOE (Romig, 1981). Retort 5 used a single vertical slot mined across its width to provide void volume, while retort 6 was a scale-up of the retort 3 design with three horizontal rooms (Figure 37). The latter design again afforded the superior oil yield (Ricketts, 1980).

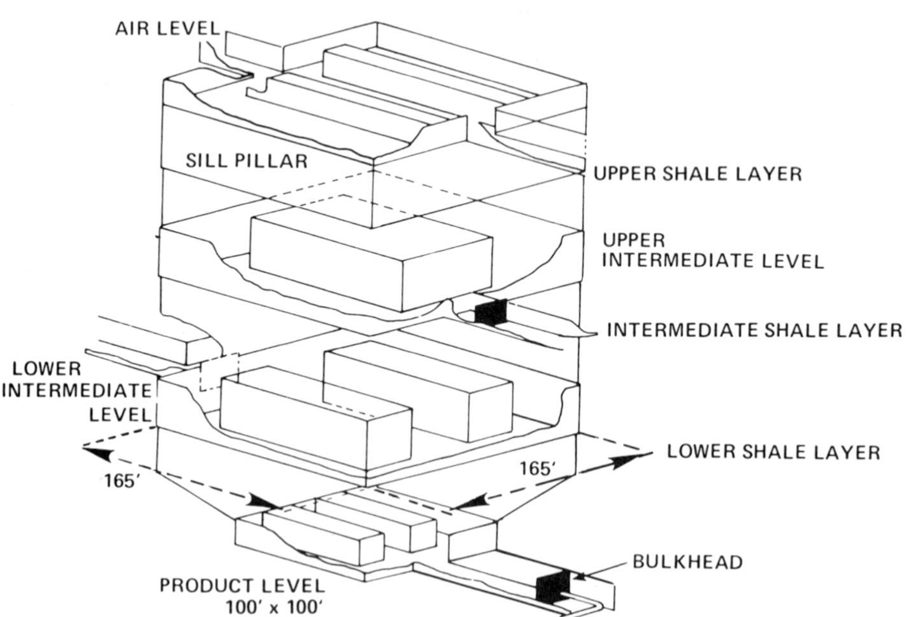

Figure 37 Isometric view of the Oxy VMIS retort 6 before blasting (Ricketts, 1980; reprinted with the permission of the Colorado School of Mines).

Retorts 7 and 8 were constructed under the second phase of the Occidental–DOE cooperative agreement. These were identical, commercial sized retorts, measuring 165 × 165 ft across and 241 ft tall. They were constructed side by side, using the retort 6 design, and burned at the same time in order to obtain reproducibility data not available from single retort tests (Figure 38). Retorts 7 and 8 afforded even higher oil yields than retort 6 (Table 25).

The distance between retorts 7 and 8 was 160 ft. To assess the effect of a shorter interretort distance, the partial height retort 8X was constructed 50 ft behind retort 8. Retort 8 had the 165 × 165 ft cross-section of retorts 7 and 8, but was only 63 ft tall. It was

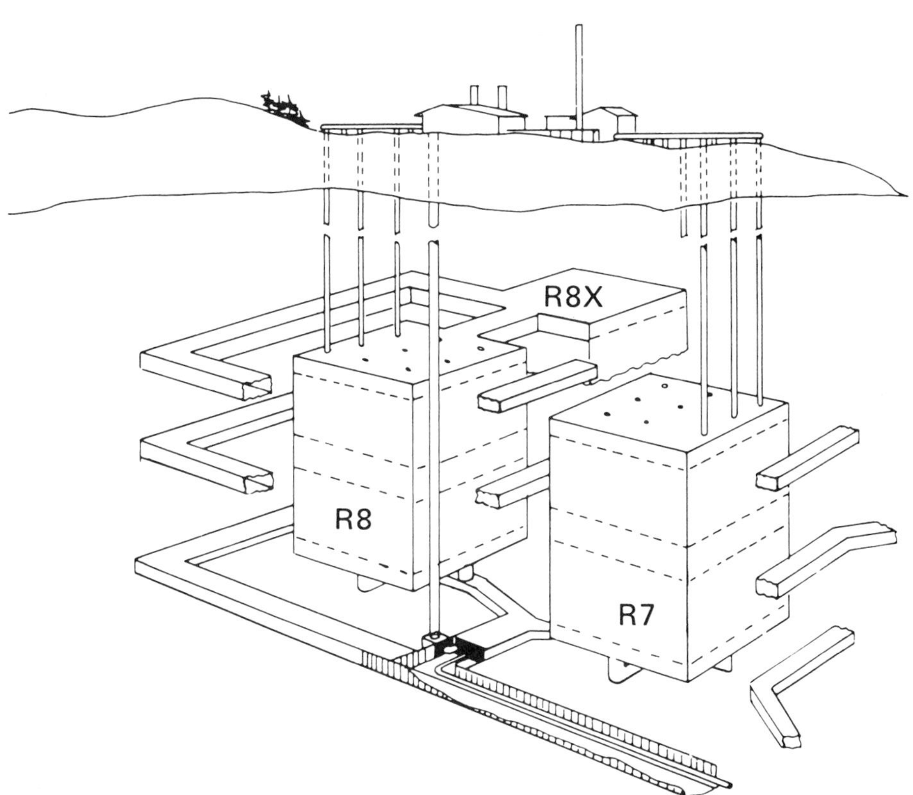

Figure 38 Isometric view of Oxy VMIS retorts 7 and 8, with partial height retort 8X in the background (Ricketts, 1982; reprinted with the permission of the Colorado School of Mines).

Table 25 Summary of the Retort Operating Parameters and Results for Occidental Retorts 7 and 8

Parameter	Retort 7	Retort 8	Total
Superficial gas velocity, ft/min	0.54	0.54	
Air/steam ratio, avg.	80/20	80/20	
Fischer assay of target rubble, gal/ton	17.3	16.9	17.1
Oil-in-place in target rubble, bbl	143,200	140,100	283,300
Produced oil, bbl	92,326	106,200	198,526
Yield, % of Fischer assay	64.5	75.8	70.1
Void rock mined, tons	111,604	114,042	225,646
Yield, bbl oil/ton of rock mined	0.83	0.93	0.88

Source: After Stevens and Zahradnik, 1983.

blasted first, so that blasting of retort 8 could provide information about the deformation, stresses, and fracturing that would occur at the shorter distances called for in the commercial development plan. No specific problems associated with the 50-ft interretort distance have been mentioned, and retort 8 performance does not seem to have suffered. However, rock structure is highly site-specific and it would be unwise to assume that such a short distance would be acceptable in any other site.

D. Rio Blanco Modified In Situ Process

The Rio Blanco Oil Shale Company (RBOSC) was originally a 50/50 joint venture of Gulf Oil Corporation and Standard Oil Company (Indiana) and is now a division of Amoco Corporation. It was formed after winning a bid in January 1974 for Federal Oil Shale Lease C-a. Site preparation for modified in situ (MIS) development on tract C-a began in late 1977. Two retorts were designed, constructed, then successfully rubblized and burned to demonstrate the RBOSC-MIS technology. The MIS retorting phase was completed during the first part of 1982. Following successful completion of the RBOSC effort, an extensive study of above-ground retorting based on Lurgi technology was made at the Gulf Oil Research Center at Harmarville,

Pennsylvania. Because tract C-a is suitable for open-pit mining, current RBOSC interest has shifted to the above-ground technology. The tract C-a operation has now been suspended until RBOSC can obtain off-tract land for disposal of overburden and spent shale. During suspension, RBOSC is maintaining the leasehold and continuing environmental monitoring work, while proceeding with internal oil shale research and development programs at the Amoco Research Centers.

Figure 39 is an isometric view showing the mine and the two experimental retorts: retort 0 is to the right of the much larger retort 1 (Berry et al., 1982). At 400 ft tall, retort 1 is the tallest in situ retort ever built.

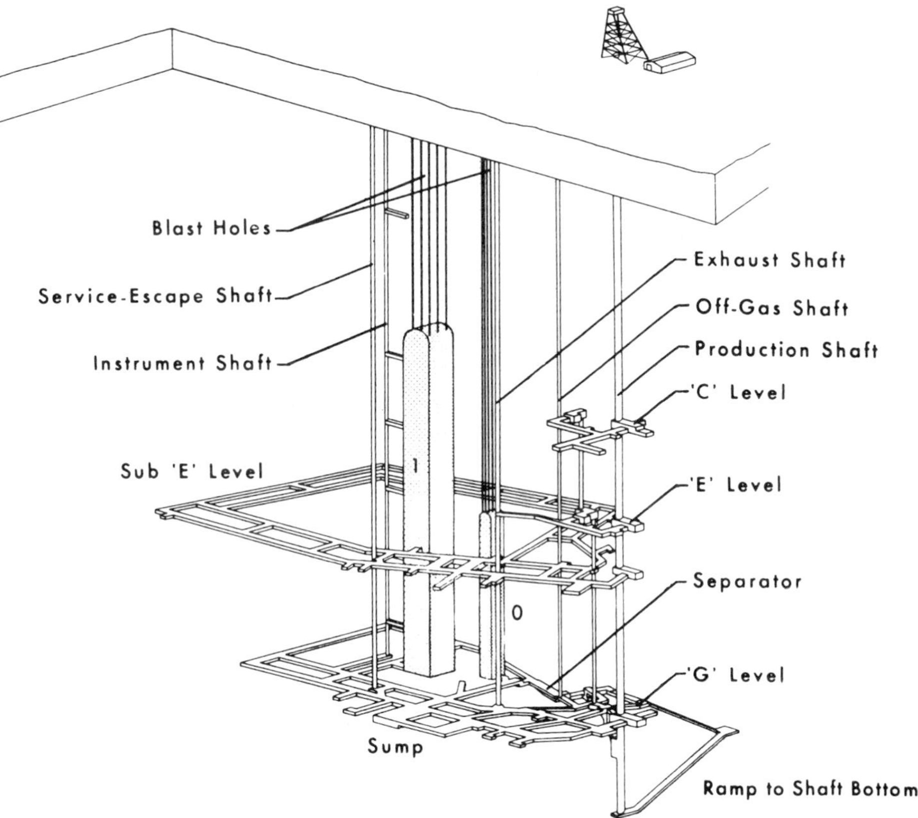

Figure 39 Rio Blanco Oil Shale Company's experimental retort 0 is to the right of the much larger retort 1 in this isometric view, which also shows the layout of the three-level mine (Berry et al., 1982; reprinted with the permission of the Colorado School of Mines).

Key features of the RBOSC mine include the production shaft which was constructed first using conventional shaft-sinking equipment, the exhaust shaft which was then raise-bored to the surface to establish a conventional ventilation system for the mine, and the service/escape shaft which was also raise-bored to the surface. A separate circuit for supplying air and steam to the retorts through the surface-drilled blastholes and for piping off-gases back to the surface through the off-gas shaft was sealed off from the working areas of the mine. A unique feature of the RBOSC design is the underground oil-water separator that is a mined cavity that minimizes the capital cost of large above-ground tankage.

The main production level is the G level, 850 ft below the surface. The G level provided access to the bottoms of the retorts before bulkheads were installed to seal off the mine from the retorts.

Retort 1 extends through the upper aquifer of the Piceance Basin in which the pressure is several hundred feet of water. Initially a ring of surface wells was drilled to dewater the area around the production shaft; later the sub-E level was mined and drain holes were drilled into the aquifer. After water production rates as high as 3600 gal/min, water rates stabilized near 1100 gal/min and only 10–30 gal/min was entering retort 1 at its ignition.

Retort 0 was not instrumented. In contrast, retort 1 was heavily instrumented with over 150 thermocouples and many pressure taps drilled into the rubble to provide a clear picture of conditions inside the operating retort.

The RBOSC proprietary rubblizing procedure that affords high void volume, and therefore high permeability, is felt to be the single factor most responsible for the success of the process. The first step in this procedure is to drill blast holes from the surface to the full depth of the planned retort and undercut a room that will serve as a product drain (Figure 40). In step 2, the lower part of each blast hole is loaded with an explosive that is detonated to blast a segment of roof down into the room at the bottom. Part of the rubble is mucked, leaving the rubble at its angle of repose. The load, blast, and muck sequence is repeated until the desired void space is obtained. In step 3, the remaining roof rock is rubblized without mucking until the entire retort is constructed with only a small free space at the top to serve as a feed gas distributor.

The RBOSC rubblizing procedure was found to give a relatively small average particle size with a random size distribution that is conducive to high, even, and maintainable permeability. This was confirmed by extensive cold flow tests with freon tracer gases. These tests indicated an overall sweep efficiency of 85% for both retorts. Sweep efficiency in retort 1 could be investigated in more detail because of its instrumentation. In this case, it was found that sweep efficiency in the top two-thirds was 100%, while that in the lower third was only 60%. Leak rates were measured under stable pressure conditions and void volumes were calculated from pressure changes with the effluent valves closed. Cold flow pressure drop as a function of flow rate could be calculated from the Ergun equation using the experimental void volume and a calculated effective mean particle size.

Ignition of the retorts was achieved within 28 hr using gas-fired burners lowered downhill. Durability of the burners, a problem in some other in situ studies, was demonstrated by repeated shutdown/reignition cycles. It was found that the higher pressure drop in ignited rubble tended to direct flow to unignited rubble and thereby to promote even spreading of the flame front.

Air and steam were fed to the ignited retort through the same boreholes used for blasting (Figure 41). Products flowed to the underground separator, from which oil and water were pumped separately to the surface. The philosophy for operation was to maximize the flame front advance within the safe operating limits of the available equipment. Consequently, the advance rate decreased with time due to the increasing pressure drop from the top of the bed to the front. Even though some problems that limited advance (air leakage, sulfur emission limits) were encountered, the front advance averaged about 3 ft/day and were considerably higher during the early stages (Figure 42).

Oil production, which totaled 24,444 bbl for retort 1, includes liquid oil from the separator room, the C_6+ condensables from the exit gas, and any oil mist found as an aerosol in the exit gas. The first liquid oil appeared about 15 days after ignition, and thereafter oil production tended to mirror the flame front advance (Figure 42).

Oil recovery in the RBOSC tests was very good, i.e., 68% of Fischer assay from each retort (Table 26). The maximum yield value in Table 26 was calculated using a computer model from Lawrence Livermore National Laboratory which takes into account the shale grade, particle size, air/steam ratio, and front advance rate. The maximum calculation assumes 100% sweep efficiency. However, it will be recalled that efficiency was only about 85% for these retorts. Thus, the experimental values are in reasonable accord with expected behavior.

In summary, the Rio Blanco effort led to a simple design that approaches laboratory results in oil recovery efficiency. The high void-volume retorts are suitable for MIS retorting at high burn rates. The entire retort can be developed from a single mine level (e.g., the G level); no access to the upper part of the retort is needed for ignition. Safe and very effective ignition was achieved using downhole burners. In addition, it has been demonstrated that successful retorting can be achieved in the dewatered upper aquifer. Even though current tract C-a interest has shifted, the success of the RBOSC-MIS

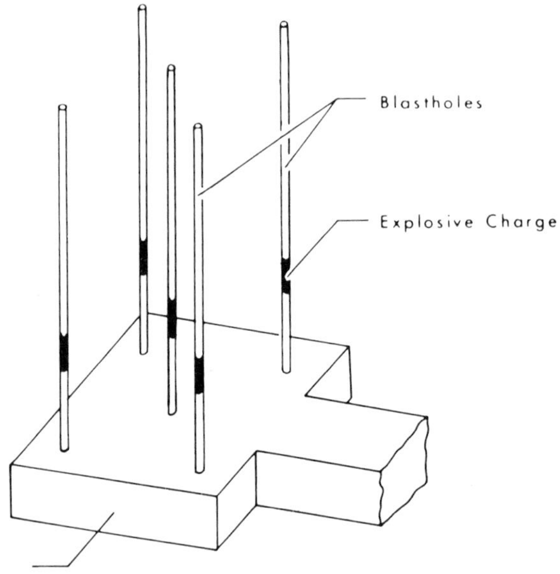

Figure 40 Construction of RBOSC's high-void retort involves three steps (Berry et al., 1982; reprinted with the permission of the Colorado School of Mines).

effort suggest that the technology may be applicable to other deposits situated even deeper beneath the surface than that at tract C-a.

E. Equity Oil/Arco BX Process

In the leached zones of the Piceance Basin, dissolution of water-soluble minerals has led to a relatively rich oil shale of high natural permeability. The Equity Oil BX project took advantage of this permeability for in situ retorting, using injected superheated steam to heat the oil shale (Jacobs et al., 1980). The project was carried out on a 1000-acre fee property jointly owned by Equity Oil and Atlantic Richfield (ARCO) and located in Rio Blanco County, Colorado, about midway between tracts C-a (Rio Blanco) and C-b (Occidental). At the project site, the leached zone averages 540 ft thick with shale assaying 24 gal/ton.

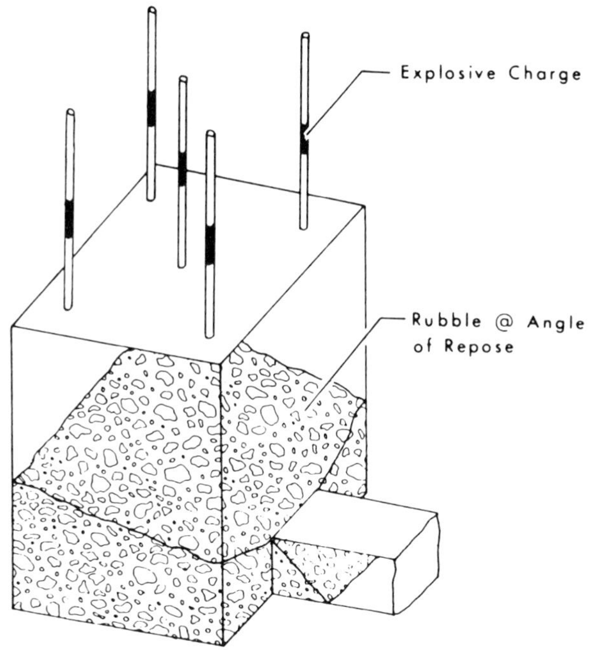
Explosive Charge
Rubble @ Angle of Repose
Step 2

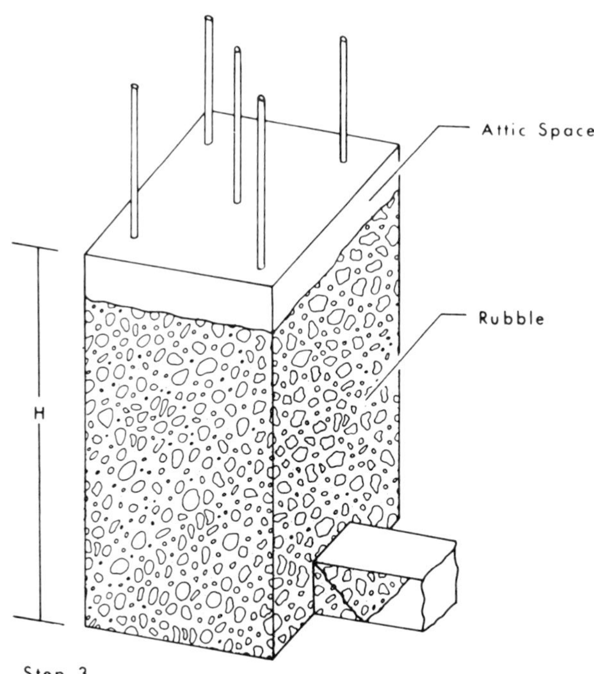
Attic Space
Rubble
H
Step 3

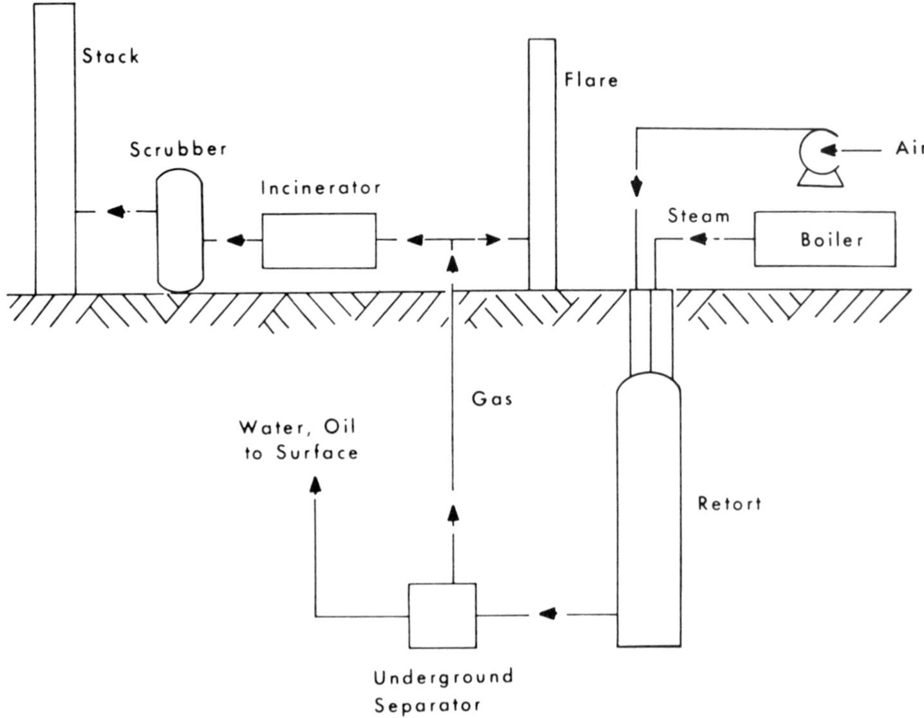

Figure 41 Schematic representation of the overall RBOSC retorting process showing the ancillary equipment on the surface (Berry et al., 1982).

The project comprised a pattern of eight injection wells, five production wells, and three observation wells covering an area of 0.7 acre (Figure 43). Initial plans were to inject steam at 540°C (1000°F), 1500 psig, and a total pattern rate of 974,000 lb steam per day (2784 bbl/day water). This was to be maintained over a 2-year period to produce a total of 650,000 bbl of shale oil (100% of Fischer assay). Because of equipment limitations and mechanical problems, these injection targets were not met. As a result, very little shale oil was actually produced. The initial oil observation was made only after a year of steam injection and 18 months of steam injection yielded 46 bbl of crude shale oil.

Not only the quantity but also the quality of permeability seems to be important. The permeability of the leached zone shale is not

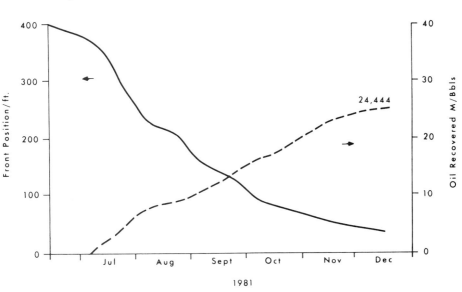

Figure 42 Flame front advance and oil production in RBOSC retort 1 (Berry et al., 1983; reprinted with the permission of the Colorado School of Mines).

Table 26 Oil Yields and Operating Parameters for Rio Blanco Modified In Situ Retorts 0 and 1

Parameter	Retort 0	Retort 1
Air/stream ratio	50/50	70/30
Front advance rate, avg. ft/day	2.7	3.0
Actual oil yield, bbl	1,876	24,444
Fischer assay of available rubble, gal/ton	17.3	21.6
Maximum calculated oil yield, % FA	70	79
Actual oil yield, % FA	68	68

Source: Data of Berry et al., 1982.

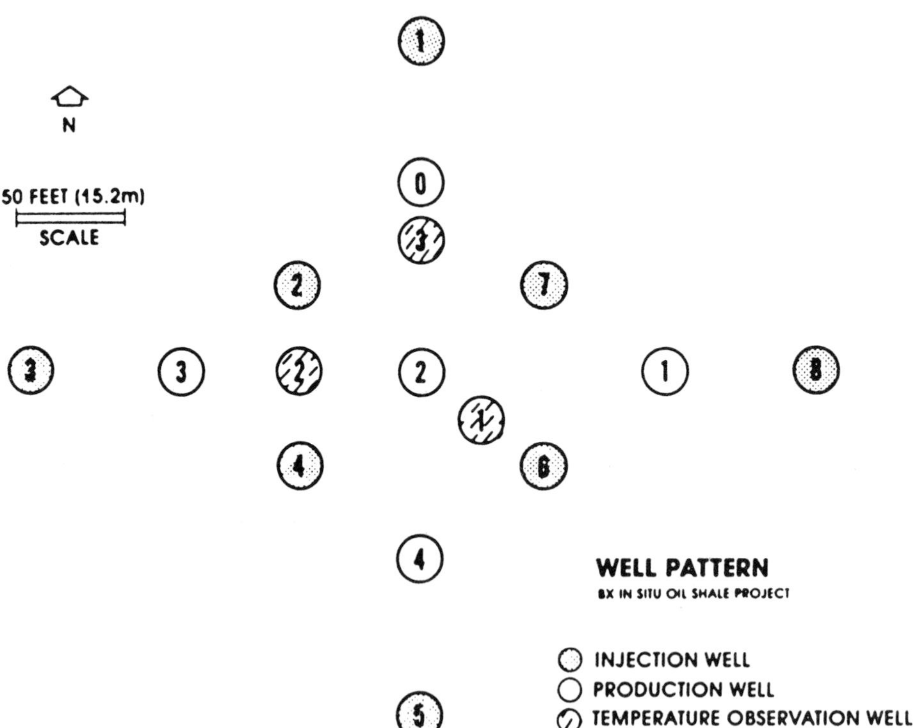

Figure 43 The Equity Oil/ARCO BX project included 16 wells drilled in a pattern covering 0.7 acre (Dougan and Docktor, 1981; reprinted with the permission of the Colorado School of Mines).

uniform; it is appreciably lower in the vertical direction than in the horizontal direction (bedding plane). The effects of this anisotropic permeability are readily apparent in Figure 44 which shows that efficient heating was confined to the two steam injection zones. These results do not preclude using the Equity Oil steam injection scheme, but does mean that more complicated and expensive injection equipment would be needed for distribution of the steam into multiple injection zones to provide uniform heating of the shale formation.

In summary, the BX project attempt to take advantage of the leached zone's natural permeability for in situ retorting by superheated steam injection did not result in high oil recoveries. In large part, the BX results were traced to equipment limitations. However, the results also showed clearly that the anisotropy of leached zone permeability is a factor to be reckoned with in any future work along these lines.

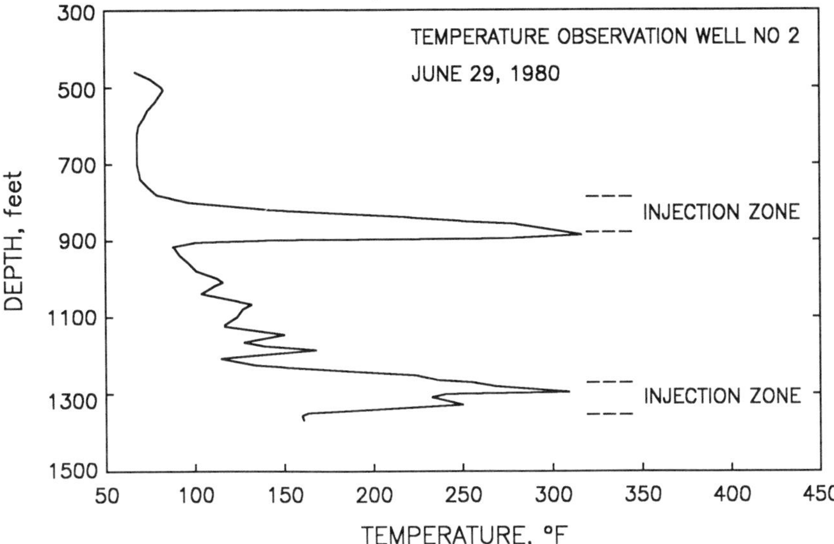

Figure 44 The temperature profile versus depth for BX observation well no. 2, located near the center of the pattern, shows that efficient heating was confined to the two steam injection zones (Dougan and Docktor, 1981; reprinted with the permission of the Colorado School of Mines).

30

Shale Oil Refining

I	Composition	1005
II	Upgrading	1007
	A. Dewatering and Solids Removal	1010
	B. Arsenic and Iron Removal	1013
	C. Noncatalytic Processes	1013
	D. Catalytic Processes	1016

I. COMPOSITION

Crude shale oil, sometimes termed *retort oil*, is the liquid oil condensed from the effluent in oil shale retorting. Crude shale oil typically contains appreciable amounts of water and solids, as well as having an irrepressible tendency to form sediments. As a result, it must be upgraded to a *synthetic crude oil* (*syncrude*) before being suitable for pipelining or substitution for petroleum crude as a refinery feedstock. However, shale oils are sufficiently different from petroleum crudes (Chap. 3) that processing shale oil presents some unusual problems.

Shale oils, especially those from Green River oil shale, have particularly high nitrogen contents—typically 1.7–2.2 wt % versus 0.2–0.3 wt % for a typical petroleum (Chap. 3, Sec. I). In many other shale oils (including those from eastern U.S. shales) nitrogen contents are lower than in the Green River shale oils, but still higher than those typical of petroleums. Because retorted shale oils are produced by a thermal cracking process, olefin and diolefin contents are high. It is the presence of these olefins and diolefins in conjunction with high nitrogen contents that gives crude shale oils their characteristic instability toward sediment formation. The sulfur contents of shale oils vary widely but are generally lower than those of high-sulfur petroleum crudes and tar sand bitumens (Chap. 3, Sec. I, and Chap. 14, Sec. I).

For raw shale oils produced from the Utah part of the Green River formation (Figure 1), the sulfur content is relatively constant in all the fractions, while nitrogen is concentrated in the higher boiling fractions (see also Chap. 4).

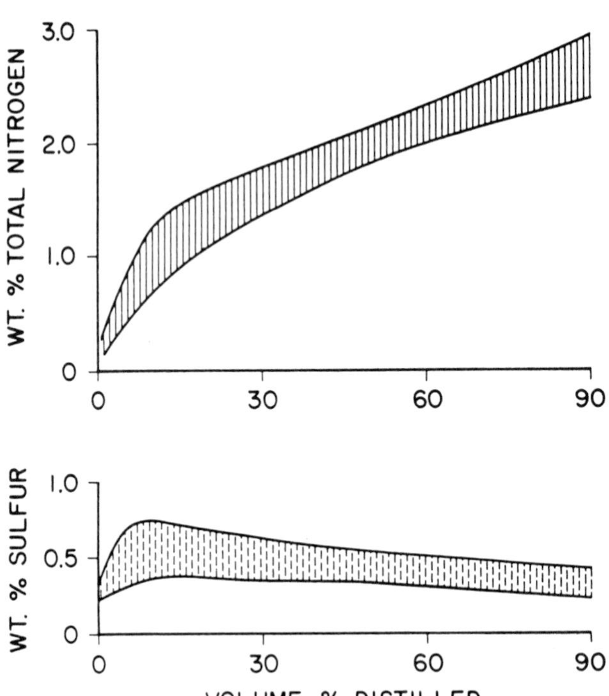

Figure 1 Nitrogen and sulfur concentrations in raw Green River shale oil as a function of distillation range. The shaded band in each case shows the range of data from oils produced by the Union-B and Paraho-DH (directly heated) retorts (Lovell, 1978; reprinted with the permission of the Colorado School of Mines).

In addition to the olefins and diolefins mentioned above, the Green River shale oils contain appreciable amounts of aromatics, polar aromatics, and pentane insolubles (asphaltenes) (Chap. 2, Sec. I.C., and Chap. 4, Sec. II). The distribution of compound types as a function of distillation range shows (Figure 2) that the concentration of polar aromatics and pentane insolubles in the higher boiling fractions parallels the nitrogen concentration in these fractions. Oxygen contents are higher than those typically found in petroleum (Chap. 3, Sec. I), but lower than those of crude coal liquids (Chap. 23). Crude shale oils also contain appreciable amounts of soluble arsenic, iron, and nickel that cannot be removed by filtration (Table 1).

II. UPGRADING

Upgrading, or partial refining, to improve the properties of a crude shale oil may be carried out with different objectives, depending on the intended use for the product:

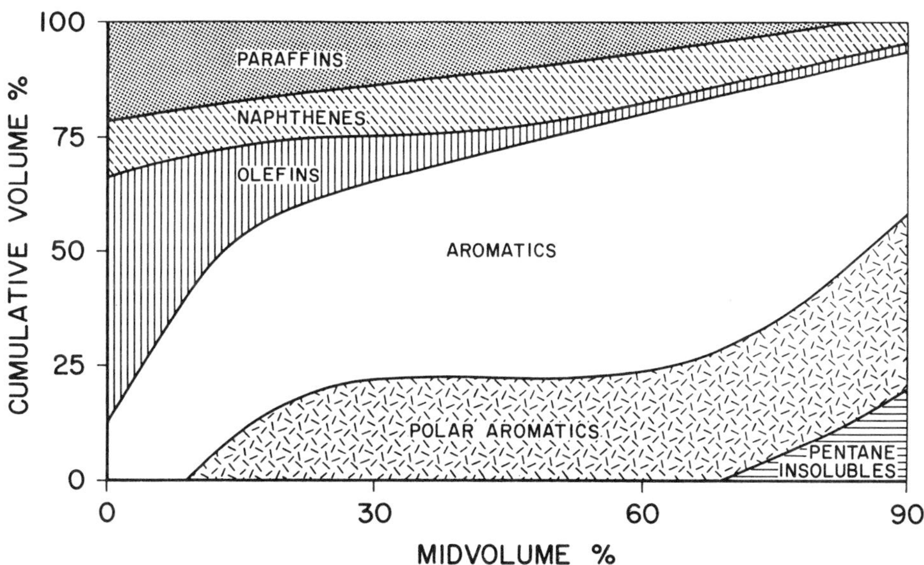

Figure 2 Distribution of compound types in crude Green River shale oil as a function of distillation range (Lovell, 1978; reprinted with the permission of the Colorado School of Mines).

Table 1 Compositions of Some Crude Shale Oils

	Paraho-DH[a] Colorado	Occidental MIS[b] Colorado	Dravo[c] Kentucky	CSR[d] Julia Creek
Gravity, °API	20.4	22.9	12.0	10.1
Pour point, °F	85	65	15	-11
Viscosity at 100°F (38°C)	213 SSU	—	—	15.9 cSt
122°F (50°C)	44.9 SSU	21.94 cSt	—	—
210°F (99°C)	—	5.27 cSt	—	—
BS&W, wt %	0.05	0.2	—	—
Conradson carbon, wt %	2.98	1.36	—	6.9
Elemental Analysis				
Carbon, wt %	83.68	84.85	83.19	83.0
Hydrogen, wt %	11.17	12.27	9.84	9.5
Nitrogen, wt %	2.02	1.51	1.14	1.3
Oxygen, wt %	1.38	0.65	1.97	1.3
Sulfur, wt %	0.70	0.64	3.86	4.9
Ash, wt %	—	0.014	0.092	—
Metals				
Arsenic, ppm	28	27.5	100	26
Iron, ppm	53	45	110	55
Nickel, ppm	2.4	6.7	—	3

Vanadium, ppm	0.17		0.42		n.d.		153
Sodium, ppm	0.3		11		n.d.		—
Hydrocarbon type							
Saturates, wt %	—		—		11.4	C_{ar} =	51.6%
Aromatics, wt %	—		—		54.7	H_{ar} =	12.4%
Polars, wt %	—		—		14.6	H_{ol} =	3.2%
Asphaltenes (C_7), wt %	0.89		0.34		—		—
Bromine number	—		23.6		—		57
Distillation (ASTM D1160),	°F	°C	°F	°C	°F	°C	
IBP	90	32	376	191	—		—
5%	255	124	467	242	439	226	—
10%	324	162	570	299	477	247	—
30%	454	234	670	354	598	314	—
50%	572	300	712	378	714	379	—
70%	675	357	820	438	836	447	—
90%	834	445	953	512	981	527	—
95%	895	479	—		—		—
EP	972	522	—		—		—
% recovered	97		87		—		—

[a] Robinson and Evin, 1983.
[b] Sikonia et al., 1977.
[c] Weichman, 1976.
[d] Atkins et al., 1986.

Stabilization to produce a pipelinable oil that can be transported to a distant refinery
More complete upgrading to produce a premium refinery feedstock with low nitrogen, low sulfur, and essentially no residuum
Upgrading to produce chemical feedstock streams
Complete refining of the crude shale oil or selected fractions to produce finished end products (e.g., gasoline, diesel, jet fuel)

It is difficult to generalize regarding shale oil processing. Not only do the shale oil properties vary; refineries vary widely. For example, there are about 300 fluid catalytic cracking (FCC) units in free world refineries—and these use more than 260 different cracking catalysts! Therefore, several of the reported large-scale studies have been selected to illustrate the major features of shale oil upgrading and refining. These studies have generally used one of three approaches:

Thermal conversion (visbreaking, coking) followed by hydrotreating
Hydrotreating followed by fluid catalytic cracking
Hydrotreating followed by hydrocracking

Much of the recent information about oil shale upgrading and refining has come from studies sponsored by the U.S. Department of Defense and Department of Energy, and carried out under contract by petroleum refiners (Table 2).

A. Dewatering and Solids Removal

Crude shale oil usually contains emulsified water and suspended solids. Therefore, the first step in upgrading is usually dewatering/ desalting.

Sikonia and his coworkers at UOP found that a conventional two-stage electric desalter reduced the water content of a crude Occidental MIS shale oil to 0.05 wt % but was not able to achieve the 10 ppm toluene-insoluble solids level desired for hydrotreating (Sikonia et al., 1983). The 100–500 ppm solids left in the feed necessitated special measures to prevent or alleviate fouling and pressure drop problems in fixed-bed reactors.

Sullivan and Stangeland (1979) described dewatering of a crude shale oil produced by the Paraho semiworks retort (indirectly heated mode). As received, the crude shale oil was an emulsion containing 6 wt % water and about 0.5 wt % fine solids. Heating to 77°C (170°F) broke the emulsion; most of the water and suspended solids settled

Table 2 Shale Oil Upgrading and Refining Studies Sponsored by the U.S. Department of Defense (DOD) and Department of Energy (DOE)

Processing program	Crude shale oil source	Upgrading contractor
• 1975		
Shale I (DOD)	Paraho I	Gary Western (Bartick et al., 1975).
• 1975 - 1976		
Aviation turbine fuels from synthetic crudes (DOD)		Exxon Research & Engineering (Taylor et al., 1978)
• 1978		
Shale II (DOD)	Paraho II	Sohio (Toledo) (Robinson, 1979)
Advanced catalytic processes (DOE)	Paraho	Chevron (Sullivan and Stangeland, 1979)
• 1978 - 1981		
Process methods	Paraho II/Oxy no. 6	Ashland (Moore et al., 1983)
Process methods	Paraho II/Oxy no. 6	Suntech (Schwedock et al., 1983)
Process methods	Paraho II/Oxy no. 6	Amoco (Tait and Hensley, 1982)
• 1979 - 1981		
Catalyst development	Occidental retort no. 6	Amoco (Tait et al., 1983)
• 1980		
Shale III	Geokinetics	HRI–Suntech

out after standing at 77°C (170°F) for 6 hr. Passage through a 15 μm filter removed little additional material; however, a 0.45-μm filter retained 252 ppm fine solids (Table 3).

Table 3 Properties of the Dewatered Paraho-IH[a] Crude Shale Oil Upgraded at Chevron's Salt Lake City Refinery

Gravity, °API	20.2	
Pour point, °F (°C)	90 (32)	
Total nitrogen, wt %	2.18	
Sulfur, wt %	0.66	
Arsenic, ppm	28	
Iron, ppm[b]	70	
Carbon, wt %	84.30	
Hydrogen, wt %	11.29	
Atomic H/C ratio	1.60	
Oxygen, wt %	1.16	
Chloride, ppm	<0.2	
Ash, wt % (ASTM D-486)	0.03	
Filter residue ash (0.45 μm filter):		
Total solids, ppm	252	
Ash, ppm	194	
BS&W (sediment water), vol %	0.1	
Bromine number	51	
Average molecular weight	326	
Viscosity, cSt		
122°F (50°C)	25.45	
210°F (99°C)	5.54	
Acid neutralization number, mg KOH/g	2.3	
Base neutralization number, mg KOH/g	38	
pH	9.2	
Maleic anhydride number, mg/g	40.6	
Heptane asphaltenes (includes any fines), wt%	0.17	
ASTM D1160 Distillation		
IBP/5	386/456°F	(197/235°C)
10/30	508/659°F	(264/348°C)
50	776°F	(413°C)
70/90	871/995°F	(466/535°C)
EP	1022°F	(550°C)
% overhead (excludes trap)	94	
% in trap	1	
% in flask	5	

[a]Produced in the indirectly heated mode.
[b]This iron was not removed by filtration through a 0.45-μm filter.
Source: Data of Sullivan and Stangeland, 1979.

B. Arsenic and Iron Removal

If not removed, the arsenic and iron in shale oil would poison and foul the supported catalysts used in hydrotreating. Because these materials are soluble, they cannot be removed by filtration. Several methods have been used specifically to remove arsenic and iron. Other methods involve hydrotreating; these also lower sulfur, olefin, and diolefin contents and thereby make the upgraded product less prone to gum formation.

Workers at Atlantic Richfield (ARCO) found that contacting the crude shale oil with hydrogen at 427°C (800°F) and 1400 psig (visbreaking conditions) resulted in precipitation of arsenic and selenium, and also lowered the pour point of the oil (Curtin, 1977). A crude shale oil containing 17.4 ppm arsenic and having a pour point of 24°C (75°F) was processed to obtain an upgraded oil with 2 ppm arsenic and a pour point of -34°C (-30°F).

Dhondt (1979) described Union Oil's treatment to remove solids and arsenic. Two-stage water washing removes suspended solids, which are recycled. Arsenic is removed by a proprietary, nickel-containing absorbent at 288–343°C (550–649°F) in the presence of hydrogen at moderate pressure (Young, 1977; Alley et al., 1984). It is claimed that this absorbent will lower arsenic content from 500 ppm to less than 1 ppm, and that the absorbent can pick up arsenic to about 80% of its own weight.

For commercial scale processing (73,000 bbl) of crude Paraho shale oil, Sohio workers installed a guard bed containing layers of high-surface 1/16-in. alumina extrudate loaded between layers of low-surface 1/4- and 1/2-in. alumina balls (Figures 3 and 4). No active metal was deposited on the guard bed packings and the bed was only operated at 215°C (420°F). As a result, little or no arsenic was removed by the guard bed. However, iron was removed as iron pyrite which remained in the guard bed. Also deposited in the guard bed was a black pitch material, whose formation may have been caused by inadvertent recycle of sulfuric acid to the shale feed tank during startup of the acid treater used as a final polishing step for the jet and diesel fuels. Buildup of pyrite and pitch was sufficient to cause bed plugging after 24 days on-stream.

C. Noncatalytic Processes

Thermal conversions, coking, and visbreaking are conceptually simple, noncatalytic methods for lowering the high pour point and viscosity of raw shale oils, in order to make the oil more suitable for

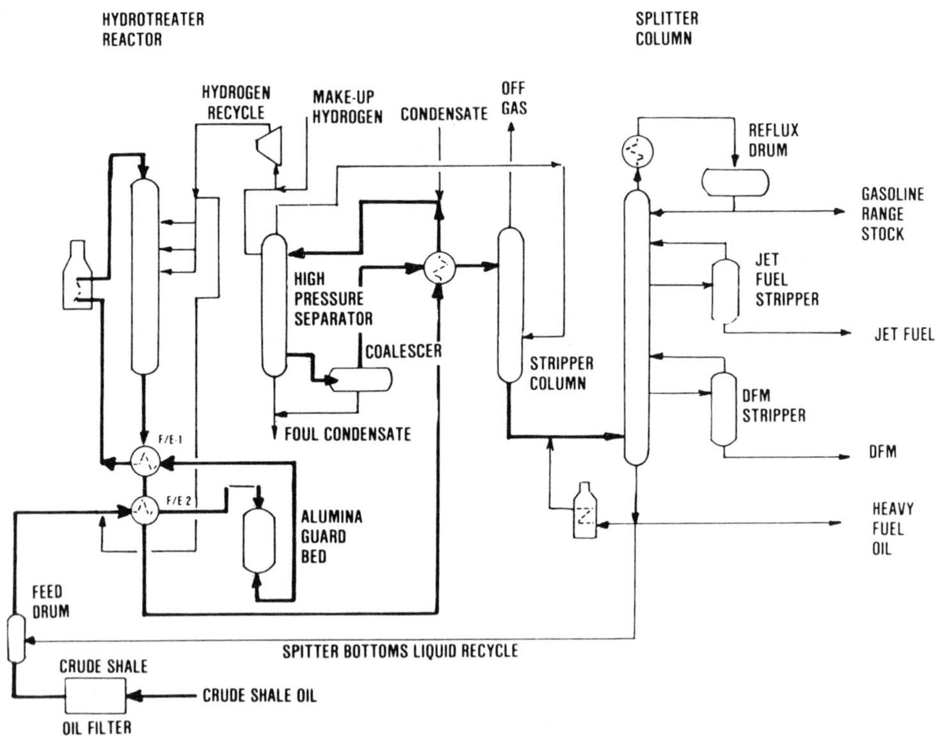

Figure 3 Sohio hydrotreating unit used to process 73,000 bbl of crude Paraho shale oil into military transportation fuels (Robinson and Evin, 1983; reprinted with the permission of Butterworth Publishers).

hydrotreating that is needed to remove nitrogen and sulfur. Coking also separates suspended solids.

Visbreaking (Chap. 6, Sec. I.B.1) is a mild thermal treatment that lowers viscosity and pour point to make the shale oil pipelinable. As noted above, it also precipitates arsenic but does little to reduce the contents of nitrogen, sulfur, or olefins. In visbreaking, the oil is heated to 420–480°C (788–896°F) for a short time (seconds to minutes) under hydrogen, during which some product is cracked to gas, along with the desired cracking. Capital costs for visbreaking are low, but energy consumption is high for the benefits obtained. As a result, visbreaking is not a preferred method for upgrading.

Delayed coking (Chap. 6, Sec. I.B.2) followed by hydrotreating was used in the upgrading of 8505 bbl of crude Paraho oil shale at

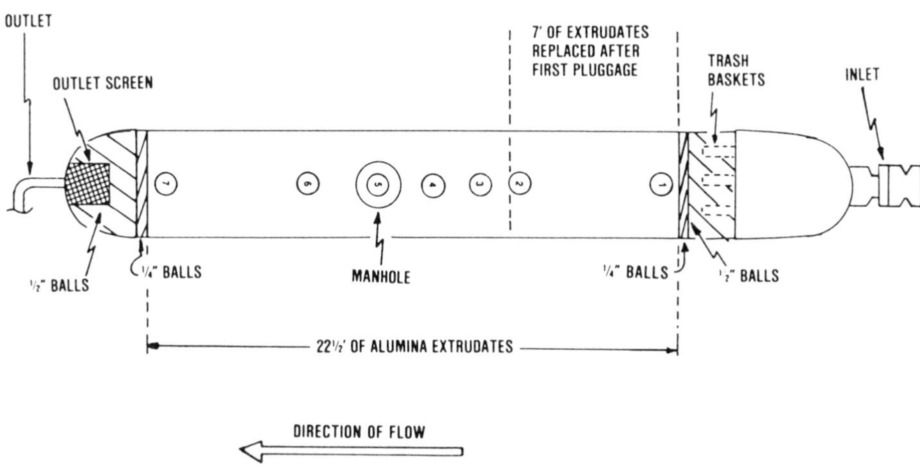

Figure 4 The guard bed used in Sohio's shale oil hydrotreating study was packed with alumina extrudate and alumina balls (Robinson and Evin, 1983; reprinted with the permission of Butterworth Publishers).

the Gary Western Refinery (Bartick et al., 1975). Hawk and coworkers (1964) used coking, followed by severe hydrotreating, to produce experimental quantities of military jet and diesel fuels in the hydrogenation pilot plant at USBM's Bruceton, PA facility. Delayed coking was studied at the bench and pilot plant scales at Chevron (Sullivan et al., 1978) and a bench scale study was carried out at Phillips Petroleum (Montgomery, 1968). Delayed coking was also used to process about 3400 bbl of Occidental MIS crude shale oil at Chevron's Salt Lake City refinery, but in this case the shale oil (13–19%) was coprocessed with the refinery's normal petroleum residuum (Sullivan et al., 1978). In delayed coking, the oil is heated to 480°C (896°F) and fed to one of two coke drums that are typically 10 ft in diameter and 100 ft tall. Pressure in the drums is sufficient to prevent vaporization; hence coking reactions take place in the liquid phase. Liquid is continuously withdrawn and coking is allowed to proceed until the coke drum is nearly full. Feed is then switched to the second drum and the first is cooled and emptied. Because no catalyst is involved, coking is very tolerant of metals and solids. However, in the Gary Western test about 20% of the shale oil feed was converted to low-value gas and nearly 30% was converted to coke. Thus, the yield of high-value transportation fuels amounted to only one-half the shale oil fed to the coker. Moreover, because of its high impurity content the shale-derived coke was not suitable for making carbon electrodes and could only be used for fuel. In the USBM and Phillips studies, considerably lower coke makes were achieved, but the coke makes of 27 and 13%, respectively, still represented a substantial loss of valuable oil. Consequently, delayed coking cannot be regarded as a preferred method for shale oil upgrading. Nevertheless, the Gary Western, USBM, and Phillips refining tests did demonstrate that shale oil can be processed into transportation fuels, using conventional refining technology with suitable adjustments of operating parameters.

Two more advanced coking technologies, Exxon's fluid coking and flexicoking processes, warrant mention. Although no reports were found of their application to shale oil upgrading, they have been successfully used to upgrade the heavy bitumen from Canadian tar sands as well as conventional petroleum residua (Speight, 1980).

Fluid coking (Chap. 6, Sec. I.B.2) is carried out with steam as a fluidizing gas that also serves to strip liquid products from the coke. Coke particles are continuously withdrawn from the coker and fed to a burner, where they are fluidized in air and partially burned and heated to about 620°C (1148°F). The hot coke is returned to the coker where it heats the incoming feed, leading to deposition of more coke. Only about 5% of the coke is actually burned to raise heat, so the "fluid coke" is withdrawn as a product. Compared to delayed coking, fluid coking offers the advantages of being a continuous and more rapid process, with the shorter pyrolysis time leading to higher liquid yields than delayed coking.

Flexicoking (Chap. 8, Sec. IV) combines fluid coking with gasification of the fluid coke. The product gas can be used as fuel or reformed to produce hydrogen. Coking, combustion, and gasification are all high-temperature operations with high heat demands. By integrating all three operations into a single close-coupled unit, flexicoking achieves very high thermal efficiency, while retaining the high liquid yield capability of fluid coking.

D. Catalytic Processes

Hydrotreating (Chap. 6, Sec. III) is more flexible and less destructive than coking as a way to remove nitrogen, sulfur, oxygen, arsenic, and metals. (Here we discuss the hydrotreatment of crude shale oil, not the hydrotreatment of syncrude that would be used to produce finished products for market.) One approach has been to distill the crude shale oil, then to hydrotreat the fractions. Shaw et al. (1975) suggested that a more reasonable processing approach would be to hydrotreat the whole (dewatered/desalted) shale oil and then to process at least the <481°C (<900°F) product in a conventional refinery. This approach was followed by Chevron in its 1977 pilot plant study of Paraho shale oil upgrading (Sullivan et al., 1978) and Sohio in refining 73,000 bbl of Paraho oil at Toledo (Evin and Robinson, 1979).

The problems caused in the Sohio test by pyrite fines accumulation in the guard bed were discussed above. The Sohio hydrotreater also accumulated fines but did not experience a pressure drop problem during the 24-day test period. However, catalyst activity declined significantly, as shown by the whole-product nitrogen plot (Figure 5). After completion of the shale oil test, the top 10% of the catalyst

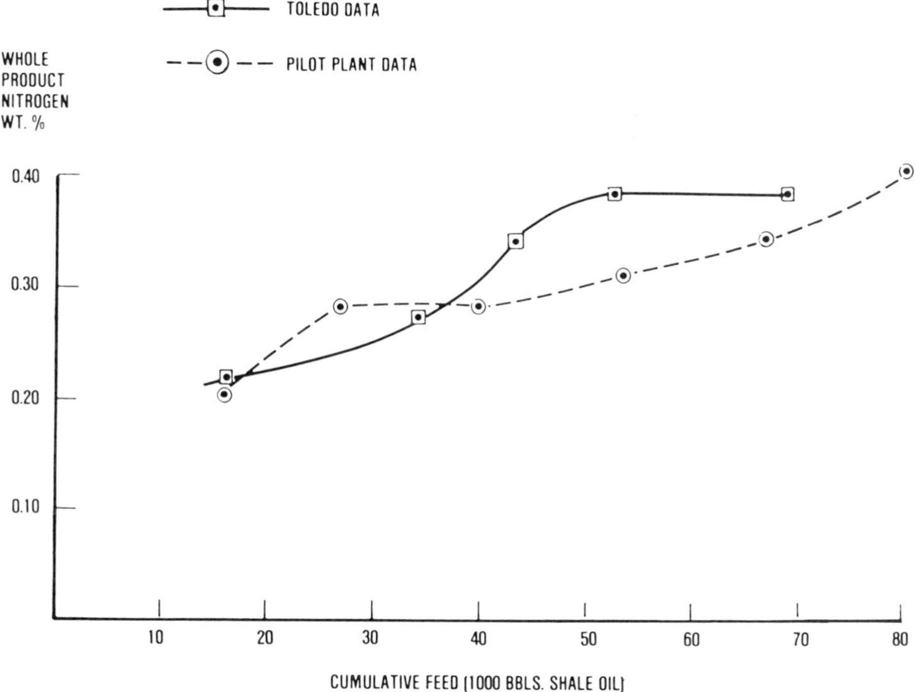

Figure 5 Increased nitrogen content of the product reflects deactivation of the catalyst during the 24 days needed to process 73,000 bbl of crude Paraho shale oil in Sohio's Toledo refinery. Data from the Sohio pilot plant study are included for comparison (data of Robinson and Evin, 1983; reprinted with the permission of Butterworth Publishers).

Table 4 Summary of Products from Sohio Refining of Crude Paraho Shale Oil

Product	Bbl	Vol %	% Nitrogen
Gasoline (including butanes)	8,473	11.96	0.067
Jet fuel: JP-5	9,546	13.73	0.220
JP-8	490		
Diesel fuel, marine (DFM)	18,939	25.90	0.430
Residual fuel oil	37,220	50.91	0.220
Total	74,939	102.50	

Source: Data of Robinson and Evin, 1983.

bed was replaced with fresh Shell 324 Ni/Mo catalyst and the reactor was placed back in service for petroleum hydrotreating. The reactor was still in service with the same catalyst 4 years later! These results demonstrate the need for an effective guard bed to remove arsenic.

Hydrogen consumption during Sohio's upgrading was about 1600 scf/bbl, yet the distillation results show that one-half of the syncrude was residual fuel oil as shown by the product summary (Table 4). Clearly, additional molecular weight reduction would be required to produce the gasoline-rich product slate typical of U.S. petroleum refineries.

Extensive hydrodenitrogenation (HDN) studies were carried out by Holmes at the Laramie Energy Technology Center (now Western Research Institute) in a small continuous unit, but under conditions that closely approximated the Sohio refinery conditions, except that Shell 514 support balls were used in the guard bed in the place of Sohio's alumina extrudate and no residuum was recycled to the hydrotreater. Adsorption chromatography on alumina and silica gel separated the starting Paraho crude shale oil and the Sohio and LETC upgraded oils into six fractions each, and the nitrogen content of each fraction was determined (Tables 5 and 6).

Fourteen nitrogen compound types were identified and quantified in the fractions, using a combination of infrared, high-resolution mass spectroscopy, and differential potentiometric titration. Because both the Sohio and LETC hydroprocessing removed about 80% of the total nitrogen, and nearly all the remaining nitrogen was characterized, the nitrogen compound distributions in the two products can be compared and correlated with hydroprocessing conditions. The extent of removal can be compared with the overall nitrogen removal of 80% to evaluate the relative susceptibility of each compound type to HDN. Nitrogen types

Table 5 Properties of Raw and Upgraded Paraho Shale Oil

		Paraho crude	Sohio hydro-processed	LETC hydro-processed
Carbon, wt %		84.0	86.5	86.1
Hydrogen, wt %		11.4	13.1	13.3
Nitrogen, wt %		2.19	0.43	0.42
Sulfur, wt %		0.66	0.02	0.05
Oxygen, wt %		2.0	0.12	0.27
H/C atomic ratio		1.63	1.82	1.85
Gravity, °API		21.4	33.8	34.9
Pour point, °F (°C)		85 (29)	85 (29)	70 (21)
Average molecular weight		311	265	295
Aromatic carbon, % C_{ar}		21	10	9
Aromatic hydrogen, H_{ar}		5.8	10.8	3.0
Simulated distillation (GC) cut point		Weight % of shale oil		
°C	°F			
38	100	nil	nil	0.3
93	199	0.2	0.2	0.8
149	300	0.7	2.8	3.8
204	399	3.7	7.5	8.2
260	500	8.9	12.9	15.0
316	601	11.9	14.0	15.3
371	700	14.8	18.3	16.4
427	801	16.3	15.7	15.0
482	900	17.7	12.4	14.2
538	1000	13.1	5.4	8.3
Residuum		12.7	10.7	2.7

Source: Robinson, 1978; Holmes, 1983.

that were more than 80% removed were relatively susceptible to HDN, while those less than 80% removed were relatively resistant to HDN.

Little difference was found in the susceptibility of different nitrogen base types to HDN; weak bases were only slightly more difficult

Table 6 Weight Distributions and Nitrogen Contents of the Fractions Obtained by Chromatography of the Hydroprocessed Shale Oils

	Paraho crude shale oil	Sohio	LETC
	Wt % of sample		
Hydrocarbon	31.7	81.6	77.6
Pyridine I	16.9	2.84	7.71
Pyrrole	3.56	0.48	2.43
Pyrrole/arylamine	5.47	2.54	1.03
Pyridine II	19.2	2.88	2.91
Amide/pyridine III	20.8	0.45	1.09
Total sample recovery	97.6	91[a]	93[a]
	Nitrogen, wt % of fraction		
Hydrocarbon	0.003	0.0006	0.0035
Pyridine I	1.22	4.51	1.68
Pyrrole	1.10	0.34	3.62
Pyrrole/arylamine	3.92	4.65	4.21
Pyridine II	4.11	5.79	3.89
Amide/pyridine III	4.53	5.99	3.29
Recovery, % of total N	100	102	98

[a]Low recovery in these cases is probably due to evaporation of the lighter components from the hydrocarbon fraction during removal of chromatographic solvent.
Source: Data of Holmes, 1983; Holmes and Thompson, 1981.

to remove than very weak or nonbasic species (Figure 6). Within experimental error, removal of nitrogen bases was the same for both hydroprocessing conditions.

Significant differences were, however, observed in the distributions of the individual compound types and ascribed to differences in HDN susceptibility and hydroprocessing conditions (Figure 7). With the exceptions of very weak bases of unknown structure (G) and *N*-alkylcarbazoles (H), the susceptibility of nitrogen compounds to HDN was

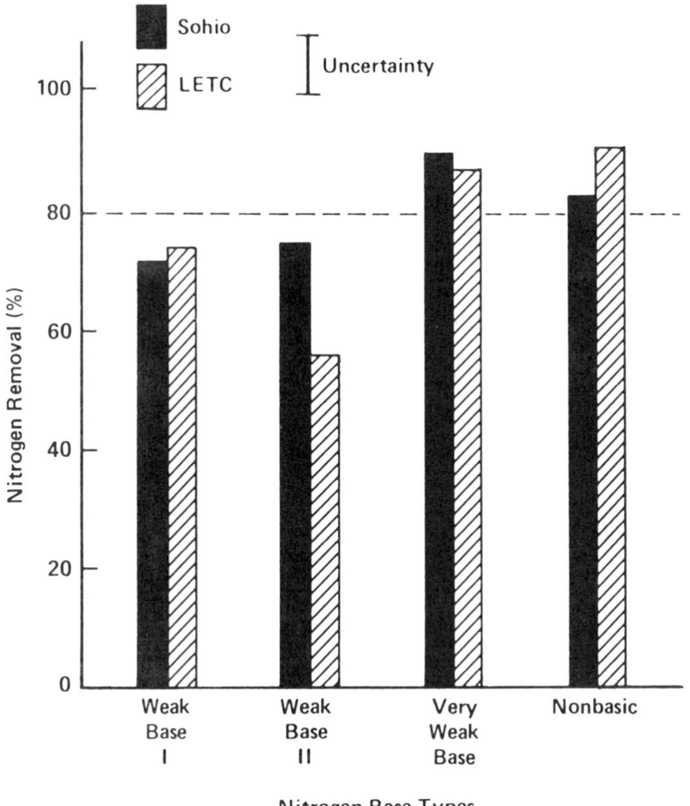

Figure 6 Removal of nitrogen bases was the same for both Sohio and LETC hydroprocessing conditions. Weakly basic nitrogen was only slightly more difficult to remove than very weak and nonbasic nitrogen (Holmes, 1983; reprinted with the permission of the Colorado School of Mines).

similar under both the Sohio and LETC conditions. It is most difficult to remove hindered alkylpyridines/quinolines (A), basic alkylpyrroles/indoles (F), and nonbasic alkylpyrroles/indoles/carbazoles/benzocarbazoles (I). The less hindered alkylpyridines/quinolines/acridines were removed in amounts proportional to the overall nitrogen removal, while the weak base II compounds of unknown structure (D) were only slightly more difficult to remove. The easiest nitrogen species to remove were the nonhindered alkylpyridines/quinolines/hydropyridines (C), alkylhydroxypyridines (E), and alkylcarboxamides/diazaaromatics

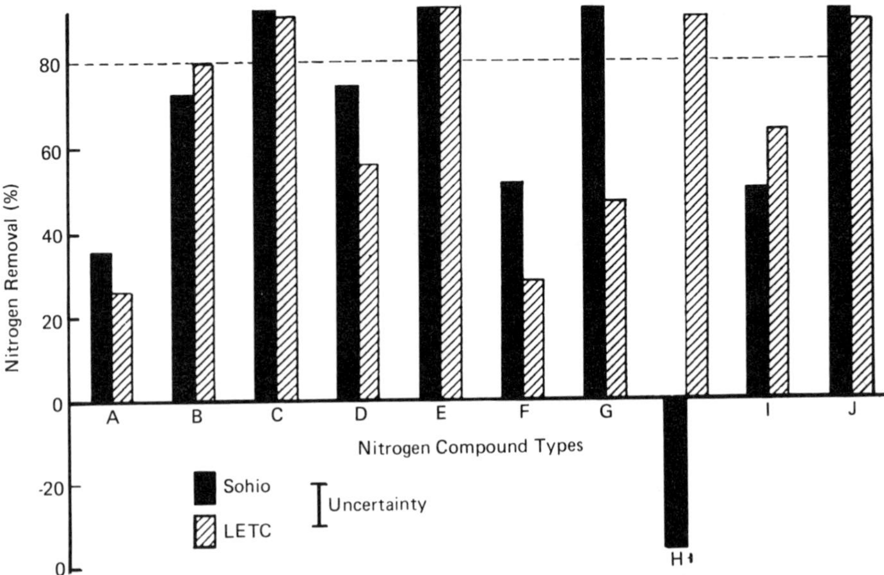

Figure 7 Removal of nitrogen compound types by hydroprocessing. Compound types are identified in the text (Holmes, 1983; reprinted with the permission of the Colorado School of Mines).

(J). The very weakly basic compounds of unknown structure (G) were apparently removed more effectively under the LETC conditions and the reason is not understood, but the amounts were small. The *build-up* of *N*-alkylcarbazoles during Sohio processing was ascribed to residuum recycle, since these heavy compounds become concentrated in the residuum.

The shale oil upgrading process developed by Gulf Oil is often described as two-stage hydrotreating but would be more properly termed a three-stage hydrotreating process (Figure 8). Data have been reported on upgrading of both western (Paraho, Union-B) and eastern (Dravo/Kentucky shale) shale oils (Lyzinski and Jones, 1984).

The pretreating section of the Gulf process is designed to remove arsenic, iron, trace metals, and residual solids, and to stabilize the oil before it enters the main hydrotreater. The first pretreating reactor is designed to remove most of the arsenic, iron, and solids with minimal hydrogen uptake. It operates at low hydrogen pressure with a low-cost disposable catalyst, and is duplicated so that one reactor can be on-stream while the other is being recharged with fresh catalyst. The second pretreating reactor stabilizes the oil and removes a

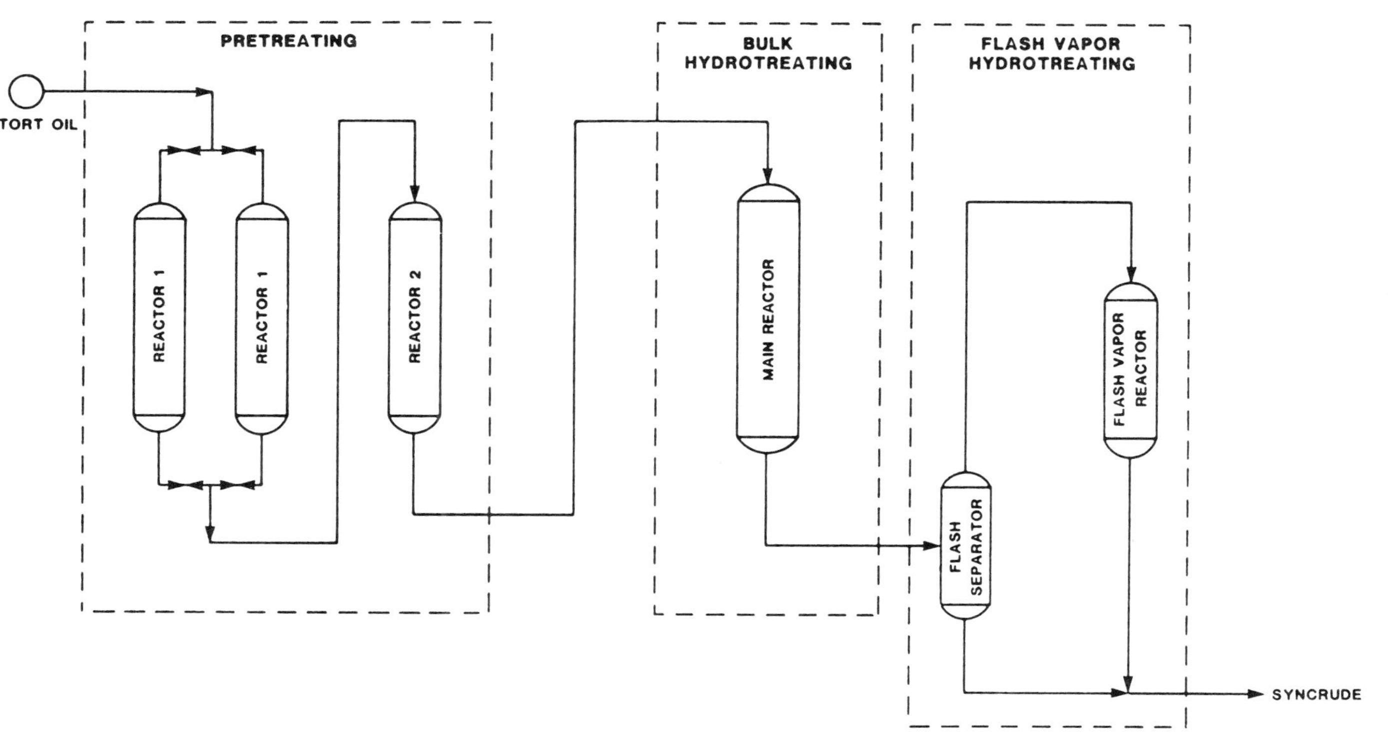

Figure 8 Flow diagram of the Gulf shale oil upgrading process (Jones et al., 1984; reprinted with the permission of the Colorado School of Mines).

substantial part of the sulfur. The main hydrotreating reactor contains a catalyst selected for high hydrodenitrogenation activity and produces an effluent that typically contains 750 ppm nitrogen. This effluent is fed to the flash separator, from which the <343°C (<650°F) vapors enter a vapor hydrotreater. Using about 15% of the total hydrogen, this final stage yields <191°C (<375°F) naphtha that typically contains <4 ppm nitrogen, and hence can be used as reformer feed and 191–343°C (375–650°F) distillate that contains 10–150 ppm nitrogen. The >343°C (>650°F) separator bottom typically contains 1000–2000 ppm nitrogen and is a good cracking feedstock.

Fluid catalytic cracking (FCC) (Chap. 2, Sec. I) of shale oils has been studied by Moore and his coworkers (1983) at Ashland Petroleum. Pilot plant FCC studies were made by Chevron (Sullivan et al., 1978). Sikonia and coworkers (1977) at UOP carried out small-scale studies of hydrotreating as a pretreatment for FCC. Both bench scale and pilot plant FCC studies of shale oil were carried out at Gulf.

In the FCC process, relatively small (40–80 μm) catalyst particles are fluidized by upflowing shale oil vapors in the reactor vessel, which is maintained at 510–538°C (950–1000°F) (Figure 9). As the cracked

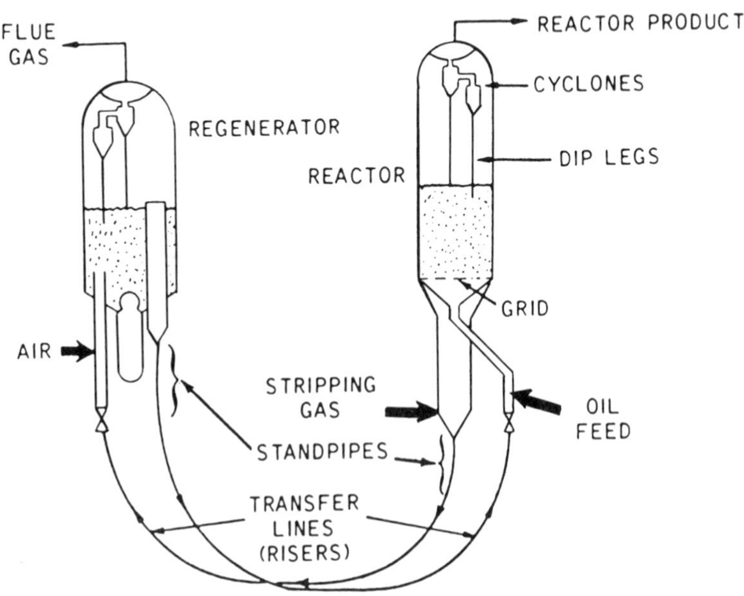

Figure 9 Fluid catalytic cracking unit.

vapors exit the reactor, they pass through cyclones where the catalyst particles are separated from the products and recycled. Along with the desired cracking reaction, some coke is deposited on the catalyst surface. Therefore, catalyst is continuously withdrawn from the cracking reactor and transported by an air stream into a combustor where the coke is burned. When returned to the cracking reactor the hot, regenerated catalyst supplies heat for vaporization of the feed.

Shale oils are rich in high molecular weight, waxy paraffinic material. Thermal cracking lowers molecular weight but yields straight-chain products of low octane number. Fluid catalytic cracking not only lowers molecular weight but also causes isomerization to produce branched products with higher octane numbers. As a result, the so-called cat-cracked shale naphtha is a more desirable feedstock for hydrotreating to make gasoline blend stock than is the naphtha for thermal cracking or coking of shale oil.

Hydrotreatment of the raw shale oil to remove basic nitrogen that would poison the acidic cracking catalyst and also promote unwanted coking is required as an FCC pretreatment (Sikonia et al., 1977). Workers at Gulf found that hydrotreated syncrudes from Union-B (Colorado shale) and Dravo (Kentucky shale) retorts performed well as FCC feeds in a MAT test, and the atmospheric tower bottoms from a Paraho syncrude gave good results in an FCC pilot plant test (Weichman, 1976).

Under an Air Force contract, Tait, Miller, and Hensley at Amoco investigated all three of the upgrading routes outlined at the start of this section (Tait et al., 1983). Of these, hydrotreating, followed by hydrocracking, offered the most flexibility and was the only approach to efficiently maximize jet fuel production. This is an important consideration, since western U.S. shale oils characteristically have high contents of straight-chain, waxy paraffins and low contents of the aromatics that are undesirable in jet fuel. (Aromatics set the smoke point in jet fuel.) Coal liquids, on the other hand, are characteristically highly aromatic and well suited for processing into gasoline blend stocks. Thus, its molecular structure makes shale oil a prime source for jet fuel.

As part of the Air Force contract, the Amoco workers developed catalysts that are capable of direct upgrading of a whole sheld oil into high yields of JP-4 jet fuel boiling range material in a single-stage process. In contrast to the multistep processes discussed above, the Amoco catalyst has the ability to sequentially saturate, dinitrogenate, and crack the whole shale oil in the presence of contaminants such as ammonia, water, and basic organic nitrogen compounds, while maintaining high selectivity toward the jet fuel boiling range materials.

The catalyst development studies were carried out on an Occidental MIS crude shale oil that contained 1.32 wt % nitrogen, 0.64 wt % sulfur, 1.33 wt % oxygen, 26 ppm arsenic, and about 60 ppm iron. Although it was recognized than an effective guard bed would be required for commercial operation, none was used in the development study, nor was the oil pretreated to remove arsenic or iron.

Screening of existing catalysts revealed that a proprietary catalyst containing cobalt, chromium, and molybdenum could effectively remove nitrogen and effect moderate conversion of the feed into the JP-4 boiling range. By screening catalysts in which the metals loading was systematically varied, it was found that optimal metal oxide loadings were 1.5% cobalt oxide, 10% chromium oxide, and 15% molybdenum oxide. The high loading of chromium oxide actually lowered denitrogenation activity but was necessary to impart high-temperature stability.

A study of different supports revealed that 20% silica/alumina was better than alumina and that 50% ultrastable molecular sieve/alumina

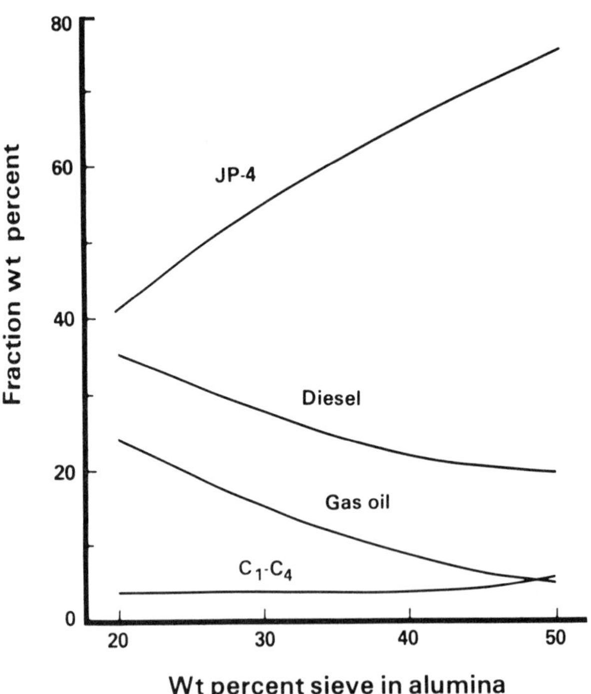

Figure 10 Effect of ultrastable molecular sieve concentration in the support on the product distribution from single-stage hydrotreating-hydrocracking of Occidental MIS crude shale oil (data of Tait and Hensley, 1982).

was more active yet. Next a series of eight supports containing 50% ultrastable sieve but having varying surface area (SA), pore volume (PV), and average pore diameter were prepared. These variations were achieved by modifying the alumina component, since modification of the sieves themselves could destroy their inherent cracking activity. Evaluation of catalysts prepared from these supports revealed that optimum activity was obtained with a catalyst having a narrow pore size distribution and APD of about 7 nm (Figures 10 and 11).

Activity testing was plagued by several upsets yet the optimized catalyst proved very durable. Based on the temperature response factor calculated from data obtained during the activity tests, an expected catalyst life of about 4.5 months was estimated for a constant JP-4 yield of 75% and reactor temperatures increasing from 413 to

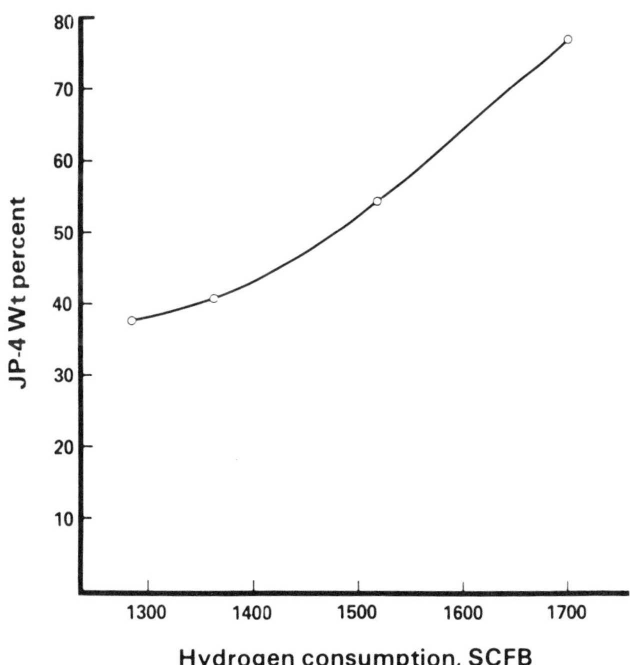

Figure 11 Correlation of JP-4 jet fuel yields with total hydrogen consumption for catalysts containing 0, 20, 30, and 50% ultrastable molecular sieve (left to right) in the support (data of Tait et al., 1982).

Table 7 Properties of Occidental MIS Crude Shale Oil Feed, Composite Product, and Jet Fuel Fractions from the Hydrocracked Shale Oil

Property	Occidental MIS crude shale oil feed	Composite product		JP-4		JP-8	
Gravity, °API	12.8	39		49.4 (49–57)		43.4 (37–51)	
Weight, %		100		76		61	
Pour point, °F (°C)	60 (16)	-5 (-21)		-85 (-65)		-40 (-40)	
Aromatics, vol %	—	—		16.0 (25.0)		18.0 (25.0)	
Olefins, vol %	—	—		1.0 (5.0)		2.5 (5.0)	
Nitrogen, ppm	31,200	1.1		0.7		1.1	
Carbon, wt %		85.82		85.99		86.10	
Hydrogen, wt %	(Atomic H/C = 1.67)	14.17		14.00 (13.6)		13.86 (13.6)	
Sulfur, wt %	0.64	—		—		—	
Oxygen, wt %	1.33	—		—		—	
Distillation, D2887		°F	°C	°F	°C	°F	°C
IBP, °F		-47	-44	22	-6	250	121
10%		203	95	190	88	322	161
20%		268	131	238	114	353	178
30%		321	161	276	136	390	199
40%		372	189	312	156	413	212
50%		410	210	346	174	436	224
60%		446	230	377	192	461	238
70%		487	253	408	209	489	254
80%		547	286	440	227	520	271
90%		624	323	480	245	564	300
EP, °F		789	421	553	289	622	328

Source: Data of Tait et al., 1983. (Co/Cr/Mo on alumina, 1800 psi H_2, 0.55 LHSV, 790°F, H_2 consumed = 1400 scf/bbl.

427°C (775–800°F). Because of the upsets encountered during testing, this was judged to be a minimum life expectancy and a catalyst life of 6 months was considered probable.

Separate distillations of the product were carried out to produce JP-4 and JP-8 jet fuels. The analytical data indicate that the samples would meet all specifications, with perhaps one exception (Table 7). The pour point of -40°C (-40°F) for the JP-8 fraction is somewhat higher than the JP-8 specification of -50°C (-58°F). A slight lowering of the boiling point range would lower the pour point to give a product that meets all JP-8 specifications.

The results demonstrated that the single-catalyst system was capable of hydrocracking a whole shale oil containing large amounts of nitrogen. However, analysis of kinetic data from the studies above led to the development of a dual-catalyst system that affords the same high product quality, plus higher throughput (Tait et al., 1983). Using the dual-catalyst system with recycle of >271°C (>520°F) distillate (no. 2 fuel oil) afforded a 91 wt % (108 vol%) yield of JP-4 boiling range material as the only liquid product. With recycle operation, the gas make increased slightly, based on fresh feed, from 6.1 to 7.9 wt %. As a result, the hydrogen consumption also increased on a fresh feed basis from 1800 scf/bbl to 2060 scf/bbl. However, because of the higher yield obtained, the hydrogen consumption dropped on a JP-4 basis from 2320 scf/bbl of JP-4 to 1920 scf/bbl for the recycle operations.

Thus, the Amoco work led to novel catalyst systems capable of directly upgrading a whole shale oil into high yields of jet fuels in a single step. First a single-catalyst hydrocracking system was developed. Then an even more active dual-catalyst system was developed with a catalyst specifically designed for high denitrogenation activity at high nitrogen content preceding the hydrocracking catalyst. The development of these catalysts is significant, since their use could eliminate the need for complex, multistep, multicatalyst processing of whole shale oils.

31

Chemicals from Shale Oil

I	Introduction	1031
II	General Chemicals	1032
III	Asphalt	1035

I. INTRODUCTION

The attractiveness of oil shale to the oil industry is, in the simplest sense, due to its composition (Sinor, 1988). The organic constituents of oil shale, on the other hand, have hydrogen-to-carbon atomic ratios similar to those found in petroleum crudes and have only small proportions of organic oxygen. Furthermore, liquid products (shale oil) obtained by thermal means from oil shale typically have an atomic hydrogen/carbon ratio of about 1.35.

In spite of the potential for added-value products, most oil shale companies have tended to concentrate on producing a single product stream. In short, there has been the tendency to compete with petroleum by producing a crude oil substitute. Unfortunately, whenever a shale oil industry was established strictly to compete head to head with petroleum, the industry became susceptible to the periodic downward fluctuations in petroleum prices.

In a few instances outside of Scotland, the industry was large enough or possessed enough technical capability to undertake the development of alternative product slates. The USSR industry has long viewed liquid shale oil products not only as an alternative fuel but also as a chemical feedstock. This is at least partly due to two factors: (1) shale retorting was established for the major purpose of producing gas and the tars were only a coproduct or byproduct; (2) the richness of the resource allowed direct combustion in boilers and power plants where otherwise an oil fuel might have been needed. Thus, the Russian shale oil industry has considered petrochemical and other nonfuel product applications to a far greater extent than in any other country (Figure 1).

II. GENERAL CHEMICALS

Shale oil contains a large variety of hydrocarbon compounds (Table 1) including paraffins, cycloparaffins, olefins, aromatics, furans,

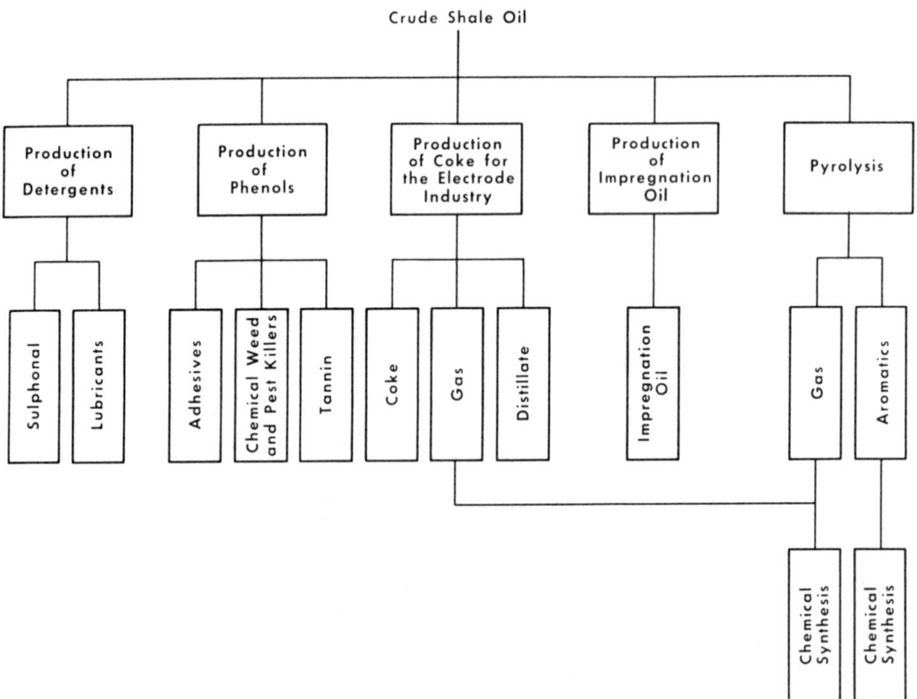

Figure 1 Schematic representation of the Russian concept of chemicals and other products from oil shale.

Table 1 Major Compound Types in Shale Oil Fractions

Saturates	*Polars—Acids*
n-Paraffins	Phenols
Isoprenoid paraffins	Carbazoles
Cycloparaffins	Pyrroles
	Aliphatic nitriles
Olefins	
α-Olefins	Hydroxyindans/tetralins
Internal olefins	Carbazoles
Aromatics	*Polars—Bases*
Benzenes	Nitriles
Styrenes	2-Ketones
Indans/tetralins	
	Acetophenones
Indans, tetralins	Pyridines
Indenes	Quinolines
Naphthalenes	Tetrahydroquinolines
Biphenyls	Tetrahydrocarbazoles
Acenaphthenes	
Benzothiophenes	
Benzofurans	
Naphthalenes	
Biphenyls	
Fluorenes	
Phenanthrenes	
Pyrenes	
Phenanthrenes	
Chrysenes	
Benzoanthracenes	

thiophenes, hydroxyaromatics, pyrroles, and pyridines. Minor constituents such as thiophenofurans, hydroxynitrogen compounds, ketones, aldehydes, acids, amides, nitriles, and dinitrogen compounds are also present. Alkyl substituents of the compounds can range from one to fifty carbon atom chains.

However, the concentrations of individual constituents can vary substantially and is dependent on the retorting process used. In general, the higher the temperature to which the oil has been exposed, the higher the concentration of aromatics and the lower the concentration of saturates and olefins (Figure 2). In most cases, the simplest member of a homologous series is the predominant compound.

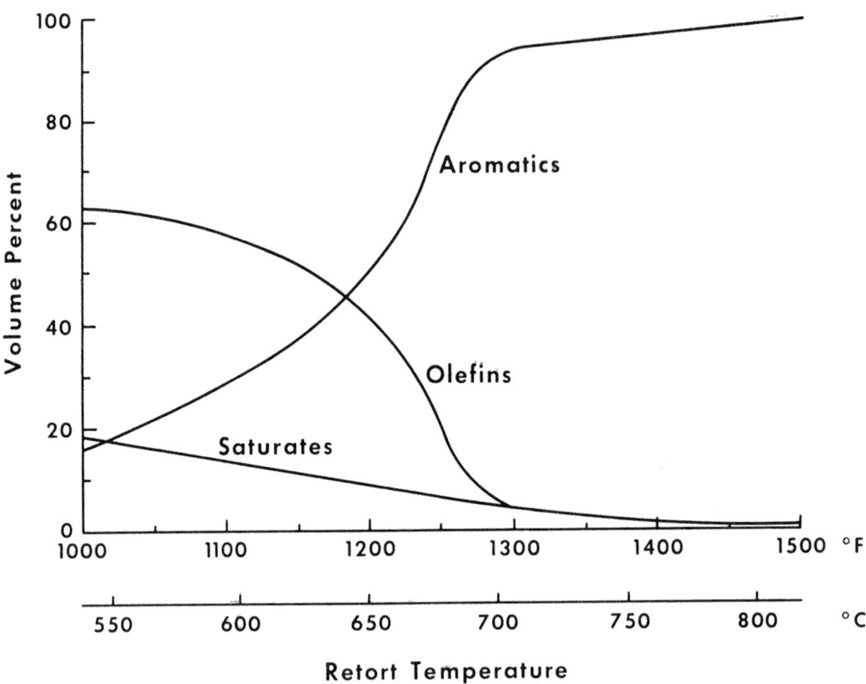

Figure 2 Variation of shale oil composition with the temperature of the retort.

In addition to being a potential source of hydrocarbon products, oil shale can also be regarded as a source of other products. For example, McKay and Blanche (1987) showed that large yields of carboxylic acids can be obtained by supercritical fluid extraction of oil shale with carbon monoxide and water, methanol and water, or toluene. Total yield of organic liquids with this procedure was higher than is obtained by retorting. Yields of as much as 90% of the organic matter (140% of Fischer assay) were obtained. Whereas the normal alkanes recovered showed odd-even predominance in carbon numbers, the normal carboxylic acids showed even-odd predominance.

To accomplish the isolation of the acids, oil shale is treated with methanol and water at 400°C (752°F) for 1 hr, and then extracted with common solvents such as methylene chloride. At higher temperatures, pyrolysis reactions increase and the product composition changes dramatically. The products obtained in the first 15 min were almost entirely carboxylic acids amounting to about 25% of the total oragnic matter in the oil shale.

It should be noted that retorted oil shale contains only trace amounts of dicarboxylic acids in the 370–535°C (698–995°F) distillate, where they may be found in petroleum in either large or small quantities depending on the geologic history of the petroleum. The carboxylic acids decarboxylate during the retorting processes used to produce shale oil.

III. ASPHALT

Approximately 2 million miles, or 93% of all surfaced roads in the United States are paved with asphalt. The United States uses approximately 21 million tons of asphalt and asphalt-related paving materials annually and spends in excess of $10 billion each year in the construction and maintenance of asphalt roadways. In addition, The Road Information Program (TRIP) in Washington, D.C. has reported that "bouncing over rough and broken pavements is costing the U.S. motorist more than $28 billion annually in added vehicle operating costs. This total includes $21.7 billion for wasted fuel and another $6.5 billion spent on repairs to tires, brakes and steering and suspension systems." These figures are staggering, representing about $191 a year per motorist.

In recent years there has been an increasing trend in the U.S. refining industry to process the heavy end of the crude oil barrel into lighter distillate and transportation fuels at the expense of producing petroleum binders for asphalt. This recent and marked increase in "bottom of the barrel" processing has resulted in periodic shortages of domestically produced asphalt binders and a greater reliance on imported petroleum feedstocks for the production of asphalt binders. There is increasing concern that some asphalts produced from these new crudes or untried crude blends are causing an increasing number of pavement problems—problems which are exacerbated by the significant increase in traffic frequency and increased truck weights in recent years. These factors are combining to create greater stresses on, and an increasing rate of failure of, our nation's roads.

Moisture damage and binder embrittlement are major causes of pavement failure. Moisture damage results from a weakening or disruption of the asphalt-aggregate bond, leading to a loss of the structural strength of the pavement. This can produce pavement stress symptoms such as rutting, raveling, flushing, and cracking that may eventually lead to pavement breakup. In addition to inadequacies in pavement design and construction, moisture damage typically is enhanced by such factors as deficiencies in the asphalt, moisture-sensitive aggregates, environmental conditions, and increased traffic loading. Binder embrittlement, on the other hand, results when the flow properties of the asphalt binder deteriorate to such an extent that, under the influence of physical or thermal stresses, the binder, and thus the pavement, fractures.

The residuum from vacuum distillation of crude shale oil can produce materials with consistencies in the asphalt range with yields on the order of 10--20% depending on the retorting process used.

As part of their continuing asphalt program, researchers at Western Research Institute (Plancher and Petersen, 1984) discovered that certain types of compounds present in shale oil cause a significant reduction in moixture damage and a potential reduction in binder embrittlement when added to asphalt. This work has been extended to investigate the strength of the nitrogen compound-aggregate bond and its resistance to water-induced weakening. One of the major causes of asphalt pavement failure is moisture-induced damage. Moisture and repeated freeze-thaw cycles tend to break apart asphalt pavement.

A comparison of the number of freeze-thaw cycles in the water susceptibility test required to induce failure in petroleum versus shale oil asphalt briquets clearly showed the superiority of shale oil asphalt (Table 2). The shale oil asphalt briquet showed no signs of failure after 100-plus cycles.

Asphalt-aggregate mixtures were subjected to tests that are commonly used in analyzing paving mixtures. These tests included evaluations of the resilient modulus, tensile strength, and water susceptibility. From the test results, it was concluded that

- Shale oil asphalt can be produced by conventional methods in acceptable grades for highway paving mixtures.
- Adhesive properties of shale oil asphalt compare favorably with those of petroleum asphalts.
- Paving mixtures containing shale oil asphalts appear to be quite resistant to damage by water.

There seems little doubt that the heavy 10–15% fraction of crude shale oil could be used to make a variety of paving grade asphalts, either alone or blended with petroleum asphalt. In Europe blends of petroleum asphalt and coal tar pitches are used in road construction; blends of shale oil and petroleum asphalt should have less of a variation in properties and could be produced to meet all existing specifications.

The oil shale challenge is clear. Sooner or later, we will need to make use of the world's abundant oil shale resource–if not for fuel, then for chemical feedstock–to replace dwindling petroleum supplies.

Table 2 Evaluation of Briquettes by the Water Susceptibility Test

Asphalt + aggregate[a]	Cycles-to-failure
Petroleum[b]	1–7
Shale oil	100

[a]Prepared using 5% (wt/wt) asphalt binder and 95% (wt/wt) aggregate.
[b]Four typical petroleum asphalts obtained from the Federal Highway Administration.

The timing will probably be dictated more by political considerations than by purely technical factors. As a result, our need to use oil shale may come upon us suddenly, giving no time for the development of new technology. We must be ready. Technology is now available that could be used to provide liquid fuels and feedstock from oil shale but is expensive. A rapid shift to such technology could have dire and far-reaching consequences. Continued research and development is needed to provide further improvements in oil shale science and technology, so that when the need arises we can use oil shale economically and in an environmentally acceptable manner.

References to Part IV

Abelson, P. H., T. C. Hoering, and P. L. Parker. 1964. In *Advances in Organic Geochemistry—1963* (U. Colombo and G. D. Hobson, eds.), Pergamon Press, London, p. 169.

Adams, R. C., C. E. Banks, W. S. Bradley, L. L. Brannick, and W. G. Christian. 1976a. Report No. NTIS-PB-274435. National Technical Information Service, Oak Ridge, Tennessee.

Adams, R. C., C. E. Banks, W. S. Bradley, L. L. Brannick and W. G. Christian. 1976b. Report No. NTIS-PB-275520. National Technical Information Service, Oak Ridge, Tennessee.

Alley, S. K., D. P. McArthur, and K. E. Whitehead. 1984. *Proc. 11th World Petroleum Congress*, John Wiley and Sons, Chichester, *4*: 221.

Allred, V. D. 1966. *Chem. Eng. Prog. 62*:55.

Allred, V. D. (ed.). 1982. *Oil Shale Processing Technology*. The Center for Professional Advancement, New Brunswick, New Jersey.

Allred, V. D., and G. I. Nielson. 1965. *Chem. Prog. Symp. Series 160*:54.

Ambles, A., M. V. Djuricic, L. J. Djordevic, and D. Votorovic. 1983. In *Advances in Organic Geochemistry—1981* (M. Bjoroy, ed.), John Wiley and Sons, Chichester, p. 554.

Atkins, A. R., C. J. Fookes, A. Muridan, and L. Stephenson. 1986. *Proc. 3rd Australian Workshop on Oil Shale*, p. 229.

Audeh, C. A. 1984. *Preprints. Div. Petrol. Chem.* Am. Chem. Soc., Washington, D.C., *29*(1):19.

Baas-Becking, L. G. M., I. R. Kaplan, and D. Moore. 1960. *J. Geol. 68*:243.

Bader, R. G., B. W. Hood, and J. B. Smith. 1960. *Geochim. Cosmochim. Acta 19*:4.

Baird, R. A. 1986. *Am. Assoc. Petroleum Geologists Bulletin 70*:111 and references cited therein.
Barakat, A. O., and T. F. Yen. 1987a. *Preprints. Div. Fuel Chem.* Am. Chem. Soc., Washington, D. C., *32*(1):43.
Barakat, A. O., and T. F. Yen. 1987b. *Fuel 66*:587.
Barakat, A. O., and T. F. Yen. 1988. *Energy and Fuels 2*:105.
Bartick, H., K. Kunchal, D. Switzer, R. Bowen, and R. Edwards. 1975. Contract No. N00014-75-C-0055. Office of Naval Research, Washington, D.C.
Baughman, G. L. 1981. *Synthetic Fuels Data Handbook*, 2nd ed., Cameron Engineers, Denver, Colorado.
Bauman, R. F., W. N. Mitchell, J. M. Eakman, and R. J. Koveal. 1987. *Proc. 4th Australian Workshop on Oil Shale*, p. 198.
Behar, F., and M. Vandenbroucke. 1986. *Rev. Inst. Francais du Petrole. 41*:173.
Bekri, O., H. Baba-Habib, C. Y. Cha, and M. C. Edelman. 1983. *Proc. 16th Oil Shale Symposium* (J. H. Gary, ed.), Colorado School of Mines Press, Golden, Colorado, p. 345.
Berg, C. 1950. U.S. Patent 2,501,153.
Berry, K. L., R. L. Hutson, J. S. Sterrett, and J. C. Knepper. 1982. *Proc. 15th Oil Shale Symposium* (J. H. Gary, ed.), Colorado School of Mines Press, Golden, Colorado, p. 385.
Bock, J., P. P. McCall, M. L. Robbins, and M. Siskin. 1984a. U.S. Patent 4,458,757.
Bock, J., P. P. McCall, M. L. Robbins, and M. Siskin. 1984b. U.S. Patent 4,461,696.
Bordovsky, O. K. 1965. *Marine Geology 3*:3.
Botto, R. E., R. Wilson, R. Hayatsu, R. L. McBeth, R. G. Scott, and R. E. Winans. 1985. *Preprints. Div. Fuel Chem.* Am. Chem. Soc., Washington, D.C., *30*(4):187.
Bradley, W. H. 1929. Prof. Paper No. 158-E. U.S. Geological Survey, Washington, D.C.
Bradley, W. H. 1931. Prof. Paper No. 168. U.S. Geological Survey, Washington, D.C.
Bradley, W. H. 1970. *Geol. Soc. Am. Bull. 81*:985.
Bradley, W. H. 1973. *Geol. Soc. Am. Bull. 84*:1121.
Braun, R. L., and A. K. Burnham. 1986. *Fuel 65*:218.
Bray, E. E., and E. D, Evans. 1961. *Geochim. Cosmochim. Acta 22*:2.
Britton, K. 1985. In *Mechanics of Oil Shale* (K. P. Chong and J. W. Smith, eds.), Elsevier, New York.
Brooks, J. 1981. *Organic Maturation Studies and Fossil Fuel Exploration*. Academic Press, London.
Brooks, J., and D. Welte (eds.). 1984. *Advances in Petroleum Geochemistry*. Academic Press, London.
Broughton, A. C. 1920. *Trans. Proc. Roy. Soc. South Aust. 44*: 386.

Bruni, C. E. 1968. *Proc. United Nations Oil Shale Symposium.*
Burlingame, A. L., and B. R. Simoneit. 1968. *Science 160*:531.
Burlingame, A. L., and B. R. Simoneit. 1969. *Nature 222*:741.
Burlingame, A. L., P. A. Haug, H. K. Schnoes, and B. R. Simoneit. 1969a. In *Advances in Organic Geochemistry—1968* (P. A. Schenck and I. Havenaar, eds.), Pergamon Press, Oxford, p. 85.
Burlingame, A. L., P. C. Wszolek, and B. R. Simoneit. 1969b. In *Advances in Organic Geochemistry—1968* (P. A. Schenck and I. Havenaar, eds.), Pergamon Press, Oxford.
Burnham, A. K., and M. F. Singleton. 1983. In *Geochemistry and Chemistry of Oil Shales* (F. P. Miknis and J. F. McKay, eds.), Symposium Series No. 230. Am. Chem. Soc., Washington, D.C., p. 335.
Camp, D. W. 1987. *Proc. 20th Oil Shale Symposium* (J. H. Gary, ed.), Colorado School of Mines Press, Golden, Colorado, p. 130.
Campbell, J. H., G. H. Koskinas, and N. D. Stout. 1978. *Fuel 57*: 372.
Cane, R. F. 1976. In *Oil Shale* (T. F. Yen and G. V. Chilingarian, eds.), Elsevier, Amsterdam.
Cane, R. F. 1979. *Proc. 12th Oil Shale Symposium* (J. H. Gary, ed.), Colorado School of Mines Press, Golden, Colorado, p. 17.
Carley, J. F. 1975. Report No. UOPKK-75-28. Lawrence Livermore National Laboratory.
Cattell, R. A., B. Guthrie, and L. W. Schramm. 1951. In *Oil Shale and Cannel Coal* (G. Sell, ed.), Institute of Petroleum, London, p. 345.
Coal Miner. 1978. *3*(3):77.
Costa Neto, C. 1983. In *Geochemistry and Chemistry of Oil Shales* (F. P. Miknis and J. F. McKay, eds.), Symposium Series No. 230. Am. Chem. Soc., Washington, D.C., p. 13.
Curtin, D. J. 1977. U.S. Patent 4,029,571.
Dancy, T. E., and V. Giedroyc. 1950. *J. Inst. Petroleum 36*:607.
Datta, R. S., and C. A. Salotti. 1982. *Proc. Eastern Oil Shale Symposium.* U.S. Department of Energy, Washington, D.C.
Davis, J. D., and A. E. Galloway. 1928. *Ind. Eng. Chem. 20*:612.
Debyser, J., and G. Deroo. 1969. *Ref. Inst. Francais du Petrole. 24*(1):21.
Degens, E. T. 1963. *Proc. 6th International Congress on Sedimentology*, Amsterdam.
Degens, E. T., and K. Mopper. 1975. In *Chemical Oceanography* (I. R. Riley and R. Chester, eds.), Academic Press, New York.
Degens, E. T., A. Prashnowsky, K. O. Emery, and J. Pimenta. 1961. *J. Neues Jahrb. Geol. und Palaentol. Montsch. 8*:413.
Degens, E. T., J. H. Reuter, and K. N. F. Shaw. 1964. *Geochim. Cosmochim. Acta 28*:45.
Demaison, G. P., and G. T. Moore. 1980. *Am. Assoc. Petrol. Geol. Bull. 64*:1179.

Demaison, G. P., A. J. J. Holck, R. W. Jones, and G. T. Moore. 1984. *Proc. 11th World Petroleum Congress* 2:17.
Dereppe, J. M., and C. Moreaux. 1987. *Fuel* 66:1008.
Dhondt, R. O. 1979. *Proc. Symposium on Synthetic Fuels from Oil Shale and Tar Sands*, Institute of Gas Technology, Chicago, p. 439.
Dick, R. D., C. Young, and W. L. Fourney. 1984. *Proc. 17th Oil Shale Symposium* (J. H. Gary, ed.), Colorado School of Mines Press, Golden, Colorado, p. 225.
Dineen, G. U. 1968. Symposium on the Development and Utilization of Oil Shale Resources, Tallinin.
Djuricic, M. V., R. C. Murphy, D. Vitorovic, and K. Biemann. 1971. *Geochim. Cosmochim. Acta* 35:1201.
Docktor, L. 1972. *Proc. Annual Meeting, AIME.*
Doilov, S. K., V. M. Efimov, R. E. Ioonas, N. A. Nazinin, E. E. Piil, K. E. Raad, I. K. Roox, N. D. Serebtyannikov, J. Y. Shaganov, L. S. Ananiev, and A. S. Voikov. 1977. U.S. Patent 4,007,093.
Donovan, R. N., and J. Scott. 1980. *J. Geol.* 16:35.
Dougan, P. M., and L. Docktor. 1981. *Proc. 14th Oil Shale Symposium* (J. H. Gary, ed.), Colorado School of Mines Press, Golden, Colorado, p. 118.
Duir, J. H., R. F. Deering, and H. R. Jackson. 1977. *Hydrocarbon Processing*. May:147.
Duir, J. H., C. F. Griswold, and B. A. Christolini. 1983. *Chem. Eng. Prog.* February:45.
Duncan, E. 1977. *Proc. 7th World Petroleum Congress*, Elsevier, Amsterdam, 3:659.
Duncan, D. C., and V. E. Swanson. 1965. Circular No. 523. U.S. Geological Survey, Washington, D.C.
Dundas, R.C., and R. T. Howes. 1923. U.S. Patent 1,469,628.
Durand, B. 1980. *Kerogen*, Editions Technip, Paris.
Durand, B., and G. Nicase. 1980. In *Kerogen* (B. Durand, ed.), Editions Technip, Paris, p. 35.
Earnest, C. M. 1982. *Thermochim. Acta* 58:271.
Earnest, C. M. 1983. *Thermochim. Acta* 60:171.
East, J. H., and E. D. Gardner. 1964. Bulletin No. 611. U.S. Bureau of Mines, Washington, D.C.
Eastman, R. L., and R. A. Quinn. 1960. *J. Am. Chem. Soc.* 82:4249.
Eele, M., and T. Hancock. 1964. U.K. Patent 330.
Eitel, G. L. 1985. *Proc. 18th Oil Shale Symposium* (J. H. Gary, ed.), Colorado School of Mines Press, Golden, Colorado, p. 216.
Eitel, G. L., and G. Domaidy. 1985. *Proc. 18th Oil Shale Symposium* (J. H. Gary, ed.), Colorado School of Mines Press, Golden, Colorado, p. 216.
Espitalie, J., J. L. LaPorte, M. Madec, P. LePlat, J. Paulet, and A. Boutfeu. 1977. *Ref. Inst. Francais du Petrole.* 32:23.

Evin, C. G., and E. T. Robinson. 1979. Contract No. N00014-79-C-0061. Navy Energy and Natural Resources R&D Office, Washington, D.C.
Ewert, W. M., and J. Scinta. 1983. *Proc. Symposium on Synthetic Fuels from Oil Shale and Tar Sands*, Institute of Gas Technology, Chicago, p. 395.
Farris, C. B. 1980. *Mining Engineering 32*(1):aa.
Faulkner, B. P., M. H. Weinecke, and R. F. Cnare. 1983. *Proc. Symposium on Synthetic Fuels from Oil Shale and Tar Sands*, Institute of Gas Technology, Chicago, p. 545.
Fester, J. I., and W. E. Robinson. 1964. *Anal. Chem. 36*:1392.
Fester, J. I., and W. E. Robinson. 1966. In *Coal Science*, Advances in Chemistry Series No. 55. Am. Chem. Soc., Washington, D.C., p. 22.
Fischer, F., and H. Schrader. 1920. *Angew. Chem. 23*:172.
Forbes, F., F. W. Kinsey, and L. J. Colaianni. 1985. *Proc. Second Australian Workshop on Oil Shale*, p. 182.
Forsman, J. P. 1963. In *Organic Geochemistry* (I. A. Breger, ed.), Pergamon Press, Oxford, p. 148.
Forsman, J. P., and J. M. Hunt. 1958. In *Habitat of Oil*. Am. Assoc. Petroleum Geologists, Tulsa, p. 747.
Frost, I. C., and K. E. Stanfield. 1950. *Anal. Chem. 22*:491.
Garrels, R. M., and C. L. Christ. 1965. *Solutions, Minerals and Equilibria*, Harper and Row, New York.
Garrels, R. M., and F. T. Mackenzie. 1971. *Evolution of Sedimentary Rocks*, W. W. Norton, New York.
Gavin, J. M. 1924. *Oil Shale*, Government Printing Office, Washington, D.C.
Goodfellow, M., and M. T. Atwood. 1974. *Colorado School of Mines Quart. 69*(2):205.
Greaves, M. J. 1981. *Proc. Symposium on Synthetic Fuels from Oil Shale and Tar Sands*, Institute of Gas Technology, Chicago, p. 323.
Gulf Oil/Standard Oil (Indiana). 1976. Rio Blanco Oil Shale Project Report.
Gwyn, J. E., S. C. Roberts, G. P. Hinds, D. E. Hardesty, and G. L. Johnson. 1980a. *Proc. 13th Oil Shale Symposium* (J. H. Gary, ed.), Colorado School of Mines Press, Golden, Colorado, p. 35.
Gwyn, J. E., S. C. Roberts, D. E. Hardesty, G. L. Johnson, and G. P. Hinds. 1980b. *Preprints. Div. Fuel Chem.* Am. Chem. Soc., Washington, D.C., *25*(3):59.
Gwyn, J. E., S. C. Roberts, D. E. Hardesty, G. L. Johnson, and G. P. Hinds. 1981. In *Oil Shale, Tar Sands and Related Materials* (H. C. Stauffer, ed.), Symposium Series No. 163. Am. Chem. Soc., Washington, D.C., p. 167.
Haddadin, R. A., and F. A. Mizyed. 1974. *Ind. Eng. Chem. Process Des. Dev. 13*:332.

Hagaman, E. W., F. M. Schell, and D. C. Cronauer. 1984. *Fuel 63*:915.
Hall, P. B., and A. G. Douglas. 1983. In *Advances in Organic Geochemistry—1981* (M. Bjoroy, ed.), John Wiley and Sons, Chichester, p. 576.
Hall, P. L., D. F. R. Midner, and R. L. Borst. 1983. *Appl. Phys. Lett. 43*:252.
Harack, A. E. 1970. *Proc. Annual Meeting AIME.*
Harack, A. E. 1971. Technical Report No. 30. U.S. Bureau of Mines, Washington, D.C.
Hartley, F. L., and C. S. Brinegar. 1959. *Proc. 5th World Petrol. Cong. II*:37.
Hawk, C. O., M. D. Schlesinger, H. H. Ginsberg, and R. W. Hiteshue. 1964. Report of Investigations No. 6548. U.S. Bureau of Mines, Washington, D.C.
Heistand, R. N. 1976. *Energy Sources 2*:397.
Herrell, A. Y., and C. Arnold. 1976. *Thermochim. Acta 17*:165.
Hills, I. R., E. V. Whitehead, D. E. Anders, J. J. Cummins, and W. E. Robinson. 1966. *Chem. Commun.* p. 752.
Himus, G. W. 1951. In *Oil Shale and Cannel Coal* (G. Sell, ed.), Institute of Petroleum, London, *2*:112.
Himus, G. W., and G. C. Basak. 1949. *Fuel 28*:57.
Hoering, T. C., and P. H. Abelson. 1965. *Carnegie Inst. Washington Yearbook 64*:218.
Holmes, S. A. 1983. In *Shale Oil Upgrading and Refining*, Butterworths, Woburn, MA, p. 159.
Holmes, S. A., and L. F. Thompson. 1981. *Proc. 14th Oil Shale Symposium* (J. H. Gary, ed.), Colorado School of Mines Press, Golden, Colorado, p. 235.
Hoskins, W. N., R. P. Upadhyay, J. D. Bills, C. R. Sandberg, and F. D. Wright. 1976. Report No. NTIS-PB-262525. National Technical Information Service, Oak Ridge, Tennessee.
Hubbard, A. B. 1965. Report of Investigations No. 6676. U.S. Bureau of Mines, Washington, D.C.
Hubbard, A. B., and J. I. Fester. 1958. *Ind. Eng. Chem.* 3:147.
Hubbard, A. B., and W. E. Robinson. 1950. Report of Investigations No. 4744. U.S. Bureau of Mines, Washington, D.C.
Hubbard, A. B., H. N. Smith, H. H. Heady, and W. E. Robinson. 1952. Report of Investigations No. 5725. U.S. Bureau of Mines, Washington, D.C.
Hunt, J. M. 1979. *Petroleum Geochemistry and Geology*, W. H. Freeman, San Francisco.
Hutton, A. C., A. J. Kantsler, and A. C. Cook. 1980. *Austr. Petrol. Explor. Assoc. 20*:44.
Hutton, A. C., J. Korth, J. Ellis, P. Crisp, and J. D. Saxby. 1987a. *Proc. 20th Oil Shale Symposium*, Colorado School of Mines Press, Golden, Colorado, p. 10.

Hutton, A. C., J. D. Saxby, and J. Ellis. 1987b. *Proc. 20th Oil Shale Symposium* (J. H. Gary, ed.), Colorado School of Mines Press, Golden, Colorado, p. 24.

Jacobs, H. R., M. J. Marzenelli, K. S. Udell, and R. M. Dougan. 1980. *Proc. 13th Oil Shale Symposium* (J. H. Gary, ed.), Colorado School of Mines Press, Golden, Colorado, p. 62.

Jaffe, F. C. 1962. *Colorado School of Mines Industrial Bulletin* 5(2):1.

Jee, C. K., J. D. White, S. K. Bhatia, and D. Nicholoson. 1977. Review and Analysis of Oil Shale Technologies. Volume 1. Oil Shale Deposits, Mining and Environmental Concerns. Report No. NTIS-FE-2343-6. National Technical Information Service, Oak Ridge, Tennessee.

Johnson, D. R., N. B. Young, and J. W. Smith. 1978. Report No. LERC/RI-77/6. Laramie Energy Research Center, U.S. Department of Energy, Washington, D.C.

Jones, J. B., and R. N. Heistand. 1979. *Proc. 12th Oil Shale Symposium* (J. H. Gary, ed.), Colorado School of Mines Press, Golden, Colorado, p. 184.

Jones, J. B., and A. A. Reeves. 1971. U.S. Patent 3,736,247.

Jones, W., D. Lyzinski, J. B. Miller, A. V. Cugini, and F. J. Antezana. 1984. *Proc. 17th Oil Shale Symposium* (J. H. Gary, ed.), Colorado School of Mines Press, Golden, Colorado, p. 133.

Kauppila, T. A. 1982a. In *Oil Shale Processing Technology* (V. D. Allred, ed.), The Center for Professional Advancement, New Brunswick, New Jersey, p. 1.

Kauppila, T. A. 1982b. In *Oil Shale Processing Technology* (V. D. Allred, ed.), The Center for Professional Advancement, New Brunswick, New Jersey, p. 23.

Kimble, B. J., J. R. Maxwell, R. P. Philp, G. Eglinton, P. Albrecht, A. Ensminger, P. Arpino, and G. Ourisson. 1964. *Geochim. Cosmochim. Acta 38*:1165.

Koons, C. B., G. W. Jamieson, and L. S. Cierszko. 1965. *Am. Assoc. Petrol. Geol. Bull. 49*:301.

Krumbein, W. C., and R. M. Garrels. 1965. *J. Geol. 60*:1.

Kurapei, G. A., and Y. E. Cheshko. 1977. *Ugol. 12*:23.

Lappi, R. L., D. I. Carey, A. H. Pelofsky, E. R. Bates, and J. F. Martin. 1982. *Proc. 15th Oil Shale Symposium* (J. H. Gary, ed.), Colorado School of Mines Press, Golden, Colorado, p. 231.

Lau, F. S., D. M. Rue, D. V. Punmani, and R. C. Rex. 1987. *Proc. Symposium on Eastern Oil Shale*, University of Kentucky.

Lekas, M. A. 1981. *Proc. 14th Oil Shale Symposium* (J. H. Gary, ed.), Colorado School of Mines Press, Golden, Colorado, p. 146.

Lekas, M. A. 1985. Report No. DE-FC20-78LC10787. U.S. Department of Energy, Washington, D.C.

Leo, R. F., and P. L. Parker. 1966. *Science 152*:649.

Linden, H. R. 1972. U.S. Patent 3,703,052.

Lovell, P. F. 1978. *Proc. 11th Oil Shale Symposium* (J. H. Gary, ed.), Colorado School of Mines Press, Golden, Colorado, p. 184.
Luts, K. 1928. *Brennstoff-Chem.* 9:217.
Lyzinski, D., and W. Jones. 1984. *Proc. Eastern Oil Shale Symposium*, Institute for Mining and Minerals Research, Lexington, Kentucky, p. 289.
Maciel, G. E., and L. W. Dennis. 1981. *Organic Geochem.* 3:105.
Maciel, G. E., V. J. Bartuska, and F. P. Miknis. 1978. *Fuel* 57:505.
Maciel, G. E., V. J. Bartuska, and F. P. Miknis. 1979. *Fuel* 58:155.
Mapstone, G. E. 1951. In *Oil Shale and Cannel Coal* (G. Sell, ed.), Institute of Petroleum, London, 2:489.
Mashin, V. N., N. D. Serebryannikov, and T. A. Purre. 1980. *Proc. 10th World Petroleum Congress* 3:311.
McCarthy, H. E., and C. Y. Cha. 1976. *Proc. 9th Oil Shale Symposium* (J. H. Gary, ed.), Colorado School of Mines Press, Golden, Colorado, p. 85.
McDonald, R. E. 1972. In *Geologic Atlas of the Rocky Mountain Region*, Rocky Mountain Association of Geologists, Denver, Colorado, p. 243.
McKay, J. F., and M. S. Blanche. 1987. *Energy and Fuels* 1:525.
McKee, J. M., and S. K. Kunchal. 1976. *Proc. 9th Oil Shale Symposium* (J. H. Gary, ed.), Colorado School of Mines Press, Golden, Colorado, p. 49.
Menzel, D. W. 1974. In *The Sea-Marine Chemistry* (D. Goldberg, ed.), John Wiley and Sons, New York.
Miknis, F. P., and J. F. McKay (eds.). 1983. *Geochemistry and Chemistry of Oil Shales*. Symposium Series No. 230, Am. Chem. Soc., Washington, D.C.
Miknis, F. P., G. E. Maciel, and V. J. Bartuska. 1979. *Organic Geochem.* 1:169.
Miknis, F. P., J. W. Smith, E. K. Maughan, and G. E. Maciel. 1982. *Am. Assoc. Petroleum Geologists Bull.* 66:1396.
Milton, C. 1971. *Contribution to Geology*, University of Wyoming, Laramie, Wyoming, *10*(1):57.
Montgomery, D. P. 1968. *Ind. Eng. Chem. Prod. Res. Dev.* 7:274.
Moore, H. F., W. A. Sutton, F. H. Turrill, R. P. Long, C. A. Johnson, and W. P. Hettinger. 1983. In *Shale Oil Upgrading and Refining* (S. A. Newman, ed.), Butterworths, Woburn, MA, p. 223.
Moore, L. R. 1969. In *Organic Geochemistry* (G. Eglinton and M. T. J. Murphy, eds.), Springer-Verlag, Berlin, p. 265.
Mraw, S. C., and C. J. Keweshan. 1986. *Fuel* 65:54.
Mraw, S. C., L. J. Heidman, S. C. Hwang, and C. Tsonopoulos. 1984. *Ind. Eng. Chem. Process. Des. Dev.* 23:577.
Mraz, T., J. DuBow, and K. Rajeshwar. 1983. *Fuel* 62:1215.
Muntean, J. V., L. M. Stock, and R. E. Botto. 1988. *Energy and Fuels* 2:108.

Nicolaus, R. A. 1960. *Rass. Med. Sper. (Suppl. 2)* 7:1.
Oil and Gas Journal. 1982a. May 10, p. 86.
Oil and Gas Journal. 1982b. March 22, p. 215.
Parks, T. J., L. T. Lynch, and D. S. Webster. 1987. *Fuel 66*:338 and references cited therein.
Parrish, R. L. 1986. Report No. SAND84-0818. Sandia National Laboratory.
Peterson, E. J., and W. D. Spall. 1983. Report No. LA-9722-MS. Los Alamos National Laboratory.
Pettijohn, F. J. 1957. *Sedimentary Rocks, 2nd ed.*, Harper and Row, New York.
Pforzheimer, H. 1981. *Proc. Symposium on Synthetic Fuels from Oil Shale and Tar Sands*. Institute of Gas Technology, Chicago, p. 315.
Piper, E. M., and O. C. Ivo. 1986a. *Proc. 19th Oil Shale Symposium* (J. H. Gary, ed.), Colorado School of Mines Press, Golden, Colorado, p. 98.
Piper, E. M., and O. C. Ivo. 1986b. *Proc. Eastern Oil Shale Symposium*, U.S. Department of Energy, Washington, D.C.
Plancher, H., and J. C. Petersen. 1984. *Preprints. Div. Petrol. Chem.* American Chemical Society, Washington, D.C. *29*(1):229 and references cited therein.
Platt's Oilgram News Service. 1985. *63*(206):4.
Platt's Oilgram News Service. 1987. *65*(121):5.
Potter, O. E., C. Fryer, G. Christiansen, P. Cunico, G. Swaine, K. C. Tam, R. Preslmaier, and C. M. Hoskin. 1984. *Proc. 2nd Australian Workshop on Oil Shale*, p. 188.
Prien, C. H. 1976. In *Oil Shale* (T. F. Yen and G. V. Chilingarian, ed.), Elsevier, Amsterdam, p. 235.
Punwani, D. V., F. S. Lau, M. J. Roberts, W. C. S. Hu, and R. C. Rex. 1986. *Proc. 19th Oil Shale Symposium* (J. H. Gary, ed.), Colorado School of Mines Press, Golden, Colorado, p. 82.
Quass, F. W. 1938. *J. Inst. Petrol. 25*:813.
Rajeshwar, K. 1981. *Thermochim. Acta 45*:253.
Rajeshwar, K. 1983. *Thermochim. Acta 63*:97.
Rajeshwar, K., R. Nottenburg, and J. DuBow. 1979. *J. Mater. Sci. 14*:2025.
Rajeshwar, K., R. J. Rosenvold, and J. B. DuBow. 1983. *Thermochim. Acta 66*:373.
Rammler, R. W. 1982. In *Oil Shale Processing Technology* (V. D. Allred, ed.), The Center for Professional Advancement, New Brunswick, New Jersey, p. 83.
Repeta, D. J., and R. B. Gagosian. 1983. In *Advances in Organic Geochemistry-1981* (M. Bjoroy, ed.), John Wiley and Sons, Chichester.
Resing, H. A., A. N. Garroway, and R. N. Hazlett. 1978. *Fuel 57*:540.
Resource Sciences Corporation, 1975. Tulsa, Oklahoma.

Rex, R. C., H. Feldkirchner, J. C. Janka, and F. C. Schora. 1983. *Proc. Symposium on Synthetic Fuels from Oil Shale and Tar Sands*, Institute of Gas Technology, Chicago, 1983.
Rezende, J. 1982. In *Oil Shale Processing Technology* (V. D. Allred, ed.), The Center for Professional Advancement, New Brunswick, New Jersey, p. 121.
Richardson, J. H., E. B. Huss, L. L. Ott, J. E. Clarkson, M. O. Bishop, J. R. Taylor, L. J. Gregory, and C. J. Morris. 1982. Report No. UCID-19548. Lawrence Livermore National Laboratory.
Ricketts, T. E. 1980. *Proc. 13th Oil Shale Symposium* (J. H. Gary, ed.), Colorado School of Mines Press, Golden, Colorado, p. 46.
Ricketts, T. E. 1982. *Proc. 15th Oil Shale Symposium* (J. H. Gary, ed.), Colorado School of Mines Press, Golden, Colorado, p. 341.
Robie, R. A., B. S. Hemingway, and J. R. Fisher. 1979. Bulletin No. 1452. U.S. Geological Survey, Washington, D.C.
Robinson, E. T. 1979. *Proc. 12th Oil Shale Symposium* (J. H. Gary, ed.), Colorado School of Mines Press, Golden, Colorado, p. 195.
Robinson, E. T., and C. G. Evin. 1983. In *Shale Oil Upgrading and Refining* (S. A. Newman, ed.), Butterworths, Woburn, MA, p. 49.
Robinson, W. E. 1969. In *Organic Geochemistry* (G. Eglinton and M. T. J. Murphy, eds.), Springer-Verlag, Berlin, p. 181.
Robinson, W. E. 1976. In *Oil Shale* (T. R. Yen and G. V. Chilingarian, eds.). Elsevier, Amsterdam, p. 61 and references cited therein.
Robinson, W. E., and G. U. Dineen. 1967. *Proc. 7th World Petroleum Congress*, Elsevier, Amsterdam, p. 669.
Robinson, W. E., H. H. Heady, and A. B. Hubbard. 1953. *Ind. Eng. Chem. 45*:788.
Robinson, W. E., D. L. Lawlor, J. J. Cummins, and J. I. Fester. 1963. Report of Investigations No. 6166. U.S. Bureau of Mines, Washington, D.C.
Robinson, W. E., J. J. Cummins, and G. U. Dineen. 1965. *Geochim. Cosmochim. Acta 29*:249.
Romig, B. A. 1981. *Proc. 14th Oil Shale Symposium* (J. H. Gary, ed.), Colorado School of Mines Press, Golden, Colorado, p. 91.
Rosenfeld, W. D. 1948. *Arch. Biochem. 16*:263.
Rosenvold, R. J., and J. Rajeshwar. 1982. *Thermochim. Acta 57*:1
Ruark, J. R. 1956. Report of Investigations No. 5279. U.S. Bureau of Mines, Washington, D.C.
Ruark, J. R., H. W. Sohns, and H. W. Carpenter. 1971a. Report of Investigations No. 7303. U.S. Bureau of Mines, Washington, D.C.
Ruark, J. R., H. W. Sohns, and H. W. Carpenter. 1971b. Report of investigations No. 7540. U.S. Bureau of Mines, Washington, D.C.
Russell, P. L. 1986. *Proc. 19th Oil Shale Symposium* (J. H. Gary, ed.), Colorado School of Mines Press, Golden, Colorado, p. 94.

Salotti, C. A., and Datta, R. S. 1983. *Proc. 16th Oil Shale Symposium* (J. H. Gary, ed.), Colorado School of Mines Press, Golden, Coloardo, p. 394.

Saxby, J. D. 1976. In *Oil Shale* (T. F. Yen and G. V. Chilingarian, eds.), Elsevier, Amsterdam, p. 103.

Schmalfeld, I. P. 1975. *Proc. 8th Oil Shale Symposium* (J. H. Gary, ed.), Colorado School of Mines Press, Golden, Colorado, p. 129.

Schmidt-Collerus, J. J., and C. H. Prien. 1974. *Preprints. Div. Fuel Chem.* Am. Chem. Soc., Washington, D.C., *19*(2):100.

Schmitz, H. H., and A. Tolle. 1987. *Erdoel, Kohle, Erdgas. Petrochem. 40*:245.

Schonbrunner, J. 1940. *J. Biochem. Z. 304*:26.

Schwedock, J. P., H. E. Reif, and A. Macris. 1983. In *Shale Oil Upgrading and Refining* (S. A. Newman, ed.), Butterworths, Woburn, MA, p. 193.

Scouten, C. G., M. Siskin, K. D. Rose, T. Aczel, S. G. Colgrove, and R. E. Pabst. 1987. *Proc. 4th Australian Workshop on Oil Shale.*

Sellers, J. B., G. R. Haworth, and P. G. Zambas. 1972. *Trans. Soc. Min. Eng. 252*:222.

Sethi, N. K., D. M. Grant, and R. J. Pugmire. 1987. *J. Magnet. Reson. 71*:476.

Shaw, R. J. 1947. Report of Investigations No. 4151. U.S. Bureau of Mines, Washington, D.C.

Sikonia, J. G., J. G. Board, J. R. Wilcox, and L. Hilfman. 1983. In *Shale Oil Upgrading and Refining* (S. A. Newman, ed.), Butterworths, Woburn, MA, p. 29.

Silbernagel, B. C., L. A. Gebhard, M. Siskin, and G. Brons. 1987. *Energy and Fuels 1*:501.

Silliman, B. 1869. *Am. J. Sci. Arts 4*(8):86.

Simoneit, B. R., and A. L. Burlingame. 1974. In *Advances in Organic Geochemistry—1973* (B. Tissot and F. Bienner, eds.), Editions Technip, Paris, p. 191.

Singleton, M. F., G. J. Koskinas, A. K. Burnham, and J. N. Raley. 1986. Report No. UCRL-53272. Lawrence Livermore National Laboratory.

Sinor, J. E. 1988. *Proc. International Conference on Oil Shale and Shale Oil.* Chemical Industry Press, Beijing, China, p. 159.

Siskin, M., G. Brons, and J. F. Payack. 1987a. *Preprints. Div. Petrol. Chem.* Am. Chem. Soc., Washington, D.C., *32*(1):75.

Siskin, M., G. Brons, and J. F. Payack. 1987b. *Energy and Fuels 1*:100.

Siwinski, R. E. 1987. *Proc. 4th Australian Workshop on Oil Shale*, p. 72.

Smith, J. W. 1956. *Ind. Eng. Chem. 48*:441.

Smith, J. W. 1969. Report of Investigations No. 7248. U.S. Bureau of Mines, Washington, D.C.

Smith, J. W. 1980. *Oil Shale Resources of the United States*, Mineral and Energy Resources Series 23(6), Colorado School of Mines Press, Golden, Colorado.

Smith, J. W. 1983. In *Geochemistry and Chemistry of Oil Shales* (F. P. Miknis and J. F. McKay, eds.), Symposium Series No. 230, Am. Chem. Soc., Washington, D.C., p. 225.

Smith, J. W., and D. R. Johnson. 1969. *Proc. 2nd International Conference on Thermal Analysis*, 1251.

Soderquist, A., D. J. Burton, R. J. Pugmire, A. J. Beeler, D. M. Grant, B. Durand, and A. Y. Huc. 1987. *Energy and Fuels 1*:50.

Sohns, H. W., L. E. Mitchell, R. J. Cox, W. I. Barnet, and W. I. R. Murphy. 1951. *Ind. Eng. Chem. 43*:33.

Sorokin, J. I. 1971. *Rev. Ges. Hydrobiol. 56*:1.

Sousa, J. J. F., N. N. Vugman, and A. S. Mangrich. 1987. *Chem. Geol. 63*:17.

Speight, J. G. 1980. *The Chemistry and Technology of Petroleum*, Marcel Dekker, New York.

Speight, J. G., and R. G. Vawter. 1988. Private communication relating to the actual location of, and a visit to, this plant in the summer of 1987.

Stanfield, K. E., and I. C. Frost. 1949. Report of Investigations No. 4477. U.S. Bureau of Mines, Washington, D.C.

Stauffer, H. C. (ed.). 1981. *Oil Shale, Tar Sands and Related Materials*. Symposium Series No. 163, Am. Chem. Soc., Washington, D.C.

Steuart, D. R. 1912. In *The Oil Shales of the Lothians*, Part III. Geol. Survey Memoirs of Scotland. H. M. Stationery Office, London.

Stevens, A. L., and R. Zahradnik. 1983. *Proc. 16th Oil Shale Symposium* (J. H. Gary, ed.), Colorado School of Mines Press, Golden, Colorado, p. 267.

Stewart, R. 1964. *Oxidation Mechanisms*, W. A. Benjamin, New York.

Stewart, R. 1965. In *Oxidation in Organic Chemistry* (K. B. Wiberg, ed.), Academic Press, New York, p. 1.

Stewart, R., and R. Van der Linden. 1960. *Disc. Faraday Soc.* p. 211.

Strickland, J. D. H. 1965. In *Chemical Oceanography* (J. P. Riley and G. Skirrow, eds.), Academic Press, New York, *I*(12):478.

Sullivan, R. F., and B. E. Strangeland. 1979. In *Refining of Synthetic Crudes* (M. L. Gorbaty and B. L. Harney, eds.), Advances in Chemistry Series No. 179. Am. Chem. Soc., Washington, D.C.

Sullivan, R. F., B. E. Stangeland, and H. A. Frumkin. 1978. *Proc. 43rd Midyear Meeting, American Petroleum Institute*, Washington, D.C.

Suuberg E. M., J. Sherman, and W. D. Lilly. 1987. *Fuel 66*:1176.

Tait, A. M., and A. L. Hensley. 1982. *Preprints. Div. Fuel Chem.* Am. Chem. Soc., Washington, D.C., *27*(2), p. 187.

Tait, A. M., J. T. Miller, and A. L. Hensley. 1982. Paper 16d. *Proc. Summer Meeting, American Institute of Chemical Engineers.*
Tait, A. M., J. T. Miller, and A. L. Hensley. 1983. In *Shale Oil Upgrading and Refining* (S. A. Newman, ed.), Butterworths, Woburn, MA, p. 73.
Tappan, H., and A. R. Loeblich. 1970. In *Symp. Palynology of the Late Cretaceous and Early Tertiary* (R. M. Kosanke and S. T. Cross, eds.), Geology Society of America, p. 247.
Thakur, D. S., and H. E. Nuttall. 1987. *Ind. Eng. Chem. Res. 26*: 1351.
Thorne, H. M. 1964. Information Circular No. 8216. U.S. Bureau of Mines, Washington, D.C.
Tisot, P. R., and H. W. Sohns. 1971. Report of Investigations No. 7576. U.S. Bureau of Mines, Washington, D.C.
Tissot, B. P., and J. Espitalie. 1975. *Rev. Inst. Francais du Petrole. 30*:743.
Tissot, B. P., and W. I. R. Murphy. 1963. Report of Investigations No. 6284. U.S. Bureau of Mines, Washington, D.C.
Tissot, B. P., and D. H. Welte. 1984. *Petroleum Formation and Occurrence*, 2nd ed., Springer-Verlag, Berlin.
Trask, P. D. 1932. *Origin and Environments of Source Sediments of Petroleum*, Gulf, Houston.
Trent, R. H., and R. K. Dunham. 1977. *Proc. 10th Oil Shale Symposium* (J. H. Gary, ed.), Colorado School of Mines Press, Golden, Colorado, p. 89.
U.S. Department of Energy, 1977. *Oil Shales and Tar Sands: A Bibliography*, National Technical Information Service, Oak Ridge, Tennessee.
Vadovic, C. J. 1983. In *Geochemistry and Chemistry of Oil Shales* (F. P. Miknis and J. F. McKay, eds.), Symposium Series No. 230. Am. Chem. Soc., Washington, D.C., p. 385.
Virgona, J. E. 1986. *Proc. 19th Oil Shale Symposium* (J. H. Gary, ed.), Colorado School of Mines Press, Golden, Colorado, p. 161.
Vitorovic, D., A. Ambles, M. Djordevic, and S. Bajc. 1987. *Preprints. Div. Petroleum Chem.* Am. Chem. Soc., Washington, D.C., *32*(1):37.
Voetter, H., I. Poll, and H. M. H. van Wechem. 1987. *Proc. 20th Oil Shale Symposium* (J. H. Gary, ed.), Colorado School of Mines Press, Golden, Colorado, p. 122.
Votorovic, D. 1980. In *Kerogen* (B. Durand, ed.), Editions Technip, Paris, p. 301.
Wallman, P. H., P. W. Tamm, and B. G. Spars. 1980. *Preprints. Div. Fuel Chem.* Am. Chem. Soc., Washington, D.C., *25*(3):70.
Wallman, P. H., P. W. Tamm, and B. G. Spars. 1981. In *Oil Shale, Tar Sands and Related Materials*. Symposium Series No. 163. Am. Chem. Soc., Washington, D.C., p. 93.

Wang, N. C., K. E. Teo, and H. Andersen. 1977. *Can. J. Chem. 55*: 4112.
Ward, C. R. 1984. *Coal Geology and Coal Technology*, Blackwell, Melbourne.
Watson, G. H., E. A. Ziemba, P. Bissery, D. Namy, R. L. Griffis, and D. E. Nicholson. 1982. *Proc. 15th Oil Shale Symposium* (J. H. Gary, ed.), Colorado School of Mines Press, Golden, Colorado, p. 397.
Watts, C. D., and J. R. Maxwell. 1977. *Geochim. Cosmochim. Acta 41*:493.
Watts, C. D., J. R. Maxwell, and H. Kjosen. 1977. In *Advances in Organic Geochemistry—1975* (R. Campos and J. Goni, eds.), Enadamisa, Madrid, p. 391.
Weichman, B. 1976. *Proc. 9th Oil Shale Symposium* (J. H. Gary, ed.), Colorado School of Mines Press, Golden, Colorado, p. 25.
Whitcombe, J. A., and R. G. Vawter. 1982. In *Oil Shale Processing Technology* (V. D. Allred, ed.), The Center for Professional Advancement, New Brunswick, New Jersey, p. 153.
Whitworth, K. 1979a. *World Coal* 5(10):26.
Whitworth, K. 1979b. *World Coal* 5(10):34.
Wilkins, E. S., H. E. Nuttall, and D. S. Thaker. 1981. *Proc. 2nd World Congress on Chemical Engineering.*
Williams, P. F. V. 1982. In *Petroanalysis-81* (B. G. Crump, ed.), Institute of Petroleum, London, p. 326.
Williams, P. F. V. 1983. *Fuel 62*:756.
Williams, P. F. V. 1985. *Fuel 64*:540.
Williams, P. F. V. 1987. *Fuel 66*:86.
Williams, P. F. V, and A. G. Douglas. 1981. In *Organic Maturation Studies and Petroleum Exploration* (J. Brooks, ed.), Academic Press, London, p. 255.
Williams, P. F. V., and A. G. Douglas. 1985. *Fuel 64*:1062.
Willmon, G. J. 1987. *Proc. 12th World Petroleum Congress*. John Wiley and Sons, New York, *4*:3.
Wise, R. L., R. C. Miller, and H. W. Sohns. 1971. Report of Investigations No. 7482. U.S. Bureau of Mines, Washington, D.C.
Wyllie, R. J. M. 1979. *World Mining 32*(11):100.
Yen, T. F. 1974. *Preprints. Div. Fuel Chem.* Am. Chem. Soc., Washington, D.C., *19*(2):109.
Yen, T. F. 1976a. In *Oil Shale* (T. F. Ten and G. V. Chilingarian, eds.), Elsevier, New York, p. 129.
Yen, T. F. 1976b. In *Science and Technology of Oil Shale* (T. F. Yen, ed.), Ann Arbor Science, Ann Arbor, Michigan, p. 193.
Young, D. A. 1977. U.S. Patent 4,046,674.
Young, D. K., and T. F. Yen. 1977. *Geochim. Cosmochim. Acta 41*: 1411.

Young, D. K., S. Shih, and T. F. Yen. 1974. *Preprints. Div. Fuel Chem.* Am. Chem. Soc., Washington, D.C., *19*(2):169.
Young, D. K., S. Shih, and T. F. Yen. 1976. In *Science and Technology of Oil Shale* (T. F. Yen, ed.), Ann Arbor Science, Ann Arbor, Michigan, p. 65.
ZoBell, C. E. 1944. API Project 34A—Annual Report for 1944. American Petroleum Institute, Washington, D.C.

Part V

Natural Gas

32

Origin, Occurrence, and Recovery

I	Introduction	1055
II	Origin	1056
III	Occurrence	1059
	A. Conventional Reservoirs	1059
	B. Unconventional Reservoirs	1060
	C. Reserves	1065
IV	Recovery	1072
V	Transportation	1083

I. INTRODUCTION

Natural gas has been known for many centuries but its initial use was probably more for religious purposes than as a fuel. For example, gas wells were an important aspect of religious life in ancient Persia because of the importance of fire in the religion of the Persians. In classical times these wells were often flared and must have been awe inspiring, to say the least (Lockhart, 1939; Forbes, 1964). There is also the possibility that the "voices of the gods" recorded by the ancients were actually natural gas forcing its way through fissures in the earth's surface (Scheil and Gauthier, 1909; Schroder, 1920). Gas wells were also known in Europe in the middle ages and were reputed to eject oil from the wells; e.g., the phenomena observed at the site near to the town of Mineo in Sicily (Forbes, 1958). There are many other such documentations for which it can be surmised that the combustible material, or the source of the noises in the earth, was actually natural gas (Forbes, 1958, 1964).

Just as petroleum was used in antiquity (Chap. 1, Sec. III), natural gas (see also Chap. 1, Sec. IV.C) was also known in antiquity. However, the use of petroleum has been relatively well documented because of its use in warfare and as a mastic for walls and roads (Table 4 in Chap. 1). The use of natural gas in antiquity is somewhat less well documented although historical records do indicate that its use (for other than religous purposes) dates back to about 250 A.D. when it was used as a fuel in China. The gas was obtained from shallow wells and was distributed through a piping system constructed from hollow bamboo stems.

There is other fragmentary evidence for the use of natural gas in certain old texts but the use is usually inferred since the gas is not named specifically. However, it is known that natural gas was used on a small scale for heating and lighting in northern Italy during the early seventeenth century. From this it might be conjectured that natural gas found some use from the seventeenth century to the present day—recognizing that gas from coal would be a strong competitor.

Natural gas was first discovered in the United States in Fredonia, New York, in 1821. In the years following this discovery, natural gas usage was restricted to the local environs since the technology for storage and transportation (bamboo pipes notwithstanding) was not well developed and, at that time, natural gas had little or no commercial value. In fact, in the 1930s, when petroleum refining was commencing an expansion in technology (Chaps. 5 and 6) that is still continuing, natural gas was not considered to be a major fuel source and was only produced as an unwanted byproduct of crude oil production.

The principal gaseous fuel source at that time (i.e., the 1930s) was the gas produced by the surface gasification of coal. In fact, each town of any size had a plant for the gasification of coal (hence, the use of the term "town gas"). Most of the natural gas produced at the petroleum fields was vented to the air or burned in a flare stack; only a small amount of the natural gas from the petroleum fields was pipelined to industrial areas for commercial use. It was only in the years after World War II that natural gas became a popular fuel commodity leading to the recognition it has currently.

II. ORIGIN

Many theories have been proposed for the origin of petroleum and natural gas but the diversity of the precursors and the variation in the prevailing physical conditions make it difficult to fully explain the origin of the petroleum and gas in any given reservoir (Chap. 8, Sec. II). The most widely accepted theory utilizes the concept of a wide range of organic natural product precursors (Chap. 1, Sec. II) and advocates that the hydrocarbons were generated from the organic detritus under the influence of pressure and temperature over geologic time (Chap. 1, Sec. II.E). Variations in the character of the organic precursors and the prevailing physical conditions, such as temperature and pressure, play an important role in the generation of oil and/or gas.

There are those theories which promote the formation of natural gas and waxy crude oils from the remains of terrestrial plants while on the other hand, the nonwaxy crude oils were produced from the organic remains of aquatic organisms. Interesting as though this theory may be, it is difficult to equate the formation of any one particular type of crude oil with any group of specific precursors. The transportation of organic detritus by rivers or by tidal forces is a major unknown as are the prevailing conditions that bring about the chemical changes necessary to convert the detritus to crude oil constituents.

There is also the theory that, because rivers do appear to have played a role in transporting terrestrial matter to the sea, formations which include ancient river deltas and ancient beaches are favorable places for gas to exist. It has been suggested that the deepest sediments are rich in organic matter that is of terrestrial origin and these sediments are overlain by marine sediments rich in matter of aquatic origin. Thus, a vertical sequence has been envisaged in which the gas-generating materials are at the bottom of the source rock and the oil-generating materials at the top.

Attractive as though these two particular theories may seem, there is in fact no universally accepted theory that satisfactorily explains all aspects of the formation of petroleum and natural gas (Gold, 1985). Nevertheless, it is generally accepted that temperature and pressure conditions have played a role in gas formation (Figure 1).

In a more general sense, natural gas (methane) is considered to originate in three principal ways: (1) the thermogenic process; (2) the biogenic process; and (3) the abiogenic process. The thermogenic process is the slow process of the decomposition of organic material that occurs in sedimentary basins and usually requires some degree of heat. The biogenic process involves the formation of methane by the action of living organisms (bacteria) on organic materials (Chap. 1, Sec. II.C; see also Gold, 1984). The abiogenic process, unlike the other two processes, does not require the presence of organic matter as the starting material (Gold and Soter, 1982, 1986). Such gas is believed to have been present in the deeper parts of the earth since the formation of the earth. Some of this gas could also have diffused to surface formations in the intervening time.

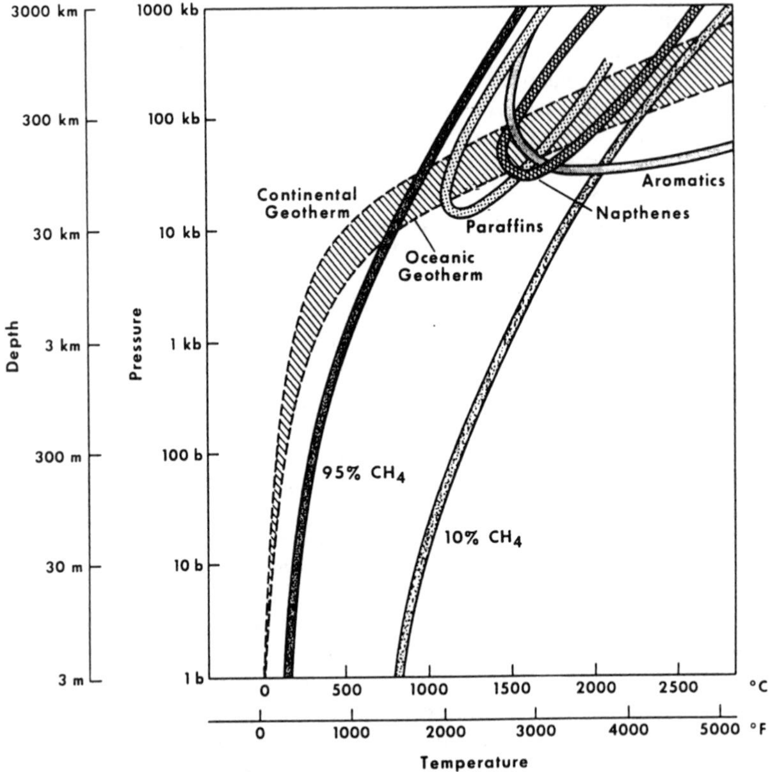

Figure 1 Generalized relationship of temperature and pressure and their influence on petroleum and natural gas precursors (Gold, 1985).

It is generally believed that most of the natural gas used today is of thermogenic and biogenic origin, and the abiogenic origin of natural gas is still considered to be highly speculative. Nevertheless, there is the belief that the abiogenic gas is that gas which could also be classified as coming from "unconventional reservoirs" (Chap. 34, Sec. II). However, such a classification of gas resources leaves much to be desired.

Once formed, it is generally believed that the direction of mobility of the hydrocarbons in the earth is in an upward direction (i.e., toward the surface). It is more than likely that there are exceptions to this general rule. For example, the hydrocarbons can also be envisaged as moving in a downward or sideways direction from their place of formation (source rock) to their place of accumulation (reservoir rock). Irrespective of the direction of movement of hydrocarbons from the source rock, the movement causes displacement of some of the seawater that originally filled the pore spaces of the sedimentary rock. This movement of the hydrocarbons is inhibited when the oil and gas reach an impervious rock that traps or seals the reservoir.

III. OCCURRENCE

Like petroleum, natural gas is located in the earth in reservoirs but, just as "conventional" petroleum reservoirs can vary considerably in character from tar sand reservoirs (Chap. 1, Sec. II.D, and Chap. 12), natural gas reservoirs can also vary considerably. For general purposes, natural gas reservoirs can be conveniently classified as conventional and nonconventional. The latter include formations such as tight sands, tight shales, geopressured aquifers, coal beds, deep sources, and gas hydrates (Meyer, 1977; Gibbons, 1985).

A. Conventional Reservoirs

There are many different types of geologic structures that are capable of forming reservoirs for the accumulation of oil and gas (Figure 2; see also Chap. 1, Sec. II.D). The depth of the reservoirs is variable and, in the case of natural gas, this may be a secondary factor in gas accumulation and storage. However, for a petroleum reservoir, depth can play a major role in preservation or conversion of a particular type of petroleum. For example, it has been noted that methane is stable at depths in excess of 40,000 ft (Barker and Kemp, 1980) and that the amount of methane surviving the pressure and, to some extent, the temperature effects that go with depth (Chap. 1, Sec. II.F) is more a function of reservoir lithology. Clean sandstones appear to be more favorable for methane preservation than carbonates (Barker and Kemp, 1980).

This concept raises the distinct possibility that natural gas may be found at depths of 15,000–30,000 ft that have not yet been explored (this chapter, Sec. III.B).

The sedimentary reservoir rocks, in which petroleum and natural gas are found, are usually composed of sandstone, limestone, or dolomite. Some of the reservoirs may be the "gas-only" type whereas an oil reservoir will usually have natural gas associated with it (either as free gas or as gas in solution in the oil; Chap. 1, Sec. IV.E). Notable exceptions to the petroleum-gas relationship are those heavy oil

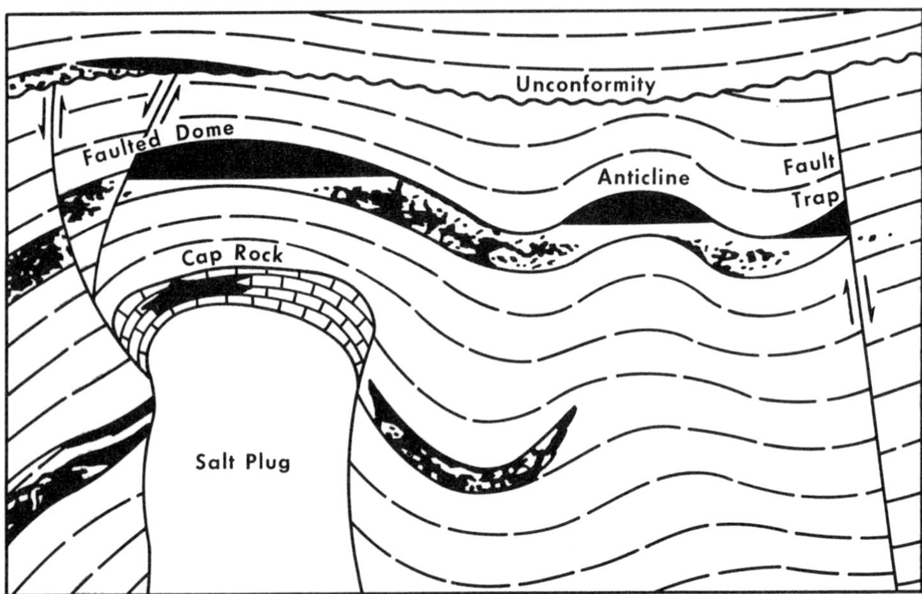

Figure 2 Composite representation of the different types of traps for petroleum and natural gas.

reservoirs, such as tar sand reservoirs (or deposits), where the oil is immobile under reservoir conditions (Chap. 14, Sec. III).

Just as oil can vary in composition depending on the placement of the well (and the depth of the well) in the reservoir, each well in the natural gas reservoir may also produce gas with a different composition. In addition, the composition of the gas from each individual well is also likely to change as the reservoir is depleted. Thus, production equipment may need to be changed as the well ages to compensate for any changes in the composition of the gas.

B. Unconventional Reservoirs

So far the emphasis has been on those subsurface reservoirs that are equivalent to the "typical" petroleum reservoir (Figure 2; see also Chap. 1, Sec. II.D). There are, however, in addition to the conventional sandstone and limestone reservoirs, other sources of natural gas which, because of their differences to those sources noted above, are classified as "unconventional" sources. These sources include tight sands, tight shales, geopressure aquifers, coal, deep sources, and gas hydrates (Nederlof, 1988).

1. Tight Sands

Tight sands (also variously called tight gas sands or tight gas) are those formations where natural gas occurs in rock formations of extremely low permeability. Geologically, the tight formations are very similar to the accumulations of "conventional" formations but the main feature of tight sand resources is the slow rate at which the gas can be produced from the formation. Substantial amounts of gas occur within such formations where the porosities are in the range 5–15% and the permeability is extremely low (0.001–1.0 millidarcy) as well as having irreducible water saturations in the range 50–70%.

Production wells in these formations generally need stimulation (such as by rock fracturing around the well bore) to increase the gas flow rates. There are many such formations in the United States (Figure 3) that contain natural gas which cannot be produced using conventional production methods (Haas et al., 1987; METC, 1988a).

The need to develop fracturing techniques to produce gas economically from tight sands cannot be overemphasized and has been the subject of many investigations (METC, 1988a). Three major techniques have been proposed: nuclear explosives, chemical explosives, and massive hydraulic fracturing (MHF).

Nuclear detonation has also been proposed to increase petroleum recovery (Atkinson and Johansen, 1964) and to recover hydrocarbon liquids from tar sand deposits (Watkins and Anderson, 1964) and oil shale deposits (Watkins and Anderson, 1964; Lekas et al., 1967), but has not been put to practice. Nuclear detonation has also been proposed for the release of gas from tight reservoirs (Figure 4; Gevertz et al., 1965; Holzer, 1968). There is always the fear that such a detonation may have unforeseen long-term consequences (such as radioactive products) that would constitute a definite danger to the user.

Chemical explosives appear to be effective only in areas where natural fractures exist, but this is not usually the case for tight sands. Massive hydraulic fracturing appears to offer the most promise (Veatch and Moschovidis, 1986). The process usually requires injection of a fracturing fluid (water) at high pressures over a prolonged period of time to induce a fracture. This is then followed by the injection of a fluid which contains "propping agents" (glass beads or sand) to prevent closure of the fracture when the fluids flow back into the well bore after cessation of the pumping operation.

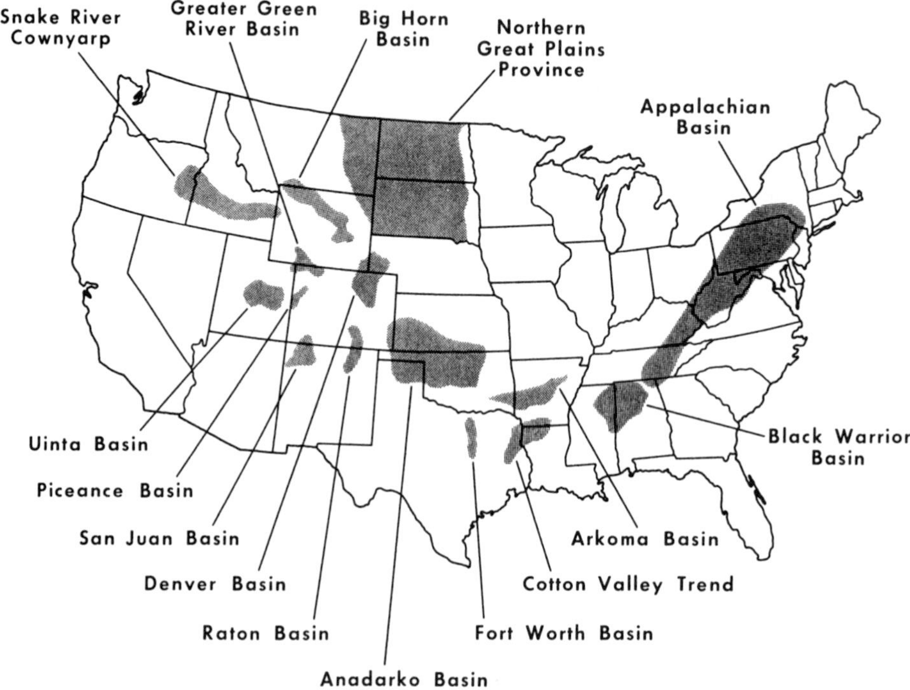

Figure 3 Location of the major tight gas formations in the United States.

2. Tight Shales

Natural gas found in tight shale beds is also referred to as Devonian shale gas and is found in the pore spaces of the shale and/or adsorbed on the shale. The shales are finely laminated deposits that are generally rich in organic matter (Chap. 25, Sec. III) but, unfortunately, only have a permeability on the order of 1 millidarcy or less. The mineral content of shales is quite diverse although common mineral constituents are quartz, with some kaolinite, pyrite, and feldspar.

The Devonian shales in the Appalachian, Illinois, and Michigan basins (Figure 5) are such gas-containing formations and can yield a gas with a heating value as high as 1250 btu/cf. These tight shales are an attractive source of gas and may contribute very significantly to gas production in the coming years. Although a commercial means

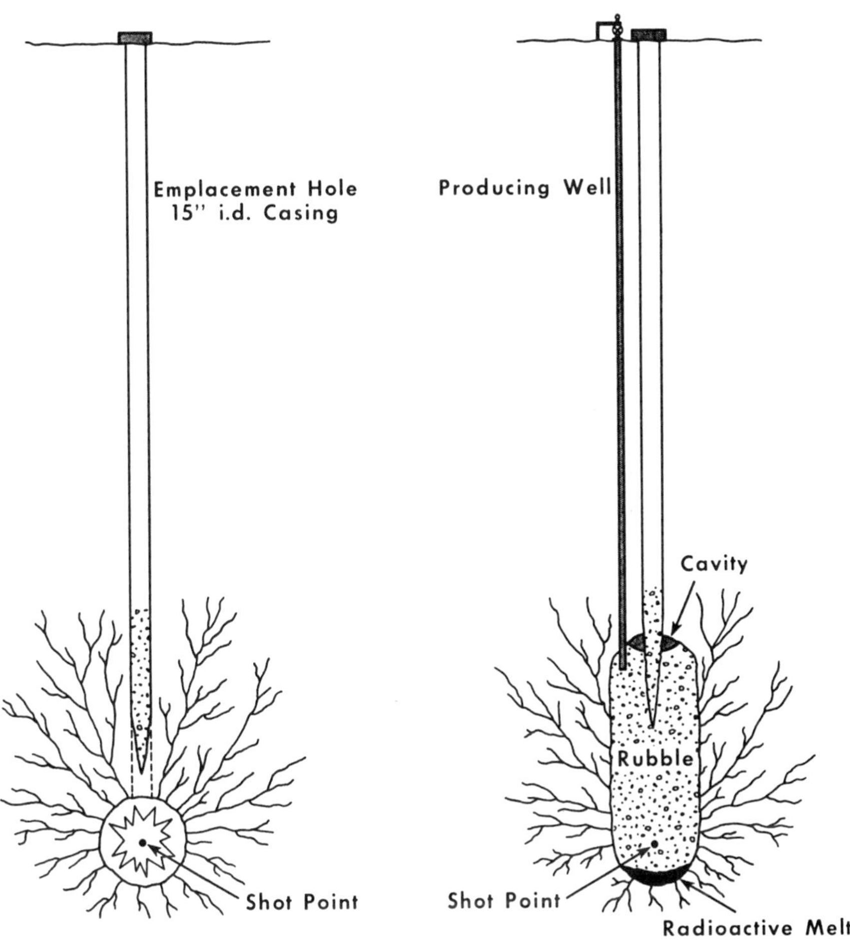

Figure 4 Schematic representation of the use of nuclear detonation to release gas from tight formations. For the release of oil from tar sands, the cavity would be below the formation for drainage purposes.

of obtaining the gas has not yet been fully defined, it is believed that massive hydraulic fracturing may be the answer to producing the gas from these tight shale formations.

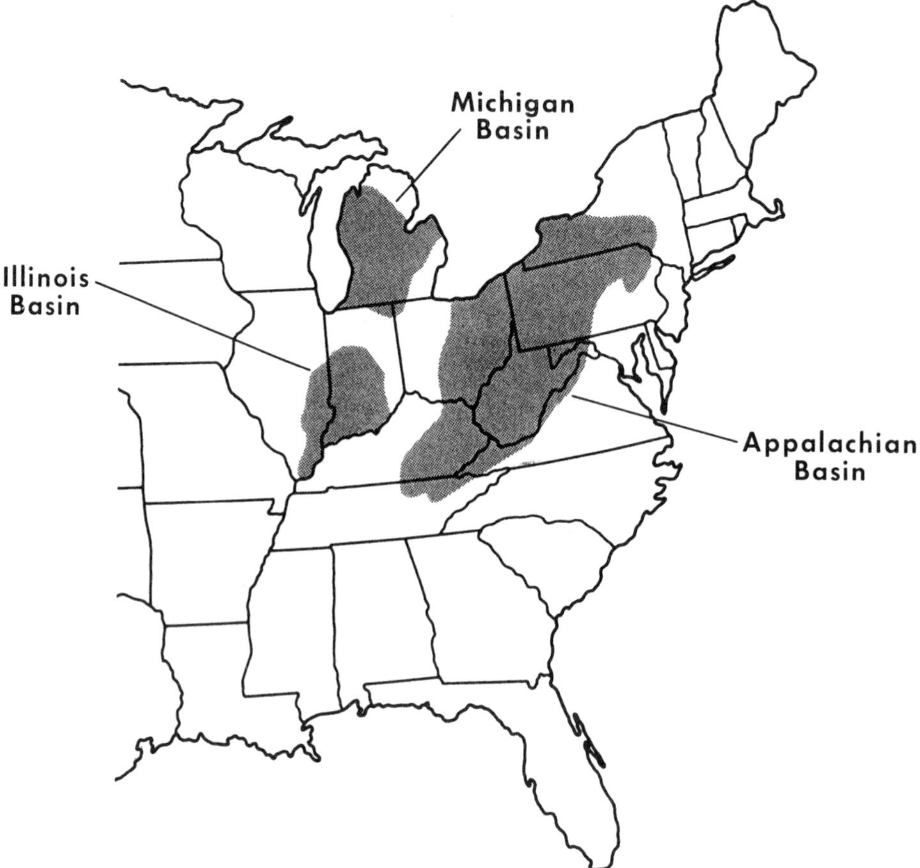

Figure 5 Location of the gas-bearing shale deposits in the United States.

3. Geopressured Aquifers

The brine in geopressured aquifers, which can form as a result of rapid subsidence, can contain up to 40 cf of natural gas per barrel of water. Such geopressured aquifers in the United States occur in a "locale" that extends onshore and offshore from Texas to Florida along the Gulf of Mexico. An estimated 1700 Tcf (1 Tcf = 1 trillion cubic feet = 1×10^{12} cubic feet) of gas reserves (unproven at this time) are estimated to be in this region.

4. Coal Seams (Coal Beds)

Methane is associated with many coal seams and is a common occurrence in coal mines. It is the dreaded "firedamp" which has been the cause of many explosions with the accompanying loss of life (Speight, 1983). The gas is occluded in the pores of the coal under pressure and is gradually released during mining operations.

Such gas in seams at depths less than 3000 ft has been estimated to be 260 Tcf in the United States but it has been estimated that practical constraints may allow production of less than 40 Tscf (Figures 6 and 7). Gaseous products are also obtained from coal during the various thermal processes that have been used/advocated for the production of gaseous fuels from coal (Chap. 22) as well as by the various thermal processes that have been advocated for the production of liquid fuels from coal (Chap. 23).

5. Deep Sources

Deep source gas is natural gas that existed deep within the earth and accumulated both as natural gas in conventional or unconventional reservoirs (Gold and Soter, 1980). There is also the possibility that deep source gas could exist in traps under "basement" rocks. The evidence for the existence of such natural gas is mainly speculative but, if proven, would imply that substantial amounts of natural gas may exist beneath existing shallow gas fields.

6. Gas Hydrates

Gas hydrates are icelike complexes (clathrates) of gas and water that form under prescribed conditions of temperature and pressure (Trofimuk et al., 1972; Holder et al., 1982). The hydrates are often found under water (at depths greater than 100 ft) and under the permafrost, and represent a potentially huge resource (Holder et al., 1983, 1984). There has also been speculation that there may even be free natural gas trapped under the hydrate resource.

C. Reserves

The reserves of natural gas are often classified as follows:

(1) Proven reserves. These are reserves of gas that are actually found (proven) by drilling. The estimates have a high degree of

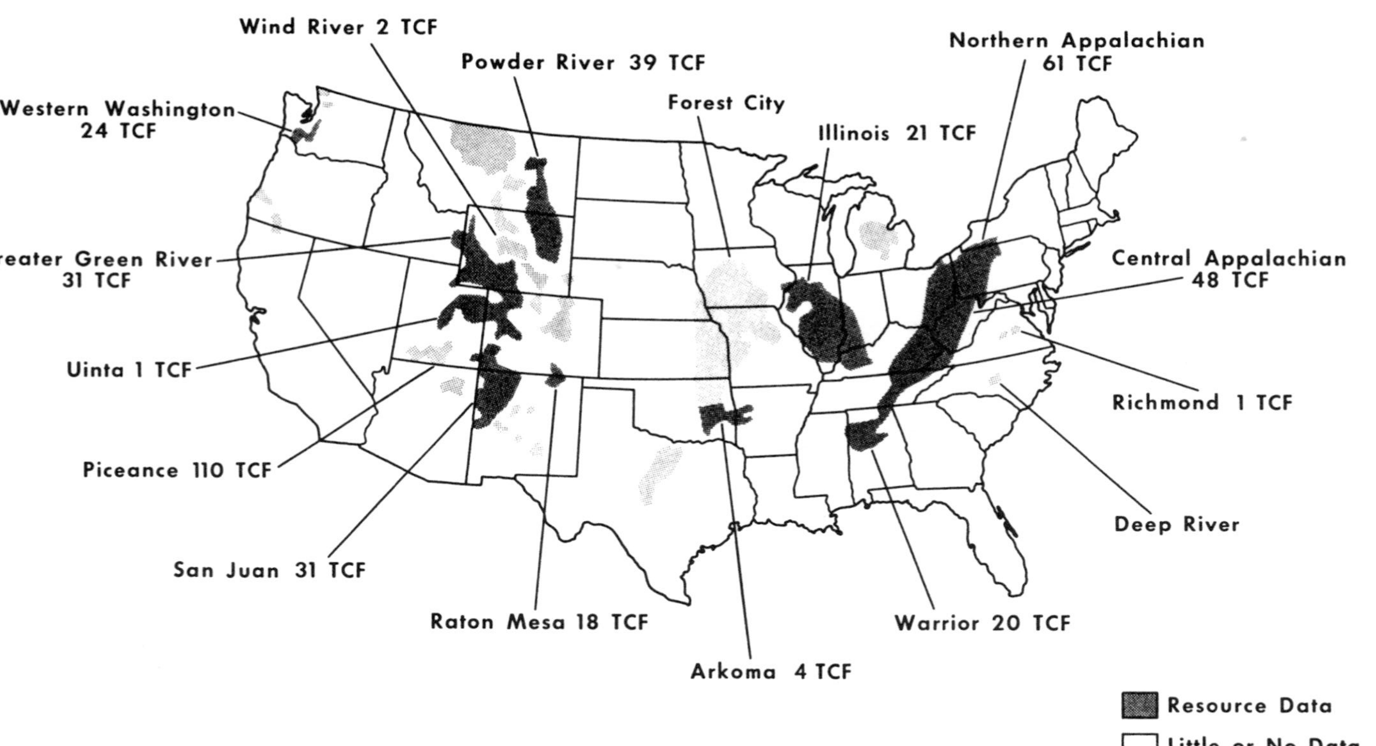

Figure 6 Estimation of the coal bed methane resources and production in the United States.

accuracy and are frequently updated as the recovery operation proceeds or are updated by reservoir characteristics using, for example, production data, pressure transient analysis, and reservoir modeling.

(2) Inferred reserves. This term is also commonly used in addition to, or in place of, potential reserves (see "Potential reserves" below). The inferred reserves are regarded as having a higher degree of accuracy than the potential reserves and the term is applied to those reserves that are estimated based on an improved understanding of reservoir frameworks (see "Proven reserves" above). The term also usually includes those reserves that can be recovered by further development of recovery technologies.

(3) Potential reserves. These are the additional resources of gas believed to exist in the earth. The data are estimated (usually from geologic evidence) but have not been substantiated by drilling operatons.

(4) Undiscovered reserves. One major issue in the estimation of the natural gas resource base is the all-too-frequent use of the term "undiscovered" gas resources. Caution is advised when using such data as a means of estimating natural gas reserves. The data are very speculative and are regarded by many energy scientists to contribute little besides unbridled optimism. The differences between the data obtained from these various estimates can be considerable, but it must be remembered that any data regarding reserves of natural gas (and, for that matter, any other fuel or mineral resource) will always be subject to questions about the degree of certainty (Figure 8).

There are three important items that outweigh the pure guesswork of "undiscovered" natural gas resources: (1) the actual discoveries of new fields; (2) the development of improved recovery technologies for already known reserves of gas in-place; and (3) the estimates of the resource base that are derived from known reservoir properties where the whole of the reservoir is not explored (McFarlane et al., 1981).

The proven reserves of natural gas are on the order of 3615.2 Tcf (1 Tcf = 1×10^{12} cf) of which some 285 Tcf exists in the United States and Canada (Table 1 and Figure 9; Grow, 1980; *BP Statistical Review of World Energy*, 1987; *BP Review of World Gas*, 1987). A more recent update for the United States (METC, 1988b) includes a breakdown by each gas-producing state for dry natural gas reserves (Table 2) which is, as expected, a somewhat lower estimate than the estimate of the total resource base. In addition, certain natural gas reserves in the United States can be inferred (see above for "inferred reserves") from what is already known from various exploration programs. If these data are taken into account, the natural gas resource base of the United States can be expanded to some 1118 Tcf and, furthermore, recovery estimates can be made on the basis of gas price (Table 3).

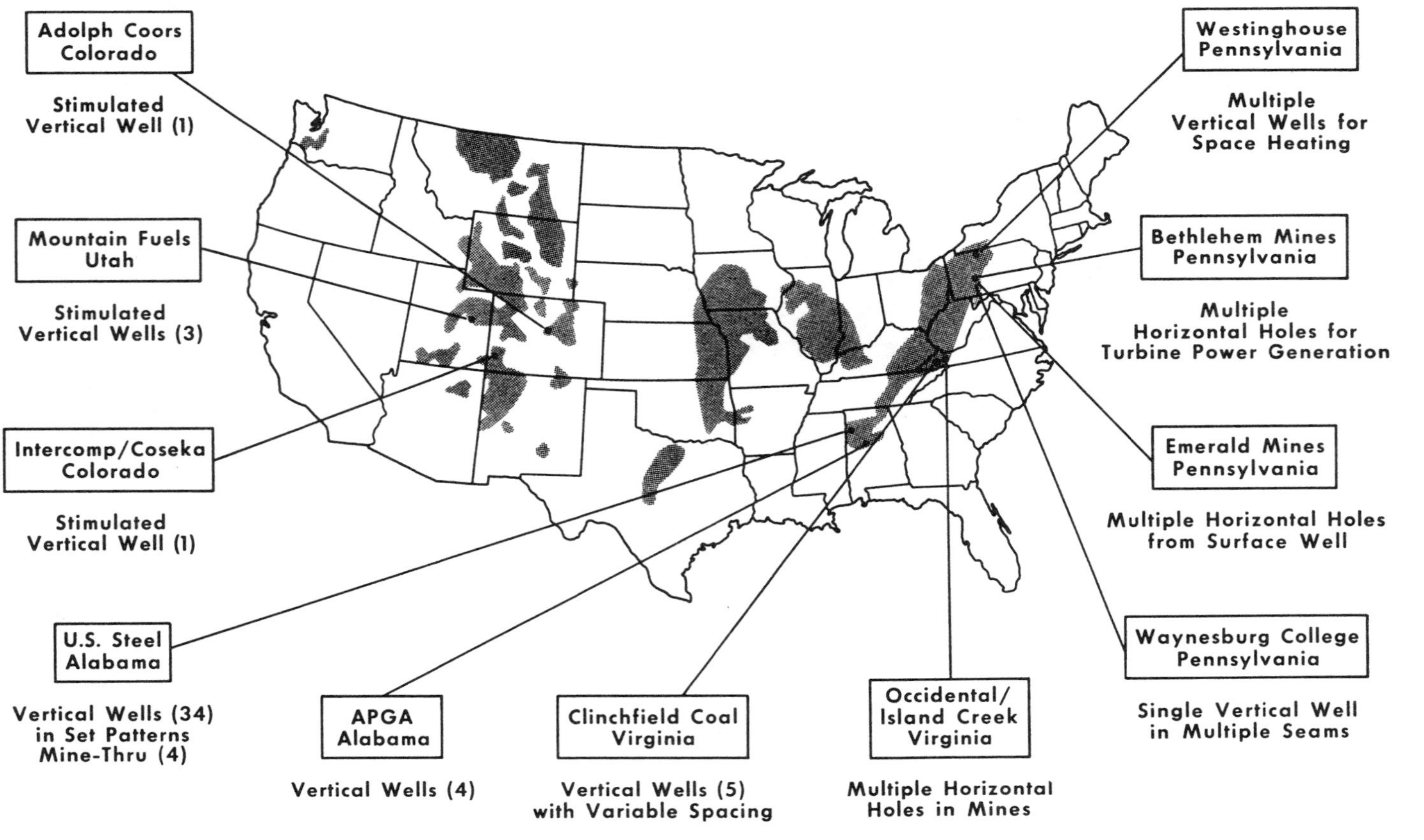

Figure 7 Locations of the major tests for coal bed methane drainage carried out by the U.S. Department of Energy.

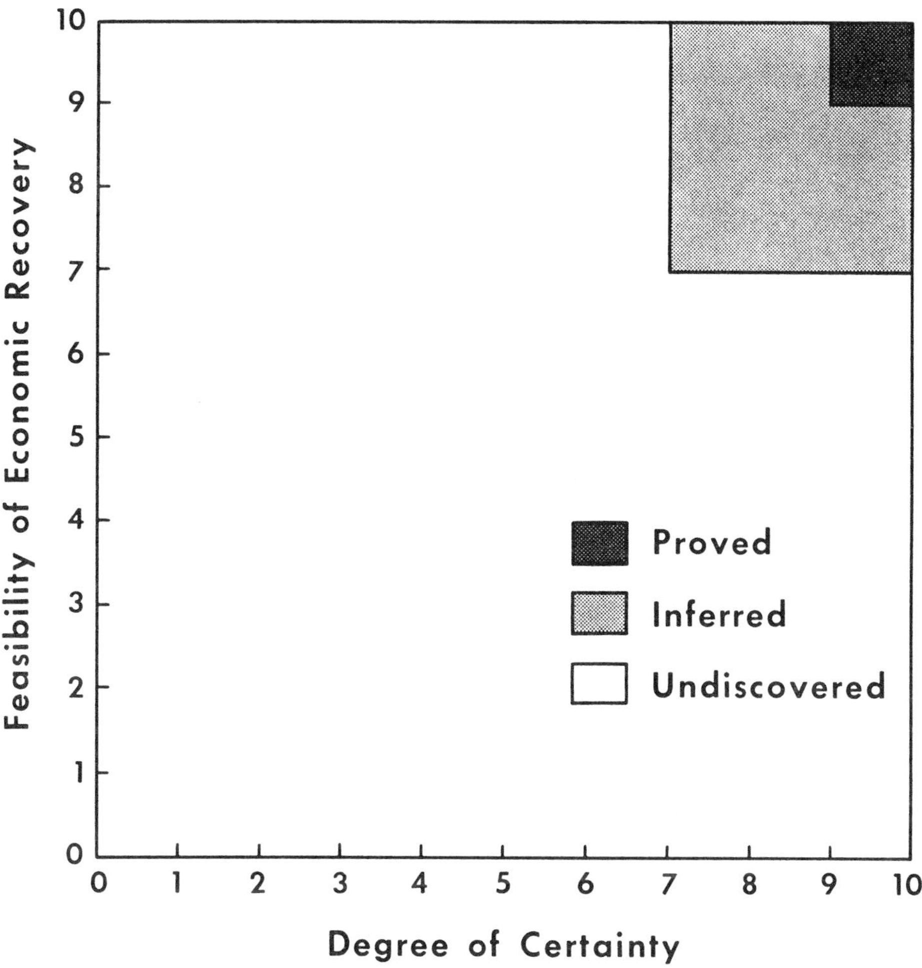

Figure 8 Representation of the degree of accuracy of resource estimation.

It should also be remembered that the total gas resource base (like any fossil fuel, or mineral, resource base) is dictated by economics (Figure 10; see also Nederlof, 1988). Therefore, when resource data are quoted, some attention must be given to the cost of recovering those resources. Most important, the economics must also include a cost factor that reflects the willingness to secure total, or a specific degree of, energy independence.

Table 1 Estimation of World Natural Gas Reserves

	Trillion cubic feet	Trillion cubic meters	Share of total (%)	R/P ratio
North America				
USA	185.4	5.2	5.1	11.6
Canada	99.6	2.8	2.8	40.0
Total North America	285.0	8.0	7.9	15.5
Latin America				
Argentina	23.0	0.7	0.6	46.5
Ecuador	4.1	0.1	0.1	a
Mexico	76.5	2.2	2.1	84.0
Trinidad	10.4	0.3	0.3	82.8
Venezuela	59.0	1.7	1.6	95.8
Others	17.1	0.5	0.5	33.7
Total Latin America	190.1	5.5	5.2	71.6
Western Europe				
Netherlands	70.4	2.0	1.9	30.6
Norway	103.2	2.9	2.9	a
United Kingdom	22.4	0.6	0.6	14.7
West Germany	6.5	0.2	0.2	14.9
Others	16.3	0.5	0.5	18.2
Total Western Europe	218.8	6.2	6.1	35.9
Middle East				
Abu Dhabi	90.0	2.5	2.5	a
Bahrain	7.0	0.2	0.2	44.7
Dubai	4.4	0.1	0.1	a
Iran	450.0	12.7	12.4	a
Iraq	28.0	0.8	0.8	a
Kuwait	35.0	1.0	1.0	a
Qatar	152.0	4.3	4.2	a
Saudi Arabia	124.0	3.5	3.4	a
Others	34.9	1.0	1.0	a
Total Middle East	925.3	26.1	25.6	a
Africa				
Algeria	106.0	3.0	2.9	85.5
Egypt	8.9	0.3	0.2	44.5
Gabon	0.5	b	b	a
Libya	21.2	0.6	0.6	a
Nigeria	47.0	1.3	1.3	a
Others	17.9	0.5	0.5	a
Total Africa	201.5	5.7	5.5	a

(continued)

Table 1 (Cont.)

	Trillion cubic feet	Trillion cubic meters	Share of total (%)	R/P ratio
Asia and Australasia				
Japan	1.1	b	b	14.6
Brunei	7.1	0.2	0.2	25.7
Indonesia	49.4	1.4	1.4	36.8
Malaysia	49.4	1.4	1.4	a
Other South East Asia	18.3	0.5	0.5	a
Bangladesh	12.7	0.4	0.4	a
India	17.6	0.5	0.5	60.4
Pakistan	18.7	0.5	0.5	48.1
Australia	18.7	0.5	0.5	35.5
New Zealand	5.7	0.2	0.2	42.7
Total Asia and Australasia	198.7	5.6	5.6	54.4
Total NCW	2019.4	57.1	55.9	58.7
Centrally-planned economies (CPEs)				
China	30.0	0.8	0.8	64.8
USSR	1550.0	43.9	42.9	64.0
Others	15.8	0.4	0.4	6.4
Total CPEs	1595.8	45.1	44.1	58.8
Total world	3615.2	102.2	100.0	58.7
Of which OPEC	1193.2	33.7	33.0	a

[a]Over 100 years.
[b]Less than 0.05.

Proved reserves of natural gas are generally taken to be those quantities which geological and engineering information indicate with reasonable certainty can be recovered in the future from known reservoirs under existing economic and operating conditions.

Reserves/Production (R/P) ratio. If the natural gas reserves remaining at the end of any year are divided by the production in that year, the result is the length of time that those remaining reserves would last if production were to continue at the then current level.

Source of data. The estimates contained in this table are those published by the *Oil and Gas Journal* in its "Worldwide Oil" issue of 29th December 1986. United Kingdom reserves data are taken from the Department of Energy's *Brown Book* published in May 1987.

Trillion equals one million million (10^{12}).

1 trillion cubic feet of natural gas = 26 million tonnes oil (approx).

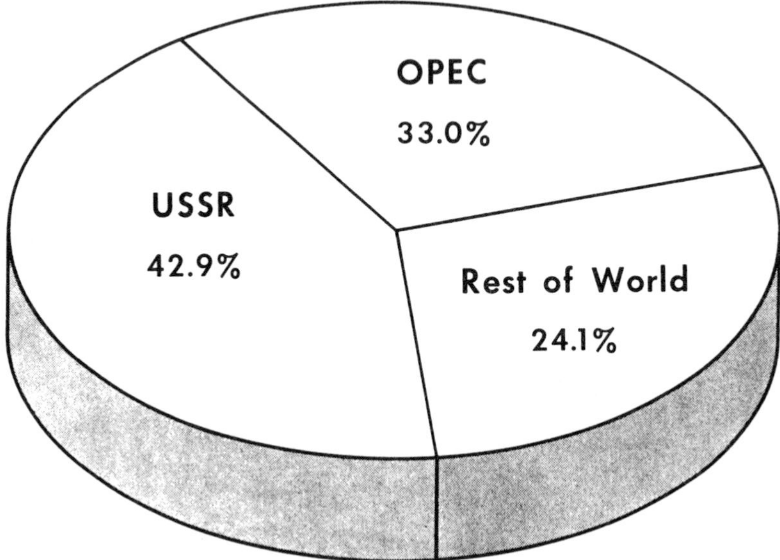

Figure 9 Schematic representation of the distribution of world natural gas reserves.

IV. RECOVERY

Further to the definition of natural gas reserves, it is essential to understand the various exploration techniques used to discover the gas. As might be expected, the type of exploration techniques that are employed depend on the nature of the site. In other words, and as for many current environmental operations, the recovery techniques applied to a specific site are dictated by the nature of the site and must be considered site-specific.

For example, in areas where little is known about the subsurface, preliminary reconaissance techniques will be necessary to identify potential reservoir systems that require further investigation. Techniques for reconaissance include satellite and high-altitude imagery, magnetic and gravity surveys, all of which have been employed to make inferences about the subsurface structure.

Once an area has been selected for further investigation, more detailed methods (such as the seismic reflection method) are brought into play. Drilling is the final stage of the exploratory program and is, in fact, the only method by which a natural gas reservoir can be conclusively identified. However, in keeping with the concept of site

Table 2 Distribution of Dry Natural Gas Reserves in the United States (Excluding Alaska and Hawaii)

Major producing states	1968 Production (Tcf)	12/31/86 Proved reserves (Tcf)	% Production (cumulative)	% Reserves (cumulative)
(1) TX (onshore + state offshore)	4.620	40.574	30	26
(2) Federal Gulf of Mex. (TX, LA, AL)	3.965	32.898	26 (56)	21 (47)
(3) Oklahoma	1.658	16.685	11 (67)	10 (57)
(4) LA (onshore + state offshore	1.741	12.930	11 (78)	8 (65)
(5) New Mexico	0.628	11.808	4 (82)	7 (72)
(6) Kansas	0.461	10.509	3 (85)	7 (79)
(7) Wyoming	0.402	9.756	3 (88)	6 (85)
Total			88	85
Total lower 48 + federal offshore	15.286	158.922		

specificity, in some areas drilling may be the only option as a means to commence the project. The risk involved in the drilling depends on what is known about the subsurface at the site. A classification scheme (Figure 11) has been devised so that exploratory wells may be categorized according to the relationship of the site to known petroleum reservoirs.

Once the drilling operations have conclusively established that a reservoir exists and the resources contained therein are sufficient to warrant economical recovery, a production and processing system is constructed.

A "typical" gas production and processing system (Figure 12) may be considered but it must be remembered that the concept of site specificity may dictate the need for an "atypical" gas production and processing system. Nevertheless, the system, like a petroleum refinery

Table 3 Estimation of Recoverable Natural Gas Resources (and Relationship to Gas Price) in the United States

	Technically recoverable gas (Tcf)[a]	Recoverable gas by price[b] <$3/ Mcf	Recoverable gas by price[b] $3–5/ Mcf
Lower 48 (conventional)			
Proved reserves, 12/31/86, onshore and offshore	159	159	–
Inferred reserves/probable resources, 12/31/86, onshore	85	85	–
Inferred reserves, 12/31/86, offshore	23	23	–
Extended reserve growth in nonassociated fields, onshore	119	56	18
Gas resources associated with oil reserve growth[c]	61	30	11
Undiscovered onshore resources	219	88	59
Undiscovered offshore resources[d]	134	54	28
Subtotal	800	495	116
Lower 48 (unconventional)			
Gas in low permeability reservoirs	180	70	49
Coal bed methane	48	8	4
Shale gas	31	10	5
Subtotal	1059	583	174
Alaska			
Alaska reserves	33	7[e]	0
Alaska inferred reserves (Cook Inlet area)	3	3	0
Alaska undiscovered, onshore and offshore	93	2[e]	2[e]
Subtotal	1188	595	176
Total	1188	595	176

[a]Volumes of gas judged recoverable with existing technology.
[b]Volumes of gas (Tcf) judged recoverable with existing technology by review panel at wellhead prices shown (1987$).
[c]Judged at oil prices of <$24 bbl and $24–40 bbl.
[d]Outer continental shelf.
[e]Component in southern Alaska.

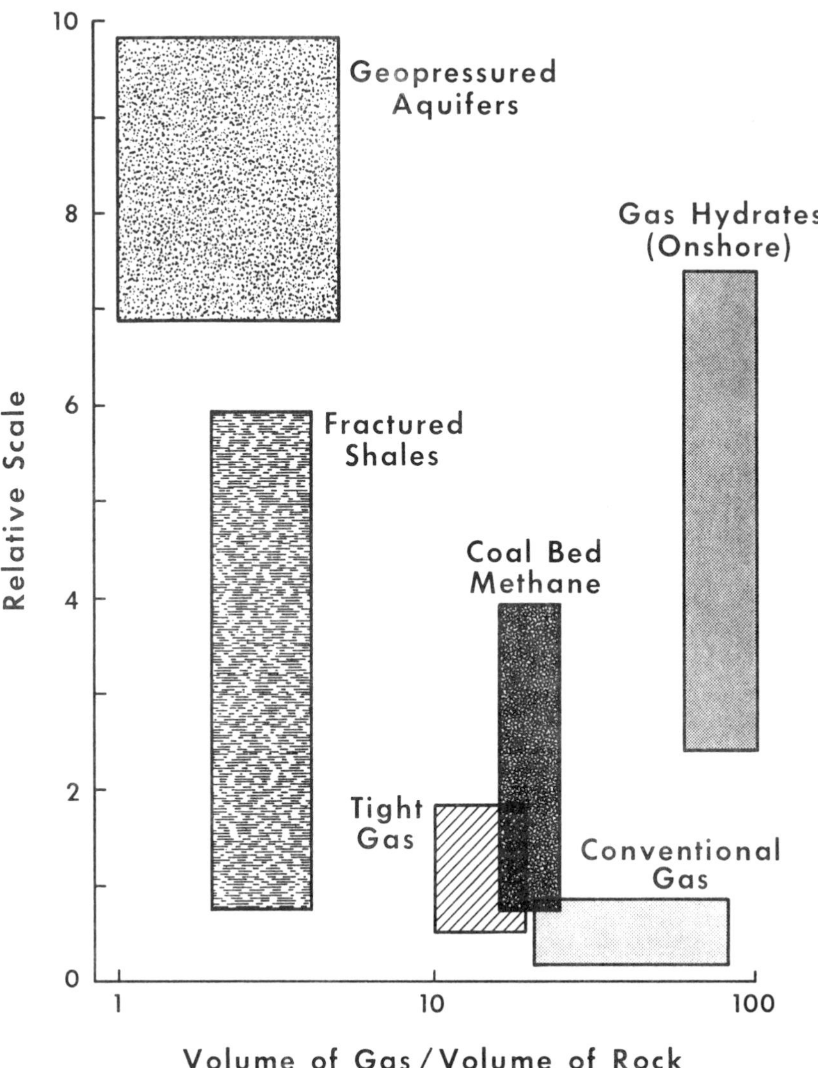

Figure 10 Relative costs of the recovery of different natural gas resources.

(Chap. 5), is an integrated collection of different units that will ultimately produce a purified gas product.

A gas production and processing system will consist of a variety of units, such as

<table>
<tr><th colspan="4" rowspan="2">Objective of Drilling</th><th rowspan="2">Initial Classification When Drilling is Started</th><th colspan="4">Final Classification After Completion or Abandonment</th></tr>
<tr><th colspan="2">Successful</th><th colspan="2">Unsuccessful</th></tr>
<tr><td colspan="4">Drilling for a New Field on a Structure or in an Environment Never Before Productive</td><td>1. New-Field Wildcat</td><td colspan="2">New-Field Discovery Wildcat</td><td colspan="2">Dry New-Field Wildcat</td></tr>
<tr><td rowspan="3">Drilling for a New Pool on a Structure or in a Geological Environment Already Productive</td><td rowspan="3">New Pool Tests</td><td colspan="2">Drilling Outside Limits of a Proved Area of Pool</td><td>2. New-Pool (Pay) Wildcat</td><td rowspan="3">New-Pool Discovery Wells (Sometimes Extension Wells)</td><td>New-Pool Discovery Wildcat</td><td rowspan="3">Dry New-Pool Tests</td><td>Dry New-Pool Wildcat</td></tr>
<tr><td rowspan="2">Drilling Inside Limits of Proved Area of Pool</td><td>For a New Pool Above Deepest Proven Pool</td><td>3. Deeper Pool (Pay) Test</td><td>Deeper Pool Discovery Well</td><td>Dry Deeper Pool Test</td></tr>
<tr><td>For a New Pool Below Deepest Proven Pool</td><td>4. Shallower Pool (Pay) Test</td><td>Shallower Pool Discovery Well</td><td>Dry Shallower Pool Test</td></tr>
<tr><td colspan="4">Drilling for Long Extension of a Partly Developed Pool</td><td>5. Outpost or Extension Test</td><td colspan="2">Extension Well (Sometimes a New-Pool Discovery Well)</td><td colspan="2">Dry Outpost or Dry Extension Test</td></tr>
<tr><td colspan="4">Drilling to Exploit or Develop a Hydrocarbon Accumulation Discovered by Previous Drilling</td><td>6. Development Well</td><td colspan="2">Development Well</td><td colspan="2">Dry Development Well</td></tr>
</table>

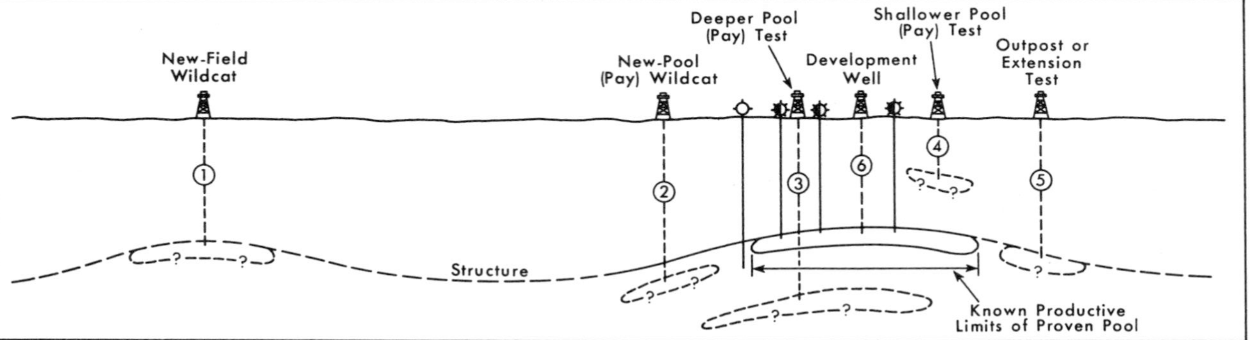

Source: Lahee classification of wells, as applied by the Committee on Statistics of Drilling of the American Association of Petroleum Geologists, and the American Petroleum Institute. Developed by Frederic H. Lahee in 1944.

Figure 11 American Association of Petroleum Geologists (AAPG) and American Petroleum Institute (API) systems of well classification.

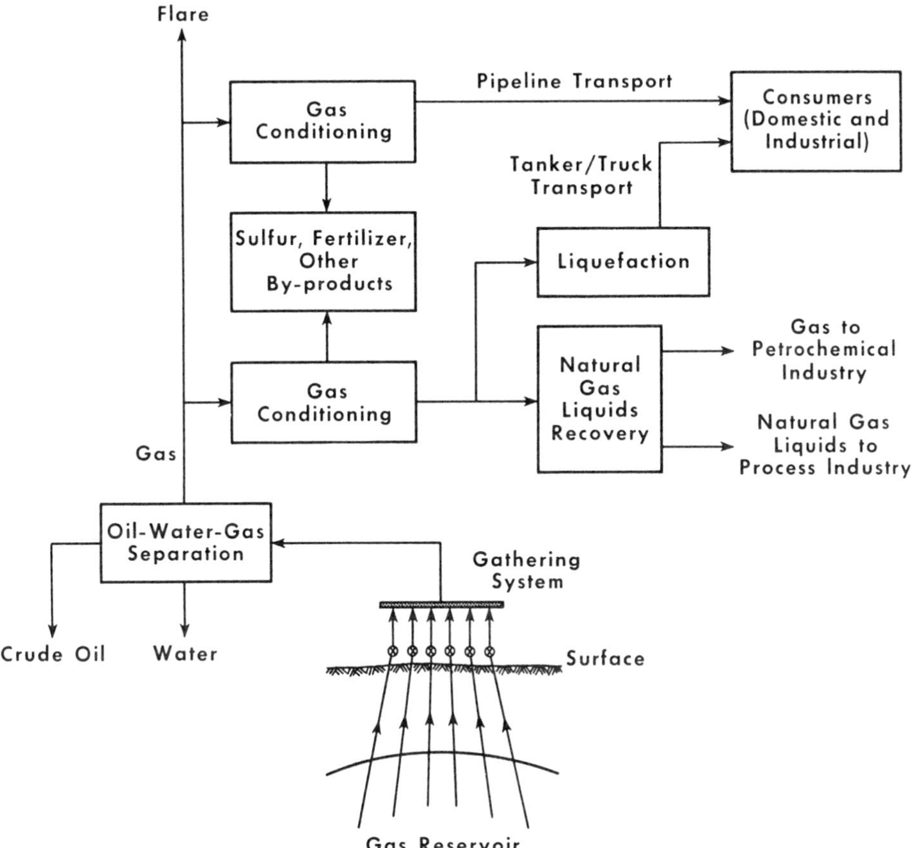

Figure 12 Schematic representation of a gas production, collection, and processing system.

1. The reservoir system for the production of gas from the reservoir
2. The gas-gathering system that is used to collect gas from several wells for separation and processing
3. The separation system for separation of the gas, oil, and water from the gas-gathering system
4. The gas-conditioning system for contaminant removal
5. The natural gas liquids recovery system
6. The gas compression/liquefaction system to prepare the gas for transportation by sea, rail, and road transport
7. The flow module for the pipeline transport of gas to consumers

All of these systems are interrelated but, for the purposes of this text, the first two systems are included as part of the production system while the last five systems are included under the processing systems (Chap. 35).

The manner in which the natural gas is produced from the reservoir depends on the properties of the reservoir rock and whether or not the gas is associated with petroleum in the reservoir. In general terms, the gas in the reservoir rock will migrate to the producing well because of the pressure differential between the reservoir and the well (Figure 13). The rate at which the gas migrates depends on the permeability of the reservoir rock; low permeability rocks may be fractured to yield better access to the well either by explosives or by hydraulic methods.

Production can continue as long as there is adequate pressure within the reservoir to produce the gas. The reservoir pressure will usually decrease as the natural gas is extracted from the reservoir and an additional method of recovery may have to be employed. For example, water may be injected into the reservoir to displace the gas from the pores of the reservoir rock (Figure 14). Such an operation helps maintain reservoir pressure during the recovery operation and will improve the recoverability of the natural gas.

Once the gas is produced from a well, it is mixed with the gas from other nearby wells and processed to obtain the residue gas that will be sold to the consumer. There are also other options for the gas; for example, when gas occurs in association with oil, it may be reinjected into the reservoir to maintain pressure for optimal recovery of the petroleum. On the other hand, the natural gas may be reinjected into the reservoir for storage if the market demand is low or if pipeline facilities for transport are not available.

Generally, the gas is gathered at low pressure and must be compressed to the processing pressure, which is usually the residue gas sales pressure (Figure 15). If the distance between the wells and the gas collection system is short, it is often an advantage to locate compressors at the plant where they can be attended for more efficient operation. For greater distances, field compressors designed for unattended operation should be used.

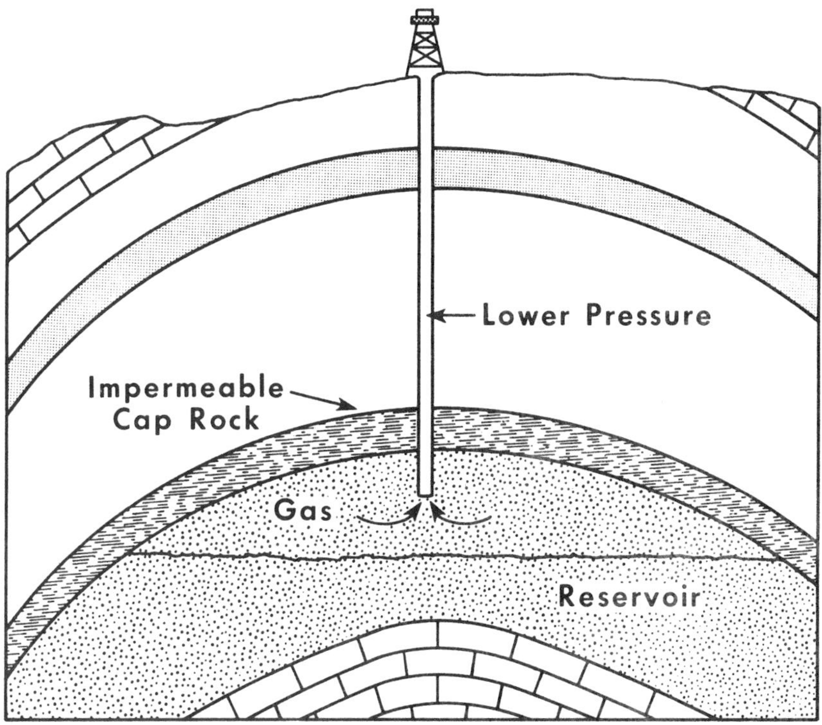

Figure 13 Schematic representation of the gas drive mechanism of recovery.

For a true gas field rather than where the gas is produced from an oil field, the gathering lines are usually of a size suitable to deliver the gas at utilization pressure. However, as the wellhead pressure declines, compression is necessary to maintain the flow of gas. In the later stages of a field's operation when the production volume of the natural gas is on the decline and gathering lines may be oversized, it may be possible to bypass the field compressors.

Hydrate formation in gas-gathering lines can be a serious problem, particularly if high pressures, low temperatures, and long gathering line distances are involved. The gas hydrates, which have a crystalline (clathrate) structure and can form at temperatures above the freezing point of water, will tend to deposit and plug lines and valves.

In order to combat gas hydrate formation in the gas-gathering lines, it is necessary to implement any one of the following: (1) temperature and pressure regimes at which hydrates cannot form, such as the use of wellhead gas heaters to heat the gas entering the gathering system;

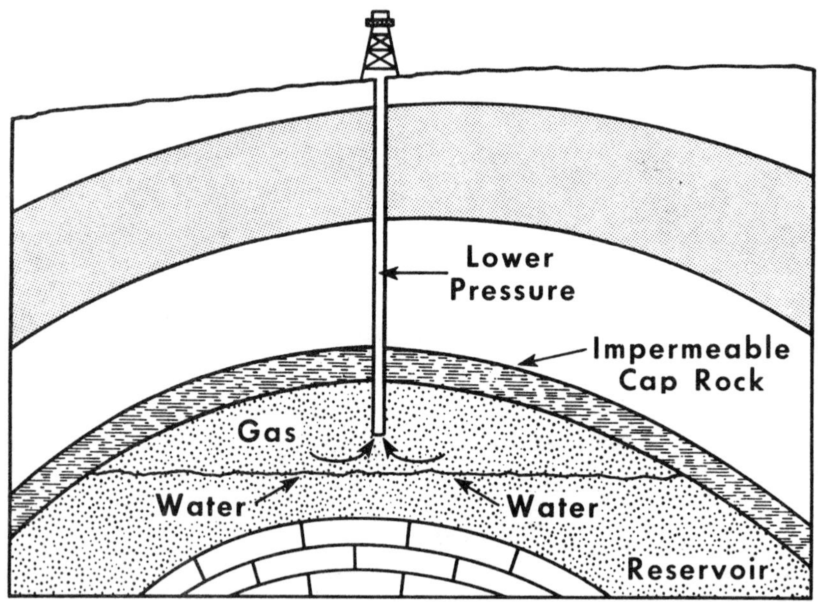

Figure 14 Schematic representation of the gas/water drive mechanism of recovery.

(2) remove the water from the natural gas at the source before it enters the gas-gathering line; (3) inject hydrate inhibitors (such as methanol or ethylene glycol) into the gas-gathering lines; (4) reinject produced gas back in the formation from which it is produced or inject additional gas from other formations into the formation to supplement the gas produced from a given oil zone (Kennedy, 1972).

Such techniques will ensure maximum recovery by maintaining the field pressure and/or preserve the gas that cannot be sold because an outlet is not available since venting may not be permitted due to local environmental regulations (Moran et al., 1986). Injection pressures of 3000–4000 psi are often employed for either of the above situations.

As noted above, underground storage of natural gas (Tek, 1987) may be necessary because of low market demand. The most common method for underground storage is to use a previously producing gas or petroleum field. The gas is pumped into the old wells by means of compressors similar to those employed to move the gas throughout the pipeline system (Kumar, 1987). The natural gas is usually stored under the same pressure conditions that originally existed in the field. Such a protocol will preserve the integrity of the formations that make

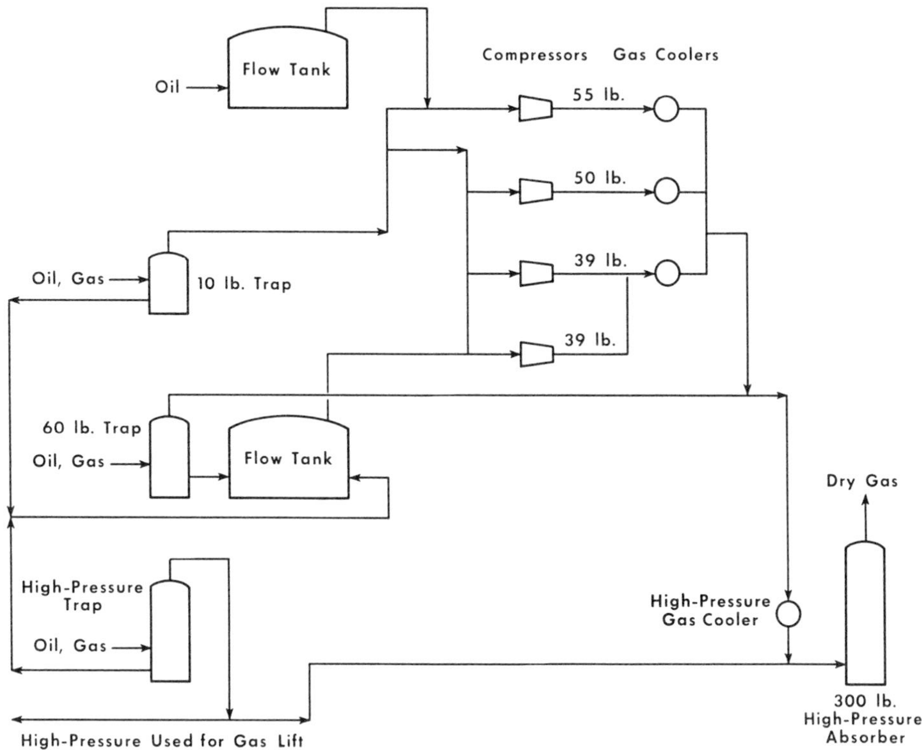

Figure 15 Schematic representation of a gas-gathering system for a petroleum field (see also Figure 12).

up the field and reduce the potential for gas loss due to formation disturbance that may occur at a higher pressure.

Other underground storage areas employ salt domes and aquifers. Salt domes are found throughout the world and generally give the appearance of a placement plug in which the adjacent strata are intruded by the salt and may be ruptured and sheared (Chap. 1, Sec. II.D; see also Pettijohn, 1957; Fairbridge, 1972; Hamblin, 1975). In many instances, salt recovery (usually by water injection, collection of the brine, and evaporation to yield the salt) leaves cavities in the earth which are suitable for the storage of the natural gas through a series of injection wells.

On the other hand, an aquifer is a lithologic unit (or even a combination of such units) that is porous in nature thereby having greater water transmissability than neighboring lithologic units (Fairbridge,

1972. The aquifer is therefore capable of storing and transmitting water. The gas is introduced (by means of wells) under pressure into the aquifer and, as the pressure increases, the gas will displace the water and commence to fill the pore spaces of the aquifer.

Another method for storage of gas is to liquefy the gas (Figure 16) and place the liquid in storage containers. However, special care must be taken to ensure that the containers are safe and leak-free (Considine, 1977).

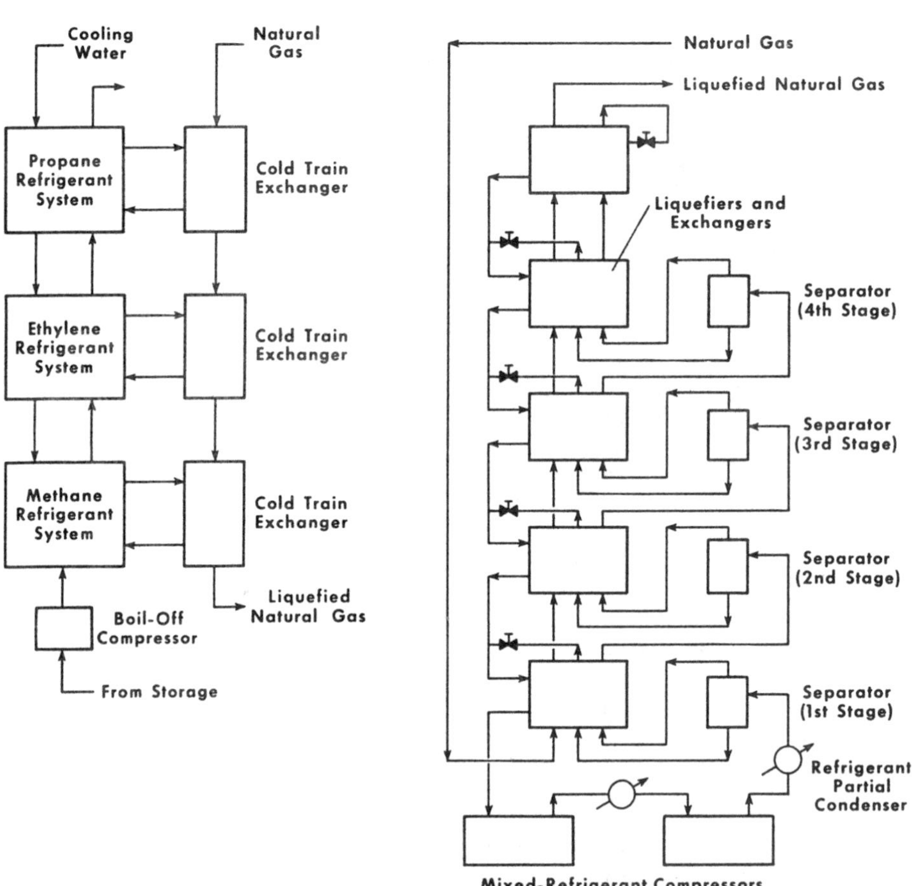

Figure 16 Schematic representation of different methods of gas liquefaction: (left) cascade system; (middle) single-pressure system; and (right) propane system.

V. TRANSPORTATION

The means by which natural gas will be transported depends on several factors: (1) the physical characteristics of the gas to be transported—whether it be in the gaseous or liquid phase; (2) the distance over which the gas will be moved; (3) features such as the geologic and geographic characteristics of the terrain, including land and sea operations; (4) the complexity of the distribution systems; and (5) any environmental regulations that are directly associated with the mode of transportation.

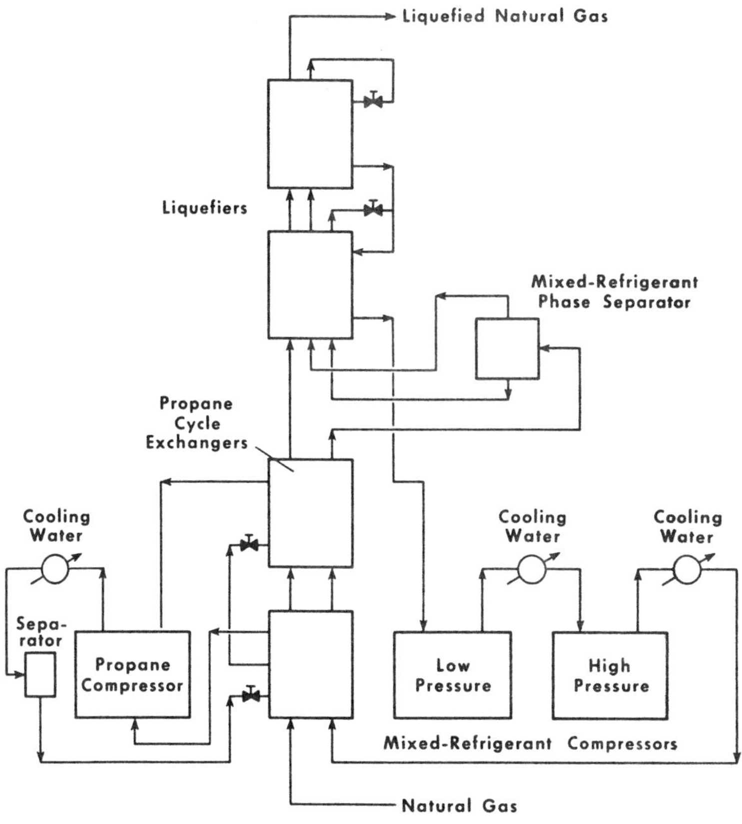

In the latter case, factors such as the possibility of pipeline rupture as well as the effect of the pipeline itself on the econosystems are two such factors that need to be addressed. In general and aside from any economic factors, it is possible to construct, and put in place, a system that is capable of transporting natural gas in the gaseous or liquid phase that allows system flexibility.

There are many such pipeline systems throughout the world and the United States (Considine, 1977). However, natural gas pipeline cmopanies must meet environmental and legal standards (Moran et al., 1986). Economic standards are also a necessity--it would be extremely foolhardy (and economically suicidal) for a company to construct several pipelines when one such system would suffice. Construction of a pipeline system involves not only environmental and legal considerations, but also compliance with the permitting regulations of the local, state, and/or federal authorities.

In general, many of the available pipeline systems use pipe material up to 36 in. in diameter (although lately larger diameter pipe has become more favorable) and sections of pipe may be up to 40 ft long. Protective coatings are usually applied to the pipe to prevent corrosion of the pipe from outside influences.

The gas pressure in long-distance pipelines may vary up to 5000 psi but pressures up to 1500 psi are more usual. To complete the pipeline, it is necessary to install a variety of valves/regulators that can be opened or closed to adjust the flow of gas and also to shut down a section of the system where an unexpected rupture may be caused by natural events (such as weather) or even by unnatural events (such as sabotage). Most of the valves/regulators in the pipeline system can now be operated by remote control so that, in the event of a rupture, the system can be closed down. This is especially valuable where it may take a repair crew considerable time to reach the site of the breakdown.

The trend in recent years has been to expand the pipeline system into marine environments where the pipeline is actually under a body of water. This has arisen mainly because of the tendency of petroleum and natural gas companies to expand their exploration programs to the sea. Lines are now laid in marine locations where depths exceed 500 ft and cover distances of several hundred miles to the shore. Excellent examples of such operations include the drilling operations in the Texas gulf and in the North Sea.

One early concern with the laying of pipelines under a body of water arose because of the bouyancy of the pipe and the subsequent need to place the pipe in a permanent position on the floor of the lake or sea bed. In such instances, the negative bouyancy of the pipe can be overcome by the use of a weighted coating (e.g., concrete) on the pipe. Other factors such as laying the pipe without too much stress

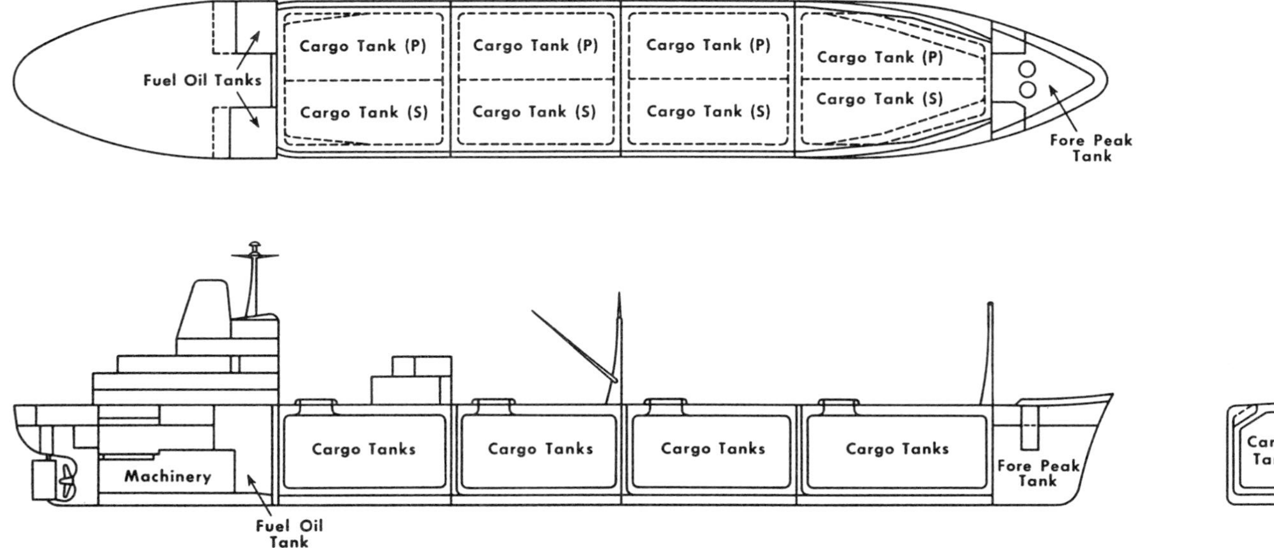

Figure 17 Various sectional views of a tanker equipped for carrying liquefied petroleum gas.

(which could otherwise induce a delayed rupture) as well as the anchoring and positioning of the pipe on the sea bed are major issues that need to be addressed.

Just as petroleum can be transported by seagoing tankers (Chap. 1, Sec. VII), natural gas is also transported by seagoing vessels. The gas is either transported under pressure at ambient temperatures (e.g., propane and butanes) or at atmospheric pressures but with the cargo under refrigeration (e.g., liquefied petroleum gas). For safety reasons, petroleum tankers are constructed with several independent tanks so the rupture of one tank will not necessarily drain the whole ship—unless it is a severe bow-to-stern (or stern-to-bow) rupture. Similarly, gas tankers also contain several separate tanks (Figure 17).

33
Definitions and Terminology

Natural gas is defined as the gaseous mixture that is associated with petroleum reservoirs (Chap. 32, Sec. III.A) and coal seams (Chap. 32, Sec. III.B) and the decay of organic material. Natural gas is predominantly methane (see below), but it should not be confused with the gaseous products from the destructive distillation or carbonization of wood and coal (Table 1); such gaseous products are manufactured gases.

Natural gas is a naturally occurring mixture of combustible hydrocarbon compounds but it does contain nonhydrocarbon compounds (Table 2). The gas occurs in the porous rock of the earth's crust either alone or with accumulations of petroleum (Chap. 1, Sec. III, and Chap. 32, Sec. III). In the latter case, the gas forms the gas cap (Figure 1; see also Chap. 1, Secs. II.D and IV.E), which is the mass of gas trapped between the liquid petroleum and the impervious cap rock of the petroleum reservoir. When the pressure in the reservoir is sufficiently high, the natural gas may be dissolved in the petroleum and is released upon penetration of the reservoir as a result of drilling operations.

Natural gas, after processing and purification, is a homogeneous mixture that is odorless; however, odor-generating additives are added during processing to enable the detection of gas leaks. It is one of the more stable flammable gases (Curry, 1981) but it is flammable within the limits of a 5–15% mixture with air (hydrogen sulfide is flammable within 4–46% in air at a much lower ignition temperature). In general, natural gas has an energy content of 1000 btu/scf and is very often priced in terms of its energy content rather than mass or volume.

Table 1 Generalized Classification of Methane-Containing Gases

1. Natural Gas

   Associated with petroleum oil deposits, coal seams, or the decay of organic matter.

2. Manufactured Gases

(a) From wood–by distillation or carbonization–wood gas.
(b) From peat–by distillation or carbonization–peat gas.
(c) From coal–by carbonization–coal gas;
–by gasification (1) in air–producer gas,
(2) in air and steam–water gas,
(3) in oxygen and steam–Lurgi gas.
–by hydrogenation.
(d) From petroleum and oil shale–by cracking–refinery gas;
–by hydrogenation–oil gas;
–by water gas reaction–oil gas;
–by partial oxidation–oil gas.

In addition to composition and thermal content (btu/scf), natural gas can also be characterized on the basis of the mode of the natural gas which is found in reservoirs where there is no or, at best, minimal amounts of crude oil.

Thus, there is "nonassociated" natural gas which is found in reservoirs in which there is no or, at best, minimal amounts of crude oil. Nonassociated gas is usually richer in methane but is markedly leaner in terms of the higher molecular weight hydrocarbons and condensate materials.

Conversely, there is also "associated" or "dissolved" natural gas which occurs either as free gas or as gas in solution in the crude oil. Gas that occurs as a solution with the crude petroleum is dissolved gas whereas the gas that exists in contact with the crude petroleum ("gas cap"; Figure 1; see also Chap. 1, Sec. IV.E) is associated gas. Associated gas is usually leaner in methane than the nonassociated gas but will be richer in the higher molecular weight constituents.

Another product is "gas condensate," which contains relatively high amounts of the higher molecular weight liquid hydrocarbons. These hydrocarbons may occur in the gas phase in the reservoir.

The most preferential type of natural gas is the nonassociated gas. Such gas can be produced at high pressure whereas associated, or dissolved, gas must be separated from petroleum at lower separator

Table 2 Generalized Composition of Natural Gas from a Petroleum Well

Category	Component	Amount (%)
Paraffinic	Methane (CH_4)	70–98
	Ethane (C_2H_6)	1–10
	Propane (C_3H_8)	Trace–5
	Butane (C_4H_{10})	Trace–2
	Pentane (C_5H_{12})	Trace–1
	Hexane (C_6H_{14})	Trace–0.5
	Heptane and higher (C_7+)	None–trace
Cyclic	Cyclopropane (C_3H_6)	Traces
	Cyclohexane (C_6H_{12})	Traces
Aromatic	Benzene (C_6H_6), others	Traces
Nonhydrocarbon	Nitrogen (N_2)	Trace–15
	Carbon dioxide (CO_2)	Trace–1
	Hydrogen sulfide (H_2S)	Trace occasionally
	Helium (He)	Trace–5
	Other sulfur and nitrogen compounds	Trace occasionally
	Water (H_2O)	Trace–5

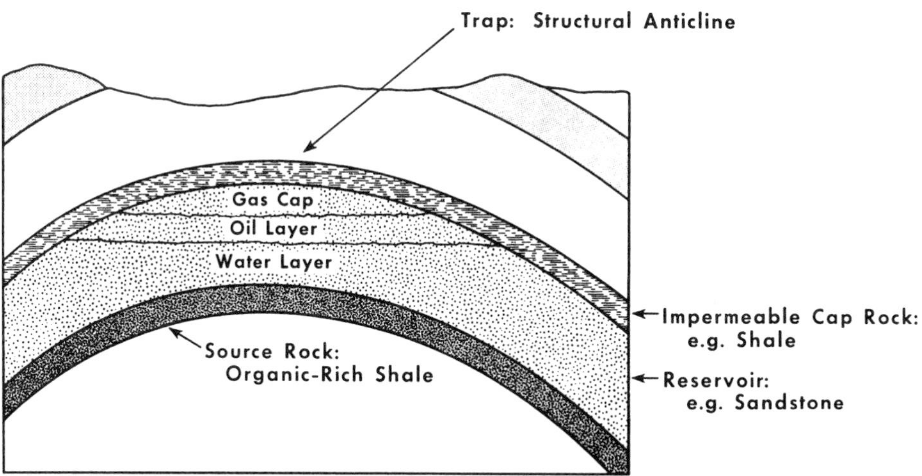

Figure 1 Illustration of a petroleum reservoir to show the gas cap.

pressures, which usually involves increased expenditure for compression. Thus, it is not surprising that such gas (under conditions that are not economically favorable) will often be flared or vented.

The nonhydrocarbon constituents of natural gas can be classified as two types of materials: (1) diluents such as nitrogen, carbon dioxide, and water vapor; (2) contaminants such as hydrogen sulfide and/or other sulfur compounds. The diluents are noncombustible gases that reduce the heating value of the gas and are on occasion used as "fillers" when it is necessary to reduce the heat content of the gas. On the other hand, the contaminants are detrimental to production and transportation equipment in addition to being obnoxious pollutants. Thus, the primary reason for gas processing (Chaps. 35 and 36) is to remove the unwanted constituents of natural gas.

The major diluents/contaminants of natural gas are (1) acid gas, which is predominantly hydrogen sulfide although carbon dioxide does occur to a lesser extent; (2) water, which includes all entrained free water or water in condensed form; (3) liquids in the gas such as higher boiling hydrocarbons as well as pump lube oil, scrubber oil, and, on occasion, methanol; and (4) any solid matter that may be present such as fine silica (sand) and scaling from the pipe (Curry, 1981).

Natural gas from different wells will vary widely in composition and analysis (Table 3) and the proportion of nonhydrocarbon constituents can vary over a very wide range. Thus, one particular natural gas field could require different production, processing, and handling protocols from those of another field.

As a result of the variances in the composition of natural gas from different locations, there are several general definitions that have been applied to the different products. Thus, natural gas can be (1) "lean" gas in which methane is the major constituent; (2) "wet" gas, which contains considerable amounts of the higher molecular weight hydrocarbons; (3) "sour" gas, which contains hydrogen sulfide; (4) "sweet" gas, which contains very little, if any, hydrogen sulfide; (5) "residue" gas, which is natural gas from which the higher molecular weight hydrocarbons have been extracted; and (6) "casinghead" gas, which is derived from petroleum but is separated at the separation facility at the wellhead.

There are many gas fields (often referred to as "dry" fields) in which liquids do not occur with the gas and the only processing required is dehydration or perhaps heating value adjustment. Other fields (often referred to as "condensate" fields) are those fields in which a liquid product or condensate is produced with the gas. In these fields, liquids often condense out from the gas (retrograde condensation) as the pressure is reduced (Katz et al., 1959). Liquid also condenses out of the gas in the formation as the pore pressure drops and may not completely vaporize before well abandonment pressure is reached. Cycling plants are often installed to prevent this loss of product.

Table 3 Variations of Natural Gas Composition with Source

Component	Type of gas field			Natural gas separated from crude oil		
				Ventura[a]		
	Dry gas, Los Medanos[a] (mole %)	Sour gas, Jumping Pound[b] (mole %)	Gas condensate, Paloma[a] (mole %)	400 lb (mole %)	50 lb (mole %)	Vapor (mole %)
Hydrogen sulfide	0	3.3	0	0	0	0
Carbon dioxide	0	6.7	0.68	0.30	0.68	0.81
Nitrogen and air	0.8	0	0	0	–	2.16
Methane	95.8	84.0	74.55	89.57	81.81	69.08
Ethane	2.9	3.6	8.28	4.65	5.84	5.07
Propane	0.4	1.0	4.74	3.60	6.46	8.76
Isobutane	0.1	0.3	0.89	0.52	0.92	2.14
n-Butane	Trace	0.4	1.93	0.90	2.26	5.20
Isopentane	0		0.75	0.19	0.50	1.42
n-Pentane	0		0.63	0.12	0.48	1.41
Hexane	0	0.7	1.25			
Heptane	0			0.15	1.05	4.13
Octane	0		6.30			
Nonane	0					
	100.0	100.0	100.00	100.00	100.00	100.00

[a]California.
[b]Canada.

The produced gas is processed to remove the higher molecular weight hydrocarbons and any liquid products. Any residue gas, rather than being sold, is injected to maintain reservoir pressure. When the reservoir has been swept of the higher molecular weight materials so that retrograde condensation can no longer occur, the field can be taken to full production.

As already noted, gas is also produced with crude oil (Chap. 1, Sec. IV.E) and such gas is usually rich in recoverable hydrocarbon liquids. Thus, construction of a gas-processing plant in conjunction with the petroleum recovery operations may be economically justifiable even at relatively low gas production rates.

A typical gas-processing plant produces residue gas and a variety of products such as ethane, liquefied petroleum gas (LPG), and "natural" gasoline (Table 4). Originally, the gas-processing plants were used to remove the gasoline components to be used as a blending stock

Table 4 Composition of (Natural) Gasoline from Natural Gas

Reid vapor pressure	Ventura gasoline plant			Ten-section gasoline plant
	38 psia	60 psia	100 psia	22 psia
Ethane	Trace	0.5	0.7	0
Propane	1.1	16.0	43.8	0
Isobutane	19.0	16.0	10.7	0.2
n-Butane	41.0	34.7	23.0	22.7
Isopentane	13.2	11.2	7.4	24.1
n-Pentane	11.3	9.5	6.3	21.0
Hexane	6.8	5.7	3.8	12.6
Heptane	5.3	4.4	2.9	13.7
Octane	1.2	1.0	0.7	4.1
Nonane	1.1	1.0	0.7	1.2
Decane	Trace	Trace	Trace	0.4
	100.0	100.0	100.0	100.0

for motor gasoline; hence, the term "gasoline" plant was often inappropriately applied to the gas-processing plant. Other fuel needs then caused a shift of focus to the liquefied petroleum gas (propane, butanes, and/or mixtures thereof) as well as the gasoline constituents. More recently, the extraction of ethane for petrochemical feedstocks (Chap. 11) has become an extremely important aspect of gas-processing operations.

34

Composition and Properties

I	Hydrocarbon Content	1096
II	Water Content	1096
III	Acid Gas Content	1097
IV	Natural Gas Liquids	1097

Natural gas contains constituents other than methane; therefore, knowledge of the occurrence and concentration of these constituents in natural gas is a prerequisite to any step in the processing sequence (Nonhebel, 1964; Lowenheim and Moran, 1975; Kohl and Resienfeld, 1985; Trusell, 1985; Willis, 1986).

For example, many wells produce natural gas that contains hydrogen sulfide. Natural gases containing hydrogen sulfide are, like petroleum products that contain hydrogen sulfide or mercaptans (Chap. 7), termed "sour" whereas natural gases that are free of hydrogen sulfide are called "sweet."

The hydrogen sulfide concentration in natural gas will usually vary from barely detectable quantities to more than 0.30% (3000 ppm). Other sulfur derivatives are not usually present in significant quantities and may occur only in trace amounts. Thus, a sulfur removal process must be very precise since natural gas contains only a small quantity of sulfur-containing compounds that must be reduced several orders of magnitude. Most consumers of natural gas require less than 4 ppm in the gas.

A characteristic feature of natural gas that contains hydrogen sulfide is the presence of carbon dioxide (generally in the range of 1–4% by volume). In cases where the natural gas does not contain hydrogen sulfide, there is usually a relative lack of carbon dioxide as well.

Sales gas specifications for natural gas include one or more of the following: water content, hydrocarbon content, heating value, specific gravity, acid gas content, temperature, and pressure. As with any property measurement, the value of any specification depends on the availability of reliable test methods to determine the specific property (Sharples and Panhill, 1985; ASTM, 1987; Kumar, 1987).

I. HYDROCARBON CONTENT

Hydrocarbon content of natural gas is usually obtained indirectly by measurement of heating value (ASTM D1826) or specific gravity (ASTM D1070; ASTM D3588). However, it must be remembered that the composition of natural gas can vary widely (Chap. 33) although, since natural gas is a multicomponent system, neither property can be changed significantly.

In some instances, the hydrocarbon dew points may be specified or limits placed on gas enrichment with reference to specific components. Of particular importance in this respect are the hexanes and higher molecular weight hydrocarbons which may condense in the gas-gathering and/or distribution systems. If significant amounts of carbon dioxide or nitrogen are present in the natural gas, neither gravity nor heating value alone will indicate hydrocarbon content. If both of these properties are measured, the presence of either carbon dioxide or nitrogen will be reflected in a higher specific gravity and lower heating value.

II. WATER CONTENT

The water content of natural gas is usually expressed as pounds of water per million cubic feet of gas or by use of dew point temperature and pressure (ASTM D1142). The two methods have a defininte relationship as shown by curves of water content as a function of saturation temperature and pressure (NGPSA, 1966). Common specifications are 1-, 4-, or 7-lb gas (i.e., lb water/Mscf gas) depending on the conditions to which the gas will be exposed.

III. ACID GAS CONTENT

Hydrogen sulfide and carbon dioxide are the acid gases associated with natural gas. They are termed acid gases because solutions of these gases in water are acidic in nature.

Besides emitting a foul odor at low concentrations, hydrogen sulfide is deadly poisonous. At concentrations above 600 ppm it can be fatal in a matter of minutes and has a toxicity comparable to that of hydrogen cyanide. Thus, it cannot be tolerated in gas to be used as domestic fuel. In addition, hydrogen sulfide is corrosive to all metals normally associated with gas-transporting, processing, and handling systems (although it is less corrosive to stainless steel), and may lead to premature failure of most such systems. On combustion, it forms sulfur dioxide:

$$2H_2S + 3O_2 = 2H_2O + 2SO_2$$

which is usually highly toxic and corrosive. Hydrogen sulfide and other sulfur compounds can also cause catalyst poisoning in refinery processes (Chap. 6).

Carbon dioxide has no heating value and its removal may be required in some instances (where acidic properties are of a lesser issue) to increase the energy content (btu/scf) of the gas. For gas being sent to cryogenic plants, removal of carbon dioxide is necessary to prevent solidification of the carbon dioxide in the plant.

Both hydrogen sulfide and carbon dioxide promote hydrate formation, and the presence of carbon dioxide may be less desirable for this reason. However, if none of these situations is encountered, there may be no need to remove the carbon dioxide.

Acid gas content is specified according to the particular impurity. The usual acid gas specification is for hydrogen sulfide (ASTM D2420; ASTM D2725; ASTM D4048). Sulfur compounds may also be present in natural gas (ASTM D3031) but may not often present a major problem if they are present only in small amounts since mercaptans are added as a warning odorant for natural gas. Carbon dioxide content may also be specified (ASTM D1945; ASTM D1946); an upper limit is commonly 5% by volume. There are also a few reported cases of carbonyl sulfide (also called carboxysulfide; COS) in natural gas.

IV. NATURAL GAS LIQUIDS

Natural gas is usually considered to be methane (CH_4) and, hence, natural gas liquids are the higher molecular weight hydrocarbons.

They are not true liquids in the sense that they are usually not in liquid form at ambient temperature and pressure. Thus, natural gas liquids are defined as (1) ethane, (2) liquefied petroleum gas, or (3) natural gasoline.

The liquefied petroleum gas is usually composed of propane (C_3H_8), butanes (C_4H_{10}), and/or mixtures thereof; small amounts of ethane and pentane may also be present as impurities. On the other hand, the natural gasoline (like refinery gasoline; Chap. 5) consists mostly of pentane (C_5H_{12}) and higher molecular weight hydrocarbons. The term "natural gasoline" has also, on occasion in the gas industry, been applied to mixtures of liquefied petroleum gas, pentanes, and higher molecular weight hydrocarbons. Caution should be taken not to confuse the term "natural gasoline" with the term "straight-run gasoline" (often also incorrectly referred to as natural gasoline), which is the gasoline that is distilled unchanged from petroleum (Chap. 5).

There are also standards for the liquid content of natural gas that are usually set by mutual agreement between the buyer and the seller, but such specifications do vary widely and can only be given approximate limits. For example, ethane may have a maximum methane content of 1.5% by volume and a maximum carbon dioxide content of 0.28% by volume. On the other hand, propane will be specified to have a minimum of 95% propane by volume, a maximum of 1–2% butane, and a maximum vapor pressure which limits ethane content. For butane, the percentage of one of the butane isomers is usually specified along with the maximum amounts of propane and pentane.

Other properties that may be specified are vapor pressure, specific gravity, corrosivity, dryness, and sulfur content. The specifications for the propane-butane mixtures will have limits on the amount of the nonhydrocarbons and, in addition, the maximum isopentane content is usually stated.

Natural gasoline may be sold on the basis of vapor pressure or on the basis of actual composition, which is determined from the Reid vapor pressure–composition curves prepared for each product source (ASTM D323).

35

Processing: General Concepts

I	Introduction	1099
II	Water Removal	1100
III	Nitrogen Removal	1103
IV	Acid Gas Removal	1103
V	Liquids Removal	1107
VI	Enrichment	1110
VII	Fractionation	1112

I. INTRODUCTION

The processes that have been developed to accomplish gas purification vary from a simple once-through wash operation to complex multistep recycle systems. In many cases, the process complexities arise because of the need for recovery of the materials used to remove the contaminants or even recovery of the contaminants in the original, or altered, form.

There are many variables in gas treatment and the precise area of application of a given process is difficult to define. Several factors must be considered: (1) the types and concentrations of contaminants in the gas; (2) the degree of contaminant removal desired; (3) the selectivity of acid gas removal required; (4) the temperature, pressure, volume, and composition of the gas to be processed; (5) the carbon dioxide/hydrogen sulfide ratio in the gas; (6) the desirability of sulfur recovery due to process economics or environmental issues.

In addition to hydrogen sulfide and carbon dioxide, natural gas may contain other contaminants such as mercaptans and carbonyl sulfide (Chap. 33). The presence of these impurities may eliminate some of the sweetening processes since some processes will remove large amounts of acid gas but not to a sufficiently low concentration. On the other hand, there are processes not designed to remove (or incapable of removing) large amounts of acid gases but capable of removing the acid gas impurities to very low levels when the acid gases are present in low to medium concentrations in the natural gas.

Process selectivity indicates the preference with which the process will remove one acid gas component relative to (or in preference to) another. For example, some processes remove both hydrogen sulfide and carbon dioxide whereas other processes are designed to remove hydrogen sulfide only. It is important to consider the process selectivity for, say, hydrogen sulfide removal compared to carbon dioxide removal which will ensure minimal concentrations of these components in the product; hence the need for consideration of the carbon dioxide/hydrogen sulfide ratio in the natural gas.

The processing of natural gas involves the use of several different types of processes but there is always overlap between the various processing concepts. In addition, the terminology used for natural gas processing can often be confusing and/or misleading because of the overlap.

For convenience in this text, natural gas processing is subdivided into two chapters. The first (this chapter) deals with the general processing concepts. The second (next chapter) deals with the description of specific processes. Cross-referencing is employed so that the reader need not miss any particular aspect of the processing operations.

II. WATER REMOVAL

Water is the most common impurity in natural gas, and its removal is necessary to prevent condensation and the formation of ice or gas hydrates. Water in the liquid phase will cause corrosion or erosion problems in pipelines and equipment, particularly when carbon dioxide and hydrogen sulfide are present in the gas. The simplest method of water removal (refrigeration or cryogenic separation; Chap. 36, Sec. I) is to cool the natural gas to a temperature equal to or (more preferentially) below dew point.

In a majority of cases, cooling alone is insufficient and for the most part impractical in field operations. Other, more convenient water removal options use (1) hygroscopic liquids (e.g., di- or triethylene glycol) and (2) solid adsorbents (e.g., alumina, silica gel, and molecular sieves). Ethylene glycol can be directly injected into the gas stream in refrigeration-type plants (see Figure 1).

The use of a hygroscopic fluid, such as ethylene glycol, for natural gas dehydration is a relatively simple operation (Figure 2). The overhead stream from the regenerator is cooled with air fins at the top of the column or by an internal coil through which the feed flows. The countercurrent vapor-liquid contact between the gas and the glycol produces a dew point of the outlet stream that is a function of the contact temperature and the residual water content of the stripped or lean glycol.

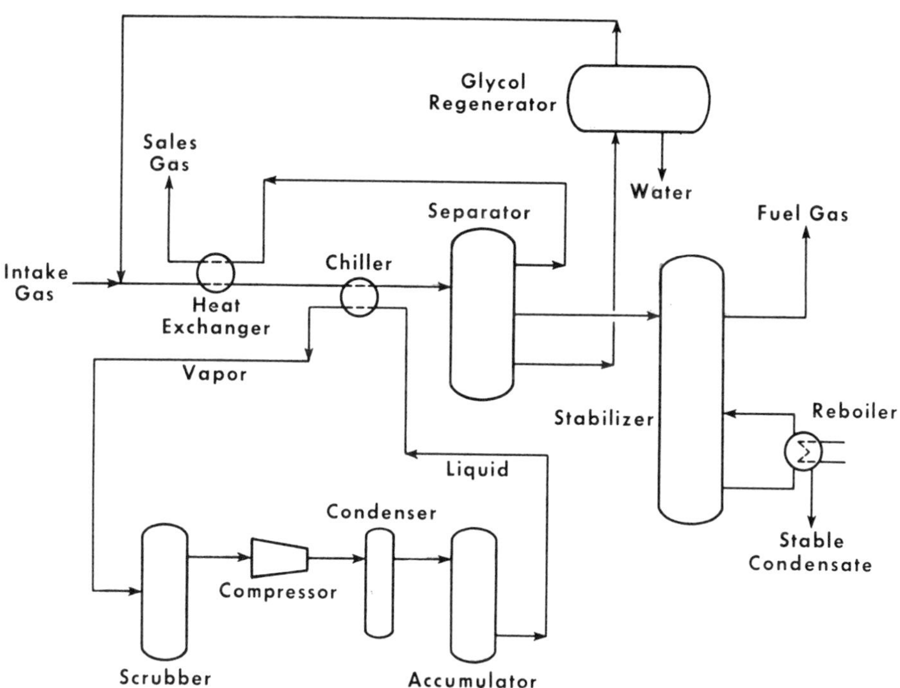

Figure 1 Schematic representation of a refrigeration process.

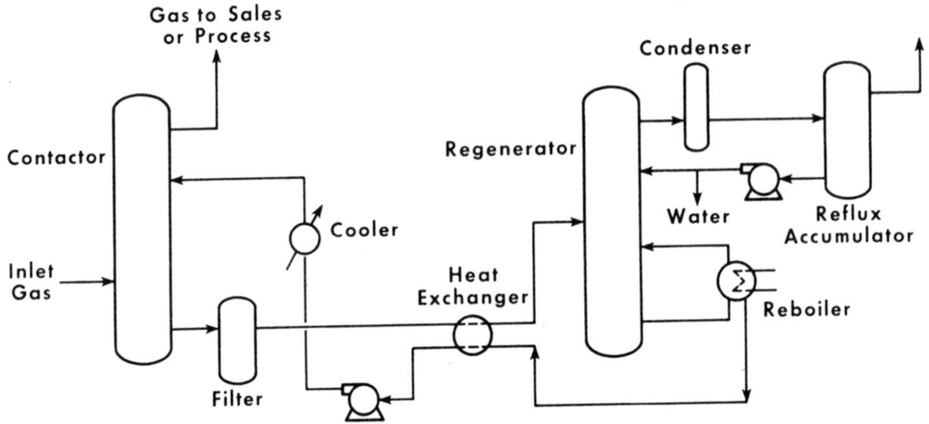

Figure 2 Schematic representation of a fluid process for gas treatment.

The regeneration (stripping) of the glycol is limited by temperature; diethylene glycol and triethylene glycol will decompose at, or prior to, their respective boiling points. Techniques such as stripping of hot triethylene glycol with dry gas (e.g., heavy hydrocarbon vapors—the Drizo process) or vacuum distillation are recommended.

Some adsorbent processes for water removal (Figure 3) employ a two-bed adsorbent treater; while one bed is removing water from the gas, the other is undergoing alternate heating and cooling. On occasion, a three-bed system is used; one bed is adsorbing, one is being heated, and one is being cooled. An additional advantage of the three-bed system is the facile conversion to a two-bed system so that the third bed can be maintained or replaced, thereby ensuring continuity of the operations and reducing the risk of a costly plant shutdown.

Silica gel and alumina have good capacities for water adsorption (up to 8% by weight). Bauxite (crude alumina) will adsorb up to 6% by weight water and molecular sieves will adsorb up to 15% by weight water. Silica is usually selected for dehydration of sour natural gas because of its high tolerance to hydrogen sulfide and to protect molecular sieve beds from plugging by sulfur. Alumina "guard beds" (which will serve as protectors by the act of attrition; see also Chap. 36, Sec. III.B and Speight, 1981) may be placed ahead of the molecular sieves to remove the sulfur compounds. Downflow reactors are commonly used for adsorption processes with an upward flow regeneration of the adsorbent and cooling the same direction as adsorption.

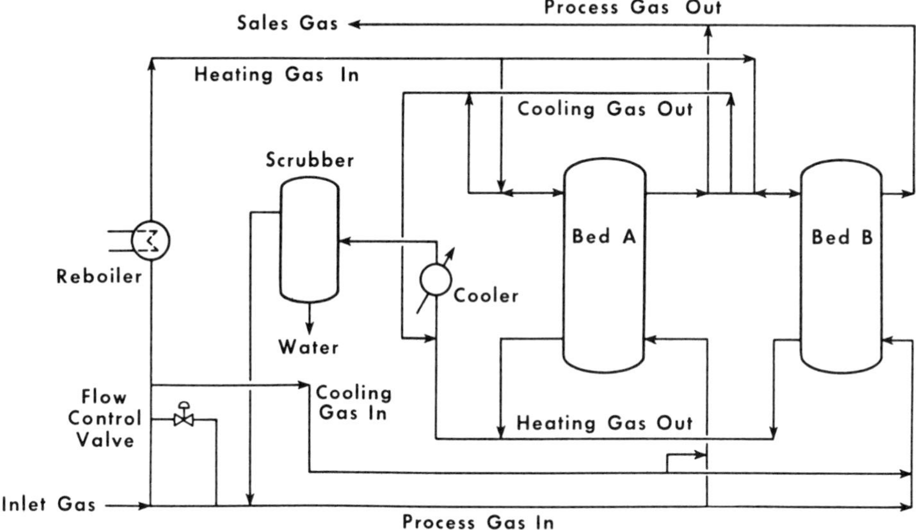

Figure 3 Schematic representation of a simple adsorption process for gas treatment.

III. NITROGEN REMOVAL

Nitrogen may often occur in sufficient quantities in natural gas to lower the heating value of the gas. Thus, several plants for the removal of nitrogen from the natural gas have been built and it must be recognized that nitrogen removal requires liquefaction and fractionation of the entire gas stream, which may affect process economics. In many cases, the nitrogen-containing natural gas is blended with a gas having a higher heating value and sold at a reduced price depending on the thermal value (btu/scf).

IV. ACID GAS REMOVAL

The removal of acid gases from natural gas streams can be generally classified in two categories: (1) chemical absorption processes and (2) physical adsorption processes. Several such processes fit into these categories (Table 1) but only the more important will be addressed in this text.

Table 1 Simplified Classification of the Various Processes for Acid Gas Removal

Chemical absorption (chemical solvent processes)	Physical absorption (physical solvent processes)
Alkanolamines	
MEA	Selexol
SNPA:DEA (DEA)	Rectisol
UCAP (TEA)	Sulfinol[a]
Selectamine (MDEA)	
Econamine (DGA)	
ADIP (DIPA)	
Alkaline salt solutions	
Hot potassium carbonate	
Catacarb	
Benfield	
Giammarco–Vetrocoke	
Nonregenerable	
Caustic	

[a]A combined physical/chemical solvent process.

Treatment of natural gas to remove the acid gas constituents (hydrogen sulfide and carbon dioxide) is most often accomplished by contact of the natural gas with an alkaline solution (Figure 2; see also Chap. 7, Sec. II.F). The most commonly used treating solutions are aqueous solutions of the ethanolamines or alkali carbonates, although a considerable number of other treating agents have been developed in recent years. Comparison of the general features of the scrubbing processes (Table 2) does give an indication of the limitations of the particular types of processes.

Most of these newer treating agents rely on physical absorption and chemical reaction. When only carbon dioxide is to be removed in large quantities, or when only partial removal is necessary, a hot carbonate solution (Chap. 36, Sec. II.E) or one of the physical solvents is the most economical selection.

Hydrogen sulfide may be removed solely by the use of several processes (Table 3). The most well-known hydrogen sulfide removal process is based on the reaction of hydrogen sulfide with iron oxide (often

Table 2 Comparison of Some of the Features of the Various Acid Gas Removal Processes

Feature	Chemical adsorption: Amine processes	Chemical adsorption: Carbonate processes	Physical absorption
Absorbents	MEA, DEA, DGA, MDEA	K_2CO_3, K_2CO_3 + MEA K_2CO_3 + DEA, K_2CO_3 + arsenic trioxide	Selexol, Purisol, Rectisol
Operating pressure, psi	Insensitive to pressure	>200	250–1000
Operating temp, °F	100–400	200–250	Ambient temperature
Recovery of absorbents	Reboiled stripping	Stripping	Flashing, reboiled, or steam stripping
Utility cost	High	Medium	Low-medium
Selectivity, H_2S, CO_2	Selective for some amines (MDEA)	May be selective	Selective for H_2S
Effect of O_2 in the feed	Formation of degradation products	None	Sulfur precipitation at low temperature
COS and CS_2 removal	MEA: not removed DEA: slightly removed DGA: removed	Converted to CO_2 and H_2S and removed	Removed
Operating problems	Solution degradation; foaming; corrosion	Column instability; erosion; corrosion	Absorption of heavy hydrocarbons

also called the iron sponge process or the dry-box method; Figure 4; see also Chap. 36, Sec. III.A) in which the gas is passed through a bed of wood chips impregnated with iron oxide. The bed is maintained in a moist state by circulation of water or a solution of soda ash.

Table 3 Comparison of the Various Processes for Hydrogen Sulfide Removal

Process	Sorbent	Temp.
Stretford	Water solution of $NaCO_3$ and anthraquinone disulfonic acid (ADA) with activation of sodium metavanadate	50–100°F 10–40°C
Giammarco–Vetrocoke	K_2CO_3 activated with basic arsenic compounds	50–212°F 10–100°C
Iron sponge	Iron oxide	Slightly above ambient
Activated carbon	Carbon	Ambient

The hydrogen sulfide reacts with the iron oxide to form iron sulfide:

$$Fe_2O_3 + 3H_2S = Fe_2S_3 + 2H_2O$$

which in turn is regenerated by passage of air through the bed:

$$2Fe_2S_3 + 3O_2 = 2Fe_2O_3 + 6S$$

which converts the iron sulfide to elemental sulfur and iron oxide.

The method is suitable only for small to moderate quantities of hydrogen sulfide. Approximately 90% of the hydrogen sulfide can be removed per bed, but bed clogging by elemental sulfur does occur and the bed must then be discarded. The use of several beds in series is not usually economical.

Removal of larger amounts of hydrogen sulfide from the natural gas requires a continuous process such as the ferrox process (Figure 5) or the Stretford process (Figure 6). The ferrox process is based on the same chemistry as the iron oxide process except that it is fluid and continuous. The Stretford process employs a solution containing vanadium salts and anthraquinone disulfonic acid (Maddox, 1974; see also Chap. 7, Sec. II.F and Chap. 36, Sec. II.F).

Most hydrogen sulfide removal processes return the hydrogen sulfide unchanged, but if the quantity involved does not justify installation of a sulfur recovery plant (usually a Claus plant; Figure 7), it is necessary to select a process which produces elemental sulfur directly.

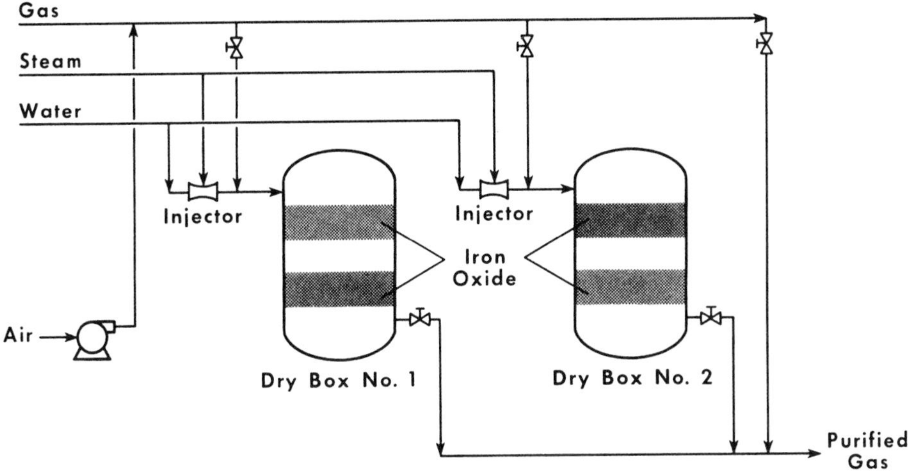

Figure 4 Schematic representation of the iron oxide process.

V. LIQUIDS REMOVAL

Recovery of liquid hydrocarbons can be justified either because it is necessary to make the gas saleable or because economics dictates this course of action. The justification for building a liquids recovery (or a liquids removal) plant depends on the price differential

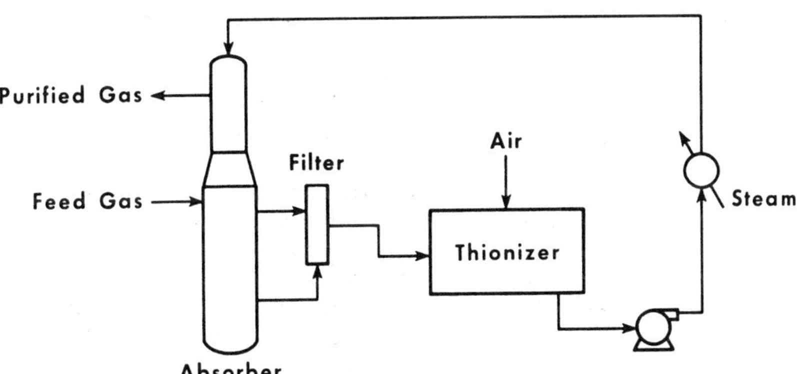

Figure 5 Schematic representation of the ferrox process.

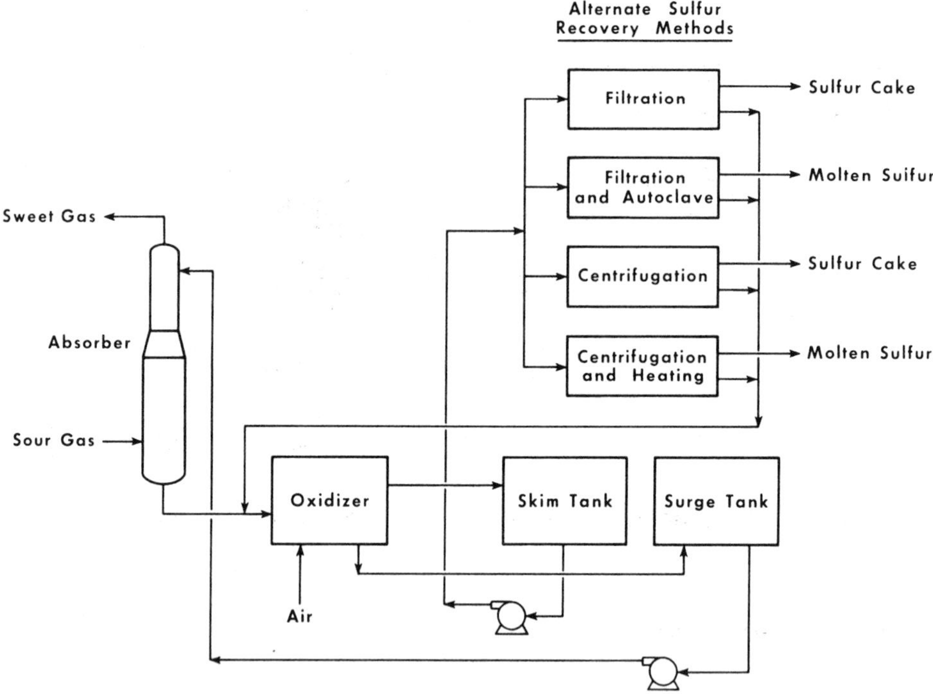

Figure 6 Simplified representation of the Stretford process.

between the enriched gas (containing the higher molecular weight hydrocarbons) and lean gas (with the added value of the extracted liquid).

If saleability of the gas is the only reason for processing, removal of liquids can be achieved as a field operation using either crude oil enrichment (this chapter, next section), adsorption (Chap. 36, Sec. II), or refrigeration processes (Chap. 36, Sec. I).

Most of the plants for liquids separation recover a substantial portion of the propane and essentially all of the butanes and higher molecular weight hydrocarbons. While these products can be conveniently transported using the more conventional transportation methods (pipeline, road and sea; Chap. 32, Sec. V), ethane recovery depends on the availability of a product pipeline, although small amounts of ethane can be moved by road or rail when mixed with higher molecular weight hydrocarbons.

a)

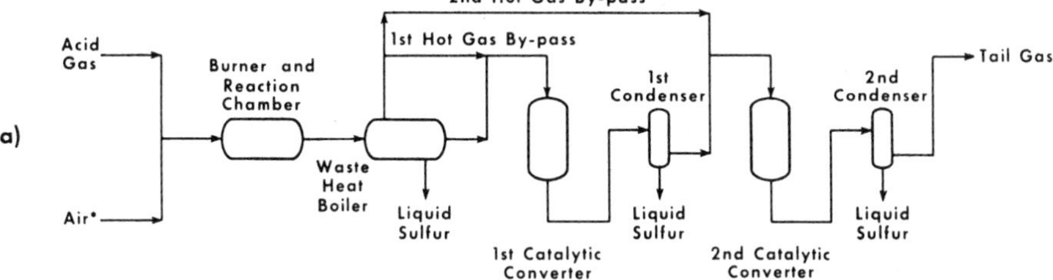

* Sufficient Air is Added to Burn 1/3 of Total H_2S to SO_2 and All Hydrocarbon to CO_2.

b)

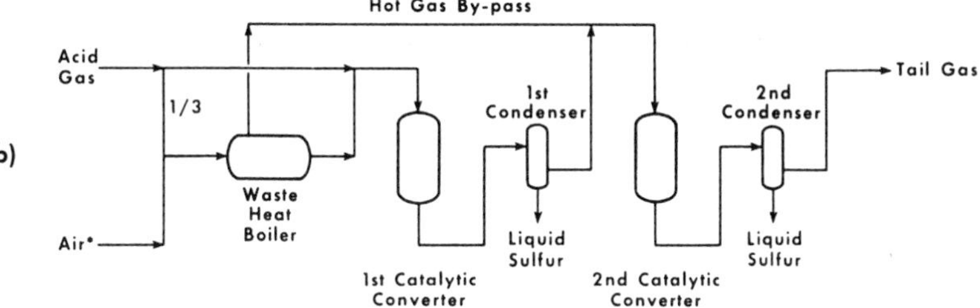

* Sufficient Air is Added to Burn All H_2S to SO_2 and All Hydrocarbon to CO_2 in 1/3 of Acid Gas.

c)

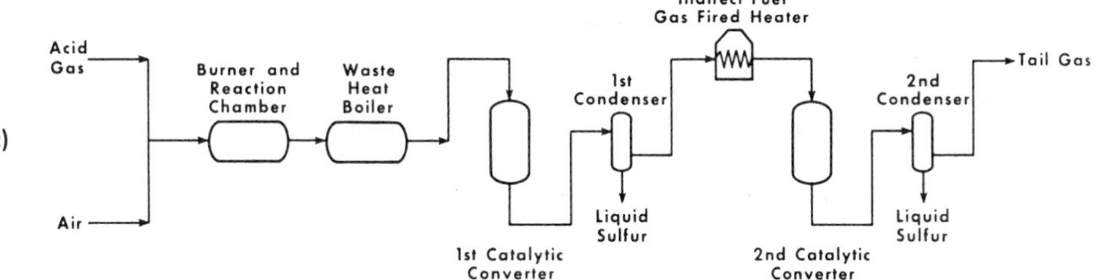

Figure 7 Representations of the various options for the Claus process: (a) once-through; (b) split-stream; and (c) process with two sulfur condensers and indirect heater.

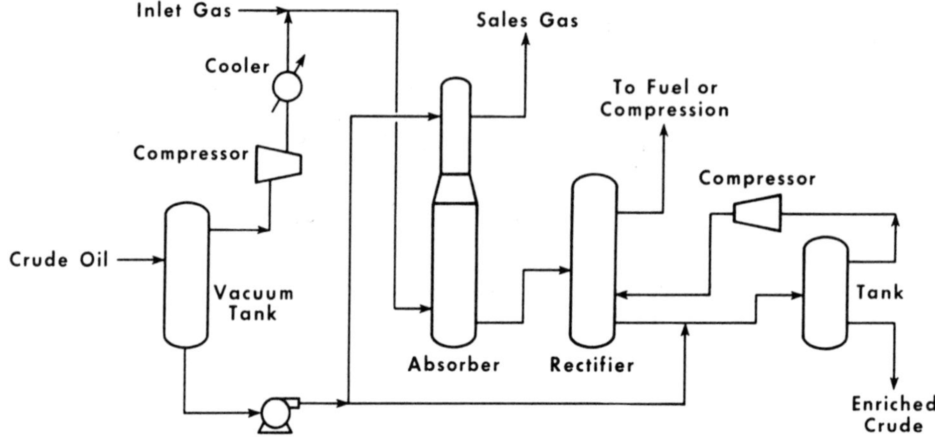

Figure 8 Schematic representation of the crude oil enrichment process.

VI. ENRICHMENT

The purpose of crude enrichment is to produce natural gas for sales and an enriched tank oil. The tank oil contains more light hydrocarbon liquids than natural petroleum and the residue gas is drier (leaner; i.e., has lesser amounts of the higher molecular weight hydrocarbons). Therefore, the process concept is essentially the separation of hydrocarbon liquids from the methane to produce a lean, dry gas.

Crude oil enrichment is used where there is no separate market for light hydrocarbon liquids, or where the increase in API gravity (Chap. 3, Sec. III.A) of the crude will provide a substantial increase in the price per unit volume as well as volume of the stock tank oil. A very convenient method of enrichment involves manipulation of the number and the operating pressures of the gas-oil separators (traps). However, it must be recognized that alteration or manipulation of the separator pressure will affect the gas compression operation as well as influence other processing steps.

One method of removing light ends involves the use of a pressure reduction (vacuum) system (Figure 8). Generally, stripping of light ends is achieved at low pressure, after which the pressure of the stripped crude oil is elevated so that the oil will act as an absorbent. The crude oil, which becomes enriched by this procedure, is then

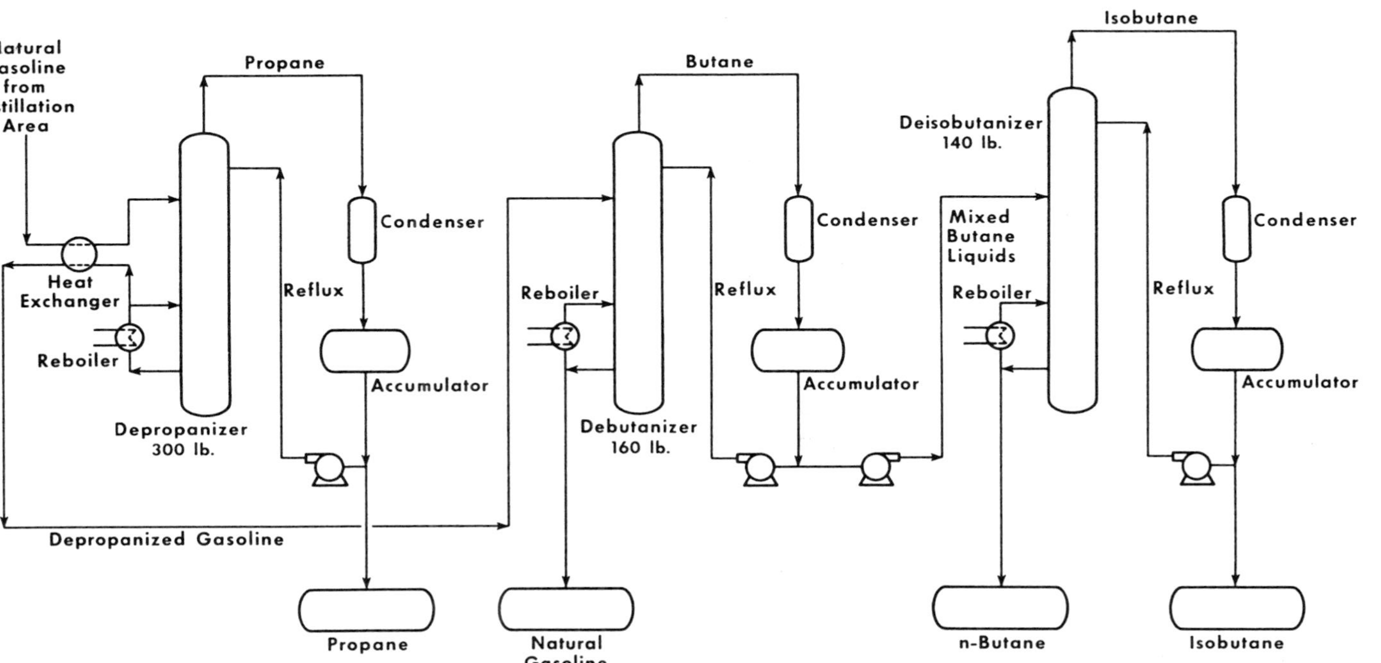

Figure 9 Schematic representation of a fractionation process.

reduced to atmospheric pressure in stages or using fractionation (rectification).

VII. FRACTIONATION

Fractionation processes are very similar to those which are classified as "liquids removal" processes (Sec. V, this chapter) but often appear to be more specific in terms of the objectives; hence the need to place the fractionation processes into a separate category. The fractionation processes are those processes used (1) to remove the more significant product stream first or (2) to remove any unwanted light ends from the heavier liquid products.

In the general practice of natural gas processing, the first unit will be a deethanizer followed by a depropanizer, then by a debutanizer, and, finally, a butane fractionator (Figure 9; see also Chap. 36, Sec. IV.C). Thus, each column can operate at a successively lower pressure thereby allowing the different gas streams to flow from column to column by virtue of the pressure gradient without necessarily using pumps.

36

Processing: Specific Processes

I Refrigeration Processes 1114
II Absorption Processes 1115
A. Oil Absorption Process 1115
B. Water Wash (Aquasorption) Process 1118
C. Selexol Process 1120
D. Alkanolamine Processes 1120
E. Carbonate Processes 1128
F. Holmes–Stretford Process 1137
G. Rectisol Process 1142
III Adsorption Processes 1142
A. Iron Oxide (Iron Sponge) Process 1145
B. Molecular Sieve Processes 1146
IV Chemical Conversion Processes 1149

As noted in Chap. 35, a description of the natural gas processing is divided between two chapters. Chapter 35 describes the general concepts for natural gas processing, and this chapter describes individual processes by which the general concepts can be implemented.

There are many processes for the removal of contaminant materials from natural gas (Kohl and Riesenfeld, 1985) and the processes are basically members of the clean-up type described in Chap. 35. Indeed, there are so many process variations that each process has its own particular niche in natural gas processing. It is not the intent to reproduce all of the processes here but to give selected examples of specific processes.

I. REFRIGERATION PROCESSES

Refrigeration (cryogenic) processes (Figure 1 in Chap. 35) can be used to upgrade low-btu gases to produce acceptable-btu products. These processes are essentially a means of separating the nonhydrocarbon contaminants from the hydrocarbon constituents of the natural gas.

Refrigeration processes generally use mechanical or compression-type refrigeration to reduce the temperature whereupon the basic separation—phase separation of the crystallized or solid product—occurs. The most common refrigerants in current use are propane and ammonia, but it is also advisable to inject ethylene glycol into the system at points where icing, or formation of gas hydrates, can occur. The glycol can be recovered from the main separator and regenerated (see, for example, Figure 2 in Chap, 35).

A process that is often used when liquid recovery (Chap. 35, Secs. V and VI) can be economically justified involves the use of a turbo-expander to produce the necessary refrigeration. Very low temperatures and high recovery of light components, such as ethane and propane, can be attained. In the process (Figure 1), the natural gas is first dehydrated with a molecular sieve followed by cooling. The separated liquid containing most of the heavy fractions is then demethanized and the cold gases are expanded through a turbine which produces the desired cooling for the process. The expander outlet is a two-phase stream that is fed to the top of the demethanizer column, which serves as a separator in which (1) the liquid is used as the column reflux and the separator vapors combined with vapors stripped in the demethanizer are exchanged with the feed gas; (2) the heated gas, which is partially recompressed by the expander compressor, is further recompressed to the desired distribution pressure in a separate compressor.

Vapor rectification processes (Figure 2) also employ mechanical refrigeration but the basic separation is accomplished using a column. Refrigeration is applied to the overhead product to produce reflux and also to partially condense the feed. If the recovery of lighter liquids is sufficiently high, the reboiler may be placed on the vapor rectifier and the stabilizer may not be necessary.

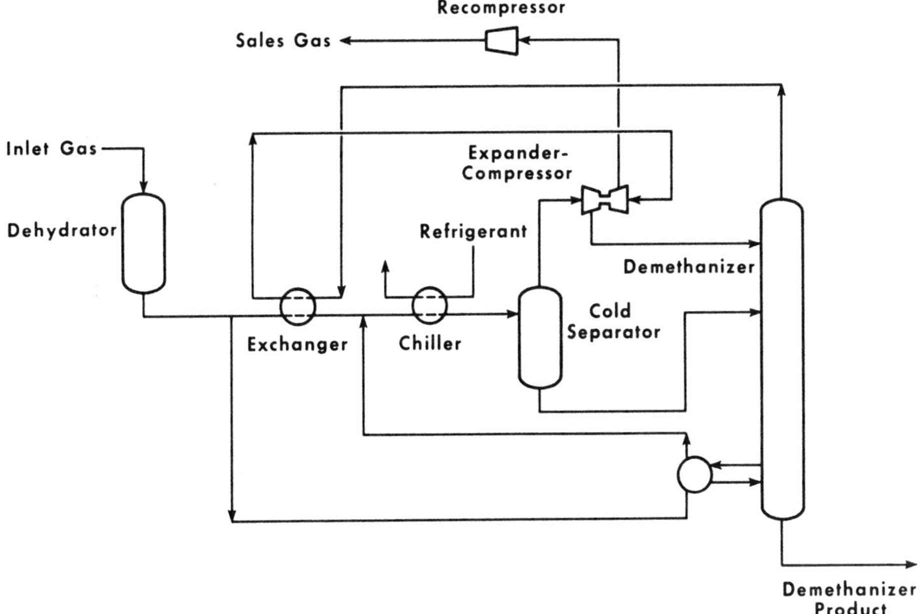

Figure 1 Schematic representation of the turboexpander section of a gas refrigeration process.

II. ABSORPTION PROCESSES

A. Oil Absorption Process

Until the early 1970s, most hydrocarbon recovery plants used the concept of oil absorption and, although many of the operating plants currently in use still employ this process, very few oil absorption units are included in newly constructed plants. Nevertheless, many older plants still use the oil absorption principle and the concept will be described in some detail here.

The oil absorption process (Figure 3) involves the countercurrent contact of the lean (or stripped) oil with the incoming wet gas (Figure 4), with the temperature and pressure conditions programmed to maximize the dissolution of the liquifiable components in the oil. The plant may also be of a dual nature insofar as refrigeration may also

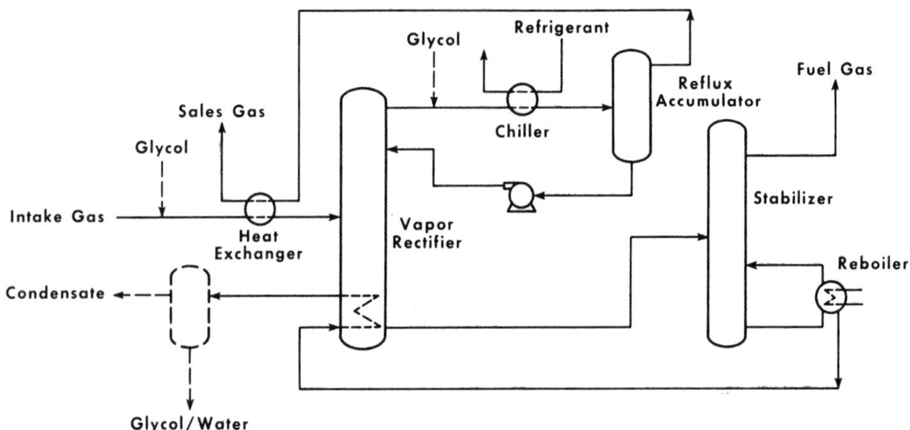

Figure 2 Schematic representation of a rectification process.

be used to obtain lower temperatures. The remainder of the plant involves (1) separation of light ends from the oil; (2) separation of absorbed materials from the oil; (3) removal of light ends from the raw product; and (4) separation of the raw product into various finished products.

The removal of any light ends may also be necessary and can be achieved using one or more additional steps. For example, methane and, in some plants, ethane can be removed in the rich-oil rectifier by pressure reduction and heating. Following rich-oil rectification, the absorbed material is removed from the oil in a stripper or a still.

If a heavy absorption oil is employed, stripping is usually achieved by preheating the oil followed by countercurrent contact with steam. Most processes use a compromise between these two operations and some plants use two stills in series: (1) a high-pressure still to condense the light ends and (2) a low-pressure still to ensure good stripping of the heavier gasoline fractions. If the oil is not stripped efficiently, the lighter components remaining in the oil will be vaporized in the absorber and lost in the residue gas stream.

The absorption processes offer reasonable selectivity for acid gas removal. In addition, the solvent used is generally recovered in good yield by flashing the rich solvent in flash tanks at successively lower pressures which requires little or no heat. Most solvents currently in use have a relatively high solubility for the higher molecular weight hydrocarbons—particularly the unsaturated and aromatic components which, because of chemical interactions, may be responsible for yielding a product that is contaminated with sulfur. Thus, for sour gases containing these particular hydrocarbons, care must be taken during

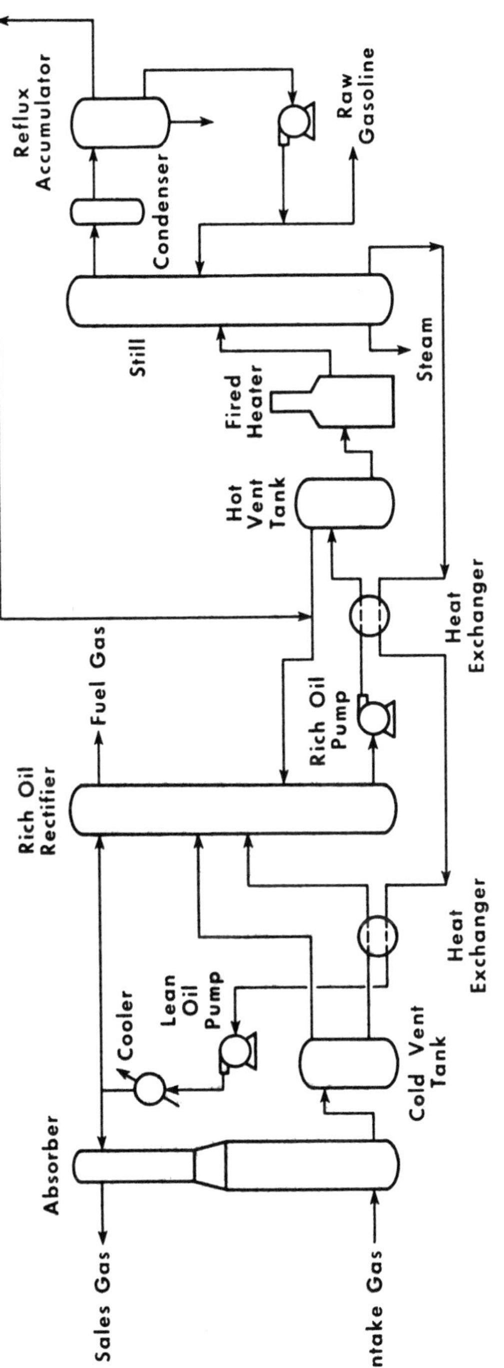

Figure 3 Schematic representation of the oil absorption process.

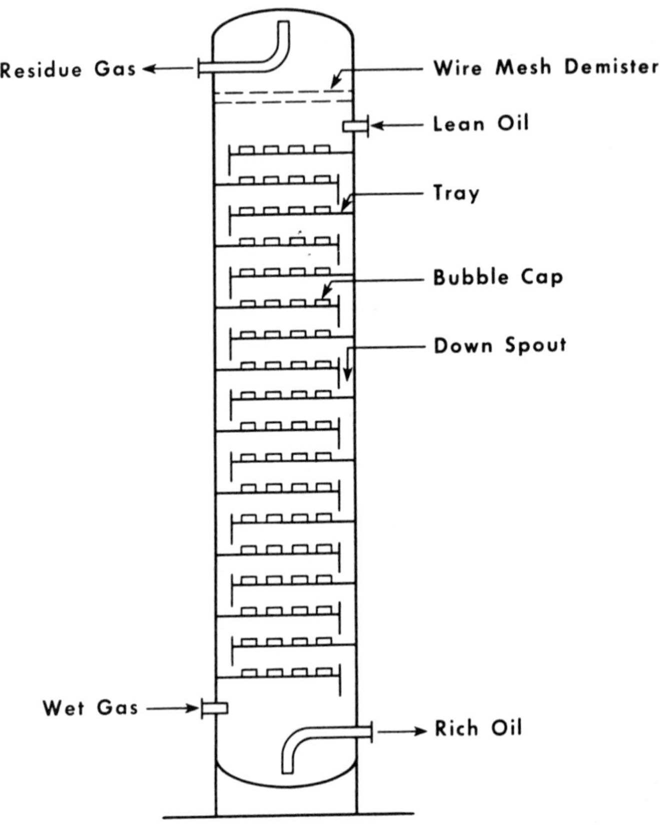

Figure 4 Schematic representation of a countercurrent bubble tray contact tower.

the regeneration step to prevent their entry into the acid gas stream that is to be sent to a sulfur recovery unit.

B. Water Wash (Aquasorption) Process

The water wash process is effective for natural gas that has a high acid gas content (including a high hydrogen sulfide/carbon dioxide ratio) that is also under high pressure. In this particular type of process, sour natural gas is passed, in an upward direction, through a contactor in which the gas flows countercurrent to the water (Figure 5). The partially sweetened gas is then passed on for further treatment (e.g., to an amine unit; see Sec. II.D).

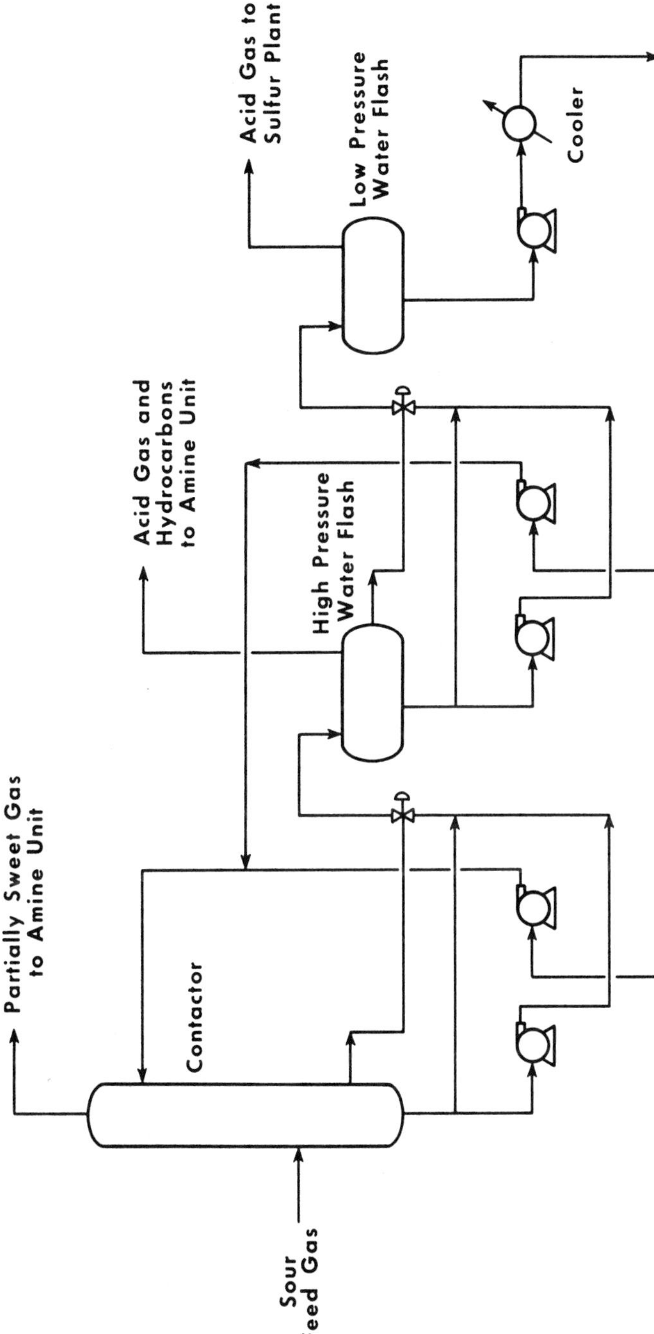

Figure 5 Schematic representation of the water-wash process.

The rich water solution from the bottom of the tower is sent to a pressurized flash tank for recovery of dissolved hydrocarbons. The water is then repressurized before being sent to a lower pressure flash tank where all of the acid gas is removed and water obtained for recycle.

C. Selexol Process

The Selexol process uses the dimethyl ether of polyethylene glycol as a solvent since the solubilities of hydrogen sulfide, carbon dioxide, and other acid gas components in this solvent are directly proportional to the partial pressures of these components. Different Selexol-based processes have been designed and used successfully for a wide range of hydrogen sulfide/carbon dioxide ratios (Hegwer and Harris, 1970).

The basic process configuration (Figure 6) has a somewhat different configuration from that employed (Figure 7) when there are low hydrogen sulfide/carbon dioxide ratios in the gas streams. As an example of the latter (i.e., a gas stream having a low hydrogen sulfide/carbon dioxide ratio), the sour natural gas is dehydrated, cooled, and sent to the absorber where it is contacted (countercurrent) with the Selexol solvent. Rich selexol from the bottom of the absorber is sent, via a surge tank to remove entrained gas that is recycled back into the absorber, to a high-pressure flash unit where most of the absorbed methane and part of the carbon dioxide is released. There is also a second flash unit where most of the vapor released is carbon dioxide which is vented. Finally, the Selexol is sent to the low-pressure flash unit where hydrogen sulfide and any remaining carbon dioxide are flashed off as the vapor stream and vented to the atmosphere.

D. Alkanolamine Processes

The alkanolamine processes are the most prominent and widely used processes for hydrogen sulfide and carbon dioxide removal. Some of the commonly used alkanolamines (Table 1) are ethanolamine (also called monoethanolamine; MEA), diethanolamine (DEA), triethanolamine (TEA), hydroxyethanolamine (usually called diglycolamine, DGA), diisopropanolamine (DIPA), and methyldiethanolamine (MDEA). The basic process configuration is quite simple (Figure 8; see also Figure 2 in Chap. 35) but there are many variations that have been designed for a specific improvement in the process operation.

Ethanolamine has the highest acid gas removal capacity and the lowest molecular weight among the olamines (Dingman and Moore, 1968; Maddox, 1982). Therefore, it offers the highest removal capacity on a unit weight or unit volume basis, from which lower solution circulation

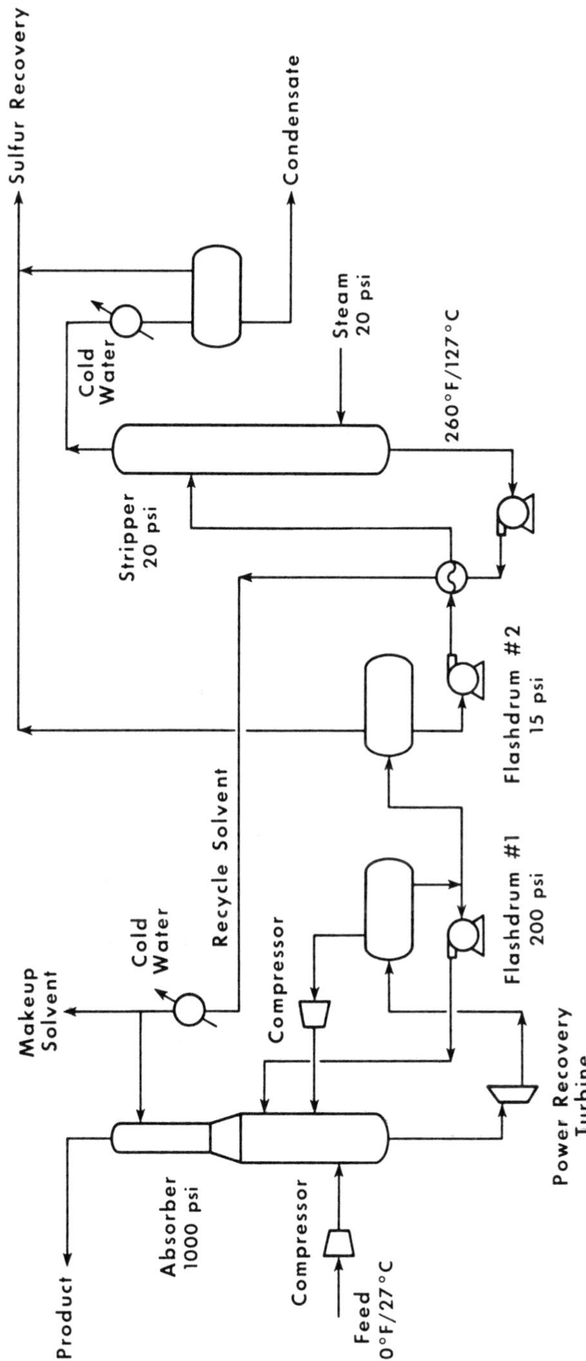

Figure 6 Generalized configuration for the Selexol process.

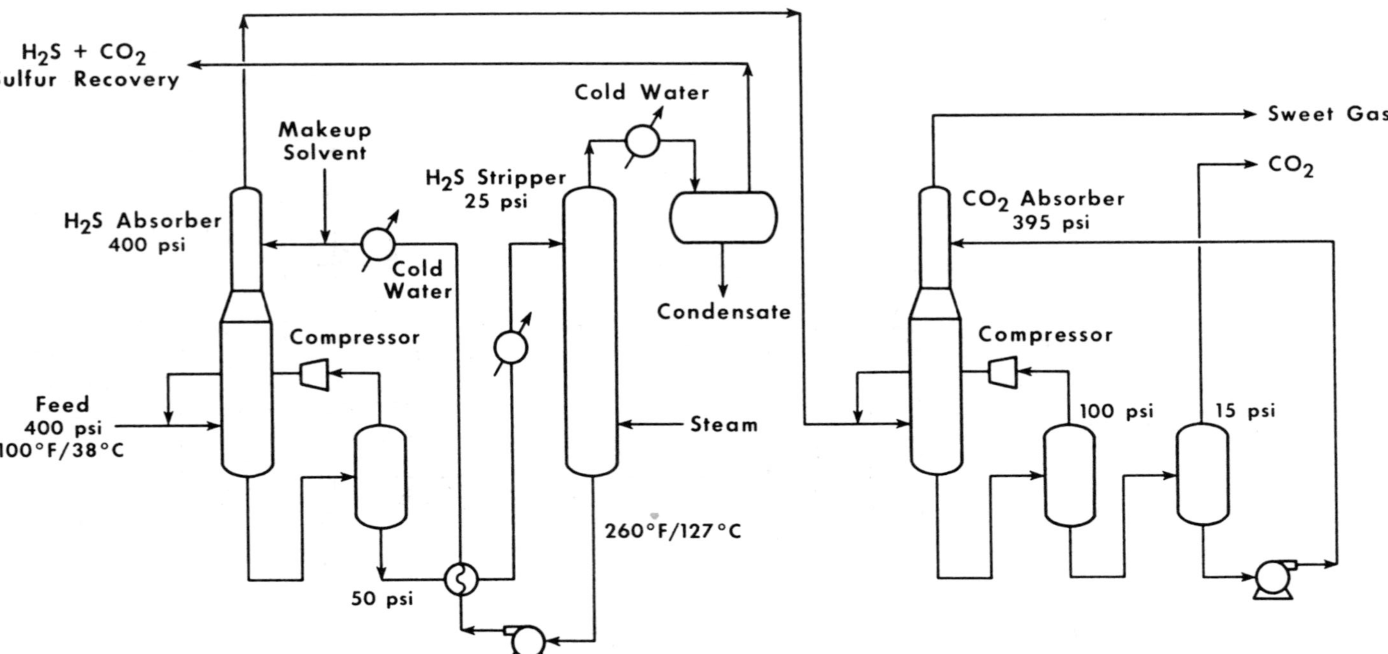

Figure 7 Schematic representation of the Selexol process for gases having low H_2S/CO_2 ratios.

Table 1 Characteristics of the Various Olamines Used in Gas Treating Units[a]

Name		Chem. formula	Mol. wt.	Vap. press. at 100°F (mm Hg)	Rel. capacity (%)
Ethanolamine (monoethanol- amine)	MEA	$HOC_2H_4NH_2$	61	1.05	100
Diethanolamine	DEA	$(HOC_2H_4)_2NH$	105	0.058	58
Triethanolamine	TEA	$(HOC_2H_4)_3N$	148	0.0063	41
Hydroxyethanol- amine	DGA	$H(OC_2H_4)_2NH_2$	105	0.160	58
Diisopropanol- amine	DIPA	$(HOC_3H_6)_2NH$	133	0.010	46
Methyl diethanol- amine	MDEA	$(HOC_2H_4)_2NCH_3$	119	0.0061	51

[a]See also Maddox, 1982.

rates in a sweetening plant are implied. It is also chemically stable but will undergo some (usually minimal) degradation (see below).

Diethanolamine is similar to ethanolamine except that it is not as reactive and therefore removal of hydrogen sulfide to meet pipeline specifications may be difficult. The reactions of diethanolamine with carbonyl sulfide and carbon disulfide are slower than the reactions of ethanolamine and lead to different products causing less filtration problems. The lower vapor pressure of the diethanolamine leads to lower vaporization losses.

Triethanolamine has been almost totally replaced by ethanolamine and diethanolamine primarily because of the lower reactivity of the triethanolamine which results in a relatively lower removal of hydrogen sulfide. Diglycolamine has the same reactivity as diethanolamine which, with a relatively lower vapor pressure, has resulted in use in recent years. Diisopropanolamine is also used to treat gas to pipeline specifications and is used in the sulfinol process (Figure 9) as well as in the ADIP process, which employs relatively concentrated solutions of the alkanolamine (diisopropanolamine) solvent (Klein, 1970). This solvent can remove carbonyl sulfide and is selective for hydrogen sulfide removal in preference to carbon dioxide removal.

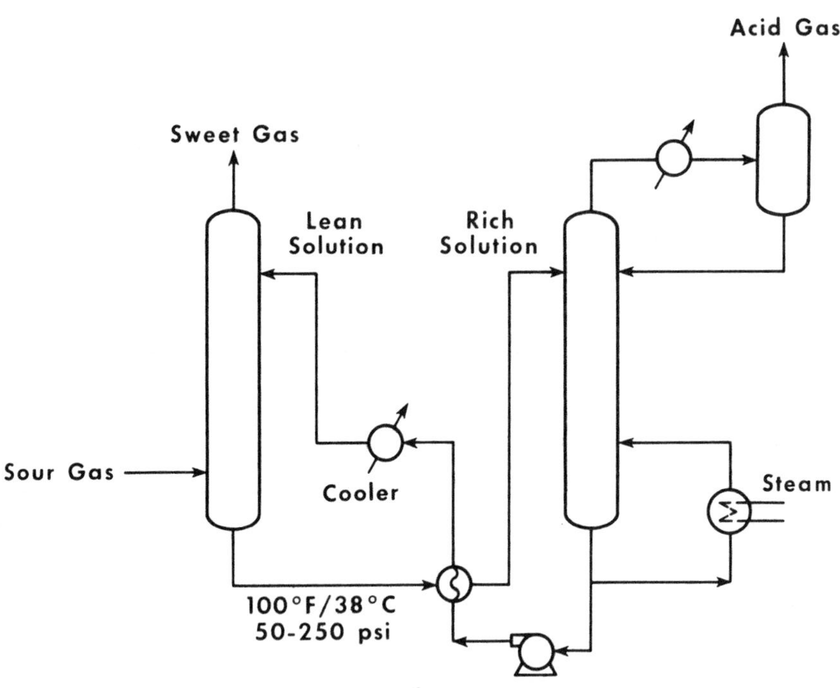

Figure 8 Simplified representation of an amine gas treatment unit.

Methyl diethanolamine has received renewed attention because it has good selectivity for hydrogen sulfide, so that any carbon dioxide remains in the gas phase.

In the olamine processes, aqueous solutions are employed. For ethanolamine, an aqueous solution having approximately 15% (by weight) of the olamine in water is usually appropriate whereas for diethanolamine a concentration of 20–30% is normally employed. For the higher molecular weight olamines, concentrations as high as 70% by weight may be necessary. Mixtures of glycol and an amine have also been used for simultaneous dehydration and desulfurization. Generally, a solution containing 10–30% by weight ethanolamine, 45–85% by weight triethylene glycol, and up to 25% by weight water is used.

The chemistry of natural gas clean-up by ethanolamine involves the following reactions:

$$RNH_2 + H_2S + RNH_4S$$

$$RNH_2 + H_2O + CO_2 + RNH_3HCO_3$$

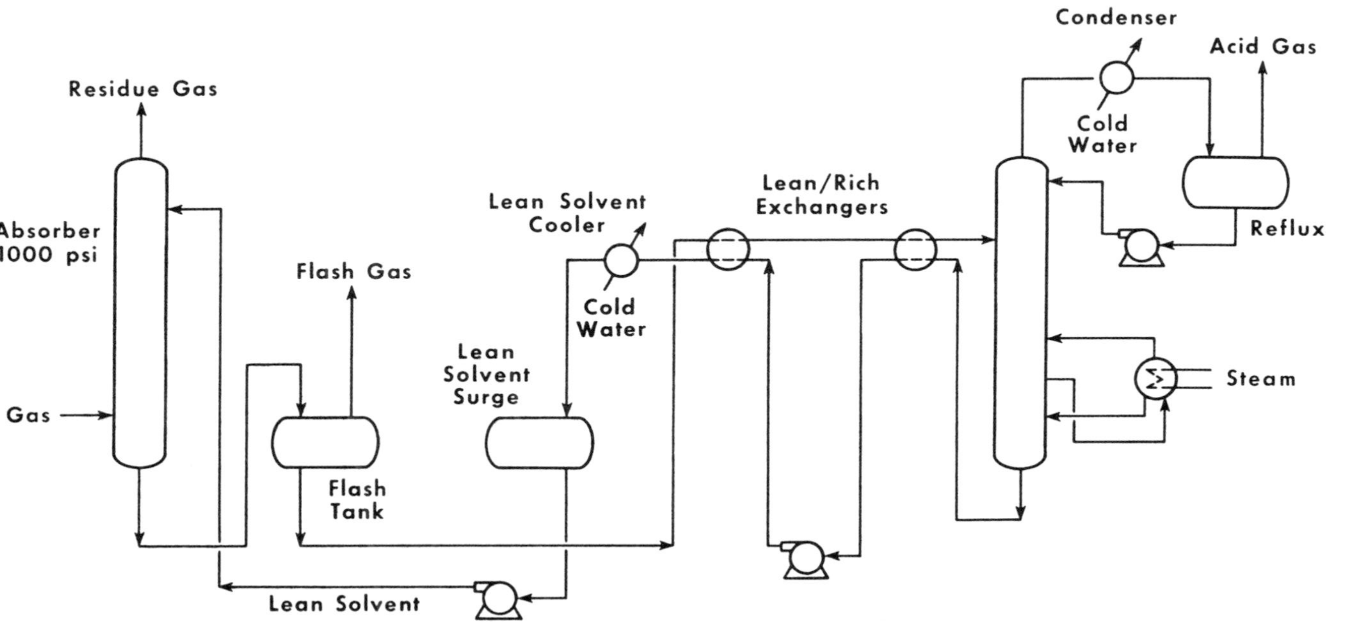

Figure 9 Schematic representation of the sulfinol process.

or for diethanolamine, the chemistry is (Edgar, 1983):

$$H_2S + 2R_2NH = (R_2HN_2)_2S$$

$$(R_2NH_2)_2S + H_2S + 2R_2NH_2HS$$

$$CO_2 + H_2O + 2R_2NH = (R_2NH_2)_2CO_3$$

$$(R_2NH_2)_2CO_3 + CO_2 + H_2O = 2R_2NH_2HCO_3$$

Ethanolamine does undergo relatively high solution losses because of the relatively high vapor pressure (which causes greater vaporization losses) and it reacts irreversibly with carbonyl sulfide and carbon disulfide. The reactions of ethanolamine with carbonyl sulfide are analogous to the reactions of ethanolamine with carbon dioxide, which may also cause some loss of the olamine:

$$HOCH_2CH_2NH_2 + CO_2 \longrightarrow \underset{\text{(ring: } CH_2-CH_2-NH-C(=O)-O\text{)}}{} + H_2O$$

Monoethanolamine **Oxazolidone-2**

$$\text{Oxazolidone-2} + HOCH_2CH_2NH_2 \longrightarrow HOCH_2CH_2N\text{(ring: } CH_2-CH_2-NH-C(=O)\text{)} + H_2O$$

Oxazolidone-2 **1-(2-Hydroxyethyl)-imidazolidone-2**

$$HOCH_2CH_2N\text{(ring: } CH_2-CH_2-NH-C(=O)\text{)} + H_2O \longrightarrow HOCH_2CH_2NHCH_2CH_2NH_2 + CO_2$$

1-(2-Hydroxyethyl)-imidazolidone-2 **N-(2-Hydroxyethyl)-ethylenediamine**

except that the reactions occur readily at ambient temperature (Pearce et al., 1961). The product slate for the reaction of carbonyl sulfide with ethanolamine is similar to the product slate for the reaction of carbon dioxide with ethanolamine except that diethanolurea ($HOCH_2CH_2NHCONHCH_2CH_2OH$) is also formed.

Diethanolamine does not appear to lose any of its activity due to the presence of carbonyl sulfide in gas streams (Figure 10). The overall losses of ethanolamine due to the presence of carbonyl sulfide in gas streams can be substantially reduced by the addition of strong

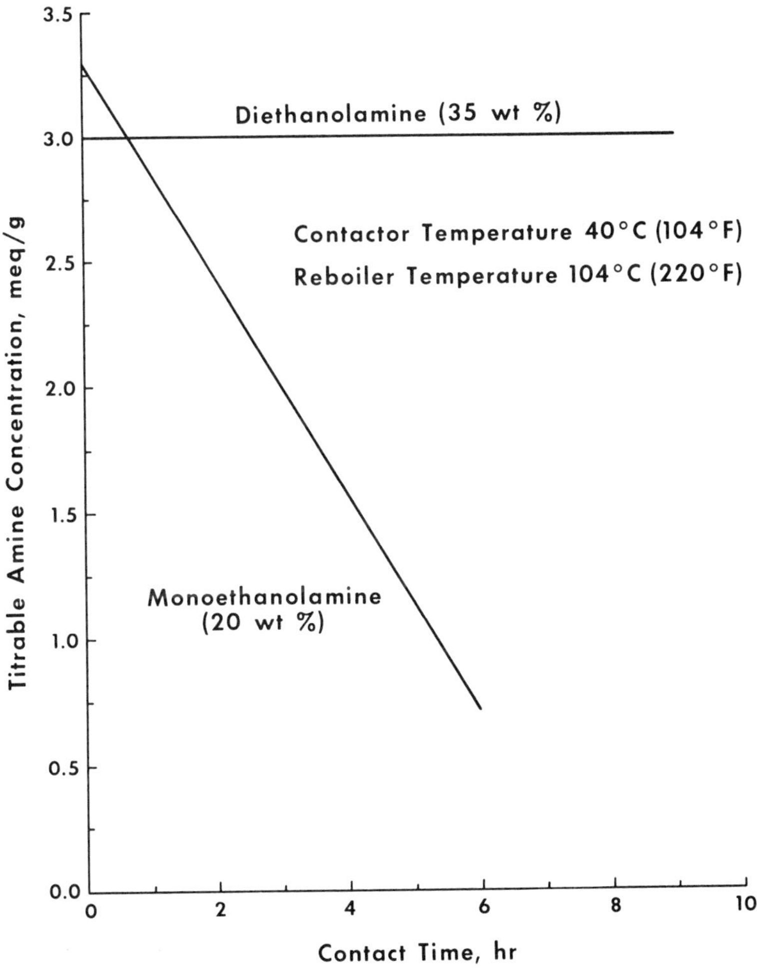

Figure 10 Illustration of the effect of carbonyl sulfide on the process life of ethanolamine and on diethanolamine (Kohl and Riesenfeld, 1985).

alkalis (e.g., sodium hydroxide or sodium carbonate) to the ethanolamine solutions (Kohl and Riesenfeld, 1985).

In a typical process (Figure 11), sour natural gas is sent upward through the contactor tower, countercurrent to the flow of monoethanolamine, and the rich solution from the bottom of the contactor is sent to a flash tank where absorbed low-boiling hydrocarbons in solution are vented. The flash tank also serves as a sediment accumulator and provisions must be made for sediment removal. After the enriched solution has been rejuvenated, the gases released at this stage are sent to the flare stack and the liquid that is accumulated by the reflux is sent to the regenerative system.

Lean ethanolamine that has accumulated at the bottom of the stripper is continuously recirculated through the reboiler. It is possible to remove up to 90% of the acid gases within the first three trays at the bottom of the absorber. The reactions are exothermic and a raise in temperature must be anticipated in this region of the absorber.

There are many variations of the typical process configuration (Figure 11), including (1) the location of the filtering system; (2) use of a packed column instead of bubble-cap; (3) use of valve-type traps in the contactor and the stripper; or (4) use of a side-stream reclaimer.

E. Carbonate Processes

The hot carbonate process (also referred to as the hot-pot process) uses an aqueous solution of potassium carbonate and often a highly concentrated solution to improve process performance. The configuration of the conventional hot carbonate process (Figure 12) is relatively straightforward but, as with the olamine processes, there have been many modifications to this process as it evolved.

The chemistry of the carbonate processes involves reaction of the potassium carbonate with the hydrogen sulfide and with the carbon dioxide in the gas stream:

$$H_2S + K_2CO_3 = KHS + KHCO_3$$

$$CO_2 + K_2CO_3 + H_2O = 2KHCO_3$$

An elevated temperature is required to ensure that the potassium carbonate and the reaction products (potassium bicarbonate and potassium bisulfide) remain in solution. The process requires a relatively high partial pressure of carbon dioxide and cannot be used for gas streams that contain only hydrogen sulfide. There are also limitations on the extent of carbon dioxide and hydrogen sulfide removal. It is not always possible to bring the natural gas to pipeline specifications with this process.

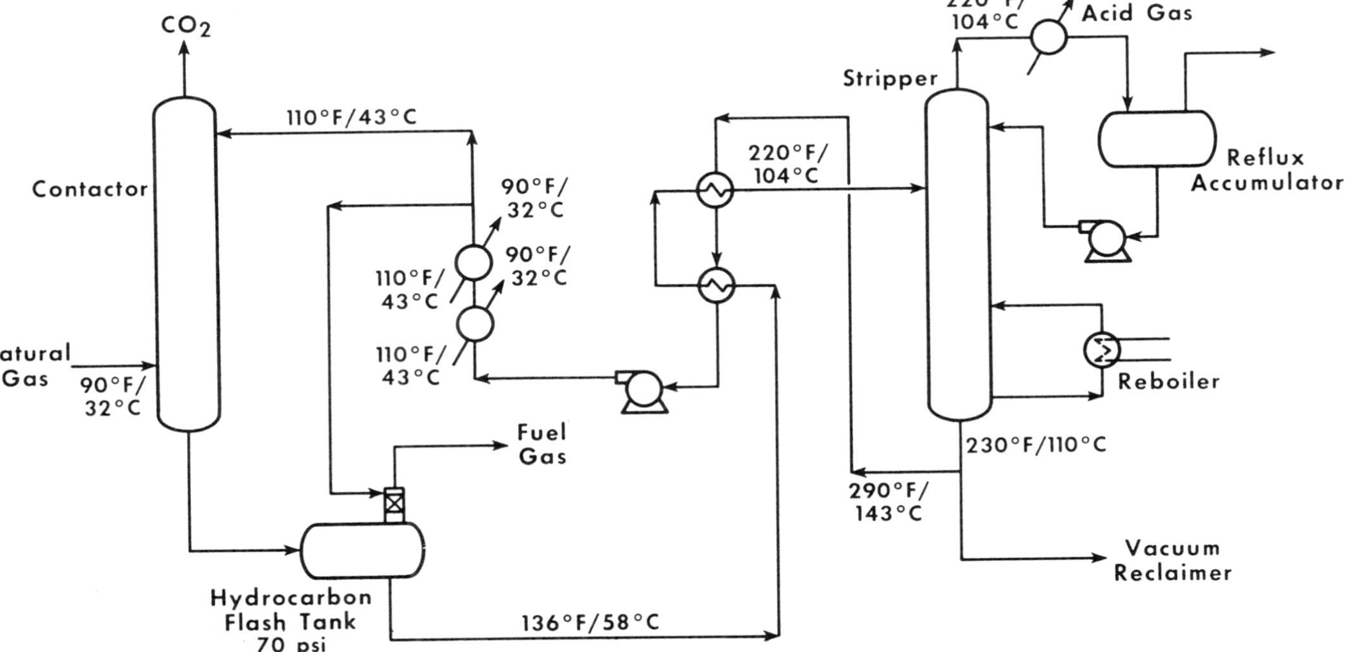

Figure 11 Schematic representation of an ethanolamine/hydroxyethanolamine (diglycolamine) process.

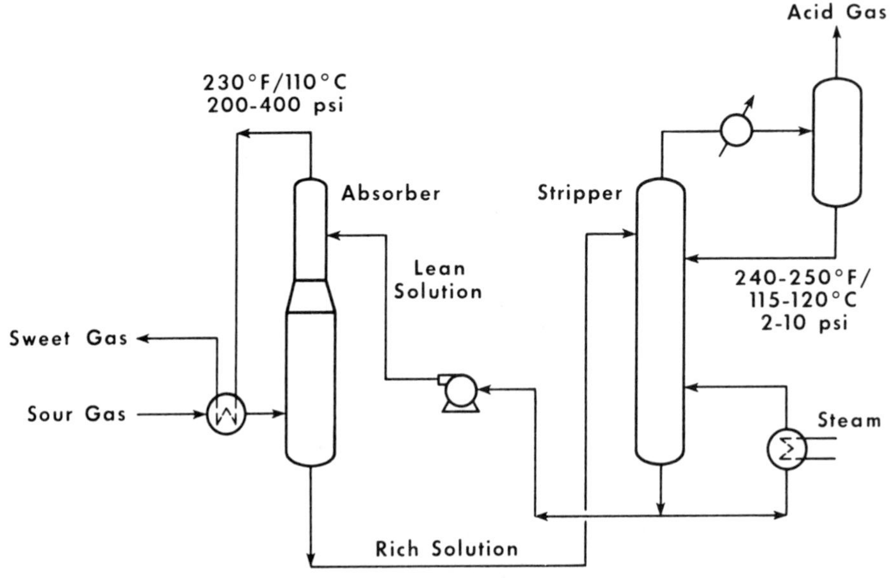

Figure 12 Schematic representation of the hot carbonate process.

Among carbonate processes, those containing an activator to increase the activity of the hot potassium carbonate solution are more popular, e.g., (1) the Benfield process; (2) the Catacarb process; and (3) the Giammarco–Vetrocoke process (see also Chap. 7, Sec. II.F).

The Benfield process has gained general acceptance for acid gas removal from gas streams; however, it is not one process but rather a collection of processes with each one tailored for a particular niche (Bartoo, 1985). The differing process configurations under the Benfield name (Figures 13 and 14) employ conventional packed or trayed towers for the countercurrent contact of liquid and gas and are configured for varying degrees of gas purification. In summary, the Benfield process is versatile and has a broad range of application with more than 500 units currently in operation.

The Catacarb process also employs a solution of potassium carbonate for the removal of hydrogen sulfide and carbon dioxide from gas streams. Several catalysts and corrosion inhibitors are used in the process but the choice depends on the composition of the gas to be treated (Gangriwala and Chao, 1985). The process is versatile and, like the Benfield process, has found general application in gas treatment operations. The simplest version of the process is a single-stage unit (Figure 15), which is used when high purity is not required in

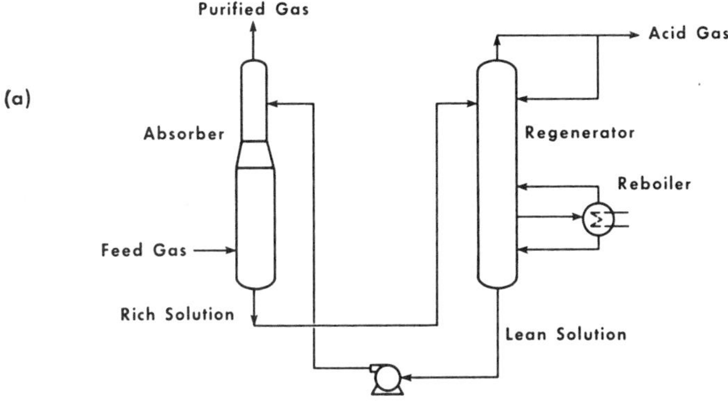

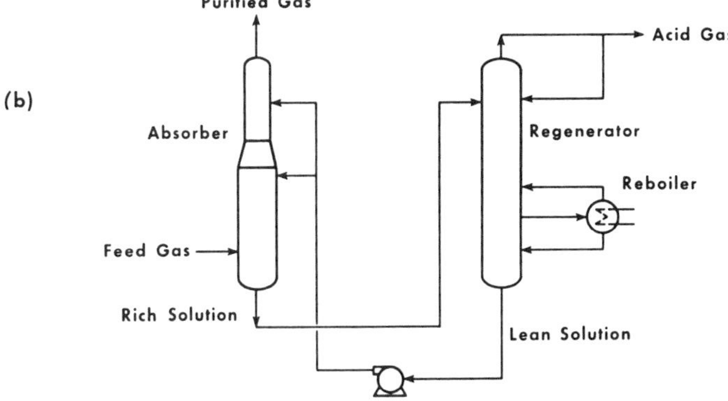

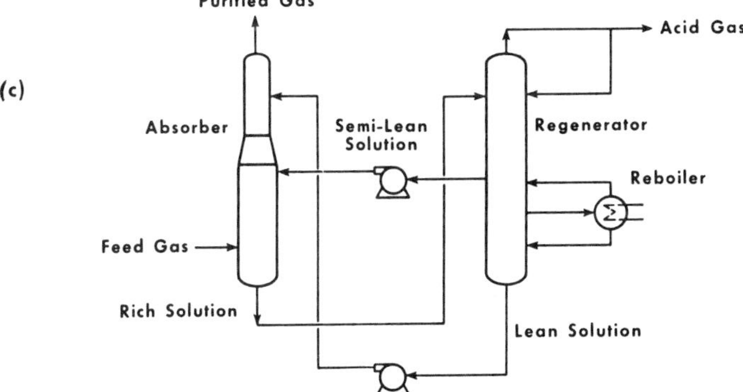

Figure 13 Schematic representation of the various configurations of the Benfield process: (a) single-stage; (b) split-flow; and (c) two-stage regenerator.

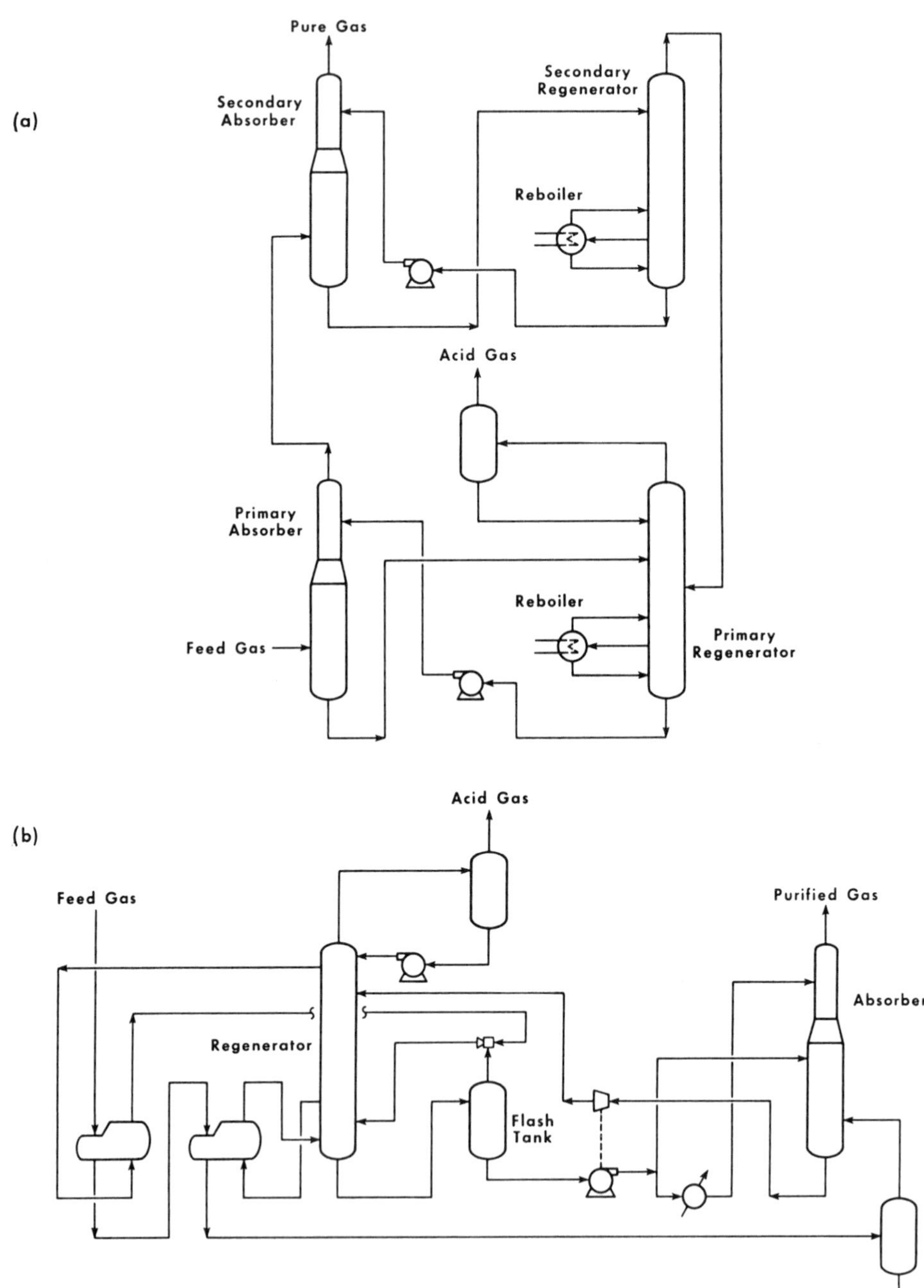

Figure 14 Schematic representation of other options for the Benfield process: (a) HiPure unit and (b) LoHeat unit.

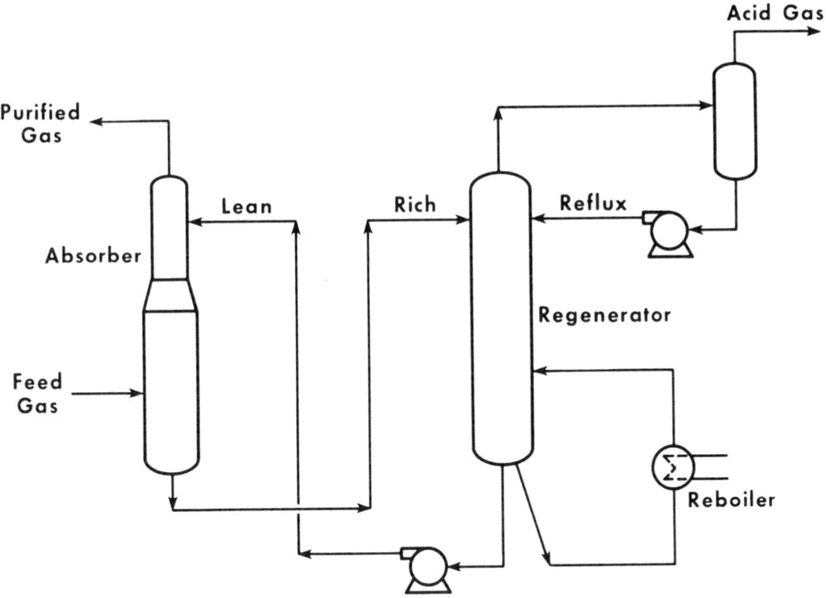

Figure 15 Schematic representation of the Catacarb single-stage process.

the treated gas. A two-stage design (Figure 16) is most efficient in terms of a higher purity product and other optional designs (Figure 17) are used to significantly reduce the process heat requirements. There are more than 100 Catacarb units currently in operation.

The Giammarco–Vetrocoke process (Figures 18 and 19) has found different applications. For example, there is one version for the removal of carbon dioxide while there is another version for hydrogen sulfide removal and yet another process version for the removal of both of these gases. As for most conventional carbonate processes, impurities such as carbon disulfide, mercaptans, and carbonyl sulfide have no detrimental effects on the solution (Jenett, 1962). The Giammarco–Vetrocoke process that is specifically designed for hydrogen sulfide removal produces elemental sulfur of high purity as the by-product and the process also has the ability to reduce the hydrogen sulfide content to less than 1 ppm. The process is reputed to be unable to handle gas streams with hydrogen sulfide concentrations greater than 1.5% (Jenett, 1962).

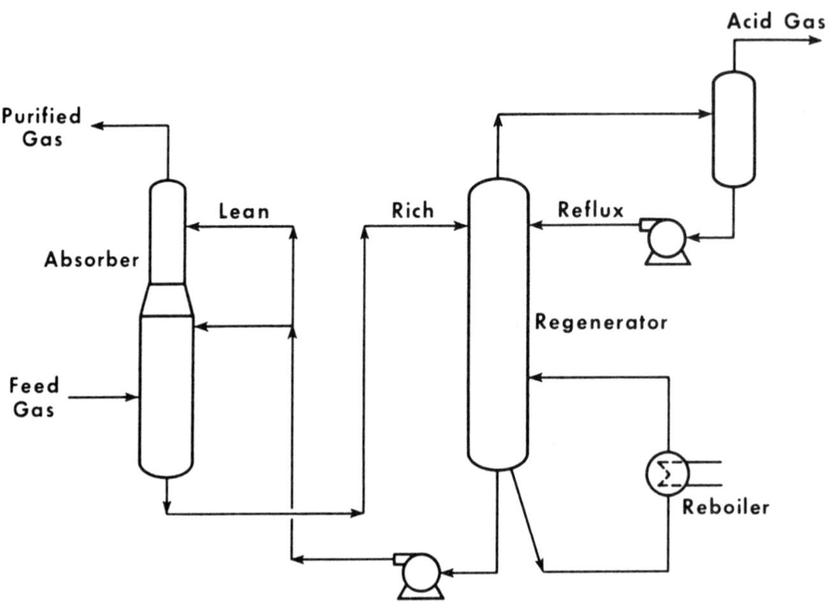

Figure 16 Schematic representation of the Catacarb split-cooled process.

There are also process configurations which use a combination of olamine and carbonate technology. For example, the diethanolamine process has been used in conjunction with the carbonate process for gas clean-up (Figure 20).

Another process in active use is the alkazid process (Figure 21), of which there are three different process variations with differing degrees of efficiency (Figure 22). This process is not obviously a carbonate process but is included here because of the similarities to the carbonate processes by virtue of the use of salts of alkali metals and organic radicals.

The alkazid DIK process uses the potassium salt of diethylglycine or dimethylglycine for the selective removal of hydrogen sulfide from gases that contain both hydrogen sulfide and carbon dioxide. Alkazid M uses sodium alanine and is effective in removing both hydrogen sulfide and carbon dioxide. The third process variation, alkazid S, uses a sodium phenolate mixture to remove contaminants such as carbon disulfide, mercaptans, and hydrogen cyanide.

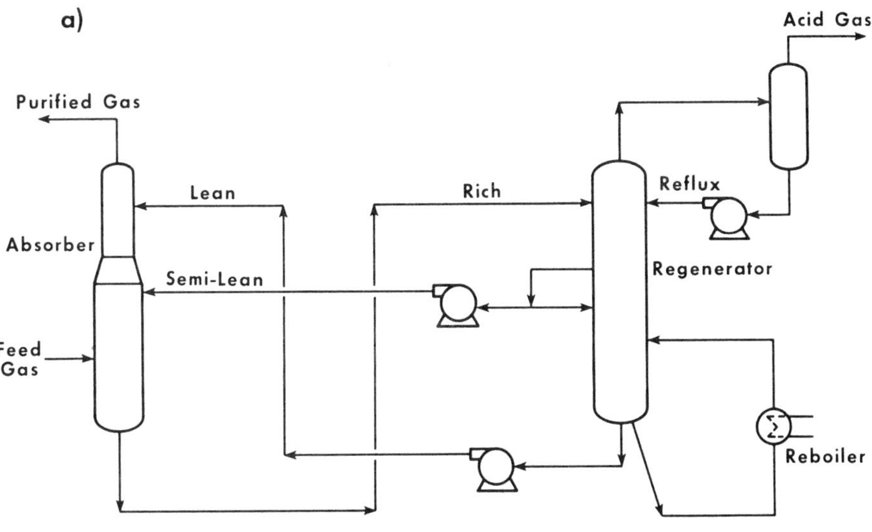

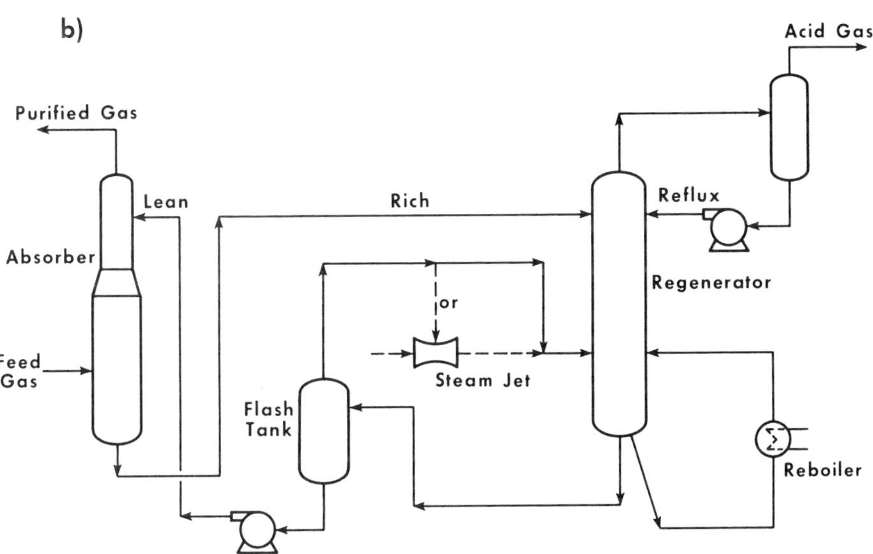

Figure 17 Schematic representation of other options for the Catacarb process: (a) two-stage unit and (b) low-heat unit.

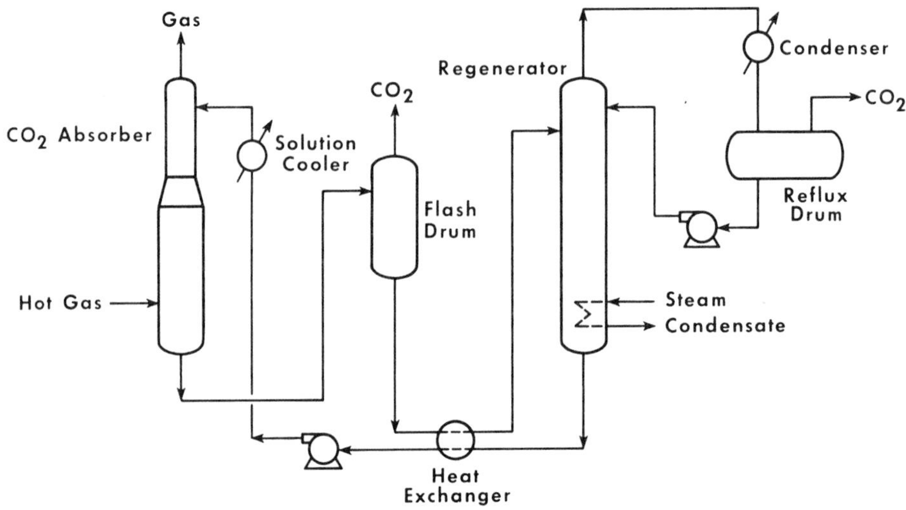

Figure 18 Schematic representation of the Giammarco–Vetrocoke process for carbon dioxide removal with steam regeneration.

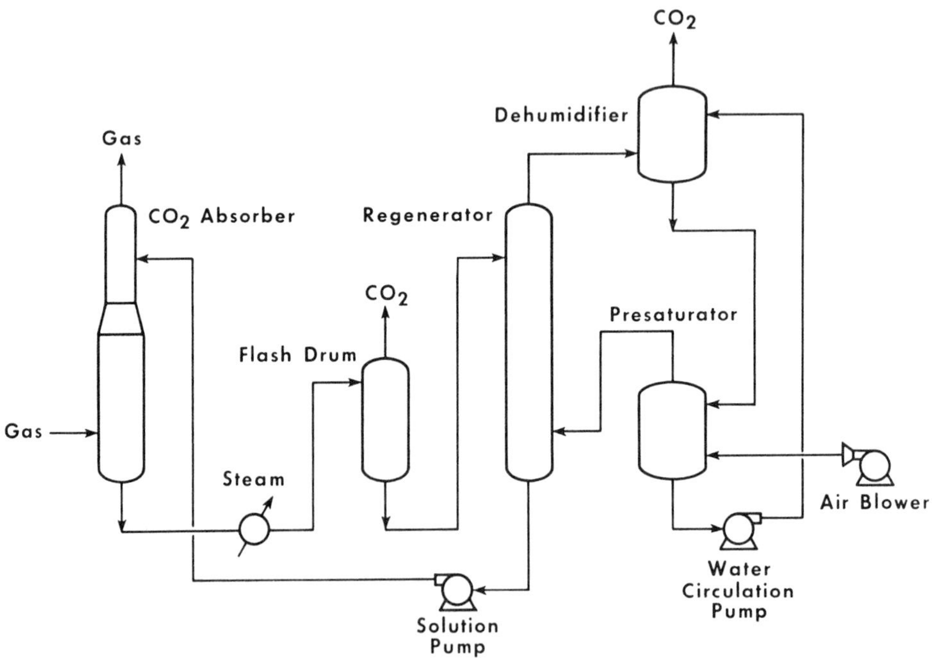

Figure 19 Schematic representation of the Giammarco–Vetrocoke process for carbon dioxide removal with air regeneration.

F. Holmes–Stretford Process

The Holmes–Stretford process is a modification of the Stretford process (Chap. 7, Sec. II.F) and is used to remove hydrogen sulfide by converting it to elemental sulfur of extremely high purity. The process is selective for hydrogen sulfide and can reduce the hydrogen sulfide concentration to as low as 1 ppm. However, the carbon dioxide concentration remains virtually unchanged during the application of this process (Nicklin et al., 1973; Moyes and Wilkinson, 1974; Ouwerkerk, 1978; Vasan, 1978).

The process (Figure 23) uses an aqueous solution containing sodium carbonate and bicarbonate (in the ratio of approximately 1:3) resulting in a pH of about 8.5–9.5, and the sodium salts of 2.6- and 2.7-anthraquinone disulfonic acid:

O, SO_3H, HO_3S, O

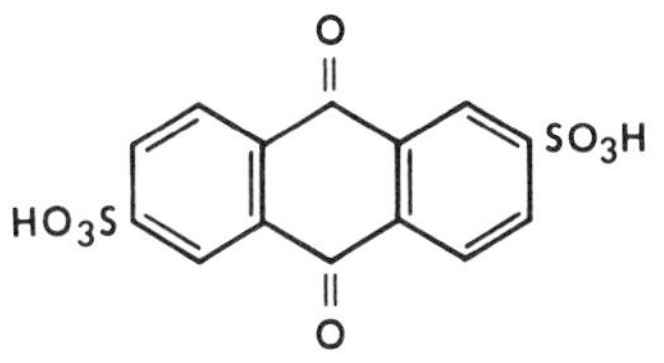

Anthraquinone-2,6-Disulfonic Acid

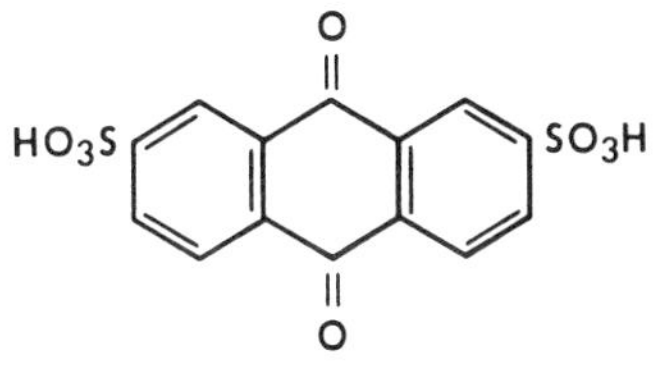

O, HO_3S, SO_3H, O

Anthraquinone-2,7-Disulfonic Acid

Several possible additives have been tested to increase the solution capacity for hydrogen sulfide and the rate of conversion of hydrosulfide to elemental sulfur. Alkaline vanadates have been found to be excellent additives for reducing hydrosulfide to sulfur, with a simultaneous valence change of vanadium from 5+ to 4+. In the presence of the anthraquinone disulfonic acid, the vanadate solution can be regenerated to a 5+ valence state.

The postulated reaction mechanism involves several steps:

(1) Absorption of the hydrogen sulfide by the alkali:

$$Na_2CO_3 + H_2S = NaHS + NaHCO_3$$

(2) Reduction of the anthraquinone disulfonic acid by addition of hydrosulfide to a carbonyl group and liberation of elemental sulfur from

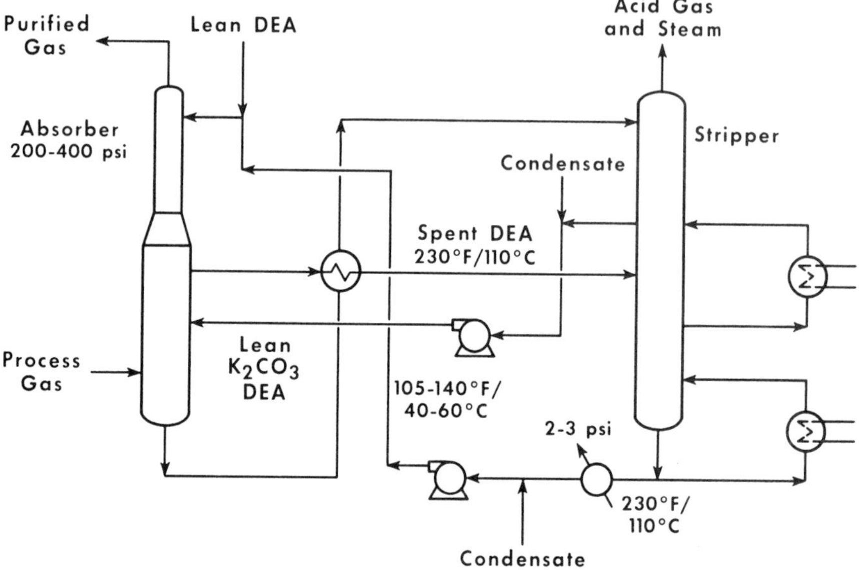

Figure 20 Schematic representation of a combined diethanolamine carbonate process.

reduced anthraquinone disulfonic acid (ADA is anthraquinone disulfonic acid and ADA-H_2 is reduced anthraquinone disulfonic acid):

$$2ADA + 2NaHS + H_2O = \underset{\text{reduced}}{2ADA\text{-}H_2} + 2NaOH + 2S$$

(3) Reoxidation of the reduced anthraquinone disulfonic acid by air:

$$2ADA\text{-}H_2 + O_2 = 2ADA + H_2O$$

(4) Reoxygenation of the alkaline solution, which also provides dissolved oxygen for the conversion of the reduced anthraquinone disulfonic acid to the anthraquinone disulfonic acid (step 3 above).

The overall chemistry of the Holmes–Stretford process is the atmospheric oxidation of hydrogen sulfide to elemental sulfur:

$$2H_2S + O_2 = 2H_2O + 2S$$

The absorption rate of the hydrogen sulfide in solution to produce the sodium bisulfide and the sodium bicarbonate is greatly aided by a

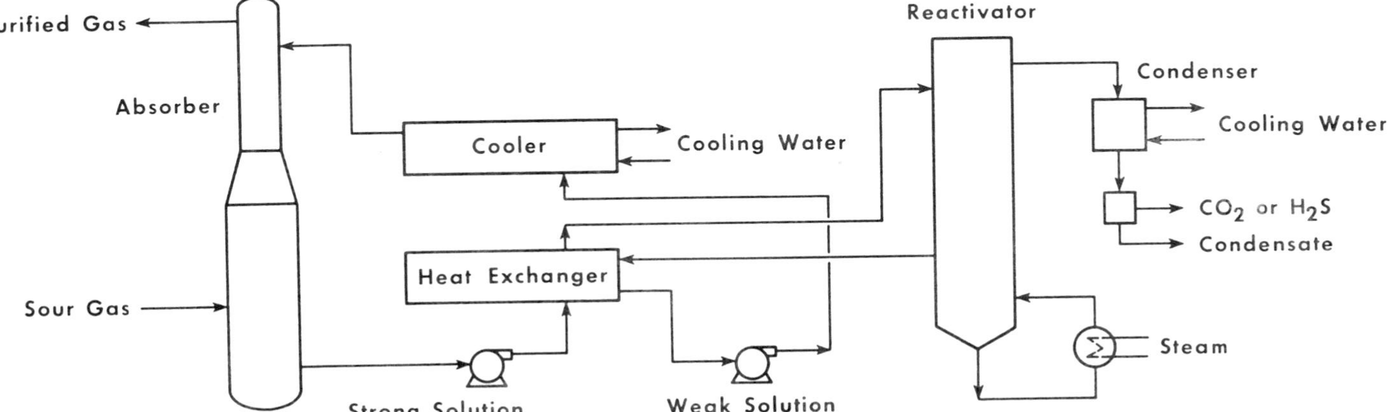

Figure 21 Schematic representation of the alkazid process.

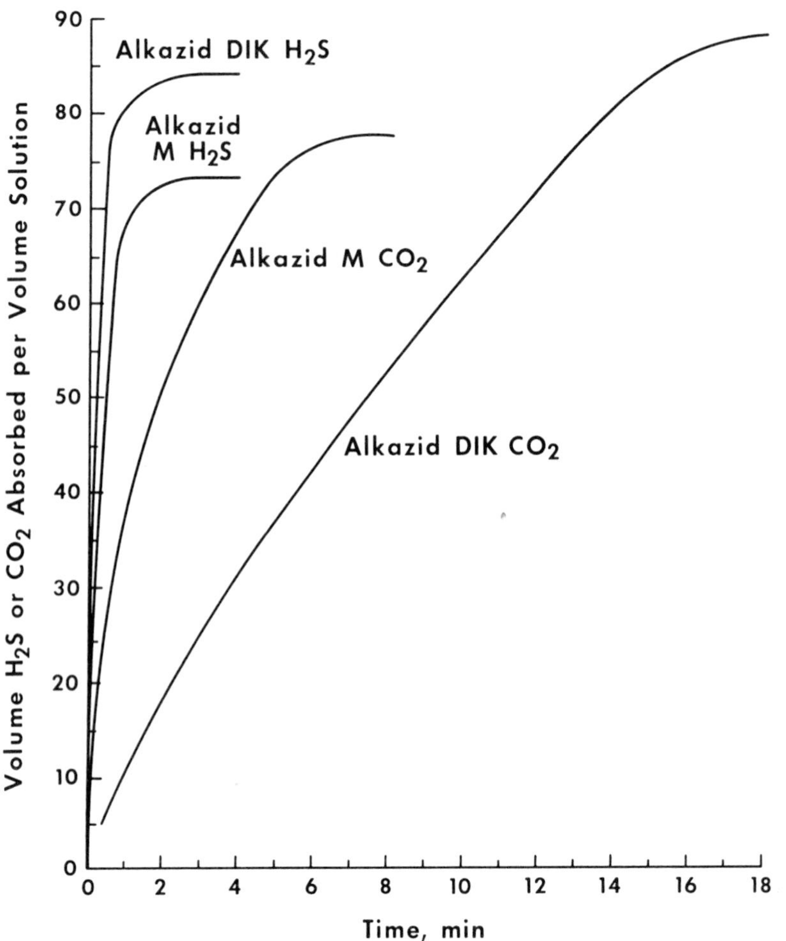

Figure 22 Illustration of the efficiency of the various alkazid process options.

high pH, whereas the conversion the sodium bisulfide to elemental sulfur is adversely affected by pH values above 9.5. Therefore a pH range of 8.5–9.5 is preferred (Nicklin and Hughes, 1977).

According to the chemistry outlined above, it might be assumed that the chemicals could be used indefinitely with only minimal replenishments for losses that occur in the absorber or within the sulfur recovery unit. However, side reactions produce dissolved solids, e.g.:

$$2NaHS + 2O_2 = Na_2S_2O_3 + H_2O$$

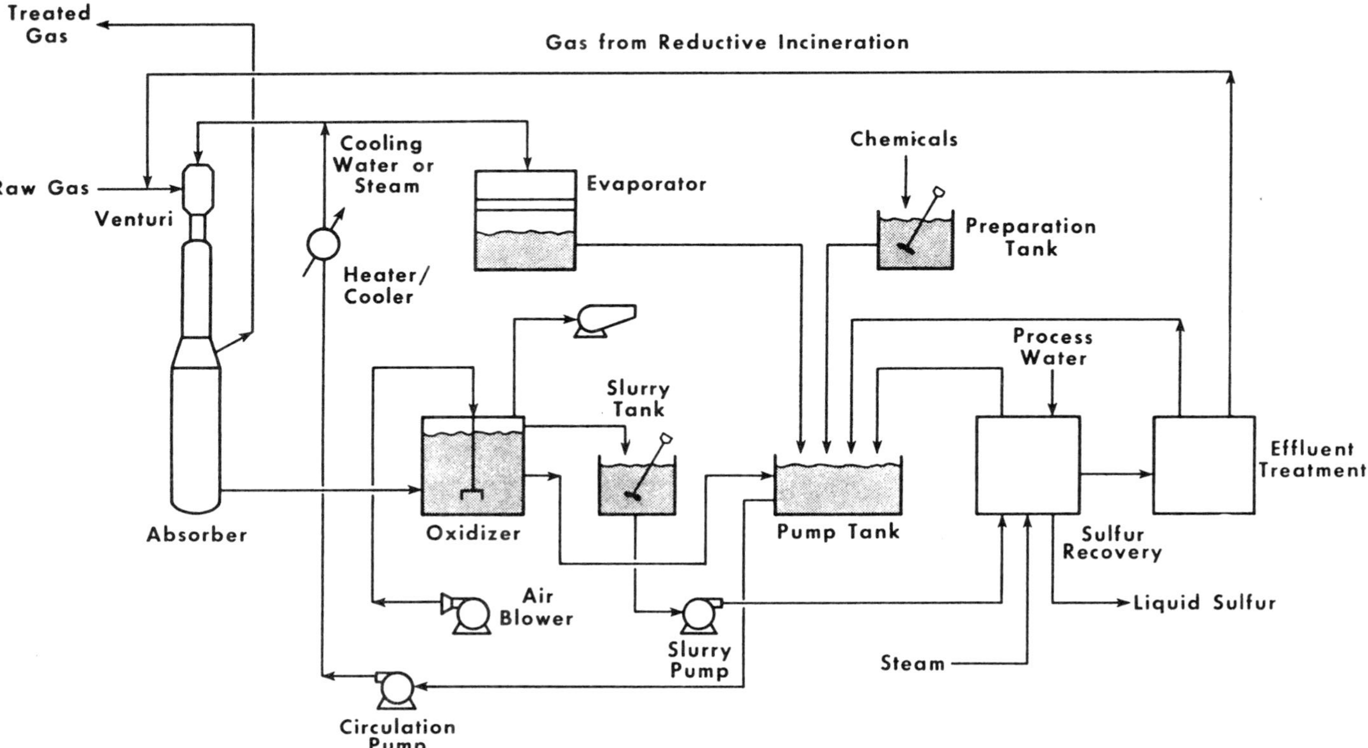

Figure 23 Schematic representation of the Holmes–Stretford process.

which increase in concentration until some of the solution must be discarded. The formation of sodium thiosulfate is probably due to some lack of reaction between the sodium bisulfide and the vanadate (because of insufficient time in the absorber) and, thus, the bisulfide is carried to the oxidizer where a reaction with oxygen occurs. These byproducts can also be formed in the contactor if the natural gas contains oxygen; high temperature and high pH also promote the thiosulfate formation.

The effluent stream from the Stretford process, containing sodium thiosulfate and in some cases sodium thiocyanate, must be treated prior to discharge. The Holmes modification involved the introduction of four alternate methods of handling the effluents: (1) evaporation or spray drying; (2) biological degradation; (3) oxidative combustion; and (4) reductive incineration. The reductive incineration concept has gained the widest acceptance since it results in, essentially, a zero effluent discharge; the process thermally degrades the sodium thiosulfate liquor to hydrogen sulfide, carbon dioxide, and a liquid stream which contains reduced vanadium salts, all of which are recycled.

G. Rectisol Process

The Rectisol process is a gas absorption process designed primarily to clean the gases from a coal gasifier. The process uses organic solvents, such as methanol, at temperatures between -60°C and -1°C (-80°F and 30°F) and involves countercurrent contact of the gas with the solvent in a trayed absorption column (Figure 24). The spent solvent is regenerated by stripping or by reboiling and the lean solvent is then recycled to the top of the absorption column.

III. ADSORPTION PROCESSES

The adsorption processes utilize the concept of contaminant removal by concentration on the surface of a solid material (Chap. 4, Sec. III). The commercial adsorbents are generally granular solids which have been prepared to have a large surface area per unit weight. These materials are frequently used in fixed beds (Chap. 6, Sec. II.A) for purification and dehydration of the natural gas. In most fixed-bed adsorption systems, the adsorption is exothermic and the bed temperature can be raised to extreme process limits if the heat evolution is not controlled.

One of the most common uses for the adsorption process in the natural gas industry is dehydration (Chap. 35, Sec. II). Although

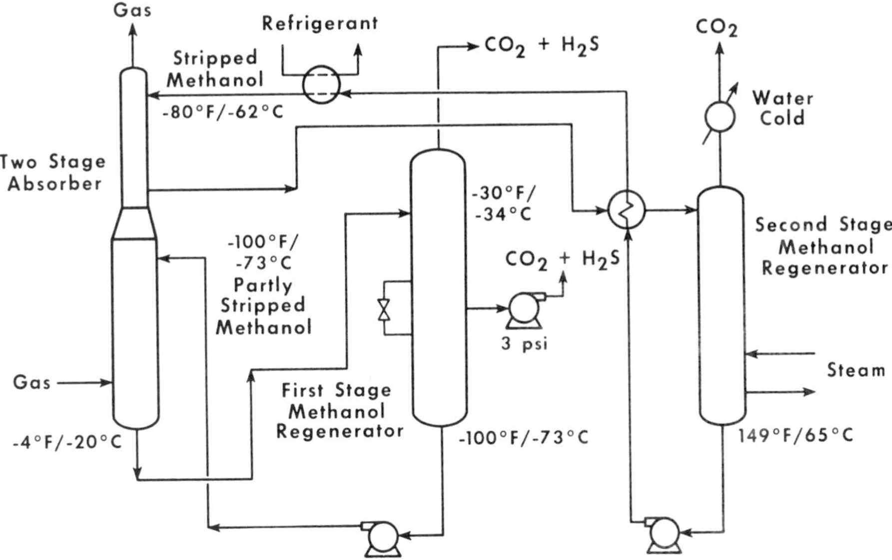

Figure 24 Schematic representation of the Rectisol process.

adsorption can be achieved using many different solid adsorbents, the great majority of dehydration adsorbents are based on silica, alumina (including bauxite), carbon, and molecular sieves. The silica- and alumina-based materials are used primarily for dehydration whereas the carbon and molecular sieves have expanded usage in the adsorption of organic materials also.

These processes are best applied to gas streams that have only moderate concentrations of hydrogen sulfide and where the carbon dioxide is not required to be removed from the gas stream. These process types are not as widely used as the liquid absorption processes but they do offer advantages such as simplicity, high selectivity for hydrogen sulfide, and process efficiency.

In a general sense (molecular sieves being the exception), the equipment used for the adsorption process is essentially the same for each adsorbent and the adsorbents themselves are interchangeable. Thus, in the simplest configuration (Figure 25), a plant for the dehydration of natural gas will consist of two contact reactors filled with the adsorbent but with only one reactor on-stream at any given time (see Chap. 6, Sec. I.B for the in-parallel twin-reactor concept used in the delayed coking operation). This allows one bed to be regenerated while not in use. Regeneration is usually accomplished by the

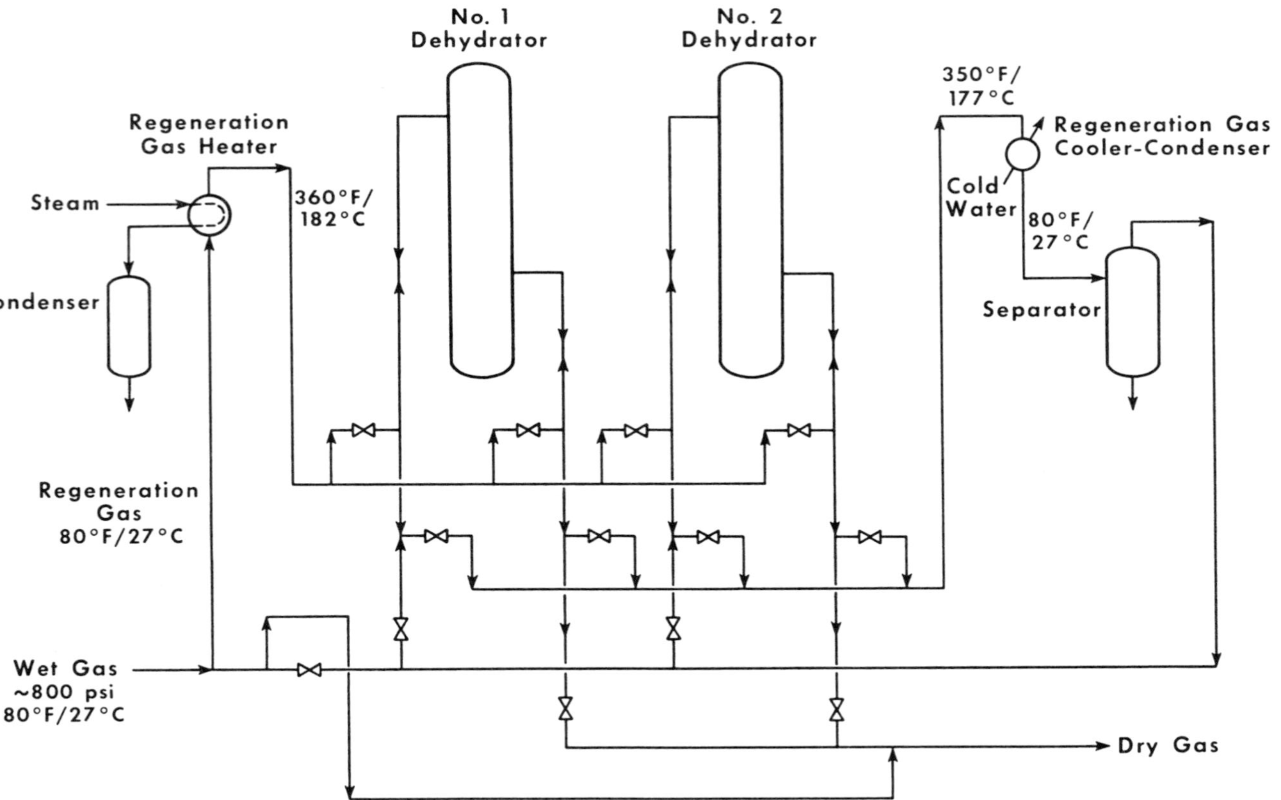

Figure 25 Schematic representation of a natural gas dehydration process.

passage of a hot gas through the bed. The regenerated bed is then brought on-stream and the used bed is regenerated.

The complete cycle is repeated at periodic intervals depending on the nature of the natural gas passing through the unit and is, because of the twin-reactor concept, a continuous process. There has been some development of adsorption-dehydration units in which the adsorbent moves from the adsorption zone to the regeneration zone in a similar manner to the movement of catalyst during petroleum processing (Chap. 6, Sec. II.A).

There is a further process option for gas that is relatively lean which involves the use of adsorption units for liquid recovery similar to that used for dehydration (Figure 3 and Sec. II in Chap. 35). The difference is that dehydration usually requires approximately 8 hr and any adsorbed hydrocarbons are displaced from the bed with water whereas hydrocarbon recovery requires 30–60 min per cycle and most of the adsorbed hydrocarbons are retained.

A. Iron Oxide (Iron Sponge) Process

In this process (Figure 26; see also Chap. 35, Sec. IV), the sour gas is passed through a bed of wood chips have have been impregnated with hydrated ferric oxide that has a high affinity for hydrogen sulfide whereupon the hydrogen sulfide is converted to ferric sulfide:

$$Fe_2O_3 + 3H_2S = Fe_2S_3 + 3H_2O$$

and the iron oxide is regenerated from the iron sulfide by passing oxygen/air over the bed:

$$2Fe_2S_3 + 3O_2 = 2Fe_2O_3 + 6S$$

The process operates in a batch-type reaction-regeneration cycle and offers the advantages of simplicity and excellent selectivity for hydrogen sulfide removal. However, bed regeneration can be difficult and expensive. In addition, sulfur will eventually cover most of the surface of the ferric oxide particles and further regeneration becomes impossible. A continuous regeneration process has also been developed (Figure 27) where small amounts of oxygen or air are added along with the sour gas at the inlet. This latter process gives an improved performance, generating a higher removal efficiency as well as better regeneration (Hollings, 1952; Moore, 1956; Taylor, 1956; Duckworth and Geddes, 1965; Kohl and Riesenfeld, 1985).

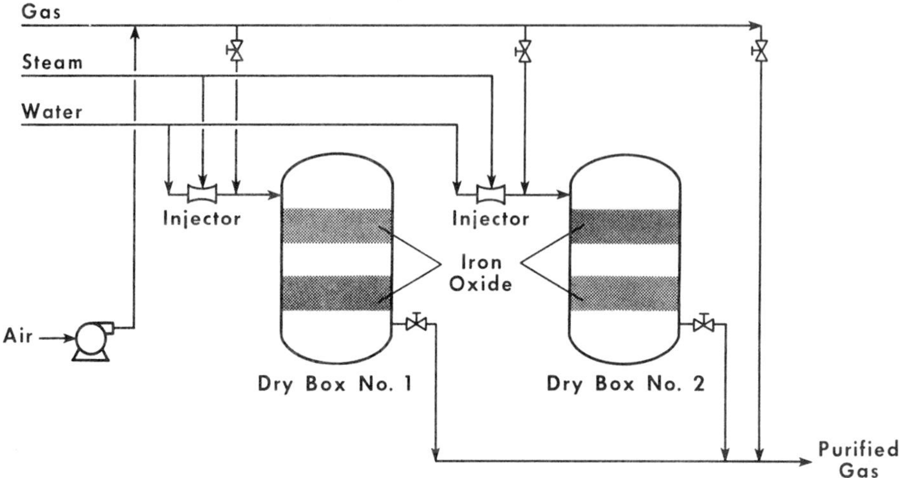

Figure 26 Schematic representation of the iron oxide (iron sponge) process.

B. Molecular Sieve Processes

Molecular sieves are synthetic forms of crystalline sodium-calcium alumina silicates which are porous in structure and have a very large surface area. The pores are uniform throughout the material and several grades of molecular sieves are available with each grade corresponding to a very narrow range of pore sizes.

Molecular sieves remove contaminants from natural gas through a combination of size selectivity ("sieving") and the physical adsorption process. Because of the narrow pore sizes, molecular sieves can discriminate among the adsorbates on the basis of molecular size. The sieves also possess highly localized polar charges on their surface that act as adsorption sites for polar materials. Therefore, small molecules that are polar (or which are polarizable) and which could conceivably pass through the pores of the sieve are also removed from the gas stream by the sieve.

Molecular sieves are highly selective for the removal of hydrogen sulfide (as well as other sulfur compounds) from natural gas and offer a continuously high absorption efficiency. They are also an effective means of water removal and thus offer a process for the simultaneous dehydration and desulfurization of natural gas. Gas that has an excessively high water content may, however, require upstream dehydration (Rushton and Hayes, 1961).

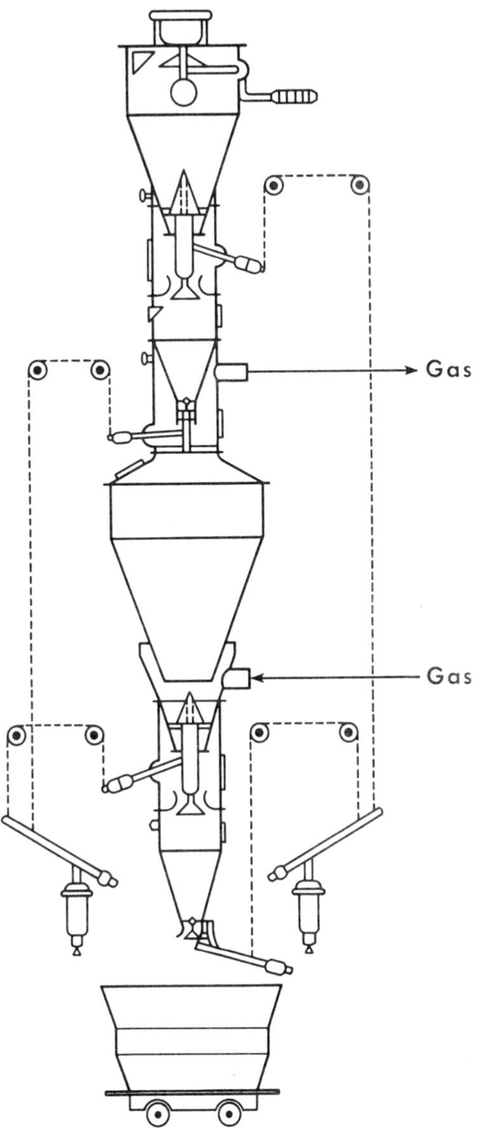

Figure 27 Schematic representation of a continuous iron oxide process.

The molecular sieve process (Figure 28) is similar to the iron oxide process. The bed is regenerated by passing a portion of the heated clean gas over the bed and as the temperature of the bed increases,

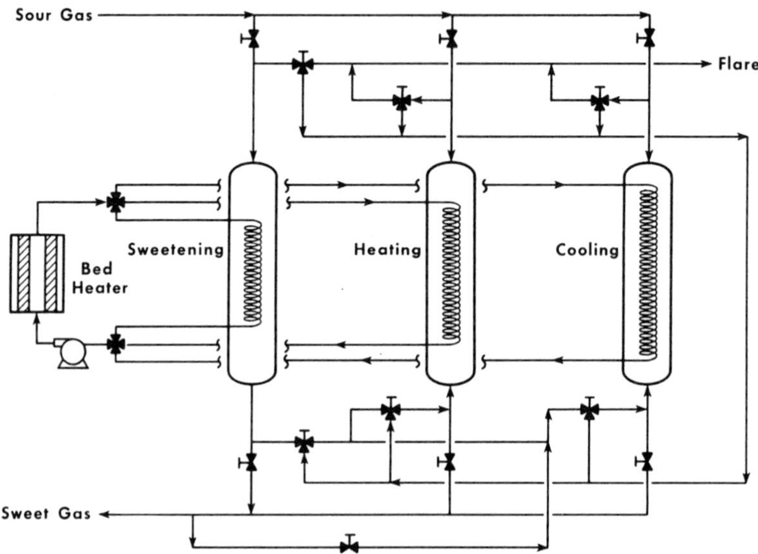

Figure 28 Schematic representation of a molecular sieve process.

it releases the adsorbed hydrogen sulfide into the regeneration gas stream. The sour effluent regeneration gas is sent to a flare stack and up to 2% of the gas treated can be lost in the regeneration process (Rushton and Hayes, 1961). A portion of the natural gas may also be lost by the adsorption of hydrogen components by the sieve.

In this process, unsaturated hydrocarbon components such as olefins and aromatics tend to be strongly adsorbed by the molecular sieves (Conviser, 1965). The molecular sieves are susceptible to poisoning by chemicals such as glycols and require thorough gas cleaning methods prior to the adsorption step. Alternatively, the sieve can be offered some degree of protection by the use of guard beds in which a less expensive catalyst is placed in the gas stream prior to contact of the gas with the sieve thereby protecting the catalyst from poisoning. This concept is analogous to the use of guard beds or attrition catalysts in the petroleum industry (see also Chap. 35, Sec. II and Speight, 1981).

The molecular sieve adsorption concept can also be employed in conjunction with other process concepts as, for example, in the process for purifying natural gas that is rich in carbon dioxide (Figure 29). For such gas streams, carbonyl sulfide may form as a byproduct:

$$H_2S + CO_2 = COS + H_2O$$

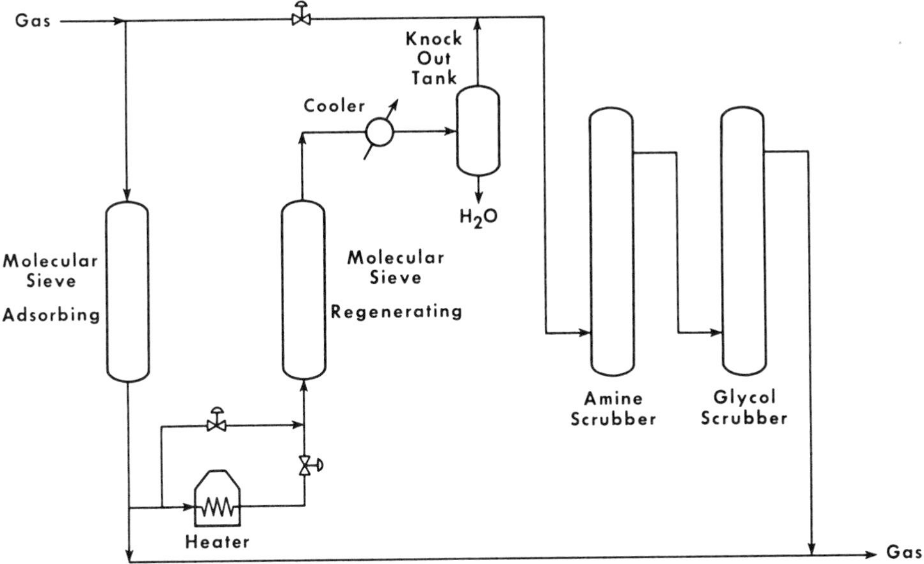

Figure 29 Schematic representation of a combined molecular sieve–absorbent process for gas cleaning.

since molecular sieves have been known to catalyze this reaction but new molecular sieves have been developed to retard the formation of carbonyl sulfide. If further clean-up of the gas stream is desired, cation is advised in the use of olamines since the presence of carbonyl sulfide in the gas stream can bring about irreversible reactions that lead to olamine loss in the process (this chapter, Sec. II.D).

IV. CHEMICAL CONVERSION PROCESSES

Processes for the removal of contaminants from natural gas that utilize the concept of chemical conversion of the contaminants usually accomplish the objectives by heterogeneous catalysis using solid catalysts in fixed-bed catalytic reactors (see also Sec. II.A in Chap. 6 and Sec. III of this chapter).

The catalytic conversion processes that are used for purification of natural gas differ from other process concepts insofar as the contaminants are (1) removed from the gas stream by a physical process and (2) converted to compounds that are not objectional. Therefore, the products may remain in the gas stream or they can be removed from the gas stream with greater ease than the original contaminants.

In this type of process, in which the reactants and the catalysts are in different phases (heterogeneous catalysis), efficient contact between the catalyst and the reactants is essential. Indeed, the reaction may be regarded as a sequence of steps in which each step is critical to the smooth operation of the process and which are (1) transfer of the reactants from the gas stream to the catalyst surface; (2) adsorption on to the surface; (3) chemical reaction on the surface of the catalyst; (4) desorption of the product from the catalyst surface; and (5) incorporation of the product into the gaseous phase (Hougen and Watson, 1946). Thus, mass transfer of the reactants and products are very important aspects of the process operation. In addition, the frequent occurrence of catalyst poisoning is another factor that can effect the efficient operation of the process. In order to diminish catalyst loss due to poisoning, the use of catalyst guard beds (see also Sec. II of this chapter and Speight, 1981) is highly recommended.

Examples of this concept include the iron oxide process (Figure 26) and the Katasulf process (Figure 30). The most common application of the iron oxide process is the complete removal of hydrogen sulfide:

$$2Fe_2O_3 + 6H_2S = 2Fe_2S_3 + 6H_2O$$

$$2Fe_2S_3 + 3O_2 = 2Fe_2O_3 + 6S$$

or

$$6H_2S + 3O_2 = 6H_2O + 6S$$

In the Katasulf process, the hydrogen sulfide in the natural gas is reacted with oxygen to form water and sulfur dioxide:

$$2H_2S + 3O_2 = 2H_2O + 2SO_2$$

The reaction is exothermic and the heat generated is used to heat the inlet gases to reaction temperature (400°C; 750°F). Catalysts used in the process are activated carbon, bauxite, and alloys of iron or nickel or copper with tungsten or vanadium or chromium. Any one of the former trio of metals reacts with the hydrogen sulfide to form the metal sulfide while any one of the latter trio of metals acts as an oxygen carrier.

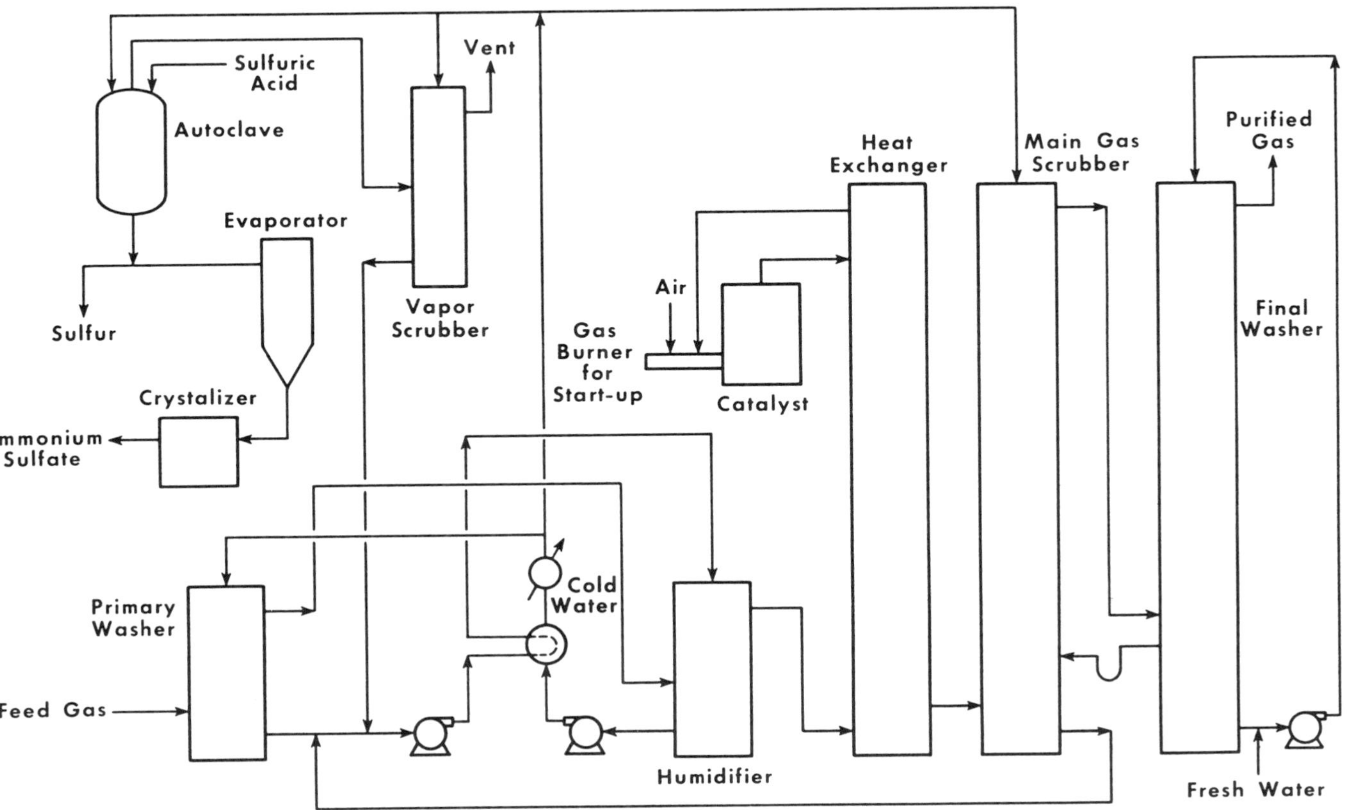

Figure 30 Schematic representation of the Katasulf process.

37

Chemicals from Natural Gas

I	Introduction	1153
II	Hydrogen	1157
III	Ammonia	1158
IV	Hydrogen Cyanide	1160
V	Carbon Dioxide	1161
VI	Carbon Disulfide	1161
VII	Chlorinated Hydrocarbons	1162
VIII	Carbon Black	1163

I. INTRODUCTION

Like the other fuel sources, petroleum and coal, natural gas is a major raw material for many chemical processes, and the potential number of chemicals that can be produced from natural gas is almost without limit.

Natural gas can be used as a source of hydrocarbons (ethane, propane, etc.) that have a higher molecular weight than methane and are important chemical intermediates (Chap. 11). Perhaps what is more important in the current context is that natural gas can be converted to a wide variety of chemicals (Figure 1; Table 1) but natural gas is also a source of synthesis gas. This leads to a wide variety of chemicals (Figure 2) which involve C_1 chemistry (i.e., the chemistry of methane and other one-carbon compounds; Sasma and Hedman, 1984).

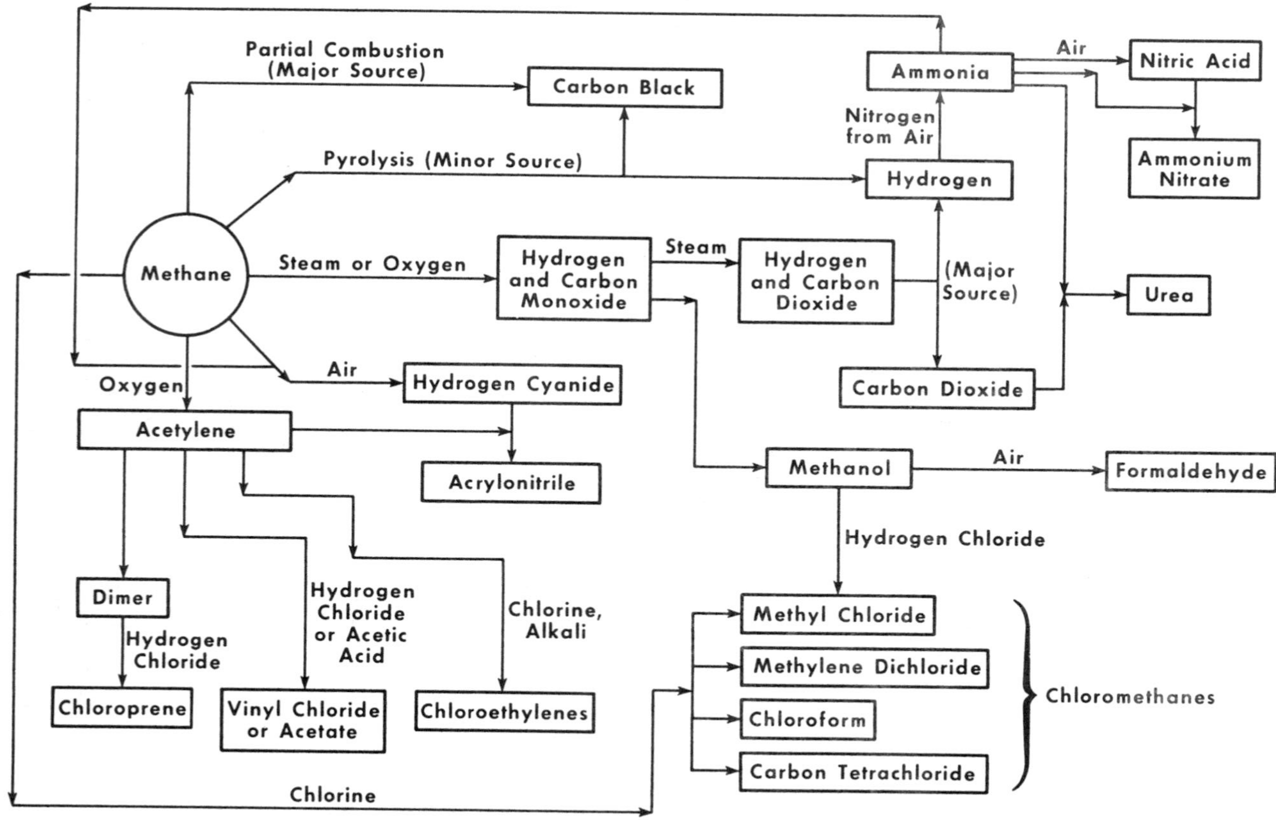

Figure 1 Schematic representation of the production of chemicals from natural gas (methane).

Table 1 Examples of the Chemicals That Can Be Produced from Natural Gas (Methane)

Basic derivatives and sources	Uses
Ammonia	Agricultural chemicals (as ammonia, salts, urea) Fibers, plastics Industrial explosives
Carbon black	Rubber compounding Printing ink, paint
Methanol	Formaldehyde (mainly for resins) Methyl esters (polyester fibers), amines, and other chemicals Solvents
Chloromethanes	Chlorofluorcarbons for refrigerants, aerosols, solvents, cleaners, grain fumigant
Hydrogen cyanide	Acrylonitrile Adiponitrile Methyl methacrylate

In this aspect, the use of natural gas for chemicals is very similar to the chemistry employed in the synthesis of chemicals from the gasification products of coal (Payne, 1987; see also Chap. 22). In this section emphasis will be placed on those materials that can be prepared directly from natural gas (methane), but it must be recognized that there are many other options for the formation of chemical intermediates and chemicals from natural gas by indirect routes, i.e., where other compounds are prepared from the natural gas which are then used as further sources of petrochemical products and liquid fuels (Figures 1 and 2; see also Chap. 11).

The preparation of chemicals and chemical intermediates from natural gas should not be restricted to those described below but should be regarded as some of the building blocks of the petrochemical industry. In summary, natural gas can be a very important source of petrochemical intermediates and solvents (Lowenheim and Moran, 1975; Sasma and Hedman, 1984; *Hydrocarbon Processing*, 1987).

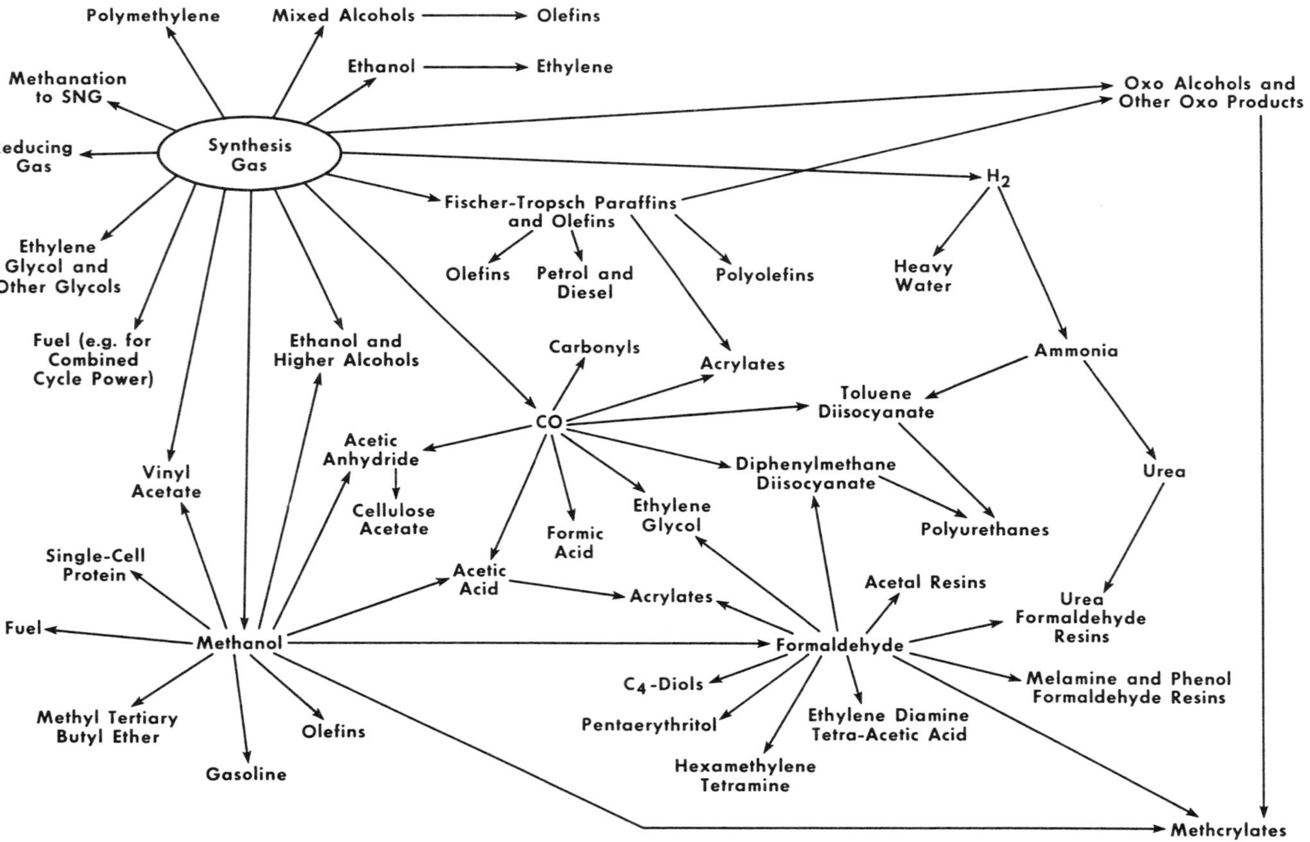

Figure 2 Other routes to chemicals from natural gas (methane) via synthesis gas (gas).

II. HYDROGEN

Although hydrogen is also one of the products when natural gas is converted to ammonia, it is appropriate to consider hydrogen separately because of the importance of this element to the petroleum and petrochemical industry. Several processes have been described for the synthesis of hydrogen (Chap. 6, Sec. IX), but it is pertinent to describe the production of hydrogen here also.

In the process (Figure 3), the hydrogen is made as a mixture with carbon monoxide (synthesis gas, also called syngas) from natural gas by heating a mixture of natural gas with oxygen and (if desired) steam at high temperature (>1000°C; >1830°F) after which the products are rapidly quenched:

$$2CH_4 + O_2 = 4H_2 + 2CO$$

The gases are then scrubbed to remove any entrained carbon followed by passage to a shift converter where the carbon monoxide is reacted with steam to produce carbon dioxide and more hydrogen:

$$CO + H_2O = CO_2 + H_2$$

The hydrogen can be purified by absorption of the carbon dioxide in a hot carbonate solution (Chap. 36, Sec. II.D) or carbon oxides may be removed by conversion to methane through passage over a nickel catalyst at 315–425°C (600–795°F).

The production of synthesis gas (syngas) opens the way to a host of different chemicals (Wender, 1987). For example, the production of methyl alcohol (methanol) from hydrogen and carbon monoxide mixtures:

$$CO + 2H_2 = CH_3OH$$

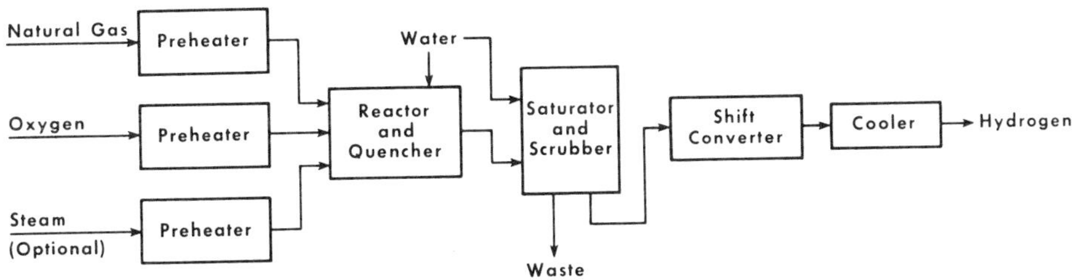

Figure 3 Schematic representation of the preparation of hydrogen from natural gas.

leads to a whole range of chemicals such as methyl chloride, methyl formate, methyl methacrylate, and methylamines:

$$CH_3OH + HCl = CH_3Cl + H_2O$$

$$CH_3OH + CO = HCO_2CH_3$$

$$CH_3OH + CH_2{:}C(CH_3)CO_2H = CH_2{:}C(CH_3)CO_2CH_3 + H_2O$$

$$CH_3OH + NH_3 = CH_3NH_2 + H_2O$$

$$CH_3OH + CH_3NH_2 = (CH_3)_2NH + H_2O$$

$$CH_3OH + (CH_3)_2NH = (CH_3)_3N + H_2O$$

III. AMMONIA

Natural gas is a particularly important feedstock for the synthesis of ammonia (Figure 4). In this process, the natural gas is preheated, passed over a bauxite catalyst to remove the sulfur (Chap. 36, Sec. III), and then treated with steam in a reforming furnace to produce hydrogen. A nickel catalyst is employed and a 70% conversion of methane to carbon monoxide and hydrogen is often realized:

$$CH_4 + H_2O = CO + 2H_2$$

The pressures in the reforming chamber are usually on the order of 500 psi with temperatures up to 950°C (1740°F).

The partially reformed gas is then led to a combustion furnace where sufficient air is added to give the 3:1 molar ratio of hydrogen and nitrogen required for the synthesis of the ammonia:

$$3H_2 + N_2 = 2NH_3$$

The reaction mixture of hydrogen and nitrogen is then compressed to approximately 4500 psi and the temperature is raised to 475°C (890°F)—although temperatures over the range of 400–600°C (750–1110°F) have also been employed. The reactor contains an iron oxide catalyst that has been promoted by small amounts of calcium, magnesium, and aluminum oxides. The gases leaving this reactor are cooled to -20°C (-4°F) where the ammonia commences to liquefy. The conversion per pass may be up to 25% and, with recirculation, the overall yield approaches 90%.

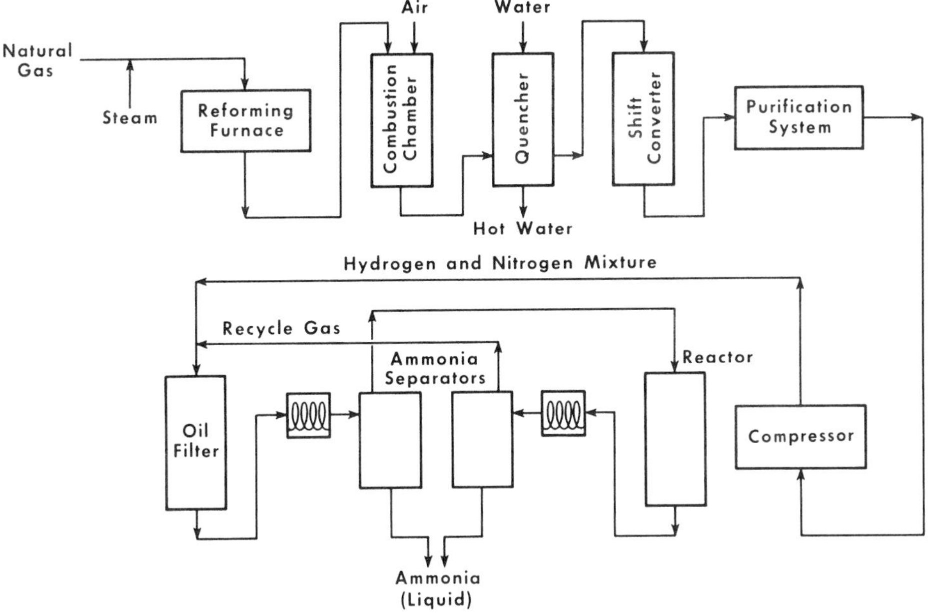

Figure 4 Schematic representation of the preparation of ammonia from natural gas.

The combustion chamber also contains a nickel catalyst so that the reforming reaction can be completed in the chamber. As the gases pass out of the combustion chamber, they are quenched (water) to 425°C (795°F), then fed to a shift converter where the amount of carbon monoxide in the mixture is reduced to <1% by use of an iron oxide catalyst:

$$CO + H_2O = CO_2 + H_2$$

The carbon dioxide is removed by absorption (Chap. 36, Sec. II).

More modern processes use a partial oxidation unit (using 95% oxygen and a large excess of natural gas) in which reaction occurs to produce the carbon monoxide and hydrogen:

$$2CH_4 + 3O_2 = 2CO + 4H_2O$$

whereupon the methane reacts with the water vapor to produce the carbon monoxide and hydrogen:

$$CH_4 + H_2O = CO + 3H_2$$

These effluent gases are then processed as already described using a shift converter and an ethanolamine unit.

The production of ammonia from natural gas opens the way to the production of ammonium nitrate by the neutralization of niric acid with gaseous ammonia:

$$HNO_3 + NH_3 = NH_4NO_3$$

IV. HYDROGEN CYANIDE

Hydrogen cyanide is another important chemical intermediate that is synthesized by the reaction of natural gas with ammonia and air (the Andrussow process; Figure 5) in which the gaseous mix is passed over a platinum (or platinum-rhodium) catalyst at elevated temperature (>1000°C; >1830°F).

$$2NH_3 + 3O_2 + 2CH_4 = 2HCN + 6O_2$$

It is essential that the hydrogen cyanide be cooled as quickly as possible after formation to prevent any side reactions leading to product loss.

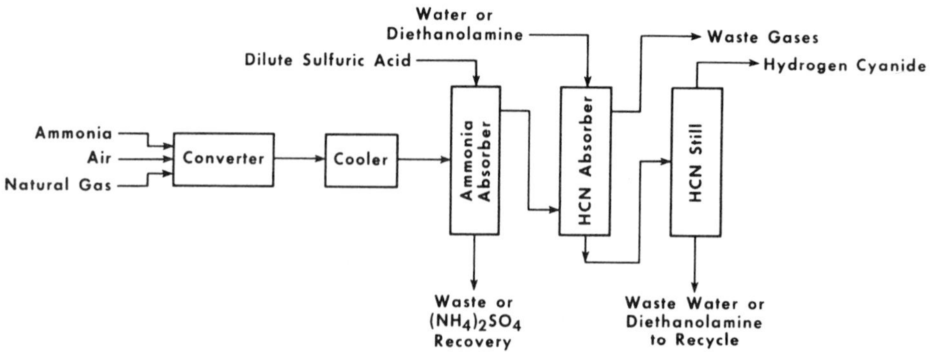

Figure 5 Schematic representation of the preparation of hydrogen cyanide from natural gas.

In another version of the process, air may be omitted (the Degussa process) and the hydrogen cyanide is formed by direct reaction of the ammonia with the natural gas:

$$NH_3 + CH_4 = HCN + 3H_2$$

In yet another variation of the process (the Fluohomic process), ammonia is reacted with higher molecular weight hydrocarbons (which can also be separated from natural gas) but usually at a much higher reaction temperature (around 1500°C; 2730°F):

$$3NH_3 + C_3H_8 = 3HCN + 7H_2$$

No catalyst is usually required other than a fluidized bed of petroleum coke.

V. CARBON DIOXIDE

Carbon dioxide may be classified as a byproduct of the ammonia synthesis (Chap. 6, Sec. III) in which natural gas is decomposed in the presence of steam to produce hydrogen and carbon dioxide:

$$CH_4 + 2H_2O = 4H_2 + CO_2$$

In addition to its use as a coolant, carbon dioxide is also used as an intermediate for chemicals such as urea:

$$CO_2 + 2NH_3 = \underset{\text{ammonium carbamate}}{NH_2COONH_4} = NH_2CONH_2 + H_2O$$

VI. CARBON DISULFIDE

Carbon disulfide is an important solvent and can be conveniently manufactured from natural gas and sulfur:

$$CH_4 + 4S = CS_2 + 2H_2S$$

In the process (Figure 6), natural gas and vaporized sulfur are sent to a reactor at 675°C (1245°F) which is packed with a catalyst such as clay or alumina. The methane is usually in a 5–10% excess and will give a near-quantitative (90–95%) conversion of the sulfur.

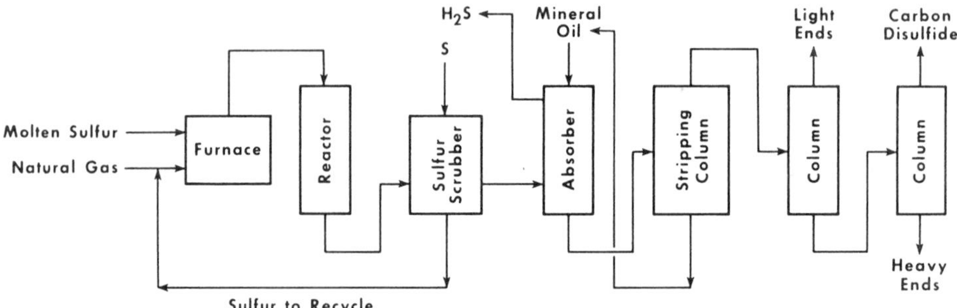

Figure 6 Schematic representation of the preparation of carbon disulfide from natural gas.

The products also contain small amounts of mercaptans (R-SH) as well as di- and polysulfides. The desulfurized gases are fed to an absorption unit where carbon disulfide is removed from the stream by contact with oil. The oil/carbon disulfide mix is then sent to a stripper where the carbon disulfide is removed as overhead and sent to a distillation unit. Two or more steps are used to purify the carbon disulfide—one step to remove any light ends and another step to separate the carbon disulfide from any higher boiling sulfur-containing products.

VII. CHLORINATED HYDROCARBONS

The reaction between natural gas and chlorine in the presence of light or a catalyst may be controlled to yield predominantly methyl chloride with smaller yields of methylene dichloride, chloroform, and carbon tetrachloride. It is more usual that the required product is carbon tetrachloride, so that most of the lesser chlorinated products are recycled to full chlorination.

In the process (Figure 7), high-purity methane (from natural gas) is preheated with the requisite amount of chlorine and passed through a reactor (fitted with mercury arc lamps) at 350–370°C (660–700°F). Although the amount of chlorine can be adjusted to obtain predominantly one product, in actual practice all four products are formed and the three lower chlorinated materials are recycled to extinction if the carbon tetrachloride is the product of choice:

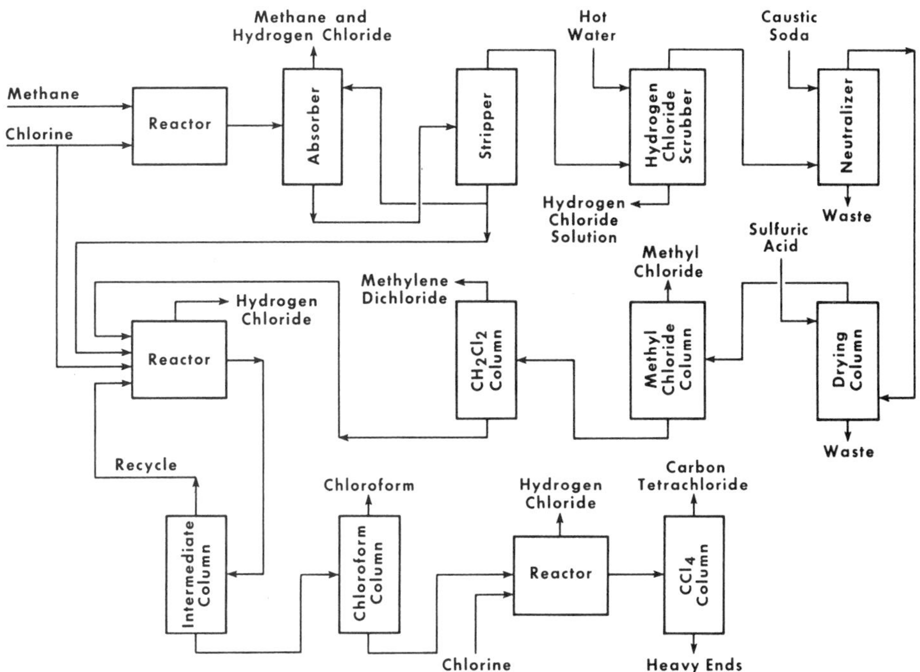

Figure 7 Schematic representation of the preparation of chlorinated hydrocarbons from natural gas.

$$CH_4 + Cl_2 = CH_3Cl + HCl$$

$$CH_3Cl + Cl_2 = CH_2Cl_2 + HCl$$

$$CH_2Cl_2 + Cl_2 = CHCl_3 + HCl$$

$$CHCl_3 + Cl_2 = CCl_4 + HCl$$

VIII. CARBON BLACK

Natural gas is also a source of carbon black; in the process (Figure 8), the natural gas is introduced into a furnace at 870°C (1600°F) whereupon the gas is decomposed to carbon and hydrogen:

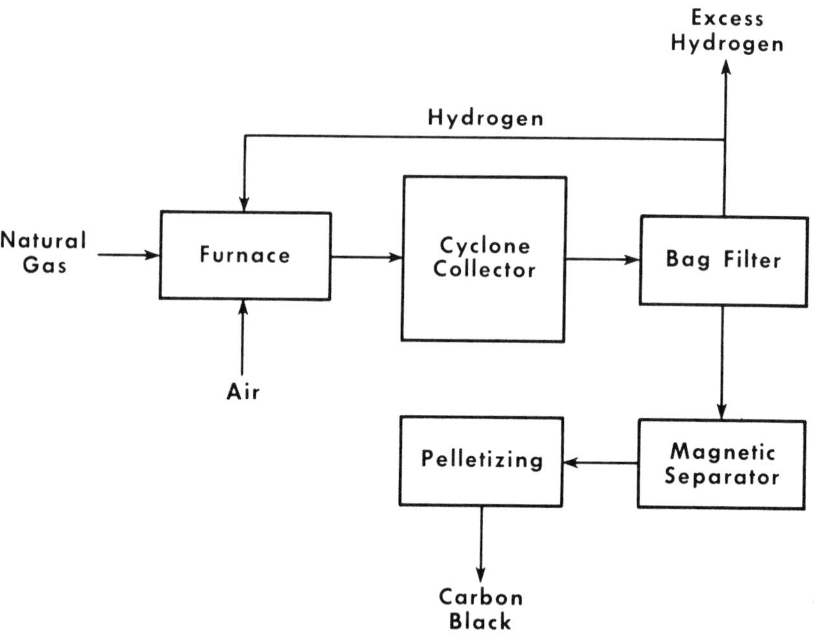

Figure 8 Schematic representation of the preparation of carbon black from natural gas.

$$CH_4 = C + 2H_2$$

Some of the carbon may remain on the interior of the furnace but the remainder is carried by the hydrogen stream through a water cooler into a bag filter and thence to storage hoppers.

A modification of this process has also been used with refinery waste gases to produce hydrogen for ammonia synthesis. It should also be remembered that any process which is capable of generating hydrogen is also useful for refinery operations where hydrogen is required for the many hydroprocesses employed in the refinery (Chap. 6, Secs. III and IX).

References to Part V

ASTM. 1987. American Society for Testing and Materials, Race Street, Philadelphia, Pennsylvania.

Atkinson, C. H., and R. T. Johansen. 1964. Report of Investigations No. 6494, U.S. Bureau of Mines, Washington, D.C.

Barker, C., and M. K. Kemp. 1980. Conference on Natural Gas Resource Development in Mid-Continent Basins: Production and Exploration Techniques. University of Tulsa, Tulsa, Oklahoma, March 11–12.

Bartoo, R. K. 1985. In *Acid and Sour Gas Treating Processes* (S. A. Newman, ed.), Gulf Publishing, Houston, Texas.

BP Review of World Gas. 1987. British Petroleum, London, September.

BP Statistical Review of World Energy. 1987. British Petroleum, London, June.

Considine, D. M. 1977. *Energy Technology Handbook*, McGraw-Hill, New York.

Conviser, S. A. 1965. *Oil and Gas J. 63*(49):130.

Curry, R. N. 1981. *Fundamentals of Natural Gas Conditioning*, PennWell Publishing, Tulsa, Oklahoma.

Dingman, J. C., and T. F. Moore. 1968. *Hydrocarbon Processing 47*(7):138.

Duckworth, G. L., and J. H. Geddes. 1965. *Oil and Gas J. 63*(37):94.

Edgar, T. F. 1983. *Coal Processing and Pollution Control*, Gulf Publishing, Houston, Texas.

Fairbridge, R. W. 1972. *Encyclopedia of Geochemistry and Environmental Sciences*. Encyclopedia of Earth Sciences Series, Vol. IVA, Dowden, Hutchinson and Ross, Stroudsberg, Pennsylvania.

Forbes, R. J. 1958. *Studies in Early Petroleum History*, E. J. Brill, Leiden, Netherlands.

Forbes, R. J. 1964. *Studies in Ancient Technology*, Vol. I, E. J. Brill, Leiden, Netherlands.

Gangriwala, H. A., and I-M. Chao. 1985. In *Acid and Sour Gas Treating Processes* (S. A. Newman, ed.), Gulf Publishing, Houston, Texas.

Gevertz, H., R. F. Lemon, W. T. Hollis, M. A. Lekas, D. C. Ward, and C. H. Atkinson. 1965. Project Gasbuggy. U.S. Atomic Energy Commission, Washington, D.C.

Gibbons, J. H. 1985. *U.S. Natural Gas Availability*, Office of Technology Assessment, Congress of the United States, Washington, D.C.

Gold, T. 1984. *Sci. Am. 251*(5):6.

Gold, T. 1985. *Ann. Rev. Energy 10*:53.

Gold, T., and S. Soter. 1980. *Sci. Am. 242*(6):154.

Gold, T., and S. Soter. 1982. *Energy Explor. Exploit. 1*(1):89.

Gold, T., and S. Soter. 1986. *Chem. Eng. News 64*(16):1.

Grow, G. C. 1980. Proc. Mid-Year Meeting, American Institute of Chemical Engineers, Philadelphia, Pennsylvania, June.

Haas, M. R., J. P. Brashear, and F. Morra. 1987. *J. Petroleum Technology (January) 29*:77.

Hamblin, W. K. 1975. *The Earth's Dynamic Systems*, Burgess Publishing, Minneapolis, Minnesota.

Hegwer, A. M., and R. A. Harris. 1970. *Hydrocarbon Processing 49*(4):103.

Holder, G. D., P. F. Angert, V. T. John, and S. L. Yen. 1982. *J. Petroleum Technol. 34*:1127.

Holder, G. D., P. F. Angert, and V. Pereira. 1983. In *Natural Gas Hydrates*, Butterworths, Boston, Massachusetts.

Holder, G. D., V. A. Kamath, and S. P. Godbole. 1984. *Ann. Rev. Energy 9*:427.

Hollings, H. 1952. *Inst. Gas Eng. Commun.* p. 407.

Holzer, F. 1968. Report No. UCRL-50386. U.S. Atomic Energy Commission, Washington, D.C.

Hougen, O. A., and K. M. Watson. 1947. *Chemical Process Principles*, Vol. 3, John Wiley and Sons, New York.

Hydrocarbon Processing. 1987. *Petrochemical Handbook*. November.

Jenett, E. 1962. *Oil and Gas J. 60*(18):72.

Katz, D. L., D. Cornell, R. Kobayashi, F. H. Poettmenn, J. A. Vary, J. R. Elenbass, and C. F. Weinaug. 1959. *Handbook of Natural Gas Engineering*, McGraw-Hill, New York.

Kennedy, J. L. 1972. *Oil and Gas J. 70*(9):69.

Klein, J. P. 1970. *Oil and Gas Int. 10*(9):109.

Kohl, A. L., and F. C. Riesenfeld. 1985. *Gas Purification*, Gulf Publishing, Houston, Texas.

Kumar, S. 1987. *Gas Production Engineering*, Gulf Publishing, Houston, Texas.
Lekas, M. A., B. G. Bray, H. C. Carpenter, H. H. Aronson, G. U. Dineen, and J. D. Downen. 1967. Project Bronco. U.S. Atomic Energy Commission, Washington, D.C.
Lockhart, L. 1939. *J. Inst. Petroleum 25*:1.
Lowenheim, F. A., and M. K. Moran. 1975. *Industrial Chemicals*, John Wiley and Sons, New York.
Maddox, R. N. 1974. *Gas and Liquid Sweetening*, Campbell Publishing, Norman, Oklahoma.
Maddox, R. N. 1982. *Gas Conditioning and Processing*, Vol. 4, *Gas and Liquid Sweetening*, Campbell Publishing, Norman, Oklahoma.
McFarlane, R. C., D. J. Grave, and J. Riccio. 1981. Report No. DOE/BC/10003-22. U.S. Department of Energy, Washington, D.C.
METC. 1988a. Technology Status Report. DOE/METC-88/0259. Morgantown Energy Technology Center, Morgantown, West Virginia.
METC. 1988b. An Assessment of the Natural Gas Resource Base of the United States. Report No. DOE/W-31109-H1. Morgantown Energy Technology Center, Morgantown, West Virginia.
Meyer, R. F. 1977. *Proc. Conference on the Future Supply of Nature-made Petroleum and Gas*, United Nations Institute for Training and Research, Laxenburg, Austria, July 5–16, Pergamon Press, New York.
Moore, D. B. 1956. *Gas World 143*:153.
Moran, J. M., M. D. Morgan, and J. H. Wiersma. 1986. *Introduction to Environmental Science*, W. H. Freeman, New York.
Moyes, A. J., and J. S. Wilkinson. 1974. *The Chemical Engineer 282*(Feb):84.
Nederlof, M. H. 1988. *Ann. Rev. Energy 13*:95.
NGPSA. 1966. *Engineering Data Book*, Natural Gas Processors Suppliers Association, Tulsa, Oklahoma.
Nicklin, T., and D. Hughes. 1977. U.S. Patent 4,049,776.
Nicklin, T., F. C. Riesenfeld, and R. P. Vaell. 1973. *Proc. 12th World Gas Conference*, June, Nice.
Nonhebel, G. 1964. *Gas Purification Processes*, George Newnes Ltd., London.
Ouwerkerk, C. 1978. *Hydrocarbon Processing 57*(4):89.
Payne, K. R. *Chemicals from Coal: New Processes*, John Wiley and Sons, New York.
Pearce, R. L., J. L. Arnold, and C. K. Hall. 1961. *Hydrocarbon Processing 40*(8):121.
Pettijohn, F. J. 1957. *Sedimentary Rocks*, Harper and Brothers, New York.
Rushton, D. W., and W. Hayes. 1961. *Oil and Gas J. 59*(38):102.
Sasma, M. E., and B. A. Hedman. 1984. *Proc. International Gas Research Conference*, Gas Research Institute, Chicago, Illinois.

Scheil, V., and A. Gauthioer. 1909. *Annales de Tukulti Ninip II*. Paris.
Schroder, O. 1920. *Keilschriftetexte aus Assur vershiedenen xiv*. Leipzig.
Sharples, R. J., and W. Panhill. 1985. *Oil and Gas J.* *83*(26):47.
Speight, J. G. 1981. *The Desulfurization of Heavy Oils and Residua*, Marcel Dekker, New York.
Speight, J. G. 1983. *The Chemistry and Technology of Coal*, Marcel Dekker, New York.
Taylor, D. K. 1956. *Oil and Gas J.* *54*(48):147.
Tek, M. R. 1987. *Underground Storage of Natural Gas*, Vol. 3, Gulf Publishing, Houston, Texas.
Trofimuk, A. A., N. V. Cherskiy, Y. F. Makagon, and V. P. Tsarev. 1972. *Int. Geol. Rev.* *15*:1042.
Trusell, F. C. 1985. *Anal. Chem.* *57*(5):191R.
Vasan, S. 1978. *Oil and Gas J.* *76*(1):78.
Veatch, R. W., and Z. A. Mocchovidis. 1986. *Proc. International Meeting on Petroleum Engineering*, Society of Petroleum Engineers of AIME, Dallas, Texas *2*:421.
Watkins, J. W., and C. C. Anderson. 1964. Information Circular No. 8219. U.S. Bureau of Mines, Washington, D.C.
Wender, I. A. 1987. In *Chemicals from Coal: New Processes* (K. R. Payne, ed.), John Wiley and Sons, New York.
Willis, P. A. 1986. *Trends Anal. Chem.* *5*(3):63.

Index

Abiogenic formation of natural gas, 1057
Absorption processes for gas cleaning, 1105, 1115
Acetals, 305
Acetone, 292
Acid gas
natural gas, 1097
removal, 669, 1103, 1104, 1105, 1120, 1123
Acid sludge, 274
Acid treating, 207, 208
Acrylates, 303
Acrylonitrile, 301
Adiponitrile, 301
Adsorption fractionation, 130
alumina, 138, 608
attapulgus clay (attapulgite), 138
clay, 138
coal liquids, 607
Fuller's earth, 138
silica, 138, 608
Adsorption processes for gas cleaning, 669, 1102, 1106, 1142
Albertite, 57
Alginite, 545
Alkali leaching coal cleaning, 536
Alkaline flooding, 36, 39, 40
Alkanolamine gas cleaning processes, 669, 1104, 1120, 1123
chemistry, 1124, 1126
Alkazid gas cleaning process, 225, 1134, 1140
Alkylation, 189
catalysts, 257
chemistry, 287
commercial processes, 190
feedstocks, 287
nobel metal catalysts, 191
Allis-Chalmers oil shale retorting process, 944
Allogenic mineral matter, 455
Ames coal cleaning process, 537
Amines, aliphatic, 300
Amine washing, 196, 1105
Amino acids, 11
Aminoalcohols, 300
Ammonia from natural gas, 1158
Ammonium sulfate from coal gas, 651

Aniline point, 114
Anoxic zone, 828
Anthraxylon, 560
Anticline, 14, 25
API gravity, 63, 69, 100
 asphaltics, 105
 carbon residue, 104
 nitrogen content, 80
 sulfur content, 77
 viscosity, 104
Aquasorption gas cleaning process, 1118
Aquifer for gas storage, 1081
Aromatic hydrocarbons
 coal liquids, 628
 coal tar, 768
 petroleum, 72, 74
 shale oil, 852, 1032, 1033
Arsenic removal from shale oil, 1013
ART process, 245
Ash in coal, 456, 581, 646
Asphalt, 18, 271
 blending, 271
 cutback, 271, 272
 emulsions, 271
 hard, 272
 native, 53, 59
 production, 62, 272
 propane, 59, 270
 shale oil, 1035
 soft, 272
 solvent, 59
 straight run, 59
 wurtzulite, 59
Asphalt base crude oil, 65
Asphaltenes, 15, 53, 55, 62, 124
 association, 236
 average structure, 230
 carbon residue, 143
 catalytic hydrodesulfurization, 143
 coal liquids, 144, 608
 coking chemistry, 230, 232
 definition, 230
[Asphaltenes]
 effects in refining, 230, 236, 241
 kerogen, 910
 natural product, 232
 shale oil, 992
 yield vs solvent, 409
Asphaltic crudes
 API gravity, 105
 metals content, 7, 323
Asphaltics content, 408
Asphaltite, 53, 56, 57
Asphaltoid, 53, 57
Assay
 Fischer, 804, 840, 841, 848, 849, 874
 modified Fischer, 840
 other methods, 843
 Tosco, 844
Associated natural gas, 1088
ASTM system of coal classification, 561
Auger mining, 519
Authentic mineral matter, 455
Azeotropic distillation, 163

Banded coal, 556, 561
Barrel, definition, 154
Basement rock, 13
Battelle hydrothermal coal cleaning process, 537
Battery, 155
Baume, 63, 100
Bauxite treating, 210
Belknap coal cleaning process, 524
Beneficiation, 520
 alkali leaching, 536
 Ames process, 537
 Battelle hydrothermal process, 537
 chemical methods, 533
 cyclone process, 527
 dynawhirlpool process, 528
 Humphrey's spiral concentrator, 532

[Beneficiation]
jib method, 531
JPL process, 538
Ledgement process, 537
Meyer (TRW) process, 533
microwave method, 540
oxydesulfurization, 536
PETC process, 537
physical methods, 521
sand-flotation process, 525
TRW process (*see* Meyer process)
Benfield process, 669
Bentonite, 252
Benzoporphyrins, 85, 88, 94
Bergius process, 737, 740
Biogenic formation of natural gas,
Bitumen, 16, 229, 544, 847, 851, 852, 853
API gravity, 321, 351
carbon residue, 323, 410
definition, 322
elemental composition (*see* ultimate composition)
fractional composition, 388, 395, 842, 843, 844
hot water process, 345
mining recovery, 338, 419, 423
molecular weight, 410
nonmining recovery, 351, 352
oil shale, 815, 847, 850, 880
pour point, 408
primary conversion, 413
properties, 321, 323, 389, 426
quality, 363
recovery, 335, 339, 343, 349, 426
solubility, 403
specific gravity, 389
sulfur content, 321
tar sand, 3, 53, 321, 384

[Bitumen]
terminology, 53, 56
thermal sensitivity, 400
transportation, 364
ultimate composition, 323, 387, 395, 848
upgrading, 411
viscosity, 396
volatility, 397
Bituminous rock, 53, 57
Bituminous sand, 53
Boghead coal, 560
Boie equation, 867
Bone coal, 559
Borehole logging, 25
Boron oxide catalysts, 252
Bright coal, 562
Brown coal, 570, 574
Bubble tower, 157, 1120
Bucket wheel excavators, 339
Bulk acid polymerization, 195
Bulk chemicals from coal, 770
Bulldozers, 339
Burton process, 168
Butadiene, 302
Butane extraction, 163
Butyl alcohol, 294

Caking properties of coal, 666
Calcining, 252
Canmet hydrocracking process, 247, 417
Cannel coal, 839
Cap rock, 13, 31
Caprolactam, 302
Carbenes, 53, 62, 124
Carbohydrates, 10, 11, 831
Carboids, 53, 62, 124
Carbon, fixed, 57, 577
Carbonate gas cleaning processes, 1105, 1128
Carbon black from natural gas, 1163
Carbon dioxide flooding, 40, 42

Carbon dioxide from natural gas, 1161
Carbon disulfide from natural gas, 1161
Carbon cycle, 799, 800
Carbon inventory, 801
Carbonization of coal, 651, 764
 high temperature, 651
 low temperature, 652
Carbon number, boiling point, 2
Carbon rejection, 239, 243, 246
Carbon residue, 237
 API gravity, 104
 asphaltenes, 114, 143
 Conradson, 111, 115
 nitrogen, 114
 petroleum, 323
 Ramsbottom, 111, 115
 sulfur, 114
 tar sand bitumen, 323, 410
 viscosity, 114
Carotenoids, 7, 830
Cascade sulfuric acid process, 191
Casinghead gas, 1089
Catacarb process, 1104, 1130
Catalysts
 alkylation, 257
 boron oxide, 252
 calcining, 252
 cobalt-molybdenum, 253
 composition, 252
 cracking, 251
 poisoning, 76
 hydroprocessing, 253, 253, 1025, 1026
 insulator, 252
 isomerization, 256
 large pore, 239
 metal poisons, 258
 noble metal, 191
 polymerization, 257
 properties, 252
 reforming, 255
 silica-alumina, 251
[Catalysts]
 silica-magnesia, 251
 thorium oxide, 252
 treating, 257
 zeolite, 254
 zirconium oxide, 251
Catalyst treating
 Demet process, 258
 Met-X process, 259
Catalytic cracking
 catalysts, 251
 commercial processes, 174
 fixed-bed processes, 175
 fluid-bed processes, 176, 1024
 heavy feedstocks, 247
 moving-bed processes, 178
 process parameters, 175, 248, 249
Catalytic gasification of coal, 678
 catalyst deactivation, 697
 catalyst volatilization, 696
 channeling activity, 686
 gas phase poisons, 690
 included solid poisons, 693
 physics, 679
 sintering, 686
 wetting properties, 684
Catalytic hydrogenation of coal, 737
Catalytic liquefaction of coal processes, 740
Catalytic reforming
 commercial processes, 184
 fixed-bed processes, 185
 hydroforming, 185
Caustic (lye) treating, 205
CCL coal liquefaction process, 739, 740
Celcon, 305
Cellulose, 10
Ceresins, 58
Chance coal cleaning process, 524
Characterization factor, 69
Char hydrogenation, 671
Chemical flooding, 35, 36
Chemical fractionation, sulfuric acid, 139

Chemicals
coal, 763, 768, 770, 773
natural gas, 1153, 1154, 1155
petroleum, 277, 280
shale oil, 1031
Chevron olefin process, 298
Chevron (STB) oil shale retorting process, 938
Chlorinated hydrocarbons from natural gas, 1162
Chlorins, 85, 86, 235
Chlorophyll, 86, 93
Christmas tree, 28, 29
Clarain, 561, 562
Classification
bitumen, 64, 817
coal, 64, 543, 555, 561, 563, 568
heavy oil, 64
kerogen, 64, 820, 826
natural gas, 64, 1088
oil shale, 839
organic facies, 827
petroleum, 63, 64
well, 1076
Clastic dyke, 506
Claus process, 1109
Clay treating, 209
Clay vein, 508
Clean coke coal liquefaction process, 738
Cleaning
coal (*see also* beneficiation) 521
gas, 221, 1100, 1103, 1107
petroleum, 152
shale oil, 1010, 1013
Cleat, 513
Coal, 3
analysis, 575
ash, 456, 646
beneficiation, 520
caking properties, 666
cannel, 839
carbonization, 651, 764
chemicals from, 763
[Coal]
classification, 555
cleaning (*see also* beneficiation) 521
coking, 648
combustion, 655
consumption, 484
definition, 438
density, 551, 554
environmental aspects of utilization, 758
exports, 484, 490
floatability, 530
formation, 441, 446
gasification, 661
geologic age, 445
hydrogen content, 550
ignition, 655
liquefaction, 735
liquids, 605
macerals, 545
macroscopic classification, 555
mild gasification, 716
mineral matter, 453, 455, 456, 575, 585
moisture, 467, 474
occurrence, 471
petrography, 439
petrology, 439
production, 507
pyrolysis, 717, 718
rank, 448, 449, 543, 554
recovery, 506
reserves, 324, 471, 473, 497, 500
structure, 585
sulfur, 473, 484, 488, 578
supply, 493
terminology, 543, 556, 557
thermal properties, 646
trace elements, 463, 583
transportation, 540
ultimate analysis, 578
underground gasification, 726
washing, 526
Coal ash, 456, 460
Coal bed (seam) gas, 1065

Coal conversion pollutants, 759
Coal gas
ammonium sulfate from, 651
Coalification, 458, 451
Coal liquids
aromatic groups, 628
chromatographic separation, 607
gas chromatography, 615
heteroatomic groups, 638
HPLC, 608
luminescence spectroscopy, 627
mass spectrometry, 625
molecular weight, 624
nmr spectroscopy, 631
size exclusion chromatography (SEC), 612
solvent fractionation, 605
structural analysis, 639
supercritical fluid chromatography (SFC), 619
thin layer chromatography (TLC), 615
titration, 621
ultraviolet spectroscopy, 627
Coal petrography, 439
Coal petrology, 439
Coal producing periods, 441
Coal rank, 448, 449, 543
elemental composition, 451
variation with depth, 450
Coal seam (bed) gas, 1065
Coal tar, 61, 764
COED coal liquefaction process, 738
COFCAW process, 361
Coke, 273
Coking
chemistry, 230
coal, 648
delayed, 171, 239, 243, 415
flexicoking, 239, 244, 1016
fluid, 172, 244, 416, 1016
heavy feedstocks, 247
process parameters, 248, 249
Coking coals, 648
Collinite, 545
Colloidal fuel combustion system, 659
Combustion, 42, 44
chemical aspects, 657
colloidal fuel system, 659
effects of coal properties, 655
elemental composition of coal, 656
entrained flow, 658
fixed-bed, 658
fluid-bed, 658
forward, 358
grindability of coal, 656
ignified fuel system, 659
magnetohydrodynamic system, 660
mineral matter in coal, 656
reverse, 358
submerged system, 660
superslagging system, 660
systems, 657
Combustion engineering coal gasification process, 673
Composition
bitumen, 388, 395, 842, 843, 844
coal, 451, 574
kerogen, 848, 856, 885
natural gas, 31, 1090, 1091, 1095
oil shale,
petroleum, 72, 145, 146, 147, 148 323
shale oil, 1005, 1006, 1034
tar sand bitumen, 321, 323, 388, 395
Compound type classification, 66
Concretion, 513
Conductivity
coal, 647
petroleum, 117
Consol synthetic fuel process, 742
Continuous contact filtration, 210
Contour mining, 519
Conventional reservoirs, 1061
Cool water coal gasification process, 675

Copper, 6
amine solutions, 199
pyrophosphate, 193
salts, 213
sweetening, 213
Correlation index, 66
Costeam process, 742
Countercurrent contact tower, 217, 1118
Cracking
catalysts, 251
catalytic, 174
chemistry, 167, 282
fixed-bed processes, 175
fluid-bed processes, 176
moving-bed processes, 178
process summary, 170
thermal, 166
Crude oil enrichment process, 1110
Cryogenic gas cleaning processes, 1114
Curie point pyrolysis of coal, 600
Cutback asphalt, 271
Cutout, 514
Cyclic steam injection, 42, 44, 354
Cyclone coal cleaning process, 527
Cyclopentadiene, 302
Cycloversion process, 256

Deasphalting, 215
duosol process, 217
gas, 15
process, 215
propane, 216
solvent, 215, 240
Debutanizer, 161
Deep source gas, 1065
Dehydration process for gas cleaning, 1143
Delayed coking
heavy oils, 415
[Delayed coking]
petroleum, 171, 239, 243
shale oil, 1014
Delrin, 305
Demetallization, 238, 240
Demet process, 258
Demonstrated reserve base (coal), 475
Density, 67, 99
Depentanizer, 161
Depropanizer, 161, 192
Desalting, 81, 153
Destructive hydrogenation, 178
Desulfurization, 143
Dewatering
natural gas, 1100, 1142
petroleum, 47, 152
refinery gas, 224, 226
refinery liquids, 226
shale oil, 1012
Dewaxing, 215, 218
catalytic, 220, 221
solvent, 219
Diesel fuel oil, 266
Dimethylamine, 300
Dipping strata, 513
Direct heating of tar sand, 348, 350
Directional drilling, 26
Distillate fuel oil, 266
Distillation, 17, 112
atmospheric pressure, 122, 151, 156
azeotropic, 163
debutanizer, 161
depropanizer, 161
extractive, 163
flash, 166, 271
fractions, 151
history, 153
light end removal, 159
plates, 157
reduced pressure, 122, 151, 158
rerunning, 159
stabilization, 159
stripping, 159

[Distillation]
superfractionation, 162
trays, 157
vacuum, 122, 151, 158
Doctor (sodium plumbite) treating, 211
Donor solvent process
Exxon, 741, 742, 744
heavy oil, 246
Dow coal liquefaction process, 737, 740
Draglines, 339
Dravo oil shale retorting process, 945
Drift access, 515
Drill bit, 26, 29
Drill collars, 26
Drilling, 26
bits, 26
cable tool method, 26, 27
casing head, 28
collars, 26
directional, 26
rotary method, 26, 28
mud, 26, 27
Drilling mud, 26, 27
Dry fields, 1089
Dry gas, 1073, 1089
Dulong equation, 865
Duosol process, 217
Durain, 559, 560
Dynawhirlpool coal cleaning process, 528

Eastman chemicals-from-coal process, 773
Economics of bitumen recovery, 419, 423
Elaterite, 57
Elemental composition (*see* Ultimate composition)
Emulsion breaking, 152
Enhanced oil recovery, 35
alkaline flooding, 36, 39, 40

[Enhanced oil recovery]
carbon dioxide flooding, 40, 42
chemical methods, 35, 36
criteria, 36
cyclic steam injection, 42, 44, 354
huff and puff method, 44
hydrocarbon flooding, 41
in situ combustion, 42, 44
micellar flooding, 36
microemulsion flooding, 36
miscible methods, 35, 39
polymer flooding, 36, 37, 38
product quality, 46, 48
steam injection, 42, 43, 44
surfactant flooding, 36, 38, 39
thermal methods, 35, 42
water flooding, 36, 37
Entrained flow reactor
coal gasification, 658, 673, 676
Environmental aspects
coal conversion pollutants, 759
coal utilization, 758
tar sand recovery, 367
EOR (*see* Enhanced oil recovery)
Epigenetic mineral matter, 455
EPRI coal liquefaction process, 739
Equity Oil/Arco oil shale retorting process, 739
Ethanolamine, 291
Ethyl alcohol, 291
Ethyl Corporation process, 299
Ethylene glycol, 290
Ethyl ether, 291
Etioporphyrins, 85, 87
Eureka Cracking process, 248
Evaluation of petroleum feedstocks, 118, 120
Exinite, 545
Exploration
borehole logging, 25
gravimeter, 25
magnetometer, 25

[Exploration]
 seismograph, 25
 subsurface, 25
Extractive distillation, 163
Exxon Donor Solvent (EDS) process, 741, 742, 744
Exxon olefin process, 297
Exxon (ESR) oil shale retorting process, 937

Facies, 827, 846
Fat oil, 162
Fats, 10
Fatty acids, 7
Fault trap, 14
 normal, 514
 reverse, 514
Faulting, 513
Feedstock evaluation, data use, 118, 120
Ferrox process, 1106
Fields (natural gas)
 condensate, 1089
 dry, 1089
Finishing processes, 180
Fischer assay, 856, 874
 effect of hydrogen, 973
 limitations, 842
 modified, 804, 840
 standard, 840, 841
 temperature profile, 842
 Young's modulus, 874
Fischer-Tropsch process, 749, 751, 752, 769
Fixed-bed processes
 catalytic cracking, 175
 combustion, 658
 reforming, 185
Fixed carbon
 coal, 453, 646, 577
 petroleum, 57
Flash distillation, 166, 271
Flash point, 109
Flexicoking
 petroleum, 239, 244
 shale oil, 1016
Flexicracking, 247
Floatability index, 530
Fluid-bed processes
 catalytic cracking, 176, 1024
 coal combustion, 658
 coal gasification, 673
 reforming, 187
Fluid catalytic cracking
 petroleum, 176
 shale oil, 1024
Fluid coking
 heavy oil, 416
 petroleum, 172
 shale oil, 1016
Fluid gas treating process, 1102
Folded strata, 515
Foots oil, 270
Formaldehyde, 288
Formed coke, 652
Fractionation, 121
 adsorption, 130, 609
 chemical, 139
 distillation, 122
 miscellaneous methods, 134, 608, 612, 619
 natural gas, 1111
 SAPA method, 130
 SARA method, 130, 132, 133
 solvent treatment, 124, 605
 thiourea, 137
 urea, 137
 USBM-API method, 131
 use of data, 140
Fractionation process for natural gas, 1111
Frases coal cleaning process, 524
Fuel oil
 diesel, 266
 distillate, 266
 furnace, 266
 heavy, 267
 residual, 266

Furnace fuel oil, 266
Fusain, 559, 560
Fusinite, 545

Galoter oil shale retorting process, 966
Gas cap, 13, 31, 1090
Gas chromatography, 615
Gas cleaning, 221
 absorption processes, 1115
 acid gas removal, 1103, 1104, 1105
 adsorption processes, 669, 1102, 1105, 1142
 alkanolamine processes, 669, 1120
 alkazid process, 1134, 1140
 Benfield process, 669, 1104, 1130
 carbonate processes, 1105, 1128
 Catacarb process, 1104, 1130
 chemical processes, 669, 1149
 chemistry, 222, 223
 Claus process, 1109
 crude oil enrichment process, 1110
 cryogenic processes, 1114
 dehydration process, 1143
 enrichment, 1110
 ferrox process, 1106
 fluid process, 1101
 fractionation process, 1111
 Giammarco-Vetrocoke process, 225, 1104, 1106, 1133
 Girbotol process, 223
 Holmes-Stretford process, 1137, 1141
 hydrogen sulfide removal, 1106
 iron oxide (sponge) process, 669, 1106, 1145, 1147, 1150
 liquids removal, 1107, 1112
 Katasulf process, 1150
[Gas cleaning]
 molecular sieve processes, 1146, 1148
 nitrogen removal, 1103
 phosphate process, 224
 Rectisol process, 669, 1142
 refrigeration process, 1101, 1114
 Selexol process, 1120
 Stretford process, 1106, 1108
 water removal, 1100, 1142
 water wash process, 1118
Gas combustion oil shale retorting process, 916
Gas condensate, 1088
Gas deasphalting, 15
Gas drive, 30, 33, 1079, 1081
Gas gathering system, 1081
Gas hydrates, 1065
Gasification of coal, 661
 catalytic, 678
 chemistry, 663, 664
 cool water process, 675
 gas composition, 673
 methane formation, 665
 mild, 716
 primary, 667
 products, 662
 reactors, 670, 674, 676
 secondary, 667
 second generation technologies, 676
Gas injection, 32
Gas liquefaction, 1082
Gas oil, 3
 heavy, 2, 151, 253
 light, 2, 151
 vacuum, 2, 151
Gasoline, 2, 3, 151, 263, 1092
 aviation, 265
 blending, 263
 components, 264
 polymer, 194
Gas storage
 aquifers, 1081
 salt domes, 1081

Geokinetics in situ oil shale retorting process, 985
Geologic history, 442
Geopressured aquifers, 1064
Giammarco-Vetrocoke process, 225, 1064, 1104, 1106, 1133
Gilsonite, 57
Girbotol process, 223
Glance pitch, 57
Global carbon inventory, 801
Glycerides, 10
Glycerin, 293
Glycogen, 10
Grahamite, 57
Gravimeter, 25
Gravity
 API, 63
 specific, 67, 99
Gravity drive, 30
Grease, 268, 269
 buttery, 269
 fibrous, 269
 smooth, 269
 soaps, 268, 269
 thickening agents, 268
Green acids, 275
Grindability of coal, 658
Guard reactors, 239
Gulf shale oil upgrading process, 1022

Halloysite, 252
Hard coal, 566
H-Coal process, 739, 740
Heat content of coal, 453
Heavy crude oil, 229
 properties and development, 426
 refinery options, 242
Heavy fuel oil, 267
Heavy gas oil, 2, 151, 253
Heavy naphtha, 151
Heavy oil, 3, 53, 55
 ART process, 245
 asphaltenes and refining, 230
 carbon rejection, 243
 catalysts, 246
 delayed coking, 243
 flexicoking, 244
 fluid coking, 244
 hydrogen addition, 239, 243
 options for refining, 242, 247
 porphyrins and refining, 233
 process summary, 248, 249
 properties, 101
 refining, 229
Heavy Oil Cracking (HOC) process, 248
Hematins, 6
Hemes, 6
Hexamethylenediamine, 304
High-btu gas, 662
High performance liquid chromatography (HPLC), 608
High-temperature coal tar, 764
H-Oil process, 181, 247
Holmes-Stretford process, 1137, 1141
 chemistry, 1137, 1138
Hot carbonate gas cleaning process, 1128
Hot-pot gas cleaning process, 1128
Hot water extraction of bitumen, 344
Houdriflow moving-bed process, 179
Houdry fixed-bed process, 176
Huff and puff method, 44
Humates, 831
Humic coal, 558
Humphrey's spiral concentrator, 534
Hydrate formation in gas lines, 1079
Hydrocarbon flooding, 41
Hydrocarbon gasification, 199

Hydrocarbons
natural gas, 1096
petroleum, 73
shale oil, 1032, 1033
Hydrocracking, 178
Canmet process, 247
catalysts, 254
process parameters, 179, 249
Hydrocyclone, 533
Hydrodemetallization, 234, 246
Hydrodenitrogenation, 178, 179, 1018
Hydrodesulfurization, 143, 205, 234
Hydrofining, 181
Hydrogasification of char, 669
Hydrogen addition, 239, 243, 244, 246
Hydrogenation
char, 669
coal, 638, 737
petroleum, 178
shale oil, 1016
Hydrogen, cyanide from natural gas, 1160
Hydrogen donor diluent visbreaking, 246, 417
Hydrogenolysis, 178
Hydrogen production, 184, 195
commercial processes, 196, 1157
from natural gas, 1157
hydrocarbon gasification, 199
Hypro process, 201
steam-methane reforming, 196
steam-naphtha reforming, 198
synthesis gas generation, 198
Hydrogen sulfide, 1095
removal from natural gas, 1106
Hydroprocessing, 178
catalysts, 253, 1025, 1026
commercial processes, 180
heavy feedstocks, 247
H-Oil, 181
hydrocracking, 178, 248, 249
hydrofining, 181
[Hydroprocessing]
hydrogenolysis, 178
hydrotreating, 248, 249, 253
hydrovisbreaking, 248, 249
shale oil, 1016, 1022
Hydropyrolysis of coal, 723
parameters, 724
processes, 738
Hydrotreating, 248, 249, 252
Hydrovisbreaking, 248, 249
Hyperforming, 186
Hypothetical resources (coal), 473
Hypro process, 201
Hytort oil shale retorting process, 971

Identified resources (coal), 473
Igneous intrusion, 514
Igneous rock, 13
Ignified fuel combustion system, 659
Ignition of coal, 655, 656
Impsonite, 57
Indicated resources (coal), 474
Indirect liquefaction of coal, 749
Inertinite, 545
Inferrred reserves
coal, 475
natural gas, 1067
In situ combustion (EOR), 42, 44
In situ gasification of coal (*see* Underground gasification of coal)
In situ recovery (tar sands)
combustion, 359
cyclic steam injection, 354
emulsification, 358
modified in situ, 362
projects, 352
steam drive, 356
In situ retorting of oil shale, 913, 980
Geokinetics process, 985
Occidental process, 985

[In situ retorting of oil shale]
Rio Blanco modified in situ (MIS) process, 994
US DOE process, 981
vertical modified in situ (VMIS) process, 985
Insulator catalysts, 252
Interfacial tension, 116
Interfoots oil, 270
International System of coal classification, 558, 563
Iron, 6
Iron oxide (sponge) gas cleaning process, 1145, 1147
chemistry, 1150
Iron removal from shale oil, 1013
Isoforming process, 256
Isomerization, 187
catalysts, 256
commercial processes, 188
fixed-bed process, 188
Isoprene, 302
Isopropyl alcohol, 292

Jig coal cleaning method, 531
JPL coal cleaning process, 548

Katasulf process, 1150
Kerogen, 544
aliphatic carbon, 857
analyses, 879, 885
aromatic carbon, 857
assay oil yield, 856, 862
characterization, 859
depolymerization, 890
H/C ratio, 825, 856, 857
heteroatom groups, 910
hydrogenolysis, 893
isolation, 878
kinetics of retorting, 866, 867
maturation, 820
micropyrolysis, 891
models, 895
[Kerogen]
N/C ratio 856
oxidation, 883
oxygen, 824, 880
pyrolysis, 891
structure, 877, 879, 895-910
thermal decomposition, 866
Type I, 821, 845, 846
Type II, 822, 845, 846
Type III, 822, 845, 846
types, 820
Kerosene, 2, 3, 68, 151, 266
Kieselguhr, 194
Kinematic viscosity, 105
Kiviter oil shale retorting process, 966
Koppers coke oven, 654
Koppers-Totzek coal gasification process, 753

Lacustrine environment, 821
Lacustrine oil shales, 831
LC-Fining process, 247
Lean gas, 32, 1089
Lean oil, 162
Ledgement coal cleaning process, 537
Lens trap, 14
Light end removal, 159
composition, 160
Light gas oil, 2, 151
Light naphtha, 151
Lignin, 11
hydroxyl groups, 11
methoxyl groups, 11
phenylpropane units, 11
Lipids, 12, 832
Liptinite, 548
Liptobiolithic coal, 559
Liquefaction (melting), 115
Liquefaction of coal
Bergius process, 737, 740
catalytic hydrogenation, 737
carbonization, 764
CCL process, 740

[Liquefaction of coal]
chemistry, 736
clean coke process, 738
COED process, 738
Consol process, 742
Costeam process, 742
Dow process, 737, 740
EPRI process, 739
Exxon process, 742, 746
H-Coal process, 740
history, 736
indirect, 746
Lurgi-Ruhrgas process, 737
multistage process, 740
Occidental process, 737, 738
processes, 736
pyrolysis, 736
Schroeder process, 737, 740
Shell-Koppers process, 753
SRC process, 742
SRL process, 742, 745
Synthoil process, 740
Toscoal process, 738
Union Carbide process, 738
University of Utah process, 740
zinc chloride catalysis, 737, 740
Liquefied petroleum gas (LPG), 226, 1092, 1098
Lithology
oil shale, 805
tar sand, 375
Longwall mine, 516
Low-btu gas, 662
Low-temperature coal tar, 764
Lubricating oil, 2, 151, 267
dewaxing, 218
viscosity, 108
Lurgi coal gasification process, 673, 753
Lurgi gas, 1088
Lurgi pressure gasifier, 662
Lurgi-Ruhrgas oil shale retorting process, 921
Lurgi-Ruhrgas process, 350, 737, 738
Lurgi-Spulgas retort, 653
Lye (caustic) treating, 205

Macerals, 544, 545, 546, 548
analysis, 552
composition, 550, 556
properties, 549
pyrolysis, 650
reactivity, 553
Macrinite, 545
Macroscopic classification of coal, 555
Magnetohydrodynamic combustion of coal, 660
Magnetometer, 25
Mahogany acids, 275
Maltenes, 63, 124
Marine shales, 830
Mark I ignified coal gasifier, 672
Measured resources (coal), 473
Medium-btu gas, 662
MEK (methyl ethyl ketone), 294
Metalloporphyrin, 93, 235
Metals
coal, 449
nonporphyrin, 96, 235
petroleum, 80, 111, 147, 323
shale oil, 990, 1008, 1013
tar sand bitumen, 323, 394
Metamorphic rock, 13
Methanation, 670
Methanol, 288
Methanol from coal, 754
Methanol-to-gasoline (MTG) process, 754, 755, 756, 757
Met-X process, 259
Meyer (TRW) coal cleaning process, 533
Micellar flooding, 35, 39
Micrinite, 545
Microfossils, 821
Microlithotype, 547

Microwave coal cleaning method, 540
Mineralized coal, 557
Mineral matter in coal, 453, 456, 575, 585
allogenic, 455
authentic, 455
combustion, 656
composition, 585
content, 575
epigenetic, 455
external, 455
internal, 455
primary, 455
secondary, 455
syngenetic, 455
Mineral wax, 53, 55
Mine
drift access, 515
room and pillar, 515
shaft access, 515
slope access, 515
Minerology
coal, 458, 462
oil shale, 801, 803, 806, 808, 810, 811
tar sand, 378, 392
Mining
coal, 515
contour, 519
oil shale, 830
open pit, 519
petroleum, 56
strip, 519
tar sand, 338, 378
Mobil MTG process, 755
Modified Fischer assay, 840
Modified in situ process
heavy oil, 362
oil shale, 994
Moisture in coal, 453, 467, 474, 646
Molecular sieve gas cleaning processes, 1146, 1148
Molecular weight
boiling point relationship, 123
[Molecular weight]
coal liquids, 624
composition, 146
Montmorillonite, 252
Moving-bed processes
catalytic cracking, 178
gasification of coal, 673, 676
reforming, 186
Moving grate oil shale retorting processes, 943
Allis-Chalmers roller grate process, 944
Dravo circular travelling grate process, 945
Superior Oil/Davy McKee circular grate retort, 948
Multistage coal liquefaction process, 740

Naphtha, 265
National Coal Board system of coal classification, 561
Native asphalt, 53, 56
Natural gas, 30, 1055
acid gas removal, 1103
associated, 1088
chemicals from, 1153, 1154, 1155
composition, 31, 1090, 1091, 1095
constituents, 1089
conventional reservoirs, 1059
dry, 32
historical use, 1055
hydrocarbon content, 1096
hydrogen sulfide, 1106
lean, 32, 1089
nitrogen removal, 1103
nonassociated, 1088
occurrence, 1059
origin, 1056
processing, 1099
production, 1077, 1078
properties, 1095
recovery, 1072

[Natural gas]
reserves, 803, 1065, 1070, 1073, 1074
residue, 32, 1089
sour, 32, 1089
sweet, 32, 1089
terminology, 1087
transportation, 1083
unconventional reservoirs, 1058, 1060
underground storage, 1080
water content, 1096
water removal, 1100, 1142
Natural gas liquids, 1097
Nickel, 6
catalyst, 1159
Nitroalkanes, 298
Nitrobenzene, 300
Nitrogen
API gravity, 80
coal, 457
coal tar, 766
composition, 147
kerogen, 910
petroleum, 79, 323
removal from natural gas, 1103
shale oil, 1006, 1017, 1020, 1021, 1022, 1033
tar sand bitumen, 323
Nobel metal catalysts, 191
Nomenclature
bitumen, 53, 56
coal, 543, 556, 557
natural gas, 1087
oil shale, 798
petroleum, 51, 151
shale oil, 799
tar sand, 322
Nonassociated natural gas, 1088
Nonbanded coal, 557
Noncoking coals, 648
Nonmining recovery, 351
Nonporphyrin metals, 235
Nonregenerative caustic treating, 206
NTU oil shale retorting process, 912
Nylon, 304

Occidental coal liquefaction process, 737, 739
Occidental in situ oil shale retorting process, 985
Oil absorption processes for gas cleaning, 1115
Oil gas, 1088
Oil sand (*see* Tar sand)
Oils fraction, 53, 55, 63
Oil shale, 3
assay, 840, 848, 856, 862
bitumen, 815, 847, 850
definition, 798
heat of retorting, 870
history, 796
lacustrine, 831
lithology, 805
marine, 828
minerology, 801, 805
mining, 832, 836
occurrence, 832
origin, 799
permeability, 875, 878
porosity, 875
production, 797, 798
properties, 804, 834, 846, 917, 924
recovery, 832
resources, 324, 795, 801, 802
retorting, 866, 912
sedimentation, 801
size reduction, 837
source materials, 815, 821, 823, 829
specific gravity, 862
terminology, 798
Oil shale retorting processes
Allis-Chalmers roller grate process, 944
Dravo circular travelling grate process, 945

[Oil shale retorting processes]
Equity Oil/Arco BX process, 998
Galoter process, 969
Hytort process, 971
Kiviter process, 966
Paraho process, 948
Superior Oil/Davy McKee process, 948
Unocal process, 958
Olefin, 281
Olefin manufacture
Chevron process, 298
Ethyl Corporation process, 299
Exxon process, 297
Shell process, 296
OPEC, 20
Open pit mining, 519
Organic facies, 827, 845, 846
Organic matter
coal, 440
oil shale, 815
petroleum, 5-12
preservation, 823
Otisca coal cleaning process, 524
Outcrops, 21
Oxidation
gas cleaning, 223
partial, 198, 199
Oxidative treating, 211
Oxo process, 294
Oxydesulfurization coal cleaning, 536, 537
Oxygen
coal, 581, 602
coal tar, 764
kerogen, 823, 824, 882, 884, 910
petroleum, 79, 323
shale oil, 1033
tar sand bitumen, 323
Ozokerite (ozocerite), 55, 58

Paraffin wax, 267, 270
Paraho oil shale retorting process, 948
Partial oxidation, 198, 199
Parting, 514
Peat
conversion into coal, 458
formation, 447
plant precursors, 444
Peat gas, 1088
Permeability
gas reservoirs, 32, 1058, 1060
oil shale, 875, 878
petroleum reservoirs, 13
tar sand, 351, 378, 382
PETC coal cleaning process, 537
Petrochemicals, 261
chemistry, 282
feedstocks, 287
historical, 281
hydrocarbons, 287
intermediates, 1155
oxygenates, 288
sources, 278, 280
steam cracking, 285
Petrography, 441
Petrolatum, 58
Petrolenes, 63
Petroleum
accumulation in sediments, 12
asphalt base, 65
catalysis in sediment, 16
classification, 63
cracking in sediment, 16
composition, 3, 55
definition, 278
distillation, 17, 109, 149
elemental composition (*see* ultimate composition)
exploration, 21
flash point, 109
fractions, 2
genesis, 3
imports, 24
mining, 56
nomenclature, 151
origin, 3
paraffin base, 65
pretreatment, 152

[Petroleum]
producers, 20
production, 30
products, 262
properties, 3, 101
recovery, 21
reserves, 22, 324
source beds, 5
source material, 5
terminology, 51
transformation, 15
transportation, 20, 21, 46, 50
treatment, 203
use, 16, 17, 18
ultimate composition, 72, 323
volatility, 109, 405
water removal, 47, 152
Petrology, 441
Petrosix oil shale retorting process, 940
Phenols from coal, 772
Phosphoric acid polymerization, 193, 194
Pipeline
natural gas, 1084
petroleum, 47
synthetic crude oil, 364
Pipe still, 157
Pitch, 53, 56, 58, 61
Plates, distillation, 157
Pollutants
coal conversion, 761
tar sand recovery, 367
Polyaromatics, 302
Polybutylene terephthalate, 305
Polycarbonate, 305
Polyenes, 7
Polyethylene terephthalate, 305
Polymer gasoline, 194
Polyphenylene oxide resin, 305
Polysaccharides, 10
Polymer flooding, 36, 37, 38
Polymerization, 193
bulk acid, 195
catalysts, 257
commercial processes, 193
[Polymerization]
copper pyrophosphate, 193
phosphoric acid catalyst, 193, 194
sulfuric acid, 193
thermal, 193
Polyolefins, 302
Porosity
oil shale, 877
petroleum reservoir, 13
tar sand, 351, 378, 382
Porphine, tetraphenyl, 87
Porphyrins, 5, 6
benzo-, 85, 88, 94
coal, 96
effects in refining, 233
etio-, 85, 87
free-base, 93
iron, 94
metallo-, 93
nickel, 94
occurrence, 94
oil shale, 96
precursors, 93
properties, 83
tetrahydrobenzo-, 85, 88, 94
vanadyl, 87, 88, 91, 94
Potential reserves, 1067
Power shovels, 339
Preasphaltenes, 606
Pretreatment (*see* Cleaning)
Primary conversion of heavy oils, 246, 413
Primary gasification of coal, 667
Primary mineral matter, 455
Primary recovery, 21
Prime coking coal, 648
Producer gas, 1088
Production
coal, 507
natural gas, 1077, 1078
oil shale, 797, 798
petroleum, 20
tar sand bitumen, 343, 348, 351
Products, petroleum, 262
Propane asphalt, 59, 270

Propane deasphalting, 216
Propylene glycol, 291, 293
Proteins, 11, 829
Proven reserves, 1065
Proximate analysis of coal, 575
Pseudovitrinite, 545
Pyritic sulfur in coal, 579, 580
Pyrobitumen, 57
Pyrolysis of coal, 647, 717, 718, 720, 722, 737, 764

Rank (coal), 448, 449, 543
 elemental composition, 451
 variation with density, 554
 variation with depth, 450
Reboiler, 158
Recoverability factor, 475
Recovery
 coal, 506
 enhanced, 34
 environmental aspects, 367
 natural gas, 1072, 1075
 oil shale, 832
 primary, 21, 30, 34
 secondary, 32, 34
 tar sand bitumen, 339, 343, 419, 423
Rectification process for gas cleaning, 1114
Rectisol process, 669, 1142
Reduced crude, 271
Reef trap, 14
Refinery gas, 1088
Refining
 carbon rejection, 243
 chemistry, 98
 hydrogen addition, 243
 nonprocessing facilities, 149
 processes, 149, 150
 shale oil, 1005
Reflectance, 553
Reforming
 catalysts, 255
 catalytic, 184
 steam-methane, 196
[Reforming]
 steam-naphtha, 198
 thermal, 183
Refractive index, 117
Refrigeration gas cleaning process, 1101, 1114
Rerunning, 159
Reserves
 coal, 324, 471, 473, 497, 500, 801, 803
 inferred, 1067
 natural gas, 803, 1065, 1070, 1073, 1074
 oil shale, 324, 795, 802, 803
 petroleum, 22, 801, 803
 potential, 1067
 proven, 1065
 shale oil, 795
 tar sand, 320, 324, 325, 329, 801
 undiscovered, 1067
Reserves/production ratio, 23, 1071
Reservoir
 anatomy, 21
 anticline, 14, 25
 coal seams (beds), 1065
 conventional, 1059
 deep sources, 1065
 fault, 14
 gas cap, 21
 gas hydrates, 1065
 geopressured aquifers, 1064
 pressure, 32, 34
 reef, 14
 salt dome, 14, 1081
 tight gas, 32, 1061
 tight sands, 1061
 tight shales, 1062
 types, 14
 unconventional, 1060
 wedgeout, 14
Reservoir rock, 12
 igneous, 13
 limestone, 13
 metamorphic, 13
 sand, 13

Residfining process, 247
Residual oil, 266
Residue gas, 32, 1089
Residuum, 2, 3, 53, 151
 atmospheric, 58
 production, 59
 properties, 54, 60, 101, 119
 vacuum, 58
Residuum hydroprocessing, 245
Resinite, 545
Resins, 53, 55, 63
Rio Blanco oil shale retorting process, 994
Road oils, 271
Rock asphalt, 57
Rock asphaltite, 58
Rock asphaltoid, 58
Rocks
 reservoir, 5, 1059
 sedimentary, 3, 438, 801, 1059
Room and pillar mine, 515
Rotating kiln coal gasification process, 676

Salt dome, 14
 gas storage, 1081
Sand-flotation coal cleaning process, 515
SAPA fractionation, 131
SARA fractionation, 131, 132, 133
SASOL process, 670, 749, 750, 751
Saybolt viscosity, 105
Schroeder coal liquefaction process, 737, 740
Sclerotinite, 545
Scrapers, 339
Secondary conversion of heavy oils, 417
Secondary gasification of coal, 668
Secondary mineral matter, 455
Sediment
 bacterial decomposition, 5, 1055
[Sediment]
 biochemical transformation, 5, 1055
 phase equilibria, 16
 pressure, 5, 16, 1058
 temperature, 16, 1058
 transformations in, 5, 15
 water content, 5
Sedimentary rocks, 3, 438
Seismograph, 25
Selexol process, 1120
Semifusinite, 545, 650
Semisplint coal, 560
SFS, 105
Shaft access, 515
Shale oil
 aliphatic carbon, 857
 aromatic carbon, 857, 1019
 asphalt, 1035
 bitumen content, 847
 chemicals, 1031
 composition, 1007, 1018, 1020, 1028, 1034
 elemental composition (*see* ultimate composition)
 H/C ratio, 856
 hydrocarbons, 852
 metals, 990, 1008
 N/C ratio, 856, 1019
 nitrogen, 1006, 1017, 1020, 1021, 1022, 1033
 properties, 855, 917, 932, 941, 943, 954, 962, 972, 976, 990, 1008, 1010, 1012, 1019, 1028
 refining, 1005
 sulfur, 1006
 terminology, 799
 ultimate composition, 847, 917, 932, 941, 974, 976, 990, 1005, 1006, 1008, 1010, 1019, 1029
 upgrading, 1007
Shale oil refining, 1005
Shale oil upgrading
 arsenic removal, 1013

[Shale oil upgrading]
catalytic processes, 1016
delayed coking, 1014
dewatering, 1010
flexicoking, 1016
fluid coking, 1016
Gulf process, 1022
iron removal, 1013
noncatalytic, 1013
visbreaking, 1014
solids removal, 1010
Shell-Koppers coal gasification process, 753
Shell olefin process, 296
Shell (SPHER) oil shale retorting process, 931
Shell (SSRP) oil shale retorting process, 934
Shell still, 168
Sidestream products, 158
Silica-alumina catalysts, 251
Silica-magnesia catalysts, 251
Sill, 514
Size exclusion chromatography (SEC), 612
Slack wax, 270
Slope access, 515
Soaker, 169
Sodium plumbite (doctor) treating, 211
Solidification, 115
Solids removal
coal, 521, 533
petroleum, 152
shale oil, 1010
tar sand bitumen, 343
Solvent asphalt, 59
Solvent deasphalting, 215, 240
Solvent dewaxing, 218, 219
Solvent extraction of coal
processes, 741
supercritical fluid, 745
Solvent fractionation
coal liquids, 605
dilution effects, 129
solvent type, 124
[Solvent fractionation]
temperature effects, 127, 129
Solvent Refined Coal (SRC-I and SRC-II) process, 741, 742, 743
Solvent Refined Lignite (SRL) process, 742, 745
Solvents, 162, 265
Solvent treating, 215
Source beds, 5
Source material, 5-12, 440,
Sour gas, 32,
Sour petroleum, 204
Specific gravity, 3, 67, 69, 99
Specific heat
coal, 646
petroleum, 117
Spent lye, 205, 206
Splint coal, 560
Split, 514
Stabilization, 159
Starches, 10
Steam cracking, 285
Steam drive process, 356
Steam flooding operations, 357
Steam injection, 42, 43, 44
Steam-methane reforming, 196
Steam-naphtha reforming, 198
Steam regenerative caustic treating, 206
Steam stimulation, 56
Sterols, 11
Still, 154
battery, 155
shell, 168
Stopes-Heerlen coal classification system, 560
Stove oil, 2, 151, 266, 267
Straight run
asphalt, 59
gasoline, 182
Stratified lake, 814
Stratigraphic trap, 329
Stretford process, 1106, 1108
Strip mining, 519

Stripping, 159
Structural studies of coal
acetylation, 589
alkylation, 586
coal liquids, 639
Curie point pyrolysis, 600
depolymerization, 586
hypothetical structures, 603
infrared spectroscopy, 597
mass spectroscopy, 598
nmr spectroscopy, 592
oxidation, 586
pyrolytic methods, 590
X-ray photoelectron spectroscopy, 601
Structural trap, 329
Structure
coal, 585
petroleum asphaltenes, 230
tar sand, 375, 380
Submerged coal combustion process, 660
Subsurface exploration, 25
Sulfatic sulfur in coal, 579, 580
Sulfonic acids, 275
Sulfur
coal, 473, 486, 490, 580, 581, 582
kerogen, 910
natural gas, 1090
petroleum, 77, 78, 100, 106, 147, 323
shale oil, 1006
tar sand bitumen, 323
Sulfuric acid
polymerization, 193
treating, 209
Supercritical fluid chromatography (SFC), 619
Supercritical fluid extraction of coal, 745
Supercritical gas extraction of coal, 747, 748
Superfractionation, 162
Superior Oil/Davy McKee circular grate retort, 948
Superslagging coal combustion process, 660
Surface retorting of oil shale, 912, 913
Chevron (STB) process, 938
Exxon (ESR) process, 937
gas combustion process, 916
Lurgi-Ruhrgas process, 921
moving grate processes, 943
NTU process, 912
Petrosix process, 940
Shell (SPHER) process, 931
Shell (SSRP) process, 934
Tosco II process, 925
Surface tension, 116
Surfactant flooding, 36, 38, 39
SUS, 105
Sweet gas, 32, 1089
Sweet petroleum, 204
Syngenetic mineral matter, 455
Synthesis gas, 198
chemicals from, 1157
Synthetic crude oil
composition, 368, 1005, 1006
properties, 368, 416
Synthoil coal liquefaction process, 739, 740

Tailings
composition, 346
pond, 346, 347
sludge, 346
Tanker transportation
natural gas, 1085
petroleum, 47
Tar
coal, 61, 764
petroleum, 53, 55, 61
Tar sand
bitumen content, 321, 383, 384
commercial history, 342
definition, 322, 326
deposits, 319, 329, 327
direct heating, 348
economics, 420

[Tar sand]
entrapment, 329
environmental aspects, 367
location, 320, 341
minerals, 378, 392
mining, 338
nonmining operations, 351
oil wet, 375
permeability, 382, 391
physical structure, 375
porosity, 382, 383, 391
properties, 336, 376, 380, 393, 426
recovery, 335, 343
reserves, 320, 324, 325
resource comparisons, 336
structure, 375, 380
tailings pond, 346
terminology, 322
traps, 329
water wet, 375
Tar sand bitumen, 229
Telinite, 545
Terminology
bitumen, 53, 56
coal, 543, 556, 557
natural gas, 1087
oil shale, 798
petroleum, 51, 151
shale oil, 799
tar sand, 322
Terpenes, 7
Tetrahydrobenzoporphyrins, 85, 88, 94
Texaco process for coal gasification, 673, 752
Thermal conductivity
coal, 647
tar sand bitumen, 410
Thermal cracking, 166
Burton process, 168
chemistry, 167
coking, 171
commercial processes, 169
history, 167
tube and tank process, 169
[Thermal cracking]
visbreaking, 170
Thermal polymerization, 193
Thermal reforming, 183
Thermogenic formation of natural gas, 1057
Thiessen-Bureau of Mines classification of coal, 557
Thin layer chromatography (TLC), 615
Thiourea, 137
Thorium oxide catalysts, 252
Tight sands, 1061
chemical explosives, 1061
nuclear detonation, 1061
Tight shales, 1062
Toscoal coal liquefaction process, 738
Tosco material balance assay, 844
Tosco II oil shale retorting process, 925
Tower
diameter, 159
distillation, 158
Trace elements
coal, 457, 463, 583
petroleum, 82
shale oil, 1013
Transformation
catalysis, 16
cracking, 16
gas deasphalting, 15
in situ, 15
weathering action, 14
Transition zone
water-oil, 13
Transportation
coal, 540
natural gas, 1083
petroleum, 20, 46, 50
tar sand bitumen, 364
Traps (*see* Reservoir)
Trays
atmospheric tower, 159
distillation, 157
liquefied gas tower, 158

[Trays]
petroleum fractionation, 158
vacuum tower, 159
Treating processes
acid, 207
bauxite, 210
caustic (lye), 205
clay, 209
continuous contact filtration, 210
copper sweetening, 213
deasphalting, 215
dewaxing, 218
doctor (sodium plumbite) method, 211
oxidative, 211
solvent, 215
sulfuric acid, 209
TRW (Meyer) coal cleaning process, 533
Tube and tank process, 169

Ultimate composition
coal, 578
petroleum, 72, 323
shale oil, 847 (*see also* Shale oil, ultimate composition)
tar sand bitumen, 323, 387, 394, 395
Unconventional reservoirs, 1058, 1060
Underground gasification of coal, 726
borehole method, 728
CRIP method, 733
methods, 732
percolation method, 730
steam method, 729
Underground retorting of oil shale (*see* In situ retorting of oil shale)
Undiscovered reserves, 1067
Union Carbide coal liquefaction process, 738
University of Utah coal liquefaction process, 740
Unocal oil shale retorting process, 954
Unsaturated fats, 10
Upgrading
coal (*see* Beneficiation)
natural gas, 1099 et seq.
petroleum, 165 et seq.
shale oil, 1007
tar sand bitumen, 411
Urea, 137
USBM-API fractionation, 131, 133
US DOE in situ oil shale retorting process, 981
US Geological Survey system of coal classification, 559

Vacuum distillation, 122, 151, 158
Vacuum flashing, 166
Vacuum gas oil, 2, 151
Vacuum residuum,
Vanadium, 6, 7, 237
van Dyke tower, 155
Vapor rectification process, 1114
Vertical modified in situ (VMIS) oil shale retorting, 985
Visbreaking
heavy feedstocks, 247
hydrogen donor diluent, 246, 417
petroleum, 170
shale oil, 1014
process parameters, 248, 249
Viscosimeter scales
API gravity, 104
interrelationship, 107
viscosity, 69, 100
index, 108
interrelationship, 107
kinematic, 105
lubricating oils, 108
Saybolt, 105
SFS, 105
SUS, 105